Principles of Zoology

Principles of Zoology

Edited by
Richard M. Renneboog, MSc

SALEM PRESS

A Division of EBSCO Information Services, Inc.
Ipswich, Massachusetts

GREY HOUSE PUBLISHING

Cover photo: Tiger. Image by THEPALMER (via iStock).

Copyright © 2019, by Salem Press, A Division of EBSCO Information Services, Inc., and Grey House Publishing, Inc.

All rights reserved. No part of this work may be used or reproduced in any manner whatsoever or transmitted in any form or by any means, electronic or mechanical, including photocopy, recording, or any information storage and retrieval system, without written permission from the copyright owner.

For information contact Grey House Publishing/Salem Press, 4919 Route 22, PO Box 56, Amenia, NY 12501

Principles of Zoology, published by Grey House Publishing, Inc., Amenia, NY, under exclusive license from EBSCO Information Services, Inc.

∞ *The paper used in these volumes conforms to the American National Standard for Permanence of Paper for Printed Library Materials, Z39.48 1992 (R2009).*

**Publisher's Cataloging-In-Publication Data
(Prepared by The Donohue Group, Inc.)**

Names: Renneboog, Richard, editor.
Title: Principles of zoology / editor, Richard Renneboog, MSc.
Description: [First edition]. | Ipswich, Massachusetts : Salem Press, a division of EBSCO Information Services, Inc. ; Amenia, NY : Grey House Publishing, [2019] | Series: Principles of | Includes bibliographical references and index.
Identifiers: ISBN 9781642653175
Subjects: LCSH: Zoology.
Classification: LCC QL45.2 .P75 2019 | DDC 590--dc23

PRINTED IN THE UNITED STATES OF AMERICA

Contents

Publisher's Note ix
List of Contributors xi
Editor's Introduction xv

Adaptive Radiation 1
Aging 4
Amphibians 6
Animal Adaptations 11
Animal Aggregation 15
Animal Bioluminescence 19
Animal Cells 21
Animal Courtship 25
Animal Demographics 29
Animal Development: Evolutionary Perspective 34
Animal Domestication 38
Animal Embryology 42
Animal Emotions 44
Animal Evolution: Historical Perspective 47
Animal Growth 51
Animal Habituation and Sensitization 55
Animal Immune Systems 59
Animal Instincts 62
Animal Intelligence 66
Animal Kingdom 69
Animal Life of Swamps and Marshes 73
Animal Life Spans 75
Animal Locomotion 79
Animal Mating 83
Animal Migration 86
Animal Physiology 90
Animal Reproduction 92
Animal Respiration and Low Oxygen 96
Apes to Hominids 101
Arachnids 105
Arthropods 108
Asexual Reproduction 111

Beaks and Bills 114
Biodiversity 116
Biological Rhythms and Behavior 121
Biology 125
Birds 127
Birth 133
Bone and Cartilage 135
Brain 137
Breeding Programs 142

Camouflage 146
Cannibalism 148
Carnivores 150
Cell Determination and Differentiation 153
Circulatory Systems of Invertebrates 157
Circulatory Systems of Vertebrates 162
Claws, Nails, and Hooves 165
Cloning of Extinct or
    Endangered Species 167
Coevolution 170
Communication 175
Communities 179
Competition 183
Convergent and Divergent Evolution 188
Copulation 191
Crustaceans 194

Death and Dying 197
Digestion 199
Digestive Tract 204
Dinosaurs 209
Displays 214

Ecological Niches 217
Ecology 221
Ecosystems 226
Embryonic Development 231
Endangered Species 235
Endocrine Systems of Invertebrates 240
Endocrine Systems of Vertebrates 243
Endoskeletons 247
Estivation 251
Estrus 253
Evolution: Animal Life 255
Evolutionary Origin of Sex Differences 259
Excretory System 262
Exoskeletons 266
Extinction 270
Extinctions and Evolutionary Explosions 274

Felidae 280
Fertilization 283
Fins and Flippers 285
Fish 288
Food Chains and Food Webs 293

Contents

Gametogenesis	297
Gas Exchange	300
Gene Flow	305
Genetic Mutations	308
Genetics	311
Grooming	316
Habitats and Biomes	319
Hardy-Weinberg Law of Genetic Equilibrium	323
Hearing	327
Herbivores	330
Herds	333
Hermaphroditism	335
Heterochrony	337
Hibernation	339
Home Building	342
Homeosis	344
Homo Sapiens and Human Diversification	348
Hormones and Behavior	355
Hybrid Zones	359
Hydrostatic Skeletons	363
Infanticide	366
Ingestion in Animals	368
Insects	373
Invertebrates	381
Isolating Mechanisms In Evolution	384
Jellyfish	388
Lactation	390
Language	393
Learning	397
Mammalian Social Systems	403
Marine Animals	407
Marine Biology	411
Mark, Release, and Recapture Methods	414
Marsupials	418
Metabolic Rates in Animals	422
Metamorphosis	425
Mimicry	429
Molting and Shedding	431
Morphogenesis	434
Multicellularity	437
Muscles in Invertebrates	443
Muscles in Vertebrates	448

Nervous Systems of Vertebrates	453
Nesting	457
Neutral Mutations and Evolutionary Clocks	460
Nocturnal Animals	464
Nonrandom Mating, Genetic Drift, and Mutation	466
Nutrient Requirements of Animals	469
Offspring Care	472
Omnivores	475
Osmoregulation	476
Packs	481
Pair-bonding	483
pH Maintenance in Animals	485
Phylogeny	489
Placental Mammals	493
Plant and Animal Interactions	496
Poisonous Animals	501
Pollution Effects on Animal Life	503
Population Analysis	508
Population Genetics	511
Predation	515
Pregnancy and Prenatal Development in Animals	519
Prehistoric Animals	523
Primates	529
Protozoa	535
Punctuated Equilibrium and Continuous Evolution	539
Regeneration	543
Reproductive Strategies in Animals	547
Reproductive System of Female Animals	551
Reproductive System of Male Mammals	554
Reptiles	558
Respiratory Systems in Animals	562
Rodents	566
Ruminants	568
Savannas and Animal Life	570
Scavengers	571
Sense Organs	574
Sexual Development	578
Shells	581
Sleep	584
Smell	586
Social Hierarchies	589
Symbiosis	591

Tails 596	Water Balance in Vertebrates 631
Teeth, Fangs, and Tusks 598	Wildlife Management 635
Tentacles 601	Wings 637
Territoriality and Aggression 603	
Thermoregulation 608	Zoology 642
Tool Use 613	Zoos 644
Ungulates 616	**Appendixes 651**
Urban and Suburban Wildlife 618	
	Branches of Zoology 653
Vertebrates 624	Key Figures in Conservation and Zoology 655
Vocalizations 626	Index 661

Publisher's Note

Salem Press is pleased to add *Principles of Zoology* as the sixteenth title in the *Principles of series* that includes *Pharmacology, Ecology, Physics, Astronomy, Computer Science, Sustainability, Biology, Scientific Research, Sustainability, Biotechnology, Programming & Coding, Climatology,* and *Robotics & Artificial Intelligence.* This new resource introduces students and researchers to the fundamentals of modern agriculture using easy-to-understand language for a solid background and a deeper understanding and appreciation of this important and evolving subject. All of the entries are arranged in an A to Z order, making it easy to find the topic of interest.

Entries related to basic principles and concepts include the following:

- Fields of Study related to the topic;
- Abstract that provides brief, concrete summary of the topic and its significance or application in pharmacology;
- Key Concepts important to a proper understanding of the topic;
- Text that gives an explanation of the background and significance of the topic to zoology by describing developments such as Animal Development, Animal Emotions, Biological Rhythms and Behavior, Convergent and Divergent Evolution, Extinctions and Evolutionary Explosions, Learning, Nesting, Pollution Effects on Animal Life, Prehistoric Animals, and Urban and Suburban Wildlife;
- Illustrations that clarify difficult concepts via models, diagrams, and charts of such key topics as the Animal Physiology, Coevolution, Embryonic Development, and Osmoregulation; and
- Further Reading lists that relate to the entry.

The book includes helpful appendixes as another valuable resource, including the following:

- Branches of Zoology, showing areas of specialization with the field;
- Key Figures in Conservation and Zoology; and
- Index.

Salem Press and Grey House Publishing extend their appreciation to all involved in the development and production of this work. This reference work begins with a comprehensive introduction to pharmacology, written by volume editor Richard Renneboog, MSc. The entries have been written by experts in the field. Their names and affiliations follow this note.

Principles of Zoology, as well as all Salem Press reference books, is available in print and as an e-book. Please visit www.salempress.com for more information.

List of Contributors

Emily Alward, JD
Independent scholar

Michele Arduengo, PhD
Promega Corporation

Iona C. Baldridge, EdD
Lubbock Christian University

Erika L. Barthelmess, PhD
St. Lawrence University

Richard Beckwitt
Claremont McKenna College

Milton Berman
Yale University

Catherine M. Bristow
Michigan State University

William R. Bromer
College of St. Francis

Alan Brown, MD, MPH
Columbia University

J. H. U. Brown, PhD
University of Maryland

Robert A. Browne, PhD
Syracuse University

John T. Burns
Author, *Cycles in Humans and Nature*

Mary E. Carey, PhD
University of Texas at Austin

Kerry L. Cheesman, PhD
Capital University

Victor W. Chen, MD
Canton, Georgia

Richard W. Cheney, Jr.
Las Vegas, Nevada

David L. Chesemore
University of Alaska

Dale L. Clayton, PhD
La Sierra University

Sneed B. Collard, MA
Missoula, Montana

Jaime Stanley Colomé
San Luis Obispo, California

Alan D. Copsey
Seattle University School of Law

Greg Cronin, PhD
University of Colorado, Denver

James F. Crow, PhD
University of Wisconsin–Madison

Charles R. Crumly, PhD
University of California, Berkeley

George Dale
American ichthyologist

Dennis R. Dean, PhD
Virginia Tech

Peter L. deFur, PhD
Virginia Commonwealth University

Philip J. Dziuk, PhD
University of Illinois

David K. Elliott, PhD
Northern Arizona University

Jessica O. Ellison
Arizona State University

List of Contributors

John W. Engel, PhD
University of Hawai'i at Manoa

Jim Fowler
Host of Wild Kingdom

Frances C. Garb, PhD
University of Massachusetts, Lowell

Soraya Ghayourmanesh, PhD
Columbia University

Pamela J. W. Gore, PhD
Georgia State University

Dalton. R. Gossett, PhD
Louisiana State University, Shreveport

Susan E. Hamilton, MA
Rutgers University

Jean S. Helgeson
Author

Peter B. Heller, PhD
Harvard University

Robert E. Herrington, PhD
Georgia Southwestern State University

Carl W. Hoagstrom
Science editor

David Wason Hollar, Jr., PhD
Pfeiffer University

David E. Hornung, PhD
St. Lawrence University

Katherine H. Houp, PhD
Ohio North University

Richard D. Howard, PhD
Purdue University

Lawrence E. Hurd, PhD
Washington and Lee University

Mary Hurd
Independent scholar

M. A. Q. Khan, PhD
University of Illinois, Chicago

Andrew C. Skinner
Independent scholar

Vernon N. Kisling, Jr.
Editor, *Zoo and Aquarium History-Ancient Animal Collections to Zoological Gardens*

Kenneth M. Klemow, PhD
Wilkes University

Robert T. Klose, PhD
University of Maine

Kari L. Lavalli, PhD
Boston University

Walter Lener
Author, *The American Biology Teacher*

Douglas B. Light, PhD
Lake Forest College

Robert Lovely, MA
University of Wisconsin–Madison

Kristie Macrakis, PhD
Ivan Allen College

Paul Madden, PhD candidate
Boston College

David Mailman, PhD
University of Houston

Nancy Farm Männikkö, PhD
National Park Service

Patricia A. Marsteller, PhD
Emory University

Sarah Crawford Martinelli, PhD
Southern Connecticut State University

Kristen L. Mauk, PhD, DNP, RN
Colorado Christian University

Linda Mealey, PhD
University of Texas, Austin

John S. Mecham, PhD
University of Texas

Roman J. Miller, PhD
Eastern Mennonite University

Eli C. Minkoff, PhD
Worcester State University

Thomas C. Moon, PhD
uOttawa

Michele Morek
National Catholic Reporter

Rodney C. Mowbray, PhD
University of Wisconsin–La Crosse

Donald J. Nash
Colorado State University

Edward N. Nelson, PhD
University of Chicago

Bryan Ness, PhD
Pacific Union College

John G. New
Independent scholar

Robert L. Patterson, PhD
University of California, Santa Barbara

Robert W. Paul, PhD
St. Mary's College of Maryland

Joseph G. Pelliccia, PhD
Bates College

George R. Plitnik, PhD
Frostburg State University

Robert Powell, PhD
Avila University

Frank J. Prerost, PhD
Midwestern University

Donald R. Prothero, PhD
Columbia University

Judith O. Rebach
Independent scholar

James L. Robinson, PhD
University of Illinois Urbana-Champaign

Lisa M. Sardinia, PhD, JD
Pacific University

Donna Janet Schroeder, PhD
University North Carolina, Chapel Hill

Geri Seitchik, PhD
La Salle University

Jon P. Shoemaker, PhD
Author

R. Baird Shuman, PhD
University of Pennsylvania

Sanford S. Singer, PhD
University of Dayton, Ohio

Susan R. Singer, PhD
Carleton College

Roger Smith, MS
Teton Science School

Dwight G. Smith, PhD
Southern Connecticut State University

Michael Steele, PhD
Wilkes University

Daniel F. Stiffler, PhD
California State Polytechnic University

List of Contributors

Frederick M. Surowiec
Education Specialist at Harvard Museum of Natural History

Sue Tarjan
University of California-Santa Cruz

Samuel F. Tarsitano, PhD
Worcester State University

David Thorndill, PhD
Essex Community College

Leslie V. Tischauser, PhD
Prairie State College

Adrian Treves, PhD
University of Wisconsin-Madison

Robert C. Tyler, PhD
University of Maryland

John V. Urbas
Independent scholar

Sarah Vordtriede, PhD
Columbia College

Marcia Watson-Whitmyre, PhD
University of Delaware

Robert W. Yost, PhD
Purdue School of Science

Samuel I. Zeveloff, PhD
Weber State University

Ming Y. Zheng, PhD
Gordon College

Ling-Yi Zhou
Independent scholar

Editor's Introduction

The ancient Greeks had a word that meant, roughly, "an animal thing that is alive." They called it ζωον (*zoön, zo-oy*), or ζωοι if there was more than one. They had another word that meant *study* or *learn*. That word was λογος (*logos*). We know that word today as the origin of the word *logic*, and as the second half of portmanteau words like *psychology* and *geology* (study of the mind and study of the physical world, respectively). So what, then, is meant by *zoology*? Nothing more, or less, than "study of living things." It is a science as varied as the myriad living things of an animal nature this world contains in the present, that it has contained in the past, and that it will contain in the future.

The subject matter of zoology is not limited to just the number and kinds of animal life on Earth. It includes all possible aspects of what defines any particular animal species and its role in the natural world. What gives an animal species its identity, both as a species and as a unique individual within that species? Where does it live, and how does it live where it lives? What are its relationships to the other animals in its environment, and to the environment itself? What does it eat? What eats it? How does it reproduce?... and so on. So many questions in such a small word—*zoology*!

As a starting point, it must be understood that this planet Earth is a closed system as an environment. There are oceans, landmasses, and an atmosphere above them, all of which make up the biosphere of the entire planet. That biosphere contains all life on Earth, in what amounts to a paper-thin layer enveloping the planet. All known living things live only within this fantastically thin environment. However, it is a dynamic thing in a constant state of change, if for no other reason than that the planet itself is a dynamic thing in a constant state of change. Changes to the environment may occur very quickly on a local scale. They can also occur very slowly, over millions and even billions of years, and at any rate in between those two extremes. Each change, no matter how fast or slow it takes place, renders a difference in the overall nature of the world environment.

With just a quick glance about, it will be very obvious that the planetary environment is not uniform in structure. From the largest scale to the smallest, Earth's environment is made up of a seemingly infinite number and variety of ecologies and environmental niches. Exploration identifies new environmental niches, and hundreds of new species within them, every year. The dynamic nature of the planet's biosphere dictates that environmental change has been a continuous process since Earth was first formed. Evidence of those earlier environments and the living creatures—the ζωοι—they contained at that time is the subject matter of paleontology. This is the branch of zoology that deals primarily with creatures and environments that no longer exist. The ultimate goal of the paleontologist is to find those traces that remain of the animal life that existed in the past, and from them assign the specific identities of those long-ago animal life forms. Paleontology includes a number of other disciplines such as paleogeology, paleobotany, and paleoecology.

As well as understanding those animal remains, paleontology seeks to understand the nature of their contemporary environments and how those animals interacted with it. Was the animal an herbivore, a carnivore, or both? If it was herbivorous, what were the plants that it ate, and are there traces of those plants to be found that would support such an assignment? If it was carnivorous, was it an active hunter, either solitary or in packs, or was it a scavenger? How did either of those possibilities function in that time and place? These and many more questions face the paleontologist, especially concerning the physiology and anatomy of the living creatures.

As a branch of zoology, paleontology is the jigsaw puzzle with the most pieces to put in place, made all the more challenging because so many of the puzzle pieces are lost and missing. It would be an impossible puzzle to assemble even as much as it has been, were it not for the fact that environmental change over time has been a continuous process and evolution has produced animals in the present day that fulfill the same environmental roles as their prehistoric counterparts. Observing animal behavior in the present day not only provides the zoologist with vital information about what makes up the identity of those animals, it also allows the paleontologist to extrapolate and project those behaviors onto the long-gone creatures of the past, in a sense breathing life into their fossilized remains. It also provides a basis for zoologists to surmise how they might evolve into the animals of the future.

Observing animal behavior in the present provides indications of how animals in the past may have behaved in their particular environmental niches. For example, observing how the elephant, the rhinoceros, the giraffe and the hippopotamus move in and interact with their respective environments provides insight into how such creatures as the sauropods moved in and interacted with their environments. Similarly, observing the way the ostrich, the emu and other birds walk, run and fly provides insight into the way the tyrannosaurs, raptors, and pteranodons walked, ran and soared in their time. The pack and herd behavior of lions, wolves, hyenas, musk-oxen, bison and deer also indicate how prehistoric pack hunters and herd animals may have behaved long ago. Most importantly, observing the behavior of animals in the present day provides insight and understanding of how all animal (and plant) life on Earth is interconnected and of humanity's place within that network since the dawn of humankind. Accordingly, paleoanthropology, anthropology, environmental sciences and ecological studies are also important aspects of zoology.

The primary aspect of zoology, however, is the ζωοι, the animal life. That posits the question of what is an animal. Animals are living things, but so are plants. Animals are composed of cells that contain the biochemical processes of life, but so are plants. It might seem that the only difference between the two is that animals can move about in their environments of their own volition, and plants cannot. It takes a much closer look to identify the fundamental differences between animals and plants. All living things, whether animal or plant in nature, are composed of cells, in each of which is a nucleus containing the deoxyribonucleic acid—the DNA—that determines every aspect of the identity of the living entity. There is, however, a fundamental difference in the structures of animal cells and plant cells. Plant cells have structurally rigid cell walls due to the materials from which they are made. Animal cells, on the other hand, have flexible cell walls consisting of a "lipid bilayer."

The lipid bilayer of an animal cell is composed of two spheroidal layers of molecules called phospholipids, the phosphate esters of fatty acids. A phospholipid has a long non-polar hydrocarbon tail extending from the highly polar phosphate group at its head. The lipid tails of the inner layer point outward from the interior of the cell, and the lipid tails of the other layer point inward. The two sets of lipid tails intertwine strongly, resulting in the stable configuration of the animal cell wall. The lipid- or fat-filled space between the two surfaces is highly hydrophobic (*water-fearing*). The outside and the inside of the cell are thus hydrophilic (*water-loving*) due to the phosphate groups that form the inner and outer surfaces of the cell wall. This structure allows the cells to have a strong adhesion with each other, both because of the interaction of the phosphate layers and through the intermediacy of various connective tissues. The animal cell wall structure also allows for the passage of materials through the animal cell wall in a significantly more versatile manner than is the case for the more rigid walls of plant cells.

It is this lipid bilayer structure that has allowed the formation of animal life and all animal species throughout evolutionary history. It is also this relatively simple, yet highly versatile, structure that is the starting point for defining what is an animal. Therefore, microbiology, biochemistry, and their many related branches of chemistry, physics, and biology are all significant aspects of the science of zoology.

In its simplest form, animal life consists of creatures whose entire physical being is just one single cell, yet within that one cell are all of the tens of thousands of biochemical reactions that maintain life. These are the microbes that are the subject of branches of zoology called bacteriology and microbiology. This single-celled form is undoubtedly the original form of both animal and plant life on Earth. Over the vast expanse of evolutionary time, these evolved into a multitude of ever larger and more complex animal and plant species. This includes not only all animal species in the present, but also the many millions that came before and are known today only by their fossilized remains. The body structures of all multicellular animals contain smaller structures—organs—composed of cells that perform different specific functions in support of the living being. Different cells produce different biochemicals that function to maintain the viability of the organism, regulating such processes as glucose metabolism, electrolyte balance, and the reproductive system.

The miracle of all this cellular differentiation and specialization is that the molecular mechanism by which animals reproduce results in the formation of a new being of that particular species and not something else. In the nucleus of each and every cell of the animal's body, no matter what type of cell it is and regardless of its function, there is the same DNA that will reproduce a complete new being of that particular species and only that species. More precisely, that DNA encodes every feature of the individual that carries it. Understanding how that works is the focus of a number of different branches of zoology, including biochemistry, genetics, reproductive and developmental biology, and endocrinology, all of which are supported by an extensive list of other science and technology fields or practices.

The cells in animals larger than the single-celled creatures are differentiated and carry out different functions. As the animal develops from its embryonic state, some cells become skin and other membranes. Others become a brain, heart, lungs, and other organs. Still others become bone, blood and other connective tissues. Each species has its unique form and structure according to the DNA by which it is encoded. Despite that, it is interesting to note that within the incredible variety of those forms and structures, the body forms of all animal organisms on Earth have the same bilateral symmetry. They have a left side and a right side that are effectively mirror images of each other across a longitudinal plane that bisects the body. These aspects of zoology are the realm of animal anatomy and physiology, biomechanics and evolutionary biology. How the DNA is passed down through generation after generation of descendants within a species is the realm of genetics, reproductive biology, and the social sciences of anthropology and sociology.

All animals, including humans, inhabit a multitude of environmental niches and habitats in Earth's biosphere, and outside of it if one includes the International Space Station. They interact intimately with the physical nature and structural complexities of their respective environments, as well as with the other creatures that share a particular environment. In so doing, they play a significant role in determining the nature of their environments, just as those environments play an equally significant role in determining the nature of the creatures they support. In evolutionary biology, this is the basic working principle of adaptive radiation. Understanding these relationships is the purpose of the study of ecology and the environmental sciences, as well as of population dynamics and evolutionary biology. It is worth noting that this aspect of zoology also touches, however lightly, several of the social sciences such as anthropology, sociology, psychology, economics, the practice of agriculture, and agronomics.

The *Animalia*—the ζωοι—have a distinct place in the circle of life. This cycle is the mechanism by which all life on Earth has continued since life first arose on the planet. It involves all living things, starting with the primary producers that capture the energy of sunlight and store it in the chemical bonds of the glucose molecules formed during photosynthesis. The plants become the food for herbivores and omnivores, which in turn become food for carnivores and other omnivores. Eventually, their lives also end and they become food for the decomposers, the fungi and bacteria that return the components of the dead bodies to Earth to be used once more by the primary producers. The cycle is continuous and functions wherever life exists. However, the cycle exists only because there is reproduction and death. Understanding this cycle encompasses a broad range of scientific disciplines, including reproductive biology, genetics, embryology, endocrinology, evolutionary biology, animal physiology, veterinary sciences, anthropology, sociology, the practice of medicine—particularly obstetrics and gynecology, and pediatric medicine—gerontology, geriatrics, and thanatology (the study of death and the process of dying). All of these are involved in the science of zoology as it regards the viability of the individual members of a species, as well as the long-term viability of the species itself.

The subject matter of zoology is not limited to the number and kinds of animal life on Earth. By extension, the practices of the science of zoology must also apply to any forms of living creatures, or their remains, that might be found on other worlds as exploration extends into space. Given that all known life forms are based on the chemical properties and interactions of a very limited number of chemical elements, particularly carbon, and that those elements can only combine in specific ways that are strictly governed by their electronic properties, then it is entirely logical and reasonable to assume that where ever in the universe those elements exist in the proper conditions for the necessary length of time, life forms—ζωοι—have come into existence.

Elsewhere in the universe, the χηνοζωοι (*xeno-zo-oy*; any non-terrestrial alien living thing of an animal nature) would also develop, reproduce, and evolve in accord with their environments. There would not, however, be zoology in those places. It should be understood that zoology is a strictly human activity and does not exist where there are no humans to carry it out. Zoology, and the basic concept of zoology, means absolutely nothing to any of the non-human living things, the ζωοι that are the subject of the science of zoology. Zoology, as the study of the kingdom *Animalia*, has only one real purpose, which is for humans to understand the nature of animal life and the place of humans within that context.

—*Richard M. Renneboog, MSc*

A

Adaptive Radiation

FIELDS OF STUDY

Biology; Genetics; Environmental Sciences; Ecology; Botany

ABSTRACT

Many different forms of evolutionary adaptations may occur among animals that started with a common ancestor. In this way, evolutionary divergences can take place, and the occupation of a variety of ecological niches is made possible. The ability to adapt is not shared by all species. Therefore, in many instances, either evolutionary divergence has been modest or the species involved has become extinct.

KEY CONCEPTS

allele: an alternative form of a gene that is located at the same position on a chromosome

evolution: the process of change through time by which the characteristics of a species of plant or animal are altered by adaptation to the environment and produce a new species distinct from the parent species

fossil: any recognizable remains of an organism preserved in the earth's crust; it may be a footprint, bones, or even feces

gene: the biological unit of heredity, which is composed of DNA and is located on a chromosome

genotype: the total genetic composition of an organism

habitat: the place where an organism normally lives or where individuals of a population live

natural selection: the process of evolution whereby organisms that are the best adapted are the most successful in reproducing and therefore in passing along their genotypes to successive generations

niche: the role of an organism in an ecological community—its unique way of life and its relationship to other biotic and abiotic factors

phenotype: the visible expression of the genetic makeup of an individual

species: a taxonomic subdivision of a genus, containing populations of similar organisms that interbreed and that usually do not interbreed with other species

ADAPTATION

In 1898, Henry F. Osborn developed the concept of adaptive radiation. According to Osborn, many different forms of evolutionary adaptations may occur among animals that started with a common ancestor. In this way, evolutionary divergences can take place, and the occupation of a variety of ecological niches is made possible according to the adaptive nature of the invading species. As may be seen with certain forms of animal life, however, the ability to adapt is not shared by all species. Therefore, in many instances, either evolutionary divergence has been modest or the species involved has become extinct.

THE PRINCIPLES OF NATURAL SELECTION

In order to understand how adaptive radiation operates, it is necessary to become familiar with the principles of natural selection. The concept of natural selection, frequently expressed as "survival of the fittest," is at the core of Charles Darwin's theory of evolution. Darwin did not mean to suggest that there was a physical struggle among organisms in order to survive. Instead, he meant that organisms compete for food, space, shelter, water, and other things necessary for existence. Rather, he meant only that those organisms best adapted for a particular habitat will have the greatest ability to survive in a particular environment. According to the concept of natural selection, all organisms of a given species will show variation in color, size, physiology, and many other characteristics; in nature, all organisms produce more offspring than can survive, so the offspring must therefore compete for the limited environmental resources. Organisms that are the best adapted (most fit) to compete are most likely to live to reproduce and pass their successful traits on to their offspring. The others, which are less fit, will be most likely to die without reproducing. When different parts of an animal population are faced with slightly different environments, they will diverge from one another and in time will become different enough to form new species, typically unable to interbreed. Natural selection also has the effect of producing different patterns of evolution. It may bring about widely different phenotypes (variable characteristics) in closely related animals, for example, or similar phenotypes in distantly related organisms. The organisms themselves may also become forces of selection through their interrelationships with other species.

The process of adaptive radiation illustrates how natural selection operates. The most frequently cited example is the evolution of Darwin's finches on the Galápagos Islands, off the west coast of South America. The islands were formed from volcanic lava about one million years ago. At first they were devoid of life, but bit by bit, several species of plants and animals arrived at them from the South American mainland. Since the nearest island is about 950 kilometers from the coast of Ecuador, it is anybody's guess how the different species arrived. It has been suggested that the birds may have been carried to the islands by strong winds, since finches are not known for their lengthy flights. Other organisms may have been carried by floating debris. In any event, the islands became populated. The mainland ancestor of the finches is not known, but it was no doubt a nonspecialized finch (a finch is about the size of a sparrow). Since there were no other birds with which to compete on the islands, the original population of finches began to adapt to the various unoccupied niches. The early offshoots of the original population were modified again and again as adaptations continued. This process resulted in the evolution of fourteen species of finches. The main feature that makes each species different is the size of their beaks, which have adapted for the various types of available foods. Today the finches live on fifteen different islands. Some of the species are found in the same area (sympatric), while others occur in different areas (allopatric). The most noteworthy example of an adaptation to a particular niche

Darwin's finches or Galapagos finches. Darwin, 1845. Journal of researches into the natural history and geology of the countries visited during the voyage of H.M.S. Beagle round the world, under the Command of Capt. Fitz Roy, R.N. 2d edition.

is the woodpecker finch. A true woodpecker has an extremely long tongue that it uses to probe for insects. Since the woodpecker finch does not have a long tongue, it has learned to use a cactus spine for insect probing, and it can therefore occupy a niche normally filled by true woodpeckers.

A more recent example of adaptive radiation in its early stages has taken place in an original population of brown bears. The brown bear can be found throughout the Northern Hemisphere, ranging from the deciduous forests up into the tundra. During one of the glacier periods, a small population of the brown bear was separated from the main group; according to fossil evidence, this small population, under selection pressure from the Arctic environment, evolved into the polar bear. Although brown bears are classified as carnivores, their diets are mostly vegetarian, with occasional fish and small animals eaten as supplements. On the other hand, the polar bear is mostly carnivorous. Besides its white coat, the polar bear is different from the brown bear in many ways, including its streamlined head and shoulders and the stiff bristles that cover the soles of its feet, which provide traction and insulation, enabling it to walk on ice.

EVOLUTION

All the genes of any population of living organisms at any given time make up its gene pool, and the ratio of alternative characteristics (alleles) in the gene pool can change because of selection pressures during the passage of time. As the ratio of alleles changes, evolution occurs. Evolution may be a random change, or it may occur because of the directive influences of natural selection. In the former case, occasional and unpredictable permanent random changes called mutations take place in the DNA molecules that compose the genes. These mutations also may be selected for by the environment or selected against by the environment. It is simply an accident if the newly mutated genes help the organism to become better adapted to its particular habitat niche. Genes may not change or become mutated through several generations (the Hardy-Weinberg law), but may change in terms of survival value if the environment changes or the species population is subjected to new mutations or natural selection.

The relative numbers of one form of allele decrease in a divergent population, while the relative numbers of a different gene increase. This progressive change is all-important in the evolutionary process that takes place between the origin of a new gene by random mutation and the replacement of the original form of the gene by descendants having the newer, better-adapted form of the gene. The result in the long term is that enough of the DNA changes, either slowly or rapidly, through divergent populations or organisms, that the new generations have become so different from the original population that they are considered new species. Many times in earth's history, a single parental population has given rise not to one or two new species but to an entire family of species. The rapid multiplication of related species, each with its unique specializations that fit it for a particular ecological niche, is called adaptive radiation, or divergent evolution.

STUDYING ADAPTIVE RADIATION

Not all scientific information is gained by experimentation: A considerable portion of science is descriptive and is based upon observation. In determining that adaptive radiation has occurred and is indeed taking place among living species, much supporting evidence has come from the study of fossils and from observations of the structural, physiological, and behavioral adaptations of current animals. Clearly, wide-scale experimentation would be out of the question. No matter how well an experiment may be designed to test the concept of adaptive radiation, the scientist could not be around thousands or millions of years from now to gather the data. Therefore, scientific observation of animal remains is the best method.

Based upon scientific observations, it has been well established that the phenomenon known as adaptive radiation is a general feature of the evolution of most organisms. Studies of the morphological features of fossilized remains help determine relationships among prehistoric animals and enable the scientist to trace adaptive radiations from a more primitive ancestral stock. In order to establish time intervals, techniques such as radioactive carbon dating, potassium-argon dating, and fluorine dating have been used.

Zoologists have also made use of the uneven distribution of blood groups (A, B, AB, and O) among different groups of animals. As more blood subgroups were discovered, they became useful in helping chart migrations and indicating relationships between species.

THE EVIDENCE AND ITS IMPLICATIONS

Adaptive radiation as an important aspect of evolution means that modern organisms have attained their diversity in form and behavior through hereditary modifications after having been separated from ancestral populations. Adaptive radiation, therefore, is attributable to the genetic changes in isolated groups of organisms or, more specifically, to a change in the relative frequency of their genes from one generation to the next that eventually results in the formation of new species.

Evidence in several areas supports the concept of adaptive radiation as an important aspect of evolution: the fossil record (the most direct evidence), biogeographic distribution of organisms, comparative anatomy and embryology, homologous and analogous structures, vestigial organs, and comparative biochemistry. Regarding comparative biochemistry, scientists agree that blood group similarities confirm evolutionary relationships among the nonhuman primates. It has been shown that the blood of higher primates, such as orangutans and chimpanzees, is closer to human blood than to that of the more primitive monkeys.

—*Jon P. Shoemaker*

FURTHER READING

Curtis, Helena, and N. Sue Barnes. *Invitation to Biology*. 5th ed. Worth, 1994.

Darwin, Charles R. *The Illustrated Origin of Species*. Abridged and introduced by Richard E. Leakey. Hill & Wang, 1979.

Dobzhansky, Theodosius. *Genetics of the Evolutionary Process*. Columbia University Press, 1970.

Farrell, Brian D., and Charles Mitter. "Adaptive Radiation in Insects and Plants: Time and Opportunity." *American Zoologist* 34, no. 1 (February 1994): 57–70.

Givnish, Thomas J., and Kenneth J. Sytsma, eds. *Molecular Evolution and Adaptive Radiation*. Cambridge University Press, 1997.

Hickman, Cleveland P., Larry S. Roberts, and Frances M. Hickman. *Integrated Principles of Zoology*. 11th ed. McGraw-Hill, 2001.

Hou, Lianhai, Larry D. Martin, Zhonghe Zhou, and Alan Feduccia. "Early Adaptive Radiation of Birds: Evidence from Fossils from Northeastern China." *Science* 274, no. 5290 (November 1996): 1164–1168.

Kimball, John W. *Biology*. 6th ed. Wm. C. Brown, 1994.

Rainey, Paul B., and Michael Travisano. "Adaptive Radiation in a Heterogenous Environment." *Nature* 394 (July 2, 1998): 69–72.

Wahrheit, Kenneth I., Jonathan D. Forman, Jonathan B. Losos, and Donald B. Miles. "Morphological Diversification and Adaptive Radiation: A Comparison of Two Diverse Lizard Clades." *Evolution* 53 (August 1999): 1226–1234.

Wallace, Robert A., Robert J. Ferl, and Gerald P. Sanders. *Biology: The Science of Life*. 4th ed. HarperCollins, 1996.

AGING

FIELDS OF STUDY

Medicine; Biology; Geriatrics; Thanatology

ABSTRACT

When change is reversible or self-maintaining, the effects of aging are often not observable. However, in animals some change is not reversible. The changes in the cells of the body accumulate over time and result in a steady downward trend. The end point of this trend is the death of the organism. Aging occurs within body systems as a result of unseen changes at the molecular and cellular levels. Although the mechanisms through which aging occurs may be understood, the causes are less clear.

KEY CONCEPTS

aging: a process common to all living organisms, eventually resulting in death or conclusion of the life cycle
cognition: ability to perceive or understand
death: the cessation of all body and brain functions
function: ability, capacity, performance
life span: length of life from birth to death
longevity: length of life

CHANGE AND AGING

Progressive and irreversible change has been called the single common property of all aging systems. When change is reversible or self-maintaining, such as one would see in a forest, for example, the effects of aging are often not observable. Growth of the forest is evident, but with the right conditions, trees within the forest may grow for hundreds of years in the absence of disease. Certain conditions of the forest system help to regenerate, renew, and reverse changes that happen within that system.

However, in animals some change is not reversible. The changes in the cells of the body accumulate over time and result in a steady downward trend. The end point of this trend is the death of the organism. Aging is a normal part of the life cycle. This is known to be true because aging changes within populations are rather predictable. The changes associated with aging that are seen in all animal species may occur for similar reasons. These may include chemical aging, extracellular aging, intracellular aging, and aging of cells.

Aging occurs within body systems as a result of unseen changes at the molecular and cellular levels. Although the mechanisms through which aging

Before hatching Newly hatched larvæ hanging on to water-weed With external gills External gills are covered over and are absorbed Limbless larva about a month old with internal gills Tadpole with hind-legs, about two months old With the fore-limbs emergin.

occurs may be understood, the causes are less clear. The fact remains that due to changes in chemical balances such as those of hormones, and to the dying of cells within the body, each of the bodily systems shows deterioration over time.

Changes that occur in domestic animals over the life span can be similar to those that occur in humans. Dogs experience the graying of their hair, a decrease in vision, and a slowing of movement with age. They also experience cataract formation, arthritis, skin problems, cancer, and diabetes. Certain breeds of animals may demonstrate a tendency toward specific illnesses or diseases. For example, German shepherds often develop hip problems, and collies commonly develop progressive arthritis that may seriously inhibit mobility by around ten years of age.

COMMON EFFECTS OF AGING

There are many variations in the effects of aging among the species of animals. The life span of animals may range from a few days (among insects) to thirty years or more, with great variation depending upon many factors. Animals that live in captivity, as pets or in zoos where they are sheltered from the effects of predation, disease, and adverse climate, also tend to live significantly longer than animals in the wild.

Very little research has been done on the aging of most animal species. The reasons for this include the difficulty of observing animals over a long period of time in their natural habitat. Aging in monkeys has been studied more than that in other animals because of the notion that aging patterns may closely reflect those of humans.

Aging monkeys show changes in their circulatory systems similar to those found in humans: There is notable atherosclerosis and arteriosclerosis, or hardening of the arteries. The heart pumps less effectively, and vessels show buildup of plaque. These changes often result in cardiac problems, including heart attacks. The respiratory system also shows a decrease in elasticity. Senile emphysema has been noted. The kidneys show signs of atrophy and sclerosis in aged monkeys. The kidneys of humans may lose up to half of the functioning nephrons with advanced age and thus become less effective in filtering waste products from the body.

Physical function or capacity tends to decline with age. This is largely due to the atrophy of muscles, which is more common as the body gets older. The joints tend to become stiffer and less mobile. Range of motion may be restricted. Changes in bone density may lead to loss of teeth, osteoporosis, and subsequent fractures. Tooth loss and osteoporosis have been documented in monkeys over the age of twenty years. Pictures of such older monkeys reveal a stooped posture, with shoulders hunched forward, similar to the kyphosis observed in many older human women.

Physical function among animals has been less studied than that in humans, but certain physiological characteristics are similar. For example, survival times after severe physical injury with blood loss and trauma decreases in both humans and animals as age increases. Male monkeys do not lose reproductive capabilities until toward the end of the life span, while females have a more restricted period of time to bear offspring. Fertility among all females tends to decline with age after its peak.

The immune system functions less effectively as age increases. This leaves the body more susceptible to a range of illnesses and diseases. Neoplasms, or tumors, are most common among mammals as they age. An impaired immune system allows various types of tumors or cancers to spread more rapidly in the older body. Response to stress and ability to adapt to stressors also decline with age. For example, older mice become less able to adapt to cold temperatures.

Social roles and behaviors among animals may also change with age. Longitudinal studies on animals in the wild are scarce, so only generalities may be speculated upon. Even studies done within controlled laboratory settings yield only broad suggestions, since numbers of animals available for study are limited. Males generally tend to dominate the females in both physical strength and social ranks. Some nonhuman primates show different characteristics with advanced age. That is, some monkeys and baboons allow older males to remain part of the social group, while other species support the male leader in the group only as long as the female harem supports him, whether younger or older. Individual monkeys in stable groups have been observed to resort less frequently to aggressive behavior to maintain their status within the group.

CAUSES OF DEATH

Among nonhuman primates, the leading cause of spontaneous death is digestive problems. Older animals that die do not always show advanced signs of tissue aging. Since much less research has been done on aging among animals than among humans, data about causes of death are rare. However, it appears that there is an increased probability of dying from trivial illnesses, perhaps due to decreased resistance factors, as animals age.

Predator-prey relationships among animals are particularly significant as causes of death. Thus, the effect of the environment on animal aging and death requires more investigation. Do animals age more quickly if they are objects of prey? Do animals relate to stress in ways similar to those of people, thus showing signs of wear and tear that are seen with premature aging under stress? Are there risk factors among animals that affect their life span? These are some of the questions that remain to be answered on the topic of aging among animals.

—*Kristen L. Mauk*

FURTHER READING

Bowden, Douglas M. *Aging in Nonhuman Primates.* Van Nostrand Reinhold, 1979.
Kohn, Robert R. *Principles of Mammalian Aging.* Prentice-Hall, 1971.
Schmidt-Nielsen, Knut. *Animal Physiology: Adaptation and Environment.* Cambridge University Press, 1997.
Slater, P. J. B. *Essentials of Animal Behavior.* Cambridge University Press, 1999.
Slobodkin, Lawrence B. *Growth and Regulation of Animal Populations.* Holt, Rinehart and Winston, 1961.

AMPHIBIANS

FIELDS OF STUDY

Biology; Taxonomy; Ecology; Environmental Studies

ABSTRACT

The term amphibian *is derived from the Greek word* amphibios, *which means "to live two lives." The majority of amphibians spend the first part of their lives as aquatic, gill-breathing larvae and then transform into terrestrial adults. The larval stage can be as short as a few weeks or as long as several years. Completion of the larval stage is triggered by hormonal events that initiate some dramatic developmental processes, collectively termed metamorphosis.*

KEY CONCEPTS

adaptive radiation: rapid speciation that occurs as the result of a particular group being able to exploit a new resource

convergence: a phenomenon in which two forms that are not closely related evolve structures that appear similar

disjunct distribution: a geographic distribution pattern in which two closely related groups are separated by large areas that are devoid of either group

metamorphosis: the complex developmental process of morphological change in which larval amphibians are transformed into adults

neoteny: the retention of larval features by adults, a process that has played a major role in the evolution of amphibians

phylogeny: the determination of the evolutionary history of a particular group of organisms

AMPHIBIAN LIFE

The term *amphibian* is derived from the Greek word *amphibios,* which means "to live two lives." The majority of amphibians spend the first part of their lives as aquatic, gill-breathing larvae and then transform into terrestrial adults. The larval stage can be as short as a few weeks or as long as several years. Completion of the larval stage is triggered by hormonal events that initiate some dramatic developmental processes, collectively termed metamorphosis.

As adults, most amphibians seek out aquatic environments in which to deposit their eggs. These can range from fast-flowing mountain streams to ephemeral roadside ditches. Most male frogs have species-specific mating calls that serve both to attract females and to prevent interbreeding. Frogs reproduce by external fertilization: the male typically grasps the female and encourages her to deposit her eggs, which he promptly fertilizes. Normally, both parents abandon the eggs, but some variations of this pattern exist. In contrast, most salamanders practice internal fertilization, accomplished after the male has performed a type of species-specific courtship dance that culminates with the deposition of a packet of sperm cells, called a spermatophore. The female squats on the spermatophore, transferring the spermatozoa to a specialized holding structure called a spermatheca. The spermatozoa can be used to fertilize her eggs up to several months after mating.

EVOLUTION OF AMPHIBIANS

Amphibians were the first vertebrates to possess adaptations that allowed them to spend considerable periods of time out of the water. The earliest fossil amphibians, members of the order Ichthyostegalia, appear in the geologic record during the Devonian period, about 320 million years ago. Among experts, there is a general consensus that the ancestor of the amphibians is to be found in the primitive, lobe-finned fish (class Osteichthyes, subclass Sarcopterygii). This conclusion is based on a detailed analysis of the comparative anatomy of hard body parts that fossilized, such as the vertebrae, shoulder girdles, teeth, and skulls. Characteristics that the first amphibians shared with their fish ancestors include internal nasal openings (nares); a strange, hinged skull; and a distinctive tooth structure, in which the enamel is folded into intricate patterns.

The environmental conditions that led to the abandonment of aquatic habitats in favor of a more terrestrial existence remain a major topic of discussion.

One scenario envisions the early amphibian ancestors in an environment that was gradually becoming more and more arid. To survive, it would have been advantageous to be able to crawl on land for short distances to escape drying pools in favor of more permanent bodies of water. Those that could migrate would have survived in higher numbers than those that lacked this adaptation. A second scenario suggests that heavy predation pressures from the jawed, carnivorous fish that are known to have been abundant in the shallow freshwater lakes of the time may have selected for individuals that could leave the water, even if only briefly. A third scenario depicts competition for food as strong in the aquatic environments and much weaker on land, where several groups of invertebrates were known to be abundant.

TAXONOMY OF FOSSIL AMPHIBIANS

The taxonomy of fossil amphibians is confusing because of a series of problems. First, on the whole, the range of morphological variations within the group has been conservative. Striking differences of taxonomic value (which would have to fossilize) are not numerous. Second, many of the skeletal elements that are important in determining the relationships of living amphibians are composed of cartilage rather than bone, and this material does not fossilize well. Because of this limitation, bony elements such as the vertebrae and skull have played a major role in determining amphibian phylogeny. The molecular techniques that have greatly assisted modern taxonomists, such as electrophoresis, immunology, karyotyping, and deoxyribonucleic acid (DNA) sequencing, cannot be used on fossilized materials.

The class Amphibia is further divided into orders. The number of orders recognized by various authorities ranges from eight to thirteen. All but three of these are extinct. One extinct order is Ichthyostegalia, which includes the earliest recorded fossil amphibians. Most were small, with elongate bodies and weakly developed limbs. Many were almost assuredly aquatic, but at least some were capable of spending extended periods of time out of the water. By the time of the Devonian period, when they first appear in the fossil record, they were already diverse, with several different genera and species present.

Another extinct group is the order Temnospondyli, which was abundant during the late Permian period and persisted until the end of the Triassic period. Most in this order were of moderate size (0.5 to 1.0 meter in body length), with low, stout profiles and flattened skulls. Some were highly aquatic and had short, weak appendages. One group, the clade Trematosauria, was marine, and they are apparently the only amphibians to have been successful at invading the oceans.

The order Anthracosauria appears in the fossil record during the Carboniferous period and was extinct by the end of the Permian. Common names of members of this group include the seymouriamorphs and the embolomeres. This order contains a mixture of terrestrial and aquatic amphibians. From an evolutionary standpoint, Anthracosauria is important because it gave rise to the ancestors of the reptiles.

Members of the order Aistopoda were eel-like, aquatic amphibians with elongate, limbless bodies. They are characterized by a large number of vertebrae—more than one hundred—that are clearly divisible into cervical, trunk, and caudal regions.

The order Nectridea consists of fully aquatic, salamander-like amphibians that persisted during the Carboniferous period. Appendages were weak or absent, and most fossils indicate body forms that were flattened dorsiventrally. They probably persisted by slowly crawling about the bottoms of ponds and lakes, where they preyed on unwitting animals that crossed their paths.

Amphibians of the order Microsauria were a diverse group of elongate, weak-limbed amphibians. Fossils of this group are fairly abundant in habitats that were swamplike during the Carboniferous period. The remarkable physical similarities between some microsaurs and some of the earliest reptiles are considered by most authorities to be the result of convergence of body form rather than an indication of a true evolutionary relationship.

The order Proanura consists of a single froglike fossil that dates from the Triassic period on the island of Madagascar. The skull is distinctly froglike, but a tail is present, and the hind limbs have not been modified for jumping.

The taxonomy of the living members of the class Amphibia is still under considerable debate. A central question revolves around whether all currently living amphibians are of monophyletic origin, meaning they derive from a single common ancestor, or whether they are of polyphyletic origin, meaning they arose independently from two or more separate stocks of fish. Based on the presence of unique features, such as pedicellate teeth, a distinctive part of the inner ear, and specialized eye receptors, most experts have concluded that all living amphibians are monophyletic and can be grouped in the superorder Lissamphibia. As recognized, living amphibians are placed in the orders Caudata, Anura, and Gymnophiona.

SALAMANDERS

Salamanders are grouped in the order Caudata. They are distributed over temperate parts of Europe, Asia, and North and South America. There are approximately 550 recognized species, grouped into eight different families. Eastern North America has the greatest overall diversity, with seven of the eight described families represented there. One family, Plethodontidae, invaded South America and underwent a period of such tremendous speciation that approximately 60 percent of the living species of salamanders are members of this family.

Almost everyone is quick to recognize a frog based on its appearance, but the same cannot be said for salamanders. The families Proteidae, Cryptobranchidae, Sirenidae, and Amphiumidae are entirely aquatic. The family Cryptobranchidae is disjunctly distributed and occurs today only in eastern Asia and eastern North America. It contains the genus *Andrias*, which includes the largest living salamanders; one species, *Andrias davidianus*, has been known to reach lengths as great as 1.8 meters. Members of the family Proteidae are today isolated in Europe and eastern North America. They are commonly called "water dogs" and are frequently dissected in comparative anatomy classes. Members of the families Sirenidae and Amphiumidae are restricted to the southeastern United States, where they are called sirens and amphiumas, respectively. Sirens have external gills and two front legs, while amphiumas have minute front and back legs and lack external gills.

The family Ambystomatidae is entirely North American in its distribution. Most species are highly secretive and are only encountered under objects or intercepted as they migrate to breeding ponds during spring rains.

The family Hynobildae is exclusively Asian in its present distribution. Reproduction in this family is considered to be primitive, in that the female lays eggs that are enclosed in a loose sac and are subsequently fertilized externally by the male.

The family Salamandridae is widely distributed in Europe, North Africa, Asia, and North America, with the greatest diversity in the Eastern Hemisphere. Many species have developed highly toxic skin secretions to protect themselves from predators; human fatalities have been recorded from eating only one salamander. Many species advertise their toxicity by being very brightly and distinctly colored, a phenomenon known as aposematic coloration, while members of other, less toxic families also display these color patterns for protection from predators, a process called mimicry.

Members of the family Plethodontidae all share the unique feature of being lungless; respiration is accomplished by diffusion across their moist skins. Many species have abandoned laying eggs in water in favor of damp, terrestrial nests. Females guard the eggs until they hatch as miniatures of the adult, having completed their abbreviated metamorphosis while still in the egg.

CAECILIANS

The order Gymnophiona consists of a highly specialized group of wormlike, limbless amphibians called caecilians. They inhabit tropical regions of North and South America, Asia, and Africa. Most are terrestrial burrowers and are rarely observed. Some primitive forms have dermal scales embedded in their skin. All caecilians possess a unique sensory organ called a tentacle. Fertilization is internal, and male caecilians possess a copulatory organ that is derived from the cloaca. In the majority of caecilian species, the females retain their eggs in the oviduct and give birth to fully developed young. Fossil caecilians are extremely rare.

FROGS

The order Anura is composed of tailless amphibians called frogs. Their hind legs are typically modified for jumping, their presacral vertebrae are usually eight in number, and their postsacral vertebrae are fused to form a coccyx. Frogs occur on all continents except Antarctica but reach their greatest diversity in the tropics of South America, Africa, and Asia. They have successfully invaded deserts, rivers, cold mountain streams, and arboreal vegetation. Several families have undergone tremendous adaptive radiation in the tropics, so that today almost 80 percent of living amphibians are anurans.

Families that are widely distributed include Bufonidae (toads), Hylidae (tree frogs), Microhylidae (mostly small frogs), and Ranidae (true frogs). These families are almost worldwide in their geographic distributions, although Australia lacks Bufonidae, Ranidae, and Microhylidae and Africa lacks Hylidae. Toads often have dry, warty skins containing numerous glands that produce noxious, protective secretions. Tree frogs have expanded disks on the tips of their toes that have allowed them to occupy arboreal habitats unavailable to many other families.

The families Leptodactylidae, Brachycephalidae, Rhinodermatidae, Pseudidae, Centrolenidae, and Dendrobatidae reach their greatest abundance in Central and South America. The leptodactylids are a diverse assemblage of nearly seven hundred species. Many of these lay eggs in specially constructed foam nests; other species have taken this a step further and deposit their eggs in damp, terrestrial locations, thereby avoiding aquatic predators almost completely. The dendrobatids are often small, brightly colored frogs that have been given the common name of poison arrow frogs because of their extremely toxic skin secretions, which have been used by some indigenous tribes to poison the tips of hunting arrows. Members of the family Pseudidae are unique in producing very large tadpoles that metamorphose into rather small frogs. The rhinodermatids consist of only two species, but one is unique in that the larvae do not feed and are carried in the mouth of the adult until they complete metamorphosis.

The Discoglossidae are found mostly in Europe, and Pelodytidae are native to southwestern Europe and western Asia. The midwife toad (*Alytes obstetricians*), a discoglossid, has an unusual reproductive mode: after fertilizing the eggs, the male cements them on his back and carries them to and from the water until they are ready to hatch.

The Rhacophoridae are a moderate-sized family of about 180 species, distributed over southern Africa and southeast Asia. Most members have expanded terminal digits, and some even have extensive webbing between their toes, which allows them to glide between arboreal perches. Diverse reproductive tactics occur in this family. Several species use water-filled tree holes in which to deposit their eggs.

The family Myobatrachidae is a diverse group of about ninety-nine species that occur in Australia and New Zealand. One species, *Rheobatrachus silus*, has a unique reproductive mode that includes brooding eggs in the stomach of the female.

The family Leiopelmatidae is a small group of frogs that are disjunctly distributed in western North America and New Zealand. The tailed frog (*Ascaphus truei*) is the only frog to possess an intromittent organ that is used to transfer sperm to the female for internal fertilization. This organ apparently evolved in response to the swift, cold streams in which the frog lives.

BASES OF CHARACTERIZATION

As a group, amphibians are easier to characterize by the morphological features that they lack than by the unique characteristics that they possess. Missing are the scales that cover fish and reptiles (although these are not closely related structures), as well as the hair and feathers associated with mammals and birds. Amphibians' skin is relatively thin and contains numerous glands. Large amounts of water can be lost or gained via the epidermis. In many forms, the skin serves as a major organ for respiration. Amphibians are ectothermic, which means that they do not have internal physiological mechanisms for maintaining a constant body temperature. The circulatory system is closed, and the heart is composed of three chambers—two atria and one ventricle.

The taxonomic relationships of salamanders are based primarily on the arrangement of bones of the skull, which of these bones possess teeth, and the shape of the centrum of the vertebrae. Living forms are further compared by the manner of reproduction. In general, salamanders have been relatively conservative, and characteristics such as the number of chromosomes have not proved especially useful in determining phylogenetic relationships. However, modern molecular techniques, such as electrophoresis, immunology, and the use of restriction enzymes, are adding to the understanding of selected groups. With these techniques, it has been possible to show that several forms that were indistinguishable based on morphological data are in fact genetically isolated from one another and are actually distinct sibling species.

The taxonomic relationships of frogs are also based largely on differences in bony anatomy. Skull morphology, the shape of the vertebral centrum and its manner of development, and the arrangement of the bones that make up the pectoral girdle are important diagnostic characters. In living forms, molecular techniques are also shedding new light on relationships. The number of chromosomes is more variable and has more value as a diagnostic tool in frogs than in salamanders. The morphology of the larvae is another important taxonomic tool.

THE IMPORTANCE OF AMPHIBIANS

There are about seven thousand recognized species of living amphibians. This number represents only a small fraction of the number of species that have been present on the earth over the past 350 million years. In many habitats, amphibians still represent a major portion of the biomass, and because ecologists often relate a group's "worth" to its biomass, amphibians can be considered major members of many terrestrial communities, often serving as keystone species for their ecological niches. Amphibians often exhibit traits such as low mobility, fidelity to breeding sites, and species-specific behaviors that are sought by ecologists and animal behaviorists for their studies. As a whole, salamanders and frogs represent some of the most thoroughly studied vertebrates.

Areas to which amphibians have contributed a significant portion of current knowledge include the evolution of mating systems, sexual selection, reproductive isolation mechanisms, niche partitioning, and community structure. Embryologists have long used amphibians to gain a basic understanding of complex development processes.

One area of extreme interest is the apparent decline of many populations of frogs and salamanders over large geographic areas, notably in western North America. Loss of breeding sites through habitat modification, acid rain, and competition from exotic species have all contributed to their demise. Another likely cause is chytridiomycosis, an infectious and deadly skin disease caused by the chytrid fungus *Batrachochytrium dendrobatidis*. The disease was first discovered in frogs in Queensland, Australia, in 1993 and was first described by Lee Berger et al. in 1998. The specific fungus was identified by Joyce E. Longcore, Allan P. Pessier, and Donald K. Nichols in 1999.

—*Robert E. Herrington*

Further Reading

Ballinger, Royce E., and John D. Lynch. *How to Know the Amphibians and Reptiles*. Brown, 1983.

Berger, Lee, et al. "Chytridiomycosis Causes Amphibian Mortality Associated with Population Declines in the Rain Forests of Australia and Central America." *Proceedings of the National Academy of Sciences* 95.15 (1998): 9031–36.

Conant, Roger, and Joseph T. Collins. *A Field Guide to Reptiles and Amphibians of Eastern and Central North America*. 3rd ed. Houghton, 1998.

Duellman, William E. *Amphibian Species of the World: Additions and Corrections*. University of Kansas, 1993.

Duellman, William E., and Linda Trueb. *Biology of Amphibians*. 2nd ed. John Hopkins University Press, 1994.

Frost, Darrel R. *Amphibian Species of the World: A Taxonomic and Geographical Reference*. Allen, 1985.

Green, David M., and Stanley K. Sessions, eds. *Amphibian Cytogenetics and Evolution*. Academic, 1991.

Heatwole, Harold, ed. *Amphibian Biology*. Vols. 1–10. Surrey, 1994–2010.

Heatwole, Harold, and John W. Wilkinson, eds. *Amphibian Biology*. Vol. 11 pt. 3. Exeter: Pelagic, 2013.

Longcore, Joyce E., Allan P. Pessier, and Donald K. Nichols. "*Batrachochytrium dendrobatidis* gen. et sp. nov., a Chytrid Pathogenic to Amphibians." *Mycologia* 91.2 (1999): 219–27. Print.

Pough, F. Harvey, et al. *Herpetology*. 3rd ed. Prentice, 2004.

Vitt, Laurie J., and Janalee P. Caldwell. *Herpetology: An Introductory Biology of Amphibians and Reptiles*. 4th ed. Academic, 2014.

Voyles, Jamie, Erica B. Rosenblum, and Lee Berger. "Interactions between *Batrachochytrium dendrobatidis* and Its Amphibian Hosts: A Review of Pathogenesis and Immunity." *Microbes and Infection* 13.1 (2011): 25–32.

Animal Adaptations

Fields of Study

Biology; Biochemistry; Genetics; Epigenetics; Ecology; Environmental Studies

Abstract

Many of the features that are most interesting and beautiful in biology are adaptations. Adaptations are the result of long evolutionary processes in which succeeding generations of organisms become better able to live in their environments. Specialized structures, physiological processes, and behaviors are all adaptations when they allow organisms to cope successfully with the special features of their environments. Adaptations ensure that individuals in populations will reproduce and leave well-adapted offspring, thus ensuring the survival of the species.

Key Concepts

coevolution: joint evolutionary change caused by the close interaction of two or more species; each species serves as the natural selection agent for the other(s)

competition: striving for a limited resource

evolution: a process, guided by natural selection, that changes a population's genetic composition and results in adaptations

fitness: the ability of an organism to produce offspring that, in turn, can reproduce successfully; the fitness of organisms increases as a result of natural selection

natural selection: the elimination of individuals with hereditary characteristics that hinder the organism's ability to survive and reproduce, and the preservation of those with traits beneficial to survival

population: a group of individuals of the same species in a particular location

species: a group of organisms that can successfully interbreed to produce living, successfully reproducing offspring

Adaptations

Many of the features that are most interesting and beautiful in biology are adaptations. Adaptations are the result of long evolutionary processes in which succeeding generations of organisms become better able to live in their environments. Specialized structures, physiological processes, and behaviors are all adaptations when they allow organisms to cope successfully with the special features of their environments. Adaptations ensure that individuals in populations will reproduce and leave well-adapted offspring, thus ensuring the survival of the species.

MUTATION AND NATURAL SELECTION

Adaptations arise through mutations—inheritable changes in an organism's genetic material. These rare events are usually harmful, but occasionally they give specific survival advantages to the mutated organism and its offspring. When certain individuals in a population possess advantageous mutations, they are better able to cope with their specific environmental conditions and, as a result, will contribute more offspring to future generations compared with those individuals in the population that lack the mutation. Over time, the number of individuals that have the advantageous mutation will increase in the population at the expense of those that do not have it. Individuals with an advantageous mutation are said to have a higher "fitness" than those without it, because they tend to have comparatively higher survival and reproductive rates. This is natural selection.

Over very long periods of time, evolution by natural selection results in increasingly better adaptations to environmental circumstances. Natural selection is the primary mechanism of evolutionary change, and it is the force that either favors or selects against mutations. Although natural selection acts on individuals, a population gradually changes as those with adaptations become better represented in the total population. Predaceous fish, for example, which rely on speed to pursue and overtake prey, would benefit from specific adaptations that would increase their swimming speed. Therefore, mutations causing a more sleek and hydrodynamically efficient form would be beneficial to the fish predator. Such changes would be adaptations if they resulted in improved predation success, improved diet, and subsequently greater reproductive success, compared with slower members of the population. Natural selection would favor the mutations because they confer specific survival advantages to those that carry the mutations and impose limitations on those lacking these advantages. Thus, those individuals with special adaptations

The peacock's train is an adaptation related to "mate choice" that selects the more fit over the less fit. (BS Thurner Hof)

for speed would have a competitive advantage over individuals that can only swim more slowly. These attributes would be passed to their more numerous offspring and, in evolutionary time, speed and hydrodynamic efficiency would increase in the population.

ENVIRONMENT AND SURVIVAL

Although natural selection serves as the instrument of change in shaping organisms to very specific environmental features, highly specific adaptations may ultimately be a disadvantage. Adaptations that are specialized may not allow sufficient flexibility (generalization) for survival in changing environmental conditions. The degree of adaptive specialization is ultimately controlled by the nature of the environment. Environments, such as the tropics, that have predictable, uniform climates and have had long, uninterrupted periods of climatic stability are biologically complex and have high species diversity. Scientists generally believe that this diversity results, in part, from complex competition for resources and from intense predator-prey interactions. Because of these factors, many narrowly specialized adaptations have evolved when environmental stability and predictability prevail. By contrast, harsh physical environments with unpredictable or erratic climates seem to favor organisms with general adaptations, or adaptations that allow flexibility. Regardless of the environment type, organisms with both general and specific adaptations exist because both types of adaptation enhance the probability of survival under different environmental circumstances.

Structural adaptations are parts of organisms that enhance their survival ability. Camouflage, which enables organisms to hide from predators or prey; specialized mouth parts that allow organisms to feed on specific food sources; forms of appendages, such as legs, fins, or webbed toes, that allow efficient movement; protective spines that make it difficult for the organism to be eaten—these are all structural adaptations. These adaptations enhance survival because they assist individuals in dealing with the rigors of the physical environment, obtaining nourishment, competing with others, hiding, or confusing predators.

PHYSIOLOGICAL VERSUS BEHAVIORAL ADAPTATION

Metabolism is the sum of all chemical reactions taking place in an organism, whereas physiology consists of the processes involved in an organism carrying out its function. Physiological adaptations are changes in the metabolism or physiology of organisms, giving them specific advantages for a given set of environmental circumstances. Because organisms must cope with the rigors of their physical environments, physiological adaptations for temperature regulation, water conservation, varying metabolic rate, and dormancy or hibernation allow organisms to adjust to the physical environment or respond to changing environmental conditions.

Desert environments, for example, pose a special set of problems for organisms. Hot, dry environments require physiological mechanisms that enable organisms to conserve water and resist prolonged periods of high temperature. Highly efficient kidneys and other excretory organs that assist organisms in retaining water are physiological adaptations related to the metabolisms of desert organisms. The kangaroo rat is a desert rodent extremely well adapted to its habitat. Kangaroo rats do not drink, but rather can obtain all of their water from the seeds they eat. They produce highly concentrated urine and feces with very low water content.

Adaptation to a specific temperature range is also an important physiological adaptation. Organisms cannot live in environments with temperatures beyond their range of thermal tolerance, but some organisms are adapted to warmer and others to colder environments. Metabolic response to temperature is quite variable among animals, but most animals are either homeothermic (warm-blooded) or poikilothermic (cold-blooded). Homeotherms maintain constant body temperatures at specific temperature ranges. Although a homeotherm's metabolic heat production is constant when the organism is at rest and when environmental temperature is constant, strenuous exercise produces excess heat that must be dissipated into the environment, or overheating and death will result. Physiological adaptations that enable homeotherms to rid their bodies of heat are the ability to increase blood flow to the skin's surface, sweating, and panting, all of which promote heat loss to the atmosphere.

Behavioral adaptations allow organisms to respond appropriately to various environmental stimuli. Actions taken in response to various stimuli are adaptive if they enhance survival. Migrations are behavioral adaptations because they ensure adequate food supplies or the avoidance of adverse environmental conditions. Courtship rituals that help in species recognition prior to mating, reflex and startle reactions allowing for quick retreats from danger, and social behavior that fosters specialization and cooperation for group survival are behavioral adaptations.

Because organisms must also respond and adapt to an environment filled with other organisms—including potential predators and competitors—adaptations that minimize the negative effects of biological interactions are favored by natural selection. Many times the interaction between species is so close that each species strongly influences the others in the interaction and serves as the selective force causing change. Under these circumstances, species evolve together in a process called coevolution. The adaptations resulting from coevolution have a common survival value to all the species involved in the interaction. The coevolution of flowers and their pollinators is a classic example of these tight associations and their resulting adaptations.

ADAPTATION IN THEORY AND PRACTICE

Charles Darwin and Alfred Russel Wallace, the mid-nineteenth century biologists who formulated the theory of evolution by natural selection, found much of the evidence for their theory in the adaptations they observed in nature. They reasoned that organisms with similar body forms and structures were closely related evolutionarily and had common ancestors which are now extinct. The modern study of relatedness among species and the evolutionary history of organisms is called systematics, and this discipline aids in understanding evolution and adaptations.

The methods used to study adaptations are largely the same as those used to examine the theory of evolution. Evolution, however, is a slow process, and, as a result, it is extremely difficult to test the theory. Instead, evidence must be collected from the past, and closely related organisms must be examined carefully to reconstruct how adaptations may have come into being. Scientists can then speculate on how adaptations occurred and how they helped organisms to survive.

Fossils, because they are a historical record of evolutionary change, are used by scientists to reconstruct evolutionary histories. Similar structures in different living organisms with essentially the same function are used by comparative anatomists to show how adaptations for a specific mode of life arose. The fins of some ancient fish and the limbs of mammals, for example, have strikingly similar bones that have a common origin, but the appendages have been modified for locomotion in very different environments. Adaptations are also studied in relation to biogeography, the geographical distribution of organisms. On the Galápagos Islands, fourteen species of finches now known as Darwin's finches are distributed geographically on the basis of their adaptations. Although the species that gave rise to these fourteen species is extinct, the existing species and their distributions suggest how evolution proceeded and how the adaptations came about.

A classic example of recent evolutionary change and adaptation comes from England. The peppered moth, with a mottled gray color, is well adapted to resting quietly on pale tree bark, with which it blends nicely. This adaptive coloration (camouflage) enhanced the moth's survival because the moths could remain largely undetected by predators during daylight hours. Between 1850 and 1950, however, industrialization near urban centers blackened tree trunks with soot, making the gray form disadvantageous, as it stood out on the contrasting background. During this period, the gray moths began to disappear from industrial areas, but a black-colored variant, previously rare, became increasingly common in the population. These circumstances made it possible for scientists to test whether the peppered moth's camouflage was adaptive.

In a simple experiment, moths were raised in the laboratory, and equal numbers of gray and black moths were released in both industrial and unpolluted rural areas. Sometime later, only half of the gray-colored moths could be recovered from the industrial sites, while only half of the black forms could be recovered from the rural sites, compared with the total number released. These results enabled the scientists to conclude that increased predation on the gray moths in industrial areas led to a greater fitness of the black moths, so the frequency of black moths increased in the population. The reverse was true at the rural sites. This is the first well-documented case of natural selection causing evolutionary change, and it illustrates the adaptive significance of camouflage.

The various ways of examining adaptations (by evolutionary history, comparative anatomy, and biogeography) demonstrate how adaptations are structurally and functionally important. These approaches also give scientists insight into the survival benefits of various adaptations.

THE FUNCTION OF ADAPTATION

Adaptations can be general or highly specific. General adaptations define broad groups of organisms whose general lifestyle is similar. For example, mammals are homeothermic, provide care for their young, and have many other adaptations in common. At the species level, however, adaptations are more specific and give narrow definition to those organisms that are more closely related to one another. Slight variations in a single characteristic, such as bill size in the seed-eating Galápagos finches, are adaptive in that they enhance the survival of several closely related species. An understanding of how adaptations function to make species distinct also furthers the knowledge of how species are related to one another.

Why so many species exist is one of the most intriguing questions of biology. The study of adaptations offers biologists an explanation. Because there are many ways to cope with the environment, and because natural selection has guided the course of evolutionary change for billions of years, the vast variety of species existing on the earth today is simply an extremely complicated variation on the theme of survival.

—*Robert W. Paul*

FURTHER READING

Birkhead, Mike, and Tim Birkhead. *The Survival Factor*. New York: Facts on File, 1990.

Brandon, Robert N. *Adaptation and Environment*. Princeton University Press, 1990.

Gould, Stephen J. *Ever Since Darwin*. W. W. Norton, 1977.

Ricklefs, Robert E. *Ecology*. 4th ed. W. H. Freeman, 1999.

Rose, Michael R., and George V. Lauder, eds. *Adaptation*. San Diego, Calif.: Academic Press, 1996.

Weibel, Ewald R. *Symmorphosis: On Form and Function in Shaping Life*. Harvard University Press, 2000.

Whitfield, Philip. *From So Simple a Beginning: The Book of Evolution*. Macmillan, 1995.

ANIMAL AGGREGATION

FIELDS OF STUDY

Biology; Biophysics; Animal Behavior; Population Studies

ABSTRACT

Some animals spend most of their time alone because the presence of other conspecifics would interfere with the use of particular resources. These animals only come together with another solitary individual to pair for reproduction. Others form groups ranging from pairs of animals to large herds. Some animals are brought together by phenomena over which they have no control. Groups formed for whatever reason can be either temporary or permanent, and are generally theorized as being beneficial for a variety of reasons.

KEY CONCEPTS

competition: interactions among individuals that attempt to utilize the same limited resource

conspecifics: members of the same species

dilution effects: the reduction in per capita probability of death from a predator due to the presence of other group members

encounter effects: the reduction in the probability of death from a predator due to a single group of *N* members being more difficult to locate than an equal number of solitary individuals

interference: the act of impeding others from using some limited resource

predation: the act of capturing, killing and consuming another organism

resource defense: the control of a resource indirectly or directly

sociality: the tendency to form and maintain stable groups

ANIMAL ASSOCIATIONS

Some animals spend most of their time alone because the presence of other conspecifics would interfere with the use of a particular resource or a suite of resources. These animals only come together with another solitary individual to pair for reproduction. Others form groups ranging from pairs of animals to large herds. Finally, some animals are brought together by phenomena over which they have no control (winds, currents, or tides) and are simply clumped in space. Groups formed for whatever reason can be either temporary or permanent, and are generally theorized as being beneficial for a variety of reasons.

Some associations are simply the result of congregating around a common food resource. Other associations arise for specific functions, such as finding mates, caring for young, providing for a learning environment for developing young, providing protection from the elements, thermoregulation or huddling, locomotory efficiency (swimming or flight), locating and subduing food items, resource defense against other groups or competing species, division of labor, population regulation, predator vigilance, and reduced predation risk via dilution, confusion, encounter, or group defense effects. While potential benefits can be many, aggregation can also result in distinct disadvantages for individuals making up the group. Such disadvantages include increased competition for resources (such as mates, food, or shelter), increased risk of disease and spread of parasites, and interference in reproductive behaviors.

REPRODUCTION AND REARING YOUNG

Living in groups can increase an individual's chance of finding a mate, but it often results in increased aggression between males who must compete for females. Because all females may come into estrus within a short time period, grouping together at specific mating territories ensures that all females will be mated. In such species, courtship rituals are often common. These rituals serve a dual purpose: They provide the male with information about the sexual receptivity of the female and they allow the female to assess the quality of the male prior to pairing with him. In some cases, the rituals also serve to bind the pair together for a breeding season, or in some species for longer periods of time. In other cases, males simply congregate at display grounds, attract, and court females. Females leave the display grounds after mating to nest elsewhere, while the males remain to court other females. This latter grouping strategy is known as *lekking*.

In some species, which both aggregate during mating seasons and form pair bonds, both male and female can care for the young. In some cases, a mating pair may have helpers at the nest—other members of the species (usually offspring) which aid the parents in raising the young. Helpers at the nest greatly contribute to the breeding success of the parental birds and gain experience themselves in rearing young. They can then use their experience to be successful parents in the next breeding season. Other examples, such as lions and elephants, include kin groups (generally sisters or a mother and her daughters) that help to raise whatever offspring are present in the group.

Colonial nesting and thus synchronous egg-laying produces offspring in large numbers, who are vulnerable to predators for only a short period of time. In this way, each parent lessens the chance that their offspring will be the ones taken by any predator (dilution effect). However, colonial nesting also presents the possibility that offspring may grow more slowly or run the risk of parasitism from fleas and mites. Some evidence exists that offspring of bank swallows gained weight more slowly if they were from large colonies versus small colonies, suggesting that large colonies were depleting their resources more rapidly. Nests from large colonies were also more often infested with fleas than those in small colonies.

Rearing young in a group gives the young opportunities to learn from more than one adult. It can also provide them with practice in tasks that later prove important when the offspring are on their own. Cooperative hunting can provide young with the opportunity to learn hunting skills from their elders. Generally, this benefit results in longer-lived species that produce only a few young per year.

SURVIVAL

Cold temperatures can cause physiological problems, particularly for animals that are ectothermic (relying upon the environment for heat). During the day, such animals can bask in the sun and maintain high body temperatures. At night, however, they may have difficulty staying warm, so many huddle together in large groups. This type of behavior allows ectotherms to continue to digest their food (a process requiring heat).

Ectothermic species are not the only ones that huddle—many birds and mammals do so as well. Some birds flock together on trees on particularly cold nights to reduce the surface area of their body that is exposed to the elements. Voles, which are normally asocial, huddle in groups during the winter to keep warm. Marmots can huddle underground in groups of twenty for up to seven months to avoid freezing temperatures in the mountain regions they occupy.

Lowered individual predation risk is theorized as a primary benefit leading to the evolution of gregarious behavior, particularly in the absence of kinship between group members. Three likely outcomes of grouping that can reduce rates of attack are the dilution effect, the confusion effect, and the encounter effect. In the dilution effect, the probability of a particular individual being killed or injured is reduced by the presence of other group members that might be attacked first. This helps to explain why ostriches lay eggs in communal clutches that only the first laying female incubates. The first female is not necessarily acting in an altruistic manner, because she is diluting the chance that predators will find her eggs and eat them or, after the chicks hatch, that predators will capture her chicks instead of those of another female. Individuals in a group may also benefit by putting other animals between themselves and the predator (the selfish herd effect). Grouping provides the opportunity to decrease the area of danger around each individual. If individuals within the group are acting in a selfish herd manner, the groups formed tend to be tightly clumped, as all individuals attempt to put other individuals on all sides around them.

Confusion effects benefit group members because predators may have difficulty in fixing upon a particular individual for attack. The time it takes for the predator to discern a particular individual from a mass of surrounding individuals may be enough for the entire group to scatter and to further confuse the predator.

In contrast, the encounter effect results when it is more difficult for a predator to find a single group of prey than an equal number of scattered individuals due to the apparent rarity of the grouped individuals. However, the actual result of aggregation may be increased risk of detection if the group becomes a more conspicuous entity, which is detrimental to the individuals making up the group, rather than beneficial.

Groups can also benefit by collective defense, which is seen in animals possessing weaponry such as horns or other piercing appendages. Commonly these groups will form a line (phalanx) or a rosette (circular structure) with weapons pointed outward toward the attacker(s). Common examples of these defensive groups include musk oxen, elephants, and spiny lobsters. The use of weapons in these defensive lines may enhance the probability of survival above that of mere dilution, since each defending animal is capable of inflicting damage on a predator.

An individual has to partition its time between foraging and being vigilant for potential predators. However, if that individual is within a group, it can spend more timing on foraging and less time on vigilance because its scan frequency for predators can decrease proportionate to the number of members of the group. Furthermore, because of the many eyes within the group, detection of predators with subsequent alerting of the danger is enhanced. Some species go so far as to alert other group members to danger by alarm calling. This may have evolved for several reasons: The caller benefits because it knows where the predator is and can position itself appropriately within the group to avoid being a target; the caller may enhance the probability of becoming the target, but at the same time, it reduces the probability that its kin will be targets (this works in kin groups); and/or the caller may attract attention to itself at the time of the call, but if it survives, it is entitled to a payback at a later date from some other individual in the group doing the same.

COMMUNAL FORAGING AND HUNTING

Frequently, food being sought for exploitation is clumped in space. As a result, the animals that feed upon this food are also clumped, and this promotes aggregation into group structures, such as herds. Because animals can observe others of their species feeding, they can follow successful foragers to feeding groups and group for no other reason. This phenomenon may allow such foragers to exploit food resources in a more systematic way. Some evidence suggests that grasses clipped by herbivore herds actually grow faster and are more productive than grasses not so clipped, and by proceeding from one patch to another, the herds actually allow the grasses to grow back in a systematic fashion. This allows the food time to replenish itself before the herd passes by it again, and would be less likely to happen if individuals exploited the resource, isolated in time and space.

Other animals form groups to facilitate hunting and capture success. However, some of these groupings can be highly variable in time, based on the food supply and how a kill is partitioned between group participants. For example, sociality in lions is controlled by food supply. When wildebeests or zebra are especially numerous, lions concentrate their efforts upon capturing them, but solitary hunters are successful only 15 percent of the time, while groups of five are 40 percent successful. Furthermore, groups of five lions are better able to protect their kill from scavengers than are groups of two or three lions, even though a group of two or three lions will maximize their daily intake of food per successful kill, while a group of five will only secure the minimum daily requirement of food. Despite their propensity for sociality, lions will hunt alone or in pairs when the migratory wildebeests or zebras leave and only resident gazelles are left. Again, the success rate for a solitary lion hunting a gazelle is only 15 percent, while that for a pair is 30 percent; however, a gazelle can provide the minimum daily requirement of food for only two lions.

Similarly, groups of dogs (the African wild dog, Asian dhole, dingoes, or wolves) are able to kill prey larger than themselves by cooperatively hunting. These packs are composed of kin (parents and their offspring) and cover large distances in order to hunt their prey. The individual dogs will spread out around the prey in a phalanx and then approach until one member selects a victim and runs after it. Other members of the pack will run after the prey in relays until the prey is exhausted and can be subdued.

Communal spiders build webs larger than a single individual could spin and capture prey larger than any single one could capture alone. They communally feed on the captured prey. Most spider species have a short interval after the spiderlings have hatched where they remained clumped in a group and live in a communal web. After a period of several days, the spiderlings disperse to take up a solitary life. In communal species, however, adults of the same species come together to form colonies of up to one thousand individuals.

Coordinated group hunting is also known in marine mammals, particularly killer whales (orcas) and humpback whales. Orcas live in matriarchal societies (pods) of two to twelve members and hunt other marine mammals. Single orcas will charge sea lions in the surf of a beach, while other pod members will wait offshore to ensnare any sea lions that respond to the charge by entering the water. Adults will also train juvenile orcas, in the process of play-stranding, to capture seals or sea lions by throwing a dead seal to a beached juvenile. Orcas will also attack baleen whales by surrounding them and biting and holding the whale underwater so that it will drown. Humpback whales are known for bubble feeding, where members of the pod surround a school of small fish and release bubbles while spiraling upward. This bubble net concentrates the terrified fish into a tight ball, which the whales then eat as they approach the surface and open their mouths. Groups of dolphins herd fish into shallow waters and surround them so they can be easily picked off.

Even groups of mixed species are known to cooperatively hunt. This is common in shorebirds, where one species might herd the prey while another species either dives to prevent the prey's escape from below or stabs at the prey to prevent its escape along the banks of the water. Pelicans and cormorants, as well as grebes and egrets, are known to form these associations.

Colonial species, particularly those that are sessile and need space to spread out, need to be able to defend their resources. Among invertebrates, larger colonies may have the competitive advantage in excluding newcomers from unoccupied space, as well as pushing other colonial species out of the way, so that the space once occupied can be overrun. Corals are well known for their warlike actions as the space on a reef becomes scarce—they will use their stinging cells to attack adjacent coral species in an attempt to kill the polyps making up the other colony. Dense colonies of bryozoans are much more able to withstand overgrowth by other invertebrates than are bryozoan colonies that are more spread out. Mussel beds can overrun barnacle colonies in the intertidal zone if their main predator is not present in sufficient numbers to keep the mussel population low.

DIVISION OF LABOR

In eusocial insects, colony members divide the labor amongst themselves. In worker bees, the labor done is dependent on the age of the bee. After emergence, a worker bee's first job is cleaning the hive. She then tends the brood, builds up the honeycomb, and guards the nest. Her final task is to forage for pollen and nectar. The change in her duties correlates to physiological changes in her nurse glands and wax glands.

Ant colonies comprise thousands of individuals, divided into workers, brood, and queen. In army ants, workers vary in size, and these size differences determine their role in the colony. Smaller workers spend most of their time tending to and feeding the larval broods; medium-sized workers make up the majority of the population and are responsible for making raids on other colonies and locating food. The largest workers are called soldiers because of their powerful jaws; they accompany the raiding parties but carry no food.

Naked mole rats are the only known vertebrate that lives in colonies like those of the social insects. Only one female breeds. Larger individuals remain in the colony and huddle to keep the entire colony warm. Smaller individuals are the worker caste and are responsible for nest building and foraging.

These examples serve to illustrate that grouping behavior of animals can serve a myriad of purposes, or can simply be the result of phenomena beyond the animal's control (patchy food resources, physical forces of nature). The functions that lead to cooperative activities are usually best explained by groups being composed of kin; however, cooperation can also evolve in the absence of kinship, provided the benefits to group members outweigh the costs.

—*Kari L. Lavalli*

Further Reading

Bertram, B. C. R. "Living in Groups: Predators and Prey." In *Behavioral Ecology: An Evolutionary Approach*, edited by J. R. Krebs and N. B. Davies. Sinauer Associates, 1978.

Halliday, Tim, ed. *Animal Behavior*. University of Oklahoma Press, 1994.

Hamilton, W. D. "Geometry for the Selfish Herd." *Journal of Theoretical Biology* 31 (1971): 294–311.

Janson, C. H. "Testing the Predation Hypothesis for Vertebrate Sociality: Prospects and Pitfalls." *Behavior* 135 (1998): 389–410.

Krebs, J. R., and N. B. Davies. *An Introduction to Behavioral Ecology*. 3rd ed. Blackwell Scientific, 1993.

Pulliam, H. R., and T. Caraco. "Living in Groups: Is There an Optimal Group Size?" In *Behavioral Ecology: An Evolutionary Approach*, edited by J. R. Krebs and N. B. Davies. 2nd ed. Sinauer Associates, 1984.

Animal Bioluminescence

Fields of Study

Biology; Biochemistry; Environmental Studies; Oceanography

Abstract

Bioluminescence is the visible light produced by luminous animals, plants, fungi, protists, and bacteria, that results from a biochemical reaction with oxygen. Bioluminescence is produced without heat. Most organisms use a luciferin/luciferase system. The luciferin molecules are oxidized through catalysis by an oxidizing enzyme (luciferase). The oxidized form of luciferin is in an excited electronic state that relaxes to the ground state by emitting its excess energy as light. Although bioluminescence is widespread in animals, its occurrence is sporadic.

Key Concepts

luciferase: one of a group of enzymes that catalyzes the oxidation of a luciferin
luciferin: one of a group of organic compounds that emits visible light when oxidized
melanophore: a melanin-containing cell
photophore: a light-emitting organ consisting of a lens, reflector, and light-emitting photogenic cells
symbiosis: the intimate living together of two dissimilar organisms in a mutually beneficial relationship

BIOLUMINESCENCE IN NATURE

Bioluminescence is the visible light produced by luminous animals, plants, fungi, protists, and bacteria, that results from a biochemical reaction with oxygen. Unlike incandescent light from electric light bulbs, bioluminescence is produced without accompanying heat. Bioluminescence was first described ca. 500 BCE, but the chemical mechanism of bioluminescence was not elucidated until the beginning of the twentieth century. The ability to luminesce appears to have arisen as many as thirty times through evolution. The chemical systems used by luminescent organisms

Bioluminescence in Antarctic krill - watercolor (Uwe Kils).

are similar but not exactly the same. Most organisms use a luciferin/luciferase system. The luciferin molecules are oxidized through catalysis by an oxidizing enzyme (luciferase). The oxidized form of luciferin is in an excited electronic state that relaxes to the ground state by emitting its excess energy as light.

Animals may produce light in one of three ways. The bioluminescence may be intracellular: Chemical reactions within specialized cells result in the emission of visible light. These specialized cells are often found within photophores. These light-producing organs may be arranged in symmetrical rows along the animal's body, in a single unit overhanging the mouth, or in patches under the eyes, and are connected to the nervous system. Alternatively, the bioluminescence may be extracellular: The animals secrete chemicals that react in their surroundings to produce light. The third option involves a symbiotic relationship between an animal and bioluminescent bacteria. Several species of fish and squid harbor bioluminescent bacteria in specialized light organs. The symbiotic relationship is specific—each type of fish or squid associates with a certain type of bacteria. The bacteria-filled organ is continuously luminous. The animal regulates the emission of light either with melanophores scattered over the surface of the organ or with a black membrane that may be mechanically drawn over the organ.

Although bioluminescence is widespread in animals, its occurrence is sporadic. Most of the bioluminescent animal species are invertebrates. Among the vertebrates, only fish exhibit bioluminescence; there are no known luminous amphibians, reptiles, birds, or mammals. Although bioluminescence is found in terrestrial and freshwater environments, the majority of luminous organisms are marine. Scientists estimate that 96 percent of all creatures in the deep sea possess some form of light generation.

FUNCTIONS OF BIOLUMINESCENCE

There appear to be three main uses of bioluminescence: finding or attracting prey, defense against predators, and communication. Although visible light penetrates into the ocean to one thousand meters at most, most fish living below one thousand meters possess eyes or other photoreceptors. Many deep-sea fishes have dangling luminous light organs to attract prey. Terrestrial flies have also exploited bioluminescence for predation. The glow of glowworms (fly larvae) living in caves serves to attract insect prey, which get snared in the glowworms' sticky mucous threads. Fungus gnats (carnivorous flies) attract small arthropods through light emission and capture the prey in webs of mucous and silk.

Bioluminescence can serve as a decoy or camouflage. For example, jellyfish such as comb jellies produce bright flashes to startle a predator, while siphonophores can release thousands of glowing particles into the water as a mimic of small plankton to confuse the predator. Other jellyfish produce a glowing slime that can stick to a potential predator and make it vulnerable to its predators. Many squid and some fishes possess photophores that project light downward, regardless of the orientation of the squid's body. The emitted light matches that of ambient light when viewed from below, rendering the squid invisible to both predators and prey.

The best known example of bioluminescence used as communication is in fireflies, the common name for any of a large family of luminescent beetles. Luminescent glands are located on the undersides of the rear abdominal segments. There is an exchange of flashes between males and females. Females respond to the flashes of flying males, with the result that the male eventually approaches the female for the purpose of mating. To avoid confusion between members of different types of fireflies, the signals of each species are coded in a unique temporal sequence of flashing, the timing of which is controlled by the abundant nerves in the insect's light-making organ. Females of one genus of fireflies (*Photuris*) take advantage of this by mimicking the response of females of another genus (*Photinus*) to lure *Photinus* males that the *Photuris* females then kill and eat.

Some marine animals also utilize bioluminescence for communication. For example, lantern fishes and hatchetfishes (the most abundant vertebrates on earth) possess distinct arrangements of light organs on their bodies that can serve as species- and sex-recognition patterns; female fire worms release luminescent chemicals into the water during mating, beginning one hour after sundown on the three nights following the full moon; and deep-sea dragonfish emit red light that is undetectable except by other dragonfish.

—*Lisa M. Sardinia*

Further Reading

Presnall, Judith Janda. *Animals That Glow*. Franklin Watts, 1993.

Robison, Bruce H. "Light in the Ocean's Midwaters: Bioluminescent Marine Animals." *Scientific American* 273 (July 1995): 60–64.

Silverstein, Alvin, and Virginia Silverstein. *Nature's Living Lights: Fireflies and Other Bioluminescent Creatures*. Little, Brown, 1988.

Toner, Mike. "When Squid Shine and Mushrooms Glow, Fish Twinkle, and Worms Turn into Stars." *International Wildlife* 24 (May/June 1994): 30–37.

Tweit, Susan J. "Dance of the Fireflies." *Audubon* 101 (July/August 1999): 26–30.

Animal Cells

Fields of Study

Biology; Biochemistry; Physiology

Abstract

The cells that make up the body of an animal are the most basic units of life. Living cells use nutritional sources for energy to maintain both structure and life processes such as growth and reproduction. Life requires structure, and cells have the minimal architectural design that enables them to retain life and pass it on to future generations of cells.

Key Concepts

adenosine triphosphate (ATP): a molecule produced in the cell that provides energy for cell processes

amino acid: the subunit that makes up larger molecules called proteins

cytoplasm: the living portion of the cell that is contained within the cell membrane

deoxyribonucleic acid (DNA): the molecular structure within the chromosomes that carries genetic information

differentiation: the process during development in which specialized cells acquire their characteristic structures and functions

gamete: the sex cells of an animal, each of which contains only one chromosome from each available pair of chromosomes found in normal body cells

gene: the part of the chromosome that includes the DNA and is the carrier of heredity

nucleus (pl. nuclei): a central cell structure that controls the activity of the cell because of the genetic material it contains

organelle: a subcellular structure found within the cytoplasm that has a specialized function

protein: a substance made up of conjoined amino acids, different types of which are the chief building blocks of cellular structures

CELLS AND LIFE

The cells that make up the body of an animal are the most basic units of life. Living cells use nutritional sources for energy to maintain both structure and life processes such as growth and reproduction. Life requires structure, and cells have the minimal architectural design that enables them to retain life and pass it on to future generations of cells.

Typically, cells are joined together in the animal body to form larger structures called tissues. By definition, a tissue, such as connective or muscle tissue, is an aggregate of similar cells and intercellular materials that combine to perform a common function. An organ, such as the skin or the biceps, is frequently composed of several types of tissues.

CELL STRUCTURE

There are essential structural characteristics of animal cells that enable them to maintain life. Most fundamentally, an animal cell must have a border that separates it from its environment or surroundings. In animal cells, that border is called the cell membrane. Within the membrane lies the cytoplasm (literally, the plasm of the cell); outside the membrane is the environment from which the cell must extract its nutritional needs and into which it must pass the waste products that result from the numerous chemical reactions taking place inside the cell.

Structurally, the cell membrane consists of a double layer of lipid molecules. These molecules are bipolar: the head end of each molecule is hydrophilic, meaning it has an attraction to or affinity for water molecules, while the tail end of the molecule is hydrophobic, meaning it tends to repel water molecules. In the formation of the membrane, these molecules are found parallel to one another and arranged in a double row. In the inner row, their heads face the inside of the cell; in the outer row, the heads face their environment. The hydrophobic tails form the interior of the membrane and provide an effective barrier that prevents the free passage of water and water-soluble substances. Interspersed among the lipid molecules are numerous protein molecules.

Some of these proteins, called transmembrane proteins, extend entirely through the lipid bilayer and have ends exposed to both the interior and the exterior of the cell. Others, called integral proteins, are found only on one side of the membrane and extend only partway into the lipid bilayer. Finally, some proteins, called peripheral proteins, are attached to the outside or inside surfaces of the membrane. Transmembrane proteins may function as channels or pores, allowing specific ions or molecules to pass through them. Integral proteins may also function as transmembrane carriers, as they can bind to specific products outside the cell, such as a certain amino acid, and then flip-flop across the membrane to the inner side and release the product into the cell's interior. Peripheral proteins on the outer surface of the membrane may serve as identification markers for other cells. The membrane itself functions to maintain the integrity of the cell's cytoplasm by holding essential components inside and preventing the cell from drying out.

Cytoplasm is a general term for the viscous fluid within the cell's membrane. In the cytoplasm are numerous small, specialized structures called cell organelles. An essential structure for living cells is the nucleus, which contains genetic information in the form of DNA. DNA determines a particular cell's structure and function. Other organelles in the cytoplasm are designed for specific processes, such as manufacturing energy, building structural proteins, or storing cell products prior to exporting them. Some organelles are enclosed within their own membranes, which are structurally very similar to the membrane forming the cell's outer boundary.

THE NUCLEUS AND ITS CONTENTS

The nucleus is an essential control center of the cell. It is enclosed within the nuclear envelope, a unique

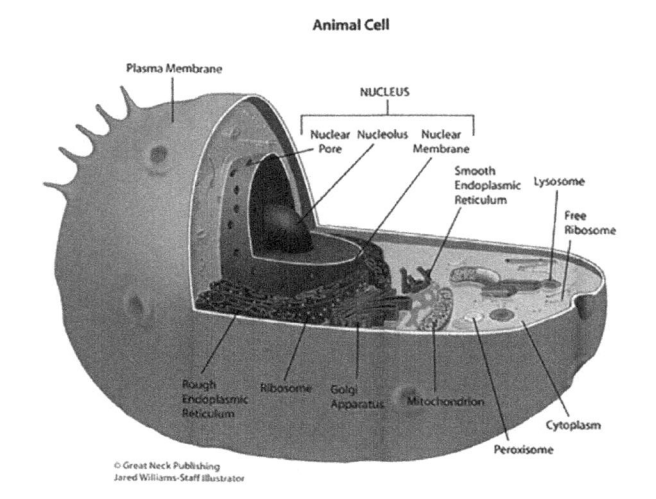

Animal Cell: A cell is the smallest living part of any kind of life. All living things are made of cells. The human body is made of billions of cells. This illustration of an animal cell shows the most important features of the cell.

membrane with relatively large pores. These pores allow large information-carrying molecules called ribonucleic acid (RNA) to pass from the nucleus to the cytoplasm. Resident within the nucleus is a dark material called the chromatin. During cell division, the chromatin material condenses into clearly observable structures called chromosomes. When the cell divides into two, each daughter cell contains an equal number of chromosomes from the original maternal cell. The chromatin material within the cell's nucleus is made up of two major types of material: DNA and associated proteins.

DNA is organized into discrete packets of information called genes, which form the backbone of the chromosomes. The genes determine the characteristics of the specific cell or organism. The DNA-associated proteins regulate the gene's expression. At times, the genes may express themselves by replicating their information into chemical messengers called RNA, which then diffuse out through the nuclear pores into the cytoplasm. Within the cytoplasm, RNA binds to an organelle called the ribosome, which produces new protein molecules. Normally, the nucleus also contains one or more dark, round structures called nucleoli, which assist in the production of the cell's ribosomes. A single nucleolus is made up of protein and RNA.

A particular cell may also contain hundreds of mitochondria, also typical organelles. Mitochondria are complex, double-membraned structures that are found throughout the cytoplasm of the cell. These oval-shaped structures contain numerous enzymes that stimulate energy-producing reactions, the net sum of which results in the formation of high-energy ATP molecules. These molecules diffuse throughout the cell to various other organelles and release their energy, thereby fueling most cell processes.

Lysosomes are cell organelles that are round in shape and are enclosed within a membrane. They contain many different enzymes that break down or digest various substances into simpler substances that the cell can use.

The endoplasmic reticulum (ER) consists of membrane-enclosed spaces in the cytoplasm. These complex membrane arrays are extensions of the outer cell membrane. Some of this membrane system is covered with ribosomes that function to produce protein. These parts of the ER are called rER, for rough ER or ribosomal ER. Other ER portions that lack ribosomes have a smooth appearance and are called smooth ER (sER).

Another cellular organelle, the Golgi apparatus, appears as a bunch of flattened bags. This organelle is usually not far distant from the ER. After the rER produces a protein product, the Golgi apparatus further processes the product and packages it for cellular export.

CELL SIZE AND FUNCTION

Animal cells have great variations in size. The smallest may be as little as four micrometers in diameter, where one micrometer equals 0.001 millimeter. The largest known cell is the single-cell ostrich egg, which is about seventy-five millimeters in diameter—more than 750,000 times larger than the smallest bacterium. Somatic cells, the cells that make up the body structures of animals, are typically more intermediate in size. The human red blood cell is about seven micrometers in diameter, while an intestinal epithelial cell is about thirty micrometers in diameter. Most animal cells are approximately the same size and typically have diameters between ten and twenty-five micrometers.

An essential characteristic for cell survival is the ratio of the cell's surface area to the volume of the cell, or its surface-to-volume ratio. Typically, cells with small diameters have large surface-to-volume ratios, while large cells have small surface-to-volume ratios.

Many of the substances that are needed for the cell's survival, such as oxygen or nutrients, enter the cell through the surface membrane by simple diffusion. The cells that have high rates of metabolism tend to be very small and have larger surface-to-volume ratios. Larger cells either have lower metabolic rates or have specialized shapes, such as numerous membrane enfoldings, to optimize diffusion of essential materials into their interiors.

In multicellular animals, cells are differentiated, meaning they have a specialized modification in their structure to enable them to perform a specific task. Thus, a striated muscle cell contains numerous myofibrils that shorten as the cell contracts, while a glandular cell may contain numerous secretory granules filled with products for export.

Animal cells can be classified on the basis of the number of chromosomes that they contain. Somatic cells, or body cells of an organism, are called diploid, because they contain the total number of chromosome pairs that is characteristic for that organism. For example, somatic cells in mosquitoes contain four pairs of chromosomes, or eight individual chromosomes. Each chromosome in the pair contains genetic information for the same genes that the other paired chromosome contains. Alternatively, sex cells or gamete cells—sperm and egg cells—contain a haploid number of chromosomes, meaning they have a single chromosome from each of the possible pairs. Thus, a human sperm cell contains a total of twenty-three individual chromosomes, while a human skin cell contains forty-six individual chromosomes.

Another way to classify animal cells is to consider the primary way that they use proteins. Some animal cells are primarily protein-secreting and manufacture much of their proteins for body use outside the cell that produces the protein. One example of this is the pancreatic acinar cell, which produces numerous digestive enzymes (proteins) and secretes them into the pancreatic duct for transport into the digestive tract, where the enzymes break down complex foods into simpler forms. Other animal cells are primarily protein-retaining cells, in which much of the manufactured protein is retained for use by the cell itself. An example of this is the keratinocyte, the prominent cell type found in the skin. This cell produces a large amount of the protein keratin, which remains stored in the cell's cytoplasm. Because of the presence of the keratin, the skin is able to maintain its protective waterproofing function. Without keratin, the skin would lose body water.

CELL SHAPES AND TYPES

Animal cells have a wide variety of shapes. Individual cells that are mobile within an aqueous environment tend to be spherical in shape. The neutrophil, a type of white blood cell, is an example of this. Cells that are mobile within tissues and migrate from one area to another often have long cytoplasmic extensions; one such cell is the macrophage, which has long, changeable, armlike processes (projections) that enable it to migrate through tissue spaces and ingest bacteria that may be found there. Other migratory cells have a tail called a flagellum. Sperm cells, for example, use their flagella to propel themselves through the female reproductive tract to reach the egg.

Some cells have branching, stationary cytoplasmic processes through which information molecules move. Often, these processes form a complex network with similar cells, such as the network of neurons in the brain. Body tissues are made of cells that have tight cell-to-cell connections between their membranes. These cells may provide a covering for or form the wall of a particular structure. Protective cells, such as skin cells, are often many layers thick.

A typical animal's body is composed of about two hundred different recognizable cell types. Most of these fall into four main categories: epithelial cells, connective cells, movement cells, and message or conveyance cells.

Epithelial cells form a continuous layer over surfaces that are external or internal to the body. Skin is an external protective tissue that is made of many such layers of cells. The inner lining of the digestive tract, for the most part, consists of a single layer of epithelial cells that absorb and secrete materials.

Connective cells provide the structural support of the animal body. Examples of these cells are fibrocytes, found in the dermal layer of the skin, and osteocytes, found in the matrix of bone.

Movement cells, responsible for body movement, are typified by the muscle cells. Muscle cells that are attached to bone and cause limb movement are called striated muscle cells. Those found in the walls of body organs such as the stomach are called smooth muscle cells; contraction of these cells causes the contents of the stomach to be mixed and stirred. Cardiac muscle cells are found in the heart, where they contract to force the movement of blood throughout the circulatory system.

Message or conveyance cells are very diverse. Branching nerve cells have long processes that carry information molecules from one part of the body to another, such as from the spinal cord to the finger. Red blood cells carry oxygen from the lungs to the body cells. Gamete cells transfer genetic information from one organism to the next generation of organisms.

STUDYING CELLS

Ever since the 1600s, when Antoni van Leeuwenhoek first used a simple magnifying lens system to study the structure of single-celled life forms, microscopes have been an important tool in cytology (the study of cells). Several types of microscopes are commonly used to study cells, whether alive or preserved.

The simplest and most common type of microscope used to study animal cells is the light (bright-field) microscope. These microscopes have magnification powers ranging from about forty to two thousand times. In their natural state, most cells are essentially colorless. In order to make microscopic viewing more effective, cells are often stained so that their structures are more readily visible.

A phase-contrast microscope is a modified light microscope that can produce visible images from transparent objects. This type of microscope is frequently used to study unstained or living cells. Its magnification range is similar to the light microscope.

A transmission electron microscope (TEM) uses electrons instead of light rays to visualize objects. A TEM passes its electron beam through very thinly sectioned cells. The density of the stained cellular structures absorbs electrons in a differential fashion so that an image corresponding to the cell's architecture can be visualized. The TEM can magnify cellular structures from 1,000 to 250,000 times. Consequently, this type of microscopy is often used to study the small subcellular organelles.

Another popular technique used to study animal cells involves growing them in cultures outside the animal's body. This technique, called *in vitro* cell culture, involves obtaining a group of living cells from an animal, usually in the form of pieces of tissue. An enzyme solution is commonly used to digest the cell-to-cell connections and produce a suspension of free individual cells. These cells are then placed in petri dishes along with a liquid medium that contains essential nutrients. Some types of animal cells, especially fibrocytes found in the connective tissues, are easily cultured with this technique. These cell cultures, if properly maintained, will grow and reproduce for generations. Experimenters can use such cell cultures to investigate how living cells function and respond to varied environmental influences.

Since cells are the basic units of life, an understanding of their function is essential for comprehending the way living organisms function. Many diseases are caused by a malfunctioning group of cells. For example, a group of cells normally undergoes an orderly sequence of growth and reproduction. At times, however, some cells become disordered and began to multiply rapidly without stopping. This may be caused by an abnormality that appears in the genetic code or by the presence of a virus that takes over the genetic controls of the cell. This situation is typical of some types of cancer.

As scientists learn more about how cells live and why they die, they will gain valuable insights into the aging process and may thereby increase the span of life. As differentiation is better understood, scientists may be able to change mature cells or even replace them if they become damaged, destroyed, or simply worn out due to age.

—*Roman J. Miller*

Further Reading

Karp, Gerald. *Cell and Molecular Biology: Concepts and Experiments*. 6th ed. Wiley, 2010.
Loewy, Ariel G., et al. *Cell Structure and Function: An Integrated Approach*. 3rd ed. Saunders, 1991.
Moyes, Christopher D., and Patricia M. Schulte. *Principles of Animal Physiology*. 2nd ed. Pearson, 2008.
Neilsen, Claus. *Animal Evolution: Interrelationships of the Living Phyla*. Oxford University Press, 2012.
Prescott, David M. *Cells: Principles of Molecular Structure and Function*. Jones, 1988.
Sadava, David E. *Cell Biology: Organelle Structure and Function*. Jones, 1993.
Telford, Ira R., and Charles F. Bridgman. *Introduction to Functional Histology*. 2nd ed. Harper, 1995.
Threadgold, Lawrence Theodore. *The Ultrastructure of the Animal Cell: International Series in Pure and Applied Biology*. Elsevier, 2013.

Animal Courtship

Fields of Study

Biology; Behavioral Studies; Environmental Studies; Veterinary Science

Abstract

Courtship is the ritualistic behavior animals carry out preceding mating. The courtship process allows the male and the female of the species to attract each other and choose a mate for reproduction. This behavior varies greatly among the species. Courtship rituals for mating can be as simple as sign stimuli, such as the emission of pheromones by certain female moths, to the complex dance of the stickleback or the bowerbird.

Key Concepts

arachnid: a class of arthropods with jointed legs and hard external skeleton that includes mites, scorpions, spiders, and ticks
estrus: the period during a female's sexual cycle when she is sexually receptive
lek: a territory used by certain animals for mating
monogamy: a mating system in which one male pairs with just one female
pheromone: a chemical produced by one animal that influences another animal of the same species
polygamy: a mating system is which an individual of one sex pairs with several of another sex

COURTSHIP AND RITUALISTIC BEHAVIOR

Courtship is the ritualistic behavior animals carry out preceding mating. The courtship process allows the male and the female of the species to attract each other and choose a mate for reproduction. This behavior varies greatly among the species. Courtship rituals for mating can be as simple as sign stimuli, such as the emission of pheromones by certain female moths, to the complex dance of the stickleback or the bowerbird.

In nature, most animals are solitary except when courting and rearing their young. For each animal the signals used to communicate with and attract a mate must be clear and unambiguous. This is to avoid any confusion in identifying the sex and species of an animal's potential mate, as the ultimate goal of nature is to encourage reproduction and survival of all species.

The procreation process in most multicellular plants and animals involves a complex form of sexual reproduction. Here unique and differentiated male and female reproductive cells called gametes unite to form a single cell known as the zygote. The zygote undergoes successive divisions to form a new multicellular organism where half the genes in the zygote come from one parent and half from the other, creating a singularly different living creature.

The sexual reproduction process requires a pair of distinct partners of opposite sexes to mate, hence the need for courtship, so that animals can find and attract prospective mates for producing offspring. It is the sexual interest of the opposite sexes that leads to mating; therefore, copulation in most animals is preceded by a period of courtship. The mating of most lower animals is governed by endocrine secretions specific to certain seasons. In female mammals, receptivity to mating is called estrus, which is operative only for short periods during the year. Cows have several such periods of estrus during the year; dogs have one or two.

As courtship is always a precursor to mating, which leads to the proliferation of the species, it is often intermixed with both greeting and aggression. Often the aggression is directed toward a rival that might wander into marked territory. What is fascinating is that quite often it is difficult to distinguish between greeting and aggression. There are many different types of courtship behavior, including shoaling, nest building, mouth or pouch brooding, parasitic mates, carnivorous mates, marking territory, pair bonding, domination, female dominance, bower building and gift offerings, communal display, exotic pirouettes, group breeding, monotremes, odor marking, color signals, sounds, and migration.

FISH COURTSHIP

The male stickleback, a fish similar to the minnow, orchestrates its mating by changing its color. During

Cicada killer wasps in courtship flight.

breeding season, the underside of the male turns bright red. The bright red not only attracts females but also instigates attacks from other males. In fact, studies have shown that during this time, just about any red object can provoke male stickleback aggression. The female's response to the red signal of the male stickleback is to advance in a posture that clearly exhibits her swollen, egg-filled underbelly. This stimulates the male to carry out a zigzag dance that in turn seduces the female into following him to the tunnel-like nest he has built. The male waits until the female struggles into the nest and immediately touches her tail with his nose and starts quivering. The vibration causes the female to release her eggs so that the male stickleback can externally fertilize them. If for some reason the male is unsuccessful in executing the last part of this intricately choreographed dance, the female will not lay her eggs. However, it is extremely interesting to note that vibrating the female with some arbitrary object, which she may even realize is not a male stickleback, works just as well to induce the release of the eggs, although, as the male did not complete the last step of the courtship ritual, he does not fertilize the eggs, but rather eats them.

AMPHIBIAN AND REPTILE COURTSHIP

Amphibians have varied and interesting courtship rituals. Generally the tailless amphibians breed communally. Amphibians such as frogs or toads will use sound to attract their mates, whereas male newts use odor to attract the females. The American tree frog (*Hyla crucifer*) sings in trios. Only the males have vocal sacs that can make sounds; the females are silent. The sounds of the males among the different species are distinctly different and versatile. The notes are delivered in different speeds and frequencies. The female has her own features of recognition. The way a female frog advertises her sex and readiness for mating is that she develops a series of granulations on her thighs, which is clearly visible to a male frog. During courtship many amphibians use their heads in different ways to rub their partners. For most frogs, the male climbing onto the female's back and aiding in the discharge of her eggs accomplishes mating. The male in turn releases his sperm after being excited by the female's movements.

In reptiles, even though the courtship behavior is not very complex, bobbing, circling, and marking of territory is noticed. Snakes such as the crotalids, including vipers and rattlesnakes, have a mutual dancelike ritual for courtship. As the snakes circle each other, they rear up facing one another and intertwine together for copulation. Most snakes of the Crotalidea family give birth to living young from eggs that hatched inside the mother's body. The colubrids, such as the boomslangs and kingsnakes, and elapids, such as the cobras, have a nuptial procedure where the male rubs his lower jaw along the back of the female in an attempt to stimulate her. It is conjectured that perhaps this behavior also stimulates him and all the other males in the matrix group. A male snake copulates by throwing coils round the female and bringing their cloacae into juxtaposition. However, sometimes they simply lie close without the retaining coils. They may use a branch or similar object to maintain a grip to hold them close together. Even though most snakes are more aggressive during mating season, few males fight among themselves. Other than among the elapids there does not seem to be much competition among the males.

This is contrary to what is noticed in crocodiles and alligators. In these reptiles, the breeding season for the males is a loud and contentious time. They will battle and resist any potential competitor with ritual gaping, lunging, and hissing. Yet the courtship behavior is serene. The male marks his territory using a secretion from his musk glands. This not only warns off unwelcome male rivals, it also attracts females. Male crocodiles and alligators confidently bellow their love calls to make their presence and location known to the females. Once a female gets paired with a male, they swim together. They progressively increase their speed and the male rubs his throat against the female's snout to mark her with his odor, and vocalizes without encumbrance before he copulates with her. The male leaves to find his own area after mating. The female goes on to build a nest above flood level using mud and vegetational debris. She covers her eggs with mud and decaying vegetation so that the sun's heat will incubate her eggs. She guards her eggs until they are hatched. She may aid the young ones to get to the water's edge by carrying them in her mouth. It has been observed that some female alligators may even stay with their young for a year or more. A female alligator is quite protective of her young and responds quickly to the sharp croaks of a young one in distress.

BIRD COURTSHIP

Birds in general have engaging and quite elaborate courtship rituals. Usually birds do not stay with the same mate all year round or from year to year. Sometimes the same pair may be together for several years, but nonetheless the pair bond, that is, the relationship of the couple, must be renewed or reinforced at the onset of each breeding season. Birds use auditory, visual, or both kinds of display in order to conclude their courtship ritual. Peacocks use intricate visual displays that involve specialized plumes. In the barbets and wrens, duetting is used for courtship and attracting the mate. The male and the female alternate their calls in such exact synchronization that it seems as though the sound is coming from one bird. This is an example of an auditory display. Some birds, such as ducks, follow an extremely stereotyped movement pattern. However, both birds must respond with the correct display or the sequence is broken. There are certain birds where a pair bond does not exist at all. The male birds gather together and display against each other, contending for the right to mate with the maximum number of females possible. This type of a gathering of males is called a lek. The bird of paradise is well known for this particular type of courtship behavior. Other examples would be manakins, sandpipers, and grouse.

The different types of bowerbirds are renowned for their extraordinary courtship behavior. These birds are commonly found in New Guinea and Australia. It is believed that the bowerbirds are close relatives of the birds of paradise of this area. For the most part, the males and females live apart from each other during the course of the year. Only during mating season do the males gather together and compete for females. Each male bird creates a clearing on the forest floor that becomes his "playing field." This is where he tries to entice and lure the females. The bird will arrange rocks, shells, colorful fruits and berries, flowers, and inanimate articles such as pieces of glass or other interesting manmade items. In some species, the males will construct different types of forms and structures. Some of these birds erect what look like upright poles of sticks around a tree trunk that are embellished with colorful flowers, mushrooms, lichens, and other objects.

The Australian male satin bowerbird is a silky blue-black bird that is approximately 20 centimeters (about 7.8 inches) long, with bright blue eyes. The male bird builds a stick mat. He then places two walls of vertical sticks down the middle. These walls may reach a height of thirty-eight centimeters (sixteen inches). The bird then mixes his own saliva with a blue or green fruit juice to form a "paint." He uses a tree bark to paint the bower, and tries to attract the grayish-green female into the bedecked bower to mate. This is one of the few known instances of birds using tools.

Another species known as the gardener or Vogelkop bowerbird makes a tentlike structure that resembles a teepee. The bird is only about 25 centimeters (10 inches) long, but the teepee can be as extensive as 1.6 meters (63 inches) across. This little teepee comes complete with a low entranceway overlooking a "garden," consisting of brightly colored flowers and objects that are diligently replenished as soon as they fade. There are about nineteen species of bowerbirds and they all have similarly elaborate courtship displays.

MAMMAL COURTSHIP

Mammals have evolved to have the most sophisticated sexual apparatus of all animals, yet in general mammals do not spend as much time in courtship as birds do. Mammals also have the most highly evolved manner of parenting. Therefore, the relationships between mates are diverse. Lower mammals, such as some rodents, practice promiscuous polygamy without any particular selection process by the female. The females are left to rear the young on their own. In bears, both the male and female are promiscuous. There are others where the male appears to have a harem. The California sea lion may have up to forty females, the Pribilof seal from forty-five to a hundred. Many deer and antelope have harems, too. Even the sedate koala may monopolize several females. These are also examples of dominance. Dominance can also extend toward other males, so that the dominant male can have all the females to himself.

Monogamy is rarer in mammals. A few canines are monogamous, but only for a particular season. Some monkeys and apes, not including chimpanzees, follow this rule. Foxes may be monogamous, and the American beaver is one rare mammal that practices monogamy for life. Even in the higher mammals, including humans, monogamy causes a decline in sexual interest for the partner and eventually sexual relations may cease if there is no stimulus of novelty. There is also the other extreme, where the animals are solitary and only meet for copulation, such as the titi monkey.

Most mammals tend to be complacent in their efforts at enticement, and rely on simpler stimuli such as secretions, color changes, and odors. For mammals, arctic foxes have an elaborate and rather graceful courtship dance. They rear up on their hind legs and face each other as though they are playing. The elephant has a most gentle approach to courtship. The female is attracted to the male by the strong scent of the two glands near his ears, which start to secrete during rut. Once the male and female have found each other, they caress one another with their trunks and intertwine these sensitive organs. They express their affection and confidence by placing the tips of their trunks in each other's mouths.

In some species, the female takes the initiative for mating. A small number of primates will do so. When the female is ready, she entices the male to attend to her. The female hedgehog will lift her tail up and lay all her spines down to solicit the male for mating.

Odor marking is very common in placental mammals. Most male mammals, even the pack animals, will mark out their territory by odor glands, urine, or feces. Some will mark the female as a sign of ownership. There are others that display color, signaling a potential partner of the opposite sex of its interest in mating. The female gelada baboon, who mostly spends her time sitting, develops a red patch on her chest similar to the one in her genitals, so that both markings can be easily observed by other baboons.

In many different animals, migration is required as part of the courtship and breeding process. Migratory rituals exist in just about all types of animals: fish, birds, and mammals. Some fish, such as the salmon or sturgeon, will travel thousands of miles in order to return to their ancestral waters to spawn. The urge is so strong that the fish will die trying to reach this place rather than give up.

Ultimately, nature's most primordial instinct is to continue the species. Every living organism feels the irrepressible compulsion to perpetuate its genes. This necessitates sexual selection, Darwin's other type of natural selection. The consequence is sexual dimorphism, including different size, color, and traits in the sexes. There are two types of sexual selection: intrasexual selection, where males contend among themselves through contests and displays for the favor of a female; and epigamic selection, where the females accept males with certain characteristics. The male may monopolize the females by practicing polygamy and thereby effecting a much more intense sexual selection for procreation purposes. Courtship behavior is about sexual selection, which eventually leads to the proliferation of the species.

—*Donald J. Nash*

Further Reading

Chinery, M. *Partners and Parents.* Crabtree, 2000.
Miller, G. *The Mating Mind: How Sexual Choice Shaped the Evolution of Human Nature.* Doubleday, 2000.
Prince, J. H. *The Universal Urge: Courtship and Mating Among Animals.* T. Nelson, 1972.
Robinson, Michael H. *Comparative Studies of the Courtship and Mating Behavior of Tropical Araneid Spiders.* Department of Entomology, Bishop Museum, 1980.
Simon, H. *The Courtship of Birds.* Dodd, Mead, 1977.

Animal Demographics

Fields of Study

Biology; Statistics; Demography; Epidemiology

Abstract

No animal lives forever. Instead, each individual has a generalized life history that begins with fertilization and then goes through embryonic development, a juvenile stage, a period in which it produces offspring, and finally death. Biologists seek to understand the processes that govern the production of new individuals and the deaths of those already present in a population. The branch of biology that deals with such phenomena is called demography.

Key Concepts

cohort: a group of organisms of the same species, and usually of the same population, that are born at about the same time
fecundity: the number of offspring produced by an individual
life table: a chart that summarizes the survivorship and reproduction of a cohort throughout its life span
mortality rate: the number of organisms in a population that die during a given time interval
natality rate: the number of individuals that are born into a population during a given time interval
population: a group of individuals of the same species that live in the same location at the same time
survivorship: the pattern of survival exhibited by a cohort throughout its life span

BIRTH, DEATH AND DEMOGRAPHY

No animal lives forever. Instead, each individual has a generalized life history that begins with fertilization and then goes through embryonic development, a juvenile stage, a period in which it produces offspring, and finally death. There are many variations on this general theme. Still, the life of each organism has two constants: a beginning and an end. Many biologists are fascinated by the births and deaths of individuals in a population and seek to understand the processes that govern the production of new individuals and the deaths of those already present. The branch of biology that deals with such phenomena is called demography.

The word "demography" is derived from Greek; *demos* means "population." For many centuries, demography was applied almost exclusively to humans as a way of keeping written rec ords of new births, marriages, deaths, and other socially relevant information. During the first half of the twentieth century, biologists gradually began to census populations of naturally occurring organisms to understand their ecology more fully. Biologists initially focused on vertebrate animals, particularly game animals and fish. Beginning in the 1960s and 1970s, invertebrate animals, plants, and microbes also became subjects of demographic studies. Studies clearly show that different species of organisms vary greatly in their demographic properties. Often, there is a clear relationship between those demographic properties and the habitat in which these organisms live.

DEMOGRAPHIC PARAMETERS

When conducting demographic studies, a demographer must gather certain types of basic information about the population. The first is the number of new organisms that appear in a given amount of time. There are two ways that an organism can enter a population: by being born into it or by immigrating from elsewhere. Demographers generally ignore immigration and concentrate instead on newborns. The number of new individuals born into a population during a specific time interval is termed the natality rate. The natality rate is often based on the number of individuals already in the population. For example, if ten newborns enter a population of a thousand individuals during a given time period, the natality rate is 0.010. A specific time interval must be expressed (days, months, years) for the natality rate to have any meaning.

A second demographic parameter is the mortality rate, which is simply the rate at which individuals are lost from the population by death. Losses that result from emigration to a different population are ignored by most demographers. Like the natality rate, the mortality rate is based on the number of individuals in the population, and it reflects losses during a certain time period. If calculated properly, the natality and mortality rates are directly comparable, and one can subtract the latter from the former to provide an index of the change in population size over time. The population increases whenever natality exceeds mortality and decreases when the reverse is true. The absolute value of the difference denotes the rate of population growth or decline.

When studying mortality, demographers determine the age at which organisms die. Theoretically, each species has a natural life span that no individuals can surpass, even under the most ideal conditions. Normally, however, few organisms reach their natural life span, because conditions are far from ideal in nature. Juveniles, young adults, and old adults can all die. When trying to understand the dynamics of a population, it makes a large difference whether the individuals are dying mainly as adults or mainly as juveniles.

PATTERNS OF SURVIVAL

Looking at it another way, demographers want to know the pattern of survival for a given population. This can best be determined by identifying a cohort, which is defined as a group of individuals that are born at about the same time. That cohort is then followed over time, and the number of survivors is counted at set time intervals. The census stops after the last member of the cohort dies. The pattern of survival exhibited by the whole cohort is called its survivorship. Ecologists have examined the survivorship patterns of a wide array of species, including vertebrate animals, invertebrates, plants, fungi, algae, and even microscopic organisms. They have also investigated organisms from a variety of habitats, including oceans, deserts, rain forests, mountain peaks, meadows, and ponds. Survivorship patterns vary tremendously.

Some species have a survivorship pattern in which the young and middle-aged individuals have a high rate of survival, but old individuals die in large numbers. Several species of organisms that live in nature, such as mountain sheep and rotifers (tiny aquatic invertebrates), exhibit this survivorship pattern. At the other extreme, many species exhibit a survivorship pattern in which mortality is heaviest among the young. Those few individuals that are fortunate enough to survive the period of heavy mortality then enjoy a high probability of surviving until the end of their natural life span. Examples of species that have this pattern include marine invertebrates such as sponges and clams, most species of fish, and parasitic worms. An intermediate pattern is also observed, in which the probability of dying stays relatively constant as the cohort gets older. American robins, gray squirrels, and hydras all display this pattern.

These survivorship patterns are usually depicted on a graph that has the age of individuals in the cohort on the x axis and the number of survivors on the y axis. Each of the three survivorship patterns gives a different curve when the number of survivors is plotted as a function of age. In the first pattern (high survival among juveniles), the curve is horizontal at first but then swings downward at the right of the graph. In the second pattern (low survival among juveniles), the curve drops at the left of the graph but then levels out to form a horizontal line. That curve resembles a backward letter J. The third survivorship pattern (constant mortality throughout the life of the cohort) gives a straight line that runs from the upper-left corner of the graph to the lower right (this is best seen when the y axis is expressed as the logarithm of the number of survivors).

In the first half of the twentieth century, demographers Raymond Pearl and Edward S. Deevey labeled each survivorship pattern: Type I is high survival among juveniles, type II is constant mortality through the life of the cohort, and type III is low survival among juveniles. That terminology became well entrenched in the biological literature by the 1950s. Few species exhibit a pure type I, II, or III pattern, however; instead, survivorship varies so that the pattern may be one type at one part of the cohort's existence and another type later on. Perhaps the most common survivorship pattern, especially among vertebrates, is composed of a type III pattern for juveniles and young adults followed by a type I pattern for older adults. This pattern can be explained biologically. Most species tend to suffer heavy juvenile mortality because of predation, starvation, cannibalism, or the inability to cope with a stressful environment. Juveniles that survive this hazardous period then become strong adults that enjoy relatively low mortality. As time passes, the adults reach old age and ultimately fall victim to disease, predation, and organ-system failure, thus causing a second downward plunge in the survivorship curve.

PATTERNS OF REPRODUCTION

Demographers are not interested only in measuring the survivorship of cohorts. They also want to understand the patterns of reproduction, especially among females. Different species show widely varying patterns of reproduction. For example, some species, such as octopuses and certain salmon, reproduce only once in their life and then die soon afterward. Others, such as humans and most birds, reproduce several or many times in their life. Species that reproduce only once accumulate energy throughout their life and essentially put all of it into producing young. Reproduction essentially exhausts them to death. Conversely, those that reproduce several times devote only a small amount of their energy into each reproductive event.

Species also vary in their fecundity, which is the number of offspring that an individual makes when it reproduces. Large mammals have low fecundity, because they produce only one or two progeny at a time. Birds, reptiles, and small mammals have higher fecundity because they typically produce a clutch or litter of several offspring. Fish, frogs, and parasitic worms have very high fecundity, producing hundreds or thousands of offspring.

A species' pattern of reproduction is often related to its survivorship. For example, a species with low fecundity or one that reproduces only once tends to have type I or type II survivorship. Conversely, a species that produces huge numbers of offspring generally shows type III survivorship. Many biologists are fascinated by this interrelationship between survivorship and reproduction. Beginning in the 1950s, some demographers proposed mathematically based explanations as to how the interrelationship might have evolved as well as the ecological conditions in which various life histories would be expected. For example, some demographers predicted that species with low fecundity and type I survival should be found in undisturbed, densely populated areas (such as a tropical rain forest). In contrast, species with high fecundity and type III survival should prevail in places that are either uncrowded or highly disturbed (such as an abandoned farm field). Ecologists have conducted field studies of both plants and animals to determine whether the patterns that actually occur in nature fit the theoretical predictions. In some cases the predictions were upheld, but in others they were found to be wrong and had to be modified.

AGE STRUCTURES AND SEX RATIOS

Another feature of a population is its age structure, which is simply the number of individuals of each age. Some populations have an age structure characterized by many juveniles and only a few adults. Two situations could account for such a pattern. First, the population could be rapidly expanding, with the adults successfully reproducing many progeny that are enjoying high survival. Second, the population could be producing many offspring that have type III survival. In this second case, the size of the population can remain constant or even decline. Other populations have a different age structure, in which the number of juveniles only slightly exceeds the number of adults. Those populations tend to remain relatively constant over time. Still other populations have an age structure in which there are relatively few juveniles and many adults. Those populations are probably declining or are about to decline because the adults are not successfully reproducing.

Since most animals are unisexual, an important demographic characteristic of a population is its sex ratio, defined as the ratio of males to females. While the ratio for birds and mammals tends to be 1:1 at conception (the fertilization of an egg), it tends to be weighted toward males at birth, because female embryos are slightly less viable. After birth, the sex ratio for mammals tends to favor females, because young males suffer higher mortality. The post-hatching ratio in birds tends to remain skewed toward males, because females devote considerable energy to producing young and suffer higher mortality. As a result, male birds must compete with one another for the opportunity to mate with the scarcer females.

THE AGE-SPECIFIC APPROACH

To understand the demography of a particular species, one must collect information about its survivorship and reproduction. The best survivor ship data are obtained when a demographer follows a group of newly born organisms (this being a cohort) over time, periodically counting the survivors until the last one dies. Although that sounds relatively straightforward, many factors complicate the collection of survivorship data; demographers must be willing to adjust their methods to fit the particular species and environmental conditions.

First, a demographer must decide how many newborns should be included in the cohort. Survivorship is usually based on one thousand newborns, but few studies follow that exact number. Instead, demographers follow a certain number of newborns and multiply or divide their data so that the cohort is expressed as one thousand newborns. For example, one may choose to follow five hundred newborns; the number of survivors is then multiplied by two. Demographers generally consider cohorts composed of fewer than one hundred newborns to be too small. Second, methods of determining survivorship are much more different for highly motile organisms, such as mammals and birds, than for more sedentary ones, such as bivalves (oysters and clams). To determine survivorship of a sedentary species, demographers often find some newborns during an initial visit to a site and then periodically revisit that site to count the number of survivors. Highly motile animals are much more difficult to census because they do not stay in one place waiting to be counted. Vertebrates and large invertebrates can be tagged, and individuals can be followed by subsequently recapturing them. Some biologists use small radio transmitters to follow highly active species. The demography of small invertebrates such as insects is best determined when there is only one generation per year and members of the population are all of the same age-class. For such species, demographers merely count the number present at periodic intervals.

Third, the frequency of the census periods varies from species to species. Short-lived species, such as insects, must be censused every week or two. Longer-lived species need be counted only once a year. Fourth, the definition of a "newborn" may be troublesome, especially for species with complex life cycles. Demographic studies usually begin with the birth of an infant. Some would argue, however, that the fetus should be included in the analysis because the starting point is really conception. Many sedentary marine invertebrates (sponges, starfish, and barnacles) have highly motile larval stages, and these should be included in the analysis for survivorship to be completely understood. Parasitic roundworms and flatworms that have numerous juvenile stages, each found inside a different host, are particularly challenging to the demographer.

THE TIME-SPECIFIC APPROACH

The survivorship of long-lived species, such as large mammals, is really impossible to determine by the methods given above. Because of their sheer longevity, one could not expect a scientist to be willing to wait decades or centuries until the last member of a cohort dies. Demographers attempt to overcome this problem by using the age distribution of organisms that are alive at one time to infer cohort survivorship. This is often termed a "horizontal" or "time-specific" approach, as opposed to the "vertical" or "age-specific" approach that requires repeated observations of a single cohort. For example, one might construct a time-specific survivorship curve for a population of fish by live-trapping a sufficiently large sample, counting the rings on the scales on each individual (which for many species is correlated with the age in years), and then determining the number of one-year-olds, two-year-olds, and so on. Typically, demographers who use age distributions to infer age-specific survivorship automatically assume that natality and mortality remain constant from year to year. That is often not the case, however, because environmental conditions often change over time. Thus, demographers must be cautious when using age distribution data to infer survivorship.

Methods for determining fecundity are relatively straightforward. Typically, fertile individuals are collected, their ages are determined, and the number of progeny (eggs or live young) are counted. Species that reproduce continually (parasitic worms) or those that reproduce several times a year (small mammals and many insects) must be observed over a period of time.

Demographers usually want to determine whether the production of new offspring (natality) balances the losses attributable to mortality. To accomplish this, they construct a life table, which is a chart with several columns and rows. Each row represents a different age of the cohort, from birth to death. The columns show the survival and fecundity of the cohort. By recalculating the survivorship and fecundity information, demographers can compute several interesting aspects of the cohort, including the life expectancy of individuals at different ages, the cohort's reproductive value (which is the number of progeny that an individual can expect to produce in the future), the length of a generation for that species, and the growth rate for the population.

USES OF DEMOGRAPHY

Demographic techniques have been applied to non-human species, particularly by wildlife managers, foresters, and ecologists. Wildlife managers seek to understand how a population is surviving and reproducing within a certain area, and therefore to determine whether it is increasing or decreasing over time. With that information, a wildlife biologist can then estimate the effect of hunting or other management practice on the population. By extension, fisheries biologists can also make use of demographic techniques to determine the growth rate of the species of interest. If the population is determined to be increasing, it can be harvested without fear of depleting the population. Alternatively, one can conduct demographic analyses to see whether certain species are being overfished.

An often unappreciated benefit of survivorship analyses is that they can help ecologists pinpoint factors that limit population growth in an area. This may be especially important in efforts to prevent rare animals and plants from becoming extinct. Once the factor is identified, the population can be appropriately managed. Increasing amounts of public and private money are allocated each year to biologists who conduct demographic studies on rare species.

—*Kenneth M. Klemow*

FURTHER READING

Begon, Michael, John L. Harper, and Colin R. Townsend. *Ecology: Individuals, Populations, and Communities.* 3rd ed. Blackwell Scientific, 1996.

———, Martin Mortimer, and David J. Thompson. *Population Ecology: A Unified Study of Animals and Plants.* 3rd ed. Blackwell Science, 1996.

Brewer, Richard. *The Science of Ecology.* 2nd ed. Saunders College Publishing, 1994.

Elseth, Gerald D., and Kandy D. Baumgardner. *Population Biology.* Van Nostrand, 1981.

Gotelli, Nicholas J. *A Primer of Ecology.* 2nd ed. Sinauer Associates, 1998.

Hutchinson, G. Evelyn. *An Introduction to Population Ecology.* Yale University Press, 1978.

Smith, Robert Leo. *Elements of Ecology.* 4th ed. Benjamin/Cummings, 2000.

Wilson, Edward O., and William Bossert. *A Primer of Population Biology.* Sinauer Associates, 1977.

Animal Development: Evolutionary Perspective

Fields of Study

Biology; Embryology; Histology; Genetics

Abstract

The idea of a relationship between individual development, or ontogeny, and the evolutionary history of a race, or phylogeny, is an old one. During the early nineteenth century, two different concepts of parallels between development and evolution arose. By the late nineteenth century, Haeckel, and others, established the biogenetic law that ontogeny recapitulates the adult stages of phylogeny. With the establishment of Mendelian genetics in the early twentieth century, the biogenetic law was largely repudiated. Most scientists interested in the relationships between ontogeny and phylogeny chiefly use comparative and theoretical methods.

Key Concepts

- **acceleration:** the appearance of an organ earlier in the development of a descendant than in the ancestor as a result of an acceleration of development
- **heterochrony:** changes in developmental timing that produce parallels between ontogeny and phylogeny; changes in the relative time of appearance and rate of development for organs already present in ancestors
- **neoteny:** either the retention of immature characteristics in the adult form or the sexual maturation of larval stages
- **ontogeny:** the successive stages during the development of an animal, primarily embryonic but also postnatal
- **paedomorphosis:** the appearance of youthful characteristics of ancestors in later ontogenetic stages of descendants
- **phylogeny:** a series of stages in the evolutionary history of species and lineages
- **recapitulation:** the repetition of phylogeny in ontogeny or of the ancestral adult stages in the embryonic stages of descendants
- **retardation:** the appearance of an organ later in the development of a descendant than in the ancestor as a result of a slowing of development

Ontogeny and Phylogeny

The idea of a relationship between individual development, or ontogeny, and the evolutionary history of a race, or phylogeny, is an old one. The concept received much attention in the nineteenth century and is often associated with the names of Karl Ernst von Baer and Ernst Haeckel, two prominent German biologists. It was Haeckel who coined the catchphrase and dominant paradigm: Ontogeny recapitulates (or repeats) phylogeny. Since Haeckel's time, however, the relations between ontogeny and phylogeny have been portrayed in a variety of ways, including the reverse notion that phylogeny is the succession of ontogenies. Research in the 1970s and 1980s on the parallels between ontogeny and phylogeny focused on the change of timing in developmental events as a mechanism for

Ernst Haeckel coined the phrase, "Ontogeny recapitulates phylogeny," meaning that in their individual development, animals repeat the major evolutionary steps taken by the species as a whole.

recapitulation and on the developmental-genetic basis of evolutionary change.

CONCEPTS OF BIOGENETIC LAW IN THE NINETEENTH CENTURY

During the early nineteenth century, two different concepts of parallels between development and evolution arose. The German J. F. Meckel and the Frenchman E. R. Serres believed that a higher animal in its embryonic development recapitulates the adult structures of animals below it on a scale of being. Baer, on the other hand, argued that no higher animal repeats an earlier adult stage but rather the embryo proceeds from undifferentiated homogeneity to differentiated heterogeneity, from the general to the specific. Von Baer published his famous and influential four laws in 1828: The more general characters of a large group of animals appear earlier in their embryos than the more special characters; from the most general forms, the less general are developed; every embryo of a given animal, instead of passing through the other forms, becomes separated from them; the embryo of a higher form never resembles any other form, only its embryo.

By the late nineteenth century, the notion of recapitulation and Baer's laws of embryonic similarity were recast in evolutionary terms. Haeckel, and others, established the biogenetic law: That is, ontogeny recapitulates the adult stages of phylogeny. It was, in a sense, an updated version of the Serres-Meckel law but differed in that the notion was valid not only for a chain of being but also for many divergent lines of descent; ancestors had evolved into complex forms and were now considered to be modified by descent. More specifically, Haeckel thought of ontogeny as a short and quick recapitulation of phylogeny caused by the physiological functions of heredity and adaptation. During its individual development, he wrote, the organic individual repeats the most important changes in form through which its forefathers passed during the slow and long course of their paleontological development. The adult stages of ancestors are repeated during the development of descendants but crowded back into earlier stages of ontogeny. Ontogeny is the abbreviated version of phylogeny. These repeated stages reflect the history of the race. Haeckel considered phylogeny to be the mechanical cause of ontogeny.

The classic example of recapitulation is the stage of development in an unhatched bird or unborn mammal when gill slits are present. Haeckel argued that gill slits in this stage represented gill slits of the adult stage of ancestral fish, which in birds and mammals were pressed back into early stages of development. This theory differed from von Baer's notion that the gill slit in the human embryo and in the adult fish represented the same stage in development. The gill slits, explained the recapitulationists, got from a large adult ancestor to a small embryo in two ways: first, terminal addition (in which stages are added to the end of an ancestral ontogeny); and second, condensation (in which development is speeded up as ancestral features are pushed back to earlier stages of the embryo). Haeckel also coined another term widely used currently in another sense: "heterochrony." He used the term to denote a displacement in time of the appearance of one organ in ontogeny before another, thus disrupting the recapitulation of phylogeny in ontogeny. Haeckel was not, however, interested primarily in mechanisms or in embryology for its own sake, but rather for the information it could provide for developing evolutionary histories.

RECAPITULATION IN THE TWENTIETH CENTURY

With the rise of mechanistic experimental embryology and with the establishment of Mendelian genetics in the early twentieth century, the biogenetic law was largely repudiated by biologists. Descriptive embryology was out of fashion, and the existence of genes made the two correlate laws to recapitulation—terminal addition and condensation—untenable. One of the most influential modifications for later work on the subject was broached in a paper by Walter Garstang in 1922, in which he reformulated the theory of recapitulation and refurbished the concept of heterochrony. Garstang argued that phylogeny does not control ontogeny but rather makes a record of the former: That is, phylogeny is a result of ontogeny. He suggested that adaptive changes in a larval stage coupled with shifts in the timing of development (heterochrony) could result in radical shifts in adult morphology.

Stephen Jay Gould resurrected the long unpopular concept of recapitulation with his book *Ontogeny and Phylogeny* (1977). In addition to recounting the historical development of the idea of recapitulation, he made an original contribution to defining and explicating the mechanism (heterochrony) involved in producing parallels between ontogeny and phylogeny. He argued that heterochrony—"changes in the relative time of appearance and rate of development for characters already present in ancestors"—was of prime evolutionary importance. He reduced Gavin de Beer's complex eight-mode analysis of heterochrony to two simplified processes: acceleration and retardation. Acceleration occurs if a character appears earlier in the ontogeny of a descendant than it did in an ancestor because of a speeding up of development. Conversely, retardation occurs if a character appears later in the ontogeny of a descendant than it did in an ancestor because of a slowing down of development. To demonstrate these concepts, Gould introduced a "clock model" in order to bring some standardization and quantification to the heterochrony concept.

He considered the primary evolutionary value of ontogeny and phylogeny to be in the immediate ecological advantages for slow or rapid maturation rather than in the long-term changes of form. Neoteny (the opposite of recapitulation) is the most important determinant of human evolution. Humans have evolved by retaining the young characters of their ancestors and have therefore achieved behavioral flexibility and their characteristic form. For example, there is a striking resemblance between some types of juvenile apes and adult humans; this similarity for the ape soon fades in its ontogeny as the jaw begins to protrude and the brain shrinks. Gould also insightfully predicted that an understanding of ontogeny and phylogeny would lead to a rapprochement between molecular and evolutionary biology.

By the 1980s, Rudolf Raff and Thomas Kaufman found this rapprochement by synthesizing embryology with genetics and evolution. Their work focuses on the developmental-genetic mechanisms that generate evolutionary change in morphology. They believe that a genetic program governs ontogeny and that the great decisions in development are made by a small number of genes that function as switches between alternate states or pathways. When these genetic switch systems are modified, evolutionary changes in morphology occur mechanistically. They argue further that regulatory genes—genes that control development by turning structural genes on and off—control the timing of development, make decisions about the fates of cells, and integrate the expression of structural genes to produce differentiated tissue. All this plays a considerable role in evolution.

DESCRIPTION VERSUS EXPERIMENTATION

Both embryology and evolution have traditionally been descriptive sciences using methods of observation and comparison. By the end of the nineteenth century, a dichotomy had arisen between the naturalistic (descriptive) and the experimentalist tradition. The naturalists' tradition viewed the organism as a whole, and morphological studies and observations of embryological development were central to their program. Experimentalists, on the other hand, focused on laboratory studies of isolated aspects of function. A mechanistic outlook was compatible with this experimental approach.

Modern embryology uses both descriptive and experimental methods. Descriptive embryology uses topographic, histological (tissue analysis), cytological (cell analysis), and electron microscope techniques supplemented by morphometric (the measurement of form) analysis. Embryos are visualized using either plastic models of developmental stages, schematic drawings, or computer simulations. Cell lineage drawings are also used with the comparative method for phylogenies.

Experimental embryology, on the other hand, uses more invasive methods of manipulating the organism. During this field of study's early period, scientists subjected amphibian embryos to various changes to their normal path of development; they were chopped into pieces, transplanted, exposed to chemicals, and spun in centrifuges. Later, fate maps came into usage in order to determine the future development of regions in the embryo. It was found that small patches of cells on the surface of the embryo could be stained, without damaging the cell, by applying small pieces of agar soaked in a vital dye. One could then follow the stained cells to their eventual position in the gastrula.

INTERDISCIPLINARY STUDIES

Evolutionary theory primarily uses paleontology (study of the fossil record) to study the evolutionary history of species, yet Gould has also used quantification (the clock model, for example), statistics, and ecology to understand the parallels between ontogeny and phylogeny. Most scientists interested in the relationships between ontogeny and phylogeny chiefly use comparative and theoretical methods. They, for example, compare structures in different animal groups or compare the adult structures of an animal with the young stage of another. If similarities exist, are the lineages similar? Are the stages in ontogenetic development similar to those of the development of the whole species?

Yet, the study of relationships between ontogeny and phylogeny is an interdisciplinary subject. Not only are methods from embryology and evolutionary theory of help, but also, increasingly, techniques are applied from molecular genetics. Haeckel's method was primarily a descriptive historical one, and he collected myriad descriptive studies of different animals. Although scientists in those days had relatively simple microscopes, they left meticulous and detailed accounts.

A fusion of embryology, evolution, and genetics involves combining different methods from each of the respective disciplines for the study of the relationship between ontogeny and phylogeny. The unifying approach has been causal-analytical, in the sense that biologists have been examining mechanisms that produce parallels between ontogeny and phylogeny as well as the developmental-genetic basis for evolutionary change. The methods are either technical or theoretical. The technical ones include the use of the electron microscope, histological, cytological, and experimental analyses; the theoretical methods include comparison, historical analysis, observations, statistics, and computer simulation.

RAMIFICATIONS BEYOND SCIENCE

The relationship between ontogeny and phylogeny is one of the most important ideas in biology and a central theme in evolutionary biology. It illuminates the evolution of ecological strategies, large-scale evolutionary change, and the biology of regulation. This scientific idea has also had far-reaching influences in areas such as anthropology, political theory, literature, child development, education, and psychology.

In the late nineteenth century, embryological development was a major part of evolutionary theory; however, that was not the case for much of the twentieth century. Although there was some interest in embryology and evolution from the 1920s to 1950s by Garstang, J. S. Huxley, de Beer, and Richard Goldschmidt, during the first three decades of the twentieth century genetics and development were among the most important and active areas in biological thought, yet there were few attempts to integrate the two areas. It is this new synthesis of evolution, embryology, and genetics that has emerged as one of the most exciting frontiers in the life sciences.

Although knowledge to be gained from a synthesis of development and evolution seems not to have any immediate practical application, it can offer greater insights into mechanisms of evolution, and a knowledge of evolution will give similar insights into mechanisms of development. A study of these relations and inter actions also enlarges humankind's understanding of the nature of the development of individuals and their relation to the larger historical panorama of the history of life.

—*Kristie Macrakis*

FURTHER READING

Bonner, J. T. *Morphogenesis: An Essay on Development.* Princeton University Press, 1952.

De Beer, Gavin. *Embryos and Ancestors.* Clarendon Press, 1951.

Gould, Stephen Jay. *Ontogeny and Phylogeny.* Harvard University Press, 1977.

———. "Ontogeny and Phylogeny—Revisited and Reunited." *BioEssays* 14, no. 4 (April 1992): 275–280.

Hall, Brian K. *Evolutionary Developmental Biology.* 2nd ed. Chapman & Hall, 1998.

Raff, Rudolf A., and Thomas C. Kaufman. *Embryos, Genes, and Evolution: The Developmental-Genetic Basis of Evolutionary Change.* Macmillan, 1983.

Schwartz, Jeffrey H. *Sudden Origins: Fossils, Genes, and the Emergence of Species.* Wiley, 1999.

Animal Domestication

Fields of Study

Archaeology; Anthropology; Biology; Genetics: Agriculture

Abstract

Although taming a wild animal, that is, acclimating it to the presence of people, is invariably a first step in domestication, the critical aspect is directing its breeding toward a functional goal. Genetic selection for desirable traits and against undesirable ones is a crucial feature of domestication. Mere propagation of a species or maintaining genetic diversity is insufficient to constitute domestication. Domestication is a synergistic relationship between humans and the animals involved.

Key Concepts

dominance social behavior: organization of a familial group around a dominant leader, whom the rest of the group follows

synergism: a mutually beneficial relationship between two (or more) different species that allows both to better survive and prosper

territorial social behavior: a social structure in which different unaffiliated familial groups within a limited geographical area are dominated by one, typically solitary, leader

TAMING VERSUS DOMESTICATION

Although taming a wild animal, that is, acclimating it to the presence of people, is invariably a first step in domestication, the critical aspect is directing its breeding toward a functional goal. Genetic selection for desirable traits and against undesirable ones is a crucial feature of domestication. Captive wild animals in zoos or circuses cannot be considered domesticated, although they may be quite tame and their breeding is controlled by humans. However, this breeding is not directed toward some functional goal. Mere propagation of a species or maintaining genetic diversity is insufficient to constitute domestication. Domestication is a synergistic relationship between humans and the animals involved. Humans benefit from the resources and services that the animals provide, such as food, fiber, shelter, clothing, work, sentinel duty, and companionship. Animals benefit by having humans protect them from harm, provide for their needs, and increase their numbers and range.

CHARACTERISTICS OF ANIMALS FAVORING DOMESTICATION

In 1865, the English naturalist Francis Galton suggested the following six physiological and behavioral characteristics that make some animals good candidates for domestication: hardiness, dominance social behavior, herd behavior, utility to humans, facile (easy) reproduction, and facile husbandry. First, by hardiness, he refers to the ability of the young to be removed from its mother and to be around humans. The guinea pig is perhaps an extreme example of tolerating removal from its mother, as it is born ready to eat solid food. Most mammals, on the other hand,

Dogs were originally domesticated to help humans hunt. Grace Goodhue Coolidge (1879–1957) with two dogs, September 1924.

initially depend on their mother's milk. Primates are poor subjects for domestication because of their helplessness at birth and their relatively long dependence on their mothers for food and nurturing.

Second, dominance social behavior is in contrast to territorial behavior and refers to one animal assuming leadership, with the rest of the group acquiescing to him or her in the hierarchy. In domestication, humans co-opt the function of the leader, and animals remain submissive even as adults. Third, herd animals are contrasted to solitary animals, or ones that disperse in response to danger. Domesticated animals are penned or otherwise restricted at various times. If they remain together in herds, they are easier to manage. Fourth, utility to humans includes their use for food, fiber, work, companionship, and even worship. Humans would not make the effort to domesticate an animal unless it had some perceived value. The purpose for domestication may change with time, however. It is likely that the initial motivation for domesticating cattle was for worship—to capture the strength and aura of these animals, which were revered and used in religious ceremonies. Work, such as pulling carts, packing, and riding, became a subsequent goal, while contemporary utility in Western societies involves meat and milk production.

The fifth characteristic is facile reproduction under confined conditions; animals with finicky reproductive behaviors and/or elaborate courtship rituals make poor candidates for domestication. Sixth, facile husbandry refers to placid disposition and versatility in terms of nutrition. Animals that are high-strung or dependent on unique foodstuffs would be weak prospects for domestication. Koalas, which eat leaves from only certain eucalyptus trees, are poor candidates. On the other hand, pigs and goats are excellent choices because they are not very fastidious in their eating habits. These six characteristics, enunciated more than a century ago, apply strongly to livestock species, but less well to dogs and cats; with regard to the latter, it has been argued that cats are not so much domesticated as merely tolerant of humans.

HISTORY OF ANIMAL DOMESTICATION

Archaeological evidence suggests that agriculture developed about ten thousand years ago, after the last ice age, at a time when the climate became warmer and more stable. Predictability of the weather is particularly crucial for plant domestication, which apparently developed synergistically with animal domestication, leading to agriculture. Domesticated animals simplified the acquisition of food and provided food storage, in the form of "walking larders"; larger animals, such as cattle, buffalo, and horses, permitted heavier work to be done and larger distances to be covered. Farming the land made it possible for humans to abandon the nomadic lifestyle of hunter-gatherers and to adopt a more sedentary lifestyle. This allowed for development of new technologies, professional specializations, and new forms of social organization.

Domestication of any animal did not occur at once, but rather over a substantial period of time, perhaps hundreds of years. Furthermore, estimating the dates for domestication is subject to considerable uncertainty and may need to be modified as new information becomes available. For some species, domestication occurred independently at more than one location. The process may have begun almost accidentally, as by raising a captured young animal after its mother had been killed and observing its behavior and response to various treatments. The domestication of an animal subsequently spread from the site of origin through trade or war.

Animal domestication occurred in various parts of the world. In the Middle East, the Fertile Crescent, stretching from Palestine to southern Turkey and down the valley of the Tigris and Euphrates Rivers, was an important site. Sheep, goats, cattle, and pigs were domesticated there by around 6000 BCE. The Indian subcontinent and east Asia were independent sites for domesticating cattle and pigs, respectively. Llamas, alpacas, and guinea pigs were domesticated in the Andes Mountains of South America. Domestication of cats occurred in Egypt and of rabbits in Europe. No native animals were domesticated in Australia, likely because none of them were suitable for domestication. It is worthy of note that few domestications have occurred in the past thousand years. It is also of interest that species suitable for domestication were not evenly distributed over the globe, which probably has had a lasting effect on the differential development of various cultures.

Archaeology, coupled with the natural history of domesticated animals and their wild relatives, has been essential in reconstructing the history of domestication. Examining skeletal remains at archaeological sites for changes in morphology and distribution by age and sex has helped scientists to deduce the extent of domestication. Lately, traditional archaeology has been supplemented by the methods of molecular biology. Examining extant breeds for their genetic relatedness has been particularly useful in distinguishing single versus dual sites of domestication.

EARLY DOMESTICATES: DOGS AND REINDEER

Dogs (*Canis familiaris*) are generally recognized as the earliest known domesticated animals. They were widespread across the Northern Hemisphere before other animals were domesticated. They were derived from wolves (*Canis lupus*), with whom they are fully interfertile. While there is evidence that dogs as a species have existed for at least 30,000 years, the earliest known domesticated dog is in a burial site in Northern Iraq that dates from 12,000 to 10,000 BCE. Other sites, dating from 10,000 to 7000 BCE, have been documented in England, Palestine, Japan, and Idaho. While it may have first occurred in China (a Chinese wolf has some of the detailed physical features of dogs), domestication probably occurred at a number of separate sites. Dogs accompanied the American Indians when they occupied the Americas in several waves prior to the end of the last ice age. Dingoes were brought to Australia by trade from Asia long after the Aborigines settled that continent 40,000 years ago. While dogs were considered a food animal, they have long been used for guarding, hunting, and companionship. Subsequently, they were developed for herding.

Reindeer (*Rangifer tarandus*) were another early domesticate, dating from around 12,000 BCE in northern Scandinavia and Russia. The reindeer has been little changed by domestication and its range has not been extended by the process. They are well-suited to their environment, but attempts to establish reindeer industries in Canada and Alaska have not been successful. Herding reindeer continues as a principal occupation of the Laplanders of Finland, Sweden, and Norway. Reindeer are used for draft (pulling loads), clothing and shelter (skins), tools (antlers), and food (meat and milk). Farming several other deer species (such as *Cervus dama* and *Cervus elaphus*) has recently gained in importance in New Zealand and western Europe, where they are raised for meat (venison) and "velvet," the new growth of antlers, the basis for traditional medicines in Asia. However, because these latter species have had little opportunity to be changed, they cannot be considered domesticated.

SHEEP, PIGS, AND CATTLE

Sheep (*Ovis aries*) were the first of the common food animals to be domesticated. They were derived from wild sheep (*Ovis orientalis*) and were first domesticated in the western Fertile Crescent around 7000 BCE. Goats (*Capra hircus*), derived from Persian wild goats (*Capra aegarus*), were first domesticated in the central Fertile Crescent slightly later, between 6000 and 7000 BCE. Sheep and goats were used for food, skins, and fiber (wool or hair). Both were later selected for milk production.

Pigs (*Sus domesticus*) probably originated at two separate sites, the central Fertile Crescent around 6000 BCE and in eastern Asia around 5000 BCE. Derived from wild pigs, they were primarily raised for meat. Despite restrictions against eating pork by Muslims and Jews, it has long been the principal meat consumed in the world. The most populous country, China, has nearly 50 percent of the world's pigs.

Cattle (*Bos taurus* and *Bos indicus*) are derived from now-extinct wild cattle (the aurochs, *Bos primigenius*) that ranged over much of Europe and Asia. They were probably domesticated at two locations independently, the western Fertile Crescent around 6000 BCE for *Bos taurus* and the Indian subcontinent around 5000 BCE for *Bos indicus*. Initially, the animals were worshiped and used in religious ceremonies. Reverence for cattle is still practiced by Hindus in India. Subsequently, they were developed for work, meat, and milk. Their hides are made into leather. Traditional cattle in Africa are derived from initial importations of *Bos taurus* and subsequent importations of male *Bos indicus*.

OTHER DOMESTICATED ANIMALS

Asiatic buffaloes (*Bubalus bubalis*) were domesticated as the water buffalo in India (3000 BCE) and as the swamp buffalo in east or southeast Asia (2000 BCE). While both were developed as draft animals, the water buffalo has also been selected as a dairy animal. Fully half of the milk production in India comes from buffaloes. In spite of its tropical origin, the Asiatic buffalo is not very heat tolerant and compensates by immersing in water or mud. Neither the African buffalo (*Syncerus caffer*) nor the American buffalo (more properly, bison, *Bison bison*) have been domesticated. Yaks (*Poephagus [Bos] grunniens*) were domesticated around 3000 BCE in Tibet or surrounding areas, where they are used as pack animals and as a source of milk, hair, hides, and, usually after an otherwise productive life, meat.

Horses (*Equus caballus*) originated from wild horses in the Caucasus Mountains around 4000 BCE. Originally used for food and skins, they were also developed for draft and, much later, for riding. Because they came to the Middle East. after the development of written language, their arrival is documented, so scholars do not need to depend solely on the archaeological record. Donkeys (*Equus asinus*) were domesticated in the Middle East or Northern Africa (3000 BCE). They are used for carrying material or people, as is the mule, an infertile cross between a horse and a donkey.

Llamas (*Lama glama*) and alpacas (*Lama pacos*) were domesticated in Peru by Incas around 4000 BCE. Llamas are from wild guanacos and alpacas from wild vicuñas, found at higher elevations. Llamas are used as pack animals, alpacas are valued for their fine wool, and both serve as sources of meat. Camels, the one-humped dromedary (*Camelus dromedarius*) and the two-humped Bactrian (*Camelus bactrianus*) were domesticated in Arabia (2000 BCE) and Central Asia (1500 BCE), respectively. Both are pack animals, with the dromedary also used for meat.

Guinea pigs (*Cavia porcellus*) were domesticated in Peru around 3000 BCE. They continue to be used as a meat animal in parts of South America. Rabbits (*Oryctlagus cuniculus*) were domesticated between 600 and 1000 CE in France. They are primarily raised for meat and fur, with angora rabbits producing a fine wool.

Cats (*Felis catus*) are the animals least changed, morphologically, by domestication. In addition, they are quite capable of surviving without human intervention. Their domestication occurred relatively late, around 2000 BCE, in Egypt, the home of the African wild cat (*Felis catus libyca*), which resembles domestic tabby cats. The early Egyptians adopted cats enthusiastically, deifying them and prohibiting their export. After conversion to Christianity, Egyptians ceased worshiping cats, which were carried to all parts of the Roman Empire and thence to the rest of the world. Cats have been used for companionship and for rodent control.

Chickens (*Gallus gallus*), along with ducks and geese (from China, the Middle East and Europe), turkeys (from North America), Muscovy ducks (from South America), and guinea fowl (from Africa) are avian species that have been domesticated. Chickens were probably derived from wild red junglefowl in southeast Asia before 2000 BCE. Cockfighting was an initial purpose for their domestication. They acquired religious significance and were also used for meat and feathers. Their selection for egg production has been a recent development. Because of improvements in breeding, feeding, and management, poultry meat production is increasing rapidly and is second to pork worldwide. In the past few centuries, two other avian species, ostriches (from Africa) and emus (from Australia), have been farmed, but it is probably incorrect to call them domesticated, as they have so far been little changed from their wild relatives.

Two insects have also been domesticated: honeybees and silkworms. Honeybees were domesticated shortly after the last ice age and were the primary source of dietary sweetener until two hundred years ago. They were also valuable for wax and venom, the latter for medicinal purposes. Silkworms, one of the ten varieties of silk-producing insects, were domesticated around 3000 BCE in China, producing fiber used in apparel.

—*James L. Robinson*

FURTHER READING

Clutton-Brock, Juliet. *A Natural History of Domesticated Mammals.* 2nd ed. Cambridge University Press, 1999.

Diamond, Jared. *Guns, Germs, and Steel: The Fates of Human Societies.* W. W. Norton, 1997.

Harris, David R., ed. *The Origins and Spread of Agriculture and Pastoralism in Eurasia.* Smithsonian Institution Press, 1996.

Ryden, Hope. *Out of the Wild: The Story of Domesticated Animals.* Lodestar Books, 1995.

Smith, Bruce D. *The Emergence of Agriculture.* Scientific American Library, 1995.

Animal Embryology

Fields of Study

Biology; Physiology; Embryology

Abstract

The formation of gametes, eggs and sperm, is usually considered the beginning of embryology. When a sperm first penetrates the egg, the polarity of the cell changes and chemicals are released by the membrane, which make it impossible for other sperm of that species to enter the same egg. Following fertilization, a period known as cleavage begins, during which cells divide rapidly. The various body organs begin to form from three germ layers that develop. Once the major organs have differentiated, the embryo matures and grows, a process usually called gestation.

Key Concepts

cleavage: cell division in the early embryo that, unlike division in adults, involves little or no growth between divisions

fertilization: the process by which the egg and sperm unite to form the zygote

gametes: the haploid cells, ova and spermatozoa, that fuse to form the diploid zygote

gastrula: the stage of development during which the endoderm (gut precursor) and the mesoderm (muscle and connective tissue precursor) are internalized

haploid: having only one of each kind of chromosome

zygote: the single cell formed when gametes from the parents (ova and sperm) unite, a one-celled embryo

HISTORICAL OVERVIEW

For thousands of years, humans have wondered how they and other organisms came to be. By 340 BCE, Aristotle had described the development of the chicken in the egg, but since most early embryos are too small to be seen by an unaided eye, his and later descriptions of development started with larger, more formed embryos. That did not change very much until the late 1600s, when development of the microscope gave a glimpse of life too small to be seen unmagnified. By the early eighteenth century, the developmental patterns of many organisms had been observed and described. There was, however, still much disagreement about how the early stages progressed. The majority of scientists believed in the theory of preformation, which said that a preformed embryo was present in the gametes. There were two main factions among the preformationists. The ovists believed that inside the egg was a tiny, fully formed organism that was stimulated to grow by the seminal fluid. Their opponents, the spermists, believed that the fully formed miniature organism was in the sperm and was nourished in its growth by the ovum. Thus, seventeenth and eighteenth century drawings of sperm and of eggs often show fully formed bodies within.

By the end of the eighteenth century, more and more scientists were deserting preformation in favor of the theory of epigenesis, first proposed by Caspar Wolff in 1789, which stated that development occurs through growth and remodeling of embryonic cells. Karl Ernst von Baer, who had published a collection of his observations and the observations of others, proposed that general features that are common to large groups of taxonomically related organisms appear earlier in development than more specialized features of individual species. After Darwin published his evolutionary theories, Müller, Haekel, and other proponents of Baer's law and of evolution proposed that the embryonic development of an organism (ontology) mirrored its evolution (phylogeny). Although this has been shown not to apply to all organisms or to all developmental sequences, it can be seen in the development of many embryos.

During the late nineteenth and early twentieth centuries, scientists' understanding of embryonic development increased dramatically as they began applying recently discovered knowledge in evolution, genetics, and cell biology to embryology. Edwin Ray Lankester and Hans Speeman were two prominent scientists who studied comparative embryonic development at that time. Also at that time, the new science of experimental embryology began as Wilhelm Roux and G. Schmidt manipulated the cells of amphibian embryos and began to discover how and why development occurred. Today, new discoveries in biology and chemistry are applied to the study of embryonic development.

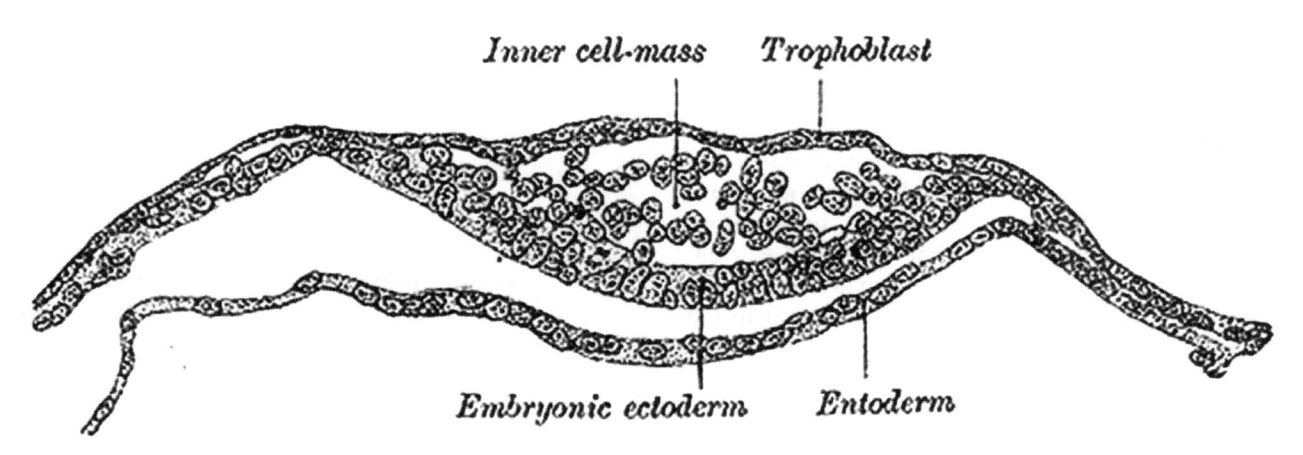

Section through embryonic disk of Vespertilio murinus.

GAMETOGENESIS

The formation of gametes, eggs and sperm, is usually considered the beginning of embryology. In sperm formation, two things need to occur, reduction of chromosomes to the haploid state and maturation of the cytoplasm. During the first part of spermatogenesis, immature cells, called spermatogonia, form four haploid cells, called spermatids, by meiosis. Spermatids then go through a maturation process in which they become streamlined and motile. They also develop an acrosome that has enzymes needed to penetrate the egg. Like sperm, eggs must become haploid and mature, but both the timing and maturation are quite different. Maturation of the cytoplasm often begins before meiosis. All the cytoplasm of the early embryo comes from the egg, so immature ova are aided by various helper cells that increase each ovum's cytoplasm and add food stores, called yolk. The amount of yolk varies considerably, from mammals that have no yolk, to birds that have huge amounts. Depending on the species, meiosis can begin at any time during cytoplasmic maturation and can be a continuous process or have one or more pauses. In sea stars and many other organisms, meiosis is complete before fertilization, while in others, such as nematodes, the egg matures fully and is released by the ovary before any meiosis begins. Sperm penetration then triggers the onset of meiosis.

FERTILIZATION AND DEVELOPMENT

Once sperm have reached the egg, the acrosomal enzymes must digest the various protective layers that surround the egg, and recognition structures on the surface of the sperm must be complementary to recognition structures on the egg cell membrane. The sperm's nucleus then enters the egg and fuses with the haploid egg nucleus. This forms a diploid cell called the zygote. Interestingly, when a sperm first penetrates the egg, the polarity of the cell changes and chemicals are released by the membrane, which make it impossible for other sperm of that species to enter the same egg.

Following fertilization, a period known as cleavage begins. During this time, cells divide rapidly with little or no growth between cell divisions. Cells become smaller and more numerous. At the end of cleavage, a structure called the blastula is formed. In some animals, such as echinoderms, amphibians, and nonvertebrate chordates, the blastula is a hollow ball of cells. In higher vertebrates, the blastula is a flat, dish-shaped structure, often called the blastodisc. In mammals, the blastula is called a blastocyst, and consists of a hollow ball of cells, called the trophoblast, and a group of internal cells, called the inner cell mass. During gastrulation, surface cells become internalized to form the three germ layers—ectoderm, mesoderm, and endoderm—that are seen in most animal embryos. A second internalization, this time of some ectodermal cells, forms the beginning of the central nervous system. After this neurulation, the various body organs begin to form from the three germ layers. As these changes progress, cells become less general and more specialized, a process called differentiation. Once the major organs have differentiated, the embryo matures and grows, a process usually called gestation. The time it takes for embryonic development varies considerably. In chickens and small rodents, the process takes about three weeks; in humans, it takes approximately nine months, while in elephants, the process can take almost two years. Some organisms emerge in very immature states that require more development. Amphibians and arthropods hatch as feeding larvae that must grow before they begin a metamorphosis that leads to the adult. Marsupials are also born at a very immature stage and must complete their embryonic development inside the mother's pouch.

—*Richard W. Cheney, Jr.*

ANIMAL EMOTIONS

FIELDS OF STUDY

Psychology; Biochemistry; Animal Behavior; Anthropology

ABSTRACT

In attempting to prove the existence and extent of emotions in animals, researchers have struggled with the question of how to identify and measure feelings in various species. For many scientists, it is nonsense to speak of animal emotions without the capacity to objectively define and measure them. Such scientists have an aversion to the nonscientific tendency to ascribe humanlike characteristics to animals. Anthropomorphism is the term used to describe this tendency.

KEY CONCEPTS

amygdala: subcortical brain structure related to emotional expression
anthropomorphism: attributing human characteristics to animal behavior
dopamine: neurotransmitter involved in movement and reward systems
field observations: observing behavior in naturalistic settings
limbic system: brain structures related to the regulation of emotions
oxytocin: hormone involved with pleasure during bonding
primary emotions: emotions related to innate motivations

FURTHER READING

Bronson, F. H. *Mammalian Reproductive Biology*. University of Chicago Press, 1989.
Carlson, B. *Patten's Foundations of Embryology*. 6th ed. McGraw-Hill, 1996.
Hartl, Daniel L., and Elizabeth W. Jones. *Essential Genetics*. 2nd ed. Jones and Bartlett, 1999.
Kumé, Matazo, and Katsuma Dan. *Invertebrate Embryology*. Translated by Jean C. Dan. NOLIT Publishing House for the US Department of Health and Human Services, 1968.

secondary emotions: emotions with a strong social component

ANTHROPOMORPHISM

In attempting to prove the existence and extent of emotions in animals, researchers have struggled with the question of how to identify and measure feelings in various species. For many scientists, it is nonsense to speak of animal emotions without the capacity to objectively define and measure them. Such scientists have an aversion to the nonscientific tendency to ascribe humanlike characteristics to animals. Anthropomorphism is the term used to describe this tendency.

DEFINING AND COMMUNICATING EMOTIONS

Defining emotions can be difficult even in humans. Psychologists view emotions as organized psychological and physiological reactions to change in one's relationship to the world. An emotion is a positive or negative transitory experience that is felt with some intensity. Emotional reactions are partly subjective experiences and partly objectively measurable patterns of behavior and physiological arousal. The subjective experiences can include how a person appraises a situation and what actions result from that appraisal. For example, when a student receives a passing grade on an extremely difficult exam, she may experience joy after appraising the situation as a success. Even with this appraisal, however,

Figure 15 from Charles Darwin's The Expression of the Emotions in Man and Animals. *Caption reads "FIG. 15.—Cat terrified at a dog. From life, by Mr. Wood."*

humans cannot decide to experience joy or some another emotion.

The subjective aspects of emotions are triggered by the thinking self and are felt as happening to the self. Objective aspects of emotions include learned and innate physiological responses and expressive displays. The expressive displays include smiles, frowns, and squinting of the eyes. The innate physiological responses are biological adjustments needed to perform the actions generated by the emotional experience. For example, if anger develops in a person, heart rate increases in order to supply additional oxygen to the muscles.

Since animals do not have the capacity of speech, any inner states cannot be expressed directly to a scientific observer. Consequently, field observations of behavior are often used to infer emotions in animals. There are problems, however, in assessing emotions through behavioral manifestations. It becomes difficult or impossible to attribute an emotion to an act with many possible motivations. If a dog chews on the shoes of an owner who is out on a date, does this indicate jealously, anger, boredom, or merely a poorly trained pet?

Historically, animals have been seen from a mechanistic perspective as being without the capacity for humanlike emotions. Behaviorism dictated that instincts and patterns of reinforcement in the environment provided the motivation for the behavior of animals. For centuries, Christian religions also promoted the idea that animals lacked humanlike emotions. The role of animals was to serve the needs of humans. The concept of "speciesism" suggested that only humans were capable of emotions because of their special place in creation.

Charles Darwin was one of the first scientists to study animal emotions and to utilize field observations to ascribe emotions to animals. In his book, *The Expression of the Emotions in Man and Animals* (1872), Darwin stressed the communicative aspects of emotion. Positive inner states were expressed through a signal for sociability, while aggressiveness indicated a desire for isolation. He believed that species developed special social signals to indicate how they would react to a social encounter. Yet the behavioristic view of animals continued to dominate the debate about animal emotions. Over a hundred years later, Jane Goodall, in her book *The Chimpanzees of Gombe: Patterns of Behavior* (1986), was criticized by the scientific community for suggesting that chimpanzees had personalities and experienced excitement and joy.

PRIMARY AND SECONDARY EMOTIONS

Today even the most critical scientists accept the fact that many animals experience a core group of emotions that are similar to those found in humans. Making the distinction between primary and secondary emotions, there exists some agreement about the basic emotions of fear and aggression. The primary emotions, such as fear, involve instinctual tendencies that are essential to survival. Fear permits escape from dangerous situations or predators. The fight-or-flight response is an instinctual pattern of behavior found in response to danger. The primary emotions, which are instinctual or hardwired into many species, can be demonstrated quite easily. When a specific stimulus is presented to an animal, a predictable response takes place. For example, if the shadow of a hawk is projected on the ground among a group of chickens, the birds will respond with "fear" and attempt to get under cover.

It is the realm of secondary emotions that creates the most controversy between those with opposing views about the extent of animal emotions. Expressions of love, grief, or jealousy may be commonplace among humans, but it is debatable whether they can be inferred in animals. Grief is commonly reported during field observations of various animals. The behaviors of elephants, chimpanzees, sea lions, and geese suggesting grief in response to the loss of a mate or offspring have been well documented. The dolphin who carries a dead baby around for several days is inferred to be experiencing both grief and love. Love has been attributed to animals such as swans or geese because of lifelong bonds that are established with a mate. Critics of these interpretations point out that animals may behave as if they are grieving or in love, yet there is no way of knowing whether this is an accurate reflection of their inner states. A central issue about the capacity of animals to experience a wide range of secondary emotions involves the ability to show self-consciousness. If an animal is able to be aware of its own inner states, it would then have the capacity to infer the mental states of others. With self-awareness comes the capacity for sympathy and empathy.

THE BIOLOGY OF EMOTIONS

Scientists examining the biology of emotions have discovered some similarities between the brains of humans and animals that help to explain the basic primary emotions. Emotions seem to arise from the parts of the brain that are located below the cortex and are part of the limbic system. These regions of the brain have remained intact across many species throughout evolution. So far, the amygdala has been identified as the central site of emotion. This almond-shaped structure is at the center of the brain. Neuroscientists have found that rats will show a pattern of fear when a particular section of the amygdala is stimulated. If the amygdala is damaged, a rat will not show the normal behavioral responses to danger, such as freezing or running away. The rat with a damaged amygdala also will not demonstrate the accompanying physiological reactions to danger, such as increased heart rate or blood pressure.

Research with humans has highlighted the amygdala's critical role in the learning of emotional associations and the recognition of emotional expressions in other individuals. Magnetic resonance imaging studies have shown that the amygdala shows activation to fearful stimuli. In humans, the brain is also involved in the control of emotional facial expressions. Smiles that occur spontaneously as a result of genuine happiness are involuntary. The extrapyramidal motor system, which depends on subcortical areas, governs involuntary smiles and fear reactions.

The chemistry of the brain also plays an important part in animal and human emotions. The neurotransmitter dopamine is released in copious amounts during periods of pleasure and excitement. Researchers have found that rats experience an increase of dopamine when engaging in activities that appear to suggest play. Research has also shown that if dopamine production is blocked in rats through the administration of a dopamine-blocking agent, the rat's play activity disappears. The effects of the hormone oxytocin have been studied in small mammals and appear to be related to sexual activity and bonding behaviors. In humans, oxytocin is released in mothers who are nursing their infants and is considered to aid in the mother-child bond. Researchers have investigated the role of oxytocin in bonding among voles. If a female vole is injected with oxytocin, the animal will quickly select a mate. When a female vole is given a drug to block oxytocin, however, mate selection never takes place.

Many scientists contend that it is illogical to believe that emotions appear suddenly in humans. If evolution takes place through the process of natural selection, the emotions found in humans would be present in early evolutionary ancestors. The similarities in brain anatomy and chemistry between animals and humans would then support the idea that some basic emotions exist in various species. Darwin believed that some facial expressions in humans are universal. These expressions are genetically determined and evolved as the most effective at telling others something about how a person is feeling. Research with infants shows the innate capacity to grimace in pain or to smile in pleasure. For the most basic emotions, people in all cultures show similar facial responses to similar emotional situations. For example, anger is linked with a facial expression recognized by almost all cultures. Perhaps it is this line of reasoning from the evolutionary context that provides the strongest support for the existence of a wide range of emotional reactions in animals.

—*Frank J. Prerost*

FURTHER READING

Bekoff, Marc. *The Smile of a Dolphin*. Discovery Books, 2000.

Griffin, Donald R. *Animal Minds*. University of Chicago P, 1992.

Marshall, Elizabeth. *The Hidden Life of Dogs*. Houghton Mifflin, 1993.

Masson, Jeffrey M. *When Elephants Weep*. Delacorte, 1995.

Mellor, D. J. "Animal Emotions, Behaviour and the Promotion of Positive Welfare States." *New Zealand Veterinary Journal* 60.1 (2012): 1–8.

Morell, Virginia. *Animal Wise: The Thoughts and Emotions of Our Fellow Creatures*. Crown, 2013.

Panksepp, Jaak. *Affective Neuroscience: The Foundation of Human and Animal Emotions*. Oxford University Press, 1998.

Sheldrake, Rupert. *Dogs That Know When Their Owners Are Coming Home*. Crown, 1999.

Animal Evolution: Historical Perspective

FIELDS OF STUDY

Evolutionary Biology; Paleontology; Genetics; Physiology; Comparative Anatomy

ABSTRACT

Evolution is the theory that biological species undergo sufficient change with time to give rise to new species. For two thousand years, however, evolution was considered an impossibility. Most scientists from Aristotle to Carolus Linnaeus in the eighteenth century insisted upon the immutability of species. Many tried to arrange all species in a single linear sequence known as the scale of being, a concept supported well into the nineteenth century. In the eighteenth century, scientists tried to reinterpret the scale of being as an evolutionary sequence, but this single-sequence idea was later replaced by Charles Darwin's concept of branching evolution. Georges Cuvier finally showed that the major groups of animals had such strikingly different anatomical structures that no possible scale of being could connect them all.

KEY CONCEPTS

adaptation: the possession by organisms of characteristics that suit them to their environment or their way of life

catastrophism: a geological theory explaining the earth's history as resulting from great cataclysms (floods, earthquakes, and the like) on a scale not now observed

Darwinism: branching evolution brought about by natural selection

essentialism (typology): the Platonic-Aristotelian belief that each species is characterized by an unchanging "essence" incapable of evolutionary change

genotype: the hereditary characteristics of an organism

Geoffroyism: an early theory of evolution in which heritable change was thought to be directly induced by the environment

Lamarckism: an early evolutionary theory in which voluntary use or disuse of organs was thought to be capable of producing heritable changes

scale of being (chain of being): an arrangement of life forms in a single linear sequence from "lower" to "higher"

uniformitarianism: a geological theory explaining Earth's history using processes that can be seen at work today

THE EVOLUTION OF EVOLUTION

Evolution is the theory that biological species undergo sufficient change with time to give rise to new species. The concept of evolution has ancient roots. Anaximander suggested in the sixth century BCE that life had originated in the seas and that humans had evolved from fish. Empedocles (fifth century BCE) and Lucretius (first century BCE), in a sense, grasped the concepts of adaptation and natural selection. They taught that bodies had originally formed from the random combination of parts, but that only harmoniously functioning combinations could survive and reproduce. Lucretius even said that the mythical centaur, half horse and half human, could never have existed because the human teeth and stomach would be incapable of chewing and digesting the kind of grassy food needed to nourish the horse's body.

For two thousand years, however, evolution was considered an impossibility. Plato's theory of forms (also called his "theory of ideas") gave rise to the notion that each species had an unchanging "essence" incapable of evolutionary change. As a result, most scientists from Aristotle to Carolus Linnaeus in the eighteenth century insisted upon the immutability of species. Many of these scientists tried to arrange all species in a single linear sequence known as the scale of being (also called the chain of being and the *scala naturae*), a concept supported well into the nineteenth century by many philosophers and theologians as well. The sequence in this scale of being was usually interpreted as a static "ladder of perfection" in God's creation, arranged from higher to lower forms. The scale had to be continuous, for any gap would detract from the perfection of God's creation.

The Comte de Buffon was perhaps the greatest naturalist of the eighteenth century, rejecting religious explanations of the state of the world in favor of scientific research.

Much exploration was devoted to searching for "missing links" in the chain, but it was generally agreed that the entire system was static and incapable of evolutionary change. Pierre-Louis Moreau de Maupertuis, in the eighteenth century, and Jean-Baptiste Lamarck were among the scientists who tried to reinterpret the scale of being as an evolutionary sequence, but this single-sequence idea was later replaced by Charles Darwin's concept of branching evolution. Georges Cuvier finally showed that the major groups of animals had such strikingly different anatomical structures that no possible scale of being could connect them all; the idea of a scale of being lost most of its scientific support as a result.

THE STRUGGLE TO CONCEPTUALIZE EVOLUTION

The theory that new biological species could arise from changes in existing species was not readily accepted at first. Linnaeus and other classical biologists emphasized the immutability of species under the Platonic-Aristotelian concept of essentialism. Those who believed in the concept of evolution realized that no such idea could gain acceptance until a suitable mechanism of evolution could be found. Many possible mechanisms were therefore proposed. Étienne Geoffroy Saint-Hilaire proposed that the environment directly induced physiological changes, which he thought would be inherited, a theory now known as Geoffroyism and carried on somewhat as one aspect of the new science of epigenetics. Lamarck proposed that there was an overall linear ascent of the scale of being but that organisms could also adapt to local environments by voluntary exercise, which would strengthen the organs used; unused organs would deteriorate. He thought that the characteristics acquired by use and disuse would be passed on to later generations, but the direct inheritance of acquired characteristics was later disproved.

Central to both these explanations was the concept of adaptation, or the possession by organisms of characteristics that suit them to their environments or to their ways of life. In eighteenth century England, the Reverend William Paley and his numerous scientific supporters believed that such adaptations could be explained only by the action of an omnipotent, benevolent God. In criticizing Lamarck, the supporters of Paley pointed out that birds migrated toward warmer climates before winter set in and that the heart of the human fetus had features that anticipated the changes of function that take place at birth. No amount of use and disuse could explain these cases of anticipation, they claimed; only an omniscient God who could foretell future events could have designed things with their future utility in mind.

The nineteenth century witnessed a number of books asserting that living species had evolved from earlier ones. Before 1859, these works were often more geological than biological in content. Most successful among them was the anonymously published *Vestiges of the Natural History of Creation* (1844), written by Robert Chambers. Books of this genre sold well but contained many flaws. They proposed no mechanism to account for evolutionary change. They supported the outmoded concept of a scale of being, often as a single sequence of evolutionary "progress." In geology, they supported the outmoded theory of catastrophism, an idea that the history of Earth had been characterized by great cataclysmic upheavals. From 1830 on, however, that theory was being replaced by the modern theory of uniformitarianism, championed by Charles Lyell. Charles Darwin read these books and knew their faults, especially their lack of a mechanism that was compatible with Lyell's geology. In his own work, Darwin carefully tried to avoid the shortcomings of these books.

DARWIN'S REVOLUTION IN BIOLOGICAL THOUGHT

Darwin brought about the greatest revolution in biological thought by proposing not only a theory of branching evolution but also a mechanism of natural selection to explain how it occurred. Much of Darwin's evidence was gathered during his voyage around the world aboard HMS *Beagle*. Darwin's stop in the Galápagos Islands and his study of tortoises and finch-like birds on these islands are usually credited with convincing him that evolution was a branching process and that adaptation to local environments was an essential part of the evolutionary process. Adaptation, he later concluded, came about through natural selection, a process in which the maladapted variations eventually died out and allowed only the more well adapted ones to survive and pass on their hereditary traits. After returning to England from his voyage, Darwin raised pigeons, consulted with various animal breeders about changes in domestic breeds, and investigated other phenomena that later enabled him to demonstrate natural selection and its power to produce evolutionary change.

Darwin's greatest contribution was that he proposed a suitable mechanism by which permanent organic change could take place. All living species, he said, were quite variable, and much of this variation was heritable. Also, most organisms produce far more eggs, sperm, seeds, or offspring than can possibly survive, and the vast majority of them die. In this process, some variations face certain death while others survive in greater or lesser proportion. Darwin called the result of this process "natural selection," the capacity of some hereditary variations (now called genotypes) to leave more viable offspring than others, with many leaving none at all. Darwin used this theory of natural selection to explain the form of branching evolution that has become generally accepted among scientists.

Darwin delayed the publication of his book for seventeen years after he wrote his first manuscript version. He might have waited even longer, except that his hand was forced. From the East Indies, another British scientist, Alfred Russel Wallace, had written a description of the very same theory and submitted it to Darwin for his comments. Darwin showed Wallace's letter to Lyell, who urged that both Darwin's and Wallace's contributions be published, along with documented evidence showing that both had arrived at the same ideas independently. Darwin's great book, *On the Origin of Species by Means of Natural Selection*, was published in 1859, and it quickly won most of the scientific community to a support of the concept of branching evolution. In his later years, Darwin also published *The Descent of Man and Selection in Relation to Sex* (1871), in which he outlined his theory of sexual selection. According to this theory, the agent that determines the composition of the next generation may often be the opposite sex. An organism may be well adapted to live, but unless it can mate and leave offspring, it will not contribute to the next or to future generations.

ACCEPTANCE OF DARWINISM IN THE TWENTIETH CENTURY

In the early 1900s, the rise of Mendelian genetics (named for botanist Gregor Mendel) initially resulted in challenges to Darwinism. Hugo de Vries proposed that evolution occurred by random mutations, which were not necessarily adaptive. This idea was subsequently rejected, and Mendelian genetics was reconciled with Darwinism during the period from 1930 to 1942. According to this modern synthetic theory of evolution, mutations initially occur at random, but natural selection eliminates most of them and alters the proportions among those that survive. Over many generations, the accumulation of heritable traits produces the kind of adaptive change that Darwin and others had described. The process of branching evolution through speciation is also an important part of the modern synthesis.

The branching of the evolutionary tree has resulted in the proliferation of species from the common ancestor of each group, a process called adaptive radiation. Ultimately, all species are believed to have descended from a single common ancestor. Because of the branching nature of the evolutionary process, no one evolutionary sequence can be singled out as representing any overall trend; rather, there have been different trends in different groups. Evolution is also an opportunistic process, in the sense that it follows the path of least resistance in each case. Instead of moving in straight lines toward a predetermined goal, evolving lineages often trace meandering or circuitous paths in which each change represents a momentary increase in adaptation. Species that cannot adapt to changing conditions die out and become extinct.

STUDYING EVOLUTION

Evolution is studied by a variety of methods. The ongoing process of evolution is studied in the field by ecologists, who examine various adaptations, including behavior and physiology as well as anatomy. These adaptations are also studied by botanists, who examine plants; zoologists, who examine animals; and various specialists, who work on particular kinds of animals or plants (for example, entomologists, who study insects). Some investigators capture specimens in the field, then bring back samples to the laboratory in order to examine chromosomes or analyze proteins using electrophoresis. Through these methods, scientists learn how the ongoing process of evolutionary change is working today within species or at the species level on time scales of only one or a few generations.

Jean-Baptiste de Lamarck is best known for his theory that acquired physical traits may be inheritable.

The long-term results of evolutionary processes are studied among living species by comparative anatomists and embryologists. Extinct organisms are studied by paleontologists, scientists who examine fossils. Biogeographers study past and present geographic distributions. All these types of scientists make comparisons among species in order to determine the sequence of events that took place in the evolutionary past. One method of reconstructing the branching sequences of evolution is to find homologies, deep-seated resemblances that reflect common ancestry. Once the sequences are established, functional analysis can be used to suggest possible adaptive reasons for any changes that took place. The sequences of evolutionary events reconstructed by these scientists represent the history of life on the earth. This history spans many species, families, and whole orders and classes, and it covers great intervals of past geologic time, measured in many millions of years.

THE HISTORICAL CONTEXT OF EVOLUTIONARY THEORY

The historical development of evolutionary theory should be viewed in two contexts: that of biological science and that of cultural history. The concept of evolution had been talked about for many years before 1859 and was usually rejected because no suitable mechanism had gained widespread acceptance. The fact that the phenomenon of natural selection was independently discovered by two Englishmen shows both that the time was ripe for the discovery and that the circumstances were right in late nineteenth century England.

Evolutionary biology is itself the context into which all the other biological sciences fit. Other biologists, including physiologists and molecular biologists, study how certain processes work, but it is evolutionists who study the reasons why these processes came to work in one way and not another. Organisms and their cells are built one way and not another because their structures have evolved in a particular direction and can only be explained as the result of an evolutionary process. Not only does each biological system need to function properly, but it also must have been able to achieve its present method of functioning as the result of a long, historical, evolutionary process in which a previous method of functioning changed into the present one. If there were two or more ways of accomplishing the same result, a particular species used one of them because found it easier to evolve one method rather than another.

Everything in biology is thus a detail in the ongoing history of life on Earth, because every living system evolves. Living organisms and the processes that make them function are all products of the evolutionary process and can be understood only in that context. As biologist Theodosius Dobzhansky once said, "Nothing in biology makes sense, except in the light of evolution."

—*Eli C. Minkoff*

FURTHER READING

Bowler, Peter J. *Evolution: The History of an Idea.* Rev. ed. University of California Press, 1989.

———. *Life's Splendid Drama: Evolutionary Biology and the Reconstruction of Life's Ancestry, 1860–1940.* University of Chicago Press, 1996.

Brandon, Robert N. *Concepts and Methods in Evolutionary Biology.* Cambridge University Press, 1996.

Darwin, Charles R. *On the Origin of Species by Means of Natural Selection: Or, the Preservation of the Favoured Races in the Struggle for Life.* John Murray, 1859.

Dobzhansky, Theodosius G. *Genetics of the Evolutionary Process.* Columbia University Press, 1970.

Gould, Stephen J. *Ever Since Darwin.* Reprint. W. W. Norton, 1992.

———. *The Flamingo's Smile.* W. W. Norton, 1985.

———. *The Panda's Thumb.* Reprint. W. W. Norton, 1992.

Grant, Verne. *The Evolutionary Process: A Critical Study of Evolutionary Theory.* 2nd ed. Columbia University Press, 1991.

Rose, Michael A. *Darwin's Spectre: Evolutionary Biology in the Modern World.* Princeton University Press, 1998.

Wills, Christopher, and Jeffrey Bada. *The Spark of Life: Darwin and the Primeval Soup.* Perseus, 2000.

ANIMAL GROWTH

FIELDS OF STUDY

Comparative Anatomy; Genetics; Physiology; Embryology

ABSTRACT

Animal development involves the slow, progressive changes that occur when the zygote, or fertilized egg undergoes mitosis, the process by which a cell divides into identical daughter cells. During development, mitosis occurs repeatedly, forming multiple generations of daughter cells. These cells increase in number and ultimately form all the cells in the body. Embryology is the study of the growth and development of an organism occurring before birth. Growth and development continue after birth and throughout adulthood. Growth ceases only at death, when the life of the individual organism is ended.

KEY CONCEPTS

differentiation: the process during development by which cells obtain their unique structure and function

fertilization: the union of two gametes (egg and sperm) to form a zygote

gamete: a functional reproductive cell (egg or sperm) produced by the adult male or female

growth: the increased body mass of an organism that results primarily from an increase in the number of body cells and secondarily from the increase in the size of individual cells

mitosis: the process of cellular division in which the nuclear material, including the genes, is distributed equally to two identical daughter cells

zygote: a fertilized egg

WITNESSING DEVELOPMENT

Animal development has been a source of wonder for centuries. Development involves the slow, progressive changes that occur when a single cell—the zygote, or fertilized egg—undergoes mitosis. Mitosis is the process by which a cell divides into identical daughter cells. During development, mitosis occurs repeatedly, forming multiple generations of daughter cells. These cells increase in number and ultimately form all the cells in the body of a multicellular animal, such as a frog, mouse, or elephant. The simple experiment of opening fertile chicken eggs to observe the embryos on successive days of their three-week incubation period illustrates the process of embryonic development. A narrow band of cells can be seen increasing in number and complexity until the body of an entire, but immature, chick is seen.

ANIMAL GROWTH AND DEVELOPMENT

An organism's growth occurs because of the increasing number of cells that form as well as because of the increasing size of individual cells. For example, a mouse increases from a single cell, the zygote, to about three billion cells during the period from fertilization to birth. Embryology is the study of the growth and development of an organism occurring before birth. Growth and development, however, continue after birth and throughout adulthood. Growth ceases only at death, when the life of the individual organism is ended. The bone marrow of human adults initiates the formation and development of millions of red blood cells every minute of life. About one gram of old skin cells is lost and replaced by new cells each day.

Development produces two major results: the formation of cellular diversity and the continuity of life. Cellular diversity, or differentiation, is the process that produces and organizes the numerous kinds of body cells. The first cell that determines an individual's unique identity, the zygote, ultimately gives rise to varying types of cells having diverse appearances and functions. Muscle cells, red blood cells, skin cells, neurons, osteocytes (bone cells), and liver cells are all examples of cells that have differentiated from a single zygote.

REPRODUCTION

Morphogenesis is the process by which differentiated cells are organized into tissues and organs. The continued formation of new individual organisms is called reproduction. The major stages of animal development include fertilization, embryology, birth, youth, adulthood—when fertilization of the next generation occurs—and death. A new individual animal is begun by the process of fertilization, when the genetic material from the sperm, produced by the father, and the egg, produced by the mother, are merged into a single cell, the zygote. Fertilization may be external, occurring in freshwater or the sea, or internal, occurring within the female's reproductive tract. While fertilization marks the beginning of a new individual, it is not literally the beginning of life, since both the sperm and egg are already alive. Rather, fertilization ensures the continuation of life through the formation of new individuals. This guarantees that the species of the organism will continue to survive in the future.

Following fertilization, the newly formed zygote undergoes embryological development consisting of cleavage, gastrulation, and organogenesis. Cleavage is a period of rapid mitotic divisions with little individual cell growth. A ball of small cells, called the morula, forms. As mitosis continues, this ball of cells hollows in the middle, forming an internal cavity called the blastocoel. Gastrulation immediately follows cleavage. During gastrulation, individual cell growth as well as initial cell differentiation occur. During this time, three distinct types of cells are formed: an internal layer called the endoderm, a middle layer, the mesoderm, and an external layer, the ectoderm. These cell types, or germ-cell layers, are the parental cells of all future cells of the body.

Cells from the ectoderm form the cells of the nervous system and skin. The mesoderm forms the cells of muscle, bone, connective tissue, and blood. The endoderm forms cells that line the inside of the digestive tract as well as the liver, pancreas, lungs, and thyroid gland. The transformation of these single germ layers into functional organs is called organogenesis. Organogenesis is an extremely complex period of embryological development. During this time, specific cells interact and respond to one another to induce growth, movement, or further differentiation; this cell-to-cell interaction is called induction. Each induction event requires an inducing cell and a responding cell.

In the formation of the brain and spinal cord, selected cells from the ectoderm form a long, thickened plate at the midline of the developing embryo. Through changes in cell shape, the outer edges of this plate fold up and fuse with each other in the middle, forming a tubular structure (a neural tube). This tubelike structure then separates from the remaining ectoderm. At the head region of the embryo, the neural tube enlarges into pockets that ultimately form brain regions.

For differentiation and development to occur, cells must be responsive to regulatory signals. Some of these signals originate within the responding cell; these signals are based in the genetic code found in the cell's own nucleus. Other signals originate outside the cell; they may include physical contact with overlying or underlying cells, specific signal molecules, such as hormones, from distant cells, or specialized structural molecules secreted by neighboring cells that map out the pathway along which a responding cell will migrate.

POSTNATAL DEVELOPMENT

Embryological development climaxes in the formation of functional organs and body systems. This period is concluded by birth (or hatching, in the case of some animals). Following birth, development normally continues. In some animals, such as frogs, newly hatched individuals undergo metamorphosis during which their body structures are dramatically altered. Newly hatched frogs (tadpoles), for example, are transformed from aquatic, legless, fishlike creatures into mature adults with legs that allow them to move freely on land.

In mammals, development and growth occur primarily after birth, as the individual progresses through the stages of infancy, childhood, adolescence, and adulthood. Mature adulthood is attained when the individual can produce his or her own gametes and participate in mating behavior.

Embryonic growth is especially impressive because the rate of cellular mitosis is so enormous. In the case of the mouse embryo, thirty-one cell generations occur during embryonic development. Thus, the zygote divides into two cells, then four, then eight, sixteen, thirty-two, and so on. This results in a newborn mouse consisting of billions of cells—produced in a period of only twenty-one days. When the newborn passes through its life stages to adulthood, its body cells may number more than sixty billion. One marine mammal, the blue whale, begins as a single zygote that is less than one millimeter in diameter and weighs only a small fraction of a gram. The resulting newborn whale (the calf) is about seven meters long and weighs two thousand kilograms: The embryonic growth represents a 200-millionfold increase in weight. Yet, for some animals, impressive growth periods also occur in the juvenile and adolescent stages of life.

In many cases, once an individual animal reaches its typical adult size, the rate of mitosis slows so that the number of new cells simply replaces the number of older, dying cells. At this maintenance stage, the individual no longer grows in overall size even though it continuously produces new cells. Since most of the cells in the mature adult have reached a final differentiated state, the function of mitosis is simply to replace the degenerating, aging cells. The slowing of the rate of cellular mitosis during this time may be attributable to the presence of specialized cell products called chalones. Chalones are thought to be local products of mature cells that inhibit further growth or mitosis.

STUDYING GROWTH

Historically, much study of animal development and growth was performed by simple observation. Aristotle, perhaps the first known embryologist, opened chick eggs during varying developmental periods. He observed and sketched what appeared to be the formation of the chick's body from a nondescript substance. With the invention of lenses and microscopes, growth and development could be studied on a cellular level. The concept of cellular differentiation arose, since investigators could see that embryonic muscle cells, for example, looked different from embryonic nerve cells. Again, much of the investigative information was descriptive in nature. Embryologists detailed the existence of the three germ-cell layers in gastrulation as well as the various tissues and primitive cells involved in organogenesis.

Experimentation as a method of investigating animal growth and development began during the nineteenth century. Lower animal species, such as the sea urchin and frog, were frequently investigated; their developmental patterns are simpler than those of mammals, their development occurs outside the maternal body, and they can be found in abundant numbers. Many of these experiments used separation or surgical techniques to isolate or regraft specific tissues or cells of interest. An attempt was made to determine how one tissue type would interact with and influence the development of another tissue type. Thus, the ideas of induction, in which some tissues affect other tissues, came into being. During this time, the descriptive and comparative observations resulting from these experimental manipulations were the major contributions of investigators.

The embryologists of the early twentieth century paid little attention to genetics. They believed that the major influences on development and growth were embryological mechanisms, although genes were thought to provide some nonessential peripheral functions. Chemical analyses of embryos attempted to establish the chemical basis for the cell-to-cell interactions that were seen during development and differentiation. During the middle portion of the twentieth century, geneticists began to investigate the role of the gene in cell function. The function of genes in the cellular manufacturing of specific proteins led to the hypothesis that each kind of cellular protein was the product of one gene. During this time, bacteria and fruit flies (*Drosophila*) were primary organisms of study because of their relatively simple genetic makeups.

In the latter part of the twentieth century, molecular biology techniques were applied to the study of development. Using techniques for transferring and replicating specific genes, researchers have greatly clarified the central importance of genes in development. Scientists came to believe that all the major developmental and differentiation influences that control cell growth are regulated through specific genes that are turned off or on.

DEVELOPMENTAL BIOLOGY

The combination of molecular biology techniques with embryological investigations has led to a new field of study—developmental biology. New methods have been developed and used. Radioactive tracer technology has allowed the investigator to label particular genes or gene products and trace their movements and influences on cell growth through several generations. Recombinant deoxyribonucleic acid (DNA) technology has allowed the isolation and replication of significant genes that are important in development. Immunochemistry uses specific proteins (antibodies) to bind to differentiating cell products and quantify them. Cell-cell hybridization allows the introduction of specific genes into the nuclei of cells in alternate differentiation pathways.

Developmental biology, with its multidisciplinary approach, is solving many of the fundamental questions of development. As scientists become better able to understand the role of genetics and cell-to-cell interactions, they gain insight into the mechanisms that control cell growth and development. Consequently, the potential to control undesirable growth or to enhance underdeveloped growth is within reach.

The problem of cell aging is also under investigation. Questions about why mature cells stop dividing and growing and what are the causes of aging constitute important areas of developmental research. While various theories have been presented, the fundamental key to cellular aging remains to be discovered. One of the most challenging areas of continuing research is the determination of how developmental patterns guide evolutionary changes. Developmental principles may provide the answer to why evolution has given rise to animal diversity. In addition, developmental biology may give scientists the information needed to predict and determine future evolutionary trends. The individual animal is a growing organism that begins as a zygote and passes through the stages of embryonic development, birth, youth, adulthood, aging, and death. Preservation of the species depends on adult individuals' producing gametes that result in the formation of a future generation of zygotes and individuals. Remarkably, each zygote contains the necessary genetic instructions to regulate the orderly processes of growth and development. Thus, animal life continues from generation to generation.

—*Roman J. Miller*

FURTHER READING

Alberts, Bruce, Dennis Bray, Julian Lewis, Martin Raff, Keith Roberts, and James D. Watson. *Molecular Biology of the Cell*. 3rd ed. Garland, 1994.

Balinsky, B. I. *An Introduction to Embryology*. Saunders College Publishing, 1981.

Baserga, Renato. *The Biology of Cell Reproduction*. Harvard University Press, 1985.

Gilbert, Scott F. *Developmental Biology*. 6th ed. Sinauer Associates, 2000.

_____, ed. *A Conceptual History of Modern Embryology*. The Johns Hopkins University Press, 1994.

Gilbert, Scott F., and A. M. Raunio. *Embryology: Contructing the Organism*. Sinauer Associates, 1997.

Hartwell, Leland H., and Ted A. Weinert. "Checkpoints: Controls That Ensure the Order of Cell Cycle Events." *Science* 246 (November 3, 1989): 629–634.

Wessels, Norman K., and Janet L. Hopson. *Biology*. Random House, 1988.

Animal Habituation and Sensitization

FIELDS OF STUDY

Psychology; Anthropology; Animal Behavior;

ABSTRACT

Habituation is a simple form of nonassociative learning. With each presentation of the habituating stimulus, the responsiveness of the organism decreases toward the zero, non response level. Habituation is important for survival of the individual. Many stimuli are continuously impinging upon it: Some are important, others are not. Important stimuli require an immediate response, but those which result in neither punishment nor reward may be safely ignored. Sensitization, a very strong response to a stimulus, involves the entire organism. Sensitization increases during the early stages of habituation but later decreases. The stronger the sensitizing stimulus and the longer the exposure to it the greater the sensitization.

KEY CONCEPTS

acetylcholine: a neurotransmitter produced by a nerve cell that enables a nerve impulse to cross a synapse and reach another nerve or muscle cell

Aplysia: a large sluglike mollusk that lives in salt water and has been used in habituation experiments; its outer covering is called the mantle

impulse: a "message" traveling within a nerve cell to another nerve cell or to a muscle cell

motor neuron: a nerve cell that causes a muscle cell to respond

neurotransmitter: a chemical substance which enables nerve impulses to cross a synapse and reach another nerve cell or muscle cell

orienting reflex: an unspecific reflex reaction caused by a change in the quantity or quality of a stimulus; it will disappear or decrease after repeated presentations of the stimulus

sensitization: an arousal or an alerting reaction which increases the likelihood that an organism will react; also, a synonym for loss of habituation with increased intensity of response

synapse: the minute space or gap between the axon of one nerve cell and the dendron of the next; also, the gap between a nerve cell and a muscle cell

HABITUATION AND TRAINING

Habituation is a simple form of nonassociative learning that has been demonstrated in organisms as diverse as protozoans, insects, *Nereis* (clam worms), birds, and humans. The habituated organism learns to ignore irrelevant, repetitive stimuli which, prior to habituation, would have produced a response. With each presentation of the habituating stimulus, the responsiveness of the organism decreases toward the zero, non response level. If habituation training continues after the zero-response level, the habituation period is prolonged. Habituation to a particular stimulus naturally and gradually disappears unless the training continues. If training is resumed after habituation has disappeared, habituation occurs more rapidly in the second training series than in the first. Habituation is important for survival of the individual. Many stimuli are continuously impinging upon it: Some are important, others are not. Important stimuli require an immediate response, but those which result in neither punishment nor reward may be safely ignored.

STIMULUS AND RESPONSE

When a new stimulus is presented, as when a sudden change in the environment occurs, the organism—be it bird, beast, or human—exhibits the "startle" or "orientation" response. In essence, it stops, looks, and listens. If the stimulus is repeated and is followed by neither reward nor punishment, the organism will pay less and less attention to it. When this happens, habituation has occurred, and the organism can now respond to and deal with other stimuli. On the other hand, if, during habituation learning, a painful consequence follows a previously nonconsequential stimulus, the organism has been sensitized to that stimulus and will respond to it even more strongly than it did before the learning sessions, whether they are occurring in the laboratory or in the field.

Young birds must learn to tell the difference between and respond differently to a falling leaf and a descending predator. A young predatory bird must learn to ignore reactions of its prey that pose no danger, reactions that the predator initially feared.

A theory known as the dual-process habituation-sensitization theory was formulated in 1966 and revised in 1973. It establishes criteria for both habituation and sensitization. Criteria for habituation (similar to those proposed by E. N. Sokolov in 1960) are that habituation will develop rapidly; the frequency of stimulation determines the degree of habituation; if stimulation stops for a period of time, habituation will disappear; the stronger the stimulus, the slower the rate of habituation; the frequency of stimulation is more important than the strength of the stimulus; rest periods between habituation series increase the degree of habituation; and the organism will generalize and therefore exhibit habituation to an entire class of similar stimuli. Stimulus generalization can be measured: If a different stimulus is used in the second habituation series, habituation occurs more rapidly; it indicates generalization.

SENSITIZATION

Sensitization, a very strong response to a very painful, injurious, or harmful stimulus, is not limited to stimulus-response circuits but involves the entire organism. After sensitization, the individual may respond more strongly to the habituating stimulus than it did prior to the start of habituation training.

There are eight assumptions about sensitization in the dual-process theory. Sensitization does not occur in stimulus-response circuits but involves the entire organism. Sensitization increases during the early stages of habituation training but later decreases. The stronger the sensitizing stimulus and the longer the exposure to it the greater the sensitization; weaker stimuli may fail to produce any sensitization. Even without any external intervention, sensitization will decrease and disappear. Increasing the frequency of sensitization stimulation causes a decrease in sensitization. Sensitization will extend to similar stimuli. Dishabituation, the loss of habituation, is an example of sensitization. Sensitization may be time-related, occurring only at certain times of the day or year.

According to the dual-process theory, the response of an organism to a stimulus will be determined by the relative strengths of habituation and sensitization. Charles Darwin, the father of evolution, observed and described habituation, although he did not use the term. He noted that the birds of the Galápagos Islands were not disturbed by the presence of the giant tortoises, *Amblyrhynchus*; they disregarded them just as the magpies in England, which Darwin called "shy" birds, disregarded cows and horses grazing nearby. Both the giant tortoises of the Galápagos Islands and the grazing horses and cows of England were stimuli that, though present, would not produce profit or loss for the birds; therefore, they could be ignored.

THE NEUROLOGY OF STIMULUS RESPONSE

Within the bodies of vertebrates is a part of the nervous system called the reticular network or reticular activating system. It has been suggested that the reticular network is largely responsible for habituation. It extends from the medulla through the midbrain to the thalamus of the forebrain. (The thalamus functions as the relay and integration center for impulses to and from the cerebrum of the forebrain.) Because it is composed of a huge number of interconnecting neurons and links all parts of the body, the reticular network functions as an evaluating, coordinating, and alarm center. It monitors incoming message impulses. Important ones are permitted to continue to the cerebral cortex, the higher brain. Messages from the cerebral cortex are coordinated and dispatched to the appropriate areas.

During sleep, many neurons of the reticular network stop functioning. Those that remain operational may inhibit response to unimportant stimuli (habituation) or cause hyperresponsiveness (sensitization). The cat that is accustomed to the sound of kitchen cabinets opening will sleep through a human's dinner being prepared (habituation) but will charge into the kitchen when she hears the sound of the cat food container opening (sensitization).

Researcher E. N. Sokolov concluded that the "orientation response" (which can be equated with sensitization) and habituation are the result of the functioning of the reticular network. According to Sokolov, habituation results in the formation of models within the reticular activating system. Incoming messages that match the model are disregarded by the organism, but those that differ trigger alerting reactions throughout the body, thus justifying the term "alerting system" as a synonym for the reticular network. Habituation to a very strong stimulus would take a long time. Repetition of this strong stimulus would cause an even stronger defensive reflex and would require an even longer habituation period.

THE ROLE OF NEUROTRANSMITTERS

Neurotransmitters are chemical messengers that enable nerve impulses to be carried across the synapse, the narrow gap between neurons. They transmit impulses from the presynaptic axon to the postsynaptic dendrite(s). E. R. Kandell, in experiments with *Aplysia* (the sea hare, a large mollusk), demonstrated that as a habituation training series continues, smaller amounts of the neurotransmitter acetylcholine are released from the axon of the presynaptic sensory neuron. On the other hand, after sensitization, this neuron released larger amounts of acetylcholine because of the presence of serotonin, a neurotransmitter secreted by a facilitory interneuron. When a sensitizing stimulus is very strong, it usually generates an impulse within the control center—a ganglion, a neuron, or the brain. The control center then transmits an impulse to a facilitory interneuron, causing the facilitory interneuron to secrete serotonin.

Increased levels of acetylcholine secretion by the sensory neuron result from two different stimuli: direct stimulation of the sensory neurons of the siphon or serotonin from the facilitory interneuron. Facilitory interneurons synapse with sensory neurons in the siphon. Serotonin discharged from facilitory interneurons causes the sensory neurons to produce and secrete more acetylcholine.

On the molecular level, the difference between habituation and adaption—the failure of the sensory neuron to respond—is very evident. The habituated sensory neuron has a neurotransmitter in its axon but is unable to secrete it and thereby enable the impulse to be transmitted across the synapse. The adapted sensory neuron, by contrast, has exhausted its current supply of neurotransmitter. Until new molecules of neurotransmitter are synthesized within the sensory neuron, none is available for release.

In 1988, Emilie A. Marcus, Thomas G. Nolen, Catherine H. Rankin, and Thomas J. Carew published the multiprocess theory to explain dishabituation and sensitization in the sea hare, *Aplysia*. On the basis of their experiments using habituated sea hares that were subjected to different stimuli, they concluded that dishabituation and sensitization do not always occur together; further, they decided, there are three factors to be considered: dishabituation, sensitization, and inhibition.

HABITUATION STUDIES

Habituation studies have utilized a wide variety of approaches, ranging from the observation of intact organisms carrying out their normal activities in their natural surroundings to the laboratory observation of individual nerve cells. With different types of studies, very different aspects of habituation and sensitization can be investigated. Surveying the animal kingdom in 1930, G. Humphrey concluded that habituation-like behavior exists at all levels of life, from the simple one-celled protozoans to the multicelled, complex mammals.

E. N. Sokolov, a compatriot of Ivan P. Pavlov, used human subjects in the laboratory. In 1960, he reported on the results of his studies, which involved sensory integration, the makeup of the orientation reflex (which he credited Pavlov with introducing in 1910), a neuronal model and its role in the orientation reflex, and the way that this neuronal model could be used to explain the conditioned reflex. Sokolov measured changes in the diameter of blood vessels in the head and finger, changes in electrical waves within the brain, and changes in electrical conductivity of the skin. By lowering the intensity of a tone to which human subjects had been habituated, Sokolov demonstrated that habituation was not the result of fatigue, because subjects responded to the lower-intensity tone with the startle or orientation reflex just as they would when a new stimulus was introduced. Sokolov concluded that the orientation response (which is related to sensitization) and habituation are the result of the functioning of the reticular network of the brain and central nervous system. Sokolov emphasized that the orientation response was produced after only the first few exposures to a particular stimulus, and it increased the discrimination ability of internal organizers. The orientation response was an alerting command. Heat, cold, electric shock, and sound were the major stimuli that he used in these studies.

E. R. Kandell used the sea hare, *Aplysia*, in his habituation-sensitization studies. *Aplysia* is a large sluglike mollusk, with a sheetlike, shell-producing body covering, the mantle. *Aplysia* has a relatively simple nervous system and an easily visible gill-withdrawal reflex. (The gill is withdrawn into the mantle shelf.) Early habituation-sensitization experiments dealt with withdrawal or absence of gill withdrawal. Later experiments measured electrical changes that occurred within the nerve cells that controlled gill movement. These were followed by studies that demonstrated that the gap (synapse) between the receptor nerve cell (sensory neuron) and the muscle-moving nerve cell (motor neuron) was the site where habituation and dishabituation occurred and that neurohormones such as acetylcholine and serotonin played essential roles in these processes. Kandell called the synapse the "seat of learning."

Charles Sherrington used spinal animals in which the connection between the brain and the spinal nerve cord had been severed. Sherrington demonstrated that habituation-sensitization could occur within the spinal nerve cord even without the participation of the brain. Pharmaceuticals have also been used in habituation-sensitization studies. Michael Davis and Sandra File used neurotransmitters such as serotonin and norepinephrine to study modification of the startle (orientation) response.

Habituation studies conducted in the laboratory enable researchers to control variables such as genetic makeup, previous experiences, diet, and the positioning of subject and stimulus; however, they lack many of the background stimuli present in the field. In her field studies of the chimpanzees of the Gombe, Jane Goodall used the principles of habituation to decrease the distance between herself and the wild chimpanzees until she was able to come close enough to touch and be accepted by them. The field-experimental approach capitalizes on the best of both laboratory and field techniques. In this approach, a representative group of organisms that are in their natural state and habitat are subjected to specific, known stimuli.

LEARNING TO SURVIVE

Habituation is necessary for survival. Many stimuli are constantly impinging upon all living things; since it is biologically impossible to respond simultaneously to all of them, those which are important must be dealt with immediately, as it may be a matter of life or death. Those which are unimportant or irrelevant must be ignored.

Cell physiologists and neurobiologists have studied the chemical and electrical changes that occur between one nerve cell and another and between nerve and muscle cells. The results of those studies have been useful in understanding and controlling these interactions as well as in providing insights for therapies. Psychologists utilize the fruits of habituation studies to understand and predict, modify, and control the behavior of intact organisms. For example, knowing that bulls serving as sperm donors habituate to one cow or model and stop discharging sperm into it, the animal psychologist can advise the semen collector to use a different cow or model or simply to move it to another place—even as close as a few yards away.

Conservationists and wildlife protectionists can apply the principles of habituation to wild animals, which must live in increasingly closer contact with one another and with humans, so that both animal and human populations can survive and thrive. For example, black-backed gulls, when establishing their nesting sites, are very territorial. Males which enter the territory of another male gull are rapidly and viciously attacked. After territorial boundaries are established, however, the males in contiguous territories soon exhibit "friendly enemy" behavior: They are tolerant of the proximity of other males that remain within their territorial boundaries. This has been observed in other birds as well as in fighting fish.

—*Walter Lener*

FURTHER READING

Alcock, John. *Animal Behavior: An Evolutionary Approach.* 6th ed. Sinauer Associates, 1998.

Alkon, Daniel L. "Learning in a Marine Snail." *Scientific American* 249 (July 1983): 70–84.

Barash, David P. *Sociobiology and Behavior.* Elsevier, 1982.

Drickamer, Lee C., Stephen H. Vessey, and Doug Meikle. *Animal Behavior: Mechanisms, Ecology, Evolution.* 4th ed. Wm. C. Brown, 1996.

Eckert, Roger, and David Randall. *Animal Physiology.* 4th ed. W. H. Freeman, 1997.

Gould, James L. *Ethology: The Mechanisms and Evolution of Behavior.* W. W. Norton, 1980.

Gould, James L., and Peter Marler. "Learning by Instinct." *Scientific American* 256 (January 1987): 74–85.

Halliday, Tim, ed. *Animal Behavior.* University of Oklahoma Press, 1994.

Klopfer, P. H., and J. P. Hailman. *An Introduction to Animal Behavior: Ethology's First Century.* Prentice-Hall, 1967.

Slater, P. J. B., and T. R. Halliday, eds. *Behavior and Evolution.* Cambridge University Press, 1994.

ANIMAL IMMUNE SYSTEMS

FIELDS OF STUDY

Immunology; Biochemistry; Biology; Bacteriology

ABSTRACT

An animal must keep its tissues from being invaded or mixed with tissues of other organisms. There are two types of protection used by animals in keeping out invaders and resisting foreign substances, nonspecific and specific defenses. In both types, the body distinguishes between cells that belong to the animal, which are "self," and anything that does not belong, or "nonself."

KEY CONCEPTS

antibody: protein produced by lymphocytes, with specificity for a particular antigen

antigen: chemical that stimulates the immune system to respond in a very specific manner

cell-mediated immunity: production of lymphocytes that specifically kill cells with foreign antigens on their surfaces

humoral immunity: production of antibodies specifically reactive against foreign antigens carried in body fluids (humors)

lymphocyte: white blood cell that produces either cell-mediated or humoral immunity in response to foreign antigens

macrophage: mature phagocytic cell that works with lymphocytes in destroying foreign antigens

PROTECTING THE "SELF"

An animal must keep itself distinct from its environment, recognizing its own tissues and keeping them from being invaded or mixed with tissues of other organisms. There are two types of protection used by animals in keeping out invaders and resisting foreign substances, nonspecific and specific defenses. In both types, the body distinguishes between cells that belong to the animal, which are "self," and anything that does not belong, or "nonself."

NONSPECIFIC DEFENSES

Even animals as primitive as sponges have the ability to recognize and maintain self-integrity. Scientists have broken apart two different sponges of the same species in a blender, intermixing the separated cells in a dish. Cells crawled away from nonself cells and toward self cells, reaggregating into clusters of organized

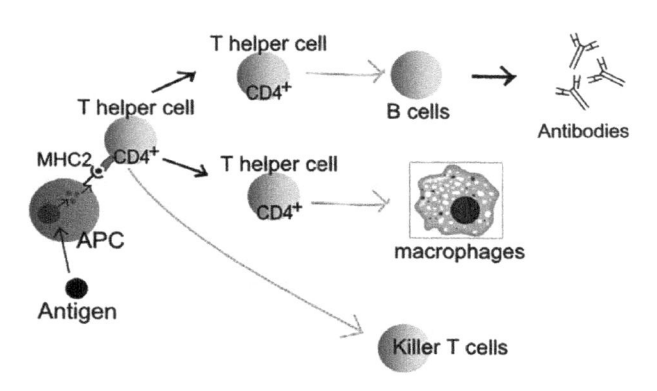

T helper cell function.

tissues containing cells of only one particular individual. Phagocytic cells that engulf and destroy foreign invaders were first identified by a scientist who had impaled a starfish larva on a thorn. He observed that, over time, large cells moved to surround the thorn, apparently trying to engulf and destroy it, recognizing it as nonself. Even earthworms have the ability to recognize and reject skin grafts from other individuals. If the graft comes from another worm of the same population, the skin is rejected in about eight months, but rejection of skin from a worm of a different population occurs in two weeks. Phagocytic cells in earthworms have immunological memory, enabling a worm to reject a second transplant from the same foreign source in only a few days.

Barriers, chemicals, and phagocytic cells are nonspecific protective mechanisms, which do not distinguish among different kinds of invaders. Tough outer coverings such as skin, hide, scales, feathers, or fur provide surface barriers. Nonspecific defenses also include secretions of mucus, sweat, tears, saliva, stomach acid, and urine, as well as body-fluid molecules, such as complement and interferon. Damaged tissues or bacterial invaders signal other cells to produce inflammation, a nonspecific response characterized by heat, redness, swelling, and pain. Cellular defenses associated with inflammation include phagocytes such as neutrophils and macrophages, which engulf and digest bacteria and debris, and natural killer cells, which destroy cancer cells or virally infected cells by poking holes in them.

IMMUNITY

The only specific defense in vertebrates is provided by the immune system, in which the component parts react against particular antigens on invaders, such as individual strains of bacteria or types of viruses. This more sophisticated protection is produced by lymphocytes that provide either cell-mediated or humoral immunity against particular antigens. Cell-mediated immunity depends on T lymphocytes (T cells) that become mature as they pass through the thymus, from which they get the "T" of their name. Humoral immunity is the function of antibodies, proteins released by B lymphocytes (B cells) that have matured and developed into plasma cells. The B lymphocytes reach maturity in the bursa of Fabricius in birds, where they were first recognized and from which they were named. Other vertebrates lack the bursa of Fabricius, and B cells mature in the bone marrow instead, but the name "B lymphocyte" still applies.

Antigen molecules are usually proteins or glycoproteins (proteins with sugars attached) that generate either an antibody response or a cellular immune response when they are foreign to the responding animal. So-called self antigens are molecules on cell membranes that identify the cells as belonging to the animal itself. An animal would not normally produce an immune response against its own antigens, but the same antigens would generate an immune response if placed in another animal to whom they were foreign. These antigens are the means by which self and nonself distinctions are made by the immune system, so the system can determine whether to ignore cells or attack them. Occasionally, self antigens, for some reason, are no longer recognized by the animal's immune system, and are attacked as if they were foreign. This causes an autoimmune disease, where the immune system destroys the body's own tissues.

Scientific understanding of how the immune system functions is largely dependent on work done using laboratory animals, including rabbits, mice, and hamsters. Laboratory mice have been highly inbred into strains where all the animals are genetically identical and their genes and antigens are well known. Studies on these mice have been essential in determining how the immune system normally works, and how it fails to work in autoimmune diseases and the inability to prevent cancer cells from proliferating.

ANTIGEN PRESENTATION AND RECEPTORS

Central to the functioning of the immune system in mammals is a system of genes called the major histocompatibility complex (MHC). These genes encode a collection of cell-surface glycoproteins that are the self antigens by which the immune system recognizes its own body cells. Class I MHC molecules are expressed on the surfaces of all nucleated cells, while Class II MHC markers are produced only by specialized cells, including cells of the thymus, B lymphocytes, macrophages, and activated T lymphocytes. Both Class I and Class II MHC molecules identify the cells bearing them as self, and these also serve as the context in which the immune system recognizes foreign antigens that are presented on the cell surface. Cells with self antigens are tolerated by the immune response of that individual animal, while cells that show foreign antigens are attacked and destroyed. Rejection of a graft or transplanted organ is reduced with more closely matched tissues, which are better tolerated by the immune system.

When bacteria evade protective barriers and chemicals to enter an animal's body, the animal's macrophages attack, engulfing and digesting the invaders. One bacterium may have thousands of different antigenic segments that can be recognized on its surface or inside the cell. Small parts of these digested cells, the individual antigens, are joined to the macrophage's newly formed MHC Class I and Class II before they are exposed on the cell surface. The foreign antigens fit into a space or pocket within the MHC molecule and are recognized by T cells that have the same MHC molecules and can respond specifically to the foreign antigen. Cytotoxic T cells (T) react to antigens held in the pocket of a Class I molecule, while helper T cells (T) respond to those presented by Class II molecules. T cells are the agents of cellular immunity, producing perforin molecules that puncture and kill cells bearing the foreign antigen against which the T cells are specific. T cells, when activated by encountering their specific antigens presented with Class II molecules, release cytokines that help to activate both T_C cells and B lymphocytes. Activated B cells divide to produce memory B cells and lymphocytes that mature into plasma cells, which secrete about two thousand antibody molecules per second over their active lifespan of four or five days.

Both T and B lymphocytes can react with their specific foreign antigens because the antigen-MHC complex binds to receptor molecules on the lymphocyte surfaces. Each clone of lymphocytes has the genetic ability to respond to a particular shape that fits its receptors. There may be millions of different receptors among the lymphocytes of a single animal, capable of binding millions of different antigens, even artificial chemicals not existing in nature. This enormous variability in response capability makes the immune system of each animal protective against many kinds of foreign invaders. Since each individual has its own set of immune responses, a population is less likely to have all its members die in an epidemic. Certain animals will be more resistant to the pathogens, so some will survive to reproduce and keep the population from extinction.

PRIMARY AND SECONDARY IMMUNE RESPONSES

When a foreign antigen is encountered by an animal for the first time, both T and B cells that can bind the antigen are activated, but not immediately. In a series of reactions, macrophages first break down the antigen-bearing cell, processing and presenting the antigen on its surface with MHC. The T cell specific to that antigen then encounters the antigen-MHC complex on the macrophage and divides to produce a clone of memory T cells and a clone of effector (activated) T cells. Activated T cells release cytokines that activate T and B cells so that they can attack the same foreign antigen. The first encounter with antigen produces a slow primary response, taking more than a week to reach peak effectiveness. During the time needed to generate this response, pathogenic bacteria or viruses can produce disease in the animal under attack. The memory T and memory B cells remain alive but inactive until the same foreign antigen is encountered again, even years later. The secondary response that results immediately when these memory cells are activated occurs so quickly that the disease process does not recur.

The importance of the immune system is seen in humans who lack its function, those with acquired immunodeficiency syndrome (AIDS). Human immunodeficiency virus (HIV) is the causative agent of AIDS, and is similar to viruses that attack other species in the same way. Most who die with AIDS really succumb to one of many opportunistic infections that cause diseases in HIV-positive individuals, but which are eradicated by the immune system in normal individuals.

—*Jean S. Helgeson*

Further Reading

Campbell, Neil A., Jane B. Reece, and Lawrence G. Mitchell. *Biology*. 5th ed. Benjamin/Cummings, 1999.
Paul, William E. *Fundamental Immunology*. 4th ed. Lippincott-Raven, 1999.
Roitt, Ivan, Jonathan Brostoff, and David Male. *Immunology*. 5th ed. C. V. Mosby, 1998.
Staines, Norman, Jonathan Brostoff, and Keith James. *Introducing Immunology*. 2nd ed. C. V. Mosby, 1993.

Animal Instincts

Fields of Study

Biology; Ethology; Psychology; Sociology; Genetics; Neurobiology

Abstract

Instincts are unlearned patterns of behavior in animals and are inherited by successive generations of the species. Ethologists call these stereotyped behaviors fixed action patterns (FAPs). Over time, such patterns are shaped by evolution, but at any given moment, a species maintains a range of instincts that are unique to its members. A number of categories of stereotyped behavioral responses exist that apply to all organic forms. The term "instinctive behavior," however, is generally reserved for animals and insects.

Key Concepts

altruism: a high degree of devotion to the interest of others that often includes self-sacrifice
ethology: a branch of biology that studies behavior
fixed action pattern (FAP): a behavior whose timing and duration are invariable for all members of a species
innate: denotes an inherited and unalterable condition or ability in an organism
neurobiology: the study of the biology of the brain
neuroethology: the study of behavior as it relates to brain functions
peptide: a chemical composed of amino acids bonded head-to-tail through amide bonds
signal: information transmitted through sound, such as bird calls, or through sight, such as body posture
stereotyped behavior: an unlearned and unchanging behavior pattern that is unique to a species

INSTINCTS AND INSTINCTIVE BEHAVIOR

Instincts are unlearned patterns of behavior in animals and are inherited by successive generations of the species. Ethologists call these stereotyped behaviors fixed action patterns (FAPs). Over time, such patterns are shaped by evolution, but at any given moment, a species maintains a range of instincts that are unique to its members. A number of categories of stereotyped behavioral responses exist that apply to all organic forms. These include the basic drives such as reproduction, feeding, and protection from predators. The term "instinctive behavior," however, is generally reserved for animals and insects, while excluding plants, bacteria, and viruses. Simple forms of stereotyped responses include a variety of reflex actions in which sensory nerve cells are affected by conditions such as heat and light. The more complex forms of responses are studied in order to understand the sequence of responses and the process of evolution that selected these patterns.

THE STUDY OF INSTINCTS

The study of instinctive behavior began in the nineteenth century with the development of clinical psychology. Scientists noticed a direct link between animal and human behavior, and since clinical experimentation of human subjects was not always possible or desirable, experiments were carried out on monkeys, mice, and guinea pigs. Experiments focused on the responses of test subjects to specific environmental conditions. An example of this type of experiment would be investigating the puzzle-solving ability of rats in a maze. The flexibility and nature of these responses were seen as part of a study of learning processes.

Meanwhile, the field of ethology began to formulate a different view of animal behavior. A European group including Konrad Lorenz, Nikolaas Tinbergen, and Karl von Frisch noticed that animals possess a specific innate capacity to perform complex activities in response to the environment. As a result, modern

Instinct

Animal Behavior: Living things perform instinctive behavior for survival. Animals exhibit instinctive behavior during migration, hibernation, pain reflex, and mating displays. These are unlearned and animals have these behaviors from their time of birth.

ethology shifted the study of animal behavior from learned responses to inherited patterns of behavior. From their beginning as "bird watchers," observing birds' courtship behavior, nest building, rearing of young, and territorial ownership, ethologists collected a body of scholarship that came to represent a respected field of study. A number of subfields have been created, ranging from neurology and genetics of behavior to ecology of species behavior. One element that links these diverse and distinctive fields of study is that of signals, language, and communications.

Charles Darwin gave the scientific world a theory of evolution, but he was also an avid observer of instinctive behavior. He observed domesticated pigeons and described how emotions are expressed by humans and animals. Other early pioneers in this field included Charles Whitman, who studied the family tree of a group of pigeons in Massachusetts, and Oskar Heinroth, who observed several species of waterfowls in Germany. Their observations led to the conclusion that members of a species not only share the same body functions and bone structures but also share behavior patterns.

The next step was to compare instinctive behavior across several closely related species. This work was accomplished between 1940 and 1960 by Lorenz and Tinbergen. Lorenz chose as his subject the courtship sequences of mallards, teals, and gadwalls. Recognition signals among ducks are a specific sequence of behaviors that include tail shaking, head flicking, and whistling. The exact sequence for different species is different, yet the components are the same. Through these studies, it is possible to conclude that courtship sequences originated in a single ancestral form, and crossbreeding between duck species produced variations in the pattern. Tinbergen arrived at a similar conclusion with his work on the calls, body postures, and movements of gulls. His study of the signaling behavior of more than fifteen species of gulls showed that their system of signals is similar. As a result of their work, Lorenz and Tinbergen provided a scientific basis both for the biological source of behavior and for an evolutionary source of instinctive reaction.

THE NATURE VERSUS NURTURE DEBATE

Although few scientists of the time had objected to the notion that genetics contributed to instinctive behavior, there was an ongoing debate as to the extent of the genetic contribution. The scientific bias was on the side of learned behavior patterns in such areas as language, signals, and body postures. It seemed clear that higher animals learn how to transfer information among themselves. One example was that an isolated songbird, denied access to parental teaching, does not instinctively know the songs of its species. Consequently, the genetic component may play only a small part in the overall development of behavior. In this debate, the contribution of Karl von Frisch shifted the balance to the side of inheritance. Von Frisch showed that honeybees communicate both the direction and the distance of a source of nectar through a sequence of dance patterns. A scouting bee returning to the hive can provide exact information to other bees. While it was argued that the larger brain sizes of higher animals contributed substantially to learning a system of communication, it was difficult to make the same argument for insects.

The debate on nature versus nurture produced extensive efforts to determine the extent to which instinctive behavior is shaped by inheritance or by learning. In the study of sea gull chicks, Jack Hailman created a series of experiments that added evidence to the side of learned behavior. The parent of a sea gull chick can elicit pecking from the chick either by pointing its bill downward or by swinging its bill from side to side. The chick will respond not only to a parent's action, however, but also to a model that has the shape of the parent. The initial feeding behavior of a newly hatched chick is a "hit and miss" affair. With growth, the chick responds more precisely to the figure of the parent and acquires greater coordination. Thus, the initial instinct is for the chick to peck at a variety of motions, but with maturity a learning component takes place to create greater discrimination.

SEQUENCES OF BEHAVIOR

During the 1950s and 1960s, researchers in instinctive behavior studied a number of animals and insects. Scientists began to understand that instincts are not a simple response to the environment but are a complex sequence of behavior. Scientists also began to find evidence suggesting that natural selection acts on behavior patterns as well as on an organism's biological makeup. The result of this research came to be categorized into three broad groups of instinctive behavior. One was the response of a simple instinct on the part of a single organ to some stimulus. An example is a nerve cell responding to light that triggers a reflex response. Other reflex reactions include locomotion and movement. Another type of stereotyped behavior was called fixed action patterns (FAPs). For example, the courting behavior of ducks can be classified as FAPs, wherein the pattern and timing of the responses are invariable for all members of the species. Similar FAPs are found in spiders, crabs, and a number of other insects and lower animals. A final group of instinctive behavior was described as modifiable action patterns (MAPs) and included fixed patterns that could be modified by the environment or by learning. For example, in species of birds, the core of nest-building behavior is fixed, but actual nest building depends on the availability of preferred building materials, which can be altered by location and setting.

With the deciphering of the genetic code during the 1950s, research in instinctive behavior shifted toward the genetic basis of innate behavior. Research has shown that one group of chemicals called neuropeptides are produced by specific genes. Within the brain, these peptides have the ability to govern stereotyped behavior. For example, a specific peptide (angiotensin II), when injected into vertebrates, causes spontaneous drinking activity. Research on neuroactive chemicals that influence behavior patterns is still in its infancy; research on brain functions and neuron pathways has only begun. In the future, biochemists expect to isolate the link between the genetic code and brain function.

There remains one area of stereotyped behavior that has puzzled scientists since the time of Darwin: the question of altruistic behavior. Scientists have wondered how to explain behaviors such as ants drowning themselves in a stream so that others can cross over them or a parent animal risking its life so that offspring can survive. W. D. Hamilton, among others, began to use probability models, which show that cooperative behavior, which may risk the lives of individuals, results in greater survival of the rest of the group. Altruistic behavior not only is "for the good of the species" but also provides the greater probability that large numbers of that group will reproduce and therefore gain an advantage.

THE FIELDWORK OF ETHOLOGY

The early students of ethology were often called "bird-watchers" because they began their work by observing the behavior of birds. Fieldwork involves repeated observations of a subject over long periods of time. Eventually, a sequence of behavior emerges, and then it is possible to read the language of the behavior. For example, Lorenz observed that the courtship sequence of the mallard duck involved some ten segmented parts, such as bill-shake, head-flick, tail-shake, and grunt-whistle.

In certain instances, fieldwork with insects and lower animals offers the possibility of direct experimentation. In his attempt to translate the dancing motion of bees, von Frisch made the food source available. Consequently, he was able to vary the distance of the food source, change its location, and alter the quantity of food. Each change in the variables produced some variation in the dance—perhaps a new "phrase" added to the language. In other areas of research, the test subject can be modified for direct experimentation. When William Keeton attempted to explain how pigeons found their way home, he used contact lenses to cover the eyes of the pigeon to block out the position of the sun. He also created secondary magnetic fields around the pigeon to test the subject's sensitivity to the earth's magnetic field.

While all studies of stereotyped behavior begin with observation, either in the field or in controlled settings, further exploration usually requires laboratory research. In Jack Hailman's study of the learning component of sea gull chicks, he constructed models of gull parents and correlated pecking accuracy with growth. He also modified features of the model sea gull to study possible changes in responses from the chicks. As the search for the causes of instinctive behavior moves further into the organism, the methodology follows—into areas of brain function (neuroethology), the chemistry of innate behavior, and the genetic component of behavior. Investigating the source of egg-laying behavior of a species of a large marine snail (*Aplysia*), Richard Scheller and Richard Axel found three genes that produce a number of peptides that govern this behavior.

INSTINCTS AND GENES

Instincts are a part of all living organisms, and observable instinctive responses are only a small part of an intricate pattern. The genetic makeup of an organism dictates specific unlearned patterns of responses and variations that, in turn, determine a favorable selection of individuals within a species. Clearly, instincts in sexual selection, reproduction, food gathering, and other basic needs are critical for the survival of members of a species. In higher animals, instinctive activities are often overshadowed (and sometimes disguised) by learned patterns of responses. For example, in dogs, the pulling back of facial muscles and the showing of teeth is a response to fear and attack. In humans, laughter is a similar response to surprise, embarrassment, and uneasiness. Because of social adaptation, however, laughter takes on additional behavioral conventions.

Until the early part of the twentieth century, instincts were thought to be learned responses to specific situations. Consequently, if aggression were learned, then it could be modified, changed, and unlearned. With the establishment of a genetic and biochemical foundation for instincts, research in stereotyped behavior has become part of a heated debate. In 1975, Edward O. Wilson published *Sociobiology: The New Synthesis.* This highly technical work found a surprisingly large audience; in it, Wilson attempted to place all social behavior on a biological basis. Although the work emphasized animal behavior, Wilson implied that all human history was also part of evolutionary biology and that his work would synthesize all the social sciences with biology. Since instincts such as aggression, selection of sexual partners, and care of the young play a prominent part in cultural activities, Wilson seems to suggest that in the future, the study of society will be grounded in neurobiology and sociobiology.

—*Victor W. Chen*

FURTHER READING

Bateson, P. P. G., Peter H. Klopfer, and Nicholas S. Thompson, eds. *Behavior and Evolution.* Plenum Press, 1993.

Frisch, Karl von. "Dialects in the Language of Bees." *Scientific American* 130 (August 1962): 2–8.

Goodall, Jane. *In the Shadow of Man.* Houghton Mifflin, 1971.

Lorenz, Konrad Z. "The Evolution of Behavior." *Scientific American* 199 (December 1959): 67–78.

Scheller, Richard H., and Richard Axel. "How Genes Control an Innate Behavior." *Scientific American* 250 (March 1984): 54–63.

Smith, John Maynard. "The Evolution of Behavior." *Scientific American* 239 (September 1978): 176–192.

Animal Intelligence

Fields of Study

Biology; Ethology; Psychology; Linguistics

Abstract

Both the general public and scientific community have long been intrigued with questions about how animals think and what are they thinking. Published reports of animal cognition increased dramatically in the last half of the twentieth century. Cognitive ethology is a relatively new discipline that studies animal intelligence. Traditionally, attitudes about animal intelligence can be sorted into those that place animals on a continuum with humans and those that see animals as distinct from humans.

Key Concepts

anthropomorphism: attributing human characteristics or states of mind to animals

cognition: transformation and elaboration of sensory input

cognitive ethology: scientific study of animal intelligence

lexigrams: symbols associated with objects or places in keyboard communication experiments with primates

protogrammar: word coined to signify the early foundation for grammar development found in primates

recapitulation: stages of human development reappearing in different animal species

FLASHES OF BRILLIANCE

Both the general public and scientific community have long been intrigued with questions about how animals think and what are they thinking. Published reports of animal cognition increased dramatically in the last half of the twentieth century. Chimpanzees in the Ivory Coast have demonstrated extensive use of rocks as tools in cracking nuts. These primates have also been reported to hide undesirable expressions from their faces and act as if blind or deaf. Vervets have been found to use an elaborate system of alarm calls that seem to function as words. Parrots can demonstrate the ability to count, and birds exhibit the capacity to make and use tools to gather food. Dolphins apparently understand and follow simple commands. Primates have been trained to use signs in a symbolic fashion, communicating their needs, desires, and thoughts.

THEORIES OF COGNITIVE ETHOLOGY

Cognitive ethology is a relatively new discipline that studies animal intelligence. Donald Griffin is considered to have founded this branch of study through the publication of *Animal Thinking* (1984) and *Animal Minds* (1992). Since the appearance of his books, numerous instances of animal intelligence have been gathered from observation and experimentation.

Traditionally, attitudes about animal intelligence can be sorted into those that place animals on a continuum with humans and those that see animals as distinct from humans. From the former perspective, animal behavior is readily interpreted as a definite sign of various cognitive skills and special abilities along a continuum of development. From a discontinuity perspective, humans alone are considered to possess the higher cognitive skill of reasoning. The higher cognitive abilities are considered to be a uniquely human capacity that sets them apart from the lower animals, who are controlled by instinct.

Charles Darwin, in *The Descent of Man* (1871), defended the idea of the intelligence of animals existing on a continuum with humans. Since animals and humans have a common ancestry, animals would have the fundamental capacities for rational choice, reflection, and insight. Darwin concluded that the differences between the minds of humans and animals were of degree rather than of kind. Following Darwin's proclamation, a number of anecdotal studies concerning animal intelligence appeared that suggested extensive cognitive ability in animals. Unfortunately, many of the examples illustrated anthropomorphism. This is the process whereby humanlike characteristics are attributed to animal behavior.

Some interpretations of Darwin's statement created a distorted view about animal evolution that persisted long into the twentieth century. The idea that life on earth represents a chain of progress from inferior to superior forms began to influence the view of animal intelligence. The theory that ontogeny recapitulates phylogeny also became popular in the early years of the twentieth century. This theory, which does not have any scientific support, suggested that the advancement of life forms corresponded to the stages of development for humans. This stepladder approach to animal intelligence led to a ranking of animals compared to the developmental stages of human infants and children.

This approach to animal intelligence is flawed because it relies on the notion that some animals are more highly evolved than others are. Evolution does not have a single point of greatest evolution. The branches of the evolutionary tree have culminated with many different species occupying special niches. Thus, the "degree" of a species' evolution depends on the extent to which it successfully occupies its niche.

ANIMALS WHO MIGHT THINK

In addition to this tendency to attribute states of mind to animals that are found in humans, there were a number of cases of labeling trained behavior in animals as signs of reasoning skills. One of the most famous examples was the case of the horse, Clever Hans, in the early 1900s. Wilhelm von Osten owned a horse that demonstrated extensive arithmetic skills. When von Osten presented a written arithmetic problem to Hans, the horse would tap out the answer with his forefoot. Clever Hans also appeared adept at telling time and answered questions about sociopolitical events by nodding or shaking his head yes or no. The horse's abilities suggested to many individuals the similarity between animal and human minds. Eventually, the Prussian Academy of Sciences discovered that Hans was not answering the questions by means of any reasoning skills, but was an astute observer of the behavior of his owner and those around him. When questions were posed to Hans, cues were provided unconsciously to the horse about the correct answer. Since horses have evolved to ascertain subtle visual cues from others in their herd, Hans was able to form a number of cued associations which led to a reward. The owner of Clever Hans was not attempting to perpetrate fraud. He believed in the possibility that a horse could have reasoning ability, but von Osten was not sophisticated in how he tested for the skills. The inadvertent cueing of an animal to respond in a certain fashion is one of the major confounding factors found in the investigation of animal intelligence.

The case of Clever Hans illustrates two other problems that confound reports concerning the level of intelligence in animals. First is the problem of anthropomorphism. People develop an emotional bond with animals and interpret behavior in order to enhance the closeness they feel to them. The second problem concerns the methods used to measure intelligence. The classic case of Köhler's chimpanzees illustrates this problem.

In the early part of the twentieth century, Wolfgang Köhler assessed the reasoning ability of chimpanzees to obtain food outside of an enclosure. After a rake was left in the enclosure, food was placed out of reach of the caged chimpanzees. The chimpanzees were able to use the rake to bring food to the cage. Köhler concluded that the animals had insight into the nature of the problem and used reasoning to achieve a solution. A further study, requiring the fitting together of two sticks in order to reach the food, also supported Köhler's conclusions. However, later experimentation has revealed that chimpanzees without a history of playing with sticks could not solve the problem. Apparently, in order to solve the problem, the chimpanzees needed an extensive history of playing with sticks, which enabled them to learn how sticks could be used at a later time. In solving the problem, they were using an instinctual tendency to play with sticks and scraping them over the ground.

PRIMATES AND SIGN LANGUAGE

A contemporary example of the problem of measurement can be provided with the case of Washoe, the first chimpanzee to be taught sign language. Because of physical inability to vocalize human speech, chimpanzees were taught sign language as a mode of communication with humans. Soon Washoe and another signing chimpanzee, Nim Chimsky, were reported to have spontaneously created novel sentences through their signing. For example, Washoe was reported to have signed the combination "water" and "bird" after seeing a swan. Being a novel combination of signs, the trainers of Washoe explained the behavior as creative insight. Unfortunately, Washoe had also shown repeated signing of meaningless combinations, leading to the conclusion that a significant pairing of signs would eventually appear not because of the primate's cognitive reasoning but as a result of chance. Inevitably, these early attempts to demonstrate animal intelligence were widely discredited as exaggeration or self-delusion on the part of the animal's trainers, and this animal language research from the late 1970s fell into disrepute.

In order avoid the ambiguities of sign language, later researchers used keyboards that related symbols to a variety of objects, people, and places. Much of this research has taken place at the Language Research Center at Georgia State University in Atlanta under the guidance of Dr. Sue Savage-Rumbaugh. In the first experiments, two chimpanzees, Austin and Sherman, were familiarized with a system of symbols or lexigrams. Each was abstract and arbitrarily associated with an object, person, place, or situation. Eventually Austin and Sherman learned to communicate with symbols illustrated on a keyboard. For example, an experiment was devised where one chimpanzee was shown where food was being deposited in a certain container while the other had control of a tool to open the container. With the keyboard present, the chimpanzees were able to communicate with one another to use the tool on the correct container.

Soon a bonobo, Kanzi, became the star pupil of this technique and learned a vocabulary of two hundred symbols. Kanzi eventually showed the capacity to construct rudimentary sentences that were generated spontaneously. The bonobos and chimpanzees trained using the keyboards appear to be exhibiting a protogrammar. This is a term to indicate the beginnings of grammar, roughly equivalent to the verbal skills seen in a human child about two to three years old.

In the late 1990s, two other chimpanzees, Panbanisha and Panzee, surpassed the capacities evidenced by Kanzi. Panbanisha was reported to understand complex sentences and use the keyboard to communicate spontaneously with the outside world. Panzee reportedly demonstrated lasting recall of hidden objects and could communicate that knowledge to blinded confederates.

In the early twentieth century, the center investigated cooperation, delayed gratification, metacognition (that is, consciousness of one's thinking), numeracy, planning and recall, and responses to inequity, among various primate species, including capuchin and rhesus monkeys. Researchers have concluded that apes, like humans, learn better from observation than direct instruction and gain new competencies without reinforcement, an area of behavior that researcher Michael Beran termed "emergent" learning. Also, remarkably, researcher Sarah Brosnan and her colleagues found that capuchins object when they observe others receiving better treatment, suggesting an evolutionary basis for ethics and morals.

Although the results have been impressive, critics of the center's activities remain. The question remains whether the animals are demonstrating extremely effective training, reacting to subtle cuing, or exhibiting some level of abstract reasoning. Other perennial issues are the need to define what exactly constitutes human language and whether linguistic performance or comprehension should be the prime indicator of cognition.

—*Frank J. Prerost*

FURTHER READING

Budiansky, Stephen. *If a Lion Could Talk*. Free Press, 1998.

"History." *Language Research Center*, Georgia State University, 11 Aug. 2011, www2.gsu.edu/~wwwlrc/3476.html. Accessed 31 Jan. 2018.

Johnson, George. "Chimp Talk Debate: Is It Really Language?" *The New York Times,* 6 June 1995, www.nytimes.com/1995/06/06/science/chimp-talk-debate-is-it-really-language.html. Accessed 31 Jan. 2018.

Moss, Cynthia. *Elephant Memories*. William Morrow, 1988.

Page, George. *Inside the Animal Mind*. Doubleday, 1999.

Savage-Rumbaugh, Sue. *Kanzi: The Ape at the Brink of the Human Mind.* John Wiley & Sons, 1994.

Savage-Rumbaugh, Sue, Stuart G. Shanker, and Talbot J. Taylor. *Apes, Language, and the Human Mind.* Oxford University Press, 1998.

Animal Kingdom

Fields of Study

Biology; Physiology; Zoology

Abstract

Human perception of the animal kingdom tends to focus on relatively large vertebrates. However, large vertebrates account for just a tiny fraction of the animal world. Over 97 percent of animal species are invertebrates. Insects and arthropods make up the vast majority of animal species and a huge percentage of the individual animals on earth. Most other animal phyla are also far more diverse and numerous than vertebrates.

Key Concepts

class: the taxonomic category composed of related genera; closely related classes form a phylum or division

invertebrates: animals lacking a backbone

phylogeny: the evolutionary history of a group of species

phylum: the taxonomic category of animals and animal-like protists that is contained within a kingdom and consists of related classes

species: a group of animals capable of interbreeding under normal natural conditions; the smallest major taxonomic category

taxonomy: the science by which organisms are classified into hierarchically arranged categories that reflect their evolutionary relationship

vertebrates: animals with a backbone or vertebral column

WHAT ARE ANIMALS

Human perception of the animal kingdom tends to focus on relatively large vertebrates. However, these large vertebrates are true minorities, accounting for just a tiny fraction of the animal world. Over 97 percent of animal species are invertebrates, the earliest animals to emerge. Insects and arthropods make up the vast majority of animal species and a huge percentage of the individual animals on earth. Most other animal phyla are also far more diverse and numerous than vertebrates. All vertebrates together constitute only part of a single phylum, Chordata. In simple terms, the small and boneless creatures called invertebrates dominate the animal kingdom. They live bountifully in diverse habitats: in pond muck, on ocean bottoms, in treetops, beneath leaf litters, and in many other environments.

Animals are easy to identify but difficult to define due to the diversity and complexity of all creatures in this kingdom. The best approach relies on a set of common characteristics that distinguish animals from individuals of other kingdoms, a field called systematics. First, animals are multicellular (made up of many cells). Second, animals are heterotrophic, obtaining nutrients and energy by consuming other organisms. Third, animals are usually capable of sexual reproduction, although other reproductive

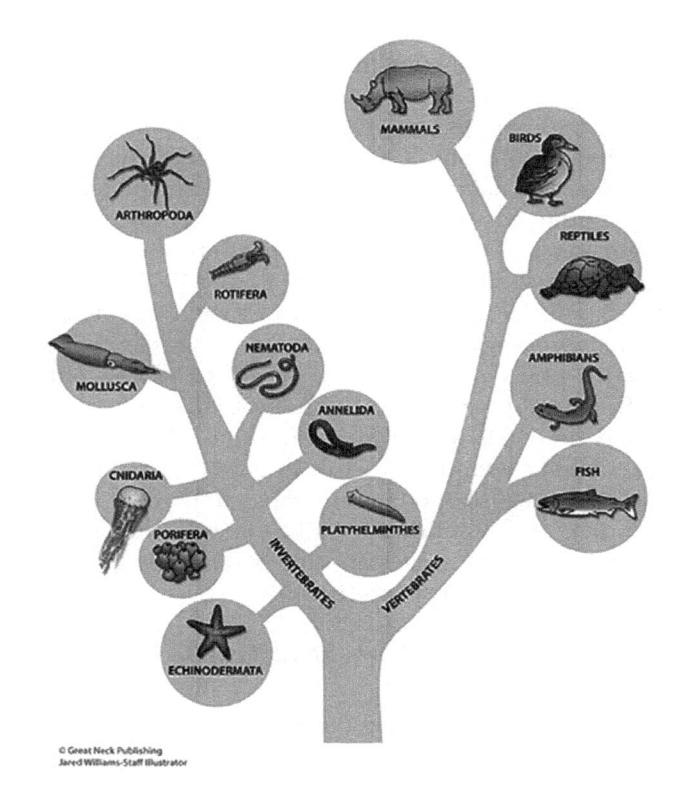

Animal Kingdom: In science, a kingdom is a large group of similar living things. This tree shows the two major divisions of the animal kingdom: Invertebrates and Vertebrates.

styles may exist. Fourth, animal cells contain no cell wall. Fifth, animals are mobile during at least some stage of their lives. Finally, animals are usually capable of rapidly responding to external stimuli through their nerve cells, muscle, or contractile tissue. These six characteristics taken together distinguish animals from other living creatures.

Based upon evolutionary theories, animal phyla show trends toward increasing cellular organization and complexity. In the most ancient phylum of animals, sponges, individual cells may have specialized functions but act independently, hence are not organized into tissues or organs. Cnidarians (jellyfish and their relatives), the phylum most closely related to sponges, have well-defined tissues that coordinate movement and sensory information. Flatworms, the next phylum to emerge, have organs and organ systems, such as a reproductive system. Organ systems are also found in all the remaining, more recently emerged animal phyla. The trend toward increasing complexity goes beyond the level of cellular organization and specialization. It includes the presence and type of symmetry in body plan, the degree of development in sensory organs and brain, the presence and type of body cavity, the presence of body segmentation, and the structure of the digestive system. The members of the latest phylum, including vertebrate animals such as seals, whales, horses, and humans, also exhibit a trend toward increasing size and sophistication of the brain.

Based upon these traits, animals can be grouped into approximately twenty-seven phyla. The nine major phyla include, from simple to more complex, Porifera (sponges), Cnidaria (hydra, anemones, and jellyfish), Platyhelminthes (flatworms), Nematoda (roundworms), Annelida (segmented worms), Arthropoda (insects, arachnids, and crustaceans), Mollusca (snails, clams, and squid), Echinodermata (sea stars and sea urchins), and Chordata (primarily vertebrates).

THE SPONGES, HYDRA, ANEMONES, AND JELLYFISH

Sponges (phylum Porifera) are the simplest multicellular animals that lack true tissues and organs. They resemble colonies in which single-celled organisms live together for mutual benefit. However, individual sponge cells are able to survive and function independently. All sponges, whether single-celled or colony-like, have a similar body plan. The body is perforated by numerous tiny pores, through which water enters, and by fewer large holes, through which water is expelled. Water travels within the sponge through canals where oxygen is extracted and microorganisms are filtered into cells for digestion. Some sponges can grow more than a meter in height. So far, more than five thousand species of sponges have been identified, all of which are aquatic and most of which are marine.

The phylum Cnidaria is composed of hydra, anemones, and jellyfish. Clearly more complex than sponges, cnidarians have distinct tissues, including contractile tissue that acts like muscle and nerve net that spreads through the body and controls movement and feeding behavior. However, they lack true organs and a brain. Their beautiful and diverse body shapes are variations of two basic body plans: tentacled and jellyfish-like. Tentacles attach to rocks and reach upward for grasping, stinging, and immobilizing prey. A jellyfish-like body can easily be carried by ocean currents. Cnidarians are radially symmetrical, with body parts arranged in a circle around the mouth and digestive cavity. All cnidarians are predators, but none hunt actively. They rely upon their tentacles to grasp small animals floundering by chance into contact with them. Once stimulated by contact, special cells called cnidocytes explosively inject poisonous or sticky darts into prey. The immobilized prey is forced through an elastic mouth into a digestive sac. The undigested food is expelled through the mouth. Cnidarians may reproduce sexually or asexually. Of the nine thousand or more species in this phylum, all are aquatic and most are marine. One of these, the corals, is of particular ecological importance.

DIVERSE FORMS OF WORMS

Flatworms (phylum Platyhelminthes) are more complex than cnidarians, yet are the simplest organisms with well-developed organs. Their bilaterally symmetrical bodies are an adaptation to active movement, as found in other, more complex organisms. Their sense organs, consisting of light-detecting eyespots and cells responsive to chemical and tactile stimuli, inform their bodies whether to feed, forge onward, or retreat. When the flatworm encounters smaller animals, it sucks its prey through a muscular tube called the pharynx, located in the middle of the body. Compared with more complex organisms, however, flatworms lack both respiratory and circulatory systems. They can produce sexually or asexually. Most flatworms are hermaphroditic, possessing both male and female sex organs within one body. Examples of flatworms include parasitic tapeworms and flukes.

Roundworms (phylum Nematoda) reside in nearly every habitat on earth. Of an estimated 500,000 species, only 10,000 have been named. They are largely microscopic, although some may reach a meter in length. They have a rather simple body plan, with a tubular gut that runs from mouth to anus. A fluid-filled hydrostatic skeleton provides support and a framework against which muscles can act. They also have a tough but flexible cuticle on the outside of the body and a simple brain that processes and transmits information. They do not have circulatory or respiratory systems. Most nematodes reproduce sexually, with the male fertilizing the female by injecting sperm inside her body. Nematodes play a crucial role in breaking down organic matter in ecosystems. Some are also parasites to humans or other animals, such as hookworms that infect human feet, *Trichinella* worms that cause trichinosis, and heartworms that attack dogs' hearts.

The prominent feature of the phylum Annelida is segmentation of the body into a series of repeating units; hence they are called segmented worms. Each body compartment is controlled by separate muscles, collectively capable of far greater complexity of movement than in other worms. A well-developed closed circulatory system distributes gases and nutrients throughout the body. Primitive hearts, in essence short, expanded segments of specialized blood vessels, can contract rhythmically. A simple brain located in the head plus nerve cords along the length of the body and within each segment control movement and other activities. Among the nine thousand or so species identified, the best known examples are the earthworm and its relatives, and leeches. However, the largest annelids, the polychaetes, live primarily in the ocean.

THE ARTHROPODS, MOLLUSKS, AND ECHINODERMS

The phylum Arthropoda comprises insects, spiders, and crustaceans. By any standard, whether number of individuals or number of species, arthropods are the most dominant animals on earth. A mere 10 percent of animals described in this phylum constitutes one million species, including insects (class Insecta), spiders and their relatives (class Arachnida), and crabs, shrimp, and their relatives (class Crustacea). The enormous success of arthropods is due to several adaptational features. The exoskeleton allows precision movement; segmentation generates specialized and more effective organ systems; these, in turn, allow higher efficiency in gas exchange, circulation, and information processing. Most arthropods have well-developed sensory systems, including compound eyes and acute chemical and tactile senses. Of the three classes, insects are the most diverse and abundant, accounting for 850,000 species identified. Insects usually have three pairs of legs plus two pairs of wings. Their ability to fly helps them escape from predators and find widely dispersed food. Insects normally go through radical changes in body form through metamorphosis, from egg to larva to pupa and finally to winged adults that mate and lay eggs.

Spiders and scorpions are examples of the class Arachnida. They typically have eight walking legs and are mostly carnivores, living on either a liquid diet of blood (ticks and mosquitoes) or predigested prey (scorpions). Simple eyes equipped with a single lens are extremely sensitive to movement, which helps in catching prey or escaping from predators. There are about fifty thousand species of arachnids. Crab, shrimp, crayfish, and their relatives make up the class Crustacea, comprising roughly thirty thousand species. They are largely aquatic, with a wide variation in size. Except for two pairs of sensory antenna and mostly compound eyes, they are highly variable in body form.

As their name suggests, members of the phylum Mollusca—snails, clams, and squid—have a moist, muscular body supported by a hydrostatic skeleton. Some have a shell of calcium carbonate to protect their body; others escape predation by moving swiftly or by being distasteful if caught. They have an open circulatory system. Their nerve systems are more advanced than those of arthropods in that more nerves are concentrated in the brain. Reproduction is sexual; some species have separate sexes, and others are hermaphroditic. Together, there are five thousand species identified, among which clams, octopuses, oysters, scallops, snails, and squid are the most familiar.

Sea stars, sea urchins, and sea cucumbers compose the phylum Echinodermata. These animals are mostly marine, and adults have radial symmetry and lack a head and distinct brain. They have very simple nervous systems, and hence move very slowly on numerous, tiny, tube feet. They feed on algae or small particles sifted from sand or water. Most species reproduce by releasing sperm and eggs into the water, where larvae develop upon fertilization. Another distinct feature of echinoderms is their endoskeleton, a hard shell of calcium carbonate enclosed by an outer skin.

PHYLUM CHORDATA: THE TUNICATES, LANCELETS, AND VERTEBRATES

Animals of this phylum exhibit tremendous diversity in form and size. They include small sea squirts and lancelets (invertebrates), and birds, fish, amphibians, reptiles, and mammals (vertebrates). Members of this phylum possess four characteristics at some stage of their lives: a notochord—a stiff yet flexible rod that extends the length of the body and provides an attachment site for muscles; a dorsal, hollow nerve cord at the anterior end of the notochord that becomes a brain; specialized respiratory openings called pharyngeal gill slits; and a tail that extends past the anus. There are only two classes of invertebrates in Chordata, lancelets and tunicates, both of which are small marine animals. Lancelets reside mainly in the sandy sea bottom and live by filtering tiny food particles from the water. Sea squirts, a member of the tunicates, send out a forceful jet of water in response to touch or danger. Their filter-feeding, saclike bodies move slowly via contraction.

Vertebrates are the most conspicuous animals on earth. Their backbones and other adaptations have contributed to their success. There are seven major classes of vertebrates.

Jawless fishes (Agnatha) were the earliest vertebrates to arise in the sea. Two examples are hagfishes and lampreys. The colorful hagfishes are strictly marine, living in communal burrows in mud, feeding on polychaete worms. Lampreys live in both fresh and salt water. Some lampreys are parasitic, attaching to fish with suckerlike mouths lined with rasping teeth. They live on blood and body fluids sucked from their hosts. Cartilaginous fishes (Chondrichthyes) are skillful predators, and include sharks, skates, and rays. Their skeletons are made up exclusively of cartilage, void of bone. Many shark species have several rows of razor-sharp teeth, with back rows moving forward as front teeth are lost to action or aging. Most sharks, as most skates and rays, are shy and retiring creatures that do not attack humans. A few species, however, can be deadly when irritated. Bony fishes (Osteichthyes), spread over a wide range of aquatic habitats, are the most diverse and abundant vertebrates on earth. As suggested by their name, bones rather than cartilage make up their skeletons. Of seventeen thousand species identified, all bony fishes have bladders that help them float effortlessly. Some have lungs and modified fins that work as legs, which help them to survive periodic drying in freshwater habitats.

Amphibians (Amphibia) live a double life between aquatic and terrestrial habitats. They represent the transition of life from water to land. Some adaptations, such as lungs, a three-chambered heart, and moist skin, help them live a temporary land life. However, other traits, requiring water for fertilization and juvenile development, restrict the range of amphibian habitats on land. Their double life and permeable skin have made amphibians particularly vulnerable to pollutants and environmental fouling. About 2,500 species have been identified, including frogs, toads, and salamanders. The seven thousand species of reptiles (Reptilia) identified have bodies of diverse forms. Turtles, snakes, lizards, alligators, and crocodiles are all reptiles, as well as the huge and now-extinct dinosaurs. Reptiles have more efficient lungs than amphibians, a tough, scaly skin that resists water loss and protects the body, a mechanism of internal fertilization, and a shelled egg.

The diversity of birds (Aves) is revealed through nine thousand species, including the delicate hummingbird, the endangered spotted owl, and the largest bird, the ostrich. Their ability to soar gracefully in the air depends on many anatomical and physiological traits. These features include a light body with hollow bones, light wings with feathers that also provide protection and insulation, reduced reproductive organs during nonbreeding periods, a single ovary in female birds, acute eyesight, and a delicate nervous system that facilitates the extraordinary coordination and balance needed for flight. Birds, which are warm-blooded, also have four-chambered hearts that help to maintain high body temperature and a high metabolic rate, crucial for flight.

The last vertebrate class, mammals (Mammalia), is represented by some 4,500 species. In addition to being warm-blooded with high metabolic rates, mammals normally possess hair, produce milk for their offspring, assume a remarkable diversity in form, and possess more highly developed brains than any other class. The bat, cheetah, elephant, mole, monkey, seal, and whale exemplify the radiation of mammals into nearly all habitats, with their bodies finely adapted to their lifestyles.

—*Ming Y. Zheng*

FURTHER READING

Blaustein, A. "Amphibians in a Bad Light." *Natural History* 103 (October 1994): 32–39.

Brusca, R. C., and G. J. Brusca. *Invertebrates*. Sinauer Associates, 1990.

Buchsbaum, Ralph, et al. *Animals Without Backbones: An Introduction to the Invertebrates*. University of Chicago Press, 2013.

Fairbairn, Daphne J. *Odd Couples: Extraordinary Differences Between the Sexes in the Animal Kingdom*. Princeton UP, 2013.

McMenamin, M. A., and D. L. McMenamin. *The Emergence of Animals: The Cambrian Breakthrough*. Columbia University Press, 1990.

Morell, V. "Life on a Grain of Sand." *Discover* 16 (April 1995): 78–86.

Rennie, J. "Living Together." *Scientific American* 266 (January 1992): 122.

ANIMAL LIFE OF SWAMPS AND MARSHES

FIELDS OF STUDY

Limnology; Biology; Ecology; Environmental Studies

ABSTRACT

Swamps and marshes differ from ponds and lakes in two primary ways: they are shallower, with seasonally fluctuating water levels, and they have little or no open water, being dominated by emergent vegetation. The predominant vegetation in marshes comprises grasses, sedges, and rushes. Marshes tend to have shallower water than swamps, and are more apt to dry out completely during the drier part of the year. The predominant vegetation in swamps comprises trees and shrubs. Not only is the water generally deeper in swamps, it is often more permanent, although water levels still tend to fluctuate seasonally.

KEY CONCEPTS

consumer food chain: a simplified description of the grazing and predator/prey relationships within an ecosystem

emergent vegetation: aquatic vegetation that grows tall enough to be visible above the water

invertebrate: a simple animal lacking a backbone

primary consumer: an organism that gets its nourishment from eating primary producers, which are mostly green plants and algae

primary productivity: production of biomass mainly by green plants

THE NATURE OF SWAMPS AND MARSHES

Swamps and marshes differ from ponds and lakes in two primary ways: they are shallower, with seasonally fluctuating water levels, and they have little or no open water, being dominated by emergent vegetation. The predominant vegetation in marshes comprises grasses, sedges, and rushes. Marshes tend to have shallower water than swamps, and are more apt to dry out completely during the drier part of the year. The predominant vegetation in swamps comprises trees and shrubs. Not only is the water generally deeper in swamps, it is often more permanent, although water levels still tend to fluctuate seasonally.

Although swamps and marshes are usually thought of as freshwater wetlands, saline swamps and marshes also exist, either in desertlike environments where evaporation rates are high, or in coastal areas. Coastal swamps and marshes tend to fluctuate in depth and salinity as the tides change. Coastal swamps are often

Wetland.

referred to as mangrove swamps or mangals. Coastal marshes are often called estuaries. Most of the following information will focus on freshwater swamps and marshes.

INSECTS AND OTHER INVERTEBRATES

The smallest of the invertebrates are primary consumers, who are at the base of the consumer food chain in wetlands. When algae is abundant in the water, primary consumers also proliferate. One of the most common of these is the water flea (*Daphnia*), which is just barely visible to the naked eye. When viewed under the microscope, the reason for its name becomes apparent, as it looks remarkably like a flea. Many species of water flea exist, varying greatly in size, head shape, swimming appendages, and other traits. Water fleas and other small invertebrates are preyed upon by larger invertebrates, tadpoles, and small fish.

The bulk of the invertebrates are insects, in either their adult or juvenile forms. Many flying insects, such as mosquitoes and dragonflies, spend their early life in the water. Dragonfly nymphs are especially vicious predators, preying upon almost anything small enough for them to grab in their strong jaws, including small fish. Adult dragonflies and their cousins, the damselflies, are commonly seen flying around marshes and swamps. Other insects, such as water striders, backswimmers, water boatmen, and diving beetles, are also abundant in the still waters, making their entire lives in or on the water. Even a few spiders (which are arachnids, not insects) have adapted to the aquatic way of life, able to stay underwater for extended periods by trapping bubbles of air next to their bodies.

Another common invertebrate group is the shellfish. Freshwater clams live buried in the sediment and filter food out of the water, while mussels can form dense assemblages on rocks and other debris in the water. A few clams and mussels from tropical and semitropical parts of the world have inadvertently been introduced to some temperate wetlands with devastating effects. Snails can also be found in many marshes and swamps, where they feed on algae growing on rocks and on the submerged stems and leaves of plants.

FISH, REPTILES, AND AMPHIBIANS

The occurrence of fish is most often associated with water depth. In shallow, seasonal marshes or swamps, fish are often absent. In deeper marshes and swamps, they can be abundant and provide food for other animals, especially birds. These fish can range from the unique, bottom-dwelling catfish to more active fish such as perch and bass. Many smaller, less noticeable species also occur, some of which are near extinction due to loss of unique habitat.

The most common reptile in marshes and swamps is the turtle. Most turtles are predators, although some feed primarily on plant material. The snapping turtle has a reputation for eating almost anything, plant or animal. Although turtles can be observed swimming in the water, they are most often seen sunning themselves on warm rocks just in reach of the water, where they can quickly escape from potential predators. Among the more dangerous reptiles of swamps and marshes are the alligators, crocodiles, and caimans. These are exclusively predatory, and although generally not very aggressive toward humans, they do attack on occasion. Snakes are also predators. Garter snakes grab their prey using their mouth and gradually swallow it. The cottonmouth is a poisonous viper that injects venom into its prey to incapacitate it before swallowing. Although snakes are more often thought of as terrestrial, they are expert swimmers.

Amphibians include salamanders and frogs, both of which spend their early life entirely in the water, where their eggs are laid. Frog eggs develop into tadpoles with tails and no legs; as they develop, the tail disappears and legs start to grow. Salamander larvae have legs from the beginning and they never lose their tail. When young, frogs and salamanders feed mostly on algae and small invertebrates in the water. As adults, frogs typically eat insects, and salamanders eat a variety of invertebrates, including worms and grubs.

BIRDS AND MAMMALS

Near the top of the food chain, a large variety of birds either live in swamps and marshes or come there to hunt. Common residents include ducks, geese, herons, and egrets. In these shallower waters, "puddling" ducks predominate, characterized by feeding behavior where they reach for vegetation on the bottom with only their rumps protruding from the surface. Herons and egrets frequently wade slowly and deliberately, searching for frogs or fish to quickly grab and eat. Many songbirds take advantage of the habitat; some, such as the marsh wren, even make their nests in the bulrushes near the margin of the water. Predatory birds such as hawks, eagles, falcons, owls, and osprey come to prey on other birds, snakes, fish, or smaller mammals.

Various mammals also take advantage of the aquatic bounty. Mink and otters freely swim in search of fish. Others, like raccoons, tend to hunt for food on the margins. Rodents of various kinds eat greens and seeds, abundant because of the water supply, and foxes and coyotes hunt the rodents. Other mammals, such as deer, come for the water, and sometimes eat the tender vegetation as well. Beaver may even be responsible for the development of a new swamp or marsh by building a dam across a creek.

HUMAN DESTRUCTION OF SWAMPS AND MARSHES

In spite of the great richness of life present in swamps and marshes, human society, in general, views these ecosystems as unsightly and useless. Consequently, many wetlands have been drained to make way for farmland, roads, or other developments. Because so much has been lost, environmental laws in the United States now prohibit further wetland destruction, unless new wetlands are formed to replace those that are lost. These laws are a recognition of the ecological importance of swamps and marshes.

The primary productivity of swamps and marshes is only surpassed by tropical rain forests and algal beds and reefs. Shallow water and ample light allow rich plant and algae growth, which supports the rich variety of organisms found here. Swamps and marshes are especially important to migrating waterfowl, who need them for their abundant food supplies. As wetlands have been lost, waterfowl numbers have been reduced. Their high productivity also makes them important in absorbing excess carbon dioxide from the atmosphere, thus reducing global warming.

—*Bryan Ness*

FURTHER READING

Bransilver, Connie, Larry W. Richardson, Jane Goodall, and Stuart D. Strahl. *Florida's Unsung Wilderness: The Swamps.* Westcliffe, 2000.

Scheffer, M. *Ecology of Shallow Lakes.* Chapman and Hall, 1998.

Streever, Bill. *Bringing Back the Wetlands.* Sainty & Associates, 1999.

Tiner, Ralph W. *In Search of Swampland: A Wetland Sourcebook and Field Guide.* Rutgers University Press, 1998.

Vileisis, Ann. *Discovering the Unknown Landscape: A History of America's Wetlands.* Island Press, 1999.

Wetzel, Robert G. *Limnology: Lake and River Ecosystems.* 3rd ed. San Diego, Calif.: Academic Press, 2001.

ANIMAL LIFE SPANS

FIELDS OF STUDY

Biology; Gerontology; Veterinary Science; Anthropology; Zoology

ABSTRACT

Life span has two common meanings, often confused. Popularly, the term can refer to the longevity of an individual, but in biology it is more abstract, a characteristic of the entire species rather than individual members. In this

sense, life span is the maximum time an individual can live, given its environment and heredity, and life expectancy is the amount of time remaining it at any point during its life span. The great variety within the animal kingdom complicates the definition of life span.

KEY CONCEPTS

life cycle: the sequence of development beginning with a certain event in an organism's life (such as the fertilization of a gamete), and ending with the same event in the next generation

life expectancy: the probable length of life remaining to an organism based upon the average life span of the population to which it belongs

life span: the maximum time between birth and death for the members of a species

metabolism: the biochemical action by which energy is stored and used in the body to maintain life

mortality rate: the percentage of a population dying in a year

LIFE SPAN AND LIFE EXPECTANCY

Life span has two common meanings, often confused. Popularly, the term can refer to the longevity of an individual, but in biology it is more abstract, a characteristic of the entire species rather than individual members. In this sense, life span is the maximum time an individual can live, given its environment and heredity, and life expectancy is the amount of time remaining it at any point during its life span.

The great variety within the animal kingdom complicates the definition of life span. For some species, life span is essentially the same as its life cycle, the return to the same developmental stage from one generation to the next. Salmon, for instance, hatch from eggs in small streams, migrate to the sea where they reach maturity, then struggle up rivers to return to the site of their hatching in order to produce more eggs. After completing the reproduction cycle, both males and females die. For most animal species, however, an individual may produce several generations of offspring before dying. In the case of humans, individuals can live long after their fertility ends.

It is often difficult to measure the life span for specific species, and in fact, a time period is seldom definitive for all species members. Rather, scientists recognize that mortality, the percentage of individuals that die each year (or each day in some cases), increases for a population of organisms until at some age it reaches 100 percent, and all individuals in a generation are dead. Finding the life span of laboratory animals is fairly simple: A population is given the best possible living conditions, and observers wait for the last individual to die. That, presumably, is the optimal life span for the species. Much the same procedure can determine the life spans of pets and animals in zoos, except that the population to be observed is much smaller, sometimes only a single individual. Likewise, the records of thoroughbred domestic animals, born and raised in captivity, provide evidence for the maximum species life span. Most animals live in the wild, however, where investigations face a great variety of conditions and anecdotal evidence can be misleading. For example, biologists long thought that bowhead whales lived only about fifty or sixty years, but in the late 1990s, various new kinds of historical and biochemical evidence identified bowheads that lived well into their second century. Moreover, species whose members can, if need be, go into dormancy show a large variation in individuals' apparent longevity, even when all members pass through a single life cycle.

In general, however, the life span variation among species falls into a narrow range of time. The shortest life spans, which last a single life cycle, can be a matter of days, while the longest last more than two hundred years. Humans enjoy the greatest longevity among primates; scientists estimate the theoretical maximum human life span to be from 130 to 150 years, more than double that of the species with the next greatest endurance, gorillas, and chimpanzees. However, several kinds of invertebrates live more than two centuries.

LIFE SPAN LIMITERS AND EXTENDERS

While each species has a theoretical maximum life span, few if any individual animals reach it. Three general influences limit longevity: environmental pressures, variations in physiological processes, and heredity.

Most domesticated species have longer life spans, frequently two times longer, than their wild relatives, and wild animals in captivity often live longer than in their natural habitat. The gray squirrel, for example, lives for three to six years in the wild, but from fifteen to twenty years in a zoo. The reason is clearly the safer, healthier, less stressful environment. Predators are one of the biggest threats in the wild, as are fluctuations in climate that affect the availability of food and shelter. Natural calamities such as hurricanes, wild fires, and earthquakes also take their toll.

Disease and chance injury also kill off many organisms, but even if disease is absent, many physiological processes in the body appear to degenerate or stop with age. Biochemists find that individual cells age and die. Cross-connections among connective tissue, such as ligaments and cartilage, gradually reduce the body's flexibility and inhibit motion. Chemical plaques build up in brain tissue, hindering the electrochemical connection among neurons. Highly reactive oxidants, the ionized molecular byproducts of cellular metabolism, also build up with age and degrade the operation of cells' mitochondria, the generators of chemical energy, as well as other cellular organelles. Moreover, laboratory tests reveal that cells can divide only a certain number of times—about fifty for some human cells—and then they die, a limit called the Hayflick finite doubling potential phenomenon. In connection with this finding, geneticists discovered that the lengths of deoxyribonucleic acid (DNA) at the tips of chromosomes, called telomeres, shorten with age and appear to play a role in cell dysfunction and death.

Other genetic factors help determine the age limit of cells and the entire organism. Research in the late 1990s with mice and roundworms uncovered genes that regulate cell life, including one kind that causes cells to suicide if their DNA or internal structure is too damaged to function properly or if the cells turn cancerous. Loss of function by genes damaged during cell division (mitosis) or by environmental toxins can impair a cell's ability to maintain metabolism, or set loose functionless or even harmful proteins and enzymes into the blood stream and lymph system—all of which can bring on illness. Organisms that escape such effects still may face genetic disease. Most human populations, for instance, now have a longer life expectancy than ever before; because of it, neurodegenerative diseases (such as Alzheimer's disease), cardiovascular dysfunction, and immunological disorders, caused or made possible by genes, grow ever more common. Because many of these life-shortening maladies were rare or nonexistent before the twentieth century, natural selection has not had a chance to cull them from the human genome. Similar diseases crop up in domestic animals and wild animals in captivity.

Late twentieth century research also identified several ways to extend life. A sharply reduced diet extends the lives of mice and fruit flies beyond their normal life span (although for the brown trout, a richer diet is life-extending), and lower than normal temperature has the same effect on fruit flies, fish, and lizards. Discovery of the relation of telomere length to cell death and isolation of genes that order cells to suicide allowed scientists to bioengineer animals with cells whose DNA self-repaired the telomeres and genes that failed to trigger cell-suicide; the result was individuals that in some cases had life spans twice the normal length or more. However, the most pervasive life-extending method is the development of social life—colonies, herds, packs—in which individuals work together to protect, shelter, and feed themselves and rear their offspring. In the case of humans (and perhaps some whales and primates), culture and intelligence permit the species to pass on recently acquired information from one generation to the next and even to alter the environment to lengthen collective and individual life spans.

SIZE AND LIFE SPAN

Biologists noted long ago that large animals live longer than small animals, but why this should be true was a mystery. In 1883, Max Rubner proposed that the relation had to do with metabolism. Large animals have a smaller skin surface to body mass ratio than smaller animals; accordingly, large animals lose heat more slowly and so can maintain body functions with less energy burned per unit of mass; in other words, a slower metabolism. Scientists since found that individuals of different animal species all use about the same amount of chemical energy during their lives, twenty-five to forty million calories per pound per lifetime. (There are significant exceptions: Humans consume about eighty million calories per pound.) Because small animals must burn energy at a higher rate, the argument holds, they physically wear out faster.

In 1932, Max Kleiber derived a mathematical relationship for Rubner's proposal. According to Kleiber's law, also known as the quarter-power scaling law, as mass rises, pulse rate decreases by the one-fourth power. So elephants, which have 104 times the mass of chickens, have a pulse rate one tenth as fast. Scientists suggest that the relation results from the geometry of circulatory systems and point out that the quarter-power scaling law is pervasive in nature, but the underlying reason for it remains unknown. In any case, plenty of exceptions to the mass-life span correlation exist. With a life span of about one hundred years, box turtles outlive fellow reptiles, for example, and humans outlive all mammals (with the possible exception of some whale species) regardless of size. Exceptions also occur among domestic species living sheltered lives: Cats have longer life spans than dogs.

THEORIES OF LIFE SPAN

A 1995 review of data from earlier animal studies suggested that heredity accounts for about 35 percent of the variation in life spans among invertebrates and mammals; 65 percent comes from unshared environmental influences. Nonetheless, theorists in the life sciences continue to debate the relative influence of genetic and other biochemical factors on the one hand and environment factors on the other hand. The debate derives from the premise that life spans are the product of the natural selection that ensured species' reproductive success. The proposals fall into three categories: random damage (stochastic) theory, programmed self-destruction theory, and ecological theory.

Random damage theories emphasize the wear and tear on the body that accumulates with metabolic action. It is the source of damage that differs from one theory to another. One holds that the buildup of metabolically produced antioxidants is the key factor, a spinoff of the long-standing conjecture that the faster an animal's metabolism is, the shorter its life span. A second theory focuses on proteins that change over time until their effect on the body alters for the worse, especially when the proteins are involved in cellular repair. There is, for example, the altered connective tissue that causes the cross-connections stiffening tendons and ligaments. Another such change is the glycosylation of proteins or nucleic acids, in which a carbohydrate is added. Glycosylation is involved in such age-related disorders as cataracts, vascular degeneration among diabetics, and possibly atherosclerosis A third theory points to the buildup of toxins inside cells, and a fourth concerns the potential problems that come from errors in metabolism or viral infection which slowly impair or kill cells. Fifth, the somatic mutations theory proposes that chance mutations accumulate in a person's nuclear or mitochondrial genome and induce cell death or produce proteins and enzymes that have aging effects.

Programmed death theories hold, as the name suggests, that a species' genetic heritage includes a built-in timer or damage sensor. Telomeres shorten as DNA ages until the genes at the end of chromosomes are unprotected and subject to deterioration during the splitting and gene crossover of mitosis. The genes then lose their ability to produce essential biochemicals, whose absence harms the body or leaves it defenseless against damage from infection or injury. Damage sensors can include the genes that instruct cancerous or malfunctioning cells to die. Although such genes clearly are a means to check the spread of disease, their cumulative effect may be harmful. Furthermore, scientists discovered genes that produce much more of, or less of, their metabolic products as cells age, which also contributes to the overall aging of the body.

The ecological theory draws conclusions about life span from a species' role in its environment. Small animals have faster metabolisms and live shorter lives, it is argued, because they are not likely to escape predators for very long. Therefore, they evolved to mature and reproduce rapidly. Large animals typically have more defenses against predators and can afford to take life slowly. Moreover, animals that evolve defensive armor, spines, or poison also avoid predation and live longer than related species that do not. Finally, species that evolve mechanisms to withstand environmental stress, as from extreme temperatures or food scarcity, also have long life spans.

The theories assume that the life span for individuals within a species serves the survival of the entire species. Yet even a species' days on earth are numbered. Environmental change can slowly squeeze them from their habitats, a catastrophe may wipe them out indiscriminately, or they may evolve into a new species. Scientists estimate that the average life span for a multicellular species lasts from one to fifteen million years. That average is stretched by several notable exceptions in the animal kingdom—such living fossils as crocodiles (140 million years old), horseshoe crabs

(200 million), cockroaches (250 million), coelacanths (a type of fish, 400 million), and certain mollusks of the genus *Neopilina* (500 million).

—Lisa M. Sardinia

FURTHER READING
Austad, Steven. *Why We Age.* John Wiley & Sons, 1997.
Bova, Ben. *Immortality.* Avon Books, 1998.
Finch, Caleb B., and Rudolph E. Tanzi. "The Genetics of Aging." *Science* 278, no. 5337 (October 17, 1997): 104–109.
Furlow, Bryant, and Tara Armijo-Prewitt. "Fly Now, Die Later." *New Scientist* 164, no. 2209 (October 23, 1999): 32–35.
MacKenzie, Dana. "New Clues to Why Size Equals Destiny." *Science* 284, no. 5420 (April 6, 1999): 1607–1608.
Margulis, Lynn, and Karlene V. Schwartz. *The Five Kingdoms: An Illustrated Guide to the Phyla of Life on Earth.* 3rd ed. W. H. Freeman, 1998.
Walford, Roy L. *Maximum Life Span.* W. W. Norton, 1983.

ANIMAL LOCOMOTION

FIELDS OF STUDY

Biophysics; Kinesiology; Biomechanics

ABSTRACT
A great many living animals, from unicellular microbes to the multicellular higher organisms, have developed the ability to move from one place to another as needed. This ability for self-directed movement is called locomotion. Locomotion is under conscious (or voluntary) control. In essence, it serves to allow living creatures to move in chosen directions, toward food, away from danger, toward mates, and so on. Locomotion is essential for the optimization of the lives of animals.

KEY CONCEPTS

airfoil: the wing of a flying animal or airplane that the provides lift and/or thrust needed for flight
angle of attack: the angle at which an airfoil meets the air passing it
eukaryote: a higher organism, whose cells have their genetic material in a membrane-bound nucleus and possess other membrane-bound organelles
lift: the force enabling flight; it occurs when an airfoil makes air passing it move so as to lower air pressure above the airfoil to values below that beneath the airfoil
locomotion: the ability of an organism to move from one place to another as needed
mitochondrion: a subcellular organelle that converts foods to carbon dioxide, water, and energy
sessile: an organism that is not capable of moving from its point of origin
striated muscle: voluntary or skeletal muscle, capable of conscious enervation

THE NEED TO MOVE
Being a sessile animal, unable to move from place to place, makes life difficult for a wide variety of reasons. For example, if food is not available in a very limited area, death may come very quickly. As a result, a great many living animals, from unicellular microbes to the multicellular higher organisms, have developed the ability to move from one place to another as needed. This ability for self-directed movement is called locomotion.

Locomotion is under conscious (or voluntary) control. In essence, it serves to allow living creatures to move in chosen directions, toward food, away from danger, toward mates, and so on. Locomotion is essential for the optimization of the lives of animals. This can be seen in two ways: Most living animals are capable of locomotion; it is particularly well developed in large, higher organisms.

THE PHYSIOLOGY OF LOCOMOTION
Many unicellular animals, such as protozoa, carry out locomotion via hairlike organelles called cilia and flagella, or by amoeba-like motion. Their movement is not complex. Multicellular animals, crawl, walk, run, swim, glide, or fly. The basis for these much more complex activities is cooperative operation

Dachshund leaping from a log at the beach.

of their skeletons and muscles, along with other processes needed to suit each organism to its environment. In addition, nerve impulse transmission coordinates locomotion.

The skeleton and the muscles act as the levers that can move to produce body motion in chosen directions and as the machines that produce the motive ability required for locomotion. Skeletons are of two types. The external, bony, jointed exoskeletons of arthropods (insects and crustaceans) surround them like medieval suits of armor. This design limits the dexterity of arthropods and the complexity of the motion possible for them. In contrast, the bony endoskeletons of vertebrates are much more complex and produce the ability for much more complicated and dexterous locomotion.

The muscle involved in locomotion can be exemplified by vertebrate striated skeletal muscle. This type of muscle is called striated because its fibers appear to be banded when viewed under light microscopes. Such muscle is under the conscious (voluntary) control of any higher organism containing it and wishing to move toward a desired place. Striated muscle is either red or white. Red striated muscle has a very good blood supply—the basis for its color—and obtains energy in the course of oxidizing foods into carbon dioxide and water. This occurs in its many mitochondria, subcellular organelles that carry out oxidative, energy-getting processes. White striated muscle has less of a blood supply and less mitochondria. It obtains most of its energy by converting food to lactic acid, without much use of oxidation or mitochondria.

All locomotion uses the contraction and relaxation of muscle or related systems, which can be thought of as muscle waves. In the unicellular eukaryotes (such as protozoa), cilia and flagella are used. These structures are cell organelles that are whiplike and identical in their makeup. However, their size and number per cell differ. One protozoan type may have on its surface hundreds of cilia, each ten to twenty microns long. Another species can have one or two flagella up to 250 microns long.

Both cilia and flagella cause movement by producing a wavelike motion transmitted all through the organelle, outward from its base. Their action uses organized protein microtubules in a network, interconnected by arms and spokes reminiscent of fibrils, and crosslinks of striated muscle. The mechanism of their action also uses a sliding mechanism initiated by a protein called dynein. Cilia and flagella are also found in multicellular eukaryotes, including humans. For example, flagella propel human sperm and are found in human lungs, where they mediate the removal of dust and dirt from the airways.

Amoeboid motion also occurs in protozoa. Here, their protoplasm flows in the direction of desired motion, forming pseudopods, followed by forward locomotion of the cell. This type of locomotion seems to depend on cytoplasm transitioning between fluid endoplasm and gel-like ectoplasm. Pseudopod formation occurs due to the action of microfilaments of actin, one of the contractile proteins also found in striated muscle. Thus, the contraction mechanism may be similar to that in striated muscle.

The various forms of locomotion include crawling, walking and running, swimming, gliding, and flying. These locomotive processes all use muscle waves based on actin and myosin fibrils. However, the complete systems involved become more and more complex due to increases of the needs of locomotion, crawling on earth's surface, walking or running on land, swimming in lakes and oceans, and gliding or flying in the air. The increases in complexity are mostly due to the need for the involvement of more and more muscles.

CRAWLING, WALKING, AND RUNNING

Crawling is movement on land without legs, due to waves of muscular activity traveling along a crawler's body. Two crawling creatures are snakes and snails.

When a snake crawls, it curves its body into bends around stones and ground irregularities. These bends travel backward along its body in muscle waves. The waves push against the stones and irregularities and move the snake forward. Snails, in contrast, usually crawl on a long foot covered with mucus, using waves of muscle contraction that travel along the foot. Snail locomotion depends on their interesting mucus, which resists gentle pressure as if it were a solid, but when pressed hard behaves like a liquid. The muscle waves of snail feet are designed so that foot portions moving forward exert pressure enough to liquefy the mucus, while parts that move backward do not.

All living organisms capable of walking or running have skeletons because the processes are intricate and require the use of a relatively large number of muscles in a complex fashion. Walking is locomotion at a slow pace, when there is no need for speedy movement. Running enables organisms to move much more quickly. It is done when predators threaten, or if there is an unthreatening reason to travel more quickly than usual, for instance, to obtain some choice food or a mate.

The complexity of walking and running is made clear by the coordination required. For example, when four-footed animals walk, at least one forefoot and one hind foot are kept on the ground at all times. In walking bipeds, at least one foot is on the ground at all times, and at intervals, both feet are on the ground. In running, there are times when both feet leave the ground simultaneously. Four-footed mammals walk slowly, but often attain higher speeds by trotting, cantering, or galloping. In each of these gaits, there are differences in the operation of the legs. For example, in a gallop, both forefeet are set down at one instant and both hind feet are set down at the next instant. This is not due to chance; rather, it allows the back muscles to contribute to the work done and spread the overall effort through the body.

Another aspect of the differences between walking and running is the interaction of the feet with the ground. Some animals, including humans, walk or run with the entire bottom of the foot on the ground; this is known as plantigrade locomotion. In contrast, most quadrupeds carry out these actions with only their toes on the ground, known as digitigrade locomotion. Such differences in ways in which vertebrates walk and run add to the complexity of the processes and the number of muscles used. The rate at which an animal moves also depends on its foot size. The larger the foot area that touches the ground while walking or running, the slower the animal. The complexity of running activity adds to total speed. For example, cheetahs can run at a speed of sixty-five to seventy miles per hour, partly by adding to their gait a series of leaps at chosen intervals. The added leaps help to produce the ability to move three times as fast as exceptional human athletes can run one hundred meter dashes.

SWIMMING, GLIDING, AND FLYING

Organisms that swim in water range from tiny ciliated or flagellate microbes to blue whales that are over thirty meters (one hundred feet) long. Their locomotive operations are complex and use oarlike rowing, hydrofoil motion of propellers, body undulation, and jet propulsion. Oar and propeller actions have some similarities, but oars push in the direction of a desired movement, while propellers work at right angles to blade motion. Propeller action is due to lift, caused by hydrofoil action in the water. Propeller-mediated animal locomotion is more complex than walking or running, especially in the case of whale flukes. These hydrofoils, in addition to complex motion, must be tilted at angles which provide more lift than drag, under widely varying conditions.

In animals that swim by undulation, movements are made like the crawling of a snake. Here, the resistance to motion of poorly compressible water makes it easier for the swimmer to slide forward along its long axis than sideways along its short axis. The squid exemplifies those organisms that move by jet propulsion. A squid does this by drawing water into its mantle and squirting it out forcefully from an organ called a funnel. The squirt is aimed forward when a squid wants to swim backward, or backward when it seeks to swim forward. A closely related species, the nautilus, uses the same method of propulsion, but is only capable of moving backwards.

One of the problems associated with locomotion by swimming is that most tissues in swimming organisms are more dense than water. Such organisms would sink, lacking ways to raise body buoyancy. These needs add to the complexity associated with swimming, compared with walking or running on land. Ways in which to counter sinking include swimming with all fins extended like airplane wings, as sharks do, to add buoyancy by increasing the surface area of the body. Also, swimming animals, such as the bony fish, adjust their densities to match that of the water via gas-filled swim bladders in their body cavities.

Only insects, birds, and bats are capable of locomotion by true flight. However, some other animals glide for short distances. Best known are flying squirrels, which glide from tree to tree. Also, flying fish use their large pectorals as wings of a sort. They leap out of the oceans with the fins spread and glide through the air for up to half a minute. As is the case with whale flukes in water, a glider supported by lift on its wings is slowed and pulled down by drag. Thus, most gliding organisms do not stay in the air for long Their time suspended aloft depends upon maintaining the appropriate angle of attack of their wings. They can adjust glide speed by changing these angles of attack. Increasing this angle slows them and decreasing it allows them to speed up.

Soaring, or gliding for long distances, is possible only for animals aloft in quickly rising air. Few animals other than birds and butterflies have wings capable of sustained gliding. The needed air currents are found mostly over hillsides or coastal cliffs and in thermals, which are currents of hot air rising from ground heated by the Sun. Vultures soaring to scan for carrion can travel for distances up to fifty miles. Butterflies, gulls, and albatrosses also soar for long distances. Long-distance soaring is very often used on migration trips, to conserve energy for other activities.

Ways to increase glide or soar distance include tilting the body to suit different angles of attack and moving wings forward or backward. These actions provide streamlining and move the center of gravity of the body. Gliding and soaring animals make turns by giving one wing a higher angle of attack than the other. Furthermore, gliding is usually done with legs retracted, to increase streamlining. The legs are lowered like the landing gear of airplanes when a braking effect is required. Clearly, locomotion in the air is even more difficult than swimming, involving much more complex interactions of the muscles and the skeleton.

Power, needed for most sustained flight, is obtained when flying animals flap their wings. There are two main animal flight types, high-speed flight and hovering flight. In the high-speed flight of birds, bats, and insects, the body moves forward at the same speed as the wings. The wings beat up and down as the animal moves forward through the air. Each down stroke produces lift and propels the flying animal forward and supports it in the air. The angle of attack of the upstroke is increased, so as to produce very little lift. This is because lift, acting upward at right angles to the wing's path, would slow high-speed flight.

In hovering flight, the body is stationary and the wings move very quickly, as when hummingbirds feed. Animals use two hovering techniques. Most often, their wings are kept horizontal and beat straight backward and straight forward. They turn upside down for each backstroke, adjusting the angle of attack so lift is obtained. The hum of hummingbirds is due to the frequencies with which their wings beat (twenty-five to fifty-five cycles per second). Insects buzz, whine, or drone, according to the sound produced by the higher frequencies at which their wings beat. In some cases, as in pigeons, hovering is accomplished when the wings clap together at the end of an upstroke. The clap enhances airflow into the space formed as they separate and so enhances lift. Initial wing-clap hovering in pigeons causes the sound first made when a flock flies away together.

—*Sanford S. Singer*

FURTHER READING

Audesirk, Gerald, and Teresa Audesirk. *Biology: Life on Earth*. 5th ed. Prentice Hall, 1999.

Dantzler, William H., ed. *Comparative Physiology*. New York: Oxford University Press, 1997.

Dickenson, Michael H., Claire T. Farley, Robert J. Full, M. A. R. Koehk, Rodger Kram, and Steven Lehman. "How Animals Move." *Science* 238, no. 5463 (April 7, 2000): 100–106.

Gamlin, Linda, and Gail Vines, eds. *The Evolution of Life*. Oxford University Press, 1987.

Hickman, Pamela, and Pat Stephens. *Animals in Motion: How Animals Swim, Jump, Slither, and Glide*. Kids Can, 2000.

Animal Mating

Fields of Study
Biology; Population Science; Animal Behavior

Abstract
Whether it is relatively simple or highly complex, the ultimate goal of courtship is to bring together two animals of different sexes of the same species to bring about successful mating and reproduction. Although they may not always be as obvious, mating patterns among animals also are quite diverse. Reproduction is the process by which all living organisms produce offspring. The need to propagate is a cardinal necessity for the preservation of the species.

Key Concepts

asexual reproduction: reproduction without the union of male and female sex cells

dioecious: having two separate sexes, namely male and female

estrus: the period of the sexual cycle during which a female is sexually receptive

gamete: a sex cell, either male or female

hermaphrodite: an animal with both male and female sex organs

monogamy: a mating system in which one male and one female comprise the main breeding unit

polygamy: a mating system in which a single adult of one sex mates with several members of the opposite sex.

COURTSHIP AND REPRODUCTION

As one observes the diverse and elaborate movements and behaviors seen in courtship displayed by all types of animals, it is easy to lose sight of what courtship is meant to accomplish. Whether it is relatively simple or highly complex, the ultimate goal of courtship is to bring together two animals of different sexes of the same species to bring about successful mating and reproduction. Although they may not always be as obvious, mating patterns among animals also are quite diverse.

Reproduction, one of nature's most fundamental and essential functions, is the process by which all living organisms produce offspring. The need to propagate is a cardinal necessity for the preservation of the species. Each living organism has its own unique way of accomplishing this requirement. The higher the animal, the more intricate the process. In one-celled organisms reproduction is usually asexual, where only one living entity is required for procreation. The one-celled animal simply splits in two, losing its original identity and thereby creating two new organisms that are characteristically exact duplicates of the parent. The process is known as fission. Some single-celled entities reproduce sexually, where two similar organisms fuse, exchange nuclear materials, and then break apart, after which each organism reproduces by fission. This form of unicellular reproduction is known as conjugation. Sometimes after conjugation, the participating organisms do not reproduce. It appears that the process is merely to revitalize the organisms. This is the most primitive method of sexual reproduction. The new organisms that are produced are from two distinct parents, having definite genetic characteristics of their own.

The procreation process in most multicellular animals involves a more complex form of sexual reproduction. Here, unique and differentiated male and female reproductive cells called gametes unite to form a single cell known as the zygote. The zygote undergoes successive divisions to form a new multicellular organism, where half the genes in the zygote come from one parent and half from the other, creating a singularly different living creature.

Since most species of animals have sexual reproduction, it must offer some advantages. Sexual reproduction results in maintaining high levels of genetic variability within a population. Genetic variability produces variability in behavior, structure, and physiology and provides a species with greater flexibility in meeting changing environmental conditions. There must also be some long-term evolutionary advantages.

A dictionary definition of mating encompasses the idea of individuals coming together to form a pair, with the implication of doing so to produce offspring. If an emphasis is placed on the latter part of the definition, then it is seen that mating may involve all degrees and types of relationships and interactions among animals. In animals that have sexual reproduction, the union of male and female sperm and eggs, also known as gametes or sex cells, may be accomplished by either external or internal fertilization. Simple or elaborate courtship rituals ready animals for mating and fertilization. Several factors may influence reproduction. Most animals have a distinct time of the year, the breeding season, when reproduction is possible. Depending on the species, the breeding season may be the spring, summer, fall, or winter. In other species, including the human and some primates, breeding may occur throughout the year. The breeding season starts with the onset of courtship and is finished when the last offspring are weaned. The same species may have different breeding seasons depending on where they live. In temperate regions of the world, there may be much variation in the breeding in different years. Any factor that affects the health of animals may modify their breeding season, including food availability, temperature, light, and population density. In general, the breeding season usually is coordinated to maximize the likelihood of survival of the young.

MATING SYSTEMS

The means by which males and females are brought together in courtship and, ultimately, copulation is achieved is by some type of mating system or pattern. Systems vary widely throughout the animal kingdom, but there are several general groupings. (A reading of the literature in cultural anthropology shows that virtually all of these mating systems can be found in different human cultures, as well.) Mating systems typically are classified under three headings: promiscuity, monogamy and polygamy. It also is useful to delineate subcategories within some of these.

Promiscuity, as it defines a mating system, means that no pair bond is formed between individuals. A technical term for this system is polybranchygamy. which literally means "many brief matings." In this system, any male may mate with any female and no one individual has exclusive rights over any individuals of the opposite sex. Males and females may copulate with from one to many of the opposite sex. Promiscuity is found more often in males than in females, and this observation is probably related to the fact that in such species males have a far lesser investment in offspring than do females. The system is found in a small percentage of bird species and many species of mammals. Promiscuity is demonstrated well in those species that form leks. A lek is an area or territory used for communal courtship displays and mating by certain species. Leks have been observed among grouse, some African antelope, a species of bat, and some insects. Leks are used solely for mating. Females are attracted to the leks by elaborate courtship displays by the males, and mate with one or a small number of the males. Usually, only a select few of the males perform most of the matings. In one study of grouse, less than 10 percent of the males carried out more than 75 percent of all copulations. The African antelope, the Uganda kob, is an example of a mammal that utilizes a lek. Here, also, a small percentage of males breed most of the adult females. The system is very effective, as nearly all of the females produce offspring.

A second major type of mating system is monogamy. It is a system in which a pair-bond is formed between one female and one male. The pair-bond may exist for only one breeding season (annual monogamy) or it may persist for one or more breeding seasons (perennial monogamy). Monogamy is very common in birds, with a large majority of species showing it. Swans and eagles are examples of species showing perennial monogamy, and sparrows and warblers are examples of species showing annual monogamy. Although it has been thought that many birds practice true monogamy, evidence has accumulated that in monogamous birds there are frequent matings of the males with females outside of the primary pair-bond relationship. These matings are known as extra-pair-bond copulations, and offer advantages to the males if they result in successful fertilization of additional eggs. Monogamy seems to occur when both the male and female have nearly identical roles in the rearing of the young.

Monogamy is far less common in mammals than it is in birds, but there are some good examples. If monogamy is likely to occur where there is equal parental investment by both males and females, it might be predicted there is less need for male mammals to practice monogamy, since only the females nourish the young with milk. Among mammals, monogamy can be observed in gibbons, foxes, wolves, beaver, red fox, and even among some small rodents. Why should there be monogamy among any mammals? It would be advantageous for males to be monogamous if by doing so they increased the likelihood of survival of their offspring. Males could do so if they helped to feed the young and helped to defend the territory against predators. Monogamy can also be looked at from the perspective of the female, a so-called female-enforced monogamy, which may take place if females can gain sole benefit from the male's efforts without having to share them with other females.

The third major type of mating system is polygamy. In polygamy, an individual of one sex forms a pair-bond with several members of the opposite sex. The two major subtypes of polygamy are polyandry, where one female mates with more than one male, and polygyny, where one male mates with more than one female. Within both polygyny and polyandry there may be either a serial type or a simultaneous type. In the former case, one male or one female bonds with several members of the opposite sex but only one at a time. In the latter case, one male or one female bonds with several individuals of the opposite sex at the same time. Simultaneous polygyny is often referred to as harem polygyny. Although it appears to resemble promiscuity, it differs from promiscuity in that a pair-bond is formed even though it may be temporary. In altricial birds, in which the young are born in a helpless state, it is often observed that polygyny is present in habitats where food is unevenly distributed and one male is able to provide food for more than one female. Polygyny also may result in cases where there is a lack of availability of suitable territories for breeding and where there may be a pressure of heavy predation.

HERMAPHRODITISM

Although most species of animals are dioecious, meaning that they have two separate sexes, male and female, some individuals and even whole species are monoecious and have both sexes in one individual. Monoecious individuals are also called hermaphrodites. Although hermaphroditism is unusual in humans and most mammals, in some species it is the rule. There are many invertebrates in which the same individual produces both eggs and sperm. Some examples of hermaphroditic species include garden snails, free-living flatworms, the common earthworm, and some fish. It might seem that the hermaphroditic condition is a beneficial and efficient one, since any single individual is capable of producing and delivering both eggs and sperm. However, there is at least one major drawback to the system, the possibility of self-fertilization. Self-fertilization does not lead to as much genetic variability in the offspring compared to what might be expected if there was cross-fertilization. In species in which hermaphroditism is the rule, there are a number of processes at work which make self-fertilization unlikely, if not impossible. Some animals produce eggs at one time and sperm at a different time, making it impossible to self-fertilize. In animals in which the sex organs mature at different times, a condition known as protandry, an individual alternates between being different sexes, functioning as a female or male first and then becoming the other sex at a later time. In some species, including the earthworm, two individuals come together to engage in a mutual copulation. The two worms are held together by a secretion produced by both of them. The togetherness allows sufficient time for sperm from one worm to travel to the other worm and fertilize that worm's eggs and vice versa. Although in most animals the act of copulation is quite short, in the earthworm it may last three hours.

It would be remiss not to mention briefly an example of parthenogenesis in invertebrates. There are a few species of fish and lizards in which only females are known and in which the offspring are produced from eggs without fertilization by sperm. The Amazon moly, a fish, goes through the acts of courtship and mating with a male, but it is with males of other species, and sperm do not fertilize any of her eggs. The eggs develop parthenogenetically and produce another generation, apparently only of females.

Even a brief discussion of mating in animals reveals the complex and diverse methods that are used to produce another generation of the rich diversity of animal life on earth.

—*Donald J. Nash*

FURTHER READING

Choe, Jae, and Bernard Crespi. *The Evolution of Mating Systems in Insects and Arachnids.* Cambridge University Press, 1997.

Daly, Martin, and Margo Wilson. *Sex, Evolution, and Behavior: Adaptations for Reproduction.* 2nd ed. Wadsworth, 1990.

Jacobs, Merle. *Mr. Darwin Misread Miss Peacock's Mind: A New Look at Mate Selection in Light of Lessons from Nature.* Nature Books, 1999.

Wallen, Kim, and Jill Schneider. *Reproduction in Context: Social and Environmental Influences on Reproduction.* MIT Press, 2000.

Animal Migration

Fields of Study

Ecology; Ethology; Biology; Biogeography; Population Studies

Abstract

Migration is a general term employed by ecologists and ethologists to describe the nearly simultaneous movement of many individuals or entire populations of animals to or between different habitats. As defined, migrations do not include local excursions made by individuals or small groups of animals in search of food, to mark territorial boundaries, or to explore surrounding environments. Nomads are migrants whose populations follow those of their primary food sources. In contrast to migrations made by populations and excursions made by individuals, the spreading or movement of animals away from others is known as dispersal.

Key Concepts

biological clock: an inherent sense of timing regulating certain types of behavioral activity such as migration

clock sense: an inherent awareness of time or time intervals used, for example, to compensate for celestial movements in navigation

dispersal: the spreading apart of individuals away from one another and away from a place; includes a directional component when passive animals are moved by winds or currents

habitat: a specific, recognizable geographical region in which a particular kind of organism lives

migrant: a group or species of animal that moves from one habitat or geographical region to another

navigation: to follow or control the course of movement from the place of origin to a specific destination

nomads: migrants without a specific habitat; wanderers

orientation: an inherent sense of geographical location or place in time

population: a group of individuals of the same species geographically located in a given habitat at the same time

MIGRATIONS, EXCURSIONS, AND DISPERSALS

Migration is a general term employed by ecologists and ethologists to describe the nearly simultaneous movement of many individuals or entire populations of animals to or between different habitats. As defined, migrations do not include local excursions made by individuals or small groups of animals in search of food, to mark territorial boundaries, or to explore surrounding environments.

Nomads are migrants whose populations follow those of their primary food sources. Such animals (the American bison, for example) do not have fixed home ranges and wander in search of suitable forage. Some scientists view nomadic movements as a form of extended foraging behavior rather than as a special case of migration. In either context, the important point is that populations change habitats in response to changing conditions.

In contrast to migrations made by populations and excursions made by individuals, the spreading or movement of animals away from others is known as dispersal. Examples of dispersal include the drift of plankton in currents and the departure of subadult animals from the home range of their parents. In numerous species (sea turtles, rattlesnakes, and salmon, for example), dispersed members of a population may return to the place of origin after a variable interval of time.

MEANS AND REASONS

Some migratory species can orient themselves and know where they are in time and space. Many birds and mammals, for example, have an inherent sense of the direction, distance, and location of distant habitats. Orientation and travel along unfamiliar routes from one place or habitat to another is called navigation. Navigators use environmental and sensory information to reach distant geographical locations, and many of them do so with a remarkably accurate sense of timing. Homing pigeons are perhaps the best-studied animal navigators. These birds are able not only to discover where they are when released but also to return to their home loft from distant geographical locations.

An aggregation of migratory Pantala flavescens *dragonflies, known as globe skimmers, in Coorg, India. (Shyamal)*

Much has been learned about how animals successfully navigate over long distances from the pioneering studies of Archie Carr. Carr proposed that green sea turtles successfully find their widely separated nesting and feeding beaches by means of an inherent clock sense, map sense, and compass sense. His investigations and those of many others continue to stimulate great interest in the physiology and ecology of navigating species and in the environmental cues to which they respond. Sensory biologists, biophysicists, and engineers have incorporated knowledge of how animals detect and use environmental information to develop new and more accurate navigational systems for human use.

Animals use a variety of cues to locate their positions and appropriate travel paths. Most species have been found to use more than one type of information (sequentially, alternatively, or simultaneously) to navigate. Among the animals known to navigate are birds (the best-studied group), lobsters, bees, tortoises, bats, marine and terrestrial mammals, fish, brittle starfishes, newts, toads, and insects. Included among the orientation guideposts that one or more of these groups may use are the positions of the sun and stars, magnetic fields, ultraviolet light, tidal fluctuations caused by the changing positions of the moon and sun, atmospheric pressure variations, infrasounds (very low frequency sounds), polarized light (on overcast days), environmental odors, shoreline configurations, water currents, and visual landmarks. Celestial cues also require a time sense, or an internal clock, to compensate for movements of the animal relative to changing positions of celestial objects in the sky. In addition to an absolute dependence on environmental cues, young or inexperienced members of some species may learn navigational routes from experienced individuals, such as their parents, or other experienced individuals in the population. Visual mapping remembered from exploratory excursions may also play a role in enhancing the navigational abilities of some birds, fish, mammals, and other animals.

The different categories of animal movements, however, are perhaps not so important as the reasons animals migrate and the important biological consequences of the phenomenon. As a general principle, migrations are adaptive behavioral responses to changes in ecological conditions. Populations benefit in some way by regularly or episodically moving from one habitat to another.

An example of the adaptive value of migratory behavior is illustrated by movement of a population from a habitat where food, water, space, nesting materials, or other resources have become scarce (often a seasonal phenomenon) to an area where resources are more abundant. Relocation to a new habitat (or to the same type of habitat in a different geographical area) may reduce intraspecific or interspecific competition, may reduce death rates, and may increase overall fitness in the population. These benefits may result in an increase in reproduction in the population. Reproductive success, then, is the significant benefit and the only biological criterion used to evaluate population fitness.

PROGRAMMED AND EPISODIC MOVEMENTS

While many factors are believed to initiate migratory events, most fall into one of two general categories.

The first and largest category may be called programmed movements. Such migrations usually occur at predictable intervals and are important characteristics of a species' lifestyle or life cycle. Programmed migrations are not, in general, density-dependent. Movements are not caused by overcrowding or other stresses resulting from an excessive number of individuals in the population.

The lifestyle of a majority of drifting animals whose entire lives are spent in the water column, for example, includes a vertical migration from deep water during the day to surface waters at night. Thus, plankton exhibit a circadian rhythm (activity occurring during twenty-four-hour intervals) in their movements. An abundance of food at or near the surface, and escape from deep-water predators, are among the possible reasons for these migrations. Daily vertical movements of plankton are probably initiated by changes in light intensity at depth, and the animals follow light levels as they move toward the surface with the sinking sun. It is interesting to note that zooplankton living in polar waters during the winter-long night do not migrate.

Monarch butterflies and many large, vertebrate animals, such as herring, albatross, wildebeests, and temperate-latitude bats, migrate from one foraging area to another, or from breeding to foraging habitats, on a seasonal or annual basis. Annual migrations usually coincide with seasonal variation. Changes in day length, temperature, or the abundance of preferred food items associated with seasonal change may stimulate mass movements directly, or indirectly, through hormonal or other physiological changes that are correlated with seasonal environmental change. The onset of migration in many vertebrates is evidenced by an increase in restlessness that seems, in human terms, to be anticipatory.

In addition to their daily vertical migrations (lifestyle movements), the life cycles of marine zooplankton involve migrations, and it is convenient to use them as examples. As discussed, most adult animal plankton are found at depth during the day and near the surface at night. In contrast, zooplankton eggs and larvae remain in surface waters both day and night. As the young stages grow, molt, and change their shapes and food sources, they begin to migrate vertically. The extent of vertical migrations gradually increases throughout the developmental period, and as adults, these animals assume the migratory patterns of their parents. Patterns of movement that change during growth and development are examples of ontogenetic, or life-cycle, migrations.

The second large category of migratory behavior includes episodic, density-dependent population movement. Such migrations are often associated with, or caused by, adverse environmental changes (effect) that may be caused by overlarge populations (cause). Local resources are adequate to support a limited number of individuals (called the "carrying capacity" of the environment), but once that number has been exceeded, the population must either move or perish. Unfortunately, migration to escape unfavorable conditions may be unsuccessful, as another suitable habitat may not be encountered. Migrations caused by overpopulation or environmental degradation are common. Pollution and habitat destruction by humankind's activities are increasingly the cause of degraded environments, and in such cases, it is reasonable to conclude that humans have reduced the carrying capacity of many animal habitats. Familiar examples of density-dependent migrations are those of lemmings, locusts, and humans.

STUDYING MIGRATION

Methods used by scientists to study the mass movements of animals are quite varied and depend on the investigator's research interests and on the kinds of organisms being investigated. Environmental or physiological factors that initiate migrations may be of interest to sensory biologists and physiological ecologists; knowledge of variation in population distributions is important to biogeographers and wildlife biologists; and migrations in predator-prey relationships, competition, pollution, and life-history strategies are important aspects of classical ecological studies.

In addition to the specific aspect of migration being studied, the particular group of animals under investigation (moths, eels, elephants, snails) requires that different methods be used. Some of the approaches used in migration-related research illustrate how information and answers are obtained by scientists.

Arctic terns migrate from their breeding grounds in the Arctic to the Antarctic pack ice each year. The knowledge that these birds make a twenty-thousand-mile annual round-trip comes from the simplest and most practical method: direct observation of the birds (or their absence) at either end of the trip. Direct observation by ornithologists of the birds in flight can establish what route they take and whether they pause to rest or feed en route. Many birds have also been tracked using radar or by observations of their silhouettes passing in front of the moon at night. Birds are often banded (a loose ring containing coded information is placed on one leg) to determine the frequency of migration and how many round-trips an average individual makes during its lifetime. From this information, estimates of longevity, survival rates, and nesting or feeding site preferences can be made.

Factors that initiate migratory behavior in terns and in other birds can often be determined by ecologists able to relate environmental conditions (changes in temperature, day length, and the like) to the timing of migrations. Physiological ecologists study hormonal or other physiological changes that co-occur with environmental changes. Elevated testosterone levels, for example, may signal the onset of migratory behavior.

How Arctic terns orient and navigate along their migratory routes is usually studied by means of laboratory-conducted behavioral experiments. Birds are exposed to various combinations of stimuli (magnetic fields, planetarium-like celestial fields, light levels), and their orientation, activity levels, and physiological states are measured. Experiments involving surgical or chemical manipulation of known sensory systems are sometimes conducted to compare behavioral reactions to experimental stimuli. In such experiments, the birds (or other test animals) are rarely harmed.

Tags of several types are used to study migrations in a wide variety of animals, including birds, bees, starfishes, reptiles, mammals, fish, snails, and many others. Tags may be transmitting collars (located by direction-finding radio receivers); plastic or metal devices attached to ears, fins, or flippers; or even numbers, painted on the hard exoskeleton of bees and other insects. Additional types of tagging (or identifying) include radioactive implants and microchips that can be read by computerized digitizers; the use of brands and tattoos; and, of great interest, the use of biological tags. Parasites known to occur in only one population of migrants (nematode parasites of herring, for example) provide an interesting illustration of how the distribution of one species can be used to provide information about another.

THE IMPORTANCE OF MIGRATION

The causes, frequency, and extent of animal migration are so diverse that several definitions for the phenomenon have been proposed. None of these has been accepted by all scientists who study animal movements, however, and it is sometimes difficult to interpret what is meant when the term "migration" is used. Most researchers have adopted a broad compromise to include all but trivial population movements that involve some degree of habitat change.

It is important to recognize that few populations of animals are static; even sessile animals (such as oysters and barnacles) undergo developmental habitat changes, which are referred to as ontogenetic migrations. Aside from certain tropical and evergreen forest areas where migrations are relatively uncommon, a significant number of both aquatic and terrestrial species move from one habitat to another at some time during their lives. In the face of environmental change, including natural events such as seasonal variation and changes caused by resource limitations and environmental degradation, animals must either move, perish, or escape by means of drastic population reduction or by becoming inactive until conditions become more favorable (hibernation, arrested development and dormancy, and diapause in insects are examples of behavioral-ecological inactivity). Migration is the most common behavioral reaction to unfavorable environmental change exhibited by animals.

One cannot understand the biology of migrators until their distribution and habitats throughout life are known. The patterns of animal movements are fascinating, and it is useful to summarize some of the major differences between them. First, many species travel repeatedly during their lives between two habitats, on a daily basis (as plankton and chimney swifts do) or on an annual basis (as frogs and elks do). Second, some species migrate from one habitat (usually suitable for young stages) to another (usually the adult habitat) only once during their lives (for example, salmon, eels, damselflies, and most zooplankton, which live on the bottom as adults). Third, some species (many butterflies, for example) are born and mature in one geographical area (England, for example), migrate as adults to a distant geographical area (Spain, for example), and produce offspring that mature in the second area. These migrations take place between generations. In a fourth pattern, one may include the seasonal swarming of social arthropods such as termites, fire ants, and bees. A fifth but ill-defined pattern is discernible, exemplified by locust "plagues," irruptive emigration in lemmings and certain other rodents, and some mass migrations by humankind, as caused by war, famine, fear, politics, or disease. These are episodic and often, if not primarily, caused by severe population stress or catastrophic environmental change.

—*Sneed B. Collard*

Further Reading

Able, Kenneth P., ed. *Gatherings of Angels: Migrating Birds and Their Ecology.* Comstock, 1999.

Aidley, David, ed. *Animal Migration.* Cambridge University Press, 1981.

Begon, Michael, John Harper, and Colin Townsend. *Ecology: Individuals, Populations, and Communities.* 3rd ed. Sinauer Associates, 1996.

Dingle, Hugh. *Migration: The Biology of Life on the Move.* Oxford University Press, 1996.

Eisner, Thomas, and Edward Wilson, eds. *Animal Behavior: Readings from "Scientific American."* W. H. Freeman, 1975.

Newberry, Andrew, ed. *Life in the Sea: Readings from "Scientific American."* W. H. Freeman, 1982.

Pyle, Robert Michael. *Chasing Monarchs: Migrating with Butterflies of Passage.* Houghton Mifflin, 1999.

Rankin, Mary, ed. *Migration: Mechanisms and Adaptive Significance.* Marine Science Institute, University of Texas at Austin, 1985.

Reader's Digest Association. *The Wildlife Year.* Author, 1993.

Schone, Hermann. *Spatial Orientation: The Spatial Control of Behavior in Animals and Man.* Translated by Camilla Strausfeld. Princeton University Press, 1984.

Animal Physiology

Fields of Study

Physiology; Kinesiology; Veterinary Medicine; Biochemistry; Biomechanics

Abstract

How animal bodies work is the subject of physiology. Early Greek philosophers such as Aristotle used deductive reasoning (the use of observations, logic, and intuition) to explain function. William Harvey (1578–1657) was among the first and most successful in employing inductive reasoning (experimentation) to explain body function. Inductive reasoning uses the scientific method to study unsolved problems.

Key Concepts

artery: a blood vessel that carries blood from the heart to the body tissues

capillary: a small vessel that connects arteries to veins; this is where respiratory gases, nutrients, and wastes are exchanged between blood and tissues

glomerulus: a specialized capillary in the kidneys that filters blood

nephron: a tubular structure in the kidneys that extracts filtrate, and reabsorbs nutrients and other valuable substances and secretes wastes

neuron: an individual nerve cell; nerves are made up of many neurons bundled together

vein: a blood vessel that returns blood to the heart

THE SCIENCE OF PHYSIOLOGY

How animal bodies work is the subject of physiology. Early Greek philosophers such as Aristotle used deductive reasoning (the use of observations, logic, and intuition) to explain function. William Harvey (1578–1657) is generally considered the father of modern physiology, as he was among the first and most successful in employing inductive reasoning (experimentation) to explain body function. Inductive reasoning uses the scientific method to study unsolved problems. This approach starts with a hypothesis. Experiments are designed to test the hypothesis. Data are collected using quantified measurements; physiologists never simply say heart rate increased in response to the experimental treatment. The result is always expressed in quantified terms; for example, one might say that heart rate increased from thirty to forty-five beats per minute. Once the data are collected, they are analyzed and interpreted in terms of the original hypothesis. The interpretation usually leads to new hypotheses which are then tested with new experiments.

THE STUDY OF SYSTEMS

Classically, physiology has been divided into several organ systems. These include the neuromuscular system, the cardiovascular system, the pulmonary system (lungs), the renal system (kidneys), the digestive system, and the endocrine system. This nomenclature is based on mammalian systems and has had to be modified to include such alternative systems as gills (branchial system) in aquatic animals.

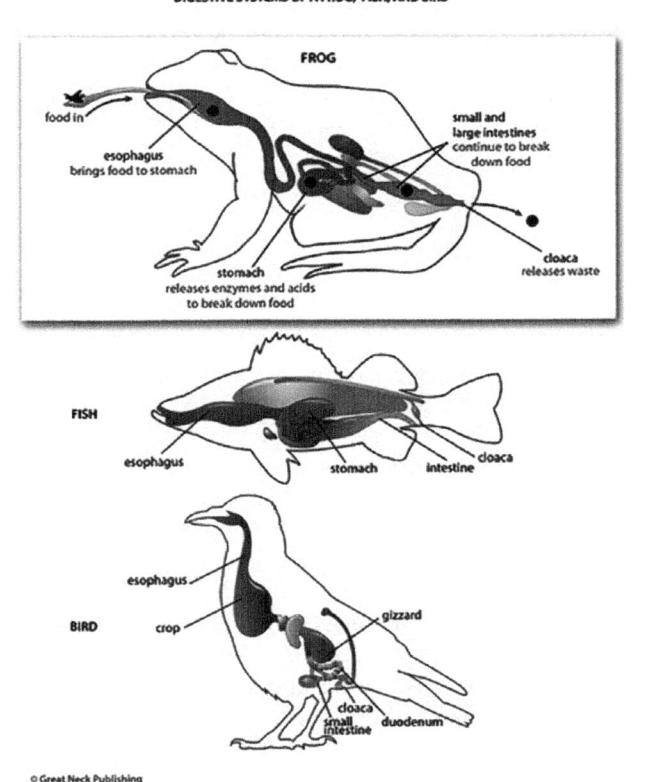

Animal Physiology: Animal physiology is the study of how animals' bodies work or function.

Physiology has been characterized as a synthetic science because it involves a synthesis of biology, chemistry, mathematics, and physics to describe body functions. The objective is to explain a given physiological function in terms which obey the laws of chemistry and physics. This often involves model building. A physiological model is a construct of hypotheses, which can be qualitative or quantitative. Harvey proposed a model in which the blood circulates from the heart to arteries, and then to capillaries which connect to veins to return the blood to the heart. He then did experiments to prove this. Quantitative or mathematical models are used to describe many functions using mathematical constructs and equations.

Physiology is studied at the molecular, cellular, tissue, organ, system, and organismic levels. There are several branches of physiology. Mammalian, fish, and insect physiologists restrict their studies to a certain group. General physiology seeks to describe functions that are common to all life forms, such as cell membrane function. Comparative physiology seeks to examine how different groups of animals accomplish similar goals while living in completely different circumstances, for example, in aquatic environments exchanging oxygen and carbon dioxide with water versus living on land and exchanging these respiratory gases with air.

Comparative studies are important for several reasons. First, the acquisition of each new piece of specific knowledge raises questions about its broader applicability. Once it is known that land animals need oxygen in air to breathe, the question arises: If a fish spends its entire life under water, why does it not drown? The answer is that it uses gills to extract oxygen efficiently from water (with a notoriously scarce oxygen supply) and to excrete carbon dioxide. Comparative physiology also helps to better understand evolution. The evolution of the vertebrate cardiovascular system from fish with two-chambered hearts, to amphibians and most reptiles with three-chambered hearts, to birds and mammals with four-chambered hearts, has attracted much interest in how the circulation of oxygen-rich blood is kept separate from oxygen-poor blood. Finally, comparative physiology can be used to study simple physiological systems in primitive animals such as invertebrates to help explain more complex systems in more advanced animals such as mammals.

GAINING KNOWLEDGE THROUGH EXPERIMENTATION

Our knowledge of how information is carried by neurons (the individual fibers of nerves) began with studies of squid neurons. Squid have giant neurons; they are so large that physiologists were able to insert electrodes inside them to discover the electrical events that produce nervous impulses and thus information transfer. Beyond this, the neurons are so large that it is possible to remove their contents and substitute artificial solutions to see how this alters neural impulse production. This is how it was discovered that the ions sodium and potassium are responsible for neural impulses. This research started in the 1920s, and developing technology has shown that mammalian neural function operates essentially the same way as squid neural function. Another example of such use of comparative physiology has led to the understanding of kidney function. In the early twentieth century, it was observed that kidney tubules in amphibians such as frogs and salamanders are large enough to allow samples of nephric tubular fluid to be removed and analyzed. Such studies led to the discovery that the initial event in urine formation is filtration of blood plasma in glomerular capillaries and that the final urine composition results from selective reabsorption and filtration of water and solutes in nephrons.

Physiology is a very interesting field in which much remains to be learned about the internal functions of animal bodies. Since it also applies to humans and human endeavors, physiology also plays in to how humans exist and go about their daily business, in terms of reducing physical stresses, ergonomics, and other such physiological factors.

—*Daniel F. Stiffler*

FURTHER READING

Boron, W., and E. Boulpaep. Medical Physiology. C. V. Mosby, 2001.

Dukes, H. H., M. J. Swenson, and W. O. Reese, eds. Dukes' Physiology of Domestic Animals. 11th ed. Comstock, 1993.

Prosser, C. L. Comparative Animal Physiology. 2 vols. 4th ed. John Wiley & Sons, 1991.

Randall, D., W. Burggren, and K. French. Eckert Animal Physiology. 4th ed. W. H. Freeman, 1997.

Schmidt-Nielsen, K. Animal Physiology. 5th ed. Cambridge University Press, 1997.

ANIMAL REPRODUCTION

FIELDS OF STUDY

Reproductive Biology; Animal Behavior Studies; Sociology

ABSTRACT

The ability to reproduce is central to the existence of any organism. Simple one-celled organisms reproduce asexually by duplicating their genetic material and dividing in half. For reproduction in sexual species the participation of two individuals is essential. Mating partners have a common interest: production of successful offspring. They also have conflicting interests. Appreciating the diversity of reproductive behaviors that occur in different organisms requires understanding of the reproductive conflicts that exist between mates even more than their mutual reproductive interests.

KEY CONCEPTS

gamete: a sexual reproductive cell that must fuse with another cell to produce an offspring: a sperm or egg

mate choice: the tendency of members of one sex to mate with particular members of the other sex

mate competition: competition among members of one sex for mating opportunities with members of the opposite sex

natural selection: the process that occurs when inherited physical or behavioral differences among individuals cause some individuals to leave more offspring than others

reproductive success: the number of offspring produced by one individual relative to other individuals in the same population

sexual dimorphism: an observable difference between males and females in morphology, physiology, and behavior

sexual selection: the process that occurs when inherited physical or behavioral differences among individuals cause some individuals to obtain more matings than others

THE BEGINNINGS OF ANIMAL REPRODUCTION

The ability to reproduce is central to the existence of any organism. Simple one-celled organisms reproduce asexually by duplicating their genetic material and dividing in half. For reproduction in sexual species, the participation of two individuals is essential. Mating partners have a common interest: production of successful offspring. They also have conflicting interests. Appreciating the diversity of reproductive behaviors that occur in different organisms requires understanding of the reproductive conflicts that exist between mates even more than their mutual reproductive interests.

Early sexual species most likely consisted of individuals that produced gametes of similar size. Except that one set of individuals produced smaller gametes that would be called sperm and another set produced larger gametes called eggs, males and females did not exist. Yet, reproduction was sexual. Through time it is thought that distinct sexes evolved because some individuals obtained reproductive advantages by producing smaller than average gametes whose greater motility increased their likelihood of fertilizing other gametes, while other individuals obtained reproductive advantages by producing larger than average gametes whose greater stores of nutrients increased the survival prospects of their young. Such evolution of gamete dimorphism would then lead to the many specializations in appearance and behavior of present-day males and females.

REPRODUCTIVE SUCCESS

For species whose parents do not provide care for their young, the maximum number of offspring that

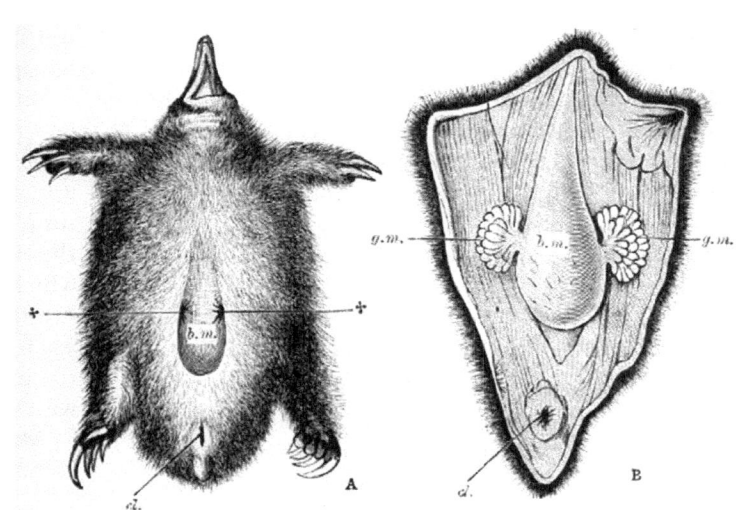

Sketch of a echidna female reproductive organs.

an individual female can produce is determined by the number of eggs that she can manufacture. For species in which parents do provide care, the number of young a female can produce might be limited more by the number of young she can raise than by the number of eggs she can make. For both these types of species, females that copulate with more than one male produce roughly the same number of young each season as females that mate with only one male. Thus, the quality of an individual mate, or the resources of paternal care that he can provide, should affect the reproductive success of females more than the number of mates they can obtain. As a result, females of most species are expected to be selective about which male fertilizes their eggs.

In contrast, the reproductive success of males of most species is not limited by the number of sperm they can produce. In species lacking parental care, males that mate with the most females usually leave the most offspring. In parental species, the reproductive consequences of mating with more than one female are slightly more complex. If one parent can provide sufficient care to raise young to independence, males that mate with multiple females usually leave more offspring than males that mate with only one female. If both parents are necessary to raise young, males that mate with only one female should leave more offspring than males with multiple mates. Given the reproductive advantages of mating with multiple females and a tendency for breeding males to outnumber breeding females, competition for mating opportunities is more pronounced in males than in females for many species. Furthermore, males should usually be less selective in mate choice than females.

MATE COMPETITION AND MATE CHOICE

Charles Darwin realized that the struggle to obtain mates could be an important process affecting the evolution of organisms. He deemed this process "sexual selection" to distinguish it from the struggle for existence, or "natural selection," and suggested two components of sexual selection: mate competition and mate choice. The importance of mate competition is rarely questioned by biologists because it often involves frequent and conspicuous aggressive interactions and distinctive weapons (for example, antlers of various ungulate species, horns of bovids, and enlarged canine teeth of primates). Mate choice is another matter. This exceedingly subtle behavior is difficult to document in any species. Although females of most species are demonstrably more reluctant to mate than males and are remarkably good at rejecting the mating attempts of males of other species as well as those of closely related males of their own species, few studies can convincingly show that females actively choose particular types of males as mates. The lack of direct evidence for the pervasiveness of mate choice in nature has caused many biologists to doubt its significance ever since Darwin first suggested it. Mate choice is also often misunderstood. The decision process it embodies is not thought to result from conscious deliberation; rather, individuals might simply react more favorably to some members of the opposite sex than to others.

FIGHTING FOR MATES

Similarities and differences in the details of mate competition and mate choice of different species can be exemplified by comparing three North American vertebrates: bullfrogs, sage grouse, and elephant seals. Male bullfrogs fight each other for small areas in ponds that females use as egg deposition sites. The wrestling matches that occur between males do not involve weapons as such, but being larger than an opponent almost always confers success. During nighttime choruses, females move among the territorial males, apparently assessing features of the male and his territory. Pairing begins when a female approaches a particular calling male and touches him. The male clasps the female, and within an hour, the female releases up to twenty thousand eggs, which the male fertilizes externally. Neither the male nor the female provides parental care for their young. Each year, roughly half of the males in a population obtain mates, and the most successful male may mate with six or seven different females.

Male elephant seals can weigh as much as three thousand kilograms (3.6 tons) and are highly aggressive. Rather than fight for territories, they fight directly for groups of females, called harems, that haul out on land to give birth. Males that monopolize large groups of females might mate with ninety or more females each year. One study revealed that less than 10 percent of the males sire all the pups in a breeding colony. Success in male competition not only involves being large but also involves having formidable canine teeth to use as weapons. During a fight, males rear up to half their length, slam themselves against the opponent, and bite him on the neck. Skin on the chests of males is highly cornified; these "shields" provide some protection against such onslaughts, but injury is still common. Mate choice by females is limited to vocalizing before and during copulation. If the male attempting to mate with her is a subordinate individual, the dominant male quickly responds to the call and attacks the copulating male.

Male sage grouse often congregate, or "lek," in traditional areas where they display and fight to control small territories. The territories function as courtship sites and places to copulate. Males provide neither resources nor parental care for their young; yet, females initiate mating and appear to be highly selective in mate choice. Near unanimity in preferred mates by the females in the population results in only a few males obtaining all the matings.

In all three of these species, many males compete for mates by fighting for territories; however, some males employ different tactics to obtain mates. Small, young bullfrog males remain silent near large, calling males and attempt to intercept any female attracted by the calling male. Small, young elephant seal males lurk about on the surf and grab females as they leave land to feed in the ocean. These males force females to copulate. Some sage grouse males attempt copulation with females away from the display arena. In another lekking bird species, the ruff, two types of males exist: territorial males and satellite males. The latter males are nonaggressive and appear to capitalize on the ability of territorial males to attract females. Unlike the case of bullfrog males, however, genetic differences of male ruffs produce the striking differences in plumage and behavior, rather than their ages.

Mate choice by females is well developed in bullfrogs and sage grouse. The benefit that female bullfrogs obtain by choosing particular males is relatively straightforward: Chosen males tend to control superior egg deposition sites that increase offspring survival. Benefits that female sage grouse obtain from mate choice are unknown. Mate choice in such lekking species continues to pose a significant question for biologists.

COMPETITION AND REPRODUCTIVE SUCCESS

Differential success of males in mate competition and mate attraction may translate to large differences in reproductive success. In contrast, variation in reproductive success among females is usually low because most females mate and produce at least some offspring. The relative amount of variation in reproductive success within each sex can influence the evolution of sexual traits. When only a few adults produce most of the offspring in a population, genes affecting the traits that underscore their success will be passed on to their offspring and quickly become the predominant characteristics of future generations. In contrast, if the most successful individual produces only slightly more offspring than other individuals, genes from all these parents will be present in roughly similar numbers in subsequent generations.

A consequence of greater variation in reproductive success of males relative to females is the evolution of elaborate sexual characteristics that are expressed only in males. These traits can be morphological, physiological, or behavioral. The extent of sexual dimorphism is predicted to be related to the relative variation in reproductive success in the sexes. Thus, species in which one or a few males sire most of the offspring produced in the population would tend to be species with considerable phenotypic differences between the sexes.

FIELD RESEARCH AND LABORATORY STUDIES

Studies of the reproductive behavior of organisms usually involve observation and experimentation of male and female interactions in nature or the laboratory. Early studies were mostly observational and cataloged the most typical behavior patterns observed in each sex. These studies ignored differences in behaviors among individuals. Because such differences can have significant consequences in terms of reproductive success, more recent studies usually involve marking males and females for individual recognition and recording various features of their morphology, behavior, and reproductive success.

Quantifying reproductive success in nature is a difficult task, and various methods have been used for different organisms. For all studies, the identity of each individual must be known. For some species, researchers can assess only the number of copulations that individuals obtain; for other species, they can count the number of young born; in yet others, they can determine the number of young that survive to independence or even sexual maturation.

Laboratory and field experimentation has been used to study a plethora of questions concerning the acquisition, function, and evolutionary significance of a variety of reproductive behaviors. Early studies of bird song investigated not only the role of song in attracting mates but also how individuals acquired their species-typical song. Choice experiments on female frogs using either naturally calling males or playbacks of recorded male calls revealed the call characteristics that females use in species and mate recognition. Crossing different species that varied in reproductive behaviors and noting the characteristics of their hybrid offspring provided some insights into the genetic basis of various behaviors. Staging aggressive interactions between males that differ physically in some regard demonstrated the significance of various male characteristics.

Researchers investigate general trends in reproductive behaviors by using the comparative method. This method usually works best when fairly closely related taxa (for example, species within the same genera or family) are considered. Using the comparative method, researchers can look for the relationships between the degree of sexual dimorphism in some characteristic and sex-specific differences in reproductive success variation, or the ecological and social conditions that affect male behaviors such as territoriality or female behaviors such as the amount of maternal care provided to young.

A thorough understanding of reproductive behaviors requires the creative use of all three methods of investigation: observations of individuals in nature to document normal behavioral patterns, precise experimentation on behaviors under controlled conditions to understand mechanisms of behavior and the stimuli that produce them, and comparison of trends in closely related species to gain insights into the evolutionary history of behavioral traits.

Reproduction is one of the defining attributes of life itself, and for most organisms, reproduction is sexual. Yet, despite the universality of sexual reproduction, the behavior patterns associated with it in different organisms are exceedingly diverse. For solitary organisms, sexual reproduction may be the only form of social behavior. For highly social species, individual interactions can be much more complex but still usually influenced by sex in some manner. Biologists seek to find some order to the variety of reproductive behaviors observed in different organisms. A unifying theme for this diversity is an evolutionary one: How do the sexes differ in maximizing the number of offspring they can produce? More specifically: How do males or females maximize the number of offspring they produce given the behavior patterns of the other sex and the various ecological factors that affect them? Thus, current research on reproductive behaviors has gone well beyond the point of merely describing what animals do to reproduce to determining why they do what they do.

—*Richard D. Howard*

Further Reading

Alcock, John. *Animal Behavior: An Evolutionary Approach.* 7th ed. Sinauer Associates, 2001.

Clutton-Brock, Tim H., F. E. Guinness, and S. D. Albon. *Red Deer: Behavior and Ecology of Two Sexes.* University of Chicago Press, 1982.

Ferraris, Joan D., and Stephen R. Palumbi, eds. *Molecular Zoology: Advances, Strategies, and Protocols.* Wiley-Liss, 1996.

Gould, James L., and Carol Grant Gould. *Sexual Selection.* 2nd ed. Scientific American Library, 1997.

Halliday, Tim. *Sexual Strategy.* University of Chicago Press, 1980.

Krebs, John R., and Nicholas B. Davis. *An Introduction to Behavioural Ecology.* 4th ed. Sinauer Associates, 1997.

Short, R. V., and E. Balaban, eds. *The Differences Between the Sexes.* Cambridge University Press, 1994.

Wiley, R. Haven. "Lek Mating System of the Sage Grouse." *Scientific American* 238 (May, 1978): 114–125.

Animal Respiration and Low Oxygen

Fields of Study

Physiology; Biochemistry; Metabolic Studies

Abstract

Oxygen is required by animal cells in order to produce energy used for growing, moving around, or simply maintaining normal body functions. Adaptation to low oxygen refers to a number of different changes in metabolism or body function, or both, that animals use to survive low-oxygen conditions. Low-oxygen conditions mean a reduction in the amount of oxygen available in relation to the need or demand for oxygen by the cells, or tissues. Low oxygen, or hypoxia, can result from either a decrease in the supply at constant demand, or an increase in demand at a constant supply.

Key Concepts

hyperventilation: an increase in the flow of air or water past the site of gas exchange (lung, gill, or skin)

hypoxia: from two Latin words, *hypo* and *oxia*, meaning "low oxygen"

metabolism: the sum of all of the reactions that take place in an animal allowing it to move, grow, and carry out body functions

respiratory pigment: a protein that "supercharges" the body fluid (blood) with oxygen; the oxygen can bind to the pigment and then be released

respiratory surface: the gill, lung, or skin site at which oxygen is taken up from the air or water into the animal, with the release of carbon dioxide at the same time and site

systemic: referring to a group of organs that function in a coordinated and controlled manner to accomplish some end, such as respiration

ventilation: the movement, often by pumping, of air or water to the site of gas exchange; commonly thought of as breathing

HYPOXIA

Adaptation to low oxygen refers to a number of different changes in metabolism or body function, or both, that animals use to survive low-oxygen conditions. Low-oxygen conditions mean a reduction in the amount of oxygen available in relation to the need or demand for oxygen by the cells, or tissues. Low oxygen, or hypoxia, therefore, can result from either a decrease in the supply at constant demand, or an increase in demand at a constant supply. The former, a reduction in oxygen supply, is the focus of the present discussion. Low oxygen resulting from increased oxygen demand usually is referred to as tissue hypoxia and is discussed only briefly here.

Oxygen is required by animal cells in order to produce energy used for growing, moving around, or simply maintaining normal body functions. At times when less oxygen is available, animals must either move to some place where there is sufficient oxygen, or change some internal function or process. A change in internal function or process is an adaptation that allows the animal to live with less oxygen or that will be a means of keeping the supply of oxygen to the tissues great enough to meet the needs of the cell.

External or environmental hypoxia results from one of two conditions: a greater utilization of oxygen by plants and animals than can be renewed by natural processes, or a lower density of air (at high elevations). It is necessary that animals cope with a decrease in oxygen in some way, in part because of the consequences of a decrease in internal oxygen supply. If oxygen supply to the tissues and cells falls,

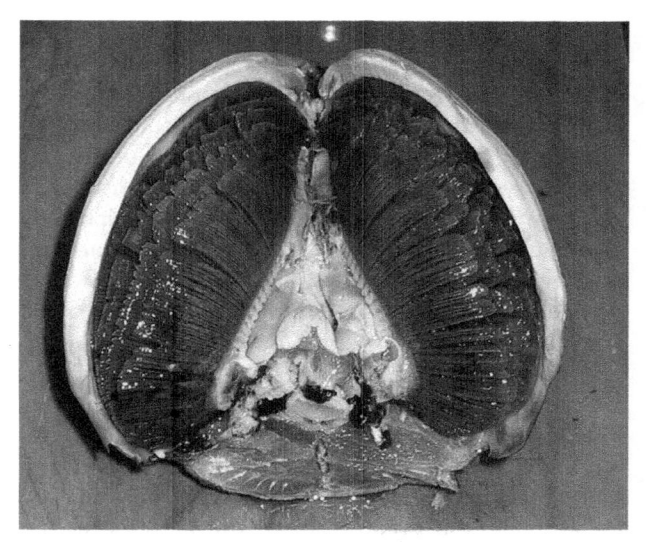

Tuna gills inside of the head. The fish head is lying on its back, the view is towards the mouth.

then the functions that require oxygen will fail or at least be reduced. The functions that fail include the "maintenance functions" of a cell, apart from growing or producing specialized chemicals. The types of low-oxygen conditions that have received the most attention from researchers are high altitude, diving by air breathers, and oxygen depletion in water. Some of these are temporary; others, such as high altitude, can last for a lifetime for animals that do not migrate.

INCREASE IN VENTILATION

One common adaptation to hypoxia is an increase in ventilation, the amount of air or water that the animal breathes. This increase is referred to as hyperventilation, and it makes up for the reduced amount of oxygen in the air or water by breathing a greater volume in a given span of time. This response is especially common in mammals that move to high altitude, fish and crabs, and some other water-breathing animals. A simple mathematical example will show how the response is effective. If an animal normally breathes one liter of air per minute and removes half the oxygen (air contains 209 milliliters of oxygen per liter), then it is taking in 104.5 milliliters of oxygen per minute. If the amount of oxygen in the air falls to only 104.5 milliliters per liter, and the animal still uses half of that (52.25 milliliters), then it must breathe two liters of air per minute to keep taking up 104.5 milliliters of oxygen per minute. Such an increase in ventilation can be accomplished by an increase in the frequency of the respiratory pump, or by increasing the volume of water or air that is moved with each "breath." For this response to occur, the nervous system must sense the reduction in oxygen and provide a nerve impulse to the brain, which then stimulates the ventilatory pump(s) to increase activity.

It may seem that this adaptation is all that would be needed for animals to survive hypoxia, but there are some limitations to this adaptation. First, hyperventilation causes increased muscular activity and an increase in oxygen used to move the respiratory muscles. The greater ventilation volume is a benefit, but the cost is a greater demand for oxygen. For animals that breathe air, the increase is rather small, but for animals that breathe water, the increase in the muscular activity causes a substantial increase in the oxygen used to pump the water, so that the "cost" may be greater than the "benefit" when the oxygen falls to low levels. Another problem exists for air breathers. Air breathers are in danger of losing water in the air that is exhaled (desert animals, such as camels, have elaborate mechanisms to conserve this respiratory water loss). Hyperventilation increases the water loss and requires the animal to drink more water. A final problem for both air and water breathers is that carbon dioxide is lost from the same respiratory surface where oxygen is taken up. Hyperventilation thus increases the loss of carbon dioxide, changing the chemical balance of the body as a whole.

BLOOD FLOW AND OXYGEN DELIVERY

In response to an internal hypoxia, many animals also exhibit an increase in the flow of blood to the tissues. This response is similar to that described for the ventilation system. An increased rate of flow compensates for a smaller amount of oxygen delivered for a given volume of blood (or respiratory medium in the case of ventilation). As with ventilation, blood flow can be elevated by increasing heart rate or by increasing the volume of blood pumped with each beat. There are numerous limitations to the effectiveness of this response, and it is only short term. The limitations center on the critical role of blood flow and blood pressure in the function of other systemic body functions. An excellent example is how kidney filtration rate increases with blood pressure.

Long-term adaptations to low oxygen often increase the ability of systemic respiratory functions to maintain oxygen delivery to the tissues. In the case of internal oxygen transport, this can be accomplished by increasing the amount of oxygen carried by the blood. A higher concentration of respiratory pigment accomplishes this, increasing either the number of red blood cells or the concentration of respiratory pigment in the blood. This adaptation requires the synthesis of new proteins and possibly new cells. Not surprisingly, many days or even weeks may be needed to increase respiratory pigment levels. Another way in which oxygen transport by the respiratory pigment may be improved is by increasing the concentration of a chemical that affects oxygen binding. This adaptation requires a change in metabolism and is discussed below.

METABOLIC ALTERATION

One final type of adaptation to low oxygen is an alteration in the basic metabolism of the animal. Metabolic changes can take one of several forms. First, a simple reduction in metabolism will lower the need and demand for oxygen by the cells. To be effective, this must occur before the oxygen has been exhausted, so as not to impair normal functions. A few animals show this type of adaptation, which is thought to result from the metabolic reactions being limited by the availability of oxygen. Second, the chemical reactions involved in metabolic pathways (a series of chemical reactions) may be altered in low-oxygen conditions so that different reactions take place to maintain energy production. The nature of these adaptations is that an alternative metabolic pathway requires different enzymes and perhaps different chemicals in the reactions. Last, a metabolic adaptation may yield a product that has an enhancing effect on oxygen transport. An enhancement of oxygen transport occurs when certain chemicals increase the ability of the respiratory pigment to bind oxygen or cause the respiratory pigment to bind oxygen at lower oxygen levels; this is called an increase in oxygen affinity. The change in metabolism at low oxygen thus results in an improvement in the supply of oxygen to the tissues. This response is seen in both vertebrates and invertebrates.

An excellent example of an animal that shows nearly all of the adaptations to low oxygen is the blue crab, the common commercial crab found throughout the Gulf Coast of North America and on the East Coast from Florida to New York. To compensate for low oxygen, the blue crab increases the flow of water over the gills, thereby keeping the amount of oxygen that actually passes over the gills nearly constant. Blue crabs also increase the heart rate, thereby increasing the rate of blood flow in the gills and to the muscles and organs of the body. This increase helps maintain the oxygen supply. If the period of hypoxia is brief, only a few hours, then these reactions may be all that is required for the animal to survive. If, however, the hypoxia continues for days or even weeks, then other responses come into play. There are changes in metabolism and in the way in which oxygen is transported in the body. Metabolism actually decreases so that less oxygen is needed by the animal. When that happens, then there must be some activity, such as swimming, that the animal gives up for lack of energy. The other change is an improvement in the way oxygen is transported to the tissues by the respiratory pigment, hemocyanin—a certain kind of protein, dissolved in the blood, that binds oxygen at the gills and can release the oxygen at the tissues where it is used by the cells. This improvement takes the form of increasing the level of a chemical in the blood that changes the binding of oxygen to hemocyanin. The hemocyanin then works as well when the animal is in hypoxic water. In addition, the crab can, and does, make the hemocyanin in a new form, so that it works better in the hypoxic conditions.

UNDERSTANDING THE RESPIRATORY AND CIRCULATORY SYSTEMS

Adaptation to low oxygen (either high demand or reduced supply) has been studied with the idea of understanding the functional capabilities of the respiratory and circulatory systems that supply oxygen to the tissues. Many different experimental protocols and procedures are used to assess the balance between oxygen uptake and oxygen demand when the external supply of oxygen is limited and demand remains constant. One approach to the study of low-oxygen conditions has been to compare animals that live at sea level in high-oxygen habitats with those living in habitats in which oxygen levels are low. A comparison of water breathers and air breathers is, strictly speaking, within the realm of consideration.

Water holds much less oxygen than air, and is therefore a low-oxygen condition. Freshwater at room temperature contains about 0.8 milliliter of oxygen per 100 milliliters of water; there are 20.9 milliliters of oxygen in the same volume of air. To obtain the same amount of oxygen, an animal must thus take all the oxygen from either 2,600 milliliters of water or 100 milliliters of air. Consequently, air breathers have much lower ventilation rates, at the same temperature, than do water breathers of the same size. The lower ventilation rate of air breathers is considered functional by reducing the loss of water from the respiratory surface.

Such adaptation is principally evolutionary and involves the transition from water breathing to air breathing in the evolutionary transition to land. There are a great many morphological as well as physiological consequences of this transition.

Adaptation to short-term hypoxia has been studied under controlled conditions in the laboratory in a variety of animals. Short-term conditions may mean anything from a few hours to weeks or even months. The length of the low-oxygen exposure generally depends on the animal used, its ability to withstand low oxygen, and the nature of the inquiry into the responses. Some clams, for example, are able to live in the absence of oxygen for several weeks. These experiments require careful monitoring of the animal and the conditions to ensure that oxygen neither rises nor falls too low and that the animals will survive.

A method that has been used to study adaptation to low oxygen in mammals is to conduct field studies in which the subjects are temporarily moved to high elevations. Mountain-climbing expeditions have been involved in some of these experiments in areas throughout the world. Additionally, experimental stations have been established at certain locations for the purpose of conducting these research projects. In this way, medical researchers are able to bring in appropriate equipment and supplies necessary to make complex and precise measurements.

All approaches to the study of low-oxygen adaptation require measurements of respiratory function or metabolic processes, or both. These measurements assess the uptake and transport of oxygen and the transport and excretion of carbon dioxide. The specific measurements are of the rate of oxygen uptake, ventilation volume and rate, blood flow, heart rate, oxygen transport properties of the blood, and oxygen uptake. In long-term monitoring studies of free-ranging animals, the animals are frequently fitted with implanted electrodes and blood sampling tubes. In this way, measurements can be made routinely over long periods without disturbing the animals.

A common measure of respiratory function is the total amount of oxygen used by an animal in a given period of time. The rate of excretion of carbon dioxide is another measure of overall function. The ratio of carbon dioxide loss to oxygen uptake is used to determine the nature of the metabolic pathways at a given time. Different metabolic pathways have characteristic oxygen uptake and carbon dioxide excretion ratios, and these are used in a predictive or diagnostic fashion. Some of the methods do not impair normal activity and can be used in low-oxygen experiments. The technique of placing small animals in respiratory chambers is used in these types of experimentation. Measurement of single organ function is used more often with larger animals, such as humans.

EVOLUTION, METABOLISM, AND ECOLOGY

Adaptation to low oxygen has been studied to understand three concepts better: evolutionary changes associated with oxygen availability, cell metabolism, and ecology of hypoxic habitats. Literally every aspect of oxygen uptake, transport, and utilization has received some attention.

Some of the evolutionary changes during the transition from water to land and from low to high altitudes have been studied as problems related to low oxygen. Results indicate that air breathers have lower ventilation rates than do water breathers of the same size. The lower ventilation rates are possible because of the higher oxygen levels, but they result in higher internal carbon dioxide levels. Animals such as insects, reptiles, and mammals have respiratory structures that are internalized and are inpocketings of the body wall. This arrangement aids in water conservation and helps keep the respiratory surfaces moist.

Just as important as research on the transition to land has been the information gained about the evolution of life in high-oxygen environments as compared to the low-oxygen conditions that are believed to have occurred in the ancient oceans.

From this research, it is clear that the major advances in respiratory systems are present in invertebrates and probably evolved quite early in the history of life on earth. Marine worms possess closed circulatory systems, respiratory pigments, red blood cells, special gas exchange structures, gills, and alternate metabolic pathways.

Biologists interested in metabolism and the factors that cause metabolic rate to change have examined the relationship between metabolic rate and other physiological functions. Specifically, oxygen supply, carbon dioxide removal, and glucose supply have been examined because all three are directly involved in aerobic metabolism. Imposing a limitation on external oxygen supply has therefore been used as an experimental tool to probe the limits and capabilities of cellular metabolism.

One of the observations that biologists have made over the years is that animals tend to find a way to live in places that are in any way habitable, and they tend to adapt to occupy new habitats. Understanding the physiological mechanisms required or used in adaptations to low-oxygen habitats, such as stagnant pools of water, has allowed explanation of some evolutionary changes.

There are several habitat types that undergo hypoxia routinely, and the utilization of the natural resources of those habitats, as well as the effective preservation of the habitats, generally dictates that attention be paid to the effects on the animals. One of the bodies of water that undergoes low-oxygen conditions is the Chesapeake Bay, and the effects on the animals there have been extensively studied.

—*Peter L. deFur*

FURTHER READING

Bicudo, J. Eduardo P. W., ed. *The Vertebrate Gas Transport Cascade: Adaptations to Environment and Mode of Life.* CRC Press, 1993.

Bryant, Christopher, ed. *Metazoan Life Without Oxygen.* Chapman and Hall, 1991.

Dejours, Pierre. "Mount Everest and Beyond: Breathing Air." In *A Companion to Animal Physiology*, edited by C. Richard Taylor, Kjell Johansen, and Liana Bolis. Cambridge University Press, 1982.

Graham, Jeffrey B. *Air-Breathing Fishes: Evolution, Diversity, and Adaptation.* Academic Press, 1997.

Hill, R. W., and G. A. Wyse. *Animal Physiology*. Harper & Row, 1989.

Hochachka, Peter, and George Somero. *Strategies of Biochemical Adaptation*. W. B. Saunders, 1973.

Paganelli, Charles V., and Leon E. Farhi, eds. *Physiological Function in Special Environments*. Springer-Verlag, 1988.

APES TO HOMINIDS

FIELDS OF STUDY

Primatology; Paleontology; Ecology; Environmental Studies

ABSTRACT

Primates are an order of the class Mammalia. The Primate order is divided into two suborders and several infraorders within those suborders. It is now taken for granted that human ancestry—if it could be traced satisfactorily—would include forms that, on other genealogies, gave rise to lower and higher primates, Old World monkeys, and a series of now-extinct creatures that were ancestral to certain of the apes as well.

KEY CONCEPTS

apes: large, tailless, semierect anthropoid primates, including chimpanzees, gorillas, gibbons, orangutans, and their direct ancestors—but excluding man and his direct ancestors

australopithecines: nonhuman hominids, commonly regarded as ancestral to man

dryopithecines: extinct Miocene-Pliocene apes; their evolutionary significance is unclear

hominid: an anthropoid primate of the family Hominidae, including the genera *Homo* and *Australopithecus*

human: a hominid of the genus *Homo*, whether *Homo sapiens sapiens* (to which all varieties of modern man belong), earlier forms of *Homo sapiens*, or such presumably related types as *Homo erectus*

primates: placental mammals, primarily arboreal, whether anthropoid (humans, apes, and monkeys) or prosimian (lemurs, lorises, and tarsiers)

stratigraphy: in geology, a sequence of sedimentary or volcanic layers, or the study of them—indispensable for dating specimens

MEET THE PRIMATES

Primates are an order of the class Mammalia. The Primate order is divided into two suborders. Suborder Prosimii (lower primates) includes lemurs, lorises, and tarsiers. Suborder Anthropoidea (higher primates), to which monkeys, apes, and humans all belong, is divided further into infraorders: Platyrrhini (flat-nosed New World monkeys, definitely not ancestral to man) and Catarrhini (down-nosed Old World monkeys, apes, and man). The infraorder Catarrhini includes two superfamilies: Cercopithecoidea (Old World monkeys) and Hominoidea (apes and man). Within the Hominoidea, finally, are three families: Hylobatidae (lesser apes), Pongidae (great apes), and Hominidae (man). As this classification suggests, it is now taken for granted that human ancestry—if it could be traced satisfactorily—would include forms

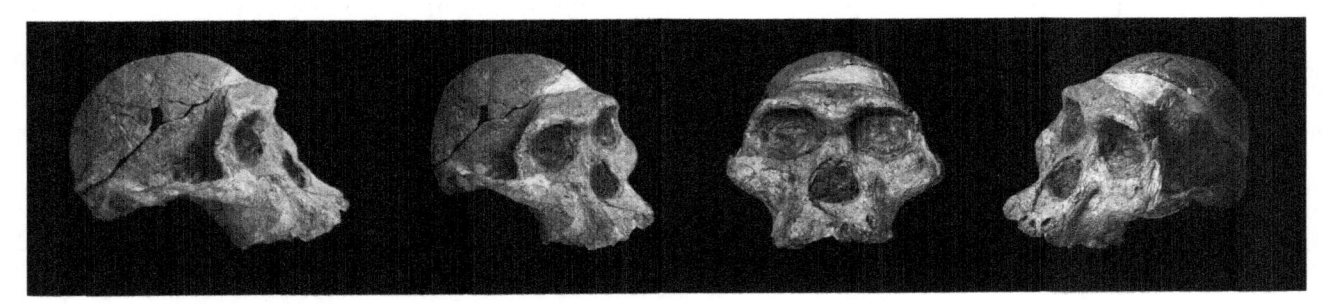

The original complete skull (without upper teeth and mandible) of a 2,1 million year old Australopithecus africanus *specimen so-called "Mrs. Ples," discovered in South Africa.*

that, on other genealogies, gave rise to lower and higher primates, Old World monkeys, and a series of now-extinct creatures that were ancestral to certain of the apes as well.

The lower primates (prosimians) first appeared about seventy million years ago. They still exist (as lemurs, lorises, and tarsiers) but have been declining for the last thirty million years, probably because of unsuccessful competition with their own descendants, the monkeys. Prosimians have five digits on each limb, but the digits have claws rather than nails, and the limbs are entirely quadrupedal. Prosimians also lack binocular vision, but they do have dentition anticipating the molar development of the higher primates.

THE HIGHER PRIMATE FOSSIL RECORD

The earliest evidence of any kind of higher primate—some tiny pieces of jaw found in Burma—dates from the Eocene epoch, about forty million years ago. Two creatures named *Amphipithecus* ("near ape") and *Pondaungia* ("found in the Pondaung Hills") have been proposed, each being a very primitive monkey or ape, but the evidence thus far is too sparse to ally these forms with any possible descendants. Some two million years later, in the Fayum Depression of Egypt (then a lush forest), *Apidium* and *Parapithecus* ("past ape") existed. Known only from jaws and teeth, they are the oldest known Old World monkeys presently recognized. Their dental pattern (arrangement of teeth), however, is the same as that of *Amphipithecus*. Other teeth from the Fayum, perhaps thirty-five to thirty million years old, have a different cusp pattern, more like an ape's (and a human's) than a monkey's. Possibly, then, *Propliopithecus* ("before more recent ape") is the earliest evidence of an ape line distinct from the monkey line.

The oldest apelike animal about which scientists know enough to regard it as a probable human ancestor is *Aegyptopithecus* ("Egyptian ape"), also found in the Fayum, in Oligocene deposits about thirty-two million years old. In addition to jaws and teeth, an almost complete skull and some postcranial bones (meaning those below the skull) have been recovered. Since the Fayum at that time consisted of dense tropical rain forest with little open space, *Aegyptopithecus* is assumed to have been an arboreal quadruped. (In 1871, before *Aegyptopithecus* was known, Charles Darwin had predicted that such a human ancestor had existed.) It is the most primitive ape yet discovered.

Proconsul ("before Consul," Consul being a chimpanzee in the London Zoo in 1933) and *Dryopithecus* ("forest ape") were either closely related to each other or identical. *Proconsul* appeared in Africa at the start of the Middle Miocene (about twenty million years ago) and was contemporary with *Dryopithecus* in Europe and Asia about fourteen million years ago. The relatively abundant fossils of these forms have been classified by some researchers into three species, together forming an extinct subfamily, the Dryopithecinae. One species in particular, *Dryopithecus major*, regularly left its remains on what were then the forested slopes of volcanoes; males grew significantly larger than females (a situation known as sexual dimorphism). In both of these respects, Dryopithecus resembled the modern gorilla, to which it may be ancestral. Like the gorilla, *Dryopithecus major* probably walked on its knuckles. In size, it was somewhat larger than a chimpanzee. The other two species, *Dryopithecus nyanzae* and *Dryopithecus africanus*, were smaller than *Dryopithecus major* and more like the chimpanzee. Outside Africa, *Dryopithecus* has been found from Spain to China.

THE ANCESTORS OF MODERN APES AND HUMANS

Limnopithecus ("lake ape"), found in deposits in Kenya and Uganda of about twenty-three to fourteen million years ago, is thought to be an earlier form of *Pliopithecus* ("more recent ape"). Its gibbonlike skulls, jaws, and teeth are plentiful in European sediments of Middle Miocene to Early Pliocene age—sixteen to ten million years ago. For some researchers, these two forms constitute a separate subfamily, the Pliopithecinae, which they consider to be part of the family Hylobatidae (lesser apes). In some respects, they resembled the modern gibbon, but other aspects of their anatomy were quite different. For example, *Pliopithecus* possessed seven lumbar vertebrae, whereas gibbons (and humans) have only five. It seems to have been primarily arboreal, swinging from branch to branch. *Pliopithecus* has been known since 1837 (in France), and since then some almost-complete skeletons have been recovered. *Sivapithecus* ("Siva's ape," Siva being a Hindu deity), found in

India and later in Africa, is a closely related form, Miocene in age. Both the dryopithecines and the pliopithecines are often regarded as the ancestors of modern apes.

Ramapithecus ("Rama's ape," Rama being another Hindu deity), found originally as a jaw fragment in India, is remarkable for its human-looking teeth. Some researchers regard it as the earliest member of the hominid line and therefore ancestral to humans. Others, however, relate *Ramapithecus* and *Sivapithecus* to modern orangutans, seeing no direct connection to man. Though *Ramapithecus* has been recovered from Late Miocene deposits in Africa and Indian and Early Pliocene ones in India (about fourteen to ten million years ago), only teeth and jaws have been found. As a result, many opinions regarding *Ramapithecus* are highly conjectural. The most striking feature of this genus, for example, is the greatly reduced size of its canine teeth, as compared with those of earlier (as well as modern) apes. Presumably, this indicates a changed diet of some sort. However, primates also use their teeth for nondietary purposes, including weaponry and display. It has therefore been suggested that the reduced tooth size of *Ramapithecus* might indicate its having begun to use other tools or weapons; if so, none has ever been found. Another conjecture has been that climatic change brought the primates down from the trees. Once on the ground, *Ramapithecus* then developed a hunter-gatherer style of sustenance that eventually included the formation of family units (male-female bonds), tool making, and a rudimentary form of language—the beginnings of culture. Unfortunately, all that is really known about *Ramapithecus* is what can be observed from a smattering of its bones.

Finally, there was *Gigantopithecus* ("giant ape"), a huge simian with protohuman teeth (clearly not ancestral to man, however) that outlasted *Dryopithecus*, *Ramapithecus*, and *Sivapithecus* to survive in Asia for almost nine million years. The largest primate that ever evolved (exactly how large is not known), it was alive in China as recently as a million years ago and perhaps more recently. There is conjecture that human experience of *Gigantopithecus* may have been the subject of the oldest folk tales of a race of giants. Known for its immense molars, *Gigantopithecus* was apparently the only successful ground-living savanna ape. It probably competed with early hominid forms and may have been exterminated by them.

In broad outline, then, these are the fossil apes. Since much of the evidence (all of it, in several cases) consists of teeth and jawbones, it is not surprising that conjecture has played a very active part in attempts to associate this evidence with the evolution of the hominids. Before 1980, there was widespread consensus among experts with regard to an evolutionary main line extending at least from *Aegyptopithecus* through *Dryopithecus* (or *Proconsul*) and *Sivapithecus* to *Ramapithecus*, the latter being regarded as the first hominid. However, portions of two *Sivapithecus* faces, recovered from Turkey in 1980 and Pakistan in 1982, impressed researchers with their orangutan-like characteristics. Since firm ties between *Sivapithecus* and *Ramapithecus* had already been established, it began to seem that the entire lineage pointed toward the orangs rather than toward man.

Another problem is that formerly accepted dating has come into question for such important branchings of the lineage as those which separated monkeys from apes and apes from man. New genetic studies having nothing to do with either fossils or stratigraphy have presented compelling (but controversial) arguments to the effect that these branchings occurred much later than hitherto believed. A third, even more serious problem is that there is virtually no pertinent fossil evidence regarding the development of simian primates into humans for a period beginning about fourteen million years ago and lasting until the appearance of the australopithecines about four million years ago. While anthropologists and biologists continue to learn more about the ancestry of modern simians, therefore, it is certainly not the case that a reliable lineage (or even a timetable) leading from other primates to humankind has been established.

THE PROBLEMS OF THEORIZING FROM FOSSILS

The study of fossil apes is a specialization within the broader field of vertebrate paleontology, or the study of fossil bones. Like all paleontologists, therefore, paleoprimatologists are necessarily concerned with fossils and their stratigraphic occurrence. Because primates still exist, however, it is also important to study the behavior of living examples. Since behavior reflects environmental conditions, it is further necessary to reconstruct the climate, flora, and fauna of the region and time in which the fossils were found.

No complete fossil ape has ever been found. Any understanding of what they may have looked like is therefore conjectural—an extrapolation from what has been recovered to what has not. Skulls are undoubtedly the most desirable evidence, but they are not the most durable of fossils. Teeth, which constitute the hardest parts of the primate body, are preserved more often than any other part. Some kinds of fossil ape are known either exclusively or primarily from their teeth and jaws. Ape teeth differ from human teeth in two significant respects: They are generally larger (the canines especially), and the cusp patterns on their molars differ. The arrangement of teeth in an ape's jaw, moreover, is angular, like a V; in a human, the arrangement is rounder, like a U. Inevitably, whenever jawbones or molar teeth are found, an attempt is made to place them somewhere on a continuum that runs between the purely simian (ape) and the purely human. This procedure not only distinguishes primitive apes from primitive humans, and one kind of fossil ape from another, but also gives rise to inevitable conjecture as to possible anticipations of the human line.

A major difficulty with evolutionary sequences based solely upon dental evidence is that the head and body of a given species have not necessarily evolved at the same rate. One may be surprisingly apelike, the other somewhat human. Even more specifically, the fact that jaws are changing does not necessarily mean that crania (or any other specific body parts) are changing also. On the whole, scientists do not yet understand the evolution of primate anatomy well enough to interpret present evidence or reconstruct missing parts with much reliability. In the absence of factual evidence, the form taken by prevailing reconstructions at any given time may owe as much to professional politics as to objective knowledge. The controversy regarding *Ramapithecus*, which (on the basis of facial bones) moved that genus from a central position at the base of the human lineage to a similar position on a separate orangutan genealogy, has been a valuable lesson in the folly of premature commitment.

ANCIENT FOSSILS AND MODERN PRIMATES

Most paleoprimatologists are to some extent modern-day primatologists also, necessarily expert in the comparative anatomy of all members of the higher primates and as knowledgeable as possible regarding their behaviors and environments. It is assumed that a changing environment (one becoming increasingly arid, for example) requires behavioral modifications and that these modifications will then create selection pressures favoring some types of anatomical variation over others; thus, one species will eventually change into (or be replaced by) another. The process is not well understood, but the primates—being so intensely studied—are often regarded as test cases for competing evolutionary theories. Some researchers stress evidence to the effect that species are always changing; others believe that species are created precipitously and then tend to endure relatively unchanged until they are abruptly superseded. All that is really known at present is that the ancient apes generally conform to a partially ascertainable progression from hypothetical earlier forms to modern-day primates—with a ten-million-year gap in between.

THE IMPORTANCE OF FOSSIL APES

The fossil apes are important for four reasons: They are an important group in their own right; they are important to the development of the mammals; they are important to the development of the primates; and, they are thought to be ancestral to humankind and therefore uniquely important among all non-human fossil genera.

One of the unique characteristics of humans is the ability to create, preserve, and transmit knowledge. *Homo sapiens* has learned to value learning, to hoard and increase knowledge in the realization that it fortifies and enhances his existence. Humans attempt to know and understand all present-day forms of life; in order to do so, however, it must also be known how these forms came into being through time. Biology, then, is inherently evolutionary and does not sharply distinguish between plants and animals of the past and those of the present.

Insofar as an understanding of life itself is the goal of biological studies, no single form of life is inherently more important than any other. From this point of view, one would say that fossil and living apes are studied for the same reason that algae, sponges, or nematodes are. A number of biologists would maintain this view. Many others, however, believe mammals—and especially primates—to be a "higher" form of life, anatomically more complex than sponges (though not necessarily more nearly perfect) and certainly capable of more complex behaviors. No mere study of anatomy, this viewpoint suggests, can sufficiently explain a primate.

The outstanding characteristic of all primates is their intelligence. One can find surprising levels of intelligence in other animals, however. Among invertebrates, such cephalopods as the squid and the octopus have the highly developed nervous systems, senses, and brains that are normally associated with mammals. Together with some birds and social insects, all the mammals are capable of surprisingly complex behavior. Nevertheless, the higher primates constitute an intellectual elite even among the mammals. Impressive as gorillas and chimpanzees can be in this respect, it is apparent that the human mind has a capacity well beyond theirs. The brains of extinct apes are seen as having been ancestral not only to those of modern apes but also to the brain of man.

Scientists are fortunate in the number of fossil ape skulls that have been found, for they make the increasing mental capacity of the higher primates easy to establish. Limb bones and other less durable parts of the skeleton are much rarer. When available, they indicate the relative lengths of arms and legs; the nature of the shoulder (a key to arboreal existence); the relation of pelvis and femur (a key to posture); and the shapes, capabilities, and functions of hands and feet. Without such evidence, scientists have only conjectures based upon the presumed place of the genus in question within a supposed evolutionary sequence. When proposed sequences differ, though they are derived from the same sparse evidence, conflicting suppositions about the evolutionary sequence are at work.

—*Samuel F. Tarsitano*

Further Reading

Berger, Lee. *In the Footsteps of Eve: The Mystery of Human Origins*. Adventure Press/National Geographic Society, 2000.

Campbell, Bernard. *Humankind Emerging*. 8th ed. Allyn & Bacon, 2000.

Conroy, Glenn C. *Primate Evolution*. W. W. Norton, 1990.

Jordan, Paul. *Neanderthal: Neanderthal Man and the Story of Human Origins*. Sutton, 1999.

Lewin, Roger. *Bones of Contention: Controversies in the Search for Human Origins*. Simon & Schuster, 1987.

Martin, R. D. *Primate Origins and Evolution: A Phylogenetic Reconstruction*. Chapman and Hall, 1990.

Szaly, Frederick S., and Eric Delson. *Evolutionary History of the Primates*. Academic Press, 1979.

Tattersall, Ian, and Jeffrey H. Schwartz. *Extinct Humans*. Westview Press, 2000.

Arachnids

Fields of Study

Biology; Zoology; Entomology; Paleontology

Abstract

Arachnids are mainly terrestrial, with some few freshwater forms found among spiders, and are found on all continents. Their forms are highly variable depending on the group or individual species. Arachnid life spans are also highly variable, but some may live for twenty-five years. All have eight legs as adults and pincerlike mouth parts that may be modified into fangs in some groups. Most members are free-living but parasites are found among the Acarina. The arachnids include some of the most venomous animals of the world.

Key Concepts

Arthropoda: animals having jointed legs, an exoskeleton, and a ventral nerve cord

book lungs: a system of blood-filled diverticula that are surrounded by air pockets located in a chamber called the atrium

chelicerae: pincerlike mouthparts used in macerating food

pedipalps: modified walking legs; these may be clawlike, as in scorpions, or have a modified structure, a palpal organ for sperm transfer, as in male spiders

podites: the parts of the jointed appendages of arachnids

sensilla: hairlike structures associated with nerves that act as mechanoreceptors and chemoreceptors

tracheas: a system of branched tubes that, in some arachnids, deliver oxygen to the blood

MEET THE ARACHNIDS

Arachnids belong to the phylum *Arthropoda*, and as such have the basic characteristics of this assemblage. These characteristics include having jointed appendages, an exoskeleton, an open circulatory system, and a ventral nerve cord. The *Arachnida* are a subgroup within the subphylum *Cheliceriformes*, those arthropods with a pair of primitively pincerlike chelicerae mouthparts. Arachnids are considered to have an aquatic origin, but most present-day forms are terrestrial in nature.

Arachnids all have the following characteristics: a body that is divided primitively into a prosoma and an opisthosoma; a pair of chelicerae used as mouth parts, a pair of pedipalps that often end in pincerlike claws and are modified for prey manipulation and sperm transfer; four pairs of walking legs in the adult (juveniles may have three pairs) originating from the prosoma; absence of antennae; simple eyes in most; coxal glands at base of the legs and Malpighian tubules extending between the hemocoels and gut tube used for excretion and osmoregulation; gut tube with diverticula; breathing accomplished by book gills (aquatic forms), book lungs, or tracheal tubes; and a dorsal heart.

The following orders comprise the *Arachnida*: *Scorpiones* (scorpions), *Uropygi* (whip-tailed scorpions), *Schizomida* (schizomids), *Amblypygi* (whip spiders), *Palpigradi, Araneae* (spiders), *Ricinuleids, Pseudoscorpionida* (false scorpions), *Solpugida* (wind scorpions), *Opiliones* (daddy longlegs), and *Acari* (ticks and mites).

SCORPIONS

The scorpions are considered to be among the earliest land animals. They are considered to be derived from water scorpions known as Eurypterids. Scorpions were established in the terrestrial environment by the Carboniferous era, but are thought to have invaded the land in the Devonian. Scorpions are mainly nocturnal and are common in desert and tropical regions. Northern climates are free of scorpions due to the extreme cold. Their bodies are

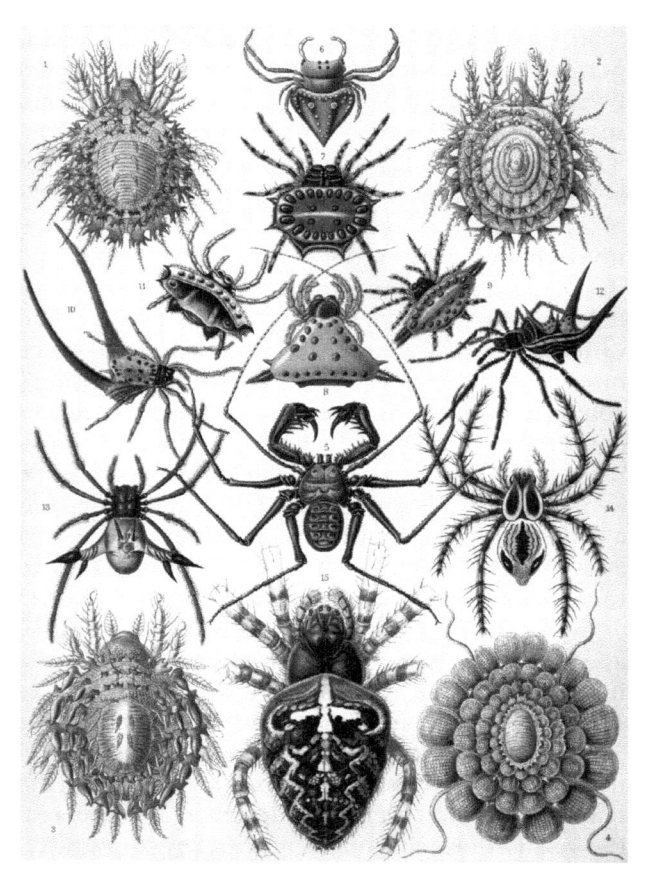

This drawing shows fifteen different species of spider. From Kunstformen der Natur *(1904), plate 66: Arachnida.*

segmented but are divided into three portions: an anterior prosoma (carapace) and an opisthosoma divided into mesosoma and a metasoma. The prosoma bears the mouth parts which are pincerlike chelicerae used to chew the prey and allow copious amounts of digestive juice to be poured externally over the prey. The next pair of appendages are the clawlike pedipalps that act to catch and hold prey. Scorpions have four pairs of walking legs, each usually with eight segments. The mesosoma bears the genital pores and a pair of comblike structures, the pectines, which are chemosensors for tracking prey as well as for digging burrows, along with the legs. Scorpions breathe through ventrally placed book lungs and thus avoid traveling in water. The tail segment or telson bears at least one stinging spine. A number of species are highly poisonous and their sting may be fatal.

UROPYGI, SCHIZOMIDA, AMBLYPYGI, AND PALPIGRADI

Uropygi, whip-tailed scorpions, look like scorpions, having chelicerae mouthparts and large pedipalps for capturing prey. Some species attain lengths of up to eight centimeters. Like scorpions, these forms are mainly nocturnal. Uropygans earn their name by elongating the telson into a whiplike structure, the flagellum. At the end of this organ are pair of glands that can spay acetic and caprylic acids at would-be predators. They range from arid climates to the tropics and subtropics of South America and southern United States. Some species have been introduced into Africa.

The *Schizomida* are small, under a centimeter in size. Like uropygans, they possess glands that spray acid, although the telson in these forms is short. The first pair of walking legs is sensory in nature. Schizomids are common in Asia, Africa, and the Americas, inhabiting leaf litter and secreting themselves under rocks and fallen trees. They are tropical and subtropical.

Amblypygi resemble whip-tailed scorpions in body form, although they lack the elongated telson. Internally they resemble spiders, but do not have fangs and use their chelicerae for tearing apart prey. The first pair of legs is sensory in nature and may be quite long in some species (up to twenty-five centimeters in length.) These whip spiders are found in tropical areas under tree bark, leaf litter, and in caves.

Palpigradi are small arachnids under three millimeters in length. About fifty-five species have been described since their discovery in 1885. Not only have they have undergone a reduction in size, but their exoskeletons are thin and colorless, facilitating the loss of respiratory organs. They have a whiplike flagellum similar to that of the *Uropygi*.

THE ARANAE

Spiders comprise a large arachnid group, with at least 35,000 known species. They inhabit all terrestrial environments, and some species have adapted to freshwater and estuarine areas. The body is normally in two parts, a prosoma and an opisthosoma. Spiders are organized into three subgroups: those with a persistent segmented opisthosoma, the *Liphistiidae*, and two groups included in the *Opisthothelae*, the mygalomorph or tarantula-like spiders, and the araneomorph spiders that are more familiar to most people. Spiders have fanglike chelicerae that inject venom and digestive enzymes into the prey that liquefy the internal organs in order that the spider may pump out the contents. Although all spiders are venomous, only a few species are dangerous to humans. Among these are the black widow and brown recluse spiders. Black widows, belonging to the *Theridiidae* or cobweb spiders, are capable of inflicting a fatal bite, owing to the neurologic aspects of their venom. Males, who are multicolored and small, are also venomous. The brown recluse spiders are fairly common throughout the southern United States. Their venom is mainly hemolytic; that is, it dissolves tissues and may create large lesions as the result of digestion of tissue. Their bite is rarely fatal. Other lethal spiders occur in Australia (the funnel web spiders) and in South America (the ctenid hunting spiders). Another major characteristic of spiders is their ability to spin silk through organs called spinnerets. The spinnerets are connected to silk glands. Silk is used for webs, cocoons for eggs and over-wintering, wrapping prey, and draglines.

Spiders have many sense organs that include hairs that sense vibration and touch. Other hairlike structures are hollow and are chemoreceptors. Slit organs that include the lyriform organs are slits in the cuticle that lead to sensory neurons. They are considered to be mechanoreceptors. Vision is variable in spiders depending upon the life habits of the spider, ranging from eight eyes arranged in two rows on the prosoma to a complete lack of eyes. The jumping and hunting spiders have the largest eyes capable of forming an image.

Spiders have interesting reproductive behaviors that range from rhythmic web tapping to flashing iridescent pedipalps. A male will pick up sperm deposited in a sperm web with a specialized structure on one of the pedipalps. If he is successful in approaching a female, he will mate with her by inserting the pedipalp tip into her seminal receptacle and beat a hasty retreat; otherwise the female may recycle him. The female will lay up to three thousand eggs sometime after the mating.

OPILIONES

Harvestmen are found throughout the world. The carapace is broadly joined with the segmented abdomen without the usual constriction seen in spiders. The legs of most are long and spindly, hence their common name of "daddy longlegs." They are most abundant in the tropics. They have pincerlike but small pedipalps that are used to feed on small invertebrates and insect eggs. They also are scavengers. For defense, they have two repugnatorial glands that give off noxious chemicals to ward off would-be predators. Unlike other arachnids, these male spiders have a penis for sperm transfer and females have an ovipositor with which they deposit their eggs in the soil.

ACARINA

Both the mites and ticks are included in this group and have the largest number of species of any arachnid taxon, possibly over a million. They may not be a monophyletic group but one that has more than one ancestor. These forms are cosmopolitan in distribution and are either free-living or parasitic on both plants and animals. Mites are the most diverse members of this assemblage. They have opted for small size with a trend toward fusing the prosoma and opisthosoma together. Mites are predatory on other arthropods, including other mites. Many others are parasitic on plants and vertebrates. Some have become aquatic and can be found in many freshwater and ocean environments where they may parasitize mollusks, crustaceans, and aquatic insects; some are suspension feeders. Mites cause various ailments that include skin mange and other irritations, feather loss in birds, as well as subcutaneous tumors. Mites can also act as vectors in disease distribution, including wheat and rye mosaic viruses. Other mites may destroy stored grain products and thus have negative economic impacts. The ticks are parasitic, blood-sucking parasites on vertebrates, although one beetle is also parasitized by this taxon. The chelicerae are modified with teeth for anchoring the tick in the skin of the host, which makes it difficult to remove. Ticks are vectors for many diseases, including protozoan, bacterial, fungal, and viral agents. Ticks transfer Rocky Mountain spotted fever as well as Lyme disease.

RICINULEI

These secretive arachnids are small, ranging from five to ten millimeters in length. There are only about thirty-five species, found in Africa, the southern United States, and Brazil. They have pincerlike pedipalps, and their third legs are modified for sperm transfer in males. There is a hood structure in front of the prosoma where brooding of the single egg of one species occurs. The reproductive habits are largely unknown. These are secretive animals living under leaf litter and in caves. They are predatory on smaller arthropods.

—*Dennis R. Dean*

FURTHER READING

Brusca, R. C., and G. J. Brusca. *Invertebrates*. Sinauer Associates, 1990.

Buchsbaum, R., M. Buchsbaum, J. Pearse, and V. Pearse. *Animals Without Backbones*. 3rd ed. University of Chicago Press, 1987.

Coddington, J. A., and H. W. Levi. "Systematics and Evolution of Spiders (Araneae)." *Annual Revue of Ecology Systematics*. 22 (1991): 565–592.

Jackson, J. A. *A Field Guide to the Spiders and Scorpions of Texas*. Gulf, 1999.

Polis, G. A., and W. D. Sissom. "Life History." In *The Biology of Scorpions*, G. A. Polis, ed. Stanford University Press, 1990.

Shultz, J. W. "Evolutionary Morphology and Phylogeny of Arachnida." *Cladistics* 6 (1990): 1–38.

ARTHROPODS

FIELDS OF STUDY

Entomology; Marine Biology; Zoology; Palaeontology

ABSTRACT

The phylum Arthropoda *is the largest phylum in the animal kingdom and consists of myriad animals belonging to several subphyla. All of these subphyla are united by a common body plan in which the body is made up of a series of repeating segments, each bearing a pair of appendages, however, arthropods have fused and modified segments for specialization. Arthropods are found in all terrestrial and freshwater habitats and on all continents. They are also located in all marine realms, including in the abyssal depths and at hydrothermal vents. Because of their widespread occurrence and their ability to occupy all niches on earth, they are considered the most successful animal phylum that has ever existed.*

KEY CONCEPTS

biramous: having two rami; antennae are bifurcated
cuticle: a noncellular, secreted body covering
ecdysis: molting; the process of removing (escaping from) the old exoskeleton
exoskeleton: a protective system of external layers and joints that allows for antagonistic action of muscles
rami: the branches of an arthropod limb or appendage
setae: hairlike organs, typically sensory in nature, arising from the cuticle
tagmatization: functional specialization of groups of segments
uniramous: having only one rami; antennae appear as a single, nonbifurcated structure

MEET THE ARTHROPODS

The phylum *Arthropoda* is the largest phylum in the animal kingdom and consists of myriad animals belonging to the subphyla *Trilobita* (extinct trilobites with biramous antennae, probably all marine), *Chelicerata* (horseshoe crabs, scorpions and pseudoscorpions, spiders, harvestmen, mites, ticks, and marine pycnogonids; all lacking antennae), *Crustacea* (all with biramous antennae), and *Uniramia* (insects, centipedes, millipedes; all with uniramous antennae). All of these subphyla are united by a common body plan similar to that of the annelids, in which the body is made up of a series of repeating segments, each bearing a pair of appendages. In contrast to annelids, however, arthropods have fused and modified segments for specialization. This fusion of groups of segments is known as tagmatization. The names for specific tagmata vary from subphylum to subphylum, but are most commonly recognized as the head, the thorax, and the abdomen. Appendages consist of articulated joints, moveable by muscles that insert onto specialized structures called apodemes that are attached to the cuticle.

Arthropods are found in all terrestrial and freshwater habitats and on all continents. They are also located in all marine realms, including in the abyssal depths and at hydrothermal vents, and are the only invertebrates to have conquered the aerial realm. Because of their widespread occurrence and their ability to occupy all niches on earth, they are considered the most successful animal phylum that has ever existed.

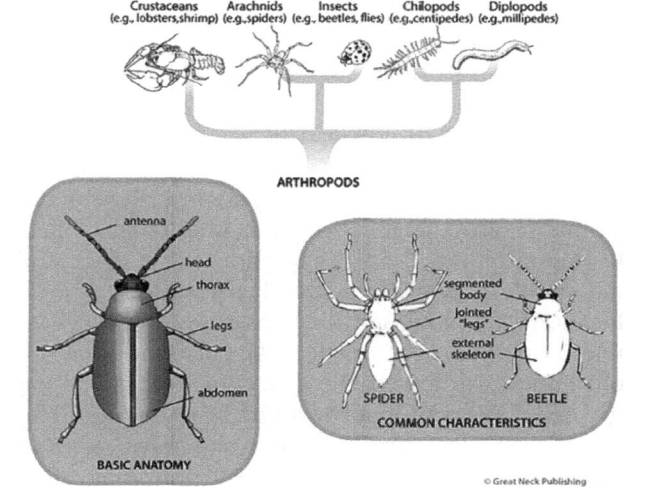

Arthropods: Arthropods are a large group of animals that have hard exterior skeletons and jointed appendages or limbs. examples are insects, lobsters, and spiders.

EXTERNAL STRUCTURE AND FUNCTION

The first and last segments of the arthropod body differ in embryological origin from all of the rest and are considered as special segments. The first is called the acron and is preoral (lies before the mouth). The last is the telson and is postanal (occurs after the anus). Increase in segment numbers takes place in a growth zone immediately in front of (anterior to) the telson. Thus, the oldest segments are toward the head, while the youngest are toward the telson. The exoskeleton of each segment consists of two thickened plates, which are connected by a membranous cuticle. The cuticle is a nonliving, secreted material consisting of three layers: an inner endocuticle, an exocuticle, and an outer epicuticle, which is usually waxy. Despite this, the cuticle has many living structures that penetrate it and project beyond its surface. Chemical and mechanical sensory organs (setae) emerge generally as hollow projections from the cuticle and are typically hairlike in structure. Gland pores also penetrate the cuticle.

The cuticle is secreted by the epidermis and becomes hardened during a process of protein tanning and can be further strengthened by depositing calcium carbonate in the endocuticle (in *Crustacea* and *Diplopoda*). However, this hardened structure restricts growth and must be shed in a process called ecdysis (commonly called molting) in order for the animal to grow larger. Ecdysis is under neural and hormonal control. In preparation for ecdysis, the old cuticle is partially digested, particularly at joint locations and along specific lines of weakness. A flexible, untanned new cuticle is laid down under the old cuticle, and then the old cuticle cracks along the weak points so that the animal can emerge. The period between successive molts is known as intermolt, and the animal in each of these periods is called an instar; during this time, cellular growth within the confines of the exoskeleton occurs. Some species of arthropods have terminal molts; that is, they molt a specific number of times and then cease all molting and growth. Other species retain the ability to molt and grow throughout their entire lives.

INTERNAL BODY PLAN

The major, tubular blood vessel (heart) is dorsal, with blood flowing toward the head rather than toward the body. The circulatory system is greatly enlarged and occupies the space of the coelomic cavity, which is vestigial. In most arthropods, the circulatory system is open, so that all internal organs are bathed directly in blood rather than fed by smaller vessels (as is common in closed systems). The heart is pierced by a series of holes (ostia) which close when the heart muscle contracts so that blood can be propelled forward. Upon relaxation of the heart muscle, the ostia open and blood reenters the heart from the sinuses that feed into the sac surrounding the heart. The blood contains clotting agents, amoebalike cells that act as phagocytic cells to attack pathogens, and oxygenating pigments.

Aquatic arthropods possess gills through which blood is circulated for gas exchange. Terrestrial forms have either book lungs or a series of cuticle-lined tubes called tracheas, which branch into smaller and smaller tubes to reach nearly all cells of the body. The tracheas open to the outside of the body via spiracles, which possess valves so that their opening and closing can be controlled to avoid water loss.

The arthropod gut has three divisions: a cuticle-lined foregut, a midgut, and a cuticle-lined hindgut. The foregut often is specialized for food storage and grinding and is separated from the midgut by a complex valve. The midgut has a large absorptive surface, which is called either the hepatopancreas or the digestive caecae. The hindgut serves as the area for water resorption and the formation of feces. Food and wastes are moved through the digestive system via muscular action due to the cuticular lining of the foregut and hindgut.

Osmoregulation and excretion of nitrogenous wastes are accomplished via one of two kinds of organs: the Malpighian tubules, or the nephridial coxal, antennae, or maxillary glands, all of which filter the blood for wastes. Marine forms excrete ammonia, while terrestrial forms conserve water through the excretion of uric acid.

The nervous system is of a ladderlike chain form, highly segmental, and ventrally located. Two longitudinal ganglia run along the midline of the animal. Paired ganglia arise in each segment and these are connected via lateral cords. The brain, composed of three pairs of ganglia, is found anteriorly above the esophagus; these ganglia then connect to the fourth set immediately below the esophagus. Sensory systems are highly evolved and include chemoreception (taste and smell), mechanoreception (vibration, touch, and deformation), and vision. Vision is accomplished via simple eyes (ocelli) and/or compound eyes, made up of a series of subunits called ommatidia. Compound eyes break up the image before it reaches the retina, and each ommatidium samples only a small part of the complete image.

Reproductive systems and strategies are highly variable. Sexes are usually separate, although hermaphroditism and parthenogenesis are known in some crustaceans and a few insects. Some arthropods display elaborate mating rituals; others do not. Some use direct copulation; others use indirect methods. Many arthropods pass through juvenile stages that are highly vulnerable to predators; hence the reproductive output of most arthropods is great. A few species, however, display direct development, where miniature adults are produced from eggs.

PHYLOGENY

Arthropods and annelids share a common ancestor, but it is unclear whether arthropods arose from an annelid ancestor or both shared an ancestor from another phylum. Some argue, based on molecular evidence, that annelids and crustaceans are not closely related at all. Traditionally, arthropods have been treated as a monophyletic group—all arose from a common arthropod ancestor. However, this idea is the subject of debate, with some arguing for a polyphyletic ancestry where each subphylum arose from a nonarthropod ancestor and then formed similar body structures via convergent evolutionary forces. Unfortunately, the fossil record does little to illuminate the phylogeny of arthropods, and current molecular evidence is contradictory.

—*Kari L. Lavalli*

Further Reading

Brusca, Richard C., and Gary J. Brusca. *Invertebrates*. Sinauer, 1990.

Dudas, Gytis, and Darren J. Obard. "Phylogeny: Are Arthropods at the Heart of Virus Evolution?" *eLife* 4 (2015): n.pag. PDF file.

Grove, David I. "Arthropods." *Tapeworms, Lice, and Prions: A Compendium of Unpleasant Infections*. Oxford University Press, 2014. 63–78.

Harrison, Frederick W., and Rainer F. Foelix, eds. *Chelicerate Arthropoda. Microscopic Anatomy of the Invertebrates*. Vol. 8. Ed. Frederick W. Harrison and Edward E. Ruppert. Wiley, 1999.

Harrison, Frederick W., and Arthur G. Humes, eds. *Crustacea. Microscopic Anatomy of the Invertebrates*. Vol. 9. Ed. Frederick W. Harrison and Edward E. Ruppert. Wiley, 1992.

Harrison, Frederick W., and Arthur G. Humes, eds. *Decapod Crustacea. Microscopic Anatomy of the Invertebrates*. Vol. 10. Ed. Frederick W. Harrison and Edward E. Ruppert. Wiley, 1992.

Harrison, Frederick W., and Michael Locke, eds. *Insecta. Microscopic Anatomy of the Invertebrates*. Vol. 11. Ed. Frederick W. Harrison and Edward E. Ruppert. Wiley, 1998.

Van Roy, Peter. "Fossils of Huge Sea Creature Shine Light on Early Arthropod Evolution." *Scientific American*. Nature America, 14 Mar. 2015. Web. 4 Jan. 2016.

Asexual Reproduction

Fields of Study

Evolutionary Biology; Genetics; Marine Biology; Biology

Abstract

Although asexual reproduction is very common in organisms such as bacteria, protists, fungi, and plants, it is rarer among animals, especially among the more complex animals. With asexual reproduction, organisms do not have to waste energy in sexual activity. However, with the exception of the rotifers, no populations of asexually reproducing organisms have very long evolutionary histories. Unlike most sexually reproducing populations with varying degrees of diversity, all members of an asexually reproducing population are identical, except for mutations that arose after the population's inception. This lowered diversity makes the population much less likely to be able to adapt to change.

Key Concepts

clone: an organism that is genetically identical to the original organism from which it was derived

diploid: having two of each chromosome; a normal state for most animals

haploid: having one of each chromosome; a normal state for animal gametes

parthenogenesis: a form of asexual reproduction where the young are derived from diploid or triploid eggs produced by the mother without any genetic input from a male

triploid: having three of each chromosome; an abnormal state which is unable to produce normal haploid gametes

ASEXUAL REPRODUCTION IN SIMPLE LIFE FORMS

Although asexual reproduction is very common in organisms such as bacteria, protists, fungi, and plants, it is rarer among animals, especially among the more complex animals. In simple animals, such as sponges (phylum Porifera), the polyps of hydra, jellyfish, corals, and sea anemones (phylum Cnidaria), and many flatworms (phylum Platyhelminthes), asexual reproduction is common. Sponges can reproduce asexually when fragments break off and become established as new individuals or when mature sponges produce gemules, overwintering buds, that are produced and released by many freshwater and a few marine sponges. Cnidarian polyps frequently reproduce by budding. The buds start as small regions of less-differentiated tissue that differentiates into a new polyp. These new polyps can separate from the original to form new individuals or can remain attached and form colonies. The largest colonies to be produced asexually by budding are those produced by the reef-building corals. In some Cnidaria, the polyps release free-floating forms of the organism called medusae (jellyfish). The medusa form of the life cycle, which was formed asexually, reproduces sexually when it matures. Many species of flatworms reproduce asexually by fragmentation. When these worms are cut into fragments, most fragments can regenerate their missing parts, and thus form several organisms from the original one. Others, like many trematode flukes, asexually reproduce by polyembryony. In this mode of reproduction, larval flukes form many juveniles of the next larval stage internally, from the mature larva's own cells. The immature larvae that are produced, which are all genetic clones of the original larva, will be released to continue the life cycle.

PARTHENOGENESIS

In higher organisms, asexual reproduction is much less frequent and far more complex. Rotifers (phylum Rotifera) are small aquatic organisms with rows of cilia around their mouths that seem to rotate as they beat. Rotifer populations are usually either mostly or entirely female. This happens because rotifers usually reproduce asexually by parthenogenesis. In this type of reproduction, the females produce diploid eggs instead of the haploid eggs that are needed for

Asexual reproduction of sea anemone. Sea anemone splits off sections of it basal discs, which then develop into new animals. At the image you could see a sea anemone with at least four "kids."

sexual reproduction. The diploid eggs have the same genes as the mother and are thus clones and mature into adults identical to their mother. In some rotifers, asexual reproduction is the only form of reproduction, and thus all of the organisms in these species are female. In other species, males are only produced during times of environmental stress, and sexual reproduction only occurs then.

Like rotifers, many populations of aphids (phylum Arthropoda), a common plant pest, are entirely or mostly female during parts of the breeding season. In spring and early summer, aphids reproduce parthenogenetically, with females producing diploid eggs that develop into adult female aphids. These eggs and the adults formed from them are clones of the original aphid. In late summer and early fall, the aphids reproduce sexually, producing haploid eggs that can be fertilized by haploid sperm from males.

Even among the very complex vertebrates, a few organisms can be found that reproduce asexually. The two most studied are the whiptail lizard, native to the deserts of the American Southwest and the northwestern part of Mexico, and the gecko, found on some tropical islands of the Pacific. These lizards exist in both sexually reproducing and parthenogenetic forms. Genetic study of their chromosomes has shown that the parthenogenetic species were first formed as diploid or sometimes triploid hybrids of two sexually reproducing forms. The hybrids could not undergo normal meiosis because different chromosomes inherited from each parent could not align properly. Thus, these lizards cannot reproduce sexually. They do, however, produce diploid or triploid eggs that can be triggered to start reproduction. The progeny that are produced are exact genetic duplicates of their mothers—clones.

With asexual reproduction, organisms do not have to waste energy in sexual activity. However, the energy savings is not without a price. With the exception of the rotifers, no populations of asexually reproducing organisms have very long histories, evolutionarily speaking. Unlike most sexually reproducing populations with varying degrees of diversity, all members of an asexually reproducing population are identical, except for mutations that arose after the population's inception. This lowered diversity makes the population much less likely to be able to adapt to change. More or less water, higher or lower temperatures, introduction of parasites, or disease could more easily wipe out the entire population.

—*Richard W. Cheney, Jr.*

FURTHER READING

Adiyodi, K. G., and Rita Adiyodi, eds. *Asexual Propagation and Reproductive Strategies*. Vol. 6, Parts A and B, in *Reproductive Biology of Invertebrates*. John Wiley & Sons, 1993, 1995.

Margulis, L., and K. Schwartz. *Five Kingdoms: An Illustrated Guide to the Phyla of Life on Earth*. 3d ed. W. H. Freeman, 1998.

Richardson, S. "The Benefits of Virgin Birth." *Discover* 17 (March 1996): 33.

Wuethrich, B. "The Asexual Life: Why Sex? Putting Theory to the Test." *Science* 281, no. 25 (September 1998): 1980–1982.

B

Beaks and Bills

Fields of Study
Ornithology; Evolutionary Biology; Ecology

Abstract
All birds have bills (or beaks), and these exhibit so much variation in their structure that they are useful as one of the distinguishing characteristics among species. Bills are toothless, except in embryos, and are covered with a horny sheath. Many of the variations in bill form are associated with specialized adaptations to different feeding habits. In addition to their obvious role in feeding behavior, bills have a function in a number of other behaviors. The specializations observed in bills are adaptations for specific types of food.

Key Concepts
adaptive radiation: the process by which many species evolve from a single ancestral species in adapting to new habitats
display: usually a visible movement or behavior used as a social signal in the context of aggression, courting, etc.
egg tooth: a hard, calcified structure on the tip of the bill of a bird embryo that is used to help the bird break its shell during hatching
polymorphism: the occurrence of two or more structurally or behaviorally different individuals within a species
sex hormones: hormones—androgens in males, estrogens in females—which are associated with sex characteristics and sexual behavior
sexual dimorphism: a difference in structure or behavior between males and females

THE ROLE OF BILLS AND BEAKS
All birds have bills (or beaks), and these exhibit so much variation in their structure that they are useful as one of the distinguishing characteristics among species. Bills are toothless, except in embryos, and are covered with a horny sheath. Many of the variations in bill form are associated with specialized adaptations to different feeding habits. In addition to their obvious role in feeding behavior, bills have a function in a number of other behaviors. Some birds use their bills in different ritualized displays, including threat displays and appeasement displays. Some bills are used for digging, for nest-building, for "sewing," for drilling, and for many other activities.

STRUCTURE OF THE BILL
Anatomically, bills are a compact, modified layer of epidermal cells formed around the bony core of the upper and lower jaws, or mandibles. The bill is normally hard and thick, but it is not rigid, and different bills can be bent and twisted in different ways and to different degrees. For their size, bills actually are relatively light, one of the many adaptations in birds for flight. Although bills do not have teeth, adaptations in some forms have functions similar to those of teeth. Some bills have sharp edges, and these may function in cutting food, for example. There also are different specializations in the bony and muscular elements that operate in conjunction with bills to compensate for the different physical stresses imposed by feeding. It is interesting to consider where birds' nostrils are located. In most species, they are found near the base of the upper mandible, but there are some notable exceptions. In that unusual bird, the kiwi, perhaps the only species of bird with a well-developed sense of smell, the nostrils are located at the tip of the bill.

Some species of birds, including puffins and pelicans, have structures or projections on the bill that are present only around the breeding season and are then shed after reproduction. The brightly colored, triangle-shaped bill of the puffin has scales on both the upper and lower mandible, which seem to serve a role during courtship.

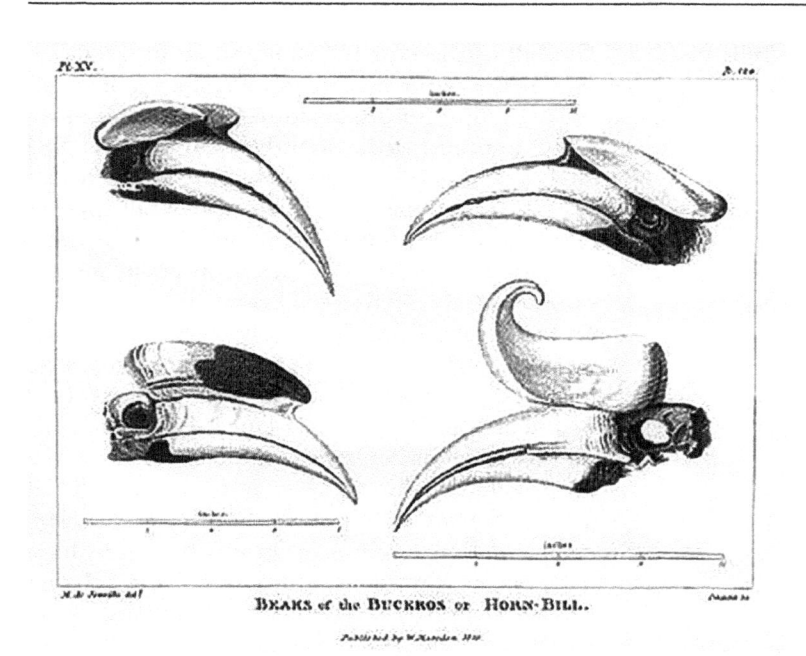

PLATE 15. BEAKS OF THE BUCEROS OR HORN-BILL. M. de Jonville delt. Swaine sc. Published by W. Marsden, 1810.

The bill of the developing embryo possesses an interesting structure: the egg tooth. The egg tooth is a hard, calcified structure on the tip of the bill of a bird embryo that is used to help the bird to crack and weaken the shell so that hatching can take place. The embryo also possesses a special muscle, the "hatching muscle" which provides the force behind the egg tooth. After hatching, the egg tooth either drops off or is absorbed.

VARIATIONS IN BILL STRUCTURE

Given the variety of foods eaten by different bird species, it is not surprising that there are so many differences in the form and size of bills. What may seem surprising, however, is the observation that closely related species may have quite different types of bills. Two good examples of this phenomenon are seen in the finches of the Galápagos Islands and in the Hawaiian honeycreepers. Both of these examples illustrate nicely the process of adaptive radiation, by which many species evolve from a single ancestral species in adapting to new habitats.

Similar to the Galápagos Islands, the islands of the Hawaiian chain are of volcanic origin and, moreover, they are even farther removed from any major landmass than are those in the Galápagos. It is thought that the ancestral form of the modern Hawaiian honeycreeper was a nectar-feeding honeycreeper type of bird that migrated to Hawaii from South America. The birds underwent a remarkable evolution, with new species formed as a result of adaptation to new feeding opportunities. Species became different not only in the shape of their bills but also in their coloration. Species were produced that were red, black, gray, yellow, and green. Beaks varied from those that were large and heavy for crushing seeds to long, thin beaks suitable for collecting nectar from flowers.

As remarkable as are differences in beaks among closely related species, even stranger are differences in bills among individuals of the same species. Bills of three different sizes are found in a finch from Africa, the black-bellied seedcracker. The bills occur in three discrete sizes, ranging from relatively small to relatively large, and the sizes appear related to the type of seeds eaten, with large bills being better suited for dealing with hard seeds and small bills being better adapted for soft seeds. Bills of intermediate size do not fare very well with either type of seed in comparison with either the small or large bills. However, the intermediates are necessary in order to have available the genetic capabilities for the other two sizes of bills. This type of selection is an example of disruptive selection, which in this case tends to produce individuals of the two extreme bill sizes.

Differences in bill structure between males and females of the same species are also known. The huia (probably extinct) is a bird in New Zealand with several unusual characteristics, the least of which is the structure of the bills in the two sexes. As is true of bills in general, the form of the bill appears related to its function. Males have a relatively straight and sharp bill, whereas females have a thin, downward curving bill which is much longer than that of the male. Males use their bills to dig open the burrows of grubs living in dead wood. After the tunnels are open, females use their bills to pull out the grubs. The different structures of the bills in males and females appear related to an interesting division of labor during feeding.

A final example of interesting variation in bills is a case in which the color of the bill changes seasonally within the same individual. The sex hormones, testosterone and estrogen, are responsible for the development of secondary sex characteristics and behavior in males and females, respectively. There are striking changes in behavior during the breeding season. The bills of male and female starlings change color, to yellow in the males and to red in the females, due to the action of these hormones during the breeding season.

MODIFICATIONS OF BILLS FOR FEEDING

As different species of birds are observed feeding, one has a sense of awe at the great diversity of dietary items and the variety of bills that are specialized to eat them. Food items include nuts, hard seeds, soft seeds, insects, larvae, roots, flowers, sap, fungi, nectar, carrion, and other types of organic matter. The specializations observed in bills are adaptations for specific types of food. In a brief discussion, it is not possible to describe the many variations seen in the size and shape of bills, but even descriptions of a few representative examples provide the basis for appreciation of the exquisite designs and functions of bills.

In carnivorous birds of prey, such as falcons and owls, the beak is used as a meat hook. The birds typically seize their prey with their feet and disarticulate the cervical vertebrae by biting into the neck. The bill has a so-called tomial tooth on each side of the upper bill just behind the curved end of the beak. There is a notch or groove on the lower mandible that fits with the tomial tooth.

Birds that catch insects in flight, such as flycatchers, have beaks that are broad and slightly hooked. Frequently, there are long bristles or feathers at the base of the bill which facilitate the catching of the moving prey.

The bill of the crossbill shows some unusual features. The tips of the beak do not meet when they are closed. The tips are displaced to the side and move past one another as the mouth is closed. The end of the lower mandible may cross to the left or to the right of the upper mandible. This type of bill appears adapted for serving as a wedge to pry open the scales of pine cones, enabling the bird to insert its tongue and to remove the seeds from inside.

An interesting specialization is seen in the beaks of tropical fruit pigeons. These birds feed on large fruits, and their bills can open not only in the vertical plane but also in a horizontal plane, like the jaws of a snake when swallowing large prey.

It is evident from these examples that bills of birds have become adapted for their feeding. There are even some examples of species whose bills have become so specialized for their specific food that the number of birds is completely dependent on the availability of that food source.

—*Donald J. Nash*

FURTHER READING

Bird, David M. *The Bird Almanac: The Ultimate Guide to Essential Facts and Figures of the World's Birds.* Firefly Books, 1999.

Castro, Isabel, and Antonia Phillips. *A Guide to the Birds of the Galápagos Islands.* Princeton University Press, 1996.

Dunn, Jon L. *National Geographic Field Guide to the Birds of North America.* 3rd ed. National Geographic Society, 1999.

Gill, Frank B. *Ornithology.* 2nd ed. W. H. Freeman, 1995.

Weiner, J. *The Beak of the Finch: A Study of Evolution in Our Time.* Alfred A. Knopf, 1994.

BIODIVERSITY

FIELDS OF STUDY

Ecology; Conservation Biology; Environmental Studies; Biogeography

ABSTRACT

The study of biodiversity is relatively new. The word was coined by Edward O. Wilson in order to explain the biological living resources and wealth lost when whole areas

of tropical rain forest were destroyed by clear-cutting and burning. Because the study of biodiversity involves the study of individual species, their loss, and their habitat, the subject is closely tied to endangered species, ecology, and conservation biology.

KEY CONCEPTS

aquaculture: the artificial growth of animals or plants that live in the water; the culture of something living in water

biodiversity: the number and kinds of animals and plants and the variations in genetic material they possess

exotic species: organisms that are not naturally found in a place but have been artificially introduced, whether by accident or intentionally

extinct: no longer found anywhere on earth

extirpated: not found in an immediate local area but found elsewhere on earth

habitat: an assemblage of plants, animals, and the physical land, water, minerals, and other elements that support a life-form; usually refers to a single species

invertebrate: an animal without a backbone; for example, worms, clams, crabs, insects, and jellyfish

species: a group or groups of interbreeding natural populations of organisms that are reproductively isolated from other such groups

subspecies: a group or groups of interbreeding organisms that are distinct and separated from similar related groups but not fully reproductively isolated

threatened species: animals or plants so few in number that they may soon be endangered and then extinct

DEFINING BIODIVERSITY

The study of biodiversity is relatively new. The word was coined by Edward O. Wilson in order to explain the biological living resources and wealth lost when whole areas of tropical rain forest were destroyed by clear-cutting and burning. Many, if not all, of these tropical rain forests were in remote areas, often in developing countries, untouched by modern civilization, and relatively unstudied by scientists. As a result, many of the plants and animals, especially insects, were unknown to scientists. Many of the species lived nowhere else on earth and, once destroyed, became extinct. Scientists realized that these tropical rain forests held a great wealth of species and that their destruction was removing large numbers, perhaps millions, of species, along with the DNA that made each species unique. Wilson referred to this loss of species as a loss of the forests' biological diversity, or biodiversity. The loss is not merely of the species that become extinct but also of the combination of highly specific genetic material in each species.

Because the study of biodiversity involves the study of individual species, their loss, and their habitat, the subject is closely tied to endangered species, ecology, and conservation biology. Endangered species are those plants and animals that are nearly extinct; threatened species are not yet at the point of being endangered but will be if measures are not taken soon. Both plants and animals, including insects, worms, clams, crayfish, and other creatures, may be considered either threatened or endangered. Specific habitats, especially those that are unique and found in only one location on earth, can also be considered threatened. Protecting individual species often means protecting the habitat the species needs for survival.

During the nineteenth century and the first part of the twentieth, some animals were hunted to extinction. Several other species were hunted to near extinction or were extirpated (removed) from large areas, largely by extensive changes to their habitats. As a result of the losses and near losses, some scientists, resource managers, and political leaders instituted programs and activities to prevent extinction and to restore species that were near extinction. Animals (and later plants) that were near extinction were officially designated as endangered and given legal protection to prevent their extinction. Eventually these efforts grew to encompass a much broader field of work by experts in many fields.

One of the more difficult aspects of protecting a single species is the importance of the habitat to the well-being of the species. This fact is one of the underlying principles driving the whole study of biodiversity. Scientists have learned that the habitat is critical for restoration of a species and that protecting the habitat is a powerful means of preventing species loss and protecting biodiversity. As a result, programs and individuals trying to protect or restore biodiversity or individual species have come into conflict with those who use or plan to use large areas of land and whole habitats for purposes such as housing or industrial development, harvesting, and aquaculture.

THREATS TO BIODIVERSITY

Scientists generally believe that three major factors cause the loss of biodiversity: resource overharvesting, habitat loss, and introduction of exotic species. Overharvesting or excessive hunting and catching caused the loss of species such as the great auk, passenger pigeon, and Steller's sea cow. Some species of whales nearly suffered the same fate. However, for the vast majority of animals and plants, overharvesting is not a major cause of endangerment or extinction. Many animals that are commonly harvested have large and dispersed populations, so harvesting stops before it causes populations to decline to the point of endangerment and extinction. This does not mean that overharvesting is not a problem, but that it is seldom the sole cause of population declines and not the most frequent cause.

A number of international controls have been implemented to prevent, limit, or discourage the killing of some of the animals that are threatened by overharvesting in many countries around the world. Many such controls ban the import or shipping of either live animals (tropical birds such as parrots) or parts of them (elephant ivory, rhinoceros horn, tiger bones, and the gall bladders of bears). The animals endangered by overharvesting—the rhinoceros, elephant, and parrot—tend to be long-lived and restricted in distribution. Their reproductive rates are low, and populations often require a large habitat. The habitats of many of these animals are threatened by human activities, which makes the effect of their being slaughtered for commercial reasons that much more pronounced. In 1997, one of the major international treaties governing commercial transactions in rare and endangered species was modified to permit some African nations to sell stocks of certain animal products that had been previously prohibited. Some experts worried that the change in the laws would again provoke illegal overharvesting.

Many animals are threatened in their native habitats by exotic species introduced into the area. These exotic species often cause biological pressures that the native species cannot handle. They may threaten native species directly by competing for food, living space, critical habitat (such as nesting space), or other necessary environmental resources. Some exotic animals are aggressive and drive away or attack the native ones. In the United States, sea lamprey were introduced into the Great Lakes and have driven populations of many native species to precipitously low levels or even extirpated them.

Exotic species may also present indirect threats by introducing diseases, parasites, or hitchhiker species that arrive with the exotic. In South Carolina and Texas, exotic shrimp are used in aquaculture, and they can carry a deadly virus that may threaten native shrimp if released into the wild by accident. Zebra mussels were carried into the Great Lakes water system in the ballast water of ocean tankers. With no natural controls, zebra mussels have multiplied to the point that they are not only a nuisance to human activities but also serious competition for native aquatic animals. Asian carp currently present a similar environmental problem in the United States and pose a threat to the Great Lakes region of both the United States and Canada.

HABITAT LOSS AND DEGRADATION

The most significant threat to biodiversity, as occurs in tropical deforestation, is habitat loss or degradation, including poisoning. Habitat loss is the elimination of a habitat, often by cutting, plowing, filling, or other construction or harvesting methods. When all the trees in a forest area are felled for wood (a practice known as clear-cutting), the habitat is lost until the trees regrow, an event that may never take place. Degradation is the elimination or lessening of some characteristic of the habitat that is critical for one or more species. Changing the temperature or speed of the water flowing in a stream can render it no longer suitable for a particular fish or mussel. Filling or draining a wetland alters the moisture content of the soil and changes its suitability for plants or animals.

One of the earliest examples of habitat degradation was the poisoning of bald eagles and peregrine falcons in the United States by the pesticide dichlorodiphenyltrichloroethane (DDT). DDT was widely used throughout the world (and still is in some countries), is highly persistent, and accumulates in animals at ever-increasing levels in a trophic system (a system in which one animal feeds on another that feeds on another, and so on). Falcons and eagles were some of the birds of prey that took in so much DDT that they were poisoned and their eggs became too thin-shelled to survive. Their populations began to decline from reproductive failure. Many species of freshwater mussels are highly sensitive to chemicals and do not survive even low levels in the water. Many of the mussel species have become rare and endangered as a result of this poisoning, and they are not the only species to suffer from environmental poisons.

Altering stream and river flow via dams, diversions, withdrawals, and dredging is another form of habitat degradation that can dramatically change the way animals and plants use the aquatic system. In river systems throughout the Pacific Northwest, salmon return from the ocean to spawn (lay eggs and reproduce) in the river where they grew from egg to fish before migrating to the open ocean. If a dam blocks the river or if the conditions in the river are no longer suitable, the salmon are not able to spawn. In salmon, the fish that spawn in a given river are distinct and genetically identifiable, almost forming a subspecies. Therefore, authorities in the Pacific Northwest have had to determine the relationship between individual spawning groups and the watershed habitat before altering streams and rivers.

The cutting of mature forests in the Pacific Northwest and in Alaska eliminated the habitat for the spotted owl. Great controversy resulted from limits, restrictions, or bans on logging in these forests, which were prized for the high quality lumber that could be made from the trees. These and similar conflicts formed the basis for many studies on scientific, legal, and policy matters concerning the Endangered Species Act in the United States.

Controversies have arisen over habitat protection programs and laws in the United States and other countries because of the restrictions and limitations imposed on development, logging, and other human activities. People on one side argue that the potential benefits from protecting a species are greater than any inconvenience or costs that might ensue, and those on the other side argue that the economic benefit from the activity that threatens the species outweighs any ecological concerns. However, in some cases, companies and governments have worked hard to preserve species believed to be of economic value.

BIODIVERSITY RESEARCH

Biodiversity is studied at the level of populations, of individuals (including molecular biology), and in the interactions between organisms and their environment. Research on biodiversity includes genetic, ecological, and behavioral studies and investigations focusing on diseases and life histories.

Much of the work on any specific species, especially large vertebrates, is conducted in the field, using techniques for observing, counting, tracking, and monitoring the population. Most invertebrates cannot be studied other than by observation and counting. One of the goals of ecological research on biodiversity is to understand the life histories and ecology of individual species. Scientists seek to explain life span, reproductive patterns and strategies, food preferences, environmental requirements, and limits. Other major goals are to describe the habitat requirements and better understand the relationship between the species and the habitat.

The management and legal aspects of various programs are included in biodiversity studies. This research is closely related to the field of conservation biology. Such study entails examining areas such as how programs work; what money is spent, by whom, how, and for what use; whether habitats or species are protected, restored, or lost; and the time required for various activities. These considerations are generally part of program evaluation, a topic that can be more thoroughly researched in the field of management science. The legal aspects of preserving biodiversity and protecting endangered species have been studied by a number of legal and policy-making organizations, including the Environmental Law Institute and Environmental Defense Fund, both of Washington, DC. The National Research Council, the operating arm of the National Academy of Sciences, has published a number of reports on the Endangered Species Act, and these usually address legal, policy, and science issues.

Modern molecular genetic techniques are applied to the study of individual species of plants and animals and their restoration. Tissue, often a blood sample, taken from a single individual is analyzed to determine the genetic relations among members of a population. Scientists may then assess the potential of the remaining individuals to act as the beginning or nucleus for a new population. Modern techniques can determine the genetic makeup of animals and whether the individuals are distantly enough related to form a breeding pair without suffering the adverse consequences of inbreeding.

Artificial breeding and culture are techniques often employed with the few remaining members of a population. These animals may be brought in from the wild and confined, as were the last remaining California condors. The animals are kept in captivity so that artificial breeding and maintenance techniques can be used to raise additional members of the population. Once the population is large enough for release or other factors are favorable, reintroduction into the wild may be attempted.

Zoological parks and botanical gardens have been involved in maintaining the few remaining specimens of some species. With the help of such facilities, animal tissue has been stored in culture, and sperm have been frozen for later use in breeding. Parks and gardens now play an active role in species propagation and husbandry through breeding, nourishment programs, studying and preventing disease, and behavioral training to reintroduce zoo animals into the wild.

INCREASING AWARENESS OF BIODIVERSITY'S IMPORTANCE

Human activities eliminated more than a few animals from the face of the earth before the second half of the twentieth century, when national and international laws were enacted to protect endangered species and the modern environmental movement began. Among the animals lost were the great auk in 1844 and the passenger pigeon in 1914. These animals were hunted to extinction under the mistaken belief that extinction could not happen, and because authorities lacked a way to limit the hunting of the animals. Despite international efforts to protect animals from extinction, in the latter part of the twentieth century, many countries allowed or encouraged the wholesale destruction of tropical rain forests, along with the animals that lived there.

In 1973, amid growing awareness of the need to prevent further extinctions, the Endangered Species Act was passed by the Congress of the United States of America. The act gave the US Fish and Wildlife Service the responsibility for enforcing the law and set procedures for determining which species should be listed as endangered. The act also established the category of threatened species to protect those that were close to becoming endangered. Subsequently, this protection was extended to the habitats on which the animals depend.

Studies by academic and government scientists revealed that most species are threatened by habitat loss or degradation, overharvesting, or exotic species. Most of the problems arise from habitat loss or degradation rather than overharvesting, which had been the cause in several of the cases that prompted passage of the Endangered Species Act. Scientifically, and eventually legally, one of the most difficult tasks is to determine habitat requirements and then ensure that the required habitat is protected. Identifying and protecting habitats became an important part of all programs aimed at protecting biodiversity. Loss or degradation of habitat became one of the criteria that federal agencies used to take legal action or make official decisions.

Many human activities such as logging, building, mining, and farming can destroy or degrade valuable habitats on which endangered species depend. As a result, endangered species protection programs and activities have been in conflict with some business interests seeking to carry out those activities that threaten species or their habitats. According to the World Wildlife Fund's *Living Planet Report 2016*, conservation efforts launched in the 1970s have done little to slow the decline of biodiversity and mitigate human damage to the global ecosystem. The report found that, between 1970 and 2012, vertebrate species population sizes have been more than cut in half, declining by 58 percent. Freshwater species experienced the greatest loss over the measured forty-two-year period, seeing an 81 percent decline in population. Marine and terrestrial species populations both declined by 36 and 38 percent respectively. Species population and biodiversity decline was greatest in South and Central America—a region rich in rainforest ecosystems—where species populations declined by an average of around 83 percent.

Efforts, activities, and programs to protect biodiversity rely on the latest techniques and methods. These include electronic tracking of large animals, molecular analysis of genetic material, and cellular studies of tissues to help determine actual or potential threats to animals. Ecologists investigate life histories, including feeding habits, environmental requirements and limits, and migration patterns.

—Peter L. deFur

FURTHER READING

Burton, John, ed. *The Atlas of Endangered Species*. 2nd ed. Macmillan, 1999.

Dhillon, Sarinder K., and Amandeep S. Sidhu. *Data Intensive Computing for Biodiversity*. Springer, 2013.

DiSilvestro, Roger L. *The Endangered Kingdom: The Struggle to Save America's Wildlife*. Wiley, 1989.

Levin, Simon A. *Encyclopedia of Biodiversity*. 2nd ed. Academic Press, 2013.

National Research Council. *Science and the Endangered Species Act*. National Academy Press, 1995.

New, T. R. *An Introduction to Invertebrate Conservation*. Oxford University Press, 1995.

Ricketts, Taylor H., et al. *Terrestrial Ecoregions of North America: A Conservation Assessment*. Island, 1999.

Rossberg, Axel G. *Food Webs and Biodiversity: Foundations, Models, Data*. Wiley, 2013.

Tudge, Colin. *The Variety of Life: The Meaning of Biodiversity*. Oxford University Press, 2000.

Wilson, Edward O., ed. *Biodiversity*. National Academy Press, 1988.

Wilson, Edward O., ed. *Half-Earth: Our Planet's Fight for Life*. Liveright, 2016.

World Wildlife Federation. *Living Planet Report 2016: Risk and Resilience in a New Era*. WWF International, 2016, www.wnf.nl/custom/LPR_2016_fullreport/. Accessed 31 Jan. 2018.

BIOLOGICAL RHYTHMS AND BEHAVIOR

FIELDS OF STUDY

Chronobiology; Environmental Studies; Endocrinology; Marine Biology

ABSTRACT

Circadian and other biological rhythms are considered a basic characteristic of life. The term "circadian" (from the Latin circa, meaning "about," and diem, "day") was coined by Franz Halberg to describe these approximately twenty-four-hour rhythms, which in time were found to exist not only in plants but also in animals and in human beings.

KEY CONCEPTS

biological rhythm: a cyclical variation in a biological process or behavior, often with a duration that is approximately daily, tidal, monthly, or yearly

circadian rhythm: a cyclical variation in a biological process or behavior that has a duration of about a day—from twenty to twenty-eight hours

endogenous: refers to rhythms that are expressions of only internal processes within the cell or organism

entrainment: the synchronization of one biological rhythm to another rhythm, such as the twenty-four-hour rhythm of a light-dark cycle

exogenous: refers to rhythms that originate outside the organism in the environment

free-running: denotes a rhythm that is not entrained to an environmental signal such as a light-dark cycle

frequency: the number of repetitions of a rhythm per unit of time, such as a heart rate of seventy beats per minute

period: the length of one complete cycle of a rhythm

photoperiodism: the responses of an organism to seasonally changing day length, that cause altered physiological states

zeitgeber: "time giver" in German, it is also referred to as a synchronizer or entraining agent

THE RHYTHM OF LIFE

Circadian and other biological rhythms have been observed and described in so many processes and behaviors in so many diverse organisms that their presence in higher plants and animals is considered a basic characteristic of life. The term "circadian" (from the Latin *circa*, meaning "about," and *diem*, "day") was coined by Franz Halberg to describe these approximately twenty-four-hour rhythms, which in time were found to exist not only in plants but also in animals and in human beings. The "circa" prefix is used also with words denoting other time periods (such as "circannual").

CIRCADIAN AND CIRCANNUAL RHYTHMS

Circadian rhythms enable animals to time precisely their daily activities. Animals are broadly classified as diurnal if they are active by day, nocturnal if they are active at night, and crepuscular if they are active at both dawn and dusk. Many species schedule their activity to start within minutes of the same time each day. Thus, the Swiss psychiatrist Auguste Henri Forel, in 1906, noticed that bees adapted to his schedule of eating breakfast on the terrace: The bees came each morning at breakfast time to feed on the jam. One of Karl von Frisch's coworkers, Ingeborg Beling, found that she could train bees to visit a feeding station every twenty-four hours but that the bees could not be trained to come every nineteen hours. Individual species of flowers produce nectar only at certain times of the day, and bees have been observed to plan their visits according to the time of nectar flow.

The activity rhythms of caged flying squirrels have been studied in detail by Patricia DeCoursey. She found that the time of the onset of activity in this nocturnal rodent was very uniform from day to day but that the time gradually drifted during the year from about 4:30 p.m. in January to 7:30 p.m. in July and then back to 4:30 p.m. by the following January. Such a pattern is called a circannual rhythm. Circannual rhythms are particularly evident in migratory birds, which show seasonal changes in both their physiology and their behavior. Many mammals have distinct reproductive seasons. Mammals that hibernate, such as ground squirrels and woodchucks, gain fat in the fall, enter hibernation, and then wake up in the spring according to a circannual rhythm.

Although the annual changes in temperature might be expected to be the environmental factor that would signal a change in season to a plant or animal, it is now known that many plants and animals respond to changes in the length of the photoperiod. This response is called photoperiodism. Photoperiodism was found first in plants in 1920, in insects in 1923, in birds in 1926, and in mammals in 1932. In a typical experiment, light was artificially added to the short days of late fall to create a longer photoperiod—similar to that characteristic of spring.

As a result, the organisms came into reproductive development months early.

Circadian rhythms have been found to play an essential role in photoperiodism. What later became known as the Bünning hypothesis postulated that a circadian rhythm was involved in the organism's mechanism which measures the length of the photoperiod. It was hypothesized that the first twelve hours of the circadian rhythm was a light-requiring phase and the last twelve hours was a dark-requiring phase. Short-day effects occurred when the light was limited to the light-requiring phase, but long-day effects occurred when light was present during the dark-requiring phase.

In this scheme light plays two roles: It is a zeitgeber to synchronize rhythms and an inducer to stimulate reproductive responses. Later experiments demonstrated that short photoperiods followed by a brief flash of light in the middle of the dark were interpreted by the organism as long photoperiods, and the organisms became reproductively developed months early. Thus, the important thing is really not how long the photoperiod is, but rather when light is present with respect to a circadian rhythm of sensitivity to light.

MARINE RHYTHMS

Some of the most dramatic examples of biological rhythms are found in marine organisms. The periods or lengths of the rhythms are rather diverse and include circadian, circatidal, circalunar, and circannual rhythms, and various combinations of them. Perhaps most famous is the rhythm of reproductive activity of the South Pacific marine worm referred to as the palolo worm. This species spawns at the last quarter of the moon in October and November (spring in the Southern Hemisphere). The worm lives buried in coral reefs and, at spawning, the last twenty-five to forty centimeters of the worm, which bears the gametes, breaks off and rises to the surface of the sea. The gametes are released into the seawater, where fertilization takes place. The spawning always occurs at daybreak. The exact timing of the spawning is an adaption that increases the chances for successful reproduction in this species.

Similarly, the California grunion, a small smelt about fifteen centimeters long, spawns in the spring at about fifteen minutes after the time of the high tide each month. During the spawnings, or "grunion runs," the fish ride the waves onto the sandy beaches, where the females burrow the posterior end of their bodies into the sand. The male curls around the female's body and releases sperm as the female lays her eggs. The fish return to the sea and the eggs continue to develop until approximately fifteen days later, when the high tide returns and uncovers the hatching young. During the grunion runs, the adult fish are caught by fishermen (legally only by hand) and are eaten. Neither the palolo worm nor the grunion has been sufficiently studied to determine what environmental factor—moonlight, gravity, magnetism, or another factor—synchronizes their rhythms so precisely.

CHRONOBIOLOGY

The broad field of the study of biological rhythms is called chronobiology. A rhythm is the cyclical repetition of a property or behavior, whether it concerns the level of body temperature, enzyme activity, or hormone level in the blood, or describes an activity of the whole animal, such as feeding patterns, daily or seasonal migrations, or seasons of reproduction. The period of the rhythm is the time it takes to complete one full cycle. This could be measured from crest to crest or trough to trough. The frequency of the rhythm refers to how many cycles occur per unit of time (such as a heart rate of seventy beats per minute). The amplitude refers to the strength of the rhythm (for example, one-half of the height of the rhythm when shown on a graph).

The properties of biological rhythms are fascinating. They are ubiquitous, innate, probably endogenous, free-running, self-sustaining, entrainable, relatively temperature-independent, and relatively unsusceptible to chemical perturbations. Biological rhythms are said to be ubiquitous because they are found everywhere—at all levels of life from cell organelles to cells, tissues, organs, whole organisms, and populations. They are found in all kinds of living things, with the possible exception of the prokaryocytes. They are said to be innate because the rhythms are not learned and are largely programmed by the genetic makeup of the organism. Biological rhythms are probably endogenous, with an oscillator inside the cells of the organism, but it should be noted that Frank A. Brown has published extensive evidence that the timing information may be exogenously derived from geophysical fluctuations.

Biological rhythms are entrainable, which means that they usually are kept in synchrony with day/night or other environmental schedules. Entrainment is maintained by an organism's responses to environmental factors called synchronizers, zeitgebers, entraining agents, or time cues. Light, temperature, noise, and feeding are some of the zeitgebers that have been identified. The rhythms are called self-sustaining because they continue in the absence of any obvious zeitgebers. Biological rhythms have been found to be relatively temperature-independent, which is important because they often function as clocks. Biological rhythms can free-run when isolated from zeitgebers. When a rhythm free-runs, its period is found to be slightly different from the entrained period.

Despite years of investigation, biological rhythms are poorly understood. The search continues to find biological bases for the rhythmic processes so commonly seen. The innermost rhythmic process is sometimes referred to as the "biological clock," since it represents the seat of the cell's or organism's timekeeping mechanism.

STUDYING BIOLOGICAL RHYTHMS

One of the earliest scientific observations of a biological rhythm was reported in 1729 by Jean Jacques d'Ortous de Mairan, a French astronomer. He made detailed observations of the reactions to constant darkness of a so-called sensitive plant that normally has its leaves unfolded during the daylight hours and folded during the night. De Mairan wondered whether the leaves respond directly to the presence of the sunlight and therefore open at dawn and close at dusk. Placing the plant in constant darkness to see how it responded, de Mairan found that the plant continued to show the rhythmic folding and unfolding of its leaves. The curious results were published in a brief report in the Proceedings of the Royal Academy of Paris.

Several years later, in 1758, Henri-Louis Duhamel repeated de Mairan's experiment and further observed that warm temperatures failed to alter the pattern of the rhythmic opening and closing of the leaves of the sensitive plant. Later studies, in the nineteenth century, revealed that in the sensitive plant (*Mimosa pudica*) the rhythmic opening and closing of leaves in constant dark completed a full cycle in 22 to 22.5 hours. It was found that plants supplied with lamps during the night and kept in darkness during the day adapted to the new schedule within a few days and unfolded their leaves only during the artificial day. Charles Darwin did experiments that convinced him that plants survived frosts more successfully when they could fold their leaves at night.

The extent to which circadian or other biological rhythms are endogenous (originate inside the organism) has been a subject of debate. Frank A. Brown spent most of his research career trying to resolve this question. Brown found that even when organisms were placed in heavy metal chambers that were airtight, it was virtually impossible to isolate an organism from its rhythmic, geophysical environment. Normally, circadian rhythms keep in synchrony with the day/night cycle (supposedly they are reset slightly each day, since they are not exactly twenty-four hours long). Brown studied in detail free-running rhythms—that is, rhythms that are found in organisms in a seemingly constant environment. When he averaged oxygen uptake data over many months, he found exact geophysical rhythms of twenty-four hours as well as exact lunar and annual rhythms in the metabolism of many different organisms, such as potatoes, carrots, hamsters, and rats. Furthermore, he showed that many animals, such as snails and flatworms, are influenced by subtle changes in the earth's magnetic field. Therefore, he concluded that the actual timing information that underlies circadian and other biological rhythms may well be exogenous and derived from the rhythmic, geophysical environment that pervades the organisms' everyday surroundings. Despite such evidence for exogenous influences, most biologists today regard biological rhythms as the product of essentially endogenous processes.

ENDOCRINE AND NERVOUS SYSTEM RHYTHMS

When looking for some basis for endogenous rhythms, researchers often investigate the endocrine and nervous systems because of their large roles in integrating and controlling biological functions. An especially interesting study has been made by Albert H. Meier. He has found that endocrine rhythms play a role in regulating the seasonal changes of physiology and behavior in the migratory white-throated sparrow and other vertebrates. By injecting birds with the hormones corticosterone and prolactin in different time relationships, he was able to induce seasonal changes. In early studies, he found that injections of prolactin either caused fat gain or fat loss, depending simply on whether the injections were given in the morning or in the afternoon. Migratory birds gain fat before they migrate and use this fat as an energy source for their flights. If the birds are given daily injections of corticosterone and prolactin four hours apart, the birds gain fat, try to fly from the south side of their cages, and do not have well-developed gonads—all characteristics of the normal fall bird. If the birds are given daily injections of corticosterone and prolactin eight hours apart, they remain lean, do not show any directed flight, and do not have well-developed gonads—traits characteristic of the normal summer bird. On the other hand, if the birds are given daily injections of cortiscosterone and prolactin twelve hours apart, the birds gain weight, try to fly out the north sides of their cages, and have gonads that will grow in response to a lengthening photoperiod—traits characteristic of the normal spring bird. Some assays, using radioactive isotopes of corticosterone and prolactin, have been made in wild populations of white-throated sparrows, and the results show that the timing of the peaks of the hormones is roughly similar to the time relationships just discussed. Further research by this group centered on the modification of brain chemistry to bring about the seasonal changes in vertebrates.

The methods used to study biological rhythms range from the simple methods used by pioneers in the field to the latest innovations in molecular biology. The field is attracting many new researchers; new discoveries are being published almost daily. Yet much remains to be done before the essential nature of biological rhythms can be understood.

IMPLICATIONS OF BIOLOGICAL RHYTHMS

There are many implications to the fact that plants and animals possess circadian and other biological rhythms. Some scientists have speculated about whether man can survive living in space vehicles that leave the geophysical environment of the earth-moon complex. Will it be necessary, they ask, to try to duplicate parts of the terrestrial geophysical environment—by, for example, installing a rhythmic magnetic field in the space vehicles?

More mundane applications of a better knowledge of rhythms are to be found in animal husbandry. The annual rhythm of the reproduction of farm animals can be manipulated to result in higher productivity. It is a standard practice to lengthen the photoperiod in the henhouse to increase egg production and minimize the winter decrease in production. Sheep are treated with the hormone melatonin, naturally produced in more abundance in the winter months, in the early fall to hasten the reproductive season. Many more benefits of a better understanding of biological rhythms await discovery.

—*John T. Burns*

FURTHER READING

Arendt, J., D. S. Minors, and J. M. Waterhouse, eds. *Biological Rhythms in Clinical Practice*. Butterworth, 1989.

Brady, John, ed. *Biological Timekeeping*. Cambridge University Press, 1982.

Cloudsley-Thompson, J. L. *Rhythmic Activity in Animal Physiology and Behaviour*. Academic Press, 1961.

Follett, Brian K., Susumu Ishii, and Asha Chandola, eds. *The Endocrine System and the Environment*. Springer-Verlag, 1985.

Moore-Ede, Martin C., Frank M. Sulzman, and Charles A. Fuller. *The Clocks That Time Us*. Harvard University Press, 1982.

NATO Scientific Affairs Division. *Rhythms in Fishes*. NATO ASI Series. Series A, Life Sciences 236. New York: Plenum, 1992.

Palmer, John D. *An Introduction to Biological Rhythms*. Academic Press, 1976.

Ward, Richie R. *The Living Clocks*. Alfred A. Knopf, 1971.

BIOLOGY

FIELDS OF STUDY

Biology; Biochemistry; Genetics; Zoology; Microbiology

ABSTRACT

Biology is an extensive field subdivided into categories based on the molecule, the cell, the organism, and the population. Molecular biology touches on biophysics and biochemistry. Cellular biology is closely related to molecular biology through understanding the functions and basic structure of the cell. Botany is the science or study of plant life. Ecology is the study of relationships between organisms and their environments. Developmental biology encompasses a number of issues, including gene regulation, genetics, and evolution. There are several other categories.

KEY CONCEPTS

biodiversity: the total and variety of all living organisms in an environment

deoxyribonucleic acid (DNA): the carrier of all an organism's genetic information

National Institutes of Health: the United States' governmental division that monitors and works to protect and improve public health

xenotransplantation: the transplantation of organs from one species to another

THE BIOLOGICAL SCIENCES

Biology is an extensive field subdivided into categories based on the molecule, the cell, the organism, and the population. Molecular biology touches on biophysics and biochemistry. It is the branch of biology that deals with the structure and development of biological systems in terms of the physics and chemistry of their molecules. Cellular biology is closely related to molecular biology through understanding the functions and basic structure of the cell. The cell is the smallest structural unit of an organism that is capable of independent functioning, consisting of one or more nuclei, cytoplasm, and organelles, all surrounded by a semipermeable membrane. Botany is the science or study of plant life. Ecology, also referred to as bionomics, is the study of relationships between organisms and their environments.

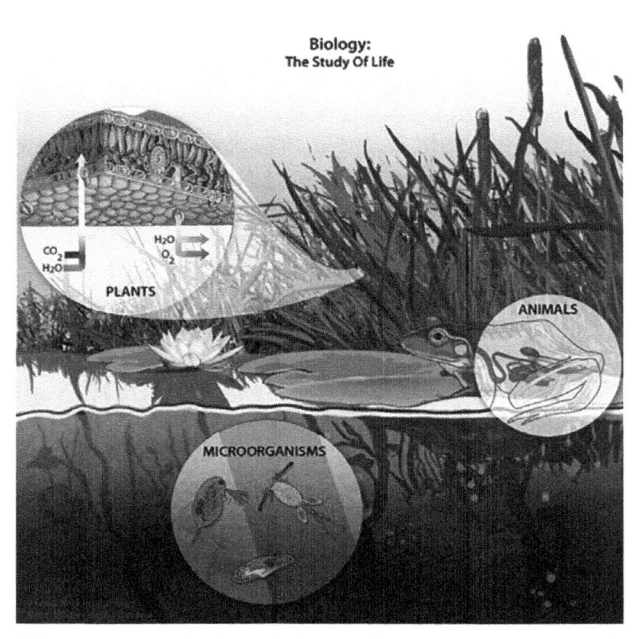

Biology encompasses the science of all living things; microscopic to macroscopic, simple to complex.

Developmental biology encompasses a number of issues, including gene regulation, genetics, and evolution. These are concepts of importance to vertebrates, invertebrates, and plants. A gene is a hereditary unit that occupies a particular location on a chromosome, determines a particular characteristic of an organism, and can undergo mutation. Genetics is the branch of biology that deals with heredity, especially the hereditary transmission and variation of inherited characteristics. Evolution is the theory that groups of organisms change with the passage of time, mainly as a result of natural selection, so that descendants differ morphologically and physiologically from their ancestors. Population genetics, the study of gene changes in populations, and ecology have been established subject areas since the 1930s. These two fields were combined in the 1960s to form a new discipline called population biology, which became established as a major subdivision of biological studies in the 1970s. Central to this field is evolutionary biology, in which the contributions of Charles Darwin are noted.

Microbiology is the branch of biology that deals with microorganisms. The study of bacteria, including their classification and the prevention of diseases that arise from bacterial infection, is the primary focus of microbiology. This branch is of interest not only among bacteriologists but also among chemists, biochemists, geneticists, pathologists, immunologists, and public health professionals. Parasitology is the study of parasites, organisms that feed on or in different organisms while contributing nothing to the survival of their hosts. Ethology is the scientific study of animal behavior. Animal behavioral studies have developed along two lines. The first of these, animal psychology, is primarily concerned with physiological psychology, and has traditionally concentrated on laboratory techniques such as conditioning. The second, ethology, had its origins in observations of animals under natural conditions, concentrating on courtship, flocking, and other social contacts. One of the important recent developments in the field is the focus on sociobiology, which is concerned with the behavior, ecology, and evolution of social animals such as bees, ants, schooling fish, flocking birds, and humans.

Ethics is the study of the general nature of morals and of specific moral choices. Bioethics addresses such issues as animal experimentation, cloning, euthanasia, gene therapy, genetic engineering, genome projects, protection of human research subjects, organ transplants, and patients' rights.

Biotechnology is the industrial use of living organisms or biological techniques developed through basic research. Biotechnology products include antibiotics, insulin, interferon, and recombinant DNA, and techniques such as waste recycling. The Office of Biotechnology Activities of the National Institutes of Health monitors scientific progress in human genetics research in order to anticipate future developments, including ethical, legal, and social concerns, in basic and clinical research involving recombinant DNA, genetic testing, and xenotransplantation (the use of organs from other species of mammals for transplants). In addition to organs donated from humans, researchers are exploring the use of partially or wholly artificial organs manufactured in the laboratory.

Biodiversity focuses on such issues as conservation, extinction and depletion from overexploitation, habitat pollution, global patterns and values of biodiversity, and endangered species protection.

Well known pioneer biologists include naturalist and explorer Sir Joseph Banks (1743–1820), naturalist and explorer Charles William Beebe (1877–1962), biochemist Gunter Blobel (b. 1936), environmentalist

Rachel Carson (1907–64), biochemist Stanley Cohen (b. 1922), biophysicist and codiscoverer of DNA, Francis Crick (1916–2004), naturalist and father of evolutionary theory Charles Darwin (1809–82), evolutionary biologist Richard Dawkins (b. 1941), marine biologist Sylvia Earle (b. 1935), bacteriologist Paul Ehrlich (b. 1932), bacteriologist and discoverer of penicillin Sir Alexander Fleming (1881–1955), microscopist Antonie van Leeuwenhoek (1632–1723), botanist and taxonomist Carolus Linnaeus (1707–23), ethologist Konrad Lorenz (1903–89), zoologist A. S. Loukashkin (1902–88), botanist and geneticist Barbara McClintock (1902–92), botanist and genetic theorist Gregor Mendel (1822–84), endocrinologist and inventor of the birth control pill Gregory Goodwin Pincus (1903–67), naturalist Alfred Russel Wallace (1823–1913), biophysicist and codiscoverer of DNA James Watson (b. 1928), Nobel Prize–winner Hamilton O. Smith (b. 1931), whose work led to the creation of recombinant DNA technology, discoverer of genetic transduction, Norton David Zimmer (1928–2012), and sociobiologist Edward O. Wilson (b. 1929).

—Mary E. Carey

FURTHER READING

Brooker, Robert J., et al. *Biology*. 3rd ed. McGraw-Hill, 2014.
Campbell, Neil A. *Biology*. 9th ed. Cummings, 2011.
Gilbert, Scott F. *Developmental Biology*. 10th ed. Sinauer, 2014.
Lavers, Chris. *Why Elephants Have Big Ears: Understanding Patterns of Life on Earth*. St. Martin's, 2001.
Meyr, Ernst. *The Growth of Biological Thought: Diversity, Evolution, and Inheritance*. Reprint. Belknap, 1985.
Watson, James D. *The Double Helix: A Personal Account of the Discovery of the Structure of DNA*. Reprint. Simon, 1998.

BIRDS

FIELDS OF STUDY

Ornithology; Biology; Zoology; Biogeography; Palaeontology

ABSTRACT

There are more than 8,600 known living species of birds. Each species has feathers; horny beaks; two hind limbs that allow them to walk, swim, or perch; and two forelimbs called wings, which often help them to fly or swim. Birds have anatomical systems similar to those of mammals, and both classes are warm-blooded, or endothermic. Along with other vertebrates, birds have a variety of physiological systems and sense organs, with unique adaptations designed to enable birds to fly.

KEY CONCEPTS

Archaeopteryx: the earliest known bird, known only from the fossil record; it lived during the Jurassic period
Aves: the class within the phylum Chordata to which all birds, and only birds, belong
crop: a specialized part of a bird's digestive system that holds and softens food
endotherm: an animal that, by its own metabolism, maintains a constant body temperature (is "warm-blooded"); birds and mammals are endotherms
gizzard: a part of a bird's stomach that uses ingested pebbles to grind up food
ornithology: the branch of biology that deals with the study of birds
thecodonts: extinct reptiles that lived during the Permian period; they were the ancestors of both dinosaurs and birds

FEATHERS

Within the animal kingdom is a class of animals that has more than 8,600 known living species, with each species having a feature found in no other class of animals: feathers. These animals, the birds (Aves), also have horny beaks; two hind limbs that allow them to walk, swim, or perch; and two forelimbs called wings, which often help them to fly or swim. Birds have anatomical systems that are quite similar to those of mammals, and both of these classes of animals are warm-blooded, or endothermic. Along with other vertebrates, birds have skeletal, muscular, circulatory, digestive, respiratory, urinary, reproductive, and nervous systems. They have an outer covering of skin and a variety of sense organs. Each system, however, has unique adaptations designed to enable birds to fly.

AVIAN PHYSIOLOGY

Feathers, found on all birds, are a lightweight body covering that insulates and protects the birds. The large, vaned feathers on the wings and tails provide most birds with the aerodynamic lift and maneuverability needed for flight. The shape, color, and size of some feathers is also important for camouflage, recognition, or behavioral displays. The skeletal systems of birds have many bones that are homologous to reptilian or mammalian bones, yet unique modifications give birds the strength, flexibility, and lightness to fly. Most birds have bones that are pneumatized, or filled with air. The supporting bones, however, are highly mineralized and very strong. Some bones that are found in reptiles and mammals are missing in birds; other bones are fused. There are also more cervical vertebrae in birds' necks, and they are unusually flexible, allowing great movement of the head and neck.

The circulatory systems of birds are very similar to those of mammals, and they have a four-chambered heart. The oxygen-carrying erythrocytes of birds are larger than those of mammals and, unlike those of mammals, have a nucleus. Heartbeat rates are a function of activity, environmental temperature, and size. A hummingbird's heart often beats more than one thousand times per minute.

The beak of a bird may serve a number of functions that include probing for, catching, crushing, tearing, and swallowing food. The toothless beak does not grind or chew food but passes it to the pharynx, next to the esophagus, and then to a storage chamber, the crop. Food eventually is passed to a glandular stomach, where digestion begins, and then to a muscular gizzard that, with the aid of a horny lining and grit or sand, grinds and digests it further. The gizzard leads to the intestine, where digestion is aided by secretions from the liver and pancreas. The feces empty into a cloacal chamber. The cloacal chamber is also the terminus of the urinary and reproductive systems. The cloaca is the only posterior opening in birds; since all sex organs are internal, including the penis and testes, the sex of a bird cannot usually be determined by examining this area.

John James Audubon was renowned for his detailed paintings of American birds.

The visual and auditory sense organs of birds are usually acutely developed. Senses of smell, taste, and touch, on the other hand, are often poorly developed. Pigeons can discriminate motions that are up to three times faster than any humans can distinguish; some hummingbirds can see ultraviolet light, which is invisible to humans. Eagles, hawks, and vultures can detect objects at remarkable distances; they have a retinal cell density that is twice that of humans. Oilbirds and some owls can locate prey in total darkness.

These anatomical systems, and the varied external sizes, shapes, and colors, as well as behaviors that include courtship patterns, migration, and vocalization, evolved during millions of years of avian evolution. Birds are thought to have evolved from reptiles, probably thecodonts, more than 150 million years ago. Several fossils of the earliest known bird, *Archaeopteryx*, have been dated at 150 million years old. The fossils clearly show feathers and wings on a crow-sized bird body that has teeth and some reptilian skeletal features.

TYPES OF BIRDS

More than nine hundred species of birds can be found only in the fossil record. Two extinct bird groups were especially common about ninety million years ago—the large, toothed, diving bird *hesperornis* and the gull-like *ichthyornis*. Two groups of huge flightless birds have become extinct within historical times: the giant moas of New Zealand (about 3 meters tall), and the elephant birds of Madagascar (about 2.7 meters tall).

Birds belong to the taxonomical class *Aves*. This class is divided further, first into orders—groups of birds that have broadly similar characteristics. Orders are divided into families, families into genera, and genera (sing. genus) into species.

FLIGHTLESS AND PERCHING BIRDS

There are several surviving orders of large, flightless, running birds. These include *Struthioniformes*, with a single living species that is the largest living bird: the 2.5-meter-tall ostrich of Africa. Somewhat similar in appearance, but with differences great enough for them to be placed in a different order, are the two species of rheas, belonging to the order *Rheiformes*. Some of these South American birds stand 1.5 meters tall. Australia has two superficially similar groups of birds in the order *Casuariiformes*: the three species of tropical forest-dwelling cassowaries, which are sometimes more than 1.5 meters tall, and the single species of the slightly taller emu, which roams the open plains.

Not far away, in New Zealand, are the kiwis. Their order, *Apterygiformes*, contains three species of the smallest of the primitive flightless birds. Some stand about thirty centimeters tall; they have no visible wings (only vestigial flaps), have feathers that look somewhat like hair, and may lay an egg that is one-quarter their body weight—proportionally the largest egg of any of the living birds.

The largest order of entirely flightless birds, with fifteen living species, is *Sphenisciformes*, the penguins. This order includes the emperor penguin of Antarctica, at 1.2 meters tall, and, along the equator, the most northern penguins, the Galápagos penguin, at a little less than 60 centimeters tall.

With rare exceptions, all other birds are capable of flight. The forty species of tinamou, however, of the order *Tinamiformes*, are weak fliers. They are thought to be most like the ancestor of all living birds. Tinamous superficially look like quail, and they are found only in Central and South America.

The largest order of birds, with 5,200 species—60 percent of all living bird species—is *Passeriformes*. These birds are collectively called perching birds because of their grasping, unwebbed toes—three in front and one behind—all at the same level. Most *Passeriformes* have anatomical features that allow them to sing very well, and they are commonly referred to as songbirds. This diverse order includes finches, cardinals, grosbeaks, buntings, sparrows, orioles, blackbirds, tanagers, vireos, waxwings, thrushes, kinglets, mimids, warblers, wrens, shrikes, nuthatches, creepers, titmice, chickadees, swallows, crows, jays, larks, flycatchers, and starlings. Lesser-known *Passeriformes* include broadbills, ovenbirds, woodcreepers, antbirds, tapacolos, manakins, pittas, birds of paradise, bowerbirds, bubuls, babblers, honey eaters, white-eyes, sunbirds, honeycreepers, waxbills, cotingas, wagtails, and flower-peckers.

AQUATIC BIRDS AND FOWL

Two bird orders are most often associated with the open oceans or the marine coastlines of continents or islands. *Procellariiformes*, with ninety living species, contains albatrosses, shearwaters, and petrels. They all have webbed feet, long, narrow wings, and nostrils consisting of raised tubes. The fifty-two species of the order *Pelecaniformes* are the only birds with all four toes connected by webs; most also have an expandable throat patch. This order includes pelicans, tropic birds, boobies, gannets, cormorants, frigate birds, and anhingas.

Many types of birds generally inhabit areas that border bodies of water. The bays, estuaries, salt marshes, swamps, freshwater lakes, and rivers of the world teem with a multitude of birds. The largest of these orders, with 313 species, is *Charadriiformes*. This order contains three fairly common groups of birds. The shorebirds, often seen probing their long bills into sand or mud for food, include sandpipers, lapwings, plovers, avocets, stilts, phalaropes, oystercatchers, and jacanas. The gull-like group includes the web-footed gulls, terns, jaegers, skuas, and skimmers. Several of the forty-three species of gulls can be very cosmopolitan and are one of the most familiar sights at seaports and coastal resorts. The third group is the web-footed alcids—auks, murres, and puffins. They are northern oceanic birds that come ashore (mostly to rocky shores and cliffs) only to breed.

The eighteen species of grebes make up the order *Podicipediformes*. These weak fliers are excellent divers and use their fleshy, lobed feet to dive for invertebrates or fish. The shallower waters of marshes and the edges of rivers and lakes are likely to have long-legged waders that belong to the order *Ciconiiformes*. These 110 species of herons, egrets, storks, ibises, bitterns, and spoonbills are usually fairly large birds with long necks and bills. They probe for mollusks, crustaceans, and fish, and sometimes for reptiles, amphibians, mammals, or birds.

The rails, gallinules, crakes, coots, and cranes are the aquatic groups of the 187 species in the order *Gruiformes*. Standing 1.5 meters tall, some of the cranes are among the tallest of the flying birds. The heaviest of all flying birds, the kori bustard, is also in this order. It weighs about twenty kilograms (approximately fifty pounds). The bustards, button quails, and several smaller families are terrestrial *Gruiformes* more often found in open plains or brushland.

The two smallest aquatic orders are *Gaviiformes*, the loons, and *Phoenicopteriformes*, the flamingos. The loons are excellent divers and breed only in or along cool northern waters. Flamingos, on the other hand, are found primarily in shallow, warmer, tropical, and temperate waters. With long legs and necks, these tall waders filter small aquatic organisms through their heavy, curved bill.

Anseriformes is an order of 151 species of semi-aquatic birds that are hunted by people throughout the world. Several species have been successfully domesticated. This order includes a small family of stoutly built South American birds, the screamers, which resemble pheasant with wing spurs. The larger family of waterfowl includes ducks, swans, and geese, which are found worldwide except for Antarctica. They all have webbed feet and swim well; some are also excellent divers. Another order which has been hunted and domesticated for centuries is *Columbiformes*. The gentle doves of this order have long been symbols of peace and, along with pigeons and sandgrouse, give the three hundred species of this order worldwide distribution, except for polar regions. The pet bird trade has been especially interested in several of the 317 species of *Psittaciformes*. The birds of this order are primarily tropical or subtropical and include parrots, cockatoos, lories, parakeets, and macaws.

The bird order that has been the most exploited, however, both by hunting and domestication for food and eggs, is *Galliformes*, the fowl-like birds. The 253 species of this order include pheasant, grouse, quail, chickens, turkeys, guinea fowls, ptarmigans, megapodes, partridges, curassows, guans, chachalacas, and the hoatzin.

PREDATOR BIRDS AND OTHER ORDERS

Man is not the only efficient predator of birds: Many of the 271 species of the order *Falconiformes* are skilled at grasping a duck off a pond, a pigeon from the air, or a sparrow from a bird feeder. The order *Falconiformes* includes hawks, eagles, falcons, kites, kestrels, caracaras, vultures, condors, buzzards, and the osprey and secretary bird. All have strong, sharp claws and bills and are powerful fliers.

Another order of powerful predators of vertebrates is *Strigiformes*, the owls. The 131 species of owls are largely nocturnal, have large eyes with binocular vision, hear exceptionally well, and fly silently with unusually soft feathers. They do not tear mammals apart as the *Falconiformes* do, but usually swallow their prey whole and later regurgitate a pellet of hair and bones.

Some 1,200 more species are included within another seven orders of birds. The order *Cuculiformes*, containing 143 species, includes the cuckoos, anis, roadrunners, and coucals. Most taxonomists also place the brightly colored touracos of Africa in this order. Most of the ninety-three species of the order *Caprimulgiformes* are either nocturnal or primarily active at twilight. This order includes night-hawks, nightjars, whippoorwills, frogmouths, potoos, and the oilbird.

Two small orders of medium-sized birds with long tails are *Coliiformes* (mousebirds) and *Trogoniformes* (trogons). While the six species of mousebirds are found only in Africa, the thirty-five species of trogons are distributed throughout tropical Africa, America, and Asia. Although the 191 species of the order *Coraciiformes* range from small to large in size, most are brightly colored and have a conspicuous bill. The kingfishers, todies, motmots, bee-eaters, rollers, hoopoes, and hornbills make up this order.

The order *Apodiformes* contains two superficially different groups of birds—the swifts and the hummingbirds. The 69 species of swifts are found in most areas of the world; the 319 species of hummingbirds live only in temperate and tropical America. Another large but more diverse order is *Piciformes*. These 374 species include woodpeckers, piculets, jacamars, puffbirds, barbets, honey guides, and toucans. Some woodpeckers are cosmopolitan and regularly frequent bird feeders.

TRACKING AND STUDYING BIRDS IN THE FIELD

Ornithologists, the scientists who study birds, have been assisted by millions of people throughout the world in their studies of bird populations and their movements. One of the largest bird projects is the Christmas bird count, sponsored by the National Audubon Society. This North American bird study has been done yearly since 1900 and involves more than forty thousand people each year. Field observers are given one day each winter to count all the birds they can find in a specified area. The information compiled about numbers of birds and species distributions is published in *American Birds*.

More than a million birds in North America are captured each year, fitted with a small aluminum band that has a unique number, and released unharmed. Some are recaptured later, sometimes years later, and perhaps hundreds or thousands of miles from the place they were banded. This technique gives information about the sizes of bird populations, mortality rates, longevity, migration routes and times, sex ratios, and age distributions. Some banders study avian parasites and sequences of feather molting. Hunters and the general public also contribute valuable information when they report dead banded birds.

Sophisticated electronic equipment has been used in many avian studies. Radio transmitters have been attached to some of the larger birds, and their individual movements have been recorded. Radar has been used to monitor both the movements of individual birds and the progress of flocks. Bird songs have been recorded and converted into graphic sonograms, which can show song differences between individuals of the same species.

Many museums have large collections of preserved or stuffed birds and bird skeletons, nests, and eggs. These may be used for anatomical or biochemical studies of birds. Taxonomists use these collections, along with studies of recently collected and live birds, to recognize species and arrange them into various classification groups.

Paleontologists studying the fossils of extinct birds have classified some nine hundred species of birds that are no longer alive. Fossils have given direct evidence about the relatedness of many bird groups; however, because they usually contain only information about bone structure, they provide limited data about species and must be interpreted carefully.

STUDYING BIRDS IN THE LAB

As well as considering the appearance and anatomical structures of birds, ornithologists now consult many areas of biology to put together the various taxonomic groups. Fieldwork has provided enormous amounts of information about the ecology, behavior, singing, breeding habits, and biogeography (location) of birds. In the nineteenth century, Charles Darwin used the structure, function, and biogeography of the finches on the Galápagos Islands, Ecuador, to work out the species of these birds. These studies contributed significantly to the development of his theory of evolution by natural selection.

Fieldwork has also helped resolve many problems that have developed in the classification of birds. In 1910, bird books listed nineteen thousand species of birds. Several thousand new kinds have been discovered since then, yet now only 8,700 species are recognized. Many studies have shown that two groups of birds that look different and were at one time considered different species are in fact able to mate and produce offspring. These two groups are then reclassified as a single species. Each original group is considered a subspecies, race, or variety, depending on how similar they are to each other. In North America, for example, the Baltimore oriole and yellow-shafted flicker are common in eastern areas, and the Bullock's oriole and red-shafted flicker are common in the west. Each type of bird looks different and can easily be recognized, so they were initially classified as four species. As towns developed in mid-America, patterns of vegetation changed, and the orioles and flickers moved with the towns until the eastern and western populations finally met. Field studies in the twentieth century showed that the two flicker types were mating and producing offspring; the two orioles were, as well. The taxonomy was adjusted to reflect this new information, and now the two flickers are considered a single species, the northern flicker. Similarly, the two orioles were combined into one species, the northern oriole.

Avian biochemistry and genetics are also areas of intensive research. Studies of the size, shape, number, and staining patterns of chromosomes have been used to show the genetic relatedness of many bird species. Just as feathers, bones, and chromosomes have been shaped by the long evolutionary history leading up to each living bird, so have molecular structures been modified over time. Bird proteins have been analyzed and compared using a variety of techniques, including spectrophotometry, electrophoresis, antibody-antigen reactions, and amino acid sequencing. Studies of avian deoxyribonucleic acid (DNA) using electrophoresis, DNA hybridization, and recombinant DNA technology have also added new insights into the evolutionary relatedness of bird species. Because DNA controls the structure of all proteins and all physical structures in a bird (or in any animal), many researchers think the comparison of DNA structure provides the best overall view of species relatedness.

POPULAR ORNITHOLOGY

There has probably been no other area of scientific inquiry that has attracted the interest of the public as much as has the study of birds. There has undoubtedly been no other area of science where the public has made such significant contributions. From the forty thousand Christmas bird count participants to the two thousand bird banders in North America, millions of interested nonscientists around the world have collected a wealth of information about birds.

Birds are often described as environmental indicators, and their population numbers and health are monitored to get an idea of overall environmental integrity. The decline in the number of migrant warblers in North America, for example, is described by some ornithologists as an indication of environmental changes in their wintering grounds in tropical America, whereas other ornithologists believe that it reflects deterioration of their breeding grounds in North America.

Conservation projects throughout the world have used bird studies as a means of focusing on when it has become necessary to fight for the protection of a particular species, subspecies, or race of birds. Major progress has been accomplished with ospreys, peregrine falcons, bald eagles, brown pelicans, trumpeter swans, whooping cranes, Kirtland's warblers, and other birds. Yet each project is time-consuming and costly and must be carefully carried out. The governmental designations "endangered species" and "threatened species" cover only very narrow, clearly defined taxonomic groups.

Birds have fascinated people—scientifically, aesthetically, and emotionally—perhaps more than any other group of animals. The vast data collected and the complex taxonomical classification of birds is something that can be passed on to generations to come. It can be hoped that an environment where these creatures can prosper will be passed on as well.

—*David Thorndill*

FURTHER READING

Austin, Oliver L., Jr. *Families of Birds*. Rev. ed. Golden Press, 1985.

"Aves." *ITIS*, Integrated Taxonomic Information System, 2013, www.itis.gov/servlet/SingleRpt/SingleRpt?search_topic=TSN&search_value=174371#null. Accessed 31 Jan. 2018.

Brown, Jerram L. *Helping Communal Breeding in Birds: Ecology and Evolution*. Princeton University Press, 2014.

Chatterjee, Sankar. *The Rise of Birds: 225 Million Years of Evolution*. 2nd ed. John Hopkins Press, 2015.

Gill, Frank B. *Ornithology*. 2nd ed. W. H. Freeman, 1995.

Gotch, A. F. *Birds: Their Latin Names Explained*. Blandford, 1981.

Hickman, Cleveland P., Larry S. Roberts, and Frances M. Hickman. *Integrated Principles of Zoology*. 11th ed. McGraw-Hill, 2001.

Margulis, Lynn, and Karlene V. Schwartz. *Five Kingdoms: An Illustrated Guide to the Phyla of Life on Earth*. 3rd ed. W. H. Freeman, 1999.

Pough, F. H., J. B. Heiser, and W. N. McFarland. *Vertebrate Life*. 5th ed. Prentice Hall, 1999.

Birth

FIELDS OF STUDY
Reproductive Biology; Genetics; Veterinary Science

ABSTRACT
Animals may be born via parturition after an internal pregnancy (a gestation period) or from an egg that hatches externally. This includes eggs that are spawned and then fertilized externally by organisms such as fish; those fertilized internally and laid in huge numbers to hatch on their own, as in snails and millipedes; or those laid in much smaller numbers and incubated by their parents, as in birds.

KEY CONCEPTS
gestation: the term of pregnancy
hormone: a substance produced by one organ of a multicellular organism and carried to another organ by the blood, which helps the second organ to function
larva: a newly hatched form of an organism that looks very different from adults of the species and must undergo metamorphosis to the adult form
metamorphosis: the form changes in a larva that turn it into the adult form
motile: able to move about spontaneously
oviparous: born from an externally incubated egg
parthenogenesis: a process whereby a female sex cell develops without fertilization, in an organism that reproduces sexually
uterus: the organ in which fertilized eggs develop during gestation
viviparous: born alive after internal gestation
zygote: a fertilized egg

WHERE DO BABIES COME FROM?

Animals are born in one of two ways. They may be born via parturition after an internal pregnancy (a gestation period). They may also be born from an egg that hatches externally. This includes eggs that are spawned and then fertilized externally by organisms such as fish; those fertilized internally and laid in huge numbers to hatch on their own, as in snails and millipedes; or those laid in much smaller numbers and incubated by their parents, as in birds.

In all animals, male and female reproductive cells (gametes) unite to form a single cell, known as a zygote. The zygote then undergoes successive cellular divisions, as well as cellular differentiation, to form a new organism. In most higher animals, individuals of a species are male or female, according to the type of reproductive cells they produce. Male reproductive cells—the sperms—are motile cells, with heads containing nuclei and tails that allow them to move. Female reproductive cells—eggs or ova—are round cells many times larger than sperms. They also contain large amounts of cytoplasm located around the nucleus.

VIVIPAROUS BIRTH

In viviparous organisms, a fertilized egg will develop into an incompletely finished miniature or miniatures of an adult of the same species. After fertilization of an egg, the zygote enters the uterus, undergoes both cell division and differentiation, and forms an embryo within the mother. In due time parturition (birth) occurs. Most viviparous organisms are mammals. Early in gestation the implanted dividing egg and the uterine wall become interconnected by a placenta, composed of both maternal and embryonic tissue. The placenta brings oxygen and nutrients to the embryo and carries away wastes. The transfer of

The mode of reproduction in birds combines internal fertilization with oviparous development. Here a Montagu's harrier chick has just hatched from its egg. (Sarinahornay)

nutrients uses the circulatory systems of both the mother and the embryo.

At the time of birth, hormonal changes cause the mother's birth canal to enlarge, the muscles of the uterus to rhythmically contract, and the embryo is expelled as a newborn. The overall process can be exemplified with the female gorilla. She menstruates monthly and can mate successfully at any time of year. Her gestation period is 9.5 months and yields one or two almost fully formed offspring. Gestation is much shorter in smaller primates and there are variations in the difficulty of parturition, related to the headfirst entry of young into the world.

Some primates—including humans—must undergo major dilation of the uterine mouth (cervix) before parturition can begin. This allows the large head of the fetus to pass out of the body safely. In species such as monkeys, in which the head of the fetus is close to the size of the cervical opening, far less dilation is needed. In other placental mammals, the position of fetuses in the uterus and the fashion of birth differ.

OVIPAROUS BIRTH

Many animals, including snails, insects, birds, lizards, and fish, lay eggs either before they are fertilized or before their young are completely developed. These organisms are termed oviparous. In the case of snails, most species are hermaphrodites. This means that each snail has both male and female sex organs. However, each individual snail usually mates with another snail of the same species, passing sperm to its partner and getting sperm from the partner. Fertilized eggs are then spawned into the water or laid on rocks or aquatic plants. The eggs hatch in two weeks to two months. Hatching is considered to be the time of birth of the young snails. In most cases, offspring hatch as miniature replicas of their parents.

In insects, eggs are laid in a wide variety of places. For example, grasshoppers lay eggs in the ground or on plants. When the offspring are born, they hatch as wingless grasshopper larvae, called nymphs. Over several months the nymphs undergo metamorphosis to adult locusts. In contrast, ants, wasps, and termites lay their eggs in special chambers in their nests (or colonies). Worker termites place eggs laid by a colony's queen into hatching chambers in "nurseries."

Termites are born as wormlike larvae when eggs hatch. The larvae undergo metamorphosis into workers, soldiers, or reproductives (kings or queens) as a result of being fed varied amounts of hormones obtained from queens.

Birds lay eggs in nests, that are located in a wide variety of locales depending on species. Adults then incubate the eggs by sitting on them. Offspring are born when they use a specialized egg tooth to break open their egg shells. In the case of lizards, the eggs are laid after they are fertilized. However, they are not cared for by parents. Large lizards such as alligators, crocodiles, and caimans lay eggs covered with hard, calcium-containing shells like those of bird eggs—reptiles and birds are distant relatives—in holes in the ground, where they hatch into offspring that look like adults. Most fish lay fertilized eggs on plants or on the bottom of the sea, lakes, or rivers, and leave their offspring to hatch on their own. These offspring then develop into adults.

OVOVIVIPAROUS BIRTH

Ovoviviparous animals produce eggs in shells like those of the oviparous organisms but the eggs are hatched within the body of the mother, or by expulsion from her body. There are numerous examples of ovoviviparous organisms among animals. They include some oysters, snails, and other gastropods, as well as numerous species of sharks, and the live-bearing tropical fish such as the guppy or swordtail. The eggs of live-bearing guppies hatch internally, just before leaving the mother's body, and the young are born alive. These young fish usually leave the body of the mother head first.

In all cases, the development of the egg or eggs of ovoviviparous species begins with internal fertilization of the female of the species. Then, the zygotes formed pass through many cycles of internal cell division and differentiation. Ultimately, each egg yields a miniature of the adult organism involved. However, there is no placenta formed and the zygote becomes the complete organism in processes that depend on a yolk sac for food and energy. Often, upon birth, the newborn organism has part of its yolk sac left and can survive for one or several days without eating.

—*Sanford S. Singer*

Further Reading

Dekkers, Midas. *Birth Day: A Celebration of Baby Animals.* W. H. Freeman, 1995.

Hayes, Karen E. N. *The Complete Book of Foaling: An Illustrated Guide for the Foaling Attendant.* Howell Book House, 1993.

Pinney, Chris C. *Veterinary Guide for Dogs, Cats, Birds, and Exotic Pets.* Tab Books, 1992.

Prine, Virginia Bender. *How Puppies Are Born: An Illustrated Guide on the Whelping and Care of Puppies.* Howell Book House, 1975.

Spaulding, C. E., and Jackie Clay. *Veterinary Guide for Animal Owners: Sheep, Poultry, Rabbits, Dogs, Cats.* Rodale Press, 1998.

Bone and Cartilage

Fields of Study

Osteology; Physiology; Biology

Abstract

Bone is the hard substance that forms the supportive framework of the bodies of all of vertebrate organisms. This framework, the skeleton, is composed of hundreds of separate bones. The bones support the bodies of vertebrates and protect their delicate internal organs, such as the brain, lungs, and liver, from injury. In addition, the muscles attached to the bones actuate them as levers to enable their function in diverse actions. Bone tissue serves as the main repository for calcium in the body, and contains the sites where the red blood cells are made.

Key Concepts

articular: pertaining to bone joints

bone: the dense, semirigid, calcified connective tissue which is the main component of the skeletons of all adult vertebrates

calcification: calcium deposition, mostly as calcium carbonate, into the cartilage and other bone-forming tissue, which facilitates its conversion into bone

cartilage: elastic, fibrous connective tissue which is the main component of fetal vertebrate skeletons, turns mostly to bone, and remains attached to the articular bone surfaces

collagen: a fibrous protein very plentiful in bone, cartilage, and other connective tissue

connective tissue: any fibrous tissue that connects or supports body organs

osteoblast: a bone cell which makes collagen and causes calcium deposition

periosteum: the fibrous membrane which covers all bones except at points of articulation, containing blood vessels and many connections to muscles

BONES OF CONTENTION

Bone is the hard substance that forms the supportive framework of the bodies of all of vertebrate organisms. This framework, the skeleton, is composed of hundreds of separate parts called bones. The bones support the bodies of vertebrates and protect their delicate internal organs, such as the brain, lungs, and liver, from injury. In addition, the muscles are attached to the bones, which act as levers to enable their function in actions as diverse as walking or swallowing. Furthermore, bone provides the calcium needs of the body, serves as the main repository for calcium storage, and contains the sites where the red blood cells are made.

Much of the bone in adult vertebrates derives from cartilage, an elastic, fibrous connective tissue which is the main component of fetal vertebrate skeletons. Such bone, for example, that of the long bones, is called cartilage bone. Cartilage is an extracellular matrix made by body cells called chondrocytes. It is surrounded by a membrane, the periosteum, and much of its firmness and elasticity arises from plentiful fibrils of the protein collagen that it contains. These fibrils and their many interconnections provide mechanical stability and very high tensile strength, while allowing nutrients to diffuse into the chondrocytes to keep them alive. The blood vessels which surround the cartilage in the periosteum

Skeleton of a Madagascar chichlid (Paretroplus polyactis) *treated with chemical dyes by American Museum of Natural History ichthyologist John S. Sparks.*

provide all of the needed nutrients and remove the cellular waste materials produced by life processes.

The cartilage-containing skeletons of newborn vertebrates become cartilage bone by ossification, a process that includes calcification, chondrocyte destruction, and replacement by bone cells, which lay down more bone. This cartilage, called hyaline cartilage, remains at the articular sites of bones. In young vertebrates, cartilage is the site for the continued growth and calcification that produces the bone lengthening required for the attainment of adult size and stature. In addition to the cartilage bone, so-called membrane bone occurs exclusively in the top portion of the skull.

Bone is thought to have developed over a half billion years ago, as shown by its presence in the fossils of fishlike carnivores of that time period. In those creatures, it seems to have been formed into interconnected external plates covering their bodies as sheaths that strengthened and protected their bodies. The existence of bone only at the surfaces of these fossils has led many scientists to suppose that the first function of bone was protection, rather than body support. Be that as it may, bone has both functions in modern organisms. It is interesting to note that many of these early organisms lacked bone in their heads. It seems possible that this lack may have led to the development of the separate mechanisms for formation of membrane bone and cartilage bone, different means to the same end.

PHYSICAL CHARACTERISTICS OF BONE

To best serve their biofunctions, bones must be very hard, strong, and rigid, but remain supple enough to stay unbroken under normal conditions. These characteristics are provided by the collagen fibrils and insoluble calcium phosphate which make up the bones. The bones must also be light enough to allow vertebrates to move easily and remain erect.

Overly heavy bones are prevented by the occurrence of two general types of bone tissue. The first of these is compact bone, the portion most familiar because it makes up the hard exterior of many bones, except at their very ends. The second bone type, cancellous bone, which appears spongy, is found at the ends of long bones and inside them. It serves to lighten the bones, acting in the same fashion as the air-filled sinuses of the skull, which diminish the overall weight of the skull without weakening it. Bones are covered on their outsides by the important fibrous membrane called the periosteum, along with cartilage. Their insides are lined by an endosteum membrane, very similar to the inner layer of the periosteum.

It is also useful to think of bones in terms of woven, lamellar, and osteonic forms. These terms indicate the relative number of cells in a bone matrix region and the arrangement of collagen fibers in the region. Collagen fibers of woven bone crisscross within the bone matrix, and its bone cells are distributed randomly. In lamellar bone the collagen fibrils are more ordered and fewer bone cells are present. Osteonic bone is also well-organized. However, its cells are found in concentric rings, with narrow channels (Haversian canals) inside them. A blood vessel passes through each canal and feeds the concentric cell rings formed around it. The bone layers form from the outside in, within the internal bone cavity. This narrows its diameter more and more. A Haversian canal and its rings develop when cancellous bone is converted into compact bone.

Bones are either "long" or "short" bones. Most long bones are located in the arms and legs. They are divided into three parts: a shaft (the diapysis), the long central part of the bone; a flared portion at each bone end (the metaphysis); and a rounded bone end (the epiphysis). The short bones, designed for flexibility, include those in the skull, spine, hands, and feet. The centers of bones—medullary cavities—are most often filled with either red or yellow bone marrow. The yellow marrow is mostly fat. Red marrow is a network of blood vessels, connective tissue, and blood-cell-making tissue. Red blood cells (erythrocytes) are made in this red marrow. Each bone has nerves that stimulate it and blood vessels that supply nutrients and take away wastes.

BONE COMPOSITION, DEVELOPMENT, AND REMODELING

Between 66 and 70 percent of bone is an inorganic mineral composite made of calcium phosphate and calcium carbonate, which is mostly hydroxyapatite. Much of the remainder of bone is the fibrous protein collagen. This mineral and protein together are called the bone matrix. Within the bone matrix are the three types of specialized cells which ensure its formation, remodeling as is needed, and continuity throughout life. The first cell type, the osteoblast, produces the bone matrix and surrounds itself with it, synthesizing collagen and stimulating mineral deposition. The second cell type, the osteocyte, is a branched cell that becomes embedded in bone matrix, is interconnected, and acts in the control of the mineral balance of the body. Finally, the osteoclast cells destroy the bone matrix whenever it is remodeled during skeleton growth or the repair of bone breaks and bone fractures.

The stepwise conversion of cartilage into bone begins when the chondrocytes of hyaline cartilage enlarge and arrange themselves in rows. This is followed by the synthesis of collagen fibers, and mineral deposition around them. Just below the inner surface of the periosteum a vascular membrane—the perichondrium—forms and supplies the osteoblasts needed for bone formation. Simultaneously, osteoclasts excavate layers through the bone layer and set the stage for the formation of additional bone.

All the bones in the bodies of the vertebrates change their sizes and shapes as these organisms pass through their lives. The processes involved are collectively called remodeling. An example of such change is the growth of the long bones in circumference as the limbs grow from puberty to adulthood. In the course of such bone growth the periosteum provides the osteoblasts required to deposit bone matrix around the bone exterior and to calcify it. At the same time the endosteum-derived osteoclasts often dissolve bone in the interior, thus enlarging the marrow cavity.

Remodeling in such cases occurs in response to biosignals including those caused by increases in the need for bone to bear additional weight or to anchor increased muscle mass. Conversely, inactivity and the lack of exercise can result in remodeling which produces diminished bone mass. The complex changes involved in bone remodeling are also controlled by vitamin D and hormones originating in the pituitary gland, the thyroid gland, and the parathyroid glands. Abnormalities in bone growth and remodeling are associated with a great many bone diseases, ranging from rickets to bone cancer.

—Sanford S. Singer

Further Reading

Alexander, R. McNeill. *Bones: The Unity of Form and Function.* Reprint. Westview, 2000.

Hukins, David W. L., ed. *Calcified Tissue.* CRC Press, 1989.

Murray, Patrick D. F. *Bones: A Study of the Development and Structure of the Vertebrate Skeleton.* Cambridge, England: Cambridge University Press, 1936. Reprint. Cambridge University Press, 1985.

Rosen, Vicki, and R. Scott Theis. *The Cellular and Molecular Basis of Bone Formation and Repair.* R. G. Landes, 1995.

Siebel, Markus J., Simon P. Robins, and John P. Bilezikian, eds. *Dynamics of Bone and Cartilage Metabolism.* Academic Press, 1999.

Vaughan, Janet Maria. *The Physiology of Bone.* 3rd ed. Clarendon Press, 1981.

Brain

Fields of Study

Physiology; Comparative Anatomy; Biology

Abstract

Using information from sensory receptors and responding to changes in the environment are generally managed by a nervous system of some sort, usually with a center where processing occurs, a brain or brainlike structure. Invertebrate nervous systems are generally very primitive and may contain only a very rudimentary brainlike structure. Some animals, however, are so structurally simple that they have no neural processing center at all.

Key Concepts

brainstem: lowest or most posterior portion of the vertebrate brain, including midbrain, pons, and medulla oblongata; controls "housekeeping" functions such as breathing and heartbeat

cell body: the central portion of a neuron, containing the nucleus, where most processing and integration of information occur

cerebellum: second largest part of the brain, manages fine muscle control and muscle memories

cerebrum: largest part of most vertebrate brains, with areas that control vocalizations, vision, hearing, smell, and taste, as well as voluntary skeletal muscle movements

cortex: thin layer of gray matter that covers surfaces of the cerebrum and cerebellum

ganglia: clustered cell bodies of neurons that may form a brain-like center in lower animals

gray matter: region of the brain or spinal cord that contains cell bodies of neurons, where information processing and storage occur

white matter: region of neural tissue that contains axons of neurons that carry electrical nerve impulses from one processing center to another

IF I ONLY HAD A BRAIN OR A BRAINLIKE STRUCTURE

Animals are multicellular organisms that obtain their nutrients by eating or ingesting other organisms, and many have locomotor abilities. Obtaining food and avoiding being eaten are behaviors enhanced by the ability of an animal to tell what is going on in its surroundings. Using information from sensory receptors and responding to changes in the environment are generally managed by a nervous system of some sort, usually with a center where processing occurs, a brain or brainlike structure. Invertebrate nervous systems are generally very primitive and may contain only a very rudimentary brainlike structure. Some animals, however, are so structurally simple that they have no neural processing center at all.

INVERTEBRATES WITH AND WITHOUT BRAINS

Sponges (phylum *Porifera*) are invertebrates with no brain or nervous system of any sort, in either the sedentary adults or the free-swimming larvae. Stimuli

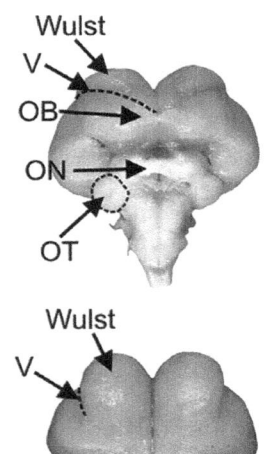

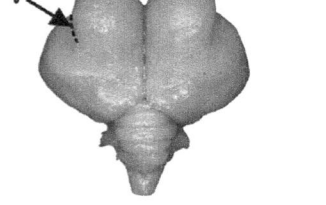

Visual processing areas of the brains of four species of birds. OT: optic tectum; ON: optic nerve; OB; olfactory bulb; V: vallecula.

received at the body surface produce responses (movements) directly, over the entire body, in these and related lower metazoan animals. Other primitive invertebrates such as hydra, jellyfish, corals, and sea anemones (phylum *Cnidaria*) have one or more nerve nets. For these radially symmetrical animals, food or danger can come from any direction in the water, and the meshlike nervous system can respond directly without a central control region. Some jellyfish also have a nerve ring that helps coordinate their movements, but no brain.

Bilaterally symmetrical invertebrates include the flatworms (phylum *Platyhelminthes*), roundworms (*Nematodes*), mollusks (*Mollusca*), segmented worms (*Annelida*), and insects and their relatives (*Arthropoda*). Most of these show cephalization, the presence of an anterior head containing the main processing center of the nervous system, specialized sensory receptors, and the mouth. Echinoderms (*Echinodermata*) such as starfish are bilaterally symmetrical as larvae, but develop radial symmetry as adults, when they lack a head. Some mollusks are not symmetrical as adults, despite the bilateral symmetry of the larvae.

Among the flatworms, some have only nerve nets like those of cnidarians, while more complex planarians, tapeworms, and flukes generally have one or more pairs of ladderlike longitudinal nerve cords with ganglia at the head. These ganglia are clusters of cell bodies of neurons, the most primitive form of a brainlike structure. Nematodes or roundworms have a nerve ring and anterior ganglia organized around the anterior digestive tract, with nerve cords extending toward the head and tail from this center. Mollusks include clams and oysters (class *Bivalva*), snails and slugs (*Gastropoda*), and octopuses and squid (*Cephalopoda*). These animals have nervous systems that vary from simple and relatively uncephalized nerve rings and nerve cords, to a more centralized system with at least four pairs of ganglia.

Octopuses and squid have the most complex nervous systems of the mollusks and are the most intelligent invertebrates. The relatively large cephalopod brain contains many clustered or fused ganglia that manage sensory information from complex eyes and produce motor instructions for extremely rapid muscular responses. Giant nerve fibers in squid are the largest neurons known in any animal, up to one millimeter in diameter in a single cell, and are able to conduct rapid impulses that allow lightning-fast movements. Extensive studies of these neurons' structure and function have provided scientific insights that are also applicable to human neurons. Gastropods and cephalopods may show extremely complex behaviors, such as homing, territoriality, and learning. An octopus can have as many as thirty functional brain centers, some of which are memory banks used for experiential learning.

Annelids such as earthworms and leeches have paired cerebral ganglia near the mouth, connected by a solid ventral nerve cord to smaller paired ganglia in each body segment. Giant nerve fibers in the nerve cord allow rapid responses to escape from threats using reflex actions and patterned behavior. Earthworms can be taught to travel a maze by simple associative learning, in which repeated stimuli become linked to a specific behavior pattern, but this learning requires many repetitions and disappears within a few days if not reinforced.

Arthropods include spiders, scorpions, ticks, and mites (class *Arachnida*), lobsters, crabs, and shrimp (*Crustacea*), and insects (*Insecta*). The nervous system in arthropods is similar to that of annelids in its segmentation, but it is much more complex, and the anterior ganglia tend to be fused into a true brain. Many arthropods have giant neurons like those of some mollusks and annelids, capable of rapid nerve impulse transmission for efficient muscle control. Insects in particular, especially ants and bees, are capable of complex learning and very intricate social behavior. Habituation allows individuals to learn to ignore repeated stimuli that do not produce harmful effects, and cockroaches and ants can learn to run mazes.

Sea urchins, sand dollars, sea stars (starfish), and sea cucumbers are echinoderms, in which the bilaterally symmetrical larvae develop a secondary radial or biradial symmetry as they mature. The resulting radial nervous system is not greatly centralized, as there is a mouth but essentially no head. The nervous system consists of a nerve ring around the mouth that is connected to radial nerves and a nerve net. Thus, behavior generally involves only localized responses to stimuli, as along one arm of a starfish.

EVOLUTIONARY DEVELOPMENT OF THE VERTEBRATE BRAIN

The location of most animals' brain or brainlike organ at the anterior or superior end of the body is important, since it places the brain at the leading end of the moving animal or at its highest point. Many sensory receptors are located in the head, and information from the eyes, ears, and nose can be rapidly received and processed if the processing center is in the same region.

In vertebrates, the nervous system is much more advanced than the primitive systems of invertebrates. The vertebrate brain is an anterior enlargement of the dorsal hollow nerve cord that develops above the notochord in all chordates. This swelling of the nerve cord allows development of a large collection of neurons that receive, process, and store information, and determine what the organism's response to that information will be. The central nervous system consists of the brain at the anterior end of the nerve cord and the spinal cord behind it, encased in a skull and vertebral column of bone or cartilage. The rest of the vertebrate nervous system is called the peripheral nervous system, with nerve fibers bundled into nerves. Clusters of the cell bodies of neurons in the central nervous system are called nuclei, while the same kind of clusters in the peripheral nervous system are called ganglia.

The components of the vertebrate embryonic brain are divided into three areas or primary vesicles, known as the forebrain (proencephalon), midbrain (mesencephalon), and hindbrain (rhombencephalon). As development occurs, the three primary vesicles form five secondary vesicles that continue to develop into the mature brain structures. The forebrain becomes subdivided into the telencephalon, which matures into the cerebrum, and the diencephalon, which contains the thalamus and hypothalamus. The midbrain does not undergo further developmental separation. The hindbrain develops into the metencephalon, which will form the pons and cerebellum, and the myelencephalon, which becomes the medulla oblongata that is connected to the spinal cord. The lower or posterior part of the brain is called the brain stem, consisting of the medulla oblongata, pons, and midbrain, which manages the most primitive functions required for life. Higher brain functions reside in the cerebrum, particularly in the outer cortex of gray matter on its surface. The cerebellum coordinates skeletal muscle or motor activities, while the diencephalon processes and sends on sensory information to the cerebrum and cerebellum, as well as being the center of autonomic or visceral motor control.

The different classes of vertebrates are grouped into subphylum Vertebrata within phylum *Chordata*, with the main classes including cartilaginous fish (*Chondrichthyes*), bony fish (*Osteichthyes*), amphibians (*Amphibia*), reptiles (*Reptilia*), birds (*Aves*), and mammals (*Mammalia*). The brains of fish and amphibians are relatively primitive as compared to those of other vertebrates, with the main control over body functions handled by the medulla oblongata, the oldest part of the vertebrate brain. Olfactory lobes for processing sensations of smell and perhaps also taste, located in the cerebrum, and optic lobes for vision, in the diencephalon, are large in comparison to other parts of the brain, and responses are generally reflexive.

Animals that lay eggs with shells, called amniotes and including reptiles and birds, are adapted to the rigorous requirements of life on land, and have larger, more complex brains than fish and amphibians. The amniote brain has a larger telencephalon and is able to process and store more information about the land environment, which is much more likely to vary than is a watery environment. In addition to having a larger telencephalon, the brain contains more gray matter that is closer to the brain surface in amniotes than in fish or amphibians.

Mammals and some reptiles have much or all of the surface of the cerebrum covered in gray matter, which forms a structure called the cerebral cortex. The evolutionarily newer portion of this cortex is called the neocortex, while the older part is called the paleocortex. The paleocortex is the control center for drive-related behaviors, such as activities associated with feeding (licking, chewing, swallowing), sexual behavior, and primitive emotions (anger, fear). The limbic system occupies the paleocortex, which is sometimes called the reptilian brain, because it is the highest brain area present in reptiles and governs nearly all their behaviors. The neocortex is a "higher" control area that is well developed even in primitive mammals, but it is most completely expressed and covers the entire cerebral surface in humans. In cetaceans (whales and dolphins) and primates, the neocortex is the center of higher learning, logical thinking, and storage of many memories. The activities of the neocortex can override the more primitive responses of the paleocortex under most conditions, but when the higher brain areas are inactive, as in alcoholic intoxication in humans or when removed surgically in experimental animals, the lower areas reassert themselves and take control, often causing inappropriate behaviors.

Mammalian brains have convolutions on the surface of the cerebrum and cerebellum, with the neural cortex following and covering every "hill" and "valley" of the convolutions. This provides a much greater surface area occupied by gray matter, especially in humans, the species in which the convolutions and cerebral cortex are most extensive. Below the gray matter surface is white matter, myelinated neuron fibers that carry information from one area of gray matter to another. Deep to this white matter are basal nuclei, gray matter centers that help regulate subconscious and involuntary control of body functions.

The gray matter of the cerebrum in birds is nearly all in the deep basal nuclei, which are relatively much larger than they are in mammals, and in an overlying gray area specific to birds called the hyperstriatum. The avian brain lacks a neocortex entirely, with no equivalent of the cerebral cortex present. The area of the basal nuclei called the corpus striatum is apparently the center for complex behavior patterns, while the hyperstriatum manages learning and memory.

Humans have been found to show lateralization of the brain, where one side of the cerebrum (left) controls language production and interpretation, while the other side (right) controls spatial awareness and artistic creativity. This lateralization is not generally seen in other vertebrates, but recently it has been observed in some birds, where memories of song patterns and migratory homing directions are located in gray matter areas on specific sides of the brain.

Because the vertebrate central nervous system develops from a dorsal hollow nerve cord, the anterior end of its hollow, fluid-filled central canal enlarges into four ventricles or spaces. These are the first and second or lateral ventricles of the cerebral hemispheres, the third ventricle within the diencephalon, and the fourth ventricle associated with the pons, medulla oblongata, and cerebellum. The midbrain retains a simple canal called the cerebral or mesencephalic aqueduct that connects the third and fourth ventricle spaces.

The fluid that fills the canal and ventricle spaces is cerebrospinal fluid (CSF), produced by filtration of fluids from the blood at specialized capillary beds called choroid plexuses within the ventricles. Besides filling the hollow spaces of the central nervous system, CSF also washes over the surfaces of the brain and spinal cord in an area below the arachnoid layer, one of the central nervous system's coverings or meninges. It provides protection against traumatic injury, delivers nutrients, removes wastes, and helps regulate neurochemicals for the central nervous system.

THE PRIMATE BRAIN

Primates, the order of animals that includes monkeys, apes, and humans, contains species that show a higher level of brain development than most other mammals. Primate brains, especially in humans, are among the largest in the animal kingdom, compared to the body size of the animal. The primate brain retains in its structure the earlier forms and functions that have developed in lower vertebrates over evolutionary time, such as the brainstem and limbic system, but higher areas give new and more complex possibilities for learning and behavior.

Because humans are upright, bipedal walkers, the human brain is at the top of the spinal cord rather than somewhat in front of it as in other primates. The human brain weighs only about 1.25 kilograms, or about 2 percent of the weight of a 75-kilogram individual, but that is still larger relative to body size than the brains of other primates, even chimpanzees. The cerebrum makes up about 87 percent of the volume of the brain, and the cerebellum occupies most of the remaining volume. The diencephalon and brainstem in primates are relatively smaller than in most other mammals, compared to the entire brain size. The cerebral cortex in humans contains only six layers of cell bodies in the gray matter on the surface of the cerebrum. Many axons extending down from these cell bodies into the underlying white matter cross-connect the neurons that receive stimuli, process information, determine responses, and store memories. Specific areas of this cerebral cortex determine the body's voluntary muscle actions, or receive and analyze sensory information from the skin, muscles, and joints. Other cortical areas process incoming information about smell, taste, vision, and hearing, and compare those sensations to previous memories or store them as new memories.

The most "human" aspect of the brain is the prefrontal cortex of the cerebrum, where logical analysis, predictions of the results of specific actions, and social interactions take place, although even in monkeys and apes the front of the brain manages social awareness and behavior. Since the primate brains of apes and monkeys are so similar to those of humans, many studies of brain function have involved experimentation on these animals, humans' closest relatives. Other mammals such as mice, rats, cats, and dogs have also served as subjects of brain studies that can be related not only to their own specific behavior, but also to how the human brain works in its various component parts. Since neurons are very similar to each other, whether they come from sea slugs, squid, or mammals, experimentation using these animals has produced insight into how all brains and nervous systems work.

—*Jean S. Helgeson*

FURTHER READING

Calvin, William H. *The Throwing Madonna: Essays on the Brain.* Updated ed. Bantam Books, 1991.

Eccles, John C. *Evolution of the Brain: Creation of the Self.* Routledge, 1989.

Falk, Dean. *Braindance.* Henry Holt, 1992.

Hickman, Cleveland P., Jr., Larry S. Roberts, and Allan Larson. *Integrated Principles of Zoology*. 9th ed. C. V. Mosby, 1993.

Marieb, Elaine N. *Human Anatomy and Physiology*. 5th ed. Benjamin/Cummings, 2001.

Mitchell, Lawrence G., John A. Mutchmor, and Warren D. Dolphin. *Zoology*. Benjamin/Cummings, 1988.

Restak, Richard M. *The Modular Brain*. Charles Scribner's Sons, 1994.

BREEDING PROGRAMS

FIELDS OF STUDY

Animal Husbandry; Genetics; Biology; Zoology; Veterinary Science

ABSTRACT

Keeping wild native animals began about 10,000 bce. While the reasons for this remain obscure, physical evidence indicates an effort to keep animals in order to meet the population's needs. It was also more convenient to have animals nearby to avoid the dangers and difficulties of hunting. As time passed, these captive animals were bred in a manner that produced traits favorable to the people keeping these animals. This process of controlled breeding became known as domestication.

KEY CONCEPTS

acclimatization: a process by which animals are adapted to new environmental conditions

animal husbandry: care and welfare of domestic animals

biotechnology: methods used to manipulate biological processes (such as reproduction)

domestication: a process by which animals are adapted biologically and behaviorally to a domestic (human) environment in order to tame and manipulate them for the benefit of humans

recombinant DNA: DNA that has been modified to contain gene sequences from different species

studbook: a record-keeping system that provides information on an animal's lineage

wildness: characteristics that define the biological and behavioral life of a species in the wild

DOMESTIC ORIGINS

Keeping wild native animals began about 10,000 BCE. While the reasons for this remain obscured because of a lack of historical documentation, the physical evidence indicates an effort to keep animals in order to meet the population's needs. It was also more convenient to have animals nearby to avoid the dangers and difficulties of hunting. As time passed, these captive animals were bred in a manner that produced traits favorable to the people keeping these animals. This process of controlled breeding became known as domestication. Originally, this was an informal affair with little control by the caretakers. Eventually, individuals keeping animals recognized the benefits derived from captive breeding and began exercising more control over the process.

Certain native species were more compatible with the human environment and more adaptable to domestication. These species had attributes favorable to this kind of controlled situation, such as the ability to adapt to new environments, social gregariousness, a dominance hierarchy that recognized humans as an alpha species, adaptable reproductive behavior, mild temperaments, and a low tendency toward flight behavior (which allowed humans to approach them). However, these attributes were not recognized initially, so there was no conscious decision to keep only certain species. Several ancient societies maintained many species in captivity. Most species, those without the favorable attributes, remained wild, or at best were only tamed. Very few species were to become domesticated, and all of these were domesticated in the early years of human agricultural development.

Various attempts were made in Europe and the European colonies to domesticate additional species during the early 1800s, but none were successfully domesticated. The primary group of domesticated animals has remained the same throughout history: the dog, cat, goat, sheep, pig, cow, horse, camel, llama, reindeer, and elephant. Those species already domesticated have been bred over many generations to improve characteristics favored by humans,

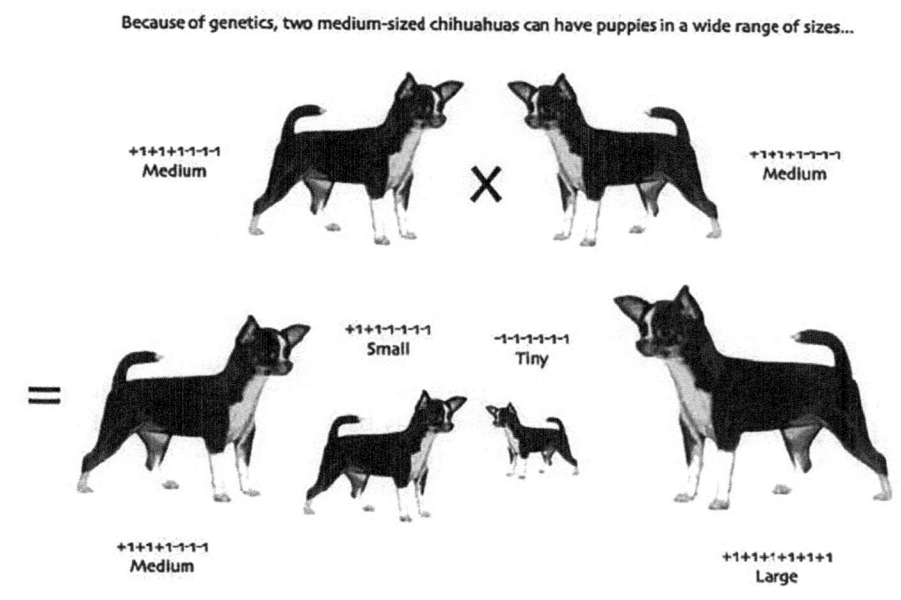

Chart of Chihuahua natural size variations when breeding a new generation.

sometimes to the point that the domesticated animal no longer resembles its wild originator.

BREEDING DOMESTICATED ANIMALS

To some extent, there were always animal breeding efforts taking place in agricultural animal herds, but for much of history they were not closely managed. This began to change in the seventeenth century, when science evolved into a reliance on observable and reproducible experimentation. That century saw improvements in livestock feeding, housing, and care, and conditions favoring the attainment of an animal's full genetic potential. Better educated and wealthy farmers were beginning to take notes and keep track of their animals, an important precursor to the breeding experiments that developed in the following century.

The eighteenth century was a time of increasing interest in agricultural improvement and experimentation. Originally, this involved breeding animals for improved adaptation to a local environment (that is, the local conditions where the farm was located). Eventually, however, this changed to breeding animals for improvements in the breed itself, without regard to its environment. Specific commercial advantages were sought, such as improved meat or milk production, or improved wool production. Eventually, as this effort at improving a breed's characteristics became widely practiced, it led to an international pedigree system, with studbooks that documented an animal's lineage.

Breed improvement and experimentation was enhanced through an exchange of information coordinated through the formation of farmers' clubs and societies. New techniques and successful experiments were publicized, and visits to breeding farms were reported upon. Journals were published that contained information previously limited to private correspondence. Monies were invested in improved breeding stock, which more widely distributed the improvements of the various breeds.

Originally, these breeding efforts were conducted on the estates of wealthy farmers at their own expense. Toward the end of the eighteenth century, the governments of England and France became interested in promoting improved husbandry practices. France established the Comité d'Agriculture in 1775, and England established the Board of Agriculture in 1793. There was also an urgent need in the European colonies to adapt European domestic animals to colonial environments, or to convert colonial species to domestication. Acclimatization societies were formed for this purpose both in Europe and in the European colonies.

These developments increased significantly in the nineteenth century. In addition to practical improvements, theoretical improvements in animal husbandry were offered as the sciences of agricultural chemistry, reproductive biology, and genetics developed. Advances in scientific knowledge aided the animal breeding efforts of farmers, and this information was better distributed once governments and universities developed agricultural departments, extension offices, and experiment stations to benefit the farmers.

Publication of *Die Organische Chemie in ihre Anwendung auf Agrikultur und Physiologie* (1840; *Organic Chemistry in Its Applications to Agriculture and Physiology*, 1840) by Justus von Liebig introduced agricultural chemistry to a wide audience and began what became known as scientific agriculture. Charles Darwin's book on evolution was published in 1859, and Gregor Mendel's work on genetics was published in 1866. Germany began a system of government-operated experiment stations in the 1870s. These combined laboratory experimentation with farm experimentation. In the United States, the US Department of Agriculture was established in 1862, as were the land grant colleges located in each of the states. Several states developed agricultural experiment stations in the late 1800s, and the federal government established a national system of experiment stations in 1887.

Implementation of federal and state programs developed in earnest during the early decades of the twentieth century. Scientific advances and a growing human population encouraged improvements in animal husbandry. Improvements in breeding techniques, improved knowledge about reproductive biology, improved veterinary care, and better housing for the animals also contributed to better animal husbandry. Eventually genetics and biotechnology began to play a major role in developing specific characteristics in each breed. After midcentury, farms decreased significantly as urban populations grew. Within this shifting demography, breeding programs gained new importance, as fewer farmers grew an ever-increasing number of domestic animals to meet the needs of a growing urban human population.

BREEDING WILD ANIMALS

Today's domestic species were once wild, but have been changed to suit human needs through the process known as domestication. Few species have been domesticated, even though attempts have been made to domesticate a wide range of species. The London Zoological Garden, the French Jardin Zoologique d'Acclimatation, and several acclimatization societies attempted to domesticate additional species in the late eighteenth century, but none were successful. The acclimatization facilities of these societies eventually closed or evolved into zoos that maintained, exhibited, and bred wild animals without changing them into tame or domestic animals.

Zoos and aquariums have always been concerned with wildlife conservation, although their effectiveness has been dependent on the era's zoological and animal husbandry knowledge, as well as the society's perceived importance of conservation. As the importance of conservation increased and the sciences related to wildlife conservation improved, zoo and aquarium conservation efforts improved. These conservation efforts included propagation programs that bred endangered species and species extinct in the wild. These propagation programs involve species studbooks, studying small populations of animals, introducing animals back into the wild, and other modern techniques.

Breeding wild animals over many generations, however, runs the risk of domesticating these animals. Of course, no such intentional domestication program has been successful. The greater risk is that the animals will become tame, will be unable to survive in the wild, and will lose their wildness. Because the breeding of wild animals is based on the animal's needs rather than human needs, propagation efforts with wild species are quite different from efforts made with domesticated animals. Nevertheless, the methods are similar.

Both wild animal breeding and domestic animal breeding require detailed studbooks in order to keep track of an animal's lineage, and the pairing of appropriate individuals is closely controlled. Many sciences provide knowledge important to the propagation programs, such as veterinary medicine, nutrition, reproductive biology, genetics, and biotechnology. Frozen tissues, artificial insemination, bioengineering, and recombinant deoxyribonucleic acid (DNA) technology play an increasingly important role in modern breeding programs.

Breeding wild animals is often more difficult than breeding domestic animals, since unusual breeding behavior is part of the reason some species are endangered. Improved knowledge about the species' social behavior and population biology needs are of assistance in the successful breeding of these difficult wild species, as is biotechnology. Frozen zoos have been established to maintain reproductive and other tissues for artificial insemination. Sometimes this artificial insemination involves the use of related surrogate species; for instance, using domestic cattle to give birth to endangered gaur. Back breeding is being attempted in order to revive extinct species; for example, breeding zebra so as to re-create the extinct quagga. As an increasing number of species become endangered and as their natural habitat continues to disappear, it is increasingly important to maintain these species through appropriate breeding programs.

—*Vernon N. Kisling, Jr.*

FURTHER READING

Clutton-Brock, Juliet. *A Natural History of Domesticated Animals*. 2nd ed. Cambridge University Press, 1999.

Hutchins, Michael, and William G. Conway. "Beyond Noah's Ark: The Evolving Role of Modern Zoological Parks and Aquariums in Field Conservation." *International Zoo Yearbook* 34 (1995): 117–130.

Peel, L., and D. E. Tribe, eds. *Domestication, Conservation, and Use of Animal Resources*. Elsevier, 1983.

Western, D., and M. Pearl, eds. *Conservation for the Twenty-first Century*. Oxford University Press, 1989.

Wiese, R. J., and Michael Hutchins. *Species Survival Plans: Strategies for Wildlife Conservation*. American Zoo and Aquarium Association, 1994.

Zeuner, Frederick E. *A History of Domesticated Animals*. Harper & Row, 1963.

C

Camouflage

Fields of Study

Biology; Biochemistry; Zoology; Animal Anatomy

Abstract

Crypsis is the art of remaining hidden. Camouflage is usually thought of as color matching: a green aphid, for example, is likely to go unnoticed while feeding on a green leaf. Background matching, or cryptic coloration, is, indeed, the most common form of camouflage, but most crypsis involves far more than matching a single color. Camouflage for most animals must be more sophisticated if it is to be useful.

Key Concepts

aposematism: use of bright, noncamouflaged colors as a warning signal to indicate toxicity or dangerousness

countershading: a form of crypsis involving dark coloration on top and light coloration on the underside

cryptic coloration: any color pattern that blends into the background

disruptive coloration: use of stripes, spots, or blotches to break up the body outline and blend into a complex background

protective mimicry: use of both color and form to mimic an inanimate feature of the environment

HIDING IN PLAIN VIEW

Crypsis is the art of remaining hidden. Camouflage is usually thought of as color matching: a green aphid, for example, is likely to go unnoticed while feeding on a green leaf. Background matching, or cryptic coloration, is, indeed, the most common form of camouflage, but most crypsis involves far more than matching a single color. Very small animals such as aphids can get away with using a single camouflage color because they are much smaller than the plants on which they spend their entire lives: They only need to match one thing. Most animals, however—even most insects—are significantly larger than aphids and are likely to spend time in more than one place. Their camouflage must be more sophisticated if it is to be useful.

If a large organism is to remain undetected, it must be camouflaged with respect to an entire scene. One way to do this is with the use of disruptive coloration, that is, the use of stripes, spots, or patches of color for camouflage. Disruptive coloration can involve large color patches, as on a pinto pony, a tabby cat, or a diamond-backed rattlesnake, or may involve tiny variations of color on each scale, feather, or hair. Many brownish or grayish mammals actually have what is called agouti coloring, with three different colors appearing on each hair.

The irregular borders of multiple color patches on an animal's body help to obscure its outline against an irregular and multicolored background, just like the blotchy greens and browns on military uniforms. An animal that has a mix of browns in its fur, feathers, skin, or scales, for example, will blend into a forest or even an open desert or tundra much better than one that is a single solid color. Even the black-and-white stripes of zebras, which seem so striking, act as a form of disruptive coloration: From far away, and especially to an animal such as a lion, which does not have good color vision, the stripes of zebras help them blend into the tall, wavy grasses of the savannah.

Countershading is another form of crypsis involving differently colored patches. Countershaded animals appear dark when viewed from above and light when viewed from underneath. Animals with countershading include orca whales with their black backs and white bellies, penguins, blue jays, bullfrogs, and weasels. Countershading works and is found as camouflage in so many kinds of animals because no matter where one lives—a desert, a forest, a meadow, or an ocean—the sun shines from above. When looking

Dead leaf mantis (Deroplatys desiccata) at Bugworld in Bristol Zoo, Bristol, England. If alarmed it lies motionless on the rainforest floor, disappearing among the real dead leaves.

up toward the sun and sky, dark things stand out and light colors blend in; when looking down toward the ground or the ocean floor, light colors stand out and dark colors blend in. Predators that are countershaded can thus approach their prey with equal stealth from either above or below; likewise, prey species that are countershaded will be equally hard to find whether a predator is searching from on high or from underneath. Countershading and other forms of disruptive coloration can occur in the same organism, so that dark spots, blotches, or stripes appear on top while paler ones appear below.

Another way of remaining undetected in a complex scene is by using protective mimicry, that is, to mimic an inanimate object in both color and form. Some insects look like thorns on plant stems; others look like leaves, twigs, or flowers. Some insects, frogs, and fishes look like rocks, lichens, or corals. Sea lions, sea dragons, and even eels can look like floating kelp or other forms of seaweed.

Some animals may not look much like the objects around them, but will disguise themselves by attaching pieces of plants or sand or other debris to their body. Some caterpillars use silk to tie bits of flowers and leaves to their body; others use saliva as a glue. Some crabs glue broken bits of shell and coral to their own exoskeleton. By using bits of local materials to camouflage itself, an animal can ensure that it matches the background. It can even change its disguise as it moves from one area into another.

Being transparent is another way to match whatever background happens to be present. Many marine invertebrates such as worms, jellyfish, and shrimp, are completely transparent. Complete transparency is less common among land animals, but some land invertebrates have transparent body parts, such as their wings, allowing them to break up the outline of their body and blend into whatever happens to be in the immediate background.

THE BEHAVIOR AND ECOLOGY OF CRYPSIS

Behavior is an important factor in the success or lack of success of any form of crypsis. For example, not even disruptive camouflage can hide something that is moving quickly with respect to its background. Because of this basic fact, predatory species that rely on speed or stamina to outrun, outswim, or outfly their prey generally have little use for camouflage. On the other hand, so-called sit-and-wait predators (such as boa constrictors or praying mantises), must be virtually perfectly camouflaged in order to remain undetected while their prey approach to within grabbing distance. In between are the stealth hunters that sneak up on their prey before making a final high speed attack; such animals must be camouflaged and slow moving when out of attack range, but do not have to be camouflaged or slow when at close range.

As with predators, prey species that rely on rapid escape maneuvers do not often bother with camouflage coloration, while prey species that cannot rely on efficient escape tactics must, instead, rely on not being seen in the first place. Prey species that can move quickly but not as quickly as their predators must detect their predators before their predators detect them, and then they must remain absolutely still until the danger has passed.

Some species use different strategies as they go through different stages in life. In many altricial species (species with dependent young that require extended parental care of the offspring), the eggs and/or young are camouflaged, even though the adults are not; the temporary spots on deer fawns and mountain lion cubs are examples. In other species, nesting or brooding females may be camouflaged while the adult males retain their gaudy plumage or attention-getting behaviors; the changing seasonal patterns of color and behavior of ducks and songbirds provide examples here. Some species may be toxic and gaudy during one stage of life, yet tasty and cryptic during another.

Finally, although camouflage is usually thought of as a visual phenomenon, crypsis is important in every sensory modality. If a prey animal is virtually invisible to its predators, but puts out a sound, a scent, or a vibration that makes it easy to locate, visual crypsis alone would be useless. For successful protection, prey species must be cryptic in whatever sensory modalities their predators use for hunting. Likewise, for successful hunting, predatory species must be cryptic in whatever sensory modalities their prey use to detect danger. For most species of both predator and prey, this means being camouflaged or blending into the background in several sensory modalities all at once.

FURTHER READING

Dettner, K., and C. Liepert. "Chemical Mimicry and Camouflage." *Annual Review of Entomology* 39 (1994): 129–154.

Ortolani, Alessia. "Spots, Stripes, Tail Tips, and Dark Eyes: Predicting the Function of Carnivore Colour Patterns Using the Comparative Method." *Biological Journal of the Linnean Society* 67, no. 4 (August 1999): 433–476.

Owen, Denis. *Camouflage and Mimicry*. University of Chicago Press, 1980.

Ramachandran, V. S., et al. "Rapid Adaptive Camouflage in Tropical Flounders." *Nature* 379, no. 6568 (1996): 815–818.

Wicksten, Mary K. "Decorator Crabs." *Scientific American* 242, no. 2 (February 1980): 146–154.

CANNIBALISM

FIELDS OF STUDY

Anthropology; Sociology; Biology; Religious Studies

ABSTRACT

Cannibalism is the process in which an organism eats another individual of its own species for food. While scientists previously thought that this behavior only occurred during stressful periods when other food sources are in short supply, research has shown that it is actually quite common in nature even under normal environmental conditions. Various forms of cannibalism have been observed in many different species, with different motivations behind the act.

KEY CONCEPTS

ecosystem: a community of organisms in relation to each other and their physical environment

endocannibalism: a form of human cannibalism in which members of a related group eat their own dead

exocannibalism: a form of human cannibalism in which unrelated humans are eaten

sexual dimorphism: consisting of different body forms according to gender, as when the females of a species are much larger than the males of the same species

HAVING SOMEONE OVER FOR DINNER

Cannibalism is the process in which an organism eats another individual of its own species for food. While scientists previously thought that this behavior only occurred during stressful periods when other food sources are in short supply, research has shown that it is actually quite common in nature even under normal environmental conditions. Various forms of cannibalism have been observed in many different species, with different motivations behind the act.

The most persuasive reason why cannibalism takes place within a species is the need to survive. Paramount to survival is the necessity to have a diet sufficient to support development and continued existence. Competition for survival begins at birth and continues throughout the life cycle of the animal. Since most animals produce more young than can survive, it sometimes occurs that the strongest of the young feeds on the weakest. Young mantids (praying mantises), black widow spiderlings, and varieties of young salamanders, for example, often feast on their brothers and sisters as soon as they are born. In some varieties of sharks, only one or two shark pups are born from the large number of eggs that the mother shark carried during gestation; the surviving shark pups consume their brothers and sisters before birth.

Another reason for cannibalism, also related to the need to survive, is the necessity to eliminate competitors within an ecosystem. Some young tiger salamanders, when living in extremely crowded conditions, develop special structures in their mouths that enable them to eat other salamanders that are their competitors. Adult male Kodiak bears often kill and eat young cubs, especially male cubs, as a means of both supplementing their diet and eliminating future competitors. Male lions and male feral cats are also known to kill and eat the cubs of another male, thus enabling them to mate with the mother of those cubs and ensure that their own offspring will survive. Male chimpanzees are also known to engage in the practice of killing and eating infants of females that they have not impregnated.

Sometimes cannibalism is related either to the lack of enough food or to a diet deficiency. Two popular household pets, guppies and gerbils, eat their young if there is not enough food available. Female gerbils also cannibalize their own or another female's litter as a means of gaining more protein in their diet. Furthermore, livestock may be forced into cannibalism by the practice of feeding them by-products of slaughtered animals to increase their protein intake; this practice is believed to have spread bovine spongiform encephalopathy, or "mad cow disease," among cattle.

Some animals appear especially capable of consuming members of their species if there is nothing else readily available. *Tyrannosaurus rex*, the famous prehistoric predator, was apparently in that category. Scientists have discovered that the North American *T. rex* may have devoured members of its own group in order to gain a fast and easy meal. South American horned frogs apparently feed on anything, including fellow horned frogs, that moves near them.

One form of cannibalism still perplexes scientists. During the mating and reproduction process, some female members of the animal kingdom kill and later consume their suitors, an act known as sexual cannibalism. It is mostly seen in insects, particularly arachnids, and often in species displaying prominent sexual dimorphism. Praying mantises and black widow spiders are perhaps best known for this practice, but not all black widows eat their mates after killing them. Scientists have discovered that if the female black widow spider is not hungry, she will not consume her dead mate. The female praying mantis, however, will always devour her mate after she has killed him. In a very few species, reverse sexual cannibalism—in which males eat females—has been observed, though rarely. Several theories have been postulated to explain sexual cannibalism, and research into the phenomenon continues.

HUMAN CANNIBALISM

While factors that explain cannibalism among animals can also be applied to humans, there are several other possibilities to examine in the case of humans. Prehistoric humans are thought to have at times engaged in cannibalism as a means of survival, and modern human cannibalism because of a natural disaster or an accident has also been recorded. There are stories of shipwrecked sailors resorting to cannibalism of their dead and even murder and cannibalism in order to survive. Perhaps two of the most famous instances of modern cannibalism forced by starvation was the 1846–1847 experience of the Donner Party in California, and the 1972 incident involving the Andes mountain crash of a plane carrying a Uruguayan soccer team. In both instances, the survivors resorted to cannibalism (but not murder) in order to withstand the peril they faced.

However, the human animal is unique in that cannibalism is also practiced as a social and religious custom not involving subsistence and survival in the normal sense. In fact, some research has suggested that the nutritional value of the human body is relatively low, providing far fewer calories than other animals. This has led many scientists to believe that hominins including early humans mostly practiced cannibalism for ritual rather than dietary purposes.

A red fire bug (Pyrrhocoris apterus) *cannibalizing another.*

Such ritual practice continued among some cultures well into the twentieth century.

One type of human cannibalism involves a genuine reverence by relatives for their dead. Called endocannibalism, this practice is based upon the belief that eating the flesh of departed relatives shows great respect and veneration of the dead. This type was practiced among the natives of islands of the southern Pacific Ocean until it was declared illegal following World War II.

Exocannibalism, or the eating of unrelated individuals, has ritualistic and religious overtones as well. Native warriors of the South Pacific, popularly referred to as headhunters, ate parts of their vanquished opponents as a means of controlling them and gaining their strength. Sixteenth century South American natives mixed the eating of captured slaves with religion, making the cannibalistic ritual into a festival. Indeed, it was actions similar to these, observed by the Spanish, that gave us the word "cannibal"—Columbus incorrectly transcribed the name of the human-eating Caribs of Cuba as *Canibalis*.

—Robert L. Patterson

Further Reading

Brown, Paula, and Donald Tuzin, eds. *The Ethnography of Cannibalism*. The Society for Psychological Anthropology, 1983.

Elgar, Mark A., and Bernard J. Crespi, eds. *Cannibalism: Ecology and Evolution Among Diverse Taxa*. Oxford University Press, 1992.

Goldman, Laurence R., ed. *The Anthropology of Cannibalism*. Bergin and Garvey, 1999.

Klitzman, Robert. *The Trembling Mountain: A Personal Account of Kuru, Cannibals, and Mad Cow Disease*. Plenum Trade, 1998.

Price, Michael. "Why Don't We Eat Each Other for Dinner? Too Few Calories, Says New Cannibalism Study." *Science*, 6 Apr. 2017, www.sciencemag.org/news/2017/04/why-don-t-we-eat-each-other-dinner-too-few-calories-says-new-cannibalism-study. Accessed 31 Jan. 2018.

Schutt, Bill. *Cannibalism: A Perfectly Natural History*. Algonquin Books of Chapel Hill, 2017.

Carnivores

Fields of Study

Zoology; Animal Physiology; Ecology; Environmental Studies

Abstract

Carnivores are a modern order of mammal that includes ten families: bears, cats, civets, dogs, hyenas, mongooses, pandas, red pandas, raccoons, and weasels. They first appear in the fossil record of the Eocene period, forty to fifty million years ago, and probably evolved from nocturnal, small, semiarboreal predators called miacids. Carnivores are recognizable by their teeth. Most are terrestrial, although the otters are aquatic. Carnivores are found on all continents except Antarctica.

Key Concepts

body mass: the average weight of females of a species, expressed in kilograms
diurnal: active mainly during the daytime
gregarious: forming groups temporarily or permanently
nocturnal: active mainly during the night
omnivore: an animal that eats both plant material and animal material

MEET THE CARNIVORES

Carnivores are a modern order of mammal that includes ten families: bears, cats, civets, dogs, hyenas, mongooses, pandas, red pandas, raccoons, and weasels. They first appear in the fossil record of the Eocene period, forty to fifty million years ago, and probably evolved from nocturnal, small, semiarboreal predators called miacids. Carnivores are recognizable by their teeth: enlarged canines, specialized for stabbing and holding prey, and carnassials, specialized for shearing flesh and skin. All carnivores eat other animals, which they capture in a variety of ways. Most are terrestrial, although the otters are aquatic. Carnivores are found on all continents except

A red fox (Vulpes vulpes) *eating a rodent—an example of a mesocarnivore. At Sunkhase National Wildlife Refuge, Maine. (Dave Small)*

Antarctica. They are recent arrivals to Australia, apparently having reached this island continent along with humans ten to forty thousand years ago.

BEARS, CATS, AND CIVETS

Bears (family *Ursidae*) are widely distributed in Eurasia and North America, but only a few forms live in tropical areas of Asia and South America (the sloth and the sun and spectacled bears). At one time, a large predatory bear of the genus *Agriotherium* lived in Africa, but currently no wild ursids exist on this continent or in Australia. Today, the largest living ursid is the polar bear, with a body mass of 320 kilograms. Many *Ursidae* share with some of the *Mustilidae* a unique reproductive physiology, called delayed implantation, in which the fertilized egg may take many months to implant in the uterus and continue its development. This may be an adaptation to hibernation and wintertime shortages of food in temperate regions. Bears in cold climates spend much of the winter hibernating in a protected den. During this time, their heart rate and metabolism slow to conserve energy. Most bears are omnivorous, but the polar bear is a specialist hunter of seals.

Cats (family *Felidae*) are distributed throughout the world, from the heights of the Himalayas (the snow leopard) to the Amazon (the jaguar). The cats are among the most carnivorous of their order and the most adaptable. The earliest felids evolved in forested areas, and most retain adaptations for tree-climbing and use of cover as concealment. Millions of years ago, much larger forms of felids existed, including the extinct saber tooth tigers whose long, bladelike upper canines were specialized for delivering killing bites to the necks of large prey, such as mammoths. Not including the many types of house cats familiar as pets, cats are a diverse family, ranging from the group-living lion (body mass 135.5 kilograms) to the solitary Geoffroy's cat (body mass 2.2 kilograms). Virtually all cats can hunt day and night, but nocturnal habits predominate. The cheetah is the only living cat that hunts exclusively by day. Often habitats will contain several species of cats, differing in size and specialized for different prey. For example, certain areas of the Amazon may contain jaguars, pumas, ocelots, and one or more forms of smaller cat, such as jaguarundi or margay.

Civets (family *Viverridae*) are restricted to tropical and subtropical areas in Africa and Asia. Civets retain many ancestral morphological features of the first carnivores. They are small (body mass range 1.2 to 13 kilograms), with adaptations for climbing trees and no specializations for pursuit or ambush of prey. Civets are nocturnal and, with one exception, arboreal. They eat both animals and fruits, although the fossa is almost exclusively predatory and the binturong eats almost only fruit. Viverrids are closely related to the mongooses.

DOGS, HYENAS, AND MONGOOSES

Dogs (family *Canidae*) are almost as widely distributed as felids, being found from the Arctic to the South American rainforest. However, canids evolved in open country habitats of North America and few have adapted to life in rain forests. The wolf was once the most widely distributed carnivore in the world. It evolved in east Asia and from there spread throughout Eurasia and the Arctic Circle, including migrating down into North America as far south as Mexico. Canids have reached their highest diversity in North American woodland-plains habitats (wolves, coyotes and foxes) and African savannahs (wild dogs and jackals). The dog family is diverse, but the domestic dog shows as much variation in body form (from tiny Chihuahua to Great Danes and mastiffs) as all of the wild dog species put together. The smallest wild dog is the North African fennec (body mass 1.5 kilograms), and the largest is the wolf (body mass 31.1 kilograms). Although some canids, such as the gray fox, can climb trees to a limited extent, the dog family is highly adapted to fast running in open ground. Canids are characterized by complex social systems, often involving cooperative care of the young by older juveniles or nonreproducing adults. Dogs hunt animals by day and night, but many forms supplement their meat diet with fruit.

Hyenas (family *Hyaenidae*) are found mainly in Africa, with one species, the striped hyena, living in southern and western Asia. However, in the Miocene and Pliocene (two to twenty-two million years ago), the hyenas were numerous, diverse, and widespread through Africa, Eurasia, and North America. Only four species of hyaenids exist today: the spotted hyena at 55.3 kilograms, the striped hyena at 35 kilograms, the brown hyena at 43.9 kilograms, and the aardwolf at 7.7 kilograms. The rise of the dog family has occurred in parallel with the decline of the hyenas. The unusual aardwolf eats termites and often digs a den in a termite mound. The other hyenas are specialized hunter-scavengers, adapted for bone-crushing with their reinforced teeth, jaws, and crania. Once considered scavengers only, field studies since the 1970s have documented the extensive hunting done by the large hyenas in Africa. All hyenas are solitary except the spotted hyena, which lives in clans. Members of clans usually disperse to hunt but also hunt in small packs, especially for large prey. Spotted hyenas are unique among carnivores because the females are larger than males and the males give the females priority of access to food and space.

Mongooses (family *Herpestidae*) are found only in Africa and warmer climates in Eurasia. They are closely related to viverrids but tend to be smaller (body mass range 0.5 to 1.5 kilograms), more terrestrial, and more often diurnal than the civets. Mongooses are renowned for their bravery in the face of snakes, but they generally hunt insects or small vertebrates, and may supplement their diet with fruits. Most mongooses are solitary, but Africa contains several forms of gregarious mongooses, such as meerkats, dwarf mongooses, and banded mongooses. These form long-lasting packs of up to thirty animals, using termite mounds or tunnels as dens during the night. It is believed that these mongooses have evolved highly social, gregarious habits because of the severe risk of predation by hawks, eagles, large carnivores, and snakes. The dwarf mongooses and meerkats show the most unusual social system, with one breeding male-female pair and multiple nonreproductive helpers feeding and protecting the young and the rest of the group. This system involves complex communication systems and a division of labor, including the use of sentinels to detect predators.

PANDAS, RACCOONS, AND WEASELS

The giant panda is the only member of its family (*Ailuropodidae*), although it is considered closely related to bears. Its habitat is restricted to the Tibetan plateau. The panda bear is large (body mass 96.8 kilograms) and specialized to eat bamboo, as well as some animal foods.

The red or lesser panda has not yet been clearly related to other carnivores. Many scientists place it alone in its own family (*Ailuridae*), while others may group it with the raccoon family (procyonids). Its size (body mass 5.7 kilograms) and omnivorous diet is consistent with procyonids, but its distribution is not. The red panda is restricted to the Tibetan plateau, and part of its diet consists of bamboo, so it has sometimes been classified with the giant panda.

Raccoons (family *Procyonidae*) are found only in the Americas. This family is rather uniform in size (body mass 0.9 to 6.7 kilograms) and in diet. Virtually all of them eat a mixture of fruit, insects, and small vertebrates. Along with the familiar, widespread raccoons, there are less well-known forms in tropical regions, including the arboreal kinkajou, distinguished by a prehensile tail that permits it to hang from branch tips in order to reach fruit at the ends. Also among the tropical procyonids is the coati, which forms large groups with complex social organization. In a coati band, adult males live alone for much of the year, while adult females and young form groups of thirty or more individuals.

Weasels (family *Mustelidae*) are found everywhere except Australia. Scientists recognize four main subfamilies. The otters (body mass 5 to 40 kilograms) are adapted to aquatic life and eat fish and shellfish. The widely distributed badgers (body mass 0.6 to 10.9 kilograms) are adapted to digging and often specialize in eating earthworms. The true skunks of North America (body mass 0.4 to 2 kilograms) are terrestrial omnivores that have specialized anal scent glands used against predators. The weasel group (body mass 0.06 to 2.3 kilograms) includes a host of long, thin forms such as mink, ferrets, and martens. Mustelids are more numerous in temperate regions than in tropical ones, although otters are found around the world. Another tropical mustelid is the tayra, a weasel-like semiarboreal predator that may attack monkeys in the trees.

In summary, the order of carnivores is very diverse in body size, habits, social organization, geographic

distribution, and basic ecology. Most members of this order are intelligent, predatory, adaptable, nocturnal, and solitary. However, among the exceptions to these general rules about carnivores are some of those species most familiar to humans: coyotes, lions, and wolves. Humans have long had a mixed view of carnivores. From the Egyptian reverence for cats and more recent romantic views of the noble wolf, positive impressions of carnivores have been countered by hatred for large predators, driven by economic concern over livestock-killing and attacks on humans.
—*Milton Berman*

FURTHER READING

Gittleman, J., ed. *Carnivore Behavior, Ecology, and Evolution.* 2 vols. Cornell University Press, 1989, 1996.

Kitchener, Andrew. *The Natural History of the Wild Cats.* Comstock Publishing Associates, 1991.

Sheldon, Jennifer W. *Wild Dogs.* Academic Press, 1992.

Sinclair, A. R. E., and Peter Arcese, eds. *Serengeti II: Dynamics, Management and Conservation of an Ecosystem.* University of Chicago Press, 1995.

CELL DETERMINATION AND DIFFERENTIATION

FIELDS OF STUDY

Embryology; Microbiology; Biochemistry; Cytobiology; Genetics

ABSTRACT

From the time a fertilized ovum begins to divide (cleavage) until it forms a complete organism, it passes through successive stages in which groups of cells become increasingly specialized. This process of cell specialization involves two key steps: determination, in which cells become committed to a certain developmental pathway, and differentiation, in which cells acquire their ultimate structure and function.

KEY CONCEPTS

commitment: the "decision" by an embryonic cell to develop in a certain way, which may be reversed if the cell is removed from its normal surroundings

embryonic induction: the point at which one embryonic tissue signals another embryonic tissue to develop in a certain way

genome: all the genes of one organism or species

morphogenesis: the development of form, including the overall form of the organism and the form of each organ and tissue

mosaic development: the process whereby early embryonic cells are determined by the cytoplasm they receive from the egg; also called determinate development

regulative development: the process whereby early embryonic cells are determined by their interactions with other cells; also called indeterminate development

restriction: reduction of the developmental potency of a cell

totipotent: the ability of a cell to develop into any kind of cell in the body

DETERMINATION, DIFFERENTIATION AND MORPHOGENESIS

From the time a fertilized ovum begins to divide (cleavage) until it forms a complete organism, it passes through successive stages in which groups of cells become increasingly specialized. This process of cell specialization involves two key steps: determination, in which cells become committed to a certain developmental pathway, and differentiation, in which cells acquire their ultimate structure and function. The term "differentiation" may also be used in a broader sense to describe the entire process of cell specialization. A third process, called morphogenesis, is needed in order to mold the embryonic cells structurally into the various tissues and organ systems of the mature organism. For example, in eye development the specialized photoreceptor cells of the retina are a product of determination and differentiation, but their organization into the overall structure of the eye is a product of morphogenesis.

THE PROCESS OF SPECIALIZATION

In animal development, the fertilized ovum (zygote) has the potential to develop into any type of cell in the body; it is said to be totipotent. As it undergoes cell division beyond the first few cleavages, the resulting cells (blastomeres) lose their totipotency. This point is reached at different stages in different species. In the sea urchin, for example, isolated cells of the two- or four-cell embryo are capable of developing into complete new adults; in other species, such as the tooth shell clam (*Dentalium*), these early cleavage cells, when isolated, do not have the potential to develop into a complete organism. In all animals, the cells of later embryos lose their ability to form complete new organisms. This loss of developmental potency of cells is called restriction and is dictated to some extent both by the type of cytoplasm with which the cell is endowed and the influence of surrounding cells. As development progresses, restricted cells become "committed" to develop into a specific tissue. Under normal conditions, these cells will always develop into this designated tissue; if, however, the cells are moved experimentally to some other part of the embryo or to another embryo, they will develop in a different way. That is why commitment is said to be reversible but determination is not.

Determination occurs when a group of cells becomes irreversibly assigned to develop into a specific tissue. Determination is the final step in restriction; beyond this point, the cells have no other developmental options. Determined cells may or may not look different from other embryonic cells, but they have changed internally so that they are committed to a particular developmental pathway. Determined cells are said to be self-perpetuating, because they can pass on heritable information about their identity and do not require stimulation by surrounding cells to develop in a certain way.

Once the fate of a cell is determined, then differentiation (also called cell differentiation or cytodifferentiation) can take place. During differentiation, the cell undergoes structural and functional changes which result in a highly specialized, mature, differentiated cell. For example, the red blood cell becomes specialized by losing its nucleus and other organelles in order to fill itself completely with the oxygen-carrying molecule hemoglobin. It also maximizes its surface area by becoming flattened and doughnut shaped. Primitive muscle cells, called myoblasts, become specialized by fusing together to form elongated multinuclear cells called myotubes. These cells further specialize by forming contractile organelles called myofibrils, composed primarily of the proteins actin and myosin.

THE ROLE OF GENES

At one time it was believed that the genetic determinants (genes) were divided up and parceled out as determination and differentiation progressed, such that each cell type received certain genes. This hypothesis was disproved in several ways, one of them being nuclear transplantation experiments, in which adult cell nuclei were transplanted into fertilized eggs whose nuclei had been removed or inactivated. A small percentage of these eggs were able to develop into normal adults, thus proving that the adult nuclei implanted in them retained all the genes necessary to form a complete organism. These and other experiments have led to the conclusion that both determination and differentiation occur because certain genes are expressed at certain times during the life history of the cell. There remains, however, the question of how genes are turned on and off to control development.

All developmental processes are believed to be controlled by genes as part of an intricate developmental program. The genes of individual cells are activated or deactivated via various signaling mechanisms in order to provide the correct cellular responses at the appropriate times. This process begins very early in development and even occurs in the egg before fertilization. Messenger ribonucleic acid (RNA) and proteins are produced by the egg and distributed unequally in the cytoplasm of different parts of the egg. When cleavage occurs, blastomeres in one part of the embryo receive cytoplasm that differs from the cytoplasm received by blastomeres in another part of the embryo; thus, some cells are endowed with one kind of messenger RNA and protein and other cells with another kind. In some species, cleavage is even unequal, in order to ensure that certain cells receive the desired cytoplasm. For example, in the nematode *Caenorhabditis elegans*, unequal cleavage results in the establishment of five different tissue types by the sixteen-cell stage. As the messenger RNA is expressed in the form of new proteins, it gives unique qualities to the cells. Some of these new proteins and the proteins made earlier in the egg may be signal molecules, which stimulate new and unique gene expression in the nucleus.

Another mechanism for turning genes on and off is called embryonic induction, which occurs when one embryonic tissue influences the development of another by releasing chemical factors called inductors. The inductors are signal molecules that instruct cells of another tissue how to develop by directly or indirectly activating certain genes. Induction is especially noticeable after the formation of distinct tissues, such as the three germ layers, ectoderm, endoderm, and mesoderm. In vertebrates, the mesoderm induces the formation of the neural tube and various other parts of the nervous system.

Another set of mechanisms, similar to embryonic induction, that controls differentiation involves the microenvironment in which the embryonic cells exist. These mechanisms include such parameters as the position of a cell in relation to other cells in the embryo; the interactions of cells with the extracellular milieu, including ions, pH, oxygen, and extracellular matrix proteins such as collagen; direct cell-to-cell contact; and the presence of specific growth and differentiation factors. The cells in one part of the developing embryo will experience a completely different set of microenvironmental influences from that of cells in another part of the embryo and consequently will be prompted to express their genes in different ways. Once the genes have been expressed, there must be a means for the daughter cells to retain the unique gene expression of the parent cell. That is most likely accomplished by proteins passed on to the daughter cells that continue to activate or deactivate the appropriate genes.

GENETIC RECOMBINATION AND TRANSCRIPTION

Although the genome is not modified extensively during embryonic development, there is evidence that certain changes occur in order to enhance differentiation. One of these mechanisms is called genetic recombination and involves breaking and rejoining deoxyribonucleic acid (DNA) at defined sites. For example, in maize, segments of DNA called transposons move around the genome and presumably take control of specific genes at certain times during development. Another example of genome modification occurs in the nematode *Ascaris*, in which parts of the chromosomes are discarded (in a process called chromosome diminution) during cleavage. The discarded chromatin is believed to be composed of extra copies of DNA sequences that are not ordinarily transcribed. The genome is also modified through the making of extra copies of essential genes (gene amplification). For example, some genes in the follicle cells surrounding the maturing oocyte of *Drosophila* are amplified about thirty times in order to code for the large amount of protein needed to make the egg chorion. As cells differentiate, the types of messenger RNA and protein they produce become increasingly selective, so that eventually each cell type has a unique pattern of gene expression. There will always, however, be common genes expressed in every cell that are needed for basic housekeeping processes, such as respiration and transport.

Selective gene transcription is carefully controlled by various mechanisms involving the blocking and unblocking of DNA. Two classes of nuclear proteins are believed to be involved in switching genes on and off: histones and nonhistones. Histones associate with DNA molecules in such a way that they block the DNA from being transcribed. The nonhistones are believed to remove or rearrange histones so that the DNA can be replicated or transcribed. Some nonhistone proteins are gene-regulatory proteins, which recognize a particular DNA sequence. The binding of these proteins with DNA can either facilitate or inhibit transcription. An additional control found only in vertebrates is methylation, by which methyl groups are added to the DNA base cytosine. In general, the inactive genes of vertebrates are more highly methylated than active genes; thus, methylation may serve to strengthen decisions involving gene expression that are made during differentiation.

Once the appropriate messenger RNA has been produced, it still must be translated into protein and the protein must be assembled and made functional. For example, hemoglobin protein (globin) translation is controlled by the presence of heme, the iron-containing portion of the hemoglobin molecule. In the absence of heme, the factor that initiates globin translation is inactivated; thus, even though the appropriate messenger RNA and other necessary ingredients are present, without heme no hemoglobin protein will be produced. Even after proteins are translated, they are subject to further regulatory mechanisms, such as assembly into functional units, activation or inactivation by various enzymes and other factors, and transport to their cellular destination.

STUDYING EMBRYOLOGY

The study of embryology was transformed from a purely descriptive science into an experimental science by investigators who developed micro surgical techniques in the 1890s. They discovered that the separated blastomeres of some early embryos, such as the sea urchin, could develop into complete normal larva (regulative or indeterminate development) and that the blastomeres of others, such as the tunicate, could form only parts of embryos (mosaic or determinate development). This led to an appreciation of the importance of nuclear-cytoplasmic interactions and the fact that egg cytoplasm distribution plays an important role in determining how certain blastomeres develop. Further microsurgical separation studies on embryos in later stages (thirty-two-cell to sixty-four-cell stages) demonstrated that the micro environment of some embryos approximates a double gradient consisting of animal pole factors in one half and vegetal pole factors in the other half. These two chemical gradients influence the cell nuclei in their respective zones and cause the cells to become progressively determined in certain ways. Thus, simply by manipulating embryonic cells, scientists were able to demonstrate the concepts of restriction and determination.

Another microsurgical technique developed in the first half of the twentieth century involves transplanting tissue from one embryo to another. When tissue that normally forms the brain is transplanted to an area of another embryo that normally forms skin, the transplanted tissue develops independently and begins to form into brain. Thus, the tissue has become irreversibly committed or determined to form a particular adult tissue. If the same transplant is done at an earlier stage, the transplanted tissue conforms to its surroundings and forms epidermis. These results indicate that tissues are capable of changing their normal fate if they are influenced by another tissue before determination. In some instances, the transplanted tissue induces the surrounding tissue to change its normal fate. Such is the case when tissue from the dorsal lip of the blastopore of an amphibian gastrula is transplanted to the lateral lip area of another gastrula. The transplanted tissue induces the formation of a second complete embryonic axis, resulting in laterally conjoined twins.

Another elementary method that has yielded a large amount of information about determination is cell marking and tracing. At first, investigators took advantage of different natural pigments that are present in certain animal embryos by following the fate of each blastomere. They discovered that each colored cytoplasm gives rise to a specific embryonic fate. For example, the yellow crescent cytoplasm of the tunicate (*Styela partita*) embryo gives rise to adult muscle cells. In other studies, cells were marked with vital dyes, carbon particles, enzymes, radioactive labels, and distinctive cells transplanted from another embryo. By tracing these labeled cells, investigators were able to ascertain their ultimate fate and when and where cell determination takes place.

One question that needed to be answered by experimental embryologists was whether the nuclei of determined and differentiated cells are irreversibly modified. That all the genetic material is present in differentiated cells could be shown by microscopic observations of the chromosomes, especially the large polytene chromosomes of larval flies, such as *Drosophila*. The only way to show whether these chromosomes were functional, however, was to transplant the nucleus of a differentiated cell into an enucleated egg and see if it could direct the development of a complete organism. The technique of nuclear transplantation (sometimes called cloning), developed in the 1950s, did indeed prove that nuclei from differentiated cells are totipotent. Success was not universal with all tissue types, however, and only a small percentage of the transplants actually succeeded, which indicates that restriction of potency does involve changes in the nucleus but not permanent modification of the genome itself.

METHODOLOGY

The study of differentiation can be approached by several methods. The simplest is to observe tissues microscopically as they differentiate. That is done most often by fixing embryos at various stages of development and observing thinly sliced sections of them with a light or electron microscope. Another method is to explant cells, tissue, or organs from embryos and observe them in culture (in vitro). By doing so, scientists can manipulate the environment of the cultured cells in order to find out precisely what conditions are necessary for differentiation to occur. In vitro culture also makes possible such techniques as synchronizing cell growth in order to study the relationship between cell division and cell differentiation, and cell fusion in order to see how the contents of one cell affect the behavior of another.

Various biochemical and molecular techniques are used to study the roles of many biological molecules in differentiation. Of particular interest are separation methods that allow scientists to isolate and identify proteins and other factors involved in the differentiation process. These molecules can be isolated by first homogenizing the cells and then separating the desired molecules by centrifugation (based on density), electrophoresis (based on electrical charge), or chromatography (based on molecular size). Once the molecules are isolated, their properties can be studied, including the biological activity. At times it is important to know if a particular sequence of DNA or RNA is present in an embryonic cell. That can be determined by a technique called hybridization, whereby single-stranded DNA is allowed to match up and adhere to a complementary strand of DNA or RNA. Usually one of the strands is radioactively labeled so that the sequence in question can be detected and measured. The usefulness of this technique has become greatly enhanced by the development of recombinant DNA technology, which allows for the construction of specific molecular probes (DNA sequences) that can be radiolabeled and used to detect cellular DNA and RNA by hybridization. One step beyond this is the technique of DNA transformation, whereby isolated genes are modified and then reintroduced into cells to determine the new properties of the altered gene when it is expressed.

Determination and differentiation are the foundations of cell diversification. There are approximately two hundred distinctly different cell types in mammals. Without cell specialization, organisms would not be able to move, breathe, think, or perform any of the many other functions necessary to sustain life.

—*Rodney C. Mowbray*

Further Reading

Alberts, Bruce, Dennis Bray, Julian Lewis, Martin Raff, Keith Roberts, and James D. Watson. *Molecular Biology of the Cell*. 3rd ed. Garland, 1994.

Bellvé, Anthony R., and Henry J. Vogel, eds. *Molecular Mechanisms in Cellular Growth and Differentiation*. Academic Press, 1991.

Carlson, Bruce M. *Patten's Foundations of Embryology*. 6th ed. McGraw-Hill, 1996.

Gilbert, Scott F. *Developmental Biology*. 6th ed. Sinauer Associates, 2000.

Hennig, W., ed. *Early Embryonic Development of Animals*. Springer-Verlag, 1992.

Ruddon, Raymond W. *Cancer Biology*. 3rd ed. Oxford University Press, 1995.

Circulatory Systems of Invertebrates

Fields of Study

Hematology; Physiology; Anatomy; Biology

Abstract

Circulatory systems are necessary when the process of diffusion no longer provides an organism with sufficient gas, nutrients, and waste exchange with its environment. Some invertebrate groups, such as sponges, coelenterates, and flatworms, have such thin body walls that diffusion can meet all their needs. For most invertebrates, however, the distance from the organism's surface to the cells in the interior is too great for diffusion to support their metabolic requirements.

Key Concepts

blood: the fluid connective tissue within blood vessels that carries raw materials to cells and carries products and wastes from them

closed circulation: a circulatory pattern in which blood is always contained within blood vessels

diffusion: the process whereby a substance moves from an area of greater concentration to one of lesser concentration, as through a cell membrane

hemolymph: the transport fluid of organisms with open circulation systems in which there is no clear distinction between blood and intercellular tissue fluid

Circulatory Systems of Invertebrates

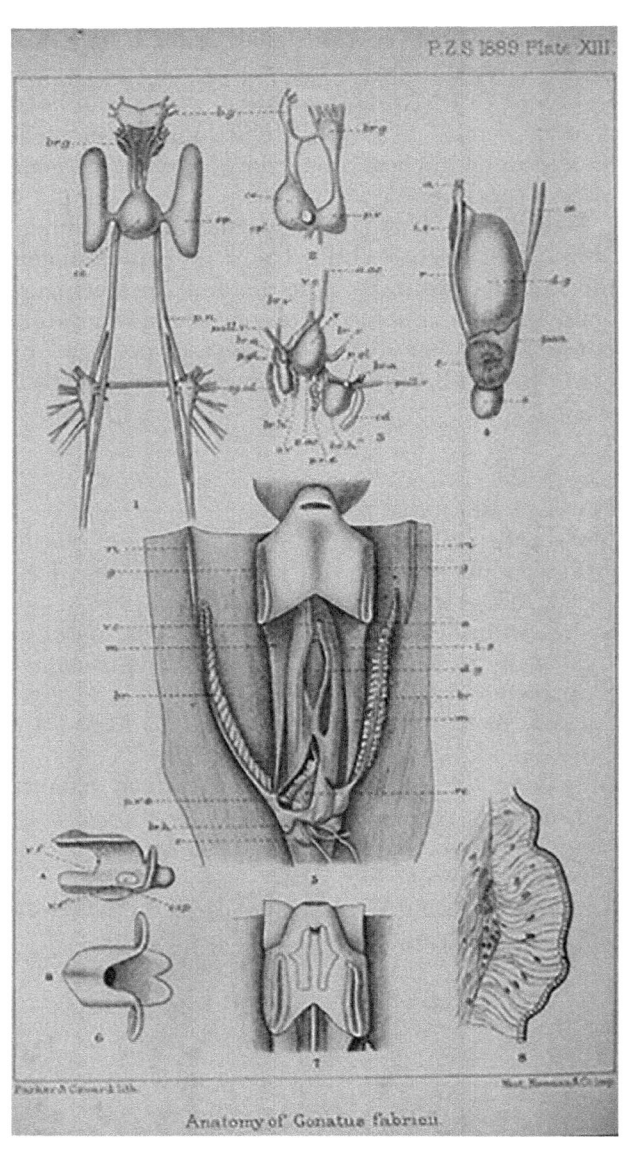

CNS and stellate ganglia from dorsal, magnification × 2; CNS without optic and stellate ganglia from dorsal, magnification × 2; Circulatory system from dorsal, unmagnified; Digestive tract of immature from lateral, magnification × 3; Mantle cavity.

lacunae: small spaces among tissue cells through which hemolymph flows in open circulatory systems

open circulation: a circulatory pattern in which the blood is not always contained within blood vessels

sinuses: larger spaces, thought to represent through channels, for hemolymph in open circulatory systems, sometimes bound by membranes

tracheal system: the respiratory system of insects and other terrestrial invertebrates; it consists of numerous air-filled tubes with branches extending into tiny channels in direct contact with body cells

THE NEED FOR A CIRCULATORY SYSTEM

Circulatory systems are necessary when the process of diffusion no longer provides an organism with sufficient gas, nutrients, and waste exchange with its environment. Some invertebrate groups, such as sponges, coelenterates, and flatworms, have such thin body walls that diffusion can meet all their needs. For most invertebrates, however, the distance from the organism's surface to the cells in the interior is too great for diffusion to support their metabolic requirements. A circulatory system is composed of three parts: the pump, the fluid that is pumped, and the vessels in which the fluid is transported. As in the design of invertebrate body plans, there is great diversity in the design of invertebrate circulatory systems.

THE COMPONENTS OF THE CIRCULATORY SYSTEM

The pump may be a simple tube lined by muscle fibers. Alternate contraction and relaxation of these muscles produces a peristaltic wave that pushes the blood along. The pump may be a heart: a localized, discrete organ whose muscle layers are the primary generators of the power that propels blood through the blood vessels. The heart can be a simple muscular enlargement, or it can be complex and multichambered, depending on the evolutionary history and the needs of the organism. Some organisms may have more than one heart.

If the pumped fluid remains within the vessels of the circulatory system, it is usually called blood. If it leaves those vessels to enter cavities surrounded by tissue cells with which it exchanges materials, it is called hemolymph. In the cavities (called lacunae if small and sinuses if large and lined with a membrane) blood mingles with intercellular fluid. Whatever it is called, the fluid is composed of water, solutes (such as salts, sugars, and other nutrients), and, in some cases, cells and formed elements. The blood may also contain a respiratory pigment that helps deliver oxygen to the cells.

FLUID TRANSPORT SYSTEMS

Blood vessels may extend only a short distance from the heart or may provide a continuous path for the transport of blood. If the blood vessels are incomplete and blood is pumped from arteries into body spaces—sinuses and lacunae—it is an open circulatory system. If the blood vessels are continuous and the blood never leaves these circulatory channels, it is a closed circulatory system. These are generalized terms for convenience; there are actually many intermediate cases. Some open systems have membrane-lined sinuses and lacunae, and exchange takes place by diffusion just as it does in the tiniest vessels of closed systems. Blood vessels are categorized according to their function. Arteries carry blood away from the heart; veins carry blood back to the heart. Closed circulatory systems have tiny tubules, called capillaries, connecting small arteries to small veins. The exchange of materials between the tissue cells and the blood takes place only in the capillaries.

A few generalities based on physical principles indicate the rules guiding fluid transport. Organisms relying on diffusion alone to circulate nutrients and eliminate wastes are limited to certain shapes and sizes: They are often only two cell layers thick. Organisms using bulk transport of fluid, on the other hand, may be as complex and as differently shaped as plants and animals are.

A fluid transport system is any system in which internal fluid movement reduces diffusion distances—either between points within an organ or between a point within an organism and the external environment. While diffusion is always used for short-distance transport, bulk flow augments diffusion for any long-distance transport. Bulk flow requires a pump and a fluid that must come into intimate contact with tissues and the environment for efficient transfer. Either the tissue layers must be thin (as in open circulatory systems) or the vessels must have a small radius (as in a closed circulatory system). The circulatory fluid should spend the majority of its time in the transfer regions (sinuses and lacunae or capillaries) and not in transit.

Such transport systems use both large and small vessels. Large vessels move fluid from one exchange site to another, and small vessels allow diffusion at the exchange sites. The total cross-sectional area of the small vessels must exceed that of the large vessels so that the flow rate in small vessels will be less than the velocity in large vessels. High-speed pumps in large vessels are preferable to low-speed pumps in the small vessels. This means that accessory hearts are only used when the "cost" of operation is not a major factor or when the pump serves some additional functional role—as when active cephalopod mollusks, such as squid and octopods, use accessory hearts to ensure adequate blood flow through their gills.

CLOSED AND OPEN CIRCULATORY SYSTEMS

Circulatory systems are traditionally divided into two categories. Closed systems are those in which the blood is always contained within distinct vessels and is physically separated from the organism's intercellular fluids. They are usually characteristic of organisms with high metabolic demands. High volumes and high pressures can be maintained in the closed vessels to aid transport and diffusion. Annelids, cephalopod mollusks, and vertebrates usually have closed circulatory systems.

Open systems are those which possess large, usually ill-defined, cavities (sinuses if bound by an endothelial layer, lacunae if not) and in which the blood is not physically separated from the intercellular fluids. Arthropods and noncephalopod mollusks have open systems.

Open circulatory systems are not always sluggish, low-pressure arrangements. Some spiders generate sufficient pressure in their open systems to use hydraulic pressure as a substitute for extensor muscles in their legs. In addition, capillaries often exist in open systems, particularly in areas such as the excretory organs and the cerebral ganglion. The major sinus in the foot of gastropod and bivalve mollusks is not a large, open cavity but a network of channels in a spongy tissue that function as capillaries. The lack of return vessels in these systems is usually a result of the fact that fluid simply has nowhere to go other than in the direction of low pressure: back to the heart. These volume constraints are sufficient to develop pressure, although this system is incompatible with high pressure and flow rates.

In the two major groups with open circulatory systems, arthropods and noncephalopod mollusks, the circulatory sinuses play an additional role: In bivalve and gastropod mollusks, the hemocoel (main body cavity) functions as a hydrostatic skeleton in locomotion and burrowing. In aquatic arthropods, it serves the same function during molting, when arthropods lose the support of the exoskeleton. In insects, the tracheal system has assumed the respiratory function, and the blood merely delivers nutrients and removes wastes. In large flying insects, the circulatory systems may also have the primary responsibility of removing heat to maintain thoracic temperatures.

PRESSURE PATTERNS IN HEART ACTION

To function effectively, the circulatory system must have a regular pattern of pressure increases that will push the blood along through the vessels. The heartbeats that accomplish this may be initiated and maintained by nerves, or they may be self-generated. If nerves initiate the contraction of the heartbeat, the heart is called neurogenic, meaning that nerve impulses generate the depolarization that results in the contraction of the heart's muscle cells. In these hearts, the heart muscle will not contract without a nerve impulse. Some species with neurogenic hearts are crustacea, horseshoe crabs, some spiders, and scorpions.

Heart muscles that continue to beat even when nervous connections are severed are called myogenic, meaning that the heart muscle contracts without external stimuli. Under these circumstances, the contraction of the muscles may occur at a different rate from that imposed by the nervous system when active. Myogenic hearts are found in mollusks and many insects.

All heart action must be modulated to respond to external and internal conditions, so even myogenic hearts usually receive some innervation. Modulation occurs through the mediation of nerves, hormones, or intrinsic controls in the heart. In the lobster, for example, nerves are crucial to maintaining the best rhythm and amplitude, but neurohormones released from a pericardial organ influence the heart action. Stretching the heart muscle, an intrinsic control, will also increase the vigor and rate of contraction.

In many cases, the structure of the heart and its suspensory ligaments contributes to the functioning of the circulatory system. Valves at the openings (ostia) to the heart prevent backflow when the heart contracts. This pulls at the ligaments. Their elasticity pulls back the walls of the heart, creating low pressure that enables the heart to fill on relaxation. In effect, the heart sucks blood from veins to refill itself for the next contraction.

In many species, the contraction of body parts contributes to the circulation of blood. Arthropods have a rigid exoskeleton; contraction in one part pushes the blood into another segment. In American lobsters, a quick flexion of the abdomen (an important locomotor movement) raises pressure in the abdomen and increases the rate of blood flow to the thorax and into the heart region.

All flow depends upon pressure differences, regardless of whether the circulatory system is open or closed, and there are two kinds of pressure. Background pressure is the pressure that prevails everywhere in the animal. Since pressure differences are responsible for flow, these pressure differences are imposed on the background pressure. If the body changes posture and increases background pressure, the blood pressure must similarly increase in order to maintain flows at the same level they were before the postural change. The blood-pressure gradient and the resistance to flow in the system affect blood flow. The low resistance in open circulatory systems probably permits relatively high rates of blood flow with relatively low pressure. The high blood-flow rate compensates for the low oxygen-carrying capacity of the blood.

Open and closed systems differ; neither is necessarily superior. The inherent weakness of the open circulatory system is that the peripheral blood flow cannot be as well controlled as that of closed circulatory systems. Yet the large sinuses are often subdivided, thereby providing discrete channels of flow, and the peripheral blood flow may be more regular than previously thought. Closed circulatory systems have a flow that is easily controlled. Flow through particular regions can be managed by using muscles to close off certain channels. Cardiac output can be distributed to meet tissue demands. In open systems, this is not possible after the blood leaves the major vessels, although muscle contractions and accessory hearts may influence peripheral flow. Whatever their patterns, the circulatory systems of invertebrates are adequately matched to their needs; they have enabled these creatures to survive and proliferate for millions of years.

STUDYING INVERTEBRATE CIRCULATION

Methods used to study invertebrate circulatory systems are varied. One basic problem is that 95 percent of all animal species are invertebrates, and many of these creatures are not known and have never been studied. The larger, more common organisms that are easiest to study have been subjected to experimentation. Many invertebrates are difficult to maintain in the laboratory, but techniques to maintain them in good health are being developed and improved. Without these culture techniques, experimenters must use recently caught subjects whose condition is doubtful.

Most knowledge of invertebrate circulatory systems is anatomical. Even a common animal must be described so that a physiologist can apply appropriate techniques to study the functioning of the heart, vessels, and blood. Descriptions are usually derived from dissection and from microscopic study. These painstaking methods have been used on known species for centuries. Larger invertebrates, such as lobsters, crabs, clams, squid, and octopods, have been studied by techniques similar to those used in vertebrate circulation physiology. A heart can be exposed and either attached to a lever that can record its contractions or attached to an electronic force transducer, which can measure the strength of contraction.

The heart is large enough to be punctured for blood samples, and hemolymph can be withdrawn from the larger sinuses. These blood samples can be analyzed for the presence and activity of cells, respiratory pigments, nutrients, and wastes, using ordinary biochemical techniques. The development of microanalytic techniques in biochemistry allows the sampling of body fluids from small insects, worms, and rare organisms that might be damaged by taking larger samples.

In addition, force transducers can be placed along blood vessels and in hemocoels to determine the pressure exerted during flow. Electronic devices can monitor the flow rate by detecting cells or the passage of a dye or magnetic substance. Radioactive tracers can be injected and their path followed. All these less invasive techniques allow the animals to survive longer and to deliver data that can be more reliably interpreted because they are from a healthy subject. The use of fewer organisms in research and the survival of rare creatures are important both to the environment and to the development of a full understanding of these complex systems.

ADVANTAGES OF CLOSED VERSUS OPEN SYSTEMS

Invertebrate circulatory systems occur in two plans, closed or open, and both styles of circulation have benefits for their users. Although open circulatory systems are usually thought to be sluggish and inefficient, they are not necessarily so. Active animals such as crustaceans and insects have open circulatory systems and metabolic rates (oxygen and nutrient demands) that equal those of the most active invertebrates, squid and octopods, which have closed circulatory systems.

Nemertean worms (*Rhynchocoela*) are a small group of inconspicuous and inconsequential worms that have an interesting circulatory system. These worms may reach thirty meters in length (although they are only a few millimeters wide), but they have a simple blood system. There may be two or three blood vessels, with connections between them, running the length of the body. The vessels have a layer of muscle in the walls. Contraction of these muscles and the main body muscles move blood—in any direction—along the vessels.

Annelids (wormlike animals) also have a closed circulatory system. The major blood vessel has pulsatile regions, often called hearts, which drive the blood forward. The pattern of blood vessels, including capillaries, is repeated in each segment of the animal, although most segments do not have "hearts." Accessory hearts occur in different segments, and the overall pattern of circulation varies greatly among species (and even within a single individual's many segments).

Arthropods and most mollusks have open circulatory systems; these invertebrates have a well-developed central heart. The heart pumps blood through an extensive arterial network, which may end in capillaries. The blood eventually leaves the blood vessels and enters lacunae. Diffusion of materials takes place in the capillaries or in the lacunae. Sinuses collect the blood for return to the heart. In these organisms, the blood follows an ill-defined path. Contractions of the body musculature affect the speed and volume of blood flow in any region.

The crustacean arthropods are a group having inactive members, which lack a heart and blood vessels entirely, and active members, which have a high level of circulatory system organization. The inactive members pump their body fluid through sinuses and lacunae using the pressure developed by muscle contraction. Decapod crustaceans, the familiar crabs, have a heart whose contraction drives blood into well-defined arteries; the return of blood to the heart from the veins occurs because of the elastic recoil of the ligaments suspending the heart in the pericardial cavity. Part of the beauty of this pattern is that only veins from the gills enter the pericardial cavity. Therefore, only oxygenated blood enters the heart and is pumped into the body. Thus, although there may be body spaces in which flow is indeterminate, the important blood flows are well controlled and can meet the needs of complex and active organisms.

—*Judith O. Rebach*

FURTHER READING

Anderson, D. T. *Atlas of Invertebrate Anatomy.* University of New South Wales Press, 1996.

LaBarbara, M., and S. Vogel. "The Design of Fluid Transport Systems in Organisms." *American Scientist* 70 (January/February 1982): 54–60.

Mader, Sylvia. *Biology: Evolution, Diversity, and the Environment.* Wm. C. Brown, 1987.

Meglitsch, Paul A., and Frederick R. Schram. *Invertebrate Zoology.* 3rd ed. Oxford University Press, 1991.

Purves, W. K., and G. H. Orians. *Life: The Science of Biology.* 6th ed. W. H. Freeman, 2001.

Schmidt-Nielsen, Knut. *Animal Physiology: Adaptation and Environment.* 5th ed. Cambridge University Press, 1998.

Starr, Cecie, and Ralph Taggart. *Biology: The Unity and Diversity of Life.* 9th ed. Brooks/Cole, 2001.

CIRCULATORY SYSTEMS OF VERTEBRATES

FIELDS OF STUDY

Hematology; Cardiology; Animal Physiology; Veterinary Medicine

ABSTRACT

Cells need a constant supply of blood. Blood transports important materials needed for metabolic, synthetic, and degradative activities, supplying energy and materials necessary for growth, repair of worn-out components of cells, reproductive activity, and other functions of the body. Among the many products that blood transports are oxygen, nutrients, metabolic wastes, heat, and hormones. The circulatory system links all tissues with one another and with the external environment to and from which many of these materials are transported.

KEY CONCEPTS

aorta: the major arterial trunk, into which the left ventricle of the heart pumps its blood for transport to the body

artery: a blood channel with thick muscular walls which transports blood from the heart to various parts of the body

atria: the two chambers of the heart, which receive venous blood from the body (via the right atrium) or oxygenated blood from the lungs (left atrium)

capillaries: the very fine vessels in various tissues, which connect arterioles with venules; it is here that the exchange between blood and the extracellular fluid takes place

cardiac output: the amount of blood ejected by the left ventricle into the aorta per minute

diastole: relaxation (filling with blood) of the heart chambers

pacemaker: a specialized group of cardiac muscle cells in the right atrium which initiates the heartbeat; also called the sinoatrial node

systole: contraction (emptying of blood) of the heart chambers

valves: specialized, thickened groups of muscle cells in the heart chambers, major arterial trunks, arterioles, and veins which prevent backflow of blood

BLOOD AND CELLS

Cells, the units of the animal body, need a constant supply of blood. Blood effects the transport of important materials needed for metabolic, synthetic, and degradative activities, supplying energy and materials necessary for growth, repair of worn-out components of cells, reproductive activity, and other functions of the body. Among the many products that blood transports through a system of closed channels are oxygen, nutrients, metabolic wastes, heat, and hormones. The circulatory system links all tissues with one another and with the external environment to and from which many of these materials are transported.

Basically, the circulatory system of vertebrates consists of two parallel systems of blood vessels: One, the arterial system, actively transports blood and its constituents from a central pumping station, the heart; the other, the venous system, more or less passively brings the blood back to the heart in a closed loop. The two systems branch again and again until they encounter all tissues in the body. In the extracellular space of tissues, the finest branches of each system, called arterioles and venules, are connected by means of a network of fine capillaries that allow the movement of blood in one direction, from the arterial into the venous system, in which valves prevent any backflow of blood. A head of pressure, generated in the heart, pumps the blood in this direction, facilitating the transport of substances as well as their movement and filtration out of the capillary membranes and into the extracellular fluid.

CIRCULATORY SYSTEMS OF FISH, AMPHIBIANS, AND REPTILES

The simplest level of organization of the circulatory system of vertebrates is seen in fish. The heart in fish consists of two chambers, an atrium (auricle) and a ventricle. The oxygen-poor, carbon-dioxide-rich blood returning from the body via a system of veins is first received by an enlarged vein, the sinus venosus, prior to entering the atrium. The atrium empties its blood into the thick-walled, muscular ventricle, which then pumps it into an enlarged artery, the conus arteriosus. The blood then passes through a major arterial trunk, the ventral aorta, going directly to the gills. The arteries in gills branch profusely and are connected via capillaries with other arteries. In the capillary bed, the blood becomes oxygenated and provides nutrients to the tissue. The oxygenated blood then flows to the head and the rest of the body, and from there returns to the heart through the venous system.

In preparation for their journey to land, ancient aquatic vertebrates had to evolve lungs for aerial breathing and had to evolve a complementary circulatory system. As demands for oxygen for a terrestrial existence increased, greater blood pressure and a new way of oxygenating blood were in order. The atrium became divided into two, the right one receiving the deoxygenated blood returning from the body and the left one receiving the oxygenated blood from the lungs (which replaced gills). The deoxygenated blood, entering the right part of the single ventricle, is pumped into the pulmonary artery, all the way to the lungs. The left part of the ventricle, receiving oxygenated blood from the left atrium, pumps it into the body. This three-chambered heart is present in amphibians and most reptiles. The oxygenated and deoxygenated bloods mix partially in the ventricle. In some amphibians, flaps and partial valves tend to prevent such mixing. Reptiles have a partition between the right and left parts of the ventricle, which is complete in alligators, crocodiles, and turtles.

CIRCULATORY SYSTEMS OF BIRDS AND MAMMALS

Later reptiles, birds, and mammals developed four-chambered hearts. This complete division of the heart into two separate right and left pumps enables birds and mammals to achieve high speeds. One pumping circuit, the pulmonary, receives blood from the body and pumps it to the lungs; the other pumping circuit, the systemic, receives oxygen-rich blood from the lungs and pumps it into the systemic circulation. Valves within the heart prevent the blood from flowing through it in the opposite direction.

The contractile tissue of the heart consists of muscle cells that receive sympathetic and parasympathetic nerve impulses. The vertebrate heart is myogenic; that is, all of its muscle cells and fibers possess an inherent capacity to contract (electrically depolarize) rhythmically; however, all these fibers are under the control of a group of specialized heart muscle cells which have a lower threshold for depolarization than other heart muscle cells: the pacemaker. In fish, amphibians, and reptiles, the pacemaker is located in the wall of the sinus venosus (the first heart chamber before the atrium). In higher vertebrates, which lack a sinus venosus, the pacemaker is found in the wall of the atrium and is called the sinoatrial node.

The wave of electrical depolarization initiated here is conducted through the atrioventricular node via a special group of fibers called the Bundle of His, which branch out into the ventricular muscle. The depolarization enters and traverses the atrioventricular node only relatively slowly but spreads down the atrioventricular bundle and its branches much more rapidly than it could travel through ordinary ventricular muscle. This regulates the sequence of contraction of the heart chambers: The atria contract first and the ventricles later, each group of muscles contracting approximately in unison.

Since the pulmonary (right) circuit is much shorter than the systemic circuit, it contains less blood volume and offers less frictional resistance to blood flow; also, the right ventricle has muscular walls that are less thick than those of the left ventricle, which has to pump large volumes of blood to the entire body via the systemic circuit. After the two ventricles are completely filled (a condition referred to as diastole), they contract simultaneously (called systole). During systole, the maximum arterial pressure is generated; during diastole (just before systole), arterial pressure decreases to a minimum. The pulmonary side of the heart contains the funnel-shaped valve between the atrium and the ventricle known as the atrioventricular valve, the right one having three flaps, or cusps (and hence named the tricuspid valve), and the left one (the bicuspid or mitral valve) having two. The free edges of these cusps hang down into the ventricular cavities and are anchored by tendonlike cords of connective tissue called chordae tendinae, each of which is attached to the ventricular wall by a lump called a papillary muscle. The pulmonary artery and the aorta originate at the base of the right and left ventricles, respectively, each having a semilunar valve at its origin. Each of these valves opens in the direction of the blood flow and prevents the backflow of blood. The ventricular contraction and the resulting turbulence in the blood produce the long, low-pitched "lub" sound that can be detected with a stethoscope. The sudden closure of the semilunar valves is similarly perceived to emit a relatively short, high-pitched "dup" sound.

BLOOD VOLUME AND BLOOD VESSELS

The volume of blood that is pumped by the heart each minute is called the minute-volume, whereby the heart beats (contractions) per minute (cardiac stroke rate) eject a typical quantity of blood per beat. This rate is altered by the body's activity and by the volume of blood returning to the heart from the veins each minute. If the venous blood volume is adequate, then increase in stroke rate can increase minute-volume. The increased stroke rate, however, involves a decrease in the ventricular filling time, and as a result, the ventricles do not fill completely. Thus, the stroke volume is decreased; at rapid heart rates, even the minute-volume may be decreased, so that it offsets the stroke rate. During systole, the ventricles do not empty completely. A small residual volume of blood remains in them. An increased venous return may cause more complete filling and emptying of the ventricles, thus increasing the cardiac output without changing the stroke rate.

The vessels at various points in the circulatory path differ anatomically and functionally. The great arteries have thick walls heavily lined with smooth muscle and contractile tissue to enable them to transport blood under pressure from the heart to peripheral tissues. The arteries become smaller and thinner-walled as they branch out toward the periphery. The systemic arteries deliver blood to the microcirculatory beds of the tissues and organs. These "capillary beds" consist of microscopic arterioles, capillaries, and venules.

The contraction (vasoconstriction) and relaxation (vasodilation) of the smooth muscles in the terminal branches of the arteries play an important role in regulating blood flow in the capillary bed. Control of the arteriole muscles is mediated by sympathetic neurotransmitters, hormones, and local effects. From the arterioles, the blood enters the capillaries, minute vessels whose walls consist of a single layer of cells, facilitating transfer of oxygen and nutrients to the tissues and the loading of metabolic waste and carbon dioxide, all via the extracellular fluid. Their density depends on the need of the particular tissue for nutrients and oxygen. The capillaries drain into small, thin-walled but muscular vessels called venules, whence the blood begins its return to the heart through the veins. The veins have elastic walls but are without muscles. The venous vasculature serves as a reservoir, storing about 60 percent of the blood.

STUDYING VERTEBRATE CIRCULATION

Circulatory systems of vertebrates have been studied since ancient times through dissection and observation of animal and human cadavers: The heart can be cut open to examine its chambers and their structures, and the body wall can be cut open from the ventral side to expose the circulatory organs. Preserved, dissected animals, including fish, amphibians, reptiles, and mammals, are available from suppliers for students of anatomy who wish to conduct their own dissections. The venous systems of these animals are dyed blue and the arterial systems are dyed red. Plastic models of the circulatory system can be purchased for classroom use.

Scientists are also interested in microcirculation, or circulation at the capillary level. One can fasten a live frog on a frog board and observe the capillaries in the frog's foot web under a microscope. The movement of the red blood cells into the capillary is observed; it is slow and intermittent. The blood flow is regulated by the central nervous system (the vasomotor center in the medulla), as well as by local conditions (such as levels of carbon dioxide, acidity, histamine, temperature, and inflammation). One can then immerse the foot in hot or cold water and observe the resulting change in blood flow. Histamine can be applied to cause vasodilation, which can be controlled by epinephrine. Drops of dilute hydrochloric acid can be applied to the foot to cause vasodilation and inflammation. Thus, it is clear from the foregoing discussion that the heart and circulatory system are of vital importance to the health of an animal.

—M. A. Q. Khan

FURTHER READING

Berne, R., and M. Levy. *Physiology*. 4th ed. C. V. Mosby, 1998.

Hill, Richard W., and Gordon Wyse. *Animal Physiology*. 2nd ed. Harper & Row, 1987.

Raven, P. H., and G. B. Johnson. *Biology*. 4th ed. McGraw-Hill, 1996.

Robinson, T. F., et al. "The Heart as a Suction Pump." *Scientific American* 254 (June 1986): 84–91.

Claws, Nails, and Hooves

Fields of Study

Animal Physiology; Biology; Marine Biology; Veterinary Science

Abstract

Claws, nails, and hooves are very often special growths of the outer skin (epidermis) of warm-blooded, vertebrate animals. The skin below a nail or claw, from which it grows, is called the matrix. If a nail, claw, or hoof part is torn off, it grows again, as long as the matrix has not been overly damaged. Nails and claws occur on the digits of fore limbs and hind limbs of vertebrates. Nails and claws are made of dead skin cells, very rich in the fibrous substance keratin. Each nail or claw consists of a root, concealed within a fold of skin at its base; a body, the exposed part attached to the skin surface; and an anterior edge overlapping the end of the digit to which it is attached.

Key Concepts

chitin: a transparent, horny substance of invertebrate exoskeletons

crustaceans: lobsters, shrimps, crabs, and barnacles

epidermis: the outer, protective layer of the skins of vertebrates

keratin: a tough fibrous substance found in hair, claws, nails, and hooves

matrix: the skin beneath a nail, involved in its growth

ungulate: a hoofed mammal

vascular papilla: a protuberance having a blood supply

WHAT ARE NAILS, CLAWS, AND HOOVES?

Claws, nails, and hooves are very often special growths of the outer skin (epidermis) of warm-blooded, vertebrate animals. Nails are considered to be claws whenever they are thicker, longer, harder, and sharper than those of humans. Nails and claws contain large amounts of keratin, the hard substance present in the skin and nails of human fingers and toes. The skin below a nail or claw, from which it grows, is called the matrix. If a nail, claw, or hoof part is torn off, it grows again, as long as the matrix has not been overly damaged. Nails and claws occur on the digits of fore limbs and hind limbs of vertebrates.

Cloven hooves of roe deer (Capreolus capreolus), *with prominent dewclaws. (Foto von Joachim Bäcker)*

Nails and claws are made of dead skin cells, very rich in the fibrous substance keratin. Each nail or claw consists of a root, concealed within a fold of skin at its base; a body, the exposed part attached to the skin surface; and an anterior edge overlapping the end of the digit to which it is attached. In clawed animals such as badgers, moles, rodents, wolves, and cats, this edge can become quite long. The matrix, the skin below the root and body of a nail or claw, is thick and covered with highly vascular papillae. The matrix color is sometimes seen through transparent horny tissue, in humans and in some other species with relatively thin nail bodies or claws. A nail or claw grows forward by a combination of continual growth of new cells both at the root and under its body.

The body of the nail or claw in the predatory and burrowing animals is much thicker than in humans. Unlike human nails, the claws of burrowing animals do not have to be trimmed; they are worn down by use. The claws of cats, from house cats to lions and tigers are retractable. This allows these animals to walk easily without wearing their claws down and to unsheathe them for the serious business of capturing and killing prey. Predatory animals, especially cats, that have retractable claws must hone them down from time to time. Cats use scratching posts—whether fallen tree limbs found in the wild or specially made carpet-covered posts provided by hopeful pet owners—for this purpose, as well as to sharpen the claws.

UNGULATE HOOVES

A hoof is a thick, hard growth at the front of a foot of a horse or other mammal in the order *Ungulata* (the hoofed mammals). The ungulates include elephants, pigs, horses, and many horned animals, such as deer and cattle. A hoof, really a "hoof-nail," is also made of keratin and it arises from the outer skin layer, or epidermis. A hoof differs physically from a nail or claw in that it is blunt and encases a toe or foot. Hooves are thought to have developed from animal toes, and ungulates are divided into those with even and odd numbers of toes. For example, a horse hoof developed from one toe. It almost completely circles the bottom of the foot. Zebras and donkeys are also animals whose hooves are single toes, not divided (or cleft). Even-toed ungulates have two or four toes. Usually, the two middle toes are developed into a cleft (divided) hoof. This vertebrate group includes deer, antelope, sheep, goats, cattle, pigs, and hippopotamuses. Hooves adhere to the earth well and give animals that have them a firm footing for pulling, running, walking on slippery or icy ground, and climbing mountains.

LIZARD AND BIRD CLAWS

Lizards, reptiles in the order *Squamata*, which also includes snakes and crocodiles, are the largest extant group of reptiles, with about three thousand species. All have dry scaly skin, clawed feet, and external ear openings. In many cases, lizard claws are used for traction on slick surfaces, or in climbing trees. They are also used by large carnivorous and omnivorous species to pick apart or tear apart food.

Most land birds have clawed feet, which help them to perch in trees. The claws are thought to have derived partly from the evolution of birds from ancient lizards. The most extensive claws seen in the birds are the talons of birds of prey (the raptors). Raptors include the night-hunting owls and the day-hunting hawks, eagles, falcons, and vultures. All are meat eaters, although the meat devoured by the small species is insects. All of the raptors have powerful bills, and all but vultures have grasping toes tipped with large, curved, sharp talons. It is believed that the weak talons of vultures are part of the reason why they eat carrion, not live prey. That is, they cannot strike live prey like the other raptors or carry it off in strong claws.

INVERTEBRATE CLAWS

Among invertebrates, insects, scorpions, and crustaceans are examples of organisms that possess claws. Examples are praying mantises and beetles, among the insects; scorpions, among the arthropods; and lobsters and crabs, among the crustaceans. In all cases, the claws are restricted to the first two legs of these invertebrates. Also, unlike those of vertebrates, the claws are not keratin-containing epidermis outgrowths. They are portions of the chitinous invertebrate exoskeletons, and therefore part of their shells.

Crustacean claws are huge, compared to animal claws. Most often, both claws are of somewhat similar size and used to capture prey. However, some crustaceans, such as hermit and fiddler crabs, have a pair of claws in which one is much larger than the other. In those cases, the larger of the claws is usually used in protection and/or mating.

Lobsters have two huge, body-length claws used to grasp prey, and to dig burrows in which they live. In "true lobster" species, the claws are especially huge. One claw is heavier than the other and has blunt teeth to crush prey. The other is somewhat smaller and has sharp teeth to tear up prey. Not all lobsters have the heavy claw on the same side. They are "right clawed" or "left clawed." During the day, a lobster stays in its burrow awaiting prey. At night it comes out to seek food, catching it and tearing it up into eating-sized chunks with its claws. Lobsters shed their shells (molt) as they grow. When they do this, their claws are soft. They cannot use them well for hunting until the new shells harden. At those times, they mostly eat carrion.

Crabs, related to lobsters, are also clawed decapods. Like other cephalopods, the crab body is sheathed in tough chitin. A crab has five pairs of walking legs, and two front legs which hold claws for feeding and defense. Male hermit and fiddler crabs have one claw that is huge compared to the other. The oversized claw in hermit crabs is used defensively, to close off the mollusk cell it uses as its abode to protect its soft, vulnerable body. Fiddler crabs also have one oversized huge claw; they use it in mating and combat.

Claws, nails and hooves are very important to animals. They are weapons used for protection and to secure food, appendages that impress potential mates, and locomotor appendages that assure safe, optimized movement around habitats. It is interesting that in organisms as far apart, evolutionarily, as lobsters and ungulates, claws and hooves are made of similar, tough insoluble materials, chitin and keratin, respectively.

—*Sanford S. Singer*

FURTHER READING

Bliss, Dorothy E. *Shrimps, Lobsters, and Crabs.* Columbia University Press, 1990.

Chapman, R. F. *The Insects: Structure and Function.* 4th ed. New York: Cambridge University Press, 1998.

Freethy, Ron. *How Birds Work: A Guide to Bird Biology.* Sterling, 1982.

Lamar, William W., Pete Carmichael, and Gail Shumway. *The World's Most Spectacular Reptiles and Amphibians.* World, 1997.

Olsen, Sandra L., ed. *Horses Through Time.* Roberts Rinehart, 1996.

Watt, Melanie. *Black Rhinos.* Raintree/Steck-Vaughn, 1998.

CLONING OF EXTINCT OR ENDANGERED SPECIES

FIELDS OF STUDY

Embryology; Palaeontology; Biology; Genetics

ABSTRACT

Conservation biologists have noted that ever since the 1600s, the rate of animal species extinction has increased significantly, primarily due to the rapid growth of the human population and the resulting destruction of animal habitats. Due to recent advances in reproductive biotechnology, the rise and decline of a species may no longer be determined solely by the forces of natural selection and the survival of the fittest. The tools of recombinant DNA technology may afford humans a role in preventing the extinction of endangered species.

KEY CONCEPTS

biodiversity: the variety of plants, animals, and habitats and the interactions among these species

biotechnology: the use of the tools of recombinant DNA technology to study or modify biological systems

cloning: the reproduction of an individual to create an offspring that is genetically identical to its parent

extinction: the dying off of all individuals of a species

gene pool: the collection of genes or genetic information in a population of individuals

nuclear transfer: the insertion of genetic material from a donor cell into a recipient cell; in reproductive technologies, the recipient cell is an egg cell from which the nucleus has been removed

EXTINCTION IS FOREVER. MAYBE.

The threat of extinction to endangered animal species is of great concern to conservation biologists as well as animal lovers everywhere. Extinction is the dying off of all members of a species. When it occurs naturally, as part of evolution in response to global climatic changes and dynamic relationships within ecosystems that result in interspecies competition, the process is termed *background extinction*.

Conservation biologists have noted that ever since the 1600s, the rate of animal species extinction has increased significantly, primarily due to the rapid growth of the human population and the resulting destruction of animal habitats. The global extinction rate is estimated to be between twenty and thirty thousand species per year, and conservation biologists have theorized that the planet is in the midst of the greatest period of animal extinction since the dinosaurs disappeared sixty-six million years ago. Such wide-scale extinctions may have profound effects on ecosystem survival, as they culminate in the irretrievable loss of thousands of species of plants and animals, some of which have never even been identified. Some of the more recent extinctions of note include the great auk and the passenger pigeon, both of which disappeared in the nineteenth and early twentieth centuries. Other species, such as the California condor, are on the verge of extinction. Mammals are not exempt from this list; in 2012, 54 percent of all primate species and subspecies were considered by the International Union for Conservation of Nature (IUCN) to be in danger of extinction. A study published in the journal *Science Advances* in January of 2017 noted that approximately 60 percent of primate species faced possible extinction.

Due to recent advances in reproductive biotechnology, however, it has become apparent that the rise and decline of a species may no longer be determined solely by the forces of natural selection and the survival of the fittest, as proclaimed by the great evolutionary biologist Charles Darwin. Rather, the tools of recombinant DNA technology may afford humans a role in preventing the extinction of endangered species. On this rapidly evolving planet, where human civilization increasingly encroaches on the domain of the natural world, the possibility of preserving endangered wildlife species via biotechnology may be the last remaining hope for treasured animal species whose numbers have dwindled or disappeared entirely. The scientific name for the technologies that enable reversing extinction is known as resurrection biology or species revivalism.

CLONING TECHNOLOGY

The basic scientific procedure that has made this hope possible is a cloning technology termed nuclear transfer. The most effective form of nuclear transfer, known as the Honolulu technique, involves the insertion of the nucleus of a somatic (body) cell from an endangered animal into an egg cell obtained from a closely related species. First, the nucleus of the endangered animal's somatic cell is removed, as is the nucleus of the egg cell. These nuclei contain the genetic material of the animals they came from. Then the nucleus of the somatic cell is inserted into the egg cell by a microinjection technique using a fine needle. The egg cell is then placed in a chemical bath to stimulate cell division. An earlier, alternative method is the Roslin technique, in which the nucleus is not removed from the somatic cell; rather, the cell is starved of nutrients so it becomes dormant, then placed next to an egg cell from which the nucleus has been removed. An electric pulse fuses the somatic cell with the egg cell, the nucleus enters the egg cell cytoplasm, and cell division begins.

Amazingly, the genetic material of a fully differentiated or mature somatic cell, such as a skin cell, becomes genetically reprogrammed once inserted into the egg cell to generate all the cells that will ultimately make up the tissues and organ systems of the fetus. Because the somatic cell and the egg cell originate from two different species, the procedure is termed interspecies nuclear transfer. This concept is critical to cloning endangered species, since often it is impossible to harvest egg cells from the threatened species, so a related but thriving species must serve as the egg donor. Once the nuclear transfer has been accomplished, the embryonic cells are implanted into the uterus of a surrogate mother, also from a related species. The result is that an animal of one species participates in the reproduction of a cloned animal of a different species.

SCIENTIFIC ADVANCES

The world's first successful cloning of an animal belonging to an endangered species by interspecies nuclear transfer was carried out by researchers at Advanced Cell Technology (ACT) in Worcester, Massachusetts, who cloned a baby gaur, an oxlike animal found in India, Indochina, and Southeast Asia. Using skin cells from the gaur as a source of genetic material, internuclear transfer by microinjection of an egg cell obtained from a cow and subsequent implantation into the uterus of a cow serving as a surrogate mother resulted in the birth of a cloned ox named Noah on January 8, 2001. Unfortunately, the cloned animal died just two days later from a common dysentery infection, apparently unrelated to the cloning procedure or its gestation in a different species.

ACT had another, more lasting success in April 2003, when two cows gave birth to cloned bantengs, a type of Asian wild cattle. The donor cells were taken from a frozen sample stored at San Diego Zoo and implanted in 2002. Although one of the bantengs was euthanized shortly after birth, the other lived for seven years, dying in April 2010. While it survived longer than any previous interspecies clone, the cloned banteng lived for less than half its species' normal life span.

The first clone of a fully extinct species was born in January 2009. Scientists at Spain's Centro de Investigación y Tecnología Agroalimentaria de Aragón (Agrifood Research and Technology Centre of Aragon) implanted DNA taken from frozen skin samples of the Pyrenean ibex, which had been declared extinct in 2000, in the eggs of domestic goats. Of the over fifty embryos that were implanted, only one resulted in a live birth. The newborn ibex lived for seven minutes before dying due to defects in its lungs.

In August 2011, a preserved baby wooly mammoth was discovered in Russia. Scientists from Russia's North-Eastern Federal University and Japan's Kinki University announced their intention to try to clone the creature, though experts have expressed skepticism about the viability of any DNA that may be found. In 2017, researchers at Harvard University in Cambridge, Massachusetts, reported that they anticipate producing a functioning wooly mammoth embryo by 2020. The researchers at Harvard claim they will use a different approach that does not rely on mammoth DNA, instead modifying an elephant genome with mammoth genes to implant into an elephant embryo to create a hybrid embryo.

FUTURE CHALLENGES

A broader question involves the potential role of cloning in maintaining species and genetic diversity. It has been argued that the selective cloning of individual members of a species will lead to a reduction in genetic diversity due to a streamlined gene pool. This may ultimately affect the survival and adaptability of a species.

Moreover, cloning does not directly address the conditions that led to the loss of species fitness, which are reflected in reproduction and survival rates. In many instances, species become endangered due to a loss of habitat; thus, survival of the species may require long-term existence in captivity. In certain cases, attempts have been made to return animals bred in captivity to their natural habitat, but it is not clear that these efforts will be successful in general. The reproductive technologies and long-term effects of captivity may have serious consequences on the survival of the species once it is reintroduced to the wild. Some successes have been noted in this area; for example, in the 1980s, the American peregrine falcon was reintroduced into the wild in eastern North America, where it had become extinct due to the widespread use of the pesticide DDT.

Other problems associated with interspecies nuclear transfer and assisted reproductive technologies in general are their high cost and low success rate. It has been estimated that the average success rate of same-species nuclear transfer is between 1 and 3 percent. The success rate for interspecies nuclear transfer is considerably less. The ACT's attempts at interspecies nuclear transfer when cloning the gaur required almost seven hundred cow eggs, about eighty of which developed as suitable donors for transplantation. About forty of these were implanted into cows, which resulted in eight pregnancies and ultimately only one live birth.

Despite these concerns, zoological parks around America have collected sperm and eggs from many types of animals in the form of "frozen zoos," which may serve as genetic repositories for producing animals of the future. In addition, the San Diego Zoo's Institute for Conservation Research and the Audubon Center for Research of Endangered Species have sponsored animal-tissue banks that could be the future source of cloned animals to safeguard species from extinction. The genetic material for the cloned bantengs was taken from the tissue bank at the San Diego Zoo.

The death of a species by extinction engenders a sense of profound loss as it brings to a close a chapter of evolutionary and biological history. It may also disrupt the delicate balance of life, as species interdependence is a fundamental component of ecological systems. As efforts to save individual species from permanent extinction by the cloning of selected members are applauded, it is important to reflect on the conditions and circumstances that have contributed to their loss.

—*Sarah Crawford Martinelli*

FURTHER READING

Bethge, Philip. "Preservation in a Petri Dish: Scientists Hope Cloning Will Save Endangered Animals." *Spiegel Online*. Spiegel Online, 8 Nov. 2012. Web. 18 Sept. 2013.

Chadwick, Douglas, and Joel Sartore. *The Company We Keep: America's Endangered Species*. National Geographic, 1996.

Corley-Smith, Graham E., and Bruce P. Brandhorst. "Preservation of Endangered Species and Populations: A Role for Genome Banking, Somatic Cell Cloning, and Androgenesis?" *Molecular Reproduction and Development* 53.3 (1999): 363–67.

Gray, Richard, and Roger Dobson. "Extinct Ibex Is Resurrected by Cloning." *Telegraph*. Telegraph Media Group, 31 Jan. 2009. Web. 18 Sept. 2013.

Lanza, Robert P., Betsy L. Dresser, and Philip Damiani. "Cloning Noah's Ark." *Scientific American* Nov. 2000: 84–89.

Myers, Norman, et al. "Biodiversity Hotspots for Conservation Priorities." *Nature* 403.6772 (2000): 853–58.

Sample, Ian. "Wooly Mammoth DNA May Lead to a Resurrection of the Ancient Beast." *Guardian*. Guardian News and Media, 30 July 2013. Web. 18 Sept. 2013.

Shapiro, Beth. *How to Clone a Mammoth*. Princeton University Press, 2015.

Tilson, Ronald, and Philip J. Nyhus, eds. *Tigers of the World: The Biology, Biopolitics, Management, and Conservation of Panthera tigris*. 2nd ed. Academic, 2010.

Wray, Britt. *Rise of the Necrofauna: The Science, Ethics, and Risks of De-Extinction*. Greystone Books, 2017.

Coevolution

FIELDS OF STUDY

Evolutionary Biology; Ecology; Environmental Studies; Animal Physiology

ABSTRACT

Coevolution is an extremely important and widespread phenomenon in the world of living things. When two or more different species experience a relationship in which any of the participating species' evolution directly affects the evolution of the other members, coevolution is taking place. This interactive type of evolution is characterized by the fact that the participant lifeforms are acting as a strong selective pressure upon one another over a period of time.

KEY CONCEPTS

antagonism: any type of interactive, interdependent relationship between two or more organisms that is destructive to one of the participants

coevolution: the interactive evolution of two or more species that results in a mutualistic or antagonistic relationship

commensalism: a type of coevolved relationship between different species that live intimately with one another without injury to any participant

parasitism: a type of coevolved relationship between different species in which one species exploits the other to its physical detriment

phytophagous: animals, also referred to as herbivorous, that feed on plants

reciprocal relationship: any type of coevolved, highly interdependent relationship between two or more species

selective pressure: evolutionary factors that favor or disfavor the genetic inheritance of various characteristics of a species

symbiosis: a type of coevolved relationship between two species in which both participants benefit; a type of mutualism

A CLOSER LOOK AT EVOLUTION

Coevolution is an extremely important and widespread phenomenon in the world of living things; it is a biological factor that is global in influence. When two or more different species experience a relationship in which any of the participating species' evolution directly affects the evolution of the other members, coevolution is taking place. This interactive type of evolution is characterized by the fact that the participant lifeforms are acting as a strong selective pressure upon one another over a period of time.

The assumption of the interdependence of all organisms is today such a commonplace and fundamental concept that it is surprising that the phenomenon of coevolution has not always enjoyed a more prominent position in evolutionary thinking. Many scientists seem to have considered sets of coevolved organisms as relatively unimportant phenomena, almost on the level of biological "curiosities." The consensus appears to have been that while numerous examples of coevolution existed in both plant and animal kingdoms, overall it was of relatively minor importance in comparison with other evolutionary phenomena, such as competition. This opinion has begun to change as researchers increasingly recognize the intrinsic and ubiquitous role that coevolution has played, and continues to play, in the evolution of life at all levels throughout earth history.

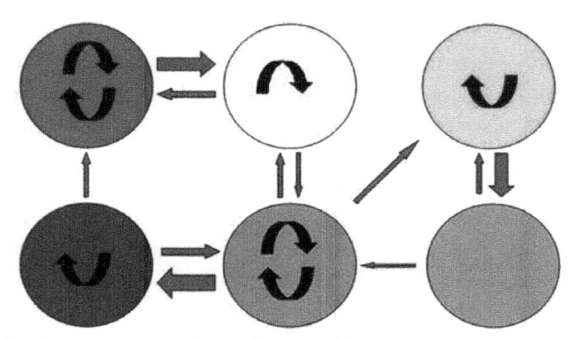

Hypothetical example of a Geographic Mosaic of Coevolution between two species: The coloured circles stand for biological communities; the arrows within the circles show interactions within local communities and represent selection on one or both (or none) species (truly reciprocal selection = hotspot). Selection among communities is generated from disparities in biotic or abiotic habitat quality. The arrows between the communities indicate gene flow (thicker arrows = more gene flow).

Graphic about the Geographic Mosaic Model of Coevolution.

GAIA

Organisms do not evolve in a biological vacuum. All organisms exist in, and have evolved within, the framework of one of a great number of delicately balanced and self-tuning biological systems or living communities termed ecosystems. Indeed, the entire planet can be regarded as one huge, incredibly complex ecosystem in which all the lesser ecosystems fit together and work together harmoniously. This planetary ecosystem has been called "Gaia" by some biologists, in reference to the ancient Greek earth goddess. In some respects, Gaia can be conceived of as actually one giant, worldwide organism. All the living communities in this huge ecosystem are products of coevolution. This phenomenon has been in effect over the vast expanses of geologic time and continues today. The only period in history when coevolution was probably not operating was at the dawn of life, billions of years ago, when the very first species of organisms appeared and had not yet established interactive communities. The importance of coevolution as a factor affecting life cannot be overstated.

Some biologists use the term coevolution in a more restricted sense to describe coevolved relationships that have developed between plants and animals, particularly between plants and animals that are herbivores or pollinators. The coevolution between plants and animals is one of the aspects of the field that has traditionally received the most attention, so this aspect of coevolution provides a useful departure point in describing the phenomenon.

COEVOLUTIONARY WARFARE

The coevolution of plants and animals, whether animals are considered strictly in their plant-eating role or also as pollinators, is abundantly represented in every terrestrial ecosystem throughout the world where flora has established itself. Moreover, the overall history of some of the multitude of present and past plant and animal relationships is displayed (although fragmentally) in the fossil record found in the earth's crust. The most elemental relationship between plants and animals is that of plants as food source. This relationship has an extremely long history, beginning with the evolution of microscopic, unicellular plants that were the earth's first autotrophs (organisms that can produce their own food from basic ingredients derived from the environment). In conjunction with the appearance of autotrophs, microscopic, unicellular heterotrophs (organisms such as animals, which must derive food from organic sources such as autotrophs) evolved to exploit the simple plants. This ancient and basic relationship has resulted in uncounted numbers of plant and animal species evolving and coevolving over billions of years of earth's history.

As both plants and animals became multicellular and more complex, more elaborate defense mechanisms evolved among plants, as did more elaborate feeding apparatuses and behavior among animals. This biological "arms race" grew ever more intense as groups of plants and animals eventually adapted to the more rigorous demands of a terrestrial existence, leaving the marine environment behind. New ecosystems developed that culminated in the world's first swamps, jungles, and forests. The plant-animal arms race engendered increasingly more sophisticated strategies of botanical defense and animal offense, and this coevolved interrelationship has continued unabated. This coevolutionary "warfare" between plants and animals has expressed itself partly through the evolution of botanical structures and chemicals that attempt either to discourage or to prevent the attentions of herbivores. These include the development of spines, barbs, thorns, bristles, and hooks on plant leaf, stem, and trunk surfaces. Cacti, holly, and rose bushes illustrate this form of plant strategy.

Another type of deterrence evolved in the form of chemical compounds that can cause a wide spectrum of negative animal responses. These compounds range in effect from producing a sensation of mild distaste, such as bitterness, to more extreme effects, such as actual poisoning of herbivore metabolisms. Plants that contain organic compounds such as tannin are examples of the chemical defensive strategy. Tannins produce several negative results in animals, including partially inactivating digestive juices and creating cumulative toxic effects that have been correlated with cancer. Plants containing tannin include trees, such as members of the oak group, and shrubs, such as those that produce the teas used as human beverages. Other plants have developed more lethal poisons that act more rapidly. Plants have also developed other strategies, such as possession of a high silica content (as found in grasses), that act to wear down the teeth of plant eaters. Animals have counteradapted to these plant defensive innovations by evolving a higher degree of resistance to plant toxins or by developing more efficient and tougher teeth with features such as harder enamel surfaces, or the capacity of grinding with batteries of teeth.

COEVOLUTIONARY ALLIANCES

Not all coevolution is characterized by having an adversarial nature; mutually beneficial relationships are also very common. Sometime during the latter part of the Mesozoic era, angiosperms, the flowering plants, evolved and replaced most of the previously dominant land plants, such as the gymnosperms and the ferns. New species of herbivores evolved to exploit these new food sources. At some point, probably during the Cretaceous period of the late Mesozoic, animals became unintentional aids in the angiosperm pollination process. As this coevolution proceeded, the first animal pollinators became more and more indispensable as partners to the plants. Eventually, highly coevolved plants and animals developed relationships of extreme interdependence exemplified by the honeybees and their coevolved flowers. This angiosperm-insect relationship is thought to have arisen in the Mesozoic era by way of beetle predation, possibly on early, magnolia-like angiosperms. The fossil record gives some support to this theory. Whatever the exact route along which plant-animal pollination partnerships coevolved, the end result was a number of plant and animal species that gained mutual benefit from the new type of relationship. Such relationships are in general termed mutualisms.

Eventually some of these plant-animal mutualisms became so intertwined that one or both participants reached a point at which they could not exist without the aid of the other. These obligatory mutualisms ultimately involved other types of animal partners besides insects. Vertebrate partners such as birds, reptiles, and mammals also became involved in mutualisms with plants. Contemporary ecosystems, such as the United States' southwestern desert, include mutualisms between aerial mammals, such as bats, and plants, such as the agave and the saguaro cactus. The bats involved are nectar drinkers and pollen eaters. They have evolved specialized feeding structures such as erectile tongues similar to those found among moths and other insects with similar lifestyles. In turn, the plants involved with the pollinating bats have evolved either reciprocal morphologies or behavior patterns to accommodate their warm-blooded visitors. For example, angiosperms coevolutionarily involved with bats have developed such specializations as bat-attractive scents, flower structures that minimize the chance of injury to bats, and petal openings timed to the nocturnal activity of bats.

SYMBIOSES, COMMENSALISMS, AND PARASITISMS

Coevolved relationships are not restricted to beneficial or nonbeneficial relationships between plants and animals. They also include an immense number of relationships between animals and other animals, and even between plants and other plants. Among these various types of coevolved situations can be found subcategories such as symbioses, commensalisms, and parasitisms. The first two involve relationships beneficial to varying degrees that feature interactions of increasing physical intimacy between or among two or more species. Parasitism involves an intimate relationship produced through coevolution in which one participant, the host, experiences serious harm or even death through exploitation by the parasite. Predation is probably the most obvious form of coevolution among higher animals such as vertebrates. Modern carnivores such as the canines and felines and their prey are a dramatic example of coevolution at work. Animal hunters over time responded to the improved defenses of their prey by evolving better senses, such as stereoscopic, three-dimensional vision, hearing with expanded range of frequency response, and more effective body structures, such as multifunctional teeth. Such teeth are termed heterodont and represent a great improvement over the simple dental array of the more primitive vertebrates, such as fish and amphibians.

Beginning with the more advanced reptiles appearing in the late stages of the Paleozoic era, teeth began to differentiate into specialized components—incisors, canines, premolars, and molars—that enhanced food acquisition and improved mastication. This, in turn, improved digestion and allowed quicker energy acquisition from food. This evolutionary advantage has reached a zenith of adaptive success among the mammals. Mammalian predators evolved fangs and efficient claws, sometimes retractable, to minimize injury and wear. Along with improved hunting senses and better dentition came increased speed from the evolution of improvements in pelvic and limb arrangements. In response to this process, vertebrate herbivores also became generally swifter or better defended, more alert, attained higher metabolic rates, and were thus better able to elude or defend against predation. Advanced predators placed an intense selective pressure on their prey herbivores, spurring ever more efficient and acutely tuned responses among the herbivore populations. Herbivores evolved either as swift forms, such as deer, or became efficiently defended, walking fortresses, such as porcupines or armadillos. Because of the pervasive effect of coevolution, the overall relationship between predator and prey has been a reciprocal one in which all participants affect one another in an interactive manner.

UNRAVELING THE INTRICACIES OF COEVOLUTION

Field research and laboratory research are pursued concurrently in the effort to unravel the intricacies of the subject of coevolution. Field research involves actual observation in nature of animals and plants, their behavior, and, especially, their interaction with other species. Special attention is given to useful clues that can be employed to establish evolutionary relationships, either presently existing or previously in effect. For example, cooperative behavior between or among several different species of animal or plant is often indicative of an established, coevolutionary relationship. If this behavior is consistent over time and can also be traced or inferred through the agency of the fossil record, more useful data are acquired concerning a possible, evolved, reciprocal relationship. Of particular importance is the confirmation of specialized physical structures that are unique to the members of the observed relationship. Examples are the specialized feeding apparatuses of pollinating animals and the specialized, accommodating flower structures of their angiosperm partners.

Such physical structures are strong evidence for the handiwork of the coevolutionary process. Direct human observation is preferable in ascertaining coevolutionary behavior; however, this is not always possible because of the rapidity of the animals involved, their habitat, their extremely small size, their preference for nocturnal activity, or their determined avoidance of humans. Consequently, electronic and mechanical aids are sometimes indispensable. These include remote-controlled still and video cameras, microscopic or telephoto lenses, infrared or ultraviolet lighting units, sonar or radar sensors, trip wires and other mechanical triggering devices, and sound recording equipment with high-gain or long-range microphones.

Laboratory research in the field of coevolution involves investigations heavily reliant on modern, sophisticated laboratory equipment and techniques. High-powered conventional, optical microscopes are employed to determine tissue and cellular structures. Scanning electron microscopes (SEMs) are employed for study of extremely small unicellular animals or plants such as planktonic organisms or extremely small organic structures. In addition to these tools of laboratory specimen observation, there are the analytical equipment and techniques used to determine the genetic codes and blood protein complexes of animals and plants to establish the degree of relatedness or divergence between various species.

MAINTAINING BALANCES

Coevolutionary studies are increasingly important in the biological sciences. One of the aims is to determine the degree of interdependence between various species, whether the relationship is between animals and plants, animals and other animals, or plants and other plants. A key factor to be determined in all these coevolved relationships is that of the nature and degree of balance attained. Although most of the biological world is forever in a state of flux, some categorical, coevolved relationships have been of long duration and can be reasonably assessed as having been in existence for tens of millions of years, such as that of flowering plants and vertebrate and invertebrate pollinators, or even hundreds of millions of years, such as the oceanic, planktonic food chain.

The degree to which these large-scale, coevolved relationships, involving entire planetary ecologies, continue to enjoy their former degree of health and well-being is of the utmost importance to human society. The present depth of understanding of the biological sciences clearly indicates the interrelatedness of all nature. Many angles of study agree that the global life system is experiencing great stress from human intervention: industrialization, urbanization, and overpopulation. It becomes increasingly urgent to know with the utmost precision all facets of the way the global life system operates, and has operated with general stability, over geological expanses of time. Every detail that contributes to this knowledge—every coevolved relationship, no matter how seemingly insignificant—adds to the total effect. This information can be used as an important resource to help maintain the stability of the entire system for ourselves and future generations.

—*Frederick M. Surowiec*

FURTHER READING

Bakker, Robert T. *The Dinosaur Heresies.* New York: William Morrow, 1986.

Barlow, Connie *The Ghosts of Evolution: Nonsensical Fruit, Missing Partners and Other Ecological Anachronisms* Basic Books, 2000.

Barth, Friedrich G. *Insects and Flowers: The Biology of a Partnership.* Translated by M. A. Biederman-Thorson. Princeton University Press, 1991.

Chaloner, William G., Peter R. Crane, and Else Marie Friis, eds. *The Origins of Angiosperms and Their Biological Consequences.* Cambridge University Press, 1987.

Clayton, Dale H., and Janice Moore, eds. *Host-Parasite Evolution: General Principles and Avian Models.* Oxford University Press, 1997.

Gould, Stephen Jay. *The Panda's Thumb.* 1980. Reprint. W. W. Norton, 1992.

Grant, Susan. *Beauty and the Beast: The Coevolution of Plants and Animals.* Charles Scribner's Sons, 1984.

Lawton, J. H., Richard Southwood, and D. R. Strong. *Insects on Plants.* Harvard University Press, 1984.

Powell, Jerry A. *Biological Interrelationships of Moths and "Yucca Schottii."* University of California Press, 1984.

Rothstein, S. I., and S. R. Robinson, eds. *Parasitic Birds and Their Hosts: Studies in Coevolution.* Oxford University Press, 1998.

Thompson, John N. *The Coevolutionary Process.* University of Chicago Press, 1994.

Communication

Fields of Study

Ornithology; Biochemistry; Animal Physiology

Abstract

A simple definition of animal communication is the transmission of information between animals by means of signals. Developing a more precise definition is difficult because of the broad array of behaviors that are considered messages or signals and the variety of contexts in which these behaviors may occur. Animal signals can be chemical, visual, auditory, tactile, or electrical. The primary means of communication used within a species will depend upon its sensory capacities and its ecology.

Key Concepts

discrete signals: signals that are always given in the same way and indicate only the presence or absence of a particular condition or state

display: a term used to indicate social signals, particularly visual signals

pheromone: a chemical substance used in communication within a species

primer pheromone: a chemical substance that affects behavior by altering physiology and is therefore not rapid in its effects

releaser: a standard signal that elicits a standard response

trophallaxis: food exchange between organisms, particularly in social insects

No Failure to Communicate

A simple definition of animal communication is the transmission of information between animals by means of signals. Developing a more precise definition is difficult because of the broad array of behaviors that are considered messages or signals and the variety of contexts in which these behaviors may occur. Animal signals can be chemical, visual, auditory, tactile, or electrical. The primary means of communication used within a species will depend upon its sensory capacities and its ecology.

Pheromones

Of the modes of communication available, chemical signals, or pheromones, are assumed to have been the earliest signals used by animals. Transmission of chemical signals is not affected by darkness or by obstacles. One special advantage is that the sender of a chemical message can leave the message behind when it moves. The persistence of the signal may also be a disadvantage when it interferes with transmission of newer information. Another disadvantage is that the transmission is relatively slow.

The speed at which a chemical message affects the recipient varies. Some messages have an immediate effect on the behavior of recipients. Alarm and sex-attractant pheromones of many insects, aggregation pheromones in cockroaches, or trail substances in ants are examples. Other chemical messages, primers, affect recipients more slowly, through changes in their physiology. Examples of primers include pheromones that control social structure in hive insects such as termites. Reproductive members of the colony secrete a substance that inhibits the development of reproductive capacity in other hive members. The chemicals important for controlling the hive are spread through grooming and food sharing (trophallaxis). Chemical communication is important not only among social and semisocial insects but also among animals, both vertebrate and invertebrate. Particularly common is the use of a pheromone to indicate that an animal is sexually receptive.

VISUAL SIGNALS

Visual communication holds forth the advantage of immediate transmission. A visual signal or display is also able to encode a large amount of information, including the location of the sender. Postures and movements of parts of an animal's body are typical elements of visual communication. Color and timing are additional means of providing information. Some visual signals are discrete; that is, the signal shows no significant variation from performance to performance. Other displays are graded so that the information content of the signal can be varied. An example of a graded display is found in many of the threat or aggressive postures of birds. Threat postures of the chaffinch vary between low-intensity and high-intensity postures. The elevation of the crest varies in ways that indicate the bird's relative readiness for combat. The song spreads of red-winged blackbirds and cowbirds show variation in intensity. In red-winged blackbirds, the red epaulets, or shoulder patches, are exposed to heighten the effect of the display. Discrete and graded signals may be used together to increase the information provided by the signal. In zebras, ears back indicates a threat and ears up indicates greeting. The intensity of either message is shown by the degree to which the mouth is held open. A widely open mouth indicates a heightened greeting or threat.

Visual displays depend upon the presence of light or the production of light. The ability to produce light, bioluminescence, is found most frequently in aquatic organisms, but its use in communication is probably best documented in fireflies, beetles belonging to the family *Lampyridae*. Firefly males advertise their presence by producing flashes of light in a species-specific pattern. Females respond with simple flashes, precisely timed, to indicate that they belong to the appropriate species. This communication system is used to advantage by females in a few predatory species of the genus *Photuris*. After females of predatory species have mated with males of their own species, they attract males of other species by mimicking the responses of the appropriate females. The males that are tricked are promptly eaten. The luminescence of fireflies does not attract a wide variety of nocturnal predators, because their bodies contain a chemical that makes them unpalatable.

An Indian palm squirrel shares its emotional feeling (love, care, fear) with its pup, which is in a cage. Photographed in June 2011, in (Batticaloa, Sri Lanka).

Visual displays are limited in the distance over which they can be used and are easily blocked by obstacles. Visual communication is important in primates, birds, and some insects, but can be dispensed with by many species that do not have the necessary sensory capacities.

AUDITORY COMMUNICATION

The limitation of visual communication is frequently offset by the coupling of visual displays with other modes of communication. Visual displays can be coupled with auditory communication, for example. There are many advantages to using sound: It can be used in the dark, and it can go around obstacles and provide directional information. Because pitch, volume, and temporal patterns of sound can be varied, extremely complex messages can be communicated. The auditory communication of many bird species has been studied intensively. Bird vocalizations are usually classified into two groups, calls and songs. Calls are usually brief sounds, whereas songs are longer, more complex, and often more suited to transmission over distances.

The call repertoire of a species serves a broad array of functions. Many young birds use both a visual signal, gaping, and calling in their food begging. Individuals that call more may receive more food. Begging calls and postures may also be used by females in some species to solicit food from mates. One call type that has been intensively studied is the alarm call. Alarm calls of many species are similar, and response is frequently interspecific (that is, interpretable by more than one species). Alarm calls are likely to be difficult to locate, a definite advantage to the individual giving the call. Calls used to gather individuals for mobbing predators are also similar in different species. Unlike alarm calls, mobbing calls provide good directional information, so that recruitment to the mobbing effort can be rapid.

Call repertoires serve birds in a great variety of contexts important for survival of the individual. Song, on the other hand, most often serves a reproductive function, that of helping a male hold a territory and attract a mate. Songs are species-specific, like the distinctive markings of a species. In some cases, songs are more distinctive than physical appearance. The chiffchaff and willow warbler were not recognized as separate species until an English naturalist named Gilbert White discovered, by examining their distinctive songs, that they are separate. The North American wood and hermit thrushes can also be distinguished more readily by song than by appearance.

Bird song can communicate not only the species of the individual singing but also information about motivational state. Most singing is done by males during the breeding season. In many species, only the male sings. In some species, females sing as well. Their songs may be similar to the songs of the males of their species or they may be distinctive. If the songs are similar to those of the males, the female may sing songs infrequently and with less volume. In some instances, the female song serves to notify her mate of her location. An interesting phenomenon found in some species is duetting, in which the male and female develop a duet. Mates may sing in alternate and perfectly timed phrases, as is done by the African boubou shrike, *Laniarius aethiopicus*. An individual shrike can recall its mate by singing the entire song alone.

Individuals in some bird species have a single song, and individuals of other species have repertoires of songs. Average repertoire size of the individual is characteristic of a species. Whether songs in repertoires are shared with neighbors or unique to the individual is also characteristic of a species. Sharing songs with neighbors permits song matching in countersinging. Cardinals and tufted titmice are species that frequently match songs in countersinging. Possible uses for matching are facilitating the recognition of intruders and indicating which neighbor has the attention of a singer. Some species of birds have dialects. The species-specific songs of one geographic region can be differentiated from the song of another geographic region. The development of dialects may be useful in maintaining local adaptations within a species, provided that females select mates of the same dialect as their fathers.

Although auditory signals of birds have received a disproportionate share of attention in the study of animal communication, auditory communication is used by a broad spectrum of animals. Crickets have species-specific songs to attract females and courtship songs to encourage an approaching female. The ears of most insects can hear only one pitch, so the temporal pattern of sound pulses is the feature by which a species can be identified. Vervet monkeys use three different alarm calls, depending upon the kind of threat present; they respond to the calls appropriately by looking up, looking down, or climbing a tree, depending upon the kind of call given.

TACTILE AND ELECTRICAL COMMUNICATION

Tactile communication differs significantly from other forms of communication in that it cannot occur over a distance. This form of communication is important in many insects, equipped as they are with antennae rich in receptors. Shortly after a termite molts, for example, it strokes the end of the abdomen of another individual with its antennae and mouthparts. The individual receiving this signal responds by extruding a fluid from its hindgut. Tactile signals are frequently used in eliciting trophallaxis (food sharing) in social insects. Tactile signals are also important in the copulatory activity of a number of vertebrates.

Additional channels of communication available in animals are electrical and surface vibration. Many modes of communication are used in combination with other modes. The channels used will depend in part on the sensory equipment of the species, its ecology, and the particular context. Most messages will be important either for the survival of the individual or the group or for the individual's ability to transmit its genes to the next generation.

COMMUNICATION STUDIES

Early study in animal communication depended primarily on careful observation of animals. This technique has been supplemented by a number of tools. The motion-picture or videotape camera permits the observer to analyze visual displays more completely. The tape recorder is a particularly versatile tool; acoustical communication can be recorded and the result used in playback experiments, in which the ethologist plays the recording in the field in order to test hypotheses about communication. Playbacks have helped scientists determine that some species of birds can discriminate between songs of neighbors and songs of strangers. Taped songs have been cut and spliced in various sequences to find out which features of a song's structure are important in species recognition.

The development of song in some species has been studied by means of isolation experiments. Birds that have been hatched and reared in the laboratory have been isolated from their species-specific song to determine whether the song needs to be learned. Some isolates are exposed to tutors (either tape recordings or living birds) at various intervals to examine the possibility of critical periods for song learning.

Sound spectrographs make it possible to produce pictures of calls and songs. Spectrographs represent song frequency on one axis and time on the other. These graphs have revealed the intricate structures of many auditory communication signals.

Information about sensory reception and about neural control of signals is an important research area. Some information in this area has come from ingenious but simple experiments. Bees have been trained to respond to color clues in association with a sugar source to determine which colors they are able to discriminate. Important knowledge in sensory research comes from determining the specific stimuli that will elicit a response in specific neurons. By using microelectrodes placed in or near the neuron, scientists can detect the presence of a response. Another technique used to determine the function of a presumed sense organ is to block or remove either the organ or the neural connections to the organ. One of the earliest experiments of this type was done in 1793 by Lazzaro Spallanzani. He found that flying bats could not avoid obstacles when their ears were tightly plugged but that the bats were able to avoid obstacles when they were blinded. In 1938, the use of an ultrasonic recorder made it possible for Donald Griffin to discover the bat's use of ultrasonic sound.

Synthetic pheromones can be produced that have the same effects on behavior as natural pheromones. Scientists can also determine at what dilution a pheromone will still evoke response. Hence, the sophisticated tools of chemistry have important applications in ethology.

TALKING TO ANIMALS

Animal communication provides a fascinating frontier for exploration. Some of the knowledge acquired has had practical applications as well. Pheromones have been used to bait traps for insect pests. In some cases, this technique is used directly as a control measure; in other instances, the traps are used to estimate population size, and other control measures are used when pest populations are high. A key advantage of pheromone use is its specificity.

Recognizing the communication signals of pets and domestic animals is often useful in their care. Knowledge of releasers, standard signals that receive standard responses from animals, is important for survival in some contexts. Knowing which signals are perceived by animals as threats allows humans to avoid triggering an attack. This knowledge also allows control of animals in less destructive ways. Recordings of alarm signals of birds are used to disperse flocks that are creating problems.

Knowledge of a species' communication repertoire is critical in the training of animals for useful work or for entertainment. Also, knowledge of their sensory capacities makes it possible for appropriate signals to be selected—particularly important as it applies to human nonverbal communication. Knowing which signals of our nonverbal communication repertoire are characteristic of the whole species is both useful and interesting.

—*Donna Janet Schroeder*

FURTHER READING

Bradbury, Jack W., and Sandra L. Vehrencamp. *Principles of Animal Communication*. 2nd ed. Sinauer, 2011.

De Waal, Frans. *Chimpanzee Politics: Power and Sex among Apes*. 25th anniversary ed. Johns Hopkins University Press, 2007.

Goodall, Jane. *In the Shadow of Man*. Rev. ed. Houghton, 2009.

Grier, James W. *Biology of Animal Behavior*. 3rd ed. McGraw-Hill, 1999.
Hart, Stephen. *The Language of Animals*. Holt, 1996.
Hauser, Marc D. *The Evolution of Communication*. MIT Press, 1996.
Roitblat, Herbert L., Louis M. Herman, and Paul E. Nachtigall, eds. *Language and Communication: Comparative Perspectives*. Erlbaum, 1993.
Wilson, Edward O. *Insect Societies*. Harvard University Press, 1971.

COMMUNITIES

FIELDS OF STUDY

Population Biology; Population Biodynamics; Ecology; Environmental Studies

ABSTRACT

An ecological community is the assemblage of species found in a given time and place. The populations that form a community interact through the processes of competition, predation, parasitism, and mutualism. The structure and function of communities are determined by the nature and strength of the population interactions within it, but these interactions are affected by the environment in which a community exists. An ecological community together with its physical environment is called an ecosystem.

KEY CONCEPTS

ecosystem: a community together with its physical environment

food chain: a pathway through which energy travels in a community

food web: the interconnections among all food chains in a community

frequency-dependent predation: predation on whichever species is most common in a community

global extinction: the loss of all members of a species

intergrade: to integrate or blend the members of different community populations along population gradients rather than as sharply divided groups

keystone species: a species that determines the structure of a community, usually by predation on its dominant competitor

local extinction: the loss of one or more populations of a species, but with at least one population of the species remaining

resilience stability: stability exhibited by a community that changes its structure when disturbed but returns to its original structure when the disturbance ends

resistance stability: stability exhibited by a community that absorbs the effects of a disturbance until it can no longer do so; then, it typically shifts permanently to an alternate structure

trophic level: a single link in a food chain; all species that obtain energy in the same way are said to be at the same trophic level

COMPONENTS OF COMMUNITIES

An ecological community is the assemblage of species found in a given time and place. The populations that form a community interact through the processes of competition, predation, parasitism, and mutualism. The structure and function of communities are determined by the nature and strength of the population interactions within it, but these interactions are affected by the environment in which a community exists. An ecological community together

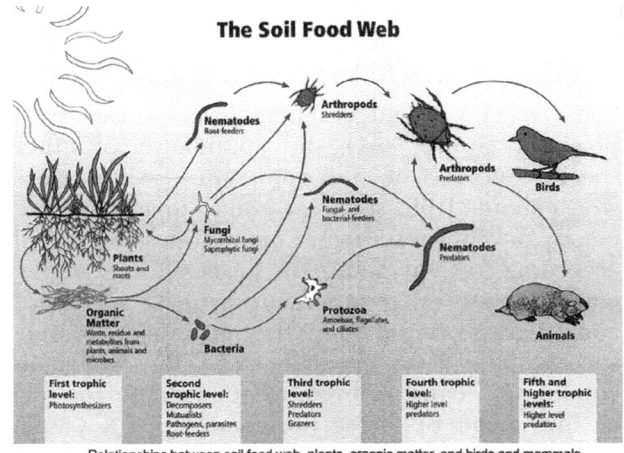

An illustration of a soil food web. (USDA)

with its physical environment is called an ecosystem. No ecological system can be studied apart from its physical environment; the structure and function of every community are determined in part by its interactions with its environment.

The species constituting a community occupy different functional roles. The most common way to characterize a community functionally is by describing the flow of energy through it. Correspondingly, communities usually contain three groups of species: those that obtain energy through photosynthesis (called producers), those that obtain energy by consuming other organisms (consumers), and those that decompose dead organisms (decomposers). The pathway through which energy travels from producer through one or more consumers and finally to decomposer is called a food chain. Each link in a food chain is called a trophic level. Interconnected food chains in a community constitute a food web. Food webs have no analogy in populations.

Very few communities are so simple that they can be readily described by a food web. Most communities are compartmentalized: A given set of producers tends to be consumed by a limited number of consumers, which in turn are preyed upon by only a few predators, and so on. Alternatively, consumers may obtain energy by specializing on one part of their prey (for example, some birds may eat only seeds of plants) but utilize a wide range of prey species. Compartmentalization is an important feature of community structure; it influences the formation, organization, and persistence of a community.

DOMINANT AND KEYSTONE SPECIES

Some species, called dominant species, can exert powerful control over the abundance of other species because of the dominant species' large size, extended life span, or ability to monopolize energy or other resources. Communities are named according to their dominant species: for example, oak-hickory forest, redwood forest, sagebrush desert, and tallgrass prairie. Some species, called keystone species, have a disproportionately large effect on community structure by preventing dominant species from monopolizing the community. Keystone species usually exert their effects through predation, while dominant species are good competitors (that is, better at obtaining and holding resources than other species).

The species that make up a community are seldom distributed uniformly across the landscape; rather, some degree of patchiness is characteristic of virtually all species. There has been conflicting evidence as to the nature of this patchiness. As one moves across an environmental gradient (for example, from wet to dry conditions or from low to high elevations), there is a corresponding change in species observed and in the type of community present. Some studies have suggested that changes in species composition usually occur along relatively sharp boundaries and that these boundaries mark the boundaries between adjacent communities. Other studies have indicated that species tend to respond individually to environmental gradients and that community boundaries are not sharply defined; rather, most communities broadly intergrade into one another.

THE NATURE OF COMMUNITY

These conflicting results have fueled a continuing debate as to the underlying nature of the community. Some communities seem to behave in a coordinated manner; for example, if a prairie is consumed by fire, it regenerates in a predictable sequence, ultimately returning to the same structure and composition it had before the fire. This coordinated response is to be expected if the species in a community have evolved together with one another. In this case, the community behaves analogously to an organism, maintaining its structure and function in the face of environmental disturbances and fluctuations (as long as the disturbances and fluctuations are not too extreme). The existence of relatively sharp boundaries between adjacent communities supports this explanation of the nature of the community.

In other communities, it appears that the response to environmental fluctuation or disturbance is determined by the evolved adaptations of the species available. There is no coordinated community response, but rather a coincidental assembly of community structure over time. Some sets of species interact together so strongly that they enter a community together, but there is no evidence of an evolved community tendency to resist or accommodate environmental change. Data support this explanation of the community as an entity formed primarily of species that happen to share similar environmental requirements.

MECHANISMS OF COMMUNITY STRUCTURE

Disagreement as to the underlying nature of communities usually reflects disagreement as to the relative importance of the underlying mechanisms that determine community structure. Interspecific competition has long been invoked as the primary agent structuring communities. Competition is certainly important in some communities, but there is insufficient evidence to indicate how widespread and important it is in determining community structure. Much of the difficulty occurs because ecologists must infer the existence of past competition from present patterns in communities. It appears that competition has been important in many vertebrate communities and in communities dominated by sessile organisms, such as plants; it does not appear to have been important in structuring communities of plant-eating insects. Furthermore, the effects of competition typically affect individuals that use identical resources, so that only a small percentage of species in a community may be experiencing significant competition at any time.

The effects of predation on community structure depend on the nature of the predation. Keystone species usually exert their influence by selectively preying on species that are competitively dominant. Predators that do not specialize on one or a few species may also have a major effect on community structure, if they attack prey in proportion to their abundance; this frequency-dependent predation prevents any prey species from achieving dominance. If a predator is too efficient, it can drive its prey to extinction, which may cause a selective predator to become extinct as well. Predation appears to be most important in determining community structure in environments that are predictable or unchanging.

NATURAL DISTURBANCES

A variable or unpredictable environment influences the structure of a community. No environment is completely uniform; longer-term or seasonal environmental fluctuations affect community structure by limiting opportunities for colonization, by causing direct mortality, and by hindering or exacerbating the effects of competition and predation. Furthermore, all communities experience at least occasional disturbance: unpredictable, seemingly random environmental changes that may be quite severe. It is useful in this regard to distinguish between disasters and catastrophes.

A disaster is an event that occurs so frequently in the life of a population that adaptation is possible; for example, fire occurs so often in tallgrass prairies that most of the plant species have become fire-adapted—they have become efficient at acquiring nutrients left in the ash and at sprouting or germinating quickly following a fire. In comparison, a catastrophe is so intense, widespread, or infrequent that a population cannot adapt to it; the eruption of Mount St. Helens in 1980, for example, was so violent and so unpredictable that the species affected could not evolve adequate responses to it.

Natural disturbances occur at a variety of scales. Small-scale disturbances may simply create small openings in a community that are filled in by other species suited to thrive in such spaces. Large disturbances are qualitatively different from small disturbances, in that large portions of a community may be destroyed, including some of the ability to recover from the disturbance. Early ecologists almost always saw disturbances as destructive and disruptive for communities. Under this assumption, most mathematical models portrayed communities as generally being in some stable state, at equilibrium; if a disturbance occurred, the community inevitably returned to the same (or some alternative) equilibrium. It later became clear, however, that natural disturbance is a part of almost all natural communities. Ecologists now recognize that few communities exhibit a stable equilibrium; instead, communities are dynamic, always responding to the last disturbance, always adjusting to the most recent environmental fluctuation.

THE LONG-TERM DYNAMICS OF COMMUNITIES

The evidence suggests that three conclusions can be drawn with regard to the long-term dynamics of communities. First, it can no longer be assumed that communities remain at equilibrium until changed by outside forces. Disturbances are so common, they occur at so many different scales and frequencies, and they so readily affect the processes of competition and predation that the community must be viewed as an entity that is constantly changing as its constituent species readjust to disturbance and to one another.

Second, communities exhibit several types of stability in the face of disturbance. A community may absorb disturbance without markedly changing, until it reaches a threshold and suddenly and rapidly shifts to a new state, called resistance stability. Alternatively, a community may change easily when disturbed but quickly return to its former state; this characteristic is called resilience stability. Resilience stability may occur over a wide range of conditions and scales of disturbance; such a system is said to be dynamically robust. On the other hand, a community that exhibits resilience only within a narrow range of conditions is said to be dynamically fragile.

Finally, there is no simple way to predict the stability of a community. At the end of the 1970s, it appeared that complex communities were generally more stable than simple communities. It appeared that stability was conferred by more intricate food webs, by more structural complexity, and by higher species diversity. On the basis of numerous field studies and theoretical models, ecologists now conclude that no such relationship exists. Both very complex communities, such as tropical rain forests, and very simple communities, such as Arctic tundra, may be very fragile when disturbed.

COMPLEX SYSTEMS

Most communities consist of thousands of species, and their complexity makes them very difficult to study. Most community ecologists specialize in taxonomically restricted subsets of communities (such as plant communities, bird communities, insect communities, or moss communities) or in functionally restricted subsets of communities (such as soil communities, tree-hole communities, pond communities, or detrivore communities).

The type of community under investigation and the questions of interest determine the appropriate methods of study. The central questions in most community studies are how many species are present and what is the abundance of each. The answers to these questions can be estimated using mark-recapture methods or any other enumeration method.

Often the aim is to compare communities (or to compare the same community at different times). A specialized parameter called similarity is used to compare and classify communities; more than two dozen measures of similarity are available. Measures of similarity are typically subjected to cluster analysis, a set of techniques that groups communities on the basis of their similarity.

Many multivariate techniques are used to search for patterns in community data. Direct gradient analysis is the simplest of these techniques; it is used to study the distribution of species along an environmental gradient. Ordination includes several methods for collapsing community data for many species in many communities along several environmental gradients onto a single graph that summarizes their relationships and patterns.

PATTERNS OF COMMUNITY RESPONSES TO DISTURBANCE

At the most basic level, destruction of a community eliminates the species comprising the community. If the community is restricted in its extent, and if its constituent species are found nowhere else, those species become extinct. If the community covers a large area or is found in several areas, local extinction of species may occur without causing global extinction.

Destruction of a community can cause unexpected changes in environmental conditions that were modified by the intact community. Even partial destruction of an extensive community can eliminate species. For example, the checkerboard pattern of clear-cutting in Douglas fir forests of the Pacific Northwest threatens the survival of the northern spotted owl, the marbled murrelet, Vaux's swift, and the red tree vole, even though fragments of the community remain. Many fragments are simply too small to support these species. A Douglas fir forest is regenerated following cutting, but this young, even-aged stand is so different from an old, mixed-age forest that it functions as a different type of community.

Altering the population of one species can affect others in a community. The black-footed ferret was once found widely throughout central North America as a predator of prairie dogs. As prairie dogs were poisoned, drowned, and shot throughout their range, the number of black-footed ferrets also declined. As of 1989, fewer than one hundred black-footed ferrets were in a captive breeding program in Wyoming in a final attempt to preserve the species.

Introducing a new species into a community can severely alter the interactions in the community. The introduction of the European rabbit into Australia led to a population explosion of rabbits, excessive predation on vegetation, and resulting declines in many native marsupials.

Finally, it appears that many communities exhibit stability thresholds; if a community is disturbed beyond its threshold, its structure is permanently changed. For example, acid deposition in lakes is initially buffered by natural processes. As acid deposition exceeds the buffering capacity of a lake, it causes insoluble aluminum in the lake bottom to become soluble, and this soluble aluminum kills aquatic organisms directly or by making them more susceptible to disease. The lesson is clear: It is far easier to disrupt or destroy natural systems (even accidentally) than it is to restore or reconstruct them.

—*Alan D. Copsey*

Further Reading

Aber, John D., and Jerry M. Melillo. *Terrestrial Ecosystems*. W. B. Saunders, 1991.

Begon, Michael, John L. Harper, and Colin R. Townsend. *Ecology: Individuals, Populations, and Communities*. 3rd ed. Blackwell Science, 1996.

Bormann, Frank H., and Gene E. Likens. "Catastrophic Disturbance and the Steady State in Northern Hardwood Forests." *American Scientist* 67 (1979): 660–669.

Goldammer, J. G., ed. *Tropical Forests in Transition: Ecology of Natural and Anthropogenic Disturbance Processes*. Birkhäuser Verlag, 1992.

Krebs, Charles J. *Ecological Methodology*. 2nd ed. Benjamin/Cummings, 1999.

———. *Ecology: The Experimental Analysis of Distribution and Abundance*. 4th ed. Benjamin/Cummings, 1994.

Pickett, S. T. A., and P. S. White, eds. *The Ecology of Natural Disturbance and Patch Dynamics*. Academic Press, 1985.

Pielou, E. C. *The Interpretation of Ecological Data: A Primer on Classification and Ordination*. John Wiley & Sons, 1984.

Competition

Fields of Study

Ecology; Environmental Studies; Population Biology; Population Dynamics

Abstract

Competition is the struggle between individuals of different species (interspecific competition) or between individuals of the same species (intraspecific competition) for food, territories, and mates in order to survive. It is a major driving force in evolution.

Key Concepts

evolution: gradual changes in organisms over time, caused by mutation and selected by the environment, resulting in better adapted organisms and new species

habitat: the type of environment in which a particular organism prefers to live, based upon various physical and chemical conditions

natural selection: the ability of an organism or species to survive, compete, and reproduce in its habitat; success is dictated by the alleles (traits) that it possesses

niche: an organism's role in its habitat environment

predation: a situation in which one animal species hunts and eats another species (examples: lynx versus hare; cheetah versus gazelle)

territoriality: a phenomenon in animal behavior whereby individual organisms occupy and defend an area from other individuals of the same or different species

threat display: a territorial behavior exhibited by animals during defense of a territory, such as charging, showing bright colors, and exaggerating body size

COMPETING TO SUCCEED

Competition is the struggle between individuals of different species (interspecific competition) or between individuals of the same species (intraspecific competition) for food, territories, and mates in order to survive. It is a major driving force in evolution, the process by which living organisms change over time, with better-adapted species surviving and less well-adapted species dying. Evolution begins with mutation, changes in the nucleotide sequence of a gene or genes, resulting in the production of slightly altered genes called alleles which encode slightly different proteins. These altered proteins are the expressed traits of an organism and may give the organism an advantage over its competitors. The organism outcompetes its rivals in the environment, and hence the environment favors the better-adapted, fitter organism, a process called natural selection. A mutation may help an organism in one environment but may hurt the same organism in a different environment (for example, albino, white or light-colored squirrels may flourish in snowy regions but may not do as well in the darker-colored environments of warm regions where they would be much more visible to predators). Mutations can be caused by chemicals called mutagens or by ionizing radiation such as ultraviolet light, X rays, and gamma radiation.

THE STRUGGLE TO SURVIVE

The science of ecology can best be defined as the experimental analysis of the distribution and abundance of organisms. Natural selection influences the distribution and abundance of organisms from place to place. The possible selecting factors include physical factors (temperature and light, for example), chemical factors such as water and salt, and species interactions. Any of these factors can influence the survivability of organisms in any particular environment. According to ecologist Charles Krebs, species interactions include four principal types: mutualism, which is the living together of two species that benefit each other (for example, humans and their pets); commensalism, which is the living together of two species that results in a distinct benefit (or number of benefits) to one species while the other remains unhurt (commensalism is shown in the relationship of birds and trees); predation, which is the hunting, killing, and eating of one species by another (examples: cats and mice; dogs and deer); and competition, which is defined as an active struggle for survival among all the species in a given environment.

This struggle involves the acquisition of various resources: food, territory, and mates. Food is an obvious target of competition. All organisms must have energy in order to conduct the cellular chemical reactions (such as respiration) that keep them alive. Photoautotrophic organisms (plants, phytoplankton, photobacteria) obtain this energy by converting sunlight, carbon dioxide, and water into sugar, a process called photosynthesis. Photoautotrophs, also called producers, compete for light and water. For example, oak and hickory trees grow taller than most pines, thereby shading out smaller species and eventually dominating a forest. All other organisms—animals, zooplankton, and fungi—are heterotrophs; they must consume other organisms to obtain energy. Heterotrophs include herbivores, carnivores, omnivores, and saprotrophs. Herbivores (plant eaters such as rabbits and cattle) obtain the sugar manufactured by plants. Carnivores (meat eaters such as cats and dogs) eat other heterotrophs in order to get the sugar that these heterotrophs received from other organisms. Omnivores, such as humans, eat plants and animals for the same reason. Saprotrophs (such as fungi and bacteria) decompose dead organisms for the same reason. Life on earth functions by intricately complex food chains in which organisms consume other organisms in order to obtain energy. Each human being is composed of molecules that were once part of other living organisms, even other humans. Ultimately, the earth's energy comes from the sun.

Competition in the animal kingdom.

COMPETITION BETWEEN AND WITHIN SPECIES

Territoriality is equally important for two reasons: An organism needs a place to live, and this place must contain adequate food and water reserves. A strong, well-adapted organism will fight and drive away weaker individuals of the same or different species in order to maintain exclusive rights to an area containing a large food and water supply. Species that are less well-adapted will be relegated to areas where food and water are scarce. The stronger species will have more food and will tend to produce more offspring, since they will easily attract mates. Being stronger or more adapted does not necessarily mean being physically stronger. A physically strong organism can be overwhelmed easily by numerous weak individuals. In general, adaptability is defined by an organism's ability to prosper in a hostile environment and leave many viable offspring.

Within a species, males attempt to attract females to their territory, or vice versa, by courtship dances and displays, often including bright colors such as red and blue and exaggerated body size. Mating displays are very similar to the threat displays used to drive away competitors, although there is no hostility involved. Generally, females are attracted to dominant males having the best, not necessarily the largest, territories.

Competition for food and territory is interspecific and intraspecific. Competition for mates is intraspecific. In an environment, the place where an organism lives (such as a eucalyptus tree or in rotting logs) is referred to as its habitat. Simultaneously, each species has its own unique niche, or occupation, in the environment (such as decomposer or carnivore). More than one species can occupy a habitat if they have different ecological niches. When two or more different species occupy the same habitat and niche, competition arises. One species will outcompete and dominate, while the losing competitors may become reduced in numbers and may be driven away from the habitat.

PECKING ORDERS

In vertebrate organisms, intraspecific competition occurs between males as a group and between females as a group. Rarely is there male-versus-female competition, except in species having high social bonding—primates, for example. Competition begins when individuals are young. During play fighting, individuals nip or peck at each other while exhibiting threat displays. Dominant individuals exert their authority, while weaker individuals submit. The net result is a very ordered ranking of individuals from top to bottom, called a dominance hierarchy or pecking order. The top individual can threaten and force into submission any individual below it. The number two individual can threaten anyone except number one, and so on. The lowest-ranked individual can threaten no one and must submit to everyone. The lowest individual will have the least food, worst territory, and fewest (if any) mates. The number one individual will have the most food, best territory, and most mates. The pecking order changes over time because of continued group competition that is shown by challenges, aging, and accidents.

Pecking orders are evident in hens. A very dominant individual will peck other hens many times but will rarely be pecked. A less dominant individual will peck less but be pecked more. A correct ranking can be obtained easily by counting the pecking rate for each hen.

In The Netherlands, male black grouse contend with one another in an area called a "lek," which may be occupied by as many as twenty males. The males establish their territories by pecking, wing-beating, and threat displays. The most dominant males occupy small territories (several hundred meters) at the center of the lek, where the food supply is greatest. Less dominant males occupy larger territories with less food reserves to the exterior of the lek. Established territories are maintained at measurable distances by crowing and flutter-jumping, with the home territory owner nearly always winning. Females, which nest in an adjoining meadow, are attracted to dominant males in the heavily contested small central territories.

A baboon troop can range in size from ten to two hundred members, but usually averages about forty. Larger, dominant males and their many female mates move centrally within the troop. Less dominant males, with fewer females, lie toward the outside of the troop. Weak individuals at the troop periphery are more susceptible to predator attacks. Dominant males exert their authority by threat displays, such as the baring of the teeth or charging; weaker males submit by presenting their hindquarters. Conflicts are usually peacefully resolved.

Female lions maintain an organized pride with a single ruling male. Young males are expelled and wander alone in the wilderness. Upon reaching adulthood, males attempt to take over a pride in order to gain access to females. If a male is successful in capturing a pride and expelling his rival, he will often kill the cubs of the pride, simultaneously eliminating his rival's descendants and stimulating the females to enter estrus for mating.

COMPETITION WITHIN NICHES

Interspecific competition occurs between different species over food and water reserves and territories. Two or more species occupying the same niche and habitat will struggle for the available resources until either one species dominates and the others are excluded from the habitat or the different species evolve into separate niches by targeting different food reserves, thus enabling all to survive in the same habitat. Numerous interspecific studies have been conducted, including crossbills, warblers, blackbirds, and insects, to mention a few.

Crossbills are small birds that live in Europe and Asia. Three crossbill species inhabit similar habitats and nearly similar niches. Each species has evolved a slightly modified beak, however, for retrieving and eating seeds from three different cone-bearing (coniferous) trees. The white-winged crossbill has a slender beak for feeding from small larch cones, the common crossbill has a thicker beak for feeding from larger spruce cones, and the parrot crossbill appropriately has a very thick beak for feeding from pine cones. The evolution of different niches has enabled these three competitors to survive.

Another example of this phenomenon is shown by five species of warblers that inhabit the coniferous forests of the American northeast. The myrtle warbler eats insects from all parts of trees up to seven meters high. The bay-breasted warbler eats insects from tree trunks six to twelve meters above the ground. The black-throated green, blackburnian, and Cape May warblers all feed near the treetops, according to elaborate studies by Robert H. MacArthur. The coexistence of five different species is probably the result of the warblers occupying different parts of the trees, with some warblers developing different feeding habits so that all survive.

G. H. Orians and G. Collier studied competitive exclusion between redwing and tricolored blackbirds. Introduction of tricolored blackbirds into redwing territories results in heavy redwing aggression, although the tricolored blackbirds nearly always prevail.

Two species of African ants, *Anoplolepis longipes* and *Oecophylla longinoda*, fight aggressively for territorial space. M. J. Way found that *Anoplolepis* prevails in sandy environments, whereas *Oecophylla* dominates in areas having thick vegetation.

Interspecific competition therefore results in the evolution of new traits and niches and the exclusion of certain species. Mathematical models of competition are based upon the work of A. J. Lotka and V. Volterra. The Lotka-Volterra equations attempt to measure competition between species for food and territory based upon the population size of each species, the density of each species within the defined area, the rate of population increase of each species, and time.

OBSERVING COMPETITION

Studies of competition between individuals of the same or different species generally follow one basic method: observation. Interactions between organisms are observed and carefully measured to determine if the situation is competition, predation, parasitism, or mutualism. More detailed analyses of environmental chemical and physical conditions are used to determine the existence of additional influences. Observations of competition between organisms involve direct visual contact in the wild, mark-recapture experiments, transplant experiments, measurements of population sizes in given areas, and competition experiments in artificial environments.

Direct visual contact involves the scientist entering the field, finding a neutral, nonthreatening position, and watching and recording the actions of the subject organisms. The observer must be familiar with the habits of the subject organism and must be keen to detect subtle cues such as facial gestures, vocalizations, colors, and patterns of movement from individual to individual. Useful instruments include binoculars, telescopes, cameras, and sound recorders. The observer must be capable of tracking individuals over long distances so that territorial boundaries and all relevant actions are recorded. The observer may have to endure long periods of time in the field under uncomfortable conditions.

Mark-recapture experiments involve the capture of many organisms, tagging them, releasing them into an area, and then recapturing them (both tagged and untagged) at a later time. Repeated collections (recaptures) over time can give the experimenter an estimate of how well the species is faring in a particular environment. This technique is used in conjunction with other experiments, including transplants and population size measurements.

In transplant experiments, individuals of a given species are marked and released into a specific environmental situation, such as a new habitat or another species' territory. The objective of the experiment is to see how well the introduced species fares in the new situation, as well as the responses of the various species which normally inhabit the area. The tricolored blackbird takeover of redwing blackbird territories is a prime example. Another example is the red wolf, a species that was extinct in the wild until several dozen captive wolves were released at the Alligator River Wildlife Refuge in eastern North Carolina. Their survival is uncertain. Accidental transplants have had disastrous results for certain species; for example, the African honeybee poses a threat to the honey industry in Latin America and the southern United States because it is aggressive and produces poorly.

Measurements of population sizes rely upon the point-quarter technique, in which numerous rectangular areas of equal size are marked in the field. The number of organisms of each species in the habitat is counted for a given area; an averaging of all areas is then made to obtain a relatively accurate measure of each population's size. In combination with mark-recapture experiments, population measurements can provide information for birthrates, death rates, immigration, and emigration over time for a given habitat.

Laboratory experiments involve confrontations between different species or individuals of the same species within an artificial environment. For example, male mouse (*Mus musculus*) territoriality can be studied by introducing an intruder into another male's home territory. Generally, the winner of the confrontation is the individual that nips its opponent more times. Usually, home court advantage prevails; the intruder is driven away. Similar studies have been performed with other mammalian, reptile, fish, insect, and bird species.

Interactions between different species are subtle and intricate. Seeing how organisms associate enables scientists to understand evolution and to model various environments. Competition is a major driving force in evolution. The stronger species outcompete weaker species for the available ecological niches. Mutations in organisms create new traits and, therefore, new organisms (more species), which are selected by the environment for adaptability.

All environments consist of a complex array of species, each dependent on the others for survival. The area in which they live is their habitat. Each species' contribution to the habitat is that species' niche. More than one species in a given habitat causes competition. Two species will struggle for available territory and food resources until either one species drives the other away or they adapt to each other and evolve different feeding habits and living arrangements. Competition can be interspecific (between individuals of different species) or intraspecific (between individuals of the same species). The environment benefits because the most adapted species survive, whereas weaker species are excluded.

—*David Wason Hollar, Jr.*

FURTHER READING

Andrewartha, H. G. *Introduction to the Study of Animal Populations.* University of Chicago Press, 1967.

Arthur, Wallace. *The Niche in Competition and Evolution.* John Wiley & Sons, 1987.

Hartl, Daniel L. *Principles of Population Genetics.* 3rd ed. Sinauer Associates, 1997.

Keddy, Paul A. *Competition.* Chapman and Hall, 1989.

Krebs, Charles J. *Ecology: The Experimental Analysis of Distribution and Abundance.* 4th ed. Benjamin/Cummings, 1994.

Lorenz, Konrad. *On Aggression.* Harcourt, Brace & World, 1963.

Raven, Peter H., and George B. Johnson. *Biology.* 4th ed. McGraw-Hill, 1996.

Wilson, Edward O. *Sociobiology: The New Synthesis.* Belknap Press of Harvard University Press, 1975.

Convergent and Divergent Evolution

Fields of Study

Evolutionary Biology; Animal Physiology; Paleontology; Biomechanics; Ecology

Abstract

Biological species have been defined as populations of organisms that are capable of successfully interbreeding (producing fertile offspring) only with other members of the same species. Members of any species possess unique sets of biological characteristics, termed "characters" that are physical expressions of a genetic code unique to members of that species. The code represents an extremely complex and thorough set of instructions for equipping an individual organism with the body and the behavioral knowledge it requires for success in the particular environment to which its species has adapted.

Key Concepts

adaptive radiation: the successful invasion by a species into a number of previously unavailable ecological niches

analogue: an individual structure shared by two or more species that is of only superficial similarity; thus, it is not indicative of a common ancestor

clade: a type of grouping of living or extinct species along lines of shared, unique structures, or homologues, indicative of a common ancestor; helpful in establishing evolutionary relationships

convergence: the evolution of a similar morphology by unrelated or only distantly related species caused by both having adapted to similar lifestyles in similar environments

divergence: the evolution of increasing morphological differences between an ancestral species and offshoot species caused by differing adaptive pressures

environmental constraints (pressures): the physical demands placed upon any species by its surroundings that ultimately determine the success or failure of its adaptations and consequently its success as a species

homologue: an individual structure shared by two or more different species that is indicative of a common ancestor

SPECIES AND NICHES IN EVOLUTION

Biological species have been defined as populations of organisms that are capable of successfully interbreeding (producing fertile offspring) only with other members of the same species. Members of any species possess unique sets of biological characteristics, termed "characters." These characters are physical expressions of a genetic code unique to members of that species. The code represents an extremely complex and thorough set of instructions for equipping an individual organism with the body and the behavioral knowledge it requires for success in the particular environment to which its species has adapted.

Because of natural selection acting upon many past generations of that species, living members are fine-tuned to a specific ecological niche, or econiche, of the greater ecosystem of which the species is a member. When conditions within the ecosystem change (a general climatic change, for example) or when other scenarios occur, such as when a smaller subpopulation of the species migrates into new, ecologically different territory or becomes isolated in some way, selective pressure is brought to bear upon members of the group or subgroup.

Random mutation is a mechanism by which selective pressure is thought to be brought about. Such mutations are changes in the genetic code that occur spontaneously in some individuals within the species in an ongoing manner. Most random mutations are insignificant phenomena with regard to the species as a whole, because most have either a neutral or negative survival value: Either they do not help the individual possessing them to survive or they are counteradaptive to an extreme degree and prove fatal. Consequently, mutations in general are not usually transmitted beyond the generation in which they occur or beyond the affected member or members. In certain scenarios, however, mutations that have a positive survival value can spread throughout the population. This is believed to be especially true when a smaller, isolated subgroup of the population is dealing with a changed or new environmental situation. Such processes are thought, for example, to have been instrumental in the evolution of groups of closely related but now morphologically distinct species found in isolated, mid-ocean island groups. These adaptive radiations of species that are monophyletic and thus share a relatively recent common ancestor are good examples of the process of evolutionary divergence at work.

THE CASE OF THE GALÁPAGOS FINCHES

In one of the best-known cases, studies have traced the presumed paths of divergence among a set of island bird species. This particular radiation produced a number of new species possessing novel adaptive morphologies evolved to exploit new econiches. This is the classic example of Darwin's finches. Darwin's finches are a group of closely related birds, numbering about fourteen species, found on the various islands of the Galápagos Archipelago, which straddles the equator. The islands are remote from any large body of land that would typically harbor similar birds: South America lies about 960 kilometers to the east across an unbroken stretch of the Pacific Ocean. In 1835, Charles Darwin, author of the highly influential work on organic evolution, *On the Origin of Species* (1859), visited the islands while employed as a naturalist on a British scientific voyage. His studies of the flora and fauna of these islands provided him with many observations that directly influenced his later writings.

Darwin's studies of the Galápagos finches convinced him, and generations of subsequent scientists, that the finches are a clear example of divergent evolution in operation. The scenario he deduced is that probably only one ancestral species arrived from South America by ocean currents or winds, established itself, and began to exploit the numerous, as yet unoccupied, econiches that the volcanic islands provided. In the relative ecological vacuum that the original finch species found among the islands, adaptive radiation occurred, resulting in the present, diverse species. The species of Darwin's finches found today on the islands exhibit a great variety of beak types, many of which are atypical of finch-type birds in general but rather are typical of birds found among totally different avian family classifications. Typical finches are noted for beaks adapted for the crushing of seeds—the diet of the usual members of the finch family, such as the familiar North American cardinal. Among the dozen or so Galápagos finches can be found a wide assortment of beaks adapted for obtaining or processing a much greater variety of diets. Darwin's finches include species with beaks and behaviors adapted for diets of insects, seeds, cacti, and other vegetal matter.

The adaptive radiation in the case of the Galápagos finches was relatively easy to work out because of the obvious environmental factors involved (the islands' remoteness and general barrenness) and the unusual variety of adaptations that the finches had made. Establishing the details of evolutionary divergence in other living ecosystems can be confusing because the numbers and types of econiches and interacting species are often far more numerous and diverse—for example, the lush and intricate ecosystem of a large, tropical rain forest such as the Amazon.

TRACING THE FOSSIL RECORD

Myriad examples of evolutionary divergence exist between both living plant and animal species throughout existing ecosystems in the modern world. The fossil record, however, also can be studied, to examine the phenomenon between extinct animal groups. This record of past life-forms preserved in the crustal rocks of Earth provides numerous examples of diverging species as organisms adapted to changing general conditions or spread into novel environments. An example is the many species of ceratopsian dinosaurs found in the latter part of the Cretaceous period of the Mesozoic era of the earth's history. Although the earlier ancestral forms appear to be bipedal and possess no significant armor, the later radiation of ceratopsians is well known by way of such impressive animals as *Triceratops*, a typical ceratopsian: a heavy, quadrupedal herbivore with a large, horned and beaked skull with a defensive, bony frill. Many variations on the basic late ceratopsian body architecture evolved through divergence. Varieties included such forms as *Pentaceratops*, *Torosaurus*, *Styracosaurus*, *Chasmosaurus*, and *Centrosaurus*, among many others. In all these later animals, the basic morphology regarding body, tail, and limbs remained the same. All forms also retained the typical massive, beaked head. What diverged were such morphological features as number and length of facial horns and length and degree of ornamentation of the frill.

One of the important things that study of such fossil forms shows is that such past examples of evolutionary divergence eloquently underscore the continuity of the evolutionary process through time down to the present living world. This continuity further reinforces the validity of basic evolutionary theory in general.

EVOLUTIONARY CONVERGENCE

A related phenomenon concerning adaptive evolution is the phenomenon of evolutionary convergence. This process can be described briefly as the evolution of similar body structures in two or more species that are only quite distantly related; they therefore come to resemble each other, sometimes to a startling degree of at least outward sameness. These sets of similar-looking but polyphyletic species frequently even display similar behavioral characteristics. All similarities found in convergence cases are believed to be attributable to the fact that the various species involved have adapted to a similar econiche within a similar ecosystem. Because in nature, form follows function and the morphology of an animal or plant is the product of environmental pressures that continuously favor the better-adapted organism, it is easy to understand how convergence can take place. Like divergence, the evidence for convergence can be traced not only among the many participants in contemporary biospace but also over the course of vast stretches of past time.

One of the classic examples of convergence is a threefold example that, conveniently, not only includes representatives from three different classes of vertebrates but also spans many millions of years of time and includes an extinct group. This is the textbook example that compares and contrasts the morphology of sharks, a type of cartilaginous fish; ichthyosaurs, an extinct type of marine reptile; and dolphins, marine mammals like whales. All three groups possess numerous member species, both fossil and alive (except for ichthyosaurs), which resemble each other in body plan and lifestyle. All three groups include species that depend, or depended, on an open ocean, fish-eating existence. Consequently, the forms of their bodies came to follow the functions dictated by their environment—sometimes termed their environmental constraints.

All three groups' general body plan began to approach a hydrodynamic ideal for a water-living animal: a streamlined fusiform, or spindle shape, efficient in passing through an aqueous medium. Besides this feature, pelagic, or open ocean-living, sharks, ichthyosaurs, and dolphins all evolved a dorsal fin to act as a vertical stabilizer for water travel. Each group also evolved a propulsive tail and a pectoral fin necessary for the demands of constant swimming and steering in water. Even more remarkable in this comprehensive example, dolphins' and ichthyosaurs' ancestors were both originally land-dwelling vertebrates that returned to the marine environment. This case presents an inclusive and persuasive argument for the reality of the phenomenon of convergent evolution.

As with the use of both living and extinct examples in the discussion of divergent evolution, the existence of fossil as well as contemporary species that display convergent morphology is convincing evidence for the process of adaptive evolution. Again, a continuity across vast stretches of time exists that connects evolutionary phenomena in a continuum.

OBSERVATION, COMPARISON, AND CLASSIFICATION

Research in the field of adaptive evolution, especially the phenomena of divergence and convergence of species, began centuries ago with the simple process of recognizing relationships in the surrounding environment between living plants and animals. The search for a unifying order to tie the complex web of animal and plant life together in some meaningful manner was for a long time a part of natural science. The modern theory of organic evolution fulfills this goal admirably in many respects. The methods used to illuminate the intricacies of evolution still encompass the type of keen, analytical observation of phenomena and reflection on their causes and effects that characterized Darwin's studies on the voyage of HMS *Beagle*. Observation, collection of specimens for comparison, classification of specimens according to a meaningful scheme, and, finally, an attempt to sort out the processes involved in a way that agrees with the dictates of strict logic are hallmarks of the scientific method at work.

Researchers investigating evolutionary divergence and convergence have powerful aids in the form of increasingly sophisticated technology. The main focus of their work is the correct interpretation of the path that various lineages took over time to arrive at known, living forms or extinct forms. In the case of living forms, technology originally developed in the field of medicine has been pressed into service to help establish relationships. For example, detailed analyses of various body tissues and fluids have been employed. Blood types have been traced with varying degrees of success, as have various proteins. Powerful optical microscopes are employed to analyze various tissue types and their structures. Since the invention of scanning electron microscopes, these more powerful instruments have further aided in probing the compositions and textures of animal and plant tissues to determine affinities among various species. In addition to these methods, very sophisticated laboratory techniques are now used to unravel and analyze deoxyribonucleic acid (DNA) strands and to try to determine the actual genetic encoding possessed by a particular organism. All these methods help establish more clearly the picture of biological relationships in regard to ancestries.

This physiological approach is obviously of limited utility with regard to fossil species. Except for such instances as the various ice age animals that were frozen in such environments as the tundra, extinct life-forms cannot be analyzed by medical means, as the original tissue has been transformed or destroyed by geological processes. In the case of most fossil forms, hard body parts such as bones and teeth (for vertebrates) and exoskeletons (in the case of invertebrates) must be analyzed in a more structural way to determine possible evolutionary relationships.

The clarification of the paths that various animal and plant lineages took during the process of their evolution further confirms the validity of basic organic evolutionary theory such as natural selection and adaptation. Study of divergent and convergent species is part of the ongoing study of living organisms that make up the functional ecosystems of which humankind is also a part. Learning more about these ecosystems and the parts that all the member species play within them is extremely important in the light of the contemporary world picture of pollution, overpopulation, and industrialization. The increased insight into how ecosystems operate from the species interaction approach is one of the positive by-products that studies of divergent and convergent evolution among species provide.

—*Frederick M. Surowiec*

Further Reading

Bowler, Peter J. *Life's Splendid Drama: Evolutionary Biology and the Reconstruction of Life's Ancestry, 1860–1940.* University of Chicago Press, 1996.

Carroll, Robert L. *Vertebrate Paleontology and Evolution.* W. H. Freeman, 1988.

Cvancara, Alan M. *Sleuthing Fossils: The Art of Investigating Past Life.* John Wiley & Sons, 1990.

Orians, Gordon H., and Otto T. Solbrig, eds. *Convergent Evolution in Warm Deserts.* Dowden, Hutchinson & Ross, 1977.

Paul, Gregory S. *Predatory Dinosaurs of the World.* Simon & Schuster, 1988.

Ridley, Mark. *Evolution.* 2nd ed. Blackwell Scientific, 1996.

Savage, R. J. G. *Mammal Evolution: An Illustrated Guide.* Facts on File, 1986.

Strickberger, Monroe W. *Evolution.* 3rd ed. Jones and Bartlett, 2000.

Copulation

Fields of Study

Animal Physiology; Animal Behavior; Reproductive Biology

Abstract

Animals have many diverse strategies to ensure that eggs and sperm are in close enough proximity for fertilization to take place. Copulation is seen in many aquatic phyla and is the rule in terrestrial phyla. In most forms of copulation, the male reproductive system has an intromittent organ, often called a penis, which deposits sperm into the female reproductive system. Once there, sperm can travel the short distance to the eggs. In some organisms, pseudocopulation is seen.

KEY CONCEPTS

amplexus: a form of pseudocopulation seen in amphibians, where the male mounts and grasps the female so that their cloacae are aligned, and eggs and sperm are released into the water in close proximity and at the same time

cloaca: a common opening for the reproductive, urinary, and digestive systems

heat: that part of the estral cycle when the female is receptive to male copulatory behavior

semen: fluid produced by the male reproductive system that contains the sperm

THE COUPLING CONSTANT

Animals have many diverse strategies to ensure that eggs and sperm are in close enough proximity for fertilization to take place. One widespread process that is seen in many different phyla is copulation. Copulation is seen in many aquatic phyla and is the rule in terrestrial phyla. In most forms of copulation, the male reproductive system has an intromittent organ, often called a penis, which deposits sperm into the female reproductive system. Once there, sperm can travel the short distance to the eggs.

In some organisms, pseudocopulation is seen. In hermaphroditic oligochaetes, such as earthworms (*Lumbricus terrestris*), two worms align in opposite directions so that their genital pores are applied to the openings of the seminal receptacles of their partners. Semen released by the genital pores flows into the seminal receptacles where it is stored. Fertilization, however, is actually external. The worms build cocoons where they lay their eggs and then deposit the stored sperm. Amphibian amplexus is also a form of pseudocopulation. Here, the male frog clasps the female in such a way that their cloacae are in close proximity. Sperm are not deposited in the female's cloaca, however. Instead, both sperm and eggs are released into the aquatic environment for external fertilization.

COPULATORY ORGANS

True copulation takes many forms. In several invertebrates, such as a few flatworms (some *Acoela*, *Rhabdocoela*, and *Polycladida*) and the bedbugs (*Cimicidae*), hypodermic injection is sometimes seen. In this form of copulation, the female has no external

Podisma pedestris *copulation.*

gonopore and the intromittent organ punctures the epidermis and deposits sperm in the underlying body tissue. This sperm must then migrate through the intercellular spaces to the female reproductive organs for fertilization of the eggs to occur. In most organisms, however, the male does not have to pierce the female's epidermis, but instead deposits sperm in an already-present opening of the female's reproductive system.

In birds and some reptiles (such as the tuatara, *Sphenodon punctatus*), the male does not have a true intromittent organ. Instead, the male must manipulate the female during mounting so that their cloacae are pressed against each other. During this "cloacal kiss," the male ejaculates sperm into the female's cloaca. In some bird species, a false penis is present in the male. These organs are not connected to the ducts of the male reproductive system and thus do not serve as intromittent organs. There is speculation that they may provide a necessary stimulation to the female during copulation.

In some fish, fins are modified for semen delivery. In guppies and their allies (*Poeciliidae*), gonopodia, modified anal fins, are used for insemination. Each gonopodium is a hollow, tubelike structure formed from the paired anal fins of the male. When mating, the male inserts his gonopodium directly into the female's gonopore. Usually, not all the sperm are used to fertilize this batch of eggs, and the rest is stored in the oviduct walls for future fertilizations. Other fish (such as the *Coodeidae*) have the anal fins modified into andropodia, which are cup-shaped structures that direct the flow of semen into the female without the andropodia actually entering the female's gonopore. Sharks (*Elasmobranchii*) have modified pelvic fins called claspers, which the male directs into the female's cloaca for insemination. Each shark has two claspers and, depending on species, either the one closer to the female or both are inserted for copulation.

Males of mammals, some reptiles, and many arthropods also have intromittent organs that deposit sperm directly into the female reproductive tract. In these copulations, by either female behavior or male manipulation, the opening of the female reproductive tract must be exposed. In many organisms, the male mounts a squatting or otherwise stationary female. Male snakes and lizards (*Squamata*) have two intromittent organs called hemipenes. Males and females line up side by side and the male uses the hemipenis closer to the female to inseminate her. Many arthropods often go through intricate body contortions to bring the male's penis in proper position for mating. This may be the common rear-mounting pattern, but can also be face-to-face or tail-to-tail. In many animal species, insertion of the penis is followed by one or more thrusting movements that lead to ejaculation.

COPULATORY BEHAVIORS

Among animals, both the lengths of time per copulation and the frequencies of copulation vary widely. When a female lion comes into heat, the male will remain near her, copulating up to one hundred times a day for periods up to ten days. Each copulation, however, lasts for just a few seconds. Other animals may copulate only once, but the copulation may be prolonged. Canid females do not usually remain stationary for mating. To remedy this, once the male mounts, his penis becomes further engorged and this effectively locks him to the female long enough to ejaculate even if she tries to get away. Other animals have hooks and barbs on their penises that may also help to lock them to a female for prolonged copulation. In some animals, prolonged copulation can last several hours. This may be a mechanism to prevent other males from fertilizing the same female.

Females can also play a role in prolonging copulation. In some water mites (*Arrenerus* sp.), the female gonopore can be opened or closed by means of chitinous plates. The smaller male inserts his intromittent organ into the female's gonopore, which then closes, trapping the male. Although sperm transfer is thought to occur in the first few minutes, the female may swim off dragging the male with her for several hours.

Copulation can be dangerous to males. In the domestic honeybee (*Apis mellifera*), a swarm of drones pursue the unmated queen. In-air copulation occurs as a drone inserts his endophallus into the queen's sting chamber. After ejaculation, a small part of the drone's phallus remains inside the queen and the drone falls to the ground and soon dies. Several more drones mate and die until the queen's spermatheca is filled. Male spiders have to be very careful when copulating. If a male does not leave the female's web immediately after depositing his sperm, the female may envenomate and then eat him. The female praying mantis (*Stagmomantis carolina*) have also been known to begin feeding on the heads of males with which they are copulating. Luckily, the headless male can continue to deposit sperm.

—*Richard W. Cheney, Jr.*

FURTHER READING

Davey, K. G. *Reproduction in the Insects*. W. H. Freeman, 1965.

Hayssen, Virginia Douglass, Ari van Tienhoven, and Ans van Tienhoven. *Asdell's Patterns of Mammalian Reproduction: A Compendium of Species-Specific Data*. Comstock, 1993.

Sieglaff, D. "Most Spectacular Mating." In *University of Florida Book of Insect Records*, edited by T. J. Walker. Gainesville: University of Florida Department of Entomolgy and Nematology, 1999.

Smith, R. L. ed. *Sperm Competition and the Evolution of Animal Mating Systems*. Academic Press, 1984.

Crustaceans

Fields of Study

Marine Biology; Oceanography; Ecology; Environmental Studies

Abstract

The majority of animals in the oceans are crustaceans, aquatic arthropods having jaws and two pairs of antennae, such as crabs, lobsters, and shrimps. Crustaceans are among the most successful of animals, dominating the oceans as the insects dominate the land. They also live in freshwater, and some even live in moist land habitats. Most crustaceans are small, but they can differ widely in body form, size, and habits. The twenty-five thousand crustacean species include lobsters up to three-quarters of a meter long and giant spider crabs with leg spans of over 3.5 meters.

Key Concepts

carapace: a hard, chitinous outer covering, such as a crustacean shell or insect exoskeleton

cephalothorax: the anterior (front) section of a crustacean, consisting of a fused head and thorax

chitin: a semitransparent, hard, horny substance forming much of crustacean shells and insect exoskeletons

decapod: any crustacean having five pairs of locomotor appendages (legs)

dorsal: at the hind (posterior) end of a living organism

hermaphrodite: an organism having both male and female reproductive systems

molting: shedding part of or all of a crustacean carapace (shell)

THE DOMINANCE OF THE *CRUSTACEAE*

The majority of animals in the oceans are crustaceans. The name indicates their class, *Crustacea*, which consists of aquatic arthropods having jaws and two pairs of antennae, such as crabs, lobsters, and shrimps. Crustaceans are among the most successful of animals, dominating the oceans as the insects dominate the land. They also live in freshwater, and some even live in moist land habitats. Most crustaceans are small, but they can differ widely in body form, size, and habits. The twenty-five thousand crustacean species include lobsters up to three quarters of a meter long and giant spider crabs with leg spans of over 3.5 meters.

PHYSICAL CHARACTERISTICS OF CRUSTACEANS

All crustaceans are covered by hard chitin coats (carapaces) acting as external skeletons (or exoskeletons), which grow as backward extensions of their heads. Carapace texture varies from bonelike to tough and leathery. The hardness of the covering depends on the amount of calcium salts in the chitin. The exoskeleton acts like armor, to protect the animal. Crustacean bodies are made up of several segments and divided into sections. Each section usually has one pair of jointed legs.

Crustacean heads are fused with several segments just behind them, forming the cephalothorax. This is followed by the abdomen. The head holds two pairs of sensory organs, jaws (mandibles), and antennae. Behind the mandibles are two other pairs of mouth parts, the maxillae. These limbs, which evolved under and in front of the mouth, are used to hold, tear, and taste food. Antennae provide the sense of touch. Each crustacean head also holds a pair of compound eyes, an unpaired eye, or both. Each eye is on a movable stalk and can be turned in any direction.

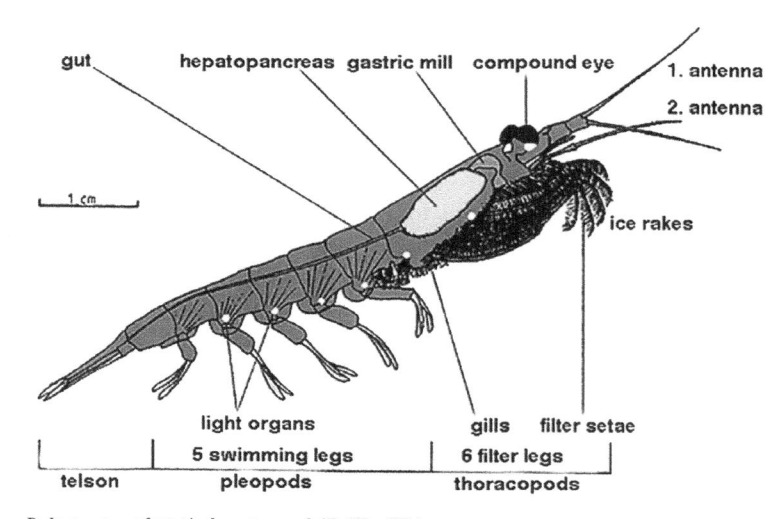

Body structure of a typical crustacean–krill. (Uwe Kils)

The cephalothorax also holds limbs used in locomotion and others used as gills in respiration. Special legs under the body are used for walking. Large legs called pincers or claws are used to catch fish, crack mollusks, dig burrows, and fight. Slow-swimming legs under the tail can be used to hold eggs. The legs at the end of the tail are flattened into a fan-shaped fin, also used in swimming.

The main body cavity is a blood circulatory system, pumped by a dorsally located heart. The crustacean intestine is a straight tube containing glands that secrete digestive fluid and absorb food. All crustaceans also have at least rudimentary brains, composed of ganglia near sense organs and below the intestine.

SHRIMP, LOBSTERS, AND CRABS

Shrimp make up two thousand crustacean species. Structurally similar to lobsters and crayfish but flattened laterally, they are green or brown. They range from insect size to 25 centimeters long, inhabiting salt water and freshwater (mostly on shallow ocean floors) and eat smaller animals and plants. Some species, krill, live deep in the oceans and are eaten by whales. Shrimps have eight pairs of appendages; the first three are used for eating and the rear five are for walking. The shrimp abdomen also contains five pairs of swimming legs and a fanlike tail.

Lobsters are marine decapods, with five pairs of thoracic locomotor appendages. They belong to the crustacean suborder *Reptantia* and are related to freshwater crayfishes. Their narrow, dark green bodies are up to three quarters of a meter in length and weigh one to seven kilograms. In each lobster, two large, body-length claws stretch forward to grasp prey, and a fan-shaped tail is used for propulsion. The three true lobster species are very important foods in North America and Europe. American and European lobsters (*Homarus americanus* and *Homarus vulgaris*) have enlarged claws. Norway lobsters (*Nephrops norvegicus*) have longer, thinner claws. In these true lobsters, one claw is heavy and has blunt teeth to crush prey. The other is smaller and has sharp teeth to tear prey up. Not all lobsters have the heavy claw on the same side. They are "right clawed" or "left clawed."

The lobster head has two pairs of antennae and its eyes are compound, on the ends of mobile stalks.

A female lobster lays 5,000 to 160,000 eggs every two years, and carries them under her tail for up to eleven months, until the young hatch. The young, initially one millimeter long, drift and swim for a month before settling on the bottom, at 2.5 centimeters long. Survivors of the next few months on the dangerous ocean bottom dig shallow burrows beneath rocks, or inhabit crevices in the ocean bed. During the day, a lobster stays inside its burrow waiting for prey. At night, it comes out to search for dead or live food. Lobsters grow by molting and may live up to fifteen years. All are primarily scavengers.

American lobsters occur only off eastern North America, from Labrador to North Carolina. Most live on the ocean bottom, at depths of three to thirty-five meters. Caught in baited lobster pots, they average lengths of twenty-five centimeters and weigh up to between two and three kilograms. Norway lobsters are most abundant near France and Spain. European lobsters are caught off Great Britain, France, Italy, Norway, and Portugal. Rock lobsters, lacking the enlarged claws of the true lobsters, are found off South Africa, Australia, New Zealand, Japan, Brazil, the United States, Mexico, and the Bahamas.

Crabs, related to lobsters and shrimp, can move sideways, burrow, and swim. They are decapods, whose smallest members, tiny pea crabs, consist of females which live in the shells of live oysters and males who live in the outside world and visit them to mate. The giant Japanese spider crabs that are found only along the southwest Pacific coast of Japan are the largest crabs, often three meters or more in circumference (including legs). Many kinds of crabs are used as human food. For example, the blue crab is the common food crab in eastern America. Two important groups, 1,500 species of *Anomura* (hermit crabs) and 4,500 species of *Brachyura* (true crabs), have similar body shapes, with small abdomens and large, broad anterior bodies, though hermit crabs have fewer walking legs. Seen most as ocean-bottom dwellers, crabs also inhabit freshwater, and some live on land.

The crab body is covered by a chitin carapace. The small abdomen under the body is most often a brood pouch. A crab has five pairs of walking legs, two sense antennae, and front legs which are pincers (claws) for feeding and defense. Crabs also have complex nervous systems, with keen compound eyes and the ability to smell and taste food. They enjoy complex mating rituals and communicate by pincer waving. Crabs often mate just after a female molts, when its shell is soft. Eggs are kept in the brood pouch and may pass through two larval stages: the initial form, zoea, does not resemble adults; the later form, megalops, does. Each time a crab molts after birth, it increases in size. Crabs may live for three to twelve years.

There are many different crab species. Some interesting crabs are the lobsterlike anomurans, called squat lobsters; sand crabs, which burrow into sand and filter suspended matter from the water; large spider crabs with long legs and slender bodies; swimming crabs (blue crabs) with paddlelike legs; and fiddler crabs, whose males each have a huge claw that they use in mating and combat. There are even land crabs, omnivores found in the tropics, that release larvae into oceans.

LIFE CYCLE OF CRUSTACEANS

Crustaceans reproduce via eggs, which usually hatch underwater. Some crustaceans, such as lobsters, carry their eggs and young on the hairs of swimming legs. The eggs of different crustaceans hatch at different stages of development. Young lobsters and crayfish look like their parents; young crabs do not. After hatching, young crustaceans grow until their shells become too tight. Then the crustacean sheds its old shell for a larger new one. The process of changing shells (molting) takes place several times during growth. The new shell is formed inside the old one, and is soft and wrinkled until exposed to the environment. When a lobster molts, its shell splits along the back and the lobster leaves through the opening. Sometimes molting accidents will occur. For example, a leg or a feeler often breaks off in the process of leaving the old shell. When the animal molts again, it grows a replacement limb. The new limb is small at first, but becomes full-sized after several molts.

Some crustaceans, such as barnacles, are hermaphrodites. All barnacles live in oceans. Their larvae are free-swimming, but adults attach to foreign objects, such as ship bottoms, wharf piles, rocks, and whales. There are five orders of barnacles. Four are parasites of shellfish. The fifth order includes stalked barnacles, originally found in warm waters. However, because barnacles attach to ships, they are found worldwide.

—*Sanford S. Singer*

FURTHER READING

Bliss, Dorothy E. *Shrimps, Lobsters, and Crabs.* Columbia University Press, 1990.

Headstrom, Richard. *All About Lobsters, Shrimps, Crabs, and Their Relatives.* Reprint. Dover Press, 1985.

Kite, L. Patricia. *Down in the Sea: The Crab.* A. Whitman, 1994.

Llamas, Andreu. *Crustaceans: Armored Omnivores.* Gareth Stevens, 1996.

Muncy, Robert J. *Species Profiles, Life Histories, and Environmental Requirements of Coastal Fishes and Invertebrates (South Atlantic): White Shrimp.* Fish and Wildlife Service, Department of the Interior, 1984.

Taylor, Herb. *The Lobster: Its Life Cycle.* Rev. ed. Pisces Press, 1984.

D

Death and Dying

Fields of Study
Animal Behavior; Biology; Thanatology; Veterinary Medicine

Abstract
The life span of all species in the animal kingdom depends upon genetic composition, environmental conditions, and the amount of energy expended throughout their lifetime. The natural life span varies from one species to another. Environmental conditions affecting the genetically determined life span and hastening death include the number and ferocity of predators, viral, bacterial and fungal disease, poisons and pollutants, changing climate, and the rise of carbon dioxide in the air.

Key Concepts
cetaceans: plant-eating marine mammals, such as whales, dolphins, and porpoises
marine mammals: part of the class of mammals that adapted to life in the sea
myocarditus: inflammation of the heart muscle
persistent organic pollutants (POPs): chemicals that remain in the environment for a very long time and can be found at long distances from where they are used or released; they are nearly all of human origin
pinnipeds: flipper-footed marine mammals, such as sea lions, fur seals, true seals, walruses
sirenians: plant-eating dugongs and manatees

THE VARIABILITY OF LIFE SPANS
The life span of all species in the animal kingdom depends upon genetic composition, environmental conditions, and the amount of energy expended throughout their lifetime. The natural life span varies from one species to another. Insects generally have the shortest lives; the adult mayfly lives only a few hours, the fruit fly lives from thirty to forty days. At the other extreme, a giant tortoise may live up to 177 years and the quahog clam can live up to 220 years. Human life expectancy has increased substantially since the beginning of the twentieth century. In 1998, it ranged from seventy-five to eighty years in the United States, Canada, Western Europe, and Australia, to fifty-five years in most African countries.

Environmental conditions affecting the genetically determined life span and hastening death include the number and ferocity of predators, viral, bacterial and fungal disease, poisons and pollutants, changing climate, and the rise of carbon dioxide in the air. Humans have contributed to the annihilation of many species and placed others close to extinction by either deliberately or accidentally destroying animal habitats and by overhunting wildlife. In 1973, the United States Congress passed the Endangered Species Act, to protect endangered animals and their habitats. The Environmental Protection Agency (EPA) monitors the fate of endangered species. If pesticide use adversely affects the habitat of an endangered species, the EPA can prohibit it.

POLLUTANTS
Maritime oil spills, which kill huge numbers of marine life and birds, affect wildlife species differently. Birds especially are sensitive internally and externally to the effects of crude oil and its refined products. If they become coated with oil and their feathers collapse and mat, the insulating properties of their feathers and down change, making them vulnerable to hypothermia. They become vulnerable to predators and can suffer from dehydration, drowning, and starvation.

Cetaceans, sirenians, and pinnipeds, who depend on air and have amphibious habits, are all susceptible to the effects of oil spills. Like birds, they can suffer from hypothermia. Due to ingesting the oil during grooming and feeding, they suffer from organ

Dead fish. Lots of footprints from the gulls that have been feeding on it.

dysfunction, congested airways, damaged lungs, gastrointestinal ulceration and hemorrhaging, and eye and skin lesions. Sea turtles are particularly vulnerable during their breeding season, as their nesting sites are on beaches and their eggs may become contaminated by the oil. Newly hatched turtles would have to move over the oiled beach to the water. Among the most deadly oil spills was the wreck in Alaska of the supertanker *Exxon Valdez*, in March 1989, which resulted in more than thirty thousand sea birds dying.

Persistent organic pollutants (POPs) have been related to many behavioral problems in birds, marine mammals, and fish, as well as in humans. Studies of humans exposed through food to POPs show a possible relationship to disruptions of the immune system. This finding has been used to explain why more seals and whales are dying and getting stranded. High levels of cancers in fish have been attributed to another class of potential POPs. Environment Canada reported that when POP levels were reduced, population declines in some birds reversed.

VIRAL AND BACTERIAL DISEASE

Like humans, nonhuman animals are vulnerable to viruses. Livestock contract highly contagious and serious diseases. Among the more commonly known are foot and mouth disease, which affects hoofed animals; scrapie, which affects sheep; and bovine spongiform encephalopathy (BSE), known also as mad cow disease, which occurs in cattle. BSE appears to jump species; humans who contract BSE can develop Creutzfeldt-Jacob disease, a fatal brain disorder. Pigs contract swine fever. Capripox occurs in sheep. In Africa, Rift Valley fever kills livestock and humans. Poultry can contract avian influenza and Newcastle disease, both of which spread rapidly, killing more than 90 percent of infected birds. Because rabies is fatal in animals and humans, many countries require a quarantine period for animals entering the country. In wildlife, trapping has been used to prevent the spread of rabies. However, usually the healthier animals are caught in traps and not the sick, who are less active, are symptomatic, and debilitated, and who are more likely to deviate from their normal behavior.

ORGANIZED ANIMAL FIGHTING, ANIMAL FARMING, AND SPORT HUNTING

Humans are voracious predators of other species, killing nonhuman animals for food, clothing, sport, scientific experimentation, and financial gain. Many animals meet their death through organized fighting: bull fighting in Spain and Mexico, and cock fighting in the United States and several Asian cultures, which usually results in the death of one or both roosters. Dog fighting, although banned in most of the United States, is still held clandestinely.

Animal agriculture is the largest food industry in the United States. Animals reared for slaughter are frequently housed in crowded conditions in large buildings, which are ideal for disease. Under natural conditions, chickens can live for as long as fifteen to twenty years. In a modern egg factory hens live about a year and a half. Each year in the United States, about two-thirds of the eighty million pigs raised for slaughter live their lives in a confinement system, as do about half of the ten million milking cows and heifers raised. When birds are debeaked and calves and pigs weaned prematurely, they can die from the shock. Slaughtering is sometimes undertaken without safeguards in place to prevent unnecessary pain.

About 7 percent of the United States' population legally hunt animals. Sport hunting of polar bears in areas of Canada eventually led to such a substantial loss of bears that the local government banned hunting in 2002. In 2001, the government of British Columbia placed a moratorium on the hunting of grizzly bears. The whale population initially decreased because of hunting. The blue whale, which once numbered 200,000, was estimated to be just 10,000 in 2001. Marine mammals also become accidentally entangled in fishing nets and collide with boats. Dolphins died at a considerable rate due to tuna fishing methods until US legislation prohibited the method and the number caught in nets was reduced dramatically. Manatees, who move slowly and sometimes sleep near the surface of the water, are particularly vulnerable to being fatally struck by motor boats.

SCIENTIFIC EXPERIMENTATION

Using animals in scientific experiments has been widely sanctioned throughout the world for testing consumer products, disease prevention and/or progression techniques, the effects of noxious agents, and psychological theories of behavior. An animal rights movement developed in the late 1970s and early 1980s to protest this use of animals, who not only died during and following the experimental procedures but were also subjected to extreme pain and injury. Industrial manufacturers and scientists were urged to find alternate methods of safety testing and conducting experiments. The Johns Hopkins Center for Alternatives to Animal Testing was founded in 1981, while In Defense of Animals (http://www.idausa.org) grew out of challenges to the University of California's research. It grew into one of the foremost animal advocacy organizations in the United States. The ethical question raised by animal rights groups is whether nonhuman animals should be treated as independent sentient beings and not as a means to human ends.

Beyond the ethical issues raised by philosophers, such as Tom Regan and Peter Singer, are the questions concerning the emotional life of animals and whether animals experience grief and have a concept of death. Marc Hauser, an animal-cognition researcher, maintains that animals, lacking a capacity for empathy, sympathy, shame, guilt, and loyalty, are without self-awareness or an awareness of what another of their species experiences, and therefore are incapable of having a deep understanding of death.

Researcher Cynthia Moss, at the Amboseli Elephant Research Project in southern Kenya, takes a different view. From her field observations, she maintains that elephants have a concept of death. They recognize one of their own carcasses or skeletons, always react to the body of a dead elephant, and have been seen putting dirt on a dead elephant's body and covering it with branches and palm fronds. Healthy elephant mothers whose young calves have died look lethargic for many days afterward, trailing behind their family. Wild animals and birds, as well as animals in captivity and animal pets, have been seen reacting to the loss of a mate or companion that can be clearly interpreted as mourning behavior and grief.

—*Susan E. Hamilton*

FURTHER READING

De Waal, Frans B. M. *Good Natured: The Origins of Right and Wrong in Humans and Other Animals.* Harvard University Press, 1996.

Gould, James L., and Carol Grant Gould. *The Animal Mind.* Scientific American Library, 1994.

Hauser, Marc D. *Wild Minds: What Animals Really Think.* Henry Holt, 2000.

Levine, Herbert M. *Animal Rights.* Steck-Vaughn, 1998.

Mason, Jim, and Peter Singer. *Animal Factories.* Harmony Books, 1990.

Moussaieff Masson, Jeffrey, and Susan McCarthy. *When Elephants Weep: The Emotional Lives of Animals.* Delacorte Press, 1995.

Regan, Tom. *The Case for Animal Rights.* University of California Press, 1983.

Singer, Peter. *Writings on an Ethical Life.* HarperCollins, 2000.

Suhowatsky, Gary. "The Role of Trapping in Wildlife Disease." http://articles.animal concerns.org/ar-voices/archive/trapping_disease.html.

Digestion

Fields of Study

Animal Physiology; Endocrinology; Veterinary Science

Abstract

The bulk of animal food consists of proteins, carbohydrates, and fats. In addition, smaller molecules such as vitamins, nucleic acids, and minerals are essential components of animal food. The selection of food and the feeding behavior of animals distinguish different animal populations, allowing them to live together in the same habitat without competing for the same resources. The organs that break down the food mechanically and chemically by the processes called digestion constitute a digestive system.

KEY CONCEPTS

enzyme: a protein that acts as a catalyst under appropriate physiological conditions to break down bonds of a large protein, fat, or carbohydrate

esophagus: the part of the oral cavity (pharynx) that transfers morsels to the stomach; it is usually a long, muscular tube with no digestive function other than transport

hormone: a chemical released into the blood for transport to a specific site, where it will perform a specific function; many hormones stimulate chemical and mechanical aspects of digestion

intestine: the part of the digestive system involved in completing the process of digestion and absorption of nutrients; usually divided into the small intestine and the large intestine, which opens to the exterior by way of the anus

mouth: the anterior part of the digestive system, used for ingesting food; it leads into the oral cavity, which opens into the esophagus

mucus: a secretion of the salivary glands and other parts of the digestive system which lubricates passages

stomach: the part of the digestive system where mechanical breakdown of food is completed and chemical digestion begins

FOOD, GLORIOUS FOOD

The bulk of animal food consists of proteins, carbohydrates, and fats. In addition, smaller molecules that make up these complex molecules—such as vitamins, nucleic acids, and minerals—are essential components of animal food. Animals obtain their food in the form of solutions, suspensions, dry particles, aggregates, and masses of particles, or whole animals and plants and their parts. It is the selection of food and the feeding behavior of animals that distinguish different animal populations and allow them to live together in the same habitat without competing for the same resources. The organs that break down the food mechanically into small particles and particles into molecules by the processes called digestion constitute what is called a digestive system. Usually the fore (anterior) part of the digestive system of animals is adapted for capturing and breaking down food (the bills and beaks of birds and the jaws of mammals are examples), and the remaining system can become specialized to store, chemically digest, and absorb the digested food and eliminate the unabsorbed food.

THE DIGESTIVE SEQUENCE

A typical functional sequence of digestion can be summarized as follows: First is the mouth, its appendages, and the oral cavity. These are involved in selecting (by taste, smell, touch), capturing, ingesting, and initial breaking down of food. Secretions here can include lubricants (mucus coming from salivary glands as well as other fluids), anticoagulants (in blood suckers), paralyzing toxins (in carnivorous coelenterates, spiders, reptiles), proteases (in cephalopod mollusks), and carbohydrases (in plant eaters). In microphages, locomotory appendages, oral tentacles with cilia, can drive currents of water containing food toward the mouth. In macrophages, locomotory appendages can be modified to capture food and ingest it. Most small aquatic animals and some large ones strain small particulate material with the help of their body surface projections (cilia, setae, bristles, legs, mucus, or nets); these microphages are called filter feeders. In vertebrates, movable jaws, and in invertebrates, hard structures or surfaces, can be used for crushing food. The mouth leads into an oral cavity whose posterior chamber is a muscular pharynx that opens into the esophagus.

Second, the muscular, tubular esophagus transfers the food, in bits, to the stomach. Sometimes, a distension in this part of the digestive system (the crop, found in cockroaches and birds, for example) is used to store food.

Third, the stomach, a muscular vessel into which the esophagus leads, mechanically mixes the food through contractions and wavelike motions, and begins the process of chemical digestion via enzymes. Sometimes the stomach is equipped with hard projections (such as the gizzards and gastric mills of birds, cockroaches, earthworms, or alligators). The lining of the stomach or of its diverticula (branches) secretes digestive enzymes and, in vertebrates, hormones and hydrochloric acid. The stomach opens into the next chamber, the intestine.

Fourth, the small intestine completes the digestive process. Its cells and the cells of its glands (the pancreas, liver, hepatic caecum) secrete digestive fluids containing enzymes and hormones that enable absorption of the resulting molecules and water into the cells of the intestine and from there into the blood. The inner lining of the intestine can be thrown into ridges and microridges, which greatly increase the surface area and thus the amount of absorption.

Fifth, the large intestine, or hindgut, reabsorbs water. The undigested food is evacuated in the form of feces through an opening to the exterior called the anus. This part of the digestive tract also stores colonies of microorganisms, especially in plant eaters, to digest cellulose, lignin, and other substances, and to provide some vitamins that the animal cannot synthesize.

CONTINUOUS AND NONCONTINUOUS FEEDERS

In animals that feed on soluble or suspended particles, called continuous feeders, digestion is a continuous process. In these animals, which include sponges, coelenterates, and flatworms, the digestive system is in the form of a tube open at one end only, and the chemical digestion of particles takes place inside each cell lining this tube. Annelids, arthropods, mollusks, and echinoderms have digestive systems that are open at both ends. These animals have developed various other systems, and the digestive system has become independent of the circulatory system. The opening of the digestive tube has allowed these animals to specialize their parts into various regions for capturing, grinding, masticating, mechanically breaking down, chemically digesting, absorbing, and eliminating their food. That, in turn, has allowed them to conduct extracellular digestion in the digestive cavity: to become discontinuous feeders.

With the evolution of a digestive tube dedicated to digestion only, animals started secreting their cellular enzymes into this cavity in response to food. This, then, constituted extracellular digestion. Extracellular digestion is present in small animals that feed on particles (microphages) or larger animals that feed on bulk food (macrophages). The animals with intracellular digestion and those microphages with extracellular digestion feed continuously and nonselectively. The evolution of a complete digestive tract, opening at both ends, and extracellular digestion have allowed evolution of larger, more active, and more advanced animals. These macrophages feed discontinuously and select their food. The time that they saved from feeding can be spent to perform other activities and evolve complex behavior patterns. Also, ingestion of a large mass of food has enabled them to obtain the bulk of their energy from this food, which provides a tremendous amount of dependable power to move and even fly.

DIGESTIVE SPECIALIZATION

The evolution of a complete digestive tube has resulted in the specialization of its parts for various digestive processes. The general pattern of functional sequence that was outlined above is evident: The digestive system is usually divided into foregut (mouth and its appendages), midgut (for chemical digestion and absorption), and hindgut (for absorption of water and elimination of undigested food). Within this general structure, however, are innumerable and complex variations in adaptation to the type of food and feeding mechanisms of different animals. Those animals that feed on solid food, for example, have appendages (such as jaws and teeth) to enable them to grind, crush, or masticate it. In addition, these may have parts of the stomach modified for storage (such as rumen in ruminants, the crops of birds) or for further grinding (the gizzards of various insects, birds, and alligators). Cows and goats, for example, have four-chambered stomachs, one of which stores colonies of bacteria. These ruminants swallow the food as a whole while grazing. Then, later, while resting, they bring the food and bacteria back to the mouth as cud to mix them together, subsequently swallowing the food. The bacteria then digest cellulose by fermentation. The microorganisms are then digested by the animal in the intestine. These animals also secrete copious amounts of saliva, which prevents abrasive damage by the solid food to the cells lining the foregut.

In fluid feeders, by contrast, the oral end is equipped with sucking apparatus containing piercing devices (as in moths, bees, flies, mosquitoes, and leeches). Some feeders on plant juices ingest large amounts of water with sugars. The last part of their foregut becomes connected with the anterior part of the hindgut, forming a filtering apparatus (as in insect leaf hoppers). Only water passes from the foregut to the hindgut, while food enters the midgut, which now does not have to process large amounts of water.

THE ROLE OF ENZYMES

The chemical breakdown of food particles takes place by means of enzyme catalysts, which are proteins that are released into the stomach and intestine (midgut) from their cells or from cells of appendages (hepatic caecum in insects, hepatopancreas in crustaceans and mollusks, and pancreas and liver in vertebrates) opening into the intestine. These enzymes are secreted in response to the entry of food in the gut. Moreover, the presence and release of specific enzymes depend on the chemical nature of the food. In plant-eating herbivores, which eat an abundance of carbohydrates (sugars), these secretions are rich in carbohydrates (carbohydrate-hydrolyzing enzymes), while in animal eaters (carnivores) protein-digesting enzymes, proteases, and fat-hydrolyzing enzymes, lipases, are predominant. In omnivores (which feed on both plants and animals), all three groups of enzymes are present. In food specialists, such as sheep blow flies (which feed on wool keratin), head lice (which feed on hair keratin), cloth moths (which feed on textile fibers), wax moths (which feed on wax), or carpet and leather beetles (which feed on keratin), the digestive fluid is rich in specific enzymes for handling one kind of food. In wood-eating termites, snails (*Helix*), and ruminant mammals, the cellulose is digested by colonies of microorganisms that are carried in parts of these animals' guts.

In addition, different enzymes are present in different stages of an animal's life cycle. For example, maggots feeding on flesh have proteases, while adult flies feeding on sugars have sucrases. The intestinal enzyme lactase, which breaks down the milk sugar lactose, is always present in land mammals at or before birth. It usually decreases after weaning. Among insects, certain leaf hoppers and moths which feed on soluble sugars (which do not require further breakdown) have no enzymes, while hoppers feeding on mesophyll cells and caterpillars actively chewing plant parts have carbohydrases and lipases. Among bees, nurses have more proteases than foragers; wax bees have no proteases; the carbohydrases are predominant in foragers, especially during midsummer, and lipase is found only in wax bees. The carnivorous turbellarians, coelenterates, cephalopod mollusks, crustaceans, scavenger insects, and starfish have more proteases and fewer carbohydrases.

In mammals, most of the enzymes (pepsinogen, trypsinogen, chymotrypsinogen, lipase) are secreted as zymogens (proenzymes) and are activated by other secretions. For example, hydrochloric acid converts pepsinogen into active pepsin in the stomach; enterokinase converts chymotrypsinogen into active chymotrypsin, which in turn activates trypsinogen to trypsin. Secretion of zymogens and their activation are precisely controlled and occur when food is present in the gut. For example, when chyme leaves the stomach, the duodenal hormone enterogastrone inhibits the release of hydrochloric acid from parietal cells so that no activation of pepsinogen occurs; otherwise, pepsin could destroy the proteins in the membranes of cells lining the gastric cavity. The digestive epithelia of animals are thereby protected from damage by physical (solid food) and chemical (enzymes, acids) sources. This digestive strategy became necessary as discontinuous feeding evolved, since the presence of food for digestion was intermittent.

STUDYING DIGESTION

A variety of observations and experiments have been performed to study the different types of digestive systems. Simple examination of the anterior (mouth) end of different animals, for example, reveals the broad range of strategies used to collect and initially break down food.

Soluble food feeders, for example, can be examined under the microscope. Observation of a microscopic slide of the head of a human tapeworm shows that it is equipped with hooks and suckers by means of which it attaches itself to the digestive tract of a person. The soluble, predigested food in the intestine needs no further breakdown and is absorbed through the flat body surface of the worm, which lacks any digestive organs. Observed under the microscope, the anterior end of a liver fluke has hooks and suckers to suck fluid; a lamprey has a round mouth and rasping tongue with which to suck the blood of its host fish; a mosquito has a piercing device to break skin and suck blood; and the mouth parts of an adult moth include a long, coiled proboscis designed to suck nectar from flowers.

Intracellular digestion of food by a variety of microorganisms can be observed in progress under the microscope. Amoebas can be starved for one or two days and then transferred to drops of a culture on a shallow depression slide. The amoebas will exhibit phagocytosis (cell eating) with the help of their "feet" (pseudopodia), surrounding the food and ingesting it. A change in the color of the Blepharisma pigment (in the food vacuole of the amoeba) can be seen—from red (indicating acidic) to neutral or colorless (indicating alkaline pH). This indicates that earlier stages of digestion are acidic; later stages, alkaline.

Paramecium can also be observed feeding on starch solution with and without a drop of iodine (which turns starch blue and inhibits feeding). If this procedure is repeated using compressed yeast in a 3 percent solution of Congo red, one can observe the direction of movement of the yeast (which has taken on the red color) as it travels in the direction of the beating of the paramecium's cilia and into the food vacuole, then as it circulates through the cytoplasm. The change in color from red to blue indicates digestion. Paramecia will also reject algae particles and ingest only yeast, indicating the presence of chemical sensory mechanisms.

Solid food eaters, which can be observed with the naked eye, reveal a variety of specially adapted parts: hard, strong mandibles for crushing leaves in the caterpillar; similar mandibles for handling solid food in the cockroach; the "Aristotle's lantern" of the sea urchin, used for grinding; the tentacles of the *Hydra*, which feeds on fine, suspended food particles; and the human jaw and teeth, with incisors, canines, and molars designed to break down a variety of food in a variety of ways.

The activity of various digestive enzymes can be determined by using appropriate substrates (the food molecules) and physiological conditions in a test tube. The source of the enzyme is the part of the digestive tract where it is produced and used. Tissue from this area is ground in a small blender or homogenizer using an appropriate buffer at about 4 degrees Celsius. The homogenate of the tissue is either used as is or fractionated using a high-speed, refrigerated centrifuge, which can fractionate cell membranes, various organelles, and cytoplasm. Then the subcell fraction, where the enzyme is located, can be used as the source of the enzyme. The enzyme is further purified by means of biochemical devices. The substrate is either natural or synthetic. The pH, temperature, and other conditions are controlled in the incubation mixture containing enzyme and substrate. Time-course aliquots (samples) are withdrawn, and the activity of the enzyme is measured by analyzing the hydrolysis product of the substrate using various spectrophotometric devices. The enzymes, from the same tissue or its subcell fraction, of animals feeding on plants, meat, or both are compared to determine how active various enzymes are in these animals. The presence of certain enzymes can be related to the chemical nature of the food.

DIGESTION AND SURVIVAL

Food selection, feeding behavior, and the structure and function of the digestive apparatus of animals form an important mechanism of survival, by which animals in a population isolate themselves from other populations to avoid competition for the same source of food. The feeding behavior depends on the type of food available (soluble, suspended, aggregates, or large organisms), and the form of the feeding apparatus (shapes and sizes of bills of birds and jaws and teeth of mammals, for example) depends on the physical nature of the food. The anatomy of the digestive system is closely adapted to the physical nature of the food, while the chemical functioning (enzymes) of the digestive systems depends on its chemical nature.

The adaptations of the digestive systems have enabled the evolution of larger and more active animals, which feed less frequently on greater bulks of food as compared with less active small animals, which may have to feed more often and even continuously.

The broad variety of different digestive systems and their enzymes has enabled animals to make the best use of the food resources available in their environment. Those animals able to exploit their environment more fully than others (such as omnivores, including humans) have a wide array of digestive enzymes that can chemically break down a wide variety of foods. They are more successful at survival than those confined to a particular type of food (food specialists with a limited ability to digest only one type of food) and, hence, are likely to survive longer as a group.

—*M. A. Q. Khan*

FURTHER READING

Baldwin, R. L. "Digestion and Metabolism of Ruminants." *Bioscience* 34, no. 4 (1989): 244–249.

Barrett, James M., Peter Abramoff, A. Krishna Kumaran, and William F. Millington. *Biology*. Prentice-Hall, 1986.

Chivers, D. J., and P. Langer, eds. *The Digestive System in Mammals: Food, Form, and Function*. Cambridge University Press, 1994.

Johnson, Leonard R., and Thomas Gerwin, eds. *Gastrointestinal Physiology*. 6th ed. C. V. Mosby, 2001.

Moog, Florence. "The Lining of the Small Intestine." *Scientific American* 245 (November 1981): 154–176.

Penry, Deborah, and Peter A. Jumars. "Modeling Animals Guts as Chemical Reactors." *American Naturalist* 129 (January 1987): 69–96.

Prosser, C. Ladd, ed. *Comparative Animal Physiology*. 4th ed. John Wiley & Sons, 1991.

Raven, P. R., and G. B. Johnson. *Biology*. 4th ed. McGraw-Hill, 1996.

Withers, Philip C. *Comparative Animal Physiology*. Saunders College Publications, 1992.

DIGESTIVE TRACT

FIELDS OF STUDY

Animal Anatomy; Animal Physiology; Endocrinology; Veterinary Science

ABSTRACT

Digestion is the process by which food is broken down into molecules that are small enough to be absorbed into the body. Digestion takes place in the digestive tract of animals. The digestive tract is a continuous tube that acts on ingested food in a sequential manner. Each part of the digestive tract is adapted to reduce the size of food particles, either mechanically or enzymatically, until they are small enough to be absorbed into the body.

KEY CONCEPTS

absorption: the movement of nutrients out of the lumen of the gut into the body

bile salts: organic compounds derived from cholesterol that are secreted by the liver into the gut lumen and that emulsify fats

digestion: the process by which larger organic nutrients are broken down to smaller molecules in the lumen of the gut

duodenum: the first part of the small intestine, where it joins the stomach

enterocytes: the cells that line the lumen of the small intestine

lumen: the central opening through the digestive tract, which is continuous from the mouth to the anus

lymphatic vessels: very thin tubes that carry water, proteins, and fats from the gut to the bloodstream

mucosa: the lining of the inner wall of the gut facing the lumen

pancreas: an organ derived from the gut that secretes digestive enzymes; it is connected to the gut by a duct through which its secretions enter the gut

plexus: a group of nerve cells and their connections to one another

sphincter: a ring of muscle that can close off a portion of the gut

WHERE THE FOOD GOES

Digestion is the process by which food is broken down into molecules that are small enough to be absorbed into the body. Digestion takes place in the digestive tract of animals. The digestive tract is a continuous tube that acts on ingested food in a sequential manner. Each part of the digestive tract is adapted to reduce the size of food particles, either mechanically or enzymatically, until they are small enough to be absorbed into the body. Consideration of the mechanisms of food intake in lower animals will illustrate the evolution of complexity as an adaptation to the changing environments of these animals.

DIGESTIVE TRACTS OF SIMPLE ANIMALS

Sponges are primitive water-dwelling animals that are attached to a fixed point in the water. They bring food into their bodies from currents of water containing particulate food passing through openings in their outer wall. These currents are created by movements of flagella on cells called choanocytes. Food enters the choanocytes by phagocytosis. Phagocytosis

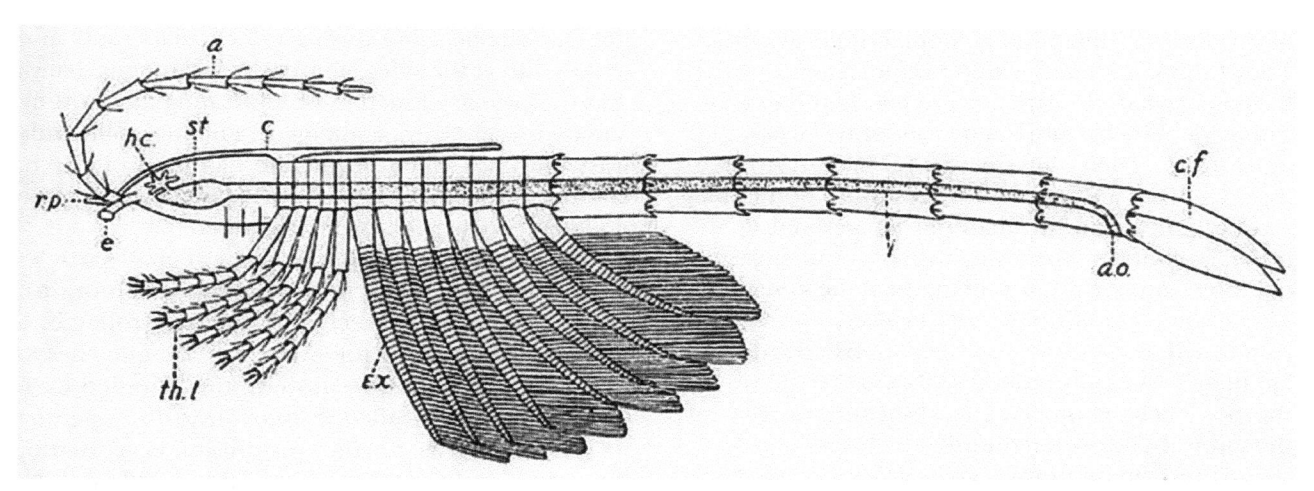

Smithsonian Miscellaneous Collections, Volume 85, Number 3, Figure 7 (on page 23):—The digestive tract of the Waptia fieldensis *Walcott. a, antennae; a.o., anal opening; c, carapace; c.f., caudal furca; e, eye; ex, exopodite; h.c., hepatic caeca.*

is the process in which a cell surrounds a particle with extensions of its cell membrane until the particle is completely surrounded and thus becomes enclosed within a small sac, or vesicle, within the cell. Intracellular enzymes then digest the food particles dissolved in the fluid into their component molecules, which then become available to the metabolic systems in the cytoplasm of the cell. Some cells, called amoebocytes, carry the food particles to other cells in the sponge by crawling through the spaces between the cells. Their travel is by amoeboid motion, in which the cell sends out an extension, called a pseudopod, and then follows it.

This method of feeding and digestion is adequate for a sponge because most of the sponge cells are in close contact with the water currents in which it lives. Thus, these cells can have direct access to food carried in the water currents. The cells that are not in close contact with water currents can be adequately supplied by the amoebocytes. The digestion, or breakdown, must be carried on inside the cells by cytoplasmic enzymes because if these enzymes were released to the extracellular surface, they would be washed away.

The coelenterates, such as jellyfishes or hydras, are more advanced than sponges and require a more elaborate digestive mechanism. These water-dwelling animals are either attached to a surface or float in the water currents. Thus, like sponges, they are dependent on food carried in the water. These animals, however, can eat live prey as well as particulate food.

They are equipped with tentacles that can reach out and trap smaller animals and paralyze them with poisoned darts called nematocysts. The tentacles then bring the food into a distinct body cavity, the gastrovascular cavity, through its one opening. The digestive cavity is, at least partially, not in direct contact with the water currents around the animal. Digestion can take place through extra cellular enzymes secreted by the cells lining this cavity. The resulting molecules are then absorbed through the cell membranes. Amoebocytes also function in these animals. These animals have limited motion through musclelike cells and this motion moves fluid within the gastrovascular cavity, thus carrying fluid to all parts of the animal.

Flatworms are more advanced than sponges and coelenterates and live in a moist, but not watery, environment. They have a distinct nervous and muscular system and can move to search for food. Their digestive tract, as that of sponges, has only a single opening. Food is pushed into this opening by the muscle action of the first part of the digestive tract, which can be protruded to the outside of the animal. Digestion is extracellular and carried to the rest of the animal through muscle contractions of the digestive tract. The digestive tract is highly branched and extends to all parts of the animal.

DIGESTIVE TRACTS OF MORE COMPLEX ANIMALS

Animals more highly evolved than flatworms, including roundworms, insects, fish, mammals, and birds, have a functionally similar digestive tract. These animals all have similar requirements, which have necessitated further, more efficient digestion. These animals are more active and thus must ingest more food. Their digestive tracts have two openings, allowing a continuous digestion: Food enters at one end and waste material is excreted at the other. In contrast, an animal with only one digestive opening cannot excrete and ingest at the same time. The greater size of these more-evolved animals also requires that digested food be absorbed into the circulatory system so that distribution to the rest of the body cells is quick. The absorbing portion of the gut is therefore surrounded by blood vessels.

Further adaptations have required sophisticated specializations of the digestive tract. These include initial chewing devices that can mechanically reduce the size of food so that it can be swallowed. Parts of the digestive system have evolved to store food until it can be efficiently digested. This adaptation allows animals to eat sporadically, when food is available, and allows time for other activities, such as hunting or hiding. Other portions of the digestive tract have become specialized to secrete powerful enzymes that sequentially break down the molecules in food to smaller and smaller molecules. Last, the terminal portions of the digestive tract retain food and extract any remaining nutritional value and eliminate the rest at a convenient time. Many of these adaptations required the formation of a space, called a coelom, between the digestive tract and the rest of the body. This space allows the gut to coil and thus become much longer than the animal, with a resulting increase in the surface area available for digestion and absorption.

The mouth, or buccal cavity, is designed for the entry of food into the digestive tract. The lips and tongue are highly sensitive to the texture and taste of food. They are capable of very precise movements because their musculature is supplied with an extensive nerve supply. The tongue can move laterally, up and down, and in and out, because it has both longitudinal and circular muscles. Movements of the jaws during chewing (mastication) cause the teeth to crush and tear food in the mouth. The teeth have an outer covering of very tough enamel, which protects them against abrasion. Some animals have teeth that grow throughout their life and replace the worn-out ends. Salivary glands in the sides or base of the jaw secrete saliva through ducts that empty into the mouth at the sides of or under the tongue. Saliva has the primary function of lubricating and wetting chewed food. Saliva contains an enzyme, called salivary amylase, which begins the digestion of starch, although the digestion is greatly slowed after the food enters the acidic stomach lumen.

After the food has been reduced to small particles and mixed with saliva, it is swallowed (deglutition). Swallowing is partly a reflex action, controlled by a center in the base of the brain. The tongue rises to the roof of the mouth, pushing the rounded mass of chewed food, called a bolus, into the opening of the esophagus. Further propulsion is created by contraction of the area between the mouth and the esophagus, called the pharynx. The esophagus is a muscular tube leading to the stomach. Contractions of muscles which encircle the esophagus cause a moving ring of contraction, called peristalsis, which propels the bolus into the stomach.

MUCOUS LAYERS

The wall of the gastrointestinal tract is similar throughout its length. The layers, from lumen outward, are the mucosa, submucosa, submucosal nerve plexus, circular muscle, myenteric nerve plexus, longitudinal muscle, and the thin connective tissue covering called the serosa. The stomach and intestines are suspended from the back wall of the abdominal cavity by a sheet of connective tissue called the mesentery. Nerves, blood vessels, and lymphatic vessels reach the gut in the mesentery.

The cavity of the gut, or lumen, is lined by a single sheet of cells called the mucosa. The mucosa contains a wide variety of cell types. Most of the mucosa is composed of a cell type which is called columnar epithelium because the cells are longer than their diameter. Mucous, or goblet, cells secrete mucus, which is the viscous slippery material that protects the cells of the gut against mechanical abrasion and chemical attack. Other cells secrete enzymes into the lumen. Hydrochloric acid is secreted by parietal cells in the stomach. Other cells in the small intestine secrete basic bicarbonate ion. These cells provide the degree of acidity or basicity appropriate to the different regions of the gut. Other cells are adapted to absorb nutrients from or to secrete fluid into the lumen of the gut.

There are also many endocrine cells in the mucosa. These cells secrete hormones into the blood when they are stimulated by nerves or by the contents of the gut. These hormones control the degree of motility or secretion of the gut and the metabolic and physiological responses of the body following feeding. Indeed, the digestive system is the largest endocrine gland in the body. These same hormones are found in the brain, where they act as neurotransmitters, and in other endocrine glands. They have numerous functions revolving around the digestive tract. Some of these hormones can increase or decrease hunger. Others prepare the body for the nutrients that will be absorbed from the digestive tract so that the nutrients can be efficiently utilized. Certain hormones can be released by different types of nutrients in the lumen of the digestive tract. Other hormones can be released through the action of nerves when food is eaten.

The layer next to the mucosa, called the submucosa, is composed of fibrous connective tissue. It provides a mechanical support for the mucosa and also contains the nerve and blood supply leading to and from the mucosa. The lymphatic vessels draining the mucosa also travel through the submucosa.

NERVE AND MUSCLE LAYERS

The next, more external layer, is a sheet of nerves, called the submucous (Meissner) plexus. These nerves send fibers inward to the mucosa and also outward to the other layers. They respond to the luminal contents and to other nerves and hormones. There are as many nerves in the gut as there are in the spinal cord. They are an intrinsic nervous system of the gut—that is, they begin and end in the gut. They are considered a separate category along with the autonomic (involuntary) and somatic (voluntary) nervous systems.

The next layer of the gut wall is a layer of visceral smooth muscle oriented circularly around the circumference of the gut. Contraction of these muscles causes a ring of contraction that may or may not move down the intestine. Next, there is another layer of nerves called the myenteric (Auerbach) plexus. Both nerve plexi are responsible for controlling and integrating the functions of the intestine. Motility of the muscles of the gut, absorption of salt, water, and nutrients, and blood flow are all regulated by these nerves. The outermost layer of the gut is composed of visceral smooth muscle oriented longitudinally along the gut. Contractions of these muscles shorten the length of the gut.

There are also rings of smooth muscle, called sphincters, which control the movement from one part of the gut to the adjacent part. These sphincters are found between the esophagus and the stomach, the stomach and small intestine, the small and large intestine, and the large intestine and the outside.

Food that enters the stomach is partially digested by the enzyme pepsin, which is secreted by the chief cells of the gastric mucosa. Pepsin begins the digestion of protein. The hydrochloric acid secreted by the parietal cells has the functions of activating the pepsin and killing bacteria. The most necessary function of the stomach is storage of food (now reduced to a semiliquid state called acid chyme, or chyme, which is pronounced *kime* rather than *chime*) and slowly propelling it into the small intestine. Additionally, the stomach secretes a substance called intrinsic factor, required for absorption of vitamin B12, which promotes red blood cell formation.

Ruminants, such as cattle and sheep, have the end of the esophagus and the beginning of the stomach modified into large chambers, called the rumen and reticulum, in which food is stored. These portions of the stomach are alkaline because of the enormous volume of basic saliva secreted by the animal. Bacterial digestion of the chyme occurs in these chambers. In addition, the contents can be regurgitated into the mouth and this cud then chewed further. After the cud is chewed and reswallowed, it bypasses the previous chambers and enters a third chamber, called the omasum, where it is churned by muscular contractions. Finally, it enters the abomasum, which is similar to the stomach of other animals.

Birds have specialized adaptations of the stomach, called the crop and gizzard. The crop is a large structure at the beginning of the stomach that stores food until it enters the stomach. The gizzard is a muscular portion of the stomach that grinds the food. This grinding by the gizzard is necessary because birds have no teeth. Frequently, birds will ingest small stones, which are stored in the gizzard and help grind the food. Groups of rounded stones found in the gut region of the fossils of large herbivorous dinosaurs indicate that the swallowing of "grinding stones" was not limited to birds.

THE SMALL INTESTINE

The stomach empties into the small intestine. The first portion of the small intestine is called the duodenum, the middle portion the jejunum, and the terminal portion the ileum. There are two large organs that are connected to the duodenum through ducts that empty into its lumen. These organs are the liver and the pancreas. The liver secretes bile salts, which are necessary to emulsify fats into small particles for absorption. Bile salts are stored in the gallbladder between meals. The gallbladder is connected, by a branch, to the duct leading from the liver to the duodenum. The pancreas secretes alkaline bicarbonate, which helps neutralize stomach acids that enter the duodenum. The pancreas also secretes many different digestive enzymes, which break down proteins, fats, carbohydrates, nucleic acids, and other large molecules. Thus, as soon as chyme enters the duodenum, it is immediately mixed with digestive enzymes and bile salts that entered the lumen from the pancreatic and bile ducts.

The chyme is mixed and propelled along the small intestine by longitudinal and circular muscle contractions. These contractions continually mix the chyme with the pancreatic enzymes and bile salts and present the digested molecules to the mucosal surface, where further digestion takes place. Most of the mucosal cells, called enterocytes, produce enzymes and absorb nutrients. Enterocytes are continuously formed in mucosal pits, called crypts. They migrate up tiny fingerlike projections, called villi, which protrude into the lumen of the gut. It takes about three days for the enterocyte to travel from the base of the crypt to the tip of the villi, and then it is sloughed into the lumen. The villi are thought to increase the surface area of the gut on which digestion and absorption take place. The enterocytes produce enzymes that are attached to the mucosal surface of the cells. These enzymes are responsible for the final stages of digestion, producing the smallest molecules, which are now in a form that can be absorbed by the intestine. Because the digestion takes place on the cells' surface, it is called contact digestion.

After molecules are in their completely digested form, they are absorbed by enterocytes, which transport them from the lumen of the gut to the circulatory or the lymphatic system. Most organic nutrients, such as amino acids, fats, and glucose, are absorbed in the first half of the small intestine, the duodenum and jejunum. Salt, water, and bile salts are absorbed primarily in the ileum. Absorption is virtually complete as long as the digestive system is functioning normally. Usually, the main problems that arise during gastrointestinal disorders are associated with malabsorption of fats. Fats require bile salts to be emulsified. Emulsification is necessary for enzymes to break down fats and also to reduce the final size of the fat microdroplet that results. If any step in this process is not functioning well, then the fats come out of suspension in the intestine and are excreted.

The final contents of the small intestine consist mostly of salts, water, indigestible fiber, and the debris from sloughed enterocytes. The small intestine empties into the large intestine, where some bacterial digestion occurs, which produces mostly small fatty acid molecules. The debris from these bacteria add to the bulk of the undigested material. Muscle contraction propels these feces through the large intestine until it is eliminated by defecation. Sphincters control the final evacuation.

STUDYING THE DIGESTIVE TRACT

The structural features of the digestive tract can be determined by classical techniques of anatomical dissection and histological examination of the cellular characteristics of the different sections of the digestive tract. The secretions and the digestive steps can be determined by sampling the luminal contents. The sampling can be done by passing a tube through the digestive tract until the end reaches the desired portion and then withdrawing a sample for biochemical analyses.

Motility can be measured by attaching a balloon to a tube passed into the digestive tract and measuring the changes in pressure from muscle contractions. Absorption can be measured by perfusing a solution of known composition from one opening in a double tube and collecting the solution remaining after it has passed through the gut lumen from a second opening.

Motility of the intestine or the presence of obstructions that prevent the passage of food along the gastrointestinal tract can be observed by X-ray techniques. A liquid substance, such as a barium suspension, which is opaque to X rays, is swallowed. A series of X rays is taken, or continuous monitoring by an X-ray camera is used. Obstructions can be visualized from the buildup of barium above the blockade. The speed of movement can be estimated to determine if the overall motility of the gastrointestinal tract is abnormal. X rays can also be used to determine directly the presence of abnormal structures such as gallstones, which form in the bile ducts, or tumors. The bile duct and gallbladder system can be visualized with X rays by administering a radioopaque dye that is secreted by the liver into the duct system.

The overall integrity of the gastrointestinal tract can be determined by ingesting inert substances of different molecular sizes and determining if they appear in the blood. Normally, only relatively small molecules can penetrate the very tight mucosal lining of the gut, unless they are nutrients of the body. The penetration of larger molecules across the mucosa indicates leaks resulting from damage to the gastrointestinal lining.

—*David Mailman*

FURTHER READING

Arms, Karen, and Pamela S. Camp. *Biology*. 4th ed. Saunders College Publishing, 1995.

Ganong, William F. *Review of Medical Physiology*. 19th ed. Appleton and Lange, 1999.

Guyton, Arthur C. *Textbook of Medical Physiology*. 10th ed. W. B. Saunders, 2000.

Johnson, Leonard R., and David H. Alpers, et al., eds. *Physiology of the Gastrointestinal Tract*. 3d ed. 2 vols. Raven Press, 1994.

Johnson, Leonard R., and Thomas Gerwin, eds. *Gastrointestinal Physiology*. 6th ed. C. V. Mosby, 2000.

Tortora, Gerard J., and Nicholas P. Anagnostakos. *Principles of Anatomy and Physiology*. 9th ed. John Wiley & Sons, 2000.

DINOSAURS

FIELDS OF STUDY

Paleontology; Paleoecology; Ornithology; Evolutionary Biology

ABSTRACT

The word "dinosaur" is the popular name for a group of land-dwelling reptilelike creatures that was active for about 165 million years. They were the dominant vertebrate animals during most of the Mesozoic era, which began 225 million years ago and ended 65 million years ago. Although dinosaurs were long considered extinct, modern research has indicated that birds are in fact living examples of dinosaurs. Among the dinosaurs were the largest animals ever known, although some of the earliest dinosaurs were very small.

KEY CONCEPTS

ankylosaurs: a group of later ornithischians characterized by heavy armor

cerotopsians: a group of later ornithischians characterized by a beaked snout and a bony frill on the back of the head

ornithischians: one of the two orders of dinosaurs; it comprises the "bird-hipped" dinosaurs

ornithopods: the early, bipedal ornithischians

saurischians: one of the two orders of dinosaurs; it comprises the "reptile-hipped" dinosaurs

sauropods: the herbivorous, quadrupedal saurischians

stegosaurs: a group of later ornithischians characterized by a row of plates down the back

thecodonts: an order of Triassic reptiles that were the ancestors of dinosaurs, birds, and crocodiles

theropods: the carnivorous, primarily bipedal saurischians

The Dinosaurs and Their Times

The word "dinosaur," which is derived from the Greek term for "terrible lizard," is the popular name for a group of land-dwelling reptilelike creatures that was active for about 165 million years. They were the dominant vertebrate animals during most of the Mesozoic era, which began 225 million years ago and ended 65 million years ago. Although dinosaurs were long considered extinct, and indeed most species were wiped out after the extinction event marking the Cretaceous-Palogene boundary, modern research has indicated that birds are in fact living examples of dinosaurs. Among the dinosaurs were the largest animals ever known, although some of the earliest dinosaurs were very small.

Imaginative depiction of the meteor impact thought to be responsible for extinction of most dinosaurs.

The Mesozoic era is divided into three periods, the Triassic, the Jurassic, and the Cretaceous, of approximately equal length. Dinosaurs first appeared in the later third of the Triassic period. Experts believe that dinosaurs developed from a group of archosauromorph reptiles such as *Marasuchus*, which was a lightly built flesh eater about 1.3 meters long. It was clearly a biped, running on its hind legs, and the long tail was presumably used as a balancing organ.

Dinosaurs are divided into two separate orders, depending on the arrangement and shape of the hip bones, which determine the way an animal walks and holds its body. The saurischians, or "reptile hips," as they are commonly called, arose in the early part of the Late Triassic; the ornithischians, or "bird hips," arose toward the end of the Triassic period.

SAURISCHIANS

The earliest dinosaurs were saurischians, which are best known from the Ischigualasto Formation of Argentina. The order *Saurischia* may be divided into two major suborders: the theropods, or "beast-footed dinosaurs," and the sauropods, or "reptile-footed dinosaurs." The theropods, which were more primitive than the sauropods, were primarily bipedal, although many of them probably used all four feet when walking or resting. The hind legs were strong and bore birdlike feet, while the forelimbs bore sharp, curved claws for seizing and holding prey. All theropods had long tails that functioned as stabilizers. The head was large, and the jaws of most of the theropods contained sharp teeth.

The theropods are divided into two major groups. A basal group, the ceratosaurs, includes such dinosaurs as *Coelophysis*, a small, agile carnivore with a long, narrow skull represented by many hundreds of specimens from the Ghost Ranch Quarry in New Mexico. However, larger dinosaurs, such as *Ceratosaurus*, are included within this group. The remaining theropods, termed tetanurans, include the Maniraptora forms, which share many advanced characteristics with birds. The largest of these is *Tyrannosaurus* rex, from the Late Cretaceous period of North America, which grew to a weight of 4,500 kilograms, a height of 6 meters, and a length of 15 meters. Theropod dinosaurs evolved feathers and gradually decreased in size over about 50 million years to give rise to birds, including the early species *Archaeopteryx*. Scientists first speculated about the connection between birds and dinosaurs as long ago as the mid-1800s, but it was only in the 1990s and after that a majority of researchers concluded that birds were indeed surviving dinosaurs rather than a separate classification.

The sauropods, which appeared slightly later in the Triassic than the theropods, have come to stand as a symbol of gigantism in land animals. They were all quadrupeds and vegetarians. They had small skulls, long necks and tails, large barrel-shaped bodies, padded feet, and large claws on the innermost toe of the forefoot and the innermost toe of the hind foot. The ancestral stock of the sauropods were the prosauropods, which were much smaller than the sauropods. Like most prosauropods, *Plateosaurus* had blunt, spatulate teeth, was an herbivore, and was quadrupedal, although it was capable of bipedal posture and gait.

The later sauropods had longer necks, and their skulls were relatively small. The limb bones became solid and pillarlike to support their great weight. This category contained many extremely large dinosaurs, including *Brachiosaurus*, which is estimated to have weighed 73,000 kilograms. Among the best known sauropods are *Brontosaurus* and *Diplodocus*, from the Late Jurassic period of North America. The group known as titanosaurs were even larger than the brachiosaurids; *Argentinosaurus* is often considered both the heaviest and longest land animal known, although the fragmentary nature of fossil evidence means that measures are only estimates and many other contenders have been suggested. Although it was once assumed that these huge beasts had to live in swamps where the water could support their great weight, it is now clear that they were terrestrial animals that used their long necks to eat from trees.

ORNITHISCHIANS

The sauropods reached their zenith in the Late Jurassic; the ornithischians replaced them as the dominant herbivores in the Cretaceous period. The expansion of this group was associated with the advent of the flowering plants during the Cretaceous period. Characteristically, a horny beak developed at the front of the mouth, and the toes ended in rounded or blunt hooves instead of claws.

The earliest ornithischians were the ornithopods. A typical example is *Hypsilophodon*, a small, swift dinosaur with a long, slender tail and long, flexible toes. The most specialized of the ornithopods were the "duck-billed dinosaurs," also known as hadrosaurs. Although they had flat beaks and no anterior teeth, the cheek region had rows of grinding teeth. The various types of duck-billed dinosaur can be distinguished by modifications of the bones associated with the nostrils. Some were molded into hollow, domelike crests, bizarre swellings of the nasal region, or long, projecting tubular structures that were possibly used to warm the air or to produce sounds. The remaining three groups of ornithischians presumably evolved from the primitive ornithopods.

The earliest of these three groups of highly specialized quadrupeds was the "plated dinosaurs," or stegosaurs, which first appeared early in the Jurassic period. This large dinosaur was more than 6 meters long. In comparison to its body size, its head was extremely small. *Stegosaurus* had an average of twenty plates arranged alternately in two parallel rows down the back. The plates were originally thought to have been used for protection, but scientists now believe that the plates could have been used for thermoregulation. *Stegosaurus* died out in the Early Cretaceous period.

The "armored dinosaurs," or ankylosaurs, are not very well known, even though their remains have been found over much of the world. Their armor consisted of a mosaic of studs over the body, spikes that protected the legs, and, in some cases, spikes on the tail. They protected themselves by crouching and drawing in their head and legs. The best-preserved example of the armored dinosaurs is an essentially intact and complete fossilized *Nodosaur* that was unearthed during excavations in the Tar Sands region of Alberta, Canada.

Some of the last dinosaurs to develop were the "horned dinosaurs," or ceratopsians. The skull was characterized by a beaked snout and a bony frill that extended from the back of the head. The ceratopsians were also distinguished from others by various patterns of horns. The skull of *Triceratops*, for example, had three sharp horns, one on the snout and one above each eye. The best known of the small ceratopsians was *Protoceratops*, which was a small, hornless dinosaur from the Gobi Desert in Mongolia. The complete fossilized remains of a *Protoceratops* locked in battle with a *Velociraptor* were unearthed in Mongolia, where they are thought to have been buried in an instant by the collapse of a sand dune during a heavy rainfall.

Studies of dinosaur eggs, nests, trackways, and bone structures have shown that smaller dinosaurs probably had a warm-blooded, or endothermic, metabolism similar to mammals. This is supported by the discovery of small theropods in China that show a covering of feathers, presumably for insulation in cold environments that could not support cold-blooded animals. Large dinosaurs, such as sauropods, may have been more efficient as ectotherms, similar to most modern reptiles. Although research has increasingly supported the theory that dinosaurs were active and endothermic rather than sluggish creatures as originally imagined, there is still much debate over the issue and it remains an active area of study.

EXTINCTION THEORIES

Several theories regarding the dinosaurs' extinction were first proposed in the late nineteenth and the early twentieth centuries. According to one popular theory, dinosaurs were wiped out because early mammals of the Cretaceous period ate their eggs. Yet the eggs of many modern reptiles have faced the same threat and have survived, primarily because reptiles lay so many eggs. Another theory suggested that the same animals ate the plants on which the dinosaurs depended. Although that is possible, virtual plagues of mammals would have been required to eradicate the dinosaurs. Some early scientists also believed that the dinosaurs became too big for their environment; that is unlikely, however, because gigantic dinosaurs had been successful for millions of years and small dinosaurs existed as well. Changes in the physical environment also occurred in the Late Mesozoic. Evidence indicates that the sea levels fell. Geologic evidence shows, though, that drastic environmental changes had occurred many times during the dinosaurs' reign without any sudden effect as detrimental as a total mass extinction.

A theory proposed in early 1979 by Luis Alvarez and Walter Alvarez suggests that the iridium that has been found in several samples of sedimentary layers between the rock of the Cretaceous and Tertiary periods came from an asteroid that struck Earth at that time near present-day Chicxulub, on the Yucatan Peninsula. Such a catastrophic event could have caused an enormous cloud of dust to circle Earth and cut off the sunlight, destroying the plants and the dinosaurs that depended on them. While according to some scientists this theory fails to explain why so many other animals, such as the mammals, managed to survive, it has long been the most generally accepted idea, sometimes in combination with other factors. Recent modeling studies and practical tests indicate that the surface effects of the global conflagration that followed the impact of the asteroid were not felt significantly just ten centimeters below the soil surface. This would suggest that many burrowing creatures could have survived the catastrophe, but would then have been faced with surviving on the devastated surface afterwards, while surface-swelling creatures would have been eliminated.

Another modern theory places the blame on the greenhouse effect. It has been argued that the reduction of the seas that occurred during the Cretaceous period caused a reduction of marine plants. As a result, the amount of carbon dioxide in the air increased, trapping heat from Earth's surface. A similar theory suggests that the eruption of a tremendous volcano produced a fatal amount of carbon dioxide. Neither theory, however, explains why other animals, especially heat-sensitive reptiles, survived.

The main alternative to the extraterrestrial catastrophist explanation is a gradual ecosystem change model. Declines in many groups of organisms that started well before the Cretaceous-Tertiary boundary are seen as having been caused by long-term climatic change, as lush tropical environments were replaced by strongly seasonal, temperate climates. The best explanation for the extinction of most dinosaur species may be a combination of the two main theories.

However, as scientists in the late 1990s and twenty-first century increasingly came to recognize the connection between dinosaurs and birds, a new consensus began to emerge that dinosaurs in fact never went fully extinct. Instead the theropods lived on as birds, surviving together with mammals and other species through whatever catastrophes or gradual changes wiped out the rest of the dinosaurs.

STUDY OF DINOSAURS

Scientists study dinosaurs mainly by examining fossils, which are animal remains that have turned to stone. If a dinosaur died near a river or in a swamp, it stood a fair chance of being preserved. Its body might sink into the mud, or floodwaters might float it downstream, where it would end up on a sandbar, on the bottom of a lake, or even in the sea. After the flesh decayed or was scavenged, the bones would be covered by sediments, such as mud or sand. Over time they bone material would become mineralized, and the weight of accumulated layers of sediment would slowly compress the remains and incorporate them into the surrounding rock: mud into shale, sand into sandstone, limey oozes into limestone or chalk.

The way a fossil is studied is determined by the category to which it belongs. The first category is petrified fossils. They may be preserved in two ways. In replacement, minerals replace the original substance of the animal after water has dissolved the soft body parts. In permineralization, minerals fill in the small air spaces in bones or shells, thereby preserving the original bone or shell. The second group of fossils is composed of natural molds that form when the bodies dissolve. Scientists make artificial casts of these molds by filling them with wax, plastic, or plaster. The third type is prints, which are molds of thin objects, such as feathers or tracks. Sometimes, even skin is preserved. Prints are formed when the soft mud in which they are made turns to stone. Scientists can determine the length and weight of the dinosaur that made a set of footprints by studying the depth, size, and distance between them.

Most fossils are found in sedimentary rock formations, which lie beneath three-fourths of the surface of Earth. The best collecting areas are places where the soil has worn away from the rocks. Areas in Colorado, Montana, Wyoming, and Alberta, Canada, have been especially rich in fossils. Most of the finds consist of no more than scraps of limb bones, odd vertebrae, loose teeth, or weathered lumps of rock with broken bone showing on the surface. Once a scientist has discovered a few fossilized fragments, he or she combs the area to find the rest of the animal. If the skeleton is embedded, it is extracted with the help of a wide variety of tools, ranging from picks and shovels to pneumatic drills. Loose fragments are glued back into place, and parts that are too soft or breakable are hardened by means of a special resin solution that is sprayed or painted on.

As the fossil is uncovered, it is encased in a block of plaster of Paris. (A more modern method uses polyurethane foam instead of plaster.) After the entire surface is covered, the fossil is rolled over, and another layer of plaster is added. After the fossil has been transported to the museum, the plaster or polyurethane coating is removed. The "development" stage involves the removal of the rock around the bones. The oldest way is by hand, using tools such as hammers and chisels; a more modern technique uses electrically powered drills similar to dentists' drills. Sandblasting and chemicals may also be employed. After the fossil is cleaned, it is ready for mounting. The bones are fastened to a steel framework that makes the skeleton appear to stand by itself.

LIFE-EARTH INTERACTION

From the dinosaurs, scientists continually learn new lessons about the physiology of such beasts, their relationship to the world in which they lived, their distribution and the bearing of that distribution on the past arrangements of the continents, various aspects of evolution, and the reasons that they became extinct. The dinosaurs played a major part in the shaping of the natural world and in turn have shaped human understanding of natural processes. For example the fossil evidence of the intermediary species *Archaeopteryx*, a primitive bird that lived during the Late Jurassic period and had a beak containing teeth, provided a powerful example of evolution. Eventually the direct transition from dinosaurs to birds would be recognized.

The disappearance of a species that seemed to rule the world for more than 100 million years brings into question the notion of a "dominant" species. Most people believe that mammals are now the dominant form of life; however, dinosaurs did not truly "rule," and neither do mammals. If one were to list the biological organisms whose influence on the planet is such that their removal would produce chaos, then that list would be headed by microorganisms so small that they can be seen only through powerful microscopes. The list would also include the green plants and the fungi. Studying dinosaurs can bring humanity's humble place in the world into perspective.

The extinction of the dinosaurs also brings into question the ability of humans to destroy the world. All species, from the simplest microorganism to the largest plant or animal, modify their immediate surroundings. They cannot avoid doing so. The success of one group, however, does not imply the failure of the groups it exploits. The complexity of individual organisms may increase, but the simpler forms do not necessarily disappear. Life continued after the demise of the dinosaurs and would probably continue to do so if humankind were destroyed.

—*Alan Brown*

FURTHER READING

Allaby, Michael, and James Lovelock. *The Great Extinction*. Martin Secker & Warburg, 1983.

Bakker, Robert T. *The Dinosaur Heresies*. William Morrow, 1986.

Benton, Michael J. *Vertebrate Paleontology*. 2nd ed. Chapman and Hall, 1997.

Charig, Alan. *A New Look at the Dinosaurs*. Avon, 1983.

Colbert, Edwin H. *Dinosaurs: An Illustrated History*. Hammond, 1983.

Currie, P. J., and Kevin Padian. *Encyclopedia of Dinosaurs*. Academic Press, 1997.

Gough, Zoe. "Dinosaurs 'Shrank' Regularly to Become Birds." *BBC*. BBC, 31 July 2014. Web. 6 Jan. 2016.

Lambert, David. *A Field Guide to Dinosaurs*. Avon, 1983.

Lucas, Spencer G. *Dinosaurs: The Textbook*. 3rd ed. McGraw-Hill, 2000.

Martin, Anthony J. *Dinosaurs Without Bones: Dinosaur Lives Revealed by Their Trace Fossils*. Pegasus, 2014.

Moore, Randy. *Dinosaurs by the Decades: A Chronology of the Dinosaur in Science and Popular Culture*. Greenwood, 2014.

Pickrell, John. *Flying Dinosaurs: How Fearsome Reptiles Became Birds*. Columbia University Press, 2014.

Witford, John Noble. *The Riddle of the Dinosaurs*. Alfred A. Knopf, 1985.

Displays

Fields of Study

Animal Behavior; Ornithology; Entomology; Biochemistry

Abstract

Some displays involve, literally and simply, the visual display of a physical feature. Physical features can indicate an individual's sex, age, and reproductive status. Many such features vary in size, shape, or color in relation to an animal's health, hormones, or social status, and are therefore referred to as status badges or signs. Often, meaningful physical features are further highlighted by behavioral displays.

Key Concepts

- **aposematic display:** use of bright, non camouflaged colors to indicate toxicity or dangerousness
- **pair-bonding:** prolonged and repeated mutual courtship display by a monogamous pair, serving to cement the pair bond and to synchronize reproductive hormones
- **pheromone:** a modified hormone that, through sense of smell, communicates information to, and has effects on, individuals other than the individual producing it
- **Principle of Antithesis:** the observation that signals communicating opposite meaning tend to be expressed using displays having opposite characteristics
- **ritualization:** an evolutionary process that formalizes the context and performance of a display so that its meaning is clear and straightforward
- **species-specific:** a behavior or trait that characterizes members of a species, is innate, and is exclusive to that species
- **status badge:** a visual feature that, based on its size or color or some other variation, indicates the social status of the bearer

WHY DO BIRDS SING?

Why do birds sing? Not because they are happy. They sing to communicate. By singing, a bird communicates its location, its species, its sex, its approximate age and size, and perhaps its current reproductive status, territory ownership, health, dominance status, and motivational state. Birdsong and other nonlinguistic forms of communication are called displays.

TYPES OF DISPLAYS

Some displays involve, literally and simply, the visual display of a physical feature. Among insects, for example, green or brown coloring is often used for camouflage. Insects that are poisonous do not need camouflage and often advertise themselves with warning colors, such as black and red, or black and orange. This is referred to as aposematic coloration or an aposematic display.

Physical features can also indicate an individual's sex, age, and reproductive status—as do peacock tails, turkey wattles, deer antlers, the canine teeth of male baboons, and the swollen genitals of estrous female chimpanzees. Many such features vary in size, shape, or color in relation to an animal's health, hormones, or social status, and are therefore referred to as status badges or signs.

Often, meaningful physical features are further highlighted by behavioral displays. A courting peacock or turkey will fan open his tail and shake it back

Many male birds have brightly coloured plumage for display. This feather is from a male Indian peafowl Pavo cristatus. (MichaelMaggs)

and forth for emphasis; a challenged buck will load plant material onto his antlers so as to exaggerate their size; an angry baboon will curl back his lip to further expose his canine teeth; and an estrous chimpanzee will approach a friendly male and assume a posture displaying her fertile state.

A particularly energetic or dramatic behavioral display not only calls attention to a physical feature, but indicates the health and vitality of the performer. The principle of honest signaling refers to the fact that large and healthy individuals tend to have brighter or more contrasting colors, make deeper-pitched and louder sounds, and produce longer, more intense performances than small or weak ones. Such differences in display quality are readily noticed by predators, potential competitors, and potential mates.

Most displays are performed by individuals and are one way: sender to receiver. Threat displays, however, may involve reciprocal signaling between two challengers or between two groups of challengers. Courtship displays also may occur in groups: In some species, males gather together to perform in what is called a lek or a lekking display. Courtship of monogamous species may include long sequences of frequently repeated, ritualized interactions in which both partners participate. Such pair-bonding displays may continue well into the breeding season and the mateship; initially serving to familiarize the pair with one another and to synchronize their hormones and breeding behavior, they may later serve as greeting displays after separation.

INTERPRETING DISPLAYS

Darwin noted that displays having opposite characteristics often signal opposite meaning. In humans, for example, a face with upturned corners of the mouth (a smile) signals friendliness, whereas a face with down turned corners of the mouth (a frown) signals displeasure. In most animals, loud, deep-pitched sounds (for example, roars and growls) indicate aggression, whereas quiet, high-pitched sounds (for example, mews and peeps) indicate anxiety or fear. Similarly, body postures exaggerating size tend to signal dominance, whereas postures minimizing size tend to signal submission. Darwin called his observation the Principle of Antithesis.

Although some rules of display can be applied across species, most displays are specialized for intraspecific (within-species) communication—male to female, parent to offspring, or dominant to subordinate—and are therefore species-specific. That is, the ability to perform and interpret a particular display (such as a particular birdsong) is generally characteristic only of individuals of a particular species and is either innate (inborn) or learned from conspecifics (individuals of the same species) during an early critical period of development.

In order that their meaning is easily and quickly conveyed, most displays also tend to be highly ritualized; that is, they are performed only in certain contexts and always in the same way. This consistency in communication prevents errors of interpretation that could be disastrous. It would be a grave mistake, for example, to interpret an aggressive signal as a sexual overture, or an alarm call (predator alert signal) as an offspring's begging call. Mistakes of interpretation are also minimized by signal redundancy; that is, messages are often conveyed simultaneously in more than one sensory modality.

DISPLAY MODALITY

Displays utilize every sensory modality. Visual displays involve the use of bright, contrasting, and sometimes changing colors; changes in body size, shape, and posture; and what ethologists call "intention movements"—brief, suggestive movements which reveal motivational state and likely future actions. Auditory displays include vocal songs and calls, as well as a variety of sounds produced by tapping, rubbing, scraping, or inflating and deflating various parts of the body. Tactile displays include aspects of social grooming, comfort contacts (such as between littermates or parents and offspring), and the seismic signaling of water-striders, elephants, frogs, and spiders which, respectively, vibrate the water, ground, plants, or web beneath them. Olfactory displays include signals from chemicals that have been wafted into the air or water, rubbed onto objects, or deposited in saliva, urine, or feces.

Olfaction (sense of smell) is the most primitive, and therefore the most common and most important, sense in the animal kingdom. Species of almost every taxonomic group use smell to signal their whereabouts and, generally, their sex and reproductive state. (Birds seem to be an exception.) Animals may also use smell to identify particular individuals, to recognize who is related to them and who is not, and to determine the relative dominance status of a conspecific.

Chemicals used in displays are called pheromones. They may be derived from waste products or hormones, acquired by ingesting certain food items, or obtained directly from plants or other animals. Some pheromones not only communicate information, but also have physical effects on their receivers.

—*Linda Mealey*

FURTHER READING

Agosta, William C. *Chemical Communication: The Language of Pheromones.* Scientific American Library, 1992.

Bailey, Winston. *Acoustic Behaviour in Insects.* Chapman and Hall, 1991.

Eibl-Eibesfeldt, I. *Ethology: The Biology of Behavior.* Translated by Erich Klinghammer. 2nd ed. Holt, Reinhart and Winston, 1975.

Guthrie, R. Dale. *Body Hot Spots: The Anatomy of Human Social Organs and Behavior.* Van Nostrand Reinhold, 1976.

Johnsgard, Paul A. *Arena Birds: Sexual Selection and Behavior.* Smithsonian Institution, 1994.

Morris, Desmond. *Animalwatching.* Crown, 1990.

Owen, Denis. *Camouflage and Mimicry.* University of Chicago, 1980.

ECOLOGICAL NICHES

FIELDS OF STUDY

Ecology; Environmental Studies; Biogeography; Animal Behavior

ABSTRACT

The idea of the niche probably had its first roots in ecology in 1910 when Roswell Johnson theorized that individuals of a particular species are only in certain places because of food supply and environmental factors that limit their distribution in an area. In 1924, Joseph Grinnel developed his concept of niche that centered on an organism's distribution having limits set on it by climatic and physical barriers. At the same time, Charles Elton's definition of niche involved the way an organism gathers food.

KEY CONCEPTS

- **community:** all the populations of plant and animal species living and interacting in a given habitat or area at a given time
- **environment:** all the external conditions that affect an organism or other specified system during its lifetime
- **food pyramid:** diagram representing organisms of a particular type that can be supported at each trophic level from a given input of solar energy in food chains and food webs
- **habitat:** place or type of place where an organism or community of organisms naturally thrives
- **organism:** any form of life
- **trophic level:** a level in a food chain or food web at which all organisms consume the same general types of food

DEFINING THE NICHE

The idea of the niche probably had its first roots in ecology in 1910. At that time, Roswell Johnson wrote that different species utilize different niches in the environment. He theorized that individuals of a particular species are only in certain places because of food supply and environmental factors that limit their distribution in an area. Later, in 1924, Joseph Grinnel developed his concept of niche that centered on an organism's distribution having limits set on it by climatic and physical barriers. At the same time, Charles Elton was defining his own idea of niche. His description of niche involved the way an organism makes its living—in particular, how it gathers food.

TROPHIC LEVELS

For many years, ecologists focused on Elton's definition and referred to niche in terms of an organism's place in the food pyramid. The food pyramid is a simplified scheme in which organisms interact with one another while obtaining food. The food pyramid is represented as a triangle, often with four horizontal divisions, each division being a different trophic level.

The base of the food pyramid is the first trophic level and contains the primary producers: photosynthetic plants. At the second trophic level are the primary consumers. These are the herbivores, such as deer and rabbits, which feed directly on the primary producers. Secondary consumers are found at the third trophic level. This third trophic level contains carnivores, such as the mountain lion. The members of the uppermost trophic level are the scavengers and decomposers, including hyenas, buzzards, fungi, and bacteria. The organisms in this trophic level break down all the nutrients (such as carbon and nitrogen) in the bodies of plants and animals and return them to the soil to be absorbed and used by plants.

It should be noted that no ecosystem actually has a simple and well-defined food pyramid. Many organisms interact with more organisms than those at the adjacent trophic levels. For example, a coyote

Two lichenes species on a rock, at Meneham, Britanny, Finistère, France. A nice exemplification of the ecological niche concept.

could be considered to belong to the third trophic level with the carnivores, but the coyote also occasionally feeds on fruits and other primary producers. Basically, all living things are dependent on the first trophic level, because it alone has the capability to convert solar energy to energy found in, for example, glucose and starch. The food pyramid takes the geometric form of a triangle to show the flow of energy through a system.

Photosynthetic plants lose 10 percent of the energy they absorb from the sun as they convert solar energy into glucose and starch. In turn, the herbivores can convert and use only 90 percent of the energy they obtain by eating plants. Hence, less energy is found at each higher trophic level. Because of this reduction in energy, fewer organisms can be supported by each higher trophic level. Consequently, the sections of the pyramid get smaller at each higher trophic level, representing the decreasing levels of energy and number of members.

INTERRELATIONSHIPS AMONG ORGANISMS

Through the years, two concepts of niche have evolved in ecology. The first is the place niche, the physical space in which an organism lives. The second is the ecological niche, and it encompasses the particular location occupied by an organism and its functional role in the community.

The functional role of a species is not limited to its placement along a food pyramid; it also includes the interactions of a species with other organisms while obtaining food. For example, the methods used to tolerate the physical factors of its environment, such as climate, water, nutrients, soils, parasites, and the like, are all part of its functional role. In other words, the ecological niche of an organism is its natural history: all the interactions and interrelationships of the species with other organisms and the environment.

The study of the interrelationships among organisms has been the focus of ecological studies since the 1960s. Before this time, researchers had focused on the food pyramid and its effect on population changes of merely a single species. One example, the classic population study of the lynx and the snowshoe hare of Canada, originally focused on the interactions of the species in the food pyramid. It was discovered that the lynx had a ten-year population cycle closely following the population cycle of its prey, the snowshoe hare. The lynx population appeared to rise, causing a decline in the population of the snowshoe hare. In the investigations that followed, however, studies diverted the focus from the food pyramid to other elements of the niche of the two species. For example, the reproductive nature of the hare provided a contradiction to the simple predator-prey explanation. The hare has a faster rate of reproduction than the lynx. It seemed impossible that the significantly lower population of lynx could effectively place sufficient predator pressure on the hare to cause its drastic decline in numbers. Therefore, it appeared that the population dynamics of the hare and lynx was regulated by more than simply a predator-prey relationship.

Later studies of the lynx and hare suggested that the peaks and dives in the two populations may also be a factor of parasites of the hare that are carried by the lynx. A rise in the lynx population increases the carriers of parasites of the hare. Therefore, it is thought that, although the hare has a much greater reproduction rate than the lynx, the population of hares will still decline because of the combination of predation by the lynx and the increased frequency of parasites of the hare. This study involved looking at more than one dimension of the ecological niche of a species and broke away from concentrating on only the interactions between organisms in the food pyramid.

NICHE OVERLAP

The goal of understanding how species interact with one another can also be better accomplished by defining the degree of niche overlap, the degree of the sharing of resources between two species. When two species use one or more of the same elements of an ecological niche, they exhibit interspecific competition. It was once believed that interspecific competition would always lead to survival of only the better competitor of the two species. That was the original concept of the principle of the competition exclusion law of ecology: No two species can utilize the same ecological niche. It was conjectured that the weaker competitor would either migrate, begin using another resource not used by the stronger competitor, or become extinct. It is now believed that the end result of two species sharing elements of ecological niches may not always be exclusion.

Ecologists theorize that similar species do, in fact, coexist, despite the sharing of elements of their ecological niches, because of character displacement, which leads to a decrease in niche overlap. Character displacement involves a change in the morphological, behavioral, or physiological state of a species without geographical isolation. Character displacement occurs as a result of natural selection arising from competition between one or more ecologically similar species. Examples might be changes in mouth sizes so that they begin to feed on different sizes of the same food type, thereby decreasing competition.

SPECIES SPECIALIZATION

The more specialized a species, the more rigid it will be in terms of its relation to its ecological niche. A species that is general in terms of its ecological niche needs will be better able to find and use an alternative for the common element of the niche. Since a highly specialized species cannot substitute whatever is being used, it cannot compete as well as the other species. Therefore, a specialized species is more likely to become extinct.

For example, a panda is a very specialized feeder, eating mainly bamboo. If a pest is introduced into the environment that destroys bamboo, the panda will probably starve, being unable to switch to another food source. On the other hand, the coyote is a generalized feeder. It has a broad variety of food types that make up its diet. If humans initiate a pest-control program, killing the population of rabbits, the coyote will not fall victim to starvation, because it can switch to feeding predominantly on rodents, insects, fruits, and domesticated animals (including cats, dogs, and chickens). Hence, species with specialized ecological niche demands (specialists) are in greater danger of extinction than those with generalized needs (generalists). Although this fundamental difference in survival can be seen between specialists and generalists, it must be noted again that exclusion is not an inevitable result of competition. Many cases of ecologically similar species coexist.

When individuals of the same species compete for the same elements of the ecological niche, it is referred to as intraspecific competition. Intraspecific competition has the opposite effect of interspecific competition: niche generalizations. In increasing populations, the first inhabitants will have access to optimal resources. The opportunity for optimal resources decreases as the population increases; hence, intraspecific competition increases. Deviant individuals using marginal resources may slowly begin to use less optimal resources that are in less demand. That can lead to an increase in the diversity of ecological niches used by the species as a whole. In other words, the species may become more generalized and exploit wider varieties of niche elements.

Representing a situation on the opposite end of the spectrum from that of two organisms competing for the same dimension of an ecological niche is the vacant niche theory. This ecological principle states that when an organism is removed from its ecological niche, space, or any other dimension of the niche, another organism of the same or similar species will reinvade.

FIELD RESEARCH

Theoretical studies of ecological niches are abstract, since humans are limited to three-dimensional diagrams, and there are more than three dimensions to an ecological niche. This multidimensionality is referred to as the n-dimensional niche. This abstract n-dimensional niche can be studied mathematically and statistically, but ecology is mainly a field science. Therefore, the focus of techniques is on those used for field research of the ecological niche.

Research that attempts to describe all the elements of the n-dimensional ecological niche would require extensive observations. Yet, ecological niches are difficult to measure not only because of the plethora of data that would have to be collected but also because of the element of change in nature. The internal and external environment of an organism is always dynamic. Nothing in life is static, even if equilibrium has been established.

These constant fluctuations create daily and seasonal changes in space and ecological niches. Therefore, because of the constant fluctuations, any merely descriptive field observations would not be reliable depictions of an organism's ecological niche. Ecologists must also resort to quantitative data of measurable features of an organism's ecological niche. For example, the temperature, pH, light intensity, algae makeup, predators, and activity level of the organism are measurable features of an ecological niche in a pond community. The difficulty is in the collection of each of the necessary measurements making up an ecological niche. The ecologist would have to limit the data to a manageable number of specific dimensions of the niche based on conjecture and basic intuition. Such limitations often lead to incomplete and disconnected measurements that can at best only partially describe a few of the dimensions of the ecological niche.

Ecologists realize that complete observations and measurements of all the dimensions of an organism's ecological niche are unattainable. The focus in understanding how a species interacts with its community centers on determining the degree of niche overlap between any two species—in other words, the level of competition for niche space and resources. Studies of this niche overlap are typically limited to dimensions that can be quantitatively measured. Yet, there is still the problem of deciding which of the dimensions are involved in the competition between the two species. Again, the ecologist must usually rely on inherent knowledge about the two species in question. Often, researchers investigating niche competition measure no more than four ecological niche dimensions to determine the niche overlap in an attempt to understand how two individuals competing for the same space, resources, or other ecological niche features can coexist.

Field methods for observations and quantitative measurements of elements of ecological niches, niche overlap, and niche competition are probably endless. To name a few, describing an organism's niche may involve fecal samples to determine its diet, fecal samples of possible predators to identify its primary predator, animal and plant species checklists of its space niche along with soil components, climatic trends, and the like. Niche competition and overlap often can be studied first in the laboratory under controlled situations. One method might involve recording the population dynamics of the species as different elements in the ecological niche are manipulated to determine which is the better competitor and what is the resource that is most responsible for limiting the population size.

NICHE AND COMMUNITY

The shift in meaning and study from merely space and trophic level placement in the food pyramid to ecological niche of n dimensions has been beneficial for the field of ecology. This focus on community ecology is obviously much more productive for the goal of ecology, the understanding of how all living organisms interact with one another, and with the nonliving elements in the environment.

Perhaps more important is the attempt to describe niches in terms of community ecology, which can be essential for some of humankind's confrontations with nature. For example, it becomes more and more apparent that synthetic chemicals are often too costly and too hazardous to continue using for control of crop pests and carriers of diseases. The goal is to control pests effectively with biological controls. Biological controls can involve the introduction of natural predators of the undesirable pest or the introduction of a virus or bacteria that eliminates the pest and is harmless to humans and wildlife.

The success of a biological control is directly proportional to the knowledge of the pest's n-dimensional ecological niche and the other organisms with which it comes in contact. A classic example of the havoc that can result from manipulations of nature without adequate ecological information is when Hawaii attempted to use biological controls to eradicate a population of snakes, which humans had accidentally introduced. The biological control used was the snake's natural predator, the mongoose. One very important dimension of the ecological niche of both species was ignored. One species was active only at night, while the other was active only during the day. Needless to say, this particular venture with a biological control was not a success.

Another relevant function of community-oriented studies of ecological niches involves endangered species. In addition to having aesthetic and potential medicinal values, an endangered organism may be a keystone species, a species on which the entire community depends. A keystone species is so integral to keeping a community healthy and functioning that if the species is obliterated, the community no longer operates properly and is not productive.

Habitat destruction has become the commonest cause of drastic population declines of endangered species. To enhance the habitat of the endangered species, it is undeniably beneficial to know what attracts a species to its particular preferred habitat. This knowledge involves the details of many of the dimensions of its ecological niche integral to its population distribution. Another common cause of endangering the survival of a species is when an introduced organism or exotic species competes for the same resources and displaces the native species. Solving such competition between native and introduced species would first involve determining niche overlap.

It is often stated that an ounce of prevention is worth a pound of cure. Thus, the researching and understanding of all the dimensions of ecological niches are integral components of preventing environmental manipulations by humankind that might lead to species extinction. Many science authorities have agreed that future research in ecology and related fields should focus on solving three main problems: species endangerment, soil erosion, and solid waste management.

This focus on research in ecology often means that studies of pristine communities, those undisturbed, will be the most helpful for future restoration projects. Although quantitative and qualitative descriptions of pristine areas seem to be unscientific at the time they are made, because there is no control or experimental group, they are often the most helpful for later investigations. For example, after a species has shown a drastic decline in its population, the information from the observations of the once-pristine area may help to uncover what niche dimension was altered, causing the significant population decrease.

—*Jessica O. Ellison*

Further Reading

Bronmark, Christopher, and Lars-Anders Hansson. *The Biology of Lakes and Ponds*. Oxford University Press, 1998.

Ehrlich, Paul R. *The Machinery of Nature*. Simon & Schuster, 1986.

Odling-Smee, F. John, Kevin N. Laland, and Marcus W. Feldman. "Niche Construction." *American Naturalist* 147, no. 4 (April 1996): 641–649.

Rayner, Alan D. M. *Degrees of Freedom: Living in Dynamic Boundaries*. World Scientific Publications, 1997.

Ricklefs, Robert E. *Ecology*. 4th ed. Chiron Press, 1999.

Shugart, Herman H. *Terrestrial Ecosystems in Changing Environments*. Cambridge University Press, 1998.

Smith, Robert L. *Ecology and Field Biology*. 6th ed. Benjamin/Cummings, 2001.

Stone, Richard. "Taking a New Look at Life Through a Functional Lens." *Science* 269, no. 5222 (July 1995): 316–318.

Ecology

Fields of Study

Ecology; Environmental Studies; Population Dynamics; Evolutionary Biology

Abstract

Ecology is the study of how organisms relate to their natural environments. The two principal concerns of ecologists are the distribution and abundance of organisms. An ecologist

views organisms as consequences of past natural selection brought about by their environments. Each organism represents an array of adaptations that can provide insight into the environmental pressures that resulted in its present form. Adaptations of organisms are also revealed by other features.

KEY CONCEPTS

adaptation: a genetic (intrinsic) feature of an organism which, through natural selection, enhances its fitness in a given environment

community: all the populations that exist in a given habitat

ecosystem: the biological community in a given habitat, combined with all the physical properties of the environment in that habitat

environment: all the forces and things external to an organism that directly affect it

fitness: the contribution of an organism to future generations; the perpetuation of its genes through reproduction

habitat: the place where an organism lives; for example, a pond or forest

natural selection: a change in the genetic makeup of a population as a result of different survival and reproduction rates (fitness) among its members

niche: the role of an organism in its environment; the sum of all factors that define its existence (temperature, energy requirements, and so on)

population: all the individuals in a habitat which are of the same species

resource: a requirement for life, such as space for living, food (for animals), or light (for plants), not including conditions such as temperature or salinity

THE NATURE OF ECOLOGY

Ecology is the study of how organisms relate to their natural environments. The two principal concerns of ecologists are the distribution and abundance of organisms: Why are animals, plants, and other organisms found where they are, and why are some common and others rare? These questions have their roots in the theory of evolution. In fact, it is difficult (and not often worthwhile) to separate modern ecological matters from the concerns of evolutionary biologists. Ecology can be divided

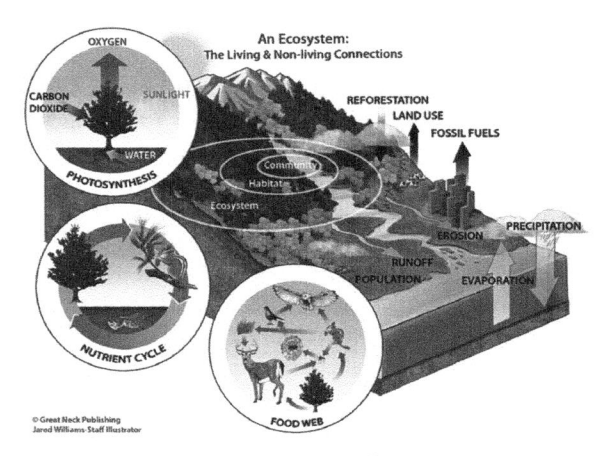

The weather, climate, and many earth processes influence natural resources such as water, nutrients, plants, soil, land, and air quality.

according to several levels of organization: the individual organism, the population, the community, and the ecosystem.

ENVIRONMENT AND NATURAL SELECTION

An ecologist views organisms as consequences of past natural selection brought about by their environments. That is, each organism represents an array of adaptations that can provide insight into the environmental pressures that resulted in its present form. Adaptations of organisms are also revealed by other features, such as the range of temperature an organism can tolerate, the amount of moisture it requires, or the variety of food it can exploit. Food and space for living are considered resources; factors such as temperature, light, and moisture are conditions which determine the rate of resource utilization. When ecologists have discovered the full range of resources and conditions necessary for an organism's existence, they have discovered its niche.

Many species, such as fish, insects and plants, have a large reproductive output. This compensates for high mortality imposed by natural selection. Other species, such as large mammals and birds, have fewer offspring. Many of these animals care for their young, thus increasing the chances that their offspring will survive to reproduce. These are two different strategies for success, based upon the principle that organisms have a finite energy budget. Energy acquired from food (animals) or sunlight (plants) must be partitioned among growth, maintenance, and reproduction. The greater the energy allocated to the care of offspring, for example, the fewer the offspring that can be produced.

The concept of an energy budget is a key to understanding evolutionary strategies of organisms, as well as the energetics of ecosystems. The amount of energy fixed and stored by an organism is called net production; this is the energy used for growth and reproduction. Net production is the difference between gross production (the amount of energy assimilated) and respiration (metabolic maintenance cost). The greater the respiration, the less energy will be left over for growth and reproduction. Endothermic animals, which physiologically regulate their body heat (mammals and birds), have a very high respiration rate relative to ectotherms (reptiles, amphibians, fish, and invertebrates), which cannot. Among endotherms, smaller animals have higher respiration rates than larger ones, because the ratio of body surface area (the area over which heat is exchanged with the environment) to volume (the size of the "furnace") decreases with increasing body size.

DEMOGRAPHY AND POPULATION REGULATION

Although single organisms can be studied with regard to adaptations, in nature most organisms exist in populations rather than as individuals. Some organisms reproduce asexually (that is, by forming clones), so that a single individual may spawn an entire population of genetically identical individuals. Populations of sexually reproducing organisms, however, have the property of genetic variability, since not all individuals are identical. That is, members of a population have slightly different niches and will therefore not all be equally capable of living in a given environment. This is the property upon which Charles Darwin's theory of natural selection depends: Because not all individuals are identical, some will have greater fitness to exist within a particular ecological niche than others of its species. Those with superior fitness will reproduce in greater numbers and therefore will contribute more genes to successive generations. In nature, many species consist of populations occupying more than a single habitat. This constitutes a buffer against extinction: If one habitat is destroyed, the species will not become extinct, because it exists in other habitats.

Two dynamic features of populations are growth and regulation. Growth is simply the difference between birth and death rates, which can be positive (growing), negative (declining), or zero (in equilibrium). Every species has a genetic capacity for exponential (continuously accelerating) increase, which will express itself to varying degrees depending on environmental conditions: A population in its ideal environment will express this capacity more nearly than one in a less favorable environment. The rate of growth of a population is affected by its age structure—the proportion of individuals of different ages. For example, a population which is growing rapidly will have a higher proportion of juvenile individuals than one which is growing more slowly.

Populations may be regulated (so that they have equal birth and death rates) by a number of factors, all of which are sensitive to changes in population size. A population may be regulated by competition among its members for the resource that is in shortest supply (limiting). The largest population that can be sustained by the available resources is called the carrying capacity of the environment. A population of rodents, for example, might be limited by its food supply such that as the population grows and food runs out, the reproductive rate declines. Thus, the effect of food on population growth depends upon the population size relative to the limiting resource. Similarly, parasites that cause disease spread faster in large, dense populations than in smaller, more diffuse ones. Predators can also regulate populations of their prey by responding to changes in prey availability. Climate and catastrophic events such as storms may severely affect populations, but their effect is not dependent upon density and is thus not considered regulatory.

INTERACTIONS BETWEEN SPECIES

Competition occurs between, as well as within, species. Two species are said to be in competition with each other if and only if they share a resource that is in short supply. If, however, they merely share a resource that is plentiful, then they are not really competing for it. Competition is thought to be a major force in determining how many species can coexist in natural communities. There are a number of alternative hypotheses, however, which involve such factors as evolutionary time, productivity (the energy base for a community), heterogeneity of the habitat, and physical harshness of the environment.

Predator-prey interactions are those in which the predator benefits from killing and consuming its prey. These differ from most parasite-host interactions in that parasites usually do not kill their hosts (a form of suicide for creatures that live inside other creatures). Similarly, most plant-eating animals (herbivores) do not kill the plants on which they feed. Many ecologists classify herbivores as parasites for this reason. There are exceptions, such as birds and rodents that eat seeds, and these can be classified as legitimate predator-prey interactions. Predators can influence the number of species in a community by affecting competition among their prey: If populations of competing species are lowered by predators so that they are below their carrying capacities, then there may be enough resources to support colonization by new species.

In many cases, the interaction between two species is mutually beneficial. Mutualism is often thought to arise as a result of closely linked evolutionary histories (coevolution) of different species. Termites harbor protozoans in their guts that produce an enzyme which can break down cellulose in wood. The protozoans thus are provided with a habitat, and termites are able to derive nourishment from wood. Some acacia trees in the tropics have hollow thorns which provide a habitat for ants. In return, the ants defend the trees from other insects which would otherwise damage or defoliate them.

Communities of organisms are composed of many populations that may interact with one another in a variety of ways: predation, competition, mutualism, parasitism, and so on. The composition of communities changes over time through the process of succession. In terrestrial communities, bare rock may be weathered and broken down by bacteria and other organisms until it becomes soil. Plants can then invade and colonize this newly formed soil, which in turn provides food and habitat for animals. The developing community goes through a series of stages, the nature of which depends on local climatic conditions, until it reaches a kind of equilibrium. In many cases this equilibrium stage, called climax, is a mature forest. Aquatic succession essentially is a process of becoming a terrestrial community. The basin of a lake, for example, will gradually be filled with silt from terrestrial runoff and accumulated dead organic material from populations of organisms within the lake itself.

ECOSYSTEMS

Ecosystems consist of several trophic levels, or levels at which energy is acquired: primary producers, consumers, and decomposers. Primary producers are green plants that capture solar energy and transform it, through the process of photosynthesis, into chemical energy. Organisms that eat plants (herbivores) or animals (carnivores) to obtain their energy are collectively called consumers. Decomposers are those consumers, such as bacteria and fungi, that obtain energy by breaking down dead bodies of plants and animals. These trophic levels are linked together into a structure called a food web, in which energy is transferred from primary producers to consumers and decomposers, until finally all is lost as heat. Each transfer of energy entails a loss (as heat) of at least 90 percent, which means that the total amount of energy available to carnivores in an ecosystem is substantially less than that available to herbivores.

As with individual organisms, ecosystems and their trophic levels have energy budgets. The net production of one trophic level is available to the next-higher trophic level as biomass (mass of biological material). Plants have higher net productivity (rates of production) than animals because their metabolic maintenance cost is lower relative to gross productivity; herbivores often have higher net productivity than predators for the same reason. For the community as a whole, net productivity is highest during early successional stages, since biomass is being added more rapidly than later on, when the community is closer to climax equilibrium.

In contrast to the unidirectional flow of energy, materials are conserved and recycled from dead organisms by decomposers to support productivity at higher trophic levels. Carbon, water, and mineral nutrients required for plant growth are cycled through various organisms within an ecosystem. Materials and energy are also exchanged among ecosystems: There is no such thing in nature as a "closed" ecosystem that is entirely self-contained.

DESCRIPTIVE, EXPERIMENTAL, AND MATHEMATICAL ECOLOGY

The science of ecology is necessarily more broadly based than most biological disciplines; consequently, there is more than one approach to it. Ecological studies fall into three categories: descriptive, experimental, and mathematical.

Descriptive ecology is concerned with describing natural history, usually in qualitative terms. The study of adaptations, for example, is descriptive in that one can measure the present "value" of an adaptive feature, but one can only conjecture as to the history of natural selection that was responsible for it. On the other hand, some patterns are discernible in nature for which hypotheses can be constructed and tested by statistical inference. For example, the spatial distribution (dispersion) of birds on an island may be random, indicating no biological interaction among them. If the birds are more evenly spaced (uniform dispersion) than predicted assuming randomness, however, then it might be inferred that the birds are competing for space; they are exhibiting territorial exclusion of one another. Such "natural experiments," as they are called, depend heavily upon the careful design of statistical tests.

Experimental ecology is no different from any other experimental discipline; hypotheses are constructed from observations of nature, controlled experiments are designed to test them, and conclusions are drawn from the results of the experiments. The basic laboratory for an ecologist is the field. Experiments in the field are difficult because it is hard to isolate and manipulate variable factors one at a time, which is a requisite for any good experiment in science. A common experiment that is performed to test for resource limitation in an organism is enhancement of that resource. If food, for example, is thought to be in short supply (implying competition), one section of the habitat is provided more food than is already present; another section is left alone as a control. If survivorship, growth, or reproductive output is higher in the enhanced portion of the habitat than in the control area, the researcher may infer that the organisms therein were food-limited. Alternatively, an ecologist might have decreased the density of organisms in one portion of the habitat, which might seem equivalent to increasing food supply for the remaining organisms, except that it represents a change in population density as well. Therefore, this second design will not allow the researcher to differentiate between the possibly separate effects of food level and simple population density on organisms in the habitat.

Mathematical ecology relies heavily upon computers to generate models of nature. A model is simply a formalized, quantitative set of hypotheses constructed from sets of assumptions of how things happen in nature. A model of population growth might contain assumptions about the age structure of a population, its genetic capacity for increase, and the average rate of resource utilization by its members. By changing these assumptions, scientists can cause the model population to behave in different ways over time. The utility of such modeling is limited to the accuracy of the assumptions employed.

Modern ecology is concerned with integrating these different approaches, all of which have in common the goal of predicting the way nature will behave in the future, based upon how it behaves in the present. Description of natural history leads to hypotheses that can be tested experimentally, which in turn may allow the construction of realistic mathematical (quantitative) models of how nature works.

ALL-ENCOMPASSING NATURE

People historically have viewed nature as an adversary. The "conquest of nature" has traditionally meant human encroachment on natural ecosystems, usually without benefit of predictive knowledge. Such environmental problems as pollution, species extinction, and overpopulation can be viewed as experiments performed on a grand scale without appropriate controls. The problem with such experiments is that the outcomes might be irreversible. A major lesson of ecology is that humans are not separate from nature; we are constrained by the same principles as are other organisms on Earth. One object of ecology, then, is to learn these principles so that they can be applied to our portion of Earth's ecosystem.

Populations that are not regulated by predators, disease, or food limitation grow exponentially. The human population, on a global scale, grows this way. All the wars and famines in history have scarcely made a dent in this growth pattern. Humankind has yet to identify its carrying capacity on a global scale, although regional famines certainly have provided insights into what happens when local carrying capacity is exceeded. The human carrying capacity needs to be defined in realistic ecological terms, and such constraints as energy, food, and space must be incorporated into the calculations. For example, knowledge of energy flow teaches that there is more energy at the bottom of a food web (producers) than at successively higher trophic levels (consumers), which means that more people could be supported as herbivores than as carnivores.

The study of disease transmission, epidemiology, relies heavily on ecological principles. Population density, rates of migration among epidemic centers, physiological tolerance of the host, and rates of evolution of disease-causing parasites are all the subjects of ecological study.

An obvious application of ecological principles is conservation. Before habitats for endangered species can be set aside, for example, their ecological requirements, such as migratory routes, breeding, and feeding habits, must be known. This also applies to the introduction (intentional or accidental) of exotic species into habitats. History is filled with examples of introduced species that caused the extinction of native species. Application of ecological knowledge in a timely fashion, therefore, might prevent species from becoming endangered in the first place.

One of the greatest challenges in ecology is the loss of habitats worldwide. This is especially true of the tropics, which contain most of Earth's species of plants and animals. Species in the tropics have narrow niches, which means that they are more restricted in range and less tolerant of change than are many temperate species. Therefore, destruction of tropical habitats, such as rain forests, leads to rapid species extinction. These species are the potential sources of many pharmaceutically valuable drugs; further, they are a genetic record of millions of years of evolutionary history. Tropical rain forests also are prime sources of oxygen and act as a buffer against carbon dioxide accumulation in the atmosphere. Ecological knowledge of global carbon cycles permits the prediction that destruction of rain forests will have a profound impact on the quality of the air.

—*Lawrence E. Hurd*

FURTHER READING

Begon, Michael, John L. Harper, and Colin R. Townsend. *Ecology: Individuals, Populations, and Communities.* 3rd ed. Blackwell, 1996.

Carson, Rachel. *Silent Spring.* 1962. 50th anniversary ed. Penguin, 2012.

Elton, Charles. *Animal Ecology.* Macmillan, 1927.

Hutchinson, G. Evelyn. *The Ecological Theater and the Evolutionary Play.* Yale University Press, 1969.

Krebs, Charles J. *The Message of Ecology.* Harper, 1987.

Molles, Manuel C., Jr. *Ecology: Concepts and Applications.* 7th ed. McGraw, 2016.

Morin, Peter Jay. *Community Ecology.* 2nd ed. Wiley, 2012.

Pianka, Eric R. *Evolutionary Ecology.* 6th ed. Benjamin/Cummings, 2000.

Ricklefs, Robert E. *Ecology.* 4th ed. Chiron, 1999.

Ecosystems

Fields of Study

Environmental Studies; Ecology; Biogeography; Botany

Abstract

The ecosystem is essentially an abstract organizing unit superimposed on the landscape to help ecologists study the form and function of the natural world. An ecosystem consists of one or more communities of interacting organisms and their physical environment. Ecosystems have no distinct boundaries; thus, the size of any particular ecosystem should be inferred from the context of the discussion. Size and boundaries are arbitrary because no ecosystem stands in complete isolation from those that surround it.

Key Concepts

biomass: the weight of organic matter, often expressed in terms of grams per square meter per year

consumers: animals, fungi, and bacteria that get energy by feeding on organic matter

food chain: an abstract chain representing the links between organisms, each of which eats and is eaten by another

food web: a network of interconnecting food chains representing the food relationships in a community

photosynthesis: the process by which green plants and algae use sunlight as energy to convert carbon dioxide and water into energy-rich compounds such as glucose
primary production: the energy assimilated by green plants and stored as organic tissue
producers: green plants and chemosynthetic organisms that can produce food from inorganic materials
respiration: the release of chemical energy to do work in plants and animals; a reversal of the photosynthetic process
trophic level: a feeding level on the pyramid of numbers, consisting of all the kinds of animals that feed at comparable levels on food chains

AN ECOSYSTEM IS BY DEFINITION

The ecosystem is essentially an abstract organizing unit superimposed on the landscape to help ecologists study the form and function of the natural world. An ecosystem consists of one or more communities of interacting organisms and their physical environment. Ecosystems have no distinct boundaries; thus, the size of any particular ecosystem should be inferred from the context of the discussion. Individual lakes, streams, or strands of trees can be described as distinct ecosystems, as can the entire North American Great Lakes region. Size and boundaries are arbitrary because no ecosystem stands in complete isolation from those that surround it. A lake ecosystem, for example, is greatly affected by the streams that flow into it and by the soils and vegetation through which these streams flow. Energy, organisms, and materials routinely migrate across whatever perimeters the ecologist may define. Thus, investigators are allowed considerable latitude in establishing the scale of the ecosystem they are studying. Whatever the scale, though, the importance of the ecosystem concept is that it forces ecologists to treat organisms not as isolated individuals or species but in the context of the structural and functional conditions of their environment.

DEVELOPMENT OF THE ECOSYSTEM CONCEPT

Antecedents to the ecosystem concept may be traced back to one of America's first ecologists, Stephen Alfred Forbes, an eminent Illinois naturalist who studied food relationships among birds, fishes, and insects. During the course of his investigations,

Rachel Carson's book Silent Spring brought the dangers facing the earth's ecosystems to public consciousness and was one of the contributing factors to the growth of environmentalism in the late twentieth century.

Forbes recognized in 1880 that full knowledge of organisms and their response to disturbances would come only from more concentrated research on their interactions with other organisms and with their inorganic (nonliving) physical surroundings. In 1887, Forbes suggested that a lake could be viewed as a discrete system for study: a microcosm. A lake could serve as a scale model of nature that would help biologists understand more general functional relationships among organisms and their environment. Forbes explained how the food supply of a single species, the largemouth bass, was dependent either directly or indirectly upon nearly all the fauna and much of the flora of the lake. Therefore, whenever even one species was subjected to disturbance from outside the microcosm, the effects would probably be felt throughout the community.

In 1927, British ecologist Charles Elton incorporated ideas introduced by Forbes and other fishery biologists into the twin concepts of the food chain and the food web. Elton defined a food chain as a series of linkages connecting basic plants, or food producers, to herbivores and their various carnivorous predators, or consumers. Elton used the term "food cycle" instead of "food web," but his diagrams reveal that his notion of a food cycle—that it is simply a network of interconnecting food chains—is consistent with the modern term.

Elton's diagrams, which traced various pathways of nitrogen through the community, paved the way for understanding the importance of the cycling of inorganic nutrients such as carbon, nitrogen, and phosphorus through ecosystems, a process that is known as biogeochemical cycling. Very simply, Elton illustrated how bacteria could make nitrogen available to algae at the base of the food chain. The nitrogen then could be incorporated into a succession of ever larger consumers until it reached the top of the chain. When the top predators died, decomposer organisms would return the nitrogen to forms that could eventually be taken up again by plants and algae at the base of the food chain, thus completing the cycle.

Elton's other key contribution to the ecosystem concept was his articulation of the pyramid of numbers, the idea that small animals in any given community are far more common than large animals. Organisms at the base of a food chain are numerous, and those at the top are relatively scarce. Each level of the pyramid supplies food for the level immediately above it—a level consisting of various species of predators that generally are larger in size and fewer in number. That level, in turn, serves as prey for a level of larger, more powerful predators, fewer still in number. A graph of this concept results in a pyramidal shape of discrete levels, which today are called trophic (feeding) levels.

ECOSYSTEMS IN TWENTIETH CENTURY THOUGHT

Although the basic concept of an ecosystem had been recognized by Forbes as early as 1880, it was not until 1935 that British ecologist Arthur G. Tansley coined the term. Though he acknowledged that ecologists were primarily interested in organisms, Tansley declared that organisms could not be separated from their physical environment, as organism and environment formed one complete system. As Forbes had pointed out half a century earlier, organisms were inseparably linked to their nonliving environments. Consequently, ecosystems came to be viewed as consisting of two fundamental parts: the biotic, or living components, and the abiotic, or nonliving components.

No one articulated this better than Raymond Lindeman, a limnologist (freshwater biologist) from Minnesota who in 1941 skillfully integrated the ideas of earlier ecological scientists when he published an elegant ecosystem study of Cedar Bog Lake. Lindeman's classic work set the stage for decades of research that centered on the ecosystem as the primary organizing unit of study in ecology. Drawing on the work of his mentor, G. Evelyn Hutchinson, and Charles Elton, Arthur Tansley, and other scientists, Lindeman explained how ecological pyramids were a necessary result of energy transfers from one trophic level to the next.

By analyzing ecosystems in this manner, Lindeman was able to answer a fundamental ecological question that had been posed fourteen years earlier by Elton: Why were the largest and most powerful animals, such as polar bears, sharks, and tigers, so rare? Elton had thought the relative scarcity of top predators was due to their lower rates of reproduction. Lindeman corrected this misconception by explaining that higher trophic levels held fewer animals not because of their reproductive rates but because of a loss of chemical energy with each step up the pyramid. It could be looked upon as a necessary condition of the second law of thermodynamics: Energy transfers yield a loss or degradation of energy. The predators of one food level could never completely extract all the energy from the level below. Some energy would always be lost to the environment through respiration, some energy would not be assimilated by the predators, and some energy simply would be lost to decomposer chains when potential prey died of nonpredatory causes. This meant that each successive trophic level had substantially less chemical energy available to it than was transferred from the one below and, therefore, could not support as many animals.

Ecologists soon expanded the principle of Elton's pyramid of numbers to model other ecosystem processes. They found, for example, that the flow of chemical energy through an ecosystem could be characterized as an energy pyramid; the biomass (the weight of organic material, as in plant or animal tissue) in a community could be plotted in a pyramid of biomass. Collectively, such pyramidal models became known as ecological pyramids.

ENERGY PRODUCTION AND TRANSMISSION

Lindeman subdivided the biotic components of ecosystems into producers, consumers, and decomposers. Producers (also known as autotrophs) produce their own food from compounds in their environment. Green plants are the main producers in terrestrial ecosystems; algae are the most common producers in aquatic ecosystems. Both plants and algae are producers that use sunlight as energy to make food from carbon dioxide and water in the process of photosynthesis. During photosynthesis, plants, algae, and certain bacteria capture the sun's energy in chlorophyll molecules. (Chlorophyll is a pigment that gives plants their green color.) This energy, in turn, is used to synthesize energy-rich compounds such as glucose, which can be used to power activities such as growth, maintenance, and reproduction or can be stored as biomass for later use. These energy-rich compounds can also be passed on in the form of biomass from one organism to another, as when animals (primary consumers) graze on plants or when decomposers break down detritus (dead organic matter).

The energy collected by green plants is called primary production because it forms the first level at the base of the ecological pyramid. Total photosynthesis is represented as gross primary production. This is the amount of the sun's energy actually assimilated by autotrophs. The rate of this production of organic tissue by photosynthesis is called primary productivity. Plants, however, need to utilize some of the energy they produce for their own growth, maintenance, and reproduction. This energy becomes available for such activities through respiration, which essentially is a chemical reversal of the process of photosynthesis. As a result, not all the energy assimilated by autotrophs is available to the consumers in the next trophic level of the pyramid. Consequently, respiration costs generally are subtracted from gross primary production to determine the net primary production, the chemical energy actually available to primary consumers.

MEASURING ECOSYSTEM PRODUCTIVITY

The carrying capacity for all the species supported by an ecosystem ultimately depends upon the system's net primary productivity. By knowing the productivity, ecologists can, for example, estimate the number of herbivores that an ecosystem can support. Consequently ecosystem ecologists have developed a variety of methods to measure the net primary productivity of different systems. Productivity is generally expressed in kilocalories per square meter per year when quantifying energy, and in grams per square meter per year when quantifying biomass.

Production in aquatic ecosystems may be measured by using the light and dark bottle method. In this technique two bottles containing samples of water and the natural phytoplankton population are suspended for twenty-four hours at a given depth in a body of water. One bottle is dark, permitting respiration but no photosynthesis by the phytoplankton. The other is clear and therefore permits both photosynthesis and respiration. The light bottle provides a measure of net production (photosynthesis minus respiration) if the quantity of oxygen is measured before and after the twenty-four-hour period. (The amount of oxygen produced by photosynthesis is proportional to the amount of organic matter fixed.) Measuring the amount of oxygen in the dark bottle before and after the run provides an estimate of respiration, since no photosynthesis can occur in the dark. Combining net production from the light bottle with total respiration from the dark bottle yields an estimate of gross primary production.

Other studies have concentrated on quantifying the rate of movement of energy and materials through ecosystems. Investigations begun in the 1940s and 1950s by the Atomic Energy Commission to track radioactive fallout were eventually diverted into studies of ecosystems that demonstrated how radionuclides moved through natural environments by means of food-chain transfers. This research confirmed the interlocking nature of all organisms linked by the food relationship and eventually yielded rates at which both organic and inorganic materials could be cycled through ecosystems. As a result of such studies, the Radiation Ecology Section at Oak Ridge National Laboratory in Tennessee became established as a principal center for systems ecology.

Ecologists sometimes extend the temporal boundaries of their studies by utilizing the methods of paleoecology (the use of fossils to study the nature of ecosystems in the past). Research of this type generally centers on the analysis of lake sediments, whose layers often hold centuries of ecosystem history embodied in the character and abundance of pollen grains, diatoms, fragments of zooplankton, and other organic microfossils.

More general trends in the methods of studying ecosystems include a continuing emphasis on quantitative methods, often using increasingly sophisticated computer modeling techniques to simulate ecosystem functions. Equally significant is a trend toward a "big science" approach, modeled on the Manhattan Project, which employed teams of investigators working on different problems related to nuclear fission in different parts of the country. The well-known international biological program, the Hubbard Brook Project in New Hampshire, and continuing projects on long-term ecological research all serve as examples of ecosystem studies that involve teams of researchers from a wide range of disciplines.

RESPONDING TO DISTURBANCE

One of the practical benefits of studying ecosystems derives from naturalist Stephen Forbes's suggestion, which he made in 1880, that the knowledge from biological research be used to predict the response of organisms to disturbance. When disturbance is caused by natural events such as droughts, floods, or fires, ecologists can use their knowledge of the structure and function of ecosystems to help resource managers plan for subsequent recolonization and succession of species.

The broad perspective of the ecosystem approach becomes particularly useful in examining the effects of certain toxic compounds because of the complexity of their interaction within the environment. The synergistic effects that sometimes occur with toxic substances can produce pronounced impacts on ecosystems already stressed by other disturbances. For example, after the atmosphere deposits mercury on the surface of a lake, the pollutant eventually settles in the sediments where bacteria make it available to organisms at the base of the food chain. The contaminant then bioaccumulates as it is passed on to organisms such as fish and fish-eating birds at higher trophic levels. Synergistic effects occur in lakes already affected by acid deposition; researchers have found that acidity somehow stimulates microbes to increase the bioavailability of the mercury. Thus, aquatic ecosystems that have become acidified through atmospheric processes may stress their flora and fauna even further by enhancing the availability of mercury from atmospheric fallout. The complexity of such interactions demands research at the ecosystem level, and ecosystem studies are prerequisite for prudent public policy actions on environmental contaminants.

—*Robert Lovely*

FURTHER READING

Begon, M., J. L. Harper, and C. R. Townsend. *Ecology: Individuals, Populations, and Communities*. Blackwell, 1996.

Bouma, Jetske A., and Pieter van Beukering. *Ecosystem Services: From Concept to Practice*. Cambridge University Press, 2015.

Clark, Tim W., A. Peyton Curlee, Steven C. Minta, and Peter M. Kareiva, eds. *Carnivores in Ecosystems: The Yellowstone Experience*. Yale University Press, 1999.

Colinvaux, Paul. *Why Big Fierce Animals Are Rare: An Ecologist's Perspective*. Princeton University Press, 1978.

Golly, Frank Benjamin. *A History of the Ecosystem Concept in Ecology: More than the Sum of the Parts*. Yale University Press, 1993.

Hagen, Joel B. *An Entangled Bank: The Origins of Ecosystem Ecology*. Rutgers University Press, 1992.

Hunter, Malcolm L., ed. *Maintaining Biodiversity in Forest Ecosystems*. Cambridge University Press, 1999.

Mustafa, Saleem, and Rossita Shapawi. *Aquaculture Ecosystems: Adaptability and Sustainability*. Wiley, 2015.

Odum, Eugene P. *Ecology and Our Endangered Life-Support Systems*. Sinauer, 1989.

Real, Leslie A., and James H. Brown. *Foundations of Ecology: Classic Papers with Commentaries*. University of Chicago Press, 1991.

Smith, Robert Leo. *Ecology and Field Biology*. 6th ed. Benjamin/Cummings, 2007.

Sullivan, Timothy J., et al. *Air Pollution and Freshwater Ecosystems: Sampling, Analysis, and Quality Assurance*. CRC, 2015.

Embryonic Development

Fields of Study

Embryology; Evolutionary Biology; Biology; Animal Physiology

Abstract

After an egg cell, or ovum, is fertilized, it divides into many smaller cells, which then undergo rearrangement and differentiation to form the embryo of a new individual. The significance of this lies in the fact that a single cell with genetic information from two parents is transformed into a multicellular structure with three germ layers that will give rise to all the organs and systems of the body.

Key Concepts

- **archenteron:** the primitive gut cavity formed by the invagination of the blastula; the cavity of the gastrula
- **blastula:** an early stage of an embryo which is shaped like a hollow ball in some animals and a small, flattened disc in others; contains a cavity called the blastocoel
- **cleavage:** the process by which the fertilized egg undergoes a series of rapid cell divisions which result in the formation of a blastula
- **gastrulation:** the transformation of a blastula into a three-layered embryo, the gastrula; initiated by invagination
- **germ layers:** the embryonic layers of cells which develop in the gastrula: ectoderm, mesoderm, and endoderm
- **invagination:** the turning of an external layer into the interior of the same structure; formation of archenteron
- **morula:** a solid ball or mass of cells resulting from early cleavage divisions of the zygote
- **neurulation:** the process by which the embryo develops a central nervous system
- **notochord:** a fibrous rod in an embryo which gives support; a structure that will later be surrounded by vertebrae
- **zygote:** the fertilized egg; the first cell of a new organism

WHERE DO BABIES COME FROM?

After an egg cell, or ovum, is fertilized, it divides into many smaller cells, which then undergo rearrangement and differentiation to form the embryo of a new individual. The fertilized egg is called a zygote, and the division of the one-celled zygote into smaller and smaller cells is called cleavage. The cellular rearrangement is known as gastrulation, and the proliferation and movement of cells into position to form the beginnings of the central nervous system is termed neurulation. The significance of these events lies in the fact that a single cell with genetic information from two parents is transformed into a multicellular structure with three germ layers that will give rise to all the organs and systems of the body.

CLEAVAGE

After fertilization, the resultant zygote undergoes many rapid cell divisions. The cleavage process results in smaller and smaller cells, called blastomeres. The cell divisions are by mitosis, which produces identical chromosomes in each new cell. When between sixteen and thirty-two cells have been formed, the structure is called a morula, from the Latin for "mulberry," which it resembles.

The morula stage is short-lived because, as soon as it is formed, processes are initiated that bring it to the next stage, known as the blastula. A cavity begins to form in the center of the morula as water flows in and pushes out the cells. The new cavity is called the blastocoel (pronounced *blast-o-seal*) and the embryonic stage the blastula. Cleavage continues until the blastula consists of hundreds of cells but is still no larger than the original zygote. The blastula is the terminal cleavage structure. The egg cell, much larger than an average cell, has been fertilized and subdivided into hundreds of normal-sized cells. The blastomeres all appear to be similar to one another, but studies have shown that the individual cells are already destined for the tissues they will become.

Cleavage and embryonic development take place inside a vessel called an egg; in some animals, such as all birds and most reptiles, the embryo is fed from an organ in the egg called the yolk (in animals like mammals, the embryo is fed directly from the mother through an umbilical cord). The principles of cleavage are the same in all vertebrate groups, but the mechanics differ according to the amount of yolk in the egg. Eggs with large amounts of yolk

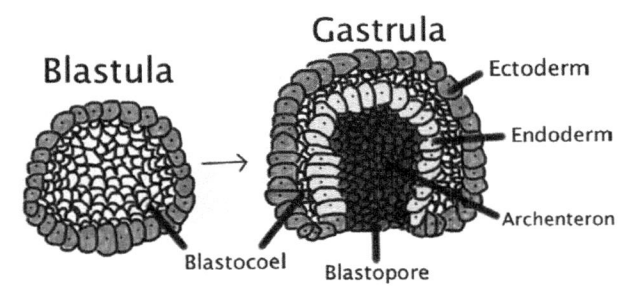

The process of gastrulation.

undergo only partial cleavage, because the yolk retards the cytoplasmic division. In birds, reptiles, and many fishes, the yolk is so dense that the cytoplasm and nucleus are crowded into a small cap or disk on one side of the cell. The cleavage divisions all occur in this small area, resulting in a flattened blastula atop the large inert yolk.

Eggs with but a moderate amount of yolk, such as amphibian eggs, are able to cleave completely. Because division proceeds more slowly through the part of the cell where yolk has accumulated, the cleavage is uneven. The cells are formed more slowly on the yolky side and are larger and fewer in number. The blastocoel is smaller and displaced to the side, with less yolk. The side with smaller blastomeres will develop into the embryo and is called the animal hemisphere. The side containing larger amounts of yolk is called the vegetal hemisphere and will provide nutrients for the embryo.

Eggs with very little yolk undergo total and equal cleavage divisions. The blastula has a large, centrally located blastocoel, and blastomeres are uniform in size. Starfish and the primitive chordate amphioxus undergo this kind of cleavage. They are often used to demonstrate the successive cleavage stages which are more easily seen in the absence of yolk. Though mammalian eggs (as from a platypus or echidna) do not have large amounts of yolk, their development is similar to that of birds. The outer layer of cells of the morula develop into a membrane, called the trophoblast, that surrounds the embryo. The embryo forms from cells in the inner region known as the inner cell mass. A large, fluid-filled blastocoel forms within the trophoblast, giving rise to the term "blastocyst" for the mammalian blastula. The inner cell mass develops atop the blastocoel as the bird embryo on the yolk.

GASTRULATION

Gastrulation is the next process in embryonic development and consists of a series of cell migrations that result in cellular rearrangement. The final gastrula will have three embryonic germ layers destined to give rise to all body structures and systems.

The first step in gastrulation is an indenting or invagination in the blastula at a spot known as the dorsal lip. Cells begin to move over the lip and drop into the interior, forming the lining of a new cavity, the archenteron, or primitive gut. Continued inward movement of cells forms a middle layer between outer cells and inner ones which have dropped in through the opening, or blastopore. The three embryonic germ layers have now been formed, and they are called ectoderm, mesoderm, and endoderm.

In animals with little egg yolk, such as the starfish, gastrulation begins when a few cells lose their adhesiveness and drop into the blastocoel. That causes a dent or depression in that area. Cells move in and deepen the depression, forming the archenteron. As the archenteron expands, the inner blastocoel shrinks and is finally obliterated. This process may be visualized as punching in the side of a hollow rubber ball with one's finger. The hole the finger makes is the blastopore; the new cavity formed by the hand represents the archenteron; and the original space inside the ball represents the blastocoel. The indentation forms two cell layers, and a third one is formed as cells continue to move in and take position between the inner and outer layers.

The outer ectoderm is destined to become epidermis and nerve tissue. The inner endoderm will form digestive glands and the lining of the digestive and respiratory systems. The middle germ layer, the mesoderm, will give rise to bone, muscle, connective tissue, and the cardiovascular and urinary systems. Additional mesoderm forms a rodlike structure known as the notochord, which lies in the roof of the archenteron. The notochord is a distinctive characteristic of chordates and gives embryonic support. The mesoderm lateral to the notochord will segregate into paired masses known as somites, each with prospective skin, bone, and nerve segments.

Gastrulation in blastulas with moderate quantities of yolk, such as amphibians have, proceeds similarly, but the archenteron is displaced toward the animal hemisphere and is filled with yolk cells. The early stages are similar to those in starfish.

Principles of Zoology

Embryonic Development

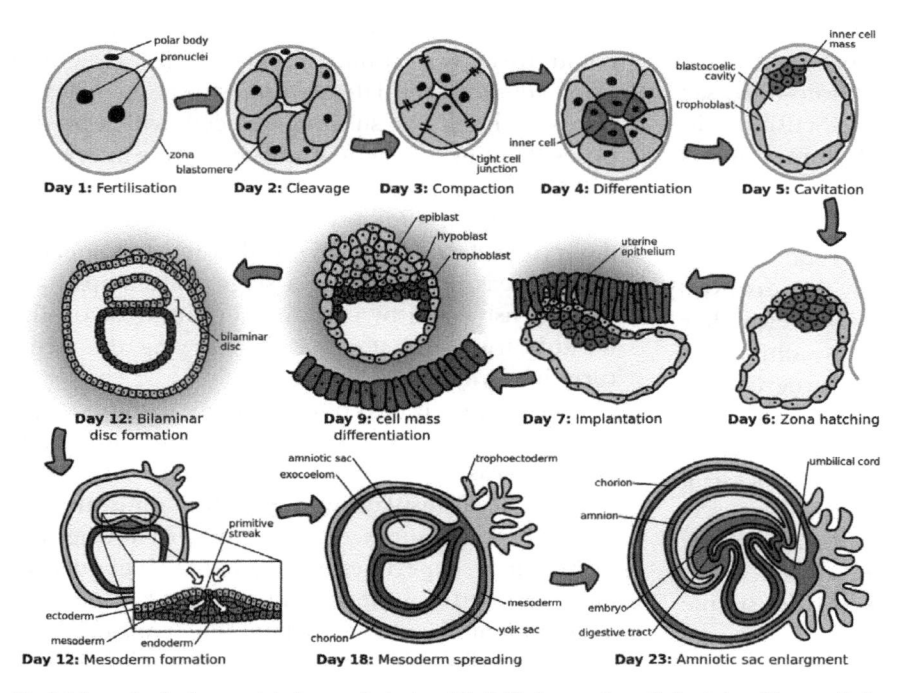

The first few weeks of embryogenesis in humans, beginning at the fertilized egg, ending with the closing of the neural tube.

Gastrulation in birds and mammals is initiated in a manner different from that in starfish and amphibians, because of the discoidal configuration of the blastula. Both groups have incomplete cleavage with embryonic development on a disklike area on one side of the egg. The upper cells of the disk separate from the lower ones, forming two layers, the epiblast and the hypoblast. After the two layers are formed, a thickening occurs in one quadrant of the blastula and soon becomes noticeable as a distinct streak, the primitive streak. The streak becomes grooved, and cells from either side begin to migrate to the groove and sink down through it. The cells then move into position between the epiblast and hypoblast. The three embryonic germ layers have been formed.

The primitive groove in the gastrula is considered homologous to the blastopore in the starfish and amphibians. After the germ layers have been established, cells continue to move in to the new cavity, the archenteron, and form a mesodermal notochord in the roof of the archenteron.

NEURULATION

Neurulation is the final stage of early embryonic development. Studies have shown that the notochord induces the neurulation process to begin. Cells just above the notochord are induced to proliferate and thicken, forming a neural plate. After the neural plate is formed, a buckling occurs in it, forming a depression known as the neural groove. Modern microscope techniques have revealed microfilaments and microtubules lying beneath the surface of the plate. Contraction of the microfilaments and elongation of microtubules appear to cause cell buckling and folding of the plate. The neural groove deepens at its cephalic end, and folds on either side continue to grow higher until they actually touch each other, forming an enclosed tube, the neural tube. At the same time that the neural tube is forming, the head is growing forward and tissue is folding beneath it so it projects forward free from the surface. Brain differentiation begins with the enlargement of the anterior end of the neural tube. The undilated caudal portion will give rise to the spinal cord. The brain forms several constrictions, so that three bulges appear. These will become the three embryonic brain divisions: the forebrain, the midbrain, and the hindbrain.

Upon completion of the three brain divisions, the embryo undergoes forward flexion of the forebrain and a lateral torsion so that the embryo comes to life with its left side on the yolk. A final caudal flexion causes the embryo to take its typical C-shaped configuration.

Extraembryonic membranes form from tissue outside the embryo to provide oxygen, nutrients, and waste storage. In birds, an outer chorion and amnion fuse to form a membrane with a large blood supply which provides for the exchange of oxygen and carbon dioxide between the embryo and the atmosphere. The allantois is a membranous sac to contain waste secretions.

In mammals, the outer chorion becomes extensively vascularized on one side and interconnects with the uterus to form the placenta. Nutrient and waste exchange between mother and baby take place in the placenta. The amnion forms a fluid-filled sac that lies closely around the embryo and cushions it. The allantois is not needed for waste storage and is not well developed.

EMBRYOLOGY

Humans have always been intrigued by the processes of gestation and birth. Aristotle questioned whether the embryo unfolds from a preformed condition and then enlarges to adult proportions or progressively differentiates from simple to complex form. Not until the eighteenth century were actual observations made of a developing embryo. The chick egg was the first to be studied, because of its large size. Early studies were descriptive, as each stage of the embryo was observed and carefully described. It was found that development does proceed from simple form to forms increasingly complex.

In the late nineteenth century, great interest developed in evolutionary theory, and comparative embryology became the focal point of studies. Clues were sought for possible evolutionary relationships between organisms. The theory emerged that embryonic stages reflect the evolutionary past of an organism.

The twentieth century has seen the explosion of experimental embryology and multiplication of knowledge. Cleavage of the large fertilized egg was first observed in the eighteenth century, but not until the late twentieth century did the mechanics begin to be understood. With improved microscope techniques, a ring of microfilaments can be seen just below the egg cell surface. These protein filaments have contractile qualities, and it was thought perhaps they lined up around the equator to contract and squeeze the cell in two. To test this hypothesis, a drug which causes microfilament subunits to break down was added to the cell culture. It was found that cell division was inhibited, suggesting that microfilaments are involved in the division process. Removal of astral rays also hindered cleavage. Each new discovery answers some questions and raises more.

Embryologists have questioned how blastomeres all formed from the same cell could differentiate into many kinds of cells and tissues. Some of the earliest experiments in embryology involved separating the first two daughter cells to demonstrate that each could form two complete individuals. How and when cells differentiate continues to be a challenge to researchers. The substance in cells which predisposes them to differentiate between one another is still not understood.

It has been discovered that each part of the embryo surface is already divided into prospective organ areas by the blastula stage. Fate maps have been constructed by marking certain areas on the blastula with vital stains and observing the structures into which they develop.

Since the early days of experimental embryology, researchers have performed all kinds of operations on embryos, marking areas and observing their movement, transplanting cells from one area to another, exchanging cell nuclei and removing portions. These experiments have led to many discoveries and better understanding of the complicated developmental process. When one considers the multitude of complex events that must take place in the development of a new individual from a single cell, it might seem impossible that the entire developmental process could occur without a slip.

Malformation usually begins during early development. Deformities may arise from inherited mistakes in the genetic code or from the harmful influence of external factors such as radiation, poor nutrition, or infection. Studies of cell migration in the embryo have led to ideas for procedures to inhibit tumor cell migration. Knowledge of normal cell development is helping to find ways to prevent abnormal cell development.

—*Katherine H. Houp*

FURTHER READING

Needham, Joseph, and Arthur Hughes. *A History of Embryology*. Rev. ed. Cambridge University Press, 2015.

Oppenheimer, Steven B., and Edward J. Carroll. *Introduction to Embryonic Development*. 4th ed. Pearson, 2004.

Schoenwolf, Gary C. *Atlas of Descriptive Embryology*. 7th ed. Pearson, 2008.

Starr, Cecie, Ralph Taggart, Christine A. Evers, and Lisa Starr. *Biology: The Unity and Diversity of Life*. 16th ed. Cengage, 2016.

Endangered Species

Fields of Study

Ecology; Environmental Studies; Biology; Agriculture; Forestry; Economics

Abstract

As of 2017, over fourteen thousand species of plants and wildlife worldwide were considered either endangered or critically endangered, with a further eleven thousand regarded as vulnerable. Fish, especially freshwater fish, are among the most immediately threatened types of animals. The destruction of species is caused in four major ways: humans hunting other species out of existence; habitats being destroyed; new species being introduced into regions and displacing native species; and nonnative plants and animals introducing diseases into environments, killing the existing species.

Key Concepts

ecosystem: a biological community and the physical environment contained in it
environment: the physical, chemical, and biological conditions of the region in which a plant or animal lives
extinction: the complete destruction of a species
habitat: the place in which a plant or animal lives
species: a group of similar individuals that can breed among themselves and produce offspring

VANISHING WILDLIFE

According to the International Union for Conservation of Nature and Natural Resources (IUCN), the foremost authority on species conservation status, as of 2017, over fourteen thousand species of plants and wildlife worldwide were considered either endangered or critically endangered, with a further eleven thousand regarded as vulnerable. In 2017, BirdLife International announced that 1,469 species of bird, or roughly one-eighth of all known living species, were threatened with extinction (meaning their conservation status is vulnerable, endangered, or critically endangered, according to IUCN categories). Fish, especially freshwater fish, are among the most immediately threatened types of animals. According to the National Oceanic and Atmospheric Association (NOAA), there are approximately 2,300 marine species that are considered either threatened or endangered (according to US government categories).

CAUSES OF SPECIES ENDANGERMENT

The destruction of species is caused in four major ways: humans hunting other species out of existence; habitats, the environments in which plants or animals grow and develop, being destroyed; new species, such as rats, cats, goats, or ground-covering plants, being introduced into regions and displacing native species; and nonnative plants and animals introducing diseases into environments, killing the existing species. For much of history, hunting was the major cause of species extinction. However, hunting has become less of a factor since governments and conservation authorities have imposed strict controls on the practice. In the second half of the twentieth century, habitat destruction and invasion by exotics (nonnative plants and animals) and the diseases they carry caused the most damage. Most biologists agree that whatever the factors involved, the rate of extinction has increased rapidly since the 1950s.

Some people have argued that the destruction of a single species of fish, bird, or flower would make little or no difference to the future of human life or the earth. They also suggest that extinctions have always taken place, even before human beings existed, and therefore are simply part of the natural process of existence. These arguments omit an important point. Individual species each inhabit a small part of an entire ecosystem, a community of plants and animals that are closely associated in a chain of survival. For example, plants absorb from the soil chemicals and minerals that are essential to their health. Animals then eat the plants—grasses, fruits, leaves, or flowers—and digest the nutrients they need for energy. Carnivorous animals then eat these plant eaters and get their energy from them. If a single species is removed from this chain, the whole ecosystem can suffer unknown and perhaps terrible consequences that cannot be reversed.

There are several tiger species: Siberian, Bengal, Sumatran, China, and Indo-Chinese. All species are listed as endangered on the U.S. Endangered and Threatened Wildlife and Plants List.

The death of an entire species constitutes a loss that cannot always be measured in economic terms. The American biologist William Beebe made the point that any species that is lost diminishes the quality of life for everyone:

> "The beauty and genius of a work of art may be reconceived, though its first material expression can be destroyed; a vanished harmony may yet inspire the composer, but when the last individual of a race of living things breathes no more, another heaven and another earth must pass before such a one can be again."

HOW SPECIES ARE LOST

Deaths of entire species grew more frequent in the 1900s. One example of this loss was the passenger pigeon, a bird so numerous in the 1820s that flocks numbered in the hundreds of millions, and John James Audubon, the famous American painter and collector, wrote that the flapping of wings on the Great Plains sounded like the roar of thunder. More than nine billion of the pigeons were alive in 1850; slightly more than sixty years later, exactly one bird, named Martha, remained—in the Cincinnati Zoo, where she had been taken in 1912. The population fell from nine billion to one in little more than half a century, then to zero when Martha died on September 1, 1917. People had found these pigeons delicious to eat and easy to kill. They formed hundreds of hunting parties, killing more than fifty thousand birds each week. No one dreamed that the passenger pigeon could ever be exterminated.

The same fate almost befell the American bison, often called the buffalo. Before the coming of railroads and white settlers in the 1860s and 1870s, the bison numbered more than one hundred million. Native Americans hunted the bison, eating their flesh and using the skins for clothing and shelter, but they killed only what they needed. The settlers, however, saw the bison as a problem that needed to be eliminated. Huge herds of bison crossed railroad tracks, forcing passenger trains to stop, and the animals interfered with farming, knocking down fences and trampling grain fields. Railroad companies and the US Army sent out hunting parties to get rid of the bison. By 1890, fewer than one thousand bison survived in a herd that had managed to escape far into northern Canada. The extermination ended only after this small herd was given protection by the Canadian government.

Stories of other near extinctions are numerous and frightening but demonstrate that action can be taken to save some if not all of the endangered species. Whales, which had been hunted commercially since the 1600s, faced possible extinction until action was taken to reduce hunting in the 1970s. Whales were easy to kill and provided oil and bone. Whale oil was the major substance burned in lamps until the electric light largely replaced oil-burning lamps in the 1880s. Europeans hunted the Atlantic whale, called the right whale because it was the "right" one to kill, into virtual extinction by the 1860s. When the right whale became too hard to find, hunters turned to the Pacific right whale and then the bowhead whale before action was taken by the world community to save remaining whales.

GREED AND IGNORANCE

Human greed has brought death or near death to many species. The desire for fur coats has killed nearly all jaguars, snow leopards, and various species of fur seals. Pribilof Island fur seals were hunted nearly into extinction in the late 1800s. A treaty between the United States, Canada, and Russia established limits on killing the species in its remote northern Pacific island habitat, but enforcement has proved difficult, and thousands of seals have been slaughtered despite international protection.

The belief that some animals are nuisances has led to the near extinction of wolves, grizzly bears, cougars, and coyotes. These predators have been poisoned and shot by the thousands and have become endangered species as a result. Attempts to kill insects with pesticides in order to control the spread of disease and improve crop yields were successful but had an unfortunate side effect: chemicals from the pesticides worked their way into ecosystems, killing millions of other forms of life. In the 1950s, dichlorodiphenyltrichloroethane, or DDT, was used to kill malaria-carrying mosquitoes, but the chemical infested the entire food chain. It entered plants that were eaten by animals and affected birds, fish, and butterflies. Pesticide poisoning also diminished the numbers of the bald eagle and peregrine falcon, which started to come back only after rigid controls on pesticides were established.

Events on the island of Madagascar, a large island in the Indian Ocean off the east coast of Africa, demonstrate most fully the deadly consequences of habitat destruction. About 180 million years ago, the island was attached to the African continent; then it was split off after a series of geological catastrophes and ended up 250 miles to the east. The split occurred just at the time mammals were emerging as a class of animals in Africa. One mammal species, the monkeylike lemur, became isolated on Madagascar and increased abundantly. Other animals caught on the island were several kinds of giant birds, including one that weighed five hundred kilograms and stood more than three meters tall. The island was isolated for millions of years, allowing hundreds of species found nowhere else in the world to evolve in the diverse island ecosystems. Madagascar had deserts, rain forests, dry forests, and seashores. About 99 percent of its reptiles, 81 percent of its plants, and 99 percent of its frogs were unique and tied specifically to the island's food chain. About two thousand years ago, the Malagasy people began to arrive on the island. They hunted, fished, and began to grow crops, destroying more than 90 percent of the forests that covered Madagascar in the process. Dozens of species died as a result, including the giant elephant bird, which was gone by 1700. Ten out of thirty-one species of lemur had died out by 1985. The loss of Madagascar's forests caused terrible erosion, which resulted in flooding and the destruction of more trees. Madagascar's entire ecological system is now threatened, and hundreds of unique species are listed as endangered. Only major restrictions on farming and habitat destruction can save these animals.

THE UNITED STATES' ENDANGERED SPECIES PRESERVATION ACT

The implications of species and habitat destruction were first described in books such as Rachel Carson's *Silent Spring* (1962). Carson, a biologist with the US Department of Interior, wrote about the effects of DDT and insecticides on birds and other animals. Her book inspired Congress to pass the Endangered Species Preservation Act in 1966, authorizing the secretary of the interior to protect certain fish and wildlife through the creation of the National Wildlife Refuge System. Congress strengthened the law in 1969 by restricting importation of threatened species and adding more domestic species to the list of those deserving protection.

The US Department of the Interior published its first list of endangered species in 1967. The list included seventy-two native species of animals, including grizzly bears, certain butterflies, bats, crocodiles, and trout. No plants were on this list. In 1973, President Richard M. Nixon signed into law the Endangered Species Act, which gave the Secretaries of the Interior and of Commerce responsibility for creating a list of endangered animals and plants, or species in immediate danger of extinction. Another list would include species threatened with extinction or those likely to become endangered in the foreseeable future. Once an animal or plant was on either of the lists, no one could kill, capture, or harm it. Penalties for violators were increased, and international or interstate trade of listed species was prohibited. Fines of up to ten thousand dollars could be imposed for knowingly violating the act, while a person could be fined up to one thousand dollars for unwittingly violating it.

A separate provision of the law mandated that federal agencies could not engage in projects that would destroy or modify a habitat critical to the survival of a threatened or endangered plant or animal. This provision became a very important tool in the battle to save species. Supporters of wildlife preservation used it to block highway and dam projects, at least until government officials could prove that construction would have no major impact on a fragile ecosystem. The law even called for affirmative measures to aid in the recovery of listed species. The Secretaries of the Interior and of Commerce were required to produce recovery plans detailing steps necessary to bring a species back to a point where it no longer needed protection. Money was appropriated for states to design recovery programs, and most states established plans of their own to deal with local crises.

A major threat to the 1973 law arose in 1978 during the Tellico Dam controversy. The Tennessee Valley Authority, a federally owned corporation, proposed building a hydroelectric dam on the Tennessee River in Loudon County, Tennessee, in the late 1970s. Shortly after plans were made public, a scientist from the University of Tennessee discovered a three-inch-long fish, the snail darter, that was unique to the area. Building the dam, a $250 million project, would destroy the snail darter's habitat and eliminate the fish. Environmentalists successfully argued in federal court that the dam had to be abandoned. The US Supreme Court supported the ruling of the lower court, arguing that when Congress passed the law, it had intended that endangered species be given the highest priority regardless of the cost or other concerns involved. However, in 1981, Congress enacted a special exemption that excluded "economically important" federal projects from the act. A federal judge then found the Tellico Dam to be without economic importance, and construction was again halted. Congressmen friendly to dam interests then added an amendment directing completion of the dam to an unrelated environmental bill, which passed, and Tellico was constructed. However, the principle of protection remained intact, and the 1973 act remained in force. The snail darter apparently survived, too, as scientists found it living in a river not far from the spot where it was originally discovered.

It is worth noting that as of 2019 the administration under President Donald J. Trump has cut back on the protections afforded by the Endangered Species Act, and reduced the mandate of the Environmental Protection Agency, causing many to question the long-term survivability of many species and habitats in the foreseeable future.

INTERNATIONAL TRADE BANS

The 1973 act also made the United States a partner in the Convention on International Trade in Endangered Species of Wild Flora and Fauna (CITES). This treaty came out of a conference in Washington, DC, that was attended by representatives from eighty nations. It created an international system for control of trade in endangered species. Enforced by the Switzerland-based International Union for the Conservation of Nature and Natural Resources (IUCN), later simply the International Union for Conservation of Nature, the convention has more than one hundred members. The IUCN maintains the IUCN Red List of Threatened Species, which sorts species into one of nine categories: extinct, extinct in the wild, critically endangered, endangered, vulnerable, near threatened, least concern, data deficient, and not evaluated. Category one consists of those in immediate danger and therefore absolutely banned from international hunting and trading. Animals and plants in categories two and three are not immediately threatened but require special export permits before they can be bought and sold because their numbers have been seriously reduced.

CITES has a major flaw, a loophole that can be exploited by any member: any nation can make a "reservation" on any listed species, exempting itself from the ban on trade. Japan has been among the most frequent users of the reservation, exempting itself from controls on, among others, the fin whale, the sei whale, the scalloped hammerhead, and the squat-headed hammerhead, all designated endangered by the IUCN. Other frequent offenders include Canada and Iceland. Unless this loophole is closed, the IUCN will be severely restricted in its ability to save threatened species.

SUCCESSES IN RESTORING ENDANGERED SPECIES

Several species in the United States—the California condor, the black-footed ferret, and the whooping crane—have been saved from extinction because of the 1973 Endangered Species Act. The road to extinction has also been reversed for the brown pelican, found in the southeastern states; the American alligator, which had been hunted almost to death in Florida; and the peregrine falcon in the eastern states. Other species, however, such as the dusky seaside sparrow, have totally disappeared, and a bird called the Guam rail, while still extant, has been declared extinct in the wild.

The northern spotted owl, found in parts of the rain forest in Oregon and Washington, has attracted a good deal of attention because of efforts to save it. The case of the owl points to the most difficult questions raised by the act: Which comes first, the welfare of the plant or animal or the economic needs of people? Each pair of spotted owls needs six to ten square miles of forest more than 250 years old in which to hunt and breed. The owls also need large hollow trees for nesting and large open fields in which to search for mice and other small animals. The majority of the suitable ecosystems are found in parts of twelve national forests in the region. At the same time, loggers in the areas need jobs. When the interests of the lumber industry and environmentalists collide, it is left to the courts to determine which interest will prevail or whether a compromise can be arranged.

Outside the United States, the future of endangered species appears much grimmer. Scientists at IUCN think several hundred thousand species will disappear by the end of the second decade of the twenty-first century. Many of these species have never even been identified or named. The most endangered habitats in the world are the tropical rain forests, which were reduced by half—nearly 3.5 million square miles—during the twentieth century. Over fifty thousand square miles are destroyed each year, mainly to provide farms and cattle ranches on land that is not suitable for long-term agriculture and that will require massive inputs of fertilizers while affording little or no protection against the loss of topsoil by erosion. Among the most threatened animal species in these forests are primates, including various species of capuchins, spider monkeys, lemurs, and macaques, as well as the western gorilla. One solution to forest destruction has been the creation of large wildlife refuges, but there are limits to the amount of land available for conservation efforts. Another solution is the establishment of more wildlife zoos. Several zoos have successful programs for saving species on the very edge of extinction. However, capacity is limited, and the very small numbers of animals in a zoo's herd create problems of interbreeding and the handing down of recessive genes.

For many species, it is too late to do very much, so scientists and biologists divide populations into three groups: those that can survive without help, those that would die whatever help was provided, and those species that might survive with help and would certainly die without it. Environmentalists focus their efforts on the plants and animals that fall into the third category. Resources are limited, however, and much work needs to be done, or extinctions will take place at a pace as yet unseen in the history of living things.

—*Leslie V. Tischauser*

FURTHER READING

BirdLife International. "Spotlight on Threatened Birds." *BirdLife International*, 2017, datazone.birdlife.org/sowb/spotthreatbirds. Accessed 31 Jan. 2018.

DiSilvestro, Roger L. *The Endangered Kingdom: The Struggle to Save America's Wildlife*. Wiley, 1989.

Doub, J. Peyton. *The Endangered Species Act: History, Implementation, Successes, and Controversies*. CRC, 2013.

"Endangered Species." *US Fish and Wildlife Service*. Dept. of the Interior, 27 Dec. 2017, www.fws.gov/endangered/. Accessed 31 Jan. 2018.

Goble, Dale D., J. Michael Scott, and Frank W. Davis. *The Endangered Species Act at Thirty*. 2 vols. Island, 2005–6.

The IUCN Red List of Threatened Species. Intl. Union for Conservation of Nature and Natural Resources, 2017, www.iucnredlist.org/. Accessed 31 Jan. 2018.

Mackay, Richard. *The Atlas of Endangered Species*. 3rd ed. University of California Press, 2009.

Olive, Andrea. *Land, Stewardship, and Legitimacy: Endangered Species Policy in Canada and the United States*. University of Toronto Press, 2014.

Randall, Jan A. *Endangered Species: A Reference Handbook*. ABC-CLIO, 2018.

Vié, Jean-Christophe, Craig Hilton-Taylor, and Simon N. Stuart, eds. *Wildlife in a Changing World: An Analysis of the 2008 IUCN Red List of Threatened Species*. Gland: IUCN, 2009. PDF file.

Wilson, Edward O. *The Diversity of Life*. Reprint. Belknap, 2010.

Endocrine Systems of Invertebrates

Fields of Study

Entomology; Animal Physiology; Animal Anatomy; Endocrinology

Abstract

The principal sources of hormones in most invertebrate phyla are neurosecretory cells. These hormonal mechanisms have been found in arthropods, annelid worms, mollusks, and echinoderms. The physiological processes that are affected are generally fundamental, long-term biological phenomena such as growth, regeneration, reproduction and development, and certain metabolic processes.

Key Concepts

budding: a form of asexual reproduction that begins as an outpocketing of the parental body, resulting in either separation from or continued connection with the parent, forming a colony

diapause: a resting phase in which metabolic activity is low and adverse conditions can be tolerated

molt: the process of replacing one exoskeleton with another

neurons: cells specialized for the conduction of electrical signals and the transmission of information (nerve cells)

neurosecretory cells: specialized neurons capable of manufacturing and releasing hormones (neurosecretions or neurosecretory hormones) and discharging them directly into circulation

photoperiod: the measure of the relative length of daylight as it relates to the potential physiological responses that exposure to daylight evokes

target organ: a specific body part that a particular hormone directly affects

trophic hormone: a hormone that stimulates another endocrine gland

ENDOCRINE SYSTEMS

The endocrine systems of many invertebrates are nearly as complicated as vertebrate endocrine systems. The principal sources of hormones in most invertebrate phyla are neurosecretory cells. These hormonal mechanisms have been found in arthropods, annelid worms, mollusks, and echinoderms. The physiological processes that are affected are generally fundamental, long-term ones that include such biological phenomena as growth, regeneration, reproduction and development, and certain metabolic processes.

The subkingdom of animals made up of all vertebrates and most invertebrates (protozoa and sponges are not included) is called the Metazoa, which is defined by the presence of nervous and endocrine systems in the animals in the group. These systems coordinate the activities of the animal so it can function as a whole. The nervous system is important in rapid communication, such as contraction of muscles for movement, while the endocrine system controls long-term processes within the body, such as the growth of organs or maintenance of appropriate metabolic concentrations. Chemical messengers released by the endocrine systems have to travel to specific target organs to exert their effects. The means of travel is the circulatory system. Because it takes time for the chemicals to accumulate to effective concentrations, they must be stable enough to remain in the body without undergoing chemical changes and without being excreted. These chemical messengers—hormones—are, then, well suited to working over long periods of time. The nervous and endocrine systems do not work independently of one another, however. It is probable that most animals' central nervous systems are strongly affected by hormones much of the time.

NEUROSECRETORY CELLS

In 1928, German chemist Ernst Scharrer hypothesized that certain nerve cells have qualities of both nerve and endocrine gland cells. These neurosecretory cells are neurons which are cellularly like gland cells, but are widespread within the invertebrate body. They receive nervous impulses, but rather than communicating through synapses with other neurons or effector cells, they terminate close to the circulatory system and release substances which travel to act on organs or upon endocrine glands. These neurosecretory substances, therefore, are themselves hormones. Neurosecretory cells are usually found in clusters within the central nervous system. Extending from the cell bodies are axons that terminate in swollen knobs associated with blood spaces. Terminals that are aggregated into a body are called neurohemal organs. Neurosecretory material is produced by the cell bodies, transported down the axons, and stored in the swollen knobs. Release is accomplished by exocytosis.

To be classified as a neurosecretory cell, three criteria must be met: The cell must have the structural features of a neuron (cell body and axonlike fibers); the axons must not synapse with other cells but end in close association with an area of body fluid (generally a blood vessel or sinus, the combination making a neurohemal organ); the neuron must contain membrane-bound vesicles within the cytoplasm. There are, in addition, two physiological criteria: Destruction of these clustered cells, or the areas or organs where they are found, produces an alteration of existing internal conditions within the organism that can be restored by replacing the removed organ or injecting an extract from it; implantation of an organ thought to be neurosecretory into a normally functioning animal brings about a change in internal state by either prompting or inhibiting the occurrence of certain events.

Neurosecretions in invertebrates may influence behavior or target another endocrine gland by trophic neurosecretions or trophic hormones. For example, in many insects neurosecretions from neurosecretory cells in the brain exert a trophic effect on the prothoracic glands, which then produce and release the hormone ecdysone that controls molting, the developmental sequence of insects. There are other examples of how hormonal release is dependent on and dictated by the nervous system. Most animals respond developmentally to environmental changes, such as seasonal variations throughout the year. If unfavorable conditions develop, the animal may compensate by going into dormancy or migrating, or may overcome the conditions by other changes in habit or physiology. Even brief fluctuations, such as a temporary shortage of food or the absence of suitable mates, may dramatically affect development. The mating act in a female insect may speed up the development of her eggs; the changing day length may control when metamorphosis begins in annelids. Stimulation of sense organs sets up nervous messages that result in changes in the amounts of circulating hormones, which generate these "new" responses.

THE ROLES OF INVERTEBRATE HORMONES

Invertebrate hormones play as many roles as there are invertebrate phyla. In the less highly organized invertebrates, endocrine glands are apparently absent; hormonal coordination depends on neurosecretions. Hormones released in the plant hydra are believed to come from the hypostomal region (the nerve ring around the oral opening) and from actively growing areas and are thought to regulate growth, regeneration, and the development of sexuality. Little is known about substances termed "wound hormones" in planarians, but their presence in wounded tissues has been inferred, even though their site of production is unclear.

All annelids possess neurosecretory cells in the brain that control growth, reproduction, and maturation. In nereids, reproductive body forms releasing eggs and sperm are controlled by at least one brain neurohormone, and normal reproductive development appears to depend on the gradual withdrawal of brain neurohormones with increasing age. Regeneration, too, is probably controlled by neurohormones.

In starfish and sea urchins, spawning of eggs is preceded by release of a "shedding hormone" found only in the radial nerves. This hormone, known as gonad-stimulating substance, also stimulates the manufacture and secretion of a second substance by the gonads called meiosis-inducing substance (MIS). MIS causes the follicle cells to pull away from the gametes so the gametes can be expelled more easily; it induces meiosis within the oocytes, and, after diffusing into the coelomic fluid, stimulates muscle contractions which cause spawning.

Many mollusks have neurons resembling neurosecretory cells that change their apparent secretory activity with conditions such as reproductive state. In a few cases, evidence has been found for neurosecretory control of reproduction, water balance, or heart function. Cephalopods such as squid, however, are one of the few classes of animals that possess endocrine glands. The cephalopod brain is connected to the optic lobes by short optic stalks bearing optic glands. As the animal matures, the size of these glands increases. These glands function in the control of reproductive development. Glands on the gills, called mesodermal bronchial glands, are also endocrine organs and are thought to function similarly to vertebrate adrenal glands.

Certainly the best-studied invertebrate system is that of insects. Insects possess discrete clusters of neurosecretory cells, well-developed neurohemal organs, and even nonneural endocrine glands. The insect endocrine system has four major components: the corpora cardiaca, a group of neurosecretory cells in the brain, the corpora allata, and the thoracic glands. The corpora cardiaca, closely associated with the heart, store and secrete hormones from the brain as well as producing their own inherent hormones. Along with the brain's neurosecretory cells, they compose the cerebral neurosecretory system. Molting is controlled by hormones called ecdysteroids produced under the brain's direction. Secretion of these ecdysteroids stimulates the release of ecdysone from the prothoracic gland.

Ecdysone, also called the molting hormone, stimulates the development of adult structures but is held in check by juvenile hormone (JH), which favors the development of juvenile characteristics. During juvenile life, JH predominates and each molt yields a larger juvenile. High levels of JH are released by the corpus allatum during early stages of life. Its major function, then, is to ensure that when molting is triggered by ecdysone secretion, the next larval stage results. When the final stage is reached, ecdysone production dramatically falls, but sufficient levels are produced to induce a molt that will result in the adult stage. Similar systems are found in the crustaceans.

Because invertebrates make up about 95 percent of the species in the animal kingdom, one might anticipate a great diversity of invertebrate endocrine mechanisms. Eventually, this expectation may be confirmed; but knowledge of endocrine systems in many invertebrate groups is, for the most part, incomplete. What is known is that in most groups of invertebrates, neurosecretory systems are distinctly more prominent than nonneural endocrine glands, which occur in very few cases.

REFINEMENT OF STUDY TECHNIQUES

Until the 1960s, the search for hormonal regulators in invertebrates was largely unsuccessful because early experiments on gonad transplantation from insects of one sex to those of the other and injection of vertebrate hormones into invertebrates yielded negative results. Strides made in the last twenty-five years are mostly the results of refinements of microscopic, operative, and analytical techniques, both chemical and physical. Arthropods have provided the most accessible material for study, and more is known about the phenomenon in crustaceans and insects than in any other group. The range of investigation is expanding, however, and neurosecretion in invertebrates is not only accepted but recognized as widespread among them.

Many problems are generally associated with determining the functions of the neurosecretory system. The classical experimental method involves removal of the suspected endocrine gland and then reimplanting it at another location in the body. If the effects of removal are reversed and normal conditions return when the organ is relocated, then a hormonal mechanism is probably involved. The problem arises, however, because removing a neurohemal organ will leave behind the cut ends that may continue to release hormones, perhaps in an uncontrolled manner. A new neurohemal organ may be rapidly regenerated so that the effects of lessened amounts of neurosecretory hormones cannot be observed. In addition, reimplanting the organ may produce several hormones in the animal in abnormal concentrations or proportions. Hormonal deficiency may not be obvious immediately because hormones stored outside the neurohemal organ may be secreted or leached out for some time after the organ's removal. Yet another problem encountered is the lack of distinct neurohemal organs; instead, scattered neurosecretory cells may be found throughout the nervous system. It is therefore difficult to determine the exact function of the mechanisms because of the virtual impossibility of removing and testing these individual cells. In these cases, the neurosecretory nature of the cells is deduced based on their structural and chemical similarity to those whose function has been already verified in other animals.

Typical of the early work on insect growth hormones was the work done in the 1930s in England by Vincent B. Wigglesworth on the metamorphosis of a bloodsucking insect named *Rhodnius*. This insect goes through five immature stages, each separated by a molt, until it reaches adulthood. During each of these stages, it engorges and stretches its abdomen by ingesting a blood meal. This filling meal apparently stimulates the release of hormones that cause molting at the end of a specified time interval following the meal. Usually, the final molt (to adulthood) occurs about twenty-eight days after the blood meal. If the insect is decapitated during the first few days after its meal, molting does not occur, even though the animal may live for several months longer. Decapitation more than eight days after the blood meal does not interfere with molting, although a headless adult is produced. Wigglesworth further showed that joining the circulatory system of a later-decapitated insect to that of an earlier-decapitated insect allows both to molt into adults. It appears obvious that some stimulus passes via the blood from one insect to the other and induces molting; it is assumed that the stimulus is a hormone that is secreted about eight days after the blood meal.

Since Wigglesworth's time, studies typical of his work have shown evidence of hormonal activity and control of many other invertebrate phyla. Experimentation with insects still outweighs all other studies, however, since they are so available and easy to work with because of their size.

COMPARING VERTEBRATE AND INVERTEBRATE HORMONES

The study of invertebrate hormones began as an attempt to draw parallels between invertebrate hormones and known vertebrate hormones. Experimenters were virtually forced to look for these similarities because of legal restrictions placed on the use of vertebrates for experimental study. The end results, however, have shown that invertebrate hormones share little with vertebrate hormones.

One of the few similar hormones, in structure at least, is prothoracicotropic hormone (PTTH), isolated from the heads of adult silkworms. (PTTH stimulates the prothoracic gland to release ecdysone, which then regulates molting and growth.) Though structurally similar to vertebrate insulin, insect insulin has no functional link with vertebrate insulin.

The interaction of neurohormones and the nervous system has been studied using the lobster, tying the release of neurohormones to its behavior. By introducing neurohormones via injection, one can induce behavioral changes, such as increased aggression. By working with these crustaceans, the apparent relationship among neurohormones, the nervous system, and behavior modification may be used in observing and controlling animal behavior.

—*Iona C. Baldridge*

FURTHER READING

Alexander, R. McNeill. *The Invertebrates.* Cambridge University Press, 1979.

Barnes, R. S. K., P. Calow, and P. J. W. Olive. *The Invertebrates: A New Synthesis.* 2nd ed. Blackwell Scientific Publications, 1993.

Hickman, Cleveland P., Larry S. Roberts, and Frances M. Hickman. *Biology of Animals.* 7th ed. MacGraw-Hill, 1998.

Hill, Richard W., and Gordon A. Wyse. *Animal Physiology.* 2nd ed. Harper & Row, 1989.

Keeton, William T., and James L. Gould. *Biological Science.* 6th ed. W. W. Norton, 1996.

Laufer, Hans, and Roger G. H. Downer, eds. *Endocrinology of Selected Invertebrate Types.* Liss, 1988.

Matsumoto, Akira, and Susumu Ishii, eds. *Atlas of Endocrine Organs: Vertebrates and Invertebrates.* Springer-Verlag, 1992.

ENDOCRINE SYSTEMS OF VERTEBRATES

FIELDS OF STUDY

Endocrinology; Reproductive Biology; Animal Physiology

ABSTRACT

The first known hormone was discovered around 1902 when William Bayliss and Ernest Starling discovered, in dogs, secretin, a hormone that stimulates pancreatic exocrine secretion in response to acid in the small intestine. Since that date, dozens of other hormones have been discovered, which control all aspects of growth, metabolism, and reproduction.

KEY CONCEPTS

feedback: in endocrinology, this usually refers to one hormone controlling the secretion of another that stimulates the first, usually in the form of negative feedback, in which the second hormone inhibits the first

gland: a tissue composed of similar cells that produce a hormone

hormone: a blood-borne chemical messenger

receptor: a protein molecule on or in a cell that responds to the hormone by binding to it and initiating a series of events that compose the response

target: cells that contain hormone receptors

ENDOCRINE SYSTEMS

Endocrine systems have been known only since the early twentieth century. The first known hormone was discovered around 1902 when William Bayliss and Ernest Starling discovered, in dogs, secretin, a hormone that stimulates pancreatic exocrine secretion in response to acid in the small intestine. Since that date, dozens of other hormones have been discovered, which control all aspects of growth, metabolism, and reproduction.

The endocrine system consists of glands that secrete chemical substances called hormones in response to specific signals. The hormones are secreted into the blood stream, where they travel to specific target cells or tissues, which contain specific receptors that allow the hormones to bind, initiating the response. Classically, hormones were thought to belong to two very different groups, the polypeptide (small protein) hormones and the steroid (cholesterol-like) hormones. It is now known that hormones can be composed of several different kinds of molecules, including fats (prostaglandins) and even gases (nitric oxide). Proteinlike hormones bind receptors found on external cell membranes to stimulate second messengers, such as cyclic adenosine monophosphate (cAMP), which activate enzymes and other cellular substances to produce a response. Steroid hormones enter target cells and bind intracellular receptors. The hormone-receptor complexes migrate to the nucleus and activate gene expression, which results in the response. This, like descriptions of many concepts in biology, is an oversimplification, and many hormones appear to work by a combination of the two mechanisms.

ENDOCRINE CONTROL SYSTEMS

Endocrine secretions are controlled by the nervous system through a complex chain of command.

Receptors around the body monitor sensory signals and alert the brain, which then relays the information to specific cells in the median eminence of the hypothalamus. For example, temperature receptors in the skin detect cold and inform the brain of potential body cooling. The brain then relays the information to cells in the hypothalamus, which secrete a molecule called thyrotropin-releasing hormone into a blood vessel called the hypothalamo-hypophysial portal vessel. This blood vessel delivers the releasing hormone to the anterior pituitary gland, which in turn secretes a hormone called thyroid-stimulating hormone (TSH), or thyrotropin, into the blood. The TSH travels to the thyroid gland to stimulate the secretion of thyroid hormones, which stimulate metabolism in liver, muscle, and other cells. Heat produced as a by-product of metabolism warms the body.

Some hormones are under dual control. Growth hormone (somatotropin) is stimulated by a releasing hormone called somatocrinin and inhibited by somatostatin. There are about seven anterior pituitary hormones that are controlled by similar mechanisms. adrenocorticotropic hormone (ACTH) is controlled by corticotropin-releasing hormone. melanocyte-stimulating hormone (MSH) and prolactin are under dual control by both releasing hormones and inhibiting hormones. The gonadotropins—follicle-stimulating hormone (FSH) and luteinizing hormone (LH)—are under the control of a single releasing hormone called gonadotropin releasing hormone. All of these control systems are subject to feedback loops which usually involve negative feedback (for example, TSH secretion being inhibited by thyroid hormone), but positive feedback loops exist (estrogen feeding back positively to stimulate LH secretion).

HORMONES CONTROLLING GROWTH, DEVELOPMENT, AND METABOLISM

The major control of growth is carried out by somatotropin (STH) from the anterior pituitary. STH does not act directly, however. Cells in the liver respond to STH to produce somatomedin, which stimulates bone growth and muscle production. Prolactin, a protein similar to STH, stimulates breast development in female mammals. In an interesting case of hormone evolution, thyroid hormone stimulates amphibian metamorphosis (tadpole to frog transition); however, in warm-blooded vertebrates, this same hormone has evolved to stimulate metabolism for the purpose of heat production in birds and mammals. Several hormones stimulate metabolism for different reasons. Epinephrine (adrenaline), in addition to elevating blood pressure, mobilizes glucose from glycogen, in response to stress. Steroid hormones, also produced in the adrenal glands, stimulate the production of glucose from noncarbohydrate molecules (gluconeogenesis). The stimulus for this is prolonged stress, for example, starvation. These glucocorticoids, such as cortisol and corticosterone, evolved early and are very important in combating stresses resulting from migration among birds and even fish.

The pancreatic hormones insulin and glucagon also effect energy metabolism. These two proteins regulate blood sugar, fat, and protein levels. After eating, insulin stimulates transport of these molecules into liver, fat, and muscle cells and then stimulates the incorporation of the simple molecules, such as glucose, amino acids, and fatty acids, into larger storage molecules, such as glycogen, protein, and fats. Glucagon has opposite actions. After a prolonged period without food intake, glucagon stimulates breakdown of complex molecules, such as glycogen and fats, into simple molecules, which are released into the blood and made available to metabolizing cells. These two hormones act independently of the pituitary and respond directly to blood-borne signals such as glucose concentration. This regulation ensures a steady delivery of nutrients to metabolizing cells in animals who only eat intermittently.

CONTROL OF WATER AND SALT BALANCE

The state of hydration and salt levels in the body are of critical importance to vertebrate animals. Dehydration has obvious severe detrimental consequences. The salt composition of body fluids is equally important. For enzymes and other proteins such as antibodies and even hormones to function properly, salt concentrations (ionic concentrations) must be maintained. For example, blood levels of sodium and potassium must be maintained at approximately 145 and 4 millimolers, respectively, in most vertebrates. These levels are lower in amphibians (100 and 2 millimolers). Water content of the body is controlled primarily by a posterior pituitary hormone called antidiuretic hormone (ADH). When the body becomes dehydrated, both concentration receptors and volume receptors in the brain trigger the secretion of ADH. This small peptide then stimulates thirst and water retention in the kidneys. In amphibians, it also stimulates water absorption by the skin and urinary bladder.

A steroid hormone produced in the adrenal glands called aldosterone stimulates the kidney and large intestine to conserve sodium. The kidneys also excrete increased amounts of potassium in response to aldosterone. Aldosterone secretion is stimulated by angiotensin II. When blood sodium levels decrease, there is a consequent loss of water and thus body fluid volume. Pressure receptors in the kidneys trigger the release of renin, which initiates a complex series of enzymatic reactions in the blood leading to the appearance of angiotensin II, which stimulates the secretion of aldosterone. When blood pressure increases, aldosterone secretion decreases, sodium excretion increases, and other hormones appear which also help eliminate sodium. Pressure receptors in the heart cause that organ to secrete atrial natriuretic peptide (ANP). ANP inhibits sodium conservation in the kidneys. Increased pressure in blood vessels activates nitric oxide synthetase, which produces nitric oxide locally. Nitric oxide increases the excretion of sodium by the kidneys.

Calcium and phosphate are also controlled by hormones. The parathyroid hormones respond directly to blood calcium concentrations. When calcium levels are low, parathyroid hormone (PTH) is secreted into the blood to stimulate three centers. In bone, PTH mobilizes calcium to elevate blood levels of this ion. Because mobilization of bone also elevates phosphate, which can be toxic at high concentrations, the kidneys become important. PTH stimulates the kidneys to increase calcium conservation and potassium excretion. PTH also stimulates uptake of calcium in the small intestine. Vitamin D enhances the action of PTH. Working antagonistically to PTH, calcitonin, produced in the thyroid gland, responds directly to high blood calcium to move this ion into bone.

DIGESTIVE HORMONES

The digestion and assimilation of food are also controlled by hormones. In meat-eating animals, beginning in the stomach, stretch of the organ and the presence of protein stimulate the secretion of gastrin into blood vessels in the wall of the stomach. The gastrin stimulates the secretion of hydrochloric acid into the lumen of the stomach to digest proteins. When the partially digested food enters the small intestine for the completion of digestion and assimilation, a slightly alkaline pH is required. The walls of the small intestine detect the acidity and secrete another pair of hormones into blood vessels. Secretin travels to the pancreas and stimulates sodium bicarbonate secretion. The sodium bicarbonate travels through the common bile duct to the small intestine, where it neutralizes the acid. Gastric inhibitory polypeptide travels to the stomach to inhibit acid secretion and stomach contractions. Another peptide, cholecystokinin-pancreozymin (CCKPZ), responds to fats and proteins in the small intestine and is subsequently secreted into the blood. This hormone travels to the gallbladder, causing it to contract and release its bile contents through the common bile duct to aid digestion of fats in the small intestine. CCKPZ also stimulates secretion of a whole host of enzymes by the pancreas. These enzymes also move through the common bile duct to the small intestine to aid in the digestion of carbohydrates, fats, and proteins. At least two other digestive hormones have been discovered that are not well understood at present. Motilin is secreted by the small intestine and stimulates stomach muscle contractions. Vasoactive intestinal polypeptide also is secreted by the small intestine and it, in turn, stimulates sodium bicarbonate secretion by the walls of the small intestine. Both hormones are of obvious benefit, but key details of their function, such as what triggers their secretion, are not clearly understood. It is important to realize that all of the hormones of the stomach and small intestine are secreted into the blood vessels in the walls of the organs, not into their lumens.

REPRODUCTIVE HORMONES

The two pituitary hormones that are involved in reproduction are called the gonadotropins, FSH and LH. These hormones are identical in males and females. The gonadal hormones differ between the two sexes. Females produce estrogens and progesterone in their ovaries. Males produce androgens (primarily testosterone) in the testes.

The mammalian menstrual cycle has two components. Both the ovarian cycle and the uterine cycle proceed simultaneously and last approximately four days in rats, sixteen days in sheep, and twenty-eight days in humans. The length and pattern of the cycle vary with species. For the sake of comparison, the human cycle is described here. The first five days of each cycle is called the menstrual period, and during this period the built-up walls of the uterus (resulting from the previous cycle) are shed and discharged through the vagina. At this time, the concentrations of FSH and LH in the blood are about the same. From the close of the menstrual period until ovulation is the follicular cycle. FSH stimulates the ovaries to begin the growth and maturation of an egg-containing follicle. This follicle produces estrogen. Estrogen feeds back negatively on FSH, causing its levels in the blood to drop. At the same time, estrogen is feeding back positively on LH, causing its levels to rise.

At the midpoint of the ovarian cycle, LH peaks and causes the now mature follicle to burst and eject an egg (ovum) into the oviduct. The ruptured follicle now becomes a corpus luteum and continues to secrete estrogen, but also begins to secrete progesterone. The estrogen, and now the progesterone, stimulate the walls of the uterus to thicken and produce glandular tubes and blood vessels. This goes on for the final half of the cycle, which is called the follicular phase in the ovaries and the proliferative phase in the uterus. If fertilization of the ovum in the oviduct fails to occur during this period, a hormone, probably a prostaglandin, builds up in the corpus luteum, causing it to stop producing estrogen and progesterone. With the loss of these two steroids, the thickened wall of the uterus is shed and the menses flows during the first five days of the next cycle.

Male reproductive endocrinology is much different. The first striking difference is that, although the pituitary hormones FSH and LH are the same, the patterns of secretion are different. Instead of the cyclic peaks found in females, males secrete constant levels of gonadotrophins. FSH stimulates sperm production and maturation in the seminiferous tubules of the testes. LH stimulates testosterone secretion by the interstitial cells of the testes. Testosterone helps FSH to stimulate sperm maturation. This androgen also stimulates such primary sex characteristics as penis and epididymal growth during puberty. The epididymis is a tubular structure that stores sperm in preparation for ejaculation. Testosterone also stimulates secondary sex characters, such as the deepening of the voice and development of muscle mass that manifest during puberty in humans.

—*Daniel F. Stiffler*

ENDOSKELETONS

FIELDS OF STUDY

Animal Physiology; Evolutionary Biology; Histology

ABSTRACT

Endoskeletons can be thought of as primarily mechanical or architectural systems. They can also be viewed as types of tissue, which involves study at the microscopic level. Endoskeletons can also be considered as products of the evolution of the endoskeletal organism. This area examines such factors as development from protochordates and the intrinsic biological advantages of the endoskeleton over the exoskeleton.

KEY CONCEPTS

appendicular skeleton: one of two main divisions of vertebrate skeletal systems, composed of the bones of the pelvic girdle, the shoulders, and the limbs

axial skeleton: the other main division of vertebrate skeletal systems, made up of the bones of the skull, the vertebral column, the ribs, and the sternum

cancellous bone: spongy bone that is composed of an open, interlacing framework of bony tissue oriented to provide maximum strength in response to normal strains and stresses

cartilage: a soft, pliable typically deep-lying tissue that constitutes the endoskeletons of primitive vertebrates, such as sharks, as well as the embryonic skeletons and jointing structures of adult higher vertebrates

compact bone: a dense type of bone, often termed lamellar bone, formed of a calcified bone matrix having a concentric ring organization

FURTHER READING

Bentley, P. J. *Comparative Vertebrate Endocrinology*. 3rd ed. Cambridge University Press, 1998.

Norman, A. W., and G. Litwack. *Hormones*. 2nd ed. Academic Press, 1997.

Norris, D. O. *Vertebrate Endocrinology*. 3rd ed. Academic Press, 1997.

Yen, S. C., R. B. Jaffe, and R. L. Barbieri. *Endocrinology: Physiology, Pathophysiology, and Clinical Management*. 4th ed. W. B. Saunders, 1999.

Haversian systems: narrow tubes surrounded by rings of bone, called lamellae, that are found within compact bones of animals having endoskeletons; the tubes contain blood vessels and bone

osteoblast: a bone-secreting cell found in vertebrates, that is instrumental in the process of ossification

TO STUDY ENDOSKELETONS

The study of endoskeletons can be approached in three basic ways. First, endoskeletons can be approached as gross structures—that is, as primarily mechanical or architectural systems. Second, they can be viewed as types of tissue—bone and cartilage—which involves study at the microscopic level. Third, endoskeletons can be considered products of the evolution of the endoskeletal organism. This area examines such factors as the organisms' development from protochordates and the intrinsic biological advantages of the endoskeleton over the exoskeleton.

THE DEVELOPMENT OF ENDOSKELETONS

The evolution of vertebrate animals, such as fishes, reptiles, and mammals, having articulating endoskeletons represents a biological quantum leap forward for the phylum *Chordata*. The development of endoskeletons permitted several significant structural advantages that allowed the higher chordates to compete successfully with invertebrates and eventually to become dominant in many varied ecosystems.

The evolution of endoskeletons allowed a greater degree of general body efficiency and organization. Coupled with a great increase in the rapidity, power, and control of movement with which endoskeletons

Starfish endoskeleton seen on X-Ray. The highly organised endoskeleton of a starfish can be seen in this x-ray image. The lines reaching up each leg are the creature's tubed feet, which operate through hydraulic pressure.

and their improved musculature endowed their possessors, endoskeletons allowed vertebrates to become the highest and fastest flyers, the swiftest runners and swimmers, and the widest ranging of animals. The organizational plan of the endoskeleton of the more highly evolved vertebrates is exemplified by that of mammals. The mammalian endoskeleton is articulated in many ways, giving it a great degree of flexibility and great range of movement. Cartilaginous joints facilitate the articulations. The endoskeleton itself is typically made of bone material composed of calcium phosphate. Other than the endoskeleton proper, minor externalizations can take the form of fingernails and toenails, claws, hooves, antlers, and horn cores, as well as teeth. In these cases, materials other than bone can sometimes originate at or near the body surface or in the skin. These expressions can be present in the form of scales or plates, though this is the exception rather than the rule among this vertebrate class. All these external expressions can be considered, in a way, forms of a limited exoskeleton.

AXIAL AND APPENDICULAR SKELETONS

Aside from such exceptions as pangolins and armadillos among mammals, and turtles and tortoises among reptiles, exoskeletal tissue does not usually constitute an architectural structure that the animal's other organs or structures depend on for support or protection. The one exception among the various bones of the vertebrate endoskeleton is the bones of the cranium, or skull. The cranial bones are a group of hard, thick bones, usually ovoid or spheroidal in general geometry, that offer extensive protection to the brain and primary sensory organs, such as the eyes and ears. The cranium is such a universal feature among vertebrates that the subphylum is sometimes referred to as the craniates. Evolutionarily lower vertebrates tend to have larger numbers of skull bones. Some fish have as many as 180 skull bones. Higher taxonomic groups have inversely lower numbers of skull bones: Amphibians and reptiles possess between fifty and ninety-five, while mammals have thirty-five or fewer. The skull itself is a member of two fundamental divisions between which the entire endoskeleton is usually subdivided; it is part of the axial skeleton. The axial skeleton also includes the bones of the vertebral system, the ribs, and the sternum. Possession of all axial features is not universal. Some vertebrates, such as the leopard frog, do not possess ribs, while others, such as the snakes, do not have sternums.

The other endoskeletal subdivision is called the appendicular skeleton, and it is made up of the bones of the pelvic girdle, shoulders, and limbs. The components of the appendicular system exhibit a great degree of variation from vertebrate group to vertebrate group and even among species, as do those of the axial system, and reflect the many different environments and lifestyles to which their respective possessors have adapted. A case in point is the numerous variations in form and length found among limb bones. The various lengths represent adaptations to such external environmental factors as the medium through which the animals move from place to place (air, water, ground surface, and subsurface) and speed. The lower and upper limb bones themselves are connected by joints. The jointing arrangements of limb bones are highly efficient mechanical developments. Two basic types of limb joint exist: the pulley joint and the ball-and-socket joint. Pulley joints are exemplified by finger and toe joints and represent great freedom of motion in one plane. Ball-and-socket joints are exemplified by shoulder or hip joints and represent freedom of universal motion. Still another joint type is a cross between these two. Such a combination of pulley and ball joints is exemplified by the elbow joint in humans.

HISTOLOGY

Another major approach to the study of the endoskeleton is its histology, or the fine details of tissues and how these tissues develop. Bone material itself is active metabolically—that is to say, it is alive. It can be considered not only an architecture along which the vertebrate body is arranged but also a complex and specialized connective tissue. As an organic material, it possesses a number of unique properties that are derived from the fact that it has evolved to perform its various duties efficiently for the size, weight, and arrangement of the materials of which it is composed. It is engineered like structurally reinforced concrete, having fibers of collagen, a tough, fibrous binding protein that is analogous to the function of steel rods.

The mineral calcium is analogous to the concrete in a building. Bone is formed in two different ways, depending on type. One process involves a means of growth of two bone types, termed lamellar bone (compact bone) and cancellous bone (spongy bone). Lamellar bone, sometimes also called membrane bone, develops through the process of ossification when certain cells called osteoblasts become bone-secreting. The osteoblastic cells, in association with numerous fibers of connective tissue cells, form a network in which layers of calcium mineral salts, called lamellae, are deposited. This network slowly builds up a plate that expands along its margins. As the plate thickens, some osteoblasts remain alive and are incorporated into the bone growth. At this point, they begin to have irregular shapes and are termed osteocytes. Spaces in which the osteocytes are sited are termed lacunae (cavities) and develop long, omnidirectional, branching processes, termed canaliculi. Neighboring canaliculi eventually link up and create a network through which life-supporting blood containing oxygen and food can reach the growing bone tissue. The canaliculi system grows such that no bone cell is more than 0.1 millimeter from a blood-carrying capillary. This overall arrangement is termed a Haversian system.

Cartilage tissue is another endoskeletal material that forms the adult skeletons of higher vertebrates, such as mammals, when they are still in early developmental stages. In mammals, this type of tissue is not formed directly, but rather by a replacement process. In mammalian embryos, most of the skeletal structure is initially laid down in the form of cartilage and then subsequently replaced by true bone. The process does not reach completion in the higher vertebrates until the animal is full-grown; in humans, this is as late as twenty-five or twenty-six years of age. Cartilage is not as hard or rigid as bone, but it is extremely tough and is resistant to forces of compression or extension. Under microscopic examination, it appears as a clear matrix which possesses numerous, embedded cells termed chondroblasts. These chondroblasts lie in fluid-filled voids termed lacunae. Chondroblasts secrete the matrix called chondrin, which surrounds the lacunae. Both the chondrin and the fluids act in an elastic manner and are resistant to compression and external shocks. Various types of cartilage have collagen fibers. The amounts of collagen fiber present determine the amount of extension that the cartilage can resist. The total effect of the cartilage's unique composition is to render it a good skeletal material for young, rapidly developing, vulnerable animals, such as mammal embryos.

VERTEBRATE EVOLUTION

The last basic approach to the study of endoskeletons is that of examining the evolutionary development of the phylum *Chordata* in general and the subphylum of vertebrates in particular. The earliest history of the chordates is only very sketchily understood; the remains of early, ancestral forms of this group made poor candidates for the fossilization process because they lacked hard body parts. A line of hypothesized evolution, therefore, has been drawn through surviving marine animals called protochordates, which presently are sessile or stationary for most of their life cycles, although before attaining this current form, they were capable of locomotion. These curious animals are considered invertebrate chordates as they possess notochords, or flexible skeletal rods that run up the long axes of their bodies. Among this group are such animals as the amphioxus and the so-called sea squirts or tunicates.

Further evolved along the path that eventually led to the present diversity of endoskeleton-possessing vertebrates are animals possessing bone matter, such as the agnathan (jawless) fishes. Later fishes evolved jaws and eventually true teeth and progressed from having cartilaginous skeletons (the class *Chondrichthyes*) to having true, bony endoskeletons (the class *Osteichthyes*). These more advanced fishes eventually gave rise to the land-pioneering class *Amphibia* and ultimately engendered the vertebrate classes of *Reptilia* (reptiles), *Aves* (birds), and *Mammalia*.

The reason that bone tissue evolved at all in the lower vertebrates is a subject that is still open to debate. Several rival theories exist; one holds that bone evolved simply as a more improved, harder material for exoskeletons superior to such material as calcium carbonate, the most common building material for invertebrate exoskeletons. Another suggests that bone evolved as a phosphate reserve as one component for energy storage and transfer for metabolic processes within the bodies of ancestral vertebrates. Still another theory postulates that bone materials such as dentine and enamel evolved originally simply as effective insulation for the electrosensory organs found in primitive, marine vertebrates. Other theories integrate versions of the theories above in complex, interactive arrangements.

ENDOSKELETAL RESEARCH
Histological research of endoskeletal tissues has in the past been the most productive approach to obtaining the large body of data on bone tissues and processes that currently exists. The majority of the data accumulated with this approach has been gained in the laboratory and has involved specialized equipment and the use of techniques tailored to produce useful information on bone cells, their composition, related tissues, and the various organic processes involved. These techniques and laboratory tools and appliances were developed laboriously over the centuries. Real progress in the field had to wait until the advent of the simple microscope. Believed to be invented by the Dutch scientist Antoni van Leeuwenhoek or one of his contemporaries in the seventeenth century, the simple optical microscope—utilizing only one lens—allowed the first close-up look at living structures at or near the cellular level.

Examination of Leeuwenhoek's original equipment has revealed that he was able to obtain the respectable magnification of as much as 250 times. His lenses thus allowed humans an entry into a world that had hitherto been barred to them: the world of the very small. Subsequent development in microscopes produced compound optical microscopes with several lenses working in series and an eventual exponential increase in magnifying power. Biologists quickly recognized that the new tool underscored the relation between structure and function of organic materials at the microscopic level. It is this critical concept that has been the key to unlocking the many secrets of the organic microstructures of living things, among them endoskeletons.

Hand in hand with the development of research using optical microscopes has been the preparation of histological sections. These are extremely thin, transparent shavings of organic tissues prepared in such a way as to facilitate microscopic examination. With the increased sophistication of the use of microscopy has come the perfected use of many different types of staining and dying. The use of stains and dyes has been selectively employed to highlight the different types of tissue being observed. A further refinement in microscopy has been the use of various lens filters, such as polarizing filters, that have added control of the light target to emphasize or de-emphasize various features.

LATE TWENTIETH CENTURY ADVANCES
After the advent of applied nuclear physics during World War II, a new technique called autoradiography appeared; this enhanced the resources available to histological research. Autoradiography involves the introduction of radioactive substances into animals; consequently, these substances are incorporated into various tissue components. The great advantage to this technique is that it can provide direct information on how long it takes for the various tissue components to be synthesized and on how long they last.

As the spectrum of isotopic labels expanded and became refined, it became possible to label and study in great detail almost every common tissue component found in animals. The study of the most intricate or delicate endoskeletal tissues thus became realistic along with the added advantage of being able to determine the durations of the metabolic processes involved in development, decay, and replacement.

A solution to the magnification limitation of the light microscope was reached when the first electron microscope (EM) was built in 1931. Further improvements were made until the 1950s saw the widespread use of more technologically advanced devices called scanning electron microscopes (SEMs), which allow observation and SEM photography, termed electron micrography, of target objects considerably less than 1 micron in diameter.

Still newer technologies, such as the use of fiber-optic probes inserted into the living bodies of animals and humans, allow benign observation of tissues in their natural state in the midst of normal processes. Fiber optics involves the transmission of light (and therefore images) through very fine, flexible glass rods by internal reflection. Fiber-optic instruments called fiberscopes allow the viewing of extremely small, and normally dark, internal structures such as skeletal tissues.

—*Frederick M. Surowiec*

FURTHER READING

Alexander, R. McNeill. *Bones: The Unity of Form and Function.* Foreword by Mark A. Norell. Macmillan, 1994.

Bilezikian, John P., Lawrence G. Raisz, and Gideon A. Rodan, eds. *Principles of Bone Biology.* Academic Press, 1996.

Carroll, Robert L. *Vertebrate Paleontology and Evolution.* W. H. Freeman, 1988.

Ham, Arthur W. *Ham's Histology.* 9th ed. J. B. Lippincott, 1987.

Hickman, Cleveland P., and Cleveland P. Hickman, Jr. *Biology of Animals.* 7th ed. McGraw-Hill, 1998.

Kardong, Kenneth V. *Vertebrates: Comparative Anatomy, Function, Evolution.* McGraw-Hill, 1998.

Kent, George C. *Comparative Anatomy of the Vertebrates.* 8th ed. McGraw-Hill, 1997.

Savage, R. J. G., and M. R. Long. *Mammal Evolution: An Illustrated Guide.* Facts on File, 1986.

ESTIVATION

FIELDS OF STUDY

Biology; Ecology; Environmental Studies; Endocrinology

ABSTRACT

During true estivation, metabolic processes including oxygen consumption, respiration, heart rate, and neurological activity decrease substantially, but the term is often used loosely to describe the activity of any animal that spends part of the warmer, drier season in a torpid state. Estivation and hibernation appear similar, although they differ physiologically. Animals most likely to estivate are mollusks, arthropods, fish, amphibians, and reptiles, although some small desert mammals estivate as well.

KEY CONCEPTS

epiphragm: covering or sealing membrane
hygroscopic: able to retain moisture
nematode: a long, cylindrical worm; some are parasitic
poikilotherm: cold-blooded or ectothermic; any organism having a body temperature that varies with its surroundings; in general, reptiles, amphibians, fish, and invertebrates
torpid: dormant, numb, sluggish in action

INVERTEBRATE ESTIVATION

A seemingly lifeless desert may be teeming with estivating life underground, waiting for seasonal rains that will awaken them to resume their life cycles. Snails, slugs, earthworms, insects, spiders, and nematodes, along with cocoons, eggs, grubs, larvae, and pupae, may all lie dormant in the soil, in building foundations and rock crevices, or under rotting logs or other vegetation. The animals are not just dormant; some of them will also be in an arrested state of sexual development called diapause. Some tropical snails can estivate for years at a time. To prepare, a snail digs a deep burrow in moist ground or under rocks. Next, it forms an epiphragm (a sealing membrane for the shell) to prevent evaporation and desiccation. Finally, its metabolism slows dramatically until it detects favorable environmental cues such as increased moisture.

LUNGFISH ESTIVATION

Among the fishes, the process of estivation is best known in the air-breathing lungfishes of Africa (*Protopterus*) and South America (*Lepidosiren*). Adult fish have paired lungs and vestigial gills; in fact, they will drown if held underwater. Both abilities—to estivate and to breathe air—have contributed to their survival in areas that experience severe seasonal droughts.

Photo of an estivating earthworm found in Norwich, Vermont, USA. The earthworm had been nestled in the nook of a rock underground. The photographer found it while digging a hole. The earthworm is shown held in a hand.

When the rivers, lakes, and marshes they inhabit dry up, lungfish dig burrows deep in the mud, leaving air passages to the surface. They curl up inside and wait out the arid conditions until rain fills up their burrows with water. *Protopterus* secretes a mucous coating that hardens, forming a tough, cocoonlike hygroscopic chamber with only one opening, connected to its mouth. When the rains come and flood the passage, the dormant fish awakens with a cough. *Lepidosiren* burrows more deeply than *Protopterus* and plugs the entrance to its air tube with perforated clay. Its burrow is somewhat larger and usually contains some water. It coats the walls with a jellylike substance to maintain moisture. During estivation, lungfish are so torpid that they make easy prey for local fisherman, who spear them in their burrows.

During estivation, energy required for reduced metabolism is provided by the breakdown of tissue protein. The waste product is urea, which is excreted in large amounts once the fish is again submerged in water. Lack of oxygen can become a problem for lungfish sealed in their burrows, but the aerobic metabolism of an estivating *Protopterus* is only 20 percent of its normal resting metabolism. Similarly, *Lepidosiren* and the swamp eel (*Synbranchus*) can survive a lack of oxygen for long periods during estivation.

ESTIVATION IN REPTILES AND AMPHIBIANS

During the summer, many reptiles and amphibians estivate. Some frogs and toads insulate themselves in cocoons composed of many layers of unshed skin. The eastern spadefoot toad digs backward into sandy soil with its spade-shaped hind feet to estivate for periods of weeks; experiments indicate that the sound of rain falling rather than moisture itself may trigger the toad's arousal. The mud turtle *Kinosternon* abandons its drying pond for a burrow where it estivates up to three months. Desert tortoises of the North American Southwest spend much of their summers estivating in burrows but emerge to drink and browse after infrequent thunderstorms. In the same region, the lizard called the chuckwalla stops eating and estivates in rock crevices, emerging every third day for about an hour at sunset.

ESTIVATION IN MAMMALS AND BIRDS

Among the few estivating mammals are small desert rodents, lemurs, and hedgehogs. Some ground squirrels remain dormant in their burrows from late summer, merging estivation with hibernation. They reappear in early spring to take advantage of new growth after winter rains. The round-tailed ground squirrel of southern Arizona (*Citellus tereticaudus*) avoids the hot, dry autumn by disappearing in August and September. Generally, the period of estivation coincides with the period most prone to scarcity of vegetation.

While a few birds may hibernate, it is unlikely that they estivate. They can migrate to more attractive regions, or, like many reptiles and most mammals, they cope with high temperatures by confining their activities to cooler parts of the day.

—*Sue Tarjan*

FURTHER READING

Badger, David. *Frogs.* Voyageur Press, 1995.
Cloudsley-Thompson, J. L. *Ecophysiology of Desert Arthropods and Reptiles.* Springer-Verlag, 1991.
Heatwole, Harold. *Energetics of Desert Invertebrates.* Springer-Verlag, 1996.
Moyle, Peter B., and Joseph J. Cech, Jr. *Fishes: An Introduction to Ichthyology.* 4th ed. Prentice Hall, 2000.
Schmidt-Nielsen, Knut. *Desert Animals: Physiological Problems of Heat and Water.* Reprint. Dover, 1979.
Warburg, Michael R. *Ecophysiology of Amphibians Inhabiting Xeric Environments.* Springer-Verlag, 1997.

Estrus

Fields of Study

Reproductive Biology; Endocrinology; Veterinary Science

Abstract

The physiological and behavioral events that make up estrus are complex and intertwined. For successful mating to occur, it is usually necessary that male and female cooperate in the act. Estrus ensures that both sexes are impelled to do so at the same time, when the female's body is ready for a pregnancy. External estrus signals vary greatly by species. In every case, they are largely produced by hormonal influences, and are keyed to evoke specific sexual response from the male of the species.

Key Concepts

copulation: mating; the insertion of the male's penis into the female's vagina to fertilize her ova

estrogens: a group of female sex hormones which regulate the estrous cycle

estrus cycle: the cycle of females' bodily changes related to reproductive potential

ova: eggs released from the females' ovaries at the height of estrus

progesterone: a female sex hormone produced chiefly in the latter half of the cycle

THE CALL TO PREGNANCY

The physiological and behavioral events that make up estrus are complex and intertwined. For successful mating to occur, it is usually necessary that male and female cooperate in the act. Estrus ensures that both sexes are impelled to do so at the same time, when the female's body is ready for a pregnancy.

During the days prior to estrus, ova in follicles ripen rapidly within the ovary. Estrogens, produced by cells in the follicle walls, increase rapidly in this phase, causing not only follicular growth but thickening of the vagina walls and the various signs of "coming into heat," and then of estrus itself. In most species, the ova spontaneously burst from their follicles at the peak of estrus and enter the oviduct, which leads to the uterus, ready to join with sperm. Females of several species, however, including the rabbit and the cat, require the added stimulus of mating for release of their ova. Once the ova are released, the ruptured follicles begin secreting progesterone in place of estrogen; this hormone prepares the uterus to support a pregnancy. This internal process is regulated through feedback connections between the ovaries and the hypothalamus in the brain. The hormonal effects that make females eager for mating also originate largely in the hypothalamus, sensitizing the nervous system to respond strongly to the presence of and stimulation by their male counterparts.

VARIATIONS BETWEEN AND WITHIN SPECIES

External estrus signals vary greatly by species. In every case, they are largely produced by hormonal influences, and are keyed to evoke specific sexual response from the male of the species. Such signals may be visual, behavioral, auditory, or scent-based, or may possibly draw on senses imperceptible to human observers.

Visual signals are often found among social mammals. These have been most observed in primates. Chimpanzee and bonobo females show a spectacular pink swelling of their genitals during estrus, which recurs every five weeks if pregnancy does not occur. Certain monkeys develop an estrus flush on their faces as well as their buttocks.

Scent and sound signals can alert males who are not initially within sight range of the receptive female. As such, they are effective for species who do not live in groups, although these signals are not confined to such species. Urine markings are among the most common scent signals. A female black rhino in estrus leaves a long trail of scent posts, which the male follows. Other scent signals are partly airborne, like those of domestic cats and dogs. Mating calls are given by many females, such as the female gibbon's ascending call, which is then answered by the male.

Behavioral changes during estrus are almost universal, as both sexes concentrate on the quest for one or several partners. The initial stages, which may suggest female coyness to an observer, are part of the courtship process. For example, the female cheetah leads several males on a headlong run across the plain, finally selecting one with whom to mate. Pet owners notice a restlessness in their dogs and cats; female cats in heat are especially likely to roam. The penultimate female signal in many species is lordosis, an arched-back posture that allows the male to mount. Mammalian species are normally either monestrous—having a single estrus period a year—or polyestrous—with several estrus periods recurring annually. The latter situation is more common.

Environmental changes can cause variations in these patterns. Such factors include climate shifts, changes in available light, nutrition, and the presence of a new male. In monestrous mammals, the estrus period usually falls at a time when resulting births will take place in the optimal season for young animals to thrive and grow. Domestication may disrupt this pattern. For instance, a female dog typically comes into heat twice a year with no regard to seasons. Her ancestor the wolf has one estrus period a year, sometime between January and April, depending upon latitude.

The length of estrus also varies by species: It lasts four to nine days in the mare, and only fourteen hours in the rat. In a few species, such as the ferret, estrus lasts for several weeks if copulation has not occurred. There are also infrequent phenomena such as "silent heats," in which the animal ovulates but the usual external signs are absent.

ESTRUS, EVOLUTION, AND ETHOLOGY

Sexual reproduction provides genetic diversity and adaptability to a species. However, in order to proceed successfully, it requires a fairly complex series of events. Estrus is central in this process. It communicates to males a female's readiness at the same time that it impels that female to mate and primes her body for pregnancy. Nature plays many variations on this basic theme, each connecting the species' estrous cycle to its environment and its whole life cycle.

Because the survival of young mammals requires maternal care, and often that of the father and/or other adult animals as well, estrus patterns are significant for social structure. In some species, estrus may help secure a pair bond, a polygynous herd, or

Gazella thomsonii, male gazelle checking female's receptivity to mating. Taken on safari in Tanzania.

bonds within a troupe or pack. It can also cause disruption, as when males battle for access to estrous females, or a new male takes over a harem and kills the young, hastening their mothers' next heat.

Humans have known of estrus ever since they first domesticated animals. Charles Darwin's *On the Origin of Species* (1859) credits estrous females' mate choices with being an evolutionary mechanism, giving many examples. Ethologists' field studies have recently described many details about species-specific estrus behavior, as well as its role in life cycles and social behavior. However, direct estrus observations are hard to make and sometimes hard to interpret, especially in wildlife.

—*Emily Alward*

FURTHER READING

Gentry, Roger L. *Behavior and Ecology of the Northern Fur Seal.* Princeton University Press, 1998.

Hrdy, Sarah Blaffer. "Why Women Lost Estrus." *Science '83* 4 (October 1983): 72–77.

Kevles, Bettyann. *Females of the Species.* Harvard University Press, 1986.

Small, M. F., and F. B. M. de Wall. "What's Love Got to Do with It?" *Discover* 13, no. 6 (June 1992): 46–51.

Young, J. Z. *The Life of Mammals: Their Anatomy and Physiology.* 2nd ed. Clarendon Press, 1975.

Evolution: Animal Life

Fields of Study

Evolutionary Biology; Paleontology; Animal Anatomy; Animal Physiology

Abstract

Current consensus holds that the so-called Big Bang occurred some fifteen billion years ago, with the sun and Earth formed about four and a half billion years ago. Subsequently, as chemical processes are assumed to have created oxygen and an organic "soup," microbial life developed out of this primordial ooze. Bacteria were the only living animal organisms for some 5.5 billion years. Eventually, millions of animals came into being. They were multicelled and were equipped with specialized body parts.

Key Concepts

gene: the basic unit of heredity
gene flow: the movement of genes from one population to another
genetic drift: change in gene frequencies in a population owing to chance
interbreeding: the mating of closely related individuals, which tends to increase the appearance of recessive genes
migration: the movement of individuals, resulting in gene flow, changing the proportions of genotypes in a population
mutation: alteration in the physical structure of the DNA, resulting in a genetic change that can be inherited
natural selection: the process of differential reproduction in which some phenotypes are better suited to life in the existing environment and thus are more likely to survive
speciation: the formation of new species as a result of geographic, physiological, anatomical, or behavioral factors
species: the basic category of biological classification representing a group of potentially or actually interbreeding natural populations which are reproductively isolated from other such groups
taxon (pl. taxa): group of related organisms at one of several levels such as the family *Canidae*, the genus *Canis*, or the species *Canis lupus*

IN THE BEGINNING

Current consensus holds that the so-called Big Bang—the high-temperature, high-density event that marked the beginning of the universe—occurred some fifteen billion years ago, with the sun and Earth formed about four and a half billion years ago. Four billion years ago, the relatively newly created sun shone with only 70 percent of its current strength. The atmosphere had no free oxygen. No bacteria, no viruses, no plants, and no animals were in existence. Subsequently, as chemical processes are assumed to have created oxygen and an organic "soup," microbial life in the form of the simplest cells without a nucleus, prokaryotes, developed out of this primordial ooze. These bacteria were the only living organisms for about two billion years. After that time, about 1.5 billion years ago, more complex cells with nuclei, eukaryotes, appeared. Thus, in all, for some 5.5 billion years, bacteria were the only existing animal organisms.

THE BEGINNINGS OF ANIMAL LIFE

Eventually, the two kingdoms, the botanical and the zoological, started to diverge and millions of animals came into being. They were multicelled and were equipped with specialized body parts. They were distinguished from plants primarily but not exclusively by their methods of feeding, locomotion, and reproduction. Most of the phyla seem to have appeared during the Precambrian period, an immense span of geological time that ended about 590 million years ago. Few fossils have remained from these prehistoric times, but the most explosive period for the development of life was the Cambrian, some 590 to 505 million years ago. In a relatively short span of ten million years, all the animal phyla currently known came into being, perhaps encouraged by an increase in oxygen in the seas, where animal life began.

Eventually, animals developed a nervous system enabling them to control their movements more appropriately, as well as sense organs to help them find suitable food. At the margins, however, the dichotomy between the botanical and the animal world has remained ambiguous, since there are

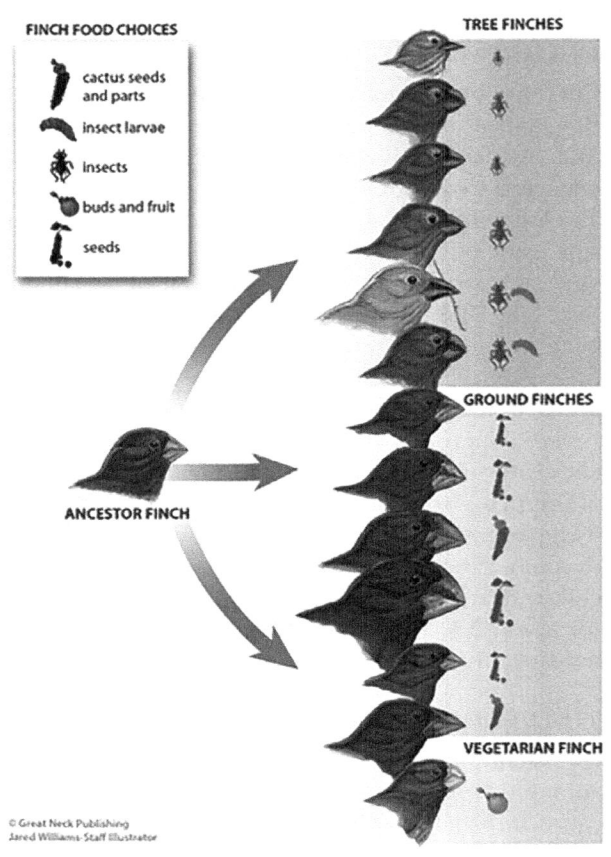

Darwin's Theory of Evolution: Darwin believed that finches' beaks evolved enabling them to survive in different ecosystems on the Galapagos Islands.

many microorganisms that defy clearcut classification. At times, these difficult cases are known as *Protista*, or Protists.

In all forms of life, including the animal kingdom, no phylum has been produced by a single evolutionary event. Nor have different animal orders appeared as a result of sudden evolutionary changes. Rather, all have come about, whether in gradual or punctuated manner, by the cumulative effect of small steps in different directions. This, at least, is the theory posited by Charles Darwin's explanation of evolution.

The time line of the animal kingdom is closely connected with this evolutionary chronology. Darwin's theory of the origin of species through natural selection, nearly universally accepted in the scientific community but at times opposed on Biblical creationist grounds and in some circles of the lay community, has it that animals, like plants, have changed since the beginning of life on earth and are still evolving today. In this view, there was no sudden creation of all species. Instead, over long periods of time, new species have evolved from isolated populations of existing species. These came to occupy new niches separate from the niches of the original species. Thus, all current species are changed descendants of others that existed previously. If there are fewer apparent links between phyla than between families further down the classification ladder, the reason is that phyla have had a longer history and so have experienced more opportunities for the elimination of intermediate forms. From an evolutionary viewpoint, the difference between species down an evolutionary line are even more recent than those between families, and so on. On average, it takes about 500,000 generations for one species to evolve into another. For species to have survived in their environments—with simultaneous changes in ecology, climate, and flora—many animal forms are now more complex and efficient than their ancestors used to be.

GENES AND EVOLUTION

Life began in the seas, so for animals to live in freshwater, let alone on dry land or in the air, many obstacles had to be overcome. The impetus to conquer these inhospitable realms came from competition, according to Darwin's theory of the survival of the fittest. Those fauna that managed to surmount the problems were the ones that underwent waves of adaptive measures and evolved into new kinds of animals.

It was only in the twentieth century that it was discovered that the characteristics of a species are passed on from parents to offspring by genes. Genes provide cells for particular features, such as webbed feet. With such characteristics inherited by offspring from parents, there is a resemblance among generations. However, at times a parent may produce quite a different offspring because of genetic change. The young, in turn, may replicate this difference in their own descendants in a process known as mutation, which may occur spontaneously for unknown reasons, but also may occur due to known causes, such as exposure to radiation. At times, mutations may be useful in allowing the species to adapt better to its environment; for instance, darker moths have a better chance of survival in a forest than lighter ones because the latter are more visible to their predators than the former. Other mutations may be harmful, such as larger size that slows down a species, making its flight from danger more difficult.

Whatever the case, within a period of 1 to 10 million years, animals may remain the same, may evolve, or may become completely extinct, as did most of the dinosaurs about 65 million years ago, after being dominant for some 350 million years.

ADAPTATION TO ENVIRONMENT

Natural selection leading to a new species may be accelerated when members of the original species move to a new environment, whether voluntarily or driven by the elements. Separated populations may develop different traits as they adapt to their new condition. Eventually, they will become sufficiently different to be unable to produce offspring with members of their original population. This process has repeated itself many times, over millions of years, and accounts for the large diversity in the animal world, not to mention the additional diversity consequent on artificial breeding by humans, widely observed among domesticated animals such as horses, cattle, and dogs.

Animals occupied new environments as species living in water moved to the land or later to the air. Thus, the step from fish to amphibian was essentially one from living in water for the whole life to living on land for the adult stage of the life cycle. The step from amphibian to reptile was one of increasingly proficient adaptation to land life at all stages of the life cycle.

Birds and mammals evolved in different directions from the reptiles, the first in adaptation to an arboreal and finally a flying life and the second as a further advance in the maintenance of an even and high body temperature—homeothermy—by combining an insulating external layer such as hair with a variety of physiological thermostats.

Events in geology, climate, and flora also determined the geographic distribution of species. Thus, marsupials are currently found almost entirely in Australasia and South America. The tiger exists only in India and Southeast Asia. The lion is restricted mainly to Africa. This pattern reflects the way in which these groups have evolved in relation to the physical world.

New animal groups evolve into many different forms, especially when they become dominant. For instance, when mammals came to occupy the dominant position, some became meat-eaters while others became vegetarians; some became smaller while others became larger; some became runners while others ended up as burrowers or flyers; still others returned to the water. This trend allowed the descendants of the original type to exploit a much greater range of environments and resources. Essentially, those species whose sense organs or brain morphology and functions improved the most ended up being dominant—primates in general and humans in particular.

TIME FRAME

The exact time of the origins of animals during earth's evolutionary history is not known because the early species were soft-bodied, at first single-celled and later multicellular life forms, that did not fossilize well. Fossils are the best material evidence of archaic times. Fossils do not appear earlier than 650 to 500 million years ago, not only because the animal life of the time was inappropriate for fossilization, but also because continued crustal shifts in the ensuing eons disturbed the very early rock formations. Accordingly, fossil evidence is unavailable for the entire early history of animals, which must consequently remain speculative. Current taxonomic interrelationships suggest the early history, and taxonomic diagrams may be regarded as presumptive evolutionary diagrams as well. However, a ball of carbon discovered in a cavity etched in a rock some 3.86 billion years old suggests that some life on earth was already possible at that time.

Knowledge is also limited by the fact that, even though over a million different species of animals have been identified, it is suspected that a similar number remain to be discovered or became extinct before such identification could be made. In the United States alone, some forty species of birds, about thirty-five species of mammals, and twenty-five other species have become extinct in the last two hundred years alone—less than a blip on earth's time scale—as a result of human activities such as the destruction of animal habitats through urbanization, the clearing of land for agricultural purposes, pollution, the introduction of new species from other parts of the world which turned out to be predatory to domestic specimens, hunting, and especially human population growth. It is widely predicted that climatic change triggered by greenhouse gases will continue, even enhance, this process, thereby endangering more animal species. Whatever the future, however, evolutionary biologists estimate that some 99 percent of all species that have ever lived on earth are now extinct.

EVOLUTION OF EXISTING AND EXTINCT HUMAN SPECIES AND AUSTRALOPITHECINES

The root of the hominid evolutionary tree is still imperfectly known. The earliest australopithecine species, *Australopithecus anamensis*, is believed to be over four million years old, by which time that branch had diverged from African ape-like ancestors. This species was followed by the *Australopithecus afarensis* nearly 3.5 million years ago. Much later came *Homo habilis*, called "skillful man" since they could presumably produce primitive tools, some two million years ago. They were followed by *Homo erectus*, "upright man," about one million years ago. Finally, *Homo sapiens*, "knowing man," emerged about 200,000 years ago. In the meantime, the australopithecine branch, after evolving through a number of intermediate species such as *A. africanus*, *A. aethiopicus*, and *A. robustus*, died out about one million years ago. To date, the earliest unearthed human-like primate fossil, that of Lucy, a three-foot-tall female discovered in Ethiopia, is about 3.2 million years old. The earliest fossil of an upright walker was discovered in the same area of Ethiopia in 2015 and is approximately 3 million years old. The oldest known fossils ever discovered, however, are those of bacteria and are in fact microscopic fossils of cells and bacteria "living" together in what was then an oxygen-free world over 3.4 billion years ago.

Modern humans are believed to have radiated out of Africa into Asia and Europe. Subsequently, cultural evolution became more prominent than biological evolution, but as modern humans evolved over the last four million years to their current condition, they developed manipulative skills, bipedalism, a change from specialized to omnivorous feeding habits, and especially, a threefold increase in cranial capacity from *H. afarensis* to *H. neanderthalensis*, together with behavior appropriate to the control of the environment.

Although humans are not the only animals capable of conceptual thought, they have refined and extended that ability until it has become their hallmark. Thus, thanks to the symbolic language of *Homo sapiens*, modern humans make possible the accumulation of experience from one generation to the next. Such cultural evolution is possessed by few, if any other animal species. It is for this reason that humans, more than other animals, have found ways to mold and change their environment according to need rather than in response to environmental demands. Because of this ability and humans' control of technology, the species has more say about their biological future than any other.

—*Peter B. Heller*

FURTHER READING

Carroll, Sean B., Jennifer K. Grenier, and Scott D. Weatherbee. *From DNA to Diversity: Molecular Genetics and the Evolution of Animal Design.* Blackwell Science, 2001.

Conway-Morris, Simon. *The Crucible of Creation: The Burgess Shale and the Rise of Animals.* Getty Center, 1999.

Ghosh, Pallab. "'First Human' Discovered in Ethiopia." *BBD.* BBC, 4 Mar. 2015. Web. 31 Dec. 2015.

Lavers, Chris. *Why Elephants Have Big Ears: Understanding Patterns of Life on Earth.* St. Martin's, 2001.

Long, John A. *The Rise of Fishes: Five Hundred Million Years of Evolution.* Johns Hopkins University Press, 1996.

"Oldest Fossils on Earth Discovered." *Science Daily.* Science Daily, 22 Aug. 2011. Web. 31 Dec. 2015.

Evolutionary Origin of Sex Differences

Fields of Study

Reproductive Biology; Ethology; Ecology; Genetics

Abstract

Those forms of reproduction in which genes are not exchanged are considered asexual. Because asexual reproduction allows numerous offspring to be produced in a short time, it is favored in situations in which a species can gain an advantage by exploiting an abundant but temporary resource. Sexual reproduction may take many forms, but all of them involve the exchange of genes. The widespread occurrence of sex, and of numerous sexual systems, shows that there must be some advantage to all the various forms of sexual reproduction, and that this advantage is sufficient to overcome the recognized advantages of asexual reproduction.

Key Concepts

anisogamy: reproduction using gametes unequal in size or motility

asexual reproduction: reproduction in which genes are not exchanged

female: an organism that produces the larger of two different types of gametes

gonochorism: sexual reproduction in which each individual is either male or female, but never both

hermaphroditism: sexual reproduction in which both male and female reproductive organs are present in the same individual, either at the same time or at different times

isogamy: reproduction in which all gametes are equal in size and motility

male: an organism that produces the smaller of two different types of gametes

parthenogenesis: asexual reproduction from unfertilized gametes, producing female offspring only

sexual dimorphism: differences in morphology between males and females

sexual reproduction: reproduction in which genes are exchanged between individuals

sexual selection: selection for reproductive success brought about by the behavioral responses of the opposite sex

ASEXUAL REPRODUCTION

The evolutionary origin of sex differences can be understood only by examining the relative benefits of sexual as compared to asexual reproduction. Those forms of reproduction in which genes are not exchanged are considered asexual. Asexual reproduction may take place from already developed body parts (vegetative reproduction) or from special reproductive tissue. In either case, however, asexual reproduction results in the rapid production of numerous individuals genetically identical to their parents. Because asexual reproduction allows numerous offspring to be produced in a short time, it is favored in situations in which a species can gain an advantage by exploiting an abundant but temporary resource, such as a newly discovered cache of food. There is also a further advantage: The individual that finds a resource that it can effectively exploit, if it can reproduce asexually, is assured that all its offspring will possess the same genotype as itself, and will thus be equally able to exploit the same resource for as long as it lasts. Despite these advantages, asexual reproduction is much less common than sexual reproduction among animals. It is a temporary stage in many species, alternating with sexual reproduction. Asexual reproduction is far more common among microorganisms such as bacteria.

FORMS OF SEXUAL REPRODUCTION

Sexual reproduction may take many forms, but all of them involve the exchange of genes. Some algae and protozoans exchange chromosomes without gametes in a process called conjugation. Most other forms of sexual reproduction use special sex cells called gametes, which exist in different "mating types." Two gametes can combine only if their mating types are different. Some simple organisms, such as the one-celled green alga *Chlamydomonas*, have gametes that are indistinguishable in size or appearance, a condition known as isogamy. Most other organisms have gametes of unequal sizes, a condition called anisogamy. Selection often intensifies the differences between gametes, producing a small, motile sperm and a much larger, immobile egg, laden with stored food (yolk).

Some sexually reproducing organisms have separate sexes, a condition called gonochorism. Individuals producing eggs are called female, while individuals producing sperm are called male. Since sperm are generally small and can be produced in great numbers, males tend to leave more offspring if they reproduce prolifically, indiscriminately, and often. Females, on the other hand, have fewer eggs to offer, and in many species they must also invest nutritional and behavioral energy in the laying of eggs and the care of the resultant offspring. Selection in these species favors females who choose their mates more carefully and take better care of their offspring.

The differing selective forces operating on the two sexes often give rise to sexual dimorphism, or differences in morphology between the sexes. Sexual dimorphism can also be reinforced by competition for reproductive success, a phenomenon first studied by Charles Darwin. Darwin called this type of competition sexual selection. It takes two basic forms—direct competition between members of the same sex, and mate choices made by members of the opposite sex.

Direct male-male competition often takes such spectacular forms as rams or stags fighting in head-to-head combat. Similar fights also occur in many other species, including a variety of turtles, birds, mammals, fishes and invertebrates. Many more species, however, engage in ritual fighting in which gestures and displays substitute for actual combat. Male baboons, for example, threaten each other in a variety of ways, including staring at each other, slapping the ground, jerking the head, or simply walking toward a rival.

Although male-male rivalry has attracted more attention in the past, female-female competition also occurs in many species. Now that more ethologists and sociobiologists are looking for evidence of such direct competition among females, it is being discovered that it is a fairly widespread occurrence which had previously escaped notice only because so few scientists suspected its existence or were interested in looking for it. Female-female competition has been found among langur monkeys, golden lion-marmosets, ichneumon wasps, and several other species.

SEXUAL SELECTION

Sexual selection in mating is selection in which reproductive success is determined at least in part by mate choice. No matter what form sexual selection may take, it results in greater reproductive success for those individuals chosen as mates, while those not chosen must try again and again if they are ever to succeed in leaving any offspring at all.

Sexual selection of this kind occurs in nearly all gonochoristic species. In some species, males will attract females by means of a visual display or by various sounds (also called calls or vocalizations). Females in such species will exercise choice by selecting among the available males. For example, male peacocks, lyre-birds, and birds of paradise will court females by showing off their elaborate tail feathers in bright gaudy displays. In other species, the females perform the display and the males do the selecting.

Sometimes, the display will include an object such as a nest constructed by one partner as an attraction to its mate. Bowerbirds, for example, construct elaborate nuptial bowers as a means of attracting their mates. These bowers, which contain a nest in the center, are sometimes adorned with attractive stones, flowers, and other brightly colored objects. In some species of animals, males and females will respond to one another by performing alternating steps; in this manner, each sex selects members of the other.

Many sexually reproducing organisms have both male organs which produce sperm and female organs which produce eggs, a condition known as hermaphroditism. Earthworms and many snails are simultaneous hermaphrodites, meaning that both male and female organs are present at the same time. Hermaphrodites often have their parts so arranged that self-fertilization is difficult or impossible. One system that guarantees cross-fertilization is serial hermaphroditism. In this system, each individual develops the organs of one sex first, then changes into the opposite sex as it matures further.

Some sexually reproducing organisms have become secondarily asexual through a process called parthenogenesis, in which gametes (eggs) develop into new individuals without fertilization. In bees and wasps, males develop parthenogenetically from unfertilized eggs, while females (with twice the chromosome number) develop from fertilized eggs.

THE COST OF SEXUAL REPRODUCTION

Sexually reproducing organisms experience a cost associated with the energy devoted to courtship behavior and to the growing of sexual parts. In addition, the act of courtship usually exposes an individual to a greater risk of predation, and the distractions of mating further increase this risk. In view of these costs, many evolutionists have wondered how sex ever evolved in the first place, or why it is so widespread. Any adaptation so complex and so costly would long ago have disappeared if the organisms possessing it were at a selective disadvantage. The widespread occurrence of sex, and of numerous sexual systems, shows that there must be some advantage to all the various forms of sexual reproduction, and that this advantage is sufficient to overcome the recognized advantages of asexual reproduction in terms of rapid proliferation with relatively low investment of energy.

The answer to this puzzle is based on the fact that asexually produced offspring are all genetically similar to the parent, while sexually produced offspring differ considerably from one another. Organisms exploiting a dependable habitat or food supply often leave more offspring if they produce numerous genetically similar offspring rapidly and asexually. On the other hand, organisms facing uncertain future conditions have a better chance of leaving more offspring if they reproduce sexually and therefore produce a more varied assortment of offspring, at least some of which might have the adaptations needed to survive in the uncertain future. Examination of those species that are capable of reproducing either way confirms this hypothesis: Whenever favorable conditions are likely to persist, they reproduce rapidly and asexually. Faced with conditions of adversity or future uncertainty, however, these same species reproduce sexually. In species that alternate between sexually produced and asexually produced generations, the asexual phases typically occur during the seasons of assured abundance, while the sexual phases are more likely to occur at the onset of harsh or uncertain conditions. Sex, in other words, is a hedge against adversity and against an uncertain future.

STUDYING SEXUAL REPRODUCTION

Most biologists who study reproduction are either ecologists, ethologists, or geneticists. Their methods include counting various kinds of offspring and measuring their genetic variability. Reproductive ecologists and ethologists also measure parental investment, or the amount of energy used by individuals of each type (and each sex) in the courting of their mates, in the production of gametes, and in caring for their young. Energy costs of this kind are generally measured by comparing the food consumption of individuals engaged in various types of activity using statistical methods of comparison among large numbers of observations.

The morphology of sex organs in various species is also studied by comparative anatomists and by specialists on particular taxonomic groups such as entomologists (who study insects), helminthologists (who study worms), malacologists (who study snails and other mollusks), and ichthyologists (who study fishes). In most hermaphroditic species, for example, the organs are so arranged as to make cross-fertilization easier and self-fertilization more difficult.

The above explanation of sexual reproduction as resulting from the greater variability among offspring facing an uncertain future is partially confirmed by studying species that can reproduce either sexually or asexually. Among these species, asexual reproduction is always favored in situations in which an individual discovers a resource (such as a habitat or a food source) too large to exploit by itself. These conditions favor individuals that can reproduce rapidly and asexually produce numerous individuals genetically similar to themselves, who then proceed to exploit the resource. Aphids, for example, produce one or several asexual generations during the spring and early summer, when plant food is abundant. In seasons or situations of great risk or uncertainty, however, the same species often reproduce sexually at somewhat greater energetic cost, leaving a wider variety of offspring but a smaller total number. Under unpredictable conditions (such as those associated with wintering in a cold, temperate climate), the greater energetic costs of reproducing sexually are more than made up by the greater genetic and ecological variability among the offspring. Sexually reproducing individuals leave more offspring (on the average) than asexual individuals under these conditions. Similarly, among hermaphroditic species, cross-fertilization results in more varied offspring than self-fertilization, and is therefore favored under such conditions.

TESTING THEORIES

The several reproductive methods studied by biologists provide a natural laboratory for the testing of several theories. Among these are theories concerned with genetic variability, natural selection, the evolution of sex, and the allocation of resources, including the theory of parental investment in the care of their offspring.

In terms of the two most general types of reproductive strategies, those species using a system called the R strategy (reproducing prolifically at small body size) may be either sexual or asexual, or may alternate between these two methods of reproduction. On the other hand, species following the K strategy (reproducing in smaller numbers at larger body size and investing time and energy in parental care) are invariably sexually reproducing and most often gonochoristic as well.

In addition to the theoretical considerations mentioned above, the study of alternative methods of reproduction gives us important insights into the reasons that our species, like other K strategists, is sexually reproducing and gonochoristic. In most species, sexual behavior is largely controlled by instincts, but learned behavior plays a major role among higher primates. Beyond what is necessary in copulation and childbirth, much of sex-specific behavior in humans is culturally defined and may differ from one society to another. This includes the norms of what behavior is appropriate (or inappropriate) for each sex and what personal qualities are considered masculine or feminine. All attempts to redefine sex roles will lead nowhere, unless one is aware of both the biological and the social underpinnings of these roles.

—*Eli C. Minkoff*

FURTHER READING

Alcock, John. *Animal Behavior: An Evolutionary Approach.* 7th ed. Sinauer Associates, 2001.
Brown, J. L. *The Evolution of Behavior.* W. W. Norton, 1975.
Campbell, Bernard, ed. *Sexual Selection and the Descent of Man, 1871–1971.* Aldine, 1972.
Campbell, Neil A. *Biology: Concepts and Connections.* 3rd ed. Benjamin/Cummings, 2000.
Clutton-Brock T. H., ed. *Reproductive Success: Studies of Individual Variation in Contrasting Breeding Systems.* University of Chicago Press, 1988.
Daly, Martin, and Margo Wilson. *Sex, Evolution, and Behavior.* 2nd ed. Wadsworth, 1983.
McGill, T. E., D. A. Dewsbury, and B. D. Sachs. *Sex and Behavior: Status and Prospectus.* Plenum, 1978.
Maynard-Smith, John. *Evolution and the Theory of Games.* Cambridge University Press, 1982.
Rosenblatt, J. S., and B. R. Komisaruk, eds. *Reproductive Behavior and Evolution.* Plenum, 1977.

EXCRETORY SYSTEM

FIELDS OF STUDY

Nephrology; Animal Physiology; Biochemistry; Endocrinology

ABSTRACT

All cells must maintain a constant internal environment with regard to specific ion concentrations, protoplasm pH, osmotic pressure, water content, and the excretion of wastes. The kidney in vertebrates and the various excretory organs found in lower animals perform this function. An animal will die more quickly from a disturbance of the composition of the internal environment of the cells than it will from an accumulation of wastes. Fortunately, the kidney performs functions that keep both of these from happening.

KEY CONCEPTS

antidiuretic hormone (ADH): a hormone produced in the hypothalamus that controls reabsorption of water in the loop of Henle

contractile vacuole: the excretory organ of several one-celled organisms

filtration: the process of diffusion of plasma from the blood to the glomerulus and nephron

glomerulus (pl. glomeruli): a capsule fitting around capillary blood vessels that receives the filtrate from the blood and passes it into the tubule

loop of Henle: a slender hairpin turn in the tubule where most adjustment of the water balance of the body occurs

Malphigian tubule: the primitive excretory organ of insects

nephridia: the primitive forms of kidneys found in worms and lower organisms

nephron: the basic excretory unit of the kidney

tubule: the long, slender part of the nephron that is the location of almost all kidney function

urea: a substance formed from by-products of protein metabolism and excreted by the kidney

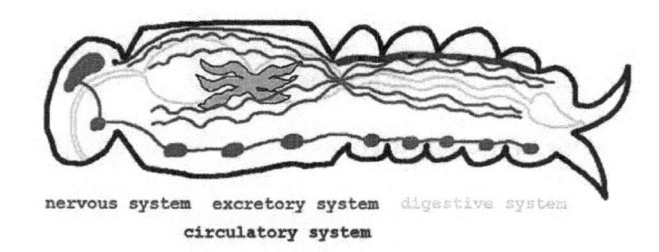

Diagram representing the digestive, circulatory, excretory and nervous systems of the grasshopper.

MAINTAINING THE CELL STATE

All cells, from the single-celled animals to the highly diversified cells of higher mammals, must maintain a constant internal environment with regard to the kinds and amounts of specific ions, the pH of the protoplasm, the osmotic pressure, the water content of the cells, and the excretion of wastes. The kidney in vertebrates and the varied types of excretory organs found in lower animals perform this function. Although the excretory function is very important, an animal will die more quickly from a disturbance of the composition of the internal environment of the cell than it will from an accumulation of wastes. Fortunately, the kidney performs functions that keep both of these from happening.

MAINTAINING FLUID HOMEOSTASIS

All cells live in a watery environment, which is maintained constant through the processes of homeostasis. Maintaining homeostasis presents different problems for different organisms. For a primate, living in the air, water loss is a constant problem, and water must be conserved. A freshwater fish, on the other hand, takes in large quantities of water that has a lower concentration of salts than its body fluids, and it must excrete the excess. An ocean fish, living in water with a much greater concentration of salts than its body fluids, must obtain water from its environment and still avoid increasing the concentration of salts in the body fluids.

A kidney is able to perform all these functions, although in each of these cases it acts in a different way. The freshwater fish must excrete large quantities of very dilute fluid in order to maintain salt conservation, whereas the saltwater fish must excrete a high concentration of salt in a very concentrated solution. The primate must be able to regulate the output of fluid as a function of water intake. Evolution has adapted the kidney of each animal to its environment.

In addition, animals survive by metabolizing foodstuffs to provide the energy for movement. One of the major metabolic processes is the breakdown of protein to produce energy for the synthesis of other proteins and the rebuilding of body structures. In the breakdown of protein, nitrogen is freed from the organic molecule and must be excreted. One of the resultant nitrogen products is ammonia, which is toxic. The ammonia is converted into a less toxic material, urea (or uric acid, in some animals), before it is excreted by the kidney. In addition, metabolism in the body usually results in the production of acids, particularly if fat is metabolized. The body functions well only within a narrow range of acidity, so unless the excess acid is removed, serious problems quickly arise. The kidney serves the function of maintaining the pH at a constant value.

THE EXCRETORY ORGANS

The organs of excretion have taken many forms. Simple single-celled organisms such as amoebas are able to form contractile vacuoles, or walled-off spaces within the cell, in which water can be stored and waste products deposited. These vacuoles are periodically transported through the cytoplasm and excreted through the external cell membrane. The size and number of vacuoles are determined by water intake and the organism's need to eliminate water as well as by the accumulation of waste materials.

As animals developed more cell types and the number of cells increased dramatically, the need to provide a constant internal environment around the cells arose. The excretory organs became, of necessity, more complex in nature. In addition to excreting waste products, they developed abilities to retain some ions and excrete others, to retain or excrete water, and to retain or excrete acids or bases to maintain a constant environment.

Many organisms have no obvious means of regulating water and salt balance, and they apparently accomplish this feat through the skin or the gut. Others have rudimentary organs of excretion. Many lower animals have nephridia, primitive versions of the kidney that excrete water and wastes and regulate ion concentration. These are simple tubes into which body fluids pass; the fluids are excreted after chemical alteration. In animals such as worms, that have segments, a pair of nephridia may be located in each segment. Some of them open into the body cavity, while others are closed. In some animals, these are well differentiated and are called flame cells. These tubular structures serve to regulate the internal environment of the body. For example, if the sodium concentration in the coelom, or internal cavity, is high, the nephridia excrete sodium; if it is low, they reabsorb it. In the insect, the organ of excretion is the Malphigian tubule, which is able to regulate ion and water exchange. The accumulated fluid is flushed into the gut, where absorption of ions and water takes place.

THE KIDNEY

In vertebrates, the kidney is the organ responsible for eliminating water and waste products. The kidney is a bean-shaped organ that receives a large blood supply from the heart. It has several auxiliary structures: the ureter, which collects fluid or urine from all the tubules; the bladder, which acts as a storage organ; and the urethra, which opens from the bladder to the outside of the body. The kidney consists of more than a million nephrons arranged symmetrically, with the lower part of each nephron pointing toward the hilus (the pole of the kidney where the ureter arises).

The kidney maintains homeostasis of the body through four basic mechanisms: filtration, reabsorption, secretion, and concentration. In filtration, a liquid portion of the blood is transferred to the tubule. There the cells proceed to reabsorb necessary materials, to secrete additional materials into the tubular liquid from the blood, and to concentrate the fluid in the tubule.

The nephron is the fundamental structure of the kidney. The nephron is a long, slender tube with different parts that are capable of secreting or reabsorbing ions, water, and other substances either to remove materials from the blood or to return materials to the blood, depending upon the needs of the body. One process it performs is the elimination of ammonia products and other waste materials.

The nephron consists of two major parts: a glomerulus and a tubule. The process of urine formation begins with filtration. The blood enters the kidney and then the glomeruli, under high pressure from the heart. The pressure forces fluid from the blood into the tubule. The amount is tremendous; in humans, every day some 180 liters (about 40 gallons of fluid) pass from the blood into the nephron. All the blood in the body passes through the nephron about thirty times per day. During this same period, seven hundred liters of blood pass through the kidneys, so only a small portion is actually filtered or transferred to the tubule. Most remarkable of all, only about 1.5 liters of urine are produced each day. The rest of this large volume is reabsorbed by the tubule.

As the fluid passes into the nephron, it enters a section of tubule at the beginning that reabsorbs much of the material needed by the body. Such things as the glucose needed for energy, amino acids for protein building, vitamins, and ions needed to maintain the correct concentration of the blood are removed and transported by the cells of the nephron back into the blood. At the same time, about 85 percent of the water of the filtrate is also transported back into the blood.

The cells of the nephron are able to secrete materials from the blood to the tubular fluid. This process is exactly the opposite of the reabsorption of substances. One of the most readily secreted substances is sodium. The cells of the body are high in potassium and low in sodium; since sodium is a constituent of every diet, the removal of excess sodium is necessary. Sodium is picked up from the blood that circulates around the nephron and is secreted into the tubule. The process of secretion also extends to other materials. The potent antibiotic penicillin was ineffective when it was first used to treat systemic infections, because it was rapidly secreted by the tubules. A high enough concentration could not be accumulated in the blood to destroy bacteria. It became useful only when a derivative that was not secreted could be found.

THE LOOP OF HENLE

The tubular fluid that has been adjusted in concentration, volume of fluid, and concentration of ions and other materials now passes into a hairpin-shaped portion of the tubule called the loop of Henle. The loop of Henle adjusts the volume of filtrate. A hormone is produced by the brain (in the hypothalamus) that is capable of altering the permeability of the cells of the loop of Henle to water. The substance, a protein hormone called antidiuretic hormone (ADH), causes the reabsorption of water from the loop. Cells of the hypothalamus respond to the concentration of particles in the blood (its osmotic pressure) and adjust the amount of water that is reabsorbed from the tubule by secreting more or less ADH as necessary to maintain a constant concentration in the blood. The range of adjustment is remarkable. The volume of urine produced can range from about 0.5 liter to more than 30 liters per day, depending upon the need. If water is administered or restricted, the water concentration of the body changes. This causes a change in the production of ADH, which in turn increases or decreases the excretion of water to return the level to normal.

Sweating also causes water loss and thus decreases urine flow. Intake of large amounts of fluid will dilute the body fluids and cause an increased urine output. There is a constant adjustment, because water is lost by breathing, through the skin, and through excretions, and the kidney must make the proper corrections. Losses have been reduced to a minimum in animals such as the kangaroo rat, which lives in the desert and must conserve water. All of its water intake is from seeds and other foods containing some water, and excretion is almost zero. The desert rat is able to concentrate urine to a level about five times that of the human.

As the urine passes from the loop of Henle, it enters upon a final adjustment in the latter portion of the tubule. Volume, concentration of material, and the like are adjusted to maintain homeostasis. Other alterations of the fluid are also made in the passage down the tubule. If the body becomes acidic, the cells of the tubule exchange sodium ions, which are neutral, for acid ions (H+), thus causing the body to lose acid. Conversely, if the body becomes basic (lacks acid or hydrogen ions), the reverse is true.

STUDYING KIDNEY FUNCTION

The kidney can be studied on many different levels. The output of the kidney, the urine, and the input, the blood, can be analyzed in order to determine function of both kidneys. In some animals, such as the frog, the individual nephrons are visible under a microscope. Through the use of the stop-flow technique, the behavior of an individual portion of the nephron can be studied. In this technique, a very small needle is introduced into one portion of the nephron, and a small drop of oil is injected. Another drop is injected in an adjacent section of the nephron, thus isolating a section between the two drops. The exchanges that occur in that portion of the nephron can then be measured by taking samples of the fluid at intervals.

The whole kidney can be studied by examination of the kidney in relation to a reference substance. For example, the chemical inulin goes through the nephron without any alteration, so its excretion can be compared with other substances: If more of another substance appears in the urine in proportion to the inulin, the substance must have been excreted from the blood into the urine and secreted. If less is present, some material must have been reabsorbed from the tubule into the blood. This is called the clearance technique, and it is widely used to predict kidney function in health and disease. For less sophisticated testing, substances—such as certain dyes—that can be taken orally and are excreted by the kidney can be used to measure the rate of excretion. Radio-opaque substances that will appear on X rays can also be used to detect overt kidney malfunction.

—*J. H. U. Brown*

FURTHER READING

Brenner, Barry M., ed. *Brenner and Rector's the Kidney*. 6th ed. W. B. Saunders, 1999.

Hainsworth, F. Reed. *Animal Physiology*. Addison-Wesley, 1981.

Smith, Homer. *From Fish to Philosopher*. Doubleday, 1961.

Starr, Cecie, and Ralph Taggart. *Biology: The Unity and Diversity of Life*. 9th ed. Brooks/Cole, 2001.

Valtin, Heinz. *Renal Function*. 3rd ed.: Little, Brown, 1995.

Exoskeletons

Fields of Study

Evolutionary Biology; Paleontology; Entomology; Ecology

Abstract

The development of such exoskeletons, in comparison with less solid structures, gave arthropods several distinct evolutionary advantages over other invertebrate phyla. As a consequence of two distinctive features—a hard, rigid exoskeleton and jointed appendages and other body parts—arthropods have become one of the most successful of all animal groups. It is by these two features that arthropods are taxonomically defined. Arthropod species account for more than three quarters of all known animal species.

Key Concepts

chitin: a cellulose-like, crystalline material that makes up 25 percent to 60 percent of the dry weight of the cuticle

cuticle: the outer arthropod exoskeleton consisting of several layers of secreted organic matter, primarily nonliving chitin

endocuticle: usually the thickest layer of the cuticle, found just outside the living epidermal cell layer and made of untanned proteins and chitin

epicuticle: the outermost and thinnest layer of the arthropod cuticle, composed mainly of the hardened protein cuticulin

epidermis: a living cellular layer that secretes the greater part of the cuticle and is responsible for dissolving and absorbing the cuticle during molting (also termed the hypodermis)

exocuticle: a thick middle layer in the cuticle made up of both chitin and rigid, tanned proteins termed sclerotin

integumentary processes: surface outgrowths from the cuticle, primarily rigid nonarticulated processes or movable articulated processes

sclerotin: a hard, horny protein constituent of the exocuticle found in arthropods such as insects; it is superficially similar to vertebrate horn or keratin

NATURAL BODY ARMOR

The evolution of invertebrate animals possessing rigid, hard exoskeletons represented a great advance for members of the phylum Arthropoda. The development of such exoskeletons—in comparison with less solid structures, such as the hydrostatic skeleton of coelenterates—gave arthropods several distinct evolutionary advantages over other invertebrate phyla. Hydrostatic skeletons, such as those possessed by sea anemones, operate by the animal's musculature being arranged in a pattern that surrounds an enclosed volume of fluid. Contraction of any one section of the muscular system creates a fluid pressure in the central cavity that is consequently transmitted in an omnidirectional manner to the rest of the body. Arthropodic exoskeletons, on the other hand, are consistently rigid and much harder because they are composed to some extent of crystalline substances. Flexibility of movement is attained by multiple jointings in the limb system and in other appendages in the body such as feeding and sensory apparatuses.

EXOSKELETONS AND ARTHROPOD SUCCESS

As a consequence of two distinctive features—a hard, rigid exoskeleton and jointed appendages and other body parts—arthropods have become one of the most successful of all animal groups; indeed, it is by these two features that they are taxonomically defined. Biologists sometimes term the enhancement of the annelid (worm) body plan of segmentation by arthropod improvement "arthropodization." Because of it, the ancestors of the present immense spectrum of arthropod species successfully adapted to myriad ecological niches in the sea, on land, and in the air. Arthropod species account for more than three quarters of all known animal species. In fact, the class Insecta, one of a number of classes within the arthropod phylum, numbers at least 700,000 known species, with new species being discovered yearly, mostly in the tropics.

The immense success of arthropods is, to a great extent, the result of the advantages provided by the composition and structure of the seemingly simple surface architecture that is the arthropod exoskeleton. This exoskeleton not only provides a substantial

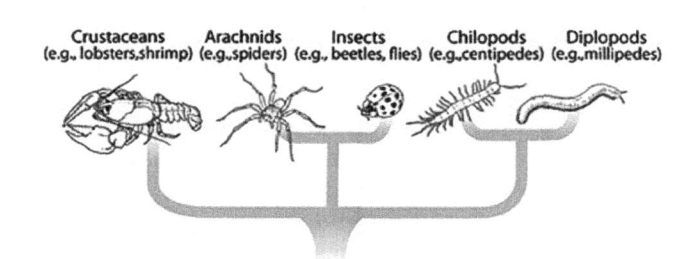

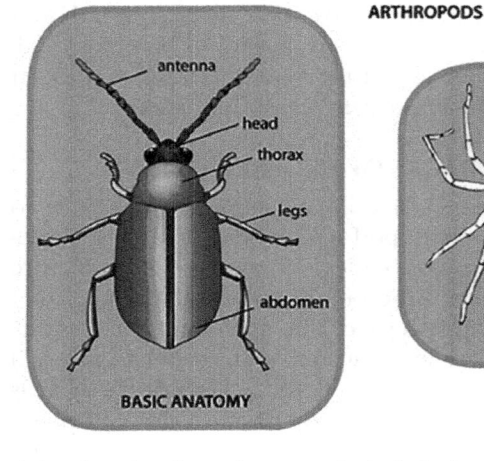

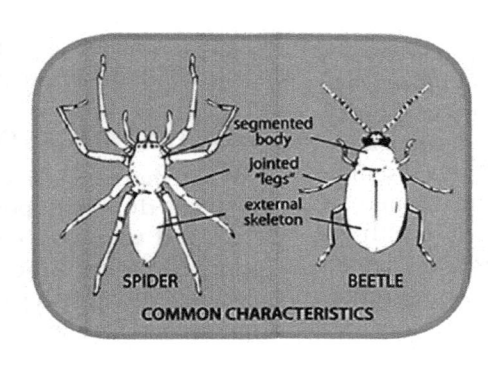

Arthropods: Arthropods are a large group of animals that have hard exterior skeletons and jointed appendages or limbs. examples are insects, lobsters, and spiders.

chemical and physical barrier between the animal and the external environment, protecting the internal organs and fluids, but also allows a degree of temperature and osmotic regulation. In addition, the exoskeleton helps deter predation, provides a solid base of attachment for an internal muscular system, and offers a good site for the location of various sense organs.

One of the most noteworthy evolutionary advantages of the exoskeleton is its service as a solid base of muscle attachment. The arthropod limbs act as a system of mechanical levers that is a much more efficient locomotive system than that of evolutionarily older and less sophisticated invertebrate locomotive systems such as that of the annelids. Because of exoskeletons and the structurally strong, jointed appendages that exoskeletons permit, arthropods possess an internal, muscular body wall broken down into separate muscles having an arrangement allowing contractions that are more localized in time and space than annelid or coelenterate muscle behavior. This more modular approach to the musculature allows arthropods to react to their environments to use energy more efficiently and with greater precision of movement and response. In fact, the inner surface of the exoskeleton acts as a limited type of endoskeleton, or inner skeleton, in that it provides good anchoring sites for muscle attachment, thus further increasing the leverage power of arthropod limbs and appendages.

THE PARTS OF THE EXOSKELETON

The exoskeleton itself can be divided into several distinct units based on function and composition. These are composed of consecutive layers that surround the animal in an arrangement similar to a medieval knight's suit of armor. Like the armor suit, the outermost exoskeleton is, in a typical arthropod, very rigid and hard; movement is possible only because both protective systems are composed of plates or body-contoured segments that incorporate narrow, flexible jointings allowing motion. The motion is usually narrowly defined in extent and direction, and it is this quality that gives both armored humans and many arthropods their often distinctively awkward and ungainly mode of movement. The larger terrestrial beetles and marine forms such as crabs and lobsters are ready examples of this. Some arthropods are nevertheless adroit and delicate in their movements, as shown by various arthropod aerialists such as the dragonflies and butterflies.

Insects can, in many ways, be considered typical arthropods and are therefore useful as models for a discussion of the exoskeleton as found among all arthropods. All the various layers of the exoskeleton, both living and nonliving, are as a whole variously termed the integument or cuticle. These layers are the skin or surface of the animal. The innermost layer

267

of the exoskeleton is termed the epidermis, or hypodermis, and is made of living cells. The epidermis is immediately external to a basement membrane that separates the epidermis from the inner body, with its organs and fluids. The epidermis is responsible for secreting the layers external to it, which are organic but actually nonliving material. From the epidermis outward, these layers in insects consist of the endocuticle, the exocuticle, and the epicuticle. (Some biologists use the term procuticle to describe the endocuticle and use epicuticle to mean both the epicuticle and the exocuticle.) Whatever precise terms may be employed, the general concept is that layers of material closer to the epidermis are more flexible and less chemically hardened than layers that are found closer to the actual exterior of the animal.

The endocuticle is usually the thickest cuticle layer and is constituted of protein mixed with a material called chitin. Chitin is a cellulose-like, crystalline material that makes up anywhere from 25 percent to 60 percent of the dry weight of the cuticle. It has many useful properties, such as resistance to concentrated alkalis and acids. In chemical composition, chitin is a nitrogenous polysaccharide. Chitin itself can be a relatively soft and flexible material that gains hardness in the outermost arthropod exoskeleton in several ways. One way is by using the presence of a material termed sclerotin. The process of hardening through the agency of sclerotin is called sclerotinization and involves a molecular change in the organization of the protein part of the cuticle. The outermost chitin found in the exocuticle of insects, for example, is thoroughly sclerotinized, which characteristically results in a darkening of the chitin. The other method by which chitin hardens is the deposition of calcium carbonate, primarily in the form of calcite. This is the process found among marine arthropods such as the Crustacea: crabs, lobsters, and shrimp, for example. This process, called calcification, occurs among Crustacea, starting in their epicuticle, or outermost exoskeletal layer, and works inward to the exocuticle and finally the endocuticle.

Besides the darkening caused by sclerotinization, coloration of the cuticle is effected in two basic ways. One is simple pigmentation caused by the presence of colored compounds found within the cuticle itself. The other is through the presence of extremely fine parallel ridges found on the epicuticle. These ridges break normal white light into its constituent wavelengths by prismatic diffraction in the same way that raindrops create rainbows. It is by this means that the effect of spectacularly iridescent rainbow hues found on many insects' wings and bodies is achieved.

Adding to the complexity of the cuticle are great numbers of sensory organs that project from or extend through the various exoskeletal layers. Prominent among these sensory structures are tactile hairs, bristles, and spines found all over the general body surface and on limb surfaces. These sensory structures, or setae, are movable and are set into thin, flexible disks on the cuticle surface itself. When one of these projections is moved, its base mechanically stimulates one or more sensory cells, setting off stimuli to which the arthropod can respond.

THE DRAWBACKS OF EXOSKELETONS

While the exoskeleton has evolved wonderfully to protect the arthropod and to enhance its locomotive and sensory abilities, its overall rigid structure presents some inherent drawbacks. Perhaps chief among these is the fact that its formidable rigidity and solidity are limitations to an individual's physical growth throughout its lifetime. Growth, in fact, is probably an arthropod's single most difficult physiological problem. This is true because once formed and hardened, an exoskeleton cannot be enlarged as the animal within enlarges with time. The physiologic solution among arthropods is the process termed ecdysis, or molting.

This process is intrinsically dangerous to the arthropod, as it leaves each individual extremely vulnerable to predation during, and immediately following, molting. It has been estimated that as much as 80 to 90 percent of arthropod mortality occurs during ecdysis. The process takes place in stages. Prior to the shedding of the old exoskeleton, a new, soft cuticle is formed beneath the old one. The new cuticle has not started along either the sclerotinization or calcification process and therefore is still soft and pliable. As the new cuticle is forming, the lower section of the old cuticle is partially dissolved by corrosive fluids secreted by cutaneous glands situated below the new cuticle. Immediately prior to the shedding, also termed casting, of the old exoskeletal cuticle, the arthropod stops feeding and absorbs more than the usual amount of water and oxygen. Its body begins to swell, and the animal makes spasmodic movements to shake off the old cuticle, most of the base of which has been removed by the corrosive process. Eventually, the old exoskeleton is effectively disconnected from the arthropod's body, and the animal extricates itself from the remains.

At this point, its new exoskeleton is soft and very pliable, and its movements are limited. Consequently, the individual is extremely vulnerable to both predation and serious damage from tearing through abrasive or sharp-edged materials in its environment. It takes some time before the animal's cuticle has hardened and thickened enough for it to resume its normal activities. In the meantime, the exoskeleton—which normally provides a great degree of protection and mobility—acts as a hindrance and danger to the arthropod.

STUDYING EXOSKELETONS

The two main approaches used to study the arthropod exoskeleton are the same types of study used by researchers in nearly all branches of the life sciences: field studies and laboratory studies. In the field approach, living arthropods are observed in nature. The specific techniques employed include both still and motion photography in various light—normal, infrared, and ultraviolet. It is important to combine the observational database with later structural analyses of specific arthropod body parts, such as exoskeletons, to ascertain how the anatomical components actually function in the natural setting. Actual physical collections are necessary for study by laboratory workers, who subject the specimens to a range of tests to determine their qualities and features in comparison with similar species and with the normal parameters that are known for previously collected members of the same species.

Specimens are often dissected—or in some cases, vivisected (disassembled while alive)—in order to record useful data such as chemical composition of exoskeletons, metabolic rates, and the estimated age of the sample. In the case of specimens raised in captivity, more precise data can be gained, as the precise age, food type, and daily or hourly intake are known with great precision.

A wide range of techniques are employed in the laboratory to analyze the structural components of exoskeletons. Among them, optical microscopy has traditionally been the primary approach. Working with dissected parts, frequently cut and chemically stained to facilitate viewing or bring out certain features selectively, researchers have used powerful microscopes capable of magnifying by a factor of many hundreds to see tiny subcomponent structures found within exoskeletal tissue. Optical microscopes have also been used to take a close look at associated cellular and noncellular organic matter, such as chitin and sclerotin. An exponential increase in magnification for study, however, has arrived with the advent of scanning electron microscopes (SEMs). These are instruments that use a beam of focused electrons to scan an object and form a three-dimensional image on a cathode-ray tube. The SEM reads both the pattern of electrons scattered by the object and the secondary electrons produced by it. This greatly enhanced ability to see smaller objects with great clarity allows scientists to see very small target sections of exoskeletal tissue, measured in microns (millionths of a meter) in circumference. Researchers can examine in minute detail the structures and interrelationships of the various layers of the arthropod cuticle.

EXOSKELETONS IN THE FOSSIL RECORD

Because the hard arthropod exoskeleton fossilizes more readily than the remains of many other animals, the evolutionary history of the phylum Arthropoda is abundantly represented in the fossil record. Much more is known about this phylum than other invertebrate phyla because of this phenomenon. Entire classes of arthropods that have left no modern descendants are known today because their substantial body armor appears in various marine strata. Examples of this are well documented in the remains of the extinct marine groups of the trilobites (similar to modern horseshoe crabs) and the eurypterids (giant "water scorpions"). In the case of the trilobites, many fossils are actually the result of cast-off exoskeletal moltings rather than the carcasses of the dead animals. This illustrates the fact that, for hundreds of millions of years, arthropods have maintained a lifestyle and evolutionary approach to physiological problems that are similar to those of modern forms. X-ray photography has been employed successfully to penetrate the hard, mineralized fossils of extinct and ancestral forms of modern arthropods, showing in good detail the internal structures of exoskeletons and other tissues. Radiographic images produced by this technique demonstrate the continuity of structure and life shared by members of this extremely successful phylum for a period extending beyond the early Cambrian period (500 million years before the present).

The exoskeleton is an evolutionary advantage shared by all arthropods. This advantage, along with their body and limb segmentation, has allowed them to move into myriad ecological niches, first in the sea and later on land, in freshwater, and finally in the air. As a phylum, arthropods are arguably the most successful of all metazoan animal phyla; they exceed all others combined in terms of the number of species, diversity, and the number of individual organisms. This ubiquity in all biomes and climates and in virtually every conceivable niche in every ecosystem makes them a force that has a constant influence on human life.

—Frederick M. Surowiec

FURTHER READING

Anderson, D. T. *Atlas of Invertebrate Anatomy*. University of New South Wales Press, 1996.

Boardman, Richard S., Alan H. Cheetham, and Albert J. Rowell, eds. *Fossil Invertebrates*. Blackwell Scientific Publications, 1987.

Clarkson, E. N. K. *Invertebrate Palaeontology and Evolution*. 4th ed. Allen & Unwin, 1998.

Ruppert, Edward E., and Robert D. Barnes. *Invertebrate Zoology*. 6th ed. Saunders College Publications, 1994.

Savazzi, Enrico, ed. *Functional Morphology of the Invertebrate Skeleton*. John Wiley & Sons, 1999.

Stachowitsch, Michael. *The Invertebrates: An Illustrated Glossary*. Illustrated by Sylvie Proidl. Wiley-Liss, 1992.

EXTINCTION

FIELDS OF STUDY

Paleontology; Evolutionary Biology; Biogeography; Geology; Environmental Studies

ABSTRACT

In 1796, Georges Cuvier demonstrated incontrovertibly that many species of once-living plants and animals had completely disappeared from Earth. Cuvier was also the first to recognize that many of the extinct species he identified had disappeared at approximately the same time—a mass extinction he attributed to catastrophic events, such as the biblical Flood. The catastrophists held the field of scientific opinion concerning extinction until Charles Lyell proclaimed the doctrine of uniformitarianism. Lyell denied that any spectacular cataclysms had occurred and maintained that no processes have affected Earth (including its flora and fauna) that are not presently observable.

KEY CONCEPTS

catastrophism: a scientific theory which postulates that the geological features of the earth and life thereon have been drastically affected by natural disasters of huge proportions in past ages

fossil: a remnant, impression, or trace of an animal or plant of a past geological age that has been preserved in the earth's crust

gene: an element of the germ plasm that controls transmission of a hereditary characteristic by specifying the structure of a particular protein or by controlling the function of other genetic material

gene pool: the whole body of genes in an interbreeding population that includes each gene at a certain frequency in relation to other genes

species: a category of biological classification ranking immediately below the genus or subgenus, comprising related organisms or populations potentially capable of interbreeding

uniformitarianism: a scientific theory that all processes which have affected the earth and living creatures thereon in the past are presently at work and observable by scientists

CATASTROPHE VERSUS UNIFORMITY

In 1796, a French naturalist, Georges Cuvier, demonstrated incontrovertibly that many species of once-living plants and animals had completely disappeared from the earth. Cuvier was also the first to recognize that many of the extinct species he identified had disappeared at approximately the same time—a mass extinction. He and his successors attributed extinctions to catastrophic events, such as the biblical Flood. The catastrophists held the field of scientific opinion concerning extinction until the publication

This species of pigeon numbered in the millions in the late 1800s. Market hunting drove the pigeon to extinction. The meat was sold to restaurants and their feathers used for hats.

of Charles Lyell's *Principles of Geology* (1830–33), which proclaimed the doctrine of uniformitarianism. Lyell maintained that no processes have affected the earth (including its flora and fauna) that are not presently observable. He denied that any spectacular cataclysms had occurred.

CLIMATE, EVOLUTION, AND EXTINCTIONS

Lyell's arguments convinced geologists, but the problem of extinctions remained: What could cause a species to die out? In 1859, Charles Darwin offered a biological explanation for extinctions that seemed to answer all questions. In *On the Origin of Species*, Darwin suggested that all life-forms engage in a perpetual struggle for survival. The best-adapted species, therefore, survive and perpetuate themselves. The less-adapted species are outcompeted and disappear. Darwin's ideas seemed to fit well with the doctrine of uniformitarianism, but the problem of mass extinctions remained. He suggested no reasons why great numbers of very different species should disappear at approximately the same time.

Although his ideas were not accepted for many years after Darwin, another scientist proposed a possible solution to the puzzle of mass extinctions as early as 1837. Louis Agassiz produced evidence that the earth has periodically undergone periods of extreme cold, with much of its surface covered by glaciers. Agassiz's glaciers are presently in existence, move very slowly, and thus fit well with the uniformitarian view. His glaciers might also explain mass extinction in evolutionary terms, since plants and animals that could not adapt to changing climate would be outcompeted by other species that could adapt, and, thus, the less-flexible species would disappear.

Uniformitarian views about extinction received a further boost in the twentieth century with the proposal of the theory of continental drift, which has evolved into the modern theory of plate tectonics. Scientists have demonstrated that the landmasses of the earth are in constant motion relative to one another and to the poles of the planet. Over millions of years, the present landmasses have occupied very different positions on the earth's surface, often drifting quite near the polar regions, causing massive climatic changes over very long periods of time. In addition, scientists have demonstrated that landmasses have subsided to be covered by the seas and have risen to create dry land from ocean floors. These massive changes would also have a profound effect on flora and fauna, creating constant competition and struggle for survival.

GENETIC FACTORS IN EXTINCTIONS

Many contemporary biologists believe that background extinctions may be the result of genetic rather than strictly climatic factors. In simple terms, extinction results from an excess of deaths over births in a given species. This excess represents, in biological terms, a reduction in the characteristics of a species that allow it to adapt to its environment. Those characteristics, biologists argue, are brought about through natural selection, a process that favors the genes that give the organism an advantage in the struggle for survival. Since natural selection has produced not only the species itself but its ancestors as well, any change in environment must work against the existing organisms. In the new process of changed environment, new gene combinations will result that will produce an organism better adapted to the changed environment. If, however, these genetic changes do not occur rapidly enough, adaptation will not take place and the species will become extinct.

Another possibility results if an organism is too well adapted to its environment. The gene pool of a species usually combines those genes that work well with one another to produce a well-adapted individual. Some biologists have observed a tendency within the gene pools of some species to resist genetic adaptation when environmental changes occur. Thus, genetic changes that could impart greater chances for survival of the individual in times of environmental turbulence do not occur. This effect also occurs when groups of genes are linked together (usually by a chromosomal inversion) in such a way as to prevent new gene combinations that might impart the ability to adjust to changed environmental conditions. This phenomenon actually retards the ability of a species to survive and may lead to extinctions.

Genetics may also explain mass extinctions. One biologist has theorized that species which live in environments that change very little over long periods of time probably have low genetic variability (a gene pool in which nonutilitarian genes have disappeared). Conversely, species in changeable environments should have much more genetic diversity, allowing them to cope with rapidly altering living conditions. If that is true, then mass extinctions might result from a rapid environmental change after a very long period of stability. When the change came, the theory goes, widespread extinctions of many species resulted. Most geneticists, however, reject this theory of mass extinctions. The species in unstable environments may actually have less genetic variability than those in stable environments. Biologists therefore have not been able to advance a plausible genetic explanation for mass extinctions.

MASS EXTINCTIONS

Geologists have identified a number of what they call "mass extinction events." The first such event known to paleontologists occurred 440 million years ago at the end of the Ordovician period, during which more than 22 percent of all families and 57 percent of all genera disappeared. Another mass extinction occurred during the Devonian period, 370 million years ago, during which more than 20 percent of all marine families disappeared. The greatest of all mass extinctions occurred 248 million years ago, at the end of the Permian period. During that event, 52 percent of all marine families, 83 percent of all genera, and a frightening 95 percent of all species became extinct. During this event, land animals and plants vanished, along with marine flora and fauna. Yet another "great dying" took place during the Triassic period, approximately 215 million years ago, when 20 percent of all marine families and 48 percent of all genera disappeared.

The mass extinction known most widely outside the scientific community occurred at the end of the Cretaceous period, about 65 million years ago. During this event, the dinosaurs vanished, along with 50 percent of all marine genera. A number of less-spectacular mass extinctions have taken place in Earth's long history, including a relatively recent one at the end of the Pleistocene epoch, which included among its victims such well-known extinct animals as the woolly mammoth and the so-called saber-toothed tiger.

Some paleobiologists and archaeologists are convinced that many of the Pleistocene extinctions were caused by the activities of human beings. They point out that the extinction of most of the large North American land mammals coincided with one theoretical date for the appearance of humans in the Western Hemisphere, approximately 11,500 years ago. According to these scientists, especially efficient human hunters were responsible for those extinctions; however, this theory seems unlikely as an explanation for all the Pleistocene extinctions. Many species in areas other than North America disappeared at the same time, most of which would have had little or no value as game. In addition, there are huge "boneyards" containing fossils of the extinct species in areas as far separated as Alaska and Florida that apparently died at the same time from causes that seem to be related to some great natural cataclysm.

ASTEROIDS AS AGENTS OF MASS EXTINCTIONS

The species that disappeared in each of these mass extinctions apparently died at approximately the same time, and scientists were at a loss to explain them until 1980. In that year, a scientific team led by Nobel laureate Luis Alvarez presented what seems to be irrefutable evidence that the extinction of the dinosaurs coincided with the collision of a huge asteroid or comet with the earth. The evidence is based on a layer of clay that separates the rock formations associated with the dinosaurs from the overlying formations, which contain fossils from the era of mammals. The clay contains large amounts of iridium and other elements that are scarce in the crust of the earth but common in asteroids and comets.

Many scientists now contend that the collision between the earth and a large extraterrestrial body would have thrown enormous quantities of dust into the atmosphere, sufficient to block out the sun's radiation for an extended period of time. The resulting subzero weather would have had devastating effects on flora and fauna, including the seemingly invincible dinosaurs. Shortly after the evidence for the collision appeared, other scientists presented evidence for an even more frightening phenomenon—periodicity of extinction events.

A scientist charting the occurrence of mass extinction events showed that they seem to occur at regularly spaced intervals, approximately every 26 to 30 million years. Almost immediately after the presentation of the evidence for periodicity, new scientific studies demonstrated more evidence that several mass extinctions other than the one during which the dinosaurs disappeared are also associated with unusually high concentrations of iridium. Taken together, these data seem to indicate that most, perhaps all, mass extinctions are caused by extraterrestrial agents and recur on a regular basis. If that is indeed the case, the implications for all the sciences, including evolutionary biology, are profound.

AN INTERDISCIPLINARY STUDY

Scientists from several different disciplines are currently studying extinctions, including mass extinctions, in a variety of ways. Many biologists believe that the most effective way to understand background extinctions is to examine those that have taken place in historic times. Geologists and paleontologists are subjecting the fossil record to a new and rigorous examination, armed with new, supersensitive techniques for ascertaining the ages of the rocks in which fossils occur in an attempt to understand mass extinctions better. These new techniques are the products of research in nuclear physics. Ecologists are particularly examining currently endangered flora and fauna, which may soon disappear. Even some astronomers are actively engaged in research into mass extinctions, scanning the heavens with powerful telescopes in search of an extraterrestrial agent that might explain the apparent periodicity of mass extinction events.

Biologists studying recent background extinctions conclude that virtually all of them are the results of the activities of humankind. The dodo and the passenger pigeon, which were harvested for food and eradicated in the 1700s; the great auk, the last known specimens of which were killed by Icelandic fishermen in 1844; the Tasmanian tiger, the last known specimen of which died in captivity in 1934 (although reports of recent sightings of the animal in the wild continue to appear sporadically); and many other species are examples of human-caused extinctions. These studies lend validity to the theory that at least some of the many extinctions during the Pleistocene resulted from the hunting activities of prehistoric peoples. More disturbing are studies which show that such modern phenomena as acid rain and ozone depletion, both results of industrialization, may be doing irreparable damage to the environment, which could result in another mass extinction event in the very near future. Indeed, some biologists and ecologists believe that such an event has already begun.

A number of geologists and paleontologists, using new dating techniques based on the rate of radioactive decay in rocks, are reassessing the ages traditionally assigned to fossils by less sophisticated techniques in the past. These studies should eventually reveal whether the mass extinction events of the remote past occurred in a very short or over a relatively longer period of time. Other geologists are searching for impact craters, to lend further credence to the theory that at least some and perhaps all mass extinctions resulted from periodic collisions between Earth and large celestial bodies.

A team of physicists (including astrophysicists), engaged in an ongoing search for a hypothetical dark companion to our sun, has postulated that the orbit of this presently undiscovered body may periodically disrupt the comet cluster on the outer fringes of the solar system, resulting in many comets being diverted into an intersection with Earth's orbit. If their search is successful, it will provide powerful substantiating evidence for the collision theory, with sobering implications. Some physicists have even proposed ways to prevent future collisions between the earth and large heavenly bodies.

The implications of background and mass extinctions are profound. If, as an overwhelming mass of evidence seems to indicate, the activities of humankind are a major cause of recent background extinctions, then those activities may soon lead to an ecological disaster of gigantic proportions. At present rates of disappearance, as many as two million species currently in existence will disappear by the middle of the twenty-first century. Unless immediate steps are taken, plant extinctions on the scale envisioned by many botanists may also cause massive climatic changes. Some ecologists and geophysicists warn that if the tropical forests disappear, the result will be the greenhouse effect. Of less immediate, but nevertheless great, concern is the theory that mass extinction events in the fossil record have resulted from collisions between the earth and asteroids or comets. The implications of the collision theory for evolutionary biology, however, are far-reaching. If the theory is correct, then the struggle for survival is not the most important feature of evolution. No matter how well adapted to its environment a species may be, survival of a collision would be largely a matter of chance. If the theory proves to be correct, biologists will need to rewrite the textbooks on evolution.

—Paul Madden

Further Reading

Archibald, J. David. *Dinosaur Extinction and the End of an Era: What the Fossils Say*. Columbia University Press, 2013.

Donovan, Stephen K., ed. *Mass Extinctions: Processes and Evidence*. Belhaven, 1989.

"The Five Worst Mass Extinctions." *Endangered Species International*. Endangered Species Internationa, 2011. Web. 31 Dec. 2015.

Leakey, Richard E., and Roger Lewin. *The Sixth Extinction: Patterns of Life and the Future of Humankind*. Doubleday, 1995.

McGhee, George R. *The Late Devonian Mass Extinction*. Columbia University Press, 1996.

Martin, Paul S., and Richard G. Klein, eds. *Quaternary Extinctions: A Prehistoric Revolution*. University of Arizona Press, 1984.

Muller, Richard. *Nemesis: The Death Star*. Weidenfeld & Nicolson, 1988.

Raup, David M. *Extinction: Bad Genes or Bad Luck?* Norton, 1991.

Stanley, Steven M. *Extinction*. Scientific American Library, 1987.

Stearns, Beverly Petersen, and Stephen C. Stearns. *Watching, from the Edge of Extinction*. Yale University Press, 1999.

Extinctions and Evolutionary Explosions

Fields of Study

Paleontology; Evolutionary Biology; Animal Anatomy; Biogeography

Abstract

Extinction of species is a continuous process, balanced by speciation events that result in the development of new species. Mass extinctions are events during which the rate of extinction rises dramatically above this background rate. Past mass extinctions were balanced by periods of explosive development that often followed, as organisms moved into vacant adaptive zones during periods of adaptive radiation. The most important of these was at the base of the Cambrian period, 544 million years ago, when all the major groups in existence originated, but other radiations occurred in the Early Triassic period and at the start of the Tertiary period.

Key Concepts

adaptive radiation: the rapid production of a new species following invasion of a new geographic region or exploitation of a new ecological opportunity

bolide: an extraterrestrial object (for example, a meteorite) that hits the earth

diversity: the number of fossil taxa (classification groups) associated with a particular place and time

Lazarus taxa: groups that apparently disappear during a mass extinction only to appear again later

mass extinction: an event in which a large number of organisms in many different taxa are eliminated

periodicity hypothesis: the proposal that mass extinctions have occurred approximately every 26 million years over the past 250 million years

Phanerozoic: an era of geologic time beginning approximately 544 million years ago at the start of the Cambrian period, when animals with mineralized skeletons became common

regression: the migration of the shoreline and associated environments toward the sea

VITA BREVIS

Extinction of species is a continuous process, and evidence of its occurrence abounds in the fossil record. It has been estimated that marine species persist for about four million years, which translates into an overall loss of about two or three species each year. This is the "background" extinction rate, and it is balanced by speciation events that result in the development of new species. Mass extinctions are events during which the rate of extinction rises dramatically above this background rate, and a number of these have been recognized in the Phanerozoic era. In each of these events, at least 40 percent of the genera of shallow marine organisms were eliminated. Using statistical methods, it has been estimated that at least 65 percent of species became extinct at each of these events, with 77 percent being eliminated at the event at the end of the Cretaceous period and 95 percent at the event at the end of the Permian period. These mass extinctions were balanced by periods of explosive development that often followed, as organisms moved into vacant adaptive zones during periods of adaptive radiation. The most important of these was at the base of the Cambrian period, 544 million years ago, when all the major groups in existence originated, but other radiations occurred in the Early Triassic period and at the start of the Tertiary period.

CAUSES OF MASS EXTINCTIONS

Attempts to explain the causes of mass extinctions have centered on terrestrial phenomena such as sea level changes, climatic changes, or volcanic activity. The sea level has shown regular fluctuations on a global level during the Phanerozoic era, and these appear to be related to the melting or formation of polar ice caps or to major tectonic events such as continental splits or the collision and uplift or subsidence of ocean ridges. Extinction events appear to be correlated mostly with periods of marine regression. During such a regression, the withdrawal of the ocean leaves a much smaller habitat for shallow marine organisms. This leads to increased crowding and competition and ultimately to an increased extinction rate. Reduction of large terrestrial vertebrates during these regressions, as happened during the events at the end of the Permian and Cretaceous periods, may be related to increased seasonality caused by the loss of the ameliorating influence of the shallow epicontinental seas.

It has also been shown that some extinctions are related to transgressive events (the spread of the sea over land areas), possibly resulting from the spread of anoxic (oxygen-poor) waters across epicontinental areas. Climatic changes seem to be correlated with eustatic events (worldwide changes in sea level), and the evidence implicating temperature as the main cause of extinctions seems weak. For example, the most important extinction event at the end of the Permian period occurred at a time of climatic amelioration marked by the disappearance of the Gondwanaland ice sheet. Volcanic activity has been presented as a possible cause of the extinctions that occurred at the end of the Cretaceous period. The Deccan Traps of northern India were erupting at that time and would have produced large quantities of volatile emissions that could have resulted in global cooling, ozone-layer depletion, and changes in ocean chemistry. However, no evidence exists as yet for the involvement of volcanic activity in other extinction events.

Although various extraterrestrial causes for mass extinction events have been suggested in the past, these ideas have gained greater credence since the publication in the early 1980s of work by Luis and Walter Alvarez, who ascribe the end-Cretaceous extinction event to the effects of the impact of a large bolide, or extraterrestrial object, perhaps ten kilometers in diameter. The impact of such a large object would have resulted in some months of darkness because of the global dust clouds generated, and this would have halted photosynthesis and resulted in the collapse of both terrestrial and marine food chains. Although cold would initially have accompanied the darkness, greenhouse effects and global warming would have followed as atmospheric gases and water vapor trapped infrared energy radiating from Earth's surface. Physical evidence for an impact rests on the presence in the period boundary layers of high concentrations of iridium and other elements generally rare at Earth's surface but abundant in asteroids. In addition, these layers often contain shocked quartz grains, otherwise found only in impact craters and at nuclear test sites, and microtectites, glassy droplets formed by impact. Although the evidence for extraterrestrial impacts having caused the other major extinction events is slight, this causal factor has been linked with the apparently regular 26-million-year periodicity exhibited by extinctions. Scientists suggest that the regular passage of an unidentified planetary body by the Oort Cloud of comets and the subsequent perturbation could result in increased asteroid impacts and extinction events on earth.

HISTORICAL MASS EXTINCTIONS

The first mass extinction event that can be recognized in the fossil record occurred in the Middle Vendian period, about 650 million years ago, when microorganisms underwent a severe decline. This event has been linked to climatic cooling related to glaciation. The extinction in the Late Ordovician period was a major event in which 22 percent of marine families became extinct. As there were two main pulses of extinction and no corresponding iridium anomaly was found, an extraterrestrial cause seems unlikely. However, sea level and temperature changes have been cited as likely causes. In addition, biologically toxic bottom waters might have been brought to the surface during periods of climatic change. The event at the end of the Devonian period had a devastating effect on brachiopods, which lost about 86 percent of genera, and on reef-building organisms such as corals. Shallow-water faunas were most severely affected; only 4 percent of species survived, although 40 percent of deeper-water species did, and cool-water faunas also survived better. This event has been linked to a significant drop in global temperatures of unknown cause.

The mass extinction event at the end of the Permian period was the most severe of the Phanerozoic era and resulted in the extinction of up to 95 percent of all marine invertebrate species. On land, amphibians and mammal-like reptiles were both badly affected, and plant diversity fell by 50 percent. No corresponding iridium anomaly was found, and the most likely explanation is climatic instability caused by continental amalgamation and the simultaneous occurrence of marine regressions. These occurrences would have disrupted food webs on a major scale. The event that occurred at the end of the Triassic period was much less severe but still involved extensive reductions in marine invertebrates and reptiles. On land, a major faunal turnover took place. Primitive amphibians, early reptile groups, and mammal-like reptiles died out and were replaced by advanced reptiles and mammals. No evidence of an impact event has been found, and the extinctions are generally correlated with widespread marine regressions.

The extinction that took place at the end of the Cretaceous period has become the most hotly debated, in large part because of the bolide impact hypothesis. Although the broad pattern of extinction among marine organisms is known, the detailed picture only encompasses microorganisms such as planktonic foraminifera and calcareous nanoplankton. Study of the ranges of these microorganisms shows that the extinctions occurred over an extended period, starting well before and finishing well after the boundary. Although much has been made of the extinction of ammonites at the end of the Cretaceous period, there are too few ammonite-bearing sections to show whether it was gradual or abrupt. On land, evidence of an increase in the population of ferns just above the boundary suggests the presence of wildfires, as ferns are usually the first plants to recolonize an area devastated by fire. However, in many sections, a return of the Cretaceous vegetation is seen above the fern increase, indicating little extinction.

POST-EXTINCTION RECOVERIES

Among the vertebrates, a picture of gradual change is seen for mammals, with drastic reductions occurring only in the marsupials. The boundary also does not seem to have been a barrier for turtles, crocodiles, lizards, and snakes, all of which came through virtually unscathed. The dinosaurs did become extinct, and much argument has centered on whether this was abrupt or occurred after a slow decline. In this context, it must be noted that there is only one area where a dinosaur-bearing sedimentary transition across the boundary can be seen, and that is in Alberta, Canada, and the northwestern United States. Records of dinosaurs in this area during the upper part of the Cretaceous period show a gradual decline in diversity, with a drop from thirty to seven genera over the last eight million years. Although explanations of the extinction of dinosaurs have ranged from mammals eating their eggs to terminal allergies caused by the rise of flowering plants to the current ideas about bolide impacts, the answer may be climate related. A major regression of the oceans occurred at this point, resulting in a drop in mean annual temperatures and an increase in seasonality. The bolide impact may have served as the death blow to taxa (animals in classification groups) that were already declining.

The main period of evolutionary expansion in the Phanerozoic era is at the base of the Cambrian period, 544 million years ago. Termed the "Cambrian explosion," it marks the development of all the modern phyla of organisms, and as many as one hundred phyla may have existed during the Cambrian period. This period seems to have lasted only about 5 million years, and the subsequent history of animal life consists mainly of variations on the anatomical themes developed during this short period of intense creativity. This period is represented in the fossil record by the remarkably well-preserved Burgess Shale fauna of British Columbia, which has been extensively described, and faunas of similar age from China and Greenland. Why the Cambrian explosion could establish all major anatomical designs so quickly is not clear. Some scientists believe that the lack of complex organisms before the explosion had left large areas of ecological space open, and when experimentation took place, particularly with the advent of hard skeletons, any novelty could find a niche. Also, the earliest multicellular organisms may have maintained a genetic flexibility that became greatly reduced as organisms became locked into stable and successful designs. Why some of the innovations were successful in the long term and others were not is unknown, as no recognized traits unite the successful taxa. It has even been suggested that success may be due to no more than the luck of the draw.

In contrast, the recoveries after the major extinctions at the end of the Permian and Cretaceous periods did not result in the development of new phyla. The earliest Triassic ecosystems were more vacant than at any time since the Cambrian period, yet no new phyla or classes appear in the Triassic period. This suggests that despite the overwhelming nature of the extinctions, the pattern was insufficient to permit major morphological innovations, in part probably because no adaptive zone was entirely vacant. Hence, despite the fact that the mass extinction at the end of the Permian period triggered an explosion in marine diversity described as the Mesozoic marine revolution, persisting species may have limited the success of broad evolutionary jumps.

READING THE FOSSIL RECORDS

All understanding of extinction events or of evolutionary explosions depends on the fossil record. The study of the diversity of organisms through time—the number of different types of organisms that occur at a particular time and place—is therefore very important. The basic data consist of compilations of extinctions of taxa plotted against similar compilations of originations of taxa. Periods when either extinction or origination was unusually high show as peaks or troughs on a graph. Unfortunately, biases in the preservation, collection, and study of fossils have conspired to obscure patterns of change in diversity.

Geological history of patterns of diversity is obscured by a variety of filters, many of which are sampling biases that cause the observed fossil record to differ from the actual history of the biosphere. The most severe bias is the loss of sedimentary rock volume and area as the age of the record increases because the volume and area correlate strongly with the diversity of organisms described from a stratigraphic interval. The quality of the record also tends to fall with increasing age because the rocks are exposed to changes that may destroy the fossils they contain. The differences in levels of representation among the paleoenvironments in the stratigraphic record also influence the composition of the fossil record; for example, shallow-marine faunas are much better represented than are terrestrial faunas.

Diversity patterns are studied at a variety of levels, from the species upward, that vary in their quality and inclusiveness. A basic problem is that many of the processes that are of interest occur at the species level or even below it, but the biases of the fossil record mean that data are best at higher levels. Diversity of shallow-marine organisms for the Phanerozoic era cannot be read directly at the species level because the record is too fragmentary. The record at the family level is much more complete because the preservation of one species in a family allows the family to be recorded. For this reason, paleodiversity studies are often conducted at the family level. However, higher taxon diversity is a poor predictor of species diversity. For example, an analysis of the mass extinction at the end of the Permian period indicates that the 17 percent reduction in marine orders and 52 percent reduction in marine families probably represent a 95 percent reduction in the number of species. Another problem with the study of fossils is that soft-bodied and poorly skeletonized groups may leave little or no record. It has generally been assumed that the ratio of heavily skeletonized to non-skeletonized species has remained approximately constant through the Phanerozoic era; however, there are no data to support this and there is some evidence that skeletons have become more robust through time in response to newly evolving predators. The net result of these biases is severe. Only 10 percent of the skeletonized marine species of the geologic past and far fewer of the soft-bodied species are known.

Despite these problems, it has been possible to show that diversity of organisms has varied in a number of ways during the Phanerozoic era. Tabulations of classes, orders, and families have been used to show that there were significant periods of increased extinction or increased evolutionary rates. One of the most important uses of these data has been the tabulation at the family level that appears to show a regular periodicity of about twenty-six million years for extinction events and that has been used to support ideas about periodic extraterrestrial events. However, although fluctuations occurred, it has also been possible to show that the number of marine orders increased rapidly to the Late Ordovician period and has remained approximately constant since then.

THE EBB AND FLOW OF LIFE ON EARTH

Mass extinctions and evolutionary explosions are the opposite faces of the pattern of diversity of organisms through time. During periods of mass extinction, the diversity of organisms on Earth has dropped drastically, and in some cases, entire lineages have been wiped out. Evolutionary explosions, on the other hand, resulted in enormous innovation, particularly at the beginning of the Cambrian period, and the development of new variations on established morphotypes (animal and plant forms and structures) later in the geologic record. Understanding the processes that caused these events is of major importance because people have reached the point where they are capable of influencing their environment in drastic ways.

Studies of extinction events have shown that they have a variety of causes, some of which appear to be environmental changes brought about by natural processes while others may be the result of extraterrestrial forces. The most severe of these extinction events occurred at the end of the Permian period, 245 million years ago, and resulted in the loss of up to 95 percent of marine invertebrate species. The cause of this extinction appears primarily to be that continents were amalgamating and oceans were retreating, which resulted in a major reduction in the habitat of shallow-marine organisms. Terrestrial habitats were also affected as the increase in continental area and loss of the ameliorating effect of extensive areas of shallow ocean brought about climatic changes. Although climatic changes are thought to be the main culprit in the majority of extinction events, some scientists believe that large bolides, or extraterrestrial bodies, struck the earth with such force as to create major changes in the environment that significantly reduced diversity. This theory has enjoyed the most popularity as the explanation for the event at the end of the Cretaceous period, during which the dinosaurs became extinct, but evidence for an extraterrestrial body's involvement in any of the other events is slight.

Whatever the cause, environmental change that results in habitat reduction is the main reason for species decline. As humans have risen to dominance over other species, the extinction rate has accelerated, and in the last half-century, this rate has climbed considerably above natural attrition as populations have increased and habitats have been altered. Although the levels of extinction have not yet reached those recorded during major extinction events of the past, some scientists believe people may be facing an ecological disaster. A better understanding of the processes surrounding past extinction events and the rebounds that followed them will help people prepare for and deal with the future.

—*David K. Elliott*

FURTHER READING

Allen, Keith C., and Derek E. Briggs, eds. *Evolution and the Fossil Record*. Smithsonian Press, 1989.

Briggs, Derek E., and Peter R. Crowther, eds. *Palaeobiology: A Synthesis*. Blackwell Scientific Publications, 1990.

Donovan, Steven K., ed. *Mass Extinctions: Processes and Evidence*. Columbia University Press, 1989.

Drury, Stephen. *Stepping Stones: Evolving the Earth and Its Life.* Oxford University Press, 1999.

Frankel, Charles. *The End of the Dinosaurs: Chicxulub Crater and Mass Extinctions.* Cambridge University Press, 1999.

Gould, Stephen J. *Wonderful Life: The Burgess Shale and the Nature of History.* W. W. Norton, 1989.

McMenamin, Mark A., and Dianna L. McMenamin. *The Emergence of Animals: The Cambrian Breakthrough.* Columbia University Press, 1989.

Officer, Charles, and Jake Page. *The Great Dinosaur Extinction Controversy.* Addison-Wesley, 1996.

Raup, David M. *Extinction: Bad Genes or Bad Luck?* W. W. Norton, 1991.

Runnegar, Bruce, and James W. Schopf, eds. *Major Events in the History of Life.* Jones and Bartlett, 1992.

Ward, Peter D., and Don Brownlee. *Rare Earth: Why Complex Life is Uncommon in the Universe.* Copernicus, 2000.

FELIDAE

FIELDS OF STUDY

Evolutionary Biology; Ecology; Veterinary Science; Paleontology

ABSTRACT

Catlike animals first appeared in fossil records approximately thirty million years ago. Cats are native to all land areas of the world except Antarctica, Australia, and some oceanic islands, inhabiting primarily forests and grassy plains. Gestation periods in large cats typically last from 3 to 3.5 months; in smaller cats, approximately 2 months. The potential longevity is probably fifteen years for most species; some individuals have lived over thirty years. Cats are very agile, have large eyes with excellent night vision; jaws adapted to seizing and gripping prey, and teeth designed for tearing and slicing flesh.

KEY CONCEPTS

carnassials: pairs of large, cross-shearing teeth on each side of the jaw
epihyal: a hyoid bone whose presence or absence determines whether a cat generally purrs or roars
hyoid bones: series of connected bones at the base of the tongue
papillae: sharp, curved projections on the tongue
vibrissae: stiff hairs, projecting as feelers from the nose and the head

THE APPEARANCE OF CATS

Catlike animals first appeared in fossil records approximately thirty million years ago. They shared typical anatomical features with later cats: long limbs ending in feet with retractable claws and skulls featuring slicing teeth and large, pointed canines. Some genera developed especially long, curved canine teeth, called "sabers." About 10 million years ago, small cats classifiable as members of the genus *Felis* appeared, and by 3.5 million years ago examples of the genus *Panthera* emerged. They did not immediately replace saber-toothed cats, whose fossils exist in deposits containing those of modern cats. The American saber-tooth, *Smilodon fatalis*, was still active toward the end of the last glaciation; some individuals were trapped in California's Rancho La Brea tar pits as late as ten thousand years ago. An estimated four-fifths of all cat species are now extinct, often having disappeared during the same period that their favorite prey species also vanished.

CLASSIFICATION

Living Felidae are usually classified into some fourteen genera containing around forty species. In 1916, R. I. Pocock, a taxonomist at the London Zoo, established the initial modern feline classification system using hyoid bones as the fundamental characteristic and the epihyal structure as distinguishing the two major cat genera originally identified: *Panthera* and *Felis*, or large cats and small cats. Today, the distinctions Pocock applied to these two genera are applied to two subfamilies of *Felidae*, the *Pantherinae* (which includes *Panthera*) and the *Felinae* (which includes *Felis*).

Pocock defined the genus *Panthera* as cats whose epihyal bone is replaced by a thin ligament; these animals normally vocalize by roaring rather than purring. Included in this genus are the large cats of Africa and Asia—the lion (*P. leo*), the tiger (*P. tigris*), the leopard (*P. pardus*), the snow leopard (*P. uncia*), and the American jaguar (*P. onca*). Today, the genus *Panthera* is joined by the genus *Neofelis*, which includes the two species of clouded leopard of Southeast Asia.

Pocock placed cats whose epihyal develops as a normal bone within the genus *Felis*. They are able to purr continuously and usually do not roar. This genus originally included some twenty-eight species

The African lion is among the largest members of the cat family. It is also the most social, living in family groups called "prides" that number as many as 23 individuals. Head rubbing among pride members is a common social behavior.

of small and medium-sized cats; it now includes only six living species (including the domestic cat, *F. catus*), with twenty-eight other species split off into eleven genera, all under the subfamily *Felinae*. For the most part these animals are small cats, including the African golden cat (*Profelis aurata*, the ocelot (*Leopardus pardalis*), and many varieties of the European and African wildcat (*Felis sylvestris*). The *Felinae* subfamily also includes medium-sized cats such as the American cougar (*Puma concolor*); three species of lynx and the bobcat occupy the *Lynx* genus; and the cheetah (*Acinonyx jubatus*) is also in this subfamily.

FELINE ANATOMY

Every cat, from the smallest domestic cat to the largest tiger, is physically equipped to become a successful predator—coat color, legs, claws, mouth, teeth, sight, hearing, and touch are all highly adapted for hunting and devouring prey.

Coat colors help cats blend into their environment while stalking prey. Most cats display a pattern of spots, stripes, or rosettes on a yellowish background, providing camouflage within forest or broken terrain. The lion's uniform coat color blends into the grassy plains where it usually hunts. Lion cubs and the young of other species developing uniform coat color as adults are born with patterned coats, indicating that this was the primitive coloration of all cat species.

Cat legs are often long and muscular, permitting short, high-speed bursts when attacking prey. Cat claws are usually retractable, pulling inward when running, but extending outward when catching or holding victims. Although cheetah claws do not fully retract, the cat's powerful muscles permit speeds of over sixty miles an hour in full pursuit. Claws and muscles make cats agile climbers who can scale trees when escaping enemies or hiding in ambush.

Cat teeth are adapted for seizing and cutting meat. Four elongated, pointed canine fangs grasp prey, and small, chisel-like incisors tear meat. The scissoring action of large carnassial teeth quickly slices meat from carcasses. Food tends to be swallowed in relatively unchewed chunks, then broken down in the digestive tract. Sharp-pointed, recurved papillae on the tongue help remove remnants of flesh from bones and are also used for drinking fluid and cleaning fur.

Many cats are nocturnal hunters, possessing sensory organs well adapted to low light. Their large eyes contain an extrasensitive reflective retinal layer, making cat eyes appear to glow in the dark, while pupils vary swiftly from fully open to tiny slits. Hearing is acute, and ears swivel easily to pinpoint sources of sound. Vibrissae, or whiskers, on nose and head permit cats accurately to locate obstacles and open paths, even when moving through darkness. The vibrissae also inform cats of the best position for gripping prey with their mouths.

FELINE BEHAVIOR

Most cats are solitary hunters leading solitary lives, joining other adults only during mating. Kittens, however, may remain with their mother for up to two years, learning how to hunt before setting off on their own. Most cats live within habitats providing little stimulus for cooperative action. Tigers stalking prey in the jungle or snow leopards living in open country with highly dispersed prey find individual hunting most efficient. Occasionally, male cheetahs join in hunting coalitions of two to four animals, but such groupings are rare.

Both solitary and social cats, such as lions, are highly territorial—clawing trees, spraying urine, or leaving uncovered feces marking area boundaries; loud roars advertise the presence of claimants. Solitary females tend to establish ranges respected by each other. Males inhabit larger territories, usually overlapping those of two or more females, but face challenges from neighboring or interloping males.

Cats use three hunting strategies: moving slowly through their home range stalking, seizing, and killing prey; setting up ambushes near burrows or climbing trees and patiently waiting to pounce upon unsuspecting victims; and inadvertently stumbling upon prey while engaged in other activities, such as searching for water. Cats prefer to kill their quarry before eating. Small animals are bitten at the nape of the neck with canine teeth, severing spinal cords; biting the throat ruptures air passages. A lion sometimes strangles an antelope, clamping its mouth over the muzzle and suffocating its victim.

Lions live in groups called prides, consisting of up to a dozen individuals who aid each other in hunting. Females and their young compose the pride's core; usually related to each other, they raise their cubs together. Two or three related adult males dominate and defend the pride, becoming the fathers of its cubs. When male cubs mature they are generally driven off, but females may become permanent members of the pride. Group hunting by females, with occasional assistance from males on a difficult kill, is an economical procedure in open terrain containing abundant large prey.

Scientists studying feral cats—domestic cats returned to the wild—found two patterns of existence. Feral cats hunting widely dispersed prey tended to be solitary, occupying separate female and male territories. Cats gathered together only at concentrated and stable food sources, such as garbage dumps and barns. In either case, a group of related females and their kittens formed the core unit; adults often aided each other raising the young. Female offspring might remain group members, but strange females were driven off. Some resident males were tolerated but faced challenges from interlopers seeking access to females. Several groups might occupy areas particularly rich in food. In all cases, resemblance to the social structure of lion prides was striking.

Adapted to widely varying environments, the *Felidae* remains one of the most successful animal families. A single species—the tiger—can be found ranging from the tropics to Siberia. However, the tiger and other feline relatives are increasingly endangered. Hunters seek many cats as trophies; the fur trade also values their striped and spotted skins, and the Asian traditional medicine market values tiger bones highly. Big cats are particularly vulnerable, as expanding human settlements constrict the large ranges needed for successful predation. Whether large cats will survive, or join the four-fifths of *Felidae* species already extinct, remains for future generations to decide.

—*Adrian Treves*

FURTHER READING
Hunter, Luke. *Wild Cats of the World.* Bloomsbury, 2015.
Kobalenko, Jerry. *Forest Cats of North America: Cougars, Bobcats, Lynx.* Firefly, 1997.
Lumpkin, Susan. *Big Cats.* Facts on File, 1993.
Lumpkin, Susan. *Small Cats.* Facts on File, 1993.
Turner, Alan. *The Big Cats and Their Fossil Relatives: An Illustrated Guide to Their Evolution and Natural History.* Columbia University Press, 1997.
Turner, Dennis C., and Patrick Bateson, eds. *The Domestic Cat: The Biology of Its Behavior.* 3rd ed. Cambridge University Press, 2014.
Vaillant, John *The Tiger. A True Story of Vengeance and Survival* Vintage Canada, 2010.
Wilson, Don E., and DeeAnn M. Reeder. *Mammal Species of the World: A Taxonomic and Geographic Reference.* Vol. 1. 3rd ed. Johns Hopkins University Press, 2005.

Fertilization

Fields of Study

Reproductive Biology; Evolutionary Biology; Embryonics

Abstract

For fertilization to occur, several things must happen: Sperm and eggs must be in close proximity, the gametes need to be compatible, the sperm must be able to penetrate the egg, and the haploid egg nucleus must combine with the haploid sperm nucleus. If any one of these is missing, fertilization will not occur.

Key Concepts

corona radiata: the layers of follicle cells that still surround the mammalian egg after ovulation
vitelline envelope: the protective layers that form around the egg while it is still in the ovary
zona pellucida: mammalian protective layer analogous to the vitelline envelope

ASSURING EGGS AND SPERM ARE IN PROXIMITY

For fertilization to occur, several things must happen: Sperm and eggs must be in close proximity, the gametes need to be compatible, the sperm must be able to penetrate the egg, and the haploid egg nucleus must combine with the haploid sperm nucleus. If any one of these is missing, fertilization will not occur.

Animals have many mechanisms to ensure that sperm and eggs are in close proximity. This can be a major concern for aquatic organisms with external fertilization, and many release gametes in the millions or even billions to assure that at least some sperm reach the appropriate eggs. To increase the chances of a meeting between same-species gametes, animals often have specialized mating behaviors. Corals are among those animals that release their gametes into the water and depend on currents to bring egg and sperm together. This is not, however, as random as it may seem. As the first coral releases its gametes, it also releases hormones that induce nearby corals of the same species to release their gametes. These also release the same chemicals with their gametes, and soon there are clouds of eggs and sperm, and the chances of a proper meeting are increased dramatically.

One species of polychaete annelid, *Eunice viridis*, or the palolo worm, has another method of assuring male and female gametes are in the same place. In this species, sexually mature worms called epitokes swarm together at the ocean's surface in response to the lunar cycle. Females then secrete a hormone that induces males to release sperm, and the sperm induce the females to shed eggs. Many fish go through elaborate courtship rituals, during which males and females release gametes at a specific point, thus assuring that egg and sperm are together. Other fish build nests where females lay eggs and males deposit sperm. Frogs and toads usually breed in the water, but the female will only release her eggs when the male is clasped to her back in amplexus. Thus, sperm are deposited on the eggs as they are being laid.

Males of other species place sperm directly in the female's reproductive tract. The male octopus has a special tentacle that is used to place one of his sperm packets in the mantle cavity of the female. Some salamander males deposit their sperm packets on the substrate during a squat dance courtship ritual. Females also do the squat dance and pick up the packet with the lips of their cloacae. In some species of water mites, females mount a special saddle-shaped extension of the males' abdomen. The male squats to deposit a sperm packet, moves ahead slightly and then squats again when the opening of the female's reproductive system is over the packet, forcing the packet into her reproductive system. Another interesting way to assure fertilization is seen in the sea horse. In these animals, a female deposits her eggs into a pouch on the male's abdomen and the male releases sperm into the pouch at the same time. The most common way to introduce sperm into a female's reproductive tract is through copulation, where the male ejaculates sperm directly into the female's reproductive tract. The motile sperm then travel to the egg. For a sperm to gain full motility, it usually must undergo a little-understood process called capacitation.

PENETRATION

Once eggs and sperm are in close proximity, the sperm must begin to penetrate the egg's protective layers. All eggs have at least one protective layer outside the cell membrane. Called the vitelline envelope in most organisms, it is synthesized in the ovary and composed primarily of polysaccharides and glycoproteins. The oviducts and uterus often secrete other protective layers around the egg. In some instances, the sperm must also penetrate these layers, for example, the jelly layers that surround sea urchin and frog eggs. In other instances, the egg is fertilized before these layers are added, as is the case with the many protective layers that surround bird and reptile eggs. A protective layer made up of cells is seen in most mammals, since the egg is released by the ovary with cells of the cumulus oophorus still attached. For the sperm to penetrate these layers, its acrosome must contain the appropriate enzymes to lyse (disintegrate) the chemicals that block its way. The acrosomal reaction must also take place in order to expose the digestive enzymes of the acrosome. This reaction depends on changes in membrane permeability to ions and subsequent changes in pH.

Once through the protective layers, the sperm makes contact with the egg's plasma membrane. If the sperm and egg are of the same species, sperm receptor molecules on the egg membrane attach to complementary molecules, called bindins, on the sperm membrane and the two membranes fuse. If the bindins on the sperm do not complement the receptors on the egg, there is no fusion and fertilization does not continue, thus preventing most interspecies crosses. However, closely related species often have bindins and receptors sufficiently alike to allow some fertilization to proceed. The products of these interspecific matings are hybrids, such as the mule.

Once the first sperm fuses with the egg, mechanisms to prevent polyspermy, the fertilization of an egg by more than one sperm, are put into place. The first block to polyspermy is common to most animals studied: a very quick and only temporary depolarization of the plasma membrane. In sea urchins, the resting membrane potential of the egg plasma membrane is approximately −70 millivolts, the inside being more negative than the outside.

Fusion of the sperm plasma membrane with the egg cell membrane causes a rapid influx of sodium ions. The positive charges neutralize negative charges in the egg until the membrane potential is raised to +10 millivolts. All this happens in less than five seconds, and lasts for about one minute before the egg cell has actively transported enough sodium out of the cell to repolarize it. While the cell is depolarized, no further sperm membranes can fuse with the egg membrane. This is often referred to as the fast or temporary block to polyspermy and seems to occur in all animals thus far studied. The fast block also sets into motion the slow or permanent block to polyspermy. The changed membrane potential of the fast block and the release of nitric oxide by the sperm allows cells to release calcium ions from storage. The initial calcium ion release causes the egg to release nitric oxide, which then increases the egg's release of calcium ions. The release of calcium ions induces the cortical reaction by which cortical granules move to the surface of the cell, fuse with the cell membrane, and empty their contents into the space between the cell membrane and the vitelline envelope. In sea urchins, the first acrosomal enzymes released break the bonds between the cell membrane and the vitelline envelope. In the presence of water, other chemicals released by the cortical granules swell, lifting the vitelline envelope away from the cell membrane. Finally, other enzymes released by the cortical granules alter the vitelline envelope, knocking off any attached sperm and causing the release of peroxide ions, which harden the envelope, making it impermeable to sperm. This impermeable barrier is renamed the fertilization membrane. The released peroxide may also provide another benefit. Any sperm that had penetrated the vitelline envelope before it hardened would be killed by the peroxide and would thus not lead to polyspermy. In other animals studied, although cortical granules do empty their contents into the perivitelline space, the permanent block to polyspermy does not seem to involve the same extensive changes to the vitelline envelope (or *zona pellucida* in mammals) that are seen in the sea urchin. In large, yolky eggs, some polyspermy does occur, but the extra sperm remain in the yolk and never reach the egg nucleus for fusion.

CELL METABOLISM AND MEIOSIS

Concomitant with the cortical reactions is an increase of metabolism in the egg, which will be necessary for nuclear fusion and cleavage. In species where the egg has not completed meiosis, it does so at this time. Which parts of the sperm enter the egg is dependent on the species. In many mammals, the entire sperm enters, while all but the tail enters in echinoderms. In other organisms, the head with the nucleus and centrioles seem to be the only things that enter. There is no evidence that any parts of the sperm other than the nucleus and centrioles are used by the zygote, and other parts that enter most probably degenerate and their components are recycled.

Studies on the mitochondria of sperm indicate that soon after entering the egg, the sperm's mitochondria are tagged by ubiquitin, the first step in breakdown and recycling. After entry, the sperm nucleus imbibes water and is converted into the male pronucleus. At the same time, the egg nucleus becomes the female pronucleus. In most animals, the male pronucleus and the female pronucleus fuse to form the diploid zygote nucleus. In some nematodes, mollusks, and annelids, however, the pronuclei remain separate until after the first cleavage division. In a few others, like the copepod *Cyclops*, the pronuclei divide separately for several cleavage divisions.

The fusion of the sperm with the egg nucleus affects many other cellular processes. One of the most interesting is the displacement of some cytoplasmic constituents. These constituents of the egg determine the fate of cells derived from the parts of the egg in which they were located and probably determine the plane of bilateral symmetry. Sperm attachment and entry often causes shifts in the position of the viscous cortical and subcortical cytoplasm, where many of the fate-determining chemicals are located.

—*Richard W. Cheney, Jr.*

FURTHER READING

Balinsky, B. *An Introduction to Embryology.* 5th ed. Saunders College Publishing, 1981.

Bronson, F. H. *Mammalian Reproductive Biology.* University of Chicago Press, 1989.

Carlson, B. *Patten's Foundations of Embryology.* 6th ed. McGraw-Hill, 1996.

Kumé, Matazo, and Katsuma Dan. *Invertebrate Embryology.* Translated by Jean C. Dan. NOLIT Publishing House for the US Department of Health and Human Services, 1968.

Fins and Flippers

Fields of Study

Animal Physiology; Ichthyology; Evolutionary Biology; Paleontology

Abstract

One of the characteristic features of aquatic vertebrate animals is the presence of single and paired appendages used for locomotion. In fishes these structures are known as fins and in marine mammals they are known as flippers or flukes (on the tail). Although these structures bear superficial resemblances to each other, there are significant differences in their anatomies, attesting their different evolutionary histories. However, they are related structures, having evolved from the same basic structures found in the earliest vertebrates.

Key Concepts

fin-fold theory: theory that fins initially evolved as long folds of tissue extending around the body

flipper: finlike structures of marine mammals that have evolved from the forelimbs of their terrestrial ancestors

four-fin system: the combined activity of paired fins in some bony fishes that makes them highly maneuverable

heterocercal: a tail in which the spine extends into the upper lobe, giving a distinctly sharklike impression

homocercal: a type of tail at which the spine ends at the base of the tail, which consists of two equal lobes

lepidotrichia: modified scales that form the supporting rays of the fins of bony fishes

pectoral and pelvic girdles: skeletal structures that form a structural base for attachment of the paired fins in fishes, connecting them to the rest of the body's skeleton

LOCOMOTION FOR AQUATIC VERTEBRATE ANIMALS

One of the characteristic features of aquatic vertebrate animals is the presence of single and paired appendages used for locomotion. In fishes, these structures are known as fins and in marine mammals they are known as flippers or flukes (on the tail). Although these structures bear superficial resemblances to each other, there are significant differences in their anatomies, attesting their different evolutionary histories. However, at a deeper level, they are related structures, having evolved from the same basic structures found in the earliest vertebrates. This is a classic example of evolutionary parallelism. Finlike structures have evolved independently in both fishes and marine mammals from a common ancestral structure.

THE EVOLUTION OF FINS

The earliest vertebrates were elongated aquatic animals, and locomotion in these animals was probably accomplished by an eel-like undulation of the body. The efficiency of this form of locomotion is decreased by unwanted motion resulting in an up-and-down (pitch) or side-to-side (yaw) seesawing and rolling around the long axis of the body. Fins probably first evolved as stabilizers to resist these motions and increase swimming efficiency. Since their appearance, however, they have also evolved other functions.

There are two theories as to how fins first evolved. The fin-fold theory suggests that the early vertebrates had two paired folds of tissue extending along the side of the body. These folds fused just behind the anus to form a single fin which extended around the tail and up onto the midline of the upper body (dorsal) surface. The theory states that several regions of these fin folds have persisted, resulting in the paired and unpaired fins in modern fishes, whereas the rest of the early fin folds have been lost. As evidence for this theory, the sand launce, *Brachiostoma* (amphioxus) is used as an example. Amphioxus is an animal that is closely related to the vertebrates, and the earliest

Fins are used by aquatic animals, such as this killer whale, to generate thrust and control the subsequent motion. (Minette Layne)

protovertebrates are believed to have resembled it. It has extensive folds that closely resemble the theoretical fin-folds of the earliest vertebrates. A different theory, the body-spine theory, states that the earliest vertebrates possessed two or more pairs of spines extending from both sides of the lower portion of the body. Fins were then formed when membranes extended from the tip of the spine to the side of the body, rather like a sail extending from the mast of a boat. Internal support structures within the fin (the endoskeleton) developed at a later date. At the current time, there is no evidence to conclusively support one theory or the other.

THE USES OF FINS

Fins in fishes are either paired, meaning there are equivalent fins on either side of the body trunk, or unpaired, meaning that there is a single fin located on the midline of the upper or lower body. The paired fins are the pectoral fins, generally located just behind the gills, and the pelvic fins, which are located behind and below the pectoral fins. The unpaired fins comprise one or more dorsal fins, located on the midline of the back, the anal fin, located on the midline of the bottom (ventral) surface of the body, behind the anus, and the caudal fin, or tail.

In fishes, the tail, or caudal fin, provides a large portion of the forward thrust required for moving through the water and has evolved as a specialized portion of the rearward end of the body trunk and spine. The spine itself is composed of repeating skeletal structures, the vertebrae, which together form a flexible column extending through the long axis of the animal. Sharks, sturgeon, and paddlefish possess a heterocercal tail, in which the vertebral column extends into the upper portion, or lobe, of the tail, giving it a distinctly asymmetric and shark-like appearance. The majority of bony fishes possess a homocercal tail, in which the spine ends at the base of the tail and in which the tail is composed of equal-sized upper and lower lobes. A few extinct species of fishes possessed hypocercal tails, in which the spine extended into the ventral lobe of the tail. These animals therefore possessed caudal fins in which the lower lobe was larger than the upper, the opposite of what is seen today in sharks.

The paired and unpaired fins of sharks are solid, broad-based, and relatively inflexible. Like all fins they possess an internal skeleton, which provides structural support as well as attachment for muscles that allow the fin to be moved. These consist of a series of cartilages, known as basals and radials, located in the base of the fins. Long rods of cartilage called ceratotrichia extend from the radials out to the edges of the fin, providing support. The principal role of these fins is to resist the yawing, pitching, and rolling motions generated during swimming. Because of the asymmetric heterocercal tail, the thrust generated in forward swimming extends downward from the upper lobe of the tail through the shark's center of gravity, pushing the animal downward. An important function of the large pectoral fins in sharks is to generate lift, in much the same manner as an airplane's wing. The lift generated by the pectoral fins counters the downward thrust and moves the animal forward through the water. The pelvic fins are specialized in male sharks and rays to form claspers, which serve as the male organs of reproduction in internal fertilization.

The fins of bony fishes are distinctly different from those seen in sharks, although the endoskeleton of the fins also consists of basal and radial bones. The fins of the majority of bony fishes, particularly the paired fins, are much narrower at their base, more flexible, and may play a more direct role in locomotion in addition to stabilizing the fish during swimming. The radials of the endoskeleton are reduced to very small structures located within the muscles at the base of the fin. These articulate with modified scales called lepidotrichia, or fin rays, which extend out to form the main structural elements of the fin. Membranous connective tissue extends between the fin rays, giving them their typical weblike appearance. The vast majority of bony fishes possess this type of fin structure, and they are known collectively as the ray-finned fishes, or Actinopterygii.

THE FINS OF BONY FISHES

A key development in the evolution of bony fishes was the appearance of a swim bladder, a gas-filled sac in the abdominal cavity that counters the fish's tendency to sink in the water and thus provides it with neutral buoyancy. With the evolution of this structure, it was no longer necessary for the pectoral fins to generate lift, and they could then be employed as brakes to stop forward motion. The pectoral fins of most bony fishes have a large surface area and a narrow base that is inserted into the body wall almost vertically, as opposed to horizontally in sharks. They are therefore admirably suited to act as brakes both singly and together, and greatly increase the ability of the fish to stop or change direction rapidly. The pelvic fins act to counterbalance any pitching or rolling motion generated by the pectorals. The combined activity of the paired pectoral and pelvic fins, often called the four fin system, provides enhanced maneuverability and control. This system is best observed in many types of coral reef butterfly fishes (Chaetodontidae), in which the pelvic fins are inserted almost directly below the pectoral fins. These fishes are highly maneuverable, able to move, turn, and stop quickly and accurately within the complex and constrained multidimensional environment of the coral reef.

A relatively small group of bony fishes possesses paired fins of a different sort. This group of fish, which includes the lungfishes and the coelacanth, is characterized by pelvic and pectoral fins that possess fleshy lobes at their base. The lobes contain muscles and skeletal elements; however, only a single basal bone articulates with the rest of the skeleton of the fish. These lobe-finned fishes (the Crossopterygii) are important because they represent descendants of the lineage that moved from the aquatic environment onto land. The pectoral and pelvic fins of these fishes gave rise to the forelimbs and hind limbs of terrestrial vertebrates, and the single bone articulating with the body skeleton is the forebear of the humerus, the bone of the upper arm.

In all fishes, the paired fins are connected to the rest of the body's support framework, the skeleton, via structures known as the pectoral and pelvic girdles. The pectoral girdle in sharks is relatively simple and consists of a large bar of cartilage that extends across between the two pectoral fins, known as the coracoid bar. Scapular processes extend above the base of the pectoral fin and connect the pectoral girdle to the skeleton. The pelvic girdle of sharks has fused into a single bar of cartilage, the puboischiac bar.

The pectoral girdle of bony fishes contains a large number of small bones including the cleithrum, supracleithrum, and the clavicle. A posttemporal bone attaches the pectoral girdle to the rear of the skull. The pelvic girdle in most bony fishes is composed of a pair of bones that act as extensions of the basal bones of the fins.

FLIPPERS

During the Tertiary period (approximately thirty-five to fifty-seven million years ago) several groups of land-dwelling mammals returned partly or completely to the aquatic environment. The best known of these are the cetaceans (whales and dolphins), but other groups include the *Sirenidae* (manatees) and the pinnipeds (seals and sea lions). These animals faced the same problems of locomotion in water as did their fish ancestors, and in the course of evolution they have evolved finlike structures called flippers and flukes. These perform the same stabilizing functions as in fishes, and bear a superficial resemblance to fins. However, closer examination of whales and dolphins demonstrates that the pectoral flippers of cetaceans are in fact modified mammalian forelimbs and contain the same bones present in the forelimbs (arms) of terrestrial mammals. Thus, the bones of the pectoral fins, which gave rise to the forelimbs of terrestrial vertebrates, have evolved back into finlike structures in these mammals as an adaptation to an aquatic lifestyle. The hind limbs, along with their bones and supporting pelvic girdle (the hip bones in land mammals) have been lost in modern dolphins and whales, although they are still present in fossil forms such as the extinct fossil whale *Basilosaurus*. These mammals also possess fleshy dorsal fins to help stabilize them during swimming and large horizontal flukes on the tail, which provide the main forward propulsive thrust to these animals as they move through the water.

—*John G. New*

FURTHER READING

Bailey, Jill. *Animal Life: Form and Function in the Animal Kingdom.* Oxford University Press, 1994.
Bone, Q., N. B. Marshall, and J. H. S. Blaxter. *Biology of Fishes.* 2nd ed. Blackie, 1995.
Moyle, Peter B., and Joseph J. Cech. *Fishes: An Introduction to Ichthyology.* 4th ed. Prentice Hall, 1999.
Riedman, Marianne. *The Pinnipeds: Seals, Sea Lions, and Walruses.* Reprint. University of California Press, 1991.

FISH

FIELDS OF STUDY

Ichthyology; Fisheries Biology; Evolutionary Biology; Oceanography; Limnology

ABSTRACT

The bony fishes constitute large and diverse superclasses of vertebrates. Like the jawless fishes and cartilaginous fishes, they are characterized by gills, fins, and a dependence on water as a medium in which to live. Unlike those fishes, however, they typically possess a skeleton made of bone. Additional features characteristic of most bony fishes include a lateral line system, scales, osmoregulation (salt balance) by means of salt retention or secretion, and a bony operculum (gill cover) over the gill openings.

KEY CONCEPTS

ctenoid scales: thin, flat, bony scales with tiny spines on the exposed rear edge, found on sunfish, perch, sea bass, and other advanced teleosts
cycloid scales: thin, flat bony scales with a smooth surface; rounded in shape, found on herrings, minnows, trout, and other primitive teleosts
ganoid scales: thick, diamond-shaped, bony scales that are covered with ganoine, a hard inorganic

substance; found on bichirs, gars, and other primitive bony fishes

Osteichthyes: the taxonomic superclass in which the bony fishes were placed; contained species related to the ancestors of higher vertebrates; now known as two superclasses, *Actinopterygii* and *Sarcopterygii.*

pectoral fins: paired fins found near the head end of the fish body; related to the forelimbs of higher vertebrates

pelvic fins: paired fins found either near the tail end of the fish body or below the pectoral fins; related to the hind limbs of higher vertebrates

swim bladder: the hydrostatic (buoyancy) organ of teleost fishes derived from the lung of more primitive bony fishes

teleosts: members of the class *Teleostei,* the most advanced of the ray-finned fishes; they compose the vast majority of living bony fish species

BONY FISHES AND JAWLESS FISHES

The superclasses *Actinopterygii* and *Sarcopterygii* (formerly known as *Osteichthyes*), or bony fishes, constitute large and diverse superclasses of vertebrates. Like the jawless fishes (*Agnatha*) and cartilaginous fishes (*Chondrichthyes*), they are characterized by gills, fins, and a dependence on water as a medium in which to live. Unlike those fishes, however, they typically possess a skeleton made of bone. Additional features characteristic of most bony fishes include a lateral line system, scales, osmoregulation (salt balance) by means of salt retention or secretion, and a bony operculum (gill cover) over the gill openings.

The fossil record of bony fishes begins nearly 400 million years ago in the early Devonian geological period, mostly in freshwater deposits. Thus there is reason to believe that bony fishes originated in freshwater habitats. Living bony fish species inhabit both freshwater habitats (58 percent of species) and marine habitats (41 percent), and some (1 percent) move between the two environments on a regular basis. This distribution does not reflect the relative proportions of these environments, since 97 percent of the earth's water is in the oceans and only 0.001 percent is in freshwater lakes, rivers, and streams (the rest is ice, groundwater, and atmospheric water). Rather, the high diversity of freshwater species is a reflection of the ease with which freshwater populations become isolated and evolve into new species.

FISH SUBCLASSES

There are six classes of bony fishes: *Dipnoi, Coelacanthi, Chondrostei, Cladistei, Holostei,* and *Teleostei.* The first two of these include a total of only eight primitive living species. Subclass *Actinopterygii* includes all the rest. The *Dipnoi,* or lungfish, are named for their possession of lungs, an ancestral characteristic suggesting that the earliest steps of bony fish evolution took place in tropical freshwaters subject to stagnation. Modern lungfishes (six species) are able to cope with such conditions by swallowing air and exchanging respiratory gases (oxygen and carbon dioxide) in the lung. Once considered closely related to terrestrial vertebrates, they are now believed to share certain similarities merely because of convergence (independent evolution of characteristics that appear similar).

The *Crossopterygii,* or fringe-finned fishes, were the dominant freshwater predators of the Devonian period. One fossil subgroup, the rhipidistians, had

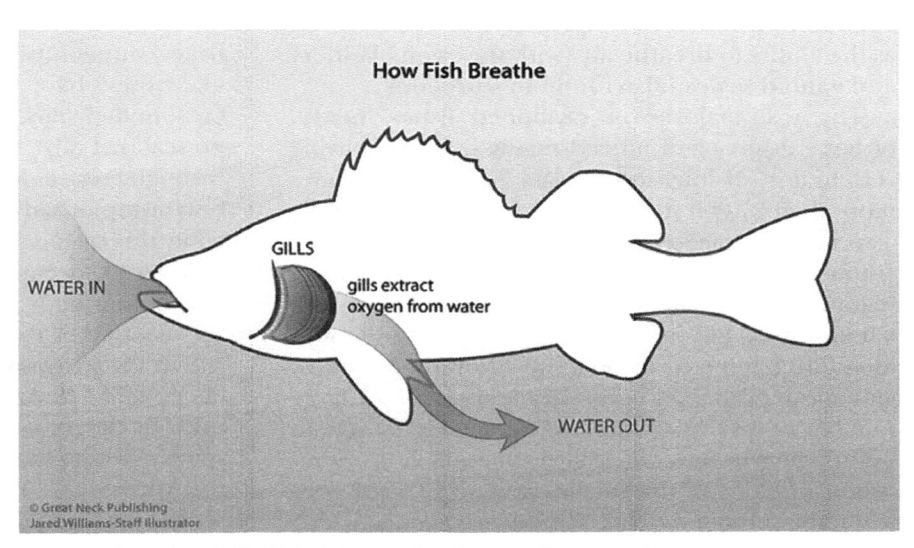

Fish breathe through lateral gills which take in oxygen from the surrounding water environment.

many features intermediate between fishes and ancestral amphibians, including tooth structure, lobed fins, and a jaw connected directly to the skull. Therefore, they are believed to represent a link between fishes and higher vertebrates. The class *Coelacanthi*, was also believed to be extinct (for 70 million years) until a coelacanth, *Latimeria chalumnae*, was taken from deep water off South Africa in 1938. This species and *L. menadoensis* are of great interest as "living fossils."

The class *Cladistei* (bichirs, birchers, lobed-finned pike, and reed fishes) include eleven living species known from swamps and rivers in tropical Africa. Though they share some characteristics with the other bony fish classes, they have some distinct features that warrant placing them in a separate class. One such feature is a dorsal fin consisting of many separate finlets, each supported by a single spine.

The *Actinopterygii*, or ray-finned fishes, comprise four classes: the *Chondrostei, Cladistei, Holostei,* and *Teleostei*. Thirty-one species of chondrosteans, which have reverted to a largely cartilaginous skeleton, include the sturgeons and paddlefishes. One species, the beluga sturgeon (*Huso huso*), source of the famous Russian caviar, may be the largest living *Actinopterygii* species. It is known to achieve a length of 8.5 meters and a weight of nearly 1,300 kilograms.

Holosteans include eight species: seven gar and one bowfin species, all known from North America. These freshwater piscivores (fish predators) are characterized by a skeleton made entirely of bone, but they do have certain other features in common with their more primitive ancestors, such as the ability to breathe air (with the swim bladder) and ganoid scales (also found in sturgeons).

The vast majority of ray-finned fishes, hence, of bony fishes—and indeed nearly half of all living vertebrates—belong to the class *Teleostei*. It includes more than 27,000 species. Among the features characteristic of teleosts are cycloid or stenoid scales (though some are scaleless), a swim bladder (lost in many bottom fishes), highly maneuverable fins, and a homocercal tail (meaning that its upper and lower lobes are symmetrical). Teleosts are represented by an amazing range of body sizes and shapes. A large number of species are quite small, enabling them to occupy niches (ways of living) unavailable to other fishes. The one of the smallest known fish is a goby from the Indian Ocean, *Trimmatom nanus*, which matures at 8 to 10 millimeters in length.

FISH SHAPES AND HABITS

There are several common body shape categories among teleosts which relate strongly to the fishes' habits. "Rover-predators" have the fusiform (streamlined) body shape that is perhaps most typically fish-like. Fins are distributed evenly around the body, and the mouth is terminal (at the end of the snout). This category includes minnows, basses, tunas, and others that typically are constantly moving—searching for and pursuing prey. "Lie-in-wait predators" tend to be more elongated, with the unpaired fins far back on the body, favoring a sudden lunge for their prey. The pike, barracuda, and needlefish typify this category.

"Bottom fishes" include a wide variety of shapes. Some are flattened for lying in close contact with the bottom (as are flatfishes such as flounder), some have flattened heads and sensory barbels (filaments with taste buds) near the mouth (as do catfishes), and some have fleshy lips for sucking food from the bottom sediment (as do suckers). A number of bottom-fish species have structures, usually modified pelvic fins, that enable them to cling to the bottom in areas with strong currents (sculpins and clingfishes have these).

"Surface-oriented fishes" tend to be small, with upward-pointing mouths, heads flattened from top to bottom, and large eyes. The mosquitofish, killifish, and flying fish belong to this category. "Deep-bodied fishes" are laterally flattened and have pectoral fins high on the body, with pelvic fins immediately below. This arrangement favors maneuverability in tight quarters such as coral reefs, thick plant beds, or dense schools of their own species. Examples include angelfishes, surgeonfishes, and freshwater sunfishes. "Eel-like fishes" have highly elongated bodies, tapering or rounded tails, and small, embedded scales (or no scales at all). They are adapted for maneuvering through crevices and holes in reefs and rocks and for burrowing in sediments. Eels, loaches, and gunnels typify this category.

Teleosts occupy habitats ranging from torrential streams high in the Himalayas to the bottom of the deepest oceanic trenches. They are found in the world's highest large lake (Lake Titicaca) and deepest lake (Lake Baikal). Some blind species live in the total darkness of underground caves. One *Tilapia* species lives in hot soda lakes in Africa at 44°C, while the Antarctic icefish *Trematomus* lives at −2°C.

The vast majority of teleost fishes, both marine and freshwater, are tropical. Southeast Asia contains the greatest number of freshwater fish species, but the Amazon and its tributaries contain almost as many (and perhaps many hundreds more, still undiscovered). Marine teleosts are most diverse in the Indo-Pacific region, especially in the area from New Guinea to Queensland, Australia. A single collection made in the Great Barrier Reef off northeastern Australia may contain one hundred or more species. Some marine teleost species have a nearly worldwide distribution, while certain freshwater species have highly restricted ranges. The Devil's Hole pupfish (*Cyprinodon diabolis*), for example, is found only in one small spring in Nevada.

Many teleosts have highly specialized associations, called symbioses, with other organisms. Some live among the stinging tentacles of sea anemones (the clownfish *Amphiprion*), within the gut of sea cucumbers (the pearlfish *Carapus*), within the mantle cavity of giant snails (the conchfish *Astrapogon*), or among the stinging tentacles of the Portuguese man-of-war jellyfish (the man-of-war fish *Nomeus*).

REPRODUCTION

Reproduction among teleosts is incredibly varied. Most species are egg-layers, often producing an enormous number of eggs. The female ocean sunfish *Mola mola* may produce up to 300 million eggs, making it the most fecund vertebrate of all. Some species are livebearers, such as the platyfishes, swordtails, and surfperches. Some species are oral brooders, incubating the eggs in the mouth of the male (as in many cardinal fishes) or of the female (as in many cichlids). In one South American cichlid species, *Symphysodon discus*, the female "nurses" its young with a whitish milklike substance secreted by the skin.

Many teleost species are hermaphroditic. A few of these are synchronous hermaphrodites (functioning as male and female at the same time), such as the hamlet *Hypoplectrus*, but many more are sequential hermaphrodites (first one sex, then the other), such as the sea bass *Serranus*. In some coral reef fishes in the wrasse family, a dominant male mates with a harem of females. If this male is removed, the largest female becomes male and takes over the missing male's behavioral and reproductive function.

In a few species, all individuals are female, as in the Amazon molly *Poecilia formosa*. It has been shown that this species is a "sexual parasite" of two related "host" species. Sperm from host males are required to activate development of Amazon molly eggs, but male and female chromosomes (genetic material) do not join, and the offspring are all genetically uniform females.

ICHTHYOLOGY

The scientific study of bony fishes dates back to Aristotle, who was the first to note, for example, that the sea bass is hermaphroditic. The "father of ichthyology" in more recent times was Peter Artedi (1705–35), whose classification system was used by Carolus Linnaeus (1707–78) in his *Systema Naturae*, which became the basis for all future classification systems.

Bony fish classification depends on the study of taxonomic features, or characters, which vary from one species, or group of species, to another. Useful characters include countable features (meristic characters), such as the number of fin supports (rays) or the number of scales in the lateral line, and measurable features (morphometric characters), such as the relative lengths of body parts. Such studies are typically done on museum specimens that are preserved in alcohol solution after fixation in formaldehyde solution. Dissecting tools, microscopes, and even X-ray machines are used for revealing meristic and morphometric characters. For studying bones, dry skeletons are sometimes prepared, or (especially for small species) specimens are "cleared and stained." This latter technique involves clearing the flesh with potassium hydroxide and staining the bones with Alizarin red stain.

Other techniques use samples of living tissue for finding taxonomic characters. Karyotyping (analysis of the chromosomes) and enzyme electrophoresis (using an electric field to separate similar proteins) are also important sources of taxonomic information.

Specimens for taxonomic studies are collected by means of netting, trapping, catching with hook and line, and spearing. Specialized techniques include electrofishing (use of an electric shocking device) for stream fishes, and ichthyocide (fish poison such as rotenone) for coral reef fishes.

Understanding the evolutionary history and classification of the Osteichthyes also depends on paleontological studies (the study of fossils). Bony fishes are well represented in the fossil record because of the superior fossilizing nature of their bony skeletons. Many fish biologists are concerned with matters other than taxonomy. Because of the economic importance of both marine and freshwater bony fishes, the science of fisheries biology (concerned with the management and exploitation of fish populations) is of great significance. Fish populations are often studied with "age and growth" techniques. Age (determined by scale analysis), length, and weight data can be used to calculate growth and mortality rates, age at maturity, and life span. Other techniques for studying fish populations involve tagging individuals (useful for making estimates of population size) and even using tiny radio transmitters that can be followed by aircraft (useful for studying fish migrations).

Ecologists and ethologists (behavioral biologists) are also active in fish studies, particularly since the invention of scuba diving, which allows direct observation of fishes in their natural habitat. An example of an important discovery made possible by scuba diving is cleaning symbiosis, common in coral reef areas. This symbiosis (an association involving members of two different species) involves a "cleaner" species (often a goby or wrasse), which feeds on the external parasites and diseased tissue of a host ("cleanee") species, which visits the cleaner for this service.

QUESTIONS STILL TO BE ANSWERED

Bony fishes are by far the most numerous of all vertebrates. They are also arguably the most diverse in terms of body form, reproductive habits, symbiotic relationships, and other characteristics. Yet much remains to be learned. Virtually every ichthyological expedition into the Amazon region, for example, returns with specimens of previously unknown species. Some ichthyologists estimate that perhaps five or ten thousand undiscovered teleosts remain in unexplored streams and remote coral reefs.

Many biological mysteries remain about even some of the most familiar species. A good example is the American eel, *Anguilla rostrata*. This predatory species spends most of its life in the rivers, streams, and lakes of eastern North America, where it is often one of the dominant species. After six to twelve years in these habitats, the adult eels swim to the ocean and apparently migrate more than five thousand kilometers to spawn in deep water in the Sargasso Sea (an area in the western Atlantic south of Bermuda).

This general location of eel spawning has been inferred from the appearance there of the tiniest eel larvae (called leptocephali, these were once considered a separate species). The larvae become larger and larger as they drift in the Gulf Stream toward the North American coast. This much has been known since 1922. The adult migration has never actually been followed, however, and no one knows exactly where, at what depth, or how they mate and spawn, nor is it known what then happens to the adults.

Despite many advances in scientific knowledge, much remains to be learned about the interrelationships, ecology, behavior, and fishery potential of the world's bony fish species.

—*George Dale*

FURTHER READING

Bond, C. E. *Biology of Fishes*. 2nd ed. W. B. Saunders, 1996.

Cailliet, G. M., M. S. Love, and A. W. Ebeling. *Fishes: A Field and Laboratory Manual on Their Structure, Identification, and Natural History*. Wadsworth, 1986.

Chiasson, Robert B. *Laboratory Anatomy of the Perch*. 4th ed. Boston: McGraw-Hill, 1991.

Friedman, Matt, and Martin D. Brazeau. "Palaeontology: A Jaw-Dropping Fossil Fish." *Nature* 502.7470 (2013): 175–77.

McClane, A. J. *McClane's Field Guide to Freshwater Fishes of North America*. Henry Holt, 1978.

Mills, D., and G. Vevers. *The Golden Encyclopedia of Freshwater Tropical Aquarium Fishes*. Golden Press, 1982.

Moyle, P. B., and J. J. Cech. *Fishes: An Introduction to Ichthyology*. 4th ed. Prentice Hall, 2000.

"Osteichthyes." *ITIS*, Integrated Taxonomic Information System, 2015, www.itis.gov/servlet/SingleRpt/SingleRpt?search_topic=TSN&search_value=161030#null. Accessed 5 Feb. 2018.

Paxton, John R., and William N. Eschmeyer. *Encyclopedia of Fishes*. 2nd ed. Academic Press, 1998.

Pough, F. H., J. B. Heiser, and W. N. McFarland. *Vertebrate Life*. 5th ed. Prentice Hall, 1999.

Seaman, William Jr., ed. *Artificial Habitats for Marine and Freshwater Fisheries*. Academic Press, 2013.

Food Chains and Food Webs

Fields of Study
Ecology; Biology; Evolutionary Biology; Biochemistry; Thermodynamics

Abstract
All activities of life are powered, whether directly or indirectly, by sunlight. The energy enters the biosphere through primary producers, which trap and convert solar energy into chemical energy, first in form of sugars and ultimately in other complex organic molecules. Energy in its chemical form is then passed along from one type of organism to another through a complex feeding relationship. Two fundamental laws underlie this function: Energy moves through ecosystems in one direction only, while nutrients cycle and recycle. Each time the energy is transferred, some of it is lost as heat. Energy needs constant replenishment from the sun.

Key Concepts

biological magnification: the increasing accumulation of a toxic substance in progressively higher feeding levels

consumer: an organism that eats other organisms

decomposers: microbes such as fungi and bacteria that digest food outside their bodies by secreting digestive enzymes into the environment

detritus feeders: an array of small and often unnoticed animals and protists that live off the refuse of other living beings

energy pyramid: a graphical representation of the energy contained in succeeding trophic levels, with maximum energy at the base (producers) and steadily diminishing amounts at higher levels

nutrient cycle: a description of the pathways of a specific nutrient (such as carbon, nitrogen, or water) through the living and nonliving portions of an ecosystem

producers: organisms that produce food for themselves as well as for nearly all other forms of life, including plants, plantlike protists, and cyanobacteria

trophic level: the categories of organisms in a community, and the position of an organism in a food chain, defined by the organism's source of energy

THE SUN IS THE SOURCE OF ENERGY FOR LIFE

All activities of life are powered, whether directly or indirectly, by a single source of energy: sunlight. The energy enters the biosphere through primary producers, which trap and convert solar energy into chemical energy, first in form of sugars and ultimately in other complex organic molecules. Energy in its chemical form is then passed along from one type of organism to another through a complex feeding relationship. Two fundamental laws underlie this function: Energy moves through ecosystems in one direction only, while nutrients cycle and recycle. Each time the energy is transferred, some of it is lost as heat. Energy needs constant replenishment from an outside source, the sun. In contrast to energy, nutrients constantly cycle and recycle in a circular flow. Nutrients enter the system from soil or water or atmosphere through primary producers (plants), and pass along to herbivores, then to carnivores. The wastes and dead bodies or body parts degraded by detritus feeders return nutrients to the ecosystem.

FEEDING RELATIONSHIPS WITHIN A COMMUNITY

Feeding relationships in a community are often defined or described through food chains and food webs. A linear feeding relationship spanning all trophic levels is called a food chain, whereas many interconnecting food chains in a community make up the food web. Obviously, different ecosystems have drastically different food chains. To illustrate who feeds on whom in a community, it is better first to examine some basic laws and structures that govern a community in general. Energy and nutrients are two common elements that sustain all communities and ecosystems. Energy enters communities through the process of photosynthesis, by which plants and other photosynthetic organisms trap a small portion of sunlight and convert it into sugars. Photosynthetic organisms, from the mighty sequoia to the zucchini and tomato plants in a garden to single-celled diatoms in the ocean, are called autotrophs or producers, because they produce food for themselves. They also produce food for nearly all other organisms, called heterotrophs or consumers.

The amount of life a community can support is determined by how much energy the producers within it can capture. Within the community, energy flows from producers, occupying the first trophic level, through several levels of consumers. The consumers that feed directly and exclusively on producers are herbivores, ranging from caterpillars to buffaloes to wheat aphids. These herbivores, also called primary consumers, form the second trophic level. Carnivores, such as the spider, eagle, fox, and birds that eat caterpillars, are meat eaters, feeding primarily on primary consumers. Carnivores are secondary consumers that form the third or higher trophic level. Some consumers, such as the black bear that eats both blueberries and salmon, occupy more than one trophic level.

A food web, however, represents many interconnecting food chains in a community, describing the actual, complex feeding relationships within a given community. A food web also reflects the feeding nature of organisms that occupy more than one trophic level. Animals such as raccoons, bears, rats, and a variety of birds are omnivores, eating at different consumer levels at different times. In addition to producers and consumers, a functional ecosystem also consists of detritus feeders and decomposers that release nutrients for reuse. The extremely diverse network of detritus feeders is made up of earthworms, mites, protists, centepedes, nematodes, worms, some crustaceans and insects, and even a few vertebrates such as vultures. Except for vultures, these organisms thrive in the garden, and compost by extracting energy stored within dead organic matter, in turn releasing it in a further decomposed state. The excretory products released serve as food for other detritus feeders and decomposers, which are primarily fungi or bacteria. Fungi and bacteria digest food outside their bodies by releasing digestive enzymes into the environment. They then absorb the nutrients they need, and leave the remaining nutrients for recycling. Without detritus feeders and decomposers, nutrients would soon be locked into organic matters, and the ecosystem will cease to be functional.

ENERGY TRANSFER AND THE NUTRIENT CYCLE

One important principle that governs the flow of energy through the ecosystem is that the energy transfer from one trophic level to the next is never efficient. The net transfer of energy between two trophic levels is roughly 10 percent. During the transfer, 90 percent of the energy is lost as heat or in other forms. Of one thousand calories stored in the producer (plants), the caterpillars that consume all plant tissues will obtain one hundred calories, a bird that eats caterpillars will extract ten calories, and when a hawk catches the bird, that portion of the energy is reduced to a mere one calorie. This inefficient energy transfer between trophic levels is called the 10% law, or an energy pyramid. This 10% law has profound impacts within an ecosystem. Plants have the most energy available to them; the most abundant animals will be those directly feeding on plants, and carnivores will always be relatively rare, especially those of large size that occupy a higher trophic level.

However, when toxic substances pass through trophic levels, the exact opposite is true. While energy diminishes in the process of flowing from lower to higher trophic levels, toxic substances progressively increase in concentration along the food chain. This phenomenon, called biological magnification, was discovered through the study of the use of the pesticide Dichlorodiphenyltrichloroethane (DDT). Tests of water samples following the use of this pesticide showed a trace amount of DDT. Tissue analyses of predatory birds in the same aquatic ecosystem, however, revealed a DDT concentration a million times greater than that in the water. Fish caught from the same waters also contained much higher DDT levels than the water, but substantially lower levels than that of the birds that consumed those fish. DDT has since been confirmed as the cause for population declines of several predatory birds, especially fish eaters such as bald eagles, brown pelicans, and cormorants. Understanding biological magnification is crucial to the prevention of widespread loss of wildlife.

Unlike energy, no mechanism or source exists to allow a constant replenishment of nutrients, other than the extremely small amount that comes from the influx of meteoritic dust each day. Essentially the same pool of nutrients has been supporting life from the beginning of life on Earth. Nutrients are elements and small molecules that form all the building blocks of life. Macronutrients are those acquired by organisms in large quantities, including water, carbon, hydrogen, oxygen, nitrogen, phosphorus, sulfur, and calcium. Micronutrients, including zinc, molybdenum, iron, selenium, and iodine, are acquired in trace quantities. Nutrients cycle from producer, to consumer, to detritus feeder and decomposer, and eventually back to producer. The major reservoirs of nutrients are in the nonliving environment, such as soil, rock, water, and atmosphere. In an undisrupted ecosystem, nutrients have cycled and recycled in a sustainable manner for thousands of years. However, human intervention, either through industry or agriculture, has created enormous problems for sustainable nutrient cycling. One example is the nitrogen cycle. Although plants may obtain nitrogen from soil, the primary source of nitrogen is the atmosphere. Nitrogen gas (mainly N_2) makes up over 70% of the air. Gaseous N_2 may be extracted by some plants or converted by lightning into usable forms that drop as nitrogen oxides in rainfall. Through industrial production and use of nitrogen fertilizers in agriculture, humans have overcharged the nitrogen cycle, causing acid rains and surface water and groundwater contamination, which pose a serious threat to natural ecosystems.

FOOD CHAINS AND FOOD WEBS IN A TERRESTRIAL ECOSYSTEM

Since different ecosystems have drastically different food chains and food webs, they can be better illustrated using examples of different communities. The first example is a land community, to provide an overview of terrestrial food chains and food webs. Plants such as maple trees and squash are the producers that occupy the first trophic level. Aphids, caterpillars, grasshoppers, and other animals that forage directly plants or plant tissues are primary consumers that occupy the second trophic level. Birds, spiders, and other insects that feed on primary consumers are secondary consumers occupying the third trophic level. Large birds, such as eagles, owls, and hawks, that eat secondary consumers are tertiary consumers making up the fourth trophic level. This food chain can go on to even higher trophic levels.

However, natural communities rarely contain well-defined groups of primary, secondary, and tertiary consumers in a linear pattern. In reality, a food web showing many interconnecting food chains in a community describes much more accurately the actual feeding relationships within a given land community. This is in part due to the omnivorous, or "eating all" nature of some animals. These animals include but are not limited to bears, rats, and raccoons. They act as primary, secondary, and even tertiary consumers at different times. Many carnivores will eat either herbivores or other carnivores, thus acting as secondary or tertiary consumers, respectively. An owl, for instance, is a secondary consumer when it eats a mouse, which feeds on plants, but a tertiary consumer when it eats a shrew, which feeds on insects. Once a shrew eats carnivorous insects, it is by itself a tertiary consumer, making the owl that feeds on the shrew a quaternary consumer that occupies the fifth trophic level. Since organisms are interlocked in such complex yet organized networks, disruption at any particular point of the food web (damage to one group of organisms) might have far-reaching effects on a whole community or ecosystem.

FOOD CHAINS AND FOOD WEBS IN A MARINE ECOSYSTEM

Coral reefs will be used as an example for marine ecosystem. Reefs are created by concerted efforts of producers—algae—and consumers—corals. In warm tropical waters, with just the right combination of bottom depth, wave action, and nutrients, specialized algae and corals build reefs from their own calcium carbonate skeletons. The reef-building corals grow best at depths of less than forty meters, where light can penetrate and allow their algal partners to photosynthesize. Algae and corals are involved in a mutualistic relationship, where algae benefit from the high nitrogen, phosphorus, and carbon dioxide levels in the coral tissues. In return, algae provide food for the coral and help produce calcium carbonate, which forms the coral skeleton.

Coral reefs provide an anchoring place for many other algae, a home for bottom-dwelling animals, and shelter and food for the most diverse collection of invertebrates and fish in the oceans. In essence, algae are the producers that occupy the first trophic level. Corals that feed on algae are primary consumers sitting at the second trophic level. Many fish (such as blue tang) that feed on corals are secondary consumers occupying the third trophic level. Larger fish, such as sharks that eat small fishes, are tertiary consumers at the fourth trophic level. A vast array of zooplanktons, invertebrates such as sponges, the poisonous blue-ringed octopus, and so on, also live in coral reef ecosystems to make extremely complex marine food webs. For example, the Great Barrier Reef in Australia is home to more than two hundred species of coral, and a single reef may harbor three thousand species of fish, invertebrates, and algae.

Similar to terrestrial ecosystems, aquatic ecosystems are also prone to human disturbance. Of all aquatic or marine ecosystems, coral reefs are probably most sensitive to certain types of disturbance, especially silt caused by soil eroding from nearby land. As silt clouds the water, light is diminished and photosynthesis reduced, hampering the growth of the corals. Furthermore, as mud accumulates, reefs may eventually become buried and the entire magnificent community of diverse organisms destroyed. Another hazard is sewage and runoff from agriculture. The dramatic rise in fertilizer in near-shore water causes eutrophication, by which excessive growth of algae blocks sunlight from the corals, deprives corals of nutrients, and suffocates corals and other organisms. A third threat to coral reefs, overfishing, is also strictly a result of human interference. It is estimated that in over eighty countries, an array of species, including mollusks, turtles, fish, crustaceans, and even corals, are being harvested much faster than they can replace themselves. Collectively, these human activities had destroyed over 30 percent of coral reefs worldwide by year 2000. Assuming no effective measure is taken to preserve or restore coral reef ecosystems, another 50 percent of reefs will disappear by year 2030. The message is clear: Once humans disturb an ecosystem, through damaging one or more species in an intricately networked food web, balance and sustainability within the whole system is affected. The price for such disruption is high and far-reaching.

—*Ming Y. Zheng*

FURTHER READING

Baskin, Y. "Ecologists Dare to Ask: How Much Does Diversity Matter?" *Science* 264 (April 8, 1994): 202–203.

Bell, R. H. V. "A Grazing Ecosystem in the Serengeti." *Scientific American* 255 (July 1971): 86–93.

Earthworks Group. *Fifty Simple Things You Can Do to Save the Earth*. Earthworks Press, 1989.

Goreau, T. F., et al. "Corals and Coral Reefs." *Scientific American* 241 (August 1979): 124–136.

Holloway, M. "Sustaining the Amazon." *Scientific American* 269 (July 1993): 90–99.

Quamman, David. *The Song of the Dodo: Island Biogeography in an Age of Extinction*. Simon & Schuster, 1997.

Gametogenesis

Fields of Study

Reproductive Biology; Developmental Biology; Molecular Biology; Biochemistry

Abstract

Sexual reproduction is the predominant mode of reproduction in animals. Sexual reproduction involves the production of gametes: the eggs and sperm. In most animals, these gametes are produced in specialized organs called gonads (ovaries and testes). In dioecious animals, the sex cells from two different individuals (one male and one female) will fuse together in a process known as fertilization to form the offspring. The advantage of sexual reproduction seems to be in its potential to produce variability in the gametes and therefore in the new organism.

Key Concepts

diploid: the number of chromosomes or the amount of genetic material normally found in the nucleus of body cells; this number is constant for a particular species of animal

gamete: a sex cell; the egg or ovum in the female and the sperm in the male

haploid: one-half of the diploid number; the number of chromosomes or the amount of genetic material found in a gamete

meiosis: reduction division of the genetic material in the nucleus to the haploid condition; it is the process used by animal cells to form the gametes

oogenesis: gamete formation in the female; it occurs in the female gonads, or ovaries

spermatogenesis: gamete formation in the male; it occurs in the male gonads, or testes

spermiogenesis: the structural and functional changes of a spermatid that lead to the formation of a mature sperm cell

NEW BEGINNINGS

Sexual reproduction is the predominant mode of reproduction in animals. Sexual reproduction involves the production of gametes: the eggs and sperm. In most animals, these gametes are produced in specialized organs called gonads (ovaries and testes). The sex cells in most animals are separate—that is, each individual animal contains either testes or ovaries, but not both. Such animals are said to be dioecious. In dioecious animals, the sex cells from two different individuals (one male and one female) will fuse together in a process known as fertilization to form the offspring. The advantage of sexual reproduction seems to be in its potential to produce variability in the gametes and therefore in the new organism.

Gametes are highly specialized cells that are adapted for reproduction. These egg and sperm cells develop by a process of gametogenesis, or gamete formation. Sperm cells are relatively small cells that are specialized for motility (movement); egg cells are larger, nonmotile cells that, in many species, contain considerable amounts of stored materials that are used in the early development of the zygote (fertilized egg).

In animals, gametogenesis consists of two major events. One involves the structural and functional changes in the formation of the gamete. The other involves the process of meiosis. Animal body cells normally contain the diploid amount of genetic material. Each species of animal has a characteristic diploid number that remains the same from generation to generation. Because fertilization involves the fusion of the egg and the sperm, bringing together each cell's set of genetic material, some mechanism must reduce the amount of genetic information in the gamete, or it would double every generation. Meiosis is a special nuclear division whereby the genetic material is reassorted and reduced to form haploid cells. Therefore, gametes are haploid, and gamete fusion during fertilization reestablishes the diploid content in the zygote.

SPERM

Sperm are highly motile cells that have reduced much of their cellular contents and are little more than a nucleus with a tail. Sperm are produced in the testes from a population of stem cells called spermatogonia. Spermatogonia are large diploid cells that reproduce by an equal division process called meiosis. Spermatogenesis is the process by which these relatively unspecialized diploid cells will become haploid cells; it is a continuous process that occurs throughout the sexually mature male's life. When a spermatogonium is ready to become sperm, it will stop dividing mitotically, enlarge, and begin the reduction division process of meiosis. These large diploid cells that begin to divide meiotically are known as primary spermatocytes.

The first step in the division process involves each primary spermatocyte dividing to form two secondary spermatocytes. Each secondary spermatocyte continues to divide, and each forms two spermatids. These spermatids are haploid cells. For each primary spermatocyte that undergoes spermatogenesis, four spermatids are formed. The spermatids are fairly ordinary cells; they must go through a process that will form them into functional sperm. The transformation process of a spermatid into a sperm is called spermiogenesis and involves several changes within the cell. The genetic material present in the nucleus begins to condense, while much of the cytoplasm and its subcellular structures are lost. The major exception to this latter event is the retention of mitochondria, cytoplasmic structures involved in energy production.

The mature sperm has three main structural subdivisions: the head, the neck (or midpiece), and the tail. All are contained within the cell's membrane. The oval head has two main parts, the haploid nucleus and the acrosome. The acrosome comes in various shapes but generally forms a cap over the sperm nucleus. The acrosome functions differently in various animals, but generally its functions are associated with the fertilization process (union and subsequent fusion of egg nucleus and sperm nucleus). Acrosomes contain powerful digestive enzymes (organic substances that speed the breakdown of specific structures and substances) that allow the sperm to reach the egg's membrane. The midpiece of the sperm contains numerous mitochondria, which provide the energy for the sperm's movement. The tail, which has the same general organization as flagella or cilia (subcellular structures used for locomotion or movement of materials), uses a whiplike action to propel the sperm forward during locomotion. The structural changes that occur during spermiogenesis are meant to streamline and pare down the sperm cell for action of a special sort and of a limited duration. The sperm's function is to "swim" to the egg, to fuse with the egg's surface, and to introduce its haploid nucleus into the egg's interior.

EGGS

The female gamete, the egg or ovum, is produced by a process known as oogenesis. This process occurs in the female gonads, the ovaries. At first glance, oogenesis and spermatogenesis appear to be very similar, but there are some striking differences. The major similarity is that both processes form gametes, which contain genetic material that has been reduced to the haploid condition. To understand oogenesis, one must consider that its goal is to produce a cell that is capable of development. The mature egg cell in all animals is large in comparison with other cells, particularly with the sperm. There are two important features of the egg that must be considered: the presence of a blueprint for development and the means to construct an embryo from that blueprint. In other words, the egg must be programmed and packaged during oogenesis. The programming refers to the information that is coded within the structure of the egg. This information includes the genetic material as well as the cytoplasmic information. Together, the nucleus and cytoplasm provide the egg with the potential to transform a simple cell into a complex preadult form. Since it is within the egg that this transformation occurs, the programming must be within the organization of the egg, and the directions for development must be within that organization. The packaging refers to the presence of all the material necessary to build embryonic structures, to nourish this developing embryo, and to provide its energy until it can obtain nourishment on its own.

As happens in spermatogenesis, the potential eggs are formed from unspecialized stem cells, in this case called oogonia. Oogonia contain the diploid amount of genetic material and divide by the process of mitosis. At some point in their life, oogonia stop dividing mitotically, enlarge, and prepare to become eggs—that is, they begin meiosis. The cell that begins this reduction division process is called the primary oocyte. Each primary oocyte divides into two cells—one large cell, the secondary oocyte, and a very small cell, the first polar body. The secondary oocyte continues the final reduction phase of meiosis and forms two cells, one large one (the ovum) and one very small one (the second polar body). The first and second polar bodies are nonfunctional byproducts of meiosis. The one functional cell, the mature egg, contains most of the cytoplasm of the primary oocyte and one-half of its genetic material. In many animals (primarily the vertebrates), all oogonia present in the ovaries enter meiosis at the same time; the initial events of oogenesis are synchronous within the animal. Oogenesis in many animals is not a continuous process, as is spermatogenesis in the male. Rather, the primary oocytes in the first stages of reduction division may remain inactivated for a long time—in some cases, for several decades. Therefore, in female animals with this format of oogenesis, a primary oocyte population is maintained, and eggs will mature as they are needed.

Thus far it appears that the egg's formation differs from the sperm's in three ways. First, in many female animals, there is a limited number of primary oocytes capable of going on to form eggs; second, this egg formation is not necessarily a continuous process; third, one primary oocyte yields one mature egg at the end of meiosis. Although these are three very important differences, there are other distinctly egg events that deal with the developmental programming and the packaging of materials in this potential gamete.

EGGS AND RNA

Little is known about the egg's storage of developmental directions or the actual programming of information, but developmental and molecular biologists are beginning to elucidate events that occur during oogenesis that are concerned with function of the egg. One such event, fairly widespread among the animal kingdom, is the formation of so-called lampbrush chromosomes during oogenesis. The chromosome's backbone unravels at many sites so that regions, composed of specific genes, loop outward from the backbone. These loops give the chromosome its distinctive lampbrushlike appearance. Large amounts of a nucleic acid known as messenger ribonucleic acid (mRNA) are being made on each loop. This mRNA is then processed and sent into the developing egg's cytoplasm, where most of it will be stored for use during early development. After fertilization, these maternal

(egg-derived) mRNAs can be used to make specific proteins necessary for the embryo.

Another event present in some developing eggs is the mass production of another type of RNA known as ribosomal RNA (rRNA). Most of this rRNA will also be stored until fertilization. After fertilization, these rRNA particles will help form cytoplasmic structures called ribosomes (the sites of protein synthesis). In addition to these egg products, many animal eggs must become filled with yolk. Yolk is the general term that covers the major storage of material in the egg.

Because the maternal proteins (yolk and other protein components) and nucleic acids (various RNAs) form the bulk of the egg cytoplasm, they have profound influences on the development of the embryo. In particular, the positions of maternal mRNAs, ribosomes, and proteins affect the organization of the embryo. It is evident, then, that the maternal genetic information and the arrangement of the products of this information provide crucial developmental information that will control much of the course of embryonic development. Therefore, the egg contributes considerably more than a haploid nucleus to the zygote.

STUDYING GAMETOGENESIS

There are several approaches to the study of gametogenesis. Early biologists employed cytological techniques (methods of preparing cells for the study of their structure and function) and microscopy to study gamete formation. These early studies were, in fact, observations of the actual events themselves. Although these early descriptive approaches gave much information about the cells involved at each stage of gamete formation, they did not provide any information about the control mechanisms for this process. Biochemical studies have contributed to the understanding of certain regulatory substances and how they function in gametogenesis. By enhancing or inhibiting the presence of these regulatory substances in the organism, investigators have been able to elucidate many of the normal events of gametogenesis.

Beginning at puberty, the hormones (substances released from endocrine glands, generally functioning to regulate specific body activity) of the hypothalamus, the pituitary gland, and the gonads interact to establish and regulate gametogenesis in the organism. gonadotropin-releasing hormone (GnRH) from the hypothalamus stimulates the release of follicle-stimulating hormone (FSH) and luteinizing hormone (LH) from the anterior portion of the pituitary gland. All three of these hormones are necessary for spermatogenesis and oogenesis.

Surgical removal of the mammalian pituitary gland (hypophysectomy) in the male leads to degeneration of the testes. Testicular function can be restored in these hypophysectomized animals by administering the hormones FSH and LH. These studies suggest that FSH and LH are necessary for normal functioning of the testes. LH appears to stimulate the release of testosterone (male hormone) by certain cells (Leydig cells) of the testes. Both testosterone and FSH are necessary for spermatogenesis, but the exact role that each of these hormones plays in male sexual physiology has yet to be determined.

Oogenesis in the female has been the subject of intense investigation. At the beginning of each ovarian cycle, from puberty to menopause, one primary oocyte present in the female's ovaries is activated to continue the process of gamete formation. Release of GnRH from the hypothalamus at the beginning of each cycle stimulates the anterior portion of the pituitary gland to release FSH, which in turn, affects the ovaries: It stimulates a primary oocyte to mature to the point that it can be released from the ovary, as a secondary oocyte, and it causes certain cells (follicle cells) in the ovary to produce estrogens, female hormones. High estrogen levels will cause the pituitary to inhibit FSH release, a negative feedback mechanism, and stimulate LH release. These estrogen-mediated events occur at approximately the middle of the ovarian cycle. LH also affects the ovaries. LH, however, is responsible for ovulation (the release of the oocyte from the ovaries) and for the formation of a cellular structure called the corpus luteum. LH also stimulates the corpus luteum to produce progesterone, another female hormone. Eventually, high levels of progesterone will inhibit LH release from the pituitary gland, and the cycle begins anew.

—*Geri Seitchik*

FURTHER READING

Epel, D. "The Program of Fertilization." *Scientific American* 237 (November 1977): 128–140.

Kinne, Rolf K. H., ed. *Oogenesis, Spermatogenesis, and Reproduction.* Karger, 1991.

Sadler, R. M. *The Reproduction of Vertebrates.* Academic Press, 1973.

Van Blerkom, Jonathan, and Pietro M. Motta, eds. *Ultrastructure of Reproduction: Gametogenesis, Fertilization, and Embryogenesis.* Kluwer, 1984.

Gas Exchange

Fields of Study

Animal Physiology; Physical Biochemistry

Abstract

Gas exchange refers to two major steps in the overall oxygen consumption and carbon dioxide excretion by the whole animal. These two steps are the movement of the respiratory medium (containing oxygen) past the site of gas exchange, known as ventilation, and the diffusion of oxygen across the gas-exchange surface into the animal.

Key Concepts

diffusion: the passive movement of a gas across a membrane from a region of high pressure to one of low pressure

epithelium: a thin layer of cells that lines a body surface, such as the lining of the lungs or the intestines

partial pressure: that part of the atmospheric pressure caused by only a single gas of many in a mixture; it is determined by how much of the gas is present in the mixture

permeability: the tendency, in this case of a membrane, to permit the movement of a gas across that membrane

respiratory medium: the water or air that contains the oxygen used by an animal to carry out biochemical reactions

ventilation: the movement of the respiratory medium to and across the site of gas exchange

IN WITH THE GOOD AIR, OUT WITH THE BAD

Gas exchange is the uptake of oxygen and the loss (or elimination or excretion) of carbon dioxide. It refers to two major steps in the overall oxygen consumption (or carbon dioxide excretion) by the whole animal. These two steps are the movement of the respiratory medium (containing oxygen) past the site of gas exchange, known as ventilation, and the diffusion of oxygen across the gas-exchange surface into the animal. The final step in diffusion of oxygen into an animal always involves diffusion from a liquid into a liquid, even in air breathers. Both ventilation and diffusion depend on the design of the structures as well as the way in which the systems and structures work. The additional steps in the whole respiratory system are the internal counterparts to ventilation and diffusion. These are perfusion, or blood flow, and diffusion from the blood to the tissues.

Basically, gas exchange takes place at a respiratory surface where the source of oxygen, the respiratory medium, is brought into contact with the surface. Oxygen diffuses into the animal, carbon dioxide diffuses out, and the spent or used respiratory medium is removed. The movement of the respiratory medium is termed ventilation. Once oxygen is in the animal, it is transported to the site of oxygen utilization, the tissues. Carbon dioxide, on the other hand, must be transported from the tissues to the site of gas exchange for excretion in gaseous form.

GAS-EXCHANGE ORGANS

Animals have three basic types of gas-exchange organs: skin, invaginations (inpocketings of the epithelium), and evaginations (outpocketings of the epithelium). All three show modifications to improve the conditions of gas exchange. Skin always permits gas exchange unless it is coated with some material that limits

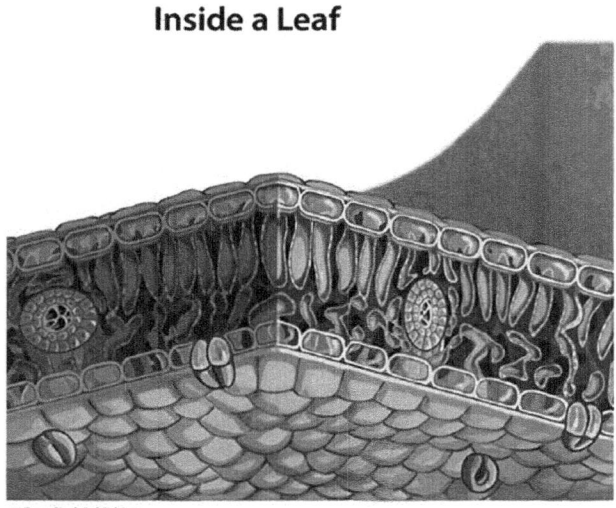

Inside a Leaf

The cross-section of a leaf is shown. The wheel-shaped structures, are veins and the darker structures are guard cells which surround each pore, or stoma. Many stoma are called stomata, and they function to exchange gases on the surface of the leaf.

diffusion. The skin of a snake or a turtle is so coated and permits very little gas exchange. The skin of a worm or an octopus, on the other hand, is quite thin and permits gas exchange quite frely. Invaginations of the external epithelium are basically what lungs and insect tracheas are, but in a highly modified condition. Evaginations of the skin are represented by the gills of aquatic animals; even when inside a cavity, as are fish and crab gills, they are still evaginations.

There is only one way that animals take up oxygen from the external medium, regardless of whether that medium is air or water. Gas must passively diffuse across the membrane that separates the animal from its environment. That membrane is a type of tissue called epithelium and is similar in nature and structure to the tissue that lines other body surfaces. The different types of epithelia are classified according to their locations and functions; those lining gills, lungs, and certain other organs of gas exchange are all known as respiratory epithelia. The respiratory epithelium not only separates the internal and external fluids but also represents a barrier to the movement of materials such as gas.

GAS DIFFUSION

Diffusion of a gas across a membrane occurs according to the laws of physics. The driving force for gas diffusion is the difference in the partial pressure of the gas across the membrane. A high external partial pressure and low internal partial pressure will provide a large difference and will enhance diffusion. Oxygen makes up 20.9 percent of the air, so that 20.9 percent of the atmospheric pressure at sea level (14.72 pounds per square inch) is attributable to oxygen (3.08 pounds per square inch). At higher altitudes, atmospheric pressure and partial pressure of oxygen are reduced.

The other factors that determine the diffusion of a gas are the thickness of the membrane across which it diffuses, the membrane's total surface area, and the nature or composition of the membrane. Obviously, a thick membrane will retard diffusion of gas because the gas must move across a greater distance. The distance the gas must diffuse is known as the diffusion distance. Additionally, the total surface area of the membrane available for diffusion has a direct effect on the rate of diffusion from one place to another. The greater the surface area, the greater the quantity of gas that can diffuse in a given time. Finally, the composition of the membrane is of critical importance in determining the diffusion of a gas. The nature of the membrane is referred to as the permeability of the membrane to the gas in question. The greater the permeability, the more easily gas diffuses. A membrane with a layer of minerals (calcium, for example) on the cells will not be as permeable as one without such a layer.

A very important point to note is that gases diffuse according to the difference in the partial pressure of the gas and not according to the concentration of the gas in the liquid. Several scientists have proved this by constructing artificial systems with two dissimilar fluids separated by a membrane. The movement of oxygen is always from high partial pressure to low and not from a high concentration to low. The reason is that pressure is a measure of molecular energy, but concentration of a gas in a liquid depends on the amount of that gas that can dissolve in the liquid—its solubility.

A FLUID-FLUID BOUNDARY

All gas exchange occurs across a fluid-fluid boundary—that is, from one liquid to another—even in air breathers. The explanation for this is that all respiratory epithelia are moist and are kept so by the cells that line the surface. If the surface were to dry out, the permeability to gases (and other materials) would be substantially reduced. Thus, in air breathers, oxygen must first dissolve in the thin fluid layer lining the surface before diffusion across the epithelium takes place.

Organs of gas exchange work as do radiators, except that a gas is exchanged instead of heat. In this system, there are two liquids, one the source and the other the sink for the transferred material, the gas. The source is the external supply, and the sink is the blood or other internal fluid. Both fluids are contained in vessels or tubes that channel and direct it, with a thin layer of epithelium between the two. In the most efficient transfer systems, both the source and the sink flow, and they flow in opposite directions. If they did not flow, then the two fluids would simply come to an equilibrium, with oxygen partial pressure the same in both of the fluids. By moving in opposite directions, each is renewed, and the difference between them is always maximized. This type of exchange system is a "countercurrent system," because the two fluids flow in opposite directions. The most efficient of the gas-exchange organs function this way.

Two other types of gas-exchange systems are based on fluids flowing in directions other than perfectly opposite to each other. Birds have a respiratory system in which blood and air do not flow in opposite directions; rather, the blood flows perpendicularly to the direction of the air flow. This is referred to as crosscurrent flow. While not as efficient as countercurrent flow, it provides for an acceptably high level of efficiency. A system in which the respiratory medium is not channeled, but the blood is instead in vessels, is known as a mixed volume system. The mammalian lung functions in this way; the air is pumped into sacs that are lined with tiny blood vessels, but the air does not flow.

VENTILATION

Ventilation of gas-exchange organs is accomplished by a pumping mechanism that brings the respiratory medium to and across the gas-exchange surface. Both water and air breathers use ventilatory pumps, but water breathers must move a much heavier and denser medium than air breathers. Pumping mechanisms may be located at the inflowing end of the system or at the outflowing end. The former are positive pressure pumps that push respiratory medium, and the latter are negative pressure pumps that pull respiratory medium into the cavity. Negative pressure is used in mammalian lungs, insect tracheas, and crab gills, while positive pressure is used in some fish that push water from the mouth into the gill chamber.

There are two basic patterns of flow of the respiratory medium through gas-exchange organs: one-way and tidal. One-way flow is found in fish, crabs, clams, and a number of other aquatic animals. Interestingly enough, the bird respiratory system also uses one-way flow. In one-way flow systems, the medium is always moving and passes over the gas exchange surface only once. In tidal-flow systems, the respiratory medium moves in and out (like the tide) through the same passages and tubes. The mammalian and insect respiratory systems both utilize tidal ventilation. The respiratory medium is not always moving, and when it is exhaled, there is some amount remaining in the cavity. The remaining respiratory medium will contain more carbon dioxide and less oxygen than fresh respiratory medium, with which it will mix upon inhalation.

MEASURING GAS EXCHANGE

The total amount of oxygen used by an animal is a gross measure of gas exchange known as oxygen uptake or oxygen consumption. Its counterpart for carbon dioxide is carbon dioxide excretion. Oxygen uptake is expressed as the amount of oxygen used per minute per kilogram of animal mass. Carbon dioxide excretion is expressed in the same terms. In theory, oxygen uptake and carbon dioxide excretion will be numerically the same, but in live animals there are several circumstances that cause the two to differ. Measurement of both rates is accomplished in similar ways. One method is the use of a respirometer and involves placing an animal in a sealed container and measuring the rate at which oxygen is depleted or carbon dioxide produced by an animal. Alternatively, respiratory medium, either air or water, is pumped through the respirometer and the oxygen or carbon dioxide measured in the inflowing and outflowing medium; the difference will be the amount used by the animal. The flow rate of the air or water must also be known for the calculations. The use of a respirometer is preferable but may not be practical for large animals, such as a horse.

In the case of animals too large to use a respirometer, oxygen uptake or carbon dioxide excretion is determined by measuring the rate of flow of the respiratory medium through the gas exchange organ and measuring the oxygen in the inspired and expired air or water. The result is the ventilation rate and the amount of oxygen extraction, the product of which is the oxygen consumption rate. This measurement is straightforward in animals with one opening for inspired and another for expired respiratory medium, such as a fish or a crab. In animals that inhale and exhale through the same organ, however, there are complications that make the measurements more difficult. Still, it is possible to measure the flow of air in and out of a lung and to collect at least some of the gas and measure either the oxygen or carbon dioxide in that air.

There is an additional advantage to the latter technique, measuring ventilation and the oxygen in the water or air. That advantage is that another measure of gas exchange is provided in these measurements. The difference between the amount of oxygen in inspired and expired water or air is the extraction (the amount taken out) and assesses the efficiency of the gas exchange organ. The efficiency is usually given as the percentage of oxygen taken out of the respiratory medium (the amount removed divided by the amount in inspired air or water). There are numerous factors that affect extraction, and measuring the efficiency provides one piece of information.

Studies of gas exchange encompass all levels of organization of animals, from the cellular to the whole animal. One of the most important levels concerns the structure of the organ and the parts of the organ. For this, it is necessary to see the spatial relationships among the parts, measure distances and areas, and count structures. The surface area, the volume, the number of structures or substructures, and the diffusion distances must all be measured. The results describe the morphology and morphometrics of the organ. Both whole, intact animals, and preserved specimens are used to make these measurements. The techniques are those used in surgery and dissection, and the results are critical to an understanding of the basic function of the respiratory organ. The electron microscope has been a powerful tool in this regard, permitting the accurate measurement of cellular-level distances, such as the diffusion distance.

MEASURING PARTIAL PRESSURE DIFFERENCE

Measuring the partial pressure difference between the inside and outside of the animal is critical because of the role that this pressure difference has in gas exchange. Partial pressure of oxygen may be measured in two ways: in the intact animal or in a sample removed from the animal and injected into an instrument. The instrument most commonly used for measuring oxygen or carbon dioxide partial pressure is an electrode that changes electrical output when oxygen diffuses across an artificial membrane into a salt solution. Some of these electrons have been miniaturized and are only four millimeters across, and they will fit in a syringe needle. Still, it is difficult to use one of these in an intact mammal. The other way to measure partial pressure of oxygen or carbon dioxide on either side of the respiratory epithelium is to withdraw a sample of the air or water on the outside or the blood from the vessels on the inside. This procedure may be routine (in animals such as fish and crabs) or somewhat difficult (as in a mammal). A small tube is threaded into the lung to withdraw the air sample.

MEASURING VENTILATION

The movement of the respiratory medium, ventilation, is an important measure in determining the rate at which oxygen is brought to the respiratory surface. The blood flow (perfusion) on the inside is the counterpart to ventilation and is equally important. Ventilation can be measured either indirectly (meaning it is calculated) or directly. Indirect determinations require measuring other functions and then calculating ventilation based on known equations. If the rate of oxygen uptake and the extraction are measured, for example, then ventilation can be calculated.

Direct measures of ventilation use an electronic sensing device to determine the flow of water or air at the site of intake or outflow of respiratory medium on the animal. A human subject can simply breathe into such an electronic or mechanical device. Nonhuman mammals are more difficult and frequently require indirect techniques.

Direct measures may be the flow rate of the respiratory medium, the frequency of breathing, the hydrostatic pressure in the respiratory chamber, or a change in shape and size of the respiratory chamber. Any of these measures can be used to monitor routine respiratory function, but all are needed to assess gas exchange completely and accurately.

It is also necessary to know the general pattern of water or air movement at the respiratory surface. To do so often requires some invasive technique and the use of an indicator, such as a dye in the respiratory medium. The movement of the medium can then be visualized to determine the pattern. In some animals, video cameras can be used to photograph flow patterns of dyed medium, particularly water.

USES OF GAS EXCHANGE STUDY

Gas exchange is studied by researchers and health practitioners both to assess basic function and to determine the source and nature of limitations of the systems of the body. These two areas may seem quite different at first; one is applied research, and the other is considered basic research. Both, however, have the same bases and use the same equations and principles. Only the animals or conditions differ.

One of the clinical applications, or contexts, in which gas exchange is studied is in respiratory distress or pulmonary (lung) disease. In these cases, the respiratory epithelium may become inflamed and thickened. This will increase the diffusion distance and retard, or limit, oxygen uptake and carbon dioxide release at the lung. Secretion of mucus by a respiratory epithelium may have a similar result for the same reasons. Mucus secretion occurs in several diseases and also takes place in fish gills when irritated by noxious chemicals in the water.

Gas exchange is also studied in diverse animals to understand evolutionary trends and pressures. Animals that live at high altitudes, for example, are constantly faced with low oxygen pressure in the air, and therefore some adjustment must be made by the animal. Scientists study the respiratory systems of these animals to determine if one of the other factors that affects gas exchange, such as diffusion distance or total surface area, is altered to compensate for the lower pressure difference.

All animals have similar basic physiological needs, including a need for oxygen to fuel the conversion of food materials into energy and other substances. Many animals have unique or specific forms or structures enabling them to survive in a particular habitat. Some of these forms affect the respiratory system—as in the differences between the respiratory surface in land animals compared with similar species that live in water. Scientists have compared the gas-exchange systems in aquatic and terrestrial species to learn more about evolutionary processes.

—*Peter L. deFur*

Further Reading

Boutilier, R. G., ed. *Vertebrate Gas Exchange: From Environment to Cell.* Springer-Verlag, 1990.

Dejours, Pierre, ed. *Principles of Comparative Respiratory Physiology.* 2nd ed. Elsevier, 1981.

Hill, R. W., and G. A. Wyse. *Animal Physiology.* Harper & Row, 1989.

Prange, Henry D. *Respiratory Physiology: Understanding Gas Exchange.* Chapman & Hall, 1996.

Rahn, H., A. Ar, and C. V. Paganelli. "How Bird Eggs Breathe." *Scientific American* 240 (February 1979): 46–55.

Raven, P. H., and G. B. Johnson. *Biology.* 4th ed. McGraw-Hill, 1996.

West, John B. *High Life: A History of High-Altitude Physiology and Medicine.*: Oxford University Press, 1998.

Gene Flow

Fields of Study

Genetics; Evolutionary Biology; Developmental Biology; Biogeography

Abstract

Genes are elements that control the transmission of a hereditary characteristic by specifying the structure of a particular protein or by controlling the function of other genetic material. Within any breeding population of a species, the exchange of genes is constant among its members. If a new gene or combination of genes appears in the population, it is rapidly dispersed among all members of the population through inbreeding. New alleles may be introduced into the gene pool of a breeding population in two ways: mutation and migration. Gene flow is integral to both processes.

Key Concepts

allele: one of a group of genes that occurs alternately at a given locus

deme: a local population of closely related living organisms

fossil: a remnant, impression, or trace of an animal or plant of a past geologic age that has been preserved in the earth's crust

gene pool: the whole body of genes in an interbreeding population that includes each gene at a certain frequency in relation to other genes

mutation: a relatively permanent change in hereditary material involving either a physical change in chromosome relations or a biochemical change in the codons that make up genes

population: a grouping of interacting individuals of the same species

speciation: the process whereby some members of a species become incapable of breeding with the majority and thus form a new species

species: a category of biological classification ranking immediately below the genus or subgenus, comprising related organisms or populations capable of interbreeding

PRE- AND POST-DARWIN SCIENCE

Prior to the nineteenth century, religious dogmatism retarded the activities of most scientists investigating the origins and nature of life by insisting on the immutability of species created by God. Despite mounting fossil evidence that many species of flora and fauna that once inhabited Earth had disappeared and that many extant species could not be found in the fossil record, pre-nineteenth century naturalists could find no viable explanation (other than divine intervention) for the disappearance of life-forms and their replacement by other forms. Then, in 1859, Charles Darwin published his epochal *On the Origin of Species,* which proposed the theory that all contemporary life-forms have evolved from simpler forms through a process he called "natural selection."

Many individuals before Darwin had proposed theories of evolution, but Darwin's became the first to be widely accepted by the scientific community. His success resulted from the careful and objective presentation of an overwhelming amount of evidence showing that species can and do change, and his concurrent promulgation of a convincing explanation of the mechanism that produces that change—natural selection. Since Darwin, scientists have modified and added new concepts to his theory, especially concerning the ways in which species change (evolve) over time. One of those new concepts, which was only dimly understood in Darwin's lifetime, is the importance of genetics in evolution, especially the concepts of migration and gene flow.

GENES AND GENE EXCHANGE

Genes are elements within the germ plasm of a living organism that control the transmission of a hereditary characteristic by specifying the structure of a particular protein or by controlling the function of other genetic material. Within any breeding population of a species, the exchange of genes is constant among its members, ensuring genetic homogeneity. If a new gene or combination of genes appears in the population, it is rapidly dispersed among all members of the population through inbreeding. New alleles may be introduced into the gene pool of a breeding population (thus contributing to the evolution of that species) in two ways: mutation and migration. Gene flow is integral to both processes.

A mutation is the appearance of a new gene or the almost total alteration of an old one. The exact causes of mutations are not completely understood, but scientists have demonstrated that they can be caused by radiation. Mutations occur constantly in every generation of every species. Most of them, however, are either minor or detrimental to the survival of the individual and thus are of little consequence. A very few mutations may prove valuable to the survival of a species and are spread to all of its members by migration and gene flow.

When immigrants from one population interbreed with members of another, an exchange of genes between the populations ensues. If the exchange is recurrent, biologists call it "gene flow." In nature, gene flow occurs on a more or less regular basis between demes, geographically isolated populations, and even closely related species. Gene flow is more common among the adjacent demes of one species. The amount of migration between such demes is high, thus ensuring that their gene pools will be similar. This sort of gene flow contributes little to the evolutionary process, since it does little to alter gene frequencies or to contribute to variation within the species. Much more significant for the evolutionary process is gene flow between two populations of a species that have not interbred for a prolonged period of time.

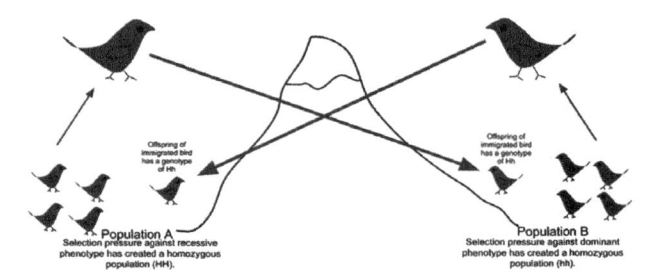

Gene flow is the transfer of alleles from one population to another population through migration of individuals. In this example, one of the birds from population A migrates to population B, which has less of the dominant alleles.

Populations of a species separated by geographical barriers often develop very dissimilar gene combinations through the process of natural selection. In isolated populations, dissimilar alleles become fixed or are present in much different frequencies. When circumstances do permit gene flow to occur between two such populations, it results in the breakdown of gene complexes and the alteration of allele frequencies, thereby reducing genetic differences in both. The degree of this homogenization process depends on the continuation of interbreeding between members of the two populations over extended periods of time.

HYBRIDIZATION

The migration of a few individuals from one breeding population to another may, in some instances, also be a significant source of genetic variation in the host population. Such migration becomes more important in the evolutionary process in direct proportion to the differences in gene frequencies—for example, the differences between distinct species. Biologists call interbreeding between members of separate species "hybridization." Hybridization usually does not lead to gene exchange or gene flow, because hybrids are not often well adapted for survival and because most are sterile. Nevertheless, hybrids are occasionally able to breed (and produce fertile offspring) with members of one or sometimes both the parent species, resulting in the exchange of a few genes or blocks of genes between two distinct species. Biologists refer to this process as "introgressive hybridization." Usually, few genes are exchanged between species in this process, and it might be more properly referred to as "gene trickle" rather than gene flow.

Introgressive hybridization may, however, add new genes and new gene combinations, or even whole chromosomes, to the genetic architecture of some species. It may thus play a role in the evolutionary process. Introgression requires the production of hybrids, a rare occurrence among highly differentiated animal species. Areas where hybridization takes place are known as contact zones or hybrid zones. These zones exist where populations overlap; in some cases of hybridization, the line between what constitutes different species and what constitutes different populations of the same species becomes difficult to draw. The significance of introgression and hybrid zones in the evolutionary process remains an area of some contention among life scientists.

Biologists often explain, at least in part, the poorly understood phenomenon of speciation through migration and gene flow—or rather, by a lack thereof. If some members of a species become geographically isolated from the rest of the species, migration and gene flow cease. The isolated population will not share in any mutations, favorable or unfavorable, nor will any mutations that occur among its own members be transmitted to the general population of the species. Over long periods of time, this genetic isolation will result in the isolated population becoming so genetically different from the parent species that its members can no longer produce fertile progeny should one of them breed with a member of the parent population. The isolated members will have become a new species, and the physical differences between them and the parent species, though initially slight, will continue to grow as more ages pass. Scientists, beginning with Darwin himself, have demonstrated that this sort of speciation, termed adaptive radiation, has occurred on the various islands of the world's oceans and seas.

STUDYING GENE FLOW

Scientists from many disciplines are currently studying migration and gene flow in a variety of ways. For decades, ornithologists and marine biologists have been placing identifying tags or markers on members of different species of birds, fishes, and marine mammals to determine the range of their migratory habits in order to understand the role of migration and subsequent gene flow in the biology of their subjects. These studies have led, and will continue to lead, to important discoveries. Most studies of migration and gene flow, however, relate to human beings.

Many of the important discoveries concerning the role of gene flow in the evolution of life come from the continuing study of the nature of genes. A gene, in cooperation with such molecules as transfer ribonucleic acid (tRNA) and related enzymes, controls the nature of an organism by specifying amino acid sequences in specific functional proteins. In recent decades, scientists have discovered that what they previously believed to be single pure enzymes are actually groups of closely related enzymes, which they have named "isoenzymes" or "isozymes." Current theory holds that isozymes can serve the needs of a cell or of an entire organism more efficiently and over a wider range of environmental extremes than can a single enzyme. Biologists theorize that isozymes developed through gene flow between populations from climatic extremes and enhance the possibility of adaptation among members of the species when the occasion arises. The combination and recombination of isozymes passed from parent to offspring are apparently determined by deoxyribonucleic acid (DNA). Investigation into the role of DNA in evolution is one of the most promising avenues to an understanding of the nature of life.

A classic example of the importance of understanding migration and gene flow in the animal kingdom is the spread of the so-called killer bees. In the 1950s, a species of ill-tempered African bee was accidentally released in South America. The African bees mated with the more docile wild bees in the area; through migration and gene flow, they transmitted their violent propensity to attack anything approaching their nests. As the African genes slowly migrated northward, they proved to be dominant.

Further research into migration and gene flow promises to provide information indispensable to the attempt to unravel the mysteries of life. Coupled with the concept of mutation, gene flow is a crucial component of evolution.

—*Paul Madden*

FURTHER READING

Ammerman, A. J., and L. L. Cavalli-Sforza. *The Neolithic Transition and the Genetics of Populations in Europe.* Princeton University Press, 1984.

Bailey, Jill. *Evolution and Genetics: The Molecules of Inheritance.* Oxford University Press, 1995.

Cavalli-Sforza, Luigi Luca. *Genes, Peoples, and Languages.* University of California Press, 2000.

Crow, J. F., and Motoo Kimura. *An Introduction to Population Genetics.* Harper & Row, 1970.

Endler, John A. *Geographic Variation, Speciation, and Clines.* Princeton University Press, 1977.

Hoffmann, Ary A., and Peter A. Parsons. *Evolutionary Genetics and Environmental Stress.* Oxford University Press, 1991.

Raup, D. M., and D. Jablonski, eds. *Patterns and Processes in the History of Life.* Springer-Verlag, 1986.

GENETIC MUTATIONS

FIELDS OF STUDY

Genetics; Developmental Biology; Biochemistry

ABSTRACT

In all living organisms, the hereditary information consists of two complementary strands of deoxyribonucleic acid, or DNA. The order of nucleotides in a strand specifies the order of the amino acids that make up proteins. Changes in the sequence of nucleotides in DNA may alter an organism's proteins, which in turn may change one or more of that organism's traits. DNA changes are called mutations, and the organisms that harbor mutations are known as mutants. The characterization of mutations and mutants is one of the best ways of discovering the function of genes. A study of mutations and mutants also has shed light on numerous genetic diseases.

KEY CONCEPTS

allele: one of many possible sequences of a gene
controlling site: a sequence of nucleotides generally fifteen to sixty nucleotides long, to which a transcriptional activator or repressor binds
gene: a sequence of one thousand to ten thousand nucleotides, which usually specifies a protein
mutation: a change in the nucleotide sequence of a gene or of a controlling site; changes in genes alter the protein, whereas changes in controlling sites determine where and how much of a protein is produced

DNA AND PROTEINS

In all living organisms, the hereditary information consists of two complementary strands of deoxyribonucleic acid, or DNA. DNA strands are constructed of subunits called nucleotides that consist of a nitrogenous base, a deoxyribose sugar, and a phosphate. Generally, DNA strands consist of millions of nucleotides attached to each other like the links of a chain. There are four different nucleotides found in DNA: adenine (A), thymine (T), guanine (G), and cytosine (C). Each nucleotide bonds only to one other type of nucleotide, known as its complementary base. The two DNA strands are held together by hydrogen bonds between complementary bases. Adenine and thymine form one complementary pair, and guanine and cytosine form the other. In other words, if there is an adenine molecule in one strand, it is hydrogen bonded to thymine in the complementary strand; if there is guanine in one strand, it is hydrogen bonded to cytosine in the other strand. Thus, the amount of A is equal to T, and the amount of G is equal to C. The order of nucleotides in a strand specifies the order of the amino acids that make up proteins.

Genetics is the study of how the information in DNA molecules is expressed and how DNA molecules account for the heredity of an organism. Changes in the sequence of nucleotides in DNA may alter an organism's proteins, which in turn may change one or more of that organism's traits. DNA changes are called mutations, and the organisms that harbor mutations are known as mutants. The original or naturally occurring version of such a trait or organism is referred to as the "wild type." The characterization of mutations and mutants has been and still is one of the best ways of discovering the function of genes and determining how organisms maintain themselves, evolve, and develop. A study of mutations and mutants also has shed light on numerous genetic diseases.

HEAT: THE CAUSE OF MOST SPONTANEOUS MUTATIONS

Most mutations are caused by the instability of the nucleotide bases. Sometimes bases hit by rapidly moving water molecules briefly alter their chemistry. These chemical changes are known as tautomeric shifts. Tautomeric shifts alter the distribution of electrons and protons in the bases, causing them to form abnormal pairings with bases in the complementary strand. For example, an abnormal adenine (A*) pairs with C, rather than T, and an abnormal guanine (G*) pairs with T rather than C.

When a DNA molecule is being replicated, spontaneous tautomer shifts can result in permanent mutations. Spontaneous mutations occur, for example, when an A in the template strand undergoes a tautomer shift (A→A*) just as the DNA polymerase reaches it. A cytosine pairs with the A* and becomes part of the new strand being synthesized by the DNA polymerase. When this new strand, with a C in it instead of a T, functions as a template, the complementary strand will have a G in it rather than an A. This type of tautomeric shift during DNA replication converts what normally would have been an A·T base pair in "granddaughter" DNA to a G·C base pair.

CHEMICALS THAT CAUSE MUTATIONS

Mutations are induced by many chemical and physical agents called mutagens. Many chemicals act as mutagens. Nitrous acid, for example, diffuses into cells and removes amino groups from DNA bases. These chemically altered bases no longer pair normally. When DNA is replicated or repaired, incorrect nucleotides are inserted opposite the chemically altered bases. Nitrous acid changes A to hypoxanthine, which pairs with C, and changes G to xanthine, which pairs with T.

Base analogues are molecules that closely resemble normal nucleotides and consequently are incorporated into DNA that is being repaired or replicated. A base analogue to thymine, such as 5-bromouracil (5BU), is efficiently incorporated into DNA. 5BU spontaneously undergoes tautomeric shifts at a high rate. The abnormal form of 5BU pairs with G rather than A. Thus, 5BU introduces many base-pair transitions in newly synthesized DNA molecules.

The most potent mutagens are alkylating agents, such as nitrosamines, methyl bromide, and ethylene oxide. These mutagens attach methyl or ethyl groups (alkyl groups) to A and G. This causes A and G to undergo tautomeric shifts at a higher-than-normal rate.

HIGH-ENERGY ELECTROMAGNETIC RADIATION AND PARTICLES

Ultraviolet (UV) light is a powerful mutagen. It generally penetrates cells but is readily absorbed by thymine and cytosine bases in DNA. When two thymines or two cytosines next to each other in a strand absorb UV light, they often react chemically with each other to form thymine or cytosine dimers that distort the DNA. These distorted regions stimulate a repair system that cuts out the dimers, as well as some DNA on either side, and replaces them with normal nucleotides. Excessive repair leads to an increased occurrence of spontaneous mutations. Sometimes a distortion in the template allows the DNA polymerase to add or to leave out nucleotides as it moves along the template during strand synthesis. This may explain how some additions and some deletions occur.

Very energetic electromagnetic radiation, such as x-rays and gamma rays, as well as high-energy particles released from radioactive atoms or arriving in the influx of cosmic ray particles from space, also induce mutations. These energetic mutagens easily penetrate cells and chemically alter many molecules in their path by stripping away electrons. Ions and radicals formed by these mutagens react with the DNA, causing bases to be released and DNA to break. DNA deletions, transpositions, and inversions may be promoted by DNA breakage.

When a gene is mutated, the protein the gene codes for generally becomes nonfunctional. In bacteria that have only one copy of each gene, traits are immediately altered by a mutation. However, in organisms that have more than one copy of a gene, a mutation in only one gene may not produce a new trait because the wild-type (normal) gene often provides enough of the essential protein. When such organisms are missing both genes, however, they may fail to develop, or they may develop in a different way.

A few mutations are beneficial to the organism that acquires them and may make the organism better adapted to its environment. These beneficial mutations may make a protein work a little better or in a different way. Some mutations are also beneficial because they create diversity in a population. Diversity promotes the survival of a population by ensuring that some organisms will survive if the environment drastically changes. A population that is too well adapted to a particular environment will not survive if there are significant changes in the environment. There have been at least five major mass extinctions during the history of life on Earth, in some cases eliminating more than 85 percent of all species. The organisms that survived these mass extinctions were much less specialized than the organisms that did not.

USEFULNESS OF MUTATIONS

Mutations have been extremely useful in the study of organisms. Mutations allow scientists to understand what a particular gene and its product do. If the mutation eliminates the gene (and product), scientists can guess what the gene does by looking at the affected organism. For example, if a mutation changes eye color, such as from red to white, the affected gene most likely has something to do with pigment synthesis or deposition of the pigment in the eye.

The study of mutations and mutant organisms has helped scientists unravel anabolic (synthetic) and catabolic (degrading) pathways, determine how parental genes combine to produce new characteristics in progeny, clarify what genes are and what they do, establish how genes are regulated, and even decipher how multicellular organisms develop and evolve.

MUTATIONS IN DEVELOPMENT AND EVOLUTION

The study of mutations and mutant organisms at the end of the twentieth century led to an understanding of how multicellular organisms develop and evolve. One of the most useful organisms in unraveling the development problem has been the small fruit fly *Drosophila melanogaster*. Thousands of mutations that affect development of this organism have been characterized. Scientists found that a hierarchy of genes are involved in development. First, maternal genes are expressed. These genes activate gap genes, which in turn activate pair-rule genes. All of these gene categories are known to be involved in regulating the expression of homeotic genes. Maternal, gap, pair-rule, and homeotic gene products all function as transcriptional activators and repressors. For example, the maternal gene product called bicoid stimulates its own synthesis, and it also inhibits the synthesis of another maternal gene product called nanos.

Maternal Genes→Gap Genes→Pair Rule Genes→Homeotic Genes

This gene hierarchy is responsible for the anterior-posterior segmentation seen in *Drosophila*. Edward B. Lewis, Christiane Nüsslein-Volhard, and Eric Wieschaus shared the 1995 Nobel Prize in Physiology or Medicine for their studies of the genes that control *Drosophila* development.

HOMEOTIC GENES

Homeotic genes are found in all multicellular organisms. Homeotic genes similar to those found in *Drosophila* control the development of segments, most visibly exemplified by the vertebrae and the bones in animals' appendages. Mutations in homeotic genes or their controlling sites affect the development of segments. Segments can be eliminated or modified by homeotic-gene-controlling site mutations.

One well-studied homeotic gene in *Drosophila* is the gene *Antennapedia* (*Antp*). Certain mutations in the controlling sites for this gene result in legs developing rather than head antennae. Another homeotic gene is *Ultrabithorax* (*Ubx*). Some mutations in the controlling sites for *Ubx* result in a second pair of wings developing where the fly would normally develop halteres, or tiny, winglike appendages that promote stable flight. Other mutations in the controlling sites for *Ubx* produce a second pair of winglike structures that are half haltere (anterior portion), half wing (posterior portion). By studying mutations and the altered traits, scientists have discovered that controlling-site mutations change when and where proteins are synthesized. For example, if a protein is to be produced in seven segments along the anterior-posterior axis of an animal, there must be at least seven different controlling sites that can respond to the different activators and repressors produced in each segment.

Numerous studies suggest that *Antp* and *Ubx* are transcriptional repressor-activators that not only repress the development of legs and wings but also stimulate the development of antennae and halteres, respectively. The study of *Drosophila* mutants is beginning to clarify how antennae and mouth parts evolved from leglike appendages and how halteres evolved from wings. The study of genes and controlling sites has led to the understanding of their role in the maintenance, development, and evolution of every organism.

—*Jaime Stanley Colomé*

FURTHER READING

Brennessel, Barbara, and Sarah Malone. "Inborn Errors of Metabolism." *Genetics & Inherited Conditions*. Ed. Jeffrey A. Knight. Salem, 2010.

Cessna, Stephen. "One Gene–One Enzyme Hypothesis." *Genetics & Inherited Conditions*. Ed. Jeffrey A. Knight. Salem, 2010.

Colomé, Jaime S. "Gene Regulation: Bacteria." *Genetics & Inherited Conditions*. Ed. Jeffrey A. Knight. Salem, 2010.

Fornari, Chet S., and Dervla Mellerick. "Homeotic Genes." *Genetics & Inherited Conditions*. Ed. Jeffrey A. Knight. Salem, 2010.

Gliboff, Sander. "Mendelian Genetics." *Genetics & Inherited Conditions*. Ed. Jeffrey A. Knight. Salem, 2010.

Kalumuck, Karen E., and Carolyn K. Beam. "Model Organism: *Drosophila melanogaster*." *Genetics & Inherited Conditions*. Ed. Jeffrey A. Knight. Salem, 2010.

Morgan, T. H. "Sex Limited Inheritance in *Drosophila*." *Science* 32.812 (1910): 120–22. Web. 6 Jan. 2016.

Sturtevant, A. H. "The Linear Arrangement of Six Sex-Linked Factors in *Drosophila*, as Shown by Their Mode of Association." *Journal of Experimental Zoology* 14.1 (1913): 43–59. Web. 6 Jan. 2016.

"Thomas H. Morgan—Biographical." *Nobel Lectures: Physiology or Medicine, 1922–1941*. Amsterdam: Elsevier, 1965. *Nobelprize.org*. Web. 6 Jan. 2016.

Thompson, James N., Jr., R. C. Woodruff, and Nicole Kosarek Stancel. "Mutation and Mutagenesis." *Genetics & Inherited Conditions*. Ed. Jeffrey A. Knight. Salem, 2010.

GENETICS

FIELDS OF STUDY

Biology; Genetics; Biochemistry; Genomics

ABSTRACT

At its most basic, genetics is the study of genes, which are sequences of DNA (deoxyribonucleic acid) that instruct cells to function through chemical processes, especially the production of proteins. Genes are passed from parents to their offspring, making them the basic units of heredity, or the passage of traits from generation to generation. Heredity and genetic variation are seen as key drivers of evolution by natural selection, making genetics an important subject in virtually every field of biology.

KEY CONCEPTS

allele: alternative forms of a single gene
chromosome: a long strand of DNA with supporting proteins, that contains many genes
deoxyribonucleic acid (DNA): the chemical polymer that is the genetic material of multicellular organisms
gene: factors in cells that are responsible for an observable characteristic of an organism
genome: all of the genetic material of an organism
genotype: the actual genetic makeup of an organism
mutation: any heritable change in the genetic material
phenotype: the observable characteristics of an organism (for example, black fur color in a cat)

GENETICS IS IN THE GENES

At its most basic, genetics is the study of genes, which are sequences of DNA (deoxyribonucleic acid) that instruct cells to function through chemical processes, especially the production of proteins. Genes are passed from parents to their offspring, making them the basic units of heredity, or the passage of traits

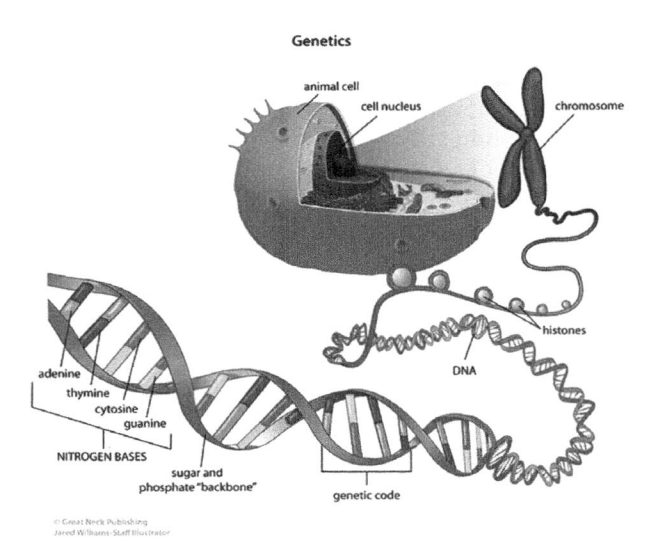

The DNA (deoxyribonucleic acid) and chromosomes within the cell's nucleus are the materials that determine an organism's genetics or hereditary traits.

from generation to generation. Heredity and genetic variation are seen as key drivers of evolution by natural selection, making genetics an important subject in virtually every field of biology. Some of the key subfields of genetics including population genetics, medical genetics, and epigenetics. Another closely related science is genomics, which focuses on complete sets of genes in organisms—genomes—rather than individual genes.

EARLY DEVELOPMENTS

While the science of genetics only developed beginning in the late nineteenth century, humans have witnessed and manipulated heredity for much longer. Before any recorded history, ancient humans chose alert pups from a litter of wolves for breeding. This practice of selectively breeding the wolves that were good companions eventually gave rise to the domesticated dog. Some of the oldest undisputed dog bones known, excavated from a twenty-thousand-year-old Alaskan settlement, demonstrate that prehistoric humans knew that traits could be passed from one generation to the next, and that selectively breeding animals (or plants) could produce an organism that possessed desired characteristics. This practice of deliberate breeding is known as artificial selection.

Humans have practiced artificial selection on numerous animals, including pigs, cattle, goats, and sheep. Homer and other Greek poets wrote about selective breeding, and part of the wealth of the ancient city of Troy was attributed to its expertise in horse breeding. Although humans had some control over the traits of domesticated animals through selective breeding, the results of matings were not always predictable, and nothing was known about the mechanism through which traits were passed from one generation to another until the mid-1800s.

MENDELIAN GENETICS

Gregor Mendel, an Austrian monk, is the undisputed father of the science of genetics. Working with garden peas, Mendel analyzed thousands of breeding experiments to describe laws that governed the inheritance of traits. Though Mendel studied a plant, Mendel's Laws of Heredity apply to all sexually reproducing organisms, including humans.

Mendel chose seven distinct traits to study in his garden peas: flower color, plant height, seed shape, seed color, pod shape, pod color, and flower position. He concluded that each of these traits was determined by a single, discrete factor, later to be called a gene. For instance, there was a gene for flower color and a gene for seed shape. Each gene had several variations, or alleles. The gene for flower color had a white allele that produced white flowers and a purple allele that produced purple flowers.

Mendel's experiments revealed that organisms have two copies of any gene for a trait. Those two copies can be identical, two purple alleles of the flower color gene, for instance; or those two alleles can be different. A pea plant could have one purple allele of the flower color gene and one white allele of the flower color gene. When an organism has two identical alleles of a gene, it is homozygous for that gene. When an organism has two different alleles of a gene, it is heterozygous for that gene.

An organism inherits one allele, or copy of a gene, from one parent and one allele from the other parent. An organism, or cell, that has two copies of all of its genetic information is called diploid. In most sexually reproducing animals, the offspring are formed when a sperm cell from the male parent fertilizes an egg from the female parent. The sperm and the egg only contain half of all the genetic information. They are said to be haploid. However, the new organism they create is diploid because it gets one copy of the genetic information from the sperm and a second copy from the egg.

Mendel's first law, or the law of segregation, states that the two copies of each gene separate during the formation of gametes (eggs and sperm), and that fertilization of the egg by the sperm is a random event. Any sperm containing any allele of a gene can fertilize any egg of the same species, regardless of the allele carried by that egg.

Mendel noted that certain alleles seemed to dominate over others. For instance, when a plant had a purple allele for flower color and a white allele for flower color, the plant always had purple flowers. Mendel called the allele that was seen in the heterozygote, in this case the purple allele, the dominant allele. The allele that was hidden or masked, he called the recessive allele. In order to show a recessive allele, an organism has to have two identical copies of a gene, both containing the same recessive allele. This is known as the homozygous recessive condition. Garden peas that have white flowers are homozygous recessive for the white allele of the flower color gene.

Homozygous recessive describes the organism's genotype, or its genetic makeup. It has two copies of the recessive allele of the gene. The observable characteristic of the organism, having white flowers, is called its phenotype.

Mendel also demonstrated that the segregation of alleles of any one gene is not dependent on the segregation of alleles of any other gene. For instance, a gamete could receive a dominant allele for an eye color gene and a recessive allele for height, or that gamete could receive the recessive alleles for both genes or the dominant alleles of both genes. This is Mendel's second law, the law of independent assortment, and it applies to any genes that are located on separate chromosomes.

Mendel's work was far ahead of its time. Although Mendel published his research in the 1800s, it was not until after his death that his work gained recognition in the scientific community. In 1900, three other scientists, each working separately on inheritance, came across Mendel's work in the course of their research. They gave him credit for his insights, and Mendel's research provided the foundation for the new discipline of genetics. An entire branch of the science is still known as Mendelian genetics.

GENES AND CHROMOSOMES

Although Mendel described the gene as the factor that was responsible for a particular trait, nothing was known about the physical makeup of a gene. (The term "gene" itself, as well as "genotype" and "phenotype" would be coined by Danish scientist Wilhelm Ludvig Johannsen in 1909.) One of the first questions scientists needed to answer was where genes are found in cells. Early studies in frogs and sea urchins indicated that the nucleus of the sperm and the nucleus of the egg combined with each other during fertilization. This observation suggested that the genetic material that determined how the fertilized egg would develop might reside in the nucleus.

As microscopes improved, scientists were able to distinguish structures within the nuclei of cells. These long, threadlike structures stained blue and were called chromosomes (Greek *chroma*, "color"). Several scientists observed that when animal and plant cells divided, the chromosomes duplicated, then separated, and each daughter cell inherited a complete set of chromosomes. The one exception to this was the cell division that produced the gametes (eggs and sperm). When an egg or a sperm cell was produced, it only contained half the number of chromosomes as the cell that produced it. If genetic information was carried on chromosomes, scientists reasoned that a sperm and an egg could each contribute half of the genetic information to the new organism at fertilization.

Some of the first evidence that chromosomes were linked to observable traits came from the studies of American graduate student Walter S. Sutton. Sutton studied grasshoppers, and his observations of chromosomes indicated that male grasshoppers always had an X and a Y chromosome, whereas female grasshoppers contained two X chromosomes. Several other scientists observed similar things in other organisms, such as fruit flies, and concluded that the physical characteristic of sex was determined by the kind of chromosomes an organism possessed.

Since chromosomes determined the trait of sex, it was possible that chromosomes contained the genes that Mendel had shown to determine physical characteristics. The first scientist to demonstrate that genes were located on chromosomes was Thomas Hunt Morgan, who showed that an eye-color gene in the fruit fly, *Drosophila melanogaster*, was located on the X chromosome.

Next, scientists wanted to know what kind of chemical molecule actually carried the genetic information. Chromosomes contain two kinds of molecules, protein and a weak acid called deoxyribonucleic acid (DNA). Experiments in the early 1930s first demonstrated that DNA is the genetic material. Oswald Avery, Colin MacLeod, and Maclyn McCarty showed that adding DNA to these bacterial cells could change their physical traits. In their experiments, they mixed a harmless strain of bacteria with DNA from bacteria that caused disease in mice. When they did this, the previously harmless bacteria changed (or transformed) into disease-causing bacteria. Two other scientists, Alfred Hershey and Martha Chase, later obtained similar results by studying a virus that infects *E. coli*.

MOLECULAR GENETICS

By the 1940s, scientists knew that genetic information was carried by genes made of DNA molecules inside cell nuclei. However, scientists did not know how the genetic information was copied accurately from one generation to the next—from one cell division to the next. Nor did scientists know how the DNA could account for the appearance of inherited changes or mutations. In order to answer these questions, scientists needed to know the precise chemical structure of DNA.

Many scientists contributed to the understanding of the structure of DNA. Erwin Chargaff obtained data that indicated that specific molecular components of the DNA molecule were always present in equal parts. These components were nitrogen-containing molecules (or nitrogenous bases). Chargaff determined that the nitrogen-containing bases adenosine and thymine were always present in a one to one ratio, and the bases guanine and cytosine were always present in a one to one ratio, no matter what species' DNA was analyzed.

Simultaneously, two scientists at Kings College in London, Rosalind Franklin and Maurice Wilkins, were attempting to make X-ray pictures of DNA molecules. Franklin obtained an X-ray film that indicated that DNA was a helical molecule. Just previous to Franklin's work, an American chemist, Linus Pauling, had made a breakthrough in solving the structure of the protein alpha helix using a model-building approach.

Two scientists working at Cambridge University in England, James D. Watson and Francis Crick, decided to use Pauling's method of model building to attempt to solve the structure of the DNA molecule. Combining the data from a variety of sources including the work of Chargaff and Wilkins, and the crucial X-ray crystallography data of Rosalind Franklin, Watson and Crick solved the structure of the DNA molecule.

Watson and Crick created a model of DNA: a double helix, like a twisted ladder. The DNA molecule was a long polymer of repeating nucleotides. Each nucleotide contained three chemical parts: a sugar, a phosphate group, and a nitrogen-containing base. The sides of the double helix ladder were formed by alternating sugars and phosphates, and the rungs were formed on the inside of the helix by specific pairings of the nitrogen-containing bases. Adenine paired with thymine to form one kind of rung. Guanine paired with cytosine to form a second kind of rung.

The order of the bases provided the information within DNA. Certain combinations of bases could form "words" that stood for parts of proteins or other molecules encoded by the DNA. The double helix could unzip like a zipper, each strand serving as a template to guide the construction of a new strand. This provided an accurate means for copying the DNA molecules from a parent cell to a daughter cell.

GENETIC ENGINEERING

The details of how DNA is passed from one generation to the next, of how mutations arise, and of how the information of DNA is actually translated into the activities of cells forms the basis of genetic research at the beginning of the twenty-first century.

One of the most important scientific discoveries that led to modern genetic technology was the discovery of a particular kind of protein, a restriction enzyme, from bacteria that cuts DNA molecules at specific sequences of bases. These restriction enzymes gave scientists the tool they needed to break DNA down into smaller pieces, eventually allowing the isolation of individual genes from the huge amount of DNA inside the nucleus of the cell.

Herbert Boyer and Stanley Cohen combined their knowledge of restriction enzymes and bacterial transformation (getting bacteria to take up DNA from the environment) to develop recombinant DNA technology and clone genes. Gene cloning involves isolating a gene of interest by using a restriction enzyme to cut it away from other DNA, and placing it in a piece of DNA called a vector that can be taken up by bacterial cells. One of the first applications of this technology was the production of human insulin. Scientists isolated the gene that encodes the information for making insulin from human DNA, cloned it into a bacterial vector, and placed the vector with the insulin gene in *E. coli*. The *E. coli* cells were able to produce large quantities of insulin. This new insulin was considerably cheaper and safer than insulin purified from human tissue.

Variations on this technique of taking a piece of DNA from one species and inserting it into the cells of another species are involved in genetic engineering of multicellular organisms. In multicellular organisms such as plants or monkeys, the DNA vector is usually a modified virus. These techniques are the basis of human gene therapy.

In the last decade of the twentieth century, entire organisms were cloned. In Scotland, Ian Wilmut and colleagues reported the first mammalian cloning of a sheep named Dolly. In Wisconsin and Japan, scientists have cloned cattle. When an organism is cloned, all of its DNA, usually contained within an intact nucleus from a cell of the adult animal, is transferred to an egg cell from which all the genetic information has been removed. The egg is then allowed to develop into a new organism. Although the new organism is young, it has the same DNA as the parent from which the nucleus was obtained.

Scientists have also developed techniques for sequencing DNA, determining the exact order and number of nitrogenous bases within the DNA of an organism's genome. In 2000, the Human Genome Project announced that the entire genome of the human had been sequenced. Many other genomes have been sequenced, including the roundworm, *C. elegans*, several plants, and even baker's yeast. The sequence of an organism gives scientists another tool in answering questions about how DNA regulates and determines the activities of cells.

By the 2010s, applied genetics had become integral in many aspects of society. For example, DNA forensic evidence is now used to convict or exonerate criminal suspects on a routine basis. The genetic engineering of food crops that are pest resistant or contain additional nutrients—known as genetically modified organisms, or GMOs—is also routine. Parents can have an embryo genetically screened for devastating genetic diseases before it is born, while those already born can undergo genetic testing for various conditions. With the cloning of entire organisms made possible, the potential cloning of a human has moved beyond the realm of science fiction. While many of these advances are clearly positive, many of them are double-edged swords, begging for informed public debate.

To many, the ethical consequences and risks of genetic engineering remain the subject of intense controversy. Some people reject human interference in natural processes on religious grounds, while others suggest that genetically modified foods may be unhealthy or have negative effects on the environment as a whole. Proponents of genetic engineering counter that humans have been manipulating genes for thousands of years—domesticated crops and animals are all technically "genetically modified"—and argue that genetics holds the potential to improve and save countless lives. Other critics do not necessarily object to genetic engineering itself, but note that the practice raises major ethical issues of accessibility and equality. They warn that the expense of genetic research and engineering threatens to give the already wealthy and powerful exclusive access to beneficial treatments and technology, while genetic testing and genetically altered "designer babies" could eventually lead to discrimination.

—*Michele Arduengo*

Further Reading

Hartwell, L., L. Hood, M. Goldberg, A. Reynolds, L. Silver, and R. Veres. *Genetics: From Genes to Genomes.* McGraw-Hill, 2000.

"Help Me Understand Genetics." *Genetics Home Reference*, US National Library of Medicine, National Institutes of Health, 30 Jan. 2018, ghr.nlm.nih.gov/primer. Accessed 1 Feb. 2018.

Lieberman, Michael, and Rick E. Ricer. *Biochemistry, Molecular Biology, and Genetics.* Wolters Kluwer, 2014.

Maczulak, Anne. *The Smart Guide to Biology.* 2nd ed. Smart Guide Publications, 2014.

Marshall, Elizabeth L. *The Human Genome Project: Cracking the Code Within Us.* Franklin Watts. 1996.

Mousseau, T. A., B. Sinervo, J. A. Endler, eds. *Adaptive Genetic Variation in the Wild.* Oxford University Press, 1999.

Petersen, Christine. *Genetics.* ABDO, 2014.

Sayre, Anne. *Rosalind Franklin and DNA.* Norton, 1975.

"Six Things Everyone Should Know About Genetics." *The American Society of Human Genetics,* 2017, www.ashg.org/education/everyone_1.shtml. Accessed 1 Feb. 2018.

Sponenberg, D. P. *Equine Color Genetics.* Iowa State University Press, 1996.

Watson, James D. *The Double Helix: A Personal Account of the Discovery of the Structure of DNA.* Ed. Gunther S. Stent. Norton, 1980.

GROOMING

FIELDS OF STUDY

Animal Behavior; Evolutionary Biology; Sociology

ABSTRACT

Grooming is health related in that the activity cares for the skin, feathers, or fur. Grooming also serves a social function. Between members of a community, grooming reduces stress, communicates and signals social status, spreads pheromones, achieves thermoregulation or pain relief, increases or decreases arousal, self-stimulates, and prevents sexually transmitted diseases. As grooming is similar through the various levels of animal taxa, it has been conjectured that grooming behavior is evolutionarily ancient.

KEY CONCEPTS

allogrooming: mutual grooming or grooming between two individuals

altruism: a behavior that increases the fitness of the recipient individual while decreasing the fitness of the performing individual

consortship: a pairing of a male and a female

pheromone: a chemical produced by one individual that influences the behavior of another individual of the group

symbiosis: a relationship between two species of organisms which is not necessarily advantageous or disadvantageous to either organism

thermoregulation: the process by which animals maintain body temperatures within a certain range

LOOK GOOD, FEEL GOOD, MAKE FRIENDS

Grooming serves a number of purposes. It is health related in that the activity cares for the skin, feathers, or fur. Grooming also serves a social function. Between members of a community, grooming reduces stress, communicates and signals social status, spreads pheromones, achieves thermoregulation or pain relief, increases or decreases arousal, self-stimulates, and prevents sexually transmitted diseases.

As grooming is similar through the various levels of animal taxa, it has been conjectured that grooming behavior is evolutionarily ancient. Most animals (mammals, birds, and insects) groom by moving their limbs over their own bodies or mouthing or licking their bodies. In some birds, sandbathing is quite common. With fish, a species with no limbs, it is not uncommon to see them rubbing or simply moving against rocks, branches, or sand, generally accomplishing what a sandbath does for a bird.

A bird preening its feathers, a cat licking its paws, or a bear brushing his back against a tree trunk are all self-grooming. This is where the animal, alone, takes care of the grooming behavior without help from another animal. However, mutual grooming is quite common. In mammalian species, this is a form a display behavior, which helps to cement the social bonds between members of the group. Yet another kind of cleaning behavior, called cleaning symbiosis, occurs between certain species of fish and shrimp. Here, one species will eat the parasites off another species. The cleaner gets food, the recipient remains debris- and parasite-free.

GROOMING AS CLEANING BEHAVIOR

Observations and experiments performed on laboratory rats showed that pregnant rats spent an increasing proportion of their time grooming their ventral surface, which includes the nipple lines and anogenital areas. When the rats were fitted with collars so they could not reach these regions, their mammary development was inhibited. The conclusion to be drawn from this is that self-stimulating grooming is a necessary part of the preparation process for nursing.

A similar conclusion was reached when looking at the function of grooming in male rats. Prepubertal males engaged in self-grooming of the anogenital area more than their female counterparts. When collars restricted the rats from grooming, their sexual development was significantly hampered. Again, the conclusion is that male rats needed to perform self-stimulating autogrooming in order to prepare for reproduction.

Another very important fact about self-grooming behavior is the appearance of a grooming pattern. For example, depending on the species, self-grooming may start around the head area and progresses downward until the entire body has been groomed. In experiments, it has been noted that a particular animal will not vary its grooming pattern. Studies on hamsters show that specific grooming tendencies evolve as the animal matures. Certain types of grooming behavior are always the result of some external stimuli, or they are the consequence of an aspect of the animal's natural behavior. A male rat invariably performs genital grooming after copulation. A dog, when aroused by some external stimuli (fear or excitement) will often groom its genitalia for de-arousal. Many animals, when afraid or nervous, will gnaw on their paws. Pigs often chew on other pigs' tails, it is thought, when they are bored or have an excess of nervous energy. The expression "licking one's wounds" originates from the observation that most animals will tend to an injury by licking the area, spreading an antiseptic found in saliva to reduce pain and decrease chances of infection. Insectivores, like solendons and shrews, spend a good deal of time grooming, using only their hind legs, to ensure thermoregulation.

Regardless of the lack of similarity between grooming species, grooming behavior is remarkably similar; reiterating that there has not been a great deal of diversification in the evolution of grooming behavior. Sea lions, seals, and walruses, regardless of whether the pelage is sparse or dense, spend much time grooming. This is typical of marine mammals. They accomplish this by a doglike scratching motion with their hind flippers and using their fore flipper to rub their head and neck while balancing on the other front flipper. They also nibble on their fur, much like dogs. It is common for these animals to rub against rocks or each other, similar to the activity of bears when they rub up against trees. Fish will also use this rubbing behavior against rocks or coral to scour their bodies free from debris.

GROOMING AS SOCIAL BEHAVIOR

Mutual grooming, also called allogrooming or allopreening, is when an animal cares for the body surface of another. Mutual grooming in animals is more a form of communication, a social act, than it is a cleansing one. Researchers have come to this conclusion because the time allocated to mutual grooming exceeds what is necessary for simple cleansing and sanitation. This behavior, especially in primates and social birds, promotes social bonding and establishes and maintains the hierarchy among members of the community.

Mutual grooming also brings attention to whether there is such a thing as animal altruism, which is the selfless delivery of service from one animal to another. Whenever one observes a parent grooming its young, the conclusion is that the motive behind the act is to tend to the health of the infant or young, which promotes the proliferation of the species. Nearly all mammals and many species of birds display this behavior. With marsupial births, the young of the Virginia opossum are in a semiembryonic state. The mother licks the embryo sac at birth so that the membrane will break. She then licks a trail from the birth canal to the pouch so that the neonate, using its developed olfactory senses, can find its way to the pouch without further aid by the mother. In altricial and semialtricial placental mammals, the young are usually born naked or with little fur. Most of these animal types, such as cats, dogs, mice, shrew, rats, and hamsters, lick the newborn to remove the birth membrane, break the umbilical cord, and eat the placenta after birth. They lick the newborn to clean the perineal region and to remove urine and feces, also aiding in thermoregulation.

When one observes animal and birds of a similar peer group performing grooming rituals, what appears to be unilateral or altruistic grooming behavior may in fact be mutual aid, in that over the long term, repayment may be expected. It is the expectation of reciprocation that establishes and maintains social status. For example, studies on nonhuman primates (mostly Old World monkeys such as rhesus macaques, stump-tailed macaques, and baboons) have shown that a competition exists among members of a group for animals considered to be excellent groomers. Alliances are formed between groomer and groomed; the groomed may provide excellent protection or be a skilled food gatherer. There is evidence of ranking among members of primate society and mutual grooming appears to assist in maintaining the social order and structure.

In primates that form consortships (Cercepethecines), grooming is important in the copulatory sequence, which is not the case in primates that do not form consortships. In general, sexual grooming serves as a male strategy to increase receptivity of an estrus female. In olive baboons and hamadryas baboons, there is a high correlation between the age of the male and the grooming of the females, implying that this may be an alternative method of securing female cooperation in mating when direct agonistic approaches fail or are no longer an option.

Equids are well-known for mutual grooming. Horses, zebras, and similar animals groom each other by pairing off and standing nose-to-nose or head-to-tail, scratching and nibbling each other's neck, back, and tail with their teeth. The nose-to-nose greeting and rubbing of noses is also typical of tapirs and rhinoceros.

INTERSPECIES GROOMING

Humans have successfully exploited the horses' need for bonding by touch and grooming. Trainers often start training horses in spring when the animals are shedding. By brushing and grooming, the trainer develops a friendship and level of trust with the animal. This is an example of interspecies bonding through grooming. This behavior is widely noticed between any combination of humans, dogs, and cats. Sometimes licking another animal or human is simply a show of affection.

Symbiotic grooming, briefly mentioned above, occurs among some species where they pair off to assist one another in the grooming process. Pilot fishes and remoras are commensal fishes (a type of symbiosis) that attach themselves to sharks and other fish. Apart from eating the remnants of the host's meal, they also feed on the external parasites that plague the host fish. Fleas and lice are symbiotic groomers as they eat harmlessly on dead skin or feathers of mammals and birds.

Grooming behavior is an integral part of existence for all higher-order animals. It is such an important part of human behavior that if proper grooming is not taught at an early age and not performed, there are negative repercussions from peers and society in general. Grooming is such an important part of human community, that in the United States alone, the annual sales of men's and women's toiletries is a multibillion dollar industry. Scientists have only recently begun to study grooming behavior closely.

—*Donald J. Nash*

FURTHER READING

Colbern, Deborah L., and Willem H. Gispen, eds. *Neural Mechanisms and Biological Significance of Grooming Behavior.* New York Academy of Sciences, 1988.

Hinde, Robert A. ed. *Primate Social Relationships: An Integrated Approach.* Blackwell Scientific Publications, 1983.

Peterson, Richard, and George A. Bartholomew. *The Natural History and Behavior of the California Sea Lion.* American Society of Mammalogists, 1967.

Poole, Trevor B. *Social Behavior in Mammals.* Chapman and Hall, 1985.

Rubenstein, Daniel I., and Richard W. Rangham, eds. *Ecological Aspects of Social Evolution: Birds and Mammals.* Princeton University Press, 1986.

H

Habitats and Biomes

Fields of Study

Ecology; Environmental Studies; Biogeography; Bioecology

Abstract

Life in the form of individual organisms composed of one or more living cells is found in a vast array of different places on Earth, each with its own distinctive types of organisms. The space in which each species lives is called its habitat. Groups of communities that are relatively self-sufficient in terms of both recycling nutrients and the flow of energy among them are called ecosystems. Some ecosystems are widely distributed across the surface of Earth and are easily recognizable as similar ecosystems, known as biomes.

Key Concepts

abiotic: the physical part of an ecosystem or biome, consisting of climate, soil, water, oxygen and carbon dioxide availability, and other physical components
biome: one of the widespread types of ecosystem on Earth, such as the Arctic tundra or the desert
biosphere: the sum of all the occupiable habitats for life on Earth
biotic: the living part of an ecosystem or biome, consisting of all organisms
community: a population of plants and animals that live together and make up the biotic part of an ecosystem
ecosystem: a relatively self-sufficient group of communities and their abiotic environment
environment: the habitat created by the interaction of the abiotic and biotic parts of an ecosystem
habitat: the specific part of the environment occupied by the individuals of a species
population: a group of all the individuals of one species
species: a group of similar organisms that are capable of interbreeding and producing fertile offspring

CALL IT HOME

Life in the form of individual organisms composed of one or more living cells is found in a vast array of different places on Earth, each with its own distinctive types of organisms. Life on Earth has been classified by scientists into units called species, whose individuals appear similar, have the same role in the environment, and breed only among themselves. The space in which each species lives is called its habitat.

HABITATS, COMMUNITIES, ECOSYSTEMS, AND BIOMES

The term "habitat" can refer to specific places with varying degrees of accuracy. For example, rainbow trout can be found in North America from Canada to Mexico, but more specifically they are found in freshwater streams and lakes with an average temperature below 20°C and a large oxygen supply. The former example describes the macrohabitat of the rainbow trout, which is a broad, easily recognized area. The latter example describes its microhabitat—the specific part of its macrohabitat in which it is found. Similarly, the macrohabitat of one species can refer to small or large areas of its habitat. The macrohabitat of rainbow trout may refer to the habitat of a local population, the entire range of the species, or (most often) an area intermediate to those extremes. While "habitat," therefore, refers to the place an organism lives, it is not a precise term unless a well-defined microhabitat is intended. The total population of each species has one or more local populations, which are all the individuals in a specific geographic area that share a common gene pool; that is, they commonly interbreed. For example, rainbow trout of two adjacent states will not normally interbreed unless they are part of local populations that are very close to one another. The entire geographic distribution of a species, its range, may be composed of many local populations.

On a larger organizational scale, there is more than only one local population of one species in any habitat. Indeed, it is natural and necessary for many species to live together in an area, each with its own micro- and macrohabitat. The habitat of each local population of each species overlaps the habitat of many others. This collective association of populations in one general area is termed a community, which may consist of thousands of species of animals, plants, fungi, bacteria, and other one-celled organisms.

Groups of communities that are relatively self-sufficient in terms of both recycling nutrients and the flow of energy among them are called ecosystems. An example of an ecosystem could be a broad region of forest community interspersed with meadows and stream communities that share a common geographic area. Some ecosystems are widely distributed across the surface of Earth and are easily recognizable as similar ecosystems known as biomes—deserts, for example. Biomes are usually named for the dominant plant types, which have very similar shapes and macrohabitats. Thus, similar types of organisms inhabit them, though not necessarily ones of the same species. These biomes are easily mapped on the continental scale and represent a broad approach to the distribution of organisms on the face of Earth. One of the more consistent biomes is the northern coniferous forest, which stretches across Canada and northern Eurasia in a latitudinal belt. Here are found needle-leafed evergreen conifer trees adapted to dry, cold, windy conditions in which the soil is frozen during the long winter.

NORTH AMERICAN BIOMES

The biomes of North America, from north to south, are the polar ice cap, the Arctic tundra, the northern coniferous forest; then, at similar middle latitudes, eastern deciduous forest, prairie grassland, or desert; and last, subtropical rain forest near the equator. Complicating factors that determine the actual distribution of the biomes are altitude, annual rainfall, topography, and major weather patterns. These latter factors, which influence the survival of the living, or biotic, parts of the biome, are called abiotic factors. These are the physical components of the environment for a community of organisms.

The polar ice cap is a hostile place with little evidence of life on the surface except for polar bears and sea mammals that depend on marine animals for food. A distinctive characteristic of the Arctic tundra, just south of the polar ice cap, is its flat topography and permafrost, or permanently frozen soil. Only the top meter or so thaws during the brief Arctic summer to support low-growing mosses, grasses, and the dominant lichens known as reindeer moss. Well-known animals found there are the caribou, musk-ox, lemming, snowy owl, and Arctic fox. Climate change, by thawing the tundra landscape to greater depths, will act to destabilize the permafrost and the overlying tundra, and thus have significant effects on the nature of the biome and its inhabitants.

The northern coniferous forest is dominated by tall conifer trees. Familiar animals include the snowshoe hare, lynx, and porcupine. This biome stretches east to west across Canada and south into the Great Lakes region of the United States. It is also found at the higher elevations of the Rocky Mountains and the western coastal mountain ranges. Its upper elevation limit is the "treeline," above which only low-growing grasses and herbaceous plants grow in an alpine tundra community similar to the Arctic tundra. In mountain ranges, the change in biomes with altitude mimics the biome changes with increasing latitude, with tundra being the highest or northernmost.

Approximately the eastern half of the United States was once covered with the eastern deciduous forest biome, named for the dominant broad-leaved trees that shed their leaves in the fall. This biome receives more than seventy-five centimeters of rainfall each year and has a rich diversity of bird species, such as the familiar warblers, chickadees, nuthatches, and woodpeckers. Familiar mammals include the white-tailed deer, cottontail rabbit, and wild turkey. The Great Plains, between the Mississippi River and the Rocky Mountains, receives twenty-five to seventy-five centimeters of rain annually to support an open grassland biome often called the prairie. The many grass species that dominate this biome once supported vast herds encompassing several millions of bison and, in the western parts, pronghorn antelope. Seasonal drought and periodic fires are common features of grasslands.

The land between the Rockies and the western coastal mountain ranges is a cold type of desert biome; three types of hotter deserts are found from western Texas west to California and south into Mexico. Deserts receive fewer than twenty-five centimeters of rainfall annually. The hot deserts are dominated by many cactus species and short, thorny shrubs and trees, whereas sagebrush, grass, and small conifer trees dominate the cold desert. These deserts have many lizard and snake species, including poisonous rattlesnakes and the Gila monster. The animals often have nocturnal habits to avoid the hot, dry daytime.

Southern Mexico and the Yucatán Peninsula are covered by evergreen, broad-leaved trees in the tropical rain-forest biome, which receives more than two hundred centimeters of rain per year. Many tree-dwelling animals, such as howler monkeys and tree frogs, spend most of their lives in the tree canopy, seldom reaching the ground.

Aquatic biomes can be broadly categorized into freshwater, marine, and estuarine biomes. Freshwater lakes, reservoirs, and other still-water environments are called lentic, in contrast to lotic, or running-water, environments. Lentic communities are often dominated by planktonic organisms, small, drifting (often transparent) microscopic algae, and the small animals which feed on them. These, in turn, support larger invertebrates and fish. Lotic environments depend more on algae that are attached to the bottoms of streams, but they support equally diverse animal communities. The marine biome is separated into coastal and pelagic, or open-water, environments, which have plant and animal communities somewhat similar to lotic and lentic freshwater environments, respectively. The estuarine biome is a mixing zone where rivers empty into the ocean. These areas have a diverse assemblage of freshwater and marine organisms.

THE BIOSPHERE

All the biomes together, both terrestrial and aquatic, constitute the biosphere, which by definition is all the places on earth where life is found. Organisms that live in a biome must interact with one another and must successfully overcome and exploit their abiotic environment. The severity and moderation of the abiotic environments determine whether life can exist in that microhabitat. Such things as minimum and maximum daily and annual temperature, humidity, solar radiation, rainfall, and wind speed directly affect which types of organisms are able to survive. Amazingly, few places on earth are so hostile that no life exists there. An example would be the boiling geyser pools at Yellowstone National Park, but even there, as the water temperature cools at the edges to about 75 degrees Celsius, bacterial colonies begin to appear. There is abundant life in the top meter or two of soil, with plant roots penetrating to twenty-two meters or more in extreme cases. Similarly, the mud and sand bottoms of lakes and oceans contain a rich diversity of life. Birds, bats, and insects exploit the airspace above land and sea up to a height of about 1,200 meters, with bacterial and fungal spores being found much higher. Thus, the biosphere generally extends about 10 to 15 meters below the surface of Earth and about 1,200 meters above it. Beyond that, conditions are too hostile. A common analogy is that if Earth were a basketball, the biosphere would constitute only the thin outer layer.

STUDYING HABITATS AND BIOMES

Abiotic habitat requirements for a local population or even for an entire species can be determined in the laboratory by testing its range of tolerance for each factor. For example, temperature can be regulated in a laboratory experiment to determine the minimum and maximum survival temperatures as well as an optimal range. The same can be done with humidity, light, shelter, and substrate type: "Substrate preference" refers to the solid or liquid matter in which an organism grows and/or moves—for example, soil or rock. The combination of all ranges of tolerance for abiotic factors should describe a population's actual or potential microhabitat within a community. Furthermore, laboratory experiments can theoretically indicate how much environmental change each population can tolerate before it begins to migrate or die.

Methods to study the interaction of populations with one another or even the interaction of individuals within one local population are much more complicated and are difficult or impossible to bring into a laboratory setting. These studies most often require collecting field data on distribution, abundance, food habits or nutrient requirements, reproduction and death rates, and behavior in order to describe the relationships between individuals and populations within a community. Later stages of these field investigations could involve experimental manipulations in which scientists purposely change one factor, then observe the population or community response. Often, natural events such as a fire, drought, or flood can provide a disturbance in lieu of a manipulation caused by man.

There are obvious limits to how much scientists should tinker with the biosphere merely to see how it works. Populations and even communities in a local area can be manipulated and observed, but it is not practical or advisable to manipulate whole ecosystems or biomes. To a limited extent, scientists can document apparent changes caused by civilization, pollution, and long-term climatic changes. This information, along with population- and community-level data, can be used to construct a mathematical model of a population or community. The model can then be used to predict the changes that would happen if a certain event were to occur. These predictions merely represent the "best guesses" of scientists, based on the knowledge available. Population ecologists often construct reasonably accurate population models that can predict population fluctuations based on changes in food supply, abiotic factors, or habitat. As models begin to encompass communities, ecosystems, and biomes, however, their knowledge bases and predictive powers decline rapidly. Perhaps the most complicating factor in building and testing these large-scale models is that natural changes seldom occur one at a time. Thus, scientists must attempt to build cumulative-effect models that are capable of incorporating multiple changes into a predicted outcome.

The biosphere, then, can be studied at different levels of organization, from the individual level through populations, communities, ecosystems, and biomes to the all-encompassing biosphere. Each level has unique relationships that require different methods of inquiry; in fact, these levels describe many of the subdisciplines within the science of ecology.

BIOSPHERE AND BIODIVERSITY

Understanding the organization of the natural world of which man is a part is essential to the continued success of humankind. By understanding the abiotic and biotic relationships within and between each ecological level, from microhabitat to biosphere, scientists can partially explain why so many species of organisms have evolved over the last 3.5 billion years. This study of biodiversity may eventually be a key to maintaining a stable biosphere, in which there would be no drastic changes in climate or community relationships. For example, compare the diversity of micro habitats and species in a natural grassland biome with the established monoculture practices of agriculture, with the latter's emphasis on one species. In a wheat field, there are fewer micro habitats available, but those few are available in abundance. This leads to an increase in the population size of "pest" species that compete with man for an abundant food resource, wheat. Understanding the microhabitat requirements of pest species can lead to the reduction of crop losses.

The goal of the study of habitats, biomes, and the biosphere is the construction of predictive models. Once scientists have a general understanding of natural ecosystem processes, mathematical models may be able to predict future changes in the environment caused by the activities of civilization or natural climatic changes. For example, they may indicate whether increased human population size or increased large-scale agriculture in or near desert biomes will lead to the spread of desertlike conditions. On a biosphere scale, they may show whether the increase in the carbon dioxide content of the atmosphere caused by man's activities will lead to global warming. Both these effects are being predicted by many scientists.

Prior to the relatively recent growth in the science of ecology and the general interest in it, man had little concern about the effects of the exploitation of natural resources such as forests, of synthetic chemical pollutants such as Dichloro-diphenyl-trichloroethane (DDT), or even of the rapid growth of the human population size. With an increasing knowledge and understanding of habitat requirements, natural community interrelationships, and cycling of nutrients and pollutants within the biosphere, however, scientists have greater predictive power concerning the effects of economic development and human population growth. Ecologists are studying the effects of changing rain forests into agricultural land, the extinction of species, the loss of the ozone layer, and many other phenomena that have the potential to change the abiotic environment and therefore affect the stability of biotic communities. These results will be incorporated into future predictive models, giving them increased accuracy.

The challenge—both to scientists and to human civilization as a whole—is to use an understanding of the biosphere to maintain a level of economic growth that is ecologically sustainable. The study of communities and ecosystems may discover ways that civilization can better adapt to its current environment rather than attempt to mold the environment to fit its own preconceived, established ideas.

—*Jim Fowler*

FURTHER READING

Allaby, Michael. *Biomes of the World.* 9 vols. Grolier International, 1999.
Bradbury, Ian K. *The Biosphere.* Belhaven Press, 1991.
Cox, George W. *Conservation Biology: Concepts and Applications.* 2nd ed. Wm. C. Brown, 1997.
Hanks, Sharon La Bonde. *Ecology and the Biosphere: Principles and Problems.* St. Lucie Press, 1996.
Luoma, Jon R. *The Hidden Forest: The Biography of an Ecosystem.* Henry Holt, 1999.
Miller, G. Tyler, Jr. *Living in the Environment.* 5th ed. Wadsworth, 1988.
Scientific American 261 (September 1989).
Sutton, Ann, and Myron Sutton. *Eastern Forests.* Alfred A. Knopf, 1986.
Wetzel, Robert G. *Limnology.* 2nd ed. W. B. Saunders, 1983.

HARDY-WEINBERG LAW OF GENETIC EQUILIBRIUM

FIELDS OF STUDY

Genetics

ABSTRACT

Population genetics is the branch of genetics that studies the behavior of genes in populations. The population is the only biological unit that can persist for a span of time greater than the life of an individual, and the population is the only biological unit that can evolve. The two main subfields of population genetics are theoretical (or mathematical) population genetics and experimental population genetics.

KEY CONCEPTS

allele: one of several alternate forms of a gene; the deoxyribonucleic acid (DNA) of a gene may exist as two or more slightly different sequences, which may result in distinct characteristics
allele frequency: the relative abundance of an allele in a population
diploid: having two chromosomes of each type
gene: a section of the DNA of a chromosome, which contains the instructions that control some characteristic of an organism
gene pool: the array of alleles for a gene available in a population; it is usually described in terms of allele or genotype frequencies
genotype: the set of alleles an individual has for a particular gene
genotype frequency: the relative abundance of a genotype in a population
haploid: having one chromosome of each type
population: the individuals of a species that live in one place and are able to interbreed
random mating: the assumption that any two individuals in a population are equally likely to mate, independent of the genotype of either individual

POPULATION GENETICS

Genetics began with the study of inheritance in families: Gregor Mendel's laws describe how the alleles of a pair of individuals are distributed among their offspring. Population genetics is the branch of genetics that studies the behavior of genes in populations. The population is the only biological unit that can persist for a span of time greater than the life of an individual, and the population is the only biological unit that can evolve. The two main subfields of population genetics are theoretical (or mathematical) population genetics, which uses formal analysis of the properties of ideal populations, and experimental population genetics, which examines the behavior of real genes in natural or laboratory populations.

Population genetics began as an attempt to extend Mendel's laws of inheritance to populations. In 1908,

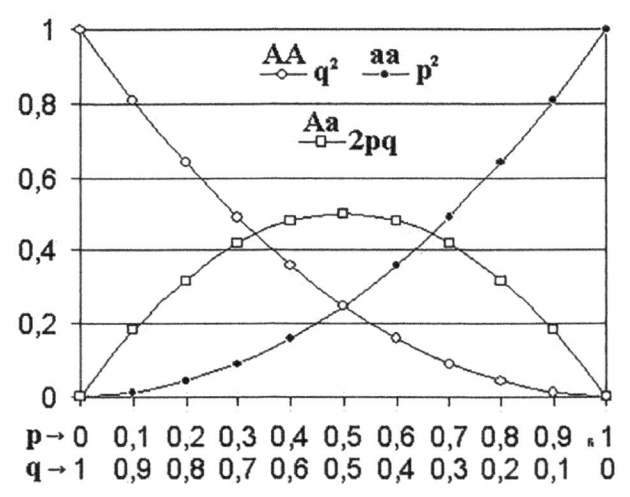

Hardy-Weinberg principle.

Godfrey H. Hardy, an English mathematician, and Wilhelm Weinberg, a German physician, each independently derived a description of the behavior of allele and genotype frequencies in an ideal population of sexually reproducing diploid organisms. Their results, now termed the Hardy-Weinberg principle, or Hardy-Weinberg equilibrium, showed that the pattern of allele and genotype frequencies in such a population followed simple rules. They also showed that, in the absence of external pressures for change, the genetic makeup of a population will remain the same, at an equilibrium. Since evolution is change in a population over time, such a population is not evolving. Modern evolutionary theory is an outgrowth of the "New Synthesis" of R. A. Fisher, J. B. S. Haldane, and Sewall Wright, which was done in the 1930s. They examined the significance of various factors that cause evolution by examining the degree to which they cause deviations from the predictions of the Hardy-Weinberg equilibrium.

ASSUMPTIONS AND PREDICTIONS

The predictions of the Hardy-Weinberg equilibrium hold if the following assumptions are true: The population is infinitely large; there is no differential movement of alleles or genotypes into or out of the population; there is no mutation (no new alleles are added to the population); there is random mating (all genotypes have an equal chance of mating with all other genotypes); and all genotypes are equally fit (have an equal chance of surviving to reproduce). Under this very restricted set of assumptions, the following two predictions are true: Allele frequencies will not change from one generation to the next, and genotype frequencies can be determined by a simple equation and will not change from one generation to the next.

The predictions of the Hardy-Weinberg equilibrium represent the working through of a simple set of algebraic equations and can be easily extended to more than two alleles of a gene. In fact, the results were so self-evident to the mathematician Hardy that he at first did not think the work was worth publishing.

If there are two alleles (A, a) for a gene present in the gene pool, let p = the frequency of the A allele and q = the frequency of the a allele. As an example, if

$p = 0.4$ (40%) and $q = 0.6$ (60%),
then $p + q = 1$,

since the two alleles are the only ones present and the sum of the frequencies (or proportions) of all the alleles in a gene pool must equal 1 (or 100 percent). The Hardy-Weinberg principle states that at equilibrium the frequency of AA individuals will be p^2 (equal to 0.16 in this example), the frequency of Aa individuals will be $2pq$, or 0.48, and the frequency of aa individuals will be q^2, or 0.36.

The basis of this equilibrium is that the individuals of one generation give rise to the next generation. Each diploid individual produces haploid gametes. An individual of genotype AA can make only a single type of gamete, carrying the A allele. Similarly, an individual of genotype aa can make only a gametes. An Aa individual, however, can make two types of gametes, A and a, with equal probability. Each individual makes an equal contribution of gametes, since all individuals are equally fit and there is random mating. Each AA individual will contribute twice as many A gametes as each Aa individual. Thus, to calculate the frequency of A gametes, add twice the number of AA individuals and the number of Aa individuals, then divide by twice the total number of individuals in the population (note that this is the same as the method to calculate allele frequencies). That means that the frequency of A gametes is equal to the frequency of A alleles in the gene pool of the parents.

The next generation is formed by gametes pairing at random (independent of the allele they carry). The likelihood of an egg joining with a sperm is the frequency of one multiplied by the frequency of the other. AA individuals are formed when an A sperm joins an A egg; the likelihood of this occurrence is

$p \times p = p^2$

(that is, $0.4 \times 0.4 = 0.16$ in the first example). In the same fashion, the likelihood of forming an aa individual is $q^2 = 0.36$. The likelihood of an A egg joining an a sperm is pq, as is the likelihood of an a egg joining an A sperm; therefore, the total likelihood of forming an Aa individual is $2pq = 0.48$. If one now calculates the allele frequencies (and hence the frequencies of the gamete types) for this generation, they are the same as before: The frequency of the A allele is

$p = (2p^2 + 2pq)/2$

(in the example, $(0.32 + 0.48)/2 = 0.4$), and the frequency of the a allele is

$q = (1 - p) = 0.6$.

The population remains at equilibrium, and neither allele nor genotype frequencies change from one generation to the next.

IDEAL VERSUS REAL CONDITIONS

The Hardy-Weinberg equilibrium is a mathematical model of the behavior of ideal organisms in an ideal world. The real world, however, does not approximate these conditions very well. It is important to examine each of the five assumptions made in the model to understand their consequences and how closely they approximate the real world.

The first assumption is infinitely large population size, which can never be true in the real world, as all real populations are finite. In a small population, chance effects on mating success over many generations can alter allele frequencies. This effect is called genetic drift. If the number of breeding adults is small enough, some genotypes will not get a chance to mate with one another, even if mate choice does not depend on genotype. As a result, the genotype ratios of the offspring would be different from the parents'. In this case, however, the gene pool of the next generation is determined by those genotypes, and the change in allele frequencies is perpetuated. If it goes on long enough, it is likely that some alleles will be lost from the population, since a rare allele has a greater chance of not being included. Once an allele is lost, it cannot be regained. How long this process takes is a function of population size. In general, the number of generations it would take to lose an allele by drift is about equal to the number of individuals in the population. Many natural populations are quite large (thousands of individuals), so that the effects of drift are not significant. Some populations, however, especially of endangered species, are very small. For example, the total population of California condors at one time numbered just 70 individuals or less, all in captivity (the California condor recovery program has since raised the numbers to 463 as of 2018, with some 270 individuals flying free in California, Arizona, Utah and Baja, Mexico).

The second assumption is that there is no differential migration, or movement of genotypes into or out of the population. Individuals that leave a population do not contribute to the next generation. If one genotype leaves more frequently than another, the allele frequencies will not equal those of the previous generation. If incoming individuals come from a population with different allele frequencies, they also alter the allele frequencies of the gene pool.

The third assumption concerns mutations. A mutation is a change in the DNA sequence of a gene—that is, the creation of a new allele. This process occurs in all natural populations, but new mutations for a particular gene occur in about one of 10,000 to 100,000 individuals per generation. Therefore, mutations do not, in themselves, play much part in determining allele or genotype frequencies. Yet, mutation is the ultimate source of all alleles and provides the variability on which evolution depends.

The fourth assumption is that there is random mating among all genotypes. This condition may be true for some genes and not for others in the same population. Another common limitation on random mating is inbreeding, the tendency to mate with a relative. Many organisms, especially those with limited ability to move, mate with nearby individuals, which are often relatives. Such individuals tend to share alleles more often than the population at large.

The final assumption is that all genotypes are equally fit. Considerable debate has focused on the question of whether two alleles or genotypes are ever equally fit. Many alleles do confer differences in fitness; it is through these variations in fitness that natural selection operates. Yet, newer techniques of molecular biology have revealed many differences in

DNA sequences that appear to have no discernible effects on fitness.

THEORETICAL AND EXPERIMENTAL GENETIC STUDIES

The field of population genetics uses the Hardy-Weinberg equations as a starting place, to investigate the genetic basis of evolutionary change. These studies have taken two major pathways: theoretical studies, using ever more sophisticated mathematical expressions of the behavior of model genes in model populations, and experimental investigations, in which the pattern of allele and genotype frequencies in real or laboratory populations is compared to the predictions of the mathematical models.

Theoretical population genetics studies have systematically explored the significance of each of the assumptions of the Hardy-Weinberg equilibrium. Mathematical models allow one to work out with precision the behavior of a simple, well-characterized system. In this way, it has been possible to estimate the effects of population size or genetic drift, various patterns of migration, differing mutation rates, inbreeding or other patterns of nonrandom mating, and many different patterns of natural selection on allele or genotype frequencies. As the models become more complex, and more closely approximate reality, the mathematics becomes more and more difficult. This field has been greatly influenced by ideas and tools originally devised for the study of theoretical physics, notably statistical mechanics. Some of the most influential workers in this field were trained as mathematicians and view the field as a branch of applied mathematics, rather than biology. As a consequence, many of the results are not easily understood by the average biologist.

Experimental population genetics tests predictions from theory and uses the results to explain patterns observed in nature. The major advances in this field have been determined, in part, by some critical advances in methodology. In order to study the behavior of genes in populations, one must be able to determine the genotype of each individual. The pattern of bands on the giant chromosomes found in the salivary glands of flies such as *Drosophila* form easily observed markers for groups of genes. Since these animals can be easily manipulated in the laboratory, as well as collected in the field, they have been the subjects of much experimental work. Using population cages, one can artificially control the population size, amount of migration, mating system, and even the selection of genotypes, and then observe how the population responds over many generations. More recently, the techniques of allozyme or isozyme electrophoresis and various methods of examining DNA sequences directly have made it possible to determine the genotype of nearly any organism for a wide variety of different genes. Armed with these tools, scientists can address directly many of the predictions from mathematical models. In any study of the genetics of a population, one of the first questions addressed is whether the population is at Hardy-Weinberg equilibrium. The nature and degree of deviation often offer a clue to the evolutionary forces that may be acting on it.

UNDERSTANDING GENOTYPES

As the cornerstone of population genetics, the Hardy-Weinberg principle pervades evolutionary thinking. The advent of techniques to examine genetic variation in natural populations has been responsible for a great resurgence of interest in evolutionary questions. One can now test directly many of the central aspects of evolutionary theory. In some cases, notably the discovery of the large amount of genetic variation in most natural populations, evolutionary biologists have been forced to reassess the significance of natural selection compared with other forces for evolutionary change.

In addition to the great theoretical significance of this mathematical model and its extensions, there are several areas in which it has been of practical use. An area in which a knowledge of population genetics is important is agriculture, in which a relatively small number of individuals are used for breeding. In fact, much of the early interest in the study of population genetics came from the need to understand the effects of inbreeding on agricultural organisms. A related example, and one of increasing concern, is the genetic status of endangered species. Such species have small populations and often exhibit a significant loss of the genetic variation that they need to adapt to a changing environment. Efforts to rescue such species, especially by breeding programs in zoos, are often hampered by an incomplete consideration of the population genetics of small populations. A third example of a practical application of population genetics is in the management of natural resources such as fisheries. Decisions about fishing limits depend on a knowledge of the extent of local populations. Patterns of allele frequencies are often the best indicator of population structure. Population genetics, by combining Mendel's laws with the concepts of population biology, gives an appreciation of the various forces that shape the evolution of the earth's inhabitants.

—*Richard Beckwitt*

FURTHER READING

Avers, Charlotte J. *Process and Pattern in Evolution.* Oxford University Press, 1989.

Ayala, Francisco J., and John A. Kiger, Jr. *Modern Genetics.* 2nd ed. Benjamin/Cummings, 1984.

Dobzhansky, Theodosius. *Genetics of the Evolutionary Process.* Columbia University Press, 1970.

Futuyma, Douglas J. *Evolutionary Biology.* 3rd ed. Sinauer Associates, 1998.

Hartl, Daniel L. *A Primer of Population Genetics.* 3rd ed. Sinauer Associates, 2000.

Nagylaki, Thomas. *Introduction to Theoretical Population Genetics.* Springer-Verlag, 1992.

Starr, Cecie, and Ralph Taggart. *Biology: The Unity and Diversity of Life.* 9th ed. Brooks/Cole, 2001.

Svirezhev, Yuri M., and Vladimir P. Passekov. *Fundamentals of Mathematical Evolutionary Genetics.* Translated by Alexey A. Voinov and Dmitrii O. Logofet. Kluwer Academic Publishers, 1990.

HEARING

FIELDS OF STUDY

Animal Physiology; Physical Biology; Neurology

ABSTRACT

In the sense organs, receptor cells are stimulated by various types of energy. These receptor cells are highly selective for specific forms of stimulus energy. Each receptor cell transduces the stimulus into a nerve impulse which travels through nerve fibers to the brain where the impulse is translated into a particular sensation. The sensory receptors for hearing are called hair cells, which are extraordinarily sensitive mechanoreceptors. Tiny filaments project from the ends of the receptor cells. The filaments bend in response to mechanical pressure, and generate the nerve impulse.

KEY CONCEPTS

auditory nerve: the cranial nerve that conducts sensory impulses from the inner ear to the brain

pitch: the frequency of sound—the higher the frequency, the greater its pitch

sound frequency: the distances between crests of sound waves measured in hertz

sound intensity: the loudness of a sound directly related to the amplitude of the sound waves measured in decibels

synapse: the functional connection between nerve cells or an effector cell, such as a sensory receptor and a nerve cell

tetrapods: vertebrates with four limbs

SENSORY RECEPTORS AND NERVE IMPULSES

Sensory perception provides the only means of communication between the external world and the nervous system. The process of sensory perception begins in the sense organs, where specially designed receptor cells are stimulated by various types of energy. These receptor cells are highly selective for specific forms of stimulus energy. For example, the photoreceptors in the eye are specific for light energy and largely ignore other forms of stimuli. Each receptor cell transduces (changes) the stimulus into an electrical charge (nerve impulse) which travels through nerve fibers to the brain, where the electrical impulse is translated into a particular sensation. The major types of sensory receptor cells are the chemo receptors (sense chemical energy), mechanoreceptors (sense mechanical energy), photoreceptors (sense light energy), thermoreceptors

Sound Waves

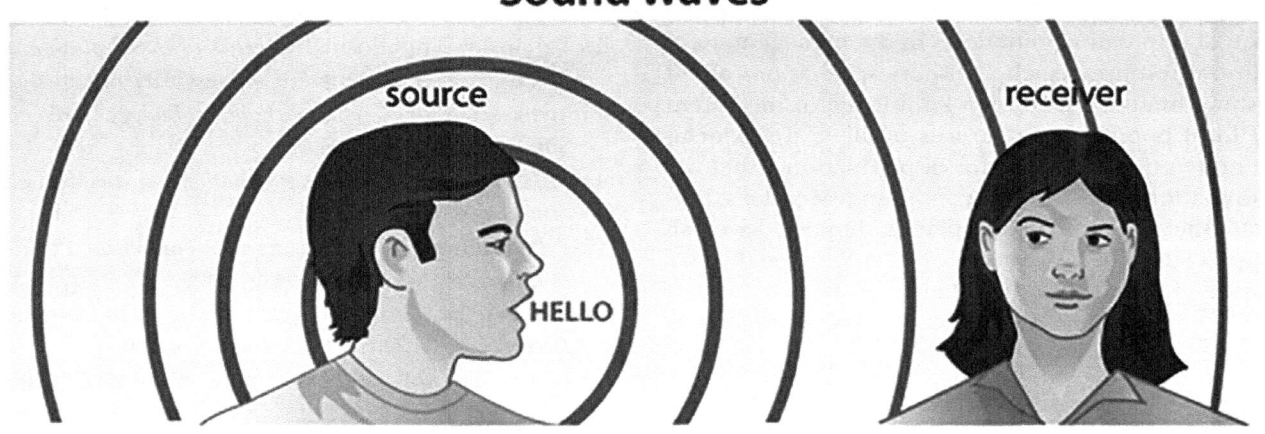

Sound waves travel through the air from a source to a receiver. The boy saying "hello" is the source, the girl is the receiver.

(sense thermal energy), and electroreceptors (sense electrical energy). The sensory receptors for the organs of hearing are called hair cells, which are extraordinarily sensitive mechanoreceptors. Tiny filaments (like hair follicles, only much smaller) called cilia project from the ends of the receptor cell. This filament bends in response to mechanical pressure, and the bending generates the nerve impulse.

Through the course of evolution, the sensory systems developed from single, independent receptor units into complex sense organs, such as the vertebrate ear, in which receptor cells are organized into a tissue associated with accessory structures. The organization of the receptor cells and the architecture of the accessory structures allow far more intricate and accurate sampling of the environment than is possible by independent, isolated receptor cells. While invertebrates possess receptor cells that sense vibrational (mechanical) energy, the true sense of hearing originated with the vertebrates.

THE VERTEBRATE EAR

All vertebrates possess a pair of membranous labyrinths (cavities lined by a membrane) embedded in the cranium, lateral to the hindbrain. This region is often referred to as the inner ear. Each labyrinth consists of three semicircular canals, a utriculus, and a sacculus, which, during development, become filled with a fluid called endolymph. The semicircular canals and utriculus are primarily associated with equilibrium, but the sacculus has evolved into an organ of hearing. In fishes, the lagena, a depression in the floor of the sacculus, has its own maculae (patches of hair cells) which respond to vibratory stimuli of relatively high frequency (sound waves). In tetrapods, the lagena has evolved into an additional fluid-filled duct of the labyrinth called the spiral duct or cochlea. With the evolution of the cochlea, the sensory region enlarged into the organ of Corti. The hair cells are located in the organ of Corti. Most of the structures of the ear assist in the transformation of sound waves (airborne vibrations) into movements of the organ of Corti, which stimulate the hair cells. These hair cells then excite the sensory neurons (nerve cells) of the auditory nerve.

Mammals are the only vertebrates to possess a true cochlea, but birds and crocodilians have a nearly straight cochlear duct that contains some of the same features, including an organ of Corti. Detection of sound in the lower vertebrates that have no cochlear ducts is carried out by hair cells associated with the utriculus and lagena. The cochlea is coiled somewhat like the shell of a snail and is divided into three longitudinal compartments. The two outer compartments, the scala tympani and the scala vestibuli, are filled with a fluid called perilymph and are connected to one another by a structure called the helicotrema. The scala media, filled with endolymph, is located between the two outer compartments and is bound by the basilar membrane and Reissner's membrane. The organ of Corti lies within the scala media and sits upon the basilar membrane. Four rows of hair cells are present in adult mammals—one inner and three outer rows. The cilia of the inner row are thought to be sensitive primarily to the velocity (speed) at which they are displaced by sound waves. The cilia of the outer rows are more sensitive and can detect the degree of deflection as well as the speed.

Sound waves are transported to the inner ear via the outer ear and middle ear. The outer ear consists of the tympanic membrane (ear drum), which is situated on the surface of the head in frogs and toads. In reptiles, birds, and mammals, the tympanic membrane is located deeper in the head at the dead end of an air-filled passageway called the outer ear canal or external auditory meatus. In mammals, there is also an outer appendage, the pinna, which collects sound waves and directs them into the outer ear canal. The tympanic membrane makes contact with the bones (ossicles) of the air-filled inner ear. In amphibians, birds, and reptiles, there is a single bone called the columella (or stapes). In mammals, there is a series of three bones. The malleus (hammer) is in contact with the tympanic membrane at one end and articulates with a second bone, the incus (anvil). The incus then articulates with a third bone, the stapes (stirrup), which connects to a structure called the oval window of the cochlea.

DETECTION OF SOUND

Sound waves striking the tympanic membrane cause it to vibrate. These vibrations are transmitted through the auditory ossicles of the middle ear and through the oval window to the perilymph. The bones of the middle ear amplify the pressure of the vibrations set up in the eardrum by airborne vibrations. Vibrations reaching the oval window pass through the cochlear fluids and the Reisner's and basilar membranes separating the cochlear compartments before dissipating their energy through the membrane-covered round window of the cochlea.

The distribution of pertubations (disturbances) within the cochlea depends on the frequencies of the vibrations entering the oval window. Very long, low frequencies travel through the perilymph of the scala vestibuli, across the helicotrema to the scala tympani, and finally toward the round window. Short wave frequencies take a shortcut from the scala vestibuli through the Reissner's membrane and the basilar membrane to the perilymph of the scala tympani. Movement of perilymph from the scala vestibuli to the scala tympani produces displacement of both Reissner's membrane and the basilar membrane. Movement of Reissner's membrane does not directly contribute to hearing, but displacement of the basilar membrane is required for pitch discrimination. Displacement of the basilar membrane into the scala tympani produces vibrations of the basilar membrane. Each region of the basilar membrane vibrates with maximum amplitude to a different sound frequency. Sounds of higher frequency (pitch) cause maximum vibrations of the basilar membrane at the apical region (closest to the stapes), while sounds of low frequency produce maximum vibrations at the distal region of the basilar membrane.

The sensory hair cells are situated on the basilar membrane with the cilia projecting into the endolymph of the cochlear duct. The cilia of the outer hair cells are embedded with the tectorial membrane located above the hair cells within the cochlear duct. Displacement of the cochlear duct by pressure waves of perilymph produces a shearing force between the basilar membrane and the tectorial membrane. This causes the cilia to bend, and the bending of the cilia produces a nerve impulse in the sensory nerve endings that synapse with the hair cells. The higher the intensity of the sound, the greater the displacement of the basilar membrane, which results in greater bending of the cilia of the hair cells. Increased bending of the cilia produces a higher frequency of nerve impulses in the fibers of the cochlear nerve that synapse with hair cells.

Since a specific region of the basilar membrane is maximally displaced by a sound of a particular frequency, those nerve cells that originate in this region will be stimulated more than nerve cells which originate in other regions of the basilar membrane. This mechanism results in a neural code for pitch discrimination. Within the brain, sensory neurons of the eighth cranial (auditory) nerve synapse with neurons in the medulla which project to the inferior colli culus of the brain. Neurons from this region of the brain project into the thalamus, which in turn sends nerve fibers to the auditory cortex of the temporal lobe of the brain. Through this pathway, neurons in different regions of the basilar membrane stimulate neurons in corresponding areas of the auditory cortex. Hence, each area of this cortex represents a different part of the basilar membrane and a different pitch.

—D. R. Gossett

FURTHER READING

Campbell, Neil A., Lawrence G. Mitchell, and Jane B. Reece. *Biology: Concepts and Connections.* 3d ed. Benjamin/Cummings, 2000.

Feldhamer, G. A., L. C. Drickamer, S. H. Vessey, and J. F. Merritt, eds. *Mammalogy: Adaptation, Diversity, and Ecology.* WCB/McGraw-Hill, 1999.

Fox, Stuart Ira. *Human Physiology.* 6th ed. WCB/McGraw-Hill, 1999.

Linzey, Donald W. *Vertebrate Biology.* McGraw-Hill, 2001.

Randall, David J., Warren Burggren, Kathleen French, and Russell Fernald. *Eckert Animal Physiology: Mechanisms and Adaptions.* 4th ed. W. H. Freeman, 1997.

HERBIVORES

FIELDS OF STUDY

Evolutionary Biology; Biology; Veterinary Sciences

ABSTRACT

Herbivores are animals whose diets consist almost entirely of plants. They have two ecological functions. First, they eat plants and keep them from overgrowing. Second, they are food for carnivores, which subsist almost entirely upon their flesh, and omnivores, which eat both plants and animals. Herbivores live on land or in oceans, lakes, and rivers. They can be insects, other arthropods, fish, birds, or mammals.

KEY CONCEPTS

- **carnivore:** any animal that eats only the flesh of other animals
- **gestation:** the term of pregnancy
- **metamorphosis:** insect development into adults, passing through two or more dissimilar growth forms
- **ruminant:** a herbivore that chews and swallows plants, which enter its stomach for partial digestion, are regurgitated, chewed again, and reenter the stomach for more digestion

FOOD FOR THOUGHT

Herbivores are animals whose diets consist almost entirely of plants. They have two ecological functions. First, they eat plants and keep them from overgrowing. Second, they are food for carnivores, which subsist almost entirely upon their flesh, and omnivores, which eat both plants and animals. Herbivores live on land or in oceans, lakes, and rivers. They can be insects, other arthropods, fish, birds, or mammals.

WILD HERBIVORES

Insects are the largest animal class, with approximately one million species. Fossils show their emergence 400 million years ago. Insects occur worldwide, from pole to pole, on land and in fresh or salt water. They are the best developed invertebrates, except for some mollusks. They mature by metamorphosis, passing through at least two dissimilar stages before adulthood. Metamorphosis can take up to twenty years or may be complete a week after an egg is laid.

Sawfly larva, Nematus ribesii, *feeding on leaf. (Daniel Mietchen)*

Many insects are herbivores. Some feed on many different plants; others depend on one specific plant species or a specific plant portion, such as leaves or stems. Relationships between insects and the plants they eat are frequently necessary for plant growth and reproduction. Among the insect herbivores are grasshoppers and social insects such as bees.

Artiodactyls are hoofed mammals, including cattle, pigs, goats, giraffes, deer, antelope, and hippopotamuses. Most are native to Africa, but many also live in the Americas, Europe, and Asia. Artiodactyls walk on two toes. Their ancestors had five, but evolution removed the first toe and the second and fifth toes are vestigial. Each support toe—the third and fourth—ends in a hoof. The hippopotamus, unique among artiodactyls, stands on four toes of equal size and width.

Artiodactyls are herbivores, lacking upper incisor and canine teeth, but pads in upper jaws help the lower teeth grind food. Many are ruminants, such as antelope, cattle, deer, goats, and giraffes. They chew and swallow vegetation, which enters the stomach for partial digestion, is regurgitated, chewed again, and reenters the stomach for more digestion. This maximizes nutrient intake from food.

Deer are hoofed ruminants whose males have solid, bony, branching antlers that are shed and regrown yearly. The deer family, approximately 40 species, occurs in Asia, Europe, the Americas, and North Africa. Deer live in woods, prairies, swamps, mountains, and tundra. Their size ranges from the 2.3 meter tall moose, to the 30 centimeter tall pudu. Deer first appear in the fossil record ten million years ago. Deer eat the twigs, leaves, bark, and buds of bushes and saplings, and grasses. Females have one or two offspring after ten-month pregnancies. Common species are the white-tailed and mule deer in the United States; wapiti in the United States, Canada, Europe, and Asia; moose in North America and Europe; and reindeer in Russia, Finland, and Alaska.

Antelope, a group of approximately 150 ruminant species, have permanent, hollow horns in both sexes. Most are African, although some are European or Asian. They eat grass, twigs, buds, leaves, and bark. There are no true antelope in the United States, where their closest relatives are pronghorns and Rocky Mountain goats (goat-antelope with both goat and antelope anatomic features). The smallest antelope, the dik-dik, is rabbit-sized. Elands, the largest antelope, are ox-sized. Unlike deer horns, antelope horns are unbranched. Most antelope run rather than fighting, and are all swift. Antelope live on plains, marshes, deserts, and forests. Females birth one or two offspring per pregnancy. Impala and gazelles, such as the springbok, are found in Africa. In Asia, Siberian saigas and goat antelope (takin) inhabit mountain ranges. Chamois goat antelope live in Europe's Alps.

Giraffes and hippos are unusual artiodactyls. Giraffes inhabit dry, tree-scattered land south of the Sahara. Their unusual features are their very long legs and necks. Males are over sixteen feet tall, including the neck. Both sexes have short, skin-covered horns. Long necks, flexible tongues, and upper lips pull leaves—their main food—from trees. Giraffes have brown blotches on buff coats and blend with tree shadows. They live for up to twenty years. They have keen senses of smell, hearing, and sight, and can run thirty-five miles per hour. Due to their two-ton weights, they live on hard ground. Giraffes rarely graze, and go for months without drinking, getting most of their water from the leaves they eat, because it is difficult for them to reach the ground or the surface of a river with their mouths. Females have one offspring after a fifteen-month gestation.

The unusual feature of hippos is that they walk on all four toes of each foot. Perhaps this is because they weigh three to four tons. Hippos are short-legged, with large heads, small eyes, small ears, and nostrils that close underwater. Huge hippo mouths hold long, sharp incisors and canines in both jaws. Hippos once lived throughout Africa. Now they are rarer, due to poaching for ivory. A hippo can be fifteen feet long and five feet high at the shoulder. Semiaquatic, hippos spend most daylight hours nearly submerged, eating aquatic plants. At night they eat land plants. Females bear one offspring at a time.

AQUATIC HERBIVORES

Fish are aquatic vertebrates, having gills, scales and fins. They include rays, lampreys, sharks, lungfish, and bony fishes. The earliest vertebrates, 500 million years ago, were fishes. They comprise over 50 percent of all vertebrates and have several propulsive fins: dorsal fins along the central back; caudal fins at tail ends; and paired pectoral and pelvic fins on sides and belly. Fish inhabit lakes, oceans, and rivers, even in Arctic and Antarctic areas. Most marine fish are tropical. The greatest diversity of freshwater species is found in African and rain-forest streams.

Fish vary in length from one centimeter to sixteen meters, and some weigh seventy-five tonnes. Many, including giant whale sharks that eat primarily plankton, are herbivores. Fish respiration uses gills, through which blood circulates. When water is taken in and expelled, oxygen enters the blood via the gills and carbon dioxide leaves. Fishes reproduce by laying eggs that are fertilized outside the body, or by internal fertilization and development with the birth of well-developed young.

DOMESTICATED ARTIODACTYLS

Bovids are domesticated artiodactyls. Most have horns and hooves. Bovid horns are spiraled, straight, tall, or L-shaped from the sides of the head. All have hooves to help them grip the ground. Most are ruminants. Their breeding habits are similar. Males fight over females and the strongest wins. Gestation, ranging from four to eleven months, yields two to three young. The young nurse for several months and then join the herd. Young males leave female herds to live with other bachelors.

Cattle are domesticated bovids, raised for meat, milk, and leather. Modern cattle come from European, African, and Asian imports. Breeding modern cattle began in Europe in the mid-1800s. Today, some three hundred breeds exist. Dairy cattle, such as Holsteins, produce copious milk. Beef cattle, such as Angus, were bred to yield meat. As of 1990, about 1.3 billion cattle were found worldwide.

Sheep are also artiodactyls. Wild sheep occur in some places, such as the North American bighorn and Mediterranean mouflon. Sheep were domesticated eleven thousand years ago from mouflon. Today, domesticated sheep have a world population of approximately 1.3 billion and inhabit most countries, being more widely distributed than any other domesticated animal. These ruminants have paired, hollow, permanent horns. Male horns are massive spirals; those of females are smaller. Adult body length is about 1.5 meters and weights are 100 to 200 kilograms. Females birth two or three young after a five-month gestation period. Sheep can live for up to twenty years.

Sheep provide wool, meat, and milk. About eight hundred domesticated breeds exist, in environments from deserts to the tropics. Those bred for wool, half the world's sheep, live in semiarid areas, are medium sized, and produce fine wool. Most are in Australia, New Zealand, and South America. Mutton-type sheep, 15 percent of the world sheep population, produce meat. Fat-tailed sheep, 25 percent of the sheep population, produce milk. In 1990, the five leading sheep countries were Australia, China, New Zealand, India, and Turkey. The United States raised less than 1 percent of the world total.

Goats are ruminants, closely related to sheep, but have shorter tails, different horn shape, and bearded males. They eat grass, branches, and leaves and breed from October to December. A five-month gestation period yields two offspring. Numerous goat breeds are domesticated worldwide for meat and milk, and as pets and burden carriers. Domesticated Angora goats yield silky mohair. Goat milk is as nutritious as cow milk and used in cheese-making.

It is clear that wild herbivores are ecologically important to food chains. This is because they eat plants, preventing their overgrowth, and they are eaten by carnivores and omnivores. Domesticated herbivores—cattle, sheep and goats, used for human sustenance—account for three to four billion living creatures. Future production of better strains of domesticated herbivores via recombinant deoxyribonucleic acid (DNA) research may cut the numbers of such animals killed to meet human needs. Appropriate species conservation should maintain the present balance of nature and sustain the number of wild herbivore species living on earth.

—*Sanford S. Singer*

FURTHER READING

Gerlach, Duane, Sally Atwater, and Judith Schnell. *Deer.* Stackpole Books, 1994.

Gullan, P. J., and P. S. Cranston. *Insects: an Outline of Entomology.* 2nd ed. Blackwell Science, 2000.

Olsen, Sandra L., ed. *Horses Through Time.* Roberts Rinehart, 1996.

Rath, Sara. *The Complete Cow.* Voyageur Press, 1990.

Shoshani, Jeheskel, and Frank Knight. *Elephants: Majestic Creatures of the Wild.* Rev. ed. Checkmark Books, 2000.

Herds

Fields of Study
Animal Behavior; Evolutionary Biology; Sociology

Abstract
Herding in animals serves several purposes, most commonly exploitation of food resources and reduction of the probability of predation via dilution effects. Sizable grazing mammals are well known for forming some of the largest and most dense herds seen. Small grazers that exploit vegetation in herds include marine and freshwater snails, tortoises and turtles, geese, and hyrax. Those grazers that feed indiscriminately tend to form herds that vary in density with season and density of foliage.

Key Concepts
carnivores: animals that eat the flesh of other animals

dilution effects: the reduction in per capita probability of death from a predator due to the presence of other group members

herbivores: animals that eat plants and show specializations of teeth and digestive tracts to do so

phalanges: the free toes of the foot; some can be modified to bear claws, hoofs, or nails

subungulates: nonhoofed mammals that support their weight on more than the terminal phalanges; some, such as elephants and hyraxes, have pads under their metatarsals, and others, such as the sirenians, have forelimbs modified into flippers

ungulates: hoofed mammals that support their weight only on the hoof-clad terminal phalanges and have teeth specialized for clipping vegetation

HERD FORMATIONS
Herding in animals serves several purposes, most commonly exploitation of food resources and reduction of the probability of predation via dilution effects. When food is spaced irregularly into patches and cannot be easily defended, individuals of a particular species will form a herd, simply as a result of coming together to feed on the same resource in the same location. Sizable grazing mammals are well known for forming some of the largest and most dense herds seen. The most familiar of these include African ungulates (wildebeests, zebra, gazelle, wild horses, rhinos, hippopotami) and subungulates (elephants, sirenians, the extinct mastadons); bison, buffalo, caribou, elk, moose, and deer in North America; kangaroo in Australia; and deer, elk, moose, antelope, wild horses, sheep, goats, pigs, boars, and peccaries in Eurasia. In addition to large mammalian herbivores, small grazers that exploit vegetation in herds include marine and freshwater snails, tortoises and turtles, geese, and hyrax.

Those grazers that feed indiscriminately tend to form herds that vary in density with season and density of foliage. Furthermore, these herds tend to vary their densities and vegetation utilization patterns with the type of vegetation exploited and its productivity potential. In seasons or areas where rainfall is common, herds tend to congregate together and exploit small patches of grass, which they clip into grazing lawns. Maintaining grazing lawns increases the productivity of the plants via increased nitrogen content and increased digestibility in freshwater, marine, and terrestrial environments. When rainfall is scarce, the herds disperse over vast areas, which allows some grasses to grow into tall meadows. The benefits to individuals in a herd from grazing a larger area and creating a grazed lawn are greater than those to a lone individual that would only be able to graze a small patch of grass and that would, presumably, be at greater risk of predation than one individual surrounded by many others. Furthermore, if the lone individual were subject to a greater risk of predation, there would be less likelihood of it being able to return to the same patch over and over again to keep it as a grazed lawn. Thus, by grouping, an individual benefits twice: once by the gain in nutrition from a grazed lawn and again from the dilution effect, where the risk of predation is diluted by the number of members forming the group.

HERDING TO AVOID PREDATION
The large herbivores that most commonly form herds are clearly visible to predators. Predation risk reduction via the dilution effect or the selfish herd effect is the other main benefit of herding. In the

Walrus are marine mammals found in the arctic regions of North America, Greenland, and northeastern Siberia. They group together in herds of both males (bulls) and females (cows). Both sexes have ivory tusks. (Joel Garlich-Miller, U.S. Fish and Wildlife Service)

dilution effect, the probability of a particular individual being killed or injured is reduced by the presence of other group members that might be attacked first. In other words, there is safety in numbers. Individuals in a group may also benefit by putting other animals between themselves and the predator (the selfish herd effect). Grouping thus provides the opportunity to decrease the area of danger around each individual. If individuals within the group are acting in a selfish herd manner, the groups formed tend to be tightly clumped, as all individuals attempt to put other individuals on all sides around them and move into a central location. Finally, formation of a herd can appear as a single, very much larger animal, making it more difficult for a predator to select one specific individual from the crowd for attack, or once an attack is initiated and herd members scatter during flight, can make it difficult for the predator to keep track of its chosen target animal.

Herd members also benefit by a reduction in the time necessary to scan for predators while foraging. An individual foraging has to split its time between consuming as much food as possible and avoiding becoming another animal's meal. Thus, a solitary individual has to be much more vigilant than does a member of a group who can rely on many other eyes and thus reduce its own scan rate. Because of this advantage and the advantage in the reduction of the rate of per capita predation, even herds that simply congregate around a resource benefit greatly.

MIGRATORY AND SOCIAL HERDS

Many large herbivores migrate in response to food availability in different seasons. Red deer, caribou, wildebeest, mountain goats, northern fur seals, and humpback whales are examples of animals that all migrate in response to seasonal changes in rainfall or food abundance. Some of these migrations are over incredible distances: Wildebeests travel about six hundred miles; northern fur seals and humpbacks can travel three to four thousand miles. During these migrations, small herds unite with bigger herds to form even larger herds. While terrestrial herbivores migrate to follow food, some marine mammals, such as gray whales and humpbacks, migrate south to calf and breed (but do not feed, living off food reserves instead) and migrate north to abundant feeding grounds.

Some herding species display social organization beyond that expected by mere association. This social organization tends to break a herd into smaller units (matrilineal groups, harems, or small territories controlled by one or several males) that are clustered within the entire herd. These species include the horses, zebras, pronghorn sheep, walruses, sea lions, seals, and elephants. Horse and zebra herds are composed of a number of small groups of females and their foals. These individual groups are overseen by a single stallion; young males leave these groups to form bachelor herds. Group members distinguish each other via a "corporate smell." Stallions generally control a group of females and will fight with challenging stallions in elaborate rearing displays. If a stallion is challenged and loses after inseminating the females in his group, the challenger will mount the females and rape them to induce them to abort. They will then come into estrus quickly and he will remate them. Stallions will groom females to cement their relationship with them; likewise, mares will groom their foals for the same purpose. Male impalas and gazelle maintain harems during the breeding season; male wildebeest do the same, but if the herd is large, the defense of the harem may be accomplished by several males, rather than one. Male elk are divided into four main categories: Primary bulls are the first to establish harems, but as they become exhausted from challenges and herding and mating females, all the while not eating, the harems are taken over by secondary bulls. Once the secondary bulls become exhausted, tertiary bulls take over. The fourth category, opportunistic bulls, only mate with females by chance.

Pinniped herds (sea lions, seals, and walruses) form breeding herds, where males establish territories. In some groups, females are herded together in harems to remain in the male's territory; in other groups, females are free to move from territory to territory. Males vigorously defend their territories against intruding males.

Sirenians, the dugong and manatee, have proved difficult to study, but dugongs often form large herds for unknown reasons. During mating, a series of males follows a receptive female to form a cluster of up to twenty animals. Males then initiate fighting to determine who will mount the female. Following mating, the main social unit is the female and her calf. A similar situation exists in the manatee, with females and their calves being the main social unit and females in estrus becoming the focus of a mating herd of males.

Herding serves important purposes for animals: utilization of a common resource (usually food) and predator reduction via a safety-in-numbers principle.

Herds can also be formed during migrations or for breeding purposes. While many herds are simply loose associations of animals, some are more highly organized into a number of social units such as harems or matrilineal groups.

—*Kari L. Lavalli*

FURTHER READING

Drickamer, L. C., S. H. Vessey, and D. Meikle. *Animal Behavior: Mechanisms, Ecology, and Evolution*. 4th ed. Wm. C. Brown, 1996.

Fryxell, J. M. "Forage Quality and Aggregation by Large Herbivores." *American Naturalist* 138, no. 2 (August 1991): 478–498.

Hamilton, W. D. "Geometry for the Selfish Herd." *Journal of Theoretical Biology* 31 (1971): 294–311.

McNaughton, S. J. "Grazing Lawns: Animals in Herds, Plant Form, and Coevolution." *American Naturalist* 124, no. 6 (December 1984): 863–886.

Reynolds, J. E., and D. K. Odell. *Manatees and Dugongs*. Facts on File, 1991.

Wallace, J. *The Rise and Fall of the Dinosaur*. Gallery Books, 1987.

Hermaphroditism

FIELDS OF STUDY

Evolutionary Biology; Reproductive Biology; Animal Physiology

ABSTRACT

Hermaphrodites are individuals that have both male and female reproductive organs—that is, they have both ovaries to produce eggs and testes to produce sperm. In most animals, hermaphroditism is a result of abnormal development and is extremely rare. In some species, however, hermaphroditism is normal.

KEY CONCEPTS

gonad: the organ that produces reproductive cells (sperm or eggs)

protandry: the condition of starting out male with the potential to become female

protogyny: the condition of starting out female with the potential to become male

sequential hermaphrodite: species or individual with the potential to change from one sex to the other

sex-limited traits: features that are only expressed in one sex

sexual dimorphism: the existence of anatomical, physiological, and behavioral differences between the two sexes of a species

simultaneous hermaphroditism: the condition of being simultaneously male and female

SEXUAL DIMORPHISM

In most species, reproduction involves sex—that is, the joining of genetic material from two individuals to create new, genetically unique offspring. In sexually reproducing species, females are those individuals which produce relatively large sex cells that are full of nutrients (eggs), and males are those individuals which produce relatively small sex cells that have little or no nutrients, but which can be produced in much greater numbers (sperm). Generally there are other differences between the sexes as well—differences in hormones, anatomy, body shape, size, color, and behavior. Such distinguishing features are referred to as sex-limited traits because they appear in only one sex; collectively, they result in sexual dimorphism—literally the two sexual forms of a single species.

Whether an individual animal develops into a male or a female depends on which genes get turned on early in development. In mammals, genes for maleness get turned on in individuals with an X and a Y chromosome, while genes for femaleness get turned on in individuals with two X chromosomes. In other animals, sex determination may depend on other factors. For example, in many reptiles, development into a male or a female depends upon the temperature of the embryo as it develops in its egg.

Hermaphrodites are individuals that have both male and female reproductive organs—that is, they have both ovaries to produce eggs and testes to produce sperm. In most animals, hermaphroditism is a result of abnormal development and is extremely rare. In some species, however, hermaphroditism is normal.

SIMULTANEOUS VERSUS SEQUENTIAL HERMAPHRODITISM

Depending on the species, hermaphroditism can be found in either simultaneous or sequential form. In species with simultaneous hermaphroditism, individuals are simultaneously both male and female; each adult has the ability to produce both sperm and eggs. Depending on the species, a single reproductive encounter between simultaneous hermaphrodites may involve both partners exchanging sperm and eggs (for example, earthworms) or may involve the partners taking turns as male and female (for example, some coral reef fishes).

On the other hand, individuals of species with sequential hermaphroditism start life as either male or female, but have the ability to change sex at some later point. In protandrous species, individuals start out as male and have the potential to later change to female. These are typically species which require large body bulk before they can produce eggs, so individuals start out as male and change to female only if they get old enough and large enough to make eggs. In protogynous species, individuals start out as female and have the potential to later change to male. These are typically species in which males must defend a harem or a territory in order to mate, so individuals start out as female and change to male only if they get large enough to fight and win.

TRIGGERS FOR SEX CHANGE

In both protandrous and protogynous species, the trigger for sex change may relate not only to body size but also to the social structure of the individual's community. For example, a large female of a protogynous species may not change to male if there is an even bigger male present who would clearly win every fight. On the other hand, a relatively small female might change to male if she is the biggest female around and the larger local males suddenly died or disappeared. Thus, body size is a relative, not an absolute trigger for sex change, and in a very few species individuals can revert back to their original sex if social circumstances change again.

When a sequential hermaphrodite changes sex, it changes not only its reproductive organ or gonad, but also all the other sexually dimorphic aspects of its anatomy, physiology, and behavior. Hormones and behavior are the first things to change; then, over a period of time, the hormonal changes induce changes in the gonad and other tissues, including the brain. As a hermaphroditic fish, for example, switches from one sex to another, it may change the size and shape of its fins, its color, its aggression level, and its sexual preferences and rituals. These miraculous changes provide visible proof of the (generally silent) presence in both sexes of sex-limited genes for both male and female attributes.

—*Linda Mealey*

FURTHER READING

Forsyth, Adrian. *A Natural History of Sex.* Charles Scribner's Sons, 1986.

Mealey, Linda. *Sex Differences: Developmental and Evolutionary Strategies.* Academic Press, 2000.

Milius, Susan. "Hermaphrodites Duel for Manhood." *Science News* 153 (February 14, 1998): 101.

Warner, R. R. "Mating Behavior and Hermaphroditism in Coral Reef Fishes." *American Scientist* 72 (1984): 128–135.

Heterochrony

Fields of Study
Evolutionary Biology; Genetics; Embryology; Animal Physiology

Abstract
Heterochronic phenomena are processes by which changes in the timing, rate, and duration of an ancestral pattern of growth and development result in changes in descendants. One of the six types of heterochrony is neoteny, in which a descendant's slower rate of development causes it to be an immature or truncated expression of its ancestor.

Key Concepts

acceleration: a faster rate of growth during ontogeny that causes a particular characteristic to appear earlier in a descendant ontogeny than it did in the ancestral ontogeny

heterochrony: any phenomenon in which there is a difference between the ancestral and descendant rate or timing of development

hypermorphosis: a phenomenon in which the rate and initiation of growth in the descendant are the same as in the ancestor but the cessation of development takes place later

ontogeny: the life history of an individual, including both its embryonic and post natal development

phylogeny: the history of a lineage of organisms, often illustrated by analogy to the branches of a tree

postdisplacement: a form of paedomor phosis in which the initiation of growth in a descendant occurs later than in the ancestor, ensues at the ancestral rate, and ceases at the ancestral point

predisplacement: a form of peramorpho sis in which the initiation of growth in a descendant occurs earlier than in the ancestor, ensues at the ancestral rate, and ceases at the ancestral point

TIME IS OF THE ESSENCE

Heterochronic phenomena are processes by which changes in the timing, rate, and duration of an ancestral pattern of growth and development result in changes in descendants. One of the six types of heterochrony is neoteny, in which a descendant's slower rate of development causes it to be an immature or truncated expression of its ancestor.

PAEDOMORPHOSIS AND PERAMORPHOSIS

The six types of heterochrony fall into two general patterns: paedomorphosis and peramorphosis. In these patterns, ancestral and descendant ontogenetic trajectories are compared. If the descendant morphology exceeds or surpasses the ancestral morphology, this is called peramorphosis. There are three types of peramorphosis: acceleration, hypermorphosis, and predisplacement. In acceleration, the beginning and end of development occurs at the same time in both the ancestral and descendant ontogenetic trajectories. The rate of development is faster in the descendant, however, so its morphology transcends that of the ancestor.

In hypermorphosis, development begins at the same time and ensues at the same rate in both the ancestor and descendant; however, the descendant continues development for a longer interval—that is, it stops growing later. In predisplacement, the descendant begins development earlier than the ancestor, and the rate of development and the time at which growth ceases remain the same. The result of both predisplacement and hypermorphosis is a longer interval of growth. In acceleration, however, the interval of growth is the same in both the ancestor and its descendant; only the rate of growth has changed.

If a descendant morphology is a truncated or an abbreviated version of ancestral morphology, this is called paedomorphosis. There are also three types of paedomorphosis: neoteny, progenesis, and postdisplacement. In progenesis, descendant development begins and proceeds at the ancestral rate but stops sooner. In postdisplacement, the ancestral rate of development and the time of cessation are the same in the descendant; however, development begins later. Thus, for post displacement and progenesis, the time interval over which the descendant morphology develops is truncated when compared to the ancestral ontogenetic trajectory. In contrast, neoteny is characterized by a slower descendant rate of development but an unchanged time interval for growth and development.

THE CONTRIBUTIONS OF VON BAER AND HAECKEL

Two scientists of the nineteenth century were especially important contributors to the early ideas concerning development and its impact on evolutionary change. Karl Ernst von Baer suggested that the features that appear early in ontogeny are those that are shared by the most organisms, whereas the features that appear later in ontogeny are those that are shared by successively smaller groups of organisms. This maxim has been called von Baer's law. Ernst Heinrich Haeckel held a different, more restricted, concept of ontogeny. He popularized the phrase "ontogeny recapitulates phylogeny." By this he meant that the ontogenetic or developmental phases through which an organism passes can be interpreted as being equal to the sequence of events that occurred during the evolutionary history of that particular organism. Haeckel's narrower concept of the relationship between ontogeny and phylogeny has been rejected by modern biologists in favor of von Baer's law.

Von Baer and Haeckel laid the foundation for ideas of heterochrony, but it was not until the later twentieth century that the subject of heterochrony was once again an area of active research. An important book by Stephen Jay Gould entitled *Ontogeny and Phylogeny* (1977) was instrumental in reopening the discussion initiated a century earlier. One of the most important things accomplished in Gould's book was the distinction that was made between different types of paedomorphosis. Frequently, in earlier works dealing with species with a juvenile appearance, no distinction was made among neoteny, post displacement, and progenesis: All three were discussed as neoteny. This confusion prevented generalization.

Organisms for which neoteny is used as an explanation for their origins include some salamanders of the family *Ambystomatidae* and the primates of the family *Hominidae* (the family that includes humans). In the latter case, *Homo sapiens* is considered a juvenilized anthropoid; the reduction in the amount of hair and the longer period of infancy can be used as evidence for this supposition.

STUDYING HETEROCHRONIC CHANGE

Two kinds of information are necessary for documenting heterochronic processes of change. Because frequent reference must be made to ancestral and descendant ontogenetic trajectories, it is of crucial importance to have available, for the organisms under study, an estimate of their phylogenetic, or evolutionary, history. This is not an easily obtained body of evidence, and much scientific debate has occurred over the procedures that should be followed in seeking to unravel phylogenetic history. Some researchers even doubted that it would ever be possible to obtain evidence sufficient for the task. Nevertheless, some small consensus has emerged. This is especially gratifying for students of heterochrony, who depend so much upon the phylogenies that anchor their conclusions.

It is also necessary to document ontogeny, development, and/or growth in either a quantitative or qualitative sense. The quantitative measurement of growth is relatively straightforward. Measurements are taken from specimens of known ages. It is most desirable that measurements be taken from the same individual specimens throughout their ontogenies. This is not always possible, nor is it always possible to obtain accurate ages for the available specimens. Another problem is that the ages of the specimens of different species might not be directly comparable. These are among the factors that complicate the otherwise simple process of measuring the sizes and shapes of specimens over time—that is, measuring in various places along an ontogenetic trajectory. Assuming that appropriate measurements have been gathered, a variety of statistical procedures can then be applied to the accumulated data. These procedures include some rather sophisticated multivariate statistics that have only become feasible with the advent of computers.

The qualitative documentation of an ontogenetic trajectory suffers from many of the same sources of complication. In this approach, ontogenies are conceptualized as a series of discrete stages, phases, events, or appearances. These sequences are determined for the organisms under examination, and the stages that bear close resemblance are sometimes considered identical. The problem with this view is that it conceives of ontogeny as being composed of static sequences;

this is a largely Haeckelian view. In fact, ontogenies are best viewed as dynamic, which makes it conceptually difficult to compare isolated parts of an indivisible ontogeny.

An unanswered question of evolutionary biology concerns the processes by which new morphologies and new organisms originate. Although many promising inroads have been made, none has been more hopeful than the idea of heterochrony. The conjecture is that small shifts in the timing, rate, and duration of ontogeny contribute to the appearance of new structures or perhaps even new organisms. This approach has so far proved both effective and promising, but it has not yet been applied to a sufficient range of organisms. Thus, it is not yet possible to tell whether heterochrony will prove to be a universally applicable approach to the study of the origin of new structures and species.

—*Charles R. Crumly*

FURTHER READING

Alberch, Pere, Stephen Jay Gould, George F. Oster, and David B. Wake. "Size and Shape in Ontogeny and Phylogeny." *Paleobiology* 5 (1979): 296–317.

Gould, Stephen Jay. *Ontogeny and Phylogeny*. Belknap Press, 1977.

Humphries, C. J., ed. *Ontogeny and Systematics*. Columbia University Press, 1988.

McKinney, Michael L., ed. *Heterochrony in Evolution: A Multidisciplinary Approach*. Plenum Press, 1988.

McKinney, Michael L., and Kenneth J. McNamara. *Heterochrony: The Evolution of Ontogeny*. Plenum Press, 1991.

McNamara, Kenneth J. *The Shapes of Time*. Johns Hopkins University Press, 1997.

_____, ed. *Evolutionary Change and Heterochrony*. John Wiley & Sons, 1995.

HIBERNATION

FIELDS OF STUDY

Animal Physiology; Endocrinology; Evolutionary Biology

ABSTRACT

During hibernation the body shuts down and requires little energy. All of its systems, including endocrine, circulation, respiration, and elimination, are reduced to the barest essentials necessary to maintain life. Animals living in this state require almost no nutrition. They draw what nourishment they need from stores of fat that they accumulate by eating a great deal immediately before hibernating. They conserve their body heat by rolling themselves into balls.

KEY CONCEPTS

anesthesiologist: a physician who administers anesthetics during surgical procedures

Celsius: a scale for measuring temperature in which freezing is zero degrees and boiling is one hundred degrees, abbreviated C

Fahrenheit: a scale for measuring temperature in which freezing is 32 degrees and boiling is 212 degrees, abbreviated F

hibernacula: the winter habitats of brown bats

synchronization: causing events to occur simultaneously

vertebrates: animals with brains and spinal cords

WHEN THE GOING GETS TOUGH, THE TOUGH GO TO SLEEP

As winter approaches, some animals enter a barely living state. Whereas the body temperature of warm-blooded animals is generally about 37°C (98.6°F), some of the larger vertebrates, such as bears, enter a restful state for several of the colder months, during which body temperatures sink to about 30°C (88°F). In another class of smaller animals, notably the brown bat, the body temperature hovers just above freezing, usually between 1° and 3°C (34° to 38°F). At these levels, the animal is barely alive. It does not bleed when cut. It breathes infrequently. Its heartbeat drops dramatically.

Groundhogs and chipmunks are found in areas where there may be sporadic periods of warm weather during the winter, at which times these animals awaken from their torpor temporarily. Their body temperatures, which will have sunk to about 10°C (50°F), increase as the temperature outside their underground lairs rises. When the temperature sinks again, these animals resume their sleep.

Animals such as bears, brown bats, some rodents, hummingbirds, whippoorwills, chipmunks, ground squirrels, skunks, and marmots have built-in mechanisms that prevent their temperatures from sinking below the levels their systems can withstand. When their temperatures approach life-threatening levels, they begin to shiver, thereby maintaining or raising their temperatures without wakening them from their slumbers.

THE WHY AND HOW OF HIBERNATION

During hibernation the body shuts down and requires little energy. All of its systems, including endocrine, circulation, respiration, and elimination, are reduced to the barest essentials necessary to maintain life. Animals living in this state require almost no nutrition. They draw what nourishment they need from stores of fat that they accumulate by eating a great deal immediately before hibernating. They conserve their body heat by rolling themselves into balls.

The mechanisms that trigger hibernation mystify scientists. Some sort of internal clock, probably responding to light and temperature, clicks in at given times, determining the beginning of hibernation and its extent. This mechanism is far from precise. Some animals that have been maintained from year to year under consistent conditions may lose their synchronization, entering hibernation possibly in spring or summer.

THE LIGHT SLEEPERS

On hearing the word "hibernation," people generally think immediately of bears, who enter periods of dormancy as winter approaches. By the time they enter their lair for their winter's sleep, they have gained considerable weight and have added to their bodies layers of fat to provide them with the nutritional reserves they will require to survive the next three or four months.

During this period of dormancy, the bear's temperature drops less than 10°F. Some scientists resist designating as hibernation the period during which such animals as bears, chipmunks, raccoons, and skunks sleep beneath ground. This period is sometimes called "winter lethargy."

Animals whose body temperatures do not drop dramatically during their dormant period may waken from their sleep several times during the winter. When the weather moderates, they often leave their lairs and scurry about seeking food. When cold weather resumes, they return to their lairs to continue their slumbers.

The eastern chipmunk is among the light sleepers. Unlike bears, it does not accumulate excess body fat to see it through its three or four dormant months. Rather, it stores food in its burrow as winter approaches, building its nest on top of the food it has gathered. It wakens frequently in winter to eat and to defecate in a section of the burrow away from the food supply. The male chipmunk usually ends its sleep late in February. Leaving the burrow, it first seeks food and water, then looks for a mate, who produces from two to five babies one month after mating.

Skunks are also among the light sleepers. When the outside temperature approaches 10°C (50°F), skunks retreat to their dens for the winter. They may take over an abandoned woodchuck's nest, but often they build their own dens below the frost line, at depths of between six and twelve feet. They line the den with dried leaves and grasses, creating a cozy nest. Although skunks are solitary in summer, they often live in groups during the winter, huddling together to keep warm.

Raccoons enter a dormant state in cold climates but are active throughout the year in milder ones, undergoing fewer body changes in winter than other true hibernators. Their body temperature drops minimally. Their heartbeat, while decreasing slightly, remains close to normal. Raccoons sleep as long as cold weather persists. They stir during warm spells but sleep again when the temperature drops.

In winter, raccoons, usually solitary dwellers, change that pattern and share their dens with other raccoons for their body warmth. Regardless of the weather, male raccoons become active late in January, which is mating season. After mating, they return to their dens for more sleep before spring.

THE HEAVY SLEEPERS

Some species undergo significant changes during dormancy that qualify them as true hibernators. In these animals, heartbeat, temperature, and respiration drop so dramatically during winter that life is barely sustained. These heavy sleepers do not respond to temporary increases in outside temperature, rather sleeping soundly through the months of their hibernation.

The most renowned of the heavy sleepers is the woodchuck, also known as the groundhog. People watch groundhogs' burrows every February 2 to determine whether the groundhog will see his shadow. If he does, legend has it, he will return to his den and there will be at least six more weeks of winter. If he does not, the prognostication is that spring weather will arrive early. The groundhog actually leaves its burrow in February seeking a mate.

Woodchucks accumulate as much body fat as possible before hibernation, sometimes achieving a weight of five kilograms. During their deep sleep, they take nourishment from their layers of fat, although they may waken at times to nibble seeds stored in their dens. By the time they emerge from hibernation, most have lost between 35 to 50 percent of their weight. The den into which woodchucks retreat when winter arrives may have a forty-foot tunnel, camouflaged at its entry, that leads into the den. The winter den is below the frost line, usually about two meters deep. The winter rooms are high in the tunnel so that they cannot flood.

When the woodchuck enters its den for hibernation, it seals off the tunnel leading into it. The outer tunnel may become the winter dwelling of skunks or rabbits. During hibernation, the woodchuck, which normally breathes about thirty-five times a minute, breathes about once in five minutes. Its body temperature drops from about 36.7°C (98°F) to 3.3°C (38°F), only slightly above freezing. Its heart, beating in warm weather at about eighty times a minute, now beats four times a minute.

Woodchucks have a layer of brown fat that builds up around their vital organs. When it is time for hibernation to end, the animal receives an instant jolt of energy from this brown fat. It begins to shiver, which gradually warms it up. Its supply of oxygen increases steadily, resulting in increased blood flow, which warms the woodchuck. It takes a few hours for it to move from its dormant state to its spring time state.

In mild climates, ground squirrels are active all year. The bodies of those that do hibernate change slowly from their summer to their winter phase. Their body temperature, around 32.2°C (90°F) in summer, drops a couple of degrees every day until it approaches 4.4° C (40°F). An internal mechanism keeps the squirrel's body temperature from dropping below that number. Brown bats enter hibernation when the air temperature stabilizes around 10°C (50°F), entering their winter quarters, or hibernacula, in swarms. In these caves, the temperature remains constant and is above freezing, which is crucial to the bats' survival, because bats subjected to lower temperatures develop fatal ice crystals in their blood.

Brown bats hang upside down in their caves in winter. Their bodies become stiff and appear to be dead. Their heartbeats drop from over five hundred beats a minute to between seven and ten beats a minute. They hibernate for three to four months, waking occasionally for water and any insects they can find to devour. In dormancy, brown bats' bodies assume the temperature of the atmosphere surrounding them.

The jumping mouse is another true hibernator. Late in October, it seals itself into its den, having gorged on all the food it can find in the weeks prior to hibernation. It curls up into a ball, placing its head between its hind legs. During hibernation, it breathes just once every fifteen minutes.

—*R. Baird Shuman*

FURTHER READING

Brimner, Larry Dane. *Animals That Hibernate.* Franklin Watts, 1991.

Busch, Phyllis S. *The Seven Sleepers: The Story of Hibernation.* Macmillan, 1985.

Carey, Cynthia, et al., eds. *Life in the Cold: Ecological, Physiological, and Molecular Mechanisms.* Westview Press, 1993.

Lyman, Charles P. *Hibernation and Torpor in Mammals and Birds.* Academic Press, 1982.

Lyman, Charles P., and Albert R. Dawe, eds. *Mammalian Hibernation: Proceedings.* Museum of Comparative Zoology, 1960.

Home Building

Fields of Study

Animal Behavior; Ecology; Environmental Studies

Abstract

The term "home" designates the area, place, or physical structure in which an animal lives, finds a safe haven, and raises its family. The nature of animal homes varies greatly. A large number of species, such as fish, whales and grazing land animals, have no fixed homes. Land animals often live in more complex homes, such as nests, dens and burrows. Some, such as chimpanzees and especially humans, are known to construct artificial shelters.

Key Concepts

artiodactyl: a hoofed mammal with an even number of toes
carnivore: an animal that eats only animal flesh
herbivore: an animal that eats only plants
mole hill: an earth mound a mole dug up in search of food.
omnivore: an animal that eats plant and animal matter
solitary: living alone

BE IT EVER SO HUMBLE...

The term "home" designates the area, place, or physical structure in which an animal lives, finds a safe haven, and raises its family. The nature of animal homes varies greatly. The very simplest example might be a spot on an ocean bottom where a marine sponge attaches to the sand and grows. Other sea animals, such as gastropods (snails) live in their shells and take their homes wherever they go. Crustaceans, such as lobsters, or land arthropods, such as scorpions, live in simple burrows in the oceans or in the earth. However, a large number of lake, ocean, and river dwelling species, such as fish and whales, have no fixed homes.

Land animals often live in more complex homes. For example, social insects such as bees, termites, and wasps inhabit nests or hives. Birds also build nests. They nest in trees, on the ground, or on rocky mountain terrain. In contrast, grazing animals such as deer and antelope live on the ground, wherever their search for food takes them on a given day. In the case of small mammals, individuals or groups often live in complex underground burrows such as mole holes. Larger carnivores and omnivores often inhabit dens that are underground burrows, as do members of the weasel family, or in caves and other natural formations, as do bears and big cats.

INSECT AND BIRD HOMES

Social insects such as bees, ants, and some wasps live in nests of differing sizes which are made of wax, paper, or dried mud. Social wasps, for example, live in spheroid paper nests that are typically about 15 centimeters in diameter, but may be considerably larger. These nests are seen in trees or under the porch roofs of human habitations. A wasp colony lasts only one year because wasps do not store food as do ants and bees. Only a few fertilized females, queens-to-be, survive the winter to begin new colonies in spring. Many solitary wasps live alone, except for breeding. Then, females build small brood nests of materials other than paper. For example, potter wasps use mud and saliva, and stone-working wasps mix small pebbles, mud, and saliva.

Termites, also social insects, are known for damaging wood homes. Most species are tropical, but some inhabit the Americas and Europe. They live in huge, long-lasting colonies that may hold millions of inhabitants. These colonies (called nests or termitaries) vary greatly. Tropical species build huge mounds with walls of soil particles and dried saliva. Inside the mounds are many chambers, passages, and good ventilation and drainage systems. Termites are often subterranean, burrowing upward into logs and wood structures from below.

Many birds, such as the commonly seen crows, robins, sparrows, doves, and other small to medium birds, nest in trees or warm places around human homes. For instance, doves may nest atop porch lights. Nests of such birds are most often made of intertwined pieces of grass, twigs, and trash discarded by humans. The nests are often abandoned yearly and in some cases are reused by other species. Also of interest are the nests of woodpeckers, which are located in holes found or manufactured in tree trunks.

Flamingos nest along shores of shallow, saltwater lagoons and lakes. The nest is a fifteen centimeter-tall mound of mud with a depression at its top. Flamingos are monogamous, and couples use the same nest over and over. In contrast, vultures usually live on bare ground under overhangs in the rock faces of mountains or in caves, building no nests, and laying eggs on the bare rock of these spartan home sites. Vultures are also monogamous and will live in a home site for up to forty years. An exception to the "bare rock" rule occurs with lammergeiers of Europe, Asia, and Africa. These bearded vultures build several nests per pair. They are conical in shape, located on rock ledges or in caves, and are used many times, in cycles, as home sites and to raise families.

MAMMAL HOMES
Mammals have a wide variety of home sites. In many cases, such sites are temporary and are simply the last place the mammal found itself each day. Creatures living in this way are usually herbivores, ranging from hippopotamuses to deer and other artiodactyls. This is because every day, these animals range over areas of several square miles or more seeking food. The other end of the home site range is seen with many omnivores and carnivores. These creatures have specific home sites or dens, which may be burrows in the ground, caves, logs, and natural crevices. Small mammals that live like this include gophers and moles. They dig burrows or tunnels with the sharp claws of their front feet. Gophers store food in chambers in the burrows. They are solitary and territorial, coming together only to breed. Females use their burrows to live in and raise young.

The homes of moles are more complex. Moles are voracious and solitary, continually burrowing in the ground for food, which includes insects, worms, slugs, snails, and spiders. They defend their homes when other moles—even of their own species—intrude. Moles only socialize when entering tunnels of females to mate.

Mole burrows or holes are close to the ground surface and may be recognized by large, central earth mounds. These "mole hills" are the earth that has been dug up in search of food. The burrows are very elaborate, holding warmly lined central nest chambers, connected galleries, bolt holes that allow escape from enemies, and many passageways.

Wolves and spotted hyenas are somewhat similar, related species. These carnivores tend to live in dens and claim large hunting territories. Wolf and hyena homes and living habits are different. Wolves live in dens or lairs that may be caves, hollow tree trunks, crevices under large fallen logs, or holes they dug in the ground. Few improvements are made in the natural dens wolves inhabit. They are shelters used for safety, protection from the elements, and for raising offspring.

Spotted (laughing) hyenas, in contrast, are much more communal. They live in clans of up to one hundred individuals and inhabit shared communities comprising many dens. In the dens they sleep, mate, and socialize. Dens may be caves on rocky ground or holes dug by individuals, as a clan grows. The individual cave or tunnel dens are most often inhabited by one individual or a female with cubs. This is because extended pairing is unusual in spotted hyenas.

Lions and tigers, the largest predatory land carnivores, often roam through large territories in search of game and inhabit dens of varying permanence. The dens are dense thickets, groups of rocks surrounded by thick underbrush, or caves whose entrances are screened by thorn bushes or dense underbrush. Very often, dens are used to birth offspring and protect the big cats from the elements.

PRIMATE HOMES
Monkeys, which live in Africa, Asia, and South and Central America, live in bands which most often inhabit trees, sheltering in the forks between branches. African baboons are large, more highly organized, ground-living primate species. They live in groups called troops, which are often found living in rocky terrain or on cliffs. In many cases, group members inhabit convenient caves.

Gorillas, the largest, strongest, rarest apes, look almost human. They inhabit West African forests from lowlands to altitudes of ten thousand feet. Gorillas live in bands of up to twenty individuals. Each band claims a territory, which may be viewed as the band's neighborhood. A band forages over several square miles each day and lacks permanent dwellings. Instead, its members build temporary shelters each night after a day of foraging for the honey, eggs, plants, berries, bark, and leaves that are their diet. When terrain and time permit, females and young sleep on temporary tree platforms made of branches and leaves. Mature males nest at the bases of these trees, to protect them.

HOMES, LIFESTYLES, AND FORMS

The homes of animals depend upon their lifestyles, habitats, and forms. Many herbivores and omnivores range widely to find sustenance. Hence, they often lay themselves down to sleep wherever the search for food takes them. Animals that live in hot, relatively dry climates often sleep out of doors. However, similar species inhabiting cool to cold climates very often build or find burrows or dens to live in. Some animals have forms and eating habits that cause them to live outdoors regardless of world location. This is typical of animals, such as zebras and reindeer, that graze daily over large areas and cannot restrict themselves to homes where they return each evening. In contrast, animals that hunt often prefer to have a safe haven where they can bring their catch home to devour in peace, while animals that live in cold climates use their homes to store food for the long winter.

—*Sanford S. Singer*

FURTHER READING

Goodman, Billie. *Animal Homes and Societies.* Little, Brown, 1991.
Robinson, W. Wright. *Animal Architects.* Blackbirch Marketing, 1999.
Shipman, Wanda. *Animal Architects: How Animals Weave, Tunnel, and Build Their Remarkable Homes.* Stackpole Books, 1994.
Whitfield, Philip, ed. *The Simon and Schuster Encyclopedia of Animals: A Who's Who of the World's Creatures.* Simon & Schuster, 1998.

HOMEOSIS

FIELDS OF STUDY

Genetics; Evolutionary Biology; Embryology

ABSTRACT

The body plans of advanced animals and plants can be viewed as a series of segments with unique identities. This is especially obvious in the annelids (segmented worms), but even in vertebrates the muscular regions and the backbone are segmented. Occasionally, one segment takes on the identity of another segment of the same organism. This is called homeosis, and it was first described in 1984.

KEY CONCEPTS

egg-polarity genes: genes whose expression in maternal cells results in products being stored in the egg in such a way as to establish polarity, such as the anterior-posterior axis
gap genes: genes expressed in the zygote that divide the anterior-posterior axis of fruit flies into several regions
homeobox: a sequence of about 180 nucleotide pairs that codes for a protein called the homeodomain, known to influence body-plan formation in numerous organisms
homeosis: a process that results in the formation of structures in the wrong place in an organism, such as a leg developing in place of a fly's antenna
homeotic selector genes: genes that determine the identity and developmental fate of segments established in fruit flies by a hierarchy of genes
imaginal disk: a small group of cells that differentiates adult fruit-fly structures after the last larval molt
pair-rule genes: segmentation genes in insects that divide the anterior-posterior axis into two-segment units
segmentation genes: genes that regulate segmentation in organisms, including gap genes, pair-rule genes, and segment-polarity genes

HOMEOTIC MUTANTS

The body plans of advanced animals and plants can be viewed as a series of segments with unique identities. This is especially obvious in the annelids (segmented worms), but even in vertebrates the muscular regions and the backbone are segmented. Occasionally, one segment takes on the identity of another segment of the same organism. This is called homeosis, and it was first described in 1984. Numerous examples of homeosis have been cited, including the antennapedia mutant of the fruit fly, which has a leg that develops in the antennal socket of the head. Much has been learned about developmental patterns in organisms, and about the evolution of these patterns, from the study of homeotic mutants.

THE GENETIC CONTROL OF BODY PLANS

The role of homeosis in elucidating genetic control of overall body plans is best illustrated in the development of the fruit fly. Early in its embryogenesis, the basic plan for the adult fly form is established, and this information is stored in imaginal disks (derived from *imago*, which is the adult form of an insect) through three larval stages and associated molts. Imaginal disks are small groups of cells that differentiate adult structures after the last larval molt. The early determination of these imaginal disks for specific developmental fates is controlled by a hierarchy of genes. This hierarchy of gene regulation has been carefully documented for the establishment of the anterior-posterior axis (a line running from the head to the abdomen) of larval and adult fruit flies. Three levels of genetic control—egg-polarity genes, segmentation genes, and homeotic selector genes—result in an adult fly with anterior head segments, three thoracic segments, and eight abdominal segments.

Egg-polarity genes are responsible for establishing the anterior-posterior axis. Mutations of these genes result in bizarre flies that lack head, thoracic, or abdominal structures. Maternal egg-polarity genes are transcribed, and the resulting ribonucleic acid (RNA) is translocated into the egg and localized at one end. This RNA is not translated into protein until after fertilization. Following translation, the proteins are dispersed unequally in the embryo, forming an anterior-posterior gradient that regulates the expression of the segmentation genes.

Segmentation genes are the second tier of the genes that establish the body plan of the fly. Within the segmentation genes, there are three levels of control—gap genes, pair-rule genes, and segment-polarity genes—resulting in progressively finer subdivisions of the anterior-posterior axis. While there is hierarchical control within the segmentation genes, genes in a given level also interact with one another. Gap genes of the embryo respond to the positional information of the gradient established by the maternal egg-polarity genes. Gap genes form boundaries that specify regional domains, and several gap genes with distinct regions of influence have been identified.

Pair-rule genes follow next in the sequence and appear to function at the level of two-segment units. Mutant pair-rule genes are responsible for flies that have half the normal number of segments. Ultimately, the larval fly body is divided into visible segments, but while the patterns are forming, genes appear to exert their influence on parasegments. A parasegment is half a segment that is "out of phase" with the visible adult segments. Parasegments include the posterior of one segment and the anterior of the adjacent segment. Developmental programming of parasegments ultimately gives rise to visibly distinct segments.

The final level of control of the segmentation genes focuses on individual segments and is controlled by the segment-polarity genes. In response to the pair-rule genes, the segment-polarity genes subdivide each segment into anterior and posterior compartments. Thus, the segmentation genes create a series of finely tuned boundaries.

HOMEOTIC SELECTOR GENES

It is the third tier of genes, the homeotic selector genes, that actually specifies segment identity. Segmentation gene mutations result in missing body parts, whereas mutations of the homeotic selector genes result in a normal number of segments, some of which have abnormal identities. Homeotic selector genes are found in two gene clusters, the antennapedia complex and the bithorax complex. Genes associated with the antennapedia complex appear to determine the fate of segments associated with the anterior body segments, while the bithorax complex is responsible for the more posterior segments, such as the abdominal segments. The pattern that arises is modulated both by interactions among the homeotic selector genes and by their interactions with the segment-polarity genes.

The genes found within the antennapedia and bithorax complexes have been identified based on mutant phenotypes that have arisen. These genes function within smaller regions of the anterior or posterior axis, much as pair-rule genes subdivide regions established by the gap genes. It is intriguing that the genes within the two complexes appear to be lined up in the same order in which they function spatially. That is, the position of an antennapedia complex gene on the chromosome relative to other antennapedia complex genes correlates with the actual position of the segments controlled by the gene. Recent analyses of homeotic mutations in beetles indicate that the homeotic genes are also physically organized in a left-to-right sequence corresponding to the location of the segments they control on the anterior-posterior axis. Unlike the fruit fly, however, the beetle has a single homeotic gene complex controlling the entire anterior-posterior axis.

The interactions between homeotic selector genes and other genes in the hierarchy may, in part, be controlled by the homeodomain protein that is coded for by the homeobox associated with many of these genes. The homeobox is a 180-nucleotide-pair sequence that is included in many homeotic selector genes and some segmentation genes, including the pair-rule gene. The homeodomain protein has a unique structure that may bind deoxyribonucleic acid (DNA) and affect its transcription. It is possible that this allows genes to regulate expression of themselves and other related genes. For example, a pair-rule gene known as *fushi tarazu* (Japanese for "not enough segments") is found in the antennapedia complex and has a homeobox, although fushi tarazu is not a homeotic selector gene. Its homeodomain can bind to the antennapedia gene and thus can regulate when this gene is turned on and off. Antennapedia, a homeotic selector gene, was first identified when a mutation of it resulted in the replacement of an antenna with a leg. Thus, the ability of the homeodomain to bind to DNA provides a way for the hierarchical control of homeotic selector genes by segmentation genes to occur.

The homeotic selector genes are ultimately linked to gene expression that leads to the development of specific structures associated with different segments. It is not known exactly how homeotic selector genes regulate segment differentiation. Although the homeotic selector genes are active early in development, they appear to be involved in programming cells for fates that are not expressed until much later.

STUDYING HOMEOSIS THROUGH MUTATIONS

Most of what is known about homeosis has been learned from studying mutations. Sometimes these mutations have arisen spontaneously; sometimes they have been induced by exposing organisms to mutagenic substances, such as chemicals or x-ray or ultraviolet radiation. The large number of mutations identified in fruit flies accounts for the wealth of information on homeosis in this organism, in sharp contrast to the limited information on humans, for whom ethical considerations prohibit mutagenesis. Mutations affecting segmentation and determination of segment identity represent defective developmental switches and provide insight into the normal developmental sequence.

Classical genetic approaches have been used with homeotic mutants to map genes to chromosomes and identify interactions between genes. For example, it can be determined whether two genes are on the same chromosome by making a series of specific matings between flies with mutations in these genes and wild-type ("normal") flies. If the genes are not on the same chromosome, offspring with one mutation will not necessarily have the other mutation. If the mutations are both on the same chromosome, they will be inherited together, except in rare situations where there is recombination between chromosomes. Geneticists use the frequency of recombination to assess how close together two genes are on a chromosome. This approach helped geneticists determine the order of genes within the antennapedia and bithorax complexes. Matings between different mutants have also established the hierarchy of genetic control among egg-polarity, segmentation, and homeotic selector genes. For example, a pair-rule mutant will have no effect on gap genes, but a gap gene mutant will affect pair-rule genes.

To visualize the results of these crosses of hierarchical mutants, researchers employed a second technique: *in situ* hybridization. To investigate the effect of gap genes on pair-rule genes, a wild-type fly embryo and one with a gap mutation affecting the middle section were exposed to radioactively labeled DNA that was a copy of the pair-rule gene fushi tarazu. The DNA hybridized (bound) to fushi tarazu RNA, and it thus labeled tissues where the fushi tarazu gene was turned on and making RNA copies of itself. Excess radioactive DNA was washed away, and a photographic emulsion that was then placed over the tissue was exposed by the bound radioactive DNA. This permitted researchers to see that the fushi tarazu gene was being expressed in the middle of the wild-type embryo, but not in the gap mutant embryo. When this experiment was repeated using radioactive gap-gene DNA in a pair-rule mutant, no effect on gap-gene expression was observed.

In situ hybridization has also provided information on how egg-polarity genes provide segmentation genes with positional information. The egg-polarity gene bicoid was identified by mutations resulting in a fly with abdominal structures but no head or thoracic structures. When radioactive DNA copies of the bicoid gene were hybridized to eggs, it was found that all the RNA was located at the anterior tip after being transferred from the mother. When this RNA was translated into protein, the protein was tagged with an antibody that identifies the bicoid protein. Tagging the protein with an antibody is similar to tagging RNA with a radioactive DNA segment; both techniques allow researchers to see how the RNA or protein is distributed in tissues. In this case, the bicoid protein formed an anterior-to-posterior gradient after the egg was fertilized, with more protein being found at the anterior end.

Homeotic genes have been isolated and used for *in situ* hybridization studies. In addition, the sequence of nucleotides in the DNA in these genes has been established. A variety of techniques is available for DNA sequencing. Generally the DNA is broken into smaller segments that can be more readily identified. Cutting the DNA yields overlapping segments, and the overall sequence can be established by piecing these overlapping fragments back together. The presence of the homeobox was established by comparing DNA sequences from different genes in flies and other organisms. It was evident, based on these types of data, that the homeobox sequence differs by only a small number of nucleotides even between very distantly related organisms.

The role of the homeobox was investigated by inserting DNA containing a homeobox from the fushi tarazu gene into bacteria in such a way that the bacteria then produced large amounts of the homeodomain protein. This protein was then tested for its ability to bind the DNA and was found to bind to specific fragments of DNA that were involved in homeosis, such as the antennapedia gene. Homeodomain protein from a mutant fushi tarazu gene was defective in its DNA binding ability. This provided evidence that the homeobox may regulate gene expression via the direct binding of the homeodomain to DNA. This is only one of numerous examples illustrating how researchers have utilized genetic mutants and molecular biology to investigate the role of homeotic genes in the development of segmented organisms.

IMPLICATIONS OF HOMEOTIC RESEARCH

Research on homeosis and homeoboxes has made significant contributions to the fields of developmental biology and evolution. Since all higher animals and plants exhibit some form of segmented development, the common link of the homeobox has intrigued scientists interested in how body plans are established. Questions concerning the evolution of segmentation patterns in animals have also arisen as more is understood about the genes affecting segmentation and how they are regulated.

Significant similarities both among the homeoboxes identified in fruit flies and among the homeoboxes of distantly related species have been found. In fruit flies, homeoboxes have been identified in homeotic selector genes, segmentation genes, and egg-polarity genes. How homeoboxes affect gene regulation and segmentation in other organisms is an exciting, and open, question. Most if not all multicellular organisms, and even some single-celled organisms, have homeoboxes, and the homeobox sequences found in most mammals are not dissimilar to those of fruit flies. Different mouse homeobox genes have been shown to be expressed in specific regions during embryogenesis. However, the expression of homeotic mutations in mammalian genes is more difficult to observe, as mammals often have more than one copy of homeotic genes, which can mask the effects of any mutations. Also, segmentation may have evolved separately in vertebrates and invertebrates, so the presence of a homeobox may or may not indicate a common developmental process for segmentation among animals.

The possibility exists that addition and modification of homeotic selector genes were responsible for the evolution of insects from segmented worms. The presence of two homeotic gene complexes in fruit flies, compared to only one in beetles, suggests that duplication of the gene complex, followed by subsequent specialization, may have allowed for greater fine-tuning of segmental identity. Evolutionary alterations of the initial homeotic gene complex may have been responsible for the addition of legs to a wormlike creature composed of similar segments, giving rise to a millipede-like creature. The reduction of all legs except for the walking legs in the thoracic region and, ultimately, the addition of wings to the thoracic region could also reflect changes in homeotic selector genes. Parts of this evolutionary journey can be reconstructed with homeotic fruit fly mutants that bear similarities to their ancestors. This is exemplified by the deletion of the antennapedia gene, which results in a wingless fly; winged insects presumably arose from non-winged insects. It is impossible, however, to create a millipede from a fly via mutations of the homeotic selector genes, so caution must be taken in speculating on the role of homeotic selector genes in insect evolution.

—Susan R. Singer

FURTHER READING

Alberts, Bruce, et al. *Molecular Biology of the Cell.* 5th ed. Garland, 2008.

Duboule, Denis, ed. *Guidebook to the Homeobox Genes.* Oxford University Press, 1994.

French, Vernon, et al., eds. *Mechanisms of Segmentation.* Research, 1988.

Garcia-Fernàndez, Jordi. "The Genesis and Evolution of Homeobox Gene Clusters." *Nature Reviews Genetics* 6.12 (2005): 881–92.

"Genes Determine Body Patterns." *Learn.Genetics.* U of Utah, n.d. Web. 19 Sept. 2013.

Gilbert, Scott F. *Developmental Biology.* 10th ed. Sinauer, 2014.

Jacobson, Brad. "Homeobox Genes and the Homeobox." *Embryo Project Encyclopedia.* Arizona State U, 10 May 2013. Web. 19 Sept. 2013.

Raff, Rudolf A., and Thomas C. Kaufman. *Embryos, Genes, and Evolution: The Developmental-Genetic Basis of Evolutionary Change.* Reprint. Indiana University Press, 1991.

Russo, V. E. A., et al. *Development: The Molecular Genetic Approach.* Springer, 1992.

Solomon, Eldra P., Linda R. Berg, and Diana W. Martin. *Biology.* 9th ed. Brooks, 2011.

Watson, James D., et al. *Molecular Biology of the Gene.* 7th ed. Pearson, 2014.

HOMO SAPIENS AND HUMAN DIVERSIFICATION

FIELDS OF STUDY

Evolutionary Biology; Paleontology; Paleobiology; Paleoanthropology

ABSTRACT

Homo sapiens sapiens—*modern humans*—*are classified taxonomically as members of the order Primates, which is part of the class Mammalia. Since humans and other members of Primates (monkeys and apes) are biologically related, scientists presume both groups to be the products of an evolutionary process similar to that which affected other divergent categories of animals. The evolution of Homo sapiens sapiens from previously existing species is also believed to account for diversification within the modern human population.*

KEY CONCEPTS

gene pool: the total collection of genes available to a species

generalized: not specifically adapted to any given environment; used to describe one group of Neanderthal humans

hominid: any living or fossil member of the taxonomic family Hominidae ("of man") possessing a human form

hominoid: referring to members of the family Hominidae and Pongidae (apes) and to the taxonomic superfamily of Hominoidae

morphology: the scientific study of body shape, form, and composition

natural selection: any environmental force that promotes reproduction of particular members of the population that carry certain genes at the expense of other members

Pleistocene epoch: the sixth of the geologic epochs of the Cenozoic era; it began about three million years ago and ended about ten thousand years ago

Würm glaciation: the fourth and last European glacial period, extending from about seventy-five thousand years ago to twenty-five thousand years ago

WISE, WISE HUMAN

All human beings on the earth today are highly adaptive animals of the genus and species *Homo sapiens sapiens* (Latin for "wise, wise human"). In terms of physical structure and physiological function, *Homo sapiens sapiens*—modern humans—are classified taxonomically as members of the order Primates, which is part of the class Mammalia. Since humans and other members of Primates (monkeys and apes) are biologically related, scientists presume both groups to be the products of an evolutionary process similar to that which affected other divergent categories of animals. The evolution of *Homo sapiens sapiens* from previously existing species is also believed to account for diversification within the modern human population such as racial differentiation (though race as it is popularly understood is entirely a social construct).

Modern humans and modern apes (the two most closely related of modern primate species) are believed to possess a common biological ancestry, or line, that diverged perhaps five or six million years ago. The scanty fossil record of this early period, in conjunction with modern genetic studies, means that the proposed evolutionary tree is continually debated and revised. For a time the fossil primate called *Ramapithecus* was thought to be the earliest known direct ancestor of modern humans, for example, but this view was discredited. By the 2010s, *Sahelanthropus tchadensis* was considered the earliest known potential hominin. The term "hominin" is generally understood to refer to modern and extinct humans as well as all immediate ancestral species, while "hominid"—which used to have that meaning, and is still used as such by some

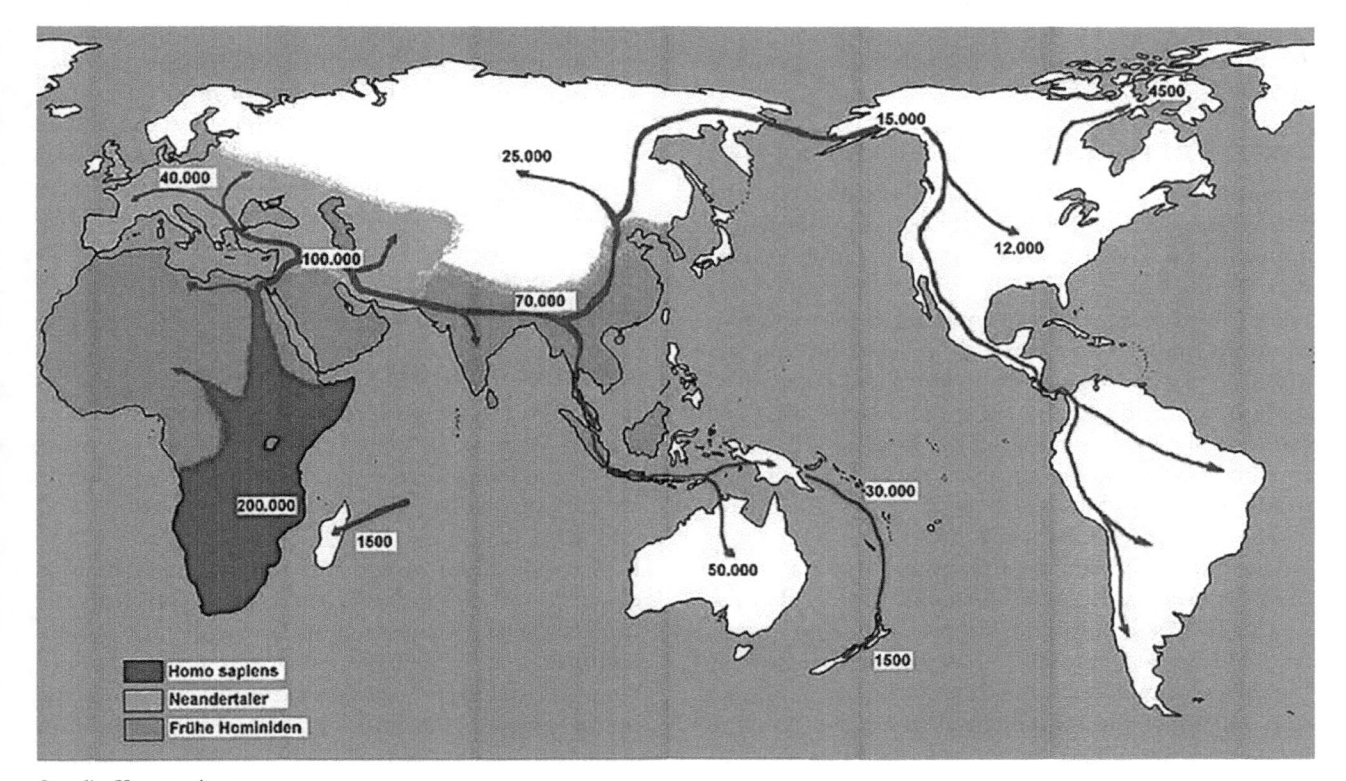

Spreading Homo sapiens.

sources—is now considered a broader category referring to hominins as well as other Great Apes (chimpanzees, orangutans, gorillas, and extinct relatives) and their ancestors.

While little is known about *Sahelanthropus tchadensis* and there is little agreement as to its classification, *Ardipithicus ramidus* is widely acknowledged as an early hominin. While still quite ape-like and likely tree-dwelling, fossils suggest *Ardipithecus* had a relatively upright posture. This points to the gradual evolution of bipedalism. Since the 1990s, the genera *Orrorin* and *Keyanthropus* have also been identified as early hominins, expanding the view of evolution in the Hominini tribe as a diverse, continually branching "bush" rather than the traditional model of a "tree" with modern humans as the culmination.

Several million years ago, during the late Pliocene epoch, early forms of the hominin *Australopithecus* appeared in Africa. They share certain characteristics with both humans and apes. Their brains are larger than those of apes but smaller than those of humans. There have been several species of *Australopithecus* identified, including *A. afarensis*, the famous example being the fossil known as Lucy. Another genus, *Paranthropus*, has also been identified as a similarly ape-like bipedal hominin.

THE EMERGENCE OF THE GENUS *HOMO*

Examples of the first undisputed members of the genus *Homo*—true human (though not *sapiens*)—appear in the fossil record about 2.1–1.5 million years ago. Samples of *Homo habilis* ("handy human") have been found in East Africa, marking the first known hominin use of tools. However, its classification has often been challenged. More widespread was *Homo erectus* ("upright human"), of which fossils have been found in China, Africa, Java, and Europe. This creature habitually walked upright, made shelters, and used sophisticated tools.

Homo erectus is also very important, since it is beleived to have been the first hominid to have used fire purposefully. It was suggested by John E. Pfeiffer, in a 1971 article entitled "When *Homo erectus* Tamed Fire, He Tamed Himself," that this first domestication of a natural force was a tremendous evolutionary step, changing the fundamental rhythms of life and human adaptability to environments. Most scholars accept the premise that *Homo erectus* was a hominid grade intermediate between the australopithecines and *Homo sapiens*.

Exactly when, where, and how advanced members of the species *Homo erectus* evolved into *Homo sapiens* are key questions in the study of human evolution, and they are questions that resist resolution. *Homo heidelbergensis* has been suggested as evolving from *H. erectus* in Africa, with some migrating out into Europe and the Middle East, where they may have diverged into Neanderthals (*Homo neanderthalensis*) and Denisovans. *H. heidelbergensis* populations remaining in Africa may then have gave rise to *Homo sapiens*, who eventually spread out of Africa in one or more waves. Genetic evidence suggests they also interbred with Neanderthals and other archaic humans, further obscuring any concept of directly progressive evolution.

However, even this general theory is unclearly documented and often contested. It might be thought that the closer one comes, in terms of time, to modern humans, the easier it would be to find the answers. In actuality, such is not the case. The ancestral line or lines leading to modern humans become hazy beginning approximately 500,000 years ago. Direct fossil evidence of the earliest members of the species *Homo sapiens* is scarce; moreover, finds of modern human fossils in the Middle East have intensified the debate about the immediate ancestry of *Homo sapiens sapiens*. Still, much evidence indicates that the middle to upper Pleistocene epoch (beginning about 350,000 years ago), known as the Paleolithic or old stone age in archaeological terms, witnessed the emergence of early *Homo sapiens*.

THE EARLIEST HOMO SAPIENS

In 1965, hominid fossil remains were found at a site named Vértesszöllös, near Budapest. They consisted of some teeth and an occipital bone (a bone at the back of the skull). The site also yielded stone tools and signs of the use of fire. Several features of the find recall *Homo erectus*, but the estimated cranial capacity of 1,400 cubic centimeters is well into the normal range for *Homo sapiens*. The age of the site was established at 350,000 years BCE. These remains have been attributed to a *sapiens-erectus* intermediate type on the grounds that the remains, and the site, show a mixture of elements reflective of the transitional hominid evolutionary process. Such an assessment placed the Vértesszöllös specimens at the root of the *Homo sapiens* evolutionary line, some 100,000 years earlier than other specimens.

A better-known example of early *Homo sapiens* comes from a gravel deposit at Swanscombe near London, England. In 1935, 1936, and 1955, three related skull pieces were unearthed that fit together perfectly to form the back of a cranial vault with an advanced (over *Homo erectus*) cranial capacity of about 1,300 cubic centimeters. This has been dated to around 275,000 to 250,000 years BCE. A more complete skull of approximately the same age (dated to the Mindel-Riss interglacial period about 250,000 years BCE) was found at Steinheim, in southern Germany, in 1933. Swanscombe's and Steinheim's advanced morphological characteristics, in combination with relatively primitive ones, such as low braincase heights, suggest that they are primitive members of the species *sapiens* and are representatives of a population intermediate between *Homo erectus* and *Homo sapiens*.

The finds at Swanscombe and Steinheim have been augmented by others from France and Italy, and especially from the Omo River region in southern Ethiopia. One Omo skull displays more mixed features (between *erectus* and *sapiens*) including flattened frontal and occipital areas, a thick but rounded vault, large mastoid processes (pointed bony processes, or projections, at the base of the skull behind the ears), and a high cranial capacity. Another skull is more fully *sapiens*, or modern in appearance. Some paleoanthropologists assert that the Omo group of fossils also helps bridge the gap between advanced *Homo erectus* and *Homo sapiens*.

NEANDERTHALS

The best-known examples of what have been considered early *Homo sapiens* come from a group of fossils known collectively as Neanderthals. Their name derives from the place where the first fossil type was discovered in 1865, the Neander Valley near Düsseldorf, Germany. Similar Neanderthal fossil types have been found at more than forty sites in Europe, North Africa, Asia, and the Middle East.

Neanderthal fossils tend to show an aggregate of distinctive characteristics that has led to their being regarded as a separate human species, *Homo neanderthalensis*. However, some scientists have considered them as a subspecies of humans, with the designation *Homo sapiens neanderthalensis*. The characteristic features of their morphology include large heads with prominent supraorbital tori (thick brow ridges), receding jaws, stout and often curved bones, and large joints.

Most important Neanderthal fossils disclose large brain capacities and, contrary to early belief, are found in sites revealing complex and sophisticated cultures. These two facts clearly separate Neanderthal humans from earlier species that exhibit some of the same morphological features. Neanderthals generally stood fully erect between 1.5 and 1.6 meters in height; they were not the stoop-shouldered brutes of early characterizations. They lived during the last glaciation (the Würm glacial stage) in Eurasia. The sites from which most examples of the Neanderthals have been recovered have commonly yielded tools of the Mousterian complex, a stone-tool industry named for the kind found at Le Moustier, France, and dating from about 90,000 to about 40,000 years BCE.

In fact, according to some researchers, two groups of Neanderthal humans seem to have existed. The first are referred to as classic Neanderthals from such sites as Germany, France, Italy, Iraq (the Shanidar 1 fossil), and the former Soviet Union. The second group, known as either generalized or progressive Neanderthalers, lived contemporaneously with, as well as later than, classic Neanderthal humans. They display a combination of modern *H. sapiens* features and typical Neanderthal characteristics (especially the prominent supraorbital torus, the forehead ridge). Included in this category for the sake of simplification are those specimens termed neanderthaloid. However, later research has suggested rethinking of these categories, including the classification of *Homo rhodesiensis* as a separate species.

CRO-MAGNONS

Neanderthals were a successful group for many thousands of years, flourishing from about 127,000 BCE to 37,000 BCE, with a wide distribution geographically. Neanderthal traces suddenly and mysteriously disappear from the fossil record, however, and they seem to have been superseded in Europe around 37,000 BCE by *Homo sapiens* with different morphology. These people became known as Cro-Magnons, so named for the Cro-Magnon cave near Les Eyzies in southwestern

France, where the first skeletons were found in 1868 and where more than one hundred skeletons have since been discovered. However, that term is largely avoided in the scientific community in favor of more specific taxonomic names, including the simple "European early modern humans (EEMH)." Indeed, Cro-Magnon skeletal anatomy is virtually the same as that of modern European and North African populations. The skull is relatively elongated, with a large cranial capacity; the brow ridges are only slightly projecting. The average height of Cro-Magnons has been estimated between 1.75 and 1.8 meters.

Cro-Magnon humans left more evidence of a highly developed culture than any of their predecessors. They made weapons and tools of bone and stone, stitched hides for clothing, and lived in freestanding shelters as well as caves. Some Cro-Magnon people produced beautiful cave paintings (they have been found in southwestern France and northern Spain) and bone carvings, and they modeled in clay. Though Cro-Magnon samples are the best-known examples of early *Homo sapiens sapiens*, mounting fossil evidence from sites outside Europe as well as genetic research from the 1980s onward suggests a much older date of origin for the emergence of modern man.

At Qafzeh, a cave near Nazareth, Israel, anatomically modern fossils classified as *Homo sapiens sapiens* were discovered in 1988 and reliably dated to 92,000 years BCE. In addition, newer fossil finds of progressive Neanderthals from Kebara Cave in Israel, taken together with earlier Neanderthal finds from the caves of et-Tabun and es-Skhul, also in Israel, make it certain that Neanderthals and modern humans coexisted for many thousands of years.

Anthropologists have puzzled over the disappearance of the Neanderthals and, more important, over where they fit in the human family tree. It appears unlikely that classic Neanderthal humans were in the direct ancestral line of modern *Homo sapiens sapiens*. Reasons for their sudden disappearance are believed to include a combination of factors: extinction because of disease, lack of adaptation to the warmer climate following glaciation, and annihilation by other human groups.

Many scholars have considered the classic Neanderthals to be a cold-adapted, specialized side branch from the modern human line that became extinct as the climate became warmer. The generalized or progressive Neanderthals are considered by some to have avoided this specialization, perhaps continuing to exist through adaptation and ultimately being absorbed by flourishing modern human populations during the late Pleistocene epoch. It is of interest to note that modern genetic studies have determined that the DNA genome of modern-day *Homo sapiens sapiens* includes genetic structures that are consistent with, and that therefore derive from, the DNA genome structures of Neanderthals. The conclusion to be drawn from this is that at some point Neanderthal and Cro-Magnon couples produced children with each other, thus introducing Neanderthal genetics into the Cro-Magnon genetic inheritance.

THE EMERGENCE OF MODERN HUMANS

Although an exact time, place, and mode of the origin of the modern human species cannot be determined, early genetic studies pointed to a date before 100,000 years BCE. Examination of mitochondrial DNA (mtDNA) from a sampling of present-day humans representing five broad geographic regions allowed researchers to propose a genetic family tree and calculate roughly (assuming a fairly constant mutation rate) a temporal origin for the modern human population. Further studies seemed to indicate that the modern human ancestral line emerged between 280,000 years and 140,000 years BCE. Later research tended to push this date back, with fossil discoveries in Morocco announced in 2017 suggesting Homo sapiens appeared as much as 300,000 years ago. In general, however, genetic evidence, in concert with fossil finds, make it plausible that a common ancestral population for *Homo sapiens sapiens* appeared in sub-Saharan Africa or the Levant (in the eastern Mediterranean region). Regional differentiation occurred, followed by radiation outward to other areas. The range of genetic and anatomical variability exhibited by fossil remains of modern humans is no greater than that known for the extant populations of modern times.

During the late Pleistocene epoch (approximately 40,000 to 11,000 years BCE) five different groups seem to have developed on the Eurasian and African landmasses. The last glaciation, approximately 30,000 to 10,000 years BCE, absorbed enough water to lower the oceans ninety meters below present levels. Emerging land bridges allowed people to move from Asia into North America,

Australia, and elsewhere. In time the major groups became subdivided into smaller ones that diverged into the human phenotypes seen today.

This view of racial diversification emphasizes the effectiveness both of geographic barriers in reducing free gene flow among varied groups of *Homo sapiens* and of environmental pressure in selecting different adaptive responses from the gene pool. These are also key factors in the entire evolutionary process by which modern humans developed over epochs into their present taxonomic position in the animal kingdom.

THE STUDY OF PALEOANTHROPLOGY AND PHYSICAL ANTHROPOLOGY

The study of human evolution is primarily the concern of the physical anthropologist and the paleoanthropologist. Evolution may be defined as change in the genetic composition of a population through time. Because evolution is thought to operate according to several principles and factors, modern human evolutionary theory is studied in the light of ideas and practices taken from different disciplines, including archaeology, biochemistry, biology, cultural anthropology, ecology, genetics, paleontology, and physics.

Early investigations into human evolution sought to establish the sequence of the human ancestral line through chronological and morphological analyses of hominid fossil remains (bones and teeth), thus placing them in their proper phylogenetic context (their natural evolutionary ordering). This has been augmented and in some cases superseded by sophisticated techniques in fossil dating and new avenues of exploration into the evolutionary process, such as genetic research.

Determination of the accurate age of a fossil is most important, since it sets the fossil in a correct stratigraphic context that allows comparison with remains from the same geologic layer or level a great distance away. Accurate dating also has helped determine the order of succession for fossils that could not be established on morphological grounds alone.

The most valuable absolute dating methods are the radioactive carbon technique, which can effectively date specimens between 60,000 years BCE and the present; the potassium-argon technique, which most easily dates material older than 350,000 years BCE; and the fission-track method, which helps bridge the gaps between other methods. These methods are based on the constant or absolute rates at which radioactive isotopes of carbon, potassium, and argon decay. When absolute dating is impossible, investigators have ascribed a relative age to fossil remains by noting the contents of the layer of rock or the deposit in which the remains were found. A layer containing remains of extinct animals is likely to be older than one containing remains of present forms.

In conjunction with dating, anatomical studies of fossil remains and comparisons with the morphological features of known hominid types, as well as comparisons to primate skeletal structures, have been primary approaches to the study of the evolutionary path of *Homo sapiens*. The species *Homo sapiens* (of which the modern human races compose a number of geographical varieties) may be defined in terms of the anatomical characteristics shared by its members. In general, these include a mean cranial capacity of about 1,400 cubic centimeters, an approximately vertical forehead, a rounded occipital (back) part of the skull, jaws and teeth of reduced size, and limb bones adapted to fully erect posture and bipedalism. Scientists assume that any skeletal remains which conform to this pattern and cannot be classified in other groups of higher primates must belong to *Homo sapiens*.

It is striking that the anatomical differences observed between *Homo erectus* and *Homo sapiens* have been confined to the skull and teeth. The limb bones thus far discovered for both are similar (though *H. erectus* appears more robust). Cranial capacity and morphology continue to be the dominant determining boundary separating sapient and presapient human species.

THE CONTRIBUTION OF OTHER SCIENCES AND SOCIAL SCIENCES

Human adaptability studies, using techniques from physiology, demographics, and population genetics, investigate all the biological characteristics of a population that are caused by such environmental stresses as altitude, temperature, and nutrition. It is believed that these normal stresses acted as genetic selectors in prehistoric times and continue to do so. Such variants as skin color and body hair are observable products of these stresses. The investigation of climatic changes during prehistoric epochs as revealed in the geologic record is important for understanding those pressures affecting the evolutionary history of humans.

Genetic studies have become indispensable to the study of human evolution. Four forces have been identified as fundamental in the evolutionary process: mutation, natural selection, gene flow, and genetic drift. Since mtDNA is inherited through the female, it is possible to calculate how much time has elapsed since the mutations that gave rise to present variations originated in prehistoric populations.

Also important to the study of the evolution of *Homo sapiens* is the examination and classification of cultural remains preserved at hominid fossil sites. Not only can the relative date of a fossil be supported, but sometimes it is also possible to reconstruct the environmental situation that may have influenced the evolutionary process operating in a population. Cultural response is an integral part of hominid adaptation, and it in turn influences natural selection. Technology changes the physical and economic environment, and economic changes alter the demographic situation. Humans continue to promote or influence their own evolution by willingly or unwillingly altering the environment to which they must adapt.

The modern methods useful for investigating the evolutionary history of *Homo sapiens* are multidisciplinary. While each of them reveals an aspect of the emergence of modern humans and complements the other methods of study, emphasis is placed on careful fieldwork, accurate dating, and comparative morphological analyses of hominid fossil remains. Increasing in importance, however, is the accumulating wealth of genetic data on human population relationships.

HUMAN EVOLUTION IN THE CONTEXT OF ANIMAL EVOLUTION

Increasing attention is being given to the biological and behavioral changes that led to the emergence of *Homo sapiens sapiens*—the last major event in human evolution. Mounting evidence continues to push backward in time the point at which modern *Homo sapiens* made its appearance in the evolutionary scheme. A clearer understanding of the evolutionary history of modern *Homo sapiens* has not only helped to define the place of the modern human species more accurately in relation to the rest of the animal kingdom, but also helped to illuminate the pressures, adaptations, and changes that have made humans what they are.

Through the pressures and process of evolution (including adaptation and natural selection), *Homo sapiens* has become one of the most successful and adaptive animals that ever lived, because it came to possess an elaborate culture (culture is based on learned behavior). The key is *Homo sapiens*' advanced mental capacity. Humans exhibit an exceptional ability to assign arbitrary descriptions to objects, concepts, and feelings and then communicate them unambiguously to others. In the late Middle Pleistocene, the hominid branch that gave rise to early *Homo sapiens* witnessed an increase in brain size, complex social organizations, continual use of fire, and perhaps even language. As to what initiated these changes, many have suggested tool use and, in turn, a hunting economy.

In a classic article published in 1960, entitled "Tools and Human Evolution," Sherwood Washburn argued that the anatomical structure of modern humans is the result of the change, in terms of natural selection, that came with the tool-using way of life. He stated that tools, hunting, fire, and an increasing brain evolved together. Washburn also argued that effective tool use led to effective bipedalism—another significant characteristic of *Homo sapiens*: humans are different from all other animals because in their use of increasingly complex tools. While these assertions have been challenged, the general connection between human evolution, intelligence, and success is strong.

The other behavioral pattern that is seen to have been of utmost importance to sapient evolution is big-game hunting. Early *Homo sapiens* was undoubtedly a big-game hunter, as were all successors until approximately 8,000 years ago. It has been argued that human intellect, interests, emotions, and basic social life are the evolutionary products of the success of hunting adaptations. Success in hunting adaptation dominated the course of human evolution for hundreds of thousands of years. The agricultural revolution and the industrial and scientific revolutions are only now releasing human beings from conditions characteristic of 99% of their evolutionary history. It has been argued that, although most humans no longer live as hunters, they are still physically hunter-gatherers. Some investigators in the field of stress biology—the study of how the human body reacts to stressful situations—suggest there may be some link between an emotional reaction, such as explosive aggression, and human evolutionary history. Again, these theories have faced important challenges, but indicate the consensus that evolutionary history plays an ongoing important role. It is clear that tools and more efficient hunting helped produce great change in hominid evolution and made humans what they are. Humans continue to be users of increasingly complex tools, such as computers, and perhaps this continued development of technology may determine the future evolutionary path of *Homo sapiens*.

—*Andrew C. Skinner*

Further Reading

Brauer, Gunter, and Fred H. Smith, eds. *Continuity or Replacement: Controversies in "Homo Sapiens" Evolution*. Balkema, 1992.

Cann, R. L., M. Stoneking, and C. A. Wilson. "Mitochondrial DNA and Human Evolution." *Nature* 325 (January 1, 1987): 31–36.

Eldredge, Niles, and Ian Tattersall. *The Myths of Human Evolution*. New York: Columbia University Press, 1982.

Gee, Henry. *The Accidental Species: Misunderstandings of Human Evolution*. University of Chicago Press, 2015.

Gibbons, Ann. "World's Oldest *Homo sapiens* Fossils Found in Morocco." *Science*, 7 June 2017, http://www.sciencemag.org/news/2017/06/world-s-oldest-homo-sapiens-fossils-found-morocco. Accessed 31 Jan. 2018.

Henke, Winfried, and Ian Tattersall. *Handbook of Peleoanthropology*. 2nd ed. Springer, 2015.

"Homo sapiens." *What Does It Mean to Be Human?* Smithsonian National Museum of Natural History, 29 June 2017, humanorigins.si.edu/evidence/human-fossils/species/homo-sapiens. Accessed 31 Jan. 2018.

Johanson, Donald, Lenora Johanson, and Blake Edgar. *Ancestors: In Search of Human Origins*. Villard Books, 1994.

Leakey, Richard E. *The Making of Mankind*. E. P. Dutton, 1981.

Lewin, Roger. *The Origin of Modern Humans*. Scientific American Library, 1998.

Mellars, Paul, ed. *The Emergence of Modern Humans: An Archaeological Perspective*. Cornell University Press, 1990.

Phenice, Terrell W. *Hominid Fossils*. Wm. C. Brown, 1973.

Tattersall, Ian. *The Fossil Trail: How We Know What We Think We Know About Human Evolution*. Oxford University Press, 1995.

Hormones and Behavior

Fields of Study

Endocrinology; Neuroendocrinology; Evolutionary Biology

Abstract

Hormones produced by developing organisms have marked effects on adult behaviors. In mammals, the fetal testis becomes active and produces testosterone, then becomes inactive before birth and remains so until puberty. The changes resulting from this brief surge of testosterone are remarkable. They include the anatomical features that distinguish male and female genitalia, changes in neural anatomy, and the sensitivity of nervous (and other) tissues to adult hormones.

Key Concepts

ablation: the technique of removing a gland to determine its function and observe what effects its removal will precipitate

androgens: masculinizing hormones, such as testosterone, responsible for male secondary (anatomical) sex characteristics and masculine behavior

behavior: an animal's movements, choices, and interactions with other animals and its environment

biological clock: a timekeeping mechanism that is "endogenous" (a part of the animal) and capable of running independently of "exogenous" timers such as day-night cycles or seasons, although the clock is normally set by them

endocrine system: a collection of glands that secrete their products into the bloodstream

estrogen: a feminizing hormone responsible for female secondary (anatomical) sex characteristics and sex-related behaviors

prolactin: a hormone responsible for secretions of milk from the mammary glands of mammals and from the crops of birds

receptor molecule: a molecule on the cell membranes of target tissues that binds to the hormone molecule and initiates the action of the hormone

THE COURTSHIP OF RING DOVES

Behavioral differences are usually attributed to two causes: differences in experience (learning) and differences in heredity (genes). Hormonal interaction involving an organism, its experience, and the highly specific behaviors that result are well illustrated by Daniel Lehrman's 1964 study of the ring dove, a small relative of the domestic pigeon.

A male ring dove begins courtship by bowing and cooing to the female when they are placed together in a cage. Toward the end of the first day of courtship, the birds choose a location and start building a nest. The doves court, nest-build, and mate, and the female lays an egg about five o'clock in the afternoon on the seventh to eleventh day. A second egg is laid about nine o'clock the next morning. The male sets for about six hours at midday, and the female occupies the nest the rest of the time. After fourteen days of incubation, the eggs hatch, and the parents feed the squabs crop milk, secreted from the lining of the adult dove's crop (enlarged gullet). At about two weeks of age the squabs (fledgling birds) begin pecking at grain, and the adults feed them less and less. With feeding chores diminished, the male begins bowing and courting, the pair start nest-building, and the cycle repeats itself.

THE HORMONAL TRIGGERS OF BEHAVIOR

The simplicity of this description belies the hormonal ferment going on beneath the placid exterior. The courtship ritual causes the production and release of estrogen and progesterone, hormones responsible for setting behavior and for the development of the oviduct. The oviduct develops from eight hundred milligrams, when the doves are placed together, to four thousand milligrams, when the first egg is laid. Birds presented with nests already containing eggs when they are first paired will build nests on top of the eggs. Even if the eggs are returned to the top of the nest by the investigator, the doves will not set until they have engaged in nest-building for five to seven days. On the other hand, if the doves are first injected with progesterone, 90 percent will set within three hours after pairing. Courtship and nest-building play a vital role in creating the hormonal conditions necessary for setting behavior. Development of the crop is, in turn, initiated when setting begins. Setting behavior causes the release of prolactin, which stimulates crop development and feeding behaviors. Unpaired doves injected with prolactin will develop crop milk and will feed squabs when exposed to them even if they have not mated. While they will feed squabs, however, they will not sit on eggs, because they have not courted, engaged in nest-building, or developed the hormonal balance necessary to support setting behaviors.

This ring dove study indicates that the behavior of one individual can alter the behavior and the hormonal balance of another individual; that an individual's behavior can alter its own hormonal balance and, hence, its own behavior; and that inanimate aspects of the environment (nest and eggs, for

INSTINCTUAL RESPONSE TO A VISUAL STIMULUS

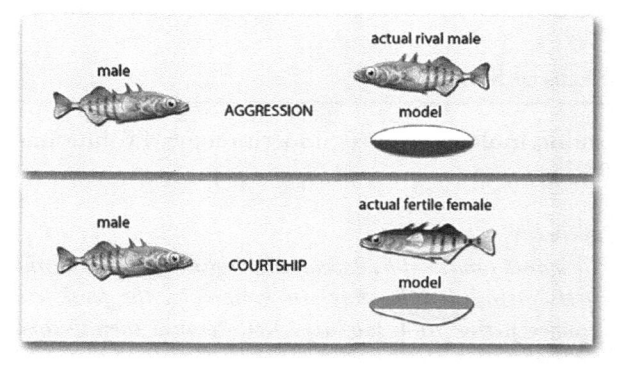

Visual Stimuli: The illustration shows the male fish's response to the rival male and fertile female fish. Instincts are immediate responses which are not learned.

example) can alter an individual's hormonal balance and, hence, its behavior. The interactions among these three factors are complex. For example, the male's behavior (courting) changes the female's behavior (nest-building), which changes her hormonal balance (estrogen and progesterone), which causes her to alter the external environment (lay eggs in the nest), which affects the pair's behavior (setting), which changes their hormonal balance (prolactin), which stimulates them to feed the squabs.

FETAL HORMONES AFFECTING ADULT BEHAVIOR

Hormones produced by developing organisms have marked effects on adult behaviors. In mammals, the fetal testis becomes active and produces testosterone, then becomes inactive before birth and remains so until puberty. The changes resulting from this brief surge of testosterone are remarkable. They include the anatomical features that distinguish male and female genitalia, changes in neural anatomy, and the sensitivity of nervous (and other) tissues to adult hormones.

Rat and mouse fetuses have several littermates, which develop side by side in a common uterus, like peas in a pod. Male mice that develop between two male embryos are more aggressive as adults than males developing between females. Similarly, females that develop between two male embryos are more aggressive as adults than females that develop between other females. In another study, females from litters that were predominantly male showed more masculine behavior (such as the mounting of other females) than females from predominantly female litters. The explanation is that testosterone produced by the male embryos' testes is absorbed into the bloodstream of sibling embryos, altering their nervous systems and hence their behaviors. In cattle, testosterone produced by a bull calf twin affects the development of his heifer twin to the extent that she is usually sterile.

Scientists have shown that pregnant rhesus monkeys treated with testosterone produced offspring that showed rougher play and more threat behavior than usual. Male rhesus monkeys experience a decrease in blood testosterone levels within six hours after losing a fight to another male and are more submissive. These studies indicate that hormones play an important role in determining male-female behavioral differences.

Many biological phenomena are repeated or change intensity over and over again throughout the life of an individual. Examples include sleep-wake cycles, menstrual cycles, and the migration cycles of birds; these are repeated approximately once a day, once a month, and once a year, respectively. The regularity of these cycles led biologists to propose a "biological clock." The golden-mantled ground squirrel avoids freezing temperatures by going into hibernation once a year. Even if these squirrels are kept in constant conditions of light and temperature to deprive them of seasonal cues, they will enter hibernation once a year. These and other data lead researchers to believe that the clock resides within the animal. Although it can be reset by environmental cues, it can also run independently of them.

HORMONES, SEASONS, AND MATING BEHAVIOR

A white-crowned sparrow, nesting in central Alaska, experiences dramatic seasonal changes and migrates to the southern United States or Mexico (more than three thousand kilometers) to avoid freezing. Central Alaska's short summer demands that the sparrow fly north as early in the spring as is safe and that it be prepared for mating and rearing chicks when it arrives. During the winter, the gonads atrophy to 1 percent or less of their breeding season weight. The bird's ability to sense the approach of spring depends on its sensing the increase in daylight. During the short winter days, the sparrow is content to stay in Arizona or Mexico, but as day length increases to fourteen or fifteen hours, the bird's hypothalamus releases hormones that stimulate the pituitary to release prolactin and gonadotropic hormones. The gonads respond by increasing in size and producing additional hormones, which stimulate the bird to begin its long migration.

When the male white-crown arrives at his breeding grounds in central Alaska, he chooses a nesting territory, attacks any male territorial intruders, and attempts to attract a mate with his constant singing. Each female chooses a mate and helps him defend the nesting site. In the next few days she feeds to gain nutrients for egg production, and her estrogen levels rise rapidly, stimulating her to solicit mating. Once the eggs are laid, the gonads of both birds begin to atrophy, estrogen and testosterone levels decline, and prolactin levels increase and stimulate feeding of the young. As the gonads atrophy, the birds become less aggressive and the male stops singing.

As the young become independent, both parents enter a "sexual refractory period," during which the gonads will not respond to artificially increased day length as they would in the spring. The birds feed voraciously, increasing body fat, which serves as fuel for the long trip south. In the next year, by early spring, the birds will have passed through the refractory period and be primed to respond to the increasing day length with a fresh hormonal flurry which will set them off on the long journey north.

Recognizing the existence of a refractory period is important. It underscores the ideas that while birds do respond to environmental conditions (day length), there is a given set of events through which the physiological machinery passes and that specific time parameters are dictated by the biological clock. White-crowned sparrows can be expected to show hormonal changes and migratory restlessness during springtime even if they had been caged and maintained in constant conditions. It is to the bird's advantage, however, to experience and recognize the seasonal changes in day length, because biological clocks tend to run a bit fast or slow. The actual measuring of day lengths allows the bird to reset that clock and arrive in Alaska at the most advantageous time for rearing a family of sparrows.

Studies of a closely related bird, the white-throated sparrow, indicate that the changes of behavior and physiology are primarily the result of two hormones: corticosterone from the adrenal cortex and prolactin from the anterior pituitary. Both hormones have daily peaks of secretion, but the timing of these daily peaks (relative to each other) changes with the seasons. If injections of these hormones are given with timing differences characteristic of specific seasons, the physiological and behavioral changes seen in the birds are characteristic of the seasons that the injections mimic.

EXPERIMENTAL ENDOCRINOLOGY

The earliest report of experimental endocrinology, in the mid-nineteenth century, demonstrated that replacements of testicular tissue would maintain comb growth and sexual behavior in castrated roosters. Techniques for determining endocrine function used today include ablation and replacement.

Ablation (removal) of endocrine tissue results in deficiency symptoms. The effects of ablation are not always unambiguous. If the testes and accessory tissues are not completely removed when a horse is castrated, for example, tissue capable of producing testosterone remains, and the consequence is an infertile gelding that behaves like a stallion.

Hormones produced by different glands can have similar physiological effects. Both the adrenal glands and the testes produce androgens (masculinizing hormones). Sexually experienced male cats do not lose their sex drive if castrated, and researchers do not have a satisfactory answer as to why this occurs. Perhaps the adrenal hormones are sufficient to maintain established feline male sexual behavior but not sufficient to initiate it in inexperienced cats. The ablation of the adrenal glands, however, has severe consequences in terms of electrolyte and blood glucose imbalances that are life threatening. Replacement of ablated endocrine tissue can reinstate normal function. If a male cat is castrated as a kitten, it will not develop normal male sexual behaviors. If, however, a normal testis is later transplanted to the abdominal cavity (or elsewhere), normal behavior will develop.

In the early 1960s, Janet Harker became convinced that the "biological clock" controlling the daily activity cycle of the cockroach was contained in the subesophageal ganglion, a patch of nervous tissue the size of a pin head resting just below the esophagus. When she ablated this ganglion, the cockroach became arrhythmic. Harker removed the legs from a normal roach and glued the roach on top of the arrhythmic roach, surgically uniting their body cavities so that the same body fluids circulated through both roaches. The arrhythmic roach ran about the cage with an activity rhythm dictated by the rhythm of hormones released into the body fluids by the legless roach on its back.

Most hormone and behavior studies involve nonhuman species and most involve sexual behaviors. Most behaviors are oriented toward perpetuating one's species. Only those individuals with behaviors conducive to rearing offspring will provide the genetic basis for behaviors represented in the next generation. It may seem that wild, unrestrained mating would be selected for, but that makes no real sense. Mate selection, shelter-seeking, feeding, maintenance of social position, and a host of other behaviors are critical to the success of one's progeny. Individuals that produce more offspring than they can feed or protect usually rear fewer than those who produce fewer to begin with. Many predator species (owls and wolves, for example) tend not to mate in years when prey is scarce. This ultimately maximizes reproduction by reserving energies that can best be spent later. This restraint is mediated by adjusting hormonal levels. Hormones have been called the ultimate arbiters of sexual behavior.

THE ENDOCRINE SYSTEM, THE NERVOUS SYSTEM, AND BEHAVIOR

The nervous system is usually thought of as the mediator of behavior, but the endocrine system is also a major player. Arguing that one system is more important would be like deciding whether height, width, or depth is more important in describing a box. It is useful, however, to discuss their differences. The nervous system has a shorter response time. Nerve impulses travel at speeds of up to 120 meters per second. Hormonal effects are much slower but are less transitory. If frightened by a false alarm, an animal may jump and run as a direct consequence of nervous system activity, but even after it recognizes that there is no real threat, it will be "keyed-up." This is a consequence of hormonal activity: Fright triggered the release of epinephrine (adrenaline) and norepinephrine from the adrenal glands. These hormones cause increased cardiac output; increased blood supply to the brain, heart, and muscles; decreased blood flow to the digestive tract; dilation of airways to breath more efficiently; and a significant increase in metabolic rate.

This is called the "fight or flight reaction," and it will affect behavior for several minutes and possibly for hours. These hormones enhance perceptions and elevate the responsiveness of the nervous system. In the final analysis, the understanding of hormones, what determines their ebb and flow, how they are affected by the environment, how hormones interact with one another, and how their levels are controlled by genetic programs of the individual and of the species is essential for the understanding of behavior.

—*Dale L. Clayton*

FURTHER READING

Adler, Norman T., ed. *Neuroendocrinology of Reproduction*. Plenum, 1981.

Alcock, John. *Animal Behavior: An Evolutionary Approach*. 6th ed. Sinauer Associates, 1998.

Becker, Jill B., S. Marc Breedlove, and David Crews. *Behavioral Endocrinology*. MIT Press, 1992.

Crews, David, ed. *Psychobiology of Reproductive Behavior: An Evolutionary Perspective*. Prentice-Hall, 1987.

Drickamer, L. C., and S. H. Vessey. *Animal Behavior: Concepts, Processes, and Methods*. 4th ed. McGraw-Hill, 1996.

Nelson, Randy Joe. *An Introduction to Behavioral Endocrinology*. 2nd ed. The Johns Hopkins University Press, 2000.

Schulkin, Jay. *Neuroendocrine Regulation of Behavior*. Cambridge University Press, 1999.

Svare, Bruce B., ed. *Hormones and Aggressive Behavior*. Plenum, 1983.

HYBRID ZONES

FIELDS OF STUDY

Evolutionary Biology; Evolutionary Genetics; Ecology

ABSTRACT

Gene flow among populations tends to increase the similarity of characters among all the demes (local populations) of a species.

Natural selection has the opposite effect: It tends to make every deme uniquely specialized for its specific habitat. Clines are one possible result of these two opposing forces; a cline is a phenomenon in which a genetic variation occurs that is caused by a difference in geographical habitat. Each species is continuously adjusting its gene pool to ensure that the species survives in the face of an environment that is continuously changing.

Key Concepts

cline: a gradual, continuous variation from one population of a species to the next that is related to differences in geography

deme: a local unit of the population of any one species

gene: the unit of molecular information, a portion of a deoxyribonucleic acid (DNA) molecule that codes for some product, such as a protein, that governs inherited traits

gene flow: the movement of genes from one part of a population to another part, or from one population to another, via gametes

gene pool: the sum total of all the genes of all the individuals in a population

hybrid: the offspring of a mating between genetically differing individuals

introgression: the assimilation of the genes of one species into the gene pool of another by successful hybridization

population: the members of a species that live in the same geographical area

species: a group of similar organisms whose members can reproduce with one another to produce fertile offspring

GENE FLOW VERSUS NATURAL SELECTION

Gene flow among populations tends to increase the similarity of characters among all the demes (local populations) of a species. Natural selection has the opposite effect: It tends to make every deme uniquely specialized for its specific habitat. Clines are one possible result of these two opposing forces; a cline is a phenomenon in which a genetic variation occurs that is caused by a difference in geographical habitat. Each species is continuously adjusting its gene pool to ensure that the species survives in the face of an environment that is continuously changing.

Comparing the characteristics of the demes of a single species usually will reveal that they are not identical. The greater the distance between the demes, the greater the differences between them will be. The grass frogs in Wisconsin differ from the grass frogs in Texas more than they differ from those in Michigan. On the average, the song sparrows of Alaska are heavier and have darker coloration than those in California. These phenomena, in which a single character shows a gradient of change across a geographical area, are called clines.

NORTH-SOUTH CLINES

Many birds and mammals exhibit north-south clines in average body size and weight, being larger and heavier in the colder climate farther north and smaller and lighter in warmer climates to the south. In the same way, many mammalian species show north-south clines in the sizes of body extremities such as tails and ears, these parts being smaller in northern demes and larger in southern demes. Increase in average body size with increasing cold is such a common observation that it has been codified as Bergmann's rule. The tendency toward shorter and smaller extremities in colder climates and longer and larger ones in warmer climates is known as Allen's rule. The trend toward lighter colors in southern climates and darker shades in northern climates has been designated Gloger's rule. The zebra, for example, shows a cline in the amount of striping on the legs. The northernmost races are fully leg-striped, and the striping diminishes toward the southern latitudes of Africa; this appears to be an example of Gloger's rule.

Another example of a cline, which does not fit any of the biogeographical rules mentioned, is the number of eggs laid per reproductive effort (the clutch size) by the European robin: This number is larger in northern Europe than it is for the same species in northern Africa. Other birds, such as the crossbill and raven, which have wide distribution in the Holarctic realm, show a clutch-size cline that reveals a larger clutch size in lower latitudes. The manifestation of such clines in clutch size is a consequence of the interplay of two different reproductive strategies that may give a species a competitive advantage in a given environment. The stability of the environment is what elicits the appropriate strategy.

In unstable environments, such as those in the temperate zone, where there may occur sudden variations in weather and extremes between seasons, a species needs to reproduce rapidly and build its numbers quickly to take advantage of the favorable warm seasons to ensure survival of the species during the harsh, unfavorable conditions of winter. This strategy is known as r strategy (r stands for the rate of increase). In the tropics, the climate is more equable throughout the year. The environment, however, can only support a limited number of individuals throughout the year. This number is called the carrying capacity. When carrying capacity is reached, competition for resources increases, and the reproductive effort is reduced to maintain the population at the carrying capacity. This is called K strategy, with K standing for carrying capacity.

In birds, clutch size tends to be inversely proportional to the climatic stability of the habitat: In temperate climates, more energy is directed to increase the reproductive rate. In the tropics, the carrying capacity is more important, resulting in a reduced reproductive rate. In the apparent contradiction of the crossbills and ravens, it may be the harshness of the habitat at higher latitudes that limits the resources available for successfully fledging a larger number of young.

GRASS FROG CLINES

The cline exhibited by the common grass frog is one of the best known of all the examples of this phenomenon. It has the greatest range, occupies the widest array of habitats, and possesses the greatest amount of morphological variability of any frog species. This variability and adaptation are not haphazard. The species includes a number of temperature-adapted demes, varying from north to south. These adaptations involve the departmental processes from egg to larva. The northernmost demes have larger eggs that develop faster at lower temperatures than those of the southernmost demes. These physiological differences are so marked that matings between individuals from the extreme ends of the cline result in abnormal larvae or offspring that are inviable (cannot survive) even at a temperature that is average for the cline region.

Leopard frogs from Vermont can interbreed readily with ones from New Jersey. Those in New Jersey can hybridize readily with those in the Carolinas, and those in turn with those in Georgia. Yet hybrids of Vermont demes and Florida demes are usually abnormal and inviable. Thus, it appears that the Vermont gene pool has been selected for a rate of development that corresponds to a lower environmental temperature. The gene pool of the Florida race has a rate of development that is slower at a higher average temperature. The mixture of the genetic makeup of the northern and southern races is so discordant that it fails to regulate characteristic rates of development at any sublethal temperature, so the resulting embryo dies before it becomes a tadpole.

There are two primary reasons why characters within a species may show clinal variation. First, if gene flow occurs between nearby demes of a population, the gene pools of demes that are close to one another will share more alleles than the gene pools of populations that are far apart. Second, environmental factors, such as annual climate, vary along gradients that can be defined longitudinally, latitudinally, or altitudinally. Because these environmental components act as selective pressures, the phenotypic characters that are best adapted to such pressures will also vary in a gradient.

HYBRIDIZATION

Hybridization is the process whereby individuals of different species produce offspring. A hybrid zone is an area occupied by interbreeding species. Partial species can and do develop on the way to becoming new species as products of hybridization. Natural hybridization and gene flow can take place between biological species no matter how sterile most of the hybrid offspring may be. As long as the mechanisms that prevent free exchange of genes between populations can be penetrated, there is the potential for a new species to develop. Because the parental species has a tendency to be replaced by the hybrid types if natural selection favors them, hybridization can be a threat to the integrity of the parental species as a distinct entity.

Hybridization between different species leads to various and unpredictable results. Any time that hybridization occurs, the isolation mechanisms of populations are overcome, forming bridging populations. Such connecting demes of hybrid origin fall into one of two general categories: hybrid swarms or introgressive demes. The formation of these types of demes reverses the process of speciation and changes the formerly distinct species into a complex mixture of highly variable individuals that are the products of the segregation and independent assortment of traits. This is the primary advantage of sexual reproduction: to produce variation in the population that is acted upon by natural selection over time. It cannot be overemphasized that hybrid swarms and introgressive demes are highly variable.

The environmental conditions that contour animal communities have endured for a very long time. In long-lived communities, every available niche has been filled by well-adapted species. When populations with new adaptive characteristics occur, there is no niche for them to occupy, so they usually die out. In contrast, when such communities are disturbed, the parity among their component species is upset, which gives new variants an opportunity to become established.

Hybrid swarms can be observed in nature by the careful investigator. The hybrid swarm forms in a disturbed habitat, where hybrid individuals backcross with the parental types to form a third population, which results from the migration of the genes of one population into the other. Such a population is designated an introgressive population. The progeny of such populations resembles the parent species, but the variations are in the direction of one parental species or the other. If introgression is extensive enough, it may eradicate the morphological and ecological distinctions of the parental types. The parental types become rarer and rarer, until they are no longer the representatives of the species.

There appear to be three reasons that first-generation hybrids occurring naturally are more likely to form offspring by backcrossing to one of the parental species than by mating with each other. Primarily, the hybrids are always rarer than the parents. Second, the parental individuals are so much more fertile than the hybrids that many more parental gametes are available than hybrid ones. Finally, backcross progeny, since they contain primarily parentally derived genes, are more likely to be well adapted to the habitat in which they originated than are the purely hybrid individuals.

INTROGRESSION

Thus, the most likely result of hybridization is backcrossing to one of the parental species. Genotypes containing the most parental genes usually have the selective advantage, and the fact that they contain a few chromosomal segments from another species gives them unique characteristics that may also be advantageous. This sequence of events—hybridization, backcrossing, and stabilization of backcross types—is known as introgression. Hybrid swarms are interesting phenomena, but they are unlikely to be of evolutionary significance except through introgression.

There are many examples of introgression among plants, but examples of introgression in animals are not common. Those that have been demonstrated are usually associated with the domestication of livestock. In the Himalayan region of Asia, there exists a relative of cattle, the yak, which is also domesticated. Many of the herds of cattle found along the western edge of the Himalayas, in central Asia, contain characteristics that clearly are derived from the gene pool of the yak. Many of these characteristics are manifested as adaptations to the harsh climatic conditions in this region.

In western Canada, there has been a modest introgression of the genes of the American bison into the gene pool of strains of range cattle. The bisonlike characters incorporated into beef cattle created a new breed called the beefalo, which exhibits such characteristics as greater body musculature, lower fat content of the flesh, and great efficiency in the utilization of range forage. A beefalo steer is ready for market in only eight months, while the same live weight is not obtained in the standard beef breed until eighteen months.

These examples serve to illustrate the concept that, as an evolutionary force, introgression is rather insignificant in natural biomes. It is almost always in the wake of human activity or the activities of their domesticated animals that the process of introgression can and does result in new combinations of gene pools from different species.

—*Edward N. Nelson*

FURTHER READING

Anderson, Edgar. *Introgressive Hybridization*. Hafner Press, 1968.

Dobzhansky, Theodosius G. *Genetics and the Origin of Species*. 1951. 3rd rev. ed. Reprint. Columbia University Press, 1982.

Kimbel, William H., and Lawrence B. Martin, eds. *Species, Species Concepts, and Primate Evolution*. Plenum Press, 1993.

Moore, J. A. "Geographic Variation of Adaptive Characters in *Rana pipiens Schreber*." *Evolution* 3 (1949): 1–24.

Hydrostatic Skeletons

Fields of Study

Hydrodynamics; Evolutionary Biology; Cellular Physiology

Abstract

Hydrostatic skeletons have evolved in a wide range of organisms including plants, protists, and animals. The basis of all hydrostatic skeletons is the material properties of water. Water cannot be compressed under biological conditions and thus can act as a support and locomotory transient skeleton. Essentially, all hydrostatic skeletons act in a similar fashion, in which water is contained in a compartment and is subjected to pressure. In this way, the compartment can become stiff and act as a skeletal support. The method used to create the pressure differs among groups.

Key Concepts

circular muscle: muscle fibers that run in a circular pattern around the body perpendicular to the long axis of the body

coelenteron: the fluid-filled gastrovascular cavity of Cnidarians

coelom: the body cavity of higher invertebrate and vertebrates, where mesodermal tissues enclose a fluid-filled space

longitudinal muscle: muscle fibers that run along the longitudinal or anterior-posterior axis of the body

pseudocoel: a fluid-filled body cavity that is bounded by mesodermal muscle on the outside and endodermal epithelium on the internal boundary

THE POWER OF WATER UNDER PRESSURE

Hydrostatic skeletons are the most primitive form of skeletal support that has evolved, because no mineralization process is necessary for their formation. Hydrostatic skeletons have evolved in a wide range of organisms including plants, protists, and animals. The basis of all hydrostatic skeletons is the material properties of water. Water cannot be compressed under biological conditions and thus can act as a support and locomotory transient skeleton. Essentially, all hydrostatic skeletons act in a similar fashion, in which water is contained in a compartment and is subjected to pressure. In this way, the compartment can become stiff and act as a skeletal support. The method used to create the pressure differs among groups.

Among the Protozoa, hydrostatic skeletons are found in the members of the phylum *Sarcomas tigophora*. Within this group are the amoeboid types that form pseudopodia, false feet that are transient structures formed by hydrostatic pressure within the cell. It is hypothesized that contractile proteins within the cell direct fluid toward the cell periphery in channels, and in so doing cause the cell membrane to bulge outward as a pseudopod. Thick pseudopodia, termed lobopodia, are found in the shell-less amoebas. In this case, there appears to be a chemical difference in the fluid used in pseudopodia formation. Plasmasol has less viscosity and is pushed into the pseudopodia; when it is distributed laterally it turns to plasmagel. In thin pseudopodia termed actinopodia, cytoplasm is forced along microtubules termed axonemes. These types of pseudopodia are normally found among foraminiferan and radiolarian amoebas.

WORM HYDROSTATIC SKELETONS

In the animal phyla, hydrostatic skeletons are developed in the *Cnidaria* and function in relationship to the gastrovascular cavity. In the polyp forms of colonial forms and especially within the *Anthozoa*, such as sea anemones, the myoepithelial cells surrounding the coelenteron make up a circular muscle band and thus are able to put pressure on the water in the coelenteron to extend the body or maintain the form of the polyp. This coelenteron extends into the tentacles of these forms and thus, via contraction of the myoepithelium, can elongate the tentacles to capture food.

Various types of flatworms comprise the phylum *Platyhelminthes*. Despite the lack of a coelom, these worms are still considered to have a hydrostatic skeleton. Since the longitudinal and circular muscles lie external to fluid-filled parenchyma tissue, the pressure exerted by this muscle extends the body by elongating this tissue. In this way, the worms can undulate in swimming or have peristaltic motion while moving along a substrate.

The nematode worms as well as other members of the *Pseudocoelomate* group have developed a fluid-filled body cavity. This cavity in most members of this group has a space with an outer boundary of mesodermal muscle and an inner boundary composed of endodermal cells. In nematodes, the muscle layer is arranged in a longitudinal pattern beneath a cuticle that is composed of layers, including fibrous collagen to maintain the shape of the worm under muscular pressure. Although less flexible than in true worms, contraction of the muscle layer can act against the pseudocoel fluid, thereby causing the body to undulate. Most nematodes need to act against a surface to have effective locomotion.

A more effective method of locomotion using a hydrostatic skeleton has been developed by annelid worms. These worms are segmented and have developed a true coelom. This means that the body cavity or coelom is bounded on all sides by mesodermally derived tissue: circular and longitudinally arranged muscles on the external boundary and membranes wrapping the gut tube. In addition, the coelom is divided in each segment into right and left halves and each segment has a membrane that separates the coelom in one segment from that of another segment. Thus, the circular musculature can constrict one side of the body while the other side is relaxed and stretched. As a result, undulation is more effective in this group than in the pseudocoelomates. When such bending is coupled with setae or segmentally arranged parapodial extensions on the body, the undulations could be used for crawling and swimming, although the latter locomotory ability is poor in some groups.

Using the hydrostatic skeleton in burrowing necessitates another evolutionary strategy involving the coelom. Here the intersegmental partitions or septa are lost or are perforated. This accomplishes the movement of coelomic fluid between segments during muscle contraction. Thus, circular muscles in posterior segments can drive fluid anteriorly, swelling and elongating the anterior portion of the animal. The posterior segments left behind can catch up with the anterior segments when the longitudinally arranged fibers contract. In this way burrowing is effected. Similar contractile wave patterns are used by terrestrial oligochaetes such as earthworms to create peristaltic-type contractions that drive the animal forward. The contraction of circular and longitudinal muscles alternate to create thick and thin areas of the body. This corresponds with elongation and subsequent contraction of the body segments. In earthworms, the segments retain their inter segmental septa. The *Hirudinida* or leeches have done away with their intersegmental septa and thus the coelomic space is continuous. Constriction of the circular muscles extends the body forward and the subsequent contraction of the longitudinal muscles will bring the rest of the body to meet it. The movement is accomplished by first attaching the posterior sucker, then elongating, attaching the anterior sucker and then pulling the rest of the body forward in the direction of the anterior sucker.

THE HYDROSTATIC SKELETONS OF ARTHROPODS, BIVALVES, AND ECHINODERMS

The development of an exoskeleton requires a change from a coelom-driven locomotion to one that uses muscles. The reduction of the coelom is a characteristic of the diverse arthropod taxon, but hydrostatic skeletons are still used in certain body areas. Flying insects, when attaining their adult state, emerge from their cocoons with folded wings. The insect must pump hemolymph into veins within the wing in order to expand them before they harden. These veins remain in the wings, and their walls and hydrostatic pressure may act to maintain the shape of the wing during flight. The ability of spiders to run fast even though they do have eight legs may lie in the hydraulic systems in their legs. Spiders have replaced extensor muscles with hydraulic spaces that, under pressure, automatically extend the legs. Thus, muscles are normally only used for flexion of the leg segments, and the legs can be moved faster than if muscles were used in both extension and flexion of the leg segments.

Although mollusks have a reduced coelom, hydrostatic skeletons are developed in certain members of this phylum. Bivalves normally burrow into the substrate and send up siphons through which water is brought in and out of the clam. In some species, water pressure is used to open and extend these siphons. In addition, the foot of the clam contains a blood sinus that, under pressure, fills with blood and expands in two ways. The foot can be extended into the substrate with subsequent swelling of the distal end. This action anchors the foot so that the clam can pull itself into the substrate. Among other mollusks, it is thought that the extension of tentacles to capture prey by squid and cuttlefishes is based upon hydraulic action of muscles on these structures.

Most echinoderms have a well-developed water vascular system derived from the coelom. This system is composed of sieve plate opening to the water. This plate is then connected via a stony tube to a ring canal. In asteroids or sea stars, this ring canal extends into the arms via radial canals. From the radial canal located in each, there extend bilaterally arranged lateral canals that enter a tube foot structure. The tube foot has two functional parts, an upper, bulblike ampulla and a lower tube foot. Contraction of the ampulla drives water into the podium, extending it and causing its tip to form a suctionlike disc that attaches to the substrate or prey. Relaxation of the ampulla withdraws water back into the ampulla, retracting the podial portion of the tube foot. In this way many echinoderms move along the ocean bottom and manipulate prey.

VERTEBRATE HYDROSTATIC SKELETONS

Like mollusks and arthropods, vertebrates have largely abandoned the coelom. Their endoskeletons have taken the place of hydrostatic skeletons. However, hydrostatic skeletons still occur, particularly in the reproductive system. The penis or hemipenes of mammals and reptiles contain spongy tissue that can engorge with blood. Venous return of the blood is largely prevented, extending the length and stiffness of the intromittent organ so that it may be inserted into the female's reproductive tract.

—*Samuel F. Tarsitano*

FURTHER READING

Barrington, E. J. W. *Invertebrate Structure and Function.* John Wiley and Sons, 1979.

Brusca, R. C., and G. J. Brusca. *Invertebrates.* Sinauer Associates, 1990.

Kristan, W. B., Jr., R. Skalak, R. J. A. Wilson, B. A. Skierczynski, J. A. Murray, F. J. Eisenhart, and T. W. Cacciatore. "Biomechanics of Hydroskeletons: Lessons Learned from Studies of Crawling in the Medicinal Leech." In *Biomechanics and Neural Control of Posture and Movement,* edited by Jack M. Winters and Patrick Crago. Springer-Verlag, 2000.

Ruppert, E. E., and R. D. Barnes. *Invertebrate Zoology.* 6th ed. Saunders College Publishing, 1994.

Infanticide

Fields of Study

Animal Behavior; Evolutionary Biology; Genetics; Sociology

Abstract

Infanticide is the intentional killing of dependent immatures by a member of the same species, or a conspecific. Victims may be as young as fertilized eggs or nearing independence from their parents. The infanticidal attackers may be strangers or relatives. The motive is always selfish competition, but sometimes the young are direct competitors and sometimes indirect competitors with the killers. There are three scientific explanations for infanticide, each of which applies to particular circumstances.

Key Concepts

conspecific: a member of the same species
genera: plural of genus, a grouping of animals above the species level
gestation: the period when the young are nourished within the mother's body; pregnancy
lactation: the period when mammal mothers produce milk to nourish their infants
ovicide: killing of fertilized eggs
raptors: predatory birds such as hawks and eagles
siblicide: infanticide committed by the siblings of the individual killed

THIS IS THE END OF THE INNOCENTS

Infanticide is the intentional killing of dependent immatures by a member of the same species, or a conspecific. Victims may be as young as fertilized eggs or nearing independence from their parents. The infanticidal attackers may be strangers or relatives. The motive is always selfish competition, but sometimes the young are direct competitors and sometimes indirect competitors with the killers. Infanticide occurs in every class of animal, but mammals and birds seem to have evolved the most pervasive and pernicious form, sexually selected infanticide. Wherever it is seen, infanticide reveals the darker side of evolutionary adaptation. There are three scientific explanations for infanticide, each of which applies to particular circumstances.

INFANTICIDE BY GENETIC RELATIVES

The first circumstance occurs when related individuals kill dependent young. For example, extreme sibling rivalry occurs in spotted hyenas (*Crocuta crocuta*) and many raptors. The unpredictability of resources or extreme competition for parental care may lead siblings to fight to the death in the den or nest. Even parents may act infanticidally. When parents have larger numbers of offspring than they can feed, the parents themselves may neglect or inflict damage on the smallest or least vigorous young, so as to reduce the number of offspring and increase each survivor's chances of success. Sibling rivalry and parental neglect or abuse of young can lead to infanticide in extreme cases when resources, such as food or parental time and energy, are in short supply. The brutal logic of natural selection shows that selfishness and lethal competition can divide even close kin. However, in most animal species, unrelated individuals are more dangerous to the young.

INFANTICIDE BY DIRECT COMPETITORS

The second circumstance in which infanticide arises is referred to as local resource competition. When immatures use resources that unrelated animals need, aggression may be severe and directed to killing the infants. All ages of infants are vulnerable, even independent juveniles. Lions (*Panthera leo*) inhabit the African savannah along with the cheetah (*Acinonyx jubata*), a direct competitor for food resources, and will unhesitatingly kill any and all juvenile cheetahs that it encounters, as well as any adult cheetah that is can. In the burying beetle (*Nicrophorus orbicollis*), male and female pairs compete for access to rotting meat in which to lay eggs. If defenders of such a resource are displaced by an intruding pair, the intruders proceed to kill and eat the eggs (ovicide) or larvae of the previous pair. Similarly, many birds, such as black-and-white casqued hornbills (*Bycanistes subcylindricus*), compete for rare tree hole nesting sites. If adults encounter another nest with eggs in such a tree hole, they will roll these eggs out or crush them and lay their own eggs. Finally, cannibalism of unrelated young has been seen in many amphibians, fish, reptiles and even in chimpanzees. Stepchildren suffer much higher rates of severe neglect and homicide than a parent's biological children. All of these cases represent extreme competition where resources are scarce and vulnerable young are eliminated or eaten by unrelated killers. The behavior patterns seen in this type of infanticide often resemble predation, are directed at young of any age, and can be performed by adults of either sex.

SEXUALLY SELECTED INFANTICIDE

The third circumstance in which infanticide occurs is motivated by sexual competition among adults. In a wide range of mammals, including over twenty genera of primates, adult males will kill unrelated infants if they can then mate with the mother. The mother of that infant does not usually reject the infanticidal male even though he has inflicted a tremendous cost on her. Typically, it takes female mammals many months—even years—to nurse young to independence. During this period of infant dependency, the mother is generally physiologically incapable of reproducing. An infanticidal male benefits if he can cut short this period so that he can fertilize the female's next egg. Therefore, a successful infanticidal male eliminates another male's offspring and advances his own reproductive career.

Sexually selected infanticide is not restricted to mammals, although the best-documented cases come from lions (*Panthera leo*), langur monkeys (*Semnopithecus entellus*), and rodents (order Sciurognathi). Some birds may also behave infanticidally. For example, the wattled jacana (*Jacana jacana*) shows a fascinating reversal. In this species, the adult females may commit ovicide and infanticide against unrelated young. Adult male jacanas incubate eggs on the nest and protect the young from predators, so females strive to monopolize the parental care donated by males. To do this, a female must eliminate the young of another female and lay her own eggs in the care of the male. Typically, the behavioral patterns seen in sexually selected infanticide differ from those in local resource competition. The young are virtually never eaten and permanent separation of mother and young is the primary goal—whereupon attacks usually cease.

EVOLUTIONARY CONSEQUENCES OF INFANTICIDE

Infanticide is widespread and may reach high frequencies in certain species, but it is not universal. Siblicide and parental neglect are found in a very restricted subset of birds and very few mammals. Local resource competition is more common, found in most classes of animals. Finally, sexually selected infanticide is most common in mammals, but even here there are many orders of mammals that never display infanticide. Species that have a very long period of lactation relative to their gestation period are most vulnerable to sexually selected infanticide. The reason for this is that infanticidal attacks are risky because of maternal defense and counterattacks by allies. Therefore, infanticidal behavior will only evolve when the males gain substantially by shortening lactational infertility of the mother.

Clearly, infants and their parents have a very strong motivation to avoid infanticide. Evolution has favored various mechanisms to reduce the risk. In general, infants avoid strangers, even in humans. In some primates, it seems that infants are born with unusual coat colorations that may impede males' efforts to determine which infants are theirs and which were fathered by other males. Parents are normally very protective of their young, both because of predators and because of the risk of attack by conspecifics. This parental protection is expressed through frequent proximity, carrying, and physical defense. One of the most fascinating consequences of a high risk of infanticide is the tendency for the father and mother to establish a long-lasting relationship whose primary benefit is protection of the young from conspecifics. This is seen in burying beetles and many primates. Alternatively, groups of mothers may cooperate in protection against infanticidal males, as in lions and langurs. Therefore, infanticide is one of the few evolutionary pressures that favors complex social relationships.

—*Adrian Treves*

FURTHER READING

Butynski, T. M. "Harem-Male Replacement and Infanticide in the Blue Monkey (*Cercopithecus mitis stuhlmanni*) in the Kibale Forest, Uganda." *American Journal of Primatology* 3 (1982): 1–22.

Hausfater, Glenn, and Sarah Blaffer Hrdy, eds. *Infanticide: Comparative and Evolutionary Perspectives.* Aldine, 1984.

Parmigiani, Stefano, and Frederik S. vom Saal. *Infanticide and Parental Care.* Harwood Academic Press, 1994.

Van Schaik, Carel P., and Charles H. Janson, eds. *Infanticide by Males and Its Implications.* Cambridge University Press, 2000.

INGESTION IN ANIMALS

FIELDS OF STUDY

Biomechanics; Developmental Biology; Biothermodynamics

ABSTRACT

Ingestion is the process of taking food into the body to satisfy nutritional and energy needs. Although basic nutritional requirements are remarkably similar for all animals, mechanisms of ingestion are exceedingly diverse. Because it is convenient to classify feeding strategies by the type of food consumed, organisms are often described as herbivores, carnivores, omnivores, and saprovores.

KEY CONCEPTS

carnivore: any organism that eats animals or animal tissues
detritus: small bits of dead matter derived from decay of plants and animals
herbivore: any animal that eats plants or plant material
invertebrate: any animal that lacks a backbone
omnivore: any animal that eats both plants and animals or their tissues
parasite: any organism that lives on or in other living organisms and obtains its food from them
plankton: microscopic plants and animals that float in water
predator: any organism that kills another living organism to eat it
protozoan: a single-celled animal-like organism
saprovore: any organism that consumes dead or decaying plant or animal matter
taxonomy: a classification scheme for organisms based primarily on structural similarities; taxonomic groups consist of genetically related animals

THE HOW AND WHY OF INGESTION

Ingestion is the process of taking food into the body to satisfy nutritional and energy needs. Although basic nutritional requirements are remarkably similar for all animals, mechanisms of ingestion are exceedingly diverse. This diversity stems from the varied nature of available food sources and the resulting behavioral adaptations and specific body forms required to procure adequate nutrition. Because it is convenient to classify feeding strategies by the type of food consumed, organisms are often described as herbivores, carnivores, omnivores, and saprovores. Although very descriptive, these terms alone are not sufficient to describe fully the feeding adaptations used by animals, especially when considering invertebrates, which make up more than 97 percent of all animal species. Accordingly, this section will expand on these ideas by further categorizing ingestion by the type, size, and consistency of food, while also describing behavioral adaptations that lead to ingestion.

SMALL PARTICLE INGESTION

Numerous animals live exclusively on a diet of very small particles that include bacteria, algae, plankton, and detritus. Although most small-particle consumers are relatively small animals, they range from

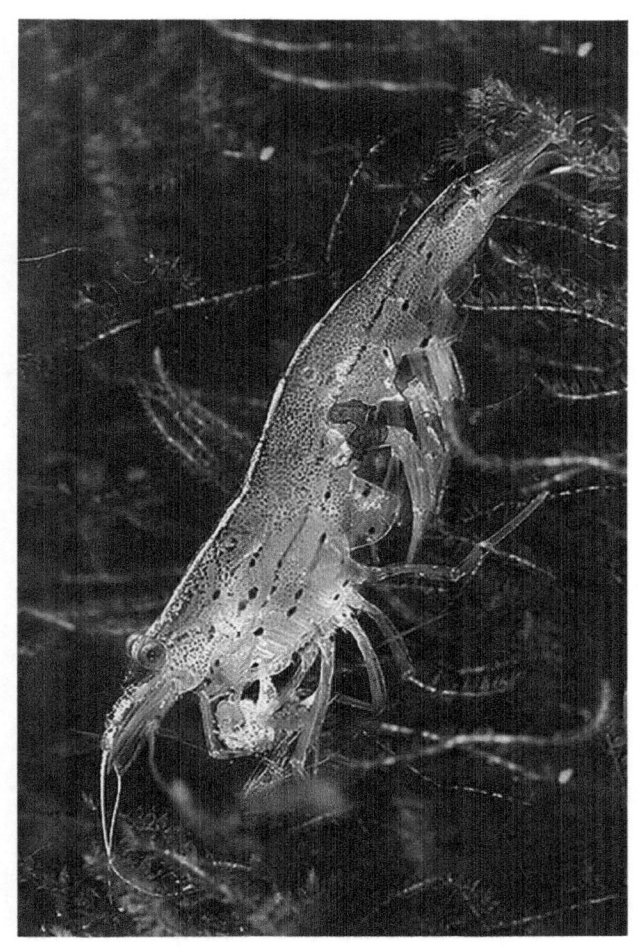

Amano or Yamatonuma shrimp (Caridina multidentata) *feeding.*

the tiniest animal-like organisms (protozoans) to the largest (whales). Many mechanisms have evolved to permit ingestion of small food particles. One common method is by endocytosis. During this process, the outer membrane of a cell surrounds a food particle and engulfs it within the cytoplasm, forming a food vacuole (a cellular organelle used for digestion). This mode of ingestion is best illustrated by protozoans. The amoeba, for example, uses pseudopodia (extensions of its cell membrane) to surround and then ingest prey. Paramecia use cilia (tiny hairlike processes on cell membranes that beat in a coordinated manner) to guide food particles into an oral region prior to ingestion by endocytosis.

Multicellular animals may also ingest small food particles by endocytosis. Sponges use flagella (motile, whiplike structures resembling long cilia) to aid in the gathering of small particles of food. The body cavity of sponges is lined with flagellated cells called choanocytes, or collar cells. The beating action of their flagella creates currents that move water through the body cavity, where food particles are removed and incorporated by the choanocytes.

Filter feeding is another strategy that is commonly used to obtain small bits of food. Most often cilia, mucus sheets, flagella, tentacles, and nets are used as filtering devices. Rotifers (small aquatic invertebrates), for example, use a special double-banded ciliary system to transport water and filter out suspended food particles, which are conducted to the mouth by a third set of cilia. In general, sessile (immobile) organisms (such as sponges, rotifers, and some oysters) are called filter feeders, because they must wait for food to come to them, then extract it from the surrounding medium.

Numerous free-moving animals also use filtering devices to obtain food. Sea cucumbers (animals related to starfish) live on the bottom of the seabed. By extending and periodically retracting sticky tentacles, they capture and ingest small food items. Clams, mussels, and some snails produce a sticky mucus that covers ciliated cells. The mucus traps fine suspended food particles, which are transported to the mouth by ciliary action. Mussels, for example, continuously pass water between their mucus-covered gill filaments for respiration, simultaneously trapping food particles. Similarly, herring and mackerel (fast-swimming fish) possess special structures called gill rakers, which act as sieves to catch plankton from water that continually passes over their gills for the process of respiration. Basking sharks and whale sharks also feed on plankton that is strained from water that enters their mouth and flows over their gills. Baleen whales use a filter consisting of a curtain of parallel filaments (called baleen) attached to the upper jaw to feed. These whales engulf a mouthful of water containing krill (small, shrimplike plankton) and use their tongue to force the water out, trapping krill in the hairlike edges of the baleen. The mouthful of krill is then swallowed. Flamingos feed in a similar manner. They use specially adapted beaks lined with filaments to strain plankton from the muddy bottoms of their aquatic habitat. Some fast-water caddis fly

369

larvae spin tiny silk nets that are used to filter small food items. The nets are then periodically gleaned. Finally, spiders may be considered filter feeders because they use their webs to "filter" flying insects from their environment.

LARGE ITEM INGESTION

In contrast with small-particle consumers, numerous animals ingest large food items. Some of these organisms are saprovores, such as earthworms. These worms consume large masses of soil and dead leaves, from which they obtain usable organic matter. Some planarians (flatworms) extend a long extensible tube (called a pharynx) from their mouth to ingest decaying material. Other examples of saprovores include millipedes, wood-eating beetles, some sea cucumbers, many roundworms, and a few snails.

An extraordinary number of animals have adapted to eat plants. Eating plants requires special structures to free plant material for ingestion. Although invertebrates lack true teeth, they have other structures to obtain plants or plant parts. Snails have a unique structure in their mouth called a radula, which acts as a miniature rasping file that scrapes plant material from surfaces and rasps through vegetation. The freed plant material is then ingested. Sea urchins scrape algae by using a highly developed oral apparatus composed of five large pointed plates. Termites use strong jaws made of chitin, the hard structural component of their external skeleton, to cut tiny chunks of wood for ingestion.

Birds use horny beaks to obtain and ingest plant material. Cardinals and grosbeaks, for example, are well adapted to hull and then consume the nutritional portion of large seeds. Herbivorous mammals use specialized teeth to obtain and chew plant material. Rodents (such as beavers, porcupines, and mice, rabbits, and hares) have chisel-like front teeth (called incisors) that are used to gnaw, slice, or pull off plant material. Other herbivorous mammals (including cows, sheep, deer, moose, elk, and giraffes) lack upper incisors and therefore use their lower incisors pressed against the roof of their mouth to pull off leaves. The ingested food is then chewed by grinding with premolars and molars.

In contrast to herbivores, many animals capture other animals and eat them. To be effective, these carnivorous predators must have appropriate behavioral adaptations to find and capture prey as well as specialized structures to seize and hold their victims. Jellyfish use tentacles that are equipped with stinging cells to grasp and subdue animals, whereas the tentacles of squid and octopuses have suction-cup-like structures to grasp and manipulate prey. The giant water bug is a carnivorous insect that hunts and captures small fish, relatively large prey for an insect. To do this, water bugs use their legs to seize and hold fish, and their piercing mouthparts to suck juices from their victims. Fish, amphibians, and reptiles have pointed teeth to seize and hold prey. It is also common for many of them to swallow their food whole. Snakes, for example, swallow whole items such as birds' eggs and small mammals. In addition, a snake's jaws are held together by elastic ligaments, permitting it to spread apart and ingest victims larger than its own head. Some large tropical snakes actually consume small pigs and deer. Chameleons and frogs swallow their prey whole, but in contrast to snakes, they use a long and sticky tongue that rapidly shoots out to capture insects. Some predatory carnivores (such as lions, tigers, bears, and dogs) have long, pointed, daggerlike teeth called canines to pierce and kill their prey. Carnivores also may have knife-like molars (called carnassials) that are used to slice flesh from bones. Carnivorous birds (such as hawks, eagles, and owls) use long, sharp talons to seize and kill small animals. Their beaks can be used to tear small pieces of food for ingestion. Further, many carnivorous animals possess specialized mechanisms for paralyzing victims. Jellyfishes, centipedes, spiders, scorpions, and some snakes possess structures that inject toxins that inhibit the nervous system of their prey. Finally, some electric eels may locate and stun their prey with electrical discharges.

LIQUID INGESTION

Surprisingly, many animals live exclusively on a liquid diet. Herbivores such as bees, butterflies, moths, hummingbirds, and some bats derive nutrition by consuming plant nectar. As a consequence of their feeding, these animals help plants reproduce by dispersing pollen. Hummingbirds have specially adapted long, narrow beaks and long tongues to suck nectar from flowers. Nectar-consuming bats also have long narrow faces and tongues. Aphids, and many other insects that consume plant sap, have highly specialized mouth parts that pierce plants and act as miniature straws to suck sap.

A number of carnivores also are adapted to consume a liquid diet. As ghoulish as it might sound, ingesting blood is the most common mechanism of feeding for these animals. Mosquitoes, for example, are equipped with a syringelike mouth part called a proboscis. Although the male sips nectar, the female mosquito uses her proboscis to pierce skin and suck blood. As with most bloodsucking animals, mosquitoes secrete an anticoagulant that prevents blood from clotting (and also makes people itch). Some flies use a similar mode of feeding, but the common housefly generally laps up food and sugary solutions. Leeches are also well adapted for bloodsucking: They have a suction-cup-like mouth that clings tenaciously to a host while their jaws make a Y-shaped cut in the skin. Further, leeches have a muscular pharynx (throat) that literally pumps blood from their host. Ticks, which are related to spiders and scorpions, have tiny heads that are designed to burrow into the skin of their host and suck blood. Unfortunately, they also are vectors for potentially serious diseases such as Lyme disease and Rocky Mountain spotted fever. Although vampire bats consume blood, they do not suck it. Instead, they lap blood with their tongue as it oozes from a shallow scrape in the skin made by their teeth. The lamprey eel, found in the waters of the lower Great Lakes, attaches itself to a host fish and uses its teeth to rasp on opening through the hosts scales and skin in order to extract blood and other liquids.

Spiders also are adapted to live on a liquid diet; however, they prey on insects that have a tough external skeleton that is not easily ingested. To feed, they first pierce the insect's exoskeleton with hollow jaws and pump in strong digestive juices that liquefy the internal contents of their prey. Later, the spider sucks the insect empty. Some young birds, such as pigeons and emperor penguins, feed on a regurgitated milklike secretion (called crop milk) that is produced by their parents' crop. In addition, all mammals begin their life as fluid feeders, ingesting milk produced by their mother.

Finally, endoparasites comprise a group of animals that consume a liquid diet by living inside other organisms. Some eat host tissues, while others rely on their host to digest food for them and, as a result, lack a digestive system. Tapeworms, for example, have a specially adapted anterior end with hooks and suckers to maintain a fixed position in their hosts' gut while they consume predigested food. In contrast, hookworms (a type of roundworm) do have a digestive system, and they use their mouth opening and toothlike structures to draw and ingest blood from the inside of their host.

STUDYING INGESTION

Much of what has been learned about ingestion has come from simple observation of feeding behavior combined with careful note-taking. In fact, observation and data collection, along with analysis and interpretation, are the most fundamental of all scientific activities. Most people have seen a robin use its beak to pull an earthworm from the ground or a cat capture a mouse. By closely watching the lifestyle and daily activities of animals, biologists have discovered what food items are eaten and how they are ingested.

The naked eye is insufficient for observing very small animals, and for this activity microscopes aid biologists. These tools have permitted observation of endocytosis of paramecia and bacteria by the amoeba as well as other feeding mechanisms of protozoans. Microscopic, inert latex beads are used to study the direction and power of feeding currents generated by cilia and flagella and the formation of food vacuoles. For example, paramecia will direct beads into their oral region by ciliary action and then engulf them by endocytosis. Dyes are also commonly used to study the direction and action of feeding currents. This method has revealed that flagellated choanocytes of sponges move water in through small pores in the sides of the sponge and out a single larger opening at the top.

Video and photographic recordings are routinely used to supplement simple observation. These procedures not only provide a permanent record of the event but also permit additional analysis to be conducted at some future time. Further, feeding mechanisms that occur very quickly are difficult to analyze with the naked eye. These events can be recorded by high-speed cinematography and later played back at a slower speed for analysis. This method of study has been used to observe the lightning-fast movement of a chameleon's tongue and the way bats catch insects with their wing membranes while in flight. Alternatively, very slow feeding events, such as endocytosis or a snake swallowing a rat, can be recorded by time-lapse photography and later viewed at a faster speed for analysis.

Mechanisms of ingestion may also be inferred by carefully analyzing the body design of an animal. For example, birds possessing beaks that have an arrangement of tightly packed vertical filaments, as found in flamingos, feed by filtering. In contrast, birds with long, pointed beaks, such as woodpeckers, probe for food in narrow places. Finally, the contents of the stomach and fecal samples from animals may be analyzed to determine the type and size of the food items that were eaten.

THE NECESSITY OF FOOD

Animals have an absolute requirement for food. Animals must ingest food items because, unlike photosynthetic organisms, they cannot manufacture all the necessary nutrients they require from raw materials. Animals require food both as a fuel source to provide energy for locomotion and metabolism and as building blocks for growth, maintenance, and repair. Obtaining sufficient food is of paramount importance for survival. Therefore, the limited availability of food is selected by animals that have the most-successful feeding strategies and body designs for procurement and ingestion of nutrients. Because different sources of nutrition are utilized, the selection pressures for obtaining food may result in vastly different feeding mechanisms among closely related animals. For example, giant water bugs, termites, and aphids are all classified as insects. Yet, they rely on different diets and therefore possess divergent methods of feeding and structures for ingestion. In addition, and conversely, selection pressures for obtaining food may result in the development of similar body structures in distantly related species (convergent evolution). Baleen whales and flamingos, for example, are classified in different taxonomic groups (mammals and birds) but have similar feeding methods and therefore similar structures for ingestion. Thus, specific feeding behaviors and structures for ingestion are primarily shaped by the nature of the food items being utilized, and the vast diversity of feeding mechanisms reflects a similar diversity in food sources.

Ultimately, the source of energy to create food comes from the Sun. Photosynthetic organisms use light to synthesize energy-rich organic compounds, such as glucose, from energy-poor inorganic compounds, such as carbon dioxide and water. One exception to this scheme occurs in certain regions of the ocean floor, near hydrothermal vents. Far removed from sunlight and organic material derived by photosynthesis, the food chain of oceanic hydrothermal vents is based upon certain bacteria that synthesize organic compounds from inorganic substances emitted by these undersea geysers.

Unfortunately, there is an unavoidable loss in usable material and energy between links in a food chain. This loss occurs because much of the energy stored in food is irreversibly lost as heat when it is used by organisms for growth, repair, and maintenance. Therefore, it is more efficient to have a lower number of links in a food chain between its base (usually plants or plankton) and its end (typically large carnivores). With that in mind, it is interesting to note that some of the largest fish (whale sharks) and mammals (whales) in the world are filter feeders of tiny organisms. These huge animals, which require a large amount of energy, taking in the equivalent of a half million calories in a single mouthful, avoid extra links in their food chain by feeding on plankton instead of other large animals.

FURTHER READING

Childress, James J., Horst Felbeck, and George N. Somero. "Symbiosis in the Deep Sea." *Scientific American* 256 (May, 1987): 115–120.

Eckert, Roger, David Randall, and George Augustine. *Animal Physiology: Mechanisms and Adaptations.* 4th ed. W. H. Freeman, 1997.

Hickman, Cleveland, Jr., Larry S. Roberts, and Frances M. Hickman. *Integrated Principles of Zoology.* 11th ed. McGraw-Hill, 2001.

Pearse, Vicki, John Pearse, Mildred Buchsbaum, and Ralph Buchsbaum. *Living Invertebrates.* Rev. 4th ed. Blackwell Scientific, 1992.

Schmidt-Nielsen, Knut. *Animal Physiology: Adaptation and Environment.* 5th ed. Cambridge University Press, 1998.

Wessels, Norman K., and Janet L. Hopson. *Biology.* Random House, 1988.

Insects

Fields of Study

Entomology; Ecology; Evolutionary Biology; Paleontology

Abstract

Insects are found in every continent except Antarctica. They are mainly terrestrial, and some are aquatic, primarily freshwater. Their lifespans are highly variable; adult mayflies live less than one week, while queen termites have been known to live for more than twenty years. Major anatomical regions of insects include the head, featuring one pair of antennae, the thorax, with three pairs of legs and up to two pairs of wings, and the abdomen, housing spiracles and genitalia; in some insects, one or both pairs of wings are modified for functions other than flight, such as protection or balance.

Key Concepts

apterous: insects without wings, such as fleas

brood: all the immature insects within a colony; these include eggs, larvae, and, in the Hymenoptera, the pupal stage

caste: one of the recognizable types of individuals within a colony, usually physically and behaviorally adapted to perform specific tasks

conspecifics: individuals of the same species

ecdysis: molting; the shedding of the exoskeleton that allows for insect growth

ectoparasite: a parasite that lives on the outside of its host's body

endoparasite: a parasite that lives within its host's body

eusocial: referring to any of the truly social species characterized by division of labor, with a sterile caste, overlapping generations, and cooperative brood care

haplodiploidy: sex determination found in the Hymenoptera, where males arise from unfertilized eggs and females from fertilized eggs

hemolymph: a liquid that serves the same function in insects as blood and lymph in humans

hexapod: six-footed, a general term for an insect

metamorphosis (complete): a transformation that occurs during the development of higher insects, in which a grublike immature form enters a resting (pupal) stage for major tissue reorganization; after pupation, the adult, which bears no resemblance to the larval form, emerges

ovipositor: egg-laying apparatus on the female abdomen, modified into a stinger in bees

pheromone: a chemical produced by one member of a species that influences the behavior or physiology of another member of the same species

phytophagous: describes insects that eat plant matter

pupa: intermediate stage between the larval and adult stages of the life cycle

spiracles: openings on the outside of the insect abdomen that lead to breathing tubes

trophallaxis: the exchange of bodily fluids between nestmates, either by regurgitation or by feeding on secreted or excreted material

vector: transmits pathogens from one host to another

THERE SURE ARE LOTS OF INSECTS IN THE WORLD

The common orders of insects include the *Coleoptera* (beetles), the *Diptera* (flies, mosquitoes), the *Hemiptera* (true bugs), the *Hymenoptera* (ants, bees, wasps), the *Buttodea* (cockroaches, termites), the *Lepidoptera* (butterflies, moths), the *Odonata* (dragonflies, damselflies), and the *Orthoptera* (crickets, grasshoppers, locusts, katydids). The history of insects dates back to the Devonian period, about 400 million years ago. Today numbering more than one million different species, insects are the most diverse class of animals

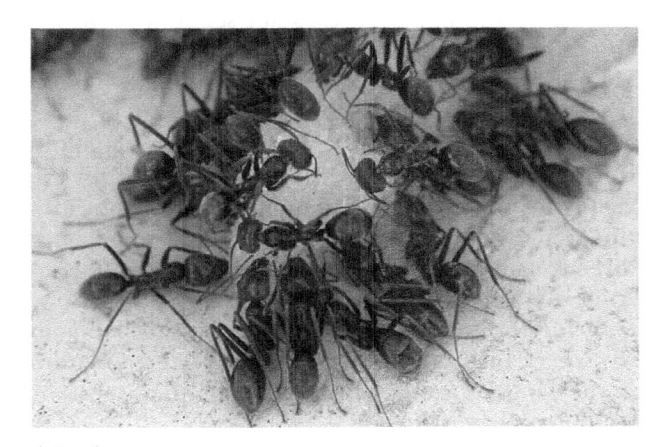

Ants eating.

on earth. Indeed, many estimates suggest there are as many as 5–10 million more extant but undescribed species. The great biological mass of insects makes them a crucial part of many ecosystems, especially as they often form basic levels of food chains. They display a wide range of morphologies and specialized evolutionary adaptations.

Insects populate almost every habitat except the deep oceans and permanently frozen land masses. They may be specialized to live underground, in live or rotting trees, in fast-flowing rivers, or stagnant puddles. Parasitic insects live attached to the outside (hair, skin) or inside (stomach lining, respiratory tract) of other animals. Crafty insects modify their habitat by building their own houses; for example, spittlebugs mix air into slimy anal secretions to form shelters; bagworms spin silk to hold leaves together around them; and some African termites glue soil particles together with saliva into colossal nests measuring six meters high and nearly four meters across.

INSECT ANATOMY

Insects are invertebrates noted for their jointed exoskeletons. Chitin makes this armor especially strong. The exoskeleton functions much like the internal bones of vertebrates, serving to protect internal organs and as sites of muscle attachment. The exoskeleton puts limits on the overall size that insects can attain and makes growth energetically costly. The insect must shed its old exoskeleton and form a functions in keeping tissues moist and transporting nutrients and hormones. There are no red blood cells or other oxygen-carrying molecules because the insect respiratory system is totally separate from the circulatory system. The circulatory system is open. Hemolymph enters a series of chambers, collectively termed the heart, through holes called ostia. Muscular pumping of the heart propels the fluid through the aorta, toward the insect's brain. In the vicinity of the brain, the hemolymph flows out into the general body cavity, or hemocoel. Hemolymph directly bathes internal organs on its way to and from the legs and wings, starting the cycle again by entering the heart ostia. There are distinct veins in the wings. Insects take advantage of fluid circulation in the wings to transfer heat, warming themselves or cooling off when needed.

THE INSECT DIET

Insects display a diverse range of diets, often highly specialized to certain ecological niches. Phytophagous insects feed on living plants; for example, caterpillars are well-known leaf eaters. Other insects feed on plant roots, shoots, flowers, stems, or fruits. When a large group of phytophagous insects occurs in the same area, the crop or forest damage that they cause can be extensive. Yet there is quite another side to the relationship between insects and plants. Insect-pollinated plants have evolved odors, shapes, and ultraviolet color patterns to attract these important visitors. When the butterfly, bee, or other pollinator stops to drink the plant's rich nectar, pollen from the male part of the flower sticks to the insect's body. When it visits another flower of the same species, it deposits

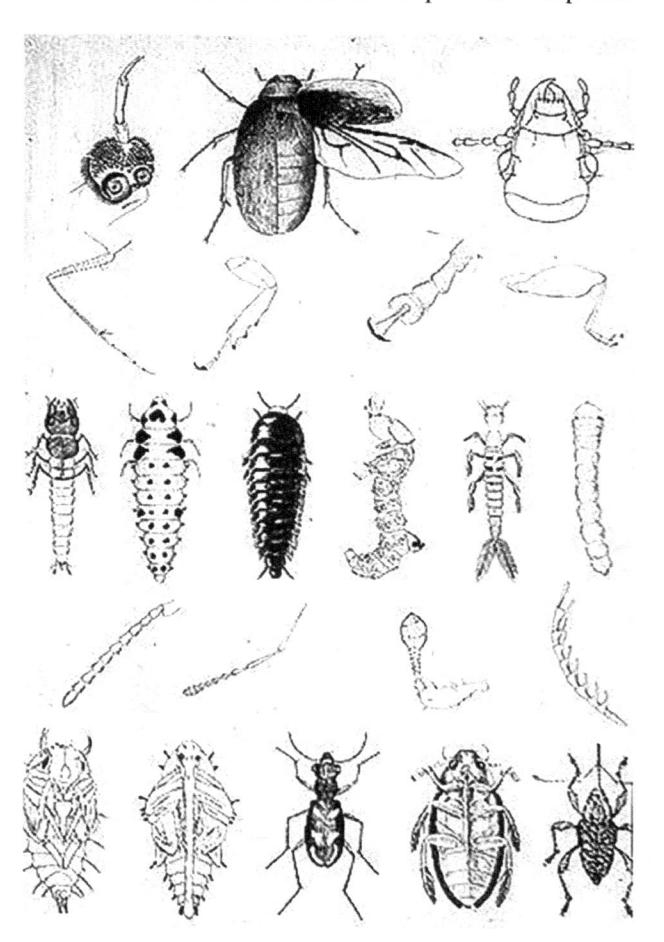

Drawings of a variety of beetles.

the pollen on the female floral organ. Without this interaction, beans, tomatoes, tea, cocoa, and many other plants could not reproduce.

Other insects are scavengers, serving an important purpose by recycling nutrients found in dead plant and animal matter. Termites are a nuisance when they infest a house but are invaluable at breaking down dead wood in nature. Dung beetles have the curious habit of forming animal feces into balls and rolling them away to feed their young. Carrion feeders assist in the decomposition of animal corpses. A succession of flies and beetles reduce a corpse to bones.

Predatory insects hunt and kill to eat. Some predators use powerful mouth parts to tear apart their prey. Other predators inject digestive enzymes into their prey, digest them externally and suck out the liquefied tissues through a specialized beak. The first pair of legs of the praying mantis are striking examples of raptorial modifications for grasping and holding struggling prey. Some insects have a varied diet, such as dragonflies, which in their aquatic nymph stage may eat crustaceans, tadpoles, and even small fish. Adult dragonflies eat bees, butterflies, and mosquitoes.

Parasitic insects typically ingest host blood, mucus, or tissues, often with minimal direct irritation or harm to the host. Trouble begins if there is a heavy parasitic burden (a large number of parasites per host), or if the parasitic insect transmits disease-causing organisms to the host. Ectoparasites live on the outside of their hosts. In this category, fleas and lice are adapted to avoid detection by the host and deter removal during normal grooming processes. Endoparasitic fly larvae live in the digestive or respiratory tract, mainly of livestock. Endoparasites of invertebrates, however, often kill the host because there is little difference in size between the parasite and host. Species of endoparasitic wasps inject their eggs into caterpillars. When the larvae hatch, they tunnel farther into the host and feed off of its tissues. Upon completion of larval development, the wasps emerge from the caterpillar, killing it in the process.

Some parasites spend their lifetimes closely associated with the host, but others only briefly visit the host. On the time continuum, lice not only form a constant, more-or-less permanent association with a single host, but many generations of lice may inhabit the same host. However, these parasites are transferred from host to host during mating, nesting, or other close contact between individuals. Unlike lice, fleas leave the host frequently between bloodmeals. On the other end of the spectrum, mosquitoes normally require multiple hosts to complete a single bloodmeal; they aim to move on before they can be slapped by hand or tail.

INSECT BEHAVIOR

Much of what insects do, they do by instinct, a genetically preprogrammed response to environmental stimuli, none the less amazing in its elegance and effectiveness. Antennae and setae, hairlike projections through the exoskeleton, serve as two of the many receptors of external stimuli. The necessity of keeping these sensory organs clean is evidenced by preening, commonly using mouth parts and legs. Major categories of environmental stimuli signal friend or foe.

Many insects locate opposite sex conspecifics by sight. Color patterns, especially in the ultraviolet (UV) spectrum attract potential mates. For example, fireflies (*Coleoptera*) use visual recognition of patterns of light flashes. In some species, females use signaling to attract males. In other species the male and female signal to each other with the proper code and response. Females of the genus *Photuris* mimic the flashing patterns of other firefly species. When the male approaches, they become prey.

Sound production and reception are also used for mate location. The sound generated by a female mosquito's beating wings in flight is picked up by the male mosquito's antennae. Male homopterans and orthopterans produce songs of courtship. Cicadas use abdominal muscles and a resonating chamber for sound production. Grasshoppers and crickets rub their wings and or legs together. Sound receptors, "eardrums" of females in these orders, are found on their forelegs or first abdominal segments. Sound production and reception vary in precise ways with temperature.

Pheromones, externally broadcast chemical signals, are another means of attracting mates. In moths, glands near the tip of the female's abdomen release pheromones that are received by the antennae of the male moths. Certain male butterflies and moths have specialized scent-producing glands on their wings. The notable pair of black patches, or androconia, on the hindwings of the male monarch butterfly secrete aphrodisiac pheromones that increase the female's receptivity to mating. Pheromones can also make already-mated females unattractive to subsequent suitors.

Members of the class *Insecta* display numerous protective behaviors and defense mechanisms when threatened. Walkingsticks are capable of losing part of a leg in the grasp of a predator. The amputated appendage continues to wiggle when severed, thus distracting the predator, while the walkingstick gets away. Some beetles play dead until the predator loses interest. Blister beetles are capable of reflex bleeding. They squeeze drops of hemolymph through joints in their exoskeleton. Their hemolymph contains cantharadine, which irritates and repels the predator. Bombardier beetles spray a noxious repellant from their anal glands for distances up to one meter.

Another means of defense is coloration. Insects may use camouflage to blend into the surroundings by resembling a leaf, twig, pebble, or flower. Some insects use bright colors to hide in plain sight. Large eyespots on wings can startle would-be predators, or at least trick them into taking a bite out of the wing rather than from the head or body. Color patterns of orange and black seem to warn vertebrate predators that the insects would not make a good meal. Some of these insects derive chemicals from their diet that make them distasteful, even harmful, to predators. Without possessing such chemicals, mimics derive protection from their resemblance to the group that does.

Social insects exhibit the most complex behavior, approaching learning. Honeybees can communicate direction of a food source, and distance from the hive, to other bees through a multipart "waggle dance." Some ants are farmers, planting and nurturing fungus gardens. Mound-building harvester ants ensure that the colony's young are kept at the optimal temperature through vertical migrations. The young are transported to top levels within the mound to warm up in the morning and evening. During the heat of the day, the young are carried to lower levels to cool off. Kidnapping of ants from other colonies to serve as workers in the home colony may not be considered a socially advanced behavior, but it is a complex one. The other major group of social insects, the termites, work together to build huge mounds, called termitaria, some reaching heights of 6 meters and diameters of over 3.5 meters. They engineer series of chimneys into the structure that can be opened and closed to regulate airflow and maintain constant temperature.

INSECT SOCIETIES

It is estimated that more than twelve thousand species of social insects exist in the world today. This number is equivalent to all the known species of birds and mammals combined. Although insect societies have reached their pinnacle in the bees, wasps, ants, and termites, many insects show intermediate degrees of social organization—providing insights regarding the probable paths of the evolution of sociality.

Many of the most robust, thriving species today owe their success in great part to benefits that they reap from living in organized groups or societies. Nowhere are the benefits of group living more clearly illustrated than among the social insects. Edward O. Wilson, one of the foremost authorities on insect societies, estimates that more than twelve thousand species of social insects exist in the world today. This number is equivalent to all the known species of birds and mammals combined. Although insect societies have reached their pinnacle in the bees, wasps, ants, and termites, many insects show intermediate degrees of social organization—providing insights regarding the probable paths of the evolution of sociality.

ANT, WASP, AND BEE SOCIETIES

Scientists estimate that eusociality has evolved at least twelve times: once in the *Isoptera*, or termites, and eleven separate times in the *Hymenoptera*, comprising ants, wasps, and bees. In addition, one group of aphids has been found which has a sterile soldier caste. Although the eusocial species represent diverse groups, they all show a high degree of social organization and possess numerous similarities, particularly with regard to division of labor, cooperative brood care, and communication among individuals. The organization of a typical ant colony is representative, with minor modifications, of all insect societies.

A newly mated queen, or reproductive female, will start a new ant colony. Alone, she digs the first nest chambers and lays the first batch of eggs. These give rise to grublike larvae, which are unable to care for themselves and must be nourished from the queen's own body reserves. When the larvae have reached full size, they undergo metamorphosis and emerge as the first generation of worker ants. These workers—all sterile females—take over all the colony maintenance duties, including foraging outside the nest for food, defending the nest, and cleaning and feeding both the new brood and the queen, which subsequently becomes essentially an egg-laying machine. For a number of generations, all eggs develop into workers and the colony grows. Often, several types of workers can be recognized. Besides the initial small workers, or minor workers, many ant species produce larger forms known as major workers, or soldiers. These are often highly modified, with large heads and jaws, well suited for defending the nest and foraging for large prey. Food may include small insects, sugary secretions of plants or sap-feeding insects, or other scavenged foods. After several years, when the colony is large enough, some of the eggs develop into larger larvae that will mature into new reproductive forms: queens and males. Males arise from unfertilized eggs, while new queens are produced in response to changes in larval nutrition and environmental factors. These sexual forms swarm out of the nest in a synchronized fashion to mate and found new colonies of their own.

With minor modifications, the same pattern occurs in bees and wasps. Workers of both bees and wasps are also always sterile females, but they differ from ants in that they normally possess functional wings and lack a fully differentiated soldier caste. Wasps, like ants, are primarily predators and scavengers; bees, however, have specialized on pollen and plant nectar as foods, transforming the latter into honey that is fed to both nestmates and brood. The bias toward females reflects a feature of the biology of the *Hymenoptera* that is believed to underlie their tendency to form complex societies. All ants, wasps, and bees have an unusual form of sex determination in which fertilized eggs give rise to females and unfertilized eggs develop into males. This type of sex determination, known as haplodiploidy, generates an asymmetry in the degree of relatedness among nestmates. As a consequence, sisters are more closely related to their sisters than they are to their own offspring or their brothers. Scientists believe that this provided an evolutionary predisposition for workers to give up their own personal reproduction in order to raise sisters—a form of natural selection known as kin selection.

TERMITE SOCIETY

The termites, or *Isoptera*, differ from the social *Hymenoptera* in a number of ways. They derive from a much more primitive group of insects and have been described as little more than "social cockroaches." Instead of the strong female bias characteristic of the ants, bees, and wasps, termites have regular sex determination; thus, workers have a fifty-fifty sex ratio. Additionally, termite development lacks complete metamorphosis. Rather, the young termites resemble adults in form from their earliest stages. As a consequence of these differences, immature forms can function as workers from an early age, and—at least among the lower termites—they regularly do so.

Termites also differ from *Hymenoptera* in their major mode of feeding. Instead of feeding on insects or flowers, all termites feed on plant material rich in cellulose. Cellulose is a structural carbohydrate held together by chemical bonds that most animals lack enzymes to digest. Termites have formed intimate evolutionary relationships with specialized microorganisms—predominantly flagellate protozoans and some spirochete bacteria—that have the enzymes necessary to degrade cellulose and release its food energy. The microorganisms live in the gut of the termite. Because these symbionts are lost with each molt, immature termites are dependent upon gaining new ones from their nestmates. They do this by feeding on fluids excreted or regurgitated by other individuals, a process known as trophallaxis. This essential exchange of materials also includes, along with food, certain nonfood substances known as pheromones.

INSECT COMMUNICATION

Pheromones, by definition, are chemicals produced by one individual of a species that affect the behavior or development of other individuals of the same species that come in contact with them. Pheromones are well documented throughout the insect world, and they play a key role in communication between members of nonsocial or subsocial species. Moth mating attractants provide a well-studied example.

Pheromones are nowhere better developed than among the social insects. They not only appear to influence caste development in the *Hymenoptera* and termites but also permit immediate communication among individuals. Among workers of the fire ant (*Solenopsis saevissima*), chemical signals have been implicated in controlling recognition of nestmates, grooming, clustering, digging, feeding, attraction or formation of aggregations, trail following, and alarm behavior. Nearly a dozen different glands have been identified which produce some chemical in the *Hymenoptera*, although the exact function of many of these chemicals remains unknown.

In addition to chemical communication, social insects may share information in at least three other ways: by tactile contact, such as stroking or grasping; by producing sounds, including buzzing of wings; and by employing visual cues. Through combinations of these senses, individuals can communicate complex information to nestmates. Indeed, social insects epitomize the development of nonhuman language. One such language, the "dance" language of bees, which was unraveled by Karl von Frisch and his students, provides one of the best-studied examples of animal behavior. In the "waggle dance," a returning forager communicates the location of a food resource by dancing on the comb in the midst of its nestmates. It can accurately indicate the direction of the flower patch by incorporating the relative angles between the sun, the hive, and the food. Information about distance, or more precisely the energy expended to reach the food source, is communicated in the length of the run. Workers following the dance are able to leave the nest and fly directly to the food source, for distances in excess of one thousand meters.

BENEFITS OF COOPERATION

Living in cooperative groups has provided social insects with opportunities not available to their solitary counterparts. Not only can more individuals cooperate in performing a given task, but also several quite different tasks may be carried out simultaneously. The benefits from such cooperation are considerable. For example, group foraging allows social insects to increase the range of foods they can exploit. By acting as a unit, species such as army ants can capture large insects and even fledgling birds.

A second benefit of group living is in nest building. Shelter is a primary need for all animals. Most solitary species use naturally occurring shelters or, at best, build simple nests. By cooperating and sharing the effort, social insects are able to build nests that are quite elaborate, containing several kinds of chambers. Wasps and bees build combs, or rows of special cells, for rearing brood and storing food. Subterranean termites can construct mounds more than six meters high, while others build intricate covered nests in trees. Mound-building ants may cover their nests with a thatch that resembles, in both form and function, the thatched roofs of old European dwellings. Colonial nesting provides two additional benefits. First, it enhances defense. By literally putting all of their eggs in one basket, social insects can centralize and share the guard duties. The effectiveness of this approach is attested by one's hesitation to stir up a hornet's nest. Nest construction also provides the potential to maintain homeostasis, the ability to regulate the environment within a desirable range.

Virtually all living creatures maintain homeostasis within their bodies, but very few animals have evolved the ability to maintain a constant external living environment. In this respect, insect societies are similar to human societies. Workers adjust their activities to maintain the living environment within optimal limits. Bees, for example, can closely regulate the internal temperature of a hive. When temperatures fall below 18°C, they begin to cluster together, forming a warm cover of living bees to protect the vulnerable brood stages. To cool the hive in hot weather, workers initially circulate air by beating their wings. If further cooling is needed, they resort to evaporative cooling by regurgitating water throughout the nest. This water evaporates with wing fanning and serves to cool the entire hive. Other social insects rely on different but equally effective methods. Some ants, and especially termites, build their nests as mounds in the ground, with different temperatures existing at different depths. The mound nests of the African termite, *Macrotermes natalensis*, are an impressive engineering feat. They are designed to regulate both temperature and air flow through complex passages and chambers, with the mound itself serving as a sophisticated cooling tower.

Finally, group living allows the coordination of the efforts of individuals to accomplish complex tasks normally restricted to the higher vertebrates. The similarities between insect societies and human society are striking. An insect society is often referred to as a superorganism, reflecting the remarkable degree of coordination between individual insects. Individual workers have been likened to cells in a body, and castes to tissues or organs that perform specialized functions. Insect societies are not immortal; however, they often persist in a single location for periods similar to the life spans of much larger animals.

The social insects have one of the most highly developed symbolic languages outside human cultures. Further, social insects have evolved complex and often mutually beneficial interactions with other species to a degree unknown except among human beings. Bees are inseparably linked with the flowers they feed upon and pollinate. Ants have actually developed agriculture of a sort with their fungus gardens and herds of tended aphids. On a more sobering note, ants are the only nonhuman animals that are known to wage war. These striking similarities with human societies have led researchers to study social insects to learn about the biological basis of social behavior and have led to the development of a new branch of science known as sociobiology.

STUDYING INSECT SOCIETY

Because of the diversity of questions that investigators have addressed regarding insect societies, many methods of scientific inquiry have been employed. In Karl von Frisch's experiments, for example, basic behavioral observations were coupled with simple but elegant experimental design to unravel the dance language of bees. The bees were raised in an observation colony. This was essentially a large hive housed between plates of glass so that an observer could watch the behavior of individual bees. Researchers followed specific workers by marking them with small numbers placed on the abdomen or thorax. Sometimes the entire observation hive was placed within a small, darkened shed to simulate more closely the conditions within a natural hive.

Bees learned to find an artificial "flower"—a glass dish filled with a sugar solution. Brightly colored backgrounds and odors such as peppermint oil were added to the sugars to provide specific cues for the bees to associate with the reward. Feeding stations were set up at fixed distances; observers could follow the exact movements of known individuals both at the feeder and at the hive. In this way, von Frisch was able to describe several types of dances (the "round dance" for near food sources, the "waggle dance" for feeding stations that were farther from the hive) and show that a returning bee could share information regarding the location and quality of a source with her nestmates. Scientists subsequently have developed robot bees that can be operated by remote control to perform different combinations of dance behaviors. This allows them to determine which parts of the dance actually convey the coded information.

The investigation of forms of chemical communication requires application of a variety of techniques. Chromatography is useful for identifying the minute amounts of chemical pheromones with which insects communicate. Chromatography (which literally means "writing with color") is particularly suitable for separating mixtures of similar materials. A solution of the mixture is allowed to flow over the surface of a porous solid material. Since each component of the mixture will flow at a slightly different rate, eventually they will become separated or spaced out on the solid material. Once the components of the pheromone have been separated and identified, their activity is assessed separately and in combination using living insects. Such bioassays allow researchers to determine exactly which fractions of the chemical generate the highest response.

Other biochemical techniques, such as electrophoresis, have been used to determine subtle behavioral differences, such as kin discrimination among hive mates. Each individual carries a complement of enzymes or proteins that catalyze biological reactions in the body. The structure of such enzymes is determined by the genetic makeup of the individual, and it varies among individuals. Because enzyme structure is inheritable, however, much as eye color is, the degree of similarities between the enzymes can be used as a measure of how closely related two individuals are. The amino acids composing the enzyme differ in their electrical charges, so different forms can be separated using the technique of electrophoresis. When a liquid containing their enzymes is subjected to an electrical field, the proteins with the highest negative charge will move farthest toward the positive pole. This provides a tool to distinguish close genetic relatives for use in conjunction with behavioral observations to test, for example, whether workers can discriminate full sisters from half-sisters, or relatives from nonrelatives, as kin selection theory would predict.

THE SUCCESS OF SOCIAL INSECTS

Social insects are among the most successful groups of animals throughout the world, especially in the tropics. Although the number of species is low when compared to all insects (twelve thousand out of more than a million species), their relative contribution to the community may be unduly large. In Peru, for example, ants may make up more than 50 percent of the individual insects collected at any site.

The study of social insects has provided scientists with new ways of looking at social behavior in all animals. Charles Darwin described the evolution of sterile workers in the social insects as the greatest obstacle to his theory of evolution by natural selection. In attempts to explain this seeming paradox, William D. Hamilton closely examined the social *Hymenoptera*, where sociality had evolved eleven separate times. Realizing that the haplodiploid form of sex determination led to sisters being more closely related to one another than they would be to their own young, Hamilton developed a far-reaching new theory of social evolution: kin selection, or selection acting on groups of closely related individuals. This theory, which provides insights into the evolution of many kinds of seemingly altruistic behaviors, arose primarily from his perceptions regarding the asymmetrical relatedness of nestmates in the social *Hymenoptera*. These insects, then, should be credited with providing the model system that has led to a subdiscipline of behavioral ecology known as sociobiology, the study of the biological basis of social behavior. Moreover, given their central roles in critical ecological processes such as nutrient cycling and pollination, it would be hard to imagine life without them.

ENVIRONMENTAL SIGNIFICANCE OF INSECTS

Insects are crucial elements of most ecosystems, and are often beneficial to humans and the environment as a whole. In addition to their roles as pollinators for many important crops, insects themselves serve as food sources to many animals, including many people around the world. Harmless insects such as dragonflies, which consume pest insects such as mosquitoes, are good candidates for biological control programs, a form of pest control that takes advantage of natural predators of the pest. Biological control shows promise as a way to reduce use of insecticides, which can have negative environmental effects.

While many people think of most insects—often colloquially called "bugs"—as nuisances, they are in fact crucial elements of most ecosystems, and indeed are often beneficial to humans and the environment as a whole. In addition to their roles as pollinators for many important crops, insects themselves serve as food sources to many animals, including many people around the world. In addition, harmless insects such as dragonflies, which consume pest insects such as mosquitoes, are good candidates for biological control programs, a form of pest control that takes advantage of natural predators of the pest. For example, ladybugs are commonly used by gardeners to control plant-damaging aphid populations. Biological control shows promise as a way to reduce the use of insecticides, which can have negative environmental effects, building up in food chains and harming other animals.

Though insecticides can be useful in controlling pest species that destroy crops or spread disease, their widespread use has also been linked to serious declines in the global insect population. Although insects are often inconspicuous and difficult to track, in the mid-2010s studies began to suggest that at least 40 percent species were in decline, with some populations dropping very steeply in a relatively short period of time. In addition to insecticide use, researchers pointed to climate change, deforestation and other habitat loss, pollution, invasive species and other factors as contributing to this trend. Scientists warned that the collapse of insect populations would have major implications on ecosystems, as insects are a vital component of food webs and fill other crucial ecological roles.

— *Catherine M. Bristow Sarah Vordtriede and Sarah Vordtriede*

Further Reading

Arnett, Ross H., Jr., and Richard L. Jacques, Jr. *Guide to Insects.* Simon & Schuster, 1981.

Borror, Donald J., Charles A. Triplehorn, and Norman F. Johnson. *An Introduction to the Study of Insects.* 6th ed. Saunders College Publishing, 1992.

Crozier, Ross H., and Pekka Pamilo. *Evolution of Social Insect Colonies: Sex Allocation and Kin Selection.* Oxford University Press, 1996.

Elzinga, Richard J. *Fundamentals of Entomology.* 5th ed. Prentice Hall, 2000.

Faegri, Knur, and Leendert Van der Pijl. *Principles of Pollination Ecology.* 3rd ed. Elsevier, 2013.

Frisch, Karl von. *The Dance Language and Orientation of Bees.* Translated by L. E. Chadwick. The Belknap Press of Harvard University Press, 1967.

Gordon, Deborah M. *Ants at Work: How an Insect Society Is Organized.* W. W. Norton, 1999.

Hoelldobler, Bert, and Edward O. Wilson. *Journey to the Ants: A Story of Scientific Exploration.* Belknap Press, 1994.

"Insecta." *ITIS,* Integrated Taxonomic Information System, 8 Nov. 2017, www.itis.gov/. Accessed 31 Jan. 2018.

Ito, Yoshiaki. *Behavior and Social Evolution of Wasps: The Communal Aggregation Hypothesis.* Oxford University Press, 1993.

Jarvis, Brooke. "The Insect Apocalypse Is Here." *The New York Times Magazine,* 27 Nov. 2018, www.nytimes.com/2018/11/27/magazine/insect-apocalypse.html. Accessed 6 Mar. 2019.

Kettle, D. S., ed. *Medical and Veterinary Entomology.* 2nd ed. CAB International, 1995.

Klowden, Marc J. *Physiological Systems in Insects.* Academic Press, 2013.

Main, Douglas. "Why Insect Populations are Plummeting—And Why It Matters." *National Geographic,* 14 Feb. 2019, www.nationalgeographic.com/animals/2019/02/why-insect-populations-are-plummeting-and-why-it-matters/. Accessed 6 Mar. 2019.

Moffett, Mark W. "Samurai Aphids: Survival Under Siege." *National Geographic* 176 (September 1989): 406–422.

O'Toole, Christopher. *Alien Empire: An Exploration of the Lives of Insects.* HarperCollins, 1995.

Prestwich, Glenn D. "The Chemical Defenses of Termites." *Scientific American* 249 (August 1983): 78–87.

Seeley, Thomas D. "How Honeybees Find a Home." *Scientific American* 247 (October 1982): 158–168.

Turpin, F. Tom. *Insect Appreciation.* 2nd ed. Entomological Society of America, 2000.

Wilson, Edward O. *The Insect Societies.* The Belknap Press of Harvard University Press, 1971.

_____. *Sociobiology: The New Synthesis.* The Belknap Press of Harvard University Press, 1975.

_____. *Success and Dominance in Ecosystems: The Case of the Social Insects.* Ecology Institute, 1990.

Invertebrates

Fields of Study

Entomology; Evolutionary Biology; Animal Physiology

Abstract

With the exception of insects, most invertebrates are aquatic, many of those being marine. Some land dwellers start life as aquatic larvae. Invertebrates follow one of three types of general body plan. Some are asymmetrical. Cylindrical organisms are radially symmetric; any cut through the center of the organism divides it in equal halves. Asymmetrical and radially symmetric animals tend to stay in one place. Animals that exhibit bilateral symmetry have right and left halves that are mirror images of each other. They are mobile and usually have a distinct head end.

Key Concepts

coelom: a true body cavity, lined by mesoderm

colony: a cluster of genetically identical individuals formed asexually from a single individual

cyst: a secreted covering that protects small invertebrates from environmental stress

gonochoristic: having separate sexes; an individual is either male or female

hydrostatic skeleton: a system in which fluid serves as the support by which muscles interact

mesoderm: a middle layer of embryonic tissue between the ectoderm and endoderm

INVERTEBRATE SHAPES

With the exception of insects, which have extensively colonized terrestrial environments, most invertebrates are aquatic, many of those being marine. Even some of the land dwellers start life as aquatic larvae. Invertebrates follow one of three types of general body plan. Some marine sponges are asymmetrical, lacking an ordered pattern to their structure. Cylindrical organisms, such as sea anemones, are radially symmetric; any cut through the center of the organism divides it in equal halves. Asymmetrical and radially symmetric animals tend to stay in one place; thus their body plan helps them to collect environmental stimuli from every direction. Animals that exhibit bilateral symmetry have right and left halves that are mirror images of each other. They are mobile and usually have a distinct head end. This area of cephalization, concentrated nerve and sensory tissues, is directed forward in their travels, giving them new information about where they are going.

INVERTEBRATE FEEDING

Among the protozoans there are a variety of diets and feeding mechanisms. Organisms in the phylum *Ciliophora* have a cytostome, a cell mouth that can be found anteriorly, laterally, or ventrally, depending on species, on these single-celled creatures. Ciliates feed primarily on bacteria, algae, and other protozoans. Members of the phylum *Amoebozoa* have a similar diet to the ciliates but, in the absence of mouths, use pseudopodia to wrap around a food item, engulfing it. *Euglena* (phylum *Euglenozoa*), a flagellate protozoan commonly used in biology laboratories, is a self-feeder (autotroph). It contains chloroplasts and uses light energy to produce sugars, photosynthesized as in plants. It is interesting to note that some euglenids can and do ingest solid food if they are exposed to darkness for too long. Another group of interesting flagellates is the hypermastigotes. Species such as *Trichonympha campanula* live in the guts of termites. *T. campanula* breaks down the high cellulose content present in the termite diet of wood products, something that the termite cannot do for itself. In return these protozoans keep some of the nutrients for themselves.

Sponges, phylum *Porifera*, do not have any organs, but they do possess specialized cell types. Choanocytes, also termed collar cells, line the inside of the sponge and capture small food particles present in the circulating water. The phylum *Cnidaria* is well known for quite another specialty. Organisms in this phylum, such as jellyfish, have specialized stinging cells called nematoblasts. One of the functions of cnidae within the nematoblasts is secretion of toxins used to paralyze and kill prey. Cnidarians are patient hunters, lying in wait until their next meal contacts a tentacle or two, triggering the toxic sting.

Tapeworms (phylum *Platyhelminthes*, class *Cestoda*) have no mouth or digestive tract. They are highly adapted to a parasitic way of life. Swimming in nutrients that their host, usually a vertebrate, is in the process of digesting, tapeworms absorb nutrients through their outer surface, and return nothing besides waste products to their host.

Many rotifers (phylum *Rotifera*) are omnivores, meaning they eat anything that will fit into their mouth. After being ingested, prey passing into the muscular pharynx encounters grinding, crushing jaws called trophi. Rotifers have a one-way digestive tract. Wastes pass out the anus rather than being expelled through the mouth, unlike many of the animals discussed thus far.

There is a wide range of feeding styles among the mollusks (phylum *Mollusca*). Some feed on plants; others feed on animals, and others feed on particles suspended in the water or mud that they themselves inhabit. There are also mollusks that are ectoparasites, living on the exterior of their host rather than inside it. An interesting molluscan feeding structure is the radula. The teeth on this tonguelike organ are replaced as they wear down or break.

The phylum *Arthropoda* is divided into two subphyla based on mouth part structure. The *Chelicerates*, including horseshoe crabs, and arachnids (spiders, mites, ticks, and scorpions) have fanglike oral appendages used to grab and shred food items. Conversely, insects and crustaceans generally have mandibles, a pair of jaws for crushing food. This basic plan can be highly modified. The specialized coiled straw proboscis of a butterfly is such an example.

Starfish (phylum *Echinodermata*) prey on large invertebrates and small fish. These echinoderms have two stomachs and can protrude the cardiac stomach through the mouth and begin to digest prey externally. This is an especially useful maneuver when feeding on particularly large prey or when eating a clam through the small opening between its shells.

RESPIRATION AND CIRCULATION

As single cells, protozoans have high body surface to volume ratio. They inhabit wet or at least moist environments and are able to take care of gas exchange through simple diffusion. Even though flatworms, phylum *Platyhelminthes*, are multicellular, they also posses a large external surface area relative to their internal volume, so they, too, rely on simple diffusion across their body surface area for gas exchange. A highly branched gastrovascular cavity allows most cells to be in contact with the digestive system, which means that nutrients do not have to circulate to remote parts of the body. Metabolic wastes generally diffuse out across the body surface.

Sponges rely on their body cavity (the spongocoel) for circulation and gas exchange. Seawater with dissolved oxygen is pulled into the spongocoel through pores in the body wall called ostia. Flagella in specialized collar cells lining the spongocoel set up an internal current to provide circulation. Water and waste products then pass out of a larger opening, called an osculum.

Cnidarians also lack specialized respiratory structures. They rely on epidermal and gastrodermal surfaces for gas exchange. The gastrodermis lines the gastrovascular, main body cavity. As its name suggests, the gastrovascular cavity functions in both gastric capacity of digestion and the vascular role of circulation.

Most aquatic mollusks have comblike gills that function in gas exchange and also in filter feeding in some. Both aquatic and terrestrial snails in the subclass *Pulmonata* have highly vascularized structures that function as modified lungs. Mollusks typically posses a heart or similar pump that circulates fluid through an open circulatory system. In an open circulatory system the conduits, or vessels, are limited, and most organs are bathed directly in the circulatory fluid. Cephalopods, such as squid and octopuses, are mollusks with closed circulatory systems. In a closed circulatory system, blood is contained in vessels. The squid actually has three hearts. The systemic heart receives oxygenated blood from the gills and sends it to the tissues. The two branchial hearts pump deoxygenated blood back to the gills.

Arthropods have open circulatory systems. The circulatory fluid, called hemolymph, enters the heart through holes, called ostia. It is then pumped through short arteries and into the body cavity. There are a number of adaptations for gas exchange among the arthropods. Spiders have book lungs consisting of multiple stacked plates, like pages in a book. Insects have trachea, tubes that connect to the outside through holes (spiracles) in the exoskeleton. The trachea transport air directly between body tissues and the environment. Rounding out the arthropods, aquatic crustaceans respire using filamentous gills.

The water vascular system (WVS) is unique to the phylum *Echinodermata*. This system of canals services thousands of tube feet, in the starfish, for example. These feet extend through the body wall and have many other functions in addition to locomotion. Due to circulating fluid within the WVS, tube feet also are involved in gas exchange, waste excretion, chemoreception, and food collection. Another echinoderm, the sea cucumber, possesses internal rather than external respiratory structures. These respiratory trees attach to the cloaca, the common collection area for the exit of digestive and reproductive systems from the body. Gases are exchanged via the pumping of seawater in and out of the cloaca.

REPRODUCTION AND DEVELOPMENT

Reproduction among the invertebrates is almost as varied as the animals themselves. Cnidarians and sponges are just two of the many phyla that reproduce asexually, offspring arising by breaking off from the parent organism. Groups such as rotifers and some arthropods alternate between asexual and sexual reproduction, depending upon environmental conditions. The term parthenogenesis is used to describe development of an egg in the absence of fertilization. Parthenogenesis tends to occur during stable, favorable conditions. Sexual reproduction, the mixing of genes from two parents through uniting of egg and sperm, produces individuals with new combinations of genes, which may adapt them for survival under stressful conditions.

There are variations in sexual reproduction as well. Some invertebrates are hermaphrodites. These animals, such as earthworms, possess both male and female reproductive systems. Some can fertilize themselves; others cannot. When hermaphrodites mate with other members of their species, each individual donates and receives sperm, resulting in twice as many offspring per mating. Gonochoristic species have separate sexes, an individual being either male or female. This is the case for most insects.

In marine invertebrates, fertilization may be external. This involves broadcasting sperm and eggs into the surrounding seawater and relying on the ocean currents to bring the two together. Internal fertilization is the rule for freshwater and terrestrial invertebrates. Sperm may be transferred directly into the female's reproductive tract through a copulatory organ such as a penis. An indirect method of sperm transfer involves a package called a spermatophore. This chemical packet commonly provides nutrients for the female and her resultant offspring.

Aquatic larvae are commonly the dispersal stage in the life history of the particular invertebrate. Many times the adults are sedentary or even sessile, anchoring themselves to the ocean floor or remaining in self-constructed burrows. The larvae, on the other hand, are planktonic, relying on the ocean currents for transportation. A similar scenario exists with spiders. The young spin silken parachutes and use the wind currents to disperse away from the web of the mother spider. However, for many terrestrial invertebrates, the young represent the feeding stage. Lepidopteran caterpillars eat voraciously, and some adults, such as the luna moth, do not eat at all, living strictly off food stores acquired as a larva. Especially with the winged insects, it is the adults who colonize new areas.

—Sarah Vordtriede

Further Reading

Barnes, Robert D. *Invertebrate Zoology*. 5th ed. Saunders College Publishing, 1987.

Campbell, Neil A., Jane B. Reece, and Lawrence G. Mitchell. *Biology*. 5th ed. Benjamin/Cummings, 1999.

Knutson, Roger M. *Fearsome Fauna: A Field Guide to the Creatures That Live in You*. W. H. Freeman, 1999.

Pechenik, Jan A. *Biology of the Invertebrates*. 4th ed. McGraw-Hill, 2000.

Robertson, Matthew, ed. *The Big Book of Bugs*. Welcome Enterprises, 1999.

Isolating Mechanisms In Evolution

Fields of Study

Evolutionary Biology; Genetics; Ecology; Bioecology

Abstract

Isolating mechanisms (reproductive isolating mechanisms) prevent interbreeding between species. The term refers to mechanisms that are genetically influenced and intrinsic. Isolating mechanisms function only between sexually reproducing species. They have no applicability to forms that reproduce only by asexual means.

Key Concepts

allozyme: one of two or more forms of an enzyme determined by different alleles of the same gene; usually analyzed by gel electrophoresis

chloroplast deoxyribonucleic acid: a circular DNA molecule found in chloroplasts; chloroplasts are cytoplasmic organelles of green plants and some protists that carry out photosynthesis

fission: the division of an organism into two or more essentially identical organisms; an asexual process

hermaphrodite: an individual with both male and female organs; functions as both male and female

hybrid: a term that in the broad sense can apply to any offspring produced by parents that differ in one or more inheritable characteristics; used to denote offspring produced by a cross between different species

mitochondrial deoxyribonucleic acid (DNA): a circular molecule of DNA found in the mitochondria; mitochondria are cytoplasmic organelles that function in oxidative respiration

nuclear ribosomal deoxyribonucleic acid (DNA): nuclear DNA that codes for the ribosomal DNAs; ribosomes are small cytoplasmic particles that function in protein synthesis

DEFINING ISOLATING MECHANISMS

Isolating mechanisms (reproductive isolating mechanisms) prevent interbreeding between species., which was first used by Theodosius Dobzhansky in 1937 in his landmark book *Genetics and the Origin of Species*, refers to mechanisms that are genetically influenced and intrinsic. Geographic isolation can prevent interbreeding between populations, but it is an extrinsic factor and therefore does not qualify as an isolating mechanism. Isolating mechanisms function only between sexually reproducing species. They have no applicability to forms that reproduce only by asexual means, as by mitotic fission, stoloniferous or vegetative reproduction, or egg development without fertilization (parthenogenesis in animals). Obligatory self-fertilization in hermaphrodites (rare in animals) is a distortion of the sexual process that produces essentially the same results as asexual reproduction. Many lower animals and protists regularly employ both asexual and sexual means of reproduction, and the significance of isolating mechanisms in such forms is essentially the same as in normal sexual species.

PREMATING MECHANISMS

Reproductive isolating mechanisms are usually classified into two main groups. Premating (prezygotic) mechanisms operate prior to mating, or the release of gametes, and, therefore, do not result in a wastage of the reproductive potential of the individual. Postmating (postzygotic) mechanisms come into play after mating, or the release of gametes, and could result in a loss of the genetic contribution of the individual to the next generation. This distinction is also important in the theoretical sense in that natural selection should favor genes that promote premating isolation; those that do not presumably would be lost more often through mismatings (assuming that hybrids are not produced, or are sterile or inferior), and this could lead to a reinforcement of premating isolation.

Ethological (behavioral) isolation is the most important category of premating isolation in animals. The selection of a mate and the mating process depends upon the response of both partners to various sensory cues, any of which may be species-specific. Although one kind of sensory stimulus may be emphasized, different cues may come into play at different stages of the pairing process. Visual signals provided by color, pattern, or method of display are often of particular importance in diurnal animals such as birds, many lizards, certain spiders, and fish. Sounds, as in male mating calls, are often important in nocturnal breeders such as crickets or frogs but are also important in birds. Mate discrimination based on chemical signals or odors (pheromones) is of fundamental importance in many different kinds of animals, especially those where visual cues or sound are not emphasized; chemical cues also are often important in aquatic animals with external fertilization. Tactile stimuli (touch) often play an important role in courtship once contact is established between the sexes. Even electrical signals appear to be utilized in some electrogenic fish.

Ecological (habitat) isolation often plays an important role. Different forms may be adapted to different habitats in the same general area and may meet only infrequently at the time of reproduction. One species of deer mouse, for example, may frequent woods, while another is found in old fields; one fish species spawns in riffles, while another spawns in still pools. This type of isolation, although frequent and widespread, is often incomplete as the different forms may come together in transitional habitats. The importance of ecological isolation, however, is attested by the fact that instances in which hybrid swarms are produced between forms that normally remain distinct have often been found to be the result of disruption of the environment, usually by humans. Mechanical isolation is a less-important type of premating isolation, but it can function in some combinations. Two related animal species, for example, may be mismatched because of differences in size, proportions, or structure of genitalia.

Finally, temporal differences often contribute to premating isolation. The commonest type of temporal isolation is seasonal isolation: Species may reproduce at different times of the year. A species of toad in the eastern United States, for example, breeds in the early spring, while a related species breeds in the late spring, with only a short period of overlap. Differences can also involve the time of day, whereby one species may mate at night and another during the day. Such differences, as in the case of ecological isolation, are often incomplete but may be an important component of premating isolation.

POSTMATING MECHANISMS

If premating mechanisms fail, postmating mechanisms can come into play. If gametes are released, there still may be a failure of fertilization (intersterility). Spermatozoa may fail to penetrate the egg, or even with penetration there may be no fusion of the egg and sperm nucleus. Fertilization failure is almost universal between remotely related species (as from different families or above) and occasionally occurs even between closely related forms.

If fertilization does take place, other postmating mechanisms may operate. The hybrid may be inviable (F1 or zygotic inviability). Embryonic development may be abnormal, and the embryo may die at some stage, or the offspring may be defective. In other cases, development may be essentially normal, but the hybrid may be ill-adapted to survive in any available habitat or cannot compete for a mate (hybrid adaptive inferiority). Even if hybrids are produced, they may be partially to totally sterile (hybrid sterility). Hybrids between closely related forms are more likely to be fertile than those between more distantly related species, but the correlation is an inexact one. The causes for hybrid sterility are complex and can involve genetic factors, differences in gene arrangements on the chromosomes that disrupt normal chromosomal pairing and segregation at meiosis, and incompatibilities between cytoplasmic factors and the chromosomes. If the hybrids are fertile and interbreed or backcross to one of the parental forms, a more subtle phenomenon known as hybrid breakdown sometimes occurs. It takes the form of reduced fertility or reduced viability in the offspring. The basis for hybrid breakdown is poorly understood but may result from an imbalance of gene complexes contributed by the two species.

It should be emphasized that in most cases of reproductive isolation that have been carefully studied, more than one kind of isolating mechanism has been found to be present. Even though one type is clearly of paramount importance, it is usually supplemented by others, and should it fail, others may come into play. In this sense, reproductive isolation can be viewed as a fail-safe system. A striking difference in the overall pattern of reproductive isolation between animals and plants, however, is the much greater importance of premating isolation in animals and the emphasis on postmating mechanisms in plants. Ethological isolation, taken together with other premating mechanisms, is highly effective in animals, and postmating factors usually function only as a last resort.

FIELD STUDIES AND EXPERIMENTAL STUDIES

Field studies have often been employed in the investigation of some types of premating isolating mechanisms. Differences in such things as breeding times, factors associated with onset of breeding activity, and differences in habitat distribution or selection of a breeding site are all subject to direct field observation. Comparative studies of courtship behavior in the field or laboratory often provide clues as to the types of sensory signals that may be important in the separation of related species.

Mating discrimination experiments carried on in the laboratory have often been employed to provide more precise information on the role played by different odors, colors, or patterns, courtship rituals, or sounds in mate selection. Certain pheromones, for example, which act as sexual attractants, have been shown to be highly species-specific in some insects. The presence or absence of certain colors or their presentation has been shown experimentally to be important in mate discrimination in vertebrates as diverse as fish, lizards, and birds. Call discrimination experiments, in which a receptive female is given a choice between recorded calls of males of her own and another species, have demonstrated the critical importance of mating call differences in reproductive isolation in frogs and toads. Synthetically generated calls have sometimes been used to pinpoint the precise call component responsible for the difference in response.

Studies on postmating isolating mechanisms have most often involved laboratory crosses in which the degree of intersterility, hybrid sterility, or hybrid inviability can be analyzed under controlled conditions. In instances in which artificial crosses are not feasible, natural hybrids sometimes occur and can be tested. The identification of natural backcross products can attest incomplete postmating, as well as premating isolation. Instances of extensive natural hybridization are of special interest and have often been subjected to particularly close scrutiny. Such cases often throw light on factors that can lead to a breakdown of reproductive isolation. Also, as natural hybridization more often occurs between marginally differentiated forms in earlier stages of speciation, new insights into the process of species formation can sometimes be obtained. Finally, such studies may yield information on the evolutionary role of hybridization, including introgressive hybridization, the leakage of genes from one species into another. Morphological analysis has long been used in such cases, and chromosomal studies are sometimes appropriate. In recent years, allozyme analysis by gel electrophoresis has become a routine tool in estimates of gene exchange, and molecular analysis of nuclear deoxyribonucleic acid (DNA), or mitochondrial DNA, have been useful. As mitochondria are normally passed on only maternally, their DNA can also be used to identify cases in which females of only one of the two species has been involved in the breakdown of reproductive isolation.

Investigations of the role of natural selection in the development and reinforcement of reproductive isolation have employed two different approaches. One has involved the measurement of geographic variation in the degree of difference in some signal character (call, color, or pattern, for example) thought to function in premating isolation between two species that have overlapping ranges. If the difference is consistently greater within the zone of overlap (reproductive character displacement), an argument can be made for the operation of reinforcement. Another approach has involved laboratory simulations, usually with the fruit fly *Drosophila*, in which some type of selective pressure is exerted against offspring produced by crosses between different stocks, and measurement is made of the frequency of mismatings through successive generations. The results of such studies to this time are contradictory, and the role of selection with regard to development of reproductive isolation requires further study.

ENHANCING REPRODUCTIVE EFFICIENCY

The efficiency of reproduction in most animals is enhanced immeasurably by premating isolating mechanisms. Clearly, in animals a random testing of potential mates without regard to type is totally unacceptable for most species in terms of reproductive capacity and time and energy resources. Premating isolation in this sense is a major factor in promoting species diversity in animal communities.

Both premating and postmating isolating mechanisms are also critical to the maintenance of species diversity in that they act to protect the genetic integrity of each form: A species cannot maintain its identity without barriers that prevent the free exchange of genes with other species. Furthermore, a species functions as the primary unit of adaptation. Every species in a community has its own unique combination of adaptive features that enable it to exploit the resources of its environment and to coexist with other species with a minimum of competition. The diversity of different species that can coexist in the same area depends upon the unique "niche" that each occupies; adaptive features that determine that niche are based on the unique genetic constitution of each species, and this genetic constitution is protected through reproductive isolation.

The development of reproductive isolating mechanisms is also critical to the formation of new species (speciation), and ultimately to the development of new organic diversity. The most widely accepted, objective, and theoretically operational concept for a sexual species is the biological species concept. Such a species can be defined as population or group of populations, members of which are potentially capable of interbreeding but which are reproductively isolated from other species. The origin of new species, therefore, depends upon the development of reproductive isolating mechanisms between populations. A major focus of research in evolutionary biology and systematics has been, and continues to be, on the various factors that influence the development of reproductive isolating mechanisms.

—*John S. Mecham*

FURTHER READING

Baker, Jeffrey J. W., and Garland E. Allen. *The Study of Biology*. 4th ed. Addison-Wesley, 1982.

Dobzhansky, Theodosius. *Genetics of the Evolutionary Process*. Columbia University Press, 1970.

Dobzhansky, Theodosius, Francisco J. Ayala, G. Ledyard Stebbins, and James W. Valentine. *Evolution*. W. H. Freeman, 1977.

Futuyma, Douglas J. *Evolutionary Biology*. 3rd ed. Sinauer Associates, 1998.

Mayr, Ernst. *Populations, Species, and Evolution*. The Belknap Press of Harvard University Press, 1970.

J

Jellyfish

Fields of Study

Ichthyology; Oceanography; Limnology; Marine Ecology

Abstract

Jellyfish species are found in all of Earth's oceans. They are almost entirely marine, although freshwater species are known. Many species appear to reproduce once a year. Polyps usually develop over a period lasting a few months but may live for several years producing clones; adult medusa forms live two to six months. They have an umbrella-like body; no head or skeleton; composed of outer epithelial layer and inner gastrodermis layer with thick elastic jellylike substance between them; gastrovascular cavity; specialized ring of epitheliomuscular cells that pulses rhythmically to propel the animal through the water; four to eight oral arms; tentacles bearing cnidoblasts containing nematocysts.

Key Concepts

ciliated: bearing short, hairlike organelles on the surface of cells, used for motility
cnidoblasts: specialized cells on the body or tentacles of jellyfish that contain nematocysts
medusa: adult umbrella- or bell-shaped forms of jellyfish, with mouth facing downward
mesoglea: gelatinous material lying between the inner and outer layers of a jellyfish
nematocyst: stinging structures containing barbs and/or poison
planula: free-swimming, ciliated jellyfish larva
polyp: immature cylindrical forms of jellyfish with mouth facing upward

MEET THE JELLYFISH

The gelatinous jellyfish are widespread in marine environments, although they are most common in tropical and subtropical regions. These ancient animals first appeared on earth over 650 million years ago. The smallest jellyfish are difficult to see without a microscope, while the largest known jellyfish is 2.5 meters in diameter; some jellyfish may have tentacles over 100 feet in length. The body plan of jellyfish is relatively primitive and contains less than 5 percent solid organic matter, the remaining bulk coming from water. They completely lack internal organs. The bell-like jellyfish bodies are composed of an outer layer of epidermis and an inner layer of gastrodermis that lines the gut. The gut has a single oral opening. Between the two layers is the mesoglea, which contains few cells and has a low metabolic rate. Four to eight oral arms are located near the oral opening and are used to transport food that has been captured by the tentacles.

Jellyfish are able to exert minimal control over their movement, being largely at the mercy of ocean currents. Jellyfish do, however, have some regulation over vertical movement. They possess a ring of muscles embedded on the underside of the bell that pulses rhythmically, pushing water out of the hollow bell. Using this jet propulsion, jellyfish can change their position in the water column, moving in response to light and prey.

FEEDING STRATEGY OF JELLYFISH

Jellyfish are simple but specialized carnivores. Although jellyfish have a low metabolic rate, they have the ability to capture large prey. These two characteristics allow jellyfish to survive in environments where prey are scarce. Jellyfish are equipped with a specialized apparatus, the cnidoblast, for defense and feeding. Cnidoblasts are found by the hundreds or thousands on the tentacles and sometimes on the body surface. Within each cnidoblast is a coiled harpoonlike nematocyst that is discharged by the presence of potential prey. The nematocyst injects poison into the prey as spines on the nematocyst anchor it to the prey. The trapped, paralyzed prey is pulled back by the tentacles and stuffed into the gastrovascular cavity to be digested. Jellyfish do not attack humans, but humans may receive stings if they encounter jellyfish. The effects of jellyfish poison on humans can range from a mild, itchy rash to death.

JELLYFISH REPRODUCTION

Most jellyfish proceed through several distinct stages in their life cycles. Male medusae produce sperm that are released from the oral opening into the oral opening of the female. The female then releases fertilized eggs which develop into slipper-shaped, solid masses of ciliated cells called planula which move through the water and eventually settle onto a solid surface. From these settled planula develop polyps that have cylindrical stalks attached to the substrate, with tentacles surrounding their mouths. At this stage, the polyps resemble sea anemones. The polyps divide and bud into tiny young jellyfish (ephyra) which are often carried far from the parent polyp by ocean currents. The ephyra develop into mature medusae over several weeks. The medusae normally live three to six months. Lack of complex physical features helps jellyfish adapt to extremes in temperature, pH, oxygen, and light. The polyps can remain dormant on the ocean floor for years awaiting favorable conditions to reproduce. As global fish populations decrease due to overfishing and environmental stressors, and other marine predators of the jellyfish decline, jellyfish gain dominance. This can lead to jellyfish blooms—massive congregations of the floating jellyfish—that further compromise fish populations and marine ecosystems.

—*Roger Smith*

FURTHER READING

Brusca, Richard C. C., and Gary J. Brusca. *Invertebrates*. Sinauer Associates, 1997.

Buchsbaum, Ralph Morris, Vicki Pearse, John Pearse, and Mildred Buchsbaum. *Animals Without Backbones*. Chicago University Press, 1987.

Gershwin, Lise-Ann. *Stung! On Jellyfish Blooms and the Future of the Ocean*. Chicago University Press, 2013.

Graham, William M., et al. "Linking Human Well-Being and Jellyfish: Ecosystem Services, Impacts, and Societal Responses." *Frontiers in Ecology and The Environment*, Nov. 2014, vol. 12, no. 9, pp. 515–23, *EBSCOhost*. Accessed 31 Jan. 2018.

Landau, Elaine. *Jellyfish*. Children's Press, 1999.

Stefoff, Rebecca. *Jellyfish*. Benchmark Books, 1997.

Twig, George C. *Jellies: The Life of Jellyfish*. Millbrook Press, 2000.

Lactation

Fields of Study

Evolutionary Biology; Reproductive Biology; Animal Physiology

Abstract

Lactation is the process by which female mammals produce milk to feed their young. The ability to produce milk is one of the defining characteristics of the class Mammalia. *All mammals, but no other animals, possess the highly specialized glands necessary for lactation. In evolutionary terms, the appearance of lactation coincides with the tendency of mammals to produce only a few offspring at a time; the provision of milk for these offspring helps to ensure their survival while removing competition between the adults and the young for food.*

Key Concepts

alveoli: the milk-producing areas within the mammary glands

colostrum: the precursor to milk that is formed in the mammary gland during pregnancy and immediately after birth of the young

ducts: the tubular structures that carry milk from the alveoli to the outside through the nipple or teat

lactation: the process of producing and delivering milk to the young; also, the time period during which milk is produced

mammary glands: the milk-producing glands found in all mammals; for example, the cow's udder contains the mammary glands

milk ejection: also known as milk letdown, this is the reflex response of the mammary gland to suckling of the nipple; the hormone oxytocin mediates this reflex

myoepithelial cells: the specialized cells within the mammary gland that surround the alveoli and contract to force milk into the ducts during milk ejection

nipple: the raised area on the surface of the skin over the mammary gland that contains the duct openings

teat: an elongated form of nipple that contains one duct opening

THE MILK OF LIFE

Lactation is the process by which female mammals produce milk to feed their young. The ability to produce milk is one of the defining characteristics of the class *Mammalia*. All mammals, but no other animals, possess the highly specialized glands necessary for lactation. Certain spiders and other insects have been observed to feed young with milk-like secretions that appear on their exoskeletons, but this is not the same as mammalian lactation. In evolutionary terms, the appearance of lactation coincides with the tendency of mammals to produce only a few offspring at a time; the provision of milk for these offspring helps to ensure their survival while removing competition between the adults and the young for food.

THE MAMMARY GLANDS

The mammary glands are the milk-producing organs. The number varies among species from two to about twenty, with a rough correlation between the number of young born and the number of glands present. The glands are located on the ventral surface of the body, either in the thoracic (in humans, for example) or abdominal region (in horses and cows) or in two lines extending almost the length of the body (in dogs and rodents). Both male and female mammals have mammary glands, because in early mammalian development the basic body plan of male and female embryos is identical. The mammary glands of males are nonfunctional, however, since they lack the hormonal stimulation necessary for lactation.

Internally, the mammary glands of all mammals follow the same basic plan, consisting of alveoli that produce milk and ducts that carry the milk to openings on the surface of the skin. The alveoli are surrounded by myoepithelial cells that contract to squeeze the milk into the ducts during suckling by the young.

Externally, considerable variation exists among the mammals in the appearance of the mammary glands and their associated openings. In the spiny anteater and platypus, the many lobes of the mammary glands each open directly to the surface of the abdominal skin through individual ducts, and the young suck the hair-covered skin to obtain the milk. In other mammals, the mammary glands are more obvious as swellings beneath the skin, with a raised area, the nipple or teat, that contains the duct openings. In some four-legged animals (cows, horses, and goats) the mammary glands are located in a baglike structure called the udder, from which are suspended the elongated teats. In humans, the nipple, which contains the openings of fifteen to twenty-five ducts, is surrounded by pigmented skin, the areola. The areola contains glands (tubercles of Montgomery) that secrete a lubricating fluid.

MILK PRODUCTION

Lactogenesis (milk production) does not begin until a female has produced young. During pregnancy, a complex of hormones prepares the mammary glands for milk production by promoting their growth and internal development. These hormones include prolactin from the mother's anterior pituitary gland, placental lactogen from the placenta within the uterus, and estrogen and progesterone, which are produced in the corpus luteum of the mother's ovary and in the placenta. Other hormones, including cortisol from the adrenal gland, thyroxine from the thyroid gland, and insulin from the pancreatic islets, may also be involved. Progesterone appears to participate in the induction of mammary development, but, paradoxically, it also prevents milk secretion during pregnancy.

Although true milk is not produced during pregnancy, a precursor to milk, colostrum, can be produced in small amounts by the mammary glands of most species. Colostrum is a sticky, yellowish, transparent liquid. Colostrum secretion continues in the first few days after birth of the young; there is then a gradual transition to production of true milk.

Milk contains water, proteins, fats, vitamins, minerals, and a unique sugar, lactose. The exact concentration of the various components varies greatly between species according to the nutritional demands of the young. The milk of seals is high in fat and other solids that contribute to rapid weight gain in the pups, a strategy that appears to be essential for their survival.

Noteworthy among the constituents of milk are antibodies produced by the mother. These antibodies help protect the newborn from disease in the period when the newborn's own immune system is immature and incapable of providing significant defense. The antibody concentration of colostrum is higher than that of true milk, and for this reason the first few days of nursing are considered the most important for immunological protection of the newborn.

The transition in production from colostrum to true milk is brought about by a change in the hormonal status of the mother. At the time of birth, the placenta is expelled from the mother's body, thus removing the source of progesterone, estrogen, placental lactogen, and other hormones. The decrease in progesterone levels is thought to be essential for the onset of lactogenesis. In addition, at the time of birth, there are changes in prolactin secretion that may play a role in initiating milk secretion.

SUCKLING

Once lactogenesis is established, a set of hormonal reflexes act to match milk production and delivery to the needs of the newborn. Suckling of the nipple involves motions similar to chewing as the infant takes the nipple between the tongue and the palate. This suckling motion stimulates nerve endings in the mother's nipple that relay signals about the stimulation back to the mother's brain. Within thirty to sixty seconds, these signals result in the release of prolactin from the mother's anterior pituitary gland and oxytocin from her posterior pituitary gland. Prolactin causes continued production of milk by the alveolar cells of the mammary glands. Oxytocin acts immediately on the myoepithelial cells of the mammary gland, causing them to contract and push milk from the alveoli into the ducts and thence through the nipple into the infant's mouth. Thus, the infant does not actually remove milk from the mammary gland by suction, but instead is responsible for promoting a hormonal reflex that results in active milk ejection, or letdown, from the mammary gland.

Because of the operation of the prolactin and oxytocin reflexes during suckling, lactation is a biological example of the principle of supply meeting demand. All that is necessary to increase milk production is to increase the suckling stimulus by nursing the young more often. Once established, lactation in some species can be sustained in this manner for years, assuming the nutritional needs of the mother are met. On the other hand, if the mother fails to nurse her offspring, the absence of the suckling stimulus will cause the mammary glands gradually to cease milk production.

The exact composition of the milk is altered as lactation continues to meet the changing nutritional needs of the growing offspring. The most extreme example of the ability of the mammary gland to change the composition of milk is seen in the kangaroo. In this animal, the newborn attaches to a teat in the mother's pouch shortly after birth and remains there for a month or more. A mother kangaroo may nurse offspring of different ages from separate teats, and each teat supplies a milk with the appropriate nutritional composition for that young.

Species vary in time spent suckling the young. The rabbit nurses her litter for only about five minutes once a day, while the rat nurses for about half an hour at a time, at intervals throughout the day. Lactation lasts about ten days in rodents, but it may persist for months in large species such as horses and cows. Continued lactation has a suppressive effect on ovulation that is thought to be attributable to interference by prolactin with the normal hormonal mechanisms that cause ovulation.

THE STUDY OF LACTATION

Although it has always been clear that milk is expelled from the mammary glands, the realization that the glands themselves actually produce the milk is a relatively recent one. Early anatomists erroneously assumed that milk must be a product of the uterus, since the uterus is involved in support and nourishment of the fetus. Thus, much of the early anatomical work attempted to show some sort of connection between the uterus and the mammary glands. It was not until the late 1800s that the light microscope clearly demonstrated that milk is formed within the mammary glands. In the twentieth century, electron microscopy showed that during pregnancy the intracellular organization of the alveoli becomes increasingly more complex as the cells become capable of milk secretion.

Various techniques for labeling compounds with radioactive or fluorescent markers have been used in conjunction with electron microscopy to examine how milk is synthesized in the alveoli. The alveoli cells extract necessary precursors from blood flowing through the mammary gland, assemble the precursors into milk components, and then secrete the constituents of milk into the mammary gland ducts. Specific routes of secretion have been identified for the major components of milk.

More recently, researchers have used cell-free systems to study the biochemical pathways involved in milk synthesis. These systems use isolated fragments of deoxyribonucleic acid (DNA), ribonucleic acid (RNA), and perhaps some cell organelles to examine the intermediate chemical steps in the synthesis of milk components. Using these techniques, researchers have been able to "watch" as complex milk proteins and constituents are assembled step by step. The knowledge of how the components of milk are assembled is leading to a fuller understanding of how the amounts of these substances in milk are hormonally regulated.

Knowledge of the hormones involved in inducing and maintaining lactation has come about through systematic assembly of information from several lines of research. Test animals can be treated with a specific hormone to determine if that hormone causes or suppresses lactation. The test animals may be males or immature females, with the goal being the duplication of the specific mix of hormones that cause lactation in the adult female. The opposite approach may also be taken: An endocrine gland can be removed from a lactating female to determine if the hormonal products of that gland are necessary for lactation. Another approach is to make careful measurements of the levels of hormones circulating in the blood as the lactational state changes; any hormone that shows a correlated change may be a good candidate for further investigation by treatment or removal from test animals. These methods have led to an understanding of the importance of prolactin, placental lactogen, oxytocin, estrogen, and progesterone in promoting lactation, but researchers still do not understand how the system is fine-tuned.

For example, considerable variation exists in the volume and quality of milk produced by different individuals—or by the same individual at different times—but these differences cannot currently be explained by any known change in hormone levels. Research is focusing not only on describing changes in circulating levels of hormones but also on elucidating the exact effects of these hormones on the biosynthetic pathways within the mammary gland.
—Marcia Watson-Whitmyre

FURTHER READING

Larson, Bruce L., and Vearl R. Smith, eds. *Lactation: A Comprehensive Treatise.* 4 vols. Academic Press, 1974.

Mepham, T. B., ed. *The Biochemistry of Lactation.* Elsevier, 1985.

Peaker, M., R. G. Vernon, and C. H. Knight, eds. *Physiological Strategies in Lactation.* Academic Press, 1984.

LANGUAGE

FIELDS OF STUDY

Animal Physiology; Developmental Biology; Sociology; Anthropology

ABSTRACT

One way to understand humans' position in the natural order of evolution is to locate the origins of language. Human language is the only animal communicative system that possesses all of the fundamental characteristics of arbitrary symbols, semanticity, grammar, productivity, duality of patterning, and displacement. Furthermore, human children are able to become effective users of such complex symbolic systems without formal teaching and within a fairly short period of time, in striking contrast to the limited expressions of animals, even after lengthy and extensive training.

KEY CONCEPTS

closed-class vocabulary: typically including the structural and functional words, such as prepositions, determiners, quantifiers, and morphological markers, closed in the sense of resisting the introduction of new members

displacement: language's power to refer to or describe things and events beyond the constraints of the here and now

grammar: the structure of a language, consisting of systematic rules to specify word formation, such as inflection, derivation, and compound words (morphology), and systematic rules to specify how words should be ordered in combination to form phrases and sentences (syntax)

open-class vocabulary: content words such as nouns, verbs, and adjectives, open in the sense of its readiness to admit new members

productivity or generativity: language's power to produce or generate an infinite number of understandable words and sentences from a finite number of symbols and rules

semanticity: meaning in language

symbol: something that stands for something else, the connection between symbol and object being arbitrary in nature

WHY DO WE SAY WHAT WE SAY?

One way to understand humans' position in the natural order of evolution is to locate the origins of language. Some scholars think that language is probably the product of the "mental mutation" of the large brain of *Homo sapiens*, because human language is the only animal communicative system that possesses all of the fundamental characteristics of arbitrary symbols, semanticity, grammar, productivity, duality of patterning, and displacement. Furthermore, human children are able to become effective users of such complex symbolic systems without formal teaching and within a fairly short period of time, in striking contrast to the limited expressions of animals, even after lengthy and extensive training. Linguist Noam Chomsky, for example, has proposed that human beings have a unique language forming capacity and human babies are innately equipped with a "Language Acquisition Device," which resides somewhere in the brain.

Many others, however, disagree with this discontinuity theory of the human language origins. Instead, they argue for an evolutionary ground (neuroanatomical, behavioral, cognitive, social, and cultural) to cultivate language formation. They believe that language is no exception from the governing of the law of evolution. For example, Philip Lieberman rejects the existence of linguistic genes or a language organ in the human brain. Rather, language had its first sprout at the intersection of the evolutionary products of neural mechanisms, communication, and cognition. These are the two main theories of language origins. The first line builds on the protolanguage theory, the second on the notion of behavior determinism in evolution.

PROTOLANGUAGE THEORY

"Protolanguage" means that utterances are not yet full language, although they serve symbolic referential and other communicative functions. The telegraphic speech of toddlers (such as "Mommy cookie!") has been used as an example. A "baby talk" was formed first, only to be shaped and refined later into a full language. It was hypothesized that modern speech could have been possible about 100,000 years ago, when the earliest *Homo sapiens* started to migrate from Africa to other places. John McCrone reasons that toolmaking and tool use, moving in troops to other regions, and collective hunting called for more group actions, which in turn promoted social interaction. These activities required joint attention and intentional communication. Using eye contact and gesture to direct attention as a means beyond reflexive behaviors to achieve joint attention might thus have been a major step toward human speech. The prolonged period for taking care of dependent human infants, a consequence of brain growth, afforded opportunities to cultivate the intimacy between mother and child. Meanwhile, social ties in a colony began to form. All these changes might have encouraged what could be called personalized noises. To meet their communicative needs, early *Homo* species were pressured to refine and stabilize their coarse communicative noises into protowords. These protowords were then passed down to the next generations. Practice of such vocalizations in turn further promoted vocal structure refinement.

The evolutionary principle of economy and efficiency was at work, so that concept categories (words) and combinatorial rules (grammar) were naturally selected, because words and grammar are far more cognitively economic and efficiently generative than mechanical one-to-one referential associations. Martin Nowak and David Krakauer, based on their mathematical and computational modeling using computer simulations, have contended that protolanguages can evolve in a nonlinguistic society. At first, signal-object associations were established, and later, combinations of sounds to form words and combinations of words to form sentences evolved into semantic and syntactic systems through natural selection to reduce mistakes in communication.

Howling wolf, in a molt stage (Canis lupus).

In addition to the supportive results of computer simulations, other empirical evidence came from studies involving primates and human children. In Leavens and Hopkins's study, 115 chimpanzees (*Pan troglodytes*) in captivity (aged three to fifty-six years), without any explicit training whatsoever, commonly employed gaze alterations and a pointing gesture in face-to-face communicative interactions with humans and among themselves. This demonstrated the presence of communicative intent and gestural precursors of language among humans' closest relatives, chimpanzees. The bonobo (*Pan paniscus*) Kanzi understands spoken English sentences and knows how to use human-designated lexigrams to announce his intention, all through laissez-faire learning without explicit teaching. The chimpanzee Washoe spontaneously taught her acquired American Sign Language (ASL) to her adopted son Loulis. Human babies are like chimpanzees in many ways. As McCrone describes it, human infants are born with the standard ape vocal plan—it is after six months that the voice box descends down into the throat. Human infants first play with vowels (coo) and then babble (combining vowels with consonants)—so vocalization comes before producing true words. They also employ gaze alterations and gestures to communicate before they say words. In the second year, they produce telegraphic speech, typically composed of two to three content words, which happens to be the mode of expression in the utterances of language-trained chimps (such as "Shirt hide" by Kanzi).

BEHAVIOR DETERMINISM

William Noble and Iain Davidson do not think that a protolanguage existed in evolution. They have also cautioned people against accepting the performance of animals in captivity and human interaction experiences as evidence to back up evolutionary arguments, because these environments are drastically different from the ecologies of the *Homo* ancestors of millions of years ago. These environments are not the same as those of the free-living primates, either. Language-trained animals' performance is like language emergence in human infants, who learn through interaction with other humans who already have language, a learning process quite different from the prehistoric origins of symbols and language from scratch.

Noble and Davidson agree that natural selection favored bipedalism, leading to neuroanatomical changes including larger brains. They do not believe, however, in biological determination of behaviors. Instead, they believe the opposite. The behavior of standing upright led to larger brains that needed to consume more energy. Meat as a good energy source had already been increased in the diet of *Homo erectus*. As meat-eaters, hominids had to run fast (either to catch prey or to escape predators). Running brought about better control of the breathing system (necessary for speaking) and adjustments to the thermoregulatory system (leading to the selection of the feature "hairlessness," which could have fostered face-to-face adult-infant interaction). In addition, hunting for meat facilitated coordinated group actions as well as tool creation. To expand food sources hominids began migrating, which further promoted groups and interaction. Thus the social context was present for the emergence of language.

In their discussion, Noble and Davidson have emphasized one important behavior responsible for the emergence of language: stone throwing. To be effective, manual control and timing control had to be achieved. As the timing control behaviors were bettered, the neuroanatomical structures improved too. These positive adjustments, together with other contextual changes, led hominids on an increasingly divergent behavioral path from their chimpanzee relatives. Better control of the forearm could develop into a pointing form. Hairlessness made it easy to carry an infant hominid in front, increasing the likelihood of adult-baby mutual observation and imitation. One such likely behavior could be arm extensions for referring. Later, better-controlled movements of forearms and fingers developed into a pointing form. Any vocalizations in company with manual gestures were first associated with the referred objects, and later became symbols, once it was realized that the sounds alone could stand for the targets themselves, even when they were out of sight.

There is consensus that the human vocal structure is a necessary condition for human speech. It has been noticed that animal vocalizations are graded in nature. A graded system contains only variations of vowel sounds but no consonants. Variations of vowels, although functional in communication, lack distinctive boundaries to mark different categories. Sue Savage-Rumbaugh and Roger Lewin have concluded that language is unlikely to emerge unless an organism can produce consonants, no matter how large its brain. How the human vocal tract acquired the ability to pronounce consonants unfortunately remains a mystery.

The evolution of language must have benefited the *Homo* species tremendously. Noble and Davidson have speculated how language could have contributed to human mentality. For example, colonization in different places would cause isolated groups to have trouble in understanding each other. Such failures and misunderstandings could contribute to the awareness of "us" vs. "them." This appreciation would lead to the realization of the possibility for a group to use their own symbolic system as a means for social control. Thus, human mentality, with language in use, is itself an evolving feature of the natural world.

It is only logical that, with language available, mental representation of the world became possible, which eventually made abstract, imaginary, retrospective, hypothetical, and metacognitive thinking a reality. No wonder these modes of thinking reflect themselves in the characteristics of human language.

ANIMAL COMMUNICATION

Animals do communicate, at least in a broad sense. Animals use vocalizations, facial expressions, gestures, body postures and movements, and even odors, to warn peers, to attract attention, to find food, to care for the young, to mark territories, and to maintain social structures. However, animal communicative systems are typically not recognized as language because of they lack the key features of a true language. Many have argued that animal communication, even among chimpanzees, is in essence instinctive and reflexive. McCrone says that these behaviors are not under conscious control, and are triggered only by an event in the immediate environment, with both parties present. Hence, chimpanzees have no true arbitrary symbols or displacement. In addition, animal communicative systems are "closed," with no combinatorial rules to create new meanings; hence they lack duality of patterning, syntax, and productivity. Edward Kako has pointed out that no animals so far, including the language-trained ones, have demonstrated the ability to understand closed-class lexical items. Despite the criticisms, animals' language-learning achievements have been acknowledged.

TALKING ANIMALS

The most famous talking parrot is Alex, an African gray parrot (*Psittacus erithacus*) trained by Irene Pepperberg. Alex is able to speak many words and phrases referring to objects, materials, actions, colors, shapes, numbers, and locations. He can answer questions that require labeling objects, classifying objects (color and substance), comparing objects ("bigger than" or "same as/different from"), and counting (from one to six).

Two Hawaii bottle-nosed dolphins (*Tursiops truncatus*), trained by Louis Herman and his colleagues, are Phoenix, with an acoustic language, and Akekamai (Ake), with a gesture-based language. They can correctly carry out commands in varied word orders (syntax) with different meanings. They have further demonstrated their semantic and grammatical knowledge by either not executing grammatically incorrect and semantically nonexecutable orders or by extracting an executable segment from an anomalous string and then completing the task according to the meaning provided in that specific syntactic structure.

Sue Savage-Rumbaugh's bonobo Kanzi is an exciting language star. Kanzi understands spoken English and uses hand signals as well as geometric symbols, and moreover, he has learned all that without explicit instruction but by mere social observation and interaction, just as a human baby learns a language. This natural learning process was successfully replicated with Kanzi's sister, Mulika.

COMPARATIVE LANGUAGE RESEARCH

Research methodology in the comparative language field has been improved greatly over the years. Designation of "language" takes into consideration the biological constraints of the species involved. Social interaction in a natural way is underscored. Possible experimenters' and trainers' biases are controlled through blind techniques such as blindfolding the eyes of the person who gives commands, using one-way mirrors and remote cameras, or separating the person who does recording and interpretation from the one who gives the command. Recent data are thus more scientifically sound than before. Yet some people are not happy with the fact that humans have imposed their dialogue on the animals. These people are now using the playback technique to decode the meaning of the signals in animals' own communicative systems. Many species have been studied in their natural ecological niches, including vervet monkeys, tigers, humpback whales, orcas or killer whales, and elephants. These animals' wild calls in nature are recorded, and then are played back to the animals to see their differentiated reactions, thus making the message decoding possible. The playback studies are very encouraging in confirming the symbolic nature of animals' natural "languages" in the wild.

It is very important to study the animal's own "language" for its own sake, for without such knowledge, commenting on nonhuman species' linguistic abilities in the frame of human language is at least prejudiced. As primatologist and psychologist Roger Foutes put it: "The best approach to science is a humble one. We are humble enough to take the animals we are studying on their own terms and allow them to tell us about themselves. Too often science takes an arrogant approach." According to him, "Someday we'll realize that the human voice is not a lone violin but part of an orchestra. We're not playing a solo; instead it's a symphony."

—*Ling-Yi Zhou*

Further Reading

Chomsky, Noam. *Aspects of the Theory of Syntax*. MIT Press, 1965.

Gardner, R. Allen, Beatrix T. Gardner, and Thomas E. Van Cantfort, eds. *Teaching Sign Language to Chimpanzees*. State University of New York Press, 1989.

Kako, Edward. "Elements of Syntax in the Systems of Three Language-Trained Animals." *Animal Learning and Behavior* 27, no. 1 (1999): 1–14.

Leavens, David A., and William D. Hopkins. "Intentional Communication by Chimpanzees: A Cross-Sectional Study of the Use of Referential Gestures." *Developmental Psychology* 34, no. 3 (1998): 813–822.

Lieberman, Philip. *The Biology and Evolution of Language*. Harvard University Press, 1984.

McCrone, John. *The Ape That Spoke: Language and the Evolution of the Human Mind*. William Morrow, 1991.

Noble, William, and Iain Davidson. *Human Evolution, Language, and Mind: A Psychological and Archaeological Inquiry*. Cambridge University Press, 1996.

Nowak, Martin A., and David C. Krakauer. "The Evolution of Language." *Proceedings of the National Academy of Sciences of the United States* 96, no. 14 (July 1999): 8028–8033.

Patterson, Francine, and Eugene Liden. *The Education of Koko*. Holt, Rinehart and Winston, 1981.

Premack, David. "Language in Chimpanzee?" *Science* 172 (1971): 808–822.

Rumbaugh, Duane M., and E. Sue Savage-Rumbaugh. "Language in Comparative Perspective." In *Animal Learning and Cognition*, edited by N. J. Mackintosh. Academic Press, 1994.

Savage-Rumbaugh, Sue, and Roger Lewin. *Kanzi: The Ape at the Brink of the Human Mind*. John Wiley & Sons, 1994.

Starr, Doug. "Calls of the Wild." *Omni* 9, no. 3 (December 1986): 52–59, 120–123.

Learning

Fields of Study

Ethology; Developmental Biology; Evolutionary Biology; Psychology

Abstract

Learning is any change or modification in behavior that is directed by previous experience and involves the nervous system but cannot be attributed to the effects of development, maturation, fatigue, or injury. These latter phenomena contribute to changes in behavior that generally do not constitute learning. Learning takes a number of forms, including habituation, sensitization, associative learning, perceptual or programmed learning, and insight.

Key Concepts

adaptation: any heritable characteristic that increases the probability that an animal will survive and reproduce in its natural environment

conditioning: the behavioral association that results from the reinforcement of a response with a stimulus

innate: any inborn characteristic or behavior that is determined and controlled largely by the genes

instinct: any behavior that is completely functional the first time it is performed

natural selection: the process of differential survival and reproduction that leads to heritable characteristics that are best suited for a particular environment

stimulus: any environmental cue that is detected by a sensory receptor and can potentially modify an animal's behavior

WHAT IS LEARNING?

Learning, as defined by ethologists, is simply any change or modification in behavior that is directed by previous experience and involves the nervous system but cannot be attributed to the effects of development, maturation, fatigue, or injury. These latter phenomena contribute to changes in behavior that generally do not constitute learning. Learning takes a number of forms, including habituation, sensitization, associative learning, perceptual or programmed learning, and insight. Each type has its own basic characteristics and adaptive significance. Habituation and sensitization are the simplest and most widespread forms of learning, and insight is the most complex and least understood form. Insight involves the ability to put two previous experiences together to solve an unrelated problem.

HABITUATION AND SENSITIZATION

Habituation and sensitization are considered the simplest forms of learning. Habituation involves a decrease in a behavioral response that results from repeated presentation of a stimulus. A young, naïve duck, for example, will exhibit an innate startle response when any hawk-shaped object is passed overhead. With repeated presentation of the hawk model, however, the intensity of the bird's reflex declines as the animal becomes habituated, or learns that the stimulus has no immediate significance. Habituation learning is common throughout the animal kingdom and has tremendous adaptive significance in that it prevents repeated response to irrelevant stimuli that could otherwise overwhelm the animal's senses and prevent it from accomplishing other critical tasks. One of the common characteristics of habituation is that after a short period, usually defined by the particular species and the stimulus in question, the animal will completely recover from the habituation experience and will again exhibit a full response to the stimulus. This, too, has important survival implications, especially for species that rely on stereotypic alarm responses for avoiding predation.

In contrast to habituation, sensitization is the increase in intensity of a response that results from the repeated presentation of a stimulus. A good example is the heightened sensitivity to even relatively soft sounds that results from the initial presentation of a loud, startling noise, such as a gunshot. Sensitization differs from habituation in important ways. First, the specific stimulus that elicits a sensitization response is different from the stimulus to which the animal becomes sensitized. Second, the underlying physiological mechanisms that control these two processes are fundamentally different.

The third and broadest category of learning is associative learning. In this type of learning, an animal makes a connection between some primary environmental stimulus (that involves either a reward or punishment) and a novel or neutral stimulus that is paired with the first stimulus.

CLASSICAL AND OPERANT CONDITIONING

The simplest form of associative learning is classical conditioning, first studied by Ivan Pavlov. Pavlov observed that when a dog is presented with food, the dog will begin to salivate. He referred to the food in this case as an unconditioned stimulus (US), and to the salivation reflex as the unconditioned response (UCR). When the unconditioned response is effectively paired with a second, novel stimulus, such as a light or bell (called the conditioned stimulus, or CS), the dog will, after several trials, associate this second stimulus (CS) with the US and begin to salivate whenever the CS is presented. The salivation reflex that occurs following presentation of the CS is termed the conditioned response (CR).

Although classically conditioned learning is often associated with the controlled experiments of psychologists, it undoubtedly occurs throughout the animal kingdom, and it may be one of the most common ways by which animals learn about their immediate environment. A good example of this is the phenomenon of taste-aversion learning. Taste-aversion learning occurs when an animal associates a specific odor or visual stimulus with an unpleasant experience resulting from the consumption of an unpalatable or poisonous food item. After even a single experience with the distasteful food, the animal will subsequently avoid ingestion, even if it means starvation. Taste-aversion learning is especially important for nonspecialist feeders that forage on a variety of foods and must periodically sample unfamiliar food items. This learning phenomenon also serves as the basis for the evolution of many warning signals in animals.

Through natural selection, many distasteful prey have evolved distinctive marks, colorations, odors, or behavioral characteristics that serve as a reminder (a CS) to predators that it is distasteful or harmful. After one negative experience with this prey, the predator learns to associate these characteristics with the sight or smell of the animal. Such characteristics have obvious survival benefits for the prey. Taste-aversion learning differs from classical conditioning in that the critical time between the CS and the US is usually much longer and in that only one trial is necessary for the learning to occur. This latter effect has important implications for animals that rely on aversion learning to avoid poisoning. Many animals, such as blue jays and rats, wait a specific length of time after ingesting a novel food item to determine whether they will become ill.

A second major form of associative learning is operant, or instrumental, conditioning. Unlike classic conditioning, in which the animal is passively involved in the learning experience, in operant conditioning the animal learns by manipulating some part of its environment. In traditional operant learning experiments the animal, for example, presses a lever or rings a bell in order to receive some reward. Because this kind of learning usually improves with practice, it is often referred to as trial-and-error learning. This kind of learning has obvious adaptive significance under natural conditions. Perhaps the best example of this is the reinforced trial-and-error learning that is necessary for many young vertebrates to perfect their feeding techniques. The naïve young of many mammals and birds greatly enhance their feeding efficiency when repeatedly allowed to manipulate their food. Similarly, many animals, including insects, use this reinforced practice to learn their way around in their habitat, home range, or territory, much the way a rat learns its way around in a maze.

PROGRAMMED LEARNING

In addition to associative learning, there are a number of types of learning that seem to involve mechanisms more complex than simple association. The most common examples include song learning (in birds) and imprinting. Ethologists often refer to these types of learning as programmed learning, since they only take place at certain times and under very restricted circumstances.

Imprinting is the process whereby a young animal develops a behavioral attachment to some other animal or object. Animals have been observed to imprint naturally on their parents, individuals of the opposite sex, food items, preferred habitats, and home streams (in the case of salmon). All such types of imprinting have two general features in common. First, the imprinting must occur during some critical period. The most familiar type of critical period is that which occurs in parental imprinting, a specific imprinting routine whereby a newborn becomes behaviorally fixed on a parent. First described by Konrad Lorenz, this type of learning requires a critical period shortly after birth, in which the young learns to recognize and follow the parent. Outside this period the learning simply cannot occur. The second characteristic common to all types of imprinting is that the young animal must be actively involved in the learning process. In fact, the strength of the imprinting seems to depend largely on the degree of this involvement.

Song learning in birds is fundamentally quite similar to imprinting in that it too requires a specific learning period. White-crowned sparrows, for example, learn their song from their fathers, usually from one to six weeks after birth. During this critical period, these young birds learn to imitate the song that is specific to their species as well as the variations and dialect characteristic of their population. When young birds are raised in isolation and prevented from hearing their own species' song, they develop an abnormal vocalization. If given the opportunity to hear a recording of a normal adult song of its species during the critical learning period, the young bird will learn to sing normally. If, on the other hand, the animal is exposed to the song of some other closely related species, the animal will not develop a normal song. This suggests that birds are somehow innately programmed to learn their species-specific songs. Thus, it seems that both imprinting and song learning are in many ways quite similar and may be controlled by the same underlying mechanisms. The tendency to classify these as complex behaviors, however, may be attributable in part to ethologists' lack of understanding of these mechanisms.

INSIGHT

Perhaps the most advanced and least understood form of learning is insight. Insight is said to differ from other forms of learning in that it is characterized by a modification in behavior that is not contingent on some particular recent experience. Instead, insight behavior involves the ability to put two independent ideas together to solve a third, unrelated, problem. Wolfgang Köhler's classic observations on learning in chimpanzees illustrate the phenomenon of insight. He observed that when a preferred food item (such as a banana) was placed out of reach of a caged chimpanzee, the animal quickly learned to use a pole as an extension of its arm to pull in the food; when the food was hung overhead, the animal would learn to stack boxes to reach the food. Examples of tool use by chimpanzees observed under natural field conditions include the use of sticks as probes for gathering insects and the use of small branches for warding off potential predators.

Although this type of problem-solving behavior seems fundamentally more complex than any other type of learning, it has been suggested that many of the specific behaviors cited as examples of insight may be nothing more than extensions of associative learning. Pigeons, for example, can be conditioned to perform certain activities that they use later in solving more complex problems. A pigeon conditioned at one time to push a box across its cage floor and at another time to climb on a box and peck at a food lever will later push and position a box under a lever so that it can peck at the lever and receive a reward. While this seems to reflect some type of problem-solving ability, it is interesting that birds that are not previously conditioned cannot solve the problem. Thus, insight may build on some form of associative learning.

Insight learning has also been invoked to explain the origin of many types of cultural learning. Cultural learning occurs when one animal in a group discovers a unique or novel behavior and the other members learn to copy the behavior through the process of observational learning. One of the classic examples of this kind of learning was observed in the blue tit, a small European bird that was observed to strip the caps off milk bottles in order to drink the cream that surfaced at the top. In relatively little time, the behavior spread and was exhibited by this species all across Western Europe. Although there is little doubt that such cultural transmission involves nothing more than the simple imitation of another animal's behavior, it is not clear whether the origin of such behaviors reflects some form of innovation.

ETHOLOGICAL AND PSYCHOLOGICAL APPROACHES

The study of behavior and learning has long been characterized by two very different methodological and philosophical approaches: those of ethology and psychology.

Ethology, the study of animal behavior, is built on several very specific assumptions and principles that clearly distinguish it from the field of psychology. First, the study of ethology involves objective, non-anthropomorphic (that is, not biased by human expectations or interpretations) descriptions and experiments of the learning process within a natural context. Konrad Lorenz, one of the founders of the field, insisted that the only way to study behavior and learning was to make objective observations under completely natural field conditions. Building on Lorenz's purely descriptive approach, Nikolaas Tinbergen conducted rigorous field experiments, similar to those that now characterize modern ethology.

The classic work of early ethologists helped demonstrate how an animal's sensory limitations and capabilities can shape its ability to learn. For example, in a series of classic learning experiments, Karl von Frisch convincingly documented the unusual visual capabilities of the honeybee. He first trained honeybees to forage at small glass dishes of sugar water and then, by attaching different visual cues to each dish, provided the animals with an opportunity to learn where to forage through the simple process of association. From these elegant (but simplistic) experiments, von Frisch found that bees locate and remember foraging sites by the use of specific colors, ultraviolet cues, and polarized light, a discovery that revolutionized how scientists view the sensory capabilities of animals.

A second important feature of ethology is that it is built on the assumption that learning depends not only on environmental experience but also on a variety of underlying physiological, developmental, and genetic factors. The work of countless neurobiologists, for example, clearly demonstrates how behavioral changes are linked to modifications in the function of nerves and neuronal pathways. By observing the response of individual nerves, neurobiologists can observe changes that occur in the nerves when an animal modifies its behavior in response to some stimulus. In a similar way, they can show how learning and behavior are affected when specific nerve fibers are experimentally cut or removed. Unfortunately, however, neurobiologists' understanding of the physiological control of learning is limited to simpler kinds of learning such as habituation and sensitization.

Like the neurobiologists, behavioral geneticists have shown that much of learning, and behavior in general, is intimately tied to internal mechanisms. The results of hybridization experiments and artificial breeding programs clearly demonstrate a strong genetic influence in learned behaviors. In fact, it has been well documented that many animals (including both invertebrates and vertebrates) are genetically programmed (or at least have a genetic predisposition) to learn only specific kinds of behaviors. Finally, the most important characteristic of ethology is that it places tremendous importance on the evolutionary history of an organism. It assumes that an animal's ability to learn is shaped largely by its evolutionary background, and it emphasizes the adaptive significance of the various types of learning.

In comparison with ethology, the field of psychology emphasizes the importance of rigorously controlled laboratory experiments in the study of learning. The most widely used methods in this field are those of classic and operant conditioning. The primary objective in these approaches is to eliminate and control as many variables as possible and thereby to remove any doubt as to the factors responsible for the behavioral changes. These approaches have met with considerable success at identifying specific external mechanisms responsible for learning. These techniques, however, tend to focus only on the input (stimulus) and output (response) of an experiment and, as a result, de-emphasize the importance of proximate mechanisms, such as physiology and genetics. In addition, these approaches generally ignore the evolutionary considerations that ethologists consider so fundamental to the study of behavior.

UNDERSTANDING THE LEARNING PROCESS

Although the approaches used to study learning vary tremendously, nearly all such studies are directed at two goals: to understand the adaptive value of learning in the animal kingdom and to understand the physiological, genetic, and psychological mechanisms that control learning. For any animal, the adaptive advantages of learning result primarily from the increase in behavioral plasticity that learning provides. This plasticity (ability to be flexible) provides the animal with a greater repertoire of responses to a given stimulus and thereby increases the chances that the animal will survive, reproduce, and pass the genes that control the learning process on to the next generation.

In comparison, the value of an innate behavior lies primarily in its ability to provide a nearly stereotypic response to a stimulus on the very first occasion on which it is encountered. Innate reflexes are especially important in situations in which there may not be a second chance for the animal to learn an appropriate response. The best examples are basic feeding responses (for example, the sucking reflex in newborn mammals) and predator-escape behaviors (alarm calls in young birds). It is a common misconception, however, that a learned behavior is attributable entirely to the animal's environment, whereas instinct is completely controlled by the genes. Many studies have demonstrated that numerous animals are genetically programmed to learn only certain behaviors. In contrast, it has been shown that instinct need not be completely fixed, but can be modified with experience. Thus, learning and instinct should not be considered two mutually exclusive events.

In addition to its evolutionary implications, the study of learning has provided considerable insight into the internal mechanisms that control and regulate behavior. These mechanisms are the cellular and physiological factors that provide the hardware necessary for learning to occur. As neurobiologists and geneticists learn more about these types of control, it is becoming increasingly evident that learning, at nearly all levels, may involve the same basic mechanisms and processes. In other words, the only difference between simple and complex behaviors may be the extent to which the learning is physically constrained by the biology of the animal. Thus, many invertebrates, by virtue of their simple body plan and specific sensory capabilities, are limited to simple learning experiences. Vertebrates, on the other hand, live longer and are not as rigorously programmed for specific kinds of behavior.

—Michael Steele

FURTHER READING

Alcock, John. *Animal Behavior: An Evolutionary Approach.* 6th ed. Sinauer Associates, 1998.

Bonner, John T. *The Evolution of Culture in Animals.* Princeton University Press, 1980.

Donahoe, John W. *Learning and Complex Behavior.* Edited by Vivian Packard Dorsel. Allyn & Bacon, 1994.

Gould, James, L. *Ethology: The Mechanisms and Evolution of Behavior.* W. W. Norton, 1982.

Grier, James W. *Biology of Animal Behavior.* 2nd ed. Times Mirror/Mosby, 1992.

Hickman, Cleveland P., Jr., Larry S. Roberts, and Frances M. Hickman. *Integrated Principles of Zoology.* Times Mirror/Mosby, 1988.

McFarland, David, ed. *The Oxford Companion to Animal Behavior.* Oxford University Press, 1987.

Mammalian Social Systems

Fields of Study

Ethology; Evolutionary Biology; Ecology; Sociology

Abstract

All levels of social organization occur in mammals. There are solitary species and herds of thousands of individuals. Between these extremes, there are many variations. No current theory accounts for the diversity of mammalian social systems, but two broad generalizations are consistently employed to explain mammalian species' social organization. These are the environmental context in which the species exists and the mammalian mode of reproduction.

Key Concepts

browser: an organism that feeds primarily on leaves and twigs of trees and shrubs rather than on grasses

carnivore: a member of the meat-eating order *Carnivora*, which includes dogs, cats, weasels, bears, and their relatives

eusocial: a social system with a single breeding female; other members of the colony are organized into specialized classes (exemplified by bees, ants, and termites)

grazer: an organism that feeds primarily on grasses

primates: members of the order Primates—monkeys, apes, and their relatives

rodent: a member of the order *Rodentia*—squirrels, rats, mice, and their relatives

savanna: a grassland with scattered trees; some ecologists restrict the term to tropical regions

territory: an area that an animal defends against other members of the species and that often contains food, shelter, and other requirements for the individual or group

ungulate: a hoofed mammal from the order *Artiodactyla* (pigs, cattle, antelope, and their relatives) or from the order *Perissodactyla* (horses, rhinoceroses, tapirs, and their relatives)

LIFESTYLES OF THE MAMMALS

All levels of social organization occur in mammals. There are solitary species, such as the mountain lion (*Felis concolor*), in which the male and female adults come together only to mate, and the female remains with her young only until they are capable of living independently. At the other numerical extreme are some of the hoofed mammals, which form herds of thousands of individuals. Other extremes might be considered in terms of specialization for social life. The most socially specialized mammal is probably the naked mole rat (*Heterocephalus glaber*) of Africa, which has a eusocial colony structure similar to that of ants, bees, and termites. Between these extremes, there are many variations. No current theory accounts for the diversity of mammalian social systems, but two broad generalizations are consistently employed to explain mammalian species' social organization. These are the environmental context in which the species exists and the mammalian mode of reproduction.

More than any other group of animals, mammals are required to form groups for at least part of their lives. Although in all sexually reproducing animals the sexes must come together to mate, mammals have an additional required association between mother and young: All species of mammal feed their young with milk from the mother's mammary glands. This group, a female and her young, is the basis for the development of mammalian social groups. In some species, the social group includes several females and their young and may involve one or more males as well.

The particular social organization adopted by a mammalian species is a response to the environmental conditions under which the species lives. The species' food supply and the distribution of that food supply are often the predominant determinants, but predators on the species are also important in determining the form of its social organization. The best way to see the effects of these factors on mammalian social structure is by example.

PRIMATE SOCIAL ORGANIZATION

The primates are the most social group of mammals. Monkeys demonstrate the importance of food supply and its distribution in determining social structure. The olive baboon (*Papio anubis*) occupies savannas, where it exists in large groups of several adult males, several adult females, and their young. Finding fifty or more animals in a group is not uncommon. Individual males do not guard or try to control specific females except when the females are sexually receptive. The group's food supply is in scattered patches, but each patch contains an abundance of food. The advantage of having many individuals searching for the scattered food is obvious: If any member finds a food-rich patch, there is plenty for all.

Predation probably also plays a role in the olive baboon's social organization. The savannas they roam have many predators and few refuges for escape. A large group is one defense against predators if hiding or climbing out of reach is not practical. Having many observers increases the chance of early detection, giving the prey time to elude the predator. A large group can also mount a more effective defense against a predator. Large groups of baboons use both of these tactics.

The hamadryas baboon (*Papio hamadryas*), on the other hand, lives in deserts in which the food supply is not only scattered but also often found in small patches. The hamadryas baboon's social structure contrasts with that of the olive baboon, perhaps because the small patches do not supply enough food to support large groups. A single adult male, one or a few adult females, and their young make up the basic group of fewer than twenty individuals. Several of these family groups travel together under certain conditions, forming a band of up to sixty animals. Within the band, however, the family groups remain intact. The male of each group herds his females, punishing them if they do not follow him. The bands are probably formed in defense against predators. They break up into family units if predators are absent. At night, hamadryas baboons sleep on cliffs, where they are less accessible to predators. Because suitable cliffs are limited, many family groups gather at these sites. Hundreds of animals may be in the sleeping troop, probably affording further protection against predators.

Though there are exceptions, forest primates consistently live in smaller groups. In many species, fewer than twenty individuals make up the social group at all times. These consist of one or a few mature males, one or a few mature females, and their offspring. The groups are more evenly distributed throughout their habitat than are groups of savanna or desert primates. In forests, the food supply is more abundant and more evenly distributed. Escape from predators is also more readily accomplished—by climbing trees or hiding in the dense cover. Under these conditions, the advantages of large groups are minimal and their disadvantages become apparent. For example, in small groups the competition for mates and food is less.

UNGULATE SOCIAL ORGANIZATION

The ungulates have all levels of social organization. African antelope demonstrate social organizations that, in some ways, parallel those of the primates. Forest antelope such as the dik-dik (*Madoqua*) and duiker (*Cephalophus*) are solitary or form small family groups, and they are evenly spaced through their environment. Many hold permanent territories containing the needs of the individual or group. They escape predators by hiding and are browsers, feeding on the leaves and twigs of trees.

Many grassland and savanna antelope, such as wildebeest (*Connochaetes*), on the other hand, occur in large herds. They outrun or present a group defense to predators and are grazers, eating the abundant grasses of their habitat. In many cases, they are also migratory, following the rains about the grasslands to find sufficient food. The social unit is a group of related females and their young. Males leave the group of females and young as they mature. They join a bachelor herd until fully mature, at which time they become solitary, and some establish territories. The large migratory herds are composed of many female/young groups, bachelor herds, and mature males. The social units are maintained in the herd. Though it may seem strange to speak of solitary males in a herd of thousands, that is their social condition. The male territories are permanent in areas that have a reliable food supply year-round, but they cannot be in regions in which the species is migratory. Under these conditions, the males set up temporary breeding territories wherever the herd is located during the breeding season.

There are parallels with primate social patterns. Large groups are formed in grasslands, and these roam widely in search of suitable food. The groups are effective as protection against predators in habitats with few hiding places. Smaller groups are found in forests, where food is more evenly dispersed and places to hide from predators are more readily found.

RODENT SOCIAL ORGANIZATION

Rodents also have all kinds of social organizations. The best known, and one of the most complex, is the social system of the black-tailed prairie dog (*Cynomys ludovicianus*). The coterie is the family unit in this case, and it consists of an adult male, several adult females, and their young. Members maintain a group territory defended against members of other coteries. Coterie members maintain and share a burrow system. Elaborate greeting rituals have developed to allow the prairie dogs of a coterie to recognize one another. Hundreds of these coteries occur together in a town. The members of these towns keep the vegetation clipped—as a result, predators can be seen from a distance. Prairie dogs warn one another with a "bark" when they observe a predator, and the burrow system affords a refuge from most predators.

The only vertebrate known to be eusocial is the naked mole rat. It occurs in hot, dry regions of Africa. The colony has a single reproductive female, a group of workers, and a group of males whose only function is to breed with the reproductive female. The workers cooperate in an energetically efficient burrowing chain when enlarging the burrow system. In this way, they are able to extend the burrow system quickly during the brief wet season. Digging is very difficult at other times of the year. The entire social system is thought to be an adaptation to a harsh environment and a sparse food supply.

CARNIVORE SOCIAL ORGANIZATION

Most carnivores are not particularly social, but some do have elaborate social organizations. Many of these are based on the efficiency of group hunting in the pursuit of large prey or on the ability of a group to defend a large food supply from scavengers. The gray wolf (*Canis lupus*) and African hunting dog (*Lycaon pictus*) are examples. In both cases, the social group, or pack, consists of a male and female pair and their offspring of several years. There are exceptions, solitary carnivores and carnivores that form temporary family units during the breeding season, such as the red fox (*Vulpes vulpes*), which hunt prey smaller than themselves. The coyote (*Canis latrans*) can switch social systems to use the food available most efficiently. It forms packs similar to those of the gray wolf when its main prey is large or when it can scavenge large animals and is solitary when the primary available prey is small.

These examples and many others show that the social groups of mammals are based on the family group. The particular social organization employed by a species is determined by the ecological situation in which it occurs. The specific aspects of the environment that seem to be most important include food abundance, food distribution, food type, and protection from predators.

FIELDWORK AND LABORATORY STUDIES

Observation has been a very important method of studying mammal societies. One of the reasons that primate and ungulate societies are so well known is that they are large and active during the day, and so are easily observable. The observer must take great pains to be inconspicuous or, in some cases, to become a part of the subject's environment. Small mammals (and sometimes larger mammals) have been kept in enclosures and observed to learn more about their social lives. The observer maps movements, records activities and interactions, and analyzes the data that result.

Simple observation is enhanced by manipulating the subjects in various ways. Individual animals can be marked, or in some cases they can be identified by natural color patterns, scars, or other marks. These marked individuals can be followed, and their behavior and interactions with other individuals observed. Radios and radioactive tracer elements are sometimes implanted in individuals, and these individuals are followed in the field. Much can be learned about a species' social behavior by following the locations of such tagged animals. In addition, they are more readily located for direct observation.

Small mammal species that are not readily observable are trapped live, marked, released, and recaptured. Mutually exclusive use of certain areas, areas used in common, and patterns of multiple captures in individual traps are some types of information from trapping that can be interpreted in terms of social behavior. Experiments are sometimes carried out in the natural context. A group or a specific individual is presented with an artificial situation, and any reactions to it are recorded, often on film or videotape.

Laboratory studies are also used to supplement the field observations. Psychological and physiological capabilities of organisms can best be studied in the controlled confines of a laboratory experiment. These data, however, must always be put back in the context of the field observations to make a meaningful contribution to the understanding of the species' social behavior.

Computer simulations and mathematical models have been used to explore the possible reactions of social systems to various environmental pressures. As with laboratory results, it is important to test predictions generated in these ways against the social system in nature before assuming their validity. Comparative studies of all the above types are of great importance. Related species, or different populations of the same species, that occur in different regions are studied and compared; these studies are tantamount to reading the data from a natural experiment.

SOCIAL ORGANIZATION AND FOOD

Mammalian societies are always organized around one or more females and their offspring. Males may also be part of the group, or they may form separate groups. The size and structure of the group are determined by the ecological setting in which it evolves. The particular ecological factors that seem to be of greatest importance in this determination are food supply, the distribution of the food, and predation (including the hiding places and escape routes available in the habitat).

Large groups occur when food is scattered in a patchy distribution. These groups are largest when the patches contain abundant food. Many organisms are more likely to find the scattered patches than is a single individual. As long as the patches have enough food for all members of the group, it is to each member's advantage to search with the group. On the other hand, if food is evenly dispersed in small units throughout the environment, the advantage of a group search is lost. Each individual will be better off searching for itself, and some strategy involving a very small social group or even solitary existence would be advantageous.

A somewhat similar argument follows for predators. If large prey are taken, a group of predators should be able to subdue the prey and protect its remains from scavengers more efficiently. If small prey are taken, solitary predators have the advantage, since the prey is easily dispatched and the predator will have it to itself. Many other factors are involved in determining the final form of a species' social organization, but the family unit and environmental context are fundamental in determination of all mammalian social structures.

Conservation of the mammal species that still exist on the earth requires knowledge of their social organization. Understanding that mammalian social organizations are responses to the environmental context in which they have evolved emphasizes the need to conserve entire ecosystems, not only the individual species that exist within them.

—Carl W. Hoagstrom

FURTHER READING

Dunbar, Robin I. M. *Primate Social Systems.* Cornell University Press, 1988.

Eisenberg, John F., and Devra G. Kleiman, eds. *Advances in the Study of Mammalian Behavior.* Special Publication 7. American Society of Mammalogists, 1983.

Gittleman, John L., ed. *Carnivore Behavior, Ecology, and Evolution.* Cornell University Press, 1989.

Immelmann, Klaus, ed. *Grzimek's Encyclopedia of Ethology.* Van Nostrand Reinhold, 1977.

Macdonald, David W. *European Mammals: Evolution and Behavior.* HarperCollins, 1995.

Nowak, Ronald M., and John L. Paradiso. *Walker's Mammals of the World.* 6th ed. 2 vols. The Johns Hopkins University Press, 1999.

Rosenblatt, Jay S., and Charles T. Snowdon, eds. *Parental Care: Evolution, Mechanisms, and Adaptive Significance.* Advances in the Study of Behavior 25. Academic Press, 1996.

Slater, P. J. B. *An Introduction to Ethology.* Reprint. Cambridge University Press, 1990.

Vaughan, Terry A. *Mammalogy.* 4th ed. Saunders College Publishing, 2000.

Wrangham, Richard W., W. C. McGrew, Frans B. M. De Waal, and Paul G. Heltne, eds. *Chimpanzee Cultures.* Harvard University Press, 1994.

Marine Animals

Fields of Study

Ichthyology; Oceanography; Ethology

Abstract

Approximately 71 percent of earth's surface is covered by salt water, and the marine environments contained therein constitute the largest and most diverse array of life on the planet. Life is believed to have originated in the oceans, and the salt water that comprises the largest constituent of the tissues of all living organisms is a vestigial reminder of the aquatic origins of life.

Key Concepts

benthos: organisms living upon, or below, the surface of the substrate that forms the ocean floor

epifauna: animals that live on the sea floor

estuary: the region where freshwater rivers empty into and mix with the marine environment of oceans or seas

infauna: animals that live in the sea floor

intertidal zone: the portion of the marine environment located between low and high tide marks

nekton: larger marine animals that have sufficient powers of locomotion to move independently of water currents

neritic zone: the shallow water areas that extend over the continental shelves up to the low tide mark

pelagic zone: portions of the marine environment that are located away from the shorelines; the open ocean environment

plankton: small animals and plants that drift with the water currents in the marine environment

WATER, WATER ALMOST EVERYWHERE

Approximately 71 percent of earth's surface is covered by salt water, and the marine environments contained therein constitute the largest and most diverse array of life on the planet. Life is believed to have originated in the oceans, and the salt water that comprises the largest constituent of the tissues of all living organisms is a vestigial reminder of the aquatic origins of life.

MARINE ZONES

The marine environment can be divided broadly into different zones, each of which supports numerous habitats. The coastal area between the high and low tide boundaries is known as the intertidal zone; beyond this is the neritic zone, relatively shallow water that extends over the continental shelves. The much deeper water that extends past the boundaries of the continental shelves is known as the oceanic zone. Open water of any depth away from the coastline is also known as the pelagic zone. The benthic zone is composed of the sediments occurring at the sea floor. Areas in which freshwater rivers empty into the saltwater oceans produce a continually mixed brackish water region known as an estuary. Estuarine zones often also include extensive wetland areas such as mud flats or salt marshes.

Zones in the marine environment are distributed vertically as well as horizontally. Life in the ocean, as on land, is ultimately supported by sunlight in most cases, used by photosynthetic plants as an energy source. Sunlight can only penetrate water to a limited depth, generally between one hundred and two hundred meters; this region is known as the photic or epipelagic zone. Below two hundred meters, there may be sufficient sunlight penetrating to permit vision, but not enough to support photosynthesis; this transitional region may extend to depths of one thousand meters and is known as the disphotic or mesopelagic zone. Below this depth, in the aphotic zone, sunlight cannot penetrate and the environment is perpetually dark, with the exception of small amounts of light produced by photoluminescent invertebrate and vertebrate animals.

The aphotic zone is typically divided into the bathypelagic zone, between seven hundred and one thousand meters as the upper range and two thousand to four thousand meters as the lower range, where the water temperature is between $4°$ and $10°C$. Beneath the bathypelagic zone, overlying the great plains of the ocean basins, is the abyssalpelagic zone, with a lower boundary of approximately six thousand meters. Finally, the deepest waters of the oceanic trenches, which extend to depths of ten thousand meters or more, constitute the hadalpelagic zone. In

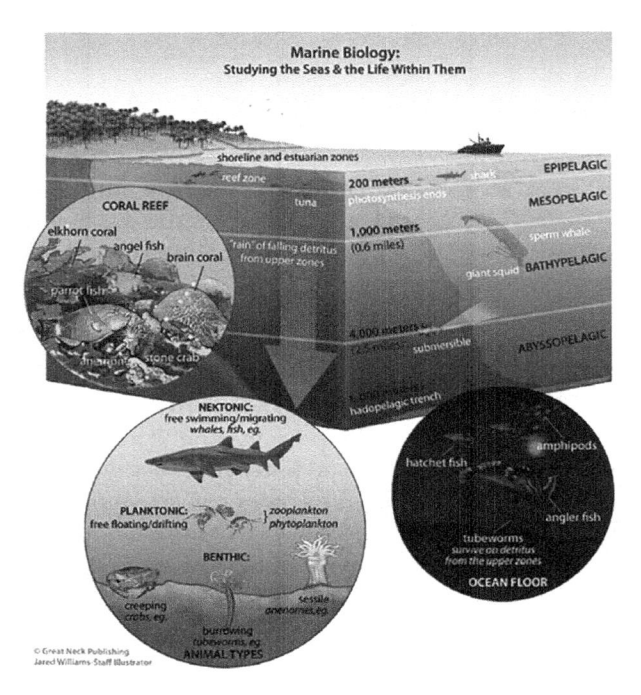

Marine biology is the study of the interactions of living things in and near the ocean ecosystem.

each of these zones, the nature and variety of marine life present is dictated by the physical characteristics of the zone. However, these zones are not absolute, but rather merge gradually into each other, and organisms may move back and forth between zones.

PLANKTON

Marine life can be divided broadly into three major categories. Those small organisms that are either free-floating or weakly swimming and which thus drift with oceanic currents are referred to as plankton. Plankton can be further divided into phytoplankton, which are plantlike and capable of photosynthesis; zooplankton, which are animal-like; and bacterioplankton, which are bacteria and blue-green algae suspended in the water column. Larger organisms that can swim more powerfully and which can thus move independently of water movements are known collectively as the nekton. Finally, organisms that are restricted to living on or in the sediments of the seafloor bottom are referred to as the benthos.

The phytoplankton, which are necessarily restricted to the photic zone, are by far the largest contributors to photosynthesis in the oceans. The phytoplankton are therefore responsible for trapping most of the solar energy obtained by the ocean (the primary productivity), which can then be transferred to other organisms when the phytoplankton are themselves ingested. The phytoplankton are composed of numerous different types of photosynthetic organisms, including diatoms, which are each encased in a unique "pillbox" shell of transparent silica, and dinoflagellates. The very rapid growth of some species of dinoflagellates in some areas results in massive concentrations or blooms that are sometimes referred to as red tides. Chemicals that are produced by red tide dinoflagellates often prove toxic to other marine organisms and can result in massive die-offs of marine life. Smaller photosynthetic plankton forms comprise the nanoplankton and also play an important role in the photosynthetic harnessing of energy in the oceans.

The zooplankton are an extremely diverse group of small animal organisms. Unlike the phytoplankton, which can make their own complex organic compounds via photosynthesis, the phytoplankton must ingest or absorb organic compounds produced by other organisms. This is accomplished by either preying upon other planktonic organisms or by feeding on the decaying remains of dead organisms. A number of zooplankton species also exist as parasites during some portion of their life cycles, living in or upon the bodies of nekton species. The largest group of zooplankton are members of the subphylum *Crustacea*, especially the copepods. These organisms typically possess a jointed exoskeleton, or shell, made of chitin, large antennae, and a number of jointed appendages. Space precludes a definitive listing of all of the zooplanktonic organisms; however, virtually all of the other groups of aquatic invertebrates are represented in the bewildering variety of the zooplankton, either in larval or adult forms. Even fishes, normally a part of the nekton, contribute to the zooplankton, both as eggs and as larval forms.

The bacterioplankton are found in all of the world's oceans. Some of these, the blue-green algae (cyanobacteria), play an important role in the photosynthetic productivity of the ocean. Bacterioplankton are usually found in greatest concentrations in surface waters, often in association with organic fragments known as particulate organic carbon, or marine snow. Bacterioplankton play an important role in renewing nutrients in the photic zones of the ocean; such renewal is important in maintaining the photosynthetic activity of the phytoplankton, upon which the rest of marine life is in turn dependent.

One of the principal problems facing plankton is maintaining their position in the water column. Since these organisms are slightly denser than the surrounding seawater, they tend to sink. Clearly this is a disadvantage, particularly since plankton typically have very limited mobility. This is especially true for the photosynthetic phytoplankton, which must remain within the photic zone in order to carry on photosynthesis. A number of strategies have evolved among planktonic species to oppose this tendency to sink. Long, spindly extensions of the body provide resistance to the flow of water. Inclusions of oils or fats (which are less dense than water) within the body provide positive buoyancy by decreasing the overall density of the plankton. Finally, some species, such as the Portuguese man-o'-war, generate balloonlike gas bladders, which provide enough buoyancy to keep them at the very surface of the epipelagic zone.

NEKTON

The nekton is composed of those larger animals that have developed locomotion to a sufficient degree that they can move independently of the ocean's water movements. Whereas the plankton are principally invertebrates, most of the nekton are vertebrates. The majority of the nekton are fishes, although reptile, bird, and mammalian species are also constituent parts. The oceanic nekton are those species which are found in the epipelagic zone of the open ocean. These include a wide variety of sharks, rays, bony fishes, sea birds, marine mammals, and a few species of reptiles. Some members of the oceanic nekton, such as blue sharks, oceanic whitetip sharks, tuna, flying fish, and swordfish, spend their entire lives in the pelagic environment; these are said to be holoepipelagic. Others, the meroepipelagic nekton, only spend a portion of their lives in the epipelagic zone, returning to coastal areas to mate, as with herring and dolphins, or returning to freshwater, as with salmon and sturgeon.

Sea birds are a special case: Although they spend much of their time flying over the epipelagic zone and nest on land, they feed in the epipelagic zone and some species may dive as deep as one hundred meters in search of prey. Some members of the nekton enter the epipelagic only at certain times in their life cycle. Eels of the family Anguillidae spend most of their lives in freshwater but return to the epipelagic zone to spawn. Additionally, at night many species of deep-water fishes migrate up into the epipelagic to feed before returning to deeper waters during the daylight hours.

The pelagic environment, unlike the terrestrial one, is profoundly three-dimensional. Nektonic animals can move both horizontally and vertically within the water column. Furthermore, since most of the pelagic environment is essentially bottomless, since there is no apparent or visible ground or substrate, the environment is basically uniform and featureless. These characteristics play an important role in the evolution of the behavior of nektonic animals. Fishes suspended in an essentially transparent and featureless medium have no shelter in which to hide from predators, nor are there any apparent landmarks to serve as directional cues for animals moving horizontally from place to place. Life in the open ocean has therefore favored adaptations for great mobility and speed with which to move across large distances and escape from predators, as well as camouflage and cryptic coloration designed to deceive potential predators or prey.

As is the case for plankton, most nektonic animals are denser than the surrounding seawater, and maintaining position in the water column is of the first importance. Most fishes possess a swim bladder, a gas-filled membranous sac within their body that opposes the tendency to sink and provides the fish with neutral buoyancy. Sharks and rays lack a swim bladder, but accumulate large concentrations of fats and oils in their liver, which also help counter the tendency to sink. Large, fast-swimming species of shark, tuna, and many billfish also rely on the generation of hydrodynamic lift to maintain vertical position in the water column. The tail and body of these fishes generate forward thrust, moving the animal through the water, and the fins, notably the pectoral fins, generate lift from the water flowing over them in a manner similar to that of an airplane's wing. Thus these animals fly through the water, but are in turn required to move continuously in order to generate lift.

All members of the nekton are carnivores, feeding on other nektonic species or upon plankton, particularly the larger zooplankton. In general, the size of the prey consumed by nekton is directly related to the size of the predator, with larger species consuming larger prey. However, the organisms that feed upon plankton, the planktivores, include a wide variety of fish species such as herring, salmon, and the whale shark, the largest extant fish species. They also include the largest marine animals of all, the baleen whales. The case of large animals feeding upon very small plankton directly addresses the need of all animals to meet their energy requirements.

For all animals, the amount of energy obtained from food consumed must necessarily exceed the energy expended in acquiring the prey. Very large animals, such as whales and whale sharks, require a great deal of energy to move their bodies through the aquatic environment, but because of their great size they are necessarily less agile than smaller forms. The amount of energy required to chase and catch these smaller animals would generally exceed the energy derived from ingesting them. Plankton, however, are relatively easy to obtain due to their very limited mobility. However, because of their small size, vast quantities of plankton must be ingested in order to meet the metabolic requirements of large marine animals. Some very large species that are not planktivores solve the energy problem by evolving behaviors for acquiring specialized diets that yield higher energy. White sharks, for example, feed on fish when young, but as they age and increase in size, marine mammals, notably seals and sea lions (pinnipeds), become a major part of their diet. Marine mammals all possess blubber, an energy-rich substance that yields much more energy than fish. Similarly, sperm whales, the largest hunting carnivores on the planet, have a diet that consists in large part of giant squid, which are hunted in the ocean depths largely using the whale's acoustic echolocation sense. Orcas (killer whales) effectively use pack hunting techniques to hunt larger whales and other marine mammals.

The deeper regions of the ocean are dominated by different types of nekton. However, even less is known about their ecology due to their relative inaccessibility. The disphotic or mesopelagic zone contains many animal species that migrate vertically into surface waters at night to feed upon the plankton there. Many of these organisms possess large, well-developed eyes and also possess light organs containing symbiotic luminescent bacteria. The majority of the fish species in this group are colored black and the invertebrates are largely red (red light penetrates water less effectively than do longer wavelengths, and these animals appear dark-colored at depth). Beneath this zone, in the bathypelagic and abyssalpelagic zones, there are many fewer organisms and much less diversity than in the shallower levels. Animals in this region are typically colorless and possess small eyes and luminescent organs. Because organisms in these deep regions are few and far between, many species have become specialized in order to maximize their advantages. Thus, deep-sea fish are characterized by large teeth and remarkably hinged jaws that allow them to consume prey much larger than might be expected from their size. Similarly, since encounters with potential mates are presumably scarce, a number of unique reproductive strategies have evolved. In the anglerfish (*Ceratius*), all of the large individuals are female and the comparatively tiny males are parasitic, permanently attaching themselves to the female. Much, however, still remains to be learned of the ecology of these deep-sea organisms.

BENTHOS

The benthos of the world's oceans consists of animals that live on the solid substrate of the water column, the ocean floor. Scientists typically divide benthic organisms into two categories, the epifauna, which live on the surface of the bottom at the sediment-water interface, and the infauna, those organisms living within the sediments. In shallow water benthic communities, members of virtually every major animal group are represented. Ecologists generally differentiate between soft bottom benthic communities (sand, silt, and mud, which comprise the majority of the benthic zone) and rocky bottom communities, which are less common proportionately. Soft bottom communities have an extensive diversity of burrowing infauna, such as polychaete worms, and mollusks, such as clams. Rocky bottom communities possess a larger proportion of epifauna, such as crustaceans and echinoderms (starfish, sea urchins, and brittle stars), living on the surface of what is essentially a two-dimensional environment. Vertical faces of the hard bottom environment, such as canyon walls or coral reefs, are often home to a wide variety of animals occupying various crannies and caves. In some parts of the world, kelp plants that are anchored to the substrate and which extend to the water surface dominate the rocky bottom substrate. In these kelp forests, large kelp plants (actually a species of brown algae) form a forestlike canopy that plays host to a wide and complex array of animals extending throughout the water column. On the deep ocean floor, the benthos is composed of representatives of virtually every major animal group: crustaceans such as amphipods, segmented polychaete worms, sea cucumbers, and brittle stars. Less common are starfish, sea lilies, anemones, and sea fans. The fishes of the deep benthos include rat tails and a number of eel species.

Estuaries, where freshwater rivers empty into marine environments, are typified by large, cyclic changes in temperature and salinity. Although estuaries have played an important role in human history as the sites of major ports, the variety and number of estuarine species tend to show less diversity of animal species due to the difficulty in adapting to the large swings in environmental conditions.

Animal life in the sea, like that on land, shows an astonishing variety of forms and behaviors, the result of natural selection. The inaccessibility and hostility of much of the world's oceans to human exploration and observation leaves much yet to be learned about the biology of marine life. Much remains to be achieved in order to obtain a useful body of knowledge concerning life in the sea.

—*John G. New*

FURTHER READING

Niesen, T. M. *The Marine Biology Coloring Book.* 2nd ed. HarperResource, 2000.

Nybakken, J. W. *Marine Biology: An Ecological Approach.* 5th ed. Benjamin/Cummings, 2001.

Robison, B. H., and J. Connor. *The Deep Sea.* Monterey Bay Aquarium Press, 1999.

Safina, C. *Song for the Blue Ocean: Encounters Along the World's Coasts and Beneath the Seas.* Henry Holt, 1998.

Marine Biology

FIELDS OF STUDY

Oceanography; Marine Biology; Ecology; Marine Chemistry

ABSTRACT

Oceans cover 71% of Earth's surface and have an average depth of 3,800 meters, which means that they represent 99% of the living space on the planet. Marine biology is the study of all the organisms that occupy this space. Despite having 99% of the planet's living space at their disposal, only about 250,000 of approximately 1.8 million described living species (14%) are marine. While the oceans may lack diversity at the species level, they are home to members of thirty-one of approximately thirty-four animal phyla, about twice the number of phyla that are represented on land or in freshwater.

KEY CONCEPTS

autotrophs: primary producers; organisms that are self-feeding; includes photosynthetic and chemo-autotrophic organisms

benthic: the area of the ocean floor; organisms associated with the sea bottom

epifauna: animals that live on the sea floor

heterotroph: consumers; organisms that must acquire energy by consuming organic material

infauna: animals that live in the sea floor

invertebrates: animals that lack backbones

littoral: the area in the intertidal zone; organisms that live in the intertidal

nekton: organisms that are strong swimmers and can move against ocean currents

pelagic: the area of open water in the oceans; organisms that occur in the water column

plankton: organisms that drift in the ocean currents because they have limited power of locomotion

THE WATERY REALM

The earliest known life forms were marine. The oceans cover 71% of Earth's surface and have an average depth of 3,800 meters, which means that they represent 99% of the living space on the planet. Marine biology is the study of all the organisms that occupy this space. Despite having 99% of the planet's living space at their disposal, only 250,000 of approximately 1.8 million described living species (14%) are marine. While the oceans may lack diversity at the species level, they are home to members of thirty-one of approximately thirty-four animal phyla, about twice the number of phyla that are represented on land or in freshwater.

Because of their vastness and humans' inability to easily visit deep waters, the oceans remain the least studied habitats on Earth. For example, scientists know that giant squid eighteen meters in length, larger than any other invertebrate, exist because they have been found in the stomachs of sperm whales and washed up on beaches. However, no human has ever seen one of these creatures living in their natural habitats until one was photographed very recently. Yet despite the comparatively small amount of study in marine biology, research in these systems has contributed greatly to the understanding of living systems.

Oceanographers divide the ocean into zones that have distinct physical characteristics, which in turn select for different organisms that are adapted to those characteristics. The benthic realm refers to the sea floor and extends from the high intertidal zone, where ocean meets terrestrial land, to depths of eleven kilometers at the oceanic trenches, the deepest parts of the ocean where the ocean floor slowly sinks back in to the interior of the earth. Organisms that live in, on, or near the ocean floor are appropriately called benthic organisms and represent 98% of all marine creatures. The pelagic realm of the ocean refers to the open ocean, basically the space between the benthos and the sea surface. Pelagic organisms, representing the remaining 2% of marine species, live in the water column. The pelagic zone can be divided into the photic and aphotic zones, a distinction that is especially important for photosynthetic organisms. The photic zone is the shallow part of the ocean that receives enough sunlight to support photosynthesis, which is about two hundred meters deep in the clearest waters, and as shallow as three meters in turbid coastal waters. The aphotic zone is where there is not enough light to support photosynthesis; it extends from the bottom of the photic zone to the ocean floor.

ORGANISMS OF THE OPEN OCEAN

Pelagic creatures can be further categorized into plankton, which drift around in the ocean currents, and nekton, which are capable of swimming against currents. Photosynthetic plankton live in the photic zone and are called phytoplankton. These algae form the base of the food web in the open ocean and are eaten by zooplankton, or heterotrophic plankton, a category that includes protists and small crustaceans such as copepods. Zooplankton are eaten by small fishes, which are eaten by larger fishes, which are eaten by sharks, birds of prey, and people. The open ocean food web is one of the longest food webs known, partly because it starts with the smallest photosynthetic organisms.

The largest animal migrations on Earth occur every day in the open oceans, in a process called diel vertical migration. Zooplankton, midwater fish, squid, and krill migrate to shallow waters at night and then return to the dark depths during the day. The main reasons for this daily migration are probably to feed and to avoid being eaten. Food densities are highest in the shallow productive waters, so these predators move to the shallows to feed at night. Because they would be susceptible to their own visually oriented predators in the well-lit shallows, they return to the dark depths during the day to avoid being eaten.

In the aphotic zone, there are only heterotrophic organisms that are supported mostly by organic material that rains down from the lit environments above. These animals live in darkness, with the exception of light produced by the animals themselves, called bioluminescence. It is common for sea creatures, especially ones at intermediate depths, to house luminescent bacteria within their tissues, which are able to produce light in order to communicate, lure prey, or illuminate the creature's bottom surface to conceal its silhouette against the dimly lit background from above. Anglerfish are deep-sea predators that attract prey near their mouths by dangling a bioluminescent lure in front of their head.

The density of organisms in the deep sea is low. Because of this low density, a long period of time can pass between meals, or between encounters with the opposite sex. To deal with the problem of infrequent meals, deep-sea creatures are often gigantic compared to their shallow-water counterparts. Their large size allows for storage of food reserves that sustain the animals between meals. Predatory fish also have large mouths and stomachs that allow them to take full advantage of any meal, regardless of its size. To overcome the problem of rare mates, the miniature males of some anglerfish have the unusual adaptation of attaching themselves to a female, where they live the rest of their lives as parasitic sperm producers.

ORGANISMS OF THE BOTTOM

Most marine species are found at the ocean floor. They occur in such familiar marine habitats as mud flats, sand flats, beaches, coral reefs, kelp forests, and the rocky intertidal. The main primary producers in benthic habitats are macroscopic seaweeds that grow attached to the bottom or microscopic algae that grow within the tissues of animals such as corals, sponges, and bryozoans. Benthic animals include mobile creatures such as fish, crabs, shrimp, snails, urchins, sea stars, and slugs. Additionally, there are numerous animals that, unlike familiar terrestrial animals, never move around as adults. These sessile animals include barnacles, sponges, oysters, mussels, corals, gorgonians, chrinoids, hydroids, and bryozoans.

Perhaps the most interesting and challenging bottom-dwelling species are found in colonies about the hydrothermal vents of the mid-oceanic ridges. The variety of creatures, both active and sessile, inhabiting those colonies is surprising, and includes various species of shrimp, tube worms and fishes. The colonies do not depend on photosynthesis as the basis of their food chains. Instead, they are based on energy derived from chemical reactions of minerals in the superheated water spewing up threw the vents.

The commonness of sessile animals in the marine benthos suggests that it is a successful way of life. These animals' lifestyle combines aspects of plant and animal lifestyles. Sessile invertebrates are plant-like in that they obtain some of their energy from sunlight (the animals themselves do not photosynthesize, but they house photosynthetic symbionts), are anchored in place, and grow in a modular fashion, like the branches of a tree. They are animallike in that they capture and digest prey and undergo embryonic development, often involving metamorphosis. In fact, nearly all benthic animals start life in the pelagic realm, drifting around as planktonic larvae and dispersing to new habitats as they develop and feed. After a few hours to several weeks of pelagic living, they sink to the ocean floor to complete life as adults.

Being stuck in one place presents special challenges for sessile animals, including food acquisition, predator avoidance, and mating. Sessile animals feed by having symbiotic algae and by filtering organic particles from passing water currents. Like plants, sessile animals use structural and chemical defenses against predators and have tremendous regenerative abilities to recover from partial predation events. Most benthic animals mate via external fertilization: sperm and eggs are spawned into the water column, and fertilization occurs outside the body of the female. Amazingly, sessile barnacles must copulate to achieve internal fertilization. These animals increase their reproductive success by being hermaphroditic, thus assuring that any neighbor is a potential mate; being gregarious to assure a high density of mates; and having a penis long enough to deliver sperm to an individual seven shell lengths away.

TROPHIC CASCADES AND KEYSTONE PREDATORS

Marine organisms live in environments that are foreign to humans, and they have lifestyles that are unique to them and different from terrestrial creatures. However, the study of marine organisms has led to advances in ecological theory that have proved to be useful in understanding terrestrial communities as well. The concept of keystone species, where a relatively rare species has a disproportionately large effect on the community structure, was discovered from research conducted on rocky intertidal habitats. Professor Robert Paine removed starfish from rocks off the coast of Washington State and observed that mussels soon crowded out seaweeds and barnacles, resulting in about a 50% reduction in species richness. He concluded that starfish are keystone predators because their predation prevented mussels from excluding less competitive species from the habitat.

Not far offshore from the rocky intertidal in kelp forests, another keystone species was shown to influence the community by means of a trophic cascade, in which the effects from a top predator "cascade" down to lower trophic levels. Sea otters, by preying on sea urchins, protect the kelps that make kelp forests, an important habitat that many marine species rely on. In the 1800s, hunters greatly reduced the number of otters by harvesting their thick furs. As a result, sea urchin populations exploded because they were relieved from predation, resulting in the decimation of kelps by the herbivorous urchins. Once otters received government protection, their numbers increased, sea urchins decreased, and kelp forests returned, at least in some areas. Interestingly, Aleutian killer whales began preying on otters in the 1980s, causing kelp forests to begin disappearing again. The reason orcas began eating otters was probably due to concurrent declines in seal and sea lion numbers, the normal prey of these killer whales. The ultimate causes of these altered food webs are uncertain, but they almost certainly are the result of human activity. Even though humans are poorly adapted for a marine existence, the evidence is mounting that humans are altering marine ecosystems in complex, novel, and unpredictable ways. The science of marine ecology is best equipped to study these effects and offer information to protect and manage ocean life.

—*Greg Cronin*

FURTHER READING

Cousteau, Jacques. *The Ocean World.* 1979. Abradale, 1985.

Dudley, Gordon H., James L. Sumich, and Virginia L. Cass-Dudley. *Laboratory and Field Investigations in Marine Life.* 10th ed. Jones, 2012.

Earle, Sylvia A. *Sea Change: A Message of the Oceans.* Putnam's, 1995.

Levinton, Jeffrey S. *Marine Biology: Function, Biodiversity, Ecology.* 4th ed. Oxford University Press, 2014.

Mladenov, Philip V. *Marine Biology: A Very Short Introduction.* Oxford University Press, 2013.

Morrissey, John F., and James L. Sumich. *Introduction to the Biology of Marine Life.* 10th ed. Jones, 2012.

Nybakken, James W., and Mark D. Bertness. *Marine Biology: An Ecological Approach.* 6th ed. Benjamin, 2005.

Ocean Drifters. Narr. Keith David. Natl. Geographic Soc., 1993. VHS.

Prager, Ellen. *Sex, Drugs, and Sea Slime: The Oceans' Oddest Creatures and Why They Matter.* University of Chicago Press, 2011.

MARK, RELEASE, AND RECAPTURE METHODS

FIELDS OF STUDY

Statistics; Population Sciences; Environmental Sciences

ABSTRACT

There are four different ways to determine the number of animals within a habitat or population. Different types of animals require different techniques for population estimation. The mark-recapture approach is often used for groups of animals whose populations are too large or too secretive for other methods. These are usually vertebrates, although mark-recapture procedures have also been used for invertebrates, such as grasshoppers.

KEY CONCEPTS

density: the number of animals present per unit of area being sampled

emigration: the movement of animals out of an area; one-way movement from a habitat type

habitat: the physical environment, usually that of soil and vegetation as well as space, in which an animal lives

home range: the space or area that an animal uses in its life activities

immigration: the movement of animals into an area; a one-way movement into a habitat type

marked: an individual animal that is identifiable by marks that may be either human-made, such as metal bands or tags, or natural, such as the pattern of a giraffe

N: a standard abbreviation for the size of an actual population; if capped with a ($\hat{n}$) it is an estimated value

population: a group of animals of the same species occupying the same physical space at the same time

recaptured: a previously marked animal that is either seen, trapped, or collected again after its initial marking

sampling: the process of collecting data, usually in such a manner that a statistically valid set of data can be acquired

TO FIND THE TOTAL

There are four different ways to determine the number of animals within a habitat or population: counting the total number of animals present, sampling part of the area (a quadrat) and extrapolating to find a total, sampling along a line-transect and measuring the distance and angle to where the animal being counted was first seen, and using the mark-recapture approach. Different types of animals require different techniques for population estimation. The mark-recapture approach is often used for groups of animals whose populations are too large or too secretive for other methods. These are usually vertebrates, although mark-recapture procedures have also been used for invertebrates, such as grasshoppers.

All mark-recapture calculations are based on how many individual animals are marked (denoted M) in the population being studied, how many animals are captured during sampling (n), and how many of the captured animals have been previously marked (m). The estimated population is commonly indicated n. The basic mathematical relationship of these data is

$n/M = n/m$, when $n = Mn/m$.

An estimate of population density can be obtained by dividing the area being sampled by the estimate of N.

MARKING

Animals may be either temporarily or permanently marked. Temporary marking may be daubing paint on an animal's body, clipping some hair off a mouse's back, or pulling off a few scales from a snake's belly. If each animal is given a unique mark, such as a number or symbol, then it is possible to determine how long the particular animal lives, its home range, and patterns of movement, such as immigration and emigration rates. Some mark-recapture calculations require that the number of times an individual animal is captured be known; "recaptured" must be separated individually for these calculations.

There are pitfalls in this census method that should not be overlooked. Several conditions must hold if mark-recapture population estimates are to be

Photo of right side view of Novisuccinea chittenangoensis. *The snail is marked with a number for monitoring the population.*

valid. The marked animals must neither lose nor gain marks. Care must be taken if natural marks, such as missing toes on a mouse, are used; additional mice losing toes would lead to an error in population estimation. Marked animals must be as subject to sampling as unmarked ones. Because of the excitement of being captured, many animals will not return to a live-trap a second time, leading to an overestimation of animal abundance. If an animal becomes easily caught and returns frequently to the trap, such as kangaroo rats often do in the desert, then this trap-happy animal produces an underestimate of population size. The marked animals must also suffer the same natural mortality as unmarked ones, and the stress of being captured and marked may cause a higher mortality rate in the animals that are marked. If this occurs, the population estimate will be too high.

The marked animals must become randomly mixed with the unmarked ones in the population, or the distribution of sampling effort must be proportional to the number of animals in different parts of the habitat being studied. If the animals are "clumped," population estimates will be either too high or too low, depending on whether the clumps of marked animals are included in the sample. Marked animals must be recognized and reported on recovery. Technicians working with the animals must be able to recognize marked animals and/or read the individual numbers per animal correctly. If marked animals are not recognized, population estimates are too high. There can only be a negligible amount of recruitment or loss to the population being sampled during the sampling period;

emigration, if occurring, should be balanced by immigration. A short time between marking the animals and collecting the additional samples for the population estimate is necessary, or the ratio of N to M changes from that existing when $n:m$ was established.

Even under the ideal conditions above, it is apparent, according to the laws of chance, that the ratio of marked to unmarked animals in the sample will not always be the same as that of marked to unmarked animals in the population; in fact, the two ratios may seldom be the same. Possibilities for sampling error can be decreased by enlarging the size of the sample. As the sample size approaches the population in size, chances for error become smaller. When the point is reached where the sample includes the entire population, there can be no error in estimation. In general, at least 50 percent of the population should be marked, and the number of marked animals in the sample should be 1.5 times the number of unmarked animals in the sample. In actual practice, it is difficult (if not impossible) to meet all these requirements. Consequently, it is often best that mark-recapture population estimates be used as measures of trends in major population fluctuations from year to year.

APPLICATIONS OF THE METHOD

There are many different formulas for utilizing the mark-recapture data to produce estimates of population size for any animals that can be marked and recaptured or observed later. The first use of the ratio of marked to unmarked animals for population estimation was for fish and ducks; the technique is usually called the Lincoln-Petersen method or index. Its formula is

$N = Mn/m,$

where N is the total estimated population, n is the number of animals sampled or captured, M represents the number of animals marked in the population before sample size n is drawn, and m equals the number of previously marked animals recovered in sample size n.

An example of how the calculations for the Lincoln-Petersen index would be made is shown by the following information. If 375 quail were banded and later, in a sample of 545, there were 85 previously banded birds recovered, therefore $N = Mn/m$, or $375 \times 545/85 = 2,404$ quail estimated to be present in the population being sampled.

When, as is the case with ring-necked pheasants, there is a variation in the capture of the sexes, caused perhaps by capturing technique, the formula can be applied to both sexes, or even to age classes, to arrive at a better estimate of the total population. For example, if 500 males and 750 females were banded before hunting season and then 360 males and 150 females were recovered after the harvest, with 150 banded males found in the 360 males checked and 50 banded females recovered in the 150 females checked, a population estimate can be made. The male population estimate would be $500 \times 360/150$, equaling 1,200 males in the population. The female population estimate would be $750 \times 150/50$, equaling 2,250 females in the population. The total population of pheasants would be estimated to be 1,200 males and 2,250 females, equaling a total of 3,450 pheasants.

The Lincoln-Petersen index differs from other mark-recapture calculations in that only two periods, the initial period when animals are marked and the second period when the sample (n) is collected, are used. If several capture periods are used, sequential formulas, such as the Schnabel method, must be utilized. In each sample taken, all unmarked animals are marked and returned to the population; marking and recapture are done concurrently. The sequential approach makes allowances for the increasing number of marked animals in the population (M). M usually increases with time, but it may decrease with known mortality or removal of marked animals from the population. All the assumptions for the Lincoln-Petersen index should also be met for the Schnabel method to produce accurate population estimates.

The Schnabel method formula for multiple sampling periods is

$N = \Sigma(C_T M_T)/R_T.$

Each line of the Schnabel method calculation corresponds to a line in the Lincoln-Petersen index calculation. C represents the number captured during sampling time one, M is the total marked, and R is the number of recaptures. The subscript T is the sample time.

An example of the Schnabel calculation can be demonstrated with the following data. For four days of trapping, the following data were obtained: day one, five animals captured, no recaptures; day two, ten animals captured, five previously marked animals in the population at the start of the second day of trapping, three previously marked animals in the day two sample; day three, fifteen animals captured, fifteen previously marked animals in the population at the start of day three of trapping, three previously marked animals in day three of trapping; day four, ten animals captured, twenty previously marked animals in the population at the start of day four, and four previously marked animals among the animals captured on day four. For the four days of trapping, then, ten animals were recaptured. From these data, the Schnabel method estimate of population density would be

$N = (5 \times 0) + (10 \times 5) + (15 \times 15) + (10 \times 20) / (0 + 3 + 3 + 4)$ with $N = 50.5$.

Another approach to the estimation of animal numbers has been developed that differs from the usual approach to mark-recapture population estimation calculations. The Eberhardt method is based not on the ratio of marked to unmarked animals but on the number of times an individual is recaptured during the recovery operations; the assumption is that this recapture frequency is related to the total population size. The relationship is believed to be a hypergeometric one. Eberhardt's formula is

$N = n / (1 - (n/t))$,

where n = number of individuals handled in recovery operations and t = the total number of captures of individuals. In this tally, t will always be greater than n unless all animals are captured only once. To use these modified mark-recapture data, individual animals must be recognizable. This calculation has been used for a number of mice studies. For example, if twenty different kangaroo rats in the Mojave Desert were captured thirty-five times, the estimated population would be $N = 20 / (1 - (20/35))$; $N = 46.67$ animals.

These represent only a few of the mark-recapture formulas available for the estimation of population size and animal density. Many other, more complicated, formulas for calculation of population estimates based on mark-recapture ratios (such as the Schumacher-Eschymeyer, DeLury, and Jolly procedures) exist, but the simplest formulas often provide the best, most usable estimates of animal numbers.

USES OF THE METHOD

The use of mark-recapture procedures allows the biologist to determine numbers of animals present in a given area. Without these numbers, the future of these animals cannot be predicted. This knowledge allows appropriate management strategies to be developed, either protecting them or providing needed control activities, such as spraying insecticides on agricultural crops before severe economic damage to the crops results. Information about animal populations is essential for determining the effects of environmental changes or human activities, such as construction, on animal communities.

The populations of different areas may be compared; population numbers between seasons or years may also be studied. The fact that certain areas have high numbers of individuals implies that these areas have good conditions for them. Wildlife managers need to know why these areas have higher numbers so that they can improve the relevant conditions in other areas. Learning this would not be possible without knowing population sizes on these respective sites. The success of management work can be judged by changes in population size.

Mark and recapture techniques are applicable to more population situations than are the other options for population estimation. The wide variety of methods available for marking animals often allows previously marked animals to be identified without actually being handled. This minimizes the stress on the marked animal by reducing human contact.

If information on how long an animal lives in the wild is needed, individual marking from population work can also serve this purpose. From their recapture points, the area used by individuals on a daily, seasonal, and yearly basis, known as the home range, can be determined. The degree of movement of individuals within the population can also be estimated. These data are economical to obtain, because the marking and capture of the animals for the population estimate also funds the cost of obtaining home range, movement, and longevity data. The rate of exploitation of the population and the rate of recruitment of new members into it can also be calculated from the ratio of marked to unmarked animals collected during the population studies. The biology of organisms in the wild cannot be adequately studied without accurate estimates of their population levels and fluctuations being known. Mark and recapture procedures are among the important scientific tools for the collection of this information.

—*David L. Chesemore*

Further Reading

Blower, J. G., L. M. Cook, and J. A. Bishop. *Estimating the Size of Animal Populations*. Allen & Unwin, 1981.

Davis, D. E. *Handbook of Census Methods for Terrestrial Vertebrates*. CRC Press, 1982.

Emery, Lee, and Richard S. Wydoski. *Marking and Tagging of Aquatic Animals: An Indexed Bibliography*. US Department of the Interior, Fish and Wildlife Service, 1987.

Krebs, C. J. *Ecological Methodology*. 2nd ed. Harper & Row, 1999.

Kunz, T. H. *Ecological and Behavioral Methods for the Study of Bats*. Smithsonian Institution Press, 1988.

Otis, D. L., K. P. Burnham, G. C. White, and D. R. Anderson. *Statistical Inference from Capture Data on Closed Animal Populations*. Wildlife Society, 1978.

Scheiner, Samuel M., and Jessica Gurevitch, eds. *Design and Analysis of Ecological Experiments*. 2d ed. Chapman and Hall, 2001.

Schemnitz, Sanford D., ed. *Wildlife Management Techniques Manual*. 4th ed. Wildlife Society, 1980.

Seber, G. A. F. *Estimation of Animal Abundance and Related Parameters*. 2nd ed. Reprint. Macmillan, 1994.

Skalski, John R., and Douglas S. Robson. *Techniques for Wildlife Investigations: Design and Analysis of Capture Data*. Academic Press, 1992.

Marsupials

Fields of Study

Animal Anatomy; Ecology; Evolutionary Biology; Animal Physiology; Zoology

Abstract

Marsupials are pouched animals that form a distinctive group within the class Mammalia. *They possess the diagnostic features of typical mammals, including high and stable body temperature, furry pelt, simple lower jaw, and mammary glands. However, there are other features that distinguish them from what are considered to be typical mammalian features.*

Key Concepts

gestation: the period of time between impregnation and birth

mammary glands: in female mammals, the organs that produce milk for the nursing of young

marsupium: the "pouch" in which infant marsupial animals are carried by their maternal parent

thylacine: a carnivorous marsupial resembling both wolf and tiger in appearance, commonly called the "Tasmanian tiger"

Marsupials Are Mammals Too

Marsupials are pouched animals that form a distinctive group within the class *Mammalia*. They possess the diagnostic features of typical mammals, including high and stable body temperature, furry pelt, simple lower jaw, and mammary glands. However, there are other features that distinguish them from what are considered to be typical mammalian features.

The kangaroo is the most commonly known marsupial, but a vast array of marsupials exist. Most marsupials are crepuscular or nocturnal, so they are often inconspicuous even in zoos. Most marsupials are found in Australia and New Zealand. Australian authorities impose strict export sanctions to protect their numerous endangered species.

There are three families of marsupials, *Didelphidae, Microbiotheriidae,* and *Caenolestidae,* that inhabit South and Central America. The opossums of North and South America are the most diverse of three families of extant marsupials outside of Australia. The American marsupials alive today are mostly small, ranging from mouse to rabbit size. These are generally either carnivorous or omnivorous, living in forests and feeding on insects. The only naturally occurring marsupial found in the United States is the Virginia opossum, *Didelphis virginiana*, which ranges across North America and into southern Canada.

For the most part, marsupials have remained curiosities for the general public. Humans have not traditionally exploited marsupials on a large scale; they have not commonly been kept as pets, the meat of larger kangaroos is mostly used only for dog and cat food, and the furs of only a few marsupials have commercial value. However, hunting as well as habitat destruction and the introduction of competing species have posed threats to some marsupials, especially in Australasia.

CLASSIFICATION AND PHYSIOLOGY

Marsupial divisions and categorizations are the subject of ongoing debate, but classifications commonly include around nineteen families, eighty-three genera, and over 270 species. Marsupials make up the infraclass *Metatheria* in the subclass *Theria*, with seven orders. There is no other group within the higher mammals that contains such a diversity of species, genera, and families as the marsupials.

Marsupials are an example of adaptive radiation. Their adaptation to their varied habitats has led to their enormous diversity of forms and the environmental niches they occupy. They are also an example of convergent evolution, as indicated by the similarities between marsupials in Australasia and placental mammals in the rest of the world. The marsupial gliders resemble the flying squirrels and lemurs, the Tasmanian tiger or thylacine was doglike, and marsupial moles resemble eutherian moles. There are many physiological similarities as well. Wombats process grasses and sedges as horses do and numbats feed on termites as anteaters do.

With few exceptions, marsupials are not conspicuous in coloration or any external physical attributes. They range in adult size from 65 kilograms to only 3 grams, though the majority of species are small, ranging in size between that of a mouse and of a small rabbit. They developed from small carnivores into herbivores the size of hippopotamuses. The larger marsupials died out only some thousands of years ago.

Marsupials exhibit a diverse range of traits. They are found in habitats as diverse as freshwater, alpine areas, hot deserts, and tropical rain forests. While kangaroos and their relatives famously spring about on their hind legs, other species climb, glide, burrow and even swim. Their diets range from purely insects to vertebrates, fungi, underground plant roots, bulbs, rhizomes and tubers, plant exudates such as saps and gums, seeds, pollen, terrestrial grasses, herbs and shrubs, and tree foliage. Because of this vast diversity it is impossible to categorize marsupials with a simple description. Instead, the physiology of marsupials must be used to categorize them.

Kangaroos.

REPRODUCTION AND DEVELOPMENT

There are three types of reproductive patterns in mammals. There are monotremes, which are egg-laying mammals, such as the platypus. There are placentals, whose embryo develops inside the uterus, and the placenta formed in the uterus provides nutrients to the developing embryo. In placentals, the offspring are born completely developed, as in humans. Finally, there are marsupials, which functionally fall in between monotremes and placentals.

Marsupials are often thought of as pouched mammals. Their embryo develops inside the uterus but, unlike placental mammals, the marsupial is born very early in its development. It completes its embryonic development outside the mother's body, attached to teats of abdominal mammary glands, which are often but not always enclosed in a pouch called the marsupium. The marsupium is formed in diverse ways,

Virginia opossum (Didelphis virginiana) *in a juniper tree in northeastern Ohio.*

ranging from the "primal pouch" (the annular skin creasing around each teat), to common marsupial walls surrounding all teats, and finally to a closed marsupium, which can be opened to the front or to the rear. The helpless embryonic form has forelimbs that are strong enough to climb from the birth canal to the mother's nipples, where it grabs on and nurses for weeks or months depending on the species.

The gestation period is eight to forty-two days, after which the young is carried in the marsupium for between thirty days and seven months. When the young are born, their eyes and ears are closed, hind limbs and tails are stumps, and they are completely hairless. Their olfactory senses are greatly developed, as are their tactile senses, allowing them to navigate their way to the marsupium.

Litter sizes range from one to twelve per birth. The young are weaned anywhere between six weeks and one year. The relationship between mother and offspring is long lasting in many species. Sexual maturity is reached between ten months and four years, depending on the species. The longer range is associated with the male koala.

While reproductive traits are the best-known distinguishing feature, other physiological features are also characteristic. Marsupials are usually woolly, with shortened forelimbs and elongated hind limbs. (In kangaroos, these physical features allow locomotion in a hopping movement only; however, at an equivalent speed, allowing for the differences in weight, a hopping kangaroo uses less energy than a running horse or dog.) In several families, second and third toes of the hind foot function as grooming claws and the first toe is always clawless, except in the shrew opossum. Vision is usually poorly developed and olfactory, tactile, and auditory senses are well developed.

BEHAVIOR

Marsupials range from pure carnivores to pure herbivores, with all the intermediate stages in between. They are usually nocturnal and crepuscular. Some species are solitary, while others live in family groups.

In all mammals, because of the milk produced by the mother, male assistance in feeding the young is less important than in birds, for example. In many marsupials, the role of the male is further reduced because the pouch takes over the functions of carrying and protecting the young and keeping it warm. A female's need for assistance in rearing young does not appear to be an important factor promoting the formation of long-lasting male-female pairs or larger social groups. The majority of marsupial species mate promiscuously. There are few examples of long-lasting bonds and they do not live in groups. Some species form monogamous pairs and harems. It is hypothesized that the lack of frequent examples of this sort is due to the lack of external pressures.

EVOLUTION

Marsupial evolutionary development is not yet clearly understood. Fossil records suggest that they may have evolved simultaneously with the placental animals about 100 million years ago, in the Cretaceous Period. Many of the oldest geological finds come from the recent Upper Cretaceous of North America, about seventy-five million years ago. Although there was some development of marsupials in North America, they later declined as placentals increased in diversity. In contrast, South America has a considerable diversity of marsupial fossil forms, indicating their persistence for more than sixty million years. Several families of living and fossil marsupials are known from South America. About two to five million years ago, a land connection between the two Americas was established again, and more placental animals reached South America, including carnivores such as the jaguar. In the face of such competition, the large carnivorous marsupials disappeared, but the small omnivores have persisted successfully to the present day. Some of them moved north to colonize in North America.

The fossil record suggests that marsupials first evolved in North America before spreading to Australasia, where they particularly thrived. The earliest marsupials found in Australia are dated from approximately twenty-three million years ago. Most modern families and forms were clearly established by that time. The relative lack of fossil records of marsupials in Asia or Africa makes the most likely route of migration from South America to Australia via Antarctica. At that time, all three southern continents were united in the land mass known as Gondwanaland. This mass of land began breaking up 135 million years ago, with South America and Antarctica still being connected until about 30 million years ago. One land mammal fossil has been found in Antarctica which is a marsupial dated to be forty million years old. The Australian plate then gradually drifted northward for another thirty million years before reaching its current latitude. This long isolation allowed the extensive development of the marsupials in Australia in the absence of competition from other placentals.

As marsupials evolved in Australia, so did the placentals in the rest of the world, filling the same ecological niches. In many cases, they adopted similar morphological solutions to ecological problems. One example is the convergent evolution of the carnivorous Tasmanian devil, a marsupial, and placental wolves of other continents. The marsupial mole is very similar in form to the placental mole. The marsupial sugar glider and the two flying squirrels of North America are also very similar.

HABITAT

The arrival of European settlers and the influx of new species—sheep, cattle, rabbits, foxes, cats, dogs, donkeys, and camels—have caused a large-scale modification of the marsupial's habitat in Australia. The first major change, however, was in the late Pleistocene, with the extinction of whole families of large terrestrial marsupials. Included in this extinction was Diprotodon, the largest browsing kangaroo. It is likely that the climatic fluctuations increased aridity and reduced the available favorable habitat. Many of the species were already under stress when humans arrived, and hunting and the use of fire by Aboriginal Australians may or may not have factored into their extinction.

European colonization, in contrast, is known to have had a direct impact on marsupial ecology. In addition to outright hunting and killing to prevent damage to crops, marsupials have been pressured by general habitat destruction and development. Perhaps even more destructive has been the introduction of foreign species that prey on or compete with marsupials. Approximately nine marsupial species have become extinct in Australia and fifteen to twenty have suffered gross reduction in range in the modern era. The most affected have been small kangaroos, bandicoots, and large carnivores such as the thylacine and native cats.

Not all the environmental changes have been unfavorable for marsupials. Many of the larger herbivores have fared well with the advent of ranching and available grazing land and watering holes already set up for stock animals. As these marsupials become competition for sheep and cattle, Australian authorities have developed programs to keep their population controlled by allowing a certain number to be shot. Most species of marsupials have little or no importance as pests and their continued existence depends largely on the maintenance of sufficient habitat to support secure populations. The control of feral foxes and cats is very important to keep predation limited.

Marsupials outside Australia have tended to fare reasonably well despite human population growth and encroachment. Indeed, some, such as the common opossum, have thrived. However, the destruction of habitats in South and Central America are considered a long-term risk to many species.

—*Donald J. Nash*

Further Reading

Dawson, Terence. *Kangaroos: The Biology of the Largest Marsupials.* Comstock, 1995.

Hume, Ian D. *Marsupial Nutrition.* Cambridge University Press, 1999.

Lee, Anthony K., and Andrew Cockburn. *Evolutionary Ecology of Marsupials.* Cambridge University Press, 1985.

"Marsupial." *Animals & Plants,* San Diego Zoo, 2018, animals.sandiegozoo.org/animals/marsupial. Accessed 31 May. 2018.

Myers, P., et al. "Metatheria: Marsupial Mammals." *Animal Diversity Web*, University of Michigan Museum of Zoology, 2018, animaldiversity.org/accounts/Metatheria/classification/. Accessed 31 May. 2018.

Paddle, Robert. *The Last Tasmanian Tiger: The History and Extinction of the Thylacine*. Cambridge University Press, 2000.

Saunders, Norman, and Lyn Hinds, eds. *Marsupial Biology: Recent Research and New Perspectives*. New South Wales University Press, 1997.

Tynedale-Biscoe, Hugh, and Marilyn B. Renfree. *Reproductive Physiology of Marsupials*. Cambridge University Press, 1987.

Metabolic Rates in Animals

Fields of Study

Biothermodynamics; Endocrinology; Biochemistry

Abstract

All living things require energy to sustain life as well as to carry out normal activities, develop, and grow. All the chemical reactions that allow energy acquisition and use are collectively called metabolism. Animals usually obtain energy by breaking down food in the presence of oxygen, so the amount of oxygen they use can be considered a measure of their rate of metabolism. Some organisms that live in oxygen-poor environments gain energy without using oxygen through anaerobic metabolism. Since all energy used by living things is eventually converted to heat, metabolic rate can be defined as the total amount of heat produced by an organism in a certain time period.

Key Concepts

adenosine triphosphate (ATP): the primary energy storage molecule in cells; links energy-producing reactions with energy-requiring reactions

anabolism: a series of chemical reactions that builds complex molecules from simpler molecules using energy from ATP

basal metabolic rate (BMR): the rate of metabolism measured when the animal is resting and has had no meals for twelve hours; used to compare different species

catabolism: a series of chemical reactions that break down complex molecules into simple components, usually yielding energy

electron transport chain: a series of electron carrier molecules found in the membrane of mitochondria; oxygen is used and ATP is made at this site

hormone: a chemical messenger molecule within organisms; acts as a regulator of cell activities

mitochondrion (pl. mitochondria): a cell organelle found in plants, fungi, animals, and protists; the site of most aerobic metabolism

specific metabolic rate: the rate of metabolism per unit body mass (calories per gram per hour)

FUELING THE ENGINE OF LIFE

All living things require energy to sustain life as well as to carry out normal activities, develop, and grow. All the chemical reactions that allow energy acquisition and use are collectively called metabolism. Animals usually obtain energy by breaking down food in the presence of oxygen, so the amount of oxygen they use can be considered a measure of their rate of metabolism. Oxygen consumption is not always a good indicator of rate of energy conversion and use. Some organisms, such as fish, that live in oxygen-poor environments gain energy without using oxygen through anaerobic metabolism. Since all energy used by living things is eventually converted to heat, metabolic rate can be defined as the total amount of heat produced by an organism in a certain time period.

Rates of metabolism measured on whole organisms can be used to study the effects of factors such as temperature, size, age, and sex on rates of energy use. Studies of metabolic rates of different species of organisms are used to determine food requirements and energy adaptations in different environments. Metabolic rate can also be measured on isolated tissues, cells, or cell organelles in order to study the different biochemical reactions that occur in tissues and cells.

FACTORS AFFECTING METABOLIC RATE

One of the most important factors that affects metabolic rate is the temperature of the organism, since within limits all chemical reactions of metabolism proceed faster at higher temperatures. The internal temperature of most invertebrate animals, fish, and amphibians is the same as the temperature of the environment in which they live. Such organisms are called poikilotherms. In poikilothermic organisms, metabolic rate increases as the environmental temperature increases. Such organisms move slowly and grow slowly when the temperature is cold, since their metabolic rate is very low at cold temperatures.

To compare the metabolic rates of different poikilotherms, one must measure their rate of metabolism under standard conditions. Standard metabolism is usually defined as the rate of energy use when the animal is resting quietly, twelve hours after the last meal, and is at a temperature of 30 degrees Celsius; however, for small invertebrates, protists, and bacteria, only temperature is usually controlled. Most reptiles, birds, and mammals can maintain their body temperature at a constant level even when the environmental temperature changes greatly. Such organisms are called homeotherms.

Birds and mammals can maintain their body temperature through internal heat production (endothermic homeothermy), while reptiles must acquire the necessary heat from their environment by changing their behavior, body posture, or coloration (ectothermic homeothermy). Most endotherms can maintain a constant body temperature over a range of temperatures (thermal neutral zone) without affecting their rate of metabolism. At temperatures outside the thermal neutral zone, metabolic rate increases to maintain constant body temperature. At colder temperatures, increased muscular activity and shivering require increased metabolic rate. Sweating and panting can increase the rate of metabolism at high temperatures. To compare the metabolic rates of endothermic homeotherms, scientists measure the basal metabolic rate (BMR), which is also referred to as the energy cost of living.

Body size (mass in grams or kilograms) is another major factor that affects basal or standard metabolic rate. A 3,800-kilogram elephant has a metabolic rate of about 1,340 kilocalories per hour, while a 2.5-kilogram cat has a rate of 8.5 kilocalories per hour, which means an elephant needs about 150 times as much food as a cat each day. A different picture emerges if one looks at energy use per kilogram of mass. For each kilogram of mass, the cat actually uses ten times the energy of a kilogram of elephant. Metabolism per unit of mass (specific metabolism) decreases as body size increases for all organisms. Since small organisms or cells have a larger surface area relative to their total volume than large ones, they can lose more heat from the surface. For two organisms of the same mass, the taller or thinner organism will have a larger surface area and higher BMR than a shorter, fatter one. More oxygen, food, and waste products can diffuse across the larger surface area; thus cell size in single-celled organisms and in different types of cells in multicellular organisms is limited by rate of energy metabolism. Small animals move faster, breathe faster, and their hearts pump faster. A mouse has a heart rate of six hundred beats per minute, while an elephant's heart rate is thirty beats per minute. Even the length of life appears to be related to the faster metabolic rate of these small creatures. Mice live only two to three years, while an elephant can live sixty years or longer.

Age and sex also influence basal metabolism. Young animals that are growing rapidly have a higher BMR than adults. As adults age, the proportion of skeletal muscle decreases and the BMR declines. Muscle tissue is metabolically very active even at rest, contributing to the higher BMR in males as opposed to females, since males have a higher proportion of body mass that is muscle. Physical or emotional stress can increase metabolic rates by increasing the catabolism of fats through the action of the hormones epinephrine and norepinephrine.

Skeletal muscle activity causes rapid short-term increases in metabolic rate. In humans, for example, a few minutes of vigorous exercise causes a twentyfold increase in the rate of metabolism, and the metabolic rate remains high for several hours. Walking, swimming, running, and flying require more energy than sitting still; however, each of these activities influences metabolic rate differently. Water is denser and has higher viscosity and resistance to movement compared to air, so more energy must be expended to swim than to walk at a given speed. Running also increases energy use, and the faster one runs, the more energy is required. Large animals, however, increase their rate of metabolism less per kilogram of mass than do small animals, so there is a metabolic advantage to large body size. Intriguingly, for the same size animal, flying is less energy-expensive than running.

MEASURING METABOLISM

Rate of metabolism can be measured as the amount of heat produced by an organism in a time period. The traditional unit of heat is the calorie; a kilocalorie is one thousand calories. The two terms are frequently confused in popular literature, although the word Calorie is traditionally used to indicate kilocalorie. In the international system of units, heat is measured in joules, and 1 calorie is equal to 4.184 joules.

Metabolic rate can be determined from the energy budget of an animal. If the total energy excreted in urine and feces is subtracted from the total energy in food eaten during a period of time, the result would be a measure of metabolic rate. The energy content of food and waste products can be determined by burning these materials in a calorimeter. The amount of heat produced is used to raise the temperature of a known amount of water. This method assumes that the organism is not growing or changing the amount of fat stored during the measurement period. It is also difficult to control metabolic activity of gut microorganisms. Although this technique is cumbersome, it may be the best way to assess energy metabolism in a normally active state for animals in their natural habitat.

More controlled measures of basal and standard metabolism can be made by isolating the animal in a calorimeter and directly measuring heat produced. This method is more accurate than the energy budget approach but still assumes that no new molecules are being produced and no activity or work is being performed. This technique is most useful for birds and small mammals that have relatively high rates of metabolism. Normal behavior and function may be altered by the confined conditions.

Indirect calorimetry is the most often used method in assessing metabolic rate in whole organisms, isolated cells, and cell components. Some factor related to energy use, such as oxygen consumed or carbon dioxide produced, is measured as an index of energy use. For aerobic metabolism, the amount of heat produced is related to oxygen use by the organism. Respirometry is the method used to monitor the oxygen used and carbon dioxide produced by an organism in a closed chamber. Oxygen consumption can be measured by absorbing carbon dioxide with soda lime and measuring the change in gas pressure in the closed system by a manometer. Oxygen electrodes can also be used to measure the decrease in oxygen concentration in water within the chamber if the animal is a water-dweller. For air-breathing animals, oxygen in the gas phase can be measured by a mass spectrometer. Carbon dioxide can also be measured by an infrared unit. In such closed systems, the gases can be monitored in the air as it enters and leaves the chamber. Respirometry can also be accomplished in open systems if the animals are fitted with breathing masks and the respired air is collected and analyzed. Oxygen consumption is a good index of metabolic rate for most animals at rest, since most of their metabolism is aerobic. Animals that live in oxygen-poor environments, such as internal parasites and mud-dwelling invertebrates, and all animals under extreme exercise often metabolize anaerobically.

To translate the amount of oxygen used into heat produced, one must know the proportion of fat, carbohydrate, and protein in the diet, since the amount of heat produced for each liter of oxygen consumed differs. In practice, it is usually assumed that only carbohydrates are being used.

Assimilating and digesting food cause large increases in metabolic rate. This increase reaches a maximum about three hours after a meal and remains above basal level for several hours in birds and mammals and up to several days in poikilotherms. Foods differ in the amount of increase in metabolic rate. Proteins, for example, cause about three times the increase in rate compared to carbohydrates or fats. The increase partially results from increased activity in cells of the digestive tract and partly from higher activity of liver and muscle cells preparing these foods for storage. Basal metabolism must then be measured after a twelve-hour fast to minimize this effect.

Since body temperature, body size, and activity affect metabolic rate of organisms, one can see that available food supplies, oxygen levels, and environmental temperatures limit the physiology of energy metabolism in different habitats. Scientists study metabolic rates in whole organisms to explain their food habits and their distributions in different habitats and to calculate energy requirements for raising animals under different conditions.

—*Patricia A. Marsteller*

Further Reading

Peters, Robert H. *The Ecological Implications of Body Size.* Cambridge University Press, 1983.

Randall, David, Warren Burggren, Kathleen French, and Russell Fernald. *Animal Physiology: Mechanisms and Adaptations.* 4th ed. W. H. Freeman, 1997.

Salway, J. G. *Metabolism at a Glance.* 2nd ed. Blackwell Scientific Publications, 1999.

Schmidt-Nielsen, Knut. *Animal Physiology: Adaptation and Environment.* 5th ed. Cambridge University Press, 1997.

Solomon, Eldra, et al. *Biology.* 3rd ed. Saunders College Publishing, 1993.

Metamorphosis

Fields of Study

Developmental Biology; Biochemistry; Ethology

Abstract

Larval creatures look nothing like the adults they will form. They are specialized primarily for growth and feeding and are unable to reproduce. At a certain time in their growth, often in response to signals from hormones or the environment, the larva undergoes a second spurt of development called metamorphosis. The word "metamorphosis" comes from the Greek word for transformation and is today defined as a major change in body form that occurs after embryonic development is completed.

Key Concepts

corpora allata: a gland in insects that synthesizes and secretes juvenile hormone (JH)

ecdysone: a hormone that triggers both molting and metamorphosis in insects as well as in many other species of animals

imaginal disk: flat sheets of cells within an insect larva; these cells will change shape during metamorphosis and form the external structures of the adult

juvenile hormone (JH): a species-specific hormone which controls whether a molt will produce a larger larva or initiate metamorphosis

larva: the reproductively immature feeding stage in the development of many species of animals, including those insects which undergo complete metamorphosis

nymph: the sexually immature feeding stage in the development of those insects which undergo incomplete metamorphosis

prothoracic gland: the gland where ecdysone is made in insects

prothoracicotropic hormone (PTTH): a hormone made in the brain of insects which stimulates the prothoracic gland to make ecdysone

pupa: the stage of insect development during which metamorphosis occurs

ANOTHER DEVELOPMENT

Unlike mammals, most animals become adults by first going through a distinctly different immature, or larval, stage of development. Larval creatures look nothing like the adults they will form. They are specialized primarily for growth and feeding and are unable to reproduce. At a certain time in their growth, often in response to signals from hormones or the environment, the larva undergoes a second spurt of development called metamorphosis.

The word "metamorphosis" comes from the Greek word for transformation and is today defined as a major change in body form that occurs after embryonic development is completed. Metamorphosis is very widespread in the animal kingdom but it has been studied most thoroughly in three different groups of organisms: the arthropods (crabs, lobsters, and insects), the amphibians (frogs and salamanders), and the echinoderms (sea urchins and starfish).

Metamorphosis affects the ways animals eat, breathe, and move; it can also change the nature of the environment in which they live. In echinoderms, the larval stage is microscopic and free swimming, while the adults are large and move very little if at all. The changes in body form thus coincide with significantly different ways of feeding and moving.

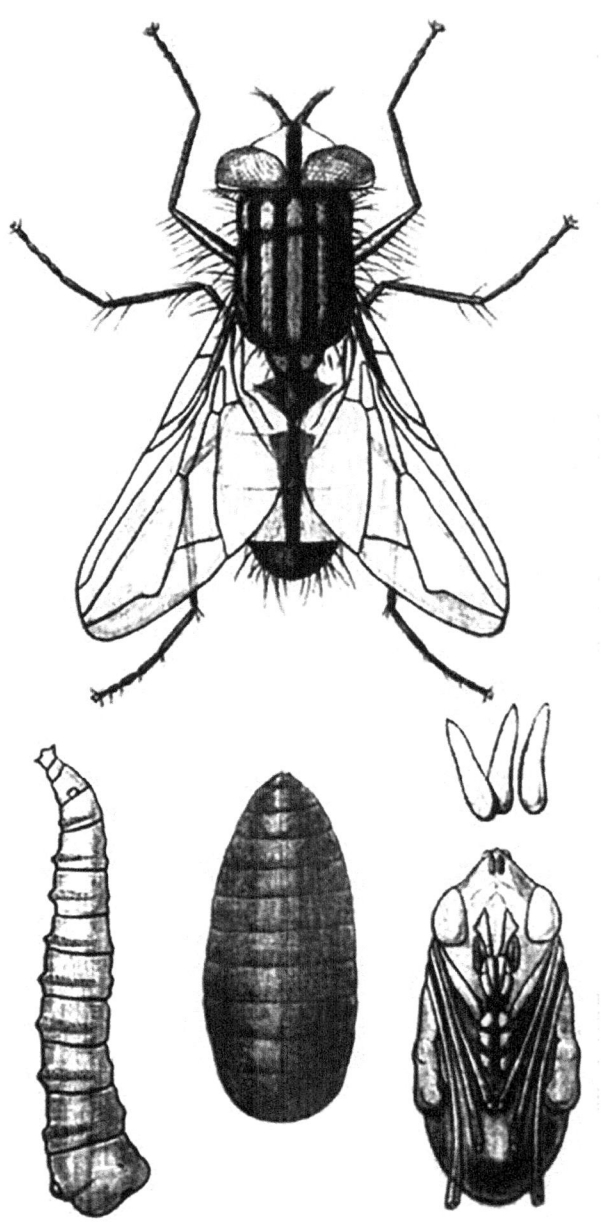

A set of drawings of the life cycle of the flesh fly.

AMPHIBIAN METAMORPHOSIS

In amphibians, metamorphosis prepares the organism for its transition from living in water to living on land. Internally, the digestive system changes to accommodate the new diet as the animal switches from consuming plants to animals. Sense organs like the eyes and ears also change to adapt to functioning predominantly in air. Finally, many of the chemical reactions which occur in the individual cells of the frog also change at this time.

Metamorphosis in amphibians is controlled by a pair of hormones. Prolactin, a protein secreted by the anterior pituitary gland, controls the rate of growth of the tadpole and suppresses metamorphosis. Thyroxine is a modified amino acid made in the thyroid gland of the tadpole and causes metamorphosis to begin. After the tadpole has grown to a certain minimal size, the thyroid gland is stimulated by environmental conditions to produce large quantities of thyroxine, which reverses the suppression exerted by prolactin and begins metamorphosis. The hormones pass through the tadpole's circulation and instruct different tissues to activate and deactivate different sets of genes that cause some tissues to degenerate, others to change, and others to grow.

These same hormones, prolactin and thyroxine, are produced by other vertebrates; nature often uses the same molecules to produce very different results in different animals. In humans, thyroxine regulates the rate of metabolism, while prolactin is crucial for milk production in nursing women. In fish, however, prolactin is crucial for keeping the cell's content of salt in balance.

INSECT METAMORPHOSIS

The hormonal control of metamorphosis in one group of animals, the insects, has yielded basic information about how genes and hormones interact to guide development in all animals. This process is understood in greater detail in insects than in any other group of animals.

Insects, like all arthropods, have a rigid exoskeleton, called a cuticle, that supports their body mass and allows the attachment of muscles for movement. Because it is external and fixed in size, however, the cuticle must be shed periodically for growth to occur. The process of insects shedding their cuticle and producing a new, larger one is called molting, and the number of molts normally occurring is regulated by the insect's genes. Many insects have developed complex behaviors to ensure that they molt in a secluded location and avoid predation. Without their hard, external covering, they are relatively defenseless.

About 10 percent of all insect species, such as the grasshoppers and true bugs, go through a process of incomplete metamorphosis. Here, the egg hatches into a juvenile form called a nymph that resembles the adult but is much smaller and is not capable of reproduction. At each molt, the nymph sheds its cuticle and grows before the newly produced cuticle hardens. When the signal for metamorphosis arrives, the molt produces an adult, often with wings, but, more important, fully capable of reproduction. Male and female adults can then mate and lay a new generation of eggs.

Complete metamorphosis, on the other hand, is a very different process that is undergone by 90% of all species of insects, including the ants, bees, flies, butterflies, and moths. Here, the larval form looks nothing like the adult. For example, the larval form of a butterfly is a caterpillar and the larval form of a fly is a wormlike maggot. Each larva undergoes a series of molts so as to be able to grow. The larva of the cecropia moth increases in mass by five thousand times during its larval development. When the trigger for metamorphosis occurs, the insect undergoes a radical change. The larva stops feeding and moving, anchors itself to a twig, leaf, or rock, and either spins a cocoon or encloses itself in its own hardened cuticle.

SILKWORM METAMORPHOSIS

The silkworm, *Bombyx mori*, undergoes complete metamorphosis, and its development begins when the egg hatches and includes five distinct larval stages. Each stage is larger than the one preceding as the larva eats, voraciously consuming several times its own body weight in food each day. At the end of the fifth larval stage, the molting event that follows is very different from the previous larval molts as the caterpillar spins a cocoon made out of silk (which is the commercially valuable product of this organism) and becomes a pupa. The next molt occurs within the pupal case when metamorphosis begins. When the adult has formed, the cocoon breaks open and the adult emerges to begin reproduction.

Metamorphosis involves the complete replacement of one body form with another. Inside the pupal case, the larval tissues break down and their molecules are reutilized in the construction of the cells and tissues of a very different looking animal, the adult. Certain groups of cells, called imaginal disks, along with the larval brain are generally the only tissues that are not broken down in this process. Opening a cocoon at this time shows that it is filled with a white, milky sap and little else as the caterpillar has been completely broken down.

The imaginal disks, round, flat sheets of cells, begin to evert or "telescope" and form the external structures characteristic of the adult cuticle. There is an imaginal disk for each eye, for each antenna, for the two or four wings, and for each of the six legs. These structures attach and the adult insect is constructed. In the well-studied fruit fly, *Drosophila*, this process takes about a week, while it can take months for other insects.

HORMONAL REGULATION OF METAMORPHOSIS

In insects, three very different hormones combine to regulate the timing of both molting and metamorphosis. Each of these three hormones is produced in a different tissue, and each has a different chemical structure and mode of function. The signal to molt originates in a small group of cells within the caterpillar's brain in response to neural or environmental signals. The hormone produced there, prothoracicotropic hormone (PTTH), is a small protein that passes through the insect's hemolymph (blood) to all parts of the body. As is true of all hormones, only certain target tissues are genetically programmed to respond to the production of this hormone. In this case, the prothoracic gland responds by producing a second hormone called ecdysone, the molting hormone. Ecdysone is a steroid, a chemical derivative of the cholesterol that the insect requires in its diet. Ecdysone is not actually the molting hormone itself, but must first undergo some minor chemical changes before it becomes active. More important, ecdysone and its derivatives are the molting hormone not only for all insects, but for all arthropods and many other animals. This chemical signal thus evolved before divergence of these organisms from a common ancestor.

Different tissues respond differently to ecdysone, but the hormone's major effect is to trigger molting by causing the hardening of the cuticle and the separation of the living cells beneath the cuticle from it. The cuticle then dries and cracks and the larva can then emerge from its old skin and grow. If PTTH production stops, the amount of ecdysone released from the prothoracic gland also falls. This happens normally in the period between molts, but can also occur during a molt. In cecropia moths, the level of PTTH drops during the pupal stage and the subsequent drop in ecdysone production causes diapause, a programmed pause in development. The pupa will remain in diapause until an environmental signal, consisting of a minimum of two weeks in the cold followed by a normal spring warming, triggers the resumption of PTTH secretion and the completion of metamorphosis.

Ecdysone acts specifically on certain groups of insect genes and not others. Many genes required for larval functions are turned off in response to ecdysone, while those genes required for molting and metamorphosis are turned on. This effect can be seen when imaginal disks are placed in a culture dish along with physiological levels of ecdysone. The disks stop growing and begin metamorphosis. Such development can even produce normal-looking legs floating free in a culture dish. The changes in shape of the disk cells can be directly attributed to the function of genes turned on by ecdysone.

The hormone ecdysone therefore not only causes metamorphosis, but also triggers certain simple molts. The third hormone involved in the control of metamorphosis is responsible for this choice, the choice to molt and grow or to undergo metamorphosis to the adult form. This compound is called juvenile hormone (JH). JH is produced in a gland called the corpora allata, and the hormone has yet a third chemical structure, a derivative of a class of molecules called terpenes. Unlike ecdysone, however, the JH of each species has a different chemical structure.

When a molt is triggered by the production of ecdysone, the type of molt that occurs will be determined by the level of JH present. During larval development, the amount of JH in the hemolymph is high and the tissues respond to ecdysone by undergoing a molt to become a larger larva. Later in development, the corpora allata stops producing JH and any new molt triggered by ecdysone causes the organism to proceed to the pupal stage and begin metamorphosis. The interaction of ecdysone and JH thus governs which type of molt will occur.

Although these hormones are crucial for controlling molting and metamorphosis, they play other roles as well. For example, adult female insects produce large quantities of ecdysone and adult males do not because ecdysone is important to egg production.

LABORATORY STUDIES OF METAMORPHOSIS

Various parts of the elaborate interplay of the three hormones that regulate metamorphosis in insects were discovered in a number of laboratories. The role of ecdysone was discovered first by Carroll Williams and Vincent Wigglesworth, who found that the prothoracic gland of the larva was responsible for producing a substance that triggered metamorphosis. They tied a fine string around the middle of a cecropia larva and observed that only the front half underwent metamorphosis. The signal was thus produced somewhere in the front half and could not reach the rear of the insect. The prothoracic glands were later discovered to be the source of the molting signal when glands transplanted to the rear half of a ligated larva caused metamorphosis to occur there as well.

Ecdysone was painstakingly purified and chemically identified by Peter Karlson's group in Germany. In a monumental effort, they extracted twenty-five milligrams of pure ecdysone from a ton of silkworms. Much of the difficulty encountered in this isolation stemmed from the lack of an easy method for identifying the presence of ecdysone in the extract they were producing.

A tedious and relatively insensitive bioassay was used. In this assay, the extract to be tested during the isolation was dissolved in an organic solvent and painted on the cuticle of an insect. If the extract contained ecdysone, the larva would begin to molt. Although time-consuming and not very reproducible, the bioassay allowed the purification of ecdysone from insects.

The role of the corpora allata and JH was found by a similar set of experiments. Wigglesworth found that if he decapitated a fourth-stage *Rhodnius* nymph and sealed the body with wax to that of a decapitated fifth-stage nymph, the insects could survive for a long period of time. This technique is called parabiosis. The mature nymph never underwent metamorphosis but rather molted in response to its own ecdysone or to ecdysone painted on its cuticle to form a larger nymph. Once again, tissue transplantation experiments showed that the hormone coming from the immature nymph that prevented metamorphosis was produced by the corpora allata. Removing the corpora allata from a larva or nymph caused either death or premature metamorphosis to the adult stage.

The role played by PTTH was also elucidated by a similar set of surgical experiments. Removing the brain of a larva prevented the production of ecdysone from the prothoracic gland so long as the removal occurred before the brain released the PTTH. After the PTTH signal was released, removing the brain had no effect on the subsequent production of ecdysone for that molt.

Today, the presence and amount of the three hormones are measured by easy, sensitive, and highly reproducible immunological techniques. In these procedures, antibodies, proteins having the chemical ability to bind tightly to a specific insect hormone, are produced. If an extract contains one of these hormones, adding its specific antibody will cause the hormone to bind to the antibody, and the amount of bound antibody can be accurately measured in a number of different ways.

These immunological procedures have allowed researchers to monitor the changing levels of the three hormones during the development of any insect and to correlate these changes with the progress of metamorphosis. Coupled with new information about the ability of hormones to turn genes on and off directly, these procedures have advanced the understanding of the mechanisms of control of metamorphosis at the molecular level.

—*Joseph G. Pelliccia*

Further Reading

Danks, H. V., ed. *Insect Life-Cycle Polymorphism: Theory, Evolution, and Ecological Consequences for Seasonality and Diapause Control.* Kluwer, 1994.

DePomerai, David. *From Gene to Animal.* 2nd ed. Cambridge University Press, 1990.

Gilbert, Lawrence I., and Earl Frieden, eds. *Metamorphosis: A Problem in Developmental Biology.* 2nd ed. Plenum, 1981.

Gilbert, Scott F. *Developmental Biology.* 6th ed. Sinauer Associates, 2000.

Koolman, Jan, ed. *Ecdysone: From Chemistry to Mode of Action.* G. Thieme Verlag, 1989.

Raven, Peter H., and George B. Johnson. *Biology.* 6th ed. Times Mirror/Mosby, 2001.

Mimicry

Fields of Study

Ethology; Evolutionary Biology; Entomology

Abstract

There are many different types of mimicry. Some mimics look like another organism; some smell like another organism; some may even feel like another organism. There are many ways that mimicking another organism could be helpful. Mimicry may help to hide an organism in plain sight or protect a harmless organism from predation when it mimics a harmful organism. It can even help predators sneak up on prey species when the predator mimics a harmless organism.

Key Concepts

adaptation: a phenotype that allows those organisms that have it a competitive advantage over those that do not have it

camouflage: patterns, colors, and/or shapes that make it difficult to differentiate an organism from its surroundings

warning coloration: the bright colors seen on many dangerous and unpalatable organisms that warn predators to stay away

ARE YOU COPYING ME?

The broadest description of mimicry is when one organism, called the operator or dupe, cannot distinguish a second organism, called the mimic, from a third organism or a part of the environment, called the model. There are many different types of mimicry. Some mimics look like another organism; some smell like another organism; some may even feel like another organism. There are many ways that mimicking another organism could be helpful. Mimicry may help to hide an organism in plain sight or protect a harmless organism from predation when it mimics a harmful organism. It can even help predators

Eyespots of foureye butterflyfish (Chaetodon capistratus) *mimic its own eyes, deflecting attacks from the vulnerable head. (Laszlo Ilyes)*

sneak up on prey species when the predator mimics a harmless organism.

In the case of hiding in plain sight, the line between camouflage and mimicry is not sharply defined. Spots or stripes that help an organism blend with the surroundings are classified as camouflage, since those patterns allow the organism to remain hidden in many areas that have mixtures of sunlight and shadow, and the organism does not look like any particular model. As an organism's appearance begins to mimic another organism more and more closely, rather than being just a general pattern, it moves toward mimicry. As in all other areas of biology, there are arguments about where camouflage ends and mimicry begins. The stripes of a tiger and the spots on a fawn are certainly camouflage. The appearance of a stick insect is more ambiguous. Its body is very thin and elongated and is colored in shades of brown and gray. Is this mimicry of a twig or just very good camouflage? Many biologists disagree. The shapes and colors of many tropical insects, especially mantids, also fall into this gray area of either extremely good camouflage or simple mimicry.

BATESIAN AND MÜLLERIAN MIMICRY

In contrast to camouflage, which hides its bearers, many species of dangerous or unpalatable animals are brightly colored. This type of color pattern, which stands out against the background, is called warning coloration. Some examples are the black and white stripes of the skunk, the yellow and black stripes of bees and wasps, red, black, and yellow stripes of the coral snake, and the bright orange of the monarch butterfly. Several species of harmless insects have the same yellow and black pattern that is seen on wasps. In addition to mimicking the coloration of the more dangerous insects, some harmless flies even mimic the wasps' flying patterns or their buzzing sound. In each case, animals that have been stung by wasps or bees avoid both the stinging insects and their mimics. This mimicry of warning coloration is called Batesian mimicry. Batesian mimicry is also seen in the mimicry of the bright red color of the unpalatable red eft stage of newts by palatable salamanders.

Sometimes two or more dangerous or unpalatable organisms look very much alike. In this case, both are acting as models and as mimics. This mimicry is called Müllerian mimicry. Müllerian mimicry is seen in monarch and viceroy butterflies. Both butterflies have in their bodies many of the chemicals found in the plants they ate as larvae. These include many unpalatable chemicals and even toxic chemicals that cause birds to vomit. If a bird eats either a monarch or a viceroy that has these chemicals, the bird usually remembers and avoids preying on either species again—a classic Müllerian mimicry. Interestingly, not all monarchs or viceroys are unpalatable. It depends on the types and concentrations of chemicals in the particular plants on which they fed as larvae. Birds that have eaten the palatable monarchs or viceroys do not reject either monarchs or viceroys when offered them as food, but birds that have eaten an unpalatable monarch or an unpalatable viceroy avoid both palatable and unpalatable members of both species. This represents both Batesian and Müllerian mimicry at work.

AGGRESSIVE MIMICRY

Mimicry by predators is called aggressive mimicry. The reef fish, called the sea swallow, is a cleaner fish, and larger fish enter the sea swallow's territory to be cleaned of parasites. The saber-toothed blenny mimics the cleaner in both appearance and precleaning behavior, but when fish come to be cleaned, the blenny instead bites off a piece of their flesh to eat. Anglerfish have small extensions on their heads that resemble worms. They use the mimic worms to lure their prey close enough to be eaten. The alligator snapping turtle's tongue and the tips of the tails of moccasins, copperheads, and other pit vipers are also wormlike and are used as lures. Certain predatory female fireflies respond to the light flashes of males of a different species with the appropriate response of the female of that species. This lures the male closer, and when the unsuspecting male is close enough to mate, the female devours him. This mimicry is quite complex, because the predatory females are able to mimic the response signals of several different species.

There are many other instances of mimicry, but the world champion mimics may be octopuses. As predators, these animals show unbelievable aggressive mimicry of other reef organisms. Octopuses can take on the color, shape, and even texture of corals, algae, and other colonial reef dwellers. As a prey species, the octopus can use the same type of mimicry for camouflage, but can also be a Batesian mimic, taking on the color and shape of many of the reef's venomous denizens.

Since in each case, being a mimic helped the organism in some way, it is not hard to understand how mimicry may have evolved. In a population where some organisms were protected by being mimics, the protected mimics were the ones most likely to mate and leave their genes for the next generation while the unprotected organisms were less likely to breed.

—*Richard W. Cheney, Jr.*

FURTHER READING

Brower, Lincoln P., ed. *Mimicry and the Evolutionary Process.* University of Chicago Press, 1988.

Ferrari, Marco. *Colors for Survival: Mimicry and Camouflage in Nature.* Thomasson-Grant, 1993.

Owen, D. *Camouflage and Mimicry.* University of Chicago Press, 1982.

Salvato, M. "Most Spectacular Batesian Mimicry." In *University of Florida Book of Insect Records*, edited by T. J. Walker. University of Florida Department of Entomolgy and Nematology, 1999. http://gnv.ifas.ufl.edu/~tjw/recbk.htm.

Wickler, Wolfgang. *Mimicry in Plants and Animals.* McGraw-Hill, 1968.

Molting and Shedding

FIELDS OF STUDY

Developmental Biology; Evolutionary Biology; Ethology; Animal Physiology

ABSTRACT

As a normal part of growth and development, some species of invertebrate and vertebrate animals undergo a process commonly called molting or shedding. Scientists term this

process ecdysis. It is common for animals to molt more frequently as larvae or juveniles and less often as they mature and become adults. Molting is also an integral part of metamorphosis. In most animals the interval between molts is called the intermolt, except insects, where it is referred to as an instar.

KEY CONCEPTS

arthropods: animals with jointed exoskeletons, including insects, crayfish, and spiders
exoskeleton: a skeleton made up of proteins and minerals, found on the outer surface of an animal
hormones: chemical messengers produced by specialized organs and cells either in the endocrine system or the nervous system; they have specific effects on different cells of the body
invertebrate: animal without an internal skeleton made up of individual bones or vertebrae
metamorphosis: a change in the physical state of an animal, such as the transformation of a tadpole into a frog or the development of a moth from a caterpillar
steroid hormone: a hormone that is made from cholesterol

PUTTING ON A NEW SKIN

As a normal part of growth and development, some species of invertebrate and vertebrate animals undergo a process commonly called molting or shedding. Scientists term this process ecdysis, which is derived from a Greek word meaning "to escape or slip out." It is common for animals to molt more frequently as larvae or juveniles and less often as they mature and become adults. Molting is also an integral part of metamorphosis. Immediately after a molt, animals may not be fully protected from the environment, and they may be more vulnerable to predators. For example, the new exoskeleton of arthropods may not be fully hardened, or some birds may lose flight feathers.

During a molt, animals may shed and replace their entire body covering or structures associated with the body surface. Molting may involve the replacement of the skin, an exoskeleton, or a cuticle in its entirety. In temperate regions, it is not unusual in the summertime to see the old exoskeleton of a cicada still clinging to a tree. Feathers, fur, or hair, which are derived from skin cells, are often shed in a cyclic fashion rather than all at once. Animals may also lose body structures during a molt. Adult male deer shed their antlers at the end of the mating season and lemmings replace their claws. In most animals the interval between molts is called the intermolt, except insects, where it is referred to as an instar.

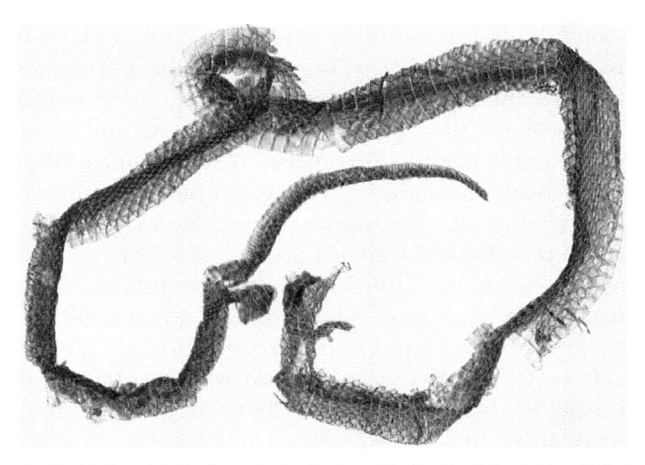

Empty skin (exuvia) of a Grass Snake (Natrix natrix; length >100 cm). Found SE of Berlin, Germany, in 1992.

THE MOLTING PROCESS

Environmental factors such as stress, temperature, and light cycle can serve as stimuli to molting in different species of animals, but the actual sequence of events leading up to shedding and replacement is strongly influenced by hormones and by specific interactions between the nervous system and the endocrine system. Hormone production is the primary function of the endocrine system. However, regions of the brain also produce hormones, which in this case are called neurohormones. One class of hormones in particular, steroid hormones, is actively involved in the regulation of molting.

Before an animal molts, a new skin or exoskeleton will have formed beneath the old one. Often the old exoskeleton splits along the midline of the posterior surface and the animal then crawls out. Different animals have specific ways to stretch the new exoskeleton before it hardens. The best way to stretch the skeleton is by increasing body size. This might be accomplished by a growth stage or by artificially increasing size. A lobster, for example, might absorb

432

enough water to increase body size by 20 percent, while other animals take up air.

MOLTING IN AMPHIBIANS AND REPTILES

Molting of an entire body surface such as the skin or exoskeleton requires the separation of the layer being shed from the underlying tissue that will become the new outer surface. The chemical and physical processes leading up to the separation will vary slightly between animal groups. Amphibians such as frogs, toads, and salamanders periodically shed the outermost surface layer of their skin. Prior to molting, mucus is secreted beneath this layer of cells. Since the mucus is secreted between the old and new surface layer, it may assist in the separation by creating a space between them. The mucus may also act as a lubricant and aid the animal in removal of the old skin. Frogs are also known to bloat themselves and increase movement to break out of the old skin. Animals will often eat the skin they have shed as a way to recycle nutrients.

Reptiles show a good deal of variability in the replacement of their scales. In fact, snakes may shed their skin several times a year. In the time immediately preceding ecdysis in snakes, a fragile zone develops. Old skin will now separate from the underlying, newly developed scales. During the molt the snake's eyes will appear cloudy, because the outer part of the eye, which is a part of the skin, is also detaching. The rattlesnake's rattle is formed from the part of the skin that remains attached to the tail at the end of each shedding cycle. Snakes shed their skin in one piece, but lizards do not. Turtles are different from both snakes and lizards. First, their scales do not overlap to form sheets and second, most turtles add new growth to existing scales and do not molt. The molting pattern in crocodiles and alligators, which shed and replace individual scales, more closely reflects events in birds than other reptiles.

MOLTING IN BIRDS

Birds shed their feathers each year and it is not unusual for some species to molt several times during the year. During the molting season, a bird's behavior may be affected in several ways, including events associated with reproduction. Molting will be influenced by the time of year and the mating season. Many female chickens often stop laying eggs when they are molting. New feathers will develop in the same area of the skin as the old ones. As the new feathers grow from the follicle they may push the old feathers out of the skin or they may be pulled out. The primary purpose of molting is to replace worn feathers with new ones, and often a pattern of feather loss will be noted. Wing feathers closest to the body are lost first and the molt progresses outward along the wings.

It is common for male birds to molt so that they can replace their duller plumage with more colorful feathers associated with attracting a mate. This is called a prenuptial molt and generally occurs in late winter or early spring. A postnuptial molt is common in both males and females. The chicks will lose their original feathers (down) when they undergo a postnatal molt to juvenile feathers. If it is a species of bird where the males have a coloration pattern different from the females, the early juvenile coloration more closely resembles the female. As the juveniles mature they will undergo successive molts to adult plumage.

The time frame for maturation can vary considerably. Eagles can take up to five years before a final molt into full adult feathers. Mature birds that depend upon flight as a method of escaping predation will not molt all of their feathers at one time. A complete molt would result in the loss of flight feathers on the wings as well as tail feathers which assist in stability and guidance during flight. Flight is less adversely affected if there is a symmetrical loss of feathers on the wings and if tail feathers are shed in groups. Birds such as ducks and geese, which are able to spend extended periods of time in the water, are able to avoid most predators by swimming or hiding in tall grass. These birds do molt all of their flight feathers at one time. Male ducks are also somewhat unusual because, during the summer months, they molt from mating colors to plumage similar in color to females. This change is called an eclipse and makes it more difficult to tell males from females.

MOLTING IN MAMMALS

When they lose hair, mammals, including humans, are molting. Like feathers, a hair grows outward from a follicle in the skin and, as new hairs grow, old hairs will be lost. Under normal conditions in humans, hair loss will be a gradual process over an individual's lifetime and it does not occur all at once. Molting in many mammals is directly influenced by length of day and interactions between the endocrine system and the nervous system. The number of molts per year varies, and many mammals molt twice a year, once in the spring and once in the fall. Foxes, however, molt once a year in the summer, and the snowshoe hare molts three times: summer, autumn, and winter. In general, it is common for the summer coat of mammals to be thinner than the winter one, and it is not unusual for the two coats to be different colors. Changing hair color provides one method of camouflage. White fur lacks pigment and blends well with snow and light surroundings. In contrast, darker shades blend better in the summer and fall. In mammals, like other animals, the transition from juvenile into adult can be associated with a molt. Deer fawns, for example, have very pronounced white spots, but when they become further developed they molt into the solid coloration of adults.

—Robert W. Yost

FURTHER READING

Gilbert, Scott F. *Developmental Biology*. 6th ed. Sinauer Associates, 2000.

Hickman, Cleveland P., Larry S. Roberts, and Allan Larson. *Biology of Animals*. 7th ed. McGraw-Hill. 1998.

Lindsey, Donald. *Vertebrate Biology*. McGraw-Hill, 2001.

Miller, Stephen A., and John B. Harley. *Zoology*. 4th ed. McGraw-Hill, 1999.

Purves, William K., David Sadava, Gordon H. Orians, and H. Craig Heller. *Life: The Science of Biology*. 6th ed. W. H. Freeman, 2000.

Morphogenesis

Fields of Study

Developmental Biology; Animal Physiology; Animal Anatomy

Abstract

Morphogenesis is the process that leads to the appearance of form in specific patterns, through the differential reproduction, growth, and movement of cells and tissues and interactions between tissues. Such movements are controlled by a variety of factors, including cell adhesion molecules (CAMs), the nature of the extracellular matrix (ECM), and the size of the cells themselves.

Key Concepts

bifurcation: the division of a Y-shaped and connected mesenchymal structure into a single proximal chondrogenic focus and two distal chondrogenic foci; this can lead to the formation of separate chondrification centers in a developing digit

chondrification: the process by which undifferentiated connective cells transform into chondrocytes (cells that make cartilage) and begin forming extracellular matrix

epithelium: the tissue that covers and lines all exposed surfaces of an organism, including internal body cavities such as the viscera and blood vessels

limb bud: thickened epithelial cells along the lateral body fold that are underlain by mesoderm, creating a paddle-shaped extension from the trunk

segmentation: the division of a structure into linearly arranged segments; it can lead to the formation of somites, or it can lead to the formation of separate chondrification centers in a developing digit

zone of polarizing activity (ZPA): a region at the posterior base of the limb bud that seems to influence the distal development of pattern in a developing limb

BECOMING

Morphogenesis is the process that leads to the appearance of form in specific patterns, through the differential reproduction, growth, and movement of cells and tissues and interactions between tissues. Such movements are controlled by a variety of factors, including cell adhesion molecules (CAMs), the nature of the extracellular matrix (ECM), and the size of the cells themselves.

Most multicellular organisms begin their lives as fertilized eggs. Although this seemingly simple cell is really quite highly organized, it is still relatively uncomplicated when compared to the structural complexity of the parent. As each egg cell divides, the daughter cells become structurally and functionally distinct. This is the beginning of the process of cell differentiation, which can lead to complex multicellular organisms. As new cells appear, grow, mature, and divide, there is a point at which the eventual fate of these cells is determined. That is, at some point the eventual location, structure, and function of the descendants of such cells are fixed. This fate determination frequently takes place very early in the development of an organism. After a particular fate is determined, it is the process of morphogenesis that allows the potential fate to be realized. Also through the process of morphogenesis, patterns of form develop that increase the structural and functional complexity of an organism.

THE RULES OF MORPHOGENESIS

Fate determination, morphogenesis, and pattern formation are not haphazard. In fact, these events are highly constrained, extremely stereotyped, and very predictable—so much so that models of morphogenesis have been formulated that rely on rules of development to establish the probabilities of particular forms developing. These rules take into account two types of conditions—initial and boundary conditions—that work together to produce pattern formation in a cascading process. Initial conditions are simply the cellular conditions that prevail in a certain part of an organism, such as cell size and number and the chemical makeup of cell membranes. Boundary conditions are the conditions that exist at boundaries of different types of tissue.

Only a few phenomena are responsible for setting the initial and boundary conditions within which a morphogenetic system operates. A variety of cellular phenomena are involved, including cell migration, cell division (producing new cell lineages), the rate of cell division, cell number, cell density, the orientation of daughter cells relative to parent cells, and even the places in which cell death occurs. Tissue-level phenomena also have an important impact on morphogenesis. In these cases, specific tissues possess certain properties and a limited ability to respond differently within the constraints allowed by those properties. For example, sheets of epithelial cells can become folded, but they seldom form a solid mass of cells. Finally, there are interactions that take place between different tissues. Among these interactions is induction, wherein one tissue will induce a specific response in an adjacent tissue. Examples of such induction are numerous and well known; they include the mesenchymal induction of dental epithelium during the formation of teeth.

LIMB MORPHOGENESIS

Although far from completely understood, one of the most thoroughly studied morphogenetic systems is the development of the tetrapod limb. (A tetrapod is a vertebrate with two pairs of limbs.) Because amphibians (such as frogs and salamanders) and chickens have large, easily obtainable eggs, they have been studied most. It will be useful to summarize the course of limb development, since it exemplifies many of the common features of morphogenesis and pattern formation.

Early in the formation of the limbs, a ridge forms along the flank of the embryo. Along this ridge, the apical ectodermal ridge (AER) develops as a thickened layer of epithelial cells. Undifferentiated connective tissue cells, called mesenchyme cells, accumulate underneath the AER and form the limb bud. These mesenchymal cells are derived from mesoderm that migrates into the limb bud following pathways of the ECM that are laced with CAMs, especially fibronectin. The AER is responsible for the formation of pattern in the proximodistal axis, which is a line from the base of a limb (the proximal end) to the outer end of the limb (the distal end).

Another focus of pattern formation also appears quite early in limb development; it is called the zone of polarizing activity (ZPA) and has influence over the anteroposterior axis of limb development. Together, the AER and the ZPA regulate a pattern of form that is shared by all tetrapods.

Under the influence of these pattern-generating centers, three types of mesenchymal chondrogenic foci will form: de novo condensations (which are unconnected), bifurcations, and segmentations. Segmentations and bifurcations both show recognizable connections, and they appear in a consistent and stereotyped fashion. Linear series of mesenchymal condensations showing connections with one another are called segmentations; a Y-shaped condensation with a single proximal focus connected to two distal foci is called a bifurcation.

In all tetrapods, a bifurcation leads to the formation of two skeletal elements in the region between the proximal bone in the arm (or leg) and the hand (or foot). In frogs and amniotes, the condensations at the base of the putative fourth digit, showing connection to the bone in the forearm (or lower leg), are the first to appear as the result of a bifurcation event. Each subsequent proximal digit condensation appears as a bifurcation. All distal condensations, after the proximal digital element forms, appear as the result of segmentation events. Thus, the development of limb elements is asymmetric and always emanates from the axis of the first digit to form chondrogenic foci.

The morphogenetic control of this pattern, which has been conserved throughout the long evolutionary history of tetrapods, is both simple and complex. It is simple in that only a few morphogenetic processes are responsible for generating the complex pattern of limb structure, so different in all the various tetrapods. It is complex in that so many phenomena can influence the ultimate structure of any particular limb. For example, the number of somites that contribute mesoderm to the limb bud mesenchyme can dictate the number of limb elements that will eventually form.

STUDYING MORPHOGENESIS

There are two general methods for studying morphogenesis. In the older—but still widely used—approach, tissues are removed from an organism in order to be examined under a microscope. Tissues are often stained (a variety of staining materials may be used) to make them more readily visible, and they are usually cut into thin slices called sections. The purpose is to determine the position of individual tissues. If carefully staged materials have been used, it is possible to get some ideas about whether cells have moved and, if so, approximately how far they have moved.

There are problems with studying morphogenesis in this way: Details can be missed, or, more worrisome, preconceived notions can bias observations. For example, a small set of sections may support a theory, but, because thin sections of tissue can vary greatly, many sections of the material under study may contravene a pet theory of development. One could simply reject the many conflicting sections, claiming them to be poorly prepared or somehow damaged material. Indeed, many sections are so rejected. With the advent of transmission electron microscopy (TEM) and scanning electron microscopy (SEM), more detailed observations could be made. New problems, however, accompany these methods. In TEM, the thinner sections which must be used increase the occurrence of variation, also increasing the chances of biased observations. SEM allows for the magnification of surfaces and a great increase in the depth of field, but it requires that materials be very carefully staged. In none of these methods is it possible actually to see individual cells move.

A second approach seeks to document cellular movements. Understanding such movements requires that individual cells be "marked." Marked cells are then grafted onto an unmarked host organism and followed. Both natural and artificial markers can be used. Pigment granules such as melanin and stained glycogen are examples of natural markers. These markers, however, can be affected by cellular activities, making it hard to follow the cells. Fortunately, there are other natural markers not so affected. For example, chick and quail cells stain differently, so the cells of one can be followed when they are transplanted into the other. One of the more frequently used artificial markers is tritiated thymidine. When it is introduced, this radioactive substance is permanently incorporated into the deoxyribonucleic acid (DNA) of the selected cells. Tritiated thymidine has several advantages: It works in any cell, it follows along with the cells, and it will be in all the offspring of the cell to which it was originally introduced. Unfortunately, as cells continue to divide, the concentrations of tritiated thymidine eventually approach undetectable levels.

Many systems have been studied using these methods, including the development of the tetrapod limb and vertebrate head, somite differentiation, and the appearance of the primary germ layers that eventually yield all other tissues.

Morphogenesis is the reflection of all the interactions that take place during the formation of a living organism and of the patterns of structure that characterize this organism. The study of morphogenesis is motivated by the desire to understand the appearance of patterns of structure. Such patterns are often conserved through evolutionary history. It is hoped that a clear picture of how these patterns are formed and regulated will show how the origin of novel morphologies is constrained. Thus, it will be possible to understand both the origin of novel forms and the maintenance of unchanging form.

—*Charles R. Crumly*

FURTHER READING

Bard, Jonathan. *Morphogenesis: The Cellular and Molecular Processes of Developmental Anatomy.* Cambridge University Press, 1990.

Bonner, John Tyler. *On Development: The Biology of Form.* Harvard University Press, 1974.

De la Cruz, María Victoria, and Roger R. Markham. *Living Morphogenesis of the Heart.* Birkhäuser, 1998.

Edelman, Gerald M., and Jean-Paul Thiery, eds. *The Cell in Contact: Adhesions and Junctions as Morphogenetic Determinants.* John Wiley & Sons, 1985.

Hinchliffe, J. R., and D. R. Johnson. *The Development of the Vertebrate Limb: An Approach Through Experiment, Genetics, and Evolution.* Clarendon Press, 1980.

Thomson, Keith S. *Morphogenesis and Evolution.* Oxford University Press, 1988.

Trinkaus, John Philip. *Cells into Organs: The Forces That Shape the Embryo.* Prentice-Hall, 1984.

Webster, Gerry, and Brian Goodwin. *Form and Transformation: Generative and Relational Principles in Biology.* Cambridge University Press, 1996.

Wessells, Norman K. *Tissue Interactions and Development.* Benjamin/Cummings, 1977.

MULTICELLULARITY

FIELDS OF STUDY

Evolutionary Biology; Paleobiology; Paleontology; Paleoecology

ABSTRACT

Multicellular organisms are those consisting of more than one cell. Three of the five kingdoms of living organisms are multicellular: the plants (Plantae), the animals (Animalia), and the fungi (Fungi). Eukaryotic cells contain a nucleus, other complex internal cell structures called organelles (such as mitochondria, which perform respiratory functions), and sometimes chloroplasts (which contain chlorophyll and perform photosynthesis). All multicellular organisms are composed of eukaryotic cells, so the eukaryotic cell must have evolved before multicellular organisms could develop.

KEY CONCEPTS

Ediacarian (Ediacaran) fauna: a diverse assemblage of fossils of soft-bodied animals that represents the oldest record of multicellular animal life on the earth

eukaryotic cell: a cell that has a nucleus with chromosomes and other complex internal structures; this is the type of cell which makes up all organisms except bacteria

fossils: the remains of ancient life preserved in sediment or rock

multicellular organisms: organisms consisting of more than one cell; there are diverse types of cells, specialized for different functions and generally organized into tissues and organs

Precambrian eon: the earliest chapter of the earth's history, covering the time interval between the formation of the earth, about 4.6 billion years ago, and the beginning of the Cambrian period, about 570 million years ago

prokaryotic cell: a primitive cell that lacks a nucleus, chromosomes, and other well-defined internal cellular structures; only members of the kingdom Monera (such as bacteria) are prokaryotic cells—all higher organisms have eukaryotic cells

PLANTS, ANIMALS AND FUNGI

Multicellular organisms are those consisting of more than one cell. Three of the five kingdoms of living organisms are multicellular: the plants (*Plantae*), the animals (*Animalia*), and the fungi (*Fungi*). The other two kingdoms consist of single-celled organisms: the bacteria (*Monera*), which have primitive prokaryotic cells, and the protists (*Protista*), which have complex eukaryotic cells. Prokaryotic cells lack a nucleus and other internal

437

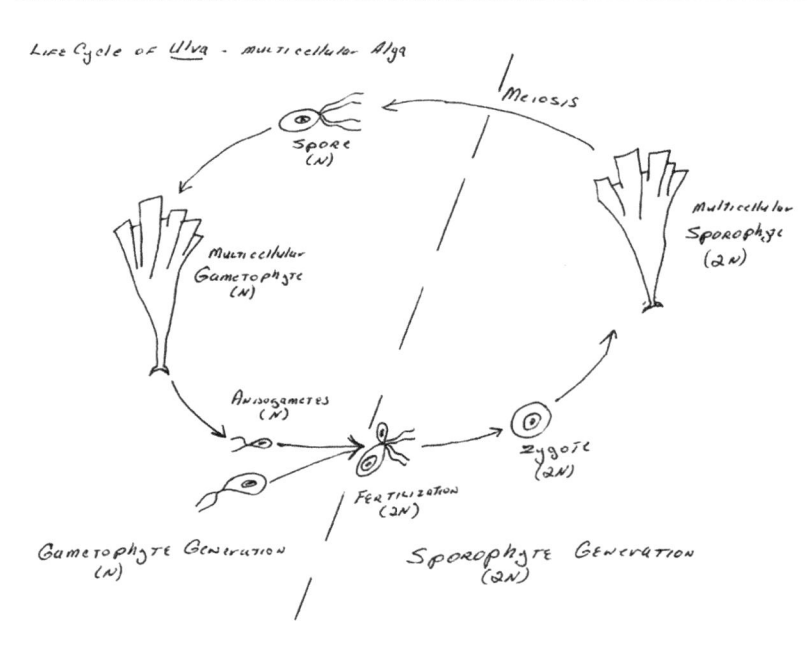

Multicellular alga alternation of generations in life cycle.

cell structures and are found today only among the bacteria. Eukaryotic cells, on the other hand, contain a nucleus, other complex internal cell structures called organelles (such as mitochondria, which perform respiratory functions), and sometimes chloroplasts (which contain chlorophyll and perform photosynthesis). All multicellular organisms are composed of eukaryotic cells, so the eukaryotic cell must have evolved before multicellular organisms could develop. It is generally accepted that simpler types of organisms evolved first, followed by more complex organisms.

THE MULTICELLULAR KINGDOMS

Plants are multicellular organisms that have chlorophyll (a green pigment used for photosynthesis), plastids (internal structures on the cell that contain chlorophyll), and a cell wall that contains cellulose. Plants are sometimes called "primary producers" because they can manufacture their own food from carbon dioxide and water through a process called photosynthesis, using sunlight for energy and producing oxygen and organic matter (carbohydrates) as by-products.

Animals are multicellular organisms that cannot produce their own food and must feed on other organisms. They are "consumers." Metazoans have many types of cells, which are organized into tissues, and groups of tissues, which form organs. There are two primary embryonic tissue layers present in all metazoans (except the sponges). These are the ectoderm (outer layer) and the endoderm (inner layer). More advanced metazoans also have a third embryonic cell layer, the mesoderm, which lies between the other two layers.

Fungi (mushrooms and their relatives) possess cell walls like plants, but unlike plants, they lack chlorophyll. Although fungi appear plantlike, they cannot produce their own food because of the absence of chlorophyll, so they must feed by ingesting organic material and therefore are consumers. Because they are neither plants nor animals, the fungi are placed in a separate kingdom (Fungi).

ADVANTAGES AND ORIGINS OF MULTICELLULARITY

Multicellularity probably evolved because it gave organisms some sort of advantage, assuring them of a greater chance of survival. Multicellularity allows organisms to become larger (which helps them to outcompete other organisms and provides a greater internal physiological stability), to have a longer life (because individual cells are replaceable), to produce more offspring (because many cells can be dedicated to reproduction), and to have a variety of body plans (which permits adaptation to various modes of life or environmental conditions). Specialization of cells for particular functions allows organisms to become more efficient.

Evidence for the origin of multicellularity comes from the fossil record, studies of the organization and biochemistry of living cells and organisms, and from studies of the embryonic and larval stages of animals. During the Archean eon (between 3.8 and 2.5 billion years ago), only single-celled, prokaryotic life (bacteria-like organisms) existed on Earth. Some of the prokaryotes were photosynthetic, including the cyanobacteria (or so-called blue-green algae). Some of these prokaryotes were colonial, with cells organized into structures such as chains, or filaments, or algal mats. These colonies differ from true multicellular organisms because they generally consist of only one type of cell rather than many types of cells. Colonies of blue-green algae formed moundlike structures called stromatolites, which were quite common during the Precambrian but are present only in a few areas today. Stromatolites appeared about 3 billion years ago but did not become abundant until about 2.3 billion years ago.

THE RISE OF EUKARYOTIC ORGANISMS

Eukaryotic organisms appeared during the Proterozoic era. The eukaryotic cell probably evolved from prokaryotic ancestors some time before about 1.4 billion years ago. The oldest convincing fossils of eukaryotic cells are generally considered to be those from the 1.3 billion-year-old Beck Spring dolomite of California. Eukaryotic fossil cells have also been found in chert from the approximately 850 million-year-old Bitter Springs formation of Australia. The earliest eukaryotes were animal-like protozoans. This evolution occurred when photosynthetic prokaryotic cyanobacteria were ingested by protozoans and then developed a symbiotic (mutually beneficial) relationship with them. The evolution of the plantlike eukaryotes probably occurred by at least 1.4 billion years ago. (This date has been suggested because primitive multicellular algae fossils are present in rocks 1.3 billion years old.)

The eukaryotic cell was a prerequisite for the development of multicellular organisms. The plantlike eukaryotes are considered to be ancestral to the multicellular algae and higher plants. The protozoans are considered to be the ancestors of the metazoans (animals). The first multicellular organisms may have been algae. Fossils that appear to be primitive multicellular algae are known from the 1.3 billion-year-old sedimentary rocks of the Belt supergroup of Montana, and the 800 million- to 900 million-year-old Little Dal group of northwestern Canada. Multicellular algae can be found living today in both freshwater and marine environments.

Fossil fungi first appear in the fossil record in the 790 million- to 1,370 million-year-old Bitter Springs formation cherts of Australia. The fossil record of fungi is poor and not well known.

THE RISE OF MULTICELLULAR ANIMALS

Multicellular animals evolved independently of the multicellular plants, probably arising from protozoan ancestors. The oldest evidence of metazoans (multicellular animals) in the geologic record is in the form of trace fossils. Trace fossils are imprints such as tracks, trails, or burrows made in sediment by moving animals. Over time, the sediment hardened into sedimentary rock as a result of compaction and cementation. The earliest trace fossils consist of simple trails and tubelike burrows. In some places, there is a succession of types of trace fossils from simple tubelike burrows in older rocks to more complex structures in younger rocks. This change suggests that the evolution and diversification of increasingly complex burrowing organisms occurred during the latter part of the Precambrian.

The oldest trace fossils are less than one billion years old, and many scientists believe that it is unlikely that any trace fossils exist in rocks much older than about 700 million years old. The trace fossils appear in the geologic record just before the first appearance of soft-bodied metazoan fossils. Structures that resemble trace fossils, however, have been reported from much older rocks. Among these questionable traces are the one-billion-year-old *Brooksella*, which resembles a jellyfish. In addition, tubelike structures from the upper Medicine Peak quartzite in Wyoming have been dated at 2 billion to 2.5 billion years, at least 1 billion years older than the oldest known metazoans. The origin of these older traces is uncertain and may be a result of inorganic processes (such as dewatering of sediment), rather than of organisms.

One possible fossil metazoan that appears to be more than 850 million years old has been reported from the Tindir group of Alaska. This fossil is less than one millimeter long and appears to be a flatworm (phylum *Platyhelminthes*). Both the age and identification of this fossil have been disputed, but if valid, it is a very important find because some biologists theorize that the earliest metazoans would have been primitive flatworms.

The oldest unquestioned metazoan fossils are the imprints of a diverse assemblage of relatively well-developed, soft-bodied marine animals. More than half of the organisms appear to be some type of cnidarian or coelenterate (related to jellyfishes), about 25 percent appear to be segmented worms (related to annelids), and a small percentage appear to be arthropods (related to insects, crabs, and lobsters). Trace fossils are also present. This assemblage of soft-bodied fossils is called the Ediacaran fauna. It was discovered in 1946 in sandstones of the Pound subgroup in the Ediacara Hills of the Flinders Range in South Australia. The exact age of the Ediacaran fauna is uncertain because there are no nearby rocks of the proper type for radiometric dating. The Ediacaran fauna is clearly Precambrian, however, judging from its position in the geologic sequence. The soft-bodied Ediacaran fossils are separated from the younger fossil shells of the Cambrian (570 million years old) by a thick section of unfossiliferous rock (up to several hundred meters thick). Since the 1950s, fossils similar to the Ediacaran fauna have been found in rocks of approximately the same age on virtually every continent on the earth (with the possible exception of Antarctica). In some of these other areas, it is possible to date radiometrically the rocks associated with the fauna. These radiometric dates indicate that the early metazoan soft-bodied fossils range from about 620 million to 700 million years old. It is likely that the metazoans evolved some time prior to 700 million years ago because these fossils represent well-developed, complex animals.

Skeletonized faunas (animals with shells or other hard parts) did not appear until approximately 580 million years ago. The skeletonized faunas are represented by microscopic scraps, cones, tubes, and plates made of calcium phosphate or a hard organic material called chitin. It is not known exactly what types of organisms produced these skeletal remains, but the tiny fossils are so diverse and complex that it is assumed that the organisms must have had a long history of evolution during the Precambrian. The origin of skeletons was advantageous to marine organisms because hard parts provide protection against predators as well as the mechanical functions of support and muscle attachment.

THEORIES OF THE ORIGIN OF MULTICELLULAR LIFE

There are a number of theories to explain the origin of multicellular life. Most of the theories are derived from studies of various types of cells and living organisms, including advanced protozoans, early developmental stages (embryos), and larval stages. Four types of cells are central to these theories, and they are grouped into two categories: motile (capable of movement) and nonmotile (not capable of movement). The motile protists include flagellate cells (those with a whiplike "tail," or flagellum) and amoeboid cells (those such as Amoeba, which move by pseudopodia, or fingerlike extensions of the cell membrane). The nonmotile stages include coccine cells (those with many nuclei, sometimes called multinucleate cells) and sporine cells (those that divide and stick together to form multicellular aggregates).

There are many theories that have been proposed to explain the origin of plants, fungi, and metazoans (animals). Formerly, it was thought that plants evolved from prokaryotic algae (cyanobacteria, or blue-green algae), but it is more likely that plants arose from a eukaryotic ancestor, such as a flagellate cell. Flagellate algae are similar to flagellate protozoans, but it is not certain whether the algae evolved from the protozoan or vice versa. The presence of plastids (such as the chloroplasts that contain chlorophyll used for photosynthesis) may be the key feature separating plants from protozoans, fungi, and animals. According to a theory proposed by Lynn Margulis, plastids evolved from prokaryotic blue-green algae that were captured by eukaryotic cells. The sporine cell is another possible ancestor of the plants. Sporine cells appear to have had the capacity to evolve beyond the colony level and to produce complex tissue-level green algae and higher plants.

Fungi used to be considered as plants that had lost (or never evolved) chlorophyll. The discovery of a single-celled stage with flagellae among the more primitive fungi, however, suggests that fungi probably evolved from protozoans.

The ancestral multicellular organisms, which gave rise to all the more-complex living animals, are all extinct. The simplest multicellular animal living today is the sponge (phylum Porifera). The sponges are not considered to be ancestors of the more complex animals because their body organization and developmental history are very different. Sponges have no tissues, mouth, or internal organs. Instead, they consist of an aggregate of flagellate and amoeboid cells (and a few other types) roughly arranged in layers. The sponges may have evolved independently from the other metazoans. Sponges are classified as a distinct side branch of the animal kingdom (*Parazoa*), with a primitive multicellular grade of organization (no tissues). The remaining multicellular animals are grouped into the *Eumetazoa*.

THEORIES OF METAZOAN ORIGIN

Several theories have been proposed to explain the origin of the metazoans. These theories can be placed into the following categories: evolution from single-celled protozoans; evolution from colonial protozoans; evolution from multinucleate coccine cells as a result of development of internal cell boundaries; and evolution from sporine cells. There are several versions of each of these theories, and there is no general agreement on which theory is best. Some researchers promote the colonial theory as the most widely accepted theory, whereas others claim no longer to take it seriously. Most experts agree that evolution of metazoans from colonial protozoans would seem to be easier than evolution directly from a single cell. Multicellularity may have arisen independently several times, in several different ways.

The colonial theory suggests that the metazoans evolved from flagellate or amoeboid protozoans that lived together in colonies, much like the modern green alga *Volvox*, which is shaped like a hollow sphere. From an original hollow spherical form, the shape of the ancestral metazoan changed as an indentation or invagination formed in the side. The indentation became larger, producing a double-walled "cup" (envision pushing one's thumb into the side of a deflated ball until that side becomes nested into the other side of the ball, forming a cuplike shape). The double-walled cup shape is referred to as a diploblastic body plan, meaning two layers of body tissue. These two layers are the ectoderm (outer layer) and the endoderm (inner layer). This process of indentation to produce a diploblastic (double-walled) form occurs in the embryos of many animals. The jellyfishes are a good example of animals with a diploblastic body plan. Nearly all groups of animals have ectoderm and endoderm (except the sponges), suggesting that nearly all groups of animals are related. Because the jellyfishes (phylum Cnidaria) have the simplest body plan, they are believed to be the most primitive. The diploblastic ancestral form has been called a gastrea. Ernst H. Haeckel, a prominent nineteenth century German biologist who studied animal embryos, believed that all bilaterally symmetrical animals evolved from a gastrea.

A second theory for the origin of metazoans suggests that the ancestral form was a bilaterally symmetrical animal resembling a flatworm. Some scientists believe that the complex organs and organ systems of metazoans are beyond the evolutionary potential of flagellate and amoeboid cells. The flatworm may have evolved from "cellularization" of a multinucleate coccine cell (formation of cell membranes around each of the nuclei) or from clumping of sporine cells. Most of the cells in the metazoans are sporine cells that stick together to form multicellular aggregates. Sporine protozoans do not exist, so it is hypothesized that sporine ancestors of the metazoans must have evolved from "pre-protozoans." These hypothetical ancestors may have been solid balls of cells resembling the early stages of many embryos. At some point, the exterior cells may have developed (or redeveloped) flagellae and become specialized for locomotion, and the interior cells may have become specialized for digestion and reproduction. Such colonies of cells would have resembled the larval (immature) form of cnidarians, called a planula larva, and, hence, they are called planuloids. Planuloids are believed to have given rise to two groups of metazoans, the cnidarians (jellyfishes and their kin) and the flatworms. The primitive flatworms are believed to have been ancestral to all other bilaterally symmetrical metazoans.

EVIDENCE FROM ROCKS

Theories to explain the origin of multicellular life have been developed by biologists as a result of studies of various types of cells and living organisms, including advanced protozoans, early developmental stages (embryos), and larval stages.

Geologists (scientists who study rocks) and paleontologists (scientists who study fossils) have a variety of techniques that they use to search for the evidence of life in Precambrian rocks (older than 570 million years). These include searching for fossil remains and chemical analysis of organic residues that are probably the breakdown products of once-living organisms.

The first step in the search for Precambrian life is to locate rocks of the proper age. Geologic maps exist for virtually all parts of the world. From an examination of these maps, it is possible to identify areas that contain rocks of the proper age. (The age of a rock is determined by radiometric dating.) Age, however, is not the only consideration. For fossil remains to be preserved, the rocks must also remain little altered from the way they were originally deposited. Metamorphism (geologic alteration caused by heat and/or pressure) has deformed many Precambrian rocks to the extent that any fossils that may have been present can no longer be recognized.

Assuming that undeformed rocks of the proper age can be located, the search begins for fossil remains. Unfortunately, most Precambrian rocks are not fossiliferous. Precambrian multicellular fossils are found in only a few places in the world. In Australia, soft-bodied Precambrian metazoan fossils are restricted to a few thin layers of sandstone in a sequence of Precambrian rock more than one thousand meters thick. In most places in the world, however, there is a thick section of unfossiliferous rock separating the Precambrian metazoan fossils from the shelly faunas in the Cambrian rocks. This unfossiliferous sequence of rock is an interval for which there is little or no information on the types of life that existed.

Before multicellular organisms appeared (prior to perhaps one billion years ago), only microscopic, single-celled organisms existed on the earth. Microscopic fossils of single-celled organisms are found by careful examination of fine-grained, dark-colored rocks such as black cherts. The black color of the rocks commonly indicates the presence of carbon, which is present in all living organisms and which may be preserved in some fossils. Very thin slices of rock are prepared and mounted on glass slides so that the organic matter can be studied. These slices of rock, called thin sections, are so thin that light can pass through them, and they are examined with a microscope. Much of the carbon in these rocks is present as amorphous (indistinct or shapeless) patches, but in some places, microscopic structures are present that appear to be the fossilized remains of single-celled organisms. Pieces of rock can also be prepared for examination using a scanning electron microscope. The search for microfossils is difficult and painstaking. Among the problems involved are the possibility of contamination by modern-day organic matter in the laboratory and the possibility that the microscopic structures may really be inorganic in origin.

Chemical tests are used to search for the products of biological activity, which may be preserved in rocks. In principle, rocks that have been influenced by biological activity should contain certain characteristic isotopic ratios. There are a number of problems inherent in searching for organic residues. Organic material may have been preserved in the rock, but it could easily have been altered subsequently by heat and pressure or by circulating fluids. In addition, circulating fluids can contaminate the rocks by introducing organic material from much younger rocks.

MULTICELLULARITY IN THE EVOLUTIONARY PROCESS

Studying the origin of multicellularity helps one to understand the conditions that led to the evolution of plant and animal life on Earth. As one begins to understand how multicellular life evolved, one may begin to wonder about why it was such a slow process. It is known that Earth formed about 4.6 billion years ago and that the first cells appeared about 3.5 billion years ago, but that the first multicellular life did not appear until approximately 1 billion years ago. In other words, it took more than 3.5 billion years for multicellular life to develop. More than three quarters of Earth's history had passed before multicellular life ever appeared.

One may also begin to wonder about the conditions that promoted the origin of multicellular life. Of all the planets in the solar system, Earth seems uniquely suited to life. Two of the most important factors involved are the presence of liquid water (which requires a specific temperature range) and the presence of an oxygen-rich atmosphere. None of the other planets in this solar system has either of these two characteristics, although there is considerable and growing evidence that certain moons of the planets Jupiter and Saturn contain huge sub-surface oceans where life and life forms of some kind may exist. Interestingly enough, Earth originally did not have liquid water or an oxygenated atmosphere. Geologic evidence suggests that Earth's early atmosphere was the result of volcanic outgassing and that it consisted of gases such as carbon dioxide, carbon monoxide, ammonia, methane, hydrogen sulfide, nitrogen, and water vapor. As the planet cooled from its original molten state, the water vapor in the atmosphere condensed to form liquid water, which fell as rain and accumulated to form the oceans, rivers, and lakes. There is abundant geologic evidence that Earth's early atmosphere lacked the free oxygen that is breathed today. In the absence of free oxygen, chemical evolution in the oceans or lakes led to the formation of organic compounds, or what has been called the "primordial soup."

The first living cells, the prokaryotes, evolved in this organics-rich water. As time passed, some of the early prokaryotic cells became photosynthetic, which allowed them not only to produce their own food from water and carbon dioxide but also to produce oxygen as a waste product. Oxygen was toxic to these early organisms. In order to survive, the cells had to develop a mechanism to adapt to the presence of increasing levels of oxygen. The buildup of oxygen led to the development of the ozone layer in the atmosphere and to the appearance of the eukaryotic cell. As the percentage of oxygen in the atmosphere increased, it is believed that some threshold level was reached, and it became possible for the environment to support multicellular organisms. That allowed a rapid diversification of life on the earth.

Hence, it appears that multicellular life on the earth appeared as a result of some prehistoric accident that resulted in global atmospheric change—the buildup of a toxic waste product (oxygen) as a result of photosynthesis by early lifeforms. One might also speculate on the possible global effects of the increasing waste products that humans are now producing. The thinning of the atmospheric ozone layer is but one manifestation of the way that life is presently changing the earth's fragile environment. Knowing that the formation of the ozone layer was probably essential to the appearance of multicellular life on the earth, it is alarming to speculate on the consequences of its destruction. Life as humans know it depends on an Earth with environmental conditions in a precarious balance.

—*Pamela J. W. Gore*

Further Reading

Ayala, Francisco J., and James W. Valentine. *Evolving: The Theory and Processes of Organic Evolution.* Benjamin/Cummings, 1979.

Bittar, E. Edward, and Neville Bittar, eds. *Evolutionary Biology.* JAI Press, 1994.

Boardman, Richard S., Alan H. Cheetham, and Albert J. Rowell, eds. *Fossil Invertebrates.* Blackwell Scientific Publications, 1987.

Cloud, Preston. *Oasis in Space: Earth History from the Beginning.* W. W. Norton, 1988.

Futuyma, Douglas J. *Evolutionary Biology.* 3rd ed. Sinauer Associates, 1998.

Minkoff, Eli C. *Evolutionary Biology.* Addison-Wesley, 1983.

Selander, Robert K., Andrew G. Clark, and Thomas S. Whittam, eds. *Evolution at the Molecular Level.* Sinauer Associates, 1991.

Stanley, Steven M. *Earth and Life Through Time.* 2nd ed. W. H. Freeman, 1989.

Muscles in Invertebrates

Fields of Study

Animal Physiology; Evolutionary Biology; Developmental Biology

Abstract

The most obvious characteristic of animals may be movement. Movement requires a mechanism. The mechanism allowing multicellular animals of all kinds to change position or posture is a muscle cell. A muscle cell, also called a muscle fiber, is specialized for contraction. A muscle is composed of many muscle fibers, and each muscle cell contains contractile molecules permitting the cell to shorten and thereby change its shape. Each muscle fiber moves by drawing its ends together or by contracting.

Key Concepts

actin: one of the two major types of contractile proteins; it forms the thin myofibrils of the sarcomere

fast muscle: muscle cells that respond quickly to nervous impulses; in invertebrates, these muscle fibers have short sarcomeres and a low ratio of thin to thick myofibrils

myosin: one of the two major contractile proteins making up the thick myofibrils

paramyosin: a structural protein associated with myosin myofibrils and thought to support them

slow muscle: muscle cells that respond slowly to nervous impulses; in invertebrates, these muscle fibers have long sarcomeres and a high ratio of thin to thick myofibrils

tropomyosin: a double-stranded protein that lies in the grooves of actin myofibrils, blocking actin from attachment to myosin

troponin: a globular protein composed of three subunits; one subunit binds calcium ions, and another draws tropomyosin away from actin, which allows myosin to form crossbridges constituting the third subunit

THE MECHANISM OF MOVEMENT

The most obvious characteristic of animals may be movement. Movement requires a mechanism. The mechanism allowing multicellular animals of all kinds to change position or posture is a muscle cell. A muscle cell, also called a muscle fiber, is specialized for contraction. A muscle is composed of many muscle fibers, and each muscle cell contains contractile molecules permitting the cell to shorten and thereby change its shape. Each muscle fiber moves by drawing its ends together or by contracting.

The contractile molecules are common to muscles of invertebrates and vertebrates. There are several kinds. Actin and myosin are the most prominent contractile molecules. Each actin molecule is rounded and forms myofibrils by assembling repetitive monomers to form a helix, a springlike shape. Two of these helices are coiled around each other like two intertwined strings of pearls.

Myosin is a more complex myofibril. Each myosin molecule is composed of a head and a tail. The tails of two myosin molecules are wound around each other helically, while the heads project from the same end but in different directions. In the presence of calcium ions, each head can swivel. If an actin molecule is exposed, the myosin head can attach to it, forming a crossbridge. Since they lie parallel to each other in the cell, when the myosin returns to its original position, it pulls the actin myofibril. The ends of the contractile unit move closer to each other, shortening the muscle cell.

Myosin tails lie parallel to each other, but at the opposite ends of the myosin myofibrils, the heads project in opposite directions. The heads lie close to the two end walls that attach the actin myofibrils. There is a small central area where only the oppositely facing tails of the myosin myofibrils are found.

A contractile unit is called a sarcomere. A sarcomere is a repeated unit within a muscle cell consisting of overlapping thick myofibrils (myosin) and thin myofibrils (actin). Since there are areas in which there is no actin and areas in which there is no myosin, some light stripes are seen crossing the myofibrils in a microscopic preparation of certain muscle fibers. Actin myofibrils connect to thick walls that mark the ends of a sarcomere. The myosin myofibrils are connected across the center of the sarcomere. During relaxation, actin is absent from the center, producing a light band. The regular pattern of light and dark areas along the length of the muscle fiber is characteristic of striated or skeletal vertebrate muscle. This vertebrate muscle type is the most studied in physiology and, therefore, is the best known.

THE SPECIFICS OF INVERTEBRATE MUSCLES

Invertebrate muscles have similar components, but the arrangement is less regular, the alignment of sarcomeres often being oblique. Actin and myosin myofibril arrangement is not readily delineated, but similar structures exist. A sarcoplasmic reticulum, similar to the endoplasmic reticulum of ordinary cells, is characteristic of vertebrate and invertebrate muscle fibers. It is arranged in a series of flattened sacs over a sarcomere's myofibrils. The sarcoplasmic reticulum can quickly absorb calcium ions, keeping their concentration around the sarcomere low. Without calcium ions, the myosin heads cannot be cocked.

Transverse tubules, or T tubules, extend along the thick end wall of each sarcomere. They conduct the nerve impulse throughout the sarcomere to initiate contraction rapidly and simultaneously. When the impulse reaches the sarcoplasmic reticulum, stored calcium ions are released. This release begins contraction. Contraction ceases only if adenosine triphosphate (ATP) is present. Because they no longer make ATP, dead animals' contracted muscles do not relax; the muscles become stiff (exhibit rigor mortis).

Other molecules are also typically found in muscle cells. Tropomyosin is a strand of molecules that lies in a groove of the helically wound actin myofibrils.

It blocks the attachment of myosin heads to actin molecules. Troponin is a three-part molecule that rests along the actin myofibril. One subunit attaches calcium ions. This changes troponin's conformation so that another subunit attracts the tropomyosin molecule, unblocking the actin myofibril so that the cocked head of myosin molecules can attach to the actin.

Paramyosin is a molecule similar to myosin found in invertebrates. It seems to act as a "filler" for myosin, keeping the myosin myofibrils aligned so that they can interact with the actin. It may be the contractile protein of some invertebrate muscle types. It is prominent in the "catch" muscles of mollusks.

MUSCLE RESPONSE RATES

Muscles may vary in their response rates even though they look the same. "Fast" muscle fibers contract rapidly in response to a nerve impulse. "Slow" muscle fibers contract several times more slowly. Although both types use the same contractile machinery, the slow muscle cells have less sarcoplasmic reticulum. This means that the calcium ions necessary to initiate contraction are not released as quickly everywhere along the contractile fibrils and are not pumped away from them as quickly as they are in the fast muscle cells.

Most invertebrate muscles are synchronous. This means that each contraction is initiated by a nerve impulse. The rate of contraction of synchronous muscles is determined by the rate of passage of nerve impulses. Even in flight, these muscles usually contract only about thirty-five times per second. Synchronous flight muscles are connected directly to the wings and are found in insects such as grasshoppers, moths, butterflies, and dragonflies.

True flies and bees, as well as beetles and true bugs, have a specialized kind of flight muscle called an asynchronous muscle. In asynchronous muscles, every contraction is not initiated by a nerve impulse. The contraction rates, up to one thousand times per second, are so rapid that nerve impulses could not be received and acted on quickly enough to produce that contraction rate. In these muscles, contraction is initiated by a nerve impulse received and transmitted by the T tubules, but there is no sarcoplasmic reticulum. Even fibrils removed from the cell and placed in a solution containing calcium ions and ATP contract and relax in an oscillatory fashion. The rates seem to be regulated by the myofibrils rather than by the calcium concentration. The T tubules seem to signal the contraction to begin and later turn off the cycling behavior of the myofibrils.

Asynchronous muscles are not directly attached to the wings of the insect. Elevator muscles are attached to the roof of the thorax. Their contraction pulls the roof of the thorax down and elevates the wings. Contraction of the wing depressor muscles pulls down the wings, but this stretches the wing elevators, stimulating them to contract and allowing the thorax roof to "pop up." Raising the thorax shortens the wing depressor muscles and terminates the active state of the depressors. They relax until stretched again by the elevation of the wings. The elevator and depressor muscle contractions follow the same sequence of stretch ® contraction ® shortening ® relaxation ® stretch, but are out of phase with each other. The frequency of wing beats depends upon the mechanical properties of the thorax and wings and not upon the frequency of nerve impulses.

Unlike vertebrate animals, there is a correlation between sarcomere length and speed of contraction. Rapidly contracting arthropod muscle fibers have short sarcomeres and relatively low ratios of thin to thick fibrils. Slowly contracting muscle fibers have long sarcomeres and high ratios of thin to thick fibrils. Intermediate types of fibers can exist within the same muscle.

VARIATIONS OF INVERTEBRATE MUSCLE TYPES

Because invertebrates are so varied, there is no generalization to which an exception cannot be found. For example, lobsters and crayfish have two separate muscle fiber types in their tail musculature. The tails are important in swimming and, particularly, in escape maneuvers. They must flex and extend rapidly to evade predators or aggressors. The bulk of the tail muscles consist of short sarcomere, rapidly contracting flexors and extensors. Thin sheets of long sarcomere—slowly contracting flexors and extensors—lie near the carapace. They are used for postural adjustments and for slow movements.

Flexors and extensors are good examples of another principle of muscle action. Muscles contract and they shorten. This shortening moves a body part.

Relaxation allows the muscle fibers to lengthen. The force needed to lengthen the relaxed muscle comes from the contraction of its antagonist, a muscle that produces movement in the opposite direction from that of the first muscle. For example, flexing the tail of a lobster or crayfish is performed by the tail flexors. Relaxation does not return the tail to its extended position. Contraction of the extensor muscles moves the tail away from the body and extends the tail again. Relaxation of the extensors may allow the tail to be less rigid in its extension, but the tail will not flex until the flexor muscles are contracted. Muscles work in antagonistic groups to produce opposite movements (flexion and extension) and to produce postural changes that allow a lobster or crayfish to maintain its position when the current changes direction or speed.

STUDYING INVERTEBRATE MUSCLES

Numerous methods are used to study the muscles of the invertebrates. Initially, the contraction patterns of whole muscles were studied by electrically stimulating muscles and their nerves. As in all biology, however, work is being done on the molecular level.

Glycerinated muscle is a preparation used to study the molecular elements necessary for contraction. Soaking muscle cells in glycerin removes the cell membrane and sarcoplasmic reticulum, but leaves the contractile fibrils intact. These can be used to determine the effect of presence or absence of ATP, calcium, magnesium, or other factors that might influence the working of the fibrils. They are used much the same as a whole muscle preparation. The ends of the fibrils are attached to one point that does not move and to another that does move when the muscle fibrils contract and relax. The movement can be recorded on paper or on an oscilloscope screen. The tension generated, or the length of shortening, can be calculated.

Microscopic analysis of the structural organization of invertebrate fibers also adds to the store of knowledge. Most invertebrate muscles do not have the striated appearance of vertebrate skeletal muscles. The myofibrils vary in their arrangements and are often difficult to discern. Microscopic analysis shows that rapidly contracting fibers have low ratios of thin to thick fibrils and slowly contracting fibers have larger ratios of thin to thick fibrils.

Muscles do not function without the coordination of the nervous system. Its contribution is different from that in vertebrates. Invertebrate muscle fibers do not exhibit the all-or-none response characteristic of vertebrates. Invertebrate muscle fibers receive innervation from several nerve fibers. A single nerve may serve many muscle fibers. Any muscle fiber is served by more than one nerve. The contraction strength of invertebrate muscle fibers depends upon the number and types of nerves sending impulses to that muscle cell at any time.

The study of the biochemical composition of myofibrils is also important. Actin and myosin are similar in all species. Vertebrate myosin occurs in two forms. One form has a higher intrinsic ATPase activity and responds quickly; the other has a slower intrinsic ATPase activity and twitches slowly. Invertebrate muscle fibrils have no such differences between myosin molecules. Paramyosin is a molecule peculiar to invertebrates. It is thought to be part of a supporting structure for the myosin tails. Some invertebrate muscles have been found not to have troponin and to have different forms of tropomyosin.

Electrophysiological studies have shown that invertebrate muscle fibers receive excitatory and inhibitory nerve fibers. Excitatory nerve fibers secrete acetylcholine and cause contraction. Inhibitory nerve fibers secrete serotonin and dopamine. The relaxation produced by these fibers is thought to be mediated by cyclic adenosine monophosphate (cAMP), a derivate of ATP. The relaxation may be a result of the breaking of crossbridges stabilized by paramyosin.

SIMPLE AND COMPLEX ARRANGEMENTS

Invertebrate neuromuscular systems are organized along the same general principles as in vertebrates: They coordinate body movements. All muscle cells shorten after receiving a stimulus, but the shortening will produce motion only if there is a skeleton, a structure to which the muscle fiber is anchored and another part to which force can be applied.

Sedentary sea anemones are relatively simple coelenterates that escape from threats by withdrawing their soft tentacles and contracting toward the substrate protected by thick body walls. Their body wall is a cylinder composed of two layers of cells separated by mesoglea, a jellylike material that allows diffusion of nutrients, wastes, and gases. Two groups of muscles are embedded in the body wall. In the outer body wall, longitudinal muscles parallel the long axis of the body. Their contraction shortens the body wall and draws it downward toward its base; its attachment is to the substrate. The circular muscles ring the inside layer of the body wall. Their contraction lengthens the body by squeezing the contents inward and narrowing the cylinder like a Chinese finger puzzle.

More complex invertebrates have more complex arrangements of muscles and skeletal parts. Their movements become more complex also. The best-known examples come from the larger and more numerous invertebrate phyla, the annelids, mollusks, and arthropods.

ANNELIDS AND MOLLUSKS

Annelids have a hydrostatic skeleton like that of the sea anemone. Their bodies are divided into discrete segments, each with a fluid-filled cavity. Circular muscles of a segment contract, pressing against the fluid in the cavity extending it. This contraction of the circular muscles stretches the longitudinal muscles and sets the tiny bristles along the side of each segment into the substrate. The contraction of the longitudinal muscles squeezes the fluid into a shorter segment which stretches the circular muscles and widens the segment. The body is pulled forward by the bristles that had been set in the substrate.

One of the few examples of the use of hydrostatic pressure in place of muscles occurs in some spiders. The hind legs of jumping spiders have flexor muscles that bring the legs toward the body. They have no extensor muscles. Rapidly increased body fluid pressure straightens the legs, causing the spider to jump forward.

Mollusks include bivalves—sedentary clams, mussels, and oysters; gastropods—the slowly moving snails and slugs; and cephalopods—some of the quickest, most intelligent, and largest invertebrates, the octopods and squid. The muscular systems of these animals are as varied as their lifestyles.

The bivalves settle as adults in one place. They are unable to move about. Whenever a predator threatens them or environmental conditions change (for example, if the tide goes out), they must close their shells to shield their delicate body tissues. "Catch" muscles protect them by closing tightly without using much energy. These obliquely striated adductor muscles can maintain their contracted state for a long period. A short train of nerve impulses initiates a contraction that may last hours or days with no further nerve impulses. The catch muscle does not stretch readily, so prolonged pressure by a predator, such as a starfish, does not lengthen the muscle. The catch muscle remains contracted until a relaxation mechanism is activated by neural impulses in separate neurons. These catch muscles contain paramyosin surrounded by myosin molecules.

ARTHROPODS

The arthropods are characterized by an exoskeleton, the joints of which allow movement. The skeleton is outside the body. It is composed of the cuticle, a hardened covering of chitin, with a thin, pliable hinge in the joints. Muscles cross the gap covered by the hinge. In arthropods, muscle tension is controlled largely or entirely by gradation of contraction within a motor unit. The degree of depolarization of these muscle fibers depends upon the frequency of impulses transmitted in motor neurons. Each fiber is a part of several motor units. In addition, inhibitory factors to arthropod muscles can prevent depolarization and, hence, contraction of the muscle (unlike vertebrate muscles). The neurons fire in short bursts to produce rapid movements.

Most rapidly contracting, short-sarcomere fibers are innervated by a phasic axon capable of producing rapid movements. The slowest, long-sarcomere muscle fibers are supplied only by the tonic axon that is active most of the time. These fibers provide for slow movements and postural adjustment. Muscle fibers with intermediate contraction times are innervated by both phasic and tonic axons. A muscle can, therefore, contract over a range of speeds and durations. An extreme example of this is the crustacean claw-opener muscle and the stretcher (extensor) of the proximal leg segment, which are both innervated by a single excitatory neuron. Separate inhibitory neurons supply each muscle so that they can be controlled independently.

In lobsters, the differentiation of the claws into a large crusher and a small cutter is correlated with muscular and neural activity. In their younger states, juvenile lobster claws have the same shape. Either claw can develop into the crusher. In the cutter claw, all the muscle fibers transform to fast fibers. In the crusher claw, all the muscle fibers become slow fibers. The change in shape and muscle type depends upon the presence of a manipulable environment and on the animal having unequal neuromuscular feedback to the central nervous system. The presence of a crusher on one side prevents the development of a crusher on the opposite side. Few generalizations describe invertebrates. Their muscles exhibit the same variety as the organisms that make up this diverse assemblage.

—*Judith O. Rebach*

FURTHER READING

Anderson, D. T. *Atlas of Invertebrate Anatomy.* University of New South Wales Press, 1996.

Hickman, C. P., Jr., L. S. Roberts, and F. M. Hickman. *Integrated Principles of Zoology.* 11th ed. McGraw-Hill, 2001.

Mader, Sylvia. *Biology: Evolution, Diversity, and the Environment.* Wm. C. Brown, 1987.

Meglitsch, Paul. *Invertebrate Zoology.* 3rd ed. Oxford University Press, 1991.

Purves, W. K., and G. H. Orians. *Life: The Science of Biology.* Sinauer Associates, 1983.

Schmidt-Nielsen, Knut. *Animal Physiology: Adaptation and Environment.* 5th ed. Cambridge University Press, 1998.

Starr, Cecie, and Ralph Taggart. *Biology: The Unity and Diversity of Life.* 9th ed. Brooks/Cole, 2001.

Wells, Martin. *Lower Animals.* McGraw-Hill, 1968.

MUSCLES IN VERTEBRATES

FIELDS OF STUDY

Animal Physiology; Developmental Biology; Kinesiology; Cardiology

ABSTRACT

The ability of vertebrates to move their bodies and many of the contents of their bodies is a feature of major importance for their survival. The movements result from the contractions and relaxations of tissues specialized for the active generation of force: the muscles. Although initially it may seem sufficient to have only one type of muscle tissue, reflection on the functional requirements makes it clear that more than one type of muscular tissue is probably necessary.

KEY CONCEPTS

involuntary: functioning automatically; not under conscious control
motor neuron: a nerve cell that transmits impulses from the central nervous system to an effector such as a muscle cell
motor unit: a motor neuron together with the muscle cells it stimulates
tissue: a group of similar cells that executes a specialized function

twitch: a rapid muscular contraction followed by relaxation that occurs in response to a single stimulus
voluntary: capable of being consciously controlled

GOT TO MOVE

The ability of vertebrates to move their bodies and many of the contents of their bodies is a feature of major importance for their survival. The movements result from the contractions and relaxations of tissues specialized for the active generation of force: the muscles. Although initially it may seem sufficient to have only one type of muscle tissue, reflection on the functional requirements makes it clear that more than one type of muscular tissue is probably necessary. For example, the movement of the limbs should be under the conscious, voluntary activation and control of the animal. Otherwise, unwanted and uncoordinated random limb movements would result, or desired and possibly vital movements would not be forthcoming. In addition, the control of the limb movements should be precise, with as wide a range as possible for the forces which can be generated.

On the other hand, consider the movement of the blood through the circulatory system. This job must be continuously performed every minute of the animal's life. It would obviously be better to have this function run automatically, without the need for conscious, voluntary activation and control. The need for precise control of the forces generated here is also not as great as for the limb movements, nor is the need for a wide range of forces as great. The flow of blood to the various organs can be much better regulated by varying the diameter of the blood vessels at or near their entrance to the organs, thereby varying the flow into the organs, while the pumping forces generated to propel the blood remain relatively constant.

For the functions of controlling blood flow into organs and the mixing and passage of food through the digestive tract, it is again most reasonable to have automatic operation of the muscles involved, without the need for conscious, voluntary activation and control. To control the huge number of such muscles consciously is an impossible task in any case. Also, these muscles need to be able to change their lengths greatly, and sometimes, to maintain a maximal contraction for very extended periods; however, very rapid actions are not as important.

The preceding considerations make it appear necessary to have three fundamentally different types of muscle tissue. In fact, vertebrates do have three types of muscle tissue, whose different characteristics match these three sets of operational requirements: skeletal muscle tissue, cardiac muscle tissue, and smooth muscle tissue.

SKELETAL MUSCLE

The skeletal muscle tissue occurs in the form of muscles that are usually attached to bones and cause movements of the skeleton. Some skeletal muscles are attached to the skin or to other skeletal muscles. They are under the animal's conscious, voluntary control. Skeletal muscle cells are long, cylindrically shaped cells with rounded ends and, when viewed under a microscope, are seen to have thousands of alternating light and dark bands oriented perpendicular to their long axis. Because of this appearance, skeletal muscle cells are said to be striated. The cells of a muscle are arranged in a parallel fashion, with their long forms mostly following the long axis of the parent muscle.

When a nerve signal stimulates a skeletal muscle cell, the muscle cell contracts relatively quickly and then, just about as quickly, relaxes (a muscle twitch).

The contraction results in a shortening and thickening of the cell, which pulls the two ends of the cell, and whatever is attached to them, toward each other. The duration of a skeletal muscle twitch varies from muscle to muscle. For example, a muscle cell from one of the muscles controlling eye movements may complete a twitch in about ten milliseconds, while a cell from the soleus muscle (found in the calf of the leg) will complete a twitch about ten times more slowly.

The strength of the overall contraction of a skeletal muscle depends on two factors: the number of individual muscle cells in the muscle which become stimulated and therefore contract (called multiple motor unit summation, or recruitment); and the frequency at which the stimulations, and therefore the contractions, occur (called temporal summation). Since most skeletal muscles are composed of hundreds of individual muscle cells, and since each of the nerve cells (motor neurons) which can stimulate the muscle makes contact with only a relatively limited number of the muscle cells (each motor neuron and the muscle cells it contacts forming a motor unit), it is possible to regulate the strength and precision of the muscle's contractions, voluntarily, over a very wide range. This is accomplished by varying the summations—multiple motor unit (number of motor units activated) and temporal (the frequency of motor unit activation)—to meet the demands of a particular situation.

When two muscle twitches occur in rapid succession, it is possible for the muscle to begin the second twitch before it has completely relaxed from the first twitch. In this case, the contraction of the second twitch adds its strength to the force developed as a result of the first contraction; therefore, the second contraction develops more force than the first contraction. The result of continuing the frequency of such twitch-evoking stimuli is a somewhat jerky, oscillating contraction called incomplete tetanus. At high frequencies of stimulation (for example, about fifty twitches per second), there is no evidence of the muscle relaxing from any one twitch before the following twitch takes place, and the muscle makes a very smooth and sustained contraction called complete tetanus. If, following an initial single twitch-evoking stimulus of a skeletal muscle cell, a second stimulus is applied too quickly, there will be no further contraction of the muscle cell, regardless of how strong the second stimulus may be. The cell's ability to respond to this second stimulus is absent until enough time has passed to allow the cell to recover its excitability. The period during which the cell's excitability is absent is referred to as its refractory period. The refractory period of skeletal muscle cells is quite short (about five milliseconds). Skeletal muscles are responsible for the movements of the skeleton (skull, limbs, fingers, toes, and trunk) and for the variety of facial expressions that humans can produce. They also permit speaking, eye movements, breathing, chewing, and swallowing.

CARDIAC MUSCLE

The second type of muscle tissue found in vertebrates is called cardiac muscle. It is the muscle tissue found in the heart. Microscopic examination of cardiac muscle cells reveals them also to possess striations similar to skeletal muscle; however, cardiac muscle is not under voluntary control. Hence, cardiac muscle is referred to as involuntary striated muscle. Cardiac muscle cells also differ from skeletal muscle in that the cardiac fibers branch and form interconnections with one another, forming a network of cells. The individual members of a network are joined by special types of cellular junctions called intercalated disks. The intercalated disks permit the excitation of a single cell, which must occur for the muscle cell to contract, to spread throughout the entire network of cardiac cells; therefore, the contraction of any one cardiac cell will result in the contraction of the entire network of cardiac cells. Thus, the heart's muscle tissue network operates as a functional unit, which is very important for the efficient development of the pressures necessary to push blood through the body's circulatory system.

Another very important characteristic of cardiac muscle is its ability to contract in a spontaneous, rhythmical fashion without the need for neural or hormonal stimuli (although both nerves and hormones are present and function as modulators of cardiac activity, also at a subconscious level). This property is termed autorhythmicity, and it accounts for the involuntary nature of cardiac muscle. Cardiac muscle twitches are about ten to fifteen times longer than those of skeletal muscle, and the refractory period (about three hundred milliseconds) is about sixty times longer than that of skeletal muscle.

The consequences of these traits are important. The long twitch maintains muscular pressure on the blood contents of the heart until most of it has been pumped out of the heart's chambers. The long refractory period prevents the development of tetanus, and allows the heart to relax between beats so that it can be refilled with blood.

SMOOTH MUSCLE

The third type of vertebrate muscle is called smooth muscle because of its lack of striations when viewed microscopically. Its basic and most important functions relate to the maintenance of stable internal conditions within vertebrate bodies. Examples of this are the regulation of blood flow to the various organs to supply them with the proper amounts of oxygen and nutrients, the maintenance of blood pressure during postural changes (such as from reclining to standing), the mixing and movement of ingested food within the digestive tract, and the directing of blood flow through or away from the skin to aid in body temperature regulation. Smooth muscle is also usually involuntary and often displays automaticity. Automaticity refers to the fact that smooth muscle often is stimulated to contract simply by being stretched, as, for example, occurs in the stomach following a meal, without the necessity of neural or hormonal stimulation. Nevertheless, involuntary nerves and hormones are both involved in regulating the actions of smooth muscle cells. The functions of smooth muscle are almost always such that contraction is the appropriate response to stretch of the organ containing the smooth muscle (as in the stomach example just mentioned).

Smooth muscle cells are very small: only about eight micrometers in diameter and between thirty and two hundred micrometers long. They are spindle-shaped, tapering toward each end. Although very small, these cells are able to contract to a length which is a much smaller fraction of their resting length than can either cardiac or skeletal muscle cells. Smooth muscle can also remain fully contracted for long time periods and consume very little energy. The speed of contraction of smooth muscle, however, is the slowest of all three muscle types.

Many smooth muscle cells are arranged as functional units which have the individual cells connected to each other by gap junctions. Gap junctions permit the passage of contraction signals from one cell to the next in the network in much the same way as it occurs in cardiac muscle. The result is that when one cell within the network begins to contract, a wave of contraction spreads throughout the entire network of smooth muscle cells.

STUDYING VERTEBRATE MUSCLE

Light and electron microscopes are frequently used to study muscle tissue. In particular, the electron microscope has made it possible to visualize the intricate and highly specialized internal structures of muscle cells, most of which are impossible to view with a light microscope, given their very small dimensions.

Many studies of muscle cells are actually studies of isolated parts of muscle cells. It is possible to break apart muscle cells and to separate many of their components from one another by the use of centrifugation. Centrifugation is the use of centrifugal force to separate objects by size and/or density. A centrifuge is a device in which samples of objects (in this case, muscle-cell structural components) that are to be separated are subjected to high centrifugal force generated by spinning the samples at great speed. The centrifugal force is the force that tends to impel an object outward from the center of rotation.

Once the various subcellular components are isolated, they can be subjected to a wide variety of biochemical tests to determine what type of molecules they contain (such as proteins, lipids, or carbohydrates), how much of each type they contain, and the exact chemical composition of these molecules (such as the amino acid sequence of a protein). In the case of muscle cells, much of the interest has centered on the large quantities of the proteins actin and myosin which they have been found to contain. Using isolated actin and myosin, scientists have learned much about their functions in muscle cells. In particular, it is now known that these two proteins are the molecules which generate the contractile forces which muscle cells are capable of producing. The precise molecular mechanisms are still unknown, but great progress has been made in discovering how actin and myosin accomplish the generation of contractile forces.

Studies of the electrical properties of muscle cells are performed using very sensitive amplifiers connected to fine glass pipette microelectrodes. These electrodes are formed from thin glass tubes which have been heated and pulled apart to produce a tapering of the wall of the tube. The final tip of the tapered tube is much smaller than a muscle cell. This makes it possible to insert the tip into the cell without killing it. When the electrodes are filled with a solution capable of conducting electrical signals, they can record the electrical activity that takes place in living, contracting, and relaxing muscle cells: their electrophysiological properties. These studies have revealed that all muscle cells possess an electrical voltage difference between their interior region and the surrounding environment. Just before a muscle cell begins to contract, this voltage difference decreases toward a zero value extremely rapidly. Combining such electrophysiological studies of muscle cells with simultaneous biochemical studies can be very rewarding.

Other techniques used in the study of muscle tissue include the observation of the relationships that various muscles have with other muscles, with nerves, and with bones, and the use of strain gauges to measure the strength of muscles when different experimental conditions occur (such as temperature changes, blood-flow variations, and pharmacological treatments). Even high speed X-ray motion pictures have been applied to the study of muscle tissue.

THE IMPORTANCE OF MUSCLE

Vertebrate muscle tissue is involved in almost every bodily function. In particular, any function requiring the movement of some body part or parts needs to have a source of motive power. For this purpose, evolution has resulted in the development of the specialized muscle tissues found in all vertebrates, which account for up to 50 percent of their total body weight.

Because of the muscle-tissue characteristics of excitability (the ability to respond to stimuli), extensibility (the ability to be stretched), contractility (the ability to shorten actively and thereby generate force for the production of work), and elasticity (the ability to return to original length following extension or contraction), vertebrates have a wide range of important capabilities available which assist them in their survival.

For the muscular system, disease states or disorders resulting from improper nutrition, injury, or toxic substances are very often life-threatening conditions. This is most obvious for disturbances of the respiratory muscles or of the muscles involved in eating and swallowing; however, muscular disorders involving the heart, the circulatory system, the digestive tract, or any major group of skeletal muscles may also prove to be life-threatening.

The most common heart problems are caused by reduction or blockage of the blood supply to the heart muscle. Reduced blood supply usually is the cause of a reduced oxygen supply. The insufficient oxygen supply weakens the heart-muscle cells, causing the condition of ischemia. Complete interruption of the blood supply to an area of cardiac muscle tissue usually results in necrosis (death) of the affected muscle cells; the condition is referred to as myocardial infarction. The dead muscle cells do not regenerate but are replaced by scar tissue, which is not contractile. This results in decreased pumping efficiency by the heart. Depending on the size and location of the dead area, the result may range from barely noticeable to sudden death.

As a consequence of some bacterial infections, it is possible for the lining of the heart muscle to become inflamed. This can then result in abnormal irritation of some of the heart-muscle cells. These cells can then disrupt the normal autorhythmicity of the heart. Irregular heartbeats and/or uncoordinated contractions among the heart-muscle cells ensues. These conditions are very serious and can lead to death.

—*John V. Urbas*

FURTHER READING

Hildebrand, Milton. *Analysis of Vertebrate Structure*. 5th ed. John Wiley & Sons, 2001.

Kardong, Kenneth V. *Vertebrates: Comparative Anatomy, Function, Evolution*. 2nd ed. Wm. C. Brown, 1998.

Keynes, R. D. *Nerve and Muscle*. 3rd ed. Cambridge University Press, 2001.

Netter, Frank H. *Musculoskeletal System: Anatomy, Physiology, and Metabolic Disorders*. CIBA-GEIGY, 1987.

Tortora, Gerard J., and Nicholas P. Anagnostakos. *Principles of Anatomy and Physiology*. 9th ed. John Wiley & Sons, 2000.

Nervous Systems of Vertebrates

Fields of Study

Animal Anatomy; Animal Physiology; Psychology

Abstract

Animals must be able to coordinate their behaviors and to maintain a relatively constant internal environment, despite fluctuations in the external environment, in order to survive and reproduce. To do so, animals must monitor their external and internal environments, integrate this sensory information, and then generate appropriate responses. The evolution of the vertebrate nervous system has provided for the efficient performance of these tasks.

Key Concepts

axon: an extension of a neuron's cell membrane that conducts nerve impulses from the neuron to the point or points of axon termination

gray matter: the part of the central nervous system primarily containing neuron cell bodies and unmyelinated axons

interneuron: a central nervous system neuron that does not extend into the peripheral nervous system and is interposed between other neurons

myelinated axon: an axon surrounded by a glistening sheath formed when a supporting cell has grown around the axon

neural integration: continuous summation of the incoming signals acting on a neuron

neurons: complete nerve cells that respond to specific internal or external environmental stimuli, integrate incoming signals, and sometimes send signals to other cells

nucleus (pl. nuclei): cluster of neuron cell bodies within the central nervous system

tract: a cordlike bundle of parallel axons within the central nervous system

white matter: the part of the central nervous system primarily containing myelinated axon tracts

THE TECHNICAL DETAILS

Animals must be able to coordinate their behaviors and to maintain a relatively constant internal environment, despite fluctuations in the external environment, in order to survive and reproduce. To do so, animals must monitor their external and internal environments, integrate this sensory information, and then generate appropriate responses. The evolution of the vertebrate nervous system has provided for the efficient performance of these tasks.

THE PATTERN OF THE VERTEBRATE NERVOUS SYSTEM

Although the various vertebrates show differences in the organization of their respective nervous systems, they all follow a similar anatomical pattern. The nervous system can be partitioned conveniently into two major divisions: the peripheral nervous system (PNS) and the central nervous system (CNS). These divisions are determined by their location and function. The CNS consists of the spinal cord and the brain. The PNS, that part of the nervous system outside the CNS, connects the CNS with the various sense organs, glands, and muscles of the body.

The PNS joins the CNS in the form of nerves, which are cordlike bundles of hundreds to thousands of individual, parallel nerve-cell (neuron) axons (long tubular extensions of the neurons) extending from the brain and spinal cord. The nerves extending from the spine are called spinal nerves, while those from the brain are called cranial nerves. The elements of the PNS include sensory neurons (for example, those in the eyes and in the tongue) and motor neurons (which activate muscles and glands, thereby causing some sort of action or change to occur). Most nerves contain both sensory and motor axons.

Thus, the PNS can be divided into two major subdivisions: sensory (or afferent) neurons and motor (or efferent) neurons. There is very little information-processing accomplished in the PNS. Instead, it relays both environmental information to the CNS (sensory function) and the CNS responses to the body's muscles and glands (motor function). Sensory neurons of the PNS are classified as somatic afferents if they carry signals from the skin, skeletal muscles, or joints of the body. Sensory neurons from the visceral organs (internal organs of the body) are called visceral afferents.

The PNS motor subdivision also has two parts. One is the somatic efferent nervous system, which carries neuron impulses from the CNS to skeletal muscles. The other is the autonomic nervous system (ANS), which carries signals from the CNS to regulate the body's internal environment by controlling the smooth muscles, the glands, and the heart. The ANS itself is subdivided into the sympathetic and parasympathetic nervous systems. These are generally both connected to any given target and cause approximately opposite effects to each other on that target (for example, slowing or increasing the heart rate).

The CNS, where essentially all information-processing occurs, has two major subdivisions: the spinal cord and the brain. Virtually all vertebrates have similarly organized spinal cords, with two distinct regions of nervous tissue: gray and white matter. Gray matter is centrally located and consists of neuron cell bodies and unmyelinated axons (bare axons without the glistening sheaths called myelin, created by supporting cells wrapping around the axons). White matter contains mostly bundles of myelinated axons (white because they have glistening myelin sheaths around them). Bundles of axons in the CNS are called nerve tracts. Within the spinal cord, these are either sensory tracts carrying impulses toward the brain, or they are motor tracts transmitting information in the opposite direction.

Interneurons are neurons positioned between two or more other neurons. They accept and integrate signals from some of the cells and then influence the others in turn. Interneurons are particularly numerous within the gray matter. In the spinal cord, they permit communication up, down, and laterally. Most axons in the cord's tracts belong to interneurons.

THE VERTEBRATE BRAIN

The brain of vertebrates is actually a continuation of the spinal cord, which undergoes regional expansions during embryonic development. The subdivisions of the brain show more variety among vertebrate species than does the spinal cord. The brain has three regions: the hindbrain, the midbrain, and the forebrain. Their structures are complex, and various systems of subdividing them exist. The major components forming the brain regions are the hindbrain's medulla oblongata, pons, and cerebellum; the midbrain's inferior and superior colliculi, tegmentum, and substantia nigra; and the forebrain's hypothalamus, thalamus, limbic system, basal ganglia, and cerebral cortex.

The hindbrain begins as a continuation of the spinal cord called the medulla oblongata. Most sensory fiber tracts of the spinal cord continue into the medulla, but it also contains clusters of neurons called nuclei. The posterior cranial nerves extend from the medulla, with most of their nuclei located there.

Also in the medulla, and extending beyond it through the pons and midbrain, is the complexly organized reticular formation. This mixture of gray and white matter is found in the central part of the brain stem but has indistinct boundaries. Essentially all sensory systems and parts of the body send impulses into the reticular formation. There are also various nuclei within its structure. Impulses from the reticular formation go to widely distributed areas of the CNS. This activity is important for maintaining a conscious state and for regulating muscle tone.

Prominent on the anterior (front) surface of the mammalian medulla oblongata are the pyramids: tracts of motor fibers originating in the forebrain and passing without interruption into the spinal cord to control muscle contraction. These tracts cross to the opposite side of the medulla before entering the spinal cord, which results in each side of the forebrain controlling muscle contraction in the opposite side of the body.

Many sensory fibers from the spinal cord terminate in two paired nuclei at the lower end of the medulla, the gracile and cuneate nuclei. Axons leaving these nuclei cross to the opposite side of the medulla and then continue as large tracts (the medial lemnisci) into the forebrain. Thus, each side of the brain gets sensory stimuli mostly from the opposite side of the body.

Immediately above the medulla is the pons. It contains major fiber pathways carrying signals through the brain stem, and a number of nuclei, including several for cranial nerves. Some pontine nuclei get impulses from the forebrain and send axons into the cerebellum, again with a majority crossing to the opposite side of the brain stem before entering the cerebellum.

On the dorsal (back) side of the medulla and pons is the cerebellum, an ancient part of the brain that varies in size among vertebrate species. The cerebellum forms a very important part of the control system for body movements, but it is not the source of motor signals. Its gray matter forms a thin layer near its surface called the cerebellar cortex and surrounds central white matter.

Vertebrates with well-developed muscular systems (for example, birds and mammals) have a large cerebellum, with several lobes and convex folding of its cortex. It is attached to the brain stem by three pairs of fiber tracts called cerebellar penduncles, which transmit signals between the left and right sides of the cerebellum and between the cerebellum and motor areas of the spinal cord, brain stem, midbrain, and forebrain. The cerebellum times the order of muscle contractions to coordinate rapid body movements.

THE MIDBRAIN AND THE FOREBRAIN

The midbrain is the second major region of the brain. The midbrain's dorsal aspect, called the tectum, is a target for some of the auditory and visual information that an animal receives. The paired inferior colliculi form the lower half of the tectum. They help to coordinate auditory reflexes to relay acoustic signals to the cerebrum. The two superior colliculi, the other half of the tectum, assist the localization in space of visual stimuli by causing appropriate eye and trunk movements. In lower vertebrates, the superior colliculi actually form the major brain target for visual signals. Connecting fiber pathways (commissures) link the individual lobes of each pair of colliculi.

The midbrain's tegmentum contains several fiber tracts carrying sensory information to the forebrain and carrying impulses among various brain-stem nuclei and the forebrain. Two cranial nerve nuclei concerned with the control of eye movements are also in the tegmentum. The reticular formation extends through the tegmentum and regulates the level of arousal. It also helps to control various stereotyped body movements, especially those involving the trunk and neck muscles. Finally, the tegmentum contains the red nucleus, which, in conjunction with the cerebellum and basal ganglia, serves to coordinate body movements. The substantia nigra functions as part of the basal ganglia to permit subconscious muscle control.

The forebrain, the final major area of the brain, differs from the lower areas in the more highly evolved functions it controls. It has a small but extremely important collection of about a dozen pairs of nuclei called the hypothalamus. These control many of the body's internal functions (such as temperature, blood pressure, water balance, and appetite) and drives (such as sexual behavior and emotions). Immediately above the hypothalamus lies the thalamus, another collection of more than thirty paired nuclei. The two thalami are the largest anterior brain-stem structures. Their ventral (front) parts relay motor signals to lower parts of the brain. The dorsal (back) parts transmit impulses from every sensory system (except olfaction, the sense of smell) to the cerebrum.

The limbic system is organized from a number of forebrain structures mostly surrounding the hypothalamus and thalamus. It determines arousal levels, emotional and sexual behavior, feeding behavior, memory formation, learning, and motivation. In general, the limbic system exchanges information with the hypothalamus and thalamus, and receives impulses from auditory, visual, and olfactory areas of the brain.

The basal ganglia function with the midbrain's tegmentum and substantia nigra, the cerebral cortex, the thalamus, and the cerebellum. These paired structures' functions are unclear, but it is known that they are important for adjusting the body's background motor activities, such as gross positioning of the trunk and limbs, before the cerebral cortex superimposes the precise final movements.

The cerebral cortex, like the cerebellum, is an ancient brain structure; however, it shows even more variation among vertebrate species than the cerebellum. It is formed into two hemispheres, which have olfactory bulbs projecting from their anterior (front) ends. The olfactory bulbs receive impulses from olfactory nerves for the sense of smell. The gray matter of the cerebral cortex is at the surface, enclosing the white matter (fiber tracts) beneath. The white matter connects various parts of the gray matter of one hemisphere with others within the same hemisphere and with corresponding parts in the opposite hemisphere. It also connects the cortical gray matter with lower brain structures. The ultimate control of voluntary motor activity resides in the motor areas of the cortex, although this control is heavily influenced by all the previously mentioned motor-control areas of the CNS.

Corresponding to each of the major senses (touch, vision, audition), there are primary sensory areas. These areas get the most direct input from their sensory organs by way of the corresponding sensory thalamic nuclei. Surrounding each primary area are association areas that receive a less direct sensory input but also more inputs from other sensory cortical areas. In general, the more intelligent an animal is, the larger are its association areas.

STUDYING THE NERVOUS SYSTEM

Many methods are used in studying vertebrate nervous systems. The level of description desired often determines the methods employed. For example, the gross structure visible to the unaided eye is usually investigated using the entire brain or spinal cord of the animal under study. It will then be photographed or drawn, sliced at various points either parallel to its long axis or across its long axis, and again photographed or drawn, until a complete series of such "sections" has been assembled.

To see finer structural details requires microscopes and very thin slices of nervous tissue (less than a millimeter in thickness). Preparation of such thin slices of this soft tissue requires that it first be either frozen or embedded in a block of paraffin wax. A special slicing device called a microtome is then used to produce the thin sections. For easy observation of different structural details (such as nuclei and fiber tracts), various chemical stains can be applied to the slices of tissue. These stains specifically color particular structural features green, blue, or some other color, thereby making them more visible.

Nervous tissue can be selectively and painlessly destroyed in an anesthetized animal by cutting a nerve trunk, inserting a fine wire electrode into the CNS and destroying tissue with electricity, or inserting a fine needle and then either injecting a chemical agent that kills nervous tissue or using a suction device to remove areas of tissue. The precision and reproducibility of wire or needle placement within the CNS is possible with a device called a stereotaxic frame. This instrument positions the animal's head and brain in an exact standard position. Then, wires or needles are inserted a certain distance away from (for example, behind, below, or to one side of) common landmarks on the skull. Stereotaxic atlases are books published by investigators for specific animals, with the exact coordinates in three-dimensional space for most CNS structures.

Following such procedures, the animal may be immediately and painlessly sacrificed, its brain or spinal cord removed, and the previously described thin slices prepared and stained. It may be necessary to allow the animal to recover from its surgical treatment, since several days or weeks must sometimes pass for the severed fiber tracts to degenerate. Then, following a painless lethal injection, the nervous tissue is prepared as above using special staining techniques, which reveal the pathways of degenerating nerve fibers in the tissue sections studied later under the microscope.

Although new techniques are constantly being developed, the preceding methods have revealed that the hundred billion neurons of the vertebrate nervous system form the most complexly organized structure known. Through this knowledge of the structure of the nervous system, it has become possible to study intelligently its functions, to diagnose its diseases, and to devise methods of treatment when it becomes damaged or diseased.

A COMPLEX STRUCTURE

The vertebrate nervous system is the most complex structure known to humankind. The human nervous system, for example, has more than a hundred billion cellular elements, and perhaps a hundred trillion points of information exchange between these elements. It is impossible to understand the details of such a structure in the same way that one can understand the structure of a radio; however, the general organizational plan can be discovered through the application of modern neuroanatomical techniques.

It is a widely accepted tenet of physiology that in order to comprehend the functioning of an organ or an organ system, it is necessary to have a critical understanding of its structure. In fact, physiology and anatomy are inseparable because function (physiology) always reflects structure (anatomy): It is impossible for an organ to perform in any other way than its architecture permits.

The nervous system is the ultimate control and communication system in the vertebrate body. Its complexity allows the vast range of vertebrate behaviors, as well as the rapid and precise regulation of the body's internal environment. The most complex vertebrate nervous systems display self-consciousness, reasoning, and language capabilities.

There are many reasons for studying the organization of vertebrate nervous systems, ranging from purely theoretical (such as determining the mechanisms of memory recall or clarifying the evolutionary relationships among vertebrate species) to very practical (such as treatments for mental illnesses, precisely defining brain death, or designing better computers).

For many, the ultimate goal is to obtain a better understanding of the relationship between the brain and the mind. It has been proposed that the mind and mental processes are emergent properties that appear when a certain degree of organizational complexity within the nervous system has been reached. The individual elements of the nervous system (the neurons) are not the constituents that think or possess consciousness. These unique capabilities are achieved as a result of the specific connections between neurons and sensory organs, and among neurons themselves.

It does not matter whether one performs an analysis of a machine or of a nervous system; it can never be expected to reveal its soul or its consciousness—if they exist. All that can be done is to admire the intelligence of its designer or the wisdom of nature.

—*John V. Urbas*

Further Reading

Butler, Ann B., and William Hodos. *Comparative Vertebrate Neuroanatomy: Evolution and Adaptation.* Wiley-Liss, 1996.

Hildebrand, Milton. *Analysis of Vertebrate Structure.* 5th ed. John Wiley & Sons, 2001.

Keynes, R. D. *Nerve and Muscle.* 3rd ed. Cambridge University Press, 2001.

Simmons, Peter J., and David Young. *Nerve Cells and Animal Behavior.* 2nd ed. Cambridge University Press, 1999.

Nesting

Fields of Study

Ornithology; Mammalogy; Ichthyology; Ethology

Abstract

Many animals build structures to house and protect themselves, their eggs, and their young. Nest sites include grasses, shrubs, and trees, but also cracks in trees, holes in the ground and banks, crevices in rocks, under the surface of the ground, within the nests of other larger animals, or even near wasp nests. The nests of birds are the most obvious and well known, but many amphibians, reptiles, fishes, social insects, and mammals also build nests of varying degrees of complexity and permanence.

Key Concepts

insectivore: any animal that feeds on insects
mallard: a species of wild duck
ornithology: the scientific study of birds
pheromones: chemicals excreted by animals into their immediate environments to identify their territorial influence to other organisms
territorial behavior: the combination of methods and actions through which an animal or group of animals protects its territory from invasion by other species
territory: any area defended by an organism or a group of organisms for purposes such as mating, nesting, roosting or feeding

HOME, SWEET HOME

Many animals build structures to house and protect themselves, their eggs, and their young. Nest sites include grasses, shrubs, and trees, but also cracks in trees, holes in the ground and banks, crevices in rocks, under the surface of the ground, within the nests of other larger animals, or even near wasp nests. The nests of birds are the most

Bird nests are most often made of plant material such as grasses, providing temporary homes for eggs and hatchlings.

obvious and well known, but many amphibians, reptiles, fishes, social insects, and mammals also build nests of varying degrees of complexity and permanence.

BIRD AND INSECT NESTS

Larger bird nests are constructed of various kinds of material, such as mud, bark, roots, twigs, hair, feathers, grass, plant fibers, shed snake skin, spider webs, lichens, or even prey remains and human-made material such as shiny, light, metallic ornaments. Many nests are cup-shaped and open, while fewer are oval, round or ball-shaped, closed, with an entrance at the side or the roof. An example of the latter is the nest of the South American ovenbird (*Furnarius*), which is often built with mud on top of a fence or another exposed surface. Others build a domed, oven-shaped nest out of plant material (the North American ovenbird) or huge nests suspended from the ends of tree branches (thorn birds). Some oceanic gulls (*Rissa tridactyla*) nest on narrow cliff ledges. Grassland and tundra owls nest on the ground or on an elevated hummock.

Some birds utilize amazing skill to construct their nests. The weavers are capable of tying knots with strips of grass or palm leaves and can prepare an exceptionally tight and compact nest. Birds of the genus *Orthotomus* are known as tailorbirds due to their ability to manufacture nests that are built in a pocket made by sewing together the edges of one or more leaves, using plant fibers. Some others (the family Contingidae) prepare very primitive and weak nests in an apparent effort to avoid the attention of predators, since they are attended minimally by the parents. Finally, woodcreepers nest in holes, while vireos weave a cup between the arms of a forked branch.

Most bird species in an area construct a unique nest in a unique location but use specific construction materials transferred from a distance. Generally, female geese and robins build their own nests, but among several other species the nest is built only by the male, who may use it as a sexual attractant in courtship. In these species, the female chooses a nest and indicates her choice by adding the nest lining.

Woodpeckers create cavities in trees, thus supplying safe nesting sites for a large number of birds. These include owls, parrots, parids, and flycatchers. In some areas, forest managers protect pileated woodpecker cavity trees and employ strategies to encourage continued production of new cavity trees where even smaller woodpecker species may find a haven for nesting.

Some insects, such as termites and carpenter bees, excavate a tunnel through solid wood. When termite colonies accidentally separate from an original nest they may migrate or march to a new nesting site and develop supplementary reproductives.

MAMMAL NESTS

Most rodents build underground residences with a central chamber, where they sleep, raise their young, and hibernate, and other chambers that serve as food storage quarters. Moles build the most complex shelters among all insectivores. The shelters have an underground nest chamber that is surrounded by concentric rings of tunnels that are interconnected by radiating ones. The presence of shallow surface tunnels allows the marking of their course and the accumulation of a large amount of earth indicates the location of the deep tunnel system. The mole uses its forefeet like a shovel to dig in a type of body movement that resembles that of a breaststroke swimmer. Shrews dig surface nests, but they also use the runs of other animals, such as rodents, and line them with plant material on which they place their offspring. Hedgehogs and solenodons construct nest chambers, usually during their breeding season, which warm them during low-temperature periods. Tree shrews build their nests of leaves and other vegetation among tree roots or in cavities of fallen timber, but they immediately desert them when they feel that their residence has been stalked or even detected by a predator.

Ground squirrels and kangaroo rats transport enough seeds and other food in their cheek pouches to last them for a whole season. Excavation of one rat den where a single five-ounce kangaroo rat resided produced nine underground storage chambers with a total of thirty-five quarts of seeds. Kangaroo rats also dig one-cubic-inch storage pits that they stuff with seeds. In one case, an area of fifty-five square feet adjacent to a den had close to nine hundred such pits. This can have a devastating effect on local crops, so-called rodent plagues. On the other hand, such underground activity tends to germinate seeds for wild grass in the arid steppe regions of Central Asia. Squirrels carry acorns and nuts to hollow trees and barns, as well as to holes in the ground. They also have an amazing memory of where the food is stored, which keeps them alive during the harsh winter. Finally, the North American pack rat or trade rat is attracted by bright, shining objects (like the magpie) and carries them home to its nest, where it stores them together with tree sticks and grass.

Cooperative breeding appears to exist in the form of communal nesting among several mammal species. Tree squirrels (*Sciuridae sciurini*), fox squirrels (*Sciurus niger*) and gray squirrels (*Sciurus carolinensis*) form kin clusters among unrelated members of their own species during all seasons, especially in winter. Gray squirrels have shown an intense female-female bond in the formation of groups.

ARTIFICIAL NESTING

Humans living in suburban and rural areas can provide the appropriate housing for cavity nesting birds, and people do have the ability to become backyard bird specialists. Artificial structures have been used extensively in the United States and Canada to increase waterfowl production, although the users of the artificial materials are not always the animals for which they were intended. During a study conducted in 1995, in which artificial nesting structures were intended for mallard use, redheads were found to be using them for normal nesting. It is believed that a combination of elevated water surfaces and subsequent limited nesting in the emergent vegetation may have made these nesting sites attractive to the redheads, which has thus provided a new alternative for nesting redheads.

Another study on the excavation and use of artificial polystyrene snags by woodpeckers was performed in Eastern Texas over a period of five years in the early 1990s. Only half of the monitored downy woodpeckers (*Picoides pubescens*) appeared to use the artificial snags for cavity excavation and later for nocturnal roosting, but not for nesting. None of the other six woodpecker species in the area excavated cavities in the artificial snags. Other animals, however, used these excavated cavities in the artificial snags. These include Carolina chickadees (*Parus carolinensis*), prothonotary warblers (*Protonotaria citrea*), southern flying squirrels (*Glaucomys volans*) and red wasps (*Polistes* sp.).

The human factor may affect, directly or indirectly, the breeding success of animals. Possible biological effects of electromagnetic fields attributed to high-voltage transmission lines are suspected to reduce the reproductive success of birds whose nests are nearby, as in the case of tree swallows. Similar effects have been postulated for the terns, gulls, and other birds that live in areas such as the Gulf Coast, where oil tankers, commercial fishing vessels, yachts, and other pleasure boats have degraded the environment. A debate and controversy occurred in the 1990s in the Pacific Northwest of the United States, where environmentalists lobbied for the preservation of the nesting habitat of the spotted owl to the detriment of the local logging industry.

—*Soraya Ghayourmanesh*

FURTHER READING

Ananthaswamy, Anil. "That Nesting Instinct." *New Scientist* 167, no. 2250 (August 5, 2000): 14.

Doherty, Paul F., and Thomas C. Grubb. "Reproductive Success of Cavity-Nesting Birds Breeding Under High-Voltage Power Lines." *The American Midland Naturalist* 140, no. 1 (July 1998): 122–128.

Dyes, John C. *Nesting Birds of the Coastal Islands: A Naturalist's Year on Galveston Bay*. University of Texas Press, 1993.

Heinrich, Bernd. "The Artistry of Birds' Nests." *Audubon* 102, no. 50 (September/October 2000): 24–31.

Labauch, Rene, and Christyna Labauch. *The Backyard Birdhouse Book: Building Nest Boxes and Creating Natural Habitats.* Storey, 1999.

Skutch, Alexander F. *Helpers at Bird Nests: A Worldwide Survey of Cooperative Breeding and Related Behavior.* University of Iowa Press, 1999.

Neutral Mutations and Evolutionary Clocks

Fields of Study

Genetics; Proteomics; Evolutionary Biology

Abstract

The hypothesis has been advanced that the vast majority of polymorphisms that occur at the molecular level are selectively neutral. Two major categories of polymorphisms are involved. The first is attributable to changes in the nucleotide sequences of deoxyribonucleic acid (DNA). The second is isozymic variation, detectable by protein electrophoresis. Isozymic variation is usually caused by changes in the amino acid composition of the protein. Since amino acids are determined by the genetic code, this type of variation also ultimately depends on changes in DNA.

Key Concepts

allele: one of two or more alternate gene forms of a single gene locus

amino acid: an organic molecule with an attached nitrogen group that is the building block of polypeptides

electrophoresis: a technique for separating molecules when they are placed in an electrical field; the separation is usually based on their charge and weight

genetic code: the three-nucleotide base sequences (codons) that specify each of the twenty types of amino acids; there can be more than one codon for a particular amino acid

nucleic acid: an organic acid chain or sequence of nucleotides, such as DNA or RNA

phylogeny: the evolutionary history of taxa, such as species or groups of species; order of descent and the relationships among the groups are depicted

polymorphic: a genotype or phenotype that occurs in more than one form in a population

DNA and Chance Changes

The hypothesis has been advanced, primarily by Motoo Kimura of the National Institute of Genetics of Japan and Masatoshi Nei of the University of Texas, that the vast majority of polymorphisms that occur at the molecular level are selectively neutral. Two major categories of polymorphisms are involved. The first is attributable to changes in the nucleotide sequences of deoxyribonucleic acid (DNA). The second is isozymic variation, detectable by protein electrophoresis. Isozymic variation is usually caused by changes in the amino acid composition of the protein. Since amino acids are determined by the genetic code, this type of variation also ultimately depends on changes in DNA. The proponents of the neutral theory admit that most of the evolution that occurs on the nonmolecular level, such as changes in morphological and behavioral traits, is attributable to natural selection. On the molecular level, however, they believe that most of the changes are caused by chance.

Selection Versus Chance

Extensive variability has been found for both DNA sequences and isozymes within the majority of natural populations. Isozymic polymorphism, in which two or more variants of an enzyme occur within the same species, ranges from approximately 15 percent in mammals to approximately 40 percent in invertebrates. Isozymic variation, even within the same individual, ranges from about 4 percent in mammals to 14 percent in some insects. Variability in DNA sequences among individuals of the same species is even higher than those found for isozymes. Proponents of the neutral hypothesis hold that these levels of variability are too high to be attributable to selection, but instead, most variability at the molecular level is attributable to chance. The result, they say, has been a large amount of enzyme and DNA variability that is selectively neutral.

They are neutral in the sense that their contributions to an organism's fitness are so small that their occurrence is attributable more to chance than to natural selection. Neutralists do not believe that most molecular mutations are neutral; they assume that most are harmful and are eliminated by natural selection. Rather, they believe that those that currently exist are adaptively equivalent. Proponents of the neutrality theory believe that changes in DNA and amino acid sequences are for the most part neutral, consisting primarily of the gradual random replacement of functionally equivalent alleles.

Although the neutral theory is able to explain much about molecular evolution, there are some issues that remain subjects of intense debate. It is known that some protein and gene variation is not neutral but, instead, under certain conditions, conveys selective advantages or disadvantages. In some organisms (for example, the Japanese macaque), there also appear to be more rare alleles than would be predicted by the neutrality hypothesis.

THE HYPOTHESIS OF A MOLECULAR "CLOCK"

If the neutrality theory of molecular evolution is correct, then changes in base sequences of DNA could act as evolutionary "clocks." This theory holds that because mutations change the DNA in all lineages of organisms at fairly steady rates over long periods of time, a clocklike relationship can be established between mutation and elapsed time. The number of base substitutions in the DNA is directly proportional to the length of time since evolutionary divergence between two or more species. The idea of molecular evolutionary clocks was forwarded in the early 1960s by Linus Pauling and Emile Zuckerkandl.

The molecular clock postulated by the neutrality hypothesis is not like an ordinary timepiece, which measures time in exact intervals. Rather, molecular changes occur as a stochastic clock, such as occurs during radioactive decay. Although there is some variability for this type of clock (it is slower or faster during some periods than others), it would be expected to keep relatively accurate time over millions of years. A potential problem arises, however, because the rate of "ticking" for the molecular clock is not the same at every position along the DNA molecule. The rate has been shown to be slow for DNA sequences that directly affect the function of a protein (for example, those at an enzyme's active site), while the rate of change has been faster for positions on the DNA that are selectively neutral, that is, where they have little or no affect on the protein's function.

In a molecular clock, the number of changes in a DNA or protein molecule are the "ticks" of the clock. The number of "ticks," in turn, estimates the extent of genetic differences between two species. With this knowledge, scientists can reconstruct phylogenies. The phylogenies are usually depicted as branching patterns, which are based upon differences in DNA base-pair sequences or amino acid sequences. They depict not only the order of descent but also the degree of relatedness.

When choosing alternative phylogenetic hypotheses, biologists usually follow the principle of Occam's razor—the simplest theory is chosen over more complex ones. Thus, the phylogenetic tree that requires the fewest mutations is preferred over those that require more mutations. By "calibrating" the molecular clock with other, independent, events, such as those obtained from the fossil record, the actual chronological times of divergence can be estimated.

For example, humans and horses differ by eighteen amino acids in the alpha chain of the blood protein hemoglobin. It has been estimated from the fossil record that humans and horses diverged from a common ancestor approximately ninety million years ago. Other evidence suggests that half the substitutions took place since the time of divergence. Since nine amino acids have changed over a ninety-million-year period, the rate of amino acid substitution would equal approximately one every ninety million years.

Since mutation rates are known to be different for different genes, the ticking rate for different genes or proteins would not necessarily be the same. For example, the rate of substitution for the genes coding for the protein histone H4 is lower than that for the genes coding for the protein gamma interferon. Yet when nucleotide substitution rates are averaged for a number of different proteins, there does appear to be a marked uniformity in the rate of molecular change over time; the ticking of a number of clocks can be averaged, leading to more accurate estimates for divergence times.

ADVANTAGES AND DISADVANTAGES OF THE THEORY

Much of the early work was done on sequence changes in proteins; however, there is a drawback to protein clocks. Their usefulness is limited because the genetic material itself is not being examined but, rather, a product coded by genes. This means that some of the changes in the genetic material may not be detected. For example, because of the redundancy of the genetic code, there could be a number of changes that occur in the nucleotide sequence of DNA that would not result in changes in the amino acid composition in a protein. Consequently, there has been great interest in directly examining the DNA itself.

Because of the advent of recombinant DNA techniques, molecular clocks can be based on changes in the genetic material. DNA-DNA hybridization also involves the comparison of DNA sequences, although on a broader scale. The DNA-DNA hybridization technique is attractive because it effectively compares very large numbers of nucleotide sites, each of which is effectively a single data point. One of the criticisms of molecular clocks is that most genes have not been found to "tick" with perfect regularity over long periods of time. During some periods, the rate of change (primarily because of mutation) may be fast, while at other periods it may be significantly slower. By comparing very large numbers of nucleotides, which represent many genes, DNA-DNA hybridization measures the average rates of change, which will produce more uniform estimates.

The concept of a molecular clock has been criticized on a number of grounds. First, it assumes that evolution in macromolecules proceeds at an approximately regular pace, whereas morphological evolution is usually recognized as occurring irregularly. It is also clear that the clock can tick at different rates among different macromolecules, whether they be proteins or DNA. Another problem is that the rate of the molecular clock varies among taxonomic groups. For example, the insulin gene has evolved much more rapidly in the evolutionary line leading to the guinea pig than in some other evolutionary lines. There are also notable differences among different parts of molecules. This variability was evident when sequences were compared among the first molecules examined in the light of the molecular clock hypothesis, notably hemoglobin molecules and cytochrome c. Another criticism is that a number of processes, known collectively as molecular drive, perturb the clock.

Recent data, however, suggest that nucleotide substitution rates in organisms as different as bacteria, flowering plants, and vertebrates are remarkably similar. For example, the average rate of substitution at "silent sites" (those at which mutation in the DNA produces no change in the amino acid encoded) is 0.7% per million years in bacteria, 0.9% for mammals, and 1% for plants. This relatively equal substitution rate across broad taxonomic categories would support the concept of the constancy of molecular clocks.

TESTING THE HYPOTHESIS

A number of types of molecular clocks have been hypothesized. In the first group are techniques that directly estimate differences in the sequences of nucleic acids that make up DNA or in amino acid sequences (which are determined by DNA). Other methods are less direct, such as DNA-DNA hybridizations and immunological techniques. All the techniques ultimately assay genetic differences caused by base pair changes in DNA. Sequence comparisons and DNA-DNA hybridization techniques are now used more extensively.

In sequence comparisons, nucleic acid replacement in DNA or amino acid replacement in proteins are compared between species. Nucleic acid substitutions can be assayed by using restriction enzymes that only recognize specific base sequences or by direct sequencing. Amino acid substitutions can be assayed by traditional biochemical techniques, through automated sequencers, or by mass spectrometers. In both types of sequence analysis, the assumption is made that the greater the number of substitutions, the greater the evolutionary distance between the species.

In DNA-DNA hybridization, DNA molecules from individuals from two species are separated into individual strands at high temperatures. The strands are then mixed at a lower temperature. This promotes joining of the strands from the different organisms. The extent of rejoining (how tightly they bond together) will be dependent on the degree of nucleotide pairing that occurs. If the nucleotide sequences between the two species are very similar, the DNA strands will bond very tightly; if there is little similarity, the bonds will be weak. The extent of bonding is measured by the temperature at which the new DNA duplex dissociates, or "melts." The higher the melting temperature, the greater the nucleotide similarity between the DNA strands from different species. The nucleotide similarity is presumed to be related to the evolutionary distance between the species.

USES OF THE THEORY

The use of molecules as clocks, in spite of their imperfect nature, has proven to be a valuable tool for inferring phylogenetic relationships among species and in estimating their times of divergence. Molecular data can be used independent of morphological and behavioral data for establishing evolutionary relationships. Similarly, divergence times estimated from the fossil record can be clarified through the use of molecular clock data. Molecular clocks have had a significant impact on evolutionary studies of organisms ranging from bacteria to humans, and molecular data have been instrumental in changing some long-held phylogenetic views. For example, the data obtained from DNA-DNA hybridizations in birds have forced a major revision in bird taxonomy.

Molecular clocks have been used to assign time scales to a large number of phylogenies. Some of the phylogenies are wide-ranging; approximate times of evolutionary divergence have been assigned to vertebrate species as diverse as sharks, newts, kangaroos, and humans. Others are more specific. Some of the best-known (and most controversial) work has been done on primates. In one set of experiments, the amino acid differences of serum albumin (a blood protein) were measured among different species of primates. By comparing the albumin from species whose divergence times were known from fossil evidence, researchers were able to "calibrate," or calculate, the mean rate of change for serum albumin. Previously, most anthropologists believed that humans and apes had diverged approximately twenty-five million years ago; the DNA-DNA hybridization data suggest a much more recent date of approximately five million years. Subsequent DNA studies have confirmed the latter estimate. This had led to a reevaluation of the primate fossil record and of the way in which primates have evolved, including humans.

Another group of researchers has used DNA-DNA hybridization data to calculate the divergence dates among different primate species. After calibrating the molecular clock with dates previously established from the fossil record, they estimated the following approximate divergence dates: Old World monkeys, 30 million years ago; gibbons, 20 million; orangutans, 15 million; gorillas, 7.7 to 11 million; and chimpanzees and humans, 5.5 to 7.7 million years ago. In contrast to earlier work, they concluded that humans and chimpanzees are genetically closer to each other than either are to gorillas. As with the serum albumin data, these new estimates of times and order of divergence led to a reexamination of primate evolution.

—*Robert A. Browne*

FURTHER READING

Avise, John C. *Molecular Markers, Natural History, and Evolution.* Chapman and Hall, 1994.

Ayala, Francisco J. "Molecular Genetics and Evolution." In *Molecular Evolution*, edited by F. J. Ayala. Sinauer Associates, 1976.

Dobzhansky, Theodosius, Francisco J. Ayala, G. Ledyard Stebbins, and James W. Valentine. *Evolution.* W. H. Freeman, 1977.

Easteal, Simon, Chris Collet, and David Betty. *The Mammalian Molecular Clock.* Springer-Verlag, 1995.

Hartl, Daniel L., and Andrew G. Clark. *Principles of Population Genetics.* 3rd ed. Sinauer Associates, 1997.

Kimura, Motoo. *The Neutral Theory of Molecular Evolution.* Cambridge University Press, 1985.

_____. *Population Genetics, Molecular Evolution, and the Neutral Theory: Selected Papers.* Edited by Naoyuki Takahata. Foreword by James F. Crow. University of Chicago Press, 1994.

Li, Wen-Hsiung. *Molecular Evolution.* Sinauer Associates, 1998.

Nei, Masatoshi. *Molecular Evolutionary Genetics.* Columbia University Press, 1987.

Strickberger, Monroe W. *Evolution.* 3rd ed. Jones and Bartlett, 2000.

Nocturnal Animals

Fields of Study

Ethology; Animal Physiology; Herpetology

Abstract

At nightfall, many animals are just beginning to waken and to function. All of these night creatures are called nocturnal animals. Their activity during the hours of darkness is usually due to a combination of factors. Most nocturnal creatures are highly adapted to living under conditions in which there is little or no light. These adaptations enable nocturnal animals to find their way through the world, locate food, and detect danger in almost total darkness. Although the diurnal species outnumber the nocturnal creatures, almost every type of creature living on earth has a nocturnal version.

Key Concepts

auditory: pertaining to hearing
diurnal: awake and functional during the daylight hours
echolocation: sonarlike determination, from sound echoes, of the positions of unseen objects
pinna (pl. pinnae): a term indicating the external ear of an animal
pit viper: a poisonous snake, such as a rattlesnake, which detects its prey via paired heat-sensing pits in its head
tympanic membrane: the eardrum

CHILDREN OF THE NIGHT

At nightfall, many animals are just beginning to waken and to function. All of these night creatures are called nocturnal animals. Folklore often brands them as evil, inimical creatures. This is not so; they are merely different from diurnal animals, which are active during the day. Their activity during the hours of darkness is usually due to a combination of factors. First, the nocturnal animals may make use at night, without having undue competition, of food sources and habitats also used by diurnal animals. Second, they may be safe from many predators who would hunt them during the daylight hours. Finally, nocturnal animals may be predators which, at night, have much less competition for

A sketch of a badger peaking out of a hole in a tree trunk.

prey, or have a better chance to capture it, than in daylight.

Most nocturnal creatures are highly adapted to living under conditions in which there is little or no light. These adaptations enable nocturnal animals to find their way through the world, locate food, and detect danger in almost total darkness. Although the diurnal species outnumber the nocturnal creatures, almost every type of creature living on earth has a nocturnal version.

REGULAR SENSES AND SPECIAL SENSES

Diurnal animals survive as long as they have a combination of senses that enables them to fit, successfully, into a daylight ecological niche. This usually includes ability to see adequately, to hear well, and to have a useful sense of smell. The ability to see, hear, and smell falls within a fairly broad range among earth's many successful diurnal animals. Furthermore, weaknesses in one sense, such as the nearsightedness of the rhinoceros, may be compensated for by enhancement of one or more of the other senses. In rhinos, both the ability to hear and to smell are very well developed.

Nocturnal animals live under conditions that are skewed far away from the normal visual range of diurnal organisms. They must operate under conditions where there is either very little light or even no light at all. For this reason, all animals that are nocturnal have at least one sense, sight, hearing, smell, taste, or touch, that is very highly developed, compared to those who are diurnal. These powerful sense adaptations enable them to survive well in the dark hours. The extent of such special sensory development varies, depending upon the organism, its needs, whether it also functions during the day, and the extent to which it is subject to predation by other organisms. In some cases, senses not seen among diurnal animals operate in nocturnal animals.

The large, carnivorous felines, such as lions, tigers, or leopards, are not threatened by many other organisms. They survive well by using a combination of keen eyesight and hearing, as well as an excellent sense of smell. To further aid these big cats, their eyes face forward, allowing very accurate judgment of distances when they hunt. They also rely quite heavily upon very acute reflexes, great strength, and the ability to run down most prey hunted.

An extreme in visual development is seen in owls, which are the ultimate nocturnal avian raptors and which function and hunt almost exclusively at night. These birds are gifted with superb vision, exceptionally sensitive hearing, and a very wide visual and aural range. For example, the night vision of many owl species is one hundred times more sensitive than that of humans. In addition, owl hearing is very acute, aided in some cases by the possession of asymmetric skulls with the two ears at different places, further enhancing their hearing. Another adaptation that optimizes owl vision and hearing is the ability to turn the neck through 270 degrees. This gives owls the widest aural and visual range of all birds. It is therefore unsurprising that owls hear even the tiniest squeak or rustle made by their prey on the ground below them, and then very efficiently locate the prey by vision.

Bats function only after the sun sets, most often feeding on insects captured while in flight. They have good hearing. However, their eyes are small and they possess relatively poor vision. Perhaps this is why they have developed a special sense, sonarlike echolocation, to pinpoint insect prey in the dark night skies and to navigate. In bat echolocation, sounds (often clicks) are emitted from the larynx or nose of the bat, depending upon species. These sounds then strike insects (and rocks or trees), and echoes bounce back to the bat's ears. The bat then uses the echoes to find its prey. Many bats have developed means to direct the sounds they make for echolocation. For example, bats with skin flaps on the nose often use them to direct sounds in the nasal passages. The excellent hearing found in bats is aided by their large, mobile, external ears. Bats are not "blind as bats." For example, they can use vision to navigate their way home, and often do. Some nocturnal birds, such as the nightjar, also use echolocation to capture their prey.

Rattlesnakes and many other snakes locate their prey in total darkness in another way that involves special sense organs. This process is most sensitive in poisonous snakes called pit vipers. These snakes have two heat-sensitive pits located on the sides of their heads. The pits help them to detect small animals which are their prey, such as rodents or birds. The detection is possible because the prey sought are warmer than their surroundings. The pit viper heat sensors are large groups of nerve endings so sensitive that they can detect the body heat of a rodent over a foot away. The heat discrimination of the pit nerve endings is huge and they can identify a temperature difference of less than one hundredth of a degree.

NOCTURNAL ANIMAL HABITATS

People generally think of nocturnal animals as living in the wild. However, many live cheek-by-jowl with humans in parks, gardens, and empty lots in urban and suburban areas. Just a few examples are badgers, raccoons, deer, and foxes. People often see them and view them as pests or problems because they raid gardens and garbage cans, and some kill pets. Another problem associated with nocturnal animals is that they can carry severe, contagious diseases (for example, foxes and raccoons may carry rabies) which can be fatal to people. Finally, many nocturnal species are on the verge of extinction. These range from insects to the big cats. A few nocturnal species, such as tigers, are finally being covered by international conservation agreements, perhaps just in time. The others should be helped to survive, too.

—*Sanford S. Singer*

FURTHER READING

Boot, Kelvin. *The Nocturnal Naturalist*. David and Charles, 1985.
Brown, Vinson. *Knowing the Outdoors in the Dark*. Stackpole, 1972.
Kappel-Smith, Diana. *Night Life: Nature from Dusk to Dawn*. Little, Brown, 1990.
Lawlor, Elizabeth. *Discover Nature at Sundown*. Stackpole, 1995.
Pettit, Theodore S. *Wildlife at Night*. G. P. Putnam's Sons, 1976.
Popper, Arthur N., and Richard R. Fay, eds. *Hearing by Bats*. Springer-Verlag, 1995.

Nonrandom Mating, Genetic Drift, and Mutation

Fields of Study

Genetics; Evolutionary Biology; Population Genetics; Sociology

Abstract

Nonrandom mating, genetic drift, and mutation are all mechanisms of genetic change in populations. These mechanisms violate the assumptions of the Hardy-Weinberg model of genetic equilibrium by increasing or decreasing the frequency of heterozygote genotypes in the population. Nonrandom mating occurs in a population whenever every individual does not have an equal chance of mating with any other member of the population. While many organisms do tend to mate randomly, there are some common patterns of nonrandom mating.

Key Concepts

allele: alternative forms of a gene for a particular trait

assortative mating: a type of nonrandom mating that occurs when individuals of certain phenotypes are more likely to mate with individuals of certain other phenotypes than would be expected by chance

gamete: a haploid sex cell that contains one allele for each gene; sperm and egg cells are gametes that fuse to form a diploid zygote

genetic variation or diversity: the total number and distribution of alleles and genotypes in a population

genotype: the complete genetic makeup of an organism, regardless of whether these genes are expressed

heterozygote: a diploid organism that has two different alleles for a particular trait

homozygote: a diploid organism that has two identical alleles for a particular trait

inbreeding: mating between relatives, an extreme form of positive assortative mating

phenotype: the expressed genetic traits of an organism

POPULATION GENETICS

Evolution is a process in which the gene frequencies of a population change over time, and nonrandom mating, genetic drift, and mutation are all mechanisms of genetic change in populations. These mechanisms violate the assumptions of the Hardy-Weinberg model of genetic equilibrium by increasing or decreasing the frequency of heterozygote genotypes in the population.

Nonrandom mating occurs in a population whenever every individual does not have an equal chance of mating with any other member of the population. While many organisms do tend to mate randomly, there are some common patterns of nonrandom mating. Often, individuals tend to mate with others nearby, or they may choose mates that are most like themselves. When individuals choose mates that are phenotypically similar, positive assortative mating has occurred. If mates look physically different, then it is negative assortative mating. Population geneticists use the term "assortative" because it means "to separate into groups," usually in a pattern that is not random. The terms "positive" and "negative" refer to the probability that mated pairs have the same phenotype more or less often than expected by chance.

Two color varieties of snow geese (*Chen hyperborea*), blue and white, are commonly found breeding in Canada, and they show positive assortative mating patterns based on color. The geese tend to mate only with birds of the same color; blue mate with blue and white with white. Since a bird's color (phenotype) is determined by the presence of a dominant blue color allele, matings between similar phenotypes are also matings between similar genotypes. Matings between similar genotypes cause the frequency of individuals that are homozygous for the blue or the white allele to be greater, and the frequency of heterozygotes to be less than if mating were random and in Hardy-Weinberg equilibrium. Negative assortative mating increases the frequency of heterozygote genotypes in the population and decreases homozygote frequency. Assortative mating does not change the frequency of the blue or white alleles in the goose population; it simply reorganizes the genetic variation and shifts the frequency of heterozygotes away from Hardy-Weinberg equilibrium frequencies.

Inbreeding is the mating of relatives and is similar to positive assortative mating because like genotypes mate and result in a high frequency of homozygotes in the population. In assortative mating, only those genes that influence mate choice become homozygous, but inbreeding increases the homozygosity of all the genes. High homozygosity means that many of the recessive alleles that were masked by the dominant allele in heterozygotes will be expressed in the phenotype. Deleterious or harmful alleles can remain hidden from selection in the heterozygote, but after one generation of inbreeding, these deleterious alleles are expressed in a homozygous condition and can substantially reduce viability below normal levels. Low viability resulting from mating of like genotypes is called inbreeding depression.

GENETIC DRIFT AND MUTATION

Genetic drift, like positive assortative mating, reduces the frequency of heterozygotes in a population, but with genetic drift, the frequency of alleles in a population changes. Nonrandom mating does not change allele frequency. Genetic drift is sometimes called random genetic drift because the mechanism of genetic change is random and attributable to chance events in small populations, such that allele frequencies tend to wander or drift. Statisticians use the term "sampling effect" to describe observed fluctuations from expected values when only a few samples are chosen, and it is easy to observe by tossing a coin. A fair coin flipped a hundred times would be expected to produce approximately fifty heads and fifty tails, plus or minus a few heads or tails. Yet, if the coin is flipped only four times, it is not too surprising to get four heads or four tails. The probability of getting either all heads or all tails on four consecutive flips is one out of eight, but the probability of getting all heads or all tails decreases to much less than one in a billion as the sample size increases from four to a hundred tosses.

Similarly, it is much easier for nonrandom events to occur in small populations than in large populations. If a population has two alleles with equal frequency for a particular trait, then the result of random mating can be simulated by tossing a coin. The frequency of each allele in the next generation would be determined by flipping the coin twice for each individual, since sexually reproducing organisms have two alleles for each trait, and counting the number of heads and tails. In a small population, only a few gametes, each containing one allele for the trait, will fuse to form zygotes. Chance events can cause the frequencies of alleles in a small population to drift randomly from generation to generation; often one allele is lost from the population.

In small populations with fewer than fifty mating pairs, alleles may be eliminated in fewer than twenty generations by random genetic drift, leaving only one allele for a particular trait in the population. Thus, all individuals would be homozygous for the remaining allele and genetically identical. Theoretically, in any finite population random genetic drift will occur, but it is usually negligible if the population size is greater than a hundred. Sometimes, disasters or disease may drastically reduce the population size, causing a bottleneck effect. The bottleneck in population size reduces genetic variability in a population because there are only a few alleles and results in random genetic drift. Many islands and new populations are established by a small group of founders that constitute a nonrandom genetic sample because they have only a fraction of the alleles from the original large population. Founder effects and bottleneck effects are phenomena that result in a loss of heterozygosity and decreased genetic variability because of the chance drift in allele frequency away from Hardy-Weinberg equilibrium values in small populations.

Mutations are any changes in the genetic material that can be passed on to offspring. Some mutations are changes at a single point in the chromosome, while at other times, pieces of the chromosome are removed, extra pieces are added, or pieces are exchanged with other chromosomes. All these changes could result in the formation of new alleles or could change one allele into a different allele. The random mistakes in the chromosomes occur at the molecular level, and only later are the changes in information or alleles translated into phenotypic differences. Thus, mutation is the ultimate source of genetic variability and is random with respect to the needs of the organism.

Most mutations are lethal and are never expressed, but nonlethal mutations provide the necessary variation for natural selection. Even though mutations are very important for evolution, they have only a small effect on allele and genotype frequencies in populations because mutation rates are relatively low. If an allele makes up 50 percent of the gene pool and mutates to another allele once for every hundred thousand gametes, it would take two thousand generations to reduce the frequency of the allele by 2 percent. The net effect of mutations is to increase genetic variability, but at a very slow rate.

STUDYING GENETIC VARIABILITY

Population geneticists use a wide variety of laboratory, field, and natural experiments to investigate genetic variability. Natural experiments are situations that have developed without a scientist intentionally designing an experiment, but conditions are such that scientists can test a theory. Researchers have used known pedigrees or ancestral histories of zoo animals and have found that mortality rates of inbred young are often two to three times higher than for noninbred young. Population geneticists use pedigrees to calculate the probability that two alleles are identical by descent; this research provides an index of the amount of inbreeding in a population.

The study of random genetic drift is usually carried out in the laboratory. Scientists often use small organisms that reproduce quickly, such as fruit flies (*Drosophila melanogaster*), to conserve space and save time. In a 1956 study of eye color conducted by Peter Buri, after only eighteen generations and sixteen fruit flies per population, more than half of the 107 populations started had only one of the two alleles for eye color.

Mutations are so rare that even fruit flies reproduce too slowly for scientists to study the effects of mutations on populations, even though much is known about the mechanism of mutation by studying *Drosophila*. Small bacterial growth chambers can hold many millions of bacterial cells, and, since they reproduce quickly, even mutations that occur in only one in a million cells can be detected. In 1955, it was found that mutation rates were very low in bacteria until caffeine was added to the growth chamber, whereupon mutation rates increased tenfold. Any chemical or type of radiation that can cause mutations is called a mutagen. Electrophoresis has also been a useful tool for the study of nonrandom mating, genetic drift, and mutations, because allele and genotype frequencies can be determined from samples of the population and unique alleles can be identified.

THE DANGERS OF INBREEDING

Most governments and religions forbid marriages between close relatives because matings between first cousins result in a 20 percent decrease in heterozygosity; for those between brothers and sisters, there is an 80 percent decrease in heterozygosity. The decrease in heterozygosity and genetic variation and increase in homozygote frequency often result in inbreeding depression because deleterious recessive alleles are expressed. All inbreeding is not undesirable; many of the prizewinning bulls and pigs at state fairs have some inbreeding in their pedigrees. Most breeds of dogs were produced by breeding close relatives so that the offspring would have particular traits.

Zookeepers and others that breed and protect rare and endangered species must continually be concerned about the negative effects of both inbreeding and genetic drift. Most zoos are lucky if they have two or three pairs of breeding adults, and total population sizes are usually very small compared to those of natural populations. These conditions mean that inbreeding may reduce the vigor of the population and genetic drift will reduce the diversity of alleles in the population, thus reducing the chances of survival for the captive species. There is hope for rare and endangered species if independent inbred lines are crossed, thus reducing the effects of inbreeding depression, and if breeding adults from other zoos or populations are traded occasionally, thus increasing the effective population size.

Mutations are the ultimate source of genetic variation and so are very important in the study of evolution, but the population-level effects of one mutation are difficult to study because of the low frequency of natural mutations. Certain nonlethal mutations may have little evolutionary impact but may be important medically because spontaneous mutations result in hemophilia or dwarfism (achondroplasia) in more than 3 out of 100,000 cases. As exposure to background radiation and chemical levels increases, mutation rates are likely to increase, as well as the incidence of mutation-related diseases.

—William R. Bromer

Further Reading

Ayala, Francisco J. *Population and Evolutionary Genetics: A Primer.* Benjamin/Cummings, 1982.

Crow, J. F. *Basic Concepts in Population, Quantitative, and Evolutionary Genetics.* W. H. Freeman, 1986.

Fisher, R. A. *The Genetical Theory of Natural Selection.* Dover, 1958.

Hartl, Daniel L. *A Primer of Population Genetics.* 3rd ed. Sinauer Associates, 2000.

Mettler, L. E., T. G. Gregg, and H. E. Schaffer. *Population Genetics and Evolution.* 2nd ed. Prentice-Hall, 1988.

Real, Leslie A., ed. *Ecological Genetics.* Princeton University Press, 1994.

Wilson, E. O., and W. H. Bossert. *A Primer of Population Biology.* Sinauer Associates, 1971.

Nutrient Requirements of Animals

Fields of Study

Biochemistry; Endocrinology; Developmental Biology

Abstract

A nutrient is any substance that serves as a source of metabolic energy, raw material for growth and repair of tissues, or general maintenance of body functions. Regardless of the source, food must provide its consumer with a sufficient amount of the essential nutrients. Animals differ widely in their specific nutritional needs, depending on the species. Within any given species, those needs may vary according to variations in body size and composition, age, sex, activity, genetic makeup, and reproductive functions.

Key Concepts

carbohydrate: an organic molecule containing only carbon, hydrogen, and oxygen in a 1:2:1 ratio; often defined as a simple sugar or any substance yielding a simple sugar upon hydrolysis

lipid: an organic molecule, such as a fat or oil, composed of carbon, hydrogen, oxygen, and sometimes phosphorus, that is nonpolar and insoluble in water

mineral: one of the many inorganic elements other than carbon, hydrogen, oxygen, and nitrogen that an organism requires for proper body function

protein: an organic molecule containing carbon, hydrogen, oxygen, nitrogen, and sulfur and composed of large polypeptides in which over a hundred amino acids are linked together

vitamin: an organic nutrient that an organism requires in very small amounts and which generally functions as a coenzyme

YOU ARE WHAT YOU EAT

Food, used to provide material for production of new tissue and the repair of old tissue as well as used as an energy source, is obtained from a variety of plant, animal, and inorganic sources. Regardless of the source, food must provide its consumer with a sufficient amount of the essential nutrients. A nutrient is any substance that serves as a source of metabolic energy, raw material for growth and repair of tissues, or general maintenance of body functions.

GENERAL NUTRITIONAL REQUIREMENTS

Animals differ widely in their specific nutritional needs, depending on the species. Within any given species, those needs may vary according to variations in body size and composition, age, sex, activity, genetic makeup, and reproductive functions. A small animal requires more food for energy per gram of body weight than does a larger animal, because the metabolic rate per unit of body weight is higher in the smaller animal. Likewise, an animal with a cool body temperature will have less energy needs and require less food than an animal with a high body temperature. An egg-producing or pregnant female will require more nutrients than a male. In order for an animal to be in a balanced nutritional state, it must consume food that will supply enough energy to supply power to all body processes, sufficient protein and amino acids to maintain a positive nitrogen balance and avoid a net loss of body protein, enough water and minerals to compensate for losses or incorporation, and those essential vitamins that are not synthesized within the body.

Activities such as walking, swimming, digesting food, or any other activity performed by an animal require fuel in the form of chemical energy. Adenosine triphosphate (ATP), the body's energy currency, is produced by the cellular oxidation of small molecules, such as sugars obtained from food. Cells usually metabolize carbohydrates or fats as fuel sources; however, when these carbon sources are in short supply, cells will utilize proteins. The energy content of food is usually measured in kilocalories, and it should be noted that the term "calories" listed on food labels is actually kilocalories (1 kilocalorie = 1000 calories). Cellular metabolism must continually produce energy to maintain the processes required for an animal to remain alive. Processes such as the circulation of blood, breathing, removing waste products from the blood, and in birds and mammals, the maintenance of body temperature, all require energy. The calories required to fuel these essential processes for a given amount of time in an animal at rest is called the basal metabolic rate (BMR). For a resting human adult, the BMR averages from thirteen hundred to eighteen hundred kilocalories per day. As physical activity increases, the BMR increases.

Energy balance requires that the number of calories consumed for body maintenance and repair and for work (metabolic and otherwise) plus the production of body heat in birds and mammals be equal to the caloric intake over a period of time. An insufficient intake of calories can be temporarily balanced by the utilization of storage fats, carbohydrates, or even protein, and will result in a loss of body weight. On the other hand, an excessive intake of calories can lead to the storage of energy sources. Animals normally store glycogen, but when the glycogen stores are full, food molecules, such as carbohydrates and protein, will be converted to fats.

NUTRIENT MOLECULES

Proteins are composed of long chains of amino acids and serve a number of important functions in all living organisms, but they are primarily used as structural components of soft tissues and as enzymes. Proteins can also be utilized as energy sources if they are broken down into amino acids. Animal tissues are composed of about twenty different amino acids. The ability to synthesize amino acids from other carbon sources, such as carbohydrates, varies among species, but few, if any, animal species can synthesize all twenty required amino acids. Those amino acids that cannot be synthesized by an animal, but are required for the synthesis of essential amino acids, are the so-called essential amino acids, and must be included in the diet. Humans, for example, require nine essential amino acids. Both plant and animal tissues can serve as protein sources, but animal protein generally contains larger quantities of the essential amino acids.

Carbohydrates are primarily used as immediate sources of chemical energy, but they can also be converted to metabolic intermediates or fats. Some carbohydrates are also structural components of larger molecules. For example, the nucleic acids deoxyribonucleic acid (DNA) and ribonucleic acid (RNA) contain the sugars deoxyribose and ribose, respectively, as an integral component of their structure. Most animals can also convert proteins and fats into carbohydrates. The principle sources of carbohydrates are the sugars, starches, and cellulose in plants and the glycogen stored in animal tissue.

Lipids are an important and essential component of all biological membranes. In addition, several animal hormones, such as the sex hormones, are lipoidal in nature. Fats and lipids are also especially suitable as concentrated energy reserves, because each gram of fat supplies twice as much energy as a gram of carbohydrate or protein and does not have to be dissolved in water. Hence, animals commonly store fat for times of caloric deficit when energy expenditure exceeds energy uptake. Some animals, such as migratory birds and hibernating mammals, store large quantities of fat to offset the times that they are not actively feeding. Lipid molecules include fatty acids, monoglycerides, triglycerides, sterols, and phospholipids.

All animals require an adequate supply of essential inorganic minerals. Carbonate salts of the metals calcium, potassium, sodium, and magnesium as well as some chloride, sulfate, and phosphate are important constituents of intra- and extracellular fluids. Calcium phosphate is present as hydroxyapatite, a crystalline material that gives hardness and rigidity to the bones of vertebrates and the shells of mollusks. Certain metals, such as copper and iron, are required for oxidation-reduction reactions and for oxygen binding and transport. The catalytic function of many enzymes requires the presence of certain metal atoms. Animals require moderate amounts of some minerals and only trace quantities of others.

Animals require a variety of vitamins, diverse and chemically unrelated organic substances. Vitamins primarily function as coenzymes for the proper catalytic activity of essential enzymes. As with amino acids, the ability to synthesize different vitamins from other carbon sources varies among species. Those essential vitamins that cannot be synthesized by the animal itself must be obtained from other sources, primarily from plants but also from dietary animal flesh or from intestinal microoganisms. Vitamin C (ascorbic acid) can be synthesized by many animals, but not by humans. Vitamins K and B_{12} are produced by intestinal bacteria in humans. Vitamins such as A, D, E, and K are fat soluble and can be stored in fat deposits within the body; however, water soluble vitamins such as vitamin C are not stored and are excreted through the urine. Hence, the water soluble vitamins must be consumed or produced continually in order to maintain adequate levels.

Although not commonly thought of as a nutrient, water is tremendously important and comprises up to 95% or more of the weight of some animal tissue. Water is replaced in most animals by drinking, ingestion with food, and to some extent, by the metabolism of carbohydrates and lipids.

—*D. R. Gossett*

FURTHER READING

Campbell, N. A., L. G. Mitchell, and J. B. Reece. *Biology: Concepts and Connections.* 3rd ed. Benjamin/Cummings, 2000.

Carr, D. E. *The Deadly Feast of Life.* Doubleday, 1971.

Fox, I. S. *Human Physiology.* 6th ed. WCB/McGraw-Hill, 1999.

Jennings, J. B. *Feeding, Digestion, and Assimilation in Animals.* 2nd ed. St. Martin's Press, 1972.

Randall, David, Warren Burggren, Kathleen French, and Russell Fernald. *Animal Physiology: Mechanisms and Adaptations.* 4th ed. W. H. Freeman, 1997.

Weindrach, R. "Caloric Restriction and Aging." *Scientific American* 274 (1996): 46–52.

Offspring Care

Fields of Study
Ethology; Sociology; Reproductive Biology

Abstract
Patterns of caring for offspring relate to several key factors. The most important factor determining the number of caretakers, as well as the quality and duration of parental care, is simply how much care the offspring need. In species in which young are born well developed and capable of surviving on their own, there is little reason for parents to provide help. In species in which young are born or hatched quite helpless with no chance of surviving on their own one or both parents are likely to stay around.

Key Concepts

alloparenting: performance of parenting duties by an individual not the parent of the offspring (though usually a relative)

altricial: the condition of being weak and relatively undeveloped at birth (or hatching) and thus dependent upon parental care for a prolonged period

cuckold: a partnered male who is helping his mate to raise offspring which are not genetically his own

nest-parasite: also called brood-parasite; an individual (or species) that lays its eggs in the nest of another individual (or species) and does no parenting at all

precocial: the condition of being strong and relatively well developed at birth (or hatching) and thus not particularly dependent upon parental care

sex-role reversal: generally used to refer to species in which the male does most of the parenting

viviparous: characterized by live birth (as opposed to egg-laying)

WHO'S MINDING THE KIDS?

In the animal kingdom, there are species in which neither parent cares for offspring, species in which one parent cares for offspring, species in which both parents care for offspring, and species in which individuals other than parents help care for offspring. Patterns of caring for offspring relate to several key factors.

PRECOCIAL SPECIES VERSUS ALTRICIAL SPECIES

The most important factor determining the number of caretakers, as well as the quality and duration of parental care, is simply how much care the offspring need. In species in which young are born well developed and capable of surviving on their own, there is little reason for parents to provide help. Hatchling fish and baby turtles, for example, are born or hatched looking just like miniature adults, and are fully capable of moving about and feeding themselves. Young of such species are referred to as precocial, and usually get no parental care at all. Young of most invertebrate species as well are capable of surviving without help, and so are typically left to fend for themselves.

On the other hand, in species in which young are born or hatched quite helpless with no chance of surviving on their own, it should not be surprising that one or both parents are likely to stay around. Young of such species are referred to as altricial. Dogs and cats are excellent examples of species with altricial offspring; so are humans. Generally speaking, mammals are more altricial than most groups of animals, and all mammalian young initially depend upon their mother for food (milk), delivered by means of lactation.

Compared to other animals, birds, too, are relatively altricial—especially raptors (birds of prey) and songbirds. When altricial birds first hatch they are featherless and unable to regulate their own body temperature; one or both parents must regularly warm the hatchlings just as they earlier warmed the eggs. Furthermore, altricial birds hatch with their eyes closed and they are not strong enough or coordinated enough to leave the nest to feed or to flee from danger.

A sketch of a mother raccoon carrying its baby.

Because baby birds need tremendous amounts of food and there are usually quite a few offspring, in most species of birds, both parents provide care.

PREY SPECIES VERSUS PREDATORY SPECIES

Although the young of birds and mammals are quite helpless compared to the young of animals in other taxonomic groups, there are variations. As a rule, prey species tend to be more precocial than predatory species and, therefore, to require less parental care.

Young of species which are herbivorous (vegetarian) but are in constant danger of being eaten by carnivorous (meat-eating) species cannot afford to be completely helpless even if they are dependent upon their mother for food. Grazing mammals, such as zebra and deer, are vulnerable to lions and wolves, so young of these species must be able to stand up and run just a few minutes after birth: they are born with open eyes, well-developed muscles, and a full coat of hair. Although mammalian young need to nurse from their mother, they are able to start grazing fairly quickly, so typically it is only the mother who provides support, and her care is not exceptionally prolonged.

Among birds the same pattern is found: Herbivorous species, such as ducks and chickens, that are preyed upon by mammalian, avian, or reptilian predators, hatch with open eyes, the ability to walk, run, or swim, and a coat of downy feathers so that even though they cannot yet fly, they can leave the nest without need of further incubation. Offspring of these species can usually eat on their own almost immediately after hatching, and therefore can often get by with only one parent to show them what to eat and to protect them until they can fly.

Compared to young of prey species, young of predatory species tend to be born in a much more altricial state, and therefore, to need more parenting. Lion cubs and wolf pups do not have open eyes or the strength and coordination to leave their nest, den, or lair until a few weeks after birth. Even after they can move about and are weaned from mothers' milk, they are still too uncoordinated to hunt successfully and must be fed by one or both parents. In fact, species such as lions and wolves, which not only have altricial young, but frequently have large litters, often recruit other adults to help raise their offspring. Amongst African (but not American) lions, related females form groups called prides that hunt together and help take care of one another's offspring. Wolves form hunting packs, and all members (both male and female) help to feed the young. This behavior is called aunting when it is done by a female relative, or more generally, alloparenting. Alloparenting is also seen among some rodents and primate species.

As with mammals, carnivorous and insectivorous birds come into the world in a more altricial state than their vegetarian brethren, and therefore need more care. In the majority of bird species, both parents cooperate to raise young and, as with some of the most altricial mammalian species, if there is a large brood, parents of some species recruit helpers. In birds, helpers are usually the parents' offspring from a previous brood or season, and are thus siblings or half-siblings of the young they are helping to raise. Species that use this extended family system are referred to as cooperative breeders—although some are less cooperative than others.

QUANTITY VERSUS QUALITY PARENTING

A third factor relevant to parenting is the number of offspring, either sequentially or in litters, that an individual produces. All individuals have a limited life span and a limited amount of energy, and during that life span they can allocate that energy either to producing a large number of offspring or to providing intensive care for a smaller number of offspring. Thus, while animals with altricial young that require intensive care do not have an option of producing huge numbers of offspring, species with precocial young do.

Most invertebrates and many vertebrates (other than birds and mammals) take the "quantity" strategy: They produce large numbers of offspring and provide no parental care at all. A large majority of the offspring of these species die before reaching adulthood, but a small percentage survive and reproduce. Extreme examples of taking a "quantity" strategy are the semelparous species, species that can only reproduce once in their lifetime. Salmon are well-known for this form of reproduction. Many spiders and insects, too, die before their only batch of young are even hatched.

At the other extreme is the strategy of having a small number of well-cared-for offspring, each of which has a high probability of survival. All altricial species are constrained to the "quality" strategy, but some precocial species opt for "quality" as well. Alligators and crocodiles are excellent mothers, protecting their eggs before they hatch, then transporting and protecting the young afterward. Some species of amphibians, fish, and even insects are devoted parents.

MATERNAL CARE VERSUS PATERNAL CARE

In mammals, if only one parent is necessary, it is always the mother who is committed to caretaking because she is the one who must provide the offspring with their first food through lactation and nursing. No other animals, however, nurse their young, so in other species, if only one parent is necessary, it does not necessarily have to be the mother who becomes the caretaker.

In single-parent species, whether it is the mother or father who becomes the caretaking parent depends, to a great extent, on parental certainty. In viviparous species (species that give birth to live young rather than lay eggs) the parent that gives birth is, for certain, one of the two genetic parents, and is therefore the parent most likely to care for the young if they need it. With rare exception, that means that mothers become the caretaking parent in viviparous species. In egg-laying (oviparous) species, parental certainty depends on whether fertilization was internal (as in birds and many invertebrates) or external (as in most fishes and amphibians). If fertilization is internal, then again, it is only the mother who is certain to be a genetic parent of the eggs she lays, and who is therefore most likely to take on the role of caretaker when only a single caretaker is needed. If, on the other hand, fertilization is external, then both parents are equally likely to be the genetic parent of any young that later hatch from eggs at the breeding site, and if parenting is needed, male and female are equally likely to become the caretaker. The result is that while most species that have external fertilization do not remain with their eggs or provide any parental care at all, in those that do, factors other than parental certainty determine which sex becomes the guardian.

In species with internal fertilization, the probability that a particular male is the father (or one of the fathers) of a particular set of offspring depends upon how many males the female mated with and when. In the few species that have been closely studied, if a mother needs help to raise her offspring, males seem to expend effort in proportion to the probability that they are actually the genetic father. This is because males that are parenting or otherwise providing resources for offspring that are not genetically their own are wasting their effort in terms of reproduction.

A male that is caring for another males' offspring is sometimes referred to as a cuckold. In some human cultures, to call a man a cuckold (or the equivalent) is a huge insult. However, in humans and other animals, when a male provides care for a female's offspring even though they are not genetically his, he is sending her a signal that he is a good provider, and perhaps he is making it more likely that she will mate with him in the future.

THE SPECTRUM OF PARENTAL BEHAVIOR

Most parents are good to most of their offspring most of the time. Parents provide food and sometimes shelter; they protect their offspring from danger and chase away predators; they may place their offspring in safe refuges; they may even teach their offspring necessary information or skills such as how to hunt, the location of productive feeding sites, or traditional migration routes.

At some point, however, there may be conflict between parents and offspring; conflict is especially common over how much care the parents provide versus how independent the offspring have become. Avian parents, for example, may displace their offspring from the nest to make way for a new brood; mammalian mothers may resort to force to wean their maturing offspring and so be able to nurse a new infant or litter. Alternatively, offspring may be sexually mature and ready to leave, but are manipulated by their parents to remain in the family acting as helpers.

The most extreme forms of parental manipulation involve neglect and what is called tolerated siblicide. In particularly bad times when resources are scarce, parents may provide care for only one or a few offspring, allowing the others to die. Even in good times, parents with very large broods or litters may neglect the smallest and weakest young. Parents of some species allow older siblings to kill, and perhaps eat, their younger siblings; in fact, this behavior is the norm in some species.

Humans tend to equate parenting with moral goodness; among other animals it equates simply to survival and reproduction. Different species provide parental care—or not—as it is needed, in order to maximize the probability that at least a few offspring will survive to maturity. In some vertebrate species parenting may be associated with intense emotions and bonding, as it is in humans, but across the animal kingdom there is no one right way to parent; what works is what works.

—Linda Mealey

FURTHER READING

Clutton-Brock, T. H. *The Evolution of Parental Care.* Princeton University Press, 1991.

Hrdy, Sarah. *Mother Nature.* Pantheon, 1999.

———, ed. "Natural-Born Mothers." *Natural History* 104, no. 12 (December 1995): 30–43.

Mealey, Linda. *Sex Differences: Developmental and Evolutionary Strategies.* Academic Press, 2000.

Sluckin, W., and M. Herbert, eds. *Parental Behavior.* Basil Blackwell, 1986.

Tallamy, Douglas W. "Child Care Among the Insects." *Scientific American,* January 1999, 72–77.

Wynne-Edwards, Katharine E., and Catherine J. Reburn. "Behavioral Endocrinology of Mammalian Fatherhood." *Trends in Ecology and Evolution* 15, no. 11 (2000): 464–468.

OMNIVORES

FIELDS OF STUDY

Evolutionary Biology; Ethology

ABSTRACT

Omnivores maximize their ability to obtain food by having digestive tracts capable of processing both plant and animal materials as food, although they are usually not capable of digesting the very tough plant material, such as grasses and leaves, that many large herbivores eat. Omnivores may also be scavengers, eating whatever carrion they may come across. Omnivores often lack the specialized food-gathering ability characteristic of pure carnivores and herbivores. Many animals often thought of as carnivores are actually omnivores, eating both plants and animals.

KEY CONCEPTS

carnivore: a flesh-eating animal
carrion: dead animals
diurnal: active during the day
herbivore: an animal that eats only plants
hermaphrodite: an organism having male and female reproductive systems
radula: a tonguelike, toothed organ for grinding food

WHAT YOU EAT IS WHAT YOU ARE

Many animals are either herbivores, who eat only plants as food, or carnivores, who eat only the flesh of other animals. The preference for one type of food or the other depends largely on the type of digestive system that the animal has, and the resources it can put into its "energy budget." Meat is generally easier to digest and requires a less complex digestive system and a relatively short intestinal tract. However, in order to get meat, carnivores have to invest a lot of time hunting their prey, and the outcome of a hunt is always uncertain. The food of herbivores is much easier to obtain, since plants do not move and all the herbivore has to do is graze on the grasses, leaves, or algae readily available around it. However, the cellulose that plants are made of is very tough to digest, and thus herbivores must have a much more complex and lengthy digestive tract than carnivores. Many herbivores are ruminants, with multipart stomachs, who have to chew and digest their food more than once in order to get adequate nutrition from it.

Carnivores and herbivores are also vulnerable to a loss of their food source. Herbivores whose digestive systems are specialized to process only one type of food will starve if that food becomes scarce due to drought or some other climatic change. Carnivores often have specialized hunting patterns that cannot be changed if the prey (usually herbivores) become scarce due to loss of their own food source.

Omnivores maximize their ability to obtain food by having digestive tracts capable of processing both plant and animal materials as food, although they are usually not capable of digesting the very tough plant material, such as grasses and leaves, that many large herbivores eat. Omnivores may also be scavengers, eating whatever carrion they may come across. Omnivores often lack the specialized food-gathering ability characteristic of pure carnivores and herbivores. Many animals often thought of as carnivores are actually omnivores, eating both plants and animals.

TYPES OF OMNIVORES

Omnivores can be found among all types of animals, living on land and in water. They include fishes, mollusks, arthropods, birds, and mammals.

Most insects are either herbivores, such as grasshoppers, or carnivores such as mantises. However some, such as yellow jacket wasps, are omnivores, eating other insects, fruit, and nectar. Omnivorous snails and slugs eat algae, leaves, lichens, insects, and decaying plant and animal matter. Their main organ for eating is called a radula, a tonguelike, toothed organ that is drawn along rocks, leaves, or plants to scrape off food; it is also used to bore holes through shells of other mollusks, to get to their flesh.

Omnivorous fish include the common carp, goldfish, catfish, eels, and minnows. Since a fish's food is often suspended in the medium through which the fish swims—water—being able to gulp up whatever comes into its mouth is an efficient way for a fish to eat. Similarly, bottom-feeders (fish that suck up material from the floor of whatever body of water they inhabit) also benefit from not needing to sort through the material before they ingest it.

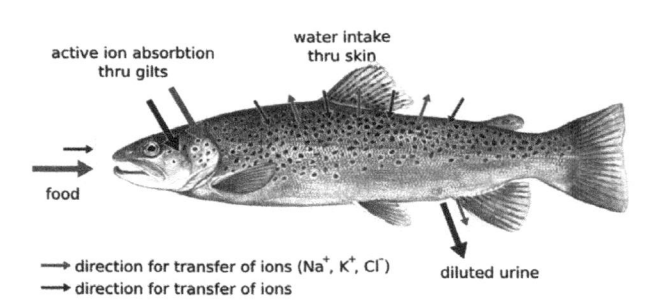

Osmoregulation by fishes.

Many birds are omnivores, such as robins, ostriches, and flamingos. The pink or red color of flamingos occurs because they eat blue-green algae and higher plants which contain the same substances that make tomatoes red. They also eat shrimp and small mollusks.

Mammal omnivores include bears, members of the weasel family, such as skunks, the raccoon family (raccoons and coatimundis), monkeys, apes, and humans. Raccoons and coatis, found only in the Americas, eat insects, crayfish, crabs, fishes, amphibians, birds, small mammals, nuts, fruits, roots, and plants. Like other omnivores, they also eat carrion. Bears eat grass, roots, fruits, insects, fishes, small or large mammals, and carrion.

—*Sanford S. Singer*

FURTHER READING

Kay, Ian. *Introduction to Animal Physiology*. Springer-Verlag, 1999.
Lauber, Patricia. *Who Eats What?* HarperTrophy, 1995.
Llamas, Andreu. *Crustaceans: Armored Omnivores*. Gareth Stevens, 1996.
McGinty, Alice B. *Omnivores in the Food Chain*. Powerkids Press, 2002.

OSMOREGULATION

FIELDS OF STUDY

Physical Biochemistry; Biochemistry

ABSTRACT

The osmotic pressure of a solution is the pressure that must be applied to a solution to prevent the movement of solvent into the solution through a semipermeable membrane. Osmoregulation, the regulation of osmotic pressure, is vital to every organism. The phenomenon collectively is called "osmosis."

KEY CONCEPTS

euryhaline: the ability of an organism to tolerate wide ranges of salinity

hyperosmotic: describes a solution with a higher osmotic pressure, one containing more osmotically active particles relative to the same volume, than the solution to which it is being compared

hypoosmotic: a solution with a lower osmotic pressure, fewer osmotically active particles relative to the same volume, than the solution to which it is being compared

isosmotic: a solution having the same osmotic pressure, the same number of osmotically active particles relative to the same volume, as the solution to which it is being compared

osmoconformer: an organism whose internal osmotic pressure approximates the osmotic pressure of its environment; such an organism is also referred to as "poikilosmotic"

osmoregulator: an organism that maintains its internal osmotic pressure despite changes in environmental osmotic pressure; such an organism is also referred to as "euryosmotic"

stenohaline: the inability of an organism to tolerate wide ranges of salinity

OSMOSIS AND OSMOTIC PRESSURE

The osmotic pressure of a solution is the measure of the tendency of pure water to pass through a semipermeable membrane to enter a solution. More precisely, the osmotic pressure of a solution is the pressure that must be applied to a solution to prevent the movement of solvent into the solution through a semipermeable membrane. Osmoregulation, the regulation of osmotic pressure, is vital to every organism. The phenomenon collectively is called "osmosis." It is the difference of hydrostatic pressure that must be created between that solution and pure water to prevent any net osmotic movement of particles in the water when the solution and pure water are separated by a semipermeable membrane. Hydrostatic pressure is a measuring device: It is a means of assessing the tendency of a solution to take on water osmotically.

Only dissolved solutes contribute to osmotic pressure. The number of individual particles determines the strength of osmotic pressure. Each particle makes a roughly equal contribution to osmotic pressure. The same number of molecules of a substance such as sodium chloride (table salt), which ionizes in water to release two ions (one sodium ion and one chlorine ion) display twice as much osmotic pressure as the same number of molecules of glucose, which retains its molecular form in water. Cells and other suspended materials do not contribute to osmotic pressure.

Several characteristics of solutions depend upon the number of particles in the solution. These are called "colligative properties." Increasing the number of particles of solute impairs the ability of the solvent to change state. The colligative properties are the freezing point, the boiling point, the osmotic pressure, and the vapor pressure. Only freezing point depression and vapor pressure are used to determine osmotic pressure.

Osmoticity refers to the osmotic pressure of solutions. Isosmotic solutions have equal osmotic pressures. A hypoosmotic solution has an osmotic pressure lower than the solution to which it is being compared; a hyperosmotic solution is one with a greater osmotic pressure than the solution to which it is being compared. These solutions can be body fluids, environmental liquids, or laboratory solutions.

The terms used to describe the changes in volume of cells exposed to solutions of differing concentrations are often confused with those comparing osmotic pressure. Changes of cell volume are described by the term "tonicity." Solutions isotonic to a cell cause no change in cell volume. Hypotonic solutions will cause the cell to swell as water diffuses into the cell; the cell may even burst. Hypertonic solutions will cause a cell to shrink as water diffuses across the cell membrane into the solution.

THE CHALLENGES OF OSMOREGULATION

This discussion reveals some of the problems that an organism encounters in the environment as concentrations of water and salts vary. Most organisms attempt to regulate both their volume and their ion content. If volume is not regulated, the chemicals within the cell will become too dilute to react or too concentrated to interact. If ions are not regulated, chemical reactions will be affected by inappropriate levels of ions, which may change the electrochemical properties of the cellular solution. Thus, there are independent challenges to volume regulation, to ion regulation, and to osmotic regulation. The homeostatic physiological responses to all three types of challenges are interconnected but distinct.

Osmoregulation is the regulation of the ratio between all dissolved particles, regardless of their chemical nature as ions or molecules, and water. All organisms are exposed to osmotic stress. Any organism incurs obligatory water losses. These occur during respiration, urination, and defecation. The organs most often thought of as participating in osmoregulation are the kidneys. They are intimately concerned with the elimination or conservation of water. Some salts are also found in urine. The proportions of the salts excreted in urine may be different from those in the body fluids because the kidney can retain required ions while eliminating less desirable ones.

Freshwater organisms are in danger of dilution, and they excrete great quantities of dilute urine. Saltwater organisms usually produce small quantities of isosmotic urine, which preferentially excretes divalent ions such as magnesium and sulfate. Other surfaces lose water, causing desiccation. The composition of the diet also influences the need for excretion of urine. Nitrogenous wastes from protein metabolism must be eliminated in urine, often as urea, which requires water for its excretion.

Carbon dioxide released during metabolism of carbohydrates and fats is eliminated by the respiratory organs. In terrestrial organisms, air leaving the lungs is usually saturated and some water is lost on expiration. The respiratory organs of aquatic organisms are gills. Their surfaces must be permeable to water. Freshwater organisms gain water through them and hypoosmotic marine organisms lose water through them.

The metabolism of carbohydrates produces what is known as metabolic water, which can be used to prevent desiccation. Metabolic water produced from the metabolism of fats is lost because of the higher rate of respiration required to supply the oxygen needed in fat oxidation.

SALT LOSS

Preformed water is present in any food. Even the driest seeds contain a small amount of water. The nutrients also always include salts. The presence of great quantities of salts may require urinary loss of water in excess of the preformed water found in the food.

Although feces may appear to be solid, they contain some water that was not absorbed in the gut. The presence of salts and other solutes in the digesta may also draw water from the hypoosmotic body fluids into the gut. One of the reasons that humans cannot drink seawater, in fact, is that the magnesium ions in ocean water increase the permeability of the gut and increase water loss, because the seawater is hyperosmotic to body fluids. More water is lost than can be gained.

Salts can be lost from the body by means other than urine formation and defecation. Marine reptiles and birds have salt glands located on the head. Since neither can produce hyperosmotic urine, these glands allow the elimination of salt with a minimum loss of water as the secretion may be four to five times as concentrated as body fluids. The cloaca of birds and the rectal glands of sharks also have the capacity to excrete salts.

One of the most fascinating mechanisms of osmoregulation is found in elasmobranch fish—the sharks, skates, and rays. Their body fluids are hyperosmotic but hypoionic to seawater. Blood salt concentrations are below those of seawater. Excess osmotic pressure is supplied by two molecules: urea and trimethylaminoxide (TMAO). Urea is toxic to most organs, but some organs resist its deleterious effects. Others would be harmed but are apparently protected by the TMAO. Retention of both urea and TMAO minimizes enzymatic disturbances by urea and allows the elasmobranchs to avoid the salt gain associated with hypoosmotic body fluids.

Organisms exposed to environmental variations have two choices: they can maintain internal constancy or homeostasis at the expense of metabolic energy, or they can allow their internal conditions to follow that of the environment. Organisms that maintain their internal osmotic pressure despite changes in external osmotic pressure are called osmoregulators. These euryosmotic organisms are protected from environmental changes. Their metabolism can continue to function, but much of the energy will be used to maintain their body fluids at the appropriate osmotic pressure.

Organisms that allow their osmotic pressure to follow that of the environment are called osmoconformers. These poikilosmotic organisms often have a limited tolerance for such changes. They are stenohaline. They may be less vigorous at salinities other than their optimal levels. The adults of such groups (for example, mollusks such as oysters and mussels) may be found in salinity extremes not tolerated by their young. These populations must be maintained by immigration of young spawned in more favorable salinity conditions.

HORMONAL REGULATION OF OSMOTIC PRESSURE

The internal osmotic pressure is affected by the hormones present in the body fluids. In invertebrates such as annelids, mollusks, and arthropods, neuroendocrine changes are seen upon changing the osmotic pressure of the environment. These changes indicate that nervous and endocrine systems are at work regulating the osmotic pressure of the organism. In most invertebrates, the biochemical nature of these hormones is unknown. Some freshwater pulmonate snails, however, produce an antidiuretic hormone and a neurosecretory factor associated with electrolyte balance. Depending upon the demands placed on them, insects such as grasshoppers and cockroaches can synthesize diuretic or antidiuretic hormones.

The best-known hormonal factors in ion regulation are studied in vertebrates. The pituitary gland produces antidiuretic hormone (ADH), which promotes water retention in terrestrial vertebrates. In fish and amphibians, ADH may induce urine formation and increase water loss through diuresis.

The adrenal gland also produces hormones that influence ion retention. In mammals, aldosterone increases reabsorption of sodium in the kidney and promotes the excretion of potassium. In nonmammalian vertebrates, extrarenal glands maintain salt and water balance by affecting the gills and intestines of fishes, the urinary bladder and skin of amphibians, and the salt glands of elasmobranchs, reptiles, and birds.

MEASURING OSMOREGULATION

Osmoregulation involves the balancing of water and solutes in the body so that the animal can continue to function. Because the presence of particles influences certain physical characteristics of the solution, these colligative properties can be used to determine osmotic pressure of solutions. Colligative properties change with increasing numbers of particles in solution: The osmotic pressure increases; the boiling point increases; the freezing point decreases; the vapor pressure decreases.

Freezing point depression and vapor pressure can be used to measure a solution's osmotic pressure. Freezing point is used most often. It works for the same reason that salt is spread on ice on sidewalks in winter. The salt lowers the freezing temperature of the water. Body fluids are much more dilute than the salt and water mixtures that melt ice, but the salinity of the ocean (approximately thirty-five parts salt for each thousand parts of solution) causes it to freeze as much as 1.6°C lower than pure water. Only marine organisms have body fluid osmotic pressures in that range. Terrestrial organisms have much less salt and therefore much lower osmotic pressures in their body fluids. Because most body fluids are so dilute, a large sample may be required to determine freezing point depression. When only small volumes of body fluid exist, the determination becomes more difficult.

Vapor pressure determinations are also used in osmometry. Usually, a small amount of the fluid being studied is tested in a capillary tube. In one ingenious method, the capillary tube is placed in a solution more concentrated than the experimental fluid. The higher osmotic pressure of the reference solution pushes a meniscus up the tube. The rate of movement of the meniscus depends upon the difference in concentrations between the experimental and the reference solutions.

Another ingenious method of using vapor pressure to determine the osmotic pressure of an experimental solution requires enough fluid to fill a depression. A glass plate with capillary tubes filled with reference solutions of known osmotic concentrations is mounted over the experimental fluid. The reference tube that exhibits no movement is at equilibrium with the experimental fluid. One of the rigors of this method is that all movement must stop.

Another method involves capturing a precise volume of experimental fluid in a capillary tube. The shape of drops of the same volume of reference solutions is compared to that of the experimental solution. Those of the same concentration will have the same shape because their vapor pressures are exerting equal force on the drop. Thermocouples are also used in vapor pressure determinations. This procedure is delicate and costly and is used infrequently. All these procedures are difficult and require patience. Now, electronic instruments analyze the constituents of solutions and allow easier calculation of the osmotic pressures of solutions than ever before.

FRESHWATER VERSUS SALTWATER ENVIRONMENTS

All organisms experience osmotic stress. There is no environment in which the osmotic pressure and the ion composition exactly match the requirements of the cells. Every organism must expend metabolic energy to maintain appropriate water and ion concentrations.

Freshwater organisms are hyperosmotic to their environment. They risk losing scarce ions through their permeable gills and in their urine and also tend to take up water through their gills or other surfaces and in their food. They face the problem of dilution of their body fluids by the environment.

Marine organisms are often isosmotic to the salt water they inhabit. They must change the concentrations of some ions, however, in order to attain this state. Magnesium is present in greater amounts in seawater than is desirable in their body fluids and must be eliminated. These organisms ion regulate even though they are not in danger of volume changes.

Marine organisms that evolved from freshwater or terrestrial ancestors are often hypoosmotic to seawater. They are in danger of desiccation as water from body fluids diffuses into the hyperosmotic ocean water. They also must regulate the types of ions that are retained and eliminated from their bodies.

Marine organisms may also be exposed to freshwater when they enter rivers, which dilute the salt content of the incoming tidal water. Under these conditions, the water is brackish—not as salty as the sea, but not as pure as freshwater. The criticality of this situation depends upon whether the organism is tolerant or intolerant of salinity changes. Organisms that can live in only a narrow range of salinities are called stenohaline. Organisms that are tolerant of wide ranges of salinities are called euryhaline.

Terrestrial organisms are always hyperosmotic to their environment, so they continually face desiccation in the air. They also must adjust the ion composition of their body fluids, because the foods that they eat may not have inorganic ions in the desired ratios and because some ions are always lost in urine.

One example of the influence of these effects concerns the interaction of oysters and the protistan parasite known as MSX. (MSX stands for "multinucleate sphere unknown," which refers to the protista *Haplosporidium nelsoni*.) The MSX organism survives in osmotic pressures greater than 0.4 osmolar. Oysters are osmoconformers which grow in saline, brackish, and nearly fresh water. At osmotic concentrations less than 0.4 osmole, oysters can survive and are unaffected by MSX. When rainfall is abnormally low, however, the salinity of brackish water increases, and oysters which were protected in low-salinity water are exposed to higher-salinity water, which allows the MSX organism to infect them.

Organisms exposed to tides may protect themselves from exposure to variations in osmotic pressure by sealing themselves off, the way snails and bivalve mollusks do. Others may move offshore to more saline waters or onshore, away from the increasing salinity. Worms that burrow in the sediments of salt water are protected from transient changes in salinity because there is little exchange of solutes with the overlying salt water.

The vertebrates adapted to their various environments by using hormones to regulate salt and water balance. Because of the differing demands of aquatic and terrestrial environments, in different groups, the same hormone may have opposite effects, but that effect is always to maintain the optimal osmotic pressure to ensure survival.

Further Reding

Brown, A. D. *Microbial Water Stress Physiology: Principles and Perspectives.* John Wiley & Sons, 1990.

Gilles, R., E. K. Hoffmann, and L. Bolis, eds. *Volume and Osmolality Control in Animal Cells.* Springer-Verlag, 1991.

Hadley, Neil F. *Water Relations of Terrestrial Arthropods.* Academic Press, 1994.

Krogh, August. *Osmotic Regulation in Aquatic Animals.* Reprint. Dover, 1965.

Prosser, C. Ladd. *Adaptational Biology: Molecules to Organisms.* John Wiley & Sons, 1986.

Smith, Homer W. "The Kidney." *Scientific American* 188 (January 1973): 40–48.

Strange, Kevin, ed. *Cellular and Molecular Physiology of Cell Volume Regulation.* CRC Press, 1994.

Packs

FIELDS OF STUDY

Ethology; Evolutionary Biology; Genetics

ABSTRACT

Packs occur whenever animals group together in highly organized social systems for traveling, hunting, feeding, and sleeping, usually with bonds of attachment between all members. These social units tend to be fairly stable in composition, in comparison to most herds. This stability results in a dominance hierarchy, in which individuals are ranked in order of the number of other animals below them. In some packs only the dominant pair breeds while the subordinates help to rear their young.

KEY CONCEPTS

- **antipredator benefits:** benefits that come from actions that protect individuals from being killed
- **carnivores:** animals that eat the flesh of other animals
- **dominance hierarchies:** ranks of individuals within a group

THE LEADER OF THE PACK

Packs occur whenever animals group together in highly organized social systems for traveling, hunting, feeding, and sleeping, usually with bonds of attachment between all members. These social units tend to be fairly stable in composition, in comparison to most herds. This stability results in a social hierarchy, commonly called a dominance hierarchy, in which individuals are ranked in order of the number of other animals below them. Dominance hierarchies are maintained by dominant animals threatening subordinates, but fights are rare. Generally, subordinates engage in appeasement behaviors or avoid the dominant individual(s) altogether.

Dominance characteristics can be linked to body size, age, or weaponry. However, in some packs, dominance is inherited. This is common in baboon and macaque monkey troops, and in hyena clans, and is dependent upon the mother's rank. Females can move up in status every time a younger sister is born; however, several subordinates can form a coalition to challenge an individual above them in status.

Advantages to dominant animals are many. Greater access to food is one common benefit, and in some packs only the dominant pair breeds while the subordinates help to rear their young. Costs to dominants are that they are frequently challenged and run a greater risk of being killed or injured.

PACKS AND PREDATION

Predatory methods differ among species that form packs. Lions, which are not capable of endurance running, rely on ambush hunting by stalking their prey and then rushing from cover. During the rushing phase, lions will target any individual that appears to be slower or weaker than others in the prey herd. This tactic, also used by wolves, requires the predator to approach the prey very closely prior to rushing. In contrast, African wild dogs are known as coursers, chasing their prey for many miles until they can either drive it into other pack members or exhaust it. Hunting success in these dogs is very high. Hyenas vary their hunting tactics based on prey species. For attacks on wildebeest, they rush the herd, run for a while, stop to choose a target, and then resume the chase. As the chase continues, more and more hyenas join, and they generally take down their prey when it turns or runs into a lake or stream. They rely upon sheer numbers to overwhelm their target.

Theoretically, the formation of packs allows individuals within the group to exploit more food

The gray wolf is a highly intelligent, social animal that lives in groups known as packs. The pack is usually comprised of a dominant male and female pair, their offspring, and other adult members.

resources than they would be able to do on their own. In some pack species, this is the case: Cooperative hunts of lions are more successful than those by solitary lions, and there is some evidence to suggest the same for African wild dogs and hyenas. However, the cooperation does not necessarily increase the amount of food available for each member of the pack; in some cases, the nutritional intake per individual decreases with an increase in the size of the pack. Nonetheless, packs may be more able to protect their kills from predators and thus may be more capable of consuming prey completely than could one individual. Packs may also be more successful at driving other packs off from a kill; lions are known for relying on the kills of hyenas for up to half of their food. Hyenas form clans of up to sixty individuals, but these clans break up when food is scarce. Similarly, lions may form larger prides when food is common, but may split into pairs or threesomes as food becomes scarce.

PACKS AND REARING OF YOUNG

If food per individual is not increased with increasing pack size, then other reasons must be present for the establishment and continuation of a pack. Suggested benefits include better defense of cubs against infanticide by outside coalitions of males that take over the pack (common in lion prides), higher reproductive success because all pack members share in feeding the offspring, and providing food for young while they grow to maturity, thereby ensuring that offspring will survive.

Rearing young in a group can benefit the young because of opportunities to learn from more than one adult. It can also provide the young with practice in certain tasks that later prove important when the offspring are on their own. Cooperative hunting can provide young with the opportunity to learn hunting skills from their elders. Generally, this benefit occurs in longer lived species that produce only a few young per year per female.

The formation of packs tends to occur in groups where kin form the nucleus of the association. Dominance hierarchies are common in such packs and serve to stabilize the relationships among group members. Benefits are then realized in terms of hunting larger prey, spreading food resources between the older and younger generation, help in rearing young from all or most members of the pack, and dissemination of skill learning from adults. In a few packs, associations can serve an antipredatory function or can allow the defense of resources that would otherwise be taken by some other species or group.

—*Kari L. Lavalli*

FURTHER READING

Drickamer, L. C., S. H. Vessey, and D. Meikle. Animal Behavior: Mechanisms, Ecology, and Evolution. 4th ed. Wm. C. Brown, 1996.

McFarland, D. Animal Behavior. 3rd ed. Longman, 1999.

Mech, L. D. The Wolf: The Ecology and Behavior of an Endangered Species. University of Minnesota Press, 1981.

Packer, C., D. Sheel, and A. E. Pussey. "Why Lions Form Groups: Food Is Not Enough." American Naturalist 136, no. 1 (July 1990): 1–19.

Schmidt, P. A., and L. D. Mech. "Wolf Pack Size and Food Acquisition." American Naturalist 150, no. 4 (October 1997): 513–517.

Whittenberg, J. I. "Group Size and Polygyny in Social Mammals." American Naturalist 115, no. 2 (February 1980): 197–222.

Pair-bonding

Fields of Study

Ethology; Reproductive Biology; Sociology

Abstract

In the animal world, there is a great variety of ways in which males and females pair or associate for reproductive purposes. Some animals pair with more than one member of the opposite sex, others pair exclusively with only one mate, some animals associate only long enough to copulate, other animals form pair bonds that last for varying lengths of time, from one reproductive period to a lifetime. While some pair-bonding has been observed in all vertebrate classes, and even in some invertebrates, it is particularly common in birds, while infrequent in fish and mammals.

Key Concepts

bond, bonding: the tie or relationship between opposite-sex partners in a pair bond

bonding behaviors: behavior patterns that establish, maintain, or strengthen the pair bond

consort pair, consortship: a temporarily bonded pair within a polygamous group

long-term pair bond: pair bonding that continues beyond a single reproductive period

monogamy: exclusive pair-bonding between one male and one female

pair, pairing: may refer to mating, sexual coupling (or copulation) or to formation of a pair bond, depending upon the context

polygamy: a mating system in which one male mates with several females (polygyny) or one female mates with several males (polyandry)

promiscuity: a mating system in which sexual partners do not form lasting pair bonds, where their relationship does not persist beyond the time needed for copulation and its preliminaries

PARTNERING UP

In the animal world, there is a great variety of ways in which males and females pair or associate for reproductive purposes (called mating systems). While some animals pair with more than one member of the opposite sex (polygamy), others pair exclusively with only one mate (monogamy). While some animals associate only long enough to copulate (promiscuity), other animals form pair bonds that last for varying lengths of time, from one reproductive period (until young leave the nest) to a lifetime. While some pair-bonding has been observed in all vertebrate classes, and even in some invertebrates (some crabs and insects), it is particularly common in birds, while infrequent in fish and mammals.

PAIR-BONDING IN BIRDS

Most bird species are monogamous, although some are polygamous, and a few may be promiscuous. In some birds, such as prairie chickens, a male and a female pair only for copulation, after which the female is on her own to nest and parent. However, most birds pair bond, remain with their mates, and cooperate in some way until their young can leave the nest and survive independently. Some birds, such as song birds, pair bond for only a single breeding season, while others, such as swans, geese, penguins, and albatrosses, exhibit pair-bonding for more than one season, sometimes for life.

The American robin, for example, is mostly monogamous while it pair-bonds and shares parenting for one breeding season. Typically, in the spring, male robins are the first to migrate north. When the females arrive about a week later, the males have already selected territories and call or

A pair of short-tailed albatrosses dance together on Midway Atoll National Wildlife Refuge. This pair later mated and hatched a chick in the 2010/2011 breeding season.

carol as though advertising for a mate. Courtship, involving three types of songs (only the males sing) and feeding, leads to mating and pair-bonding. The female builds a nest and lays her clutch of usually three or four eggs. While the female incubates the eggs, the male stands guard nearby, though he may help with the incubation as well. After the eggs hatch, both parents forage and feed the young. When the young leave the nest, they still need approximately two weeks of feeding and parental care before they are mature enough to survive independently. While the father provides this care, the mother repairs the nest and lays another clutch of eggs. American robins typically have two or three broods per year. In this case, the pair bond usually endures for the breeding season, but not through migration south for the winter and the nonbreeding season.

PAIR-BONDING IN FISH

Most fish do not form lasting pair bonds. An interesting exception, however, is the seahorse. At the beginning of the seahorse breeding season, males and females court for several days. The female produces eggs, and when they reach maturity, she deposits them into an egg pouch on the male seahorse's trunk. Then she swims away. Male seahorses typically remain in a habitat of approximately one square meter, while females range over an area that may be a hundred times as large. During the male's pregnancy, the female returns every day for a short, five- to ten-minute morning visit. During this visit, the seahorse pair exhibit social interaction and bonding behaviors reminiscent of courtship. Afterward, the female swims away until the next day, when she returns for another visit. After two to three weeks of pregnancy, the male gives birth to a few dozen or more baby seahorses during the night. When his mate returns in the morning for their daily visit, the male is ready for courtship, and the mating and reproductive cycle starts over again. Females have been observed to refuse to mate with other males during their mate's pregnancy, and so appear to be monogamous for the breeding season. In laboratory experiments, where mates were separated during pregnancy and the female interacted with another male, the original pair bond was broken. Thus, the social interaction and bonding behaviors that occur during the seahorse couple's short daily visits appear to be important factors in maintenance and longevity of pair-bonding.

PAIR-BONDING IN MAMMALS

Polygamy is the most common mating system among mammals. Within polygamous (or promiscuous) groups, such as baboons and chimpanzees, a male and female may pair temporarily and separate themselves somewhat from the group, while engaging in social grooming and sexual activity. Temporary pairings that do not endure to the end of a reproductive cycle are referred to as consort pairs or consortships in the primate literature. When pair-bonds in a polygamous system are more lasting, the social structure is sometimes called a harem.

Only a small minority (3 to 5 percent) of mammals exhibit monogamous pair-bonding. Gibbons, the smallest of the apes, are particularly interesting because they are the exceptional case, and exhibit not only long-term monogamous pair-bonding but a social organization somewhat analogous to the human nuclear family. Gibbons pair at eight to ten years of age, and have five to six offspring, spaced about three years apart, over their ten- to twenty-year reproductive lifetime. Gibbon offspring remain with their family groups until they approach or reach sexual maturity, when they may leave voluntarily or be evicted by the same-sex parent. Gibbons are territorial, and family members cooperate as needed to defend both territory and mate or family. Just prior to sunrise every morning, mated males sing solo songs that can be heard up to a kilometer away, seemingly identifying their territory as occupied. Later in the morning mated females sing their own songs, and join their mates in singing duets, which appear to publicize both territory and pair-bonding.

PAIR-BOND FORMATION AND MAINTENANCE

Formation of a pair-bond usually involves the behavior patterns of courtship and mating. Usually, the male initiates pair formation, while the female decides whether a bond is formed or not. How long a pair-bond endures depends upon various factors. Many animal pairs maintain and strengthen their relationship by continuing the bonding behaviors that were initially used in courtship. Some animals maintain close physical proximity, groom each other, communicate with movement (display) or with vocalizations (call or song), or share food, nests, and territory. Reproductive success and dependency of young also appear to maintain pair-bonding.

—*John W. Engel*

FURTHER READING

Birkhead, T. R. *Promiscuity: An Evolutionary History of Sperm Competition and Sexual Conflict.* Faber & Faber, 2000.

Black, J. M., ed. *Partnerships in Birds: The Study of Monogamy.* Oxford University Press, 1996.

Elia, I. *The Female Animal.* Henry Holt, 1988.

Lott, D. F. *Intraspecific Variation in the Social Systems of Wild Vertebrates.* Cambridge University Press, 1991.

Sussman, R. W. *Lorises, Lemurs, and Tarsiers.* Vol. 1 in *Primate Ecology and Social Structure.* Pearson, 1999.

pH Maintenance in Animals

Fields of Study

Biochemistry; Endocrinology

Abstract

The maintenance of pH is a collection of processes that regulate the hydrogen ion concentration of body fluids. The total body fluids account for about 60 percent of body mass. The total body fluids account for about 60 percent of body mass. The body fluids are divided into several compartments. Each of these compartments is regulated in order to maintain normal pH. The pH levels of the different compartments can be very different. In mammals, the extracellular pH as measured in the plasma is generally 7.4. The specialized fluids within the extracellular compartment can differ. Intracellular pH can vary from cell type to cell type but tends to be lower than extracellular pH.

Key Concepts

acidosis: a body fluid pH of less than 7.4 at 37 degrees Celsius

alkalosis: a body fluid pH greater than 7.4 at 37 degrees Celsius, the opposite of acidosis

anaerobic metabolism: metabolism in the absence of oxygen that leads to the production of lactic acid, a strong acid

partial pressure: the pressure exerted by a specific gas in a mixture of gases such as the atmosphere; it is analogous to concentration

pH: the negative logarithm of the hydrogen ion concentration, with higher hydrogen ion concentrations indicating lower pH; the pH scale goes from 0 to 14, with a pH of 7 being neutral, values below 7 indicating acidity, and values above 7 indicating alkalinity

strong acid: an acid that dissociates almost completely into its component ions; hydrochloric acid, for example, dissociates almost completely into hydrogen ions and chloride ions

weak acid: an acid that does not dissociate to a great extent; carbonic acid, for example, dissociates to produce some ions, but most of the molecules remain in their original forms

pH, THE HYDROGEN ION CONCENTRATION OF BODY FLUIDS

The maintenance of pH is a collection of processes that regulate the hydrogen ion concentration of body fluids. The total body fluids account for about 60 percent of body mass. The body fluids are divided into several compartments. The largest volume of body fluid is in the intracellular compartment, and this consists of about 40 percent of body mass. The extracellular fluid consists of interstitial fluid (about 15 percent of body mass) and several specialized compartments. The largest of the specialized compartments is the vascular space, or blood volume. About half of the blood volume is in blood cells that belong to the intracellular volume. The other half is plasma, which makes up about 5 percent of the body mass. There are several other small extracellular spaces, such as the cerebrospinal fluid and the aqueous and vitreous humors of the eyes, that comprise smaller percentages of the extracellular volume. Each of these compartments is regulated in order to maintain normal pH. The pH levels of the different compartments can be very different. In mammals, the extracellular pH as measured in the plasma is generally 7.4. The specialized fluids within the extracellular compartment can differ. For example, the cerebrospinal fluid has a pH of 7.32. Intracellular pH can vary from cell type to cell type but tends to be lower than extracellular pH. For example, skeletal muscle has an intracellular pH of 6.89, while red blood cells have a pH of 7.20.

pH MAINTENANCE

The first line of defense in the maintenance of pH is through the use of chemical buffers. Buffers are chemicals that resist pH change by absorbing hydrogen ions from acidic solutions or contributing hydrogen ions to alkaline solutions. They are made by producing a mixture of a weak acid and the salt of a weak acid. An example of a buffer is given by the bicarbonate buffer system. The components are H_2CO_3 (weak acid) and $NaHCO_3$ (salt of the weak acid). When a strong acid such as HCl is added to a solution containing this buffer pair, the H+ released from the highly dissociated HCl combines with HCO_3-; that is completely dissociated from the $NaHCO_3$ to form weakly dissociated H_2CO_3. This prevents the large pH drop that would otherwise occur because the H2CO3 is a weak acid. This buffer system is primarily extracellular. The body fluids also contain intracellular buffer systems which can resist pH changes when acids are introduced.

Acids can be introduced in several ways. Exercise and oxygen deprivation cause a buildup of lactic acid as a result of anaerobic metabolism. Metabolism of sulfur-containing amino acids causes the production of sulfuric acid, another strong acid. Both aerobic and anaerobic metabolism cause the buildup of carbon dioxide, which combines with water to form carbonic acid; even though this is a weak acid, the large accumulation of carbon dioxide can lower pH.

Acids produced by any of these means are buffered by one of the buffer systems in the body. In addition to the bicarbonate ($NaHCO_3$) buffer system described above, there are two intracellular buffers that participate in pH maintenance. In the phosphate buffer system, monosodium phosphate (H_2NaPO_4), a weak acid, dissociates to produce H+ and $HNaPO_4-$. The H+ is exchanged for Na+ by the kidneys or buffered on protein, leaving the salt disodium phosphate (HNa_2PO_4). This salt dissociates into Na+ and $HNaPO_4-$. The $HNaPO_4-$ can then absorb H+ ions which have dissociated from strong acids to form H_2NaPO_4, which is weakly dissociated.

The phosphate buffer system is the least important of the buffer systems because there is so little phosphate in the body. The most important buffer system in the body is protein. Some of the individual amino acids in proteins can act as weak acids and accept H+ ions; the amino acid histidine is the most significant of these under the temperature and pH conditions of the body. Protein buffering is primarily an intracellular phenomenon, as the protein concentration in extracellular fluid is relatively low. The intracellular protein buffer makes up about three quarters of the total body buffering capacity.

ACID-BASE DISTURBANCES

The most important extracellular buffer is the bicarbonate buffer system discussed above. This system is aided directly by a second line of defense against pH disturbances. This is called physiological buffering. Physiological buffering refers to the fact that the supply of the most important components of the bicarbonate buffer system, carbon dioxide and bicarbonate, can be controlled. The organs responsible for this control are the lungs (carbon dioxide) and kidneys (bicarbonate). For example, during exercise, the buildup of lactic acid consumes bicarbonate (HCO_3-) and lowers pH. This is called a metabolic acidosis. The low pH stimulates the rate of breathing, which in turn results in increased elimination of carbon dioxide by the lungs. This lowers the partial pressure (concentration) of carbon dioxide in the blood, reducing the amount of carbonic acid (H_2CO_3) that can dissociate to form H+. The reduction of carbon dioxide is called a respiratory alkalosis. This respiratory alkalosis compensates for the metabolic acidosis caused by the depletion of bicarbonate.

The kidneys provide the final correction for the initial metabolic acidosis. These organs accomplish this in two ways: reabsorption of bicarbonate, and secretion of H+ into the urine in exchange for Na+. When bicarbonate concentration in the extracellular fluid is reduced by a metabolic acidosis, the hormone aldosterone, produced by the adrenal glands, stimulates the secretion of H+ and reabsorption of bicarbonate. This raises both the pH and bicarbonate concentration back to normal. The increased pH relaxes stimulation of breathing, allowing breathing rate to slow and the partial pressure of carbon dioxide to return to normal.

The above is an example of one of the four basic types of acid-base disturbance: metabolic acidosis (bicarbonate concentration reduced), metabolic alkalosis (bicarbonate concentration increased), respiratory acidosis (an increase in the partial pressure of carbon dioxide), and respiratory alkalosis (a decrease in the partial pressure of carbon dioxide). As before, metabolic acidosis is compensated for by respiratory alkalosis. This is part of a general pattern in which one condition is compensated for by the opposite condition. Another example would be a respiratory acidosis, lowering the pH and stimulating bicarbonate retention and H+ excretion in the kidneys to produce a compensatory metabolic alkalosis.

BODY TEMPERATURE AND ION EXCHANGE

Temperature also affects pH. This is usually of no consequence to warm-blooded animals such as mammals and birds, which have a constant body temperature. It is, however, of great consequence to cold-blooded animals such as fish, amphibians, and reptiles, in which body temperatures vary with the environmental temperature. There is a decrease in pH by about 0.015 for each degree increase in temperature. This means that the pH at 5 degrees Celsius is 7.88, or 0.48 pH units higher than at mammalian body temperature (pH 7.40 at 37 degrees Celsius). This does not mean that cold-blooded animals do not regulate pH. While the regulation differs from that of warm-blooded animals in that pH varies, it is still regulated in the sense that it varies in a very predictable manner. By controlling partial pressures of carbon dioxide and bicarbonate ion concentrations in much the same way that mammals do, cold-blooded vertebrates maintain their body fluid pH at a constant 0.6 to 0.8 pH units higher than the pH of pure water at the same temperature. It is, thus, the relative alkalinity that is regulated, rather than the pH.

While the kidneys of mammals are intimately involved in pH maintenance, the situation in lower vertebrates can be much different. Most fish use ion exchange transport systems in their gills to regulate pH. Specialized cells in the exterior lining of the gill surfaces transport sodium ions (Na+) from the water bathing the animals into the blood in exchange for H+. Similarly, chloride ions (Cl–) are transported into the animal in exchange for HCO_3–. When fish become acidotic, they increase Na+/H+ exchange and decrease Cl–/HCO_3– exchange. This results in a net elimination of H+ and a net conservation of HCO_3– to elevate the pH of the body fluids. While the kidneys of fish do participate in pH maintenance, they seem to play only a minor role, usually about 5 percent of the total regulatory effort. The kidneys of amphibians also appear to play a minor role in pH maintenance. The major regulatory organs in these animals are the skin and urinary bladder. Aquatic salamander larvae increase Na+/H+ exchange and decrease Cl–/HCO_3– exchange across their skin to regulate internal pH. Toads, which are primarily terrestrial and thus seldom in contact with water, appear to use their urinary bladders for pH regulatory ion exchanges. Relatively little is known about the role of the reptilian kidney in pH maintenance. The alligator kidney plays a major role in acid elimination. The urinary bladder of turtles also participates in pH regulation in much the same manner as the urinary bladder of the toad. The bird kidney plays a major role in pH maintenance, and its urine pH values are similar to the pH values of mammalian urine.

STUDYING PH MAINTENANCE

The regulation of pH is usually studied by measuring the pH and the partial pressure of carbon dioxide of the blood. This measurement of the regulation of extracellular pH is a reflection of the overall acid-base status of the individual being studied. The first requirement is to be able to sample blood from an undisturbed, resting subject. This usually means an indwelling arterial cannula, through which blood from an artery can be sampled. The blood is then injected into a chamber in a blood-gas analyzer containing a glass pH electrode that measures the pH of the plasma. Additional blood is injected into another chamber that contains oxygen and carbon dioxide electrodes that measure the partial pressures of these two gases.

Bicarbonate concentration can be calculated from the mathematical relationship among pH, the partial pressure of carbon dioxide, and the bicarbonate concentration when the former two are known. When making these kinds of measurements, it is important to measure the pH and partial pressure of the blood at the same temperature as the animal. Thermostated electrodes are used for this. The calculation used to estimate bicarbonate concentration requires the use of two constants which are also temperature-dependent. It is important to use the constants that are appropriate to the temperature of the animal and the measurement conditions.

Alternatively, bicarbonate concentration can be approximated by measuring the total amount of carbon dioxide (the sum of the carbon dioxide, carbonic acid, bicarbonate, and carbonate concentrations). Under physiological conditions, more than 99% of the carbon dioxide is in the form of HCO_3^-. When strong concentrated acid (HCl) is added to a measured volume of blood or plasma, all the carbonic acid, bicarbonate, and carbonate are released as carbon dioxide. This carbon dioxide can then be measured with a carbon dioxide electrode. The resulting total carbon dioxide, minus the concentration of carbon dioxide measured before adding the HCl, is very close to being equal to the concentration of bicarbonate.

While there are a number of indirect methods for measuring intracellular pH, the best results are achieved directly with micro pH electrodes which can be used to impale single cells. These techniques must be done on isolated tissues or anesthetized, restrained animals, whose acid-base status does not, in the least, resemble that of resting, undisturbed animals. The techniques are useful for studying pH maintenance at the cellular level and useful extrapolations can be made to the whole animal.

The usual approach to the study of pH maintenance is to induce a disturbance in the pH of an experimental animal and then to follow changes in the pH, partial pressure of CO_2, and the concentration of HCO_3^- in the blood. Breathing rates and volumes of air exchanged by the lungs can be measured to determine the contribution of the lungs to pH maintenance. For example, increasing the rate of breathing will increase the rate of elimination of carbon dioxide and lower the partial pressure of this gas in the blood. Urine collection can be done to assess the contribution of the kidneys to pH maintenance, and in aquatic animals that do not rely heavily on their kidneys for pH maintenance, changes in the composition of the water in which the animals are situated can be used to assess the contribution of the gills (fish) or skin (amphibians) to pH maintenance.

Examples of experimental manipulations that can produce pH disturbances in animals include infusion of acids and bases into the blood to produce metabolic acidosis and alkalosis respectively. Alternatively, animals can be exercised to elevate lactic acid in their blood to produce a metabolic acidosis. Respiratory acidosis and alkalosis can be induced by artificially altering breathing rates. Alternatively, elevation of the partial pressure of carbon dioxide in the air or water of an experimental animal will produce a respiratory acidosis.

pH MAINTENANCE AND HOMEOSTASIS

Maintenance of pH is part of the overall phenomenon of homeostasis or constancy of the internal environment in a changing external environment. The primary reason for maintaining consistent acid-base balance is the need to keep constant the conformation (shape) of proteins. Proteins are very sensitive to changes in pH, and the usual consequence of such changes is a change in the shapes of proteins. Protein conformation is critically important to proper protein function.

All enzymes (proteins that cause necessary biochemical reactions to occur) have very specific shapes that must be maintained in order for them to function properly. Changes in pH cause alterations in those shapes and compromise the effectiveness of the enzymes. Many other proteins require certain conformational states. Cell membranes contain proteins that determine which chemicals can enter and leave the cell's interior. These proteins are called carriers if they actively transport specific substances into or out of the cell interior. Alternatively, membrane proteins can act as specific channels that allow passive movement of only specific molecules or ions across the membrane. Both carriers and channels are sensitive to pH-induced conformational changes. There are other membrane-bound proteins called receptors that respond best to hormones and other chemical messengers when the pH is maintained at the norm. Proteins in the immune system, such as antibodies, also require proper pH in order to function optimally.

Cold-blooded animals do not regulate their pH to a specific point, but instead regulate the difference between their pH and the pH of pure water at the same temperature. This relative alkalinity ensures constant protein conformation in the same way that constant pH at constant temperature ensures constant protein conformation in warm-blooded animals, because the weak acid nature of proteins causes them to be partially dissociated from H+ in the pH and temperature range experienced by animals. The conformation of protein is determined, to a large extent, by the net electrical charge (distribution of + and − charges over the protein). The net charge is influenced by the degree of dissociation of the protein and H+. As discussed above, the dissociation in protein that is important to pH maintenance is the dissociation of H+ from histidine. As long as the dissociated fraction of histidine remains constant, the protein conformation will remain constant. As long as the relative alkalinity is maintained, the fractional dissociation of histidine remains constant and thus protein conformation does not change. In reality, the regulation of mammalian pH at 7.4 (at 37°C) is one specific example of this general rule. The relative alkalinity of mammalian blood at 37°C is the same as the relative alkalinity of fish blood at 5°C or of reptile blood at 42°C.

—*Daniel F. Stiffler*

FURTHER READING

Berne, R. M., and M. N. Levey. *Principles of Physiology*. 3rd ed. C. V. Mosby, 2000.

Davenport, Horace W. *The ABC of Acid-Base Chemistry*. 6th ed. University of Chicago Press, 1974.

Heisler, N., ed. *Acid-Base Regulation in Animals*. Elsevier, 1986.

Jacquez, John A. *Respiratory Physiology*. McGraw-Hill, 1979.

Keyes, Jack L. *Fluid, Electrolyte, and Acid-Base Regulation*. Jones and Bartlett, 1999.

Lowenstein, Jerome.. Oxford University Press, 1993.

Stewart, P. A. *How to Understand Acid-Base: A Quantitative Acid-Base Primer for Biology and Medicine*. Elsevier, 1981.

Valtin, Heinz. *Renal Function: Mechanisms Preserving Fluid and Solute Balance in Health*. 3rd ed. Little, Brown, 1995.

PHYLOGENY

FIELDS OF STUDY

Evolutionary Biology; Genetics; Animal Physiology

ABSTRACT

Phylogeny traces the history of life on Earth through the study of how animals and plants have developed over time and how they are related to one another. It is similar to taxonomy, the science of classifying organisms based on their structure and functions. Phylogeny, as a system of classification, classifies the living world by similarities in ancestry rather than appearance.

KEY CONCEPTS

deoxyribonucleic acid (DNA): the genetic material of most living organisms

evolution: the process by which the variety of plant and animal life has developed over time from the most primitive to the most complex life-forms

kingdom: the highest category into which organisms are classified; there are now believed to be five kingdoms

molecule: the smallest part of a chemical compound

phylum: a group that consists of several closely related classes

species: a group of organisms that can produce offspring with each other

A SYSTEM OF CLASSIFICATION

Phylogeny traces the history of life on Earth through the study of how animals and plants have developed over time and how they are related to one another. It is similar to taxonomy, the science of classifying organisms based on their structure and functions. Taxonomists create family trees of living and extinct species in order to discover the origins and lines of descent of various forms of life. Very few family trees are complete to their fossil origins, however, because of gaps in the fossil record. The first system of classification was devised by the eighteenth century Swedish scientist Carolus Linnaeus. Linnaeus classified life-forms based on their appearance; the more they resembled each other in size, shape, and form, he believed, the more closely they were related.

The theory of evolution developed by nineteenth century naturalist Charles Darwin was based on accumulated changes over time through natural selection and led to a new way of looking at the history of life and to the development of phylogeny as a method of classification. The new system classified the living world by similarities in ancestry rather than appearance. Life histories of species were derived from the study of comparative anatomy (the search for common features in different species), embryology (the study of the development of life from the egg to birth), and biochemistry (the study of invisible chemical characteristics of cells that link species that can look very dissimilar).

Modern technology allows scientists to measure differences in deoxyribonucleic acid (DNA) molecules among species. DNA carries the genetic material of an organism and plays a central role in heredity. The degree of difference between two species helps scientists determine how much modern species have changed from their ancestors and to estimate when important events in evolution occurred. If a sufficient fossil record exists, biochemists can determine the timing of a major change that led to the development of a new characteristic such as a longer neck in giraffes or a different size eyeball in a bat.

CLASSIFYING EARTH'S SPECIES

Studies of evolutionary change suggest that anywhere from three million to twenty million species now exist, with millions more having become extinct. Earth began about 5 billion years ago, and it has been occupied by living organisms for about 3.5 billion years. For about 2.5 billion years, the planet was populated only by single-celled bacteria. The earliest forms of life most likely formed by chance during a long period of random chemical reactions in the primordial world. Random encounters between chemicals in the seas and the atmosphere can produce amino acids and proteins. Small drops of proteins, made up of carbon, hydrogen, oxygen, nitrogen, and sulfur, somehow bound together and became organisms; that is, they began to reproduce themselves, which is the quality that makes them life-forms.

Every living thing is given a two-part Latin name under a system of binomial nomenclature. This name is usually based on an evolutionary, phylogenetic relationship. The first part of the name indicates the organism's genus; the second part identifies the specific species. For example, *Acer saccharum*, *Acer nigrum*, and *Acer rubrum* are all kinds of maple trees. They share the genus *Acer*, and their species names mean sugar, black, and red, respectively. No rules govern species names, though frequently they refer to a prominent feature such as color or characteristic, such as in the case of *Homo sapiens*, the Latin name for human beings, which literally means "the thinking one."

Of the five kingdoms of life-forms, one (containing the bacteria) is made up of prokaryotes; the other four are eukaryotes. Prokaryotes do not have cell nuclei; eukaryotes do. In prokaryotes, cells are smaller and simpler in structure, and the DNA is not organized into chromosomes. Chromosomes, the threadlike structures found in the cell nucleus of all eukaryotes, determine the characteristics of individual organisms, with offspring receiving an equal number of chromosomes from each parent. In bacteria, however, reproduction takes place by a simple division of the parent into two masses that eventually separate. Bacteria do not share chromosomes with other members of the species; in other words, reproduction is asexual.

THE NON-ANIMAL KINGDOMS

All bacteria belong to the kingdom *Prokaryotae*. They are in terms of total numbers the most successful form of life in the universe. Some bacteria, which are germs, can cause disease, while some bacteria, such as antibiotics, can be used to cure disease. Bacterium-like fossils that were found in rocks from an Australian gold mine are about 3.5 billion years old. Bacteria are found in every climate and habitat, and are the first to invade and populate new habitats. Seventeen phyla of bacteria are known to exist, and none can be seen without a microscope.

The four other kingdoms of living things and the common names given to their most important phyla are *Protoctista* (algae, protozoa, slime molds), *Fungi* (mushrooms, molds, lichens), *Animalia* (sponges, jellyfish, flatworms, ribbon worms, rotifers, spiny-headed worms, parasitic nematodes, horsehair worms, mollusks, priapulid worms, spoon worms, earthworms, tongue worms, velvet worms, insects, beard worms, starfish, arrowworms, and chordates), and *Plantae* (mosses, ferns, and pine-bearing and flowering plants). Phyla are based on similarities in evolutionary development. Life-forms are divided into ninety-two phyla: There are seventeen phyla in the kingdom *Prokaryotae*; twenty-seven in *Protoctista*; five in Fungi; thirty-three in *Animalia*, and ten in *Plantae*. Each phylum is divided into classes, then orders, families, genera (the singular form is genus), and finally, species. The last two divisions are based on the most recent evolutionary differences. Some phyla have only a few genera and species, while others, such as those in *Animalia*, have millions. A species consists of a group of similar individual organisms that can breed and reproduce offspring. A genus contains similar species, but members of a genus cannot reproduce; only members of a species can.

Protoctista (also called *Protista*) means the "very first to establish." It is the kingdom made up of living forms that are neither animals, plants, nor fungi. It includes red, green, and brown algae, seaweeds, water and slime molds, and most protozoa (single-celled life-forms including amoebas). The twenty-seven phyla in this kingdom are found mostly in water habitats such as rivers, freshwater lakes, and oceans.

The kingdom *Fungi* (from a Greek word meaning "sponge") may have descended from the kingdom *Protoctista*. The oldest fossil *Fungi* are about 300 million years old. There are more than one hundred thousand identified species, including bread molds, yeasts, and mushrooms. Thousands more species are believed to exist but have yet to be identified.

THE ANIMAL KINGDOM

All plants, from mosses to giant redwoods, belong to the kingdom *Plantae* (from a Latin word for "plant"), and most animals belong to the kingdom *Animalia*. This kingdom contains all multicellular, heterotrophic, diploid organisms that develop from an egg and sperm. A heterotrophic animal takes food into the body and digests it, or breaks it down, into energy for use. A diploid organism has two sets of chromosomes, one derived from the female parent and the other from the male. The characteristic that defines all *Animalia*, however, is that they develop from a blastula, a hollow ball of cells found in animal embryos (the earliest stage of development, when the egg just begins to divide) but not in plants. The blastula forms a layer around a central open space or cavity in the embryo.

Animalia (from a Latin word meaning "soul" or "breath") is the kingdom with the largest number of life-forms, from *Placozoa*, a phylum with a single species usually found growing on aquarium walls, to the phylum *Arthropoda* (from a Greek word meaning "jointed foot") that contains more than a half million identified species, including spiders, scorpions, beetles, shrimp, lobsters, crayfish, crabs, flies, centipedes, millipedes, butterflies, moths, and all other species of insects. Seventeen of the thirty-three phyla in *Animalia* are various kinds of worms found in the shallow and deep water of lakes, rivers, and oceans and in the ground. One worm phyla, *Pentastoma*, consists of more than seventy species that live exclusively in the tongues, lungs, nostrils, and nasal sinuses of dogs, foxes, goats, horses, snakes, lizards, and crocodiles. Another phyla, *Platyhelminthes*, consists of worms that live in bat dung and other equally unusual environments. Only two of the phyla, *Arthropoda* and *Chordata* (animals with nerve cords), live entirely on land. Among the phyla, *Arthropoda* has the largest number of identified species, more than a million, and possibly millions more not yet identified. Most of the species of *Animalia* that have ever lived are now extinct. Thirty-two of the thirty-three phyla are invertebrates, which means they lack backbones.

THE HISTORY OF LIFE-FORMS

The phylogenetic method of classifying life, using animals as an example, begins with the earliest fossil records available. The earliest members of the *Animalia* kingdom evolved from members of the *Protoctista* kingdom about 565 million years ago. Exactly when and which members were the first is subject to debate; however, the first large, multicellular fossils date to this time. About 530 million years ago, there was a gigantic explosion of life-forms, including all kinds of clams, snails, arthropods, crabs, and trilobites. After a few tens of millions more years came echinoderms (starfish and sea urchins) and eventually chordates, out of which emerged fish and mammals.

The periods of explosive growth and massive extinction of life-forms are believed to have been caused by some major environmental changes, the exact nature and cause of which have not been determined. Scientists believe an explosion of new life could have been caused by an increase in atmospheric oxygen, which would allow life-forms to live out of water, and a mass extinction could have been the result of gigantic dust clouds, large enough to darken the sun's light for millions of years, stirred up by huge meteors crashing into earth. Such a critical darkening of the earth is believed to have taken place about sixty-five million years ago, plunging temperatures to near freezing and killing off thousands of species including all the dinosaurs.

THE PHYLUM CHORDATA

Despite the tremendous losses resulting from the extinction of as much as 95% of the existing species, life-forms continued to survive and evolve. The most successful of the new forms emerged from the ancestors of the phylum *Chordata* (from a Latin word for "cord"). This phylum includes all mammals, birds, amphibians, reptiles, and species with backbones (vertebrates). More than forty-five thousand species of chordates exist. Three key features are used to classify members of the phylum. A member must have a single nerve cord along its back. In mammals, this cord has developed into the spinal cord and brain. It must also have a notochord, a bony rod located between the nerve cord and the digestive tract, which supports the body and the muscles and is found in the embryo and the adult. The last requirement is that members have gill slits in the throat at some stage of development, whether in the embryo stage as is true with land animals or throughout their entire life as in fish. Gill slits show that land animals developed from sea creatures.

The phylum *Chordata* includes two superclasses, *Pisces* (fish) and *Tetrapoda* (four-limbed forms). There are two classes of living *Pisces*, *Chondrichthyes* (boneless creatures), which includes sharks, skates, and rays, and *Osteichthyes* (bony fish), a class that contains about twenty-five thousand species. The earliest fish fossils are about 500 million years old. There are four classes in *Tetrapoda*: *Amphibia*, *Reptilia*, *Aves*, and *Mammalia*.

The class *Amphibia* contains about two thousand species of frogs, toads, and salamanders. Members of this class evolved about 370 million years ago and were the first vertebrates to live on land. *Amphibia* must lay their eggs in water and must live close to water, or their soft skin will dry out and cause their deaths.

The class *Reptilia* has more than five thousand species. It includes turtles, lizards, snakes, and crocodiles. Reptiles develop from an egg, live on land, and have dry, scaly skin. The largest reptiles, the dinosaurs, died out more than sixty million years ago. Reptiles are typically cold-blooded, which means they cannot control the temperature of their blood, but there is a great deal of evidence that indicates many, if not all, species of dinosaurs were in fact warm-blooded. Reptiles breathe air through lungs, and they are not required to lay their eggs in water. They probably evolved out of an amphibian species between 300 million and 320 million years ago.

The class *Aves* contains nine thousand species of living birds, with thousands more extinct. *Aves* evolved from dinosaur species perhaps 200 million years ago. Unlike reptiles, *Aves* class members have the ability to regulate their internal temperature. They have feathers rather than scales, but they lack teeth.

The fourth class is *Mammalia*, which has more than forty-five hundred living species, including human beings. Mammals are classed into *Metatheria* (marsupials) and *Eutheria* (placentals). *Metatheria*, such as kangaroos, have external pouches in which their young are born live. *Eutheria* have vaginas through which the fully developed young pass during birth. Most mammals are members of *Eutheria*. This includes the orders of *Insectivora* (hedgehogs, shrews, and moles), *Primates* (lemurs, monkeys, apes, and human beings), *Carnivora* (dogs, cats, and bears), and *Pinnipedea* (seals and sea lions.)

THE BENEFITS OF CLASSIFICATION

Life on Earth includes millions of different types of organisms. Phylogeny is one method of making sense out of so many different life-forms. It provides a system that links organisms together on the basis of their evolutionary history of development. By grouping vast numbers of forms into related groups, phylogenetics helps bring some order to what seems like chaos. Attempts to classify life-forms have always been controversial, and there still is much disagreement in the scientific community over various types of classification. However, phylogeny is the most widely agreed upon system.

Developments in techniques used to analyze molecules and DNA in the 1980s have shed light on the evolutionary development of species and have demonstrated new links among living plants and animals. The information gleaned using these techniques can also be used to trace the ancestry of modern species back to the earliest fossil evidence available from

575 million years ago. Molecular comparisons have shown that the development of species from spores and eggs into embryos and adult stages is very similar across a wide range of phyla.

These techniques produce data that consist of long sequences of the four nucleic acids that make up the information contained in DNA. The patterns formed by the acids are very similar among related organisms. The more closely related the sequence of nucleic acids, the more closely the species are related. Closely related species differ only very slightly in the way their DNA is structured. The more species differ in their DNA structure, the more distant they are in evolutionary terms. By measuring and comparing the differences in a gene that controls basically the same function in various species, scientists can construct an evolutionary tree. Species can be placed on the tree at the points where they begin to diverge from other genus members. In this way, examinations of the phylogenetic development of organisms can be used to create a tree of life showing the close relationships among all organisms. The tree also shows that all forms of life are related because they originally came from the same source in some ancient pool where the right chemicals just happened to bump into one another, mix, and begin to produce offspring.

—Leslie V. Tischauser

Further Reading

Bakker, Robert T. *The Dinosaur Heresies.* Zebra/Kensington, 1986.

Copeland, Herbert F. *The Classification of Lower Organisms.* Pacific Books, 1956.

Desmond, Adrian J. *The Hot-Blooded Dinosaurs.* Blond & Briggs, 1975.

Gould, Stephen Jay. *Wonderful Life: The Burgess Shale and the Nature of History.* Vintage, 2000.

Margulis, Lynn, and Karlene V. Schwartz. *Five Kingdoms: An Illustrated Guide to the Phyla of Life on Earth.* 3rd ed. New York: W. H. Freeman, 1999.

Romer, A. S. *The Vertebrate Story.* University of Chicago Press, 1971.

Wilson, Edward O. *The Diversity of Life.* W. W. Norton, 1992.

Placental Mammals

Fields of Study

Developmental Biology; Endocrinology; Animal Physiology

Abstract

The placental mammals form a diverse and successful group that includes the insectivores, primates, rodents, rabbits, whales, dolphins and porpoises, carnivores, seals, aardvarks, elephants, hyraxes, manatees, uneven-toed mammals and even-toed (cloven-hoofed) mammals. Placental mammals are viviparous. This allows longer gestation, and as a result the young are born in a more advanced (precocial) state of development.

Key Concepts

chorion: the outer cellular layer of the embryo sac of reptiles, birds, and mammals; the term was coined by Aristotle

embryo: a young animal that is developing from a fertilized or activated ovum and that is contained within egg membranes or within the maternal body

endometrium: an inner, thin layer of cells overlying the muscle layer of the uterus

fetus: a mammalian embryo from the stage of its development where its main adult features can be recognized, until birth

maternal: referring to the female parent

ovum: an unfertilized egg cell

uterus: in female mammals, the organ in which the embryo develops

viviparous: producing young that are active upon birth (often referred to as live birth); the embryo is nurtured within the uterus

OVIPAROUS, VIVIPAROUS, AND OVOVIVIPAROUS MAMMALS

Monotremes are oviparous, or egg-laying mammals. Marsupials are ovoviviparous, meaning that the

493

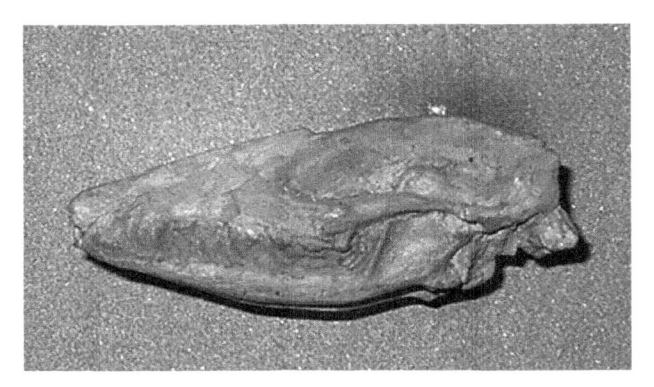

Asioryctes nemegtensis, *one of the earliest placental mammals. (Mesozoic mammals from the Gobi Desert. Museum of Evolution, Warsaw)*

egg is large and has a yolk adequate to nourish the embryo during its early development, but it remains unattached to the wall of the uterus. Gestation in marsupials is necessarily short, therefore, and the young are born in an immature, fetal stage. They make their way to the mother's pouch and continue to grow, nourished by the mother's milk. All other mammals, termed placental, are viviparous. The small egg, lacking food substance, becomes attached to the uterine wall, and the developing embryo is nourished by the mother's blood passing through a placenta. This process allows longer gestation, and as a result the young are born in a more advanced (precocial) state of development.

The placental mammals form a diverse and successful group that includes the insectivores (such as shrews, hedgehogs, and moles), bats, sloths, anteaters, armadillos, primates (to which humans belong), rodents, rabbits, whales, dolphins and porpoises, carnivores (such as cats, dogs, and bears), seals, aardvarks, elephants, hyraxes, manatees, uneven-toed mammals (such as tapirs, horses, and rhinoceroses), and even-toed (cloven-hoofed) mammals such as pigs, camels, deer, sheep, cattle, and goats.

PLACENTA STRUCTURE

The term "placenta" comes from the Latin meaning "flat cake," and is used to describe the flat structure (in most animals) which attaches the developing embryo and fetus to the wall of the uterus. The term was first used in 1559 CE, although knowledge of the placenta, at least in humans, goes far back into antiquity. Reference to it may be found in many ancient texts and drawings, including the Old Testament books of the Bible. Early Egyptians considered the placenta to be the seat of the external soul.

The placenta is the organ responsible for the transmission of materials between mother and fetus prior to birth—the only bridge between them. In many species, it has important endocrine functions, producing hormones necessary for development of the fetus or for maintenance of the pregnant state. It is essentially a product of both the developing ovum and the mother. The placenta is commonly referred to as the afterbirth, extruded from the mother's uterus at the end of the birth process.

When the ovum is released by the ovary and fertilized by sperm, it eventually comes to rest in the hollow cavity of the uterine horn (one of two chambers) or the uterus (single chamber). While it moves toward that area it begins to divide, forming a ball of cells known as a morula. As development continues, the ball becomes hollow, and it is then referred to as a blastocyst. Within this hollow blastocyst a few cells protrude into the cavity, forming a knob. These are the only cells that will eventually develop into the embryo and fetus. The rest of the cells are responsible for forming the supportive structures, including the chorion, the placenta, and the umbilical cord (which attaches the embryo to the placenta).

Among the eutherian (placental) mammals, a variety of placental configurations occurs. In many species the entire chorionic sac becomes connected to the uterine wall, and transfer of materials between the maternal and fetal compartments occurs over the whole surface. In other species, a much more specialized system develops. Here parts of the chorion (a membrane equivalent to the one that lines the shell of reptile and bird eggs) become highly specialized, establishing an intimate relationship with the uterine tissues. Thus, transfer of materials occurs only in one select region of the chorion, referred to as the placenta. It is these tissues, with their flattened, cakelike appearance, from which the name derives.

The most important feature of the placenta is the close contact between the fetal blood vessels in the placenta and the maternal blood vessels in the uterine wall. While it is a common misconception that the fetal and maternal blood mix or flow together, this is not a correct picture. What actually occurs is that the two blood systems come close to each other, at which point materials that can diffuse out of one vessel may diffuse into another. Thus, transfer of materials from mother to fetus, and vice versa, can occur. The closer the two blood pools come to each other, the better it is for transfer. Nutritional, respiratory, and excretory products are transferred.

PLACENTA TYPES

In the epithelio-chorial placenta, as found in the hoofed mammals (such as *Artiodactyla* and *Perissodactyla*), whales, and lemurs, the wall of the uterus retains its surface epithelium. The minimum separation of bloods is four cells thick (two epithelial cell layers and the endothelium of the blood vessels). Thus, it is vital to have a very large surface area to allow for adequate movement of materials. While this is a large improvement over the marsupial system, it is nonetheless 250 times less efficient at salt transfer than the placenta used by humans.

The separation between mother and fetus is reduced in carnivores and sloths. Here the chorion invades the uterine epithelium and comes in direct contact with the epithelium of the maternal blood vessels, allowing a more uniform transfer of materials. The most advanced form, showing minimal separation, is the hemo-chorial placenta, found in humans, rodents, bats, and most insectivores. Here the maternal blood vessel walls are chemically broken down and the invading chorion is now in direct contact with the maternal blood stream. Because this is so much more efficient for materials exchange, the size of the interacting surfaces can be much reduced.

In addition to the exchange of materials, the placenta plays an important role in immunology, and without the placenta, the mother's body would reject the developing embryo like any other foreign body. It is this tolerance of the embryo that separates the placental mammals from the marsupials, and allows for gestation periods to be extended. Fetuses in placental mammals receive antibodies from their mothers, thus enhancing their early immunity to disease.

The epithelio-chorial placenta, being only in contact with the uterine wall and not being invasive, is readily shed by the uterus when the fetus is born. There is no damage to the maternal tissue. The more invasive types of placenta, including the hemo-chorial type, can only be lost by separation through the uterine tissues. Thus, birth in species with hemo-chorial placentas is of necessity associated with some degree of maternal bleeding. In fact, in many hemo-chorial placentas, the blastocyst actually digests (chemically) the endometrial lining of the uterus and comes to lie completely within it. The endometrium then heals over the blastocyst, which then grows, fully surrounded by endometrium. The true placenta forms on the deep pocket of the endometrium in which the blastocyst lies. When the fetus is expelled from the uterus (that is, at birth), it ruptures the now very thin layer of stretched endometrium that covers the chorion. The placenta separates from the uterus as a result of rupture of the uterine blood vessels and tissues when the uterus contracts down after expulsion of the fetus. Bleeding between uterus and placenta produces a clot which eventually seals off the broken blood vessels and forms the basis for endometrial repair.

Most mammalian placentas, regardless of type, have some type of endocrine (hormonal) function. While the specific hormones may vary from species to species, two in particular are found in most placental mammals. Chorionic gonadotropin is a hormone secreted by the placenta which acts upon the ovary to increase progesterone synthesis. Progesterone, in return, is responsible for maintenance of pregnancy during the early phases. Placental lactogen, another hormone secreted by most placentas, acts on the mother to stimulate mammary gland development. This occurs throughout gestation, so that the mammary glands are ready for suckling by the time the offspring is born.

GESTATION PERIODS

The length of gestation varies tremendously among placental mammals. In elephants the gestation period is as long as twenty-two months. However, size alone is not the determining factor. The giant among all mammals, the blue whale, has a gestation period of only eleven months, not appreciably longer than the human (nine to ten months).

Many bats and other mammals have a delayed implantation, in which the fertilized ovum remains dormant or its development is retarded at first, thus considerably extending the gestation period and delaying birth until the optimal season of warm weather or abundant food is present. Often a placenta is not found during this period of delay, with arrest of development occurring in the blastula stage. Thus, the gestation period of the fisher, a small North American carnivore with delayed implantation, is forty-eight to fifty-one weeks, or about the same as that of the blue whale. Gestation varies from twenty-two to forty-five days in squirrels, twenty to forty days in rats and mice, two to seven months in porcupines, six months in bears, and fourteen to fifteen months in giraffes.

Gestation length is ultimately constrained by the size of skull which will fit through the maternal pelvis. Where agility, speed, or long distances of travel put a premium on the mother's athleticism, gestation length is often shortened, and the birth weight of the offspring will be low.

Animals having long gestation periods, or whose young mature slowly and are suckled for a long period of time, generally do not breed as often as others. Many species of mice breed repeatedly throughout the spring, summer, and fall, having a gestation period of about twenty days, and being mature and ready to breed by twenty-one days of age. Many others, such as bears, coyotes, and weasels, breed only once a year. Environmental conditions, and the adaptability of various species to these conditions, play a large role in breeding cycles. It is clearly advantageous for young to be born during the season of least severe weather and to be weaned when food is most abundant. Many tropical mammals breed and give birth throughout the year, whereas in temperate or cold climates young are usually born in the spring or summer.

Similar factors also influence the number of young born in each litter among different species. Their rate of growth until weaned, mortality rates, adult activity cycles, and other factors no doubt help determine the litter size as well. Many rodents have three to six young per litter; a few species of mice can have as many as eighteen. Seals, whales, and most species of bats and primates bear only a single young at one time.

—*Kerry L. Cheesman*

Further Reading

Gilbert, Scott. Developmental Biology. 6th ed. Sinauer Associates, 2000.

Harris, C. Leon. Concepts in Zoology. 2nd ed. HarperCollins, 1996.

Macdonald, David, ed. The Encyclopedia of Mammals. Facts on File, 1984.

Randall, David, Warren Burggren, and Kathleen French. Eckert Animal Physiology. 4th ed. W. H. Freeman, 1997.

Silverthorn, Dee Unglaub. Human Physiology: An Integrated Approach. 2nd ed. Prentice Hall, 2001.

Plant and Animal Interactions

Fields of Study

Ecology; Botany; Evolutionary Biology

Abstract

Ecology involves a systematic analysis of plant-animal interactions through the considerations of nutrient flow in food chains and food webs, exchange of such important gases as oxygen and carbon dioxide between plants and animals, and strategies of mutual survival between plant and animal species through the processes of pollination and seed dispersal. Ecologists study both abiotic and biotic features of such plant and animal interactions.

Key Concepts

cellular respiration: the release of energy in organisms at the cell level, primarily through the use of oxygen

chlorophyll: one of several forms of photoactive green pigments in plant cells that is necessary for photosynthesis to occur

coevolution: a mutualistic relationship between two different organisms in which, as a result of natural selection, the organisms become interdependent

cross-pollination: the transfer of pollen grains and their enclosed sperm cells from the male portion of a flower to a female portion of another flower within the same species

food chain: a diagram illustrating the movement of food materials from green plants (producers) through various levels of animals (consumers) within natural environments

natural selection: the survival of variant types of organisms as a result of adaptability to environmental stresses

pistil: a female portion of a flower that produces unfertilized egg cells

stamen: a male portion of a flower that produces pollen grains and their enclosed sperm cells

PLANTS, ANIMALS, AND ECOLOGY

Ecology represents the organized body of knowledge that deals with the interrelationships between living organisms and their nonliving environments. Increasingly, the realm of ecology involves a systematic analysis of plant-animal interactions through the considerations of nutrient flow in food chains and food webs, exchange of such important gases as oxygen and carbon dioxide between plants and animals, and strategies of mutual survival between plant and animal species through the processes of pollination and seed dispersal.

Ecologists study both abiotic and biotic features of such plant and animal interactions. The abiotic aspects of any environment consist of nonliving, physical variables, such as temperature and moisture, that determine where species can survive and reproduce. The biotic (living) environment includes all other plants, animals, and microorganisms with which a particular species interacts. Certainly, two examples of plant and animal interactions involve the continual processes of photosynthesis and cellular respiration. Green plants are often classified as ecological producers and have the unique ability to carry out both these important chemical reactions. Animals, for the most part, can act only as consumers, taking the products of photosynthesis and chemically releasing them at the cellular level to produce energy for all life activities.

PLANT-ANIMAL MUTUALISM

One topic that has captured the attention of ecologists involves the phenomenon of mutualism, in which two different species of organisms beneficially reside together in close association, usually revolving around nutritional needs. One such example demonstrating a plant and animal association is a certain species of small aquatic flatworm that absorbs microscopic green plants called algae into its tissues. The benefit to the animal is one of added food supply; the adaptation to this alga has been so complete that the flatworm does not actively feed as an adult. The algae, in turn, receive adequate supplies of nitrogen and carbon dioxide and are literally transported throughout tidal flats in marine habitats as the flatworm migrates, thus exposing the algae favorably to increased sunlight.

A similar example of mutualism has been reported by ecologists studying various types of reef-building corals, which are actually marine, colonial animals that grow single-celled green algae called zooxanthellae within their bodies. The coral organisms use the nutrients produced by these algae as additional energy supplies, enabling them to build more easily the massive coral reefs associated with tropical waters. In 1987, William B. Rudman reported a similar situation while researching the formation of such coral reefs in East African coastal waters. He discovered a type of sea slug called a nudibranch that absorbs green algae into its transparent digestive tract, producing an excellent camouflage as it moves about on the coral reefs in search of prey. In turn, the algae growing within both the coral and sea slugs receive important gases from these organisms for their own life necessities.

An example of plant-animal mutualism that has been documented as a classic example of coevolution involves the yucca plant and a species of small, white moth common throughout the southwestern United States. The concept of coevolution builds upon Charles Darwin's theories of natural selection, reported in 1859, and describes situations in which

Bumblebee pollinating an azalea.

two decidedly different organisms have evolved into a close ecological relationship characterized by compatible structures in both. Thus, coevolution is a mutualistic relationship between two different species that, as a result of natural selection, have become intimately interdependent. The yucca plant and yucca moth reflect such a relationship. The female moth collects pollen grains bearing sperm cells from the stamens of one flower on the plant and transports these pollen loads to the pistil of another flower, thereby ensuring cross-pollination and fertilization. During this process, the moth will lay her own fertilized eggs in the flower's undeveloped seed pods. The developing moth larvae have a secure residence for growth and a steady food supply. These larvae will rarely consume all the developing plant seeds; thus, both species (plant and animal) benefit.

DEFENSIVE MUTUALISM

Although these examples demonstrate the evolution of structures and secretions that reflect mutual associations between plants and animals, other interactions are not so self-supportive. Plant-eating animals, called herbivores, have always been able to consume large quantities of green plants with little fear of reprisal. Yet, some types of carnivorous plants have evolved that capture and digest small insects and crustaceans as nutritional supplements to their normal photosynthetic activities. Many of these plants grow abundantly in marshlike environments, such as bogs and swamps, where many insects congregate to reproduce. Such well-known plants as the Venus's-flytrap, sundews, butterworts, and pitcher plants have modified stems and leaves to capture and consume insects and spiders rich in protein. On a smaller scale, in freshwater ponds and lakes, a submerged green plant commonly known as the bladderwort partially satisfies its protein requirements by snaring and digesting small crustaceans, such as the water flea, within its modified leaves.

A form of ecological interaction commonly classified as mimicry can be found worldwide in diverse environments. In such situations, an animal or plant has evolved structures or behavior patterns that allow it to mimic either its surroundings or another organism as a defensive or offensive strategy. Certain types of insects, such as leafhoppers, walking sticks, praying mantids, and katydids (a type of grasshopper), often duplicate plant structures in environments ranging from the tropical rain forests to the northern coniferous forests of the United States. Such exact mimicry of their plant hosts affords these insects protection from their predators as well as camouflage that enables them to capture their own prey readily. In other examples of mimicry, some insects will absorb unpalatable plant substances in their larval stages and retain these chemicals in their adult forms, making them undesirable to birds as food sources. The monarch butterfly demonstrates this type of interaction with the milkweed plant. The viceroy butterfly has evolved colorations and markings similar to those of the monarch, thereby ensuring its own survival against bird predators. Certain species of ambush bugs and crab spiders have evolved coloration patterns that allow them to hide within flower heads of such common plants as goldenrod, enabling them to grasp more securely the bees and flies that visit these flowers.

NONSYMBIOTIC MUTUALISM

Many ecologists have been studying the phenomenon known as nonsymbiotic mutualism: different plants and animals that have coevolved morphological structures and behavior patterns by which they benefit each other without necessarily living together physically. This type of mutualism can be demonstrated in the often unusual and bizarre shapes, patterns, and colorations that more advanced flowering plants have developed to attract various insects, birds, and small mammals for pollination and seed dispersal purposes. Pollination essentially is the transfer of pollen grains (and their enclosed sperm cells) from the male portion of a flower to the egg cells within the female portion of a flower. Pollination can be accomplished by the wind, by heavy dew or rains, or by animals, and it results in the plant's sexual production of seeds that represent the next generation of new embryo plants. Accessory structures, called fruits, often form around seeds and are usually tasty and brightly marked to attract animals for seed dispersal. Although the fruits themselves become biological bribes for animals to consume, often the seeds within these fruits are not easily digested and thus pass through the animals' digestive tracts unharmed, sometimes great distances from the original plant. Other types of seed dispersal mechanisms involve the evolution of hooks, barbs, and sticky substances on seeds that enable them to be easily transported by animals' fur, feet, feathers, or beaks to new regions for possible plant colonization. Such strategies of dispersal reduce competition between the parent plant and its new seedlings for moisture, living space, and nutritional requirements.

The evolution of flowering plants and their resulting use of animals in pollination and seed dispersal probably began in dense, tropical rain forests, where pollination by the wind would be cumbersome. Because insects are the most abundant form of animal life in rain forests, strategies based upon insect transport of pollen probably originated there. Because structural specialization increases the possibility that a flower's pollen will be transferred to a plant of the same species, many plants have evolved a vast array of scents, colors, and nutritional products to attract many insects, some birds, and a few mammals. Not only does pollen include the plant's sperm cells, but it is also rich in food for these animals. Another source of animal nutrition is a substance called nectar, a sugar-rich fluid often produced in specialized structures called nectaries within the flower itself or on adjacent stems and leaves. Assorted waxes and oils are also produced by plants to ensure plant-animal interactions. As species of bees, flies, wasps, butterflies, and hawkmoths are attracted to flower heads for these nutritional rewards, they unwittingly become agents of pollination by transferring pollen from male portions of flowers (stamens) to the appropriate female portions (pistils). Some flowers have evolved distinctive, unpleasant odors reminiscent of rotting flesh or feces, thereby attracting carrion beetles and flesh flies in search of places to reproduce and deposit their own fertilized eggs. As these animals consummate their own relationships, they often become agents of pollination for the plant itself. Some tropical plants such as orchids even mimic a female bee, wasp, or beetle, so that its male counterpart will attempt to mate with it, thereby ensuring precise pollination.

Among the bird species, probably the hummingbirds are the best examples of plant pollinators. Various types of flowers with bright, red colors, tubular shapes, and strong, sweet odors have evolved in tropical and temperate regions to take advantage of the hummingbirds' long beaks and tongues as an aid to pollination.

Because most mammals, such as small rodents and bats, do not detect colors as well as bees and butterflies do, flowers that use them as pollinators do not rely upon color cues in their petals but instead focus upon the production of strong, fermenting or fruit-like odors and abundant pollen rich in protein to attract them. In certain environments, bats and mice that are primarily nocturnal have replaced day-flying insects and birds in these important interactions between plants and animals.

EXPERIMENTS WITH PLANT-ANIMAL INTERACTION

Contemporary ecologists have gone beyond the purely descriptive observations of plant-animal interactions (initially within the realm of natural history) and have designed controlled experiments that are crucial to the development of such basic concepts as coevolution. For example, the use of radioactive isotopes and the marking of pollen with dye and fluorescent material in field settings have allowed ecologists to demonstrate precise distances and patterns of pollen dispersal. Ecologists and insect physiologists have cooperatively studied how certain insects, such as bees, are sensitive to ultraviolet light. When some flowers are viewed under ultraviolet light, distinct floral patterns become evident to guide these insects to nectar pollen sources.

Through basic research, Carolyn Dickerman reported in 1986 that animal color preferences vary throughout the season. Insect pollinators, who must feed every day, will adapt to these changes by shifting their foraging behavior. Research in the field has demonstrated that some species of flowers, such as the scarlet gilia, will produce differently colored flowers to accommodate shifts in pollinator species. Early in the growing season, this plant will produce long, red, tubular-shaped flowers to attract hummingbirds. As the hummingbirds migrate, the flowers will later become lighter in hue and be pollinated primarily by nocturnal hawkmoths.

In the laboratory, ecologists and biochemists have cooperatively analyzed the chemical composition of plant secretions and products. The chemical analysis of nectar indicates great variation in composition, correlating with the type of pollinator. Flowers pollinated by beetles generally have high amino acid content. The nectar associated with hummingbird-pollinated flowers is rich in sugar. Pollen also varies widely in chemical composition within plant species. Oils and waxes are major chemical products in the pollen of plants visited primarily by bees and flies. For bat-pollinated flowers, the protein content is quite high.

Research has also successfully analyzed how certain plants have been able to develop toxins as chemical defenses against animals. These protective devices include such poisons as nicotine and rotenone that help prevent insect and small mammal attacks. A more remarkable group of protective compounds recently isolated from some plants are known as juvocimines. These chemicals actually mimic juvenile insect hormones. Insect larvae feeding on leaves containing juvocimines are prevented from undergoing their normal development into functional, breeding adults. Thus, a specific insect population that could cause extensive plant damage is locally reduced.

Ecological interactions between plants and animals are diverse and varied. These plant-animal interactions can be viewed as absolute necessities for developing food chains and food webs and for maintaining the global balances of such important gases as oxygen and carbon dioxide. The interactions can also be very precise, limited, and crucial for determining species survival or extinction. By analyzing varied plant-animal interactions, from the microscopic level to the global perspective, one can more fully appreciate all the ecological relationships that exist on the earth.

THE ECOLOGY OF INTERACTION

The ecological importance of plant-animal interactions cannot be stressed enough. Modern-day agriculture owes its existence to the activities of such insect pollinators as honeybees in regard to the production of domestic fruits, vegetables, and honey. It is becoming increasingly evident to many ecologists and forestry scientists how important certain bird species, such as blue jays and cedar waxwings, are in natural reforestation of burned and blighted areas through their seed dispersal strategies. The plant horticulture and floral industries also are developing an appreciation of specific plant-animal interactions that produce more viable natural strains of flowers and ornamental shrubbery. The study of natural chemical defenses produced by some plants against animal invasions is most promising. The renewed interest in earlier efforts to extract such plant products as nicotine, rotenone, pyrethrum, and caffeine may produce natural compounds that can be effective insecticides without the long-term, environmental hazards associated with such human-made pesticides as malathion, chlordane, and dichloro-diphenyl-trichloroethane (DDT).

Finally, humankind is realizing that it is important to understand and protect certain plant-animal interactions associated with the tropical regions of the earth; otherwise, the global balance of oxygen and carbon dioxide could be seriously disrupted. Also, these tropical areas represent the last natural environments for the continuation of important plant species that produce secretions and products that have favorable medicinal qualities for humans and domestic livestock. By maintaining these populations and understanding how certain animals interact with them, humans can be guaranteed a viable supply of beneficial plant species whose medicinal values can be duplicated within the laboratory.

—*Thomas C. Moon*

FURTHER READING

Abrahamson, Warren G., ed. *Plant-Animal Interactions.* McGraw-Hill, 1989.

Barth, Frederick G. *Insects and Flowers: The Biology of a Partnership.* Princeton University Press, 1991.

Dickerman, Carolyn. "Pollination: Strategies for Survival." *Ward's Natural Science Bulletin,* Summer, 1986, 1–4.

Howe, Henry F., and Lynne C. Westley. *Ecological Relationships of Plants and Animals.* Oxford University Press, 1990.

John, D. M., S. J. Hawkins, and J. H. Price, eds. *Plant-Animal Interactions in the Marine Benthos.* Clarendon Press, 1992.

Lanner, Ronald M. *Made for Each Other: A Symbiosis of Birds and Pines.* Oxford University Press, 1996.

Meeuse, Bastian, and Sean Morris. *The Sex Life of Flowers.* Facts on File, 1984.

Norstag, Knut, and Andrew J. Meyerriecks. *Biology.* 2nd ed. Charles E. Merrill, 1985.

Rudman, William B. "Solar-Powered Animals." *Natural History* 96 (October 1987): 50–53.

Poisonous Animals

Fields of Study

Herpetology; Entomology; Ichthyology; Evolutionary Biology

Abstract

Substances that cause disease symptoms, injure tissues, or disrupt life processes on entering the body are poisons. When ingested in large quantities, most poisons kill. Poisons may be contacted from minerals, in vegetable foods, or in animal attack. Any poison of animal origin is a venom. Venoms are delivered by biting, stinging, or other body contacts. The mechanism for development of the ability to make venom is not clear.

Key Concepts

anaphylaxis: hypersensitivity to a foreign substance, such as a venom, that causes discomfort and can even kill
arachnid: an arthropod having eight legs; a spider
arthropod: an organism with a horny, segmented external covering and jointed limbs
hemotoxin: a substance that causes blood vessel damage and hemorrhage
neurotoxin: a substance that damages the nervous system, most often nerves that control breathing and heart action
venom: a poison made by an animal

WARNING! HERE IS FOUND POISON

Substances that cause disease symptoms, injure tissues, or disrupt life processes on entering the body are poisons. When ingested in large quantities, most poisons kill. Poisons may be contacted from minerals, in vegetable foods, or in animal attack. Any poison of animal origin is a venom. Venoms are delivered by biting, stinging, or other body contacts. These animal poisons are used to capture prey or in self-defense. Often, it seems that ability to make venom arose in animals that were too small, too slow, or too weak to otherwise maintain an ecological niche. The mechanism for development of the ability to make venom is not clear.

The most familiar poisonous animals are snakes, insects, spiders, and some other arachnids. Poisonous species, however, occur throughout the animal kingdom, including a few mammals and lizards, and some fish. The severity of venom effects depends on its chemical nature, the nature of the contact mechanisms, the amount of venom delivered, and victim size. For example, all spiders are poisonous. However, their venom is usually dispensed in small amounts that do not affect humans. Hence, few spiders kill humans, though they kill prey and use venom in self-defense very effectively.

Chemically, venoms vary greatly. Snake venoms are mixtures of enzymes and toxins. Study of their effects led to the identification of hemotoxins, which cause blood vessel damage and hemorrhage; neurotoxins, which paralyze nerves controlling heart action and respiration; and clotting agents, which excessively promote or prevent blood clotting. Cobras, coral snakes, and arachnids all have neurotoxic venoms.

POISONOUS LIZARDS, ARACHNIDS, AND INSECTS

The poisonous lizards are useful to explore first, because only two species are known: Gila monsters and beaded lizards (both holoderms). They inhabit the southwestern deserts of the United States and Mexico. They do not strike like snakes; rather, they bite, hold

The venom of the black mamba (Dendroaspis polylepis) is the most rapid-acting venom of any snake species and consists mainly of highly potent neurotoxins; it also contains cardiotoxins, fasciculins, and calciseptine. (Bill Love/Blue Chameleon Ventures)

501

on, and chew to poison. Holoderm bites kill prey, but rarely kill humans. Beaded lizards grow to three feet long and Gila monsters grow to two feet long.

Most poisonous arthropods are spiders and scorpions. Both use venom to subdue and/or kill prey. As stated earlier, few spiders endanger humans because their venom is weak and is not injected in large quantities, but some species have very potent venom and harm or even kill humans. Best known of these are black widow spiders and brown recluse spiders. Though rarely lethal to humans, black widow bites cause cramps and paralysis. Brown recluse spider venom is more potent, but the brown recluse spider is less commonly encountered than the black widow spider.

There are approximately six hundred scorpion species, of sizes between one and ten inches. All have tail-end stingers. Large, tropical scorpions can kill humans, while American scorpions are smaller and less dangerous. Scorpions are more dangerous than spiders because they crawl into shoes and other places where their habitat overlaps that of humans.

Many insects, such as caterpillars, bees, wasps, hornets, and ants, use venom in self-defense or to paralyze prey to feed themselves or offspring. Caterpillars use poison spines for protection. Bees, wasps, hornets, and ants use stingers for the same purpose. The venom of insects also kills many organisms that seek to prey on them. Humans, however, are rarely killed by insect bites. Such bites are usually mildly to severely painful for a period from a few minutes to several days. However, severe anaphylaxis occurs in some cases, followed by death.

POISONOUS SNAKES

Poisonous snakes are colubrids, elapids, or vipers, depending on their anatomic characteristics. All have paired, hollow fangs in the front upper jaw. The fangs fold back against the upper palate when not used, and when a snake strikes they swing forward to inject a venom that attacks the victim's blood and tissues. The heads of poisonous snakes are scale-covered and triangular. Such snakes are found worldwide and among them are pit vipers, named for the pits on each side of the head that contain heat receptors. The pits detect warm-blooded prey, mostly rodents, in the dark. Pit vipers include rattlesnakes, moccasins, copperheads, fer-de-lance, and bushmasters.

The population and species of American and European poisonous snakes differ. In North America, twenty such snake types occur: elapid coral snakes and copperheads, sixteen rattler types, and cottonmouths (all vipers). Vipers are found everywhere but Alaska. Rattlers have the widest habitat, as shown by their abundance in the snake-rich Great Plains, Mississippi Valley, and southern Appalachia. In contrast, copperheads and cottonmouths are abundant in Appalachia and the Mississippi valley, respectively. Mexican poisonous snakes are divided into two ranges, the northern from the US-Mexican border to Mexico City, and the southern, south of Mexico City. In the north, snakes are mostly rattlers, as in the contiguous United States. Coral snakes and pit vipers are plentiful in the south. Most perilous are the five- to eight-foot fer-de-lance, whose venom kills many humans. In South America, all vipers but rattlers are tropical. Bushmasters, the largest South American vipers, and elapid coral snakes are nocturnal and rarely endanger humans. Tropical rattlers and lance-headed vipers, somewhat less nocturnal, kill many. Europe has few snakes, due to its cool climates and scarce suitable habitats. Its few vipers range almost to the Arctic Circle. Eastern Mediterranean regions hold most of the European vipers.

There are many poisonous snakes in Africa and Asia. North Africa, mostly desert, has few snakes. Central Africa's diverse poisonous snakes are colubrid, elapid, and viper types. Elapids include dangerous black mambas, twelve to fourteen feet long, and smaller cobras, which also occur in South Africa. Among diverse vipers, the most perilous are Gaboon vipers and puff adders. The Middle East, mostly desert, has few poisonous snakes. Southeast Asia has the most poisonous snakes in the world—elapids, colubrids, and vipers. This is due to snake habitats that range from semiarid areas to rain forests. The huge human population explains why this area has the world's highest incidence of snakebite and related death, due to vipers, cobras, elapids, and sea snakes. Vipers bite most often, but elapids cause a larger portion of deaths. The Far East snake population is complex and its snakebite incidence is also high. Its important poisonous snakes are pit vipers.

Australia and New Guinea have large numbers of poisonous snakes. Australia has 65% of the world's snakes, while New Guinea has 25%. Also, sea snakes occur offshore and in some rivers and lakes. However, these countries have few snakebite deaths, due to the small size and nocturnal nature of most of indigenous snakes.

POISONOUS FISH AND AMPHIBIANS

Venomous fish are dangerous to those who enter the oceans, especially fisherman who take them from their nets. The geographical distribution of these fish is like that of all other fish. The highest population density is in warm temperate or tropical waters. Numbers and varieties of poisonous fish decrease with proximity to the North and South Poles and they are most abundant in Indo-Pacific and West Indian waters.

A well-known group of poisonous fishes is the stingrays (dasyatids). They inhabit warm, shallow, sandy-to-muddy ocean waters. Dasyatids lurk almost completely buried, awaiting prey that they sting to death with barbed, venomous teeth in their tails. The tail poison is made in glands at the bases of the teeth. Small, freshwater dasyatids are found in South American rivers, such as the Amazon, hundreds of miles from the river mouths. Stingrays near Australia grow to fifteen-foot lengths. Emphasizing wide distribution of stingrays and their danger to humans is their mention in Aristotle's third century BCE writings, and the death of John Smith in 1608, killed by a stingray while exploring Chesapeake Bay.

Also well known are the venomous *Scorpaenidae* fish family, many members of which cause very painful stings. Zebra fish and stonefish are good examples. Both, like all scorpaenids, have sharp spines supporting dorsal fins. The spines, used in self-defense, have venom glands. The most deadly fish venom is that of the stonefish, which, when stepped on, can kill humans.

Poisonous animals which endanger by contact are exemplified by zebra fish and stonefish, just mentioned, and by poisonous frogs or toads. Most such frogs and toads, such as poison dart frogs, live in Africa and South America. They secrete poisons through the skin. Contact with the poisons causes effects which range from severe irritation to death in humans. The poisons frighten away or kill most predators that attempt eat them. The ecological function of poisonous animals is seen as helping to keep down the population of insects, rodents, arachnids, and small fishes. They thus contribute to maintaining the balance of nature. Poisonous land animals, such as scorpions and many poisonous snakes, are often nocturnal and add another dimension to pest control by nighttime predation.

—*Sanford S. Singer*

FURTHER READING

Aaseng, Nathan. *Poisonous Creatures*. Twenty-first Century Books, 1997.

Edström, Anders. *Venomous and Poisonous Animals*. Krieger, 1992.

Foster, Steven, and Roger Caras. *Venomous Animals and Poisonous Plants*. Houghton Mifflin, 1994.

Grice, Gordon D. *The Red Hourglass: Lives of the Predators*. Delacorte, 1998.

POLLUTION EFFECTS ON ANIMAL LIFE

FIELDS OF STUDY

Ecology; Environmental Studies; Environmental Sciences

ABSTRACT

During the last decade of the twentieth century, the environmental problems predicted by environmental scientists decades previously began to be aggravated in a variety of ways. As a result of pollution, decreases in biodiversity and the extinction of both plant and animal species have accelerated. The ignorance and inaction of ordinary citizens will lead to disastrous consequences for the environment, threatening humanity's very existence.

KEY CONCEPTS

acid rain: rainfall containing nitric or sulfuric acid, caused by the release of nitrogen oxide or sulfur dioxide into the air

chemical pollutants: harmful chemicals manufactured and released into the environment, where they frequently contaminate ecosystems

chlorofluorocarbons (CFCs): a group of very stable compounds used widely since their development in 1928 that, once risen into the stratosphere, cause ozone depletion

greenhouse effect: the process by which certain gases, such as carbon dioxide and methane, trap sunlight energy in the atmosphere as heat, resulting in global warming as more gases are released to the atmosphere by human activities

ozone layer: the ozone-enriched layer of the upper atmosphere that filters out some of the sun's ultraviolet radiation, which causes skin and other types of cancer

THE TICKING TIME BOMB

During the last decade of the twentieth century, the environmental problems predicted by environmental scientists decades previously began to be aggravated in a variety of ways. These included population explosion, food imbalances, inflation brought about by scarcity of energy resources, acid rain, toxic and hazardous wastes, water shortages, major soil erosion, the depletion of the ozone layer, and greenhouse effects. As a result of pollution, decreases in biodiversity and the extinction of both plant and animal species have accelerated. The burning and cutting of thousands of square miles of rain forest not only destroyed habitats for numerous animal species but also caused irreversible damage to ecosystems and climate. The recurring drought and famine in Africa are testimony to the negative effects of human activity.

The well-being of animals as well as humans will not be protected against the ecological consequences of human actions by remaining ignorant of those actions. In order to take effective measures to reduce pollution and protect natural resources and the environment, people must first recognize these problems. The ignorance and inaction of ordinary citizens will lead to disastrous consequences for the environment, threatening humanity's very existence.

SOURCES AND TYPES OF POLLUTION

Industrialization and the expansion of the human population have left no places on Earth undisturbed. In simple terms, human interference in natural ecosystems is the single most important source of pollution. First, heavy dependence on fossil fuels for energy, and on synthetic chemicals and materials, helped to dump millions of metric tons of nonnatural compounds and chemicals into the environment. Among such chemicals and compounds are fertilizers, insecticides, fungicides, and herbicides, and home products. Application of excess chemical

Plastic pollution covering Accra beach in Ghana, 2018. (Muntaka Chasant)

fertilizers to soil hampers natural cycling of nutrients, depletes the soil's own fertility, and destroys the habitat for thousands of small animals residing in the soil. Farm runoff carries priceless topsoil, expensive fertilizer, and animal manure into rivers and lakes, where these potential resources become pollutants. In the city, water pours from sidewalks, rooftops, and streets, picking up soot, silt, oil, heavy metals, and garbage. It races down gutters into storm sewers, and a weakly toxic soup gushes into the nearest stream or river. Many of these chemicals also seep into the ground, contaminating groundwater.

Plants and factories manufacturing these chemical products are another source of pollutants and contamination. Burning fossil fuels releases greenhouse gases, carbon oxides, and methane. Coupled with deforestation in many regions of the world, carbon dioxide concentration in the atmosphere has steadily climbed, from 290 parts per million in 1860 to more than 395 parts per million (ppm) in 2013, more than a 36 percent rise due to industrialization. That same year, atmospheric carbon dioxide surpassed 400 ppm for the first time in Earth's history. In 2016, the average concentration of carbon dioxide was 402.9 ppm, and in May 2017, it reached 409 ppm.. The resultant global warming will have far-reaching effects on plants, animals, and humans in ways still not understood.

Acid rain, a result of overcharging the atmosphere with nitric oxides and sulfur dioxide (two gases also released by the burning of fossil fuels), has increased the acidity of soil and lakes to levels many organisms cannot survive. In 1991, the US National Acidic Precipitation Assessment Program reported that 5% of New England lakes were acidic, some to such levels that they were no longer able to support fish. This happens when much of the food web that sustains the fish is destroyed. Clams, snails, crayfish, and insect larvae die first, then amphibians, and finally fish. The detrimental effect is not limited to aquatic animals. The loss of insects and their larvae and small aquatic animals has contributed to a dramatic decline in the population of black ducks that feed on them. The result is a crystal-clear lake, beautiful but dead.

Another serious problem created by the chemical industry is ozone depletion. Chlorofluorocarbon (CFC) compounds contain chlorine, fluorine, and carbon. Since their development in the 1930s, these compounds were widely used as coolants in refrigerators and air conditioners, aerosol-spray propellants, agents for producing Styrofoam, and cleansers for electronic parts. These chemicals are very stable and for decades were considered to be safe. Their stability, however, turned out to be a real problem. In gaseous form, they rise into the atmosphere, where the high energy level of ultraviolet (UV) light breaks them down, releasing chlorine atoms, which in turn catalyzes the breakdown of ozone to oxygen gas. As a result of the decline of ozone and the depletion of the ozone layer, UV radiation rose by an average of 8 to 10 percent per decade between 1979 and 1992. This depletion of the ozone layer poses a threat to humans, animals, plants, and even microorganisms.

Pollution in the form of plastic waste harms marine habitats and wildlife, such as when marine animals eat or become tangled in such debris. Plastic waste makes up 60 to 80 percent of marine debris, and the demand for plastic bags, bottles, and containers is growing. According to PlasticsEurope, the global production of plastics reached 322 million tons in 2015, a significant increase from 230 million tons a decade earlier. And, according to a report released by the Ellen MacArthur Foundation in January 2016, plastic is produced at twenty times the rate that it was in 1964. Of the plastic produced, 40% winds up in landfills and only 5% are recycled effectively.

POLLUTION EFFECTS OF CHEMICALS

The degradation of air, land, and water as a result of the release of chemical and biological wastes has wide-ranging effects on animals. On a large scale, pollution destroys habitats and leads to population crashes and even the extinction of species. Hazardous chemicals introduced into an environment sometime render it unfit for life, as in Love Canal, New York, and Times Beach, Missouri. At the individual level, pollution causes abnormalities in growth, development, and reproduction. Hazardous chemicals, introduced either intentionally (such as fertilizers, herbicides, and pesticides) or through neglect (such as industrial wastes), have a variety of detrimental, sometimes devastating effects on animals. They affect the metabolism, growth and development, reproduction, and average life span of many species.

In the 1940s, the new insecticide dichlorodiphenyltrichloroethane (DDT) was regarded as a miracle. It saved millions of lives in the tropics by killing

505

the mosquitoes that spread deadly malaria, and it saved millions more lives by destroying insect pests that fed on crops, thus increasing crop yields. This miraculous pesticide, however, turned out to be a long-lasting nemesis to the environment and many species of wildlife. In the mid-1950s, the World Health Organization (WHO) used DDT on the island of Borneo to control malaria. DDT entered food webs through a caterpillar. Wasps that fed on the caterpillar were destroyed, and then gecko lizards that ate the poisoned insects accumulated high levels of DDT in their bodies. Both geckos and the village cats that ate the geckos died of DDT poisoning. As the cats died, the local rat population exploded. The village was then threatened with an

outbreak of plague, carried by the uncontrolled rats. Meanwhile, in the United States, ecologists and wildlife biologists during the 1950s and 1960s witnessed a stunning decline in the populations of several predatory birds, especially fish eaters such as bald eagles, cormorants, ospreys, and brown pelicans. The population decline drove the brown pelican and bald eagle close to extinction. In 1973, the US Congress passed the Endangered Species Act, which banned the use of DDT, and the once-threatened species have since somewhat recovered.

Although DDT has been banned in much of the world, there is a growing concern over the effects of a number of similar compounds known as xenoestrogens or environmental estrogens. These compounds interfere with normal sex-hormone functions by mimicking the effects of the hormone estrogen or enhancing estrogen's potency. High levels of xenoestrogens, such as dioxin and polychlorinated biphenyls (PCBs), in the Great Lakes have led to a sharp decline in populations of river otters and a variety of fish-eating birds, including the newly returned bald eagles. These chemicals are also the cause of deformed offspring or eggs that never hatch. In 1980, a spill of dichlorodiphenyldichloroethylene (DDE), a common by-product of DDT, in Florida's Lake Apopka led to a 90% drop in the birth rate of the lake's alligators.

EFFECTS OF AIR POLLUTION

Air pollution leads to acid rain and the greenhouse effect, as well as damage to the ozone layer. Acid rain falls in areas that are great distances from the source of the acids, destroying forests and lakes in sensitive regions and causing fish populations to dwindle or disappear entirely. Data collected from the Adirondack lakes and Nova Scotia rivers since the 1980s clearly show declines in acid-sensitive species. Similar results were obtained by analyzing fish population and water acidity in Maine, Massachusetts, Pennsylvania, and Vermont. The effects of acid rain on other animals are indirect, either through the dwindling fish population, thus eliminating a food source, or through stunted forest growth, thus disturbing other animals' habitats.

The effect of global warming on the animal kingdom is also a serious and complex issue. As global temperature rises, ice caps in polar regions and glaciers melt, ocean waters expand in response to atmospheric warming, and the sea level elevates. The expected sea-level rise will flood coastal cities and coastal wetlands. These threatened ecosystems are habitats and breeding grounds for numerous species of birds, fish, shrimp, and crabs whose populations could be severely diminished. The Florida Everglades will virtually disappear if the sea level rises two feet. The impact of global warming on forests could be profound. The distribution of tree species is exquisitely sensitive to average annual temperature, and small changes could dramatically alter the extent and species composition of forests, and thus the population distribution of animals. A rise may make temperatures unsuitable for some species, hence reducing biodiversity. In addition, an increase in the average temperature enhances the severity and length of dry spells, which inevitably increases the incidence of wildfires.

The effect of the depleted ozone layer on animals is yet to be fully understood. It is known that the high energy level of ultraviolet (UV) radiation can damage biological molecules, including the genetic material deoxyribonucleic acid (DNA), causing mutation. In small quantities, UV light helps the skin of humans and many animals produce vitamin D, and causes tanning. However, in large doses, UV light causes sunburn and premature aging of skin, skin cancer, and cataracts, a condition in which the lens of the eye becomes cloudy. Due to UV radiation's ability to penetrate, even animals covered by hair and thick fur cannot escape from these detrimental effects. UV damage caused by depleted ozone costs US farmers billions of dollars annually in reduced crop yields. All who depend on forestry and agriculture may bear a much higher cost if the emission of pollutants that destroy ozone is not regulated soon.

POSSIBLE REMEDIES

The various types of pollution all have serious effects on the plant and animal species that share this planet. It is all too easy to document the impacts of pollution on human health and ignore its effects on the rest of the living world. Any possible remedies to alleviate these problems should start with education at an individual as well as a global level. The tasks seem to be insurmountable, and no organization or country can do it alone. It takes willingness to accept short-term inconvenience or economic sacrifice for the benefit of the long run.

Synthetic chemical pollutants that are poisoning both people and wildlife could be largely eliminated without disrupting the economy, as reported in a study published in 2000 by the Worldwatch Institute, a Washington, DC–based environmental organization. The report presents strong evidence from three sectors that are major sources of these pollutants—paper manufacturing, pesticides, and polyvinyl chloride (PVC) plastics—to show that nontoxic options are available at competitive prices. Agricultural pollution can be mitigated, significantly reduced, or virtually eliminated through the use of proper regulation and economic incentives. Farmers from Indonesia to Kenya are learning how to use less of various chemicals while boosting yields. Since 1998, all farmers in China's Yunnan Province have eliminated their use of fungicides while doubling rice yields by planting more diverse varieties of the grain.

Other efforts have already led to positive changes. In 1989, the US government established the Acid Rain Program in an effort to reduce sulfur dioxide and nitrogen oxide emissions. By 2007, emissions of sulfur dioxide had been reduced by 40 percent. Also in 1989, the Montreal Protocol went into effect; this international treaty, which as of 2012 has been signed by 197 nations, calls for the complete elimination of CFCs.

In most, if not all, cases, the question is not whether it is possible to alleviate pollution of the environment; rather, it is whether we realize the urgency and are willing to sacrifice to do it. For the common well-being of generations to come, better approaches have to be taken to preserve the environment and biodiversity.

—*Ming Y. Zheng*

FURTHER READING

Bondareff, Joan M., Maggie Carey, and Carleen Lyden-Kluss. "Plastics in the Ocean: The Environmental Plague of Our Time." *Roger Williams University Law Review*, vol. 22, no. 2, 2017, pp. 360–383. *Legal Source.* Accessed 5 Feb. 2018.

Brown, Lester R. *State of the World, 2000.* Norton, 2000.

Elliott, John E., Christine A. Bishop, and Christy A. Morrissey, eds. *Wildlife Ecotoxicology: Forensic Approaches.* Springer, 2011.

"Graphic: The Relentless Rise of Carbon Dioxide." *Global Climate Change*, NASA, 26 Jan. 2018, climate.nasa.gov/climate_resources/24/. Accessed 5 Feb. 2018.

Hill, Julia Butterfly. *The Legacy of Luna: The Story of a Tree, a Woman, and the Struggle to Save the Redwoods.* Harper, 2000.

Johnson, Arthur H. "Acid Deposition: Trends, Relationships, and Effects." *Environment* 28.4 (1986): 6+.

Kendall, Ronald J., et al., eds. *Wildlife Toxicology: Emerging Contaminant and Biodiversity Issues.* CRC, 2010.

Lindsey, Rebecca. "Climate Change: Atmospheric Carbon Dioxide." *Climate.gov*, National Oceanic and Atmospheric Administration, 17 Oct. 2017, www.climate.gov/news-features/understanding-climate/climate-change-atmospheric-carbon-dioxide. Accessed 5 Feb. 2018.

Lippmann, Morton, ed. *Environmental Toxicants: Human Exposures and Their Health Effects.* 3rd ed. Wiley, 2009.

Sampat, Payal. *Deep Trouble: The Hidden Threat of Groundwater Pollution.* Worldwatch, 2000.

Sindermann, Carl J. *Coastal Pollution: Effects on Living Resources and Humans.* Taylor, 2006.

POPULATION ANALYSIS

FIELDS OF STUDY

Statistics; Population Genetics; Ecology

ABSTRACT

A population is a group of organisms belonging to the same species that occur together in the same time and place. Populations can change over time. They increase or decrease in size, and their change in size can depend on a wide variety of factors. Population analysis is the study of biological populations, with the specific intent of understanding which factors are most important in determining population size.

KEY CONCEPTS

continuous growth: growth in a population in which reproduction takes place at any time during the year rather than during specific time intervals

density-dependent growth: growth in a population in which the per capita rates of birth and death are scaled by the total number of individuals in the population

discrete growth: growth in a population that undergoes reproduction at specific time intervals

population analysis: the study of factors that influence growth of biological populations

DETERMINING POPULATION SIZE

A population is a group of organisms belonging to the same species that occur together in the same time and place. For example, a wildlife biologist might be interested in studying the population of porcupines that inhabits a hemlock forest or the population of bark beetles that lives on a particular tree. Populations can change over time. They increase or decrease in size, and their change in size can depend on a wide variety of factors. Population analysis is the study of biological populations, with the specific intent of understanding which factors are most important in determining population size.

In order to conduct a population analysis, one must first determine whether the population of interest is best understood as discrete or continuous. A discrete population is one in which important events such as birth and death happen during specific intervals of time. A continuous population is one in which births, deaths, and other events take place continuously through time. Many discrete populations are those with nonoverlapping generations. For example, in many insect populations, the adults mate and lay eggs, after which the adults die. When the juveniles achieve adulthood, their parental generation is no longer living. In contrast, most continuous populations have overlapping generations. For instance, in antelope jackrabbits (*Lepus alleni*), females may give birth at any time during the year, and members of several generations occur together in space and time.

MODELING ANIMAL POPULATIONS

The dynamics of animal populations is affected by a wide variety of demographic factors, including the population birth rate, death rate, sex ratio, age structure, and rates of immigration and emigration. In order to understand the effects of these factors on a population, biologists use population models. A model is an abstract representation of a concrete idea. The representation created by the model boils the concrete idea into a few critical components. By building and examining population models, population analysts investigate the relative importance of different factors on the dynamics of a given population.

A basic mathematical model of population size is as follows:

$$N_{t+1} = N_t + B - D + I - E \quad \text{(equation 1)},$$

where N_{t+1} equals the population size after one time interval, N_t equals the total number of individuals in the population at the initial time, B equals the number of births, D equals the number of deaths, I equals the number of immigrants into the population, and E equals the number of emigrants leaving the population. This simple model boils population size down to just four factors, B, D, I, and E. This model is not meant to be a true or precise representation of the population; rather, it is meant to clarify the importance of the factors of birth, death, immigration, and emigration on population size. To use the same model to examine the rate of growth of a population through time, it can be rearranged as follows:

$$N_{t+1} - N_t = B - D + I - E \quad \text{(equation 2)}.$$

That is, the increase or decrease in the population size between time intervals t and t+1 is based on the number of births, deaths, immigrants, and emigrants.

When population biologists choose to focus specifically on the importance of birth and death in population dynamics, population models are simplified by temporarily ignoring the effects of immigration and emigration. In this case, the degree of change in the population between time intervals t and t+1 becomes:

$$N_{t+1} - N_t = B - D \quad \text{(equation 3)}.$$

It is usually safe to assume that the total number of births (B) and deaths (D) in a population is a function of the total number of individuals in the population at the time, N_t. For example, if there are only ten females in a population at time t, it would be impossible to have more than ten births in the population. More births and deaths are possible in larger populations. If B equals the total number of births in the population, then B is equal to the rate at which each individual in the population gives birth, times the total number of individuals in the population. Likewise, the total number of deaths, D, will be equal to the rate at which the individuals in the population might die times the total number of individuals in the population. In other words,

$$B = bN_t \text{ and } D = dN_t \quad \text{(equation 4)},$$

where b and d represent the per capita rates of birth and death, respectively.

Given this understanding of B and D, the original model becomes

$$N_{t+1} - N_t = (bN_t) - (dN_t) \text{ or } N_{t+1} - N_t = (b-d)N_t \quad \text{(equation 5)}.$$

It would be useful to find a variable which can represent per capita births and deaths at the same time. Biologists define r as the per capita rate of increase in a population, which is equal to the difference between per capita births and per capita deaths:

$$r = b - d \quad \text{(equation 6)}.$$

Thus, the equation which examines the changes in population size between time intervals t and $t+1$ becomes:

$$N_{t+1} - N_t = rN_t \quad \text{(equation 7)}.$$

A numerical example works as follows. In a population that originally had 1000 individuals, a per capita birth rate of 0.1 births per year and a per capita death rate of 0.04 deaths per year, the net change in the population size between the year t and $t+1$ would be:

$$r = 0.1 - 0.04 = 0.06$$
$$N_{t+1} - N_t = 0.06(1000) = 60$$

In other words, the population would increase by sixty individuals over the course of one year.

This model works for populations in which events take place during discrete units of time, such as a population of squirrels in which reproduction takes place at only two specific times in a single year. In contrast, many populations are continuously reproductive. That is, at any given time, any female in the population is capable of reproducing. When these conditions are met, time is viewed as being of a more fluid than discrete nature, and the population exhibits continuous growth. Models of population growth are slightly different when births and deaths are continuous rather than discrete. One way to imagine the difference between a population with continuous rather than discrete growth is to imagine a population in which each time interval is infinitesimally small. When these conditions are met, the model for population growth becomes:

$$\delta N/\delta t = rN \quad \text{(equation 8)},$$

where $\delta N/\delta t$ represents the changes in numbers in the population over very short time intervals. The per capita rate of increase (r) can now also be called the instantaneous rate of increase because the population is one with minute time intervals.

How does a population biologist select the best model? Which model is best depends on exactly what it is that a scientist is trying to understand about a population. In the first model presented above (equation 1), the different effects of birth, death, immigration, and emigration can be compared relative to one another. In the second model, the effects of immigration and emigration are ignored and the effects of birth and death are summarized into one constant called the per capita rate of increase (equations 7 and 8). If the scientist is trying to understand the cumulative effects of B, D, I, and E on the population, then equation 1 would represent a good model. On the other hand, if the scientist is trying to understand how births and deaths influence the net changes in population size, equation 7 or 8 would be a better model.

EFFECTS OF DENSITY ON POPULATION GROWTH

When dealing with a continuous rather than a discrete population, equation 8 represents the rate of population growth as a function of per capita births and per capita deaths in the population. Equation 8 represents a population that is growing exponentially without bound. In other words, regardless of the population size at any given time, the per capita rate of increase remains the same. It would be reasonable to assume that per capita rates of increase can actually change with changes in overall population size. For example, in a population of bark beetles inhabiting the trunk of a tree, many more resources are available to individual beetles when the population is small. Resources must be shared between more and more individuals as the population size increases, which can result in changes to the per capita rate of increase. A model of population growth that incorporates the effect of overall population density on the per capita rate of increase might look like:

$$\delta N/\delta t = r(1-N/K)N \quad \text{(equation 9)},$$

where K is equal to the carrying capacity, the maximum number of individuals in the population that there are adequate resources to support. The per capita rate of increase in equation 9 is not simply r by itself, but becomes $r(1-N/K)$. The per capita rate of increase is a function of rates of birth and death scaled by the population size and the carrying capacity of the habitat. If the population is very large relative to the number of individuals that the habitat can support, then $N \approx K$, and the expression $(1-N/K)$ becomes approximately equal to 0. When so, equation 9 takes the form

$$\delta N/\delta t = r(0)N = 0 \quad \text{(equation 10)}$$

and the rate of population growth is zero. In other words, the population has ceased growing. On the other hand, if the population is very small relative to the number of individuals the habitat can support, then $N <\!\!< K$ and the expression $(1-N/K)$ becomes approximately equal to 1. When so, equation 9 takes the form

$$\delta N/\delta t = r(1)N = rN \quad \text{(equation 11)}$$

and the rate of population growth remains a function of the rates of birth and death, but not the population size or carrying capacity. Thus, equation 9 represents what is called density-dependent growth.

EFFECTS OF SEX RATIO AND AGE STRUCTURE ON POPULATION GROWTH

The model set forth in equation 9 only takes into account the ways in which births, deaths, and population density relative to carrying capacity influence population growth. Sometimes it is helpful to understand how other factors such as the sex ratio and age structure in a population influence rates of growth. For example, deer hunters are not always allowed to take equal numbers of bucks and does from a population. Similarly, fishermen are often restricted in the size of fish they are allowed to keep when fishing. These wildlife management restrictions on the sex and size of animals that can be hunted arise from the fact that both age and sex can influence population growth rates. Models that incorporate the effects of age structure and population sex ratios will not be covered here. Suffice it to say that a population that consists mostly of young individuals yet to reproduce will grow more quickly than a population equal in size but consisting of mostly older individuals who have finished reproducing. Similarly, a population with a highly skewed sex ratio that has many more males than females will not grow as quickly as a population of equal size in which the numbers of males and females are equal.

Population analysis is the study of biological populations, with the specific intent of understanding which factors are most important in determining population size. Factors such as the per capita rates of birth and death, the population density, age structure, and sex ratio all contribute to population size. Understanding how these factors interact to influence population size is critical if biologists hope to manage populations of organisms at sustainable levels for hunting or fishing and if conservation biologists hope to prevent populations from going extinct.

—*Erika L. Barthelmess*

FURTHER READING

Gardali, Thomas, et al. "Demography of a Declining Population of Warbling Vireos in Coastal California." *The Condor* 102, no. 3 (August 2000): 601–609.

Hastings, Alan. *Population Biology: Concepts and Models.* Springer-Verlag, 1997.

Hedrick, Philip W. *Population Biology: The Evolution and Ecology of Populations.* Jones and Bartlett, 1984.

Johnson, Douglas H. "Population Analysis." In *Research and Management Techniques for Wildlife and Habitats*, edited by Theodore A. Bookhout. 5th ed. The Wildlife Society, 1994.

Population Genetics

Fields of Study

Classical Genetics; Developmental Genetics; Molecular Genetics

Abstract

Population genetics is the description and analysis of genetic traits and their causative genes at the population level. Population genetics uses information from classical, developmental and molecular genetics, and helps explain why populations are so variable, why some harmful traits are common, why most animals and plants reproduce sexually, how evolution works, why some animals are altruistic in a cutthroat world, and how new species arise.

Key Concepts

deoxyribonucleic acid (DNA): the chemical basis of genes
dominant: requiring only one copy of a gene for expression of the trait
gene: the unit of heredity; a short stretch of DNA encoding a specific product, usually protein
genotype: the gene makeup of an individual
inbreeding: the mating of individuals more closely related than the population average
meiosis: the two cell divisions leading to egg or sperm, during which the genes from the two parents are mixed
mutation: a sudden, unpredictable change in a gene
phenotype: a trait or combination of traits, the result of the genotype and environment
random genetic drift: the random change of gene frequencies because of chance, especially in small populations
recessive: requiring two copies of a gene for the trait to be expressed
selection: differential survival and reproduction rates of different genotypes

What Population Genetics Does

Population genetics is the description and analysis of genetic traits and their causative genes at the population level. Classical genetics deals with the rules of genetic transmission from parents to offspring, developmental genetics deals with the role of genes in development, and molecular genetics looks at the molecular basis of genetic phenomena. Population genetics uses information from all three fields and helps explain why populations are so variable, why some harmful traits are common, why most animals and plants reproduce sexually, how evolution works, why some animals are altruistic in a cutthroat world, and how new species arise.

The Sources of Variability

Simple observation demondtrates that animals are highly variable. Some dogs are big, others small; some wiry, others big boned; some long haired, others curly; some with special talents such as herding or retrieving, others with none; and some with diseases or defects, others normal. All of these are the results of various genes combined with environmental influences. Unless an animal is an identical twin, no one else shares that individual's genotype and no one ever will. Population genetics looks at variability in a population and examines its sources and the forces that maintain it.

Variability can come from genetic mutations. For example, about one child in ten thousand is born with dominant achondroplasia (short-legged dwarfism). Some children with the trait inherit the condition from an affected parent, but most have normal parents. They are therefore the result of a new mutation. Many mutations are deleterious and are eventually eliminated from the population by the lowered survival or fertility rates of those who have the mutation, but while they remain in the population, they add to its variability.

Occasionally, a seemingly harmful mutation persists, for example, the gene that causes sickle-cell anemia, a severe disease characterized by red blood cells that become sickle shaped in certain laboratory tests. The causative gene is recessive, meaning that two copies are needed to produce the anemia, but the disease is very common in some parts of Africa. The harmful, anemia-causing gene persists in the population because if a person has only one gene with the trait rather than two, that gene confers resistance to malaria, the major cause of debility in that part of the world.

Although the genes in these two examples have large and conspicuous effects, the great majority of mutations and the great bulk of genetic variability in the population are the result of a large number of genes with individually small effects, often detected only through statistics. The variability of quantitative traits such as size is due mainly to the cumulative action of many individual genes, each of which produces its small effect. The average size stays roughly constant from generation to generation because individuals who are too large or too small are at a disadvantage. However, such individuals continuously arise from new mutations.

The driving force in evolution is natural selection, that is the differential survival and fertility of different genotypes. New mutations occur continuously. Most of these are harmful, although usually only mildly so, but a small minority are beneficial. The rules of Mendelian inheritance ensure that the genes are thoroughly scrambled every generation. Natural selection acts like a sieve, retaining those genes that produce favorable phenotypes in the various combinations and rejecting others. Such a process, acting over eons of time, has produced the variety and specific adaptations that can be found throughout the animal kingdom.

THE FORCES AGAINST CHANGE

Although evolutionary progress is the result of natural selection, most selection does not accomplish any systematic change. Most selection is directed at maintaining the status quo—eliminating harmful mutations, keeping up with transitory changes in the environment, and eliminating statistical outliers (extremes of variation). Most of the time, evolutionary change is very slow.

In most populations, mating is essentially random in that mates do not choose each other because of the genes they carry. There are exceptions, of course, but for the most part, random mating can be assumed. This permits a great simplification known as the Hardy-Weinberg rule. This rule says that if the proportion of a certain gene, say A, in the population is p and of another, say a, is q, then the three genotypes AA, Aa, and aa appear in the proportions p^2, $2pq$, and q^2, respectively. (Remember that p and q are fractions between zero and one.) This is a simple application of elementary probability and the binomial theorem. Furthermore, after a few generations of random mating, genotypes at different loci also equilibrate, which means that the frequency of a composite genotype is the product of the frequencies at the constituent loci. The reason that this is so useful is that the number of genotypes is enormous, but a population can be characterized by a much smaller number of gene frequencies.

Genotypes are transient, but genes may persist unchanged for many generations. This has led the great theorists of population genetics, J. B. S. Haldane and R. A. Fisher in England and Sewall Wright in the United States, to make the primary units the frequency of individual genes and develop theories around this concept, making free use of the simple consequences of random mating. Such a gene-centered view has been described by scholar Richard Dawkins as the "selfish gene." A population can be thought of as a collection of genes, each of which is maximizing its chance of being passed on to future generations. This causes the population to become better adapted because those genes that improve adaptation have the best chance of being perpetuated.

An extension of this notion is kin selection. The concept holds that, to the extent that behavior is determined by genes, individuals should be protective of close relatives because relatives share genes. The fact that brothers and sisters share half their genes should lead a brother to be half as concerned with his sister's survival and reproduction as with his own. Evolutionists believe that altruistic behavior in various animals, including humans, is the result of kin selection. The degree of self-sacrifice to protect a close relative is proportional to the fraction of shared genes. Parents regularly make sacrifices for their children, and this is what evolutionary theory would predict.

One way in which populations depart from random mating is inbreeding, the mating of individuals more closely related than if they were randomly chosen. Related individuals share one or more ancestors; hence an inbred individual may get two copies of an ancestral gene, one through each parent. In this way, inbreeding increases the proportion of homozygotes. Because many deleterious recessive genes are hidden in the population, inbreeding can have a harmful effect by making genes homozygous. Similarly, if the population is subdivided into local units, mating mostly within themselves, these local units will be more homozygous than if the entire population mated at random. Small subpopulations will be more subject to purely random fluctuations in gene frequencies known as random genetic drifts. Therefore, subdivisions of a population often differ significantly, particularly with respect to unimportant genes.

THE ARGUMENT AGAINST AVERAGE EFFECTS

The gene-centered view of evolution is not always accepted. Some evolutionists believe that it is simplistic to view an individual as a bag of genes, each trying to perpetuate itself. They emphasize that genes often interact in complicated ways, and that a theory that deals with only average gene effects is incomplete. Modern theories of evolution take such complications into account.

This different viewpoint has led to a major controversy in evolution, one that has not yet been settled. Wright emphasized that many well-adapted phenotypes depend on genes that interact in very specific ways; two or more genes may be individually harmful but when combined produce a beneficial effect. He argued that selecting genes on the basis of average effects cannot produce such combined effects. He believed that a population subdivided into many partially isolated units provides an opportunity for such interactions. An individual subpopulation, by random drift, might chance upon such a happy gene combination, in which case, the whole population can be upgraded by migrants from this subpopulation. Whether evolutionary advance results from gene interactions in subpopulations, from mass selection in largely unstructured populations, or from a combination of both is a question that remains unresolved.

Population genetics theory, along with the techniques of molecular genetics, has greatly deepened our knowledge of historical evolution. Everyone is familiar with tree diagrams of common ancestry that show, for example, birds and mammals branching off from early ancestors. In the past these had to be constructed using external phenotypes and fossils. These techniques for measuring the relatedness of different species and determining their ancestral relations have been replaced by DNA sequencing, which produces much surer results. It has long been suspected that genes can persist for very long evolutionary periods, changing slightly to perform new, often related but sometimes quite different functions. This belief has been confirmed repeatedly by molecular analysis. The similarity of the DNA sequences between some plant and animal genes is so great as to leave no doubt that they were both derived from a common ancestral gene a billion or more years ago.

NEUTRAL MUTATION AND THE BENEFITS OF SEXUAL REPRODUCTION

Most gene mutations have very small effects, and the smaller the effect, the less likely it is to be noticed. Molecular techniques have enabled scientists to detect changes in DNA without regard to the traits they cause or whether they have any effect at all. The Japanese geneticist Motoo Kimura has advanced the idea that most evolution at the DNA level is not the result of natural selection but simply the result of mutation and chance, a concept termed neutral mutation. In vertebrates, especially in mammals, most of the DNA has no known function. The functional genes make up a very small fraction of the total DNA. Many scientists believe that most DNA evolution outside the genes—and some within—is the result of changes that are so nearly neutral as to be determined by chance. How large a role random drift plays in the evolution of changes in functional proteins is still not certain.

A few animals and a large number of plants reproduce asexually. Instead of reproducing by using eggs and sperm, the progeny are carbon copies of the parent. Asexual reproduction has obvious advantages. If females could reproduce without males, producing only female offspring like themselves, reproduction would be twice as efficient. However, despite its inherent inefficiency, sexual reproduction is the rule, undoubtedly because of the gene-scrambling process that sex produces. The ability of a species to produce and try out countless gene combinations confers an evolutionary advantage that outweighs the cost of males. Another advantage of gene scrambling is that it permits harmful mutations to be eliminated from the population in groups rather than individually.

Population genetics is also concerned with the processes by which new species arise. Scientists believe that a population somehow becomes divided into two or more isolated groups, separated perhaps by a river, mountain range, or other geographical barrier. Each group then follows its own separate evolutionary course, and the groups' dissimilar environments accentuate their differences. Eventually so many differences between the two groups evolve that they are no longer compatible. The products of interspecies crosses, or hybrids, often do not develop normally or are sterile (like the mule). Sometimes the two species do not mate because they are so different.

THEORY, OBSERVATION, AND EXPERIMENT

Population genetics involves theory, observation, and experiment. Population genetics examines how genes are influenced by mutation, selection, population size, migration, and chance. Scientists develop mathematical models that embody these theories and compare the results obtained using the models with data from laboratory experiments or field observations. These genetic models have become more and more sophisticated to take into account complex gene interactions and increasingly realistic population structures. The models are further complicated by efforts to account for random processes. Often the mathematical geneticist relies on computers to perform complex analyses and computations.

One of the simpler models, which makes the assumption that mating is random, is the Hardy-Weinberg principle. If the proportion of gene A in the population is p and that of gene a is q, then the three genotypes AA, Aa, and aa appear in the proportions p^2, $2pq$, and q^2, respectively. The proportion of Aa is $2pq$ rather than simply pq because this genotype represents two combinations, maternal A with paternal a and paternal A with maternal a. This principle can be used to predict the frequency of persons with malaria resistance from the incidence of sickle-cell anemia. If one-tenth of the genes are sickle-cell genes and the other nine-tenths are normal, the frequency of two genes coming together to produce an anemic child is 0.1×0.1, or 0.01. The frequency of those resistant to malaria, who have one normal and one sickle-cell gene, is $2 \times 0.1 \times 0.9$, or 0.18. A slight extension of the calculation (using the rates of malaria infection and death from the disease) can be used to estimate the death rate from malaria. Another mathematical model can be formed based on the molecular genetics theory of neutral mutation.

A neutral mutation, because it is not influenced by natural selection, has an expected rate of evolution that is equal to the mutation rate. Mathematical models embodying this theory are used to quantitatively predict what will happen in an experiment or what an observational study will find and act as a test of the theory. Neutral mutation theory is quite complicated and requires advanced mathematics.

Observational population genetics consists of studying animals and plants in nature. Evolution rates are inferred from the fossil record. Field observations can determine the frequency of genes in different geographical areas or environments. The frequency of self- and cross-pollination can often be observed directly. Increasingly, DNA analysis, which can detect relationships or alterations that are not visible, is being used to support field observations. For example, molecular markers have been used to determine parentage and relationship. DNA analysis revealed that certain birds that do not reproduce but care for the progeny of others are in fact close relatives, consistent with kin selection theory.

Increasingly, population genetics has begun to rely on experimentation. Plants and animals can be used to study the process of selection, but to save time and reduce costs, most laboratory experiments involve small, rapidly reproducing organisms such as the fruit fly, Drosophila. Some of the most sensitive selection experiments have involved the use of a chemostat, a container in which a steady inflow of nutrients and steady outflow of wastes and excess population permit a population to maintain a stable number of rapidly growing organisms, usually bacteria. These permit very sensitive measurements of the effects of mutation. Evolutionary studies that would require eons if studied in large animals or even mice can be completed in a very short time.

EXPLAINING, QUANTIFYING, AND PREDICTING EVOLUTION

The greatest intellectual value of population genetics has been to provide a theory of evolution that is explanatory, quantitative, and predictive. Population genetics places knowledge of mutation, gene action, selection, inbreeding, and population structure in a unified framework. It brings together Charles Darwin's theory of evolution by natural selection, Gregor Mendel's laws of inheritance, and molecular genetics to create a coherent picture of how evolution took place and is still occurring.

Population genetics has provided explanations for variability in a population, the prevalence of sexual rather than asexual reproduction, the origin of new species, and behavioral traits such as altruism. It has also provided an understanding of why some harmful diseases are found in the population. Population genetics has been used in animal and plant breeding to create rational selection programs. Using quantitative models, the results of various selection schemes can be compared and the best one chosen.

A particularly telling example of a situation in which population genetics predicted an outcome that has become painfully obvious is the development of resistance to insecticides, herbicides, and antibiotics. As people used these products more and more, the insects, weeds, and bacteria they were trying to eliminate developed resistance, and new products had to be developed to replace those rendered ineffective. The development of resistance represents evolution by natural selection that took place not over hundreds or thousands of years but in just a few years. Probably the most problematic area of resistance is antibiotics because some treatable diseases are again threatening to move beyond the ability of medicine to cure. A major challenge to ecologists, microbiologists, physicians, and population geneticists is how to deal with the increasingly difficult problem of disease-producing microorganisms that are resistant to antibiotics.

—*James F. Crow*

Further Reading

Crow, James F. *Basic Concepts in Population, Quantitative, and Evolutionary Genetics.* W. H. Freeman, 1986.
Dawkins, Richard. *The Blind Watchmaker.* W. W. Norton, 1986.
_____. *The Selfish Gene.* Rev. ed. Oxford University Press, 1999.
Falconer, Douglas S. *Introduction to Quantitative Genetics.* 4th ed. John Wiley & Sons, 1996.
Fisher, R. A. *The Genetical Theory of Natural Selection.* Rev. ed. Dover, 1958.
Haldane, J. B. S. *The Causes of Evolution.* Reprint. Cornell University Press, 1993.
Hartl, Daniel. *A Primer of Population Genetics.* 3rd ed. Sinauer Associates, 2000.
Hartl, Daniel, and Andrew Clark. *Principles of Population Genetics.* 3rd ed. Sinauer Associates, 1997.
Kimura, Motoo. *The Neutral Theory of Molecular Evolution.* Cambridge University Press, 1985.
Maynard Smith, John. *The Theory of Evolution.* Cambridge University Press, 1993.
Wright, Sewall. *Evolution and the Genetics of Populations.* 4 vols. University of Chicago Press, 1968–1978.

Predation

Fields of Study

Ethology; Biosociology; Ecology

Abstract

Predation is an interaction between two organisms in which one of them, the predator, derives nutrition by killing and eating the other, the prey. Obvious examples include lions feeding on zebras and hawks eating rodents, but predation is not limited to interactions among animals.

Key Concepts

competition: the interaction among two or more organisms of the same or different species that results when they share a limited resource

food web: the sum total of the feeding relationships (links) between trophic levels in ecosystems

functional response: the rate at which an individual predator consumes prey, dependent upon the abundance of that prey in a habitat

mimicry: the resemblance of one species (the model) to one or more other species (mimics), such that a predator cannot distinguish among them

numerical response: the abundance of predators dependent upon the abundance of prey in a habitat

population regulation: stabilization of population size by factors such as predation and competition, the relative impact of which depends on abundance of the population in a habitat

trophic level: a level at which energy is acquired in a food web—the herbivore level obtains energy from plants; the carnivore level from herbivores and other carnivores

A BROADER VIEW OF PREDATORS AND PREY

Predation is an interaction between two organisms in which one of them, the predator, derives nutrition by killing and eating the other, the prey. Obvious examples include lions feeding on zebras and hawks eating rodents, but predation is not limited to interactions among animals. Birds that feed on seeds are legitimate predators since they are killing individual organisms (embryonic plants) to derive energy. There are a number of species of carnivorous plants, such as sundews and pitcher plants, that capture and consume small animals to obtain nitrogen in habitats lacking sufficient quantities of that nutrient. Most animals that feed on plants (herbivores) do not kill the entire plant and therefore are not really predators. Exceptions to this generalization are some insects that reach infestation levels, such as gypsy moths or locusts, and can kill the plants upon which they feed. The majority of herbivore-plant associations, however, are more properly described as parasite-host interactions in which the host plant may suffer damage but does not die.

There are special cases in which parasitism and predation may be combined. One of these is the interaction between parasitoid wasps and their hosts, usually flies. Adult female parasitoids attack and inject eggs into fly pupae (the resting stage, during which fly larvae metamorphose into adults), and the larvae of the wasp consume the fly. The adult parasitoid is therefore a parasite, while the larval wasp acts as a predator.

PREDATOR-PREY INTERACTIONS

Predator-prey interactions can be divided into two considerations: the effects of prey on predators, and the effects of predators on their prey. Predators respond to changes in prey density (the number of prey organisms in the habitat) in two principal ways. The first is called numerical response, which means that predators change their numbers in response to changes in prey density. This may be accomplished by increasing or decreasing reproduction or by immigrating to or emigrating from a habitat. If prey density increases, predators may immigrate from other habitats to take advantage of this increased resource, or those predators already present may produce more offspring. When prey density decreases, the opposite will occur. Some predators, which are known as fugitive species, are specialized at finding habitats with abundant prey, migrating to them, and reproducing rapidly once they are established.

Cape May warblers are good at finding high densities of spruce budworms (a serious pest of conifers) and then converting the energy from their prey into offspring. This strategy allows the birds to persist only because the budworms are never completely wiped out; they are better at dispersing to new habitats than are the birds.

The second response of predators to changing prey density is called functional response. The rate at which predators capture and consume prey depends upon the rate at which they encounter prey, which is a function of prey density. If the predator has a choice of several prey species, it may learn to prefer one of them. If that prey is sufficiently abundant, this situation results in a phenomenon known as switching, the concentration by the predator on the preferred prey. It may entail a change in searching behavior on the part of the predator, such that former prey items will no longer be encountered as frequently.

Animals have evolved a number of defense mechanisms that reduce their probability of being eaten by predators. Spines on horned lizards, threatening displays by harmless snakes, camouflage of many cryptic animals, toxic or distasteful chemicals in insects and amphibians, and simply rapid movement—all are adaptations that may have evolved in response to natural selection by predators. A predator that can learn to prefer one prey item over another is smart enough to learn to avoid less desirable prey. That capability is the basis for a phenomenon—known as aposematism—among potential prey species which are toxic and/or distasteful to their predators. Aposematic organisms advertise their toxicity by bright coloration, making it easy for predators to learn to avoid them, which in turn saves the prey population from frequent taste-testing. Many species of insects are aposematic. Monarch butterflies are bright orange with black stripes, an easy signal to recognize. They owe their toxicity to the milkweed plant, which they eat as caterpillars. The plant contains cardiac glycosides, which are very toxic. The monarch caterpillar is immune to the poison and stores it in its body so that the adult has a high concentration of it in its wings. If a bird grabs the butterfly in flight, it is likely to get a piece of wing first, and this will teach it not to try orange butterflies in the future.

Some potential prey species that are not themselves toxic have evolved to resemble those that are; these are called Batesian mimics. Viceroy butterflies, which are not toxic, mimic monarchs very closely, so that birds cannot tell them apart. One limit to Batesian mimicry is that mimics can never get very numerous, or their predators will not get a strong enough message to leave them alone. Another kind of mimicry involves mimics that are as toxic as their models. The advantage with this type, Müllerian mimicry, is that the predator has to learn only one coloration signal, which reduces risk for both prey populations. In this relationship, the mimic population does not have to remain at low levels relative to the model population. A third type of mimicry is more insidious—aggressive mimicry, in which a predator resembles a prey or the resource of that prey in order to lure it close enough to capture. There are tropical praying mantids that closely resemble orchid flowers, thus attracting the bees upon which they prey. There are some species of fireflies that eat other species of fireflies, using the flashing light signal of their prey to lure them within range.

THE CHOICE OF PREY

What determines predator preference for prey? Since prey are a source of energy for the predator, it might be expected that predators would simply attack the largest prey they could handle. To an extent, this choice holds true for many predators, but there is a cost to be considered. The cost involves the energy a predator must expend to search for, capture, handle, and consume prey. In order to be profitable, a prey item must yield much more energy than it costs. Natural selection should favor reduction in energetic cost relative to energetic gain, the basis for optimal foraging theory. According to this theory, many predators have evolved hunting strategies to optimize the time and energy spent in searching for and capturing prey.

Some predators, such as web-building spiders and boa constrictors, ambush their prey. The low energetic cost of sit-and-wait is an advantage in environments that provide plentiful prey. If encounters with prey become less predictably reliable, however, an ambush predator may experience starvation. Spiders can lower their metabolic energy requirements when prey is unavailable, whereas more mobile predators, such as boa constrictors, can simply shift to active searching. Probably because of the likelihood of facing starvation for extended periods of time, ambush predation is more common among animals that do not expend metabolic energy to regulate their body temperatures (ectotherms) than among those that do (endotherms). Some predators, such as wolves and lions, hunt in groups. This allows them to tackle larger (more profitable) prey than if they hunted alone. Solitary hunters generally have to hunt smaller prey.

Natural communities consist of food webs, constructed of links (feeding relationships) among trophic levels. Each prey species is linked to one or more predators. Most predators in nature are generalists with respect to their prey. Spiders, snakes, hawks, lions, and wolves all feed on a variety of prey. Some of these prey are herbivores, but some are themselves predators. Praying mantids eat grasshoppers (herbivores), but they also eat spiders (carnivores) and each other. Thus, generalist predators have a bitrophic niche, in that they occupy two trophic levels at the same time.

PREDATORY RELATIONSHIPS AND POPULATION FLUCTUATION

It is an open question whether predators and prey commonly regulate each other's numbers in nature. There are many examples of cyclic changes in abundance over time, in which an increase in prey density is followed by an increase in the numbers of predators, and then the availability of prey decreases, also followed by a decrease in predators. Are predators causing their prey to fluctuate, or are prey responding to some other environmental factor, such as their own food supply? In the second case, prey may be regulated by food, and in turn, may be regulating predators, but not the reverse.

Predators can sometimes determine the number of prey species that can coexist in a habitat. If a predator feeds on a prey species that could outcompete (competitively exclude) other prey species in a habitat, it may free more resources for those other species. This relationship is known as the keystone effect. Empirical studies have indicated that the number of prey species in some communities is directly related to the intensity of predation (numerical and functional responses of predators) such that at low intensity, few species coexist because of competitive exclusion; at intermediate intensity, the diversity of the prey community is greatest; and at high intensity, diversity decreases because overgrazing begins to eliminate species. This intermediate predation hypothesis depends upon competition among prey species, which is not always the case.

STUDYING PREDATION

The central question in the study of predation is: To what extend do predators and their prey regulate one another? Most studies suggest that predators are usually food limited, but the extent to which they regulate their prey is uncertain. It is one thing to observe predators in nature and another to assess their importance to the dynamics of natural communities. Like other aspects of ecology, studies of predation can be descriptive, experimental, and/or mathematical.

At the descriptive level, characteristics of both predator and prey populations are assessed: rates of birth and mortality, age structure, environmental requirements, and behavioral traits. Qualitative and quantitative information of this type is necessary before predictions can be made about the interactions between predator and prey populations. General lack of such information in natural ecosystems is largely responsible for failures at biological control of pests and management of exploited populations.

Experimental studies of predation involve manipulation of predator and/or prey populations. A powerful method of testing the importance of predation is to exclude a predator from portions of its accustomed habitat, leaving other portions intact as experimental controls. In one such experiment, excluding starfish from marine intertidal communities of sessile invertebrates resulted in domination by mussels and exclusion of barnacles and other attached species; in the absence of the predator, one prey species was capable of competitively excluding others. This keystone effect depends upon two factors—that the prey assemblage structure is determined by competition and that the predator preferentially feeds on the species that is the best competitor in the assemblage. Clearly, not all food webs are likely to be structured in this way.

Another method of experimental manipulation is to enhance the numbers of predators in a community. For complex natural communities, both additions and exclusions of predators have revealed direct (depression of prey) and indirect (enhancement) effects. Since generalist predators are bitrophic in nature, they may interact with other carnivores in such a way as to enhance the survival of herbivores that normally would fall victim. In one experiment, adding praying mantids to an insect community resulted in a decrease in spiders and a consequent increase in aphids, normally eaten by these spiders. Such results are not uncommon and contribute to the uncertainty of prediction. Mathematical models have been constructed to depict predator-prey interactions in terms of how each population affects the growth of the other. The simplest of these models, known as the Lotka-Volterra model for the mathematicians who developed it, describes a situation in which prey and predator populations are assumed to be mutually regulating. This model, which was developed for a single prey population and single predator species, has been modified by many workers to provide more realism, but it is far from predicting many competitive situations in complex natural communities.

As with the rest of modern ecology, these different approaches must be blended in order to build a robust picture of how important predators are in natural ecosystems. This knowledge would allow more successful prediction of the outcomes of human intervention and more intelligent management of exploited populations. Predation is a key interaction in natural ecosystems; understanding the nature of this interaction is central to any understanding of nature itself.

—*Lawrence E. Hurd*

FURTHER READING

Allen, K. R. *Conservation and Management of Whales.* University of Washington Press, 1980.

Crawley, M. J., ed. *Natural Enemies: The Population Biology of Predators, Parasites, and Diseases.* Blackwell Scientific Publications, 1992.

Ehrlich, Paul R., and Anne H. Ehrlich. *Extinction: The Causes and Consequences of the Disappearance of Species.* Random House, 1981.

Fabre, Jean Henri. *The Life of the Spider.* Translated by Alexander Teixeira de Mattos. Dodd, Mead, 1916.

Gause, G. F. *The Struggle for Existence.* Reprint. Dover, 1971.

Hassell, Michael P. *Arthropod Predator-Prey Systems.* Princeton University Press, 1978.

Krebs, J. R., and N. B. Davies, eds. *Behavioral Ecology: An Evolutionary Approach.* 4th ed. Blackwell Scientific Publications, 1997.

Levy, Charles Kingston. *Evolutionary Wars: A Three-Billion Years Arms Race.* Basingstoke, 2000.

Mowat, Farley. *Never Cry Wolf.* Atlantic-Little, Brown, 1963.

PREGNANCY AND PRENATAL DEVELOPMENT IN ANIMALS

FIELDS OF STUDY

Obstetrics and Gynecology; Developmental Biology; Endocrinology

ABSTRACT

Intrauterine development is important for several reasons. The uterus provides a unique environment that not only permits development, but promotes it. A recently fertilized egg has little capability of living if not surrounded by the uterus. The pregnant female sustains a completely undifferentiated single cell in the form of a fertilized egg through the stages of embryonic differentiation to that of a fetus capable of living outside the uterus.

KEY CONCEPTS

corpus luteum (pl. corpora lutea): the structure on the ovary that is formed from the follicle after the egg has been released; it secretes progesterone

embryo: a fertilized egg as it undergoes divisions from one cell to several thousand cells, but before the individual is completely differentiated into a fetus

estrogen: a hormone secreted by the ovary and placenta for development of the uterus

fetus: a differentiated but undeveloped individual with organ systems usually identifiable as a member of a species

ovary: the female gonad that is the source of eggs to be fertilized and hormones to maintain pregnancy

oviduct: a narrow, hollow tube which takes the newly ovulated egg from the ovary, provides the site for fertilization, and transports the new embryo to the uterus.

placenta: the tissue providing contact for exchange of nutrients between the lining of the mother's uterus and the blood supply from the fetus

progesterone: the hormone, essential for maintenance of pregnancy, that is secreted by the corpus luteum

uterus: the hollow internal female organ that accommodates the embryo as it grows to term

THE IMPORTANCE OF GESTATION IN MAMMALS

Gestation is a period of intrauterine development in animals that are "viviparous," meaning "bearing live young." "Pregnant" comes from a Latin word meaning "to possess important contents." Intrauterine development is important for several reasons. The uterus provides a unique environment that not only permits development, but promotes it. A recently fertilized egg has little capability of living if not surrounded by the uterus. The pregnant female sustains a completely undifferentiated single cell in the form of a fertilized egg through the stages of embryonic differentiation to that of a fetus capable of living outside the uterus. Proper temperature is maintained, the necessary fluids and nutrients are provided, wastes are removed, attacks by microorganisms are prevented, oxygen is supplied in a form useful to the embryo, and other compounds essential for cell growth are present.

The uterus is normally a safe haven with all the accommodations needed for the new potential animal. Because gestation takes place concealed inside the body, much myth and mystery have arisen over the process. Recent advances have dispelled some of the mystery, but much is still not completely understood. The present knowledge has been obtained from research and observations on laboratory and domestic animals, supplemented by observations on humans. The mechanisms of maintenance of pregnancy, number of young, length of gestation, anatomy of the uterus, and hormones vary considerably between species, thus preventing generalizations that might apply to more than a few species.

THE COURSE OF PREGNANCY

Pregnancy begins when the recently ovulated egg is fertilized by a sperm in the upper segment of the oviduct. Depending on the species, the fertilized egg resides in the oviduct for forty to one hundred twenty hours before being transported to the uterus. At a particular time during that interval, upon the laying down of certain fundamental tissues, it is referred to as an embryo. The rate of passage is influenced by the concentration of hormones in the mother.

Normally, estrogen production from the ovary is great before fertilization, when progesterone is not produced. After ovulation, the corpus luteum forms on the ovary, estrogen declines, and progesterone rises. This sequence permits the embryo to enter the uterus when the uterus is ready to support it. Progesterone will cause the oviduct to relax and allow the embryo to enter the uterus. Administration of large amounts of progesterone by injection will cause the oviduct to relax prematurely, and embryos will enter the uterus in eight or ten hours. High concentrations of estrogen, on the other hand, will cause the embryos to be retained in the oviduct for many days. Estrogen and progesterone also have a profound influence on the nature and function of the lining of the uterus. Estrogen causes rapid growth and, in synergism with progesterone, causes proliferation of glands lining the uterus. These glands secrete a complex of the compounds necessary for embryonic growth and development, called "uterine milk," for the nourishment of the young embryo before it becomes attached to the uterus.

The embryo must in some way make its presence known to the mother's constitution so that the uterus continues to be a hospitable environment for the entire gestation. There are a great variety of ways that have evolved in different species to maintain pregnancy. The act of mating will cause a female rabbit to ovulate and maintain a pseudopregnancy for about eighteen days. Signals from the fetus take over during the next thirteen days to maintain the corpora lutea.

Pregnant lioness.

If the mating was not a fertile one, the rabbit will mate again about eighteen days after the first mating. Rats normally have recurrent periods of estrus every four or five days. Mating will delay the next estrus for about twelve days. When the mating is fertile, the fetus provides signals for maintenance for the remaining nine days of gestation. Even without mating, dogs will often become pseudopregnant. This pseudopregnancy can extend as long as the normal gestation period of about sixty-five days. In these cases of pseudopregnancy, the hormones present, changes in the uterus, and maternal behavior are very similar to a normal pregnancy that would produce live young.

In many animal species, the interval between ovulations is somewhat prolonged by the spontaneous formation of one or more corpora lutea if there is no pregnancy. Examples are: sheep, seventeen days; horse, pig, goat, cow, twenty-one days; and human, twenty-eight days. In each of these cases, the embryo must signal the mother to establish the pregnancy several days before the next expected ovulation. Embryos of some species produce estrogens that are thought not only to cause uterine growth and enlargement, so as to accommodate the growing fetus, but also to maintain the corpora lutea.

In all species, the corpus luteum is necessary to produce progesterone during at least the first one-third of gestation. In the goat, pig, rabbit, and rat, the ovaries and the corpora lutea on them are essential throughout the gestation period. Progesterone is essential throughout gestation in all species, but in the sheep and human, the fetus and placenta supply progesterone for maintenance of pregnancy after about fifty-five days, even when the ovaries are removed. In summary, estrogen and progesterone are both essential for pregnancy for a number of functions. The source of these hormones can be either or both the ovaries and placenta, with variation among species.

THE UTERINE ENVIRONMENT

Embryos arriving in the uterus must have not only sustenance in the form of an ever-changing, compatible, fluid environment, but also a place and space to develop. In species with only one young born, this may not seem to be a problem, but in litter bearers (with as many as ten per birth), it is important that each embryo have sufficient space. In the pig, the uterus is V-shaped, with the ovaries at the tips. The two uterine horns form the sides of the V, and the body of the uterus and the birth canal consisting of the cervix and vagina are at the bottom. One uterine horn may be as long as 150 to 200 centimeters. The fetuses each occupy a segment of the uterus about 30 centimeters long; thus they are arranged like peas in pods in the two horns. The embryos enter the uterus forty-eight hours after ovulation at the four-cell stage. Embryos cannot move by themselves, so the motility of the uterus helps intrauterine migration. Embryos gradually move down each uterine horn during the period from day two to day nine when they reach the body of the uterus. From day nine to day twelve, some of the embryos continue to move into the other uterine horn. Some embryos arise from one ovary implant in the uterine horn on the side of the ovary of origin while others migrate through the body of the uterus to the other side. Movement stops at day twelve, when the embryos cannot move farther.

During this period of distribution, the embryo is dividing and growing. At day six, the embryo is about thirty-two cells and is still compact, resembling a tiny raspberry or morula about two hundred microns in diameter. The embryo forms a hollow ball of cells called a blastocyst, at days seven and eight. Some of the cells of the blastocyst are destined to become the new fetus and some will develop into the placenta. At around day twelve, the embryo is twenty to thirty centimeters in length—a very thin, fragile bit of tissue. It is at this stage that each embryo occupies the place and space in the uterus that it will grow within during the remaining 104 days of gestation. At this stage also the embryo histologically signals the mother of its presence so that the pregnancy may continue. The corpora lutea form and produce progesterone for about 14 days after ovulation, regardless of the presence of embryos. There is no need for a signal before day twelve. When embryos are present at day twelve and provide an adequate signal, the corpora lutea persist, maintaining the secretion of progesterone that maintains the pregnancy and prevents recurring ovulation and a new estrus cycle. In the event of an inadequate signal (as the result of no embryos or too few), the corpora lutea regress and a new ovulation occurs.

The pig is unique in that the proportion of the uterus that is occupied determines whether the signal is adequate. If half of the uterus is not occupied by embryos at day thirteen, the signal is inadequate and the corpora lutea will regress, even though there may have been some live embryos in some segment of the uterus. The anatomy of the uterus of the rat and rabbit is such that each horn is entirely separate and embryos cannot migrate between horns. In cows, sheep, and mice, intrauterine migration occurs infrequently. The embryos of the sheep and the cow signal the corpora lutea on the ovaries adjacent to the horns in which they are implanted. If an embryo is removed and placed in the opposite horn, the signal rarely goes from the uterine horn on one side to the ovary on the opposite side. The pregnancy will cease in that case in spite of the presence of an embryo. The uterus of the horse consists primarily of the body with no pronounced horns. Embryos of the horse do not migrate great distances and have only a local effect, influencing only the ovary on one side of the uterus.

The heart of the pig embryos begins to beat at about day sixteen. The uterus begins to expand at day eighteen and by day thirty the fetuses are about two centimeters long. The fetus is now surrounded by a round ball of fluid, the amnion. It will be floating in a zero-gravity environment for the next fifty days. The pig fetus grows roughly two centimeters every ten days until term. The first thirty days of gestation are critical for establishing systems and forming organs. It is during these embryonic stages that the embryo is very sensitive to toxins. The mother, the placenta, and the fetus all strive to protect the developing fetus from the harsh external environment.

FETAL LOSS

Not all embryos and fetuses survive to term. For most species, only about 55 percent of the eggs finally end up as live fetuses. There are many potential causes for this prenatal loss. Some eggs are not fertilized because either the sperm or egg is faulty. Because the lives of an unfertilized egg and a sperm are both finite, limited to a matter of hours, the timing of the meeting of the sperm and egg at fertilization must be synchronized precisely, or a nonviable embryo will result. Even if fertilization

occurs, the egg or sperm may have inherent chromosomal defects that produce a nonviable embryo.

Any disruption of the rate of development of the embryo or in the rate of change in the uterus or composition of uterine milk will cause loss of the embryo. Synchronous, parallel development of mother's uterus and the embryo are absolutely essential for survival of the embryo. Factors such as toxic chemicals, estrogens, and severe dietary changes can greatly affect chances of survival. In some cases, the amounts of progesterone and estrogen and other hormones essential for pregnancy are slightly abnormal. In species that bear large litters, the uterine space available to each fetus is not always equal, with the result that some fetuses do not have sufficient uterine resources for survival. Considering these and other hazards, it is remarkable that the proportion of individuals that are normal at birth is as great as it is.

Not all embryos grow and implant within a few days of entering the uterus. The embryos of badgers, mink, lactating rats, seals, kangaroos, armadillos, and roe deer are among those that have a delayed implantation. The embryo develops to the early blastocyst, stops for several months until uterine conditions are favorable, and then completes the normal development. When a lactating rat with delayed implantation weans the pups, the embryos resume growth. In mink and some other species, the ratio of light to dark each day seems to influence the time of resumption of growth. The knowledge of this phenomenon goes back more than one hundred years, as indicated by the records of roe deer collected by the physician of a noble on a large estate. He had an opportunity to examine the uteri of roe deer killed for meat. He had noted the dates of mating but found no fetuses even after several months. This led to the discovery that the blastocyst's development had been arrested for a prolonged period before resumption.

THE STUDY OF GESTATION

The accumulation of knowledge on pregnancy and prenatal development has taken place over a long period. From the early superficial, gross observations of dead animals to the recent sophisticated and detailed techniques of observation of both mother and embryo or fetus, methods of exploring the changes that take place have developed. One procedure in current practice has its origin in antiquity. A cesarean section is a surgical procedure in which the fetus is brought out of the uterus through an incision in the abdominal wall and the uterus. (The term derives the name from the erroneous notion that the infant Julius Caesar was delivered by that procedure.) Currently, embryos can be removed very early and still survive. The ability to recover embryos and keep them alive outside the body for several days during the first six or seven days after fertilization has been a major step in study of embryos. The culture fluid must have many characteristics similar to uterine milk to be compatible with living embryos. Temperature, the ratio of gases, the osmotic pressure of the fluid, and chemical composition must all be near those of the normal uterine environment. Each component may be modified slightly to achieve optimum livability at different stages of development and in different species. Embryos from several species may now be frozen and preserved for many years. This creates many possibilities for studies but also creates ethical and moral dilemmas.

The transfer of embryos between mothers of the same species has provided a means of separating the maternal genetic influence from the effect of intrauterine environment on the fetus. Embryos from smaller breeds of rabbits grow larger than normal when transferred to a larger breed. Zebra foals have been born to horse mares as the result of transferring zebra embryos to the horse. The use of ultrasound to obtain an image of the fetus, the uterus, and surrounding tissue by noninvasive means has been very helpful. Now it is possible to get a frequent picture of the course of events in the uterus to monitor development of the fetus. Nuclear magnetic resonance imaging can generate a picture of the uterus and contents noninvasively. By nuclear magnetic resonance spectroscopy, chemical reactions can be measured in very small segments of the uterus and placenta. Optical systems inserted into the amniotic cavity with the fetus permit direct visualization of the fetus. Perfection of surgical techniques has permitted insertion of cannulas into blood vessels of experimental fetuses to monitor the concentration of hormones. The development of the brain of fetuses has been measured by use of minute electrodes and cannulas placed in specific points.

The concentration of hormones circulating in the mother and fetus can be measured, often and with great precision and specificity, by radioimmunoassay of very small quantities of blood. These hormones change with different stages of gestation. The normal changes and concentrations have been determined for a wide variety of animals. An interesting aspect of the emerging picture of the various species studied is the great variety of mechanisms by which pregnancy exists. No two species have the same mechanism even though they may appear to be very similar. As this rapidly growing area of knowledge expands, humans will be able to shed light on the inside of the uterus, which has been called one of the darkest places on earth.

FOSTERING OPTIMAL PREGNANCIES

Humankind has entered into an era of unprecedented ability to understand the complexities of pregnancy in animals. The efficiency of the animals related to human consumption—in production of food, fiber, recreation, and companionship—depends largely on the efficiency of pregnancy. Among cows, horses, pigs, sheep, and goats, the proportion of mated females that conceive and bear young is less than 100%. The proportion of the fertilized eggs that develop into differentiated embryos is less than 100%. The proportion of embryos that develop into full-term fetuses born alive is less than 100%. Of those fetuses born alive, less than 100% survive more than a few days because of problems that had occurred prenatally. As an example, the runt of the litter often cannot compete successfully and perishes. Often, these animals were stunted in the uterus because of inadequate space or other resources, such as energy and gas exchange. Some of the inadequacies of resources may be avoided by proper prenatal care.

Avoiding adverse effects on embryos and fetuses of all species should be humanity's aim; it may be possible to correct both maternal and fetal problems by judicious prenatal monitoring and treatment. A problem anticipated is half-solved. A sheep with twins needs different feeding and management from one with a single lamb. Prenatal monitoring can anticipate a problem, and treatment may prevent it.

—*Philip J. Dziuk*

FURTHER READING

Cole, H. H., and P. T. Cupps. *Reproduction in Domestic Animals*. 3rd ed. Academic Press, 1977.

Foxcroft, G. R., D. J. A. Cole, and B. J. Weir. *Control of Pig Reproduction II*. Journals of Reproduction and Fertility, 1985.

Gilbert, Scott F. *Developmental Biology*. 6th ed. Sinauer Associates, 2000.

Hennig, Wolfgang, ed. *Early Embryonic Development of Animals*. Springer-Verlag, 1992.

Mossman, H. W. *Vertebrate Fetal Membranes*. Rutgers University Press, 1987.

Novy, M. J., and J. A. Resko. *Fetal Endocrinology*. Academic Press, 1981.

Pasqualini, J. R., and F. A. Kincl. *Hormones and the Fetus*. Pergamon Press, 1986.

Pierrepoint, Colin G., ed. *The Endocrinology of Pregnancy and Parturition*. Alpha Omega Alpha, 1973.

Shostak, Stanley. *Embryology: An Introduction to Developmental Biology*. HarperCollins, 1991.

Wilson, J. G., and F. C. Fraser. *Handbook of Teratology*. Plenum, 1977.

PREHISTORIC ANIMALS

FIELDS OF STUDY

Paleontology; Developmental Biology; Paleoecology

ABSTRACT

Although cells with respiratory and photosynthetic organelles (mitochondria and chloroplasts) evolved about 1,500 million years ago, no trace of animals has been found in the fossil record before 1,000 million years ago. The fossil record of microscopic animals is nearly nonexistent between 1,000 and 600 million years ago. All of our knowledge of prehistoric animals comes from the fossil record that formed during the last 600 million years.

KEY CONCEPTS

- **arthropods:** the phylum of invertebrate animals having clearly segmented bodies and appendages (legs, antennae, and mouth parts) with many segments and joints
- **chordates:** the phylum of animals that have a stiff, rodlike structure called the notochord running their length; chordates include the boneless fishlike animals similar to amphioxus, as well as fish, amphibians, reptiles, birds, and mammals
- **evolution:** the science that studies the changes in astronomical bodies, the earth, and living organisms and creates explanations of how changes occur and how entities and events are related
- **geological periods:** the twelve divisions in successive layers of sedimentary and volcanic rocks, which are differentiated by the distinctive fossils present within each division; because of many recently discovered fossils and careful dating, the beginning and ending dates for these periods tend to vary somewhat among different references
- **paleontology:** the branch of geology that deals with prehistoric forms of life through the study of fossil animals, plants, and microorganisms

THE DAWN OF TIME

The fossil record indicates that eubacteria evolved as long as 3,800 million years ago, whereas a comparison of ribosomal ribonucleic acid (RNA) suggests that cells with nuclei probably diverged from archaebacteria as long as 2,800 million years ago. Although cells with respiratory and photosynthetic organelles (mitochondria and chloroplasts) evolved much later, about 1,500 million years ago, no trace of animals has been found in the fossil record before 1,000 million years ago. The fossil record of microscopic animals is nearly nonexistent between 1,000 and 600 million years ago. All of our knowledge of prehistoric animals comes from the fossil record that formed during the last 600 million years.

Near the end of the Precambrian period (540 million years ago), a few sponges, jellyfish, colonial filter feeders, wormlike animals, early mollusks, and primitive arthropods are found in the fossil record. One or more minor extinction events that eliminated up to 50 percent of all species brought the Precambrian period to an end. Animals whose body plan was based on segmentation quickly diversified into a myriad of forms.

HOW DID SO MANY ANIMALS DEVELOP SO QUICKLY?

The Cambrian period (540 to 500 million years ago) is characterized by the "sudden" appearance (in 1 to 10 million years) of a large number of morphologically different animals. One of the most innovative body plans that evolved near the end of the Precambrian period was that associated with the many wormlike animals that developed segmented bodies. Segmentation allowed for rapid evolution of all sorts of animals simply by changing the number of segments, their size and shape, as well as the

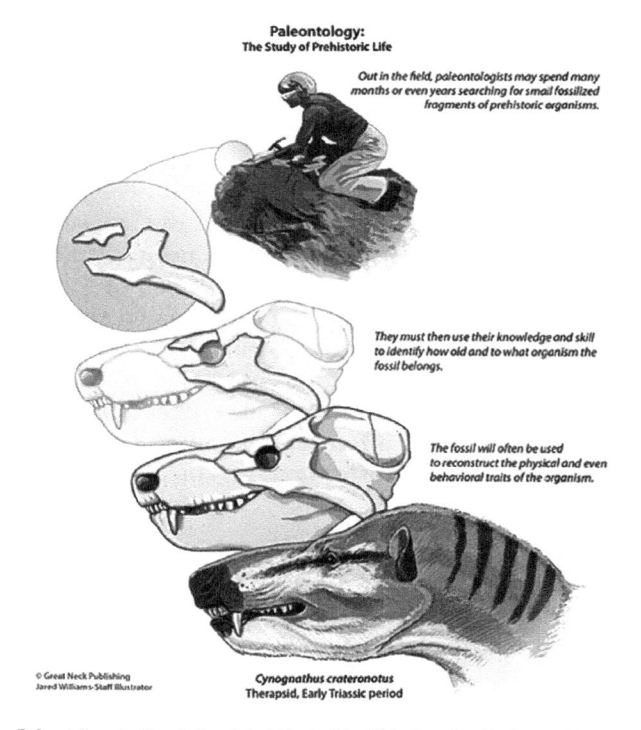

Paleontology is the study of prehistoric life. This branch of science offers an understanding of what the earth was like many years ago.

associated appendages. Evolution of homeotic genes and the genes they regulate altered not only segment number and segment characteristics but also appendages. For example, legs could be easily changed into preening, mating, and sensory appendages. During the Cambrian period, all kinds of marine worms, caterpillar-like animals, arthropods, and chordates made their appearance. Some of these caterpillar-like animals had broad legs (*Aysheaia*) and resembled velvet worms of today, whereas others had long, sharp spikes for legs (*Hallucigenia*) and are unrelated to any other known animals.

Many of the early arthropods vaguely resemble today's sow bugs, shrimp, centipedes, and horseshoe crabs. A very successful group of marine arthropods that superficially resemble sow bugs were the trilobites. A trilobite had a prominent, centrally located, segmented, concave, dorsal ridge running the full length of its body. The trilobites differentiated and evolved for over 300 million years until the few remaining forms went extinct during the Permian-Triassic mass extinction (250 million years ago) that wiped out nearly 90% of all species. A trilobite had a head, thorax-abdomen, and tail (pygidium). The segments of the head were fused but the thoracic-abdominal and tail segments were varied. Depending upon the trilobite, thoracic-abdominal segments ranged from nine (low of six) to thirteen (high of twenty-two) in number, whereas tail segments varied from six to sixteen (high of twenty-two). Legs and other specialized appendages protruded ventrally from the segments. Gills were also associated with a number of the segments. Although some of the first trilobites had "eyes," their compound nature is not clear because of their size. Most of the later trilobites developed compound eyes that closely resemble the compound eyes seen in extant insects.

One of the most unusual arthropod-like creatures of the Cambrian period was the enormous predator known as *Anomalocaris*, which ranged up to three feet in length. It had a long, oval, dorsoventrally flattened head with compound eyes on short stalks near the back of the head. Protruding from the ventral region at the front of the head were two grasping feeding appendages that for many years were mistaken for separate shrimplike organisms. Just behind the feeding appendages was a circular mouth resembling a cored pineapple slice. This mouth had been misinterpreted as a jellyfish by early discoverers. Saw-shaped "teeth" lined the inside of the mouth and most likely contracted around animals that were positioned in the mouth by the feeding appendages. A number of fossil trilobites have been found with missing chunks that appear to have been torn away by *Anamalocaris*. No arthropods of today have any sort of mouth like that of *Anamalocaris*. This creature was a specialized swimmer, propelled by winglike flaps that appear to consist of numerous overlapping flaps originating from most of the body segments.

THE EVOLUTION OF FISH

The oldest fossils of jawless fish, referred to as the *Agnatha*, are about 470 million years old. The jawless lampreys and hagfish of today are most closely related to the ancient *Agnatha* of the Ordovician period (500 to 440 million years ago). These prehistoric *Agnatha* attached to larger animals through their whorls of "teeth" and rasped the flesh of the host. Lamprey teeth are horny, sharp structures devoid of calcium, derived from the skin. Because the teeth contained enamel-like proteins, they are considered to be precursors to modern teeth, which consist of enamel over dentine.

The more advanced *Agnatha* that developed in the Silurian and Devonian periods did not survive the Devonian-Carboniferous mass extinction. Yet they are important because they suggest when the evolution of mineralized teeth, bone, body armor, eyes, and paired limbs occurred. The jawed fish first evolved in the Devonian period (410 to 360 million years ago) but did not dominate until after the Devonian-Carboniferous mass extinction. The earliest jawed fishes were sharks. The discovery of fossilized shark teeth dated to the Devonian period suggest that they evolved during this period. The earliest fossil skeletons of sharks discovered to date, however, are from the Carboniferous period (360 to 290 million years ago).

THE FIRST TERRESTRIAL VERTEBRATES

The first land-dwelling vertebrates were amphibians that evolved from freshwater, lobe-finned fishes closely related to *Eusthenopteron*, a crossopterygian that lived midway through the Devonian period (410 to 360 million years ago). Most amphibians are tetrapods that begin life in a watery environment but turn to a terrestrial life as adults. Extant amphibians include frogs, toads, salamanders, and the legless, snakelike caecilians.

An unusual prehistoric amphibian was *Ichthyostega*, a three-foot-long crocodile-like animal that lived somewhere between 350 and 370 million years ago. Its jaws were lined with two rows of teeth. Along the front was a row of large teeth and behind them was a row of densely packed, small, sharp teeth bounded by a couple of very large canine teeth. The skull was similar in some respects to the lobe-finned fish that lived during the Devonian. Its neck was only two to three vertebrae long and the pectoral bones were attached to the skull and backbone by muscle.

External gills were a prominent feature of the juvenile or larva. The adult however, lost the external gills when it underwent metamorphosis and developed primitive lungs. The aquatic juvenile amphibians depended on gills to breath, but when they metamorphosed into terrestrial adults they developed primitive lungs out of part of the swim-bladder. Air was swallowed into the intestine-swim bladder tract. A high concentration of capillaries in the wall of these simple lungs carried oxygen to the heart.

The pelvic bones were attached to the spinal chord through muscles. *Ichthyostega* had a heavy rib cage to support its weight out of water, in contrast to other contemporaneous vertebrates. In general, the early amphibians had very short leg bones compared to the primitive reptiles that evolved from them. *Ichthyostega* had seven digits on both its front and back feet, whereas *Acanthostega* had eight digits. Other ancient amphibians had variations in the number of digits. Living amphibians generally have four to five digits on the front feet and five digits on the back feet. *Ichthyostega*'s diet was mainly fish, but on land it may have consumed arthropods and smaller amphibians.

Mastodonsaurus giganteus, another crocodile-like animal, lived 230 million years ago, during the early Triassic period (250 to 205 million years ago). It was the largest amphibian that ever existed, growing to nearly thirteen feet in length. Like *Ichthyostega*, two rows of pointed teeth lined the jaws of this amphibian, making it a formidable match for the early reptiles.

A small group of amphibians lost their legs and took on the appearance of snakelike animals. Today these legless amphibians are highly segmented, making them look more like giant earth worms than snakes. One of the largest extant caecilians is *Caecilia thompsoni*, which lives in Colombia and grows to a length of five feet.

REPTILE DIVERSITY

The first reptiles evolved from amphibians during the early Carboniferous period (360 to 290 million years ago). Early reptile skeletons can be distinguished from amphibian remains by the skull and jaw characteristics, the number of neck vertebrae, the absence of external gills, the length of the leg bones, and the number of foot and toe bones.

Prehistoric amphibians had dorsoventrally flattened skulls and skull bones similar to their crossopterygian fish relatives. Amphibians had no nasal holes in their skulls because they breathed through their gills or through their mouths. Reptiles, however, had anterior nasal openings in the snout. The diapsid reptiles had two additional holes behind the eye openings for muscle attachment, whereas synapsid reptiles had one hole, and anapsid reptiles had none. The early amphibians had necks with two to three vertebrae, whereas reptiles generally had necks with more than six vertebrae. The early amphibians generally had very short front and rear leg bones and feet with six to eight digits, whereas early reptiles had front and rear feet with only five digits. Later amphibians reduced the number of digits to four in the front and five in the rear, similar to what is found in extant frogs. Early amphibians had digits that varied considerably, whereas early reptiles usually had digits with two, three, four, five, and three bones (moving from thumb to little finger).

The evolution of reptiles required the development of a new type of egg. The new egg provided a watery environment in which the embryo and juvenile could develop. An amniotic sac evolved to contain the developing reptile and its watery environment, the yolk sac developed to provide the nutrients needed for growth, and the allantoic sac was acquired to help eliminate wastes. The evolution of a tough, membranous egg covering allowed oxygen and carbon dioxide exchange but prevented water loss. Fish and amphibians have to lay their eggs in water because they never evolved the adaptations needed to lay their eggs on land. Reptiles, dinosaurs, birds, and mammals are all amniotes and can lay their eggs on land because their eggs have amniotic sacs, yolk sacs, or allantoic sacs.

Although amphibians' skin was protected by thick layers of secreted mucous, it was always in danger of drying or burning under the sun if the animal strayed too far from water. Reptiles and dinosaurs evolved thick scales to prevent drying and burning, whereas birds and mammals evolved feathers and hair to protect their skin. Recent fossil evidence demonstrates quite clearly that many dinosaurs also sported a coat of feathers.

FLYING REPTILES

The oldest flying reptiles are dated to the late Triassic period (250 to 205 million years ago). The first of these pterosaurs (flying lizards) had long, thin tails, often more than half the length of their bodies. The wings consisted of a leathery skin that was supported along its anterior margin by the pterosaurs' front legs and a highly elongated fourth digit. Digits one, two, and three were tipped with claws. The wings were attached to the body and hind legs, and the wingspan ranged from one to six feet. Their elongated jaws were lined with numerous sharp teeth that they used to grab surface fish, like some present day birds. In the late Jurassic and Cretaceous period, pterosaurs lost their long tails. Some became quite large. *Pteranodon* had a wingspan of up to twenty-five feet, whereas the largest *Quetzalcoatlus* had a wingspan of nearly forty feet. Toothless *Pteranodon* fished on the move, scooping up surface fish, whereas *Quetzalcoatlus* hunted much like the vultures of today, but swooping down on a dead dinosaur instead of the carcass of a wildebeest. Both these large flying reptiles were consummate soaring reptiles, able to take advantage of the slightest updraft of air because of their long, narrow wings.

REPTILES RETURNED TO THE SEA

Some reptiles became adapted to living most of their lives in the sea. *Nothosaurus* lived near the beginning of the Triassic period (250 to 205 million years ago) and grew up to ten feet long. Its neck and tail each accounted for a third of its length. *Nothosaurus* used its legs and webbed feet to swim and to drag itself onto beaches, where it laid its eggs.

By the end of the Triassic period, marine reptiles such as *Liopleurodon* and *Plesiosaurus* had flippers instead of legs and feet. *Liopleurodon* sometimes grew to lengths of seventy-five feet and achieved weights of nearly one hundred tons. It was twenty times heavier than *Tyrannosaurus rex* and had teeth twice as long. Long necks and tails were characteristic of the plesiosaurs, which had anywhere from twenty-eight to seventy neck vertebrae, depending upon the genus. The small mouth and teeth of *Plesiosaurus* were adapted to capturing small fish and squid.

Over a period of millions of years, ichthyosaurs evolved an extremely long, thin snout. The leg bones of these reptiles shortened and the digits were remodeled to form flippers. A dorsal and tail fin appeared. Some of these reptiles have the distinction of having the largest eyes of any animal that ever lived. Fully grown *Temnodontosaurus* had eyes with diameters of nearly eleven inches. (In comparison, adult elephants have eyes that are two inches in diameter, and those of blue whales are six inches wide.) These reptiles became so adapted to their marine lives that they could not return to land to lay eggs. Ichthyosaurs solved the problem of producing the next generation by giving birth to live babies. A fossil of *Stenopterygius* illustrates a baby exiting the reptile's birth canal much like the marine mammals of today.

All these marine reptiles had nasal openings and lungs that had to be filled with air from time to time. This need for oxygen required them to surface just like extant marine mammals. Even though the marine reptiles were highly adapted to the sea, none survived the Cretaceous-Tertiary extinction, 65 million years ago.

DINOSAURS

Dinosaurs evolved from reptiles sometime after the Permian extinction, 250 million years ago. Dinosaurs developed legs and a skeleton that supported them from beneath, in contrast to most reptiles, which supported themselves on legs that protruded from the sides of their bodies. Dinosaurs were slow to diversify during the Triassic (250 to 205 million years ago) and early Jurassic (205 to 145 million years ago) periods. Even so, many distinctive animals evolved from four major groups present near the beginning of the Jurassic.

Evolution of animals in the *Eurypoda* led to tetrapods such as *Stegosaurus*. These huge vegetarians up to thirty feet in length had large, diamond-shaped plates that protected their necks and backs from predators. It is theorized that these large dinosaurs were cold-blooded and also used the plates to rapidly cool or heat their bodies. There is considerable evidence, however, that dinosaurs were in fact hot-blooded creatures. As such, the actual function of *Stegosaurus*' dorsal plates is more enigmnatic.

Animals in the *Euornithopoda* were mostly tetrapods. *Iguanosaurus* was a large vegetarian with big, hornlike claws on its thumb and toes that ended in "hoofs." *Hadrosaurus* was a large vegetarian with a long, hollow, head crest along the top of its skull that protruded up to 1.5 feet behind the head. *Parasaurolophus* grew up to thirty feet in length and had a head crest over 5 feet long. These animals could blow air through the head crest to create various trumpeting sounds. *Triceratops* was another large vegetarian that grew up to nine feet long. It had a sharp, horny beak, a rhinoceros-like horn above and somewhat posterior to its nostrils, two long horns above its eyes, and a bony neck frill protruding from the back of its skull to protect its neck from predators. It has long been thought that *Triceratops* used its long horns in an offensive manner rather like battering rams, but it has been shown that the stress applied to the facial bones by this would easily fracture them, rendering the creature terminally helpless. *Styracosaurus*, a relative of Triceratops, grew to lengths of seventeen feet. A number of smaller horns on either side of two long horns extended the neck frill to make *Styracosaurus* appear larger and to protect its neck.

Certain *Sauropoda* evolved into some of the largest land animals that ever existed. The vegetarian tetrapods *Titanosaurus* weighed up to seventy tons, while *Supersaurus* ranged up to fifty-five tons, *Brachiosaurus* forty-five tons, and *Seismosaurus* thirty tons. Some scientists argue that these reptiles must have been cold-blooded, because they could not possibly process enough vegetation to maintain a constant temperature. In comparison, warm-blood mammals such as elephants often reach weights of six to eight tons, whereas warm-blooded blue whales may attain weights of up to ninety tons.

ARCHAEOPTERYX, THE DINOSAUR THAT FLEW

Theropod dinosaurs were warm-blooded, bipedal animals that came in a myriad of sizes and superficially resembled *Tyranosaurus rex*. Approximately 155 million years ago, a number of therapod dinosaurs were developing feathered coats. Most of the feathers were similar to down feathers that superficially resemble a tuft of hair. A tuft of hair, however, consists of hairs derived from many hair follicles, whereas a down feather is a single unit in which each hairlike barb attaches to the base of the feather. Some theropod dinosaurs also produced vaned feathers having a distinct shaft and barbs attached along the length of the shaft. Like scales, feathers develop from the differentiation of certain epithelial cells. The earliest stages of feather development resemble the first stages of reptilian scale formation. This suggests that one or more genes evolved that blocked scale development in favor of feather development.

During the late Jurassic period, many of the small, bipedal theropod dinosaurs related to the ancestor of birds evolved long arms, three-fingered hands with claws, fused clavicles (the beginnings of a wishbone), feet with three toes for running, first toes protruding from the back of the foot, and hollow bones.

The first primitive bird, *Archaeopteryx*, is dated to the late Jurassic, about 150 million years ago. Fossil remains suggest that *Archaeopteryx* had primary, secondary, and tertiary flight feathers originating from its palm, forearm, and upper arm, respectively. These flight feathers were covered at their base by other feathers known as main coverts and lesser coverts. The three clawed digits of the hand appear not to have supported primary feathers; instead they protruded three-quarters of the way along the wing.

In modern birds, the palm and first digit bones have fused and greatly diminished so that a single, small thumb bone protrudes near the wrist. This first finger supports feathers that run parallel to the front of the wing. The palm bones associated with the second and third digits have partially fused at their ends to become a bone that roughly resembles a fused miniature radius and ulna. The second digit that supports the primary feathers that extend the wing and compose the wing tip has shortened and thickened. The third finger has been reduced to a single, tiny bone.

Modern flying birds have large, keel-like sterna for attaching the muscles that move the wings up and down. Although *Archaeopteryx* had wings that suggest it was capable of performing intricate maneuvers in flight, its tiny sternum indicates it probably lacked the power to take off from the ground. It probably launched itself by running rapidly until it developed sufficient lift. Its feet are similar to those on fast-running dinosaurs. The first digit is high on the foot, suggesting that *Archaeopteryx* did not depend upon trees or high bushes to launch itself.

Archaeopteryx inherited a long, bony tail from its dinosaur ancestors. The tail supported numerous vaned feathers along its length. In modern birds, the bony tail has been greatly reduced by the loss of numerous segments at the end and the fusion of segments at the base. The short, bony tail in modern birds is called the pygostyle. Most of the tail in modern birds consists of long feathers that extend toward the rear from the pygostyle.

Archaeopteryx also inherited a head that closely resembled that of its dinosaur ancestors. The skull and mandible are light because of cavities in the bones, whereas its elongated snout, not yet a true beak, has a mouth lined with teeth. Although *Archaeopteryx* and many primitive birds of the Cretaceous period did not survive the Cretaceous-Tertiary extinction, 65 million years ago, a few did. These survivors gave rise to the myriad bird species of today.

—*Jaime Stanley Colomé*

Further Reading

Bakker, Robert T. *The Dinosaur Heresies* Zebra/Kensington, 1986

Chen, Ju-Yuan, Di-Ying Huang, and Chia-Wei Li. "An Early Cambrian Craniate-like Chordate." *Nature* 402 (December 2, 1999): 518–522.

Desmond, Adrian J. *The Hot-Blooded Dinosaurs* Blond & Briggs, 1975.

Dowswell, Paul, John Malam, Paul Mason, and Steve Parker. *The Ultimate Book of Dinosaurs.* Parragon, 2000.

Gould, Stephen Jay. *Wonderful Life: The Burgess Shale and the Nature of History.* W. W. Norton, 1989.

Hofrichter, Robert, ed. *Amphibians: The World of Frogs, Toads, Salamanders, and* Newts. Firefly Books, 2000.

Motani, Ryosuke. "Rulers of the Jurassic Seas." *Scientific American* 283, no. 6 (2000): 52–59.

Padian, Kevin, and Luis M. Chiappe. "The Origin of Birds and Their Flight." *Scientific American* 278, no. 2 (1998): 38–47.

Sereno, Paul C. "The Evolution of Dinosaurs." *Science* 284, no. 5423 (June 25, 1999): 2137–2146.

Primates

Fields of Study

Morphology; Ethology; Sociobiology; Genetics; Molecular Biology

Abstract

The primates are an order of mammals that includes monkeys, apes, lemurs, tarsiers, and humans. Anatomically, primates have many similarities to the earliest placental mammals. Primitive features retained by primates include five-fingered hands and feet with individually mobile fingers and toes, a collarbone (clavicle), a relatively simple cusp pattern in the molar teeth, and the freedom of movement within the forearm that allows the wrist to rotate without moving the elbow.

Key Concepts

arboreal: tree-dwelling, or pertaining to life in the trees

binocular vision: vision using two eyes at once, with overlapping visual fields

brachiation: a form of locomotion, also called arm-swinging, in which the body is held suspended by the arms from above

Catarrhini: a primate group including Old World monkeys, apes, and humans, with reduced tails and only two pairs of premolar teeth

clavicle: the collarbone, connecting the top of the breastbone to the shoulder

opposable: capable of rotating so that the fingerprint surface of the thumb or big toe approaches the corresponding surfaces of other fingers or toes

placentals: mammals whose unborn young are nourished within the mothers' uteri

stereoscopic vision: vision with good depth perception

visual cortex: the part of the cerebral cortex concerned with vision

visual predation: catching prey (such as insects) by sighting them visually, judging their exact position and distance, and pouncing on them

THE BRANCHES OF THE PRIMATE FAMILY TREE

The primates are an order of mammals that includes monkeys, apes, lemurs, tarsiers, and humans. The other primates are thus the nearest relatives to human beings. Anatomically, primates have many similarities to the earliest placental mammals. Primitive features retained by primates, but lost in many other mammals, include five-fingered hands and feet with individually mobile fingers and toes, a collarbone (clavicle), a relatively simple cusp pattern in the molar teeth, and the freedom of movement within the forearm that allows the wrist to rotate without moving the elbow.

Many primates live in trees; those that do not, have anatomical features showing that their ancestors were tree-dwellers. These include features directly concerned with arboreal locomotion, those concerned with vision and intelligence, and those concerned with reproduction. Arboreal locomotion, the ability to climb trees, has many direct consequences in primate anatomy. Most primates, for example, possess long, agile arms and legs. Grasping hands and feet give them the ability to hold objects and to climb by wrapping the fingers around branches. This is different from the many other arboreal animals that dig their claws into the bark. The primate grasp is aided by the development of opposable thumbs and big toes. If a primate grasps an object, the fingerprint surface of the thumb faces the corresponding surfaces of the other fingers. This is possible because the bone supporting the thumb can rotate out of the plane of the other fingers. Most primate thumbs and big toes are opposable; human feet are unusual. Individual mobility of all fingers and toes is a primitive mammalian ability that primates have kept but which many other mammals have lost through evolution. Combined with the grasping ability of hands and feet, this characteristic allows primates to manipulate objects of all sizes rather skillfully.

Primates also have hairless friction-skin on their palms and soles, which are supplied with a series of parallel ridges in complex patterns (fingerprints). These ridges provide a high-friction surface, which helps in grasping objects—or in holding branches without slipping. Claws, which could get in the way and cause injury, have generally been reduced to fingernails that rub against things and thus stay short. The clavicle (collarbone), which strengthens the shoulder region, is present in primitive mammals. Many modern mammals have lost this bone, but in primates, it has been retained and often strengthened.

PRIMATE VISION

Primates are primarily visual animals. Whereas many other animals rely strongly on smell, primates rely on sight, and particularly on color vision. One reason for this visual orientation of primates is that the

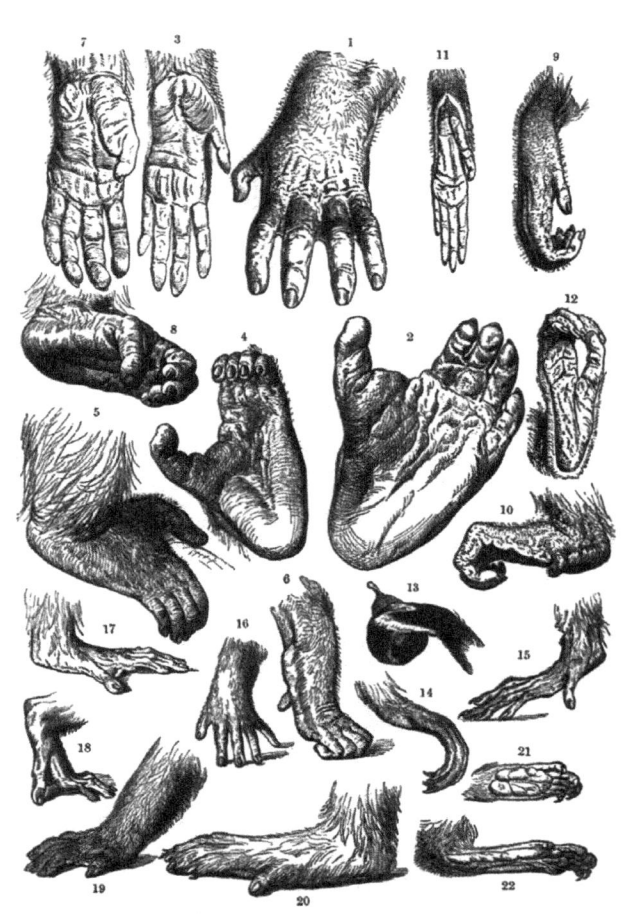

HANDS AND FEET OF APES AND MONKEYS.
1, 2, Gorilla; 3–8, Chimpanzee; 9, 10, Orang; 11, 13, Gibbon; 14, 15, Guereza; 16–18, Macaque; 19, 20, Baboon; 21, 22, Marmoset.

An 1893 drawing of the hands and feet of various primates. (Richard Lydekker)

exact position of the next branch can best be judged visually, and good depth perception is essential for avoiding a fall. Primates often hunt by visual predation, stalking an insect or other small prey, judging its exact position and distance, and then pouncing on it.

Vision in depth (stereoscopic vision) requires two eyes with overlapping visual fields (binocular vision). Each eye sees the same objects from slightly different angles; the brain combines the two images into a single three-dimensional image. Binocular vision is made possible by the forward position of the eyes. The eyes need protection in this position, and it is furnished in most primates by a bony structure called the postorbital bar. A complex organization of the brain is required for binocular vision. The cerebral cortex of the brain is therefore expanded, especially the part known as the visual cortex. The surface of the brain develops a large number of folds—primate brains are more heavily folded than any others. One characteristic fold is called the calcarine fissure, a characteristic feature of all primates.

Primates can pick up unfamiliar objects with the fingers and bring them in front of the face for closer visual inspection. This behavior requires coordination of eye, brain, and hand in a very complex interaction. It is through this behavior that young primates learn to manipulate and understand their environments and to investigate them with a characteristic curiosity. Primates rely heavily on learned behavior, including exploratory play. Primates are the most intelligent animals, and the great expansion of the cerebral cortex reflects this. Most primates grow up in social groups, where they interact in complex ways with other individuals. They have a curiosity about the world, especially about other primates.

Primate behavior is complex, and most of it is learned. The price paid for reliance on learned behavior is youthful inexperience. Infant primates have much to learn, and they have not yet had much time to learn it; they therefore depend significantly on their parents. A long and extended period of intensive parental care of offspring characterizes nearly all primates. This would not be possible if primates were born in large litters, because their mothers would be too busy caring for so many babies at once. Furthermore, infant primates are carried frequently, placing a burden on their mothers, some of whom need both arms free for climbing. Accordingly, primates are usually born one at a time. Twins and other multiple births are uncommon. There are several related features in primate reproductive systems. Female primates need only one uterus to carry their young and one pair of nipples to nourish them after birth. The right and left uteri, paired in other mammals, are fused into a single uterus called a uterus simplex. Unlike female dogs or pigs, which have many nipples in parallel rows, female primates have only a single pair, located high in the chest region.

PRIMATE CLASSIFICATION

There are many different types of primates and several ways of classifying them. The classification followed here divides primates into the suborders *Paromomyiformes*, *Haplorhini*, and *Strepsirhini*. The suborder *Paromomyiformes* contains the earliest primates, all now extinct. They lived from about fifty to eighty million years ago (from the Cretaceous period to the Eocene epoch). They are known mostly from fossilized teeth, but the genus *Plesiadapis* is known from nearly complete skeletons. All these primates were small, from roughly the size of a mouse to that of a cat. Their limbs were long and agile, with hands and feet showing the grasping ability typical of arboreal primates. The teeth, which differed among the three families of *Paromomyiformes*, were modified for a diet of mostly fruits but also some insects. A postorbital bar was not present; the eyes had not yet shifted forward. The visual fields overlapped much less than they do in modern primates, showing that these animals were not efficient visual predators. Some paleontologists have suggested excluding them from the order Primates for this reason.

SUBORDER *HAPLORHINI*

The majority of primates belong to the suborder *Haplorhini* and possess relatively large brains and relatively short faces. The noses of all *Haplorhini* are dry and have no external connection to the mouth; the upper lip goes straight across the mouth and is not divided by a groove, as it is in the *Strepsirhini*. The structures of the ear region, the placenta, and certain other anatomical features show that the *Haplorhini* are closer to the ancestral mammals than are the *Strepsirhini*. The geologic record of haplorhines extends from the Paleocene to the Recent (or Holocene) epoch.

There are three subgroups of *Haplorhini*: *Tarsioidea*, *Platyrrhini*, and *Catarrhini*. Of these, the *Tarsioidea* are the oldest and most primitive. The tarsioids are small primates with rounded heads. The eyes are rotated forward and protected from behind by a postorbital bar; visual fields overlap significantly. Tarsioids flourished during the Eocene epoch, about forty to fifty million years ago. They lived across Europe, Asia, and North America, all of which then had subtropical climates. Tarsioids disappeared from Europe and North America as climates became colder, but they survived in Asia. The living genus *Tarsius* now inhabits the East Indies from Borneo to Celebes. *Tarsius* is a big-eyed, nocturnal primate with an elongated ankle. It has a peculiar form of locomotion called vertical clinging and leaping. It clings to vertical bamboo stalks much of the time, then uses its hind legs to jump to the next perch. At least one extinct tarsioid had similar adaptations.

The *Platyrrhini* include the South and Central American primates. They are characterized by nostrils that open directly forward, three pairs of premolar teeth, and tails that are strong and sometimes assist in locomotion. Platyrrhines probably evolved from tarsioid primates that reached South America prior to Miocene times. Included in the *Platyrrhini* are the small forest primates called marmosets (*Callithricidae*). Male marmosets often have striking white or yellowish facial markings (such as tufted ears, eyebrows, and mustaches) which may be useful in mate recognition. The other *Platyrrhini* are the New World monkeys (*Cebidae*). Familiar ones include squirrel monkeys (*Saimiri*), capuchin monkeys (*Cebus*), and howler monkeys (*Alouatta*). All these monkeys are arboreal, and several of them are skilled acrobats. A few species, including the squirrel monkeys, can hang by their tails.

The *Catarrhini* include the Old World monkeys, apes, and humans. All *Catarrhini* are characterized by noses that protrude from the face with the nostrils opening downward, as human noses do. There are only two pairs of premolars. The tails are weak or absent and are useless in locomotion. Two very early catarrhine species occurred in Burma, but they are poorly known and of uncertain relationships. The oldest undoubted *Catarrhini* occur as fossils in the Fayum deposits of Egypt, which are Oligocene in age—about thirty-five million years old. Fayum primates are thought to be of tarsioid ancestry. One type, *Apidium*, has definite tarsioid resemblances. Other Fayum primates include several small apes, such as an early gibbon, a possible ancestor of Old World monkeys, and *Aegyptopithecus*.

Living *Catarrhini* include the Old World monkeys, or cercopithecoids, along with apes and humans. Familiar cercopithecoids include baboons (*Papio*), guenons and vervet monkeys (*Cercopithecus*), macaques (*Macaca*), langurs (*Presbytis* and *Pygathrix*), and colobus monkeys (*Colobus*); there are many others. The large and complex social groups of macaques and baboons have been studied repeatedly. There are dominance relationships within these primate societies, but they are usually expressed by gestures and displays instead of by fighting.

The family *Pongidae* includes the apes. The smaller, more lightly built apes are called gibbons (*Hylobates* and *Symphalangus*). These skillful acrobats are often placed in a family by themselves, the *Hylobatidae*. The more typical "great" apes include the orangutans (*Pongo*) of Asia, the gorillas and chimpanzees (*Pan*) of Africa, and several fossil apes, such as *Dryopithecus*. Modern apes have long arms and shorter legs. They exhibit a variety of locomotor patterns: arm-swinging (brachiation) in gibbons, knuckle-walking in gorillas and chimpanzees, and a four-handed type of clambering in the orangutan.

The family *Hominidae* includes humans, who walk upright and communicate using language. All living humans belong to the single species *Homo sapiens*.

SUBORDER *STREPSIRHINI*

The *Strepsirhini*, or *Lemuroidea*, include lemurs and their relatives. Once thought to be "lower" on the evolutionary scale, the lemuroids are now known to be separately specialized in many ways. Their placentas, for example, are of a peculiar type not found among other primates or among primitive mammals. The ear regions of their skulls show certain anatomical peculiarities; the same is true of the brain and the facial muscles.

All true lemurs live on the island of Madagascar, off the eastern coast of Africa. Most lemurs have the lower front teeth modified to form a comblike structure used in cleaning the fur. Other lemuroids include the small, agile galagos of mainland Africa and four types of slower-moving lorises in Africa and southern Asia. Many fossil lemuroids are known from the Eocene epoch.

GENETIC COMPARISONS

For many years, the classification of primates was based largely on anatomy, physiology, and paleontology. Now, however, the molecular structure of proteins and other types of biochemical evidence are used to make additional comparisons. Closely related primates have proteins with similar or identical sequences. For the most part, family trees based on protein sequences confirm the traditional classifications, with some exceptions; the separateness of gibbons from other apes is based largely on such biochemical evidence.

Geneticists have compared the chromosome sequences of humans and apes. They have found that the sequences of banding patterns in chimpanzee and gorilla chromosomes are nearly identical and that fewer than twenty chromosomal inversions (end-to-end changes) separate chimpanzees and humans.

TYPES OF PRIMATE STUDY

Primates are studied by different specialists, each using different methodologies. Primate anatomists and functional morphologists dissect primates and compare their structures to those of other species. Morphologists may also take numerous measurements and analyze the results statistically, often with the aid of a computer.

Primate ethologists and sociobiologists study the behavior of living primates, both in the field and in the laboratory. Field studies have been conducted among the chimpanzees in the Gombe Stream Reserve in Tanzania, the rhesus macaques of India, the Japanese macaques of Japan, the howler monkeys of Barro Colorado Island in the Panama Canal Zone, and the baboons of the East African savannas. Studies have confirmed that these primates have diverse locomotor patterns, varied diets, and complex and flexible social organizations that help them find food, avoid predators, and survive under difficult circumstances.

Molecular biologists and geneticists analyze the structures of important proteins and of deoxyribonucleic acid (DNA). This information provides important evidence of the relatedness of one group of primates to another. It was through such studies that molecular biologists were able to show that chimpanzees and gorillas are nearly identical, that humans are very closely related to these African apes, and that the gibbons of Asia are much more distantly related.

In general, the findings of molecular biology and genetics have tended to confirm the earlier findings based on comparative anatomy and paleontology, but with occasional exceptions.

Observations on living species are supplemented by studies of fossil primates. Paleontologists are always trying to connect living and extinct species into common family trees. Some molecular biologists have made estimates, based on biochemical differences, of the time of evolutionary divergence between apes (*Pongidae*) and humans (*Hominidae*). These estimates are somewhat controversial because they place the hominid-pongid split at less than five million years ago—in contrast to earlier estimates of ten million years or more made by paleontologists in the 1960s. Reinterpretations of the fossil primate *Ramapithecus* and other fossils from Asia support the biochemists' view that the hominid-pongid split took place less than five million years ago.

Many scientists with other fields of interest use nonhuman primates as experimental subjects for medical or biochemical research, including cancer research, space exploration, and the testing of new drugs. These scientists also make important additions to knowledge of how similar these primates are to humans. In particular, monkeys and apes are subject to nearly all the same diseases as are humans. Many drugs and surgical procedures designed to aid in the fight against these diseases are first tested in monkeys or apes because the physiological responses of these primates to the drugs or other medical procedures are nearly always the same as they are in humans. One case of particular importance is acquired immunodeficiency syndrome (AIDS). This disease, which attacks the immune system, does not occur in rats and mice, but it does occur in monkeys and apes. For these reasons, monkeys and apes are used extensively in tests on the AIDS virus and on possible treatments or cures for this disease.

HUMANS AND OTHER PRIMATES

Since the nonhuman primates are *Homo sapiens*' closest relatives, most studies of primates involve inevitable comparisons with humans. Most of the knowledge that is gained in studying the primates helps scientists understand the human species better. Primates, especially macaques, are commonly used in medical and behavioral research. New drugs, for example, are often tested in monkeys because the physiological reactions of these primates are likely to be more similar to those of humans than would be the case in rats or mice.

Chimpanzees are used when an even closer approximation to human anatomy or human intelligence is important to the experiment. From the use of these and other primates in medical research, knowledge of human health and human diseases has been greatly enhanced.

Scientific understanding of human behavior and other aspects of human biology is similarly enhanced by studies of other primates. Studies of sign language among apes, for example, have greatly increased the understanding of human language and the ways in which it is learned. Both gorillas and chimpanzees have successfully been taught American Sign Language (ASL), a gesture-language commonly used by deaf people. Many researchers who have worked with these apes believe that they show true language skills in their use of ASL, although some linguists disagree. Apes who use sign language can converse about past, present, and future events, faraway places, hypothetical ("what if?") situations, pictures in books, and individual preferences. These apes apparently can use language to lie, to play games or make puns, or to create definitions, such as "a banana is a long yellow fruit that tastes better than grapes."

Studies on the social organization of baboons, langurs, and other nonhuman primates have greatly increased understanding of how human social organization might have originated. Most nonhuman primates have social groups based on friendships and dominance relationships. Larger and stronger individuals tend to get their way more often, but usually by gesturing or threatening rather than by actually fighting. Even this observation, however, must be qualified, because encounters involving three or more individuals are generally very complex, and less dominant individuals can often manipulate these complex situations to their advantage.

THREAT OF EXTINCTION

A January 2017 study published in the journal *Science Advances* found that 75% of all primate species are in decline, and 60% of all primate species are threatened with complete extinction. The study, which was the largest of its kind to date and involved the skills and knowledge of over thirty primatologists, studied every primate species known to humankind. The study concluded that the decline in primate numbers and the all-out threat of extinction for many species was due in large part to habitat loss and devastation caused by human commercial development and over population, encroaching human agriculture and farmland, increased hunting by humans, and expanded mining by humans for minerals such as lithium, which is used to power modern-day batteries. The report noted that if these factors continued unabated, several primate species would become extinct by the year 2067. According to the report, every species of ape and 87% of the lemur species was threatened with extinction.

Despite the declining numbers, over 80 new primate species were identified from 2000 to 2017, with a total of 505 primate species being described as of January 2017. The most recent discovery occurred in China in January 2017 and described a new species of gibbon. Although this may seem hopeful, scientists explain the increase in the discovery of new primate species as being a direct result of human-caused deforestation, which allows researchers to access once-remote species. The rush to identify and describe new species is even greater, many feel, for fear that entire species will disappear before they are even known.

—*Eli C. Minkoff*

FURTHER READING

Ciochon, Russell L., and John G. Fleagle, eds. *Primate Evolution and Human Origins*. Benjamin/Cummings, 1985.

Estrada, Alejandro, et al. "Impending Extinction Crisis of the World's Primates: Why Primates Matter." *Science Advances*, vol. 3, no. 1, e16000946, doi: 10.1126/sciadv.16000946. Accessed 20 Feb. 2017.

Fleagle, John G. *Primate Adaptation and Evolution*. 3rd ed. Academic, 2013.

Jolly, Alison. *Evolution of Primate Behavior*. Macmillan, 1972.

Kinzey, Warren G., ed. *New World Primates: Ecology, Evolution, and Behavior*. Walter de Gruyter, 1997.

Napier, J. R., and P. H. Napier. *A Handbook of Living Primates*. Academic, 1967.

Napier, J. R., and P. H. Napier. *The Natural History of the Primates*. MIT Press, 1985.

Rowe, Noel. *The Pictorial Guide to the Living Primates*. East Hampton: Pogonias, 1996.

Sleeper, Barbara. *Primates: The Amazing World of Lemurs, Monkeys, and Apes*. San Francisco: Chronicle, 1997.

Smuts, Barbara B., et al., eds. *Primate Societies*. University of Chicago Press, 1987.

Zimmer, Carl. "Most Primate Species Threatened with Extinction, Scientists Find." *The New York Times*, 18 Jan. 2017, www.nytimes.com/2017/01/18/science/almost-two-thirds-of-primate-species-near-extinction-scientists-find.html. Accessed 20 Feb. 2017.

PROTOZOA

FIELDS OF STUDY

Microbiology; Protozoology; Cytology; Genetics; Developmental Biology

ABSTRACT

Protozoa have been known only since the seventeenth century, when Dutch naturalist Antoni van Leeuwenhoek observed them with simple microscopes of his own making. The protozoa are a vast assemblage of extremely diverse organisms whose only common characteristic is that they are all unicellular. They were previously classified as members of the animal kingdom but are now recognized as a distinct group and occupy a kingdom of their own, Protista.

KEY CONCEPTS

autotroph: any organism capable of synthesizing its own food by using either solar or chemical energy

ciliate: a protozoan that uses short, hairlike structures called cilia for locomotion

flagellate: a protozoan that uses long, whiplike structures called flagella for locomotion

heterotroph: any organism that must consume other organisms or organic substances to obtain nutrition

nucleus: a cellular organelle that coordinates the cell's activities and contains the genetic information

organelle: any of several structures found inside an individual cell, analogous to the organs of multicellular organisms

phagocytosis: obtaining food by engulfing it

pseudopods: cytoplasmic extensions of a protozoan's body, used for locomotion and the engulfing of food

unicellular: consisting of only one cell

LITTLE BEASTIES

The protozoa are a vast assemblage of extremely diverse organisms whose only common characteristic is that they are all unicellular. They were previously classified as members of the animal kingdom but are now recognized as a distinct group and occupy a kingdom of their own, *Protista*.

Protozoa have been known only since the seventeenth century, when Dutch naturalist Antoni van Leeuwenhoek observed them with simple microscopes of his own making. Leeuwenhoek's discovery was accidental; he was simply curious about the microscopic nature of such common substances as rainwater and scrapings from his mouth and teeth. His observations of small, active creatures associated with these substances were the first recorded descriptions of microscopic life. Although Leeuwenhoek made no effort to classify protozoa (which he called "little beasties"), his observations were later confirmed by scientists who recognized the importance of his discoveries.

PROTOZOAN DIVERSITY

The diversity of protozoan types cannot be overemphasized. Though some have a close evolutionary relationship, others have followed very different paths. Some protozoa are animal-like, obtaining energy by consuming food from external sources, and others resemble plants in that they contain chlorophyll and use the energy of the sun to manufacture food.

Protozoa occur in salt water and freshwater, in the soil, and inside larger organisms. In short, they are wherever moisture is present. Most are motile, although some are sessile (attached), and others, like volvox, form colonies. The great majority of protozoa are microscopic, although the very largest are just visible to the naked eye. The blood parasite *Anaplasma* is only one-sixth to one-tenth the size of a human red blood cell, while the ciliate *Spirostomum* (a freshwater protozoan that uses tiny hairlike cilia for locomotion) is up to three millimeters long and can be seen without a microscope.

An individual protozoan, because it consists of only one cell, does not contain any organs. Instead, it harbors analogous structures called organelles, which enable it to carry out its life functions. Central among these organelles is the nucleus, which contains the genes and serves as a sort of control center for cell activities. The cells of protozoa contain from one to many nuclei, depending on the species. Other organelles that distinguish the protozoa are those used for locomotion. These include whiplike flagella, hairlike cilia, and flowing extensions of the cell called pseudopodia, or false feet, which are generated in the direction of movement. Locomotory organelles are important in the classification of the protozoa. Many protozoa, especially freshwater species, possess an organelle called the contractile vacuole, which serves to pump excess water out of the cell. This organelle is especially important in freshwater protozoa, which tend to accumulate water by osmosis because their bodies contain more dissolved substances than the water in which they live. As water flows into these protozoa, the contractile vacuole absorbs it, expands, and finally contracts to flush the water through the cell membrane and into the surrounding environment.

Protozoa can be divided into two groups based on how they get their food: autotrophic and heterotrophic. Autotrophic protozoa contain chlorophyll and are able to convert solar energy into usable energy. The freshwater *Euglena* are autotrophs. Heterotrophic protozoa must capture their food, either funneling it through mouthlike openings or, like the amoeba, engulfing food particles with their pseudopods. This latter method of obtaining food is called phagocytosis.

Gas exchange in protozoa occurs through simple diffusion of gases across the cell membrane. Some protozoa live in environments such as stagnant water or the guts of animals where little oxygen is present. These protozoa are called facultative anaerobes, which means they can live with or without oxygen.

PROTOZOAN REPRODUCTION

Most protozoa reproduce asexually: a single parent produces offspring that are identical genetic copies of itself. Asexual reproduction in protozoa is accomplished in two ways, through fission and budding. In fission, the parent cell simply divides into two offspring cells of equal size. In budding, the offspring is a smaller cell that grows off the larger parent cell. Some protozoa carry out sexual reproduction, in which two parent cells release smaller sex cells called gametes. Two gametes fuse and grow into a new protozoan that is genetically different from either parent.

Conjugation is an interesting process that occurs in ciliate protozoa and may be linked to the reproductive process. It involves the temporary linking of two cells, followed by an exchange of nuclei, after which the cells separate. However, no offspring are formed. It is thought that conjugation is a means of renewing the nuclei of certain protozoa so that they may continue to reproduce asexually. Another process that many protozoa undergo is called encystment, in which the protozoa secrete a thickened sheath around themselves and enter a period of dormancy. Encystment enables protozoa to survive adverse environmental conditions such as drought or cold. When conditions become favorable again, the protozoa emerge from their cysts and resume their normal activities.

PROTOZOAN PHYLA

The kingdom *Protista* is subdivided into five groups, or phyla: *Mastigophora*, *Sarcodia*, *Sporozoa*, *Cnidospora*, and *Ciliophora*. The phylum *Mastigophora* contains the flagellates: protozoa that move by means of one or more whiplike flagella. The mastigophorans are of two types: phytoflagellates, which contain chlorophyll and are primarily autotrophic; and zooflagellates, which are primarily heterotrophic. Many zooflagellates are parasites of insects and vertebrates. The flagellates of the phylum *Mastigophora* show such a high degree of structural diversity that it is impossible to describe a typical representative. However, among the more interesting of the phylum members are the dinoflagellates, which occur in both marine and freshwater environments. They have two flagella, one of which points away from the cell and the other of which lies transversely around the cell. Many dinoflagellates contain a golden-brown photosynthetic pigment called xanthophyll and are therefore autotrophic, but others are unpigmented and are heterotrophic. These protozoa are encased in a cellulose sheath, or pellicle, which is often highly ornamented with spikes, hooks, or other prominences. Certain marine species are responsible for red tides, a population explosion that is responsible for massive fish kills and causes paralytic poisoning in humans who consume shellfish taken from red tide waters.

The protozoa in the phylum *Sarcodina* move about and engulf their food by means of pseudopodia. Amoebas belong to this phylum, as well as many marine, freshwater, and terrestrial species. The members of this phylum are the simplest protozoa and have few organelles. Some have exquisitely crafted external skeletons. The radiolarians, which are restricted to the marine environment, have beautifully sculptured exoskeletons made of silicon dioxide or strontium sulfate. The skeleton is riddled with tiny pores through which the living organism within is able to extrude thin filaments of cytoplasm for purposes of feeding, locomotion, and anchorage. Radiolarians exist in great numbers in the surface waters of the ocean and, upon dying, sink to the bottom where in some areas their skeletons form a large proportion of ocean bottom sediments.

Two phyla are closely related: the *Sporozoa* and *Cnidospora*. These are parasitic protozoa collectively referred to as sporozoans because some members of both groups go through sporelike infective stages. The phylum *Sporozoa* contains parasites that infect cells of the intestine and blood and are responsible for causing malaria in humans. These sporozoans belong to the genus *Plasmodium*, a group containing more than fifty species, four of which infect humans. All require a mosquito for transmission of the disease-causing protozoan. When the mosquito bites a person, the parasite is introduced into the blood with the saliva of the insect. From the blood, the protozoan travels to the individual's liver, where it reproduces, then escapes to invade red blood cells. Once in the red blood cells, it consumes hemoglobin, reproduces, and escapes to invade additional red blood cells. This cycle of invasion and escape is responsible for the debilitating chills and fever characteristic of malaria. The phylum *Cnidospora* contains amoeboid parasites of fishes and insects. This phylum's members are separated from those in *Sporozoa* based on differences in the nature of the infective spores.

The largest phylum is *Ciliophora*, which has more than eight thousand species. Of all the phyla, this one has the most homogeneous members as far as their evolutionary relationships are concerned. They all appear to have arisen from a common ancestor. Members of this phylum are commonly referred to as ciliates because they use cilia for locomotion and for sweeping food particles into the cytostome (cell mouth). They are also characterized by the presence of two nuclei, a small micronucleus and a larger macronucleus whose sole purpose is the synthesis of deoxyribonucleic acid (DNA, the genetic material). Although some ciliophorans are sessile, or attached, many are nimble, quick, and carnivorous and are therefore perhaps the most animal-like of all the protozoa. They reproduce either asexually by fission or through sexual conjugation, but they never release free gametes into the surrounding environment. Ciliophorans are common in fresh and salt water. Some are parasitic.

PROTOZOOLOGY

Although Antoni van Leeuwenhoek had little sense of what he was looking at when he first observed protozoa in the seventeenth century, he used the same tool that later scientists did, the microscope. Protozoology—the study of organisms too small to be observed with the naked eye—has become an area of specialization within microbiology.

Protozoa are best examined while alive, as specimens that have been killed, fixed, and stained often become distorted. Some of the smaller species can be viewed under the cover slip of a microscope slide; others, especially the larger specimens, must be viewed under less restrictive conditions: in a petri dish with a dissecting microscope or under a raised cover slip with corners that have been supported with paraffin or some other material.

One of the challenges of studying protozoa with the microscope is the speed with which they move and the tedium of keeping track of them by constantly adjusting the position of the slide. For this reason, a solution of methyl cellulose is often added to the slide preparation to increase the viscosity of the water and slow down the organisms.

As many protozoa are almost transparent, it is necessary to highlight their features by using dyes or stains that will not kill them. Such substances, called vital stains, render various structures in the protozoa visible under the microscope. For example, the vital stain Janus Green B selectively stains mitochondria, the energy-producing organelles, while neutral red has an affinity for vacuoles, the organelles that contain engulfed food particles. In studying protozoan structure, scientists use an assortment of vital stains and regular stains, which are used when protozoa are killed and preserved on a slide. Electron microscopy, used to examine the extremely fine structure of protozoa, has special stains and fixatives of its own for maintaining the structure of these organisms under the sustained electron bombardment required to illuminate the specimen.

Protozoa are ideal subjects of study for cytologists, geneticists, and developmental biologists because they reproduce asexually and therefore provide the researcher with an endless succession of clones, or genetically identical individuals. Scientists can subject these clones to various environmental conditions and observe the effects in their offspring.

The primary method of maintaining protozoa in a laboratory is a culture, which usually consists of one species of protozoa and the bacterium on which it feeds. The object is to exclude other organisms that might harm the protozoan. As toxic waste products build up in the culture dish, a small number of protozoa are transferred to a clean culture medium. This procedure is called subculturing and continues as long as the culture is maintained. There are two types of cultures: monoxenic and axenic. In monoxenic cultures, a single species of protozoa is maintained with a single species of bacteria. In axenic cultures, a single species of protozoa lives in isolation from any other organisms. The axenic culture is of value in certain types of nutritional studies.

Parasitic protozoa can be examined separate from their hosts in a laboratory but only when kept under conditions that provide them with a chemical and physical environment that will keep them alive for as long as necessary. For example, some parasitic protozoa of warm-blooded animals must be kept in a warm culture medium to prevent their being distorted by cooling.

THE BASE OF THE FOOD CHAIN

Protozoa are of first and foremost importance to ecology. These organisms make up the lowest links of the food chain that eventually leads to humans. Autotrophic protozoa are called producers because, like plants, they are able to use the energy of sunlight to manufacture food substances. Other protozoa are heterotrophs, or consumers, and must take food from their environments. Dinoflagellates are a group of autotrophs with a notorious role in the marine environment. These organisms, which are responsible for red tides, can pass their potent toxins up the food chain to cause paralytic shellfish poisoning in humans. The toxic materials they release into the ocean during periods of intense reproductive activity also lead to massive fish kills, which mean losses for the fishing industry.

Foraminifers are a type of shelled amoeba in the phylum *Sarcodina*. Some foraminifers existed millions of years ago, when Earth's fossil fuel deposits were being generated. Therefore, the presence of their fossil remains in drill cores helps indicate where oil is likely to be found.

An interesting relationship exists between protozoa and termites. Cellulolytic (cellulose-breaking) protozoa inhabit the guts of termites and are the organisms actually responsible for digestion of the wood that termites eat. Without these protozoa, the termites would not be able to feed on wood and would no longer be able to damage homes and other structures.

PROTOZOAN INFECTIONS

In people, protozoa cause numerous infections, generally limited to four sites in the body: the intestines, the genital tract, the bloodstream and tissues, and the central nervous system (brain and spinal cord). Not all parasitic protozoa cause death in humans, but they are responsible for untold misery in wide areas of the world, especially underdeveloped countries where preventive medicine is unknown and too expensive for these societies to employ.

The two most common protozoan infections are malaria and amoebic dysentery. The malarial parasite is transmitted through the bite of the *Anopheles* mosquito and eventually destroys the body's red blood cells, causing anemia and recurrent debilitating chills and fever. Malaria is a major health problem in parts of Africa, Asia, and Central and South America. More than one hundred million people are afflicted at any given time, and about one million die of the disease each year. Amoebic dysentery flourishes in overcrowded, unsanitary conditions. The disease gradually erodes the inner lining of the intestines, causing ulcers. Amoebic dysentery is entirely avoidable when food and water are handled carefully to prevent their contamination or are purified if contamination is suspected.

Other protozoa of medical importance are *Trichomonas vaginalis*, which parasitizes the genital tract; *Giardia lamblia*, a parasite of the small intestine; and *Toxoplasma gondii*, which infects the fetus and can cause congenital abnormalities.

Protozoa play an important role in the degradation of both human and industrial wastes in the environment. In sewage treatment, for example, anaerobic protozoa (those that flourish in the absence of oxygen) and bacteria play a role in the degradation of raw sewage. After they have done their work, the effluent is passed to a set of tanks with aerobic (oxygen-requiring) protozoa for further processing. Autotrophic protozoa have been used to deal with industrial wastes containing high levels of nitrates and phosphates. The settling tanks are illuminated to promote the growth of the autotrophs, which absorb and metabolize the industrial chemicals as part of their normal life processes.

—*Robert T. Klose*

Further Reading

Barnes, Robert D. *Invertebrate Zoology*. 6th ed. Saunders College Publishing, 1993.

Hegner, Robert W. *Big Fleas Have Little Fleas: Or, Who's Who Among the Protozoa*. Dover, 1968.

McKane, L., and J. Kandel. *Microbiology: Essentials and Applications*. McGraw-Hill, 1996.

Margulis, L., et al. *Illustrated Glossary of Protoctista*. Jones and Bartlett, 1993.

Morholt, E., and Paul F. Brandwein. *A Sourcebook for the Biological Sciences*. Harcourt Brace, 1980.

Roberts, Larry S., et al. *Foundations of Parasitology*. 6th ed. McGraw-Hill, 2000.

Whitten, R., and W. Pendergrass. *Carolina Protozoa and Invertebrates Manual*. Carolina Biological Supply, 1980.

Punctuated Equilibrium and Continuous Evolution

Fields of Study

Evolutionary Biology; Paleontology; Paleoecology; Population Biology

Abstract

Although Charles Darwin's most influential work was entitled On the Origin of Species (1859), in fact it did not address the problem in the title. Darwin believed that species formed by gradual transformation of existing ancestral species (known as gradualism). In this view, species are not real entities but merely arbitrary segments of continuously evolving lineages that are always in the process of change through time. Paleontologists tried to document examples of this kind of gradual evolution in fossils, but remarkably few examples were found.

Key Concepts

allopatric: populations of organisms living in different places and separated by a barrier that prevents interbreeding

gradualism: the idea that transformation from ancestor to descendant species is a process spanning millions of years

macroevolution: large-scale evolutionary processes that result in major changes in organisms

microevolution: small-scale evolutionary processes resulting from gradual substitution of genes and resulting in very subtle changes in organisms

speciation: the process by which new species arise from old species

species selection: a higher level of selection above that of natural selection is postulated to take place on the species level

stasis: the long-term stability and lack of change in fossil species, often spanning millions of years of geologic time

sympatric: populations of organisms living in the same place, not separated by a barrier that would prevent interbreeding

NOT THE "ORIGIN" OF SPECIES

Although Charles Darwin's most influential work was entitled On the Origin of Species (1859), in fact it did not address the problem in the title. Darwin was concerned with showing that evolution had occurred and that species could change, but he did not deal with the problem of how new species formed. For nearly a century, no other biologists addressed this problem either. Darwin (and many of his successors) believed that species formed by gradual transformation of existing ancestral species, and this viewpoint (known as gradualism) was deeply entrenched in the biology and paleontology books for a century. In this view, species are not real entities but merely arbitrary segments of continuously evolving lineages that are always in the process of change through time. Paleontologists tried to document examples of this kind of gradual evolution in fossils, but remarkably few examples were found.

THE ALLOPATRIC SPECIATION MODEL

By the 1950s and 1960s, however, systematists (led by Ernst Mayr) began to study species in the wild and therefore saw them in a different light. They noticed that most species do not gradually transform into new ones in the wild but instead have fairly sharp boundaries. These limits are established by their ability and willingness to interbreed with each other. Those individuals that can interbreed are members of the same species, and those that cannot are of different species. When a population is divided and separated so that formerly interbreeding individuals develop differences that prevent interbreeding, then a new species is formed.

Mayr showed that, in nature, large populations of individuals living together (sympatric conditions) interbreed freely, so that evolutionary novelties are swamped out and new species cannot arise. When a large population becomes split by some sort of barrier so that there are two different populations (allopatric conditions), however, the smaller populations become isolated and prevented from interbreeding with the main population. If these allopatric, isolated populations have some sort of unusual gene, their numbers may be small enough that this gene can spread through the whole population in a few generations, giving rise to a new species. Then, when the isolated population is reintroduced to the main population, it has developed a barrier to interbreeding, and a new species becomes established. This concept is known as the allopatric speciation model.

The allopatric speciation model was well known and accepted by most biologists by the 1960s. It predicted that species arise in a few generations from small populations on the fringe of the range of the species, not in the main body of the population. It also predicted that the new species, once it arises on the periphery, will appear suddenly in the main area as a new species in competition with its ancestor. These models of speciation also treated species as real entities, which recognize one another in nature and are stable over long periods of time once they become established. Yet, these ideas did not penetrate the thought of paleontologists for more than a decade after biologists had accepted them.

In 1972, Niles Eldredge and Stephen Jay Gould proposed that the allopatric speciation model would make very different predictions about species in the fossil record than the prevailing dogma that they must change gradually and continuously through time. In their paper, they described a model of "punctuated equilibrium." Species should arise suddenly in the fossil record (punctuation), followed by long periods of no change (equilibrium, or stasis) until they went extinct or speciated again. They challenged paleontologists to examine their biases about the fossil record and to see if in fact most fossils evolved gradually or rapidly, followed by long periods of stasis.

In the years since that paper, hundreds of studies have been done on many different groups of fossil organisms. Although some of the data were inadequate to test the hypotheses, many good studies have shown quite clearly that punctuated equilibrium describes the evolution of many multicellular organisms. The few exceptions are in the gradual evolution of size (which was specifically exempted by Eldredge and Gould) and in unicellular organisms, which have both sexual and asexual modes of reproduction. Many of the classic studies of gradualism in oysters, heart urchins, horses, and even humans have even been shown to support a model of stasis punctuated by rapid change. The model is still controversial, however, and there are still many who dispute both the model and the data that support it.

IMPLICATIONS OF PUNCTUATED EQUILIBRIUM

One of the more surprising implications of the model is that long periods of stasis are not predicted by classical evolutionary theory. In neo-Darwinian theory, species are highly flexible, capable of changing in response to environmental changes. Yet, the fossil record clearly shows that most species persist unchanged for millions of years, even when other evidence clearly shows climatic changes taking place. Instead of passively changing in response to the environment, most species stubbornly persist unchanged until they either go extinct, disappear locally, or change rapidly to some new species. They are not infinitely flexible, and no adequate mechanism has yet been proposed to explain the ability of species to maintain themselves in homeostasis in spite of environmental changes and apparent strong natural selection. Naturally, this idea intrigues paleontologists, since it suggests processes that can only be observed in the fossil record and were not predicted from studies of living organisms.

The punctuated equilibrium model has led to even more interesting ideas. If species are real, stable entities that form by speciation events and split into multiple lineages, then multiple species will be formed and compete with one another. Perhaps some species have properties (such as the ability to speciate rapidly, disperse widely, or survive extinction events) that give them advantages over other species. In this case, there might be competition and selection between species, which was called species selection by Steven Stanley in 1975. Some evolutionary biologists are convinced that species selection is a fundamentally different process from that of simple natural selection that operates on individuals. In species selection, the fundamental unit is the species; in natural selection, the fundamental unit is the individual. In species selection, new diversity is created by speciation and pruned by extinction; in natural selection, new diversity is created by mutation and eliminated by death of individuals. There are many other such parallels, but many evolutionary biologists believe that the processes are distinct. Indeed, since species are composed of populations of individuals, species selection operates on a higher level than natural selection.

If species selection is a valid description of processes occurring in nature, then it may be one of the most important elements of evolution. Most evolutionary studies in the past have concentrated on small-scale, or microevolutionary, change, such as the gradual, minute changes in fruit flies or bacteria after generations of breeding. Many evolutionary biologists are convinced, however, that microevolutionary processes are insufficient to explain the large-scale, or macroevolutionary, processes in the evolution of entirely new body plans, such as birds evolving from dinosaurs. In other words, traditional neo-Darwinism says that all evolution is merely microevolution on a larger scale, whereas some evolutionary biologists consider some changes too large for microevolution. They require different kinds of processes for macroevolution to take place. If there is a difference between natural selection (a microevolutionary process) and species selection (a macroevolutionary process), then species selection might be a mechanism for the large-scale changes in Earth's history, such as great adaptive radiations or mass extinctions. Naturally, such radical ideas are still controversial, but they are taken seriously by a growing number of paleontologists and evolutionary biologists. If they are supported by further research, then there may be some radical changes in evolutionary biology.

PATTERNS OF EVOLUTION

Determining patterns of evolution requires a very careful, detailed study of the fossil record. To establish whether organisms evolve in a punctuated or gradual mode, many criteria must be met. The taxonomy of the fossils must be well understood, and there must be large enough samples at many successive stratigraphic levels. To estimate the time spanned by the study, there must be some form of dating that allows the numerical age of each sample to be estimated. It is also important to have multiple sequences of these fossils in a number of different areas to rule out the effects of migration of different animals across a given study area. Once the appropriate samples have been selected, then the investigator should measure as many different features as possible. Too many studies in the past have looked at only one feature and therefore established very little. In particular, changes in size alone are not sufficient to establish gradualism, since these phenomena can be explained by many other means.

Finally, many studies in the past have failed because they picked one particular lineage or group and selectively ignored all the rest of the fossils in a given area. The question is no longer whether one or more cases of gradualism or punctuation occurs (they both do) but which is predominant among all the organisms in a given study area. Thus, the best studies look at the entire assemblage of fossils in a given area over a long stratigraphic interval before they try to answer the question of which tempo and mode of evolution are prevalent.

Since the 1940s, evolutionary biology has been dominated by the neo-Darwinian synthesis of genetics, systematics, and paleontology. In more recent years, many of the accepted neo-Darwinian mechanisms of evolution have been challenged from many sides. Punctuated equilibrium and species selection represent the challenge of the fossil record to neo-Darwinian gradualism and overemphasis on the power of natural selection. If fossils show rapid change and long-term stasis over millions of years, then there is no currently understood evolutionary mechanism for this sort of stability in the face of environmental selection. A more general theory of evolution may be called for, and, in more recent years, paleontologists, molecular biologists, and systematists have all been indicating that such a radical rethinking of evolutionary biology is on the way.

—*Donald R. Prothero*

FURTHER READING

Bennett, K. D. *Evolution and Ecology: The Pace of Life.* Cambridge University Press, 1997.

Eldredge, Niles. *Time Frames: The Rethinking of Darwinian Evolution and the Theory of Punctuated Equilibria.* Simon & Schuster, 1985.

Gerhart, John. Cells, *Embryos, and Evolution: Toward a Cellular and Developmental Understanding of Phenotypic Variation and Evolutionary Adaptability.* Blackwell Science, 1997.

Gould, Stephen J. "The Meaning of Punctuated Equilibria and Its Role in Validating a Hierarchical Approach to Macroevolution." In *Perspectives on Evolution,* edited by Roger Milkman. Sinauer Associates, 1982.

Gould, Stephen J., and Niles Eldredge. "Punctuated Equilibrium: The Tempo and Mode of Evolution Reconsidered." *Paleobiology* 3 (1977): 115–151.

Hoffman, Antoni. *Arguments on Evolution: A Paleontologist's Perspective.* Oxford University Press, 1988.

Levinton, Jeffrey S. *Genetics, Paleontology, and Macroevolution.* 2nd ed. New Cambridge University Press, 2000.

Mayr, Ernst. *Animal Species and Evolution.* Harvard University Press, 1963.

Moller, A. P. *Asymmetry, Developmental Stability, and Evolution.* Oxford University Press, 1997.

R

REGENERATION

FIELDS OF STUDY

Developmental Biology; Embryology; Biochemistry; Developmental Genetics

ABSTRACT

Regeneration is a process by which some organisms replace damaged or missing tissue using living cells adjacent to the affected area. The phenomenon is not well understood, but several animal systems have enabled developmental geneticists to develop strong models describing the process.

KEY CONCEPTS

chemotaxis: a process by which cells are attracted to a chemical, moving from low to high concentrations of the chemical, until the cells cluster

determination: an event in organismal development during which a particular cell becomes committed to a specific developmental pathway

differentiation: the process by which a determined cell specializes or assumes a specific function

fate map: a map of determined, but undifferentiated, tissue by which specific cell regions can be identified as giving rise to specific adult structures

imaginal disk: a determined, undifferentiated tissue in fruit fly larvae that gives rise to a specific adult structure

positional information: a concept by which differentiating cells organize themselves to produce a particular tissue type based upon cell-to-cell interactions

stem cell: a determined, undifferentiated cell that is hormonally activated and changes into a specific cell type

transdetermination: an event by which a determined, undifferentiated cell changes its determination, thereby giving rise to a different tissue type

FIX, REPAIR, OR REPLACE

Regeneration is a process by which some organisms replace damaged or missing tissue using living cells adjacent to the affected area. The phenomenon is not well understood, but several animal systems have enabled developmental geneticists to develop strong models describing the process.

The replacement of tissue is a common occurrence in fungi and plants. Regeneration can also occur in animals, although the capacity for regeneration progressively declines with increasing complexity in the animal species. Among primitive invertebrates (animal species lacking an internal skeleton), regeneration frequently occurs. A planarium can be split symmetrically into right and left halves. Each half will regenerate its missing mirror-image, resulting in two planaria, each a clone (exact genetic copy) of the other.

In higher invertebrates, regeneration occurs in echinoderms (such as starfish) and arthropods (such as insects and crustaceans). A radially symmetrical starfish can regenerate one or several of its five arms. Regeneration of appendages (limbs, wings, and antennae) occurs in insects such as cockroaches, fruit flies, and locusts. Similar processes operate in crustaceans such as lobsters, crabs, and crayfish. Limb regeneration extends even to lower vertebrate species (species having an internal skeleton) such as amphibians and reptiles, although on a very limited basis.

In amphibians, the newt can replace a lost leg. In reptiles, some lizards can lose their tails when captured by a predator, thus assisting their escape. If the lizard is young, the tail can regenerate. The tail breaks because of a breakage plane in the tail which severs upon hormonal activation. The glass-snake lizard is such a species.

PRINCIPLES OF REGENERATION

Regeneration in these organisms is based upon two principles: the symmetrical organization of cells in the organism, and the reversal of determination and differentiation in the surviving cells, termed blastema, adjacent to the missing tissue. These two factors are fundamental to the development of the organism.

In animal systems, two major body symmetries emerge—radial symmetry and bilateral symmetry. In radially symmetric organisms, including plants and primitive invertebrates such as hydra, jellyfish, and starfish, body tissues are arranged in a circular orientation about a central axis. Appendages may also be present that likewise orient in a circular pattern of cells. In bilaterally symmetrical organisms, including animal species such as planaria, arthropods, fish, amphibians, reptiles, birds, and mammals, the body is oriented into mirror-image halves about a central body plane, resulting in its having right and left equivalent structures along each half.

Body symmetry is critical for tissue regeneration because of positional information. The cells of an individual organize into a specific pattern during development. Cell-to-cell interactions and chemical messengers between cells provide the cells of a given tissue with information signals directing the cells to grow in a particular direction or pattern. The loss of a tissue portion may stimulate the remaining cells to carry out a programmed growth, that is, to complete a specified pattern.

Determination and differentiation both play an important role in the genetic basis for development. All cells of an organism contain the same genetic material; that is, they all contain the same deoxyribonucleic acid (DNA), the same genes. In a specialized organism (one having different tissue types—eyes, ears, skin, and nerves), these cells must behave differently, even though they contain the same genetic information. The process by which identical cells with the same genetic information give rise to different tissues is termed differentiation.

What causes similar cells to differentiate to form different tissues is a process called determination. Prior to differentiation, cells become determined, meaning that some genes in these cells are turned on, making certain proteins, while other genes are turned off, not making other proteins. All cells of a specific tissue have the same genes that are turned on or off and therefore make the same proteins (for example, all red blood cells manufacture hemoglobin). Cells of other tissue types have different genes that are turned on or off and make other proteins (for example, epidermis cells manufacture keratin, not hemoglobin). How cells determine and differentiate depends upon chemical signals (hormones). Hormones signal different cells that receive chemically coded information based upon their location in the organism, which is based upon the organism's symmetry.

COCKROACHES, NEWTS, AND FRUIT FLIES

Three principal animal regenerative systems have been studied: cockroach limb regeneration, newt limb regeneration, and fruit fly imaginal disk regeneration. In all animal systems, regeneration occurs primarily in younger individuals undergoing metamorphosis, which is a change in development involving considerable alterations in body size and physical appearance. Adult regeneration is incomplete or does not occur.

Severing a cockroach limb results in distal regeneration; the remaining leg part regenerates the lost portion. If the middle portion of a leg is removed and the remaining leg parts are grafted together, complete regeneration of the missing middle portion occurs precisely between the grafted parts, but grafting requires correct orientation of the body parts. If the leg parts are grafted backward, regeneration will be

A dwarf yellow-headed gecko (Lygodactylus luteopicturatus) *with regenerating tail pictured in Dar es Salaam, Tanzania. Approximately 7 cm long. (Muhammad Mahdi Karim)*

distorted, resulting in a malformed limb, sometimes with multiple leg structures sprouting from one limb. Virtually identical results have been obtained for newt limb regeneration. Furthermore, limb regeneration in the presence of certain chemicals such as retinol palmitate causes complete limb regeneration, including undamaged regions. The net result is a severely deformed limb.

For the fruit fly *Drosophila melanogaster*, the period from egg to adult is approximately ten days at 25 degrees Celsius. The period includes roughly seven days during which the organism proceeds through three larval (maggot) stages, followed by a three-day immobile pupal stage, during which metamorphosis occurs. Metamorphosis involves the replacement and modification of larval body structures with adult body structures (eyes, legs, and wings). The cells that are to become the adult structures are present, but dormant, in the larval stages. These special cells, called imaginal disks, are determined to become specific adult structures, but remain undifferentiated until activated by the hormone ecdysone during metamorphosis.

There is one imaginal disk for each future body structure (two eye imaginal disks, six leg imaginal disks, for example). Gerold Schubiger and other geneticists have determined "fate maps" for each imaginal disk. They have determined what each cell group on each disk will become in the adult. For example, the male genital disk has been mapped so that specific cell groups are associated with the formation of the specific adult structures of, for example, the heart, penis, and testes.

Experiments by Peter J. Bryant and Schubiger focused upon removing parts of leg and wing imaginal disks. The remaining surviving cells either regenerated the missing tissue or duplicated themselves. Whether regeneration or duplication occurred depended upon the amount of tissue lost and the position of the surviving imaginal disk tissue. If a small portion of a disk or a particular region of the disk was lost, then the remaining disk cells regenerated the missing part, thus producing a complete disk and ultimately a normal adult structure. If a large section of a disk or a sensitive region of it was lost, then the remaining cells duplicated a mirror-image of themselves, giving rise to a useless adult structure.

MODELS OF REGENERATION

These collective studies, especially those involving the *Drosophila* imaginal disks, have produced two comprehensive regeneration models: the gradient regeneration model and the polar coordinate model. The gradient regeneration model, proposed by Victor French, explains the regenerative capacity of a given tissue by arranging the cells of the tissue along a gradient of regenerative capacity. Cells are arranged in order of high regenerative capacity to low regenerative capacity. This high-to-low regenerative gradient is directly correlated to the positional information of each cell. Cells located proximal (near) to the main body axis have high regenerative capacity. Cells located distally (far) from the main body axis have progressively lower regenerative capacities. Removal of distal cells results in their replacement by regeneration of the proximal highly regenerative cells. The removed distal cells, which lack regenerative information because they are at the low end of the gradient, cannot regenerate the proximal cells.

The gradient regeneration model can best be visualized as a right triangle with its hypotenuse (longest side) being a downward slope. Highly regenerative cells (proximal to the main body axis) are at the top of the slope. They contain positional information for themselves plus information for all cells below them on the slope. Higher cells can replace lost lower cells (located distal to the main body axis). Distal cells low on the slope have considerably less positional information and therefore can only duplicate themselves.

The polar coordinate model, proposed by French with Peter J. Bryant and Susan V. Bryant, is a more elaborate version of the gradient model that explains not only the *Drosophila* imaginal-disk experiments but also the cockroach and newt regeneration experiments. This model is a three-dimensional gradient that covers regeneration not only in a proximate to distal direction but also from the exterior to interior. The polar coordinate model can best be visualized as a cone. The circular end of the cone represents proximal tissue, whereas the tapered tip represents distal tissue, thus simulating the proximal-to-distal regeneration gradient.

On the circular (proximal) base of the cone, imagine a bull's-eye. The outermost circle represents exterior tissue, whereas the circle center (bull's-eye) represents the most interior tissue. There is thus a three-dimensional regeneration gradient—proximal-to-distal and exterior-to-interior. The circle is furthermore subdivided clockwise into twelve regions, completing the polar coordinate model of tissue regeneration capacity.

The tissue pattern of regeneration will again favor those cells located at high gradient positions, namely proximal (cone base) and exterior (outside circle). These cells possess positional information for regeneration of lower gradient tissue. Lower gradient cells, located distally (cone tip) and interiorly (circle center), will have limited positional information and will be capable only of duplicating themselves.

For clockwise regeneration, the polar coordinate model operates by the shortest intercalation route; that is, if a small tissue section is lost, the remaining large section will regenerate the lost piece based upon positional information. If a large tissue section is lost, however, the remaining small section will lack sufficient regenerative information and will be capable only of duplicating a mirror image of itself.

The French, Bryant, and Bryant polar coordinate model is a three-dimensional regenerative capacity gradient intended to model a tissue based on cell position. The model really boils down to one principle: a large, proximal (or exterior) group of cells can regenerate missing tissue; a small, distal (or interior) group of cells cannot.

STUDYING REGENERATION

Developmental geneticists have studied regeneration in a variety of ways. Among the principal experimental techniques have been fate map determination of imaginal disks and limb regeneration, already discussed above, transdetermination of imaginal disks, and studies of simple organismal development.

Geneticists have found that under special circumstances, an imaginal disk or a portion of an imaginal disk can change its pattern of determination, that is, it transdetermines. A wing occasionally grows from an eye, for example, or a leg from a wing. Cells that are programmed to follow one developmental route follow another route instead. The cause of transdetermination is unknown, but the process does follow specific patterns. For example, a genital imaginal disk can transdetermine to form an antenna or leg, but not vice versa. An antenna disk can transdetermine to produce an eye, wing, or leg, but the wing and leg disks cannot transdetermine to an antenna.

Further regenerative studies involve the model developmental systems, including the cellular slime mold *Dictyostelium discoideum*. In the presence of adequate food, this exists as single, amoeba-like cells. If the cells are starved, they release a chemical attractant (chemotaxic) substance, cyclic AMP, that attracts the cells to one another. The resulting cellular mass moves as a single unit until the organism finds a suitable food source, upon which the cells differentiate and specialize to produce and release spores, each of which subsequently gives rise to a new amoeba-like stage. Such studies are necessary because regeneration ultimately involves changes in the determination and differentiation of cells.

FUTURE PROSPECTS

Regeneration research presents two opportunities for further development: an understanding of higher cell differentiation and growth, and prospects for replacing lost or damaged human tissues and organs. The polar coordinate model for tissue regeneration indicates that tissue replacement depends upon blastemas at the damaged area replacing tissue using positional information. Future research includes genetic and molecular studies to identify intercellular chemotaxic molecules and other information molecules that mediate cell-to-cell communication and thereby control how cells develop and grow in specified patterns. Such research will unravel important clues to cellular development and regeneration.

Scientists' understanding of cellular differentiation and growth is currently limited to more primitive species (such as *Dictyostelium discoideum* and *Drosophila melanogaster*). The most effective breakthroughs have been with the *Drosophila* imaginal disk studies and cockroach and newt limb regeneration experiments. Much more work remains to be done, particularly with respect to genetic and molecular analysis. The action of specific steroid and protein hormones on cellular growth and differentiation is a further avenue of research.

While regeneration research has been pursued for many decades, it is still in its infancy. Further research will allow the understanding of organismal development.

—*David Wason Hollar, Jr.*

FURTHER READING

Alberts, Bruce, Dennis Bray, Julian Lewis, Martin Raff, Keith Roberts, and James D. Watson. *Molecular Biology of the Cell.* 3rd ed. Garland, 1994.

Goodenough, Ursula. *Genetics.* 3rd ed. Holt, Rinehart and Winston, 1984.

Klug, William S., and Michael R. Cummings. *Concepts of Genetics.* 6th ed. Charles E. Merrill, 2000.

Lewin, Benjamin. *Genes VII.* Oxford University Press, 1999.

Sang, James H. *Genetics and Development.* Longman, 1984.

Starr, Cecie, and Ralph Taggart. *Biology: The Unity and Diversity of Life.* 9th ed. Brooks/Cole, 2001.

Wallace, Robert A., Jack L. King, and Gerald P. Sanders. *Biosphere: The Realm of Life.* 2nd ed. Scott, Foresman, 1988.

REPRODUCTIVE STRATEGIES IN ANIMALS

FIELDS OF STUDY

Population Biology; Reproductive Biology; Population Genetics

ABSTRACT

Having a reproductive strategy implies only that an organism has evolved a pattern that maximizes its success in the production of offspring. The concept of reproductive strategies is closely related to that of natural selection. Natural selection results in the more fit individuals within a population, under a given set of environmental circumstances, being more likely to pass on their genes to future generations. The organism's reproductive strategy, then, is that blend of traits enabling it to have the highest overall reproductive success.

KEY CONCEPTS

bet-hedging: a reproductive strategy in which an organism reproduces on several occasions rather than focusing efforts on a single or few reproductive events

K strategy: a reproductive strategy typified by low reproductive output; common in species living in areas having limited critical resources

litter size: the number of offspring produced per birth; also referred to as "clutch size"

population density: number of individuals per unit of area, especially when it pertains to the group's reproductive potential

r strategy: a reproductive strategy involving high reproductive output; found often in unstable or previously unoccupied areas

reproductive strategy: a set of traits that characterizes the successful reproductive habits of a group of organisms

AN AUTOMATIC STRATEGY FOR REPRODUCTIVE SUCCESS

The term "reproductive strategies" is probably something of a misnomer. A strategy implies that an organism has had conscious forethought in determining how to proceed with its reproductive events, that some planning has occurred. With the exception of humans, who can plan aspects of their parenthood, this is virtually impossible. Having a reproductive strategy implies only that an organism has evolved a pattern that maximizes its success in the production of offspring.

The concept of reproductive strategies is closely related to that of natural selection. Natural selection results in the more fit individuals within a population, under a given set of environmental circumstances, being more likely to pass on their genes to future generations. By this process, the gene pool (genetic makeup) of the population is altered over time. An organism's fitness can be assessed by evaluating two key characteristics—survival and reproductive success. The organism's reproductive strategy, then, is that blend of traits enabling it to have the highest overall reproductive success. Application of the term "reproductive strategy" has also been extended to describe patterns beyond individual organisms: the population, species—even entire groups of similar species, such as carnivorous mammals.

Examination of reproductive strategies is part of the larger study of life-history evolution, which attempts to understand why a given set of basic traits has evolved. These traits include not only those pertaining to reproduction but also those such as body size and longevity. To consider a reproductive strategy appropriately, one must view it within the context of the organism's overall life history, precisely because these traits (particularly body size) often affect reproductive traits. One should also evaluate the role that the organism's ancestry plays in these processes. A species' evolutionary history can have a profound effect on its current attributes.

TRAITS AND BEHAVIORS

A reproductive strategy consists of a collection of basic reproductive traits, including litter, or "clutch," size (the number of offspring produced per birth), the number of litters per year, the number of litters in a lifetime, and the time between litters, gestation, or pregnancy length. The age of the mother's first pregnancy is also a consideration. Another trait is the degree of development of the young at birth. In different species, mothers put varying levels of time and energy into the production of either relatively immature, or altricial, offspring or offspring that are well developed, or precocial.

Reproductive strategies also consist of behavioral elements, such as the mating system and the amount of parental care. Mating systems include monogamy (in which one male is mated to one female) and polygamy (in which an individual of one sex is mated to more than one from the other). The type of polygamy when one male mates with several females is called polygyny; the reverse is known as polyandry.

Finally, physiological events such as those involved in ovulation (what happens when the egg or eggs are shed from the ovary) may also be used to characterize a reproductive strategy. Some mammals are spontaneous ovulators. Females shed their eggs during the reproductive cycle without any physical stimulation. Other mammalian species are induced ovulators—a female ovulates only after being physically stimulated by a male during copulation. These patterns, induced and spontaneous ovulation, may be regarded as alternate reproductive strategies, each enabling a type of species to reproduce successfully under certain conditions.

The overall effectiveness of a reproductive strategy is important to consider with respect to the relative

Common house geckos (Hemidactylus frenatus) *mating. The male is biting the female to immobilize her. (Basile Morin)*

success of the offspring (even those in future generations) in leaving their own descendants. A sound reproductive strategy results in increased fitness. An organism's fitness as it affects the population's gene pool may not be adequately assessed until several generations have passed.

THE R AND K SELECTION MODEL

The model of r and K selection is the most widely cited description of how certain reproductive traits are most effective under certain environmental conditions. To appreciate this model, an understanding of elementary population dynamics is needed. At the early stages of a population's growth, the rate of addition of new individuals (designated r) tends to be slow. After a sufficient number of individuals is reached, the growth rate can increase sharply, resulting in a boom phase. In most environments, however, unrestrained growth cannot continue indefinitely. Critical resources—food, water, and protective cover—become more scarce as the environment's carrying capacity (K) is approached. Carrying capacity is the maximal population size an area can support. When the population approaches this level, growth rate slows, as individuals now have fewer resources to convert into the production of new offspring.

This pattern is defined as density-dependent population growth—the density or number of individuals per area that influences its growth. This description of population dynamics is also referred to as logistic growth and was conceived by the Belgian mathematician Pierre-François Verhulst in the early nineteenth century. It has successfully described population growth in many species.

The r and K selection model was presented by Robert H. MacArthur and Edward O. Wilson in their influential book, *The Theory of Island Biogeography* (1967). They argued that in the early phase of a population's growth, individuals should evolve traits associated with high reproductive output. This enables them to take advantage of the relatively plentiful supply of food. The evolution of such traits is called r selection, after the high population growth rates occurring during this phase. They also suggested that, as the carrying capacity was approached, individuals would be selected that could adjust their lives to the now reduced circumstances. This process is called K selection. Such individuals should be more efficient in the conversion of food into offspring, producing fewer young than those living in the population's early phase. In a sense, a shift from productive to efficient individuals occurs as the population grows.

Other biologists, most notably Eric Pianka, have extended this concept of r and K selection to entire species rather than only to individuals at different stages of a population's growth. Highly variable or unpredictable climates commonly create situations in which population size is first diminished but then grows rapidly. Species commonly occurring in such environments are referred to as r strategists. Those living in more constant, relatively predictable climates are less likely to go through such an explosive growth phase. These species are considered to be K strategists. According to this scheme, an r strategist is characterized by small body size, rapid development, high rate of population increase, early age of first reproduction, a single or few reproductive events, and many small offspring. The K strategist has the opposite qualities—large size, slower development, delayed age of first reproduction, repeated reproduction, and fewer, larger offspring.

Various combinations of r and K traits may occur in a species, and few are entirely r- or K-selected. Populations of the same species commonly occupy different habitats during their lives or across their geographic ranges. An organism might thus shift strategies in response to environmental changes—it may, however, be constrained by its phylogeny or ancestry in the degree to which its strategies are flexible.

CRITICISM OF THE MODEL

Because the r and K model of reproductive strategies seems to explain patterns observed in nature, it has become widely accepted. It has also met with considerable criticism. Charges against it include arguments that the logistic population-growth model (on which the r and K strategies model is based) is too simplistic. Another is that cases of r and K selection have not been adequately tested. Mark Boyce, an ecologist, has persuasively argued that for the r and K model to be most useful it must be viewed as a model of how population density affects life-history traits. Within this framework, also called density-dependent natural selection, the concept of r and K selection remains true to the one that MacArthur and Wilson originally proposed. Boyce suggests that the ability of r and K selection to explain reproductive strategies will have the best chance of being realized when approached in this fashion.

In addition to the r and K model, there are many other ways of describing reproductive strategies. For example, some species, such as the Chinook salmon, are semelparous: They reproduce only once before dying. The alternate is to be iteroparous—having two or more reproductive events over the organism's life. If juvenile death rates are high, an individual might be better off reproducing on several occasions rather than only once. (This reproductive strategy is referred to as "bet-hedging.") Finally, it has also been useful to evaluate reproductive strategies based on the proportion of energy that goes into reproduction relative to that devoted to all other body functions. This mode of analysis addresses such considerations as reproductive effort and resource allocation.

STUDYING REPRODUCTIVE STRATEGIES

Initially, one who studies the reproductive strategy of an organism should attempt to characterize its reproduction fully. The sample examined must be representative of the population under consideration—it should account for the variability of the traits being measured. Studies can involve any of several approaches. Short-term laboratory studies can uncover some hard-to-observe features, but there is no substitute for long-term field research. By studying an organism's reproduction in nature, a biologist has the best chance of determining how its reproduction is shaped by an environment. If the research is performed over several seasons or years, patterns of variability can be better understood. This is important in determining how the physical environment influences reproductive traits.

After data have been systematically collected, it might then be possible to characterize a reproductive strategy. Imagine that a mouse population becomes established in a previously uninhabited area and that the population has a high reproductive rate (it produces large litters). The young develop quickly and produce many young themselves. Because of this combination of circumstances, one might consider the reproductive strategy to be r-selected since the population has a high reproductive output in an unexploited area. Though the concept of r and K strategies is problematic, it still is common to typify a strategy as r- or K-selected based upon this approach.

Because a reproductive strategy needs to be seen as part of an organism's overall life history, however, other things should be measured to understand it fully. These may include the life span and population attributes such as survival patterns. Values should be taken for different age groups to characterize the population's strategy. Correlational analysis is a statistical procedure that is used to evaluate reproductive strategies. Through such a methodology, one assesses the degree of association between two variables or factors. This may involve relationships between two reproductive variables or between a reproductive and an environmental variable—for example, to determine whether there is a significant correlation between litter size and decreasing body size in mammals. If one were found to occur, the conclusion that smaller species typically have larger litters might be drawn, which is, in fact, true. Such an analysis enables the characterization of a change in reproductive strategy based on body size. Simply establishing a correlation does not prove that causation has occurred—it does not automatically mean that one factor is responsible for the expression of the other.

Multivariate statistical procedures are also used to analyze reproductive strategies. These allow the determination of how groups of reproductive traits are associated and of how they can be explained by several factors. One might determine that a certain bird species produces its greatest number of young, and that the young grow most rapidly, at northern locations having high snow levels. Such an approach is often needed in dealing with reproductive strategies—a combination of traits typically requires explanation.

REPRODUCTION AND SURVIVIAL

The characterization of an organism's reproductive strategy involves more than an understanding of reproductive traits. There is a successful process by which offspring are produced, and reproductive success is one of the two principal measures of fitness—the other is survival. Because a successful reproductive strategy ultimately results in high fitness, any discussion of these strategies bears directly on issues of natural selection and evolution.

An organism's reproductive strategy represents perhaps the most significant way in which the organism adapts to its environment. A successful reproductive strategy represents a successful mode of passing genes on to the next generation, so traits associated with a reproductive strategy are under intense natural selection pressure. If environmental conditions change, the original strategy may no longer be as successful. To the extent that an organism can shift its reproductive strategy as circumstances change, its genes will persist.

The study of reproductive strategies has helped scientists understand why certain modes of reproduction occur, based upon observations of a species itself and of its environment. An understanding of reproductive strategies may also be of some practical use. An organism's reproduction directly influences its population dynamics.

If an animal has small litters and is at an early age at first reproduction, its population should grow at a concomitantly high rate. These and other components of reproduction may strongly affect a species' population growth. A knowledge of how reproduction influences population dynamics can be important in wildlife management activities, which can range from strict preservation efforts to overseeing trophy hunting.

—*Samuel I. Zeveloff*

Further Reading

Austin, C. R., and R. V. Short, eds. *The Evolution of Reproduction.* Cambridge University Press, 1976.

Boyce, Mark S. "Restitution of r- and K-Selection as a Model of Density-Dependent Natural Selection." *Annual Review of Ecology and Systematics* 15 (1984): 427–447.

Clutton-Brock, T. H., F. E. Guiness, and S. D. Albon. *Red Deer: Behavior and Ecology of Two Sexes.* University of Chicago Press, 1982.

Ferraris, Joan D., and Stephen R. Palumbi, eds. *Molecular Zoology: Advances, Strategies, and Protocols.* Wiley-Liss, 1996.

MacArthur, Robert H., and Edward O. Wilson. *The Theory of Island Biogeography.* Princeton University Press, 1967.

Pianka, Eric R. *Evolutionary Ecology.* 6th ed. Harper & Row, 2000.

Wrangham, Richard W., W. C. McGrew, Frans B. M. De Waal, and Paul G. Heltne, eds. *Chimpanzee Cultures.* Harvard University Press, 1996.

Reproductive System of Female Animals

Fields of Study

Animal Anatomy; Animal Physiology; Reproductive Biology; Obstetrics and Gynecology; Endocrinology

Abstract

The function of the mammalian female reproductive system, in cooperation with the male reproductive system, is to produce offspring. The female must produce ova, provide the site for the combination of ova with sperm from the male (fertilization), and nourish and protect a developing fetus during pregnancy. She also must provide for the delivery of offspring from her body. These functions are carried out by a group of organs and a number of hormones. The major organs of the system include ovaries, which produce ova and hormones; uterine tubes that transport the ovum and provide the site of fertilization; the uterus; the vagina; and the external genitalia. Hormones important to the function of the female reproductive system include estrogen and progesterone, and follicle-stimulating hormone (FSH) and luteinizing hormone (LH) released under control of releasing factors produced by the hypothalamus.

Key Concepts

anterior pituitary gland: the front portion of the pituitary gland, which is attached to the base of the brain; the source of luteinizing hormone (LH) and follicle-stimulating hormone (FSH)

estrus cycle: hormonally controlled changes that make up the female reproductive cycle in most mammals; ovulation occurs during the estrus (heat) period

external genitals: the external reproductive parts of the female

gonad: the primary reproductive organ (the ovary in females and the testes in males), which produces sex cells (gametes) and sex hormones

menstrual cycle: a series of regularly occurring changes in the uterine lining of a nonpregnant primate female that prepares the lining for pregnancy

ovary: the female gonad, which produces ova and the hormones estrogen and progesterone

ovum (pl. ova): the female reproductive cell (gamete); a mature egg cell

uterus: the hollow, thick-walled organ in the pelvic region of females that is the site of menstruation, implantation, development of the fetus, and labor

Where Do Babies Come From?

The function of the mammalian female reproductive system, in cooperation with the male reproductive system, is to produce offspring. The role of the female is very complex: She must produce gametes (sex cells) called ova (singular, ovum) or eggs, provide the site for the combination of ova with sperm from the male (fertilization), and nourish and protect a developing fetus during pregnancy. She also must provide for the delivery of offspring from her body to the outside. These functions are carried out by a group of organs or structures and a number of chemicals called hormones. The major organs of the system include ovaries, which produce ova and hormones; uterine (Fallopian) tubes that transport the ovum and provide the site of fertilization; the uterus, which houses the developing offspring; the vagina, which receives the male penis and sperm during sexual intercourse and also functions as a birth canal; and the external genitals. Hormones important to the function of the female reproductive system include estrogen and progesterone from the ovaries, and follicle-stimulating hormone (FSH) and luteinizing hormone (LH) from the anterior pituitary gland. Release of FSH and LH is under control of chemicals called releasing factors, which are produced by a small region of the brain called the hypothalamus.

THE OVARIES

The paired ovaries are the primary sex organ, or gonad, of the female. They are analogous to the testes in the male reproductive system and actually develop from the same tissue. The size and shape of the ovary depend on the age and size of the female and whether the female usually has a single offspring or several at one time. Before birth, small groups of cells called follicles are formed in each ovary. In the center of each follicle is a single large cell called an oocyte, which is able to mature into an ovum. In other words, all the oocytes a female will ever produce were already in place before she was born. The ova are special cells; they are formed by a process (meiosis) that results in a cell with only half of the chromosomes found in other body cells. The only other cells that divide by this process are those that form sperm in males.

When an ovum and sperm unite, then, each cell contributes half the necessary chromosomes to make a new complete cell. This new cell, the first cell of an offspring, will have characteristics of each parent. The follicles develop in response to the hormone FSH from the anterior pituitary gland. A mature follicle releases its mature ovum through the wall of the ovary into the pelvic cavity. This process is called ovulation and is controlled by LH and FSH.

The ovary also produces the female sex hormones—estrogen and progesterone. Estrogen is produced by the maturing follicle cells. In addition to causing growth of the sex organs at puberty and stimulating growth of the uterine lining each month, estrogen is responsible for the appearance of female secondary sex characteristics. The follicle cells that are left behind following ovulation form a structure called the corpus luteum, which produces both estrogen and progesterone. The most important function of progesterone is to stimulate the lining of the uterus to complete its preparation for pregnancy.

THE ACCESSORY ORGANS

The rest of the internal structures of the reproductive system are called accessory organs. The first of these is a pair of uterine (Fallopian) tubes, or oviducts, which extend from each ovary into the uterus. They are frequently shaped like funnels, with fingerlike ends, called fimbria, that partially surround each ovary. Movements of the fimbria sweep the ovum and some attached cells into the uterine tube following ovulation. If fertilization is to take place, it will be in the uterine tube.

The uterus in most mammals consists of two horns and a body, although much variation occurs. Marsupials, mammals that have pouches, such as the opossum, have two completely separate uteri, each opening to the outside through a separate vagina. Rats, mice, and rabbits have uteri with two horns. Primates have simple uteri with no horns. The uterus has an amazing ability to expand during pregnancy. In all cases, the wall of the uterus is thick and muscular. This muscle layer, the myometrium, is able to contract rhythmically and powerfully to move the young down the birth canal and out of the mother's body during the birth process. The lining of the uterus is called the endometrium.

The uterus narrows down into a muscular, necklike region called the cervix. This structure acts like a valve to keep the opening into the uterus closed most of the time. This prevents bacteria and other harmful objects from entering. The final internal accessory structure is the vagina, a thin-walled muscular tube. The vagina surrounds the cervix of the uterus at its anterior end and extends to its opening to the outside of the body. It allows for childbirth, sexual intercourse, and, in primates, menstrual flow. The walls of the vagina normally touch one another and have deep folds that allow for stretching without damage. A thin fold of tissue called the hymen partially covers the external opening of the vagina. This structure has no function and varies considerably in different mammals.

The external structures of the female reproductive system are called the external genitalia. These include the labia majora, labia minora, and clitoris. Two thick, hair-covered folds of skin, the labia majora, protect and enclose other structures. In some mammals, two smaller hair-free folds of skin are located within the labia majora. These folds, the labia minora, are very prominent in primates but small in most other mammals and completely lacking in some. They enclose a region called the vestibule. Within the vestibule are located the clitoris, the external opening from the urinary system (the urethra), and the external opening of the vagina. The clitoris is a small structure almost covered by the anterior ends of the labia minora. It is very sensitive, being richly supplied with nerve endings and blood vessels.

SEXUAL MATURITY

Reproduction can occur only after females reach sexual maturity. In mammals, this requires the full development of the reproductive structures. The point at which maturity is attained is ultimately under the control of the hypothalamus, as it controls the release of FSH and LH. Many factors, such as attainment of a particular body weight, temperature, day length, and climate may influence the release of hormones.

After the female reaches maturity, reproductive activities are cyclic. In mammals, there are two different kinds of reproductive cycles. Most mammals have an estrus cycle in which females will mate with a male only if they are "in heat," which happens at certain restricted times. An estrus cycle is divided into stages: an inactive phase, called anestrus, which may last for days, weeks, months, or years; proestrus, during which the follicles are developing; estrus, when ovulation occurs; and metestrus, when the ova are moving into the oviduct. Females mate, and may become very aggressive about finding a mate, during estrus only. Usually ovulation is triggered by LH from the pituitary gland. In some mammals, including cats and rabbits, ovulation does not occur until the animal mates. Many females signal that they are in estrus. The signals may be chemical—a special scent which carries for a long distance, for example—or visual. Chimpanzees, for example, develop pink swollen skin on the external genitalia during estrus.

THE MENSTRUAL CYCLE

Primates have a menstrual cycle instead of an estrus cycle. The menstrual cycle is coordinated by estrogen and progesterone from the ovary. These hormones, in turn, are controlled by FSH and LH from the anterior pituitary gland, so all the functions of the reproductive structure are coordinated and synchronized. The three stages of the menstrual cycle are the menses, proliferation, and secretion stages.

In menses, the thick endometrium is sloughed off and flows out of the uterus and out of the body through the vagina. This is also called the menstrual flow or menstrual period. The menstrual fluid consists of roughly equal parts of blood and other accumulated bodily fluids. In the proliferation phase, the endometrium again grows thick. Ovulation occurs in the ovary at the end of this stage, following a sudden increase in the release of LH from the anterior pituitary gland. In the secretion stage, the endometrium becomes very thick and cushiony and prepared to nourish a developing embryo if fertilization has occurred. If fertilization did not occur, the endometrial cells die and the cycle begins again. These stages are controlled by estrogen and progesterone from the ovary. Female primates will mate throughout the entire menstrual cycle.

STUDYING THE FEMALE REPRODUCTIVE SYSTEM

Detailed examination of individual reproductive tissues is performed using a variety of very thin tissue slices, various dyes and stains, and microscopes. Frequently, preserved tissue is used. Electron microscopes have made it possible to magnify single cells, or parts of cells, several thousand times to observe minute details of structure. Fresh tissue is also examined. It is possible to freeze a small tissue sample quickly, slice it very thin, and then expose the tissue to chemicals, which can add to researchers' understanding of the function of particular cells.

The study of reproductive hormones and the understanding of their function demand the use of many different methods. Again, much information comes from nonhuman studies. The procedures vary widely but a typical laboratory experiment may involve removing the ovaries from a female rat and then injecting small amounts of estrogen or progesterone to observe the response of the endometrium. It is also possible to use chemicals that block, or inhibit, one or more specific hormones. By creating an abnormal, controlled situation and observing the results, an understanding about the role of individual hormones within a complex interrelated system can be obtained.

It is also frequently necessary to measure how much hormone is present in some bodily fluid, either for research to gain understanding of normal function or for medical diagnosis. This is very difficult, as most hormones occur in very minute concentrations. Procedures called radioimmunoassay (RIA) techniques, introduced during the late 1950s and early 1960s, represent a very important advance in the study of hormone concentrations. These procedures, which use special recognition molecules for each hormone, plus certain hormones that have been purified and made radioactive, make it possible to measure levels of hormones as low as one trillionth of a gram (a picogram).

In a system as complex in its function as the female reproductive system, it is not surprising that information is obtained from a variety of sources. Each technique has made a contribution to an understanding of the whole system.

UNDERSTANDING THE FEMALE REPRODUCTIVE SYSTEM

The reproductive system is unique among all body systems. It is the only system not called upon to function continuously for the well-being of the individual. It is nonfunctional during the early part of the female's life, then is activated by chemical messages from the anterior pituitary gland. Its primary function is not, after all, the well-being of one individual but rather the continued existence of the species. It is also unique in that it must interact with another individual, a male, in order to fulfill this function. Throughout the reproductive years, all the functions of the female reproductive systems are directed toward pregnancy.

—*Frances C. Garb*

FURTHER READING

Banks, William J. *Applied Veterinary Histology*. 3rd ed. Mosby-Year Book, 1993.

Berne, Robert M., and Matthew N. Levy. *Principles of Physiology*. 3rd ed. C. V. Mosby, 2000.

Hayssen, Virginia, Ari van Tienhoven, and Ans van Tienhoven, eds. *Asdell's Patterns of Mammalian Reproduction: A Compendium of Species-Specific Data*. Comstock, 1993.

Hickman, Cleveland P., Jr., Larry Roberts, and Frances Hickman. *Integrated Principles of Zoology*. 11th ed. Mosby College Publishing, 2001.

Rijnberk, A., and H. W. De Vries. *Medical History and Physical Examination in Companion Animals*. Translated by B. E. Belshaw. Kluwer, 1995.

REPRODUCTIVE SYSTEM OF MALE MAMMALS

FIELDS OF STUDY

Reproductive Biology; Endocrinology

ABSTRACT

The reproductive systems of all male mammals have the same basic design. The reproductive organs produce sperm and deliver it to the outside of the body. The testes are the sites of sperm production. The sperm can be regarded as packages of chromosomes that the animal passes on to his offspring. The testes of most mammals are located in the scrotum, a pouch of skin and muscle that is suspended outside the abdomen. The penis is designed to deliver sperm to the female system.

Hormones and nerves control and coordinate the functions of the reproductive organs.

KEY CONCEPTS

chromosome: a molecule of deoxyribonucleic acid (DNA) that contains a string of genes, which consist of coded information essential for all cell functions, including the creation of new life

ejaculation: the process of expelling semen from the male body

endocrine glands: glands that produce hormones and secrete them into the blood

erection: the process of enlargement and stiffening of the penis because of increased blood volume within it

fertilization: the union of a sperm with an ovum; fertilization is the first step in the creation of a new individual

gamete: a reproductive cell—sperm in the male, ovum in the female; produced in the gonads, gametes contain a set of chromosomes from the adult male or female

gonad: the organ responsible for production of gametes—the testis in the male, the ovary in the female

gonadotropin: a hormone that stimulates the gonads to produce gametes and to secrete other hormones

semen: the sperm-containing liquid that is expelled from the male body

BASIC DESIGN

The reproductive systems of all male mammals have the same basic design. The reproductive organs produce sperm and deliver it to the outside of the body. The sperm can be regarded as packages of chromosomes that the animal passes on to his offspring. Hormones and nerves control and coordinate the functions of the reproductive organs.

THE BRAIN AND REPRODUCTION

Although the brain is not usually considered to be a component of the reproductive system, part of the brain is, in fact, essential to the function of the reproductive organs because of the hormones produced there. This part of the brain is the hypothalamus, a relatively small area that acts without conscious control. The hypothalamus is located in the lower middle of the brain; it contains centers that control eating, drinking, body temperature, and other essential functions.

Hypothalamic control over reproduction in the male is primarily by way of the hormone called gonadotropin-releasing hormone (GnRH). GnRH is released from the hypothalamus to enter blood vessels that carry it to the pituitary gland, a small gland suspended just below the hypothalamus. When GnRH arrives at the pituitary, it stimulates the pituitary to produce and release two more hormones, follicle-stimulating hormone (FSH) and luteinizing hormone (LH). FSH and LH in the male are identical to hormones of the same names in the female. The names of these hormones describe their functions in the female. Like other hormones, FSH and LH are released into the blood and circulate throughout the body. FSH and LH are called gonadotropin hormones: gonadotropin means "gonad stimulating." These are the hormones that stimulate the gonads (testes in the male, ovaries in the female) to produce sperm or eggs and to secrete gonadal hormones. In the male, the gonadal hormones are primarily testosterone and related hormones. There is a chain of hormonal commands, with GnRH from the hypothalamus at the top of the chain. GnRH stimulates the pituitary to secrete FSH and LH, which in turn stimulate the testes to produce sperm and testosterone.

In addition to the chain leading from the brain to the pituitary to the testes, information is sent back to the brain from the testes, a checks-and-balances system using principles of negative feedback to ensure that the hormones are produced in the appropriate quantities. If, for example, the hormone system gets slightly out of balance, leading to too much testosterone being produced, this excess of testosterone will be sensed by the hypothalamus. It will cause a temporary shutdown of GnRH production, leading to the system's correcting itself, because then a little less testosterone will be produced. If testosterone levels fall too low, the opposite will happen: GnRH, and then FSH and LH, and then finally testosterone, will all increase, again resulting in a correction of the original aberration. The hormonal system is a delicately balanced network that ensures the proper functioning of the testes.

THE TESTES

The testes are the sites of sperm production. Within the testes are hundreds of tiny tubes, the seminiferous tubules, that are responsible for sperm production. The sperm develop gradually from round cells called spermatogonia, which are located in the walls of the seminiferous tubules. As a sperm matures, it develops a long, whiplike tail attached to an oval head. The head of the sperm contains chromosomes, the genetic information of the male that will be passed on to his offspring. The sperm of some mammals can be distinguished under the microscope by characteristic differences in their appearance.

Between the seminiferous tubules are clusters of hormone-producing cells, the interstitial or Leydig cells. The Leydig cells produce testosterone and related hormones. Testosterone is essential for proper sperm development. In addition, testosterone is responsible for the development of male body features, including, in most species, a large muscle mass, and for the growth of the reproductive organs during puberty. In some animals, testosterone is also linked to aggressive and reproductive behaviors.

The testes of most mammals are located in the scrotum, a pouch of skin and muscle that is suspended outside the abdomen. In some animals, the testes may be withdrawn into the abdomen when the animal is startled or when it is not in the breeding condition.

The function of the scrotum is to maintain the temperature of the testes at a few degrees lower than average body temperature. The capability to maintain this temperature of the scrotum is rooted in the fact that the muscles within the scrotum are responsive to temperature. Under warm conditions, the scrotum relaxes, allowing the testes to move away from the body and lose heat. In cool temperatures, the opposite occurs: The scrotum wrinkles, pulling the testes closer to the body and allowing them to stay warmer. The reduced temperature maintained by the scrotum is mandatory for the production of normal fertile sperm. Fever or other situations that raise the temperature of the scrotum can interfere with sperm production, even resulting in temporary infertility. In a few large mammals (such as elephants, whales, and dolphins), the testes are not located within a scrotum, but instead occupy a position in the abdomen. It is not known why these species apparently do not require a temperature lower than that of the body for sperm production.

THE EPIDIDYMUS AND VAS DEFERENS

Sperm are removed from the testes by a system of tubes that lead out of the body. Located next to each testis within the scrotum is the epididymis, a highly coiled tube that is directly connected to the seminiferous tubules of the testes. The epididymis serves two functions: sperm maturation and sperm storage. The epididymis is drained by a long, thin tube called the vas deferens, which carries sperm out of the scrotum through the inguinal canal into the abdomen. The inner end of the vas deferens is a widened area that may serve as a site of storage for mature sperm.

The vas deferens passes in a loop next to and under the bladder, the sac that stores urine until it can be removed from the body. Immediately beneath the bladder, the vas deferens is connected by a short tube, the ejaculatory duct, to the urethra. The urethra is the long, fairly straight tube that carries either urine from the bladder or sperm from the reproductive system. A valve located in the urethra below the bladder opens and closes to prevent sperm and urine from mixing, so that only one type of fluid is in the urethra at a time.

From their site of production in the testes, sperm pass through the epididymis, the vas deferens, the ejaculatory duct, then finally the urethra to the outside of the body. As sperm are expelled from the body along this route, they are mixed with seminal fluid to produce semen. Seminal fluid is secreted into the tubes by three sets of glands: the seminal vesicles, the prostate, and the bulbourethral (Cowper's) glands. The sperm never enter these glands; fluid is squeezed out of them into the tubes where the sperm are located.

THE PENIS

The penis is designed to deliver sperm to the female system. The penis consists of a long shaft with an enlarged head, the glans. The skin of the penis, especially the glans, is extremely sensitive to touch. In some species, the penis is withdrawn into a sheath of skin except during sexual arousal.

Internally, the penis contains the outer segment of the urethra, as well as erectile tissue. This erectile tissue is designed like a sponge. The many blood vessels in the erectile tissue are capable of greatly expanding and increasing the quantity of blood that they contain. When this happens, the erectile tissues swell, and the entire penis increases in length and width and becomes stiff. This process, called erection, is an involuntary reflex: It cannot be consciously prevented or caused. Erection can result from direct stimulation of the penis, as during sexual contact, or from erotic sights or sounds. In some animals, a bone within the penis, the baculum, assists in maintenance of the erection.

Continued sexual stimulation will eventually result in an ejaculation, with semen being forced out of the body by contractions of muscles in the fluid-producing glands and along the tube system. Ejaculation is coordinated by nerves that arise in the spinal cord. The normal volume of fluid ejaculated varies from species to species. In man, it is usually two to six milliliters; it may be up to one hundred milliliters in pigs. The ejaculate of most animals contains many millions of sperm per milliliter of fluid.

STUDYING MALE REPRODUCTION

The hormonal system that controls the male reproductive system is the subject of much research. The most straightforward type of hormonal research is simply descriptive: The scientist seeks to describe the levels of the reproductive hormones when the animal of interest is in different physiological states. The hormones can be measured in blood samples taken from the animals. Obtaining a blood sample from an experimental animal may pose difficulties: Some large animals may be difficult to restrain, and some small animals may not have veins large enough for an easy puncture. Another consideration is how often blood samples should be taken. Endocrinologists have become increasingly aware of the importance of the pattern of hormone release over time. In particular, it now appears that fluctuations in hormone levels within a time frame of minutes or hours may be critical in regulating the responses of hormone target sites. To obtain blood samples with such a high frequency, researchers usually implant a cannula into a vein of the animal; the cannula can be left in place for repeated blood sampling with very little stress to the animal.

Scientists interested in hormonal feedback may examine the roles of specific hormones by removing one of the endocrine glands from the system, and then examining the effects on the remaining hormones. For example, the testes (as the site of testosterone production) can be removed from an experimental animal. Blood samples after the surgery can then be assayed to determine the circulating levels of LH, FSH, and GnRH. The endocrine glands may be left in place, but the researcher may administer hormones either by injection or by implanting timed-release capsules containing the hormone under the skin. Then, blood samples taken from the animal will reveal how levels of hormones produced by the animal's own endocrine glands have changed as a result of the exposure to the added hormone.

A technique that is widely used to study males of seasonally breeding species is to subject the animals to carefully controlled environmental conditions. Length of exposure to light, temperature, rainfall, nutrients in the diet, and other factors can be controlled in the laboratory to determine which acts as the cue for seasonal reproduction. The status of the reproductive system can be determined by various methods. The testes can be measured: Inactive testes are usually smaller and lighter in weight. Hormone levels in the blood can be measured: Testosterone and other hormones may decrease when the animal is reproductively quiescent, or the male can be exposed to a female to determine whether he will show mating behavior.

For some types of research, the most revealing experiments may not use the entire animal (referred to as in vivo research), but will instead focus on specific organs. Living samples of organs can be maintained in the laboratory for such in vitro experimentation. For the in vitro approach, a small piece of living tissue can be removed from an animal and the cells suspended in a liquid that contains the nutrients necessary for their life. Under these isolated conditions, the scientists can investigate a number of areas such as which hormones tissues produce and the hormones that make the tissue itself respond. Organs respond optimally to a particular pattern of hormonal stimulation, and this is another important area of research. By combining the results of in vivo and in vitro experiments, scientists are able to piece together a complete picture of how the reproductive system functions.

CONTROLLING REPRODUCTION

Knowledge of how the male reproductive system functions has allowed scientists to develop technologies for controlling reproduction to enhance or curtail fertility in domestic animals. Knowledge of male reproductive physiology has been applied to the management of domestic breeding populations. Hormone measurements and sperm counts can be used to determine the optimum age at which to begin breeding young stock. Techniques for collecting and storing semen can be combined with artificial insemination of females to increase the number of offspring produced by valuable males, thus resulting in improvement of the population. These methods are particularly valuable to breeders of large animals because maintaining large numbers of males of these species (such as stallions and bulls) can be costly and difficult because of the aggressive behaviors that these males may exhibit.

The study of seasonal breeding has also been of value in agriculture. Scientists now know much about the environmental conditions that are responsible for promoting reproductive activity in many domestic species. Farmers can apply this knowledge to their breeding stock to increase production throughout the year. Another area in which reproductive studies are of vital importance is the enhancement of the breeding of captive animals that are endangered in the wild. Zoos, once considered merely spectacles for entertainment, are now seen by many as the last hope of saving many species on the verge of extinction. Knowledge of the conditions necessary for successful breeding of exotic animals will help to increase their numbers and, perhaps, to return them to the wild.

—Marcia Watson-Whitmyre

Further Reading

Carter, Carol Sue, I. Izja Lederhendler, and Brian Kirkpatrick, eds. *The Integrative Neurobiology of Affiliation.* MIT Press, 1999.

Knobil, Ernst, and Jimmy D. Neill, eds. *The Physiology of Reproduction.* 2 vols. 2nd ed. Raven Press, 1994.

Marshall Graves, J. A., R. M. Hope, and D. W. Cooper, eds. *Mammals From Pouches and Eggs: Genetics, Breeding, and Evolution of Marsupials and Monotremes.* CSIRO, 1990.

Nalbandov, A. V. *Reproductive Physiology of Mammals and Birds: The Comparative Physiology of Domestic and Laboratory Animals and Man.* W. H. Freeman, 1976.

Setchell, B. P. *The Mammalian Testis.* Cornell University Press, 1978.

Van Tienhoven, Ari. *Reproductive Physiology of Vertebrates.* 2nd ed. Cornell University Press, 1983.

Reptiles

Fields of Study

Herpetology; Paleontology; Evolutionary Biology; Developmental Biology

Abstract

Reptiles are a class of vertebrates characterized by their ability to produce cleidoic eggs. The development of this egg allowed animals to exploit terrestrial habitats. The egg and the reptiles' dry, horny scales differentiate all living reptiles from amphibians. Skeletal features and the lack of feathers and hair differentiate reptiles from birds and mammals, respectively. It is the combination of features that characterizes reptiles and distinguishing extinct forms from other closely related vertebrate groups is often dependent on a single characteristic and may not be precise.

Key Concepts

Anapsida: a group of reptiles in which the temporal region of the skull lacks openings

Chelonia (testudines): a living order of reptiles composed of turtles and tortoises

cleidoic egg: a shelled egg equipped with internal membranes that make terrestrial reproduction possible

Crocodylia: a living order of reptiles that includes crocodiles and alligators

Diapsida: a group of reptiles in which the temporal region of the skull is characterized by two openings

Euryapsida: an extinct group of reptiles in which the temporal region of the skull is characterized by a single opening situated high on the side of the skull

Rhynchocephalia: a living order of reptiles represented by a single species, the tuatara

Squamata: a living order of reptiles composed of lizards and snakes

Synapsida: an extinct group of reptiles in which the temporal region of the skull is characterized by a single opening; this group gave rise to mammals

venom: a toxic substance that must be injected in order to elicit damaging effects

DETERMINING WHAT IS A REPTILE

Reptiles are a class of vertebrates characterized by their ability to produce cleidoic eggs (which are similar to bird eggs and were their evolutionary precursors). The development of this egg, protected by an impervious shell, was a historic step, as it allowed animals to exploit terrestrial habitats. The egg and the reptiles' dry, horny scales differentiate all living reptiles from amphibians. Skeletal features (single bones for sound conduction in the middle ear and jaws composed of several bones) and the lack of feathers and hair differentiate reptiles from birds and mammals, respectively. One must realize, however, that it is the combination of features that characterizes reptiles and that, since soft tissues such as reproductive tracts and skin do not fossilize well, distinguishing extinct forms from other closely related vertebrate groups is often dependent on a single characteristic and may not be precise. For example, *Archaeopteryx*, a primitive bird, would have been classified as a reptile had not feathers been adventitiously preserved.

EARLY REPTILES, TURTLES, AND CROCODILES

Reptiles arose from amphibians roughly 315 million years ago, during the Carboniferous period. These "stem-reptiles" gave rise to all other groups. The "cheek," or temporal region, is very important in reptilian classification. In early forms, the region was solid (lacked openings). These forms are placed in the subclass *Anapsida* (without recesses or openings). Along with some of these earliest reptilian fossils are some from a distinctly different group, with a single temporal opening. These animals are placed in the subclass *Synapsida*. Referred to as "mammal-like reptiles," they gave rise to mammals prior to their extinction. During the Permian period (280–215 million years ago), another reptilian group appeared. This group, characterized by two temporal recesses, is placed in the subclass *Diapsida* (two openings), to which the majority of reptiles, living and extinct, belong. Another subclass, the *Euryapsida*, with a single opening high on the side of the skull, became extinct near the end of the Mesozoic era (approximately 175 million years ago). It included large marine (sea-dwelling) forms, such as fishlike ichthyosaurs and flat-bodied, long-necked plesiosaurs.

Anapsids include not only the stem-reptiles, but also a surviving order, the *Chelonia* or *Testudines*, composed of turtles and tortoises. Primitive turtles appear almost fully formed in the fossil record, and their relationship to the first reptiles is uncertain. Their principal feature is a shell, composed of flattened ribs fused to layers of bony tissue, usually covered by large, flat scales called scutes. The upper portion, to which vertebrae are fused, is the carapace; the underside is the plastron. They are connected by bridges. Shoulders and hips have been modified and are located within the rib cage, a unique arrangement. All lay eggs on land. The order contains almost 250 species in 75 genera and 13 families.

Although it is easy to fall into the trap of thinking that "if it has a shell, it is a turtle" and "a turtle is a turtle is a turtle," the group is quite diverse. The most common (and probably primitive) body plan is associated with marsh dwellers. Characterized by somewhat flattened shells (streamlined for locomotion through water) and webbed feet for propulsion, these surprisingly agile swimmers include the familiar sliders that drop into the water from logs or the bank when approached too closely. Though adept in water, they are no match for sea turtles; these animals have reduced, flattened shells to minimize resistance and limbs modified into paddles with which they fly through the water using movements almost identical to those of birds. Though they lay their eggs on land, they are practically helpless there. Graceful in the water, they often swim considerable distances, guided by a very effective navigational sense.

Sea turtles include the largest living reptiles, the leatherbacks, which may exceed a ton in weight. Other turtles are bottom dwellers; ambush predators or scavengers, they lie in wait or slowly crawl along bottoms of ponds and streams. They possess webbed feet and flat, often rough shells. Algae growing on their shells serves as camouflage, hiding them from prey and predators. A final body plan characterizes land turtles, which include tortoises. A firm, generally high-domed shell minimizes surface area through which water might be lost (critical in terrestrial animals) and also resists attacks by predators to which a land dweller is exposed. Some very large forms are quite long-lived; one documented record exceeds 150 years.

The term "dinosaur" is often used to describe any large, extinct reptile, including mammal-like synapsids, marine euryapsids, and diapsids such as flying reptiles and some large lizards. Used properly, however, it refers to only one group of diapsids, the *Archosauria*, or "ruling reptiles," which gave rise to birds. Other close relatives are in the living order *Crocodylia*, which now contains only twenty-two species in eight genera and three families. These animals share some very advanced features that cause some authorities to place them in a distinct class, the *Dinosauria*. All have fully partitioned hearts, allowing separate circuits for oxygenated blood to be carried to the body and deoxygenated blood to the lungs. Recent evidence indicates that many dinosaurs may have possessed birdlike capabilities for temperature regulation, allowing levels of activity beyond that of other reptiles. Modern crocodilians, some exceeding 7.5 meters in length, are quite aquatic and feed principally on fish or animals ambushed as they drink. They are restricted to tropical and subtropical zones. All are egg-layers.

LEPIDOSAURIA AND *SQUAMATA*

All other living reptiles are in the diapsid group *Lepidosauria* (scaly reptiles). The tuatara, a lizardlike reptile up to sixty centimeters long, is the only surviving member of the order *Rhynchocephalia*, a diverse assembly that coexisted with dinosaurs. Restricted to roughly thirty small islands off the coast of New Zealand and well adapted to a cool climate, it demonstrates considerable longevity (approximately 120 years) but also has a low reproductive rate. It feeds primarily on insects and eggs and the young of sea birds or other tuataras, with which it shares burrows.

Arguably the most successful reptilian group, extinct or living, is the order *Squamata*. All are equipped with efficient vomeronasal organs with which they "smell" by sampling air or substrate with their tongues. They may lay eggs or give live birth. There are approximately 3,750 species of lizards, suborder *Sauria* or *Lacertilia*, in almost 400 genera and some 16 families. They are found from north of the Arctic Circle to the southern tip of South America; such range is perhaps attributable to their exceedingly efficient capacity for thermoregulation. Some lizards at below-freezing temperatures may maintain body temperatures near 20 degrees Celsius.

It is difficult to characterize lizards because of their tremendous diversity. Some are legless—an adaptation for burrowing or living in dense grass. Others have the capacity for gliding. Feet may be modified for running (many run on their hind legs, in one instance so rapidly that the lizard can run on water for considerable distances) or climbing, with digits equipped with claws and/or adhesive pads. Teeth may be used for grasping, cutting, or crushing food. Tails may be prehensile (capable of grasping), may be used for balancing while climbing or running, may come equipped with spines or knobs for defense, may be capable of fat storage, and may even break off if grasped by a predator (often to regenerate rapidly). Some lizards are excellent swimmers; the marine iguana of the Galápagos Islands feeds primarily on seaweed. Two species are venomous. The smallest lizards are only a few centimeters long; the largest may exceed three meters.

SNAKES

Snakes, suborder *Serpentes* or *Ophidia*, are distinct from lizards in that they lack external ears, eyelids, and limbs (at least one of which most lizards have—worm lizards lack these features and have been treated as snakes or even placed into a separate suborder). Despite these constraints, snakes are quite diverse: There are almost 2,400 species in more than 400 genera and 11 families, ranging from the Arctic Circle to the southern tip of South America. Leglessness was a primitive adaptation for burrowing, but modern snakes also swim, crawl, climb, and, in one case, even glide adeptly. The elongated body form that accompanies limblessness requires that paired internal organs be arranged longitudinally, with one often degenerating. Digestive tracts are short and straight, resulting in all snakes being carnivorous, as meat is more easily digested by snakes than plant material.

Locomotion is surprisingly varied. The familiar serpentine movement works well either on land or in water, but heavy-bodied snakes often use rectilinear locomotion, pushing their bodies in straight lines by alternately raising and retracting their large belly scales. In tight quarters, snakes anchor their necks and pull their bodies forward or, alternately, push off using anchored tails (many are equipped with spines for this purpose). A few snakes, especially on loose substrates such as sand, sidewind, pushing down to prevent sliding while lifting loops of their bodies laterally.

Snakes swallow their prey whole. Jaws, which are loosely attached to the skull and to each other, alternately slide forward and pull back on food with recurved teeth. Prey may be swallowed alive, killed by constriction, or killed with venom. Venom injection may accompany a bite or may be facilitated by special fangs in the rear or front of each jaw. In vipers, the bones to which fangs are attached rotate, so that very long fangs can be folded back when not in use. Burrowing snakes tend to be slender and small, with smooth scales and rigid heads. Aquatic snakes are usually stout, with rough scales to prevent slipping through water. Arboreal snakes (climbers) are often extremely slender. Active hunters are usually more slender than ambush predators, which eat more rarely but can consume much larger items. Some snakes have temperature-sensitive pits with which they find prey in the dark. Sizes range from a few centimeters to almost ten meters.

STUDYING REPTILES

Methods used to study reptiles are determined by the nature of the particular investigation. Historical studies rely on paleontological methods. The discovery of fossils, followed by recovery, preservation, reconstruction, and analysis, leads to an understanding of the structure and function of prehistoric animals and provides information about both how they lived and conditions in which they existed. Comparisons, especially of structures, with other fossils and with animals alive today constitute much of the field of comparative anatomy and lend insights into relationships between various living and extinct forms. These studies, in turn, lead into the discipline of systematics, which attempts to reconstruct relationships and build classification schemes accordingly. The actual naming of various groups is called taxonomy. Since fossil records are typically incomplete, however, other methods must be used to establish fully the nature of relationships.

Similarities and differences between living forms may be established on the basis of detailed anatomical studies or various biochemical techniques. In the latter case, the analysis of the molecular structure of the deoxyribonucleic acid (DNA) and proteins produced by different species (or even by different populations of the same species) allows determinations of how closely related certain forms may be. These studies also have considerable evolutionary implications, providing insights not only into methods that might have resulted in evolutionary changes among reptiles but also into the processes that were responsible for the origins of birds and mammals.

One phenomenon that was first discovered in reptiles is parthenogenesis, the development of an individual from an unfertilized egg, a process that leads to all-female populations. In lizards these often result from hybridization between two species, the offspring of which are distinctive. Analysis of mitochondrial DNA, which is passed to descendants through the egg (never the sperm), allowed determination of which hybridizing parental species was maternal.

Another field of study that uses biochemical techniques and that focuses on reptiles concerns the venoms produced by some snakes and two lizard species. Knowing the composition of venoms is important in determining their effectiveness as devices to kill prey, but it is also significant in that some substances in these venoms have been shown to possess functional traits that have important medical implications.

Reptiles have also been used widely in ecological and behavioral studies. Methods include both field and laboratory techniques. Studies in natural situations often involve extensive observations and, as such, require organisms that are easily observed. Laboratory studies, in which environmental factors are often simulated and then modified, demand subjects that are small and readily maintained in captivity. Many of these studies also have physiological implications. Studies monitoring such factors as body temperatures, food intake, and foraging strategies often rely on reptiles, especially lizards, as they are ideally suited to these observational and experimental investigations. Considerable work in reproductive physiology also relies on reptiles; their eggs are accessible, and they demonstrate developmental patterns similar to those in birds and mammals. This aptitude as a subject for studies has also resulted in reptiles becoming the focus of many biogeographical studies, especially on islands, where (as in deserts) they often dominate the fauna. For many of the same reasons, environmental studies of endangered species or altered habitats frequently use reptiles as models.

REPTILES AND OTHER ANIMALS

The study of reptiles not only increases scientists' knowledge of this fascinating group of animals, but also has many other applications. Historically, reptiles were the first group of fully terrestrial vertebrates;

they dominated Earth for many millions of years and gave rise to both birds and mammals. Thus, studies of fossil forms, with additional insights from investigations of living species, provide insights into the conditions that prevailed on Earth during prehistoric times. They also lead to theories regarding relationships between animal groups and lend understanding to the origins and nature of early birds and mammals.

Studies of this type need not be restricted to arcane facts relevant only to times long past; they are also significant in understanding long-term biological processes, such as those that reflect climatic cycles and periods of mass extinction (both problems of considerable interest today). Also, since many reptilian groups are sufficiently old to predate the breakup of Pangaea (the single landmass that existed historically and which has since broken up into the continental areas of today), or at least its subsequent parts, applications can be made to the areas of biogeography and even geology.

Physiological investigations of reptiles have been invaluable in developmental and reproductive studies and in increasing knowledge of how animals interact with their environments. Lizards, especially, have been widely studied in regard to their ability to thermoregulate effectively using environmental sources of heat. These studies have numerous applications to broader investigations of homeostasis and adaptations to cold and hot environments by many animals (including man). Lizards also have been widely used as models in behavioral and ecological studies. Especially in the tropics, they are abundant, diverse, easily observed, and often remarkably well-adapted to their environment.

Like birds, much of their behavior is quite stereotypical, that is, innate and consistent. As a result, the recognition of patterns is much easier than it is in secretive mammals, for example, where the problems are further magnified by frequent modifications of instinctive mechanisms by learned behaviors. This same visibility and ease of observation lends itself to ecological investigations. Lizards have been more widely utilized in niche partitioning studies than has any other animal. Niche partitioning studies seek to investigate how limited resources are used by animal communities. They often center on the hypothesis that food habits are critical, but microhabitat preferences, activity cycles, and other aspects of environmental impact are also involved.

—*Robert Powell*

FURTHER READING

Bakker, Robert T. *The Dinosaur Heresies: New Theories Unlocking the Mystery of the Dinosaurs and Their Extinction.* Kensington, 1986.

Conant, Roger. *A Field Guide to Reptiles and Amphibians of Eastern and Central North America.* 3rd ed. Houghton Mifflin, 1998.

Desmond, Adrian J. *The Hot-Blooded Dinosaurs.* Blond & Briggs, 1975.

Halliday, Tim R., and Kraig Adler. *The Encyclopedia of Reptiles and Amphibians.* Facts on File, 1998.

Hickman, Cleveland P., Larry S. Roberts, and Frances M. Hickman. *Integrated Principles of Zoology.* 10th ed. Times Mirror/Mosby, 2001.

Kirshner, David S. *Reptiles and Amphibians.* National Geographic Society, 1996.

Spellerberg, Ian F. *Mysteries and Marvels of the Reptile World.* Scholastic, 1995.

Stebbins, Robert C. *A Field Guide to Western Reptiles and Amphibians.* 2nd ed. Houghton Mifflin, 1985.

Vitt, Laurie J., Janalee P. Caldwell, and George R. Zug. *Herpetology: An Introductory Biology of Reptiles and Amphibians.* 2nd ed. Academic Press, 2001.

RESPIRATORY SYSTEMS IN ANIMALS

FIELDS OF STUDY

Animal Anatomy; Animal Physiology; Respirology

ABSTRACT
Animals generally meet their energy needs by oxidation of food, and the respiratory system supplies the oxygen necessary for cell metabolism while removing its waste product, carbon dioxide. Oxygen is available either dissolved in water or as a component of the air, and animals have evolved special organ structures to effectively obtain oxygen from their environment.

Key Concepts

alveolus: the thin-walled, saclike lung structure where gas exchange takes place

chemoreceptor: specialized nervous tissue that senses changes in pH (hydrogen ions) and oxygen

countercurrent exchanger: the process where a medium (air or water) flowing in one direction over a tissue surface encounters blood flowing through the tissue in the opposite direction; this improves the gas diffusion by maintaining a concentration gradient

diffusion: the process by which gas molecules move from a higher to a lower concentration through a medium or across a permeable barrier; the rate at which gases cross a barrier is increased by the surface area, and gas concentration gradient is decreased by the thickness of the barrier; gas solubility determines the amount that crosses the barrier

gill: an evaginated organ structure where the membrane wall turns out and forms an elevated, protruding structure; typically used for water respiration

lung: an invaginated organ structure where the membrane wall turns in and forms a pouch or saclike structure

OXIDATION OF FOOD

Animals generally meet their energy needs by oxidation of food, and the respiratory system supplies the oxygen necessary for cell metabolism while removing its waste product, carbon dioxide. Oxygen is available either dissolved in water or as a component of the air, and animals have evolved special organ structures to effectively obtain oxygen from their environment.

ORGANS OF GAS EXCHANGE

Single-cell and simple organisms, such as flatworms and protozoa, can obtain sufficient oxygen to meet their energy demands by simple diffusion through their body surface. Some amphibians utilize gas exchange through their skin to supplement their lung respiration, but generally, larger, more complex animals require specialized organ systems with a large surface area for gas exchange and a circulatory system for distribution of oxygen to each cell. The basic mechanism, however, for gas exchange between

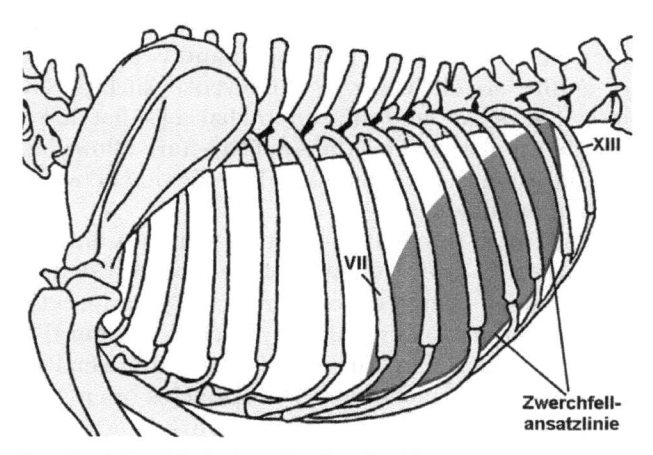

Lateral projection of the diaphragm in a dog. (Uwe Gille)

the environment and the blood and between the blood and cells is by diffusion. The three major types of gas exchange organs are the gill for water respiration, the lung for air and in some special cases water respiration, and the tracheal system of tubules for air respiration in insects.

Gills consist of several gill arches located in the operculum or gill cover on each side of the fish's head. A gill arch contains two rows of gill filaments, and each filament has a row of parallel platelike structures on its surface called lamellae. The lamellae are everted structures that rise up from the filament surface and are only a fraction of a millimeter apart. Water flows between the lamellae, and oxygen diffuses from the water into the lamellar capillary blood. The lamellar blood flows in the opposite direction of the water flow and creates a countercurrent exchanger. The countercurrent maximizes the diffusion of oxygen into the lamellar capillary blood by maintaining a diffusion gradient over its entire length.

Lungs, in contrast to gills, are invaginations, where the surface turns in and forms a hollow or saclike structure. Lungs typically are divided into two functional areas: the conducting zone and the respiratory zone. The conducting zone branches from the trachea to the bronchioles and distributes air to the respiratory zone but is not involved in gas exchange. The respiratory zone comprises the majority of the lung and contains small respiratory bronchioles and ducts that lead to the primary gas exchange area, the alveolus. The alveoli vary from simple saclike structures in a pulmonate land snail to the complex alveolar wall structure of mammals. The alveolar wall is fifty micrometers thick, or about one fiftieth the thickness of a sheet of paper, and is composed of epithelial cells covering the alveolar surface, an interstitial space, and the endothelial cells that make up the capillaries. This thin-wall structure allows for the diffusion of oxygen and carbon dioxide between the air and blood.

The insect tracheal respiratory system is unique because it is both the gas exchange and distribution system. Pairs of openings on the insect's thorax and abdomen called spiracles regulate the movement of air in and out of a system of smaller tubules called tracheoles. The spiracles open and close in a pattern that allows unidirectional flow of air through the tubule system. The tracheoles branch and extend throughout the insect's body and deliver oxygen to the cells independent of the circulatory system.

AIR AND WATER ENVIRONMENTS

Important aspects of the atmosphere for respiration are the barometric pressure and concentration of gases, temperature, and humidity. The atmospheric gases important to animals are oxygen, carbon dioxide, and nitrogen, and the atmosphere is a constant 20.95% oxygen, 0.03% carbon dioxide, and 79% nitrogen (plus other inert gases). The rate of diffusion of oxygen from the inspired air into the circulation depends on the partial pressure of the oxygen. The barometric pressure, however, decreases with increasing altitude, and this decreases the partial pressure of oxygen, which decreases the diffusion of oxygen into the blood. Thus an animal's difficulty in obtaining adequate oxygen at higher altitudes is related to the reduction in atmospheric pressure and not to a change in the %age of oxygen in the atmosphere.

The temperature and amount of water vapor or humidity in the atmosphere are variable, and during inspiration, the inspired air is warmed to body temperature and saturated with water vapor (100% humidity). The heat and moisture come from the airways and can potentially cool and dehydrate an animal. Therefore, a minimal amount of air is inspired to prevent excess heat and water loss. However, heat-stressed animals will use this respiratory heat loss or panting to cool their bodies.

Water poses several challenges for respiration compared to air: a lower oxygen content, slower gas diffusion rate, higher viscosity, and greater weight. The amount of oxygen available in the water is thirty times less than that found in air. Thus, more water has to flow over the gill surface for adequate oxygen delivery. The speed at which oxygen moves through water is ten thousand times slower than oxygen moving through air. Thus, the distance between the water and the gill surfaces can only be a fraction of a millimeter apart. In contrast, the lung gas exchange surfaces are a few millimeters apart.

Water's greater viscosity and weight compared to air require more energy to move water over the gill surface. Water-breathing animals compensate for this by having a unidirectional flow through the gill. This avoids water being moved, stopped, and then moved again in the opposite direction, which works well for air, but would be very energy costly for the heavier, more viscous water.

The gill structure depends on water to support and separate the rows of lamellar structures. Thus, when a fish is exposed to air, the gill structure collapses on itself and greatly reduces the surface area available for oxygen diffusion. Thus, the fish will suffocate if not returned to the water.

BREATHING WATER AND BREATHING AIR

Water can be moved through the gill lamellae by either opercular pumping or ram ventilation. Opercular pumping involves the movement of the mouth and opercular covering to create pressure gradients for unidirectional flow of water through the mouth, across the gill surface, and out the opercular covering (unidirectional flow). Ram ventilation takes advantage of the fish's forward speed to flow water through the mouth and gill. Opercular pumping is used from rest to slow swimming speeds, and a fish switches over to ram ventilation when swimming at faster speeds.

For air breathers, inspiration (inflating the lungs) can be accomplished by either positive-pressure or negative-pressure breathing. Positive-pressure breathing requires air pressure to inflate the lungs, which is similar to inflating a balloon or tire with a compressed air. The pressure is considered positive because it is greater than atmospheric pressure. For example, frogs use positive-pressure breathing by closing their mouths and then elevating the floor of the mouth. This compresses and pressurizes the air and forces it into the lung. The elastic lung tissue is stretched like an inflated balloon by the increased volume. The process of the air moving out of the lung is called expiration. When the frog relaxes and opens its mouth the lung elastic recoil forces the air out similar to a balloon deflating.

With negative-pressure breathing, the lung is pulled open by contraction of the diaphragm. The pressure becomes negative (below atmospheric pressure), and air flows into the lung until it equalizes with the atmospheric pressure. If additional inflation is required, such as during exercise, accessory inspiratory muscles lift the ribs to inflate the lungs further. Expiration is accomplished by the relaxation of the inspiratory muscles, and the lung elastic recoil increases airway pressure and air flows out of the lung.

Inspiration is always an active process, whereas expiration results from the passive elastic recoil of the lung tissue. However, active expiration is possible by contracting muscles that pull the ribs down and by using abdominal muscles to push the diaphragm farther into the thoracic (chest) cavity.

SETTING BREATHING RATE

In water-breathing animals, such as fish and lobsters, the level of oxygen sets the ventilation rate (volume of water moved through the gill per minute) such that as oxygen content in the water decreases, the frequency of breathing movements increases. During fast swimming, fish using ram ventilation regulate the mouth opening so that the amount of water flowing over the gills just meets tissue oxygen demand. A wider mouth opening than is necessary increases the fish's frictional drag through the water and thus decreases the energy efficiency. Carbon dioxide is highly soluble in water and easily diffuses from water-breathing animals. Thus, blood carbon dioxide levels in water-breathing animals are very low and not used to regulate respiration rate.

In air-breathing animals, the blood levels of carbon dioxide and oxygen regulate the ventilation rate (air volume moved in and out of the lungs per minute). Carbon dioxide quickly diffuses from the small capillaries in the brain circulation into the fluid surrounding the brain cells (cerebral spinal fluid). Here the carbon dioxide reacts with water and forms carbonic acid. The hydrogen ions released from the carbonic acid stimulate chemoreceptor cells that in turn stimulate the respiratory center in the medulla, located in the brain stem. Higher concentrations of carbon dioxide increase the hydrogen ion concentration and thus increase ventilation rate. Air-breathing animals primarily regulate ventilation rate by carbon dioxide produced from metabolism and not low blood oxygen levels.

However, oxygen can regulate ventilation in animals at high altitudes. Oxygen partial pressure is sensed by chemoreceptors in the aorta and the carotid artery. These peripheral chemoreceptors sense the partial pressure of oxygen in the blood plasma, and as the partial pressure of oxygen in the air decreases, such as with altitude, the partial pressure of oxygen in blood also decreases. This increases ventilation, which then compensates for the lower oxygen partial pressure. In addition to low oxygen partial pressure, the peripheral chemoreceptors are stimulated by blood acidosis. For example, lactic acid released from skeletal muscles during strenuous exercise stimulates the ventilation rate in animals and humans.

—*Robert C. Tyler*

FURTHER READING

Fish, F. E. "Biomechanics and Energetics in Aquatic and Semiaquatic Mammals: Platypus to Whale." *Physiological and Biochemical Zoology* 73, no. 6 (2000): 683–98.

Pough, F., H. Heiser, J. B. Heiser, and W. N. McFarland. *Vertebrate Life.* 5th ed. Prentice Hall, 1999.

Schmidt-Nielson, Knut. *Animal Physiology: Adaptation and Environment.* 5th ed. Cambridge University Press, 1997.

Weibel, Ewald R. *The Pathway for Oxygen.* Harvard University Press, 1984.

Willmer, Pat, Graham Stone, and Ian Johnston. *Environmental Physiology of Animals.* Blackwell Science, 2000.

Withers, P. C. *Comparative Animal Physiology.* Saunders, 1992.

Rodents

Fields of Study

Rodentology; Mammalogy; Evolutionary Biology; Ecology

Abstract

Rodents are found on every continent except Antarctica. Their habitats include, but are not limited to, mostly on land or underground, in forests, plains, and deserts; some live in a partly freshwater environment, using ponds or streams. Most smaller species live for one to three years, while large rodents survive for over ten years. All rodents possess incisor teeth designed for gnawing, which grow continuously from the roots and wear away at their tips, giving them chisel-like edges that can gnaw through very hard materials; however, they lack cuspids (tearing teeth) seen in carnivores

Key Concepts

cuspid: a tearing tooth found in the mouth of a carnivorous animal

herbivore: any animal that subsists entirely on plant foods

incisor: a cutting tooth which acts like scissors or a chisel

molar: a flat tooth found at the back of the jaw and used to grind food

omnivore: an animal which eats both plants and other animals

THE *RODENTIA*

Rodents, comprising about two thousand species, form the largest, most abundant mammal order. They are found almost everywhere on the earth. Most are ground dwellers and many rodent species dwell underground in burrows or tunnel networks of varying complexity and size. However, rodents also dwell in tree nests (squirrels) or lodges in ponds and streams (beavers), or simply run in herds (capybaras). Judging from fossil remains, rodents were widespread and plentiful fifty million years ago. It is believed that they evolved from small, insect-eating mammals, and did not develop into large species until a million years ago. The largest ancient rodents were giant, bear-sized beavers. Contemporary rodents are usually small. However, the largest modern rodents are herbivorous capybaras, which grow to approximately 100 pounds as adults.

Rodents also show remarkable diversity in their diets. These range from the vegetarian capybaras to the all-encompassing diet of omnivorous rats, which will eat meat. Rodents have many roles relative to humans. Hamsters and other small rodents are pets, capybaras are eaten as food, chinchillas are fur sources, and a few, such as rats and mice, are pests that compete with humans for their food crop supplies. The tremendous adaptability of rodents, especially rats, explains their wide geographical distribution in areas differing hugely in climate.

PHYSICAL CHARACTERISTICS OF RODENTS

Among the two thousand known rodent species, size varies widely. Some small adult mouse species weigh about a fifteenth of a pound. At the other extreme, capybaras, largest of contemporary rodents, are the size of pigs.

Regardless of size, all rodents possess pairs of large, chisel-like front teeth in both the upper and lower jaws. The roots of these incisor teeth are located far back in rodent jawbones and grow continuously. Rodents lack the tearing teeth (cuspids) of carnivores as well as several premolars. Therefore, a large space exists between their incisors and molars. This allows the incisors to operate well in gnawing. The design of rodent dentition also allows the gnawed food to be transferred easily to the molars for efficient grinding.

A rodent.

In addition, the muscles of the rodent lower jaw are arranged so as to enable its easy movement backward, forward, and laterally. This optimizes grinding of gnawed food.

Rodent incisors are different from those in other animals. Their continued growth from the root is valuable, especially because only the front surfaces of these teeth are protected by enamel, the hardest material in teeth. Thus, gnawing food causes the rear surfaces of the teeth to wear down faster than their front surfaces. This wear pattern is the basis for development of the chisel-like incisor edges. It continues as long as a rodent eats regularly, keeping the incisors sharp. Another interesting aspect of rodent mouths is that cheek fur grows inside the mouth and fills up the space between incisors and molars. This hair acts as padding and filters out food chunks too large to be swallowed comfortably.

Other than the special development of "gnawing machinery" of the mouth and teeth, rodents are anatomically unspecialized, with no other ubiquitous anatomic features. Where any special characteristic has developed in some rodents, it appears to be due to environmental need. For example, claws and front paws of burrowing rodents, such as woodchucks and moles, make them efficient diggers. In addition, gliding adaptations in some squirrels allow them to "fly" (or actually glide) from tree to tree. Furthermore, leaping rodents such as the kangaroo rat use both hind feet together to enhance leaping capability. Yet another such adaptation is the webbed feet seen in beavers.

THE LIVES OF RODENTS

Rodents, like all other mammals, are warm-blooded. They carry offspring to term in a uterus where each fetus is connected to the mother via the placenta, give birth to them, and nurse them. Depending on the rodent, the sequence of events between fertilization and the end of the nursing period takes between 5.5 weeks for a small mouse, to well over a year for large rodents. The process is easiest to describe for rats, although it is quite similar for mice and hamsters.

After fertilization, rat eggs make their way into a complex uterus which can hold eight to sixteen fetuses. There, each attaches to the uterine wall and develops, over three weeks, into a rat pup. The pups are born pink, hairless, blind, and incompletely developed. They are then nurtured by their mothers, who have the instinct of all mammals to care for their offspring. Rats breast-feed their pups for three weeks. At the end of this time, they are fully covered in hair, have full vision, and have begun to eat foods other than milk.

In another month the pups are sexually mature and can breed. This makes it clear why omnivorous wild rats pose a threat to humans. Any pair of rats can produce up to eighty offspring per year. Furthermore, within six weeks after birth, any two offspring can, and do, reproduce.

Inbred laboratory rats live for two to three years, depending on the strain. Males are much larger than females (often twice their size) and may attain body weights up to two pounds. In the wild the life expectancy of rats varies greatly. However, reports of animals living for over five years occur. Some males have been reported to be as large as small cats or dogs. Wild rats live in complex tunnels as colonies of a hundred or more animals.

Other rodents live different versions of the life of rats. Litter size, gestation time, group organization all vary. For example, the larger rodents have only a few offspring per litter, and some rodents live in tree nests (squirrels) or lodges in ponds (beavers). Life expectancies may be ten years or more, assuming death by natural causes.

DESTRUCTIVE AND BENEFICIAL RODENTS

Rats and mice interact extensively with humans in a destructive fashion. The problems involved are competition for food, and disease transmission from rodents to humans. Rats and mice, viewed as pests, are known to eat 10 to 25% of grain crops grown, harvested, and stored worldwide. This percentage varies depending upon the extent of use of rodenticides, such as warfarin, in various nations and the extent of agricultural technology. Very careful use of rodenticides is important because they are quite toxic to humans. Rats are also known to attack living fowl such as turkeys, sometimes fatally, while they roost at night in commercial growing barn operations.

Rodents are disease vectors, historically causing outbreaks of serious epidemics of the bubonic plague and tularemia. This was especially serious during the Middle Ages, when rats were responsible for the transmission of the Black Death. Currently, most sporadic outbreaks of rodent-derived infectious disease are handled by use of rodenticides to kill carriers and antibiotics to destroy rodent-borne microorganisms that infect humans. Most often it is not the rodents themselves that cause disease outbreaks. Rather, infection occurs as contaminated fleas and ticks move from rodents to humans. Rats are seen as the main disease vectors because they abound near and in human habitations. However, mice and any other infected rodents can be disease vectors.

Concerning beneficial use of rodents, one can point to the myriad rats, mice, hamsters, and guinea pigs utilized as laboratory animals in testing and developing pharmaceuticals, the identification of toxic cosmetic, paint, and food components, isolation of disease cures, and so on. This aspect of research is likely to become less common because a significant and vocal segment of the population deems it morally inappropriate to submit animals to these testing procedures.

Another benefit of rodents that is becoming morally unacceptable is harvesting rodent fur. Beaver fur was once hugely important to the world fur trade. Presently, as beaver populations are low, the use of rodents to provide fur for human use has shifted to muskrats, nutria, and chinchillas, which are valued for their attractive, luxuriant coats.

—*Sanford S. Singer*

FURTHER READING
Alderton, David. *Rodents of the World*. Facts on File, 1996.
Lacey, Eileen A., James P. Patton, and Guy N. Cameron, eds. *Life Underground: The Biology of Subterranean Rodents*. University of Chicago Press, 2000.
Stoddart, D. Michael, ed. *Ecology of Mammals*. John Wiley & Sons, 1979.
Webster, Douglas, and Molly Webster. *Comparative Vertebrate Morphology*. Academic Press, 1974.

RUMINANTS

FIELDS OF STUDY
Mammalogy; Animal Anatomy; Animal Physiology; Evolutionary Biology

ABSTRACT
Ruminants are herbivorous animals that store their food in the first chamber of the stomach, called the rumen, when it is first swallowed, then after some digestion has taken place, regurgitate it as "cud," which is chewed again and reswallowed into another chamber of the stomach for further digestion. This maximizes the amount of nutrition the animal is able to derive from hard-to-digest plant food.

KEY CONCEPTS
artiodactyl: a herbivore that walks on two toes, which have evolved into hoofs
carnivore: an animal that eats only animal flesh
esophagus: the tube through which food passes from mouth to stomach
gestation: the term of pregnancy
herbivore: an animal that eats only plants
nutrient: a nourishing food ingredient
omnivore: an animal that eats both plants and animals

DEFINING THE RUMINANTS
Ruminants are herbivorous animals that store their food in the first chamber of the stomach, called the rumen, when it is first swallowed, then after some digestion has taken place, regurgitate it as "cud," which is chewed again and reswallowed into another chamber of the stomach for further digestion. This maximizes the amount of nutrition the animal is able to derive from hard-to-digest plant food. Wild ruminants tend to eat very quickly, getting as much food mass into their rumens as possible, then retiring to places of safety where they can digest at their leisure. Ruminants include sheep, cows, camels, pronghorns, deer, goats, and antelope. They eat lichens, grass, leaves, and twigs.

The main ruminant suborders are the *Tylopoda* and *Pecora*. Tylopods have three-chambered stomachs. Examples are camels and llamas. Pecorids are sheep, goats, antelope, deer, and cattle. Most pecorids have horns or antlers. These true ruminants have four-chambered stomachs, whose compartments are called the rumen, reticulum, psalterium, and abomasum. The abomasum is most similar to the stomachs of nonruminant mammals, while the other three compartments are developments peculiar to ruminants.

Ruminants chew or grind food between their lower molars and a hard pad in the gums of the upper jaw. The rumen collects partly chewed food when it is first swallowed. The food undergoes digestion in the rumen and passes into the reticulum. There, it is softened by further digestion into cud. Then the reticulum returns the cud to the mouth for rechewing, so that it is mixed with more saliva. Swallowing the chewed cud next sends food to the third compartment, the psalterium, for more digestion. The psalterium empties into the abomasum, where food mixes with gastric juices and digestion continues. Finally, the food enters the intestine, which absorbs nutrients that are then carried through the body by means of blood.

The stomachs of ruminants also contain numerous microorganisms. These help to break down the cellulose in plants, and also protect the ruminant from the effects of any toxins used as defense mechanisms by the plants.

DOMESTICATED RUMINANTS

Cattle, sheep, goats, and reindeer are all domesticated ruminants, although there are still wild species of each. The world population of these ruminants exceeds four billion. Ruminants are useful food sources for humans because they are large mammals, providing a lot of meat, as well as milk, wool, hide, and fuel. Yet, because they eat plants, they are low on the food chain; since 90% of the energy from any food source is captured in digestion, domesticated ruminants are relatively efficient transmitters of food energy from plants to humans. The large size of these ruminants, however, means that their metabolisms are relatively slow, and thus they can afford the time it takes to digest grasses, leaves, and twigs through rumination, whereas smaller herbivores with higher metabolisms, such as rodents, must digest food more quickly and thus eat more nutrient-rich plant food, such as seeds.

SOME WILD RUMINANT SPECIES

Wild ruminants are important to food chains because they eat plants, thereby preventing plant overgrowth. They also are eaten by carnivores and omnivores. Bactrian camels, with three-chambered, tylopod stomachs, inhabit the steppes and mountains of the Gobi Desert. These two-humped camels are domesticated as food and draft animals in Afghanistan, Iran, and China. Bactrian camels subsist on a diet of grass, leaves, herbs, twigs, and other plant parts. Their humps contain stored fat. Given the extreme aridity of their native environment, ruminant digestion allows them to derive the maximum nutrition from scarce food supplies.

Chamois goats live in the mountains of Europe and southwestern Asia. Their diet consists of grass and lichens in the summer, while in winter they eat pine needles and bark. Pronghorns live in the open plains and semideserts of the North American West. They are the only living *Antilocapridae*, relatives of antelope. These true ruminants eat herbs, sagebrush, and grasses in the summer, and dig under the snow for grass and twigs in the winter. The large reindeer of northern Europe and Asia inhabit forests, grasslands, and mountains. Reindeer eat grass, moss, leaves, twigs, and lichens. Sable antelope live in southeastern African woodlands and grasslands, where they eat grass and shrub leaves and twigs.

—*Sanford S. Singer*

FURTHER READING

Church, D. C., ed. *The Ruminant Animal: Digestive Physiology and Nutrition.* Waveland Press, 1993.

Constantinescu, Gheorghe M., Brian M. Frappier, and Germain Nappert. *Guide to Regional Ruminant Anatomy Based on the Dissection of the Goat.* University of Iowa Press, 2001.

Cronje, P. B., E. A. Boomker, and P. H. Henning, eds. *Ruminant Physiology: Digestion, Metabolism, Growth, and Reproduction.* Oxford University Press, 2000.

Wilson, R. T. *Ecophysiology of the Camelidae and Desert Ruminants.* Springer Verlag, 1990.

Savannas and Animal Life

Fields of Study

Ecology; Population Biology; Mammalogy

Abstract

Savannas, or tropical grasslands, are vast open spaces on which grow a large variety of plant life. Savannas usually endure long periods of drought that are punctuated by one or two rainy seasons. Although the principal vegetation is grass, trees or tall bushes appear occasionally on the landscape or along streams.

Key Concepts

carnivores: flesh-eating animals
herbivores: plant-eating animals
marsupials: animals having a pouch on the abdomen for carrying their young
ruminants: grass-eating animals that chew again food that has been swallowed
ungulates: hoofed animals

TROPICAL GRASSLANDS

Savannas, or tropical grasslands, are vast open spaces on which grow a large variety of plant life. Savannas usually endure long periods of drought that are punctuated by one or two rainy seasons. When the rains come, large herds of animals, mostly ungulates, make their annual journey along centuries-old migration routes from the river valleys where they have spent the dry season, to the fresh grass on the savanna. Although the principal vegetation is grass, trees or tall bushes appear occasionally on the landscape or along streams.

The huge African savanna is ancient, having probably evolved about sixty-five million years ago. Other areas are newer, some having been created by humans when forests were cleared to accommodate farming. Savannas also exist in South America, largely in Venezuela, and in northern Australia.

PLANT-EATING ANIMALS

The African savanna varies from very dry regions to areas of swamp, lake, and woodland, and can support the largest variety of herbivores in the world. Many animals are capable of living together because most of them have their own specific feeding habits. The hippopotamus, reedbuck, and waterbuck remain near the water, while various gazelles prefer dry areas, receiving moisture from plants. While the zebra chooses open grassland, the wildebeest (gnu), giraffe, and antelope are equipped to forage in the bush and also the woodland by virtue of their long snouts for gathering leaves and stems. All parts of trees and shrubs provide food; while some animals feed on the tough outer parts of grass, others eat the tender, fresh foliage or leaves of wild flowers.

The herbivores found on the African savanna are also the world's largest land animals. African elephants live in grassland, bush, and forest, in mountainous country and near lakes. Every day, elephants eat vast quantities of grass, leaves, twigs and bark, sometimes destroying trees and helping to create savannas. Elephant herds incorporate smaller groups of four to twenty elephants, led by the older females.

The white rhinoceros, weighing more than three tons, is one of Africa's rarest animals. The herds are composed of small family groups of one male, one or two females, and several young. Its smaller relative, the black rhinoceros, exists more abundantly. Browsing on leaves and branches, the black rhinos are protected by their size and horns. Humans are their only real enemies. Many of the grass eaters on the African savanna are antelope, which belong to a suborder of animals called ruminants. Equipped with complex stomachs, ruminants eat food which passes from the mouth, through the several chambers of the stomach, and back again to the mouth. The slow process of rumination, or chewing the cud, provides the animal with more safety from predators as chewing can be accomplished later in a less dangerous place than grazing.

In the savanna of northern Australia are found marsupials—kangaroos, koalas, and wombats. Marsupials also live on the South American savanna, along with the armadillo. Other particular species found in these geographically isolated areas are long-legged, flightless birds—the ostrich, rhea, and the emu.

PREDATORS AND SCAVENGERS

When the sun's energy is converted by plants into food for herbivores, or primary consumers, then the predators and scavengers, or animals who live by preying upon other animals, become secondary consumers in the food chain. Big cats, such as the lion, cheetah, and leopard, are powerful animals that stalk and run down their prey. Lions function in teams, with females assuming most of the work of hunting.

African wild dogs are smaller animals that have strong jaws, sharp teeth, and a keen sense of smell. They live and hunt in packs made up of six to twenty members. Other predators, such as the hyena and the jackal, kill their prey and feed at night off the carcass of the animal; powerful, far-sighted vultures and marabou stork feed by day, each eating a different part of the carcass. Smaller scavengers—crows, ravens, rats, and insects—move in, helping to dispose of dead bodies that might carry disease organisms. Smaller predators of the African savanna include the desert lynx, which pursues the smaller antelope, and the black spotted serval, which hunts ground squirrels, large rats, and guinea fowl. Other predators are the genet and the fox. Puff adders and cobras poison their victims and are themselves attacked by the mongoose and the secretary bird. The aardvark hunts ants and termites at night.

Various species of monkeys, baboons, and vervets have adapted to living on the savanna by living mostly on the ground, trying to avoid predators. In the Australian savanna, the kangaroos are preyed upon by the dingoes, or wild dogs.

—*Mary Hurd*

FURTHER READING

Alden, Peter, et al. *National Audubon Society Field Guide to African Wildife.* Chanticleer Press, 1995.
Combes, Simon. *Great Cats: Stories and Art From a World Traveler.* Greenwich Workshop Press, 1998.
Miius, S. "When Elephants Can't Take It Anymore." *Science News* 155 (May 29, 1999): 341–342.
Sherr, Lynn. *Tall Blondes: A Book About Giraffes.* Andrews McMeel, 1997.
Silver, Donald M. *African Savanna.* McGraw-Hill, 1997.

SCAVENGERS

FIELDS OF STUDY

Ethology; Ecology; Evolutionary Biology

ABSTRACT

Many carnivorous animals do not live only by killing prey. To varying extents they eat scraps from other predator kills and the corpses of animals that died of old age, injury, or illness. Many scavengers are poorly designed for hunting. Regardless, they are ecologically important. They assure that carrion does not become breeding places for disease organisms. All scavengers have a keen sense of smell to help them to find decaying flesh. Scavengers rarely die from eating carrion, because natural selection has given them tolerance of foods that kill other animals.

KEY CONCEPTS

carrion: the corpses of dead animals
cartilage: the wing case of a beetle
chitinous: made of fibrous chitin
elytra: the wing case of a beetle

NATURE'S CLEAN-UP CREW

Carnivorous worms, bears, beetles, kites, vultures, gulls, and hyenas do not live only by killing prey. To varying extents they eat scraps from other predator kills and the corpses of animals that died of old age, injury, or illness. Many scavengers are poorly designed for hunting. Some have short talons, lack optimum teeth, or are awkward. Regardless, they are

571

Hooded crow feedeing on pigeon's carrion (scavenger).

ecologically important. They assure that carrion does not become breeding places for disease organisms.

All scavengers have a keen sense of smell to help them to find decaying flesh. They need not worry about corpses fighting back, although they do risk dangerous diseases. However, scavengers rarely die from eating carrion, because natural selection has given them tolerance of foods that kill other animals.

SCAVENGER WORMS AND BEETLES

The sea mouse is a segmented worm, related to earthworms. An undersea scavenger, it lives in shallow coastal waters worldwide. In a sea mouse, each body segment is separated from the others by chitinous tissues. On the sides of the segments are muscular protrusions having bristly hairs, used in locomotion. Finer hairs growing from the segments allow the worm to sense its surroundings. As a sea mouse crawls through the seabed, it eats carrion. The worm has a flat, three-inch wide, nine-inch long body. Scales and bristles, like gray fur, cover it and lead to the name sea mouse.

Scarab beetles are one of the twelve thousand species of colorful beetles, which grow up to six inches long. They have horns on their heads or thoraxes. Males use the horns in combat during mating. Many beetles eat carrion. Dung beetles are most often called scarabs. Some scarabs (tumble bugs, dung beetles) form dung into balls and roll it into their burrows. There it becomes food for them or for their larvae, hatched from eggs deposited on the pellets.

Ancient Egyptians worshiped tumble bugs, viewing the dung pellets as symbolizing the world and the scarab horns as symbolizing the sun's rays. Thinking that scarabs caused good fortune and immortality, they used scarab carvings as charms and replaced the hearts of the embalmed dead with carved scarabs.

BEARS, JACKALS, RACCOONS, AND HYENAS

Many mammals, such as bears, jackals, hyenas, and raccoons, are scavengers only when the need arises. Among bears, the best known scavenger is the American black bear. It averages six feet long, approximately three feet tall at the shoulder, and weighs three hundred pounds, which is small and light for a bear. This small size may make it unable to compete for choice bear fare and explain why black bears eat carrion. Given the option, black bears prefer twigs, leaves, fruit, nuts, corn, berries, fish, insects, beehives, and honey.

Raccoons, nocturnal animals, inhabit swamps or woods near water. They eat frogs, crayfish, fish, birds, eggs, fruits, nuts, rodents, insects, and carrion. Raccoons have stout, catlike bodies, masked faces, coarse yellow-gray to brown fur, body lengths up to two feet, and ringed, bushy tails, and weigh up to fifteen pounds. They are solitary, except for mating in January and February. After two-month pregnancies, females give birth to three to seven babies, which remain with them for a year. Wild raccoons live for up to seven years.

Jackals are Old World wild dog scavengers that live on plains, deserts, and prairies. Golden jackals live from North Africa to southeast Europe and India. Black-backed jackals inhabit East and South Africa. Jackals have foxlike heads, but otherwise resemble wolves. They are nocturnal creatures, holing up in dens during the day. They eat carrion and small birds and mammals, hunting in packs. Jackals live for up to fifteen years.

African and Asian hyenas are also built something like wolves. The two main species, spotted (laughing) hyenas and striped hyenas, were once thought to eat only carrion. This notion arose from observations of their carrion eating and because hyena hind legs, shorter than their front legs, make them look awkward.

African "laughing" hyenas have brown fur liberally sprinkled with spots, large heads, and jaws and teeth that are able to crush bones. These hyenas, six feet long and three feet tall at the shoulder, are less clumsy than they look. It is now thought that they are the greatest killers of zebras. Nocturnal hyena packs kill prey or eat carrion. Hyenas live in groups of about sixty members. Females have one or two cubs, after four-month pregnancies. Little is known about striped hyenas, which are smaller and less aggressive. They are tan, with dark, vertical stripes and live from East Africa to Asia. Hyenas live for up to twenty-five years.

SCAVENGER BIRDS

Kites, one bird scavenger group, are hawks found in warm parts of all continents. Their legs and feet are small and weak, so they eat carrion and small animals. Swallow-tailed kites are two feet long, with beautiful white bodies and black wings and tails. They inhabit the southern United States. The smaller, white-tailed kite inhabits the Americas, Europe, and Asia. Kites often hover in the air, searching for insects and small mammal carrion.

Vultures are larger carrion eaters. New World vultures are related to storks and live in the temperate and tropical regions of the Americas. The Old World vultures of Europe and Asia resemble New World vultures in eating habits, but are related to hawks and eagles. Unlike kites, vultures look like carrion eaters. Most are ugly and dark feathered. They lack plumage on their heads and necks, minimizing messiness from blood and gore. Vultures are predisposed to carrion eating because they have blunt claws, which are poor weapons for hunting. They are suited to finding carrion by their ability to sustain long flights and their sharp eyes. Vultures fly in flocks, except during mating, when pairs nest on cliffs or in caves. Most vultures lay three eggs among bare rocks. Parents incubate the eggs and feed their young, which can fly when six months old.

There are six New World vulture species. Turkey vultures (buzzards) live in the southern United States and northern Mexico. The largest North American land bird is a vulture, the California condor, which is up to six feet long, with a wingspan up to eleven feet. Andean condors are the largest South American vultures. Condors have black body plumage. The naked heads and necks of vultures and condors vary in color and in the presence or absence of feather ruffs and wattles.

There are fourteen Old World vulture species. Most interesting are the bearded vultures called lammergeiers, four feet long and weighing up to fifteen pounds. They inhabit mountains up to fifteen thousand feet high in Europe, Africa, and Asia. Lammergeiers build nests on ledges or in mountain caves. They have tan chest and stomach plumage, white faces, and dark brown wings and tails. Masklike black feather "beards" surround their eyes and beaks. Also of interest are Egyptian vultures, two feet long, with naked yellow heads, white body feathers, and black wings. They live in Mediterranean areas and as far east as India. Many Old World vultures are not as funereal-looking as the New World breeds.

Gulls, also scavenger birds, are pigeon-sized and long-winged. Most live on the oceans and large, inland lakes. Adults have gray plumage on their wings, backs, and heads, webbed feet, and white under parts. They are graceful fliers and swimmers and nest in large colonies on rocky islands or in marshes. Gulls eat fish, other water animals, insects, carrion, garbage, and the eggs or young of other birds.

Best known are white herring gulls. Adults are two feet long and in addition to the typical gray and white gull motifs have black wing tips, yellow bills, and flesh-colored legs. They eat fish, shellfish, garbage, and carrion. Often commercial fishermen, cleaning catches, see gulls swarming behind their boats, awaiting fish offal. Herring gulls mate and lay about six eggs. Females incubate them for three weeks and feed the offspring until, at six weeks old, they strike out on their own.

SHARKS

There are over three hundred shark species. They differ from bony fish in having skeletons made of cartilage, not bone. Many sharks eat nearly all large marine animals. They vary from forty to fifty foot long, to six-inch-long species. Most abundant in tropical and subtropical waters, nearly all sharks are viewed as aggressive carnivores.

Sharks are usually gray, having leathery skins and gills behind the head. Shark tails are not symmetrical and shark skeletons end in upper tail lobes. Shark fins and tails are rigid, not erectile, as in bony fish. Sharks also have a keen sense of smell, sensing traces of blood and homing in on their sources. Despite their great strength, sharks are mostly scavengers, eating injured fish and carrion. They also eat seals, whales, and fish.

BENEFITS OF SCAVENGING

Scavengers consume carrion, preventing its decay and the endangerment of the health of other animals. This is one of their main ecological functions. Species such as the sea mouse, the carrion beetle, scarabs, vultures, and condors choose to eat carrion. Others, such as American black bears, jackals, raccoons, kites, sharks, and gulls, given a choice much prefer catching live prey.

Also, some scavengers' perceived food sources may be based on incomplete data. For example, hyenas were dubbed scavengers based on their awkward appearance and a few observations. More careful study showed that they are more ruthless nocturnal predator than scavenger, and a match for any lion in a fight. In addition, regardless of preference in obtaining their food, scavengers such as sharks and vultures have another important ecological role in the oceans and on land, killing the injured or weak members of other species. This activity helps those species to select for strong individuals, enhancing chances of species survival.

—*Sanford S. Singer*

FURTHER READING

Ammann, Karl, and Katherine Amman. *The Hunters and the Hunted.* Bodley Head, 1989.
Earle, Olive L. *Scavengers.* William Morrow, 1973.
Evans, Arthur V. *An Inordinate Fondness for Beetles.* Reprint. University of California Press, 2000.
Parker, Steve, and Jane Parker. *The Encyclopedia of Sharks.* Firefly Books, 1999.
Wilbur, Sanford R., and Jerome A. Jackson. *Vulture Biology and Management.* University of California Press, 1983.

SENSE ORGANS

FIELDS OF STUDY

Neurobiology; Evolutionary Biology; Animal Physiology

ABSTRACT

Animals, unlike plants, must use behavior, a set of responses to internal and external events occurring in their environment, to survive. It is the need for behavior that has formed the basis of virtually all of animal evolution.

KEY CONCEPTS

adaptation: the decrease in the size of the response of a sense organ following continuous application of a constant stimulus
modality: a specific type of sensory stimulus or perception, such as taste, vision, or hearing
phasic receptors: receptors that adapt quickly to a stimulus
receptive field: the area upon or surrounding the body of an animal that, when stimulated, results in the generation of a response in the sense organ

receptor cells: sensory cells within sense organs that are directly responsible for detecting stimuli.
receptor potential: a change in the distribution of electric charge across the membrane of a receptor cell in response to the presentation of a stimulus
tonic receptors: receptors that typically show little or no adaptation to a continuously applied stimulus
transduction: the translation of a stimulus's energy into the electrical and chemical signals that are meaningful to the nervous system

ANIMAL BEHAVIOR

Plants are able to form complex organic compounds for nutrition from simple molecules such as carbon dioxide and water via the process of photosynthesis. Animals, on the other hand, rely upon obtaining complex organic compounds already formed by other organisms to meet their nutritional needs. Since such sources generally take the form of other organisms, these must be located and consumed by the animal. In short, animals, unlike plants, must use behavior, a set of responses to internal and external events occurring in their environment, to survive. It is the need for behavior that has formed the basis of virtually all of animal evolution.

The first and most vital element of behavior is the detection of events occurring both within the body of the animal and in the surrounding external environment. The role of the sense organs is to detect these events (which are called stimuli) and translate them into the complex series of electrical and chemical signals that is the language of the brain. It is important to understand that the stimuli (such as sound, light, or heat) which animals detect and which may possess behavioral significance are meaningless to the brain. Brains are capable of interpreting only those signals that are in its language of electrical impulses and chemical interactions. Thus sense organs have two vital functions: the detection of environmental stimuli and the translation of that stimulus into the language that is meaningful to the nerve cells of the brain. This latter process is called transduction, and it is the ultimate role of sensory organs.

RECEPTOR CELLS AND TRANSDUCTION

Sensory organs typically consist of several different types of cells. Receptor cells are directly responsible for transducing the stimulus into the electrical language of the nervous system. Supporting cells play a number of different roles and in some cases may themselves become receptor cells. For example, in mammalian taste buds the receptor cells routinely die after ten to fourteen days and are replaced by supporting cells that transform to become receptor cells. Accessory structures assist in the process of transduction, such as the lens of the eye. Finally, sensory nerve fibers are stimulated chemically by the receptor cells and send information concerning the presence of a stimulus into the central nervous system.

The process of transduction occurs when a stimulus interacts physically with a sense organ, causing a change in the distribution of electrical charge across the cell membrane of the receptor cell. This change in transmembrane voltage is referred to as a receptor potential, and the size of the receptor potential corresponds directly to the intensity of the stimulus applied to the receptor cell. There must therefore be a minimum intensity of the stimulus required to generate a receptor potential, referred to as the sensory threshold. In general, the sensory threshold corresponds to the smallest stimulus intensity that an animal can detect. At the same time there is a maximum receptor potential that can be generated by a receptor cell, no matter how intense the stimulus. The intensity at which this occurs is known as receptor saturation. Above this level, when the receptor is saturated, it is impossible for the animal to discriminate whether one stimulus is more intense than another. Between the upper and lower limits of threshold and saturation, a change in the intensity of the stimulus will result in a corresponding change in the magnitude of the receptor potential. This range of intensities is known as the dynamic range of the sensory organ and within this range of intensities animals can discriminate between stimuli of different intensities.

When a continual stimulus is applied, and a receptor potential is generated across the membrane of the receptor cell, the nature of the receptor potential may change with time. If the receptor potential decreases with time, even though the applied stimulus remains constant, adaptation is said to occur. If one jumps into a pool of cool water on a warm summer day, the initial sensation is that of coolness against the skin. However, this perception disappears with time as the temperature receptors in the skin adapt until the water has no perceptible temperature. There are limits to adaptation, however; immersing one's hand in very hot water does not result in an eventual disappearance of the perception of heat. Some senses exhibit sensitization, the opposite of adaptation, in which progressively less stimulus is required to elicit a sensation with increasing time. The responses to certain types of painful stimuli demonstrate sensitization.

The rates and degrees of receptor adaptation vary among different types of receptors and the specific types of information they detect. Some receptors adapt very quickly to an applied stimulus; after a short period of steadily applied stimulation the receptor potential disappears. Such receptors are said to be phasic receptors. For example, if one carefully deflects a hair on the back of one's arm with a pencil point and then holds it steadily in the deflected position, the sensation generated by the deflection quickly disappears. Other receptors show very little adaptation over time; such receptors are said to be tonic receptors. Many sorts of pain receptors are tonic, as anyone who has experienced a toothache may attest.

Phasic and tonic receptors represent the extreme ends of a continuum. Most receptors can be said to be phasic-tonic receptors, which exhibit a greater or lesser degree of adaptation to continuous stimulation. Phasic and tonic receptors send different types of information concerning the nature of the stimulus to the brain. Tonic receptors send precise information concerning the duration and the intensity of a stimulus to the central nervous system. Such information may be useful, for instance, in determining the degree to which a limb is flexed. Phasic receptors, on the other hand, relay precise information about changes in the stimulus rather than its duration. Since animals live in a dynamic world, detecting small changes in the environment caused by the presence of a predator or prey may be of first importance. Both types of information are crucial to survival.

THE SENSES

Classically, there are considered to be five senses (vision, hearing, touch, taste, and smell), but in reality there are many different senses and these can be further divided into subsenses. For example, the sense of temperature actually consists of two different sensory systems, one that detects heat and another that detects cold stimuli. Furthermore, these are both linked, at some level, with the detection of pain (intensely hot stimuli are also painful, but mild heat is not). Neurobiologists refer to a type of sense as a modality (such as taste) and the detection of variations within that modality as a quality (such as sweet versus sour). There are numerous different modalities, and more are being discovered every year. The ability of animals to detect and use many different types of complex information from their environment is a fascinating and continually unfolding story.

Because changes in electrical and chemical activity are the only language understood by the central nervous system, information arriving in the brain from different sensory systems must be kept segregated from each other to avoid confusion. Thus, any activity arriving from the eyes via the optic nerve is interpreted by the brain as "light," whether or not light is actually present. Electrical or physical stimulation of the optic nerve in the absence of actual light will also be perceived by the animal as "light." The sensory systems within the brain are thus organized into a series of labeled lines, each dedicated to a specific sensory modality. Any electrical activity in a labeled line is interpreted by the brain as the presence of that modality.

Sensory organs can generally be broadly classified by the nature of the events that they are capable of detecting and transducing. Mechanoreceptors detect mechanical forces applied directly or indirectly to the body. The sense of touch is the most familiar sense employing mechanoreceptors, but there are others, such as hearing and balance, that are equally important. Chemoreceptors detect signals that occur when chemicals of different types come in contact with sensory organs (such as in taste or smell). Electromagnetoreceptors detect energy contained in the electromagnetic spectrum. The most familiar, and one most heavily relied on in most mammals, is vision. Visual sensory organs (eyes) detect the energy contained within a limited range of frequencies within the electromagnetic spectrum commonly referred to as visible light. Other familiar electromagnetoreceptors detect heat (infrared radiation) or its lack (cold).

Other animals detect other portions of the electromagnetic spectrum and in some cases are capable of directly detecting electrical and magnetic fields. These three categories are not absolutely rigid, however. Some types of receptors such as hygroreceptors (that detect the water content of air) seem not to fall conveniently in any one category, whereas others, such as nociceptors (pain receptors) straddle several categories and may respond to mechanical, chemical, or thermal stimuli. Similarly, the sense of balance employs information from several kinds of sense organs responding to a number of discrete stimuli, such as the direction of the earth's gravitational pull, rotational acceleration of the head, and body position.

EXTERORECEPTORS AND ENDORECEPTORS

Sensory receptors of all three types may be used to detect either stimuli originating in sources outside the body (exteroreceptors) or stimuli originating within the body itself (endoreceptors). These latter receptors play an absolutely vital role in the maintenance of a constant internal chemical environment and temperature (homeostasis). If the internal environment varies outside of a narrow set of parameters, death can quickly ensue. It is the task of endoreceptors to detect fluctuations in the internal environment and signal the body's involuntary control mechanisms (the autonomic nervous system) to effect corrections. For example, the detection of a drop in the core body temperature in mammals can result in a variety of responses, including the shunting of blood away from the skin surface (to minimize loss of heat to the environment), erection of hair or fur on the skin (what humans experience as goose bumps) to trap a layer of warmed air next to the skin, and shivering, which generates heat via muscular contractions.

There are many other endoreceptors in the body, which detect stimuli such as the amount of dissolved gases in the blood (oxygen and carbon dioxide), sugars, salts, and the amount of water present in the body. Another important role of the endoreceptors is proprioception, the detection of the relative positions of the body's parts in relation to one another. It is this sense that allows an individual to touch their nose with the tip of their finger when their eyes are closed. "Muscle sense," or kinesthesia, and the detection of the amount of flexion of the joints are included as parts of the modality of proprioception.

There are very many different types of exteroreceptors, and they comprise all three general classes of receptors. All of them, however, are dedicated to detecting external events that impinge in some manner upon the body. Exteroreceptors may be scattered across the surface of the body (such as in the sense of touch) or confined within specialized structures (such as the eyes or ears). The twofold function of all of these receptors is essentially the same: Exteroreceptors relay information about the nature of a stimulus as well as its location with respect to the animal (localization). This latter purpose is crucial; it may be important to know that a given sound indicates the presence of a predator, but it is equally critical to know from whence the threat originates so that appropriate behaviors can then be generated.

All sensory organs typically possess a receptive field, that area of space on or around the body which, when stimulated, results in the generation of a response in the receptor. The size of the receptive field may vary widely among different types of receptors, and it is the receptive field size together with the density of the receptors in a given area that determines the acuity, or spatial resolution, of the sense system. For example, in humans the skin of the fingers and lips contains a very large number of tactile (touch) receptors, most of which possess very small receptive fields. This allows individuals to discern to a very high degree precisely where on the skin a stimulus is occurring. In other areas of the body, like the back of the neck, the density of receptors in the skin is much lower and it is more difficult to localize exactly where a stimulus is being applied. The density of the receptor cells (rods and cones) in the eye decreases from the central portion of the retina toward the edges. That is why visual acuity is greatest when looking directly at an object and why it is very difficult to read using one's peripheral vision.

The localization of a given stimulus is critical to the organization of an appropriate response. The location of a given sense organ on the body corresponds to a location within the central nervous system, and adjacent receptors are represented by adjacent locations within the brain. Thus, there is a sensory map of the body within the brain, and it is via this topographic organization that information about stimulus location is maintained. There are many sensory maps within the brain and they play a prominent role in the organization of behaviors.

RESPONDING TO SENSORY DATA

Sense organs may be sensitive to a wide array of different qualities and provide the animal with a general sensory scene of the surrounding environment (as in mammalian vision) or may be restricted to a narrow range of stimuli that serve as channels of communication between animals of the same species. This latter case is particularly true for special chemical senses that detect chemicals that have specific behavioral meanings for members of the same species (pheromones). Very often, the sensitivity of sensory systems lies between these two extremes, with the system showing greatest sensitivity to ranges of stimuli that have greater behavioral significance to the animal. Dolphins, for example, locate objects underwater by echolocation, emitting a high-frequency call and then listening for the returning echoes. The greatest sensitivity of the dolphin auditory system is to the range of sound frequencies that are reflected back as echoes, although dolphins can hear other sounds as well.

As animals and their behaviors have evolved, so have their sense organs, providing the animals with the competitive advantages that allow them to survive and reproduce. Furthermore, natural selection frequently results in the evolution of very similar sense organs in widely divergent animals. The eyes of mammals closely resemble those of octopuses, despite the fact that these animals are not at all closely related and their common ancestor lacked complex eyes. Such convergent evolution of sense organs has resulted from the adaptive pressures on both of these animal groups that depend strongly upon vision to organize behavior.

Sometimes the sensory systems of different animal species evolve in tandem: This coevolution of sensory systems is a direct result of the interactions of the species. Bats hunt for flying moths by echolocation, emitting ultrasonic calls and homing in on the echo reflected from the moth. Many moth species, in response, have evolved "ears," located on either side of their abdomen, that are specialized to detect the calls of hunting bats. Depending upon the intensity of the detected call (and thus, the nearness of the bat) the moths display different behaviors. If the intensity is low, indicating the bat is still at a distance, the moth will fly away from the side of the body upon which the call is loudest. If the intensity of the bat's call reaches a certain level, however, the moth will execute an erratic, fluttering crash dive toward the ground in a final attempt to escape. In an additional twist, the dogbane tiger moth emits ultrasonic pulses of its own when it detects the calls of an approaching bat. Such calls may jam the bat's echolocation by interfering with the detection of the returning echoes.

A review of all of the different types of sense organs currently known in animals and the manner in which they are used to organize and shape behaviors would fill an entire volume, and more are being continually discovered. Sense organs are, in a very real sense, the keys to our individual understanding of the world. They provide the information upon which the daily understanding of reality is entirely based.

—*John G. New*

Further Reading

Downer, J. *Supernature: The Unseen Powers of Animals.* Sterling, 2000.
Gregory, R. L. *The Oxford Companion to the Mind.* Reprint. Oxford University Press, 1998.
Halliday, Tim, ed. *The Senses and Communication.* Springer-Verlag, 1998.
McFarland, D. *The Oxford Companion to Animal Behavior.* Reprinted and corrected ed. Oxford University Press, 1987.

Sexual Development

Fields of Study

Reproductive Biology; Embryology; Developmental Biology

Abstract

Sexual reproduction in most organisms involves individuals with some obviously different physical and behavioral features. Biologically, the real difference between males and females is the type of sex cells they produce. Eggs and sperm are produced in gonads—the ovaries of females and the testes of males. The gonads of higher animals also produce sex hormones, chemical messengers that affect both embryonic and adult sexual development.

Key Concepts

androgens: the general term for a variety of male sex hormones, such as testosterone and dihydrotestosterone
genital tubercle: a small swelling or protuberance toward the front of an embryo's genital area; it is destined to become the penis tip or clitoris
genitalia: the external sex structures
gonad: the structure that produces eggs or sperm cells and sex hormones; the ovary or the testis
hermaphrodite: a single organism that produces both eggs and sperm

labial folds: the paired ridges of tissue on either side of the embryo's genital area, which become penis and scrotum in males and labia in females

Müllerian ducts: the embryonic ducts that will become the female oviducts or Fallopian tubes, uterus, and vagina

parthenogenesis: the development of an unfertilized egg

urogenital groove: a slitlike opening behind the genital tubercle that will become enclosed in the penis but remain open in females

Wolffian ducts: an embryonic duct system that becomes the internal accessory male structures that carry the sperm

THE MALES AND FEMALES OF THE SPECIES

While some lower forms of life with no recognizably different sexes exchange genetic material in a form of sexual reproduction, sexual reproduction in most organisms involves individuals with some obviously different physical and behavioral features. Biologically, the real difference between males and females is the type of sex cells they produce—whether large eggs specialized to support embryonic development or tiny sperm specialized for moving to the egg. Eggs and sperm are produced in gonads—the ovaries of females and the testes of males. The gonads of higher animals also produce sex hormones, chemical messengers that affect both embryonic and adult sexual development.

Even these basic sex distinctions are rather flexible in some organisms. Sometimes, sex is determined entirely by the environment. One kind of marine worm becomes a female unless it attaches as a larva to an adult female, whereupon it becomes a male—probably because of hormones secreted by the female. Temperature can control sex development in some animals, such as mosquitoes and amphibians. Sex may also be determined by size. Since it takes more energy to produce eggs than sperm, when food is scarce it may be more adaptive to be male. The European oyster begins adult life as a male, changes to a female as it grows larger, and reverts to being a male after shedding eggs.

In territorial animals, being a large male may be an advantage. A tropical wrasse, or "cleaner fish," travels with a harem of smaller females. If he is removed, the largest female becomes a male within a few days. Many organisms, including earthworms, snails, and some fish, are hermaphrodites—functional males/females that can fertilize themselves or exchange sperm with others. Some insects, worms, crustaceans, goldfish, whip-tailed lizards, and even turkeys lay eggs that can develop without fertilization, a process called parthenogenesis. This strategy is not sound in the long run, since it does not promote genetic diversity. It is an advantage for an organism living under good conditions, however, where an all-female population can exploit the ideal environment most efficiently.

THE ORIGIN OF SEX DIFFERENTIATION

Sex differentiation probably originated as differential growth of either the ovary or the testis, mediated in various ways by hormones or other environmental factors. Later in evolution it came under genetic control, which made the process more independent of environment and made possible the development of more complex reproductive structures and behavior.

The genetic sex of an animal is determined by the father at fertilization. In most species, females have two matching X chromosomes, males have an unmatched X and a smaller Y chromosome. If a normal egg with one X chromosome is fertilized by an X-bearing sperm, the XX embryo is genetically female. A Y-bearing sperm will produce a genetic male, XY. In butterflies, fishes, and birds, however, females have XY chromosomes and males have XX. Initially, XX and XY embryos look identical and in a sense are still sexually bipotential. Their gonads are "indifferent," that is, able to form either an ovary or a testis. Each has two sets of undifferentiated sex ducts. One set, the Wolffian ducts, will become the sperm ducts and other male structures. The other set, the Müllerian ducts, form the female oviducts, uterus, and vagina.

Soon, however, genes on the Y chromosome direct the inner part of the indifferent gonad to become a testis, which then produces the male sex hormones (androgens) and Müllerian-inhibiting substance (MIS), which control further events in male development. An androgen called testosterone causes the male duct system to persist and develop, and MIS makes the female duct system degenerate. Testosterone has other developmental effects, as indicated by the fact that in monkeys, male behavior is linked to the length of embryonic exposure to testosterone.

Without the influence of the Y chromosome an XX gonad begins to develop into an ovary. The role of female sex hormones in development is unclear, since in mammals female embryonic development can occur in the absence of female hormones. The mammalian embryo has a tendency to develop in the female direction unless specific influences prevent it. The Wolffian ducts are actually remnants of a drainage system from a temporary embryonic kidney that disappears before birth. Only the presence of male sex hormones will keep these tubes from disintegrating. The Müllerian ducts, on the other hand, tend to persist unless acted upon by the anti-Müllerian substance. In birds, the embryonic ovary is the dominant gonad, and it actively feminizes the reproductive tract. It has been suggested that the early male development in mammals is necessary to allow male differentiation in the female-hormone-rich uterine environment.

Until differentiation begins, both sexes also have the same vaguely female-looking external sex structures or genitalia. In both sexes, a small protuberance called the genital tubercle is found toward the front or belly side of the embryo. Behind the genital tubercle is a slitlike opening, the urogenital groove; it is flanked by two sets of paired folds or swellings, like a river valley paralleled by two sets of ridges on either side. In the female, the genital tubercle will form a small structure called the clitoris. The urogenital groove will remain open, forming a vestibule into which the vagina and the urethra open, which empties the urinary bladder. The folds on either side of the groove will remain relatively unchanged to form labial folds.

In the male, the genital tubercle will become the tip of the penis, and the innermost urogenital folds will fuse together to form the body of the penis; the "scar" of this joining may be seen on the underside of the penis. This fusion closes off the urogenital groove and encloses the male's urethra within the tubelike penis. The outer pair of ridges will fuse to form the scrotum, the sac that encloses the two testes, which descend into the scrotal sac before birth.

Another androgen, dihydrotestosterone, may be responsible for the development of these external male structures.

STUDYING SEXUAL DEVELOPMENT

Many sex-determining mechanisms can be studied by simple modification of the environment. For example, by varying temperature, hormone level, social-group composition, or other environmental factors it is actually possible to reverse the sexes of some invertebrates, fish, and amphibians. Castration experiments are commonly used to study the effects of hormones on the sexual development of birds and mammals. For example, castrated mammals of either sex develop in the female pattern. Since in birds only the left ovary develops, castration of hens may result in the transformation of the right gonad into a testis, with complete functional sex reversal.

Sex development can be studied with naturally occurring hormone imbalances, as in freemartin calves—sterile, masculinized females whose male twin exposed them before birth to male sex hormone. The same masculinizing effect on female fetuses can be achieved by injecting a pregnant mammal with androgens or even by growing an embryonic ovary and testis together in an organ culture outside the body. In each case, the female structures are masculinized by the male hormones.

Sex chromosome mutants in animals as diverse as fruit flies, mice, and birds can be used to study chromosomal influences in sex determination. To use a human example, there are sterile XX men with a tiny piece of the short arm of the Y chromosome attached to one X. There are also XY females who show a deletion of the same short arm of the Y. These observations have led geneticists to think that the testis-determining genes are on the short arm of the Y, since without it an individual—even one with a Y—is female.

THE EVOLUTIONARY ADVANTAGE OF SEXUAL REPRODUCTION

In spite of its great biological costs, sexual reproduction is practiced by almost every kind of living thing. It confers tremendous evolutionary advantage on a species by producing a new individual with the genetic characteristics of two parents but with unique combinations of features that may make the offspring more successful than either parent. The advantage of having separate sexes for reproduction is that it permits the development of extremely specialized reproductive organs for the very different requirements of sperm or egg production, and, when needed, intrauterine support for the embryo. Though some organisms show great sexual flexibility, higher animals and plants have tended toward sexual stability, probably because of the high cost of sex reversal for organisms with highly specialized sexual structures.

The study of sex differentiation helps advance scientific knowledge in many areas. Modes of sex determination often provide clues to evolutionary relationships among groups of organisms. In addition, the development of sex differences makes a good model system for the study of more general questions. For example, one might use the control of sexual size differences to attack the broader question of what makes mice smaller than elephants. The control of sex differentiation by environmental factors, to use another example, might provide geneticists with a way to study how genes are turned on by hormones, temperature, or other external influences.

For embryologists, the stepwise determination and differentiation of the mammalian reproductive system is an excellent general development model; it involves a genetically controlled sequence of events that includes both the preservation of one embryonic structure (the Wolffian duct) and the removal of another (the Müllerian duct). Hormonally controlled events include a wide variety of developmental sequences, from the externally visible large-scale changes involved in the shaping of the external genitalia to the biochemical differentiation that programs the brain hypothalamus for its complex control of the menstrual cycle.

—*Michele Morek*

Further Reading

Carlson, Bruce M. *Patten's Foundations of Embryology.* 6th ed. McGraw-Hill, 1996.

Hickman, Cleveland, et al. *Biology of Animals.* 7th ed. Times Mirror/Mosby, 1998.

Hopf, Alice. *Strange Sex Lives in the Animal Kingdom.* McGraw-Hill, 1981.

Naftolin, Federick, et al. *Science* 211 (March 20, 1981).

Rothwell, Norman V. *Understanding Genetics.* 4th ed. Oxford University Press, 1988.

Wrangham, Richard W., W. C. McGrew, Frans B. M. De Waal, and Paul G. Heltne, eds. *Chimpanzee Cultures.* Harvard University Press, 1996.

Shells

Fields of Study

Malacology; Conchology; Developmental Biology

Abstract

Shells are external coverings that are produced by an organism. As such, they require a process by which the constituents of the shell are deposited in a site-directed fashion. Having a matrix containing the enzyme carbonic anhydrase ensures that calcium will only precipitate at that matrix site and nowhere else. In this way, the organism can control the shape of the shell by laying down a fiber matrix, usually composed of protein, as in mollusks, or a protein-chitin mixture, as is found in arthropods. How calcium and carbonate ions are brought together varies from phylum to phylum.

Key Concepts

carapace: the exoskeleton of arthropods

carbonic anhydrase: an enzyme used in the mineralization process to interconvert carbon dioxide and bicarbonate

laminate structure: having a layered shell, as in the exoskeletons of crustaceans and the valves of clams

matrix: composed of proteins or protein-chitin polymers that act as nucleation sites for mineralization

nacreous layer: the pearl-like inner layer of molluscan shells

pore canals: sites that house the cytoplasmic extensions of the crustacean hypodermis

prismatic layer: the outer crystalline layer of the molluscan shell

SHELL FORMATION FROM CALCIUM AND CARBONATE IONS

Shells are external coverings that are produced by an organism. As such, they require a process by which the constituents of the shell are deposited in a site-directed fashion. What this means is that calcium and carbonate cannot come in contact with one another when their concentrations exceed their solubility product; otherwise, they will precipitate (form a solid). Thus, these ions must be directed to the area where the preferred precipitation is to take place. For this reason, a matrix is needed to provide a negative attractive force for calcium or other bivalent ions, such as magnesium.

Other parts of the matrix may be composed of or house the enzyme carbonic anhydrase for the reversible conversion of carbon dioxide to bicarbonate. The bicarbonate will degrade to carbonate ion that can then react with the positively charged calcium ion. Calcium can be taken out of the seawater or diet and concentrated; likewise, bicarbonate ions can be formed in the gills of some of these organisms and transported to deposition sites. However, having a matrix containing the enzyme carbonic anhydrase ensures that the calcium will only precipitate at that matrix site and nowhere else.

In this way, the organism can control the shape of the shell by laying down a fiber matrix, usually composed of protein, as in mollusks, or a protein-chitin mixture, as is found in arthropods. The dumping of bicarbonate outside of a tissue where calcium is present will cause precipitation on the tissue membranes, a consequence with detrimental effects on the cells of the tissue. How calcium and carbonate ions are brought together varies from phylum to phylum.

THE PROTOZOA

Among protists, mineralization is usually accomplished within a membrane-bound vacuole. The precipitated calcium carbonate is then exported by exocytosis to reside on the external surface of the cell. A matrix is secreted in the vacuole to facilitate the mineralization process. External calcium carbonate shells are found among the order *Sarcodina* of the phylum *Sarcomastigophora*. The shells of these amoebas become chambered with growth and are perforated by small openings, through which protrude slender pseudopodia that may be interconnected to form reticulopodia. Other sarcodines, the *radiolaria* and *heliozoa*, use silica as their shell material. In this case, too, the shell or test is perforated with puncta, or small openings, through which extend slender axopodia that will collect food.

SHELLS OF ARTHROPODS

The mineralized shell makes up the exoskeletons of the subphylum *Crustacea* and the subphylum *Merostomata*, which includes the horseshoe crabs. The mineralization process in the *Crustacea* is more complex than that of mollusks because the exoskeleton must be periodically shed and a new exoskeleton must be formed in the growth process. A percentage of the old exoskeleton is reabsorbed prior to molting and then redeposited in the new exoskeleton. Like mollusks, an epithelial, sheetlike tissue, the hypodermis, is responsible for the formation of the fiber matrix upon which is deposited calcium and carbonate. Unlike mollusks, the hypodermis forms cellular extensions that elongate as the layered exoskeleton is constructed. These cellular processes come to lie in pore canals as mineralization proceeds. The first layers laid down are the first to be mineralized, so new matrix layers are being formed while the mineralization of earlier deposited layers is proceeding.

Other differences found between the mollusks and crustaceans is the amount of protein-chitin matrix of *Crustacea* and the mainly protein matrix of the *Mollusca*. The fiber matrix of crustaceans averages about 40% by volume, a volume as high as the protein matrix of bone. Because the enzyme becomes entombed in the matrix during the mineralization process, new matrix must be laid down in stepwise fashion. This gives the exoskeleton a laminate structure that is divided up into an outer epicuticle, a middle exocuticle, and an inner endocuticle. A noncalcified membrane lies between the hypodermis and endocuticle.

Shells of arthropods are quite variable, but most have a carapace covering the dorsal part of the body (the head and the thorax). The jointed plates of the abdomen are also calcified, as are the legs and antennae. Ostracod arthropods have a clamlike shell that is hinged dorsally. Barnacles are the other crustacean group that has a different sort of exoskeleton.

Their exoskeleton is composed of plates called parietes that form a wall around the organism.

SHELLS OF MOLLUSKS

The shells of mollusks are usually arranged into two major layers: an outer prismatic layer and an inner nacreous or pearly layer. The prismatic layer is composed of vertically oriented prisms that are bounded by matrix to separate each prism from one another. The prisms are elongated and extend upward to the organic layer of the shell, termed the periostracum. The nacreous layer is composed of cross-laminated lamellae. The lamellae are oriented in different directions, much like plywood layering. This serves to increase the strength of the shell in its resistance to cracking. The amount of matrix in the shells of mollusks is quite low and may be less than 1 percent of the shell volume. The mantle tissue that lines the shell is responsible for exporting the matrix, calcium, bicarbonate, and carbonic anhydrase for shell mineralization. A space, the extrapallial space, lies between the mantle and the shell and is filled with fluid to facilitate the transfer of ions to the shell. Whereas shell is deposited along the area of the mantle, new shell is deposited along the edge of the mantle.

The shells of mollusks are found in many members of this taxon. The class *Monoplacophora* have a single cap-shaped shell covering a muscular foot. The class *Polyplacophora* have an elongated body and eight shell plates covering their dorsal surface. More familiar shells of mollusks are found among the classes *Gastropoda* and *Bivalvia*. Gastropods have a single coiled shell that may be twisted, with an opening for the protrusion of the foot and head. The opening may be covered by an operculum that, in some groups, is calcified. The coiling of the shell may keep the weight of the shell over the foot for balance. The shell is used for protection against predation and to prevent desiccation in land forms. The animal can retreat into the shell when frightened or if poor water conditions occur. Many groups lose the shell in both terrestrial and marine forms and are sluglike.

Bivalves, as the name implies, have two shells or valves that enclose the animal and are hinged dorsally with an elastic ligament. Some species of both gastropods and bivalves grow spines by the evagination and elongation of the mantle to export a matrix for mineralization. This is carried out along the edge of the mantle; a ventral slit in the spine often remains to show where the mantle was during spine formation. Spine formation appears to be a device used by gastropods and bivalves to thwart predation, although this idea has not been fully tested. Spines may make these animals difficult to grasp, for example, by the claws of crabs or lobsters. In addition, the spines may make the animal difficult to swallow or crush in the mouths of fishes. Another strategy seen by these mollusks is the thickening of the shell, especially around its lip or edge. This makes it difficult for a crab to start breaking the shell from the edge, which is usually the thinnest part of the shell as this area represents new shell deposition.

Other molluscan groups with external shells are the nautiloids and extinct ammonites. These cephalopods often have coiled shells that resemble those shells of gastropods in their external appearance. Their shells are chambered, with the animal living in the endmost chamber. Gas can be secreted into the old chambers to act as a buoyancy device. The chamber partitions add strength to the shell to withstand ocean pressure at great depths. Nautiloid shells lack a periostracum. Ammonites were a similar cephalopod that occurred in great numbers during the Mesozoic era. Their shells were coiled or straight and developed complex sutures between the separate chambers. Rib patterns developed as well.

The class *Scaphopoda* has tusklike shells that are open at both ends. The animal extends out the lower opening and the upper opening extends above the sediment in this burrowing form for water flow into and out of the mantle cavity.

—*Samuel F. Tarsitano*

FURTHER READING

Carter, J. G., ed. *Skeletal Biomineralization: Patterns, Processes, and Evolutionary Trends*. Vol. 1. Van Nostrand Reinhold, 1990.

Compére, P., and G. Geoffinet. "Elaboration and Ultrastructural Changes in the Pore Canal Systems of the Mineralized Cuticle of *Carcinus maenas* During the Molt Cycle." *Tissue and Cell* 19 (1987): 859–875.

Kamat, S., X. Su, R. Ballarini, and A. H. Heuer. "Structural Basis for the Fracture Toughness of the Shell of the Conch, *Strombus gigas*." *Nature* 405 (June, 2000): 1036–1040.

Roer, R., and R. Dillaman. "The Structure and Calcification of the Crustacean Cuticle." *American Zoologist* 24 (1984): 893–909.

Simkiss, K., and K. M. Wilbur. *Biomineralization.* Academic Press, 1989.

Vermeij, G. J. *Evolution and Escalation.* Princeton University Press, 1987.

SLEEP

FIELDS OF STUDY

Neurobiology; Neurophysiology; Ethology; Evolutionary Biology

ABSTRACT

Behaviorally, sleep can be recognized by four basic features. It generally consists of 1) a prolonged period of physical immobility during which there is 2) reduced sensitivity to environmental stimuli, and which 3) typically occurs in specific sites and postures and 4) in a twenty-four hour (circadian) pattern. Using this broad definition of sleep, almost all animals can be said to sleep.

KEY CONCEPTS

circadian rhythm: a physiological or behavioral cycle that occurs in a twenty-four hour pattern

diurnal: habitually active during the day

electroencephalogram (EEG): a chart of brain wave activity as measured by electrodes glued to the surface of the skull

nocturnal: habitually active during the night

nonrapid eye movement (NREM) sleep: sleep characterized by relaxed muscles and slow brain waves

rapid eye movement (REM) sleep: sleep characterized by fast brain waves, during which dreaming typically occurs

HOW TO TELL IF SOMETHING IS SLEEPING

Behaviorally, sleep can be recognized by four basic features. It generally consists of 1) a prolonged period of physical immobility during which there is 2) reduced sensitivity to environmental stimuli, and which 3) typically occurs in specific sites and postures and 4) in a twenty-four hour (circadian) pattern. Using this broad definition of sleep, almost all animals can be said to sleep.

THE ECOLOGY OF SLEEP

Since animals are more vulnerable to danger when sleeping than when awake, most animals sleep in sites and postures that help to maximize their safety. For an insect or a small lizard this might mean wedging into a crack in tree bark or burying themselves under leaf litter. For a snake, small bird, or mammal it might mean sleeping in a nest, burrow, or tree hollow. Some animals can adopt a particular sleep posture that helps them to blend into the background to avoid detection. Animals that cannot hide or camouflage themselves might try sleeping while semiprotected in the center of a family or larger group.

Animals can also modify their sleep sites and postures to help regulate their temperature. In cold temperatures, sleep sites and postures can be chosen so as to cover exposed skin on the face or feet; birds fluff their feathers and mammals fluff their fur to trap air like a blanket; small animals huddle together to keep warm. In particularly hot temperatures, well-chosen sleeping sites may protect an animal from direct exposure to the sun, and specific postures can be adopted to facilitate heat loss.

Sleep periods also tend to be taken at times that are most safe. Diurnal species are those that are typically active in the day and do most of their sleeping at night; nocturnal species those that are more active at night and do most of their sleeping during the day. In general, birds, reptiles, and shallow-water species tend to be active during the day while mammals and deep-water species tend to be active at night, but there are many exceptions to this generalization. Whether a particular species is primarily diurnal or primarily nocturnal depends upon many aspects of its ecology and physiology, but most exhibit some kind of circadian pattern of rest and activity.

THE PHYSIOLOGY OF SLEEP

Neurophysiologically, "sleep" can be distinguished from "rest" in vertebrate animals only by measuring

A sleeping Arctic fox with its tail wrapped around itself.

changes in brain state. Fish, amphibians, reptiles, birds, and mammals all show changes in brain waves that accompany the progressive muscle relaxation that characterizes deeper and deeper states of sleep.

During their sleep periods, fish, amphibians, and reptiles slowly progress into more and more relaxed stages of sleep, then remain in their deepest state for a prolonged period of time, eventually returning slowly back to the waking state. Birds and mammals, on the other hand, show a pattern of alternating states of sleep within each sleep period. The first state is called NREM sleep (for "nonrapid eye movement sleep"). NREM sleep is characterized by relaxation of the muscles, slowed breathing and heart rate, and slow waves in the EEG (a measure of brain activity). Alternating with periods of NREM sleep are periods of REM sleep. REM sleep is characterized by rapid eye movements, irregular heart rate and breathing, and fast waves in the EEG that look identical to brain activity while awake. Although the brain is very active during REM sleep, most muscles are deactivated, leading some people to refer to REM sleep as "paradoxical sleep." In humans, it is during REM sleep that dreaming typically occurs.

Amazingly, some birds and marine mammals can sleep on one side of their brain and body while the other side remains awake. In marine mammals, it is thought that one-sided sleeping may enable an animal to keep swimming and stay near the surface in order to breathe. In birds, one-sided sleeping is thought to be a way for a particularly vulnerable animal simultaneously to get some rest and still remain alert for predators.

ACROSS-SPECIES AND DEVELOPMENTAL PATTERNS OF SLEEP

Large animals have longer sleep periods than small animals. Large animals also sleep more deeply (are more relaxed and have fewer arousals) than small ones and, among birds and mammals, have a greater proportion of REM sleep. According to the vigilance model of sleep, this is because large animals are less vulnerable than small animals, and so can afford to be less alert. Supporting this idea is the fact that for a given size animal, species that are predators typically sleep more and sleep more deeply than animals that are prey. Cougars, for example, sleep more and sleep more deeply than the deer they hunt, while falcons sleep more and sleep more deeply than pigeons and ducks.

Although large animals tend to have longer sleep periods than small animals, small animals generally have more total sleep time than large animals because they sleep more often. It is not known whether this pattern results because small animals need more sleep or because what sleep they do get is shallower and more disrupted.

Consistent with the fact that small animals sleep more than large animals is the fact that in any particular species, young animals sleep more than adults. Not only does total daily sleep time drop as an animal ages, so does the relative percentage of time spent in REM sleep. Human babies have more total sleep time and a greater percentage of REM sleep than adults, and young adults have more sleep time and a greater percentage of REM than elderly adults. The same pattern seems to hold true for other species as well.

POSSIBLE FUNCTIONS OF SLEEP

Besides the vigilance model, there are three other models which try to explain across-species and across-age differences in sleep. One of these suggests that sleep is necessary for learning. Since large animals generally live longer than small animals, they typically have a greater capacity for learning. Likewise, for a given size animal, predatory species typically rely more on learning, while prey species rely more on instinct. (A prey animal who makes a mistake is dead, whereas a predatory animal who makes a mistake can always try again.) According to this model, larger animals and predatory animals not only can afford to sleep long and deeply, they actually need more sleep in order to process and encode information. This model also accounts for the facts that young animals sleep more than older ones (they have more to learn), and that after accounting for body size and predator/prey status, mammals sleep more than birds.

A second model suggests that sleep is necessary for visual-motor coordination. This model was originally formulated to try to explain why birds and mammals have REMs during sleep but fish, reptiles, and amphibians do not. Birds and mammals have a much more complex visual system than other vertebrates. This model also attempts to explain why young animals sleep more than older animals—their visual and motor systems are not yet fully developed—and why young animals of altricial species (those born or hatched relatively helpless) sleep more than young animals of precocial species (those born or hatched at an advanced stage of development).

A third model suggests that sleep functions as a mechanism for thermoregulation. Among warm-blooded species, small animals, having a greater surface-to-volume ratio, both lose heat and overheat more rapidly than large ones; thus, they would need to rest more frequently but for shorter periods. Young animals, according to this model, need to sleep more than older animals because their thermoregulatory abilities are not yet fully developed. Likewise, altricial animals sleep more than precocial animals because their thermoregulatory mechanisms are less well developed.

All or none of these models may be correct; although virtually all animals sleep, no one yet really knows why.

—*Linda Mealey*

FURTHER READING

Campbell, S. S., and I. Tobler. "Animal Sleep: A Review of Sleep Duration Across Phylogeny." *Neuroscience and Biobehavioral Review* 8 (1984): 269–300.

Hobson, J. Allan. *Sleep*. Scientific American Library, 1995.

Kryger, M. H., T. Roth, and W. C. Dement, eds. *Principles and Practice of Sleep Medicine*. W. B. Saunders, 1989.

Meddis, Ray. "On the Function of Sleep." *Animal Behaviour* 23 (1975): 676–690.

SMELL

FIELDS OF STUDY

Animal Physiology; Neurophysiology; Neurobiology

ABSTRACT

Responses to chemicals are fundamental at all stages in biological organization. Chemotaxis, an oriented response toward or away from chemicals, has been observed in species ranging from single-cell animals such as bacteria and protozoa to very complex multicellular animals including humans. The development of sensitivity (both positive and negative) to particular chemicals is the dominant sense in most animals.

KEY CONCEPTS

anosmia: the clinical term for the inability to detect odors

chemotaxis: an oriented response toward or away from chemicals

olfaction: the sense of smell

olfactory receptors: receptor organs which have very high sensitivity and specificity and which are "distance" chemical receptors

pheromones: species-specific compounds (odors) which, acting as chemical stimuli at a distance, have a profound effect on an animal's behavior

STOP AND SMELL THE WORLD

Responses to chemicals are fundamental at all stages in biological organization. Chemotaxis, an oriented response toward or away from chemicals, has been observed in species ranging from single-cell animals such as bacteria and protozoa to very complex multicellular animals including humans. An attraction to a chemical is referred to as positive chemotaxis whereas a rejection or repulsion is called negative chemotaxis. The development of sensitivity (both positive and negative) to particular chemicals is the dominant sense in most animals. In general, receptor organs which have very high sensitivity and specificity, and which are distance chemical receptors, are called olfactory; the receptors of moderate sensitivity, usually found in the mouth, which are associated with feeding and are stimulated by dilute solutions are called taste receptors. Smells can be delivered to the olfactory receptors through air, as is the case with terrestrial animals ranging from insects to humans. On the other hand, smells can be delivered to the olfactory receptors through water, as is the case with aquatic animals such as insects and fish.

SMELL IN INSECTS

In insects, the olfactory (smell) receptors are located on the antennae. Because of the superficial location of their receptors and, more especially, because of their suitability for electrophysiological studies, insects have contributed much basic information about the mechanisms of olfaction. The olfactory receptors of most insects are highly specialized and can detect very trace amounts of compounds that are biologically important to the animal.

Olfaction is an important sensory modality for insects, particularly in mating, egg laying, and food selection. Numerous male insects, such as moths and cockroaches, are attracted by species-specific compounds called pheromones that act as chemical stimuli at a distance. Pheromones can be thought of as a language based on the sense of smell. Pheromones are often divided into two categories. Releaser pheromones initiate specific patterns of behavior. For example, they serve as powerful sex attractants, identify territories or trails, signal danger, and bring about swarming or similar types of grouping behavior. Primer pheromones trigger physiological changes in metabolism related to sexual development, growth, or metamorphosis. These changes are usually mediated through the endocrine system.

Male silk moths and gypsy moths may be attracted from a distance of between two and four kilometers by a releaser pheromone from the scent glands of the females. Males will attempt to mate with any object that has touched the female scent gland; however, males deprived of their antennae do not even orient toward the female. Synthetic releaser pheromones are now being used in traps to attract pest insects such as the gypsy moth and the Japanese beetle.

Chemical communication in social insects is used for alarm, attraction, recruitment, and recognition of nest mates and of castes. Ants give off alarm releaser pheromones from mandibular glands and so are able to warn other ants of impending danger. Army ants deposit releaser pheromones on trails to food sources or to nest sites. Primer pheromones secreted by the queen bee cause the worker bees to cluster and swarm, and they suppress the rearing of other queens in the hive.

Mosquitoes are attracted chemically to warm-blooded animals and are sensitive to several chemicals. Carbon dioxide (the metabolic waste product excreted through the lung) attracts them and they are able to orient themselves and fly to the source of this compound. They also react positively to other mammalian body products. Most common insect repellants work by interfering with the olfactory ability of the mosquito, so that the insect can no longer follow an odor toward its source.

SMELL IN FISH

The olfactory receptors in most fish are located in olfactory sacs in a pit on the head. Chemicals are brought to the receptors while swimming or during respiratory movements.

Odors and the olfactory sense play a major role in the life of many species of fish. For example, homing in salmon is controlled mostly by the "smell" of the water in which the fish was born. By following the smell trail composed of the minerals found in the water, salmon are able to return to breed in the same stream in which they were born.

Fish can also become rapidly conditioned to odors. For example, once a pike has attacked a school of minnows, the odor of other pike in the water becomes associated with an alarm response in the minnows.

SMELL IN TERRESTRIAL VERTEBRATES

In vertebrates, the olfactory receptor cells are located in the nose along the respiratory airflow path. As a result, when air is brought into the nose either during breathing or sniffing, odorant molecules are delivered to the headspace above the mucus-coated olfactory receptors. The odor molecules then bind to hairlike cilia on the olfactory receptors, producing a signal that is transmitted to the central nervous system. Because they stimulate different receptors, different smells produce different patterns of electrical activity. These odorant-specific patterns are used by the brain in smell identification.

Olfactory receptor cells are primary receptors, with axons running directly to the brain. This makes olfactory receptor cells unique, since most other sensory cells send their signals through processing centers (called synapses) before the message is carried to the brain. In the case of the olfactory receptor cells, all the information recorded by the cell is transmitted to the central nervous system. Once in the brain, the output of the olfactory receptors is sent to the limbic system (a portion of the brain involved with memory), the endocrine system, and throughout the rest of the central nervous system. The connections to the limbic system result in the very strong association that odors have in memory recognition. In humans, smells can often trigger very vivid memories. The rest of the brain also sends messages back to the bulbs, amending the pleasure of a food aroma when the stomach is full. Unlike other neurons, olfactory receptor cells constantly replicate. As a result, after a life span of about thirty days, olfactory receptor cells are replaced.

Odors help bond mothers to their newborn babies. A mother cuddling her infant will invariably brush her nose in the baby's hair to inhale its sweet aroma. She can identify her baby by its smell as much as by its cry. Additionally, one-day-old infants of many species have been shown to be able to recognize the smell of their mothers. A mother rat licks her nipples so that her blind pups can follow the scent of her saliva to the milk. Likewise, a mother kangaroo produces a saliva trail so the newly born and blind babies can follow the trail from the uterus to the mother's pouch. Wash the nipples and eliminate the saliva trail, and the pups are lost.

Female rodents who periodically smell male urine will move more quickly into puberty than females that do not. If a pregnant female mouse smells the urine of a male of another colony, she will immediately terminate her pregnancy. Also, if the olfactory nerves of a newborn rat pup are cut, the rats will never develop sexually.

A diminished sense of smell is termed hyposmia. Hyposmia can occur following a cold or after head trauma, and humans experience some reduction in the sense of smell with age. Also, most conditions that reduce the flow of air through the nose will reduce olfactory acuity. For example, a stuffy nose as a result of an allergy, a cold, or a nasal polyp often creates hyposmia. Anosmia is the complete loss of the ability to detect airborne odorants. Head trauma and severe nasal obstructions can produce anosmia. If the cause of hyposmia or anosmia is related to a blocking in the nasal airflow passageways, then treatment with steroids and/or surgery often can restore the olfactory loss.

Human experience seems to draw a sharp contrast between taste and smell. Taste is the chemical sense related to sampling compounds that come in directed contact with the inside of the mouth whereas smell is the ability of the nose to monitor airborne chemicals, often from distant sources. However, the sensations of taste and smell are not completely independent, since smell can influence taste and vice versa. For example, a lemon smell in the nose can make distilled water appear to "taste" bitter, and a sugar solution in the mouth can affect the perception of a fruit smell such as cherry.

Much of what is usually perceived of as being a taste is really a smell. For example, with the nose blocked, it is difficult to tell coffee from bitter water or an onion from a potato. As humans chew, volatile compounds in the food are released into the air in the back of the throat. These compounds then make their way up the back of the nasal cavity, where they stimulate the olfactory receptors, producing a smell sensation that dramatically enriches the perception of the taste. This combination of smell and taste is referred to as flavor. What is often thought of as "taste" is actually a combination of smells and tastes, with additional contributions to the flavor coming from temperature and pain receptors in the nose and mouth.

—*David E. Hornung*

FURTHER READING

Association for Chemoreception Sciences. www.achems.org.

Getchell, T. V., R. L. Doty, L. M. Bartoshuck, and J. B. Snow, eds. *Smell and Taste in Health and Disease.* Raven Press, 1991.

Gibbons, Byron. "The Intimate Sense of Smell." *National Geographic* 170, no. 3 (1986): 321–361.

Vroon, Piet. *Smell: The Secret Seducer.* Translated by Paul Vincent. Farrar, Straus and Giroux, 1997.

Social Hierarchies

Fields of Study

Biosociology; Ethology; Population Biology; Evolutionary Biology

Abstract

In order to prevent constant fighting over resources, many animal species have adapted a system of dominance, or a dominance hierarchy or social hierarchy. The dominance hierarchy is a set of aggression-submission relationships among the animals of a population. With an established system of dominance, the subordinate individuals will acquiesce rather than compete with the dominant individuals for resources.

Key Concepts

- **adaption:** in evolutionary biology, any structure, physiological process, or behavior that gives an organism an advantage in survival or reproduction in comparison with other members of the same species
- **aggression:** a physical act or threat of action by one individual that reduces the freedom or genetic fitness of another
- **competition:** the active demand by two or more organisms for a common resource
- **dominance:** the physical control of some members of a group by other members, initiated and sustained by hostile behavior of a direct, subtle, or indirect nature
- **fitness:** in the genetic sense, the contribution to the next generation of one genotype in a population relative to the contributions of other genotypes
- **sociobiology:** the study of the biological basis of the social behavior of animals

WHY SOCIAL HIERARCHIES EVOLVED

All animal species strive for their share of fitness. In this struggle for reproductive success, there is often competition among the individuals that make up the population. This competition is generally for some essential resource such as food, mates, or nesting sites. In many species, the competition over resources may lead to actual fighting among the individuals. Fighting, however, can be costly to the individuals involved. The loser may suffer real injury or even death, and the winner has to expend energy and still may suffer an injury. In order to prevent constant fighting over resources, many animal species have adapted a system of dominance, or what students of sociobiology call a dominance hierarchy or social hierarchy. The dominance hierarchy is a set of aggression-submission relationships among the animals of a population. With an established system of dominance, the subordinate individuals will acquiesce rather than compete with the dominant individuals for resources.

GENERAL CHARACTERISTICS OF HIERARCHIES

The simplest possible type of hierarchy is a despotism where one individual rules over all other members of the group and no rank distinctions are made among the subordinates. Hierarchies more frequently contain multiple ranks in a more or less linear fashion. An alpha individual dominates all others, a beta individual is subordinate to the alpha but dominates all others, and so on down to the omega individual at the bottom, who is dominated by all of the others. Sometimes, the network is complicated by triangular or other circular relationships where two or three individuals might be at the same dominance level. Such relationships appear to be less stable than despotisms or linear orders.

Hierarchies are formed during the initial encounters between animals through repeated threats and fighting, but once the issue of dominance has been determined, each individual gives way to its superiors with little or no hostile exchange. Life in the group may eventually become so peaceful that the existence of ranking is hidden from the observer until some crisis occurs to force a confrontation. For example, a troop of baboons can go for hours without engaging in sufficient hostile exchanges to reveal their ranking, but in a moment of crisis such as a quarrel over food the hierarchy will suddenly be evident.

Some species are organized in absolute dominance hierarchies in which the rank orders remain constant regardless of the circumstances. Status within an absolute dominance hierarchy changes only when individuals move up or down in rank through additional interaction with their rivals. Other animal societies are arranged in relative dominance hierarchies. In these arrangements, such as with crowded domestic house cats, even the highest-ranking individuals acquiesce to subordinates when the latter approach a point that would normally be too close to their personal sleeping space.

The stable, peaceful hierarchy is often supported by status signs. In other words, the mere actions of the dominant individual advertise his dominance to the other individuals. The leading male in a wolf pack can control his subordinates without a display of excessive hostility in the great majority of cases. He advertises his dominance by the way he holds his head, ears, and tail, and the confident face-forward manner in which he approaches other members of his pack. In a similar manner, the dominant rhesus monkey advertises his status by an elaborate posture which includes elevated head and tail, lowered testicles, and slow, deliberate body movements accompanied by an unhesitating but measured scrutiny of other monkeys he encounters. Animals not only utilize visual signals to advertise dominance, but they also use acoustic and chemical signals. For example, dominant European rabbits use a mandibular secretion to mark their territory.

SPECIAL PROPERTIES OF DOMINANCE HIERARCHIES

A stable dominance hierarchy presents a potentially effective united front against strangers. Since a stranger represents a threat to the status of each individual in the group, he is treated as an outsider. When expelling an intruder, cooperation among individuals within the group reaches a maximum. Chicken producers have long been aware of this phenomenon. If a new bird is introduced to the flock, it will be subjected to attacks for many days and be forced down to the lowest status unless it is exceptionally vigorous. Most often, it will simply die with very little show of fighting back. An intruder among a flock of Canada geese will be met with the full range of threat displays and repeated mass approaches and retreats.

In some primate societies, the dominant animals use their status to stop fighting among subordinates. This behavior has been observed in rhesus and pig-tailed macaques and in spider monkeys. This behavior has been observed even in animal societies, such as squirrel monkeys, that do not exhibit dominance behavior. Because of the power of the dominant individual, relative peace is observed in animal societies organized as despotisms, such as hornets, paper wasps, bumblebees, and crowded territorial fish and lizards. Fighting increases significantly among the equally ranked subordinates as they vie for the dominant position when the dominant animal is removed.

Young males are routinely excluded from the group in a wide range of aggressively organized mammalian societies such as baboons, langur monkeys, macaques, elephant seals, and harem-keeping ungulates. At best, these young males are tolerated around the fringes of the group, but many are forced out of the group and either join bachelor herds or wander as solitary nomads. As would be expected, these young males are the most aggressive and troublesome members of the society. They compete with one another for dominance within their group and often unite into separate bands that work together to reduce the power of the dominant males. Males in the two groups show different behaviors. Among the Japanese macaques, the dominant males stay calm and aloof when introduced to a new object so as to not risk loss of their status, but the females and young males will explore new areas and examine new objects.

Nested hierarchies are often observed in some animal species. Societies that are divided into groups can display dominance both within and between the various components. For example, white-fronted geese establish a rank order of several subgroups including parents, mated pairs without young, and free juveniles. These hierarchies are superimposed over the hierarchy within each of the subgroups. In wild turkeys, brothers establish a rank order among their brotherhood, but each brotherhood competes for dominance with other brotherhoods on the display grounds prior to mating.

DOMINANTS AND SUBORDINATES

To be dominant is to have the priority of access to the essential resources of life and reproduction. In almost all cases, the superior dominant animals will displace the subordinates from food, mates, and nest sites. In the matter of obtaining food, for example, wood pigeons are flock feeders. The dominant pigeons are always found near the center of the flock when feeding and feed more quickly than the subordinate birds at the edge of the flock. The birds at the edge of the flock accumulate less food and often obtain just enough to sustain them through the night. Among sheep and reindeer, the lowest-ranking females are also the worst-fed animals and among the poorest of mothers. Baby pigs compete for teat position on the mother and once established will maintain that position until weaning. Those piglets that gain access to the most anterior teats will weigh more at weaning than those who have to settle for posterior teat positions. In gaining access to mates, one study with laboratory mice has shown that while the dominant males constituted only one third of the male population, they sired 92% of the offspring.

Life is still not all that hopeless for the subordinates. Oftentimes the loser in the battle for dominance is given a second chance, and in some of the more social species, the subordinate only has to await its turn to rise in the hierarchy. In some species, cooperation among subordinate groups, especially kin groups, can lead to the formation of a new colony and a new opportunity to establish dominance. In other species, it may well be advantageous for the subordinate to stay with the group. For example, individual baboons and macaques will not survive very long if they are away from the group's sleeping area, and they will have no opportunity to reproduce. It has been shown that even a low-ranking male eats well if he is part of a troop, and he may occasionally have the opportunity to mate. In addition, the dominant male will eventually lose prowess, and the subordinate will have a chance to move up in the dominance hierarchy.

—D. R. Gossett

Further Reading

Barash, David P. *Sociobiology and Behavior*. 2nd ed. Elsevier, 1982.

Campbell, Neil A., Lawrence G. Mitchell, and Jane B. Reece. *Biology: Concepts and Connections*. 3rd ed. Benjamin/Cummings, 2000.

Feldhamer, G. A., L. C. Drickamer, S. H. Vessey, and J. F. Merritt. *Mammalogy*. WCB/McGraw-Hill, 1999.

Ridley, Mark. *Animal Behavior: An Introduction to Behavioral Mechanisms, Development, and Ecology*. 2nd ed. Blackwell Scientific Publications, 1995.

Wilson, E. O. *Sociobiology: The Abridged Edition*. Belknap Press of Harvard University Press, 1980.

Wittenberger, James F. *Animal Social Behavior*. Duxbury Press, 1981.

Symbiosis

Fields of Study

Parasitology; Symbiotology; Evolutionary Biology

Abstract

Predator-prey relationships, competition between species for limited resources, and symbiosis are the major forms of species interactions. "Symbiosis" is a term used to describe nonaccidental, nonpredatory associations between species. Symbiosis involves many types of dependent or interdependent associations between species. Symbioses are seldom rapidly fatal to either of the associating species (symbiotes) and are often of long duration.

Key Concepts

commensalism: symbiotic associations based chiefly on some form of food sharing, which may also involve shelter, protection, or cleaning

host: by convention, the larger of two species involved in a symbiotic association

intermediate host: an animal species in which nonsexual developmental stages of some commensals and parasites occur

mutualism: a type of commensalism in which both symbiotes benefit from the association in terms of food, shelter, or protection

parasite: a symbiote that must live in intimate contact with its host to survive; a parasite may be pathogenic or beneficial to the host

parasite mix: all the individuals and species of symbiotes living in a host concurrently

reservoir host: a host species other than the one of primary interest in a given research study

symbiosis: all forms of evolved, nonaccidental, nontrivial, interspecies associations, excluding predator-prey relationships

symbiote: a species involved in any form of symbiotic association with another species

ANIMAL INTERACTIONS

Understanding the ways in which different species of animals interact in nature is one of the fundamental goals of biology. Predator-prey relationships, competition between species for limited resources, and symbiosis are the major forms of species interactions, and these have profoundly influenced the diversity and ecology of all forms of life. Significant advances have been made in understanding how organisms interact, but in studies of symbiosis (which literally means "living together") one finds the most complex, interesting, and important examples of both cooperation and exploitation known in the living world.

Symbiosis involves many types of dependent or interdependent associations between species. In contrast to predator-prey interactions, however, symbioses are seldom rapidly fatal to either of the associating species (symbiotes) and are often of long duration. With the exception of grazing animals that do not often entirely consume or destroy their plant "prey," most predators quickly kill and consume their prey. While a predator may share its prey with other individuals of the same species (clearly an example of "living together"), such intraspecific behavior is not considered to be a type of symbiosis. Fleas, some ticks, mites, mosquitoes, and other bloodsucking flies are viewed as micropredators rather than parasites.

All organisms are involved in some form of competition. The abundance and availability of environmental resources are finite, and competition for resources occurs both between members of the same species and between individuals and populations of different species. When the number of individuals in a population increases, the intensity of competition for limited food, water, shelter, space, and other resources necessary for survival and reproduction also increases. Thus, competition plays a major role in populations of free-living animals (those not inhabiting the body of other organisms) and in populations living on or in other animals. For example, both tapeworms and whales must compete for resources, and both have evolved habitat-specific adaptations to accomplish this goal. Whales compete with whales, fishes, and other predators for food; tapeworms compete with tapeworms and other symbiotes (such as roundworms) for food and space; and tapeworms and whales compete with each other for food in the whale's gut.

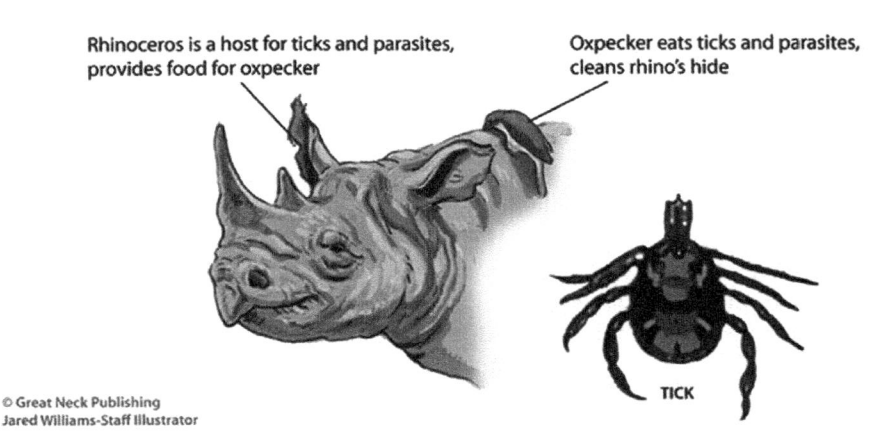

"Symbiosis" is a term used to describe nonaccidental, nonpredatory associations between species. When used by itself, the term "symbiosis" does not provide information on how or why species live together, or the biological consequences of their interactions. Recognizably different forms of symbioses all have one or more characteristics in common. All involve "living together"; most involve food sharing; many involve shelter; and some involve damage to one or both symbiotes.

HOSTS AND SYMBIOTES

Host species may be thought of as landlords. Hosts provide their symbiotes (also called symbionts) with transportation, shelter, protection, space, some form of nutrition, or some combination of these. Host species are generally larger and structurally more complex than their symbiotes, and different parts of a host's body (skin, gills, and gut, for example) may provide habitats for several different kinds of symbiotes at the same time. The three primary categories of symbiosis most commonly referred to in popular and scientific works are commensalism, mutualism, and parasitism.

Symbiotes that share a common food source are known as commensals (literally, "mess-mates"). In the usual definition of commensalism, one species (usually referred to as the commensal, although both species are commensals) is said to benefit from the relationship, while the other (usually referred to as the host) neither benefits nor is harmed by the other. Adult tapeworms which live in the intestinal tracts of vertebrate hosts provide a classic example of commensals. Adult tapeworms share the host's food, usually with little or no effect on otherwise healthy hosts. As in all species, however, too large a tapeworm population may result in excessive competition, lower fitness, or disease in both the host and the tapeworms. For example, the broad fish tapeworm of man, *Diphyllobothrium latum*, may cause a vitamin B_{12} deficiency and anemia in humans when the worm burden is high. In addition to tapeworms, many human symbiotes called "parasites" are, in fact, commensals.

External commensals (those living on the skin, fur, scales, or feathers of their hosts) are called epizoites. A good example of an epizoite is the fish louse (a distant relative of the copepod), which feeds on mucus of the skin and scales of fishes. Another type of commensalism is called phoresis (phoresy), which involves passive transportation of the commensal (phoront) by its host. Examples of phoreses include barnacles carried by whales and sea turtles, and remoras (sharksuckers), which, in the absence of sharks, may temporarily attach themselves to human swimmers. In inquilinism, the transported commensal (inquiline) shares, or more accurately, steals, food from the host, or may even eat parts of the host. Perhaps the best-known inquilines are the glass- or pearlfishes, which take refuge in the cloacae of sea cucumbers and often eat part of the host's respiratory system. A unique type of commensalism, known as symphilism, is found in certain ants and some other insects (hosts) which "farm" aphids (symphiles) and induce them to secrete a sugary substance which the ants eat.

MUTUALISM

The most diverse type of commensalism is mutualism. In some works, particularly those dealing with animal behavior, mutualism is used as a synonym of symbiosis; hence, the reader must use caution in order to determine an author's usage of these terms. As used here, mutualism is a special case of commensalism, a category of symbiosis. The relationship between mutuals may be obligatory on the part of one or both species, but it is always reciprocally beneficial, as the following examples illustrate. Some species of hermit crabs place sea anemones on their shells or claws (sea anemones are carnivores which possess stinging cells in their tentacles). Hermit crabs without anemones on their shells or claws may be more vulnerable to predators than those with an anemone partner. Hermit crabs, which shred their food in processing it, lose some of the scraps to the water, which the anemones intercept, and consume. Thus, the crab provides food to the anemone, which in turn protects its provider. Such relationships, which are species-specific, are probably the result of a long period of coevolution.

A different type of mutualism, but one having the same outcome as the crab-anemone example, is found in associations between certain clown fishes and sea anemones. Clown fishes appear to be fearless and vigorously attack intruders of any size (including scuba divers) that venture too close to "their" anemone. When threatened or attacked by predators, these small fishes dive into an anemone's stinging tentacles, where they find relative safety. Anemones apparently share in food captured by clown fishes, which have been observed to drop food on their host anemone's tentacles.

Cleaning symbiosis is another unique type of mutualism found in the marine environment. In this type of association, marine fishes and shrimp of several species "advertise" their presence by bright and distinctive color patterns or by conspicuous movements. Locations where this behavior occurs are called "cleaning stations." Instead of being consumed by predatory fishes, these carnivores approach the cleaner fish or shrimp, stop swimming, and sometimes assume unusual postures. Barracudas, groupers, and other predators often open their mouths and gill covers to permit the cleaners easy entrance and access to the teeth and gills. Cleaners feed on epizoites, ectoparasites, and necrotic tissue that they find on host fishes, to the benefit of both species. Some studies have shown that removal of cleaning symbiotes from a coral reef results in a significant decrease in the health of resident fishes.

593

PARASITISM

Parasitism is a category of symbiosis involving species associations that are very intimate and in which competitive interactions for resources may be both acute and costly. The extreme intimacy (rather than damage) between host and parasite is the chief difference between parasitism and other forms of symbiosis. Parasites often, but not always, live within the cells and tissues of their hosts, using them as a source of food. Some types of commensals also consume host tissue, but in such cases (pearl fishes and sea cucumbers, for example) significant damage to the host rarely occurs. Commensalism is associated with nutritional theft.

Some, but not all, parasites harm their hosts, by tissue destruction (consumption or mechanical damage) or toxic metabolic by-products (ammonia, for example). Commonly, however, damage to the host is primarily the result of the host's own immune response to the presence of the parasite in its body, cells, or tissues. In extreme cases, parasites may directly or indirectly cause the host's death. When the host dies, its parasites usually die as well. It follows that the vast majority of host-parasite relationships are sublethal. A number of parasites are actually beneficial or crucial to the survival of their hosts. The modern, and biologically reasonable, definition of parasitism as an intimate type of symbiosis, rather than an exclusively pathogenic association between species, promotes an ecological-evolutionary understanding of interspecies associations. Most nonmedical ecologists and symbiotologists agree that two distinct forms of intimate associations, or parasitisms (with many intermediate types), occur in nature. The most familiar are those involving decreased fitness in humans and in their domestic animals and crops.

Among animal parasites, malarial parasites, hookworms, trypanosomes, and schistosomes (blood flukes) cause death and disease in millions of people each year. The degree to which these parasites are pathogenic, however, is partly the result of preexisting conditions of ill health, malnutrition, other diseases, unsanitary living conditions, overcrowding, or lack of education and prevention. Parasites which frequently kill or prevent reproduction of their hosts do not survive in an evolutionary sense, because both the parasites and their hosts perish. Both members of intimate symbiotic relationships constantly adapt to their environments, and to each other. Over time, evolutionary selection pressures result in coadaptation (lessening of pathogenicity) or destruction or change in form of the symbiosis.

Nonpathogenic or beneficial host-parasite associations are among the most highly evolved of reciprocal interactions between species. The extreme degree of intimacy of the symbiotes (not lack of pathogenicity) distinguishes this type of parasitism from mutualism. Parasitic dinoflagellates (relatives of the algae that cause "red tides") are found in the tissues of all reef-building corals. These photosynthetic organisms use carbon dioxide and other waste products produced by corals. In turn, the dinoflagellates (*Symbiodinium microadriaticum*) provide their hosts with oxygen and nutrients that the corals cannot obtain or produce by themselves. Without parasitic dinoflagellates, reef-building corals starve to death. Similar host-parasite relationships occur in termites, which, without cellulose-digesting parasitic protozoans in their gut, would starve to death.

STUDYING SYMBIOSIS

Early studies of symbiosis focused primarily on discovery and description of commensals and parasites found in humans and their domestic food animals. Malarial parasites and some of the important trematodes (flatworms known as flukes) and nematodes (roundworms) of humans were described by the ancient Greeks, and references to the guinea worm (*Dracunculus medinensis*) are found in the Bible, where it is called "the fiery serpent." Some of the dietary conventions or laws observed in modern cultures have the side effect of preventing harmful symbioses, although it is still debated whether these proscriptions actually have their basis in early observations. It is widely known, for example, that pork products, if eaten at all, should never be consumed without thorough cooking. Swine are intermediate hosts for two very pathogenic human parasites, the trichina worm (the nematode *Trichinella spiralis*) and the bladderworm (the infective stage of the tapeworm *Taenia solium*). Much research in parasitology involves the description of symbiotes, particularly those of potential medical, veterinary, or agricultural importance.

The life cycles of many commensals and parasites are extremely complex and often involve two or more intermediate hosts living in different environments, as well as free-living developmental stages. Except for symbiotes of medical importance, relatively few complete life cycles have been worked out. Knowledge of life cycles remains as one of the most important areas of research in parasitology and is usually the phase of research following the description of new species.

Scientists have long recognized that "chemical warfare" (antibiotics, antihelminthics, insecticides) against microbial and animal parasites, and their insect and other vectors, provides only short-term solutions to the control or eradication of symbiotes of medical importance. Research attempts are being made to find ways of interrupting life cycles, sometimes with the use of parasites of other parasites. This research requires sophisticated ecological and biochemical knowledge of both the host-parasite relationship and the parasite mix. Studies of the parasite mix are ecological (parasite-parasite and host-parasite competition), immunological (host defense mechanisms and parasite avoidance strategies), and ethological (host and symbiote behavioral interactions) in nature. Investigators involved in this kind of research must be well trained in many of the biological disciplines, including epidemiology (the distribution and demographics of disease).

Immunology is the most promising modern research area in parasitology. Not only have specific diagnostic tests for the presence of cryptic (hidden or hard to find) parasites been developed, but also vaccines may be discovered that can protect people from such destructive protozoan diseases as malaria. Malaria has killed more humans than any other disease in history, and it currently causes the death of more than one million people, and lowers the quality of life for millions of others, each year.

THE INTERRELATIONSHIP OF SPECIES

All species are involved in complex interrelationships with other species that live in or on their bodies, or with which they intimately interact behaviorally or ecologically. Such interactions may play a minor role in the life and well-being of one or both of the associates, or they may be necessary for the mutual survival of both. In relatively few symbiotic relationships, one or both species may suffer damage or death. Pathogenic associations are relatively rare, because disease or death of one symbiote generally results in corresponding disease or death of the other. Such relationships, which cannot persist over evolutionarily long periods of time, may nevertheless cause catastrophic loss of life in nonadapted host populations.

Domestic animals cannot live in some parts of the world, such as the central portion of Africa, because they have little or no resistance to parasites of wild species, which are the normal hosts and are not harmed. Native species have coadapted with the parasites. This situation presents a moral dilemma to humans. In the face of human needs for space and other resources, should native animals be displaced or killed? Or should human populations proactively slow their reproductive rates? History shows that humanity has often chosen to take the former course.

The common view that animals which live in other animals are degenerate creatures that take advantage of more deserving forms of life is understandable but inaccurate. Symbiotes are highly specialized animals that do not live cost-free, or always to the detriment of their hosts. Symbiotic relationships between species have vastly increased the diversity, complexity, and beauty of the living world.

—*Sneed B. Collard*

FURTHER READING

Boothroyd, John C., and Richard Komuniecki, eds. *Molecular Approaches to Parasitology*. Wiley-Liss, 1995.

Caullery, Maurice. *Parasitism and Symbiosis*. Sidgwick and Jackson, 1952.

Limbaugh, Conrad. "Cleaning Symbiosis." *Scientific American* 205 (1961): 42–49.

Margulis, Lynn. "Symbiosis and Evolution." *Scientific American* 225 (1971): 48–57.

Margulis, Lynn, and Dorion Sagan. *Slanted Truths: Essays On Gaia, Symbiosis, and Evolution*. Copernicus, 1997.

Noble, Elmer, Glenn Noble, Gerhard Schad, and Austin MacGinnes. *Parasitology: The Biology of Animal Parasites*. 6th ed. Lea & Febiger, 1989.

Toft, Catherine Ann, Andre Aeschlimann, and Liana Bolis, eds. *Parasite-Host Associations: Coexistence or Conflict?* Oxford University Press, 1991.

Whitefield, Philip. *The Biology of Parasitism: An Introduction to the Study of Associating Organisms*. University Park, 1979.

Zann, Leon P. *Living Together in the Sea*. T. F. H., 1980.

Zinsser, Hans. *Rats, Lice, and History*. Reprint. Bantam, 2000.

TAILS

FIELDS OF STUDY

Animal Anatomy; Animal Physiology; Ethology; Evolutionary Biology

ABSTRACT

The shape, morphology, and structure of the tail vary according to the nature and behavior of the specific animal. Burrowing insectivores, like all other burrowing animals, usually have no tail. In contrast, climbing and running species have very large tails. Slow-moving animals, such as hedgehogs, have short tails, while chameleons have coiled tails.

KEY CONCEPTS

arboreal: living in trees
burrowing insectivore: an insect-eating animal that usually lives in nests formed by digging holes or tunnels in the ground
invertebrate: animals without backbones, such as insects, frogs, and snakes
prehensile tails: tails that are adapted for seizing and holding
vertebrate: animals with backbones, such as mammals
viscera: any internal body organ, such as intestines or entrails

THE MANY USES OF THE TAIL

The tail is the prolongation of the backbone, beyond the trunk of the body, of any animal, insect, or fish. The tail of the vertebrate is composed of flesh and bone but does not contain any viscera. For many aquatic animals (such as fish and amphibians) as well as animals that use water as part of their living environment (such as crocodiles, otters, and whales), the tail is fundamental to their locomotive ability. Squirrels and other arboreal animals use the tail to keep balance and as a rudder when they jump from branch to branch, while others, such as the spider monkey and the chameleon, use the tail as an extra limb to increase their mobility through the branches of the rain forest. The tail may also be used as a means of defense for porcupines, as a warning signal in rattlesnakes, as a hunting weapon in alligators and scorpions, an ornamental sexual attractant in peacocks, or even a communication tool for dogs. Most birds do not carry a tail; instead, the prolongation has been fused into the short pygostyle bone, which serves as the holder of tail feathers and assists the birds in flying.

The shape, morphology, and structure of the tail vary according to the nature and behavior of the specific animal. Burrowing insectivores, like all other burrowing animals, usually have no tail. In contrast, climbing and running species have very large tails. Slow-moving animals, such as hedgehogs, have short tails, while chameleons have coiled tails.

Tail of a scorpion.

TAILS IN INVERTEBRATES

A number of lizards have thick tails that are covered by large, spiny, hard scales. The tail is often used as a defensive measure against predators such as snakes, especially when the head and body of the lizard are wedged between rocks. Moreover, lizards are capable of shedding their tails, which wriggle in a way that may confuse their predator, thus giving them enough time to escape. Each vertebra of such a tail has a fracture line along which it splits when the tail muscles contract. A unique lizard is the Solomon Islands skink, whose prehensile tail muscles are bound both to the vertebrae and to a fibrous sheath of collagen located under the skin, thus creating a much stronger and more flexible tail. Stimulation of the nerve contraction in the severed position keeps the tail moving for several seconds after severing occurs. Normally the tail splits off in one place, but in a few lizard species, such as the glass snakes (*Ophisaurus*), tails may be broken in more than one place. The stump usually heals very quickly and a new tail regenerates, although it is not as long and not as elaborate as the original. Bioengineers in the late 1990s conducted research trying to isolate the gene that is responsible for the regeneration of the severed tails.

Amphibians such as tadpoles and salamanders also lose and, in many cases, regenerate their tail, which has a spinal cord. Tadpole tails have a stiff rod for support, called the notochord, while the salamander's tail has a backbone with vertebrae. No tail is regenerated in the salamander if the spinal cord is severed, unlike the tadpoles, where the tail is reformed regardless of the fate of the severed spinal cord.

Tails in invertebrates are used in characteristic, unique ways among the different animals. Iguanas use their tails like large oars to swim in water, while tucking its legs close to its body. When threatened, the armadillo lizard puts its tail in its mouth, rolls over, and assumes the shape of a tight ball. Day geckos use their tails as a means of support while jumping from tree to tree to avoid their main predator, the falcon. The skink has a very short tail, whose shape is very close to that of its head. This confuses its enemies, since they do not know the direction the skink is going to take while escaping.

TAILS IN VERTEBRATES

The squirrel owes its name to its tail. The name is derived from the Greek word *skia* (meaning "shadow") and *oura* ("tail"), indicating that the tail is large enough to shade the rest of the animal body. Unlike the bat, which is the only mammal that truly flies, the flying squirrel is the only vertebrate that glides. Using its strong and sturdy back feet to jump from the top of a tree, it flattens its tail and spreads the loose folds of the skin so that it can glide in air. Just before landing, the gray squirrel lowers its tail first, then quickly lifts it and lands on its hind feet. When in danger, the red squirrel attempts to scare its predator by flicking its tail while using a series of noises, such as whistling, chattering, and chirping, and stomping its hind feet.

Other nonmammals, such as snakes, crocodiles, and turtles, may lose their tails to predators or to accidents, although not voluntarily. In fact, some snakes, such as the African python *Calabaria* and the oriental venomous *Maticora*, wave their thick, colored tails toward their enemy while retreating slowly. Both male and female diamondback rattlesnakes have the ability to rattle their tail ninety times per second. During the motion, the sonic muscles pump calcium out of the myoplasm fifty times faster than the locomotor muscles do. As a result, the filaments in the sonic muscles release each other and get ready for the next contraction much more quickly than the locomotor muscles.

Parrots (*Psittaciformes*) are the most popular birds that possess colorful and widely variable tails. In some species the tail is short, square, or rounded; in others it is long and pointed, but no parrots have forked tails. Birds that fly long distances tend to have longer tails, sometimes longer than the total length of their body. Climbing parrots usually possess rounded wings and blunt tails.

Long, elaborate tails are considered by evolutionists as unusual ornaments to win mates and use in elaborate courtship rituals. Wildlife scientists believe that the more attention-getting displays also give an indication of which bird will make a good parent. In agreement with Charles Darwin's theory of sexual selection, female animals of some species develop a preference for armaments that now have purely ornamental function, while others show preference for a certain trait which males eventually have to adopt if they are to mate. Male swallows that have long tails have a much higher degree of paternity and produce more biological offspring as compared to similar birds that are short tailed, indicating a distinct positive correlation between male tail length and paternity in this species.

The function of the heterocercal tail in shark locomotion has been given two explanations. The first one suggests that as a result of the lift created by beating the tail, the net force acting on the tail is directed dorsally and anteriorly. In the so-called Thomson's model, the tail generates a net force directed through the shark's center of gravity.

Sea animals use their tails in peculiar ways. In the depths of the species-rich Amazon River, electric fish and catfish predominate. Among the unusual incidents observed, electric fish appear to eat the tails of other fish. Eels plant their tails in burrows they dug in the sand underwater and let their bodies wave in the current, while waiting for food, such as drifting tiny crustaceans, fish eggs, and plankton, to reach them.

—*Soraya Ghayourmanesh*

FURTHER READING

Adler, Tina. "Record-Breaking Muscle Reveals Its Secret." *Science News* 150 (July 27, 1996): 53.

Coates, Michael J., and Martin J. Cohn. "Fins, Limbs, and Tails: Outgrowths and Axial Patterning in Vertebrate Evolution." *BioEssays* 20, no. 5 (May, 1998): 371.

Gould, Stephen Jay. "Hooking Leviathan by Its Past: Two Tales of Tails Confirm the Theory of the Whale's Return to the Sea." *Natural History* 103 (May, 1994): 8.

Major, Peter F. "Tails of Whales and Fins, Too." *Sea Frontiers* 33 (March/April, 1987): 90–96.

Stegermann, Eileen. "Sturgeon: The King of Freshwater Fishes." *The Conservationist* 49 (August, 1994): 18–23.

Stewart, Doug. "The Importance of Being Flashy." *International Wildlife* 25 (September/October, 1995): 30–37.

Wright, Karen. "When Life Was Odd." *Discover* 18 (March, 1997): 52–57.

TEETH, FANGS, AND TUSKS

FIELDS OF STUDY

Dontology; Evolutionary Biology; Neurotoxicology; Animal Physiology

ABSTRACT

Teeth, fangs, and tusks are found in vertebrates, and are used for obtaining and masticating food and for defense. Fangs and tusks are elongated canine or incisor teeth, which serve to deliver venom (fangs) or as defensive weapons (tusks). Tusks, which are made of ivory, have been valued for making jewelry and decorative objects, which has led to the endangerment of the elephants and walruses that carry them.

KEY CONCEPTS

elapids: a snake classification which includes cobras and rattlesnakes that have short, fixed front fangs

heterodont: having two or more types of teeth, such as molars and incisors

homodont: having teeth all of the same type

incisor teeth: teeth that are located in the front of the mouth and whose function is to tear, hold, and cut the prey

toxin: any substance, such as the venom in snakes or spiders, that is toxic to an animal

Viperidae: poisonous terrestrial or semiaquatic snakes

vipers: a snake classification that includes copperheads and rattlesnakes, which have long, movable front fangs

ANIMAL DENTITION

Teeth, fangs, and tusks are found in vertebrates, and are used for obtaining and masticating food and for defense. Fangs and tusks are elongated canine or incisor teeth, which serve to deliver venom (fangs) or as defensive weapons (tusks). Tusks, which are made of ivory, have been valued for making jewelry and decorative objects, which has led to the endangerment of the elephants and walruses that carry them.

African elephants are the largest land mammal in the world, weighing several tons. They are intelligent, social animals. Elephants experienced a severe population decline in the late 1980s due to illegal killing (poaching) for their ivory tusks.

TEETH

Teeth are hard, resistant structures that are found on the jaws, as well as in or around the mouth and pharynx of vertebrates. Teeth were formed through the evolution of bony structures found in primitive fish. A tooth consists of a crown and one or more roots. The crown is the visible, functional part, while the root is attached to the tooth-bearing bone. However, many living organisms, such as birds, turtles, whales, and many insects, do not have teeth.

Although the teeth of many vertebrates have been adapted for special uses, their function is to catch and masticate food and defend against predators or enemies, as well as other specific purposes. Rodents and rabbits, for instance, have curved incisors that are deeply embedded in the jaws and grow longer with age. Several types of apes have enlarged canines for defense, while the sawfish, which is the only animal with teeth completely outside its mouth, uses its teeth to attack its prey. All perissodactyls, a group of herbivorous animal species characterized by an odd number of toes on the hind foot, including horses, rhinoceroses, and tapirs, have evolved specialized forms of teeth that are adapted for grinding. Generally, lizards are insectivores and have sharp tricuspid teeth that are adapted for grabbing and holding. Mollusk and crustacean feeders, such as the caiman lizard, have blunt, rounded teeth along the jaw margin or on the palate.

Tooth-bearing animals may be heterodont or homodont. Most mammals are heterodont and carry two or more types of teeth, such as the incisors and the molars. The purpose of the incisors is to tear and bite into the food, while the molars crush and grind the food. Cats do not have flat-crowned crushing teeth; instead only the stabbing and anchoring canine teeth cut up the food, which is then swallowed. In the case of elephants, the upper second incisors have developed into ivory tusks, which are the longest and heaviest teeth in any living animal. On the other hand, fish and most reptiles are homodonts—their teeth are all of about the same size, and their purpose is to catch prey. This is the main reason why their teeth are regularly replaced during their life span.

Snakes have teeth that curve back toward the throat. Thus, as soon as the prey is caught, it is pushed into the throat. In poisonous snakes, teeth called fangs have a canal through which the poison (venom), normally stored in glands that are in the roof of the mouth, may be ejected.

FANGS

Fangs are long, pointed teeth used by many animals primarily as a means of self-defense or for securing their prey. Many snakes use their hollow fangs as hypodermic needles to puncture their victim and, in the case of a venomous snake, to inject toxic venom. Because of the snake's muscular elasticity, the fang tips at penetration average 112% further apart than their bases at rest. The wound resulting from penetration of the flesh is called a snakebite. A nonvenomous snakebite is usually similar to a puncture wound which, when untreated, may become infected and in extreme cases may develop gangrene. That of a venomous snake is much more serious and there is a potentially lethal effect, which depends on several factors such as the size of the victim, the bite location, the quantity of venom that has been injected, the speed of venom absorption into the victim's blood circulation, and the speed with which first aid and the antidote are given.

Several venomous snakes, such as the ringhals and the black necked cobra (types of African cobras), have the ability to spit. A fine stream of venom is forced out of each fang which, instead of having to go through a straight canal that ends in a long opening, is forced through a different canal that turns sharply

599

forward to a small round opening on the front surface. Contraction of the muscle that surrounds the poison gland leads to the spitting of the venom, which is harmful to human eyes unless washed quickly. Front-fanged snakes include pit vipers such as rattlesnakes, fixed-fanged snakes such as brownsnakes and cobras, seasnakes, and true vipers (*Viperidae*). All have hollow, tubelike fangs created by the extension of the dentine across the anterior seam. Evolutionary herpetologists have postulated that this anterior seam may have been open several tens of thousands of years ago.

Only poisonous snakes possess fangs and venom glands, which are considered to be an evolutionary result of salivary glands found in primitive fish. Poisonous snakes bite the victim with their fangs and proceed to inject the venom in the wound in their effort to kill their prey. The venom contains toxins, which are chemical compounds that have the potential of attacking the blood and the nervous system with lethal consequences. The proteinaceous enzymes that are found in the venom are also used to digest the eaten animal. These snakes are generally subdivided into vipers and elapids. Vipers, which include copperheads and rattlesnakes, have long, movable front fangs, while elapids, such as cobras and rattlesnakes, have short front fangs that are fixed. Occasionally some venomous snakes have a fang on the upper jaw in the rear of the mouth, which makes them less harmful to large animals since the venom injection occurs at a much slower pace. The devouring or swallowing of the prey usually takes place only after the venom has taken full effect and the animal is dead. Generally, small snakes embed their fangs in the prey for a longer duration than larger snakes.

The analysis and determination of the composition of the hard dental tissues of the *Viperidae* has been conducted using classical microscopy, scanning electron microscopy, transmission electron microscopy, X-ray diffraction, and infrared spectroscopy. The results have shown a thin, calcified outer layer, composed of very small needlelike crystals that are randomly distributed. The calcified outer layer contains pores and collagen fibers that are incompletely mineralized, especially in the wall of the poison canal. Chemical analysis of the dentine has indicated a poorly mineralized apatite with a high level of carbonate content.

All six species that belong to the marine fish classified under the genus *Chauliodus* (order *Salminoformes*) are called viperfish. They are all small in size, the largest being the Pacific viperfish (*C. macouni*), which is no more than one foot long. They are characterized by their long fangs, which protrude from the upper and lower jaws and are used to securely grab their prey. The tigerfish of the Zambezi river, with its two rows of long and very sharp teeth, has been seen as more fearsome than the true piranha or the mako shark.

TUSKS

Evolutionary theories suggest that the tusks of both the walrus and the wild boar are enlarged canine teeth, while in the case of the pig, the lower incisor has been modified with time into an organ that is used for digging purposes. In male Indian elephants and African elephants of both sexes, the tusk is the upper incisor, which continues to grow throughout their lifetime. The female Indian elephant has either no tusks or very small ones. Male Ceylonese elephants, found in Sri Lanka, generally have no tusks, while Sumatran elephants, found in Indonesia, bear the longest tusks. Elephant tusks from Africa are typically six feet long, conical at the end, and weigh approximately fifty pounds each. Indian elephants have slightly smaller tusks. Each tusk from the largest pair, recorded and exhibited at the British Museum, is twelve feet long, has a barrel circumference of eighteen inches and weighs close to 150 pounds.

Studies by Raman Sukumar, an ecologist at the Indian Institute of Science in Bangalore, and Milind Watve, a microbiologist, indicate that male Asian elephants with longer tusks are prone to host many fewer parasites. This is in agreement with the theory of the evolutionary biologist William D. Hamilton, of the University of Oxford, who proposed in 1982 that males that carry genes resistant to parasites have the ability to be healthier and therefore live longer. At the same time, these species develop secondary sexual characteristics which enable females to select males that will produce better offspring.

Anatomically, the tusk is composed of several layers, the innermost of which grows the latest. One third of the tusk is embedded in the bone sockets of the elephant's skull. At the beginning, the head end of the tusk has a hollow cavity that becomes almost fully solid with aging. Only a narrow nerve channel runs through the center of the tusk to its end.

Ivory is a type of dentin that is the major component of the elephant tusk and is desirable worldwide for its beauty and durability. There are generally two types of ivory, soft and hard. The ivory isolated from the tusks from East African elephants is soft, while that found in the West African elephants is hard. Hard ivory is usually darker in color and is straighter than the soft type, which has a fibrous internal texture, is less brittle, and is a more opaque white. The demand for ivory has led to a large number of elephant slaughtering incidents and a dangerous decline of the African elephant population, beginning in the late nineteenth century and continuing into the twenty-first. The discovery by archaeologists in the early 1980s of tusk material in a Greek ship that was sunk around 1400 BCE revealed that ivory was a trade commodity even during the Bronze Age.

Elephants belong to the order *Proboscidea*, whose early ancestors were not larger than the average pig. It is believed that during the process of evolution, the lower jaw elongated beyond the upper and eventually turned into tusks. As a result, the nose and upper lips developed into an elongated cover to the projecting lower jaw. During the Eocene Epoch (between fifty-four and forty million years ago) the upper tusks were lost, and a downward-hooked, tusk-tipped mandible developed. The mandible and its tusks became more shovel-like during the Miocene era (twenty-six to seven million years ago). It is believed that the tusks assumed a cylindrical shape not very much later. It appears that tusks were also part of the anatomy of the woolly mammoth, as seen in the specimen that died about twenty thousand years ago and was discovered in a Siberian excavation in the 1990s.

—*Soraya Ghayourmanesh*

FURTHER READING

Bagla, Pallava. "Longer Tusks Are Healthy Signs." *Science* 276 (June 27, 1997): 1972.

Bower, Bruce. "Bronze Age Trade Surfaces from Wreck." *Science News* 126 (December 8, 1984): 359.

Chapple, Steve. "Fish with Fangs." *Sports Afield* 216 (June/July, 1996): 144.

Coppola, M., and D. E. Hogan. "When a Snake Bites." *Journal of the American Osteopathic Association* 94, no. 6 (June, 1994): 766.

Hayes, W. K. "Ontogeny of Striking, Prehandling, and Envenomation Behavior of Prairie Rattlesnakes (*Crotalus v. viridis*)." *Toxicon* 29, no. 7 (1992): 867–875.

TENTACLES

FIELDS OF STUDY

Animal Anatomy; Animal Physiology; Evolutionary Biology; Marine Biology

ABSTRACT

Tentacles are slender, leglike or armlike protrusions from the body of a living organism. They are used for protection, as organs of touch, or to capture food. They are often seen in coelenterates, an animal phylum that includes jellyfish, anemones, and coral polyps. The name coelenterate comes from the Latin for "hollow intestine."

KEY CONCEPTS

bud: protuberance used in asexual reproduction
budding: bud development into a complete organism
carrion: dead animals
funnel: an opening in a cephalopod mantle, providing oxygen and propulsion
gamete: sperm or an egg
nematocyst: poison sting cell
radula: tonguelike, toothed organ that grinds food and drills holes in shells of prey

WHAT ARE TENTACLES?

Tentacles are slender, leglike or armlike protrusions from the body of a living organism. They are used for protection, as organs of touch, or to capture food. They are often seen in coelenterates, an animal phylum that includes jellyfish, anemones, and coral polyps. The name *coelenterate* comes from the Latin for "hollow intestine."

Coelenterates are hollow tissue sacs with two layers of cells in their walls. The cells carry out digestion,

Jellyfish floating under Arctic ice.

> **JELLYFISH: POISON TENTACLES**
>
> Box jellyfish (*Chironex fleckeri*), native to Australia's north coast and Southeast Asian coastlines, are the most poisonous creatures on earth. They get their name from their box-shaped, translucent bodies, which are approximate 35 centimeters wide. Sixty tentacles, up to 5 meters long and 6 millimeters in diameter, hang from the box.
>
> The box holds an eye and a mouth. Prey—plankton, small fish, and shrimp—are caught in tentacles and pulled into the mouth. Tentacles hold huge numbers of sting cells called nematocysts. When something comes in contact with the tentacles, the nematocysts inject a poison which paralyzes and kills their prey, and can cause extreme pain and even death to large animals or humans. Severe stings will kill within four to six minutes after a human is stung.

excretion, and reproduction. There is one body opening, a mouth. Food enters it, is digested, and used by the cells. Wastes also leave through the mouth, surrounded by long, slender sense organs called tentacles, which also grab food and pull it into the mouth.

WORM AND MOLLUSK TENTACLES

Some worms have tentacles. For example, sandworms (family *Nereidae*) inhabit shallow ocean waters worldwide. Most live in sand or mud burrows. They have colorful bodies ranging from one inch to three feet long and are in the same phylum as earthworms. A sandworm body has several hundred segments. Each segment has two muscular parapodia, both attached to bundles of bristly chaetae. The first segment of a sandworm holds two light-sensitive tentacles, four eyes, and two sensory palps. The second segment has eight more tentacles called cirri. The palps and cirri help sandworms find crustaceans, small fish carrion, and other prey.

Many cephalopods have tentacles. For example, the nautilus (family *Nautilidae*) is found in South Pacific and Indian oceans. All nautilids have soft bodies with spiral, brown and white shells. Their mouths have beaks and radulae (toothed tongues) to eat carrion and crustaceans on ocean bottoms. Around the nautilid mouth are almost one hundred short tentacles that are used to sense objects, feed, and move. At one side of the arms, an opening called a funnel allows water to enter and carry oxygen to gills. A nautilid can also spray a jet of water from the funnel, to swim.

OCTOPUS AND SQUID TENTACLES

Octopuses (family *Octopodidae*) have eight slender, flexible tentacles. On the underside of each arm is a row of suckers. Sensors in suckers serve in defense, detect prey, capture it, and identify its texture, shape, and taste. Arms join at their bases into a bulb-shaped head/body.

An octopus squeezes water out of the mantle cavity and moves its tentacles to swim. It captures mollusk and crab prey by wrapping tentacles around them and using its suckers to tear them to pieces. In the center of the arm juncture is its mouth, which has a radula and a beak to continue shredding prey.

The giant squid (*Architeuthis harveyi*) inhabits the Atlantic Ocean depths. Its torpedo-shaped body is up to 5 meters in diameter and 8 meters long from head to tail. This cephalopod tapers toward its posterior and has at the tail end two fins for swimming and steering. Two eyes provide excellent vision. In front of the eyes are eight round, elongated tentacles with suckers and hooks on their undersides. Two longer tentacles lack suckers and hooks, except at the tip. The tentacles are thirty-six feet long, and from tail to tentacle tip giant squid often exceed sixty feet in length. They are earth's largest invertebrates. Squid hide near the ocean bottom and ambush or pursue prey. The two long tentacles are shot forward to seize their victims, which are passed along to the mouth. A beak in the mouth crushes and tears up prey.

Squid and octopus are eaten by humans worldwide. Nautilids, sandworms, and related organisms help maintain the balance of nature, eating carrion and helping to keep the ocean clean. Other tentacled organisms are eaten by fish that humans use as food. On the other hand, squid and octopuses eat food fish and crabs, competing with humans, and jellyfish can kill via the nematocysts in their tentacles.

—*Sanford S. Singer*

Further Reading

Hunt, James C. *Octopus and Squid*. Monterey Bay Aquarium Foundation, 1997.

León, Vicki. *A Tangle of Octopuses: Plus Cuttlefish, Nautiluses, and a Giant Squid or Two*. Silver Burdett Press, 1999.

Taylor, Leighton R., and Norbert Wu. *Jellyfish*. Lerner, 1998.

Yonge, Charles M., and T. E. Thompson. *Living Marine Molluscs*. Collins, 1976.

Territoriality and Aggression

Fields of Study

Ethology; Evolutionary Biology; Population Biology; Biosociology

Abstract

Any field or forest inhabited by animals contains countless invisible lines that demarcate territories of individuals of many different species. Most organisms appear to attend only to the territorial claims made by members of their own species. A map of individual territories for each species in the same habitat would show little consensus on the value of particular areas. Yet basic similarities exist in why and how different species are territorial.

Key Concepts

adaptive function: the reason that a characteristic evolved by means of natural selection

conspecifics: members of the same species

dominance hierarchy: a social system, usually determined by aggressive interactions, in which individuals can be ranked in terms of their access to resources or mates

home range: an area that an animal frequently uses but does not defend

population regulation: long-term stability of population size at a level that prohibits overexploitation of resources

resource-holding potential: the ability of an individual to control a needed resource relative to other members of the same species

strategy: a behavioral action that exists because natural selection favored it in the past (rather than because an individual has consciously decided to do it)

territoriality: the active defense of an area that is required for survival and/or reproduction

TERRITORIES AND SPECIES

Any field or forest inhabited by animals contains countless invisible lines that demarcate territories of individuals of many different species. Humans are oblivious to these boundaries yet have quick perception of human property lines; other animals are equally oblivious to human demarcations. Most organisms, in fact, appear to attend only to the territorial claims made by members of their own species.

If separate maps of individual territories could be obtained for each species in the same habitat and superimposed on one another, the resulting hodgepodge of boundaries would show little consensus on the value of particular areas. Yet basic similarities exist in why and how different species are territorial.

CAUSES OF TERRITORIALITY

The existence of aggression and territorial behavior in nature hardly comes as a surprise. Even casual observations at a backyard bird feeder reveal that species that are commonly perceived as friendly can be highly aggressive. The observation of birds at feeders can lead to interesting questions concerning territorial behavior. For example, bird feeders usually contain much more food than any one bird could eat: Why, then, are aggressive interactions so common? Moreover, individuals attack conspecifics more often than birds of other species, even when all are eating the same type of seeds.

Aggressive defense of superabundant resources is not expected to occur in nature; however, bird feeders are not a natural phenomenon. Perhaps the aggressive encounters that can be observed are merely artifacts of birds trying to forage in a crowded, novel situation, or perhaps bird feeders intensify aggressive interactions that occur less frequently and less conspicuously in nature. While the degree to which aggression observed at feeders mirrors reality is open to question, the observation of a greater intensity of interactions between conspecifics definitely reflects a natural phenomenon.

Members of the same species are usually more serious competitors than are members of different species because they exploit exactly the same resources; members of different species might only share a few types of resources. Despite the ecological novelty of artificial feeders, noting which individuals win and lose in such an encounter can provide valuable information on the resource-holding potential of individuals that differ in various physical attributes such as body size, bill size, or even sex. For organisms that live in dense or remote habitats, this type of information can often be obtained only by observations at artificial feeding stations.

Territorial defense can be accomplished by visual and vocal displays, chemical signals, or physical encounters. The sequence of behaviors that an individual uses is usually predictable. The first line of defense may involve vocal advertisement of territory ownership. One function of bird song is to inform potential rivals that certain areas in the habitat are taken. If song threats do not deter competitors, visual displays may be employed. If visual displays are also ineffective, then residents may chase intruders and, if necessary, attack them. This sequence of behaviors is common in territorial interactions because vocal and visual displays are energetically cheaper than fighting and involve less risk of injury to the territory owner.

It may be less obvious why fighting is a necessary component in territorial interactions for both territory owners and intruders. Without the threat of bodily injury, there is no cost to intruders that steal the resources of another individual. This would severely hamper an owner's ability to control an area. On the other hand, if intruders never physically challenge territory owners, then it would pay for all territory owners to exaggerate their ability to defend a resource. Thus, physical aggression may be essential. Animals do not frequently kill their opponents, however, so there must be something that limits violence.

Various species of animals possess formidable weapons, such as large canine teeth or antlers, that are quite capable of inflicting mortal wounds. Furthermore, a dead opponent will never challenge again. Yet fights to the death are rare in nature. When they do occur, some novel circumstance is usually involved, such as a barrier that prohibits escape of the losing individual. Restraint in normal use of weapons, however, probably does not indicate compassion among combatants. Fights to the death may simply be too costly, because they would increase the

A pair of Sichuan Takin fight.

chance that a victor would suffer some injury from a loser's last desperate attempts to survive.

FUNCTIONS OF TERRITORY

Territories can serve various functions, depending on the species. For some, the area defended is only a site where males display for mates; for others, it is a place where parents build a nest and raise their offspring; for others, it may be an all-purpose area where an owner can have exclusive access to food, nesting sites, shelter from the elements, and refuge from predators. These different territorial functions affect the area's size and the length of time an area is defended. Territories used as display sites may be only a few meters across, even for large mammal species. Territorial nest sites may be smaller still, such as the densely packed nest sites guarded by parents of many colonial seabirds. All-purpose territories are typically large relative to the body size of the organism. For example, some passerine birds defend areas that may be several hundred meters across. Although all three types of territories may be as ephemeral as the breeding season, it is not uncommon for all-purpose territories to be defended year around.

The abundance and spatial distribution of needed resources determine the economic feasibility of territoriality. On one extreme, if all required resources are present in excess throughout the habitat, territory holders should not have a reproductive advantage over nonterritory holders. At the other extreme, if critical resources are so rare that enormous areas would have to be defended, territory holders might again have no reproductive advantage over nonterritory holders. If needed resources, however, are neither superabundant nor extremely rare and are somewhat clumped in the habitat, territoriality might pay off. That is, territorial individuals might produce more offspring than nonterritorial individuals.

Studies of territoriality raise more questions than biologists can answer. Researchers investigate how large an area an individual defends and whether both sexes are equally territorial. They seek to determine whether the territories of different individuals vary according to quality. The density of conspecifics may influence territoriality; on the other hand, territoriality itself may serve to regulate population size, although evidence suggests that this is an incidental effect.

All-purpose territories vary considerably in size, depending on the resource requirements of the individuals involved and the pattern of temporal variation in resource abundance. In some organisms, individuals only defend enough area to supply their "minimum daily requirements." In others, individuals defend a somewhat larger area—one that could still support them even when resource levels drop. In still others, individuals defend territories that vary in size depending on current resource levels. For example, pied wagtails (European songbirds) defend linear territories along riverbanks that are about six hundred meters long during the winter. The emerging aquatic insects they consume are a renewable resource, but renewal rates vary considerably during the season. Rather than adjusting territory size to match the current levels of prey abundance in the habitat, wagtails maintain constant territory boundaries. This inflexibility persists even though territories that extend for only three hundred meters could adequately support an individual for about one-third of the season. In contrast, the territory size of an Australian honey eater varies widely during the winter. Nectar productivity of the flowers visited by honey eaters varies considerably during the season. By adjusting territory size to match changing resource levels, individual birds obtain a relatively constant amount of energy each day (about eighteen kilocalories).

TERRITORIAL ROLES

In some species, only males are territorial. In other species, both sexes defend territories, but males defend larger territories than females do. In some mammals in which both sexes are territorial, males are aggressive only to other males, and females are only aggressive to other females. In these species, male territories are sufficiently large to encompass the territories of several females. Presumably, these males have increased sexual access to the females within their territories. Perhaps the most curious example of sex-specific territorial behavior is observed in a number of coral reef fish, in which all individuals in the population are initially female and not territorial. As the individuals grow older and larger, some develop into males. Once male, they engage in territorial behavior.

Within a species, significant variation in territory quality exists among individuals. Studies on numerous species have demonstrated a relationship between territory quality and an individual's resource-holding potential. For example, larger individuals tend to control prime locations more often than smaller individuals. In addition, possession of higher-quality territories often results in increased reproductive success. For some species, this occurs because individuals with better territories obtain mates sooner or obtain more mates than individuals with poorer territories. In other species, possession of superior territories increases the survival chances of the owner.

As the density of conspecifics increases, the ability of individuals to control territories decreases. In some species, the territorial system may break down completely, with all individuals scrambling for their share of needed resources in a chaotic fashion. In other species, the territorial system is replaced by a dominance hierarchy. All competitors may remain in the area, but their access to resources is determined by their rank in the hierarchy. For example, elephant seal males can successfully defend areas containing from eighty to a hundred females from other males. Very dense clusters of females, however (two hundred or more), attract too many males for one male to monopolize. When this happens, one male—usually the largest male—dominates the rest and maintains disproportionate access to females.

Territoriality undeniably has an adaptive function: to increase the survival and reproductive success of individuals. Territoriality can also have several possible incidental effects, one of which was once considered to be an adaptive function: serving as a means of population regulation. The reasoning behind this hypothesis is simple. The number of territories in a habitat would limit the number of reproducing individuals in a population and would thereby prevent overpopulation that could cause a population crash. Support for this hypothesis would include demonstration that a significant number of nonbreeding adults exist in a population. Indeed, for several species, experimental removal of territory owners has revealed that "surplus" individuals quickly fill the artificially created vacancies. In most of the species studied, however, these surplus individuals are primarily males. Population growth can be curbed only by limiting the number of breeding females, not the number of breeding males. Furthermore, the population regulation argument assumes that some individuals abstain from reproduction for the good of the population. If such a population did exist, a mutant individual that never abstained from reproducing would quickly spread, and its descendants would predominate in future generations.

TERRITORIALITY IN THE FIELD

Territoriality is typically investigated in the field using an observational approach. Initial information collected includes assessing the amount of area used by each individual, how much of that area is defended from conspecifics, and exactly what is being defended. It is relatively easy to discern the spatial utilization of animals. For many species, all that is required is capturing each individual, marking it for field identification, and watching its movements. For species that range long distances, such as hawks or large mammals, and species that are nocturnal, radio telemetry is frequently used. This methodology requires putting radio transmitters on the individuals to be followed and using hand-held antennas, or antennas attached to cars or airplanes, to monitor movements. For fossorial species (animals that are adapted for digging), animal movements are often determined by repeated trapping. This method involves placing numerous baited live traps above the ground in a predetermined grid.

Knowing the spatial utilization of an animal does not document territoriality. Many types of animals repeatedly use the same regions in the habitat but do not defend these areas from conspecifics. Such "home ranges" may or may not contain areas that are defended (that is, territories). Territorial defense can be readily documented for some animals by simply observing individual interactions. These data often need to be supplemented by experiments. Behavioral interactions might only occur in part of the organism's living space because neighbors do not surround it. For these individuals, researchers play tape-recorded territorial vocalizations or place taxidermy mounts of conspecifics in different locations and note the response of the territory holder. For other species, such as fossorial rodents, direct estimates of territory size cannot be obtained because aggressive interactions cannot be observed; as a result, territory boundaries must be inferred from trapping information. Regions in which only the same individual is repeatedly trapped are likely to be areas that the individual defends. This is an indirect method, however, and can be likened to watching the shadow of an organism and guessing what it is doing.

It is often difficult to determine exactly what an animal is defending in an all-purpose territory where organisms use many different types of resources. Which resource constitutes the "reason" for territorial defense is not readily apparent. On the other hand, several resources may contribute in some complex way. For many species these things simply are not known. This uncertainty also complicates estimates of territory quality. For example, red-winged blackbirds in North America have been particularly well studied for several decades by different investigators in various parts of the species range. Males defend areas in marshes (or sometimes fields), and some males obtain significantly more mates than others. Biologists think that males defend resources that are crucial for female reproduction. Some males may be more successful at mating than others because of variation in territory quality. Yet the large number of studies done on this species has not yielded a consensus on what the important resources are, whether food, nest sites, or something else.

Theoretical investigations of territorial behavior often employ optimality theory and game theory approaches. Optimality theory considers the benefits and costs of territorial defense for an individual. Benefits and costs might be measured simply as the number of calories gained and lost, respectively. Alternatively, benefits might be measured as the number of young produced during any one season; costs might be measured as the reduction in number of future young attributable to current energy expenditures and risks of injury. For territorial behavior to evolve by means of natural selection, the benefits of territorial behavior to the individual must exceed its costs.

Game theory analyses compare the relative success of individuals using alternative behaviors (or "strategies"). For example, two opposing strategies might be "defend resources from intruders" and "steal resources as they are encountered." In the simplest case, if some individuals only defend and other individuals only steal resources, the question would be which type of individual would leave the most offspring. Yet defenders interact with other defenders as well as with thieves, and the converse holds for thieves. By considering the results of interactions within and between these two types of individuals, a game theory analysis can predict the conditions under which one strategy would "win" or "lose" and how the success of each type of individual would vary as the frequency of the other increases in the population.

A complete understanding of territoriality involves not only empirical approaches in the field but also the development of testable theoretical models. Considerable advances have been made recently merging these two methodologies. Future investigations will no doubt include experimental control over resource levels that will allow definitive tests of predictions of alternative theoretical models.

TERRITORIALITY AND AGGRESSION

The importance of investigating any biological phenomenon might be measured by its contribution to understanding nature in general and humankind in particular. By these criteria, aggression and territoriality may be among the most important topics that could ever be studied. Among animals in general, some species are highly aggressive in defending their living space, and others ignore or tolerate conspecifics in a nearly utopian manner. Some animals are territorial during only part of the annual cycle, and some only in specific areas that they inhabit; others remain aggressive at any time and in any place. Thus, a main goal for researchers is to unravel the ecological and evolutionary conditions that favor aggressive behavior and territoriality.

Aggression and territorial behavior appear to have evolved in various organisms because, in the past, aggressive and territorial individuals reproduced more successfully than nonaggressive and nonterritorial ones. An implicit assumption of behavioral biologists is that animals other than humans do not interact aggressively because of conscious reasoning, nor are they consciously aware of the long-term consequences of aggressive acts. Should these consequences be detrimental, natural selection will eliminate the individuals involved, even if this means total extinction of the species. Humans are different. They are consciously aware of their actions and of the consequences of such actions. They need only use conscious reasoning and biological knowledge of aggressive behavior to create conditions that can reduce conflict between individuals and groups.

—*Richard D. Howard*

FURTHER READING

Alcock, John. *Animal Behavior.* 7th ed. Sinauer Associates, 2001.

Allen, Colin, and Marc Bekoff. *Species of Mind: The Philosophy and Biology of Cognitive Ethology.* MIT Press, 1997.

Davies, Nicholas B., and John R. Krebs. *An Introduction to Behavioral Ecology.* 4th ed. Blackwell Scientific Publications, 1997.

Dennen, J. van der, and V. S. E. Falger, eds. *Sociobiology and Conflict: Evolutionary Perspectives on Competition, Cooperation, Violence, and Warfare.* Chapman and Hall, 1990.

Howard, Eliot. *Territory in Bird Life.* Atheneum, 1962.

Ratcliffe, Derek A. *The Peregrine Falcon.* 2nd ed. Academic Press, 1993.

Wilson, Edward O. *Sociobiology.* The Belknap Press of Harvard University Press, 1975.

THERMOREGULATION

FIELDS OF STUDY

Biochemistry; Biothermodynamics; Evolutionary Biology

ABSTRACT

Body-temperature regulation by animals is essential for life. The maintenance of life relies on the sum of all chemical reactions or metabolic activity in an organism. These reactions are facilitated by catalysts, substances not directly involved in a reaction as either a product or reagent but essential for accelerating the process or allowing the reaction to proceed under conditions compatible with life. Biological catalysts are complex temperature-sensitive proteins. Homeostasis is the maintenance of a constant internal environment, one suitable for proper enzymatic activity.

KEY CONCEPTS

conduction: a transfer of heat from one substance to another with which it is in contact

countercurrent mechanism: a heat exchange system in which heat is passed from fluid moving in

ectotherm: an animal that regulates its body temperature using external (environmental) sources of heat or means of cooling

endotherm: an animal that regulates its body temperature using internal (physiological) sources of heat or means of cooling

heliotherm: an animal that uses heat from the sun to regulate its body temperature

homeostasis: the maintenance by an animal of a constant internal environment

homeotherm: an animal that strives to maintain a constant body temperature independent of that of its environment

optimum temperature: the narrow temperature range within which the metabolic activity of an animal is most efficient

poikilotherm: an animal that does not regulate its body temperature, which will be the same as that of its environment

thermogenesis: the generation of heat in endotherms by shivering or increased oxidation of fats

BODY-TEMPERATURE REGULATION

Body-temperature regulation by animals is essential for life. The maintenance of life relies on the sum of all chemical reactions or metabolic activity in an organism. These reactions are facilitated by catalysts, substances not directly involved in a reaction as either a product or reagent but essential for accelerating the process or allowing the reaction to proceed under conditions compatible with life. For example, a reaction that in a test tube, might require exceedingly high temperatures will proceed, if catalyzed, at normal body temperatures. Biological catalysts are complex proteins called enzymes. These are fragile molecules and are quite temperature-sensitive. If exposed to excessively high or low temperatures, they will be denatured and lose their functional properties.

Homeostasis is the maintenance of a constant internal environment, one suitable for proper enzymatic activity. Homeostatic mechanisms involve three components: a sensor (or receptor) that reacts to changes in environmental conditions, a coordinator (or integrator) that responds to information

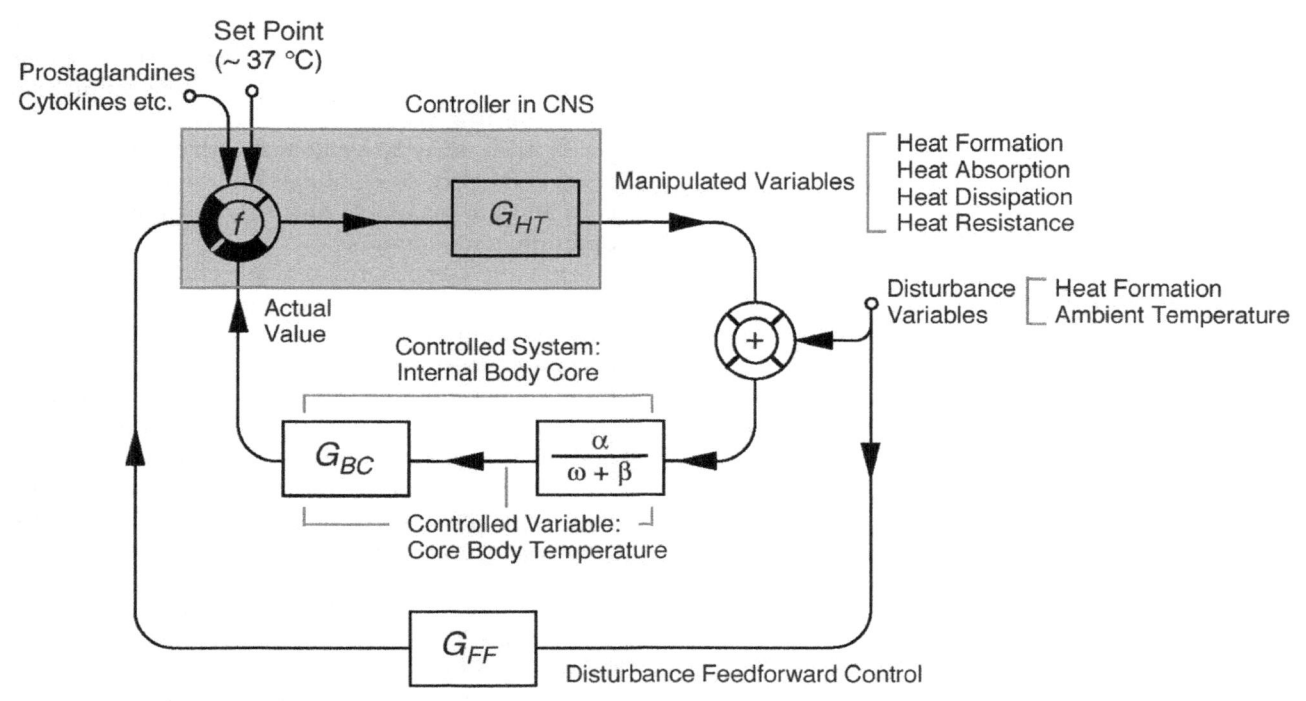

Simplified information processing structure of human thermoregulation.

from the sensor, and one or more effectors (activated by the coordinator), which elicit appropriate, regulatory responses.

Temperature sensors are scattered throughout the bodies of most animals, but those specifically associated with temperature regulation in vertebrates (animals with backbones) are found in the hypothalamic region of the brain. Coordinators are found within the brain (or its equivalent in simpler animals), again in the hypothalamus of more advanced types. Effectors may be any structure capable of affecting temperature.

Animals generally function at temperatures between 4°C and 40°C. Peak metabolic efficiencies, however, exist over a much narrower range, called the optimum temperature. This temperature varies by the animal and its habitat. Optimum temperatures often approach lethal limits, the highest temperature an animal can tolerate. This necessitates precise control of temperature in order to avoid exceeding those limits. Within lower temperature ranges, some animals can alter metabolic requirements in order to adapt to changing temperatures without sacrificing efficiency. This process, which involves complex biochemical and cellular adjustments, is called "temperature compensation."

Animals that utilize metabolic mechanisms to maintain constant, relatively high body temperatures are often referred to as being "warm-blooded." Others, whose body temperatures are not regulated or are regulated primarily by behavioral means, are called "cold-blooded." That these terms are imprecise and irrelevant becomes obvious when one considers that the temperature of a desert-dwelling "cold-blooded" lizard or insect may often exceed that of any bird or mammal. On the other hand, the core temperature of some hibernating mammals may be reduced to being anything but "warm."

Most invertebrates (animals lacking backbones) as well as many fishes, amphibians, and some reptiles, do not regulate body temperatures; they are called poikilotherms. They monitor environmental conditions, attempt to seek out areas where temperatures are suitable, and avoid those where they are not. Their temperatures are essentially identical to environmental temperatures. If excessively high temperatures are unavoidable for more than short periods, death may occur. Low temperatures are seldom fatal (unless below freezing) but will result in diminution of metabolic functions, causing the animal to become torpid, or inactive. Since these animals are vulnerable, they will seek shelter, which is why insects, for example, are rarely encountered during colder months.

ECTOTHERMS

Animals that regulate body temperatures fall into two categories. Those that utilize environmental sources of heat are called ectotherms (animals that "heat" their bodies using external sources). Those that utilize physiological temperature control mechanisms are called endotherms (animals that "heat" their bodies using internal sources). Since endotherms (birds and mammals) strive to keep temperatures constant, they may also be called homeotherms (animals that maintain constant temperatures). All regulators must invest considerable energy in the process. To minimize that expenditure, they utilize microhabitats in which regulatory mechanisms are not necessary.

Ectotherms use behavior, enhanced by physical or physiological mechanisms, to take advantage of environmental conditions. A principal source of heat for most ectotherms is sunlight; temperature regulators that rely on the sun are called heliotherms (animals that "heat" their bodies using the sun). Lizards from temperate zones (areas with moderate and/or seasonal climates) are the most efficient ectotherms and may serve as models to illustrate the process. Tropical species, which live in constant, warm environments, tend to be poor regulators.

Sunlight and heat may be assimilated directly by basking lizards or indirectly by convection from sun-heated surfaces. Basking occurs when an animal exposes itself to sunlight by seeking unshaded perches. Position and posture are critical. Lizards will orient themselves in order to expose the greatest amount of surface to the sun. This involves a position in which the animal is broad-side to the sun. Surface area is further enhanced by flattening the body dorsoventrally (top to bottom). Similarly, animals may absorb heat from the substrate. Lizards flatten themselves against a warm surface to maximize the area through which heat is assimilated.

Area is critical in elevating temperatures, either by basking or convection, but does not increase proportionately with volume as animals increase in size. Thus, large ectotherms require disproportionately more energy and time to raise their temperatures than animals with similar proportions but smaller dimensions. This explains why the first animals to emerge in the spring or early morning tend to be small. Also, since dark colors absorb more radiation (heat and light), cold animals will stimulate pigment cells and are invariably much darker than those at optimum temperatures.

That these mechanisms work effectively is illustrated by observations of active lizards at near-freezing temperatures at high elevations in the Andes of South America. When these lizards are captured, body temperatures of 31°C are recorded. In another study, lizards active at -4°C have been found to have body temperatures above 10°C. Some investigators have observed lizards, buried in sand during the night, emerging slowly, exposing only their heads. Since many lizards have large blood sinuses in their heads, it has been suggested that they can raise their body temperatures while minimizing exposure to predators. It is unlikely that this is effective, as heat gained would be rapidly lost to the substrate by conduction. Only if the ground were warmer than air and only until body temperature reached that of the ground would this mechanism be operative.

In ectotherms, cooling is a much more difficult proposition. Without access to a source of "cold," ectotherms can do little more than minimize heat absorption. Coloration is lighter to increase reflection, orientation is toward the sun, posture involves lateral (side-to-side) compression, and animals will "tiptoe," lifting themselves away from warm substrates. If these are inadequate, animals must seek shelter. Many desert-dwelling lizards exhibit activity cycles that peak twice each day (morning and evening) to avoid cold nights and hot midday periods.

ENDOTHERMS

Endotherms use physiological effectors to raise or lower temperatures. If cold, they will generate heat (thermogenesis) by rapid muscular contractions (shivering) or increased oxidation of fats. Simultaneously, devices minimizing heat loss will be implemented. These include lowered ventilation (breathing) rates; since inhaled air is warmed during passage through the respiratory tract, heat is lost with each expiration. Also, superficial blood vessels narrow (vasoconstriction), reducing flow of warm blood to the skin, from which heat is lost by convection. Attempts to insulate skin are illustrated by "goose bumps." Though ineffective in sparsely haired humans, this reaction to cold is quite effective in mammals with thick body hair or fur. Muscles attached to hair follicles contract and draw hairs into an upright position, and the ends droop, trapping dead air between matted ends and skin. A fine undercoat in many species enhances the process. Still air is an excellent barrier to heat flow. A similar device affecting feathers exists in birds.

When hot, endotherms keep muscular activity to a minimum, increase ventilation rates (panting), and expand superficial blood vessels (vasodilation). Rates of heat dissipation in some mammals are enhanced by sweating. Sweating and panting rely on evaporative cooling, the same principle involved in using radiators to prevent hot automobile engines from overheating. Endotherms adapted to hot climates produce concentrated urine and dry feces to conserve water, since much is lost in cooling.

Many of these mechanisms are surface-area related. Consequently, endotherms in hot climates, especially large species with relatively poor surface-to-volume ratios, often possess structures, such as elephant's ears, to increase area through which heat may be dissipated. On the other hand, endotherms occupying cold habitats are designed to minimize exposed surfaces. For example, arctic hares have short ears and limbs compared to the otherwise similar jackrabbits of warmer climes. In addition, cold-adapted endotherms may decrease rates of heat loss from poorly insulated appendages by means of countercurrent mechanisms. Heat from blood in arteries flowing outward into a limb is passed to venous blood returning inward to the heart. This minimizes the amount of heat carried into a limb, whose surface-to-volume ratio is very high. It also functions to warm the returning blood, which prevents cooling of the body core. The appendages themselves are very cold; portions may even be at below-freezing temperatures. Actual freezing is prevented by special fats in the extremities.

STUDYING THERMOREGULATION

Specific methods vary according to the subject, approach, and discipline in question. Anatomy (study of structure), using both micro- and macroscopic methods, often centers on surface-related phenomena. For example, studies investigating the vascularization (blood supply) of whale flukes, whose physiology is difficult to study, have indicated that these are quite capable of dissipating heat and have led to the knowledge that these animals, even in cold water, because of their large size and poor surface-to-volume ratios, have potential problems with overheating. The role of blubber was reevaluated in this light and is now recognized as being one of fat storage with little to do with insulation. Furthermore, with new technologies in electron microscopy, anatomists have been able to describe, often for the first time, the complex structural components of organs (and even cells) that are active in thermoregulation.

Physiological studies of function are of two major types. One involves measurements of activity under different thermal regimes; for example, patterns of locomotion or digestion (involving specifically neural and muscular or neural, muscular, and glandular entities, respectively) may be observed at different temperatures. Often, these include observations of performance on treadmills or of rates at which food items are processed in controlled laboratory settings. On a different scale, metabolic activity itself might be linked to temperature by measuring rates of oxygen consumption in special metabolic chambers or utilization rates of products necessary for particular chemical reactions. These types of investigations have led to the determination of optimum and lethal temperatures in many species.

A second type of physiological study deals with actual thermoregulation. The ability to monitor body temperatures continuously, even in small animals, by means of radiotelemetry has made possible whole series of experiments in which animals' thermal responses to induced or natural conditions can be evaluated. Investigations of this type have provided insights into, for example, adaptive hypothermia (significantly reduced body temperatures) in small endotherms such as bats and hummingbirds. These species drastically reduce their core temperatures when inactive in order to conserve energy otherwise rapidly lost as heat through their relatively large surface areas.

Since laboratory work often fails to simulate natural conditions adequately, observations of animals in nature have been instituted. These seek to evaluate thermoregulation in the contexts of ethology (the study of behavior) and ecology (the study of organisms' relationships with their environments). These types of studies frequently entail prolonged observations until patterns of behavior or habitat use emerge and can be quantified and evaluated. The use of rapid-reading thermometers or implanted radio-thermistors facilitates understanding of the often-subtle modifications in thermoregulatory behavior or microhabitat use characteristic of many animals. Relating recorded temperatures to changes in posture, position, orientation, activity level, and ambient temperatures of substrate and air has, for example, led to an appreciation of how efficiently some ectotherms regulate temperature and the complexity of the mechanisms involved.

APPLICATIONS OF THERMOREGULATION RESEARCH

Long restricted by concepts of "warm-blooded" versus "cold-blooded" animals, investigators did not begin in-depth explorations of thermoregulation until the twentieth century. Most early efforts grew out of medical studies dealing with dynamics of human temperature regulation, especially in the context of pathological states associated with fever or trauma-induced hypothermia. Monitoring these conditions led to an appreciation of how complex temperature regulation is and how many of the body's systems are involved. These studies, in turn, led to investigations of similar mechanisms in animals. Initially, most dealt with laboratory animals, but pioneering investigations into thermoregulation by animals in natural habitats soon opened whole new vistas. These studies were subsequently extended to "cold-blooded" species, which in turn led to an appreciation of how effective behavioral temperature regulation could be. In the 1970s, suggestions that at least some dinosaurs may have been homeotherms stimulated further interest in this field of study.

Most heat exchange with the environment occurs through skin or respiratory systems; muscular systems generate heat as a by-product of contraction; digestive and urinary systems regulate elimination of wastes, which influences retention or loss of heat-bearing water; cardiovascular systems transport heat; and nervous and endocrine systems regulate the entire complex. In addition, all cells require a proper thermal environment and may affect heat production by altering rates of oxidative metabolism. Therefore, a more complete understanding of thermoregulation has enhanced scientists' awareness of both normal and pathological functions in most body systems. Specific medical applications of these studies include induced hypothermia during surgery-related trauma and treatment of accident-related hypothermia using mechanisms first observed under natural conditions in animals.

Studies of temperature-regulating mechanisms, both behavioral and physiological, have also provided insights into relationships between animals and their environments. Thermoregulatory needs have been used to explain behavioral and ecological phenomena for which causative agents were previously unknown. From a practical perspective, this knowledge is useful in developing management tools to sustain disrupted or endangered ecosystems. Appropriate techniques must be developed with a thorough knowledge of the dynamics in any given system, and this must be based on biological criteria rather than human perceptions. For example, reforested areas have often been managed as crops, with all the attendant problems of monocultures (areas cultivated for plants of only one species). Among these is the lack of biodiversity (variety of life-forms). When efforts began to take into consideration microhabitat requirements, often related to temperature regulation, varieties of plants—many with little or no commercial value in themselves—were planted. This resulted in managed areas becoming capable of supporting many different species.

Finally, a more complete knowledge of structures related to thermoregulation has been applied by paleontologists (scientists who study fossils) to the study of dinosaurs. Long thought to be "sluggish," lizardlike ectotherms, dinosaurs are now thought by many investigators to have been more like mammals and birds in their physiological capabilities. This image is more in tune with their domination of the earth for some hundred million years.

—*Robert Powell*

Further Reading

Avery, Roger A. *Lizards: A Study in Thermoregulation.* University Park Press, 1979.

Bakker, Robert T. *The Dinosaur Heresies: New Theories Unlocking the Mystery of the Dinosaurs and Their Extinction.* Kensington, 1986.

Desmond, Adrian J. *The Hot-Blooded Dinosaurs.* Blond & Briggs, 1975.

Dukes, H. H. *Dukes' Physiology of Domestic Animals.* 11th ed. Comstock, 1993.

Gans, Carl, and F. Harvey Pough, eds. *Physiology C: Physiological Ecology.* Vol. 12 in *Biology of the Reptilia.* Academic Press, 1982.

Hickman, Cleveland P., Larry S. Roberts, and Frances M. Hickman. *Integrated Principles of Zoology.* 11th ed. McGraw Hill, 2001.

Johnston, Ian A., and Albert F. Bennett, eds. *Animals and Temperature: Phenotypic and Evolutionary Adaptation.* Cambridge University Press, 1996.

Schmidt-Nielsen, Knut. *Desert Animals: Physiological Problems of Heat and Water.* Dover, 1979.

Schmidt-Nielsen, Kurt. *How Animals Work.* Cambridge University Press, 1972.

Tool Use

Fields of Study

Ethology; Population Genetics; Evolutionary Biology

Abstract

In general, a tool is considered to be something that is not an integral part of an animal's body but is used by the animal to accomplish a specific task. It is difficult to define tools accurately. Quite often, objects taken directly from the environment, such as stones or sticks, are used as tools without further modification by the animal. Other times, the object may be modified prior to use. Tools allow the user to complete a task more easily or to accomplish a task that may not have been possible without the advantage provided by the tool.

Key Concepts

echolocation: the ability of animals to locate objects at a distance by emitting sound waves which bounce off an object and then return to the animal for analysis

ectoparasite: a parasite, such as a tick, that lives on the external surface of the host

ethology: the study of an animal's behavior in its natural habitat

insight learning: using past experiences to adapt and to solve new problems

pheromone: a hormone produced by an animal and then released into the environment

predator: an organism that kills and eats another organism, generally of a different species

primates: a group of mammals including apes, chimpanzees, monkeys, humans, lemurs, and tarsiers

WHAT MAKES A TOOL A TOOL

In general, a tool is considered to be something that is not an integral part of an animal's body but is used by the animal to accomplish a specific task. For example, a lobster may use its claw to crack open shells; however, since the claw is a normal appendage of the lobster, it is not considered to be a tool. When humans use a similar object, a nutcracker, to open shells, the nutcracker serves as a tool. It is difficult to define tools accurately. Examples of tools acceptable under the definition of one scientist may not meet the criteria set down by another investigator. Some scientists expand the definition of tool use to include specialized structures some animals use to extend their capability to locate and capture prey. These capabilities might include echolocation or sonar, electromagnetic fields, and specialized cells used for feeding such as the cnidocytes used by jelly fish. Other scientists consider products produced by an organism to be used to capture food as tools. Under this definition, a spider's web can be considered to be a tool.

Quite often, objects taken directly from the environment, such as stones or sticks, are used as tools without further modification by the animal. Other times, the object may be modified by actions such as stripping the leaves from a stick prior to use. Tools allow the user to complete a task more easily or to accomplish a task that may not have been possible without the advantage provided by the tool. The size, shape, and even texture of tools varies across the animal kingdom. Some animals use trees as tools and others use grains of sand. Some fish use spurts of water as tools. In addition to capturing or obtaining food, tools are also used in grooming, for defense, or even as protection from the elements. Thus, animals that use tools are actively interacting with and even modifying their environment.

STICKS AND STONES USED AS TOOLS

Many different species of animals, including insects, fish, birds, mammals, and primates, are known to use tools in some way during their everyday activities. While many different types of tools are used in the animal kingdom, the stick is a common and readily available tool. The use of sticks as tools has been well documented in nonhuman primates, such as chimpanzees, apes, and orangutans. Primates often use insight to solve a problem using tools and the young learn to use tools from either observing or being taught by the adults.

A classic example of insight learning leading to multiple tool use in chimpanzees was shown by Wolfgang Köhler, an early twentieth century psychologist. Chimpanzees held in captivity were offered food that had been placed beyond their normal reach. When boxes and sticks were added in the enclosure, the chimpanzees stacked the boxes, climbed them, and then used the sticks to knock down bananas that were hanging overhead. If one stick was not long enough, they would connect them together.

Orangutans and chimpanzees will strip the leaves from a stick and then use it to probe into the nest of insects such as ants or termites. When the stick is removed from the nest, the insects crawling over it can be eaten. Leaves themselves have been used by chimpanzees to gather water for drinking. Birds, too, use sticks to probe for insects and to remove them from crevices in the bark of trees. Some birds, the Galápagos woodpecker finch for example, will use their bill to trim and modify the twig before using it as a probe. Pacific island crows use their beaks to modify leaves to specific shapes, as well as sticks, before using them as probes. In the absence of sticks or leaves, some animals will use cactus spines as probes. Elephants use trees and sticks in various ways. They will rub against a tree or they may pick up a stick with their trunk to scratch. They have been observed to use tree trunks as levers and to use sticks to remove ectoparasites. When monkeys throw sticks and rocks, they are using these objects as tools for defense.

Stones are another common tool. Sea otters use stones in two different ways. Some otters will carry stones with them when they dive and use the stone as a hammer to free a tightly adhered abalone from a rock. While floating along the surface on their backs, otters use stones to crack open the shells of abalone or of bivalves such as clams, mussels, or oysters, which they also pluck from under the water. Otters may use bottles floating in the water to crack shells.

Birds use stones in a similar way. Egyptian vultures pick up stones in their beaks and use them in a pecking fashion, like a hammer, to crack open an ostrich egg. If this method fails, they will fly at the egg while clasping the stone in their talons. Mongooses also use rocks to crack eggs. Other birds, such as eagles, gulls, and crows, drop shelled animals such as turtles onto the rocks to crack their shells. Vultures are known to drop bones of prey onto rocks to crack them open and expose the marrow. Chimpanzees use stones to crack open nuts, analogous to humans using a hammer and anvil. Even spiders use stones as tools. The trap-door spider, *Stanwellia nebulosa*, uses a stone as a defensive tool. If forced to retreat when being attacked, the spider uses a stone to close off its burrow behind it.

OTHER TOOLS

In Japan, one species of crow uses a very different tool, a car. It has been reported that these crows use cars as nut crackers by placing the nut on the road and, after a car has run over it, retrieving the nut meat. If the car should miss hitting the shell, the crow may try again.

Humans are not the only species to use tools for fishing. Some green herons are known to drop objects into the water to attract fish looking for food. The herons then consume the curious fish. The archer fish uses jets of water shot from its mouth to knock insects off overhanging branches and into the water. Some scientists do not view this as a tool because the water passes along a specialized region of the mouth. However, it is similar to using a bow and arrow to subdue prey from a distance. Octopuses use water shot from their siphon system as a broom to clean the exoskeletons of eaten invertebrates from its den. An octopus may also use the jet of water to modify the size of the den. Another group of animals that uses a form of liquid tool belongs to the spider family, *Scytodidae*. These spiders shoot sticky material from modified venom glands to entangle their prey.

Spiders use their webs as tools in various ways. Those species of spiders that construct webs make them with silk produced from modified appendages called spinnerets. Webs are used to ambush animals that happen to enter into them. Some spiders strum their webs and use them as tools for communicating. Others may spin a long single strand of silk that they use as a drag line to find their way back or as a safety line to catch themselves. In some species, young spiders make silk parachutes which trap the air currents and allow them to be dispersed far from the nest. Spiders of the genus *Mastophora* spin a single thread, on the end of which is a sticky globule. By suspending the thread from one leg, the spider uses the web to "fish" for male moths, which are attracted to the sticky globule containing chemicals similar to the pheromones produced by female moths to lure males for mating.

The jellyfish and the hydra, two members of the phylum *Cnidaria*, have specialized cells, cnidocytes, concentrated on the surface of their tentacles. Inside these cells is an organelle, the nematocyst, which contains a thread. The nematocyst is stimulated to discharge when prey are near to it. This thread may have a barb on its tip that will penetrate the body surface of the prey, or it may be a lasso that wraps around the prey. The prey is then pulled into the digestive cavity of the cnidarian.

SONAR

Bats and dolphins are two good examples of animals that use echolocation to locate prey. Since sound waves can travel over great distances, the prey can be well beyond the predator's immediate area. The objects do not need to be large in order to be detected. Bats are able to locate mosquitoes. By analyzing the sound waves returning after bouncing off an object, the bat knows which objects are moving and which are stationary. The moving objects represent potential prey. Some potential prey, moths, have evolved a way to detect that they are being tracked by a bat. Thus, they are able to take evasive action and seek shelter near a stationary object such as a tree, or by landing on the ground. In a similar manner, dolphins use a series of high-frequency clicks to track fish. However, the fish, unlike the moths, are often not aware that they are being followed.

—*Robert W. Yost*

FURTHER READING

McFarland, David. *Animal Behavior*. 3rd ed. Longman Science and Technology, 1998.

McGrew, W. C. *Chimpanzee Material Culture: Implications for Human Evolution*. Cambridge University Press, 1992.

Maier, Richard. *Comparative Animal Behavior: An Evolutionary and Ecological Approach*. Allyn & Bacon, 1998.

Sherman, Paul W., and John Alcock, eds. *Exploring Animal Behavior: Readings from "American Scientist."* 2nd ed. Sinauer Associates, 1998.

U

Ungulates

Fields of Study
Evolutionary Biology; Mammalogy; Ethology; Ecology

Abstract
Ungulates are a large group of dissimilar vertebrate animals, grouped together because their outermost toe joints are encased in hooves. There are four ungulate orders. The size extremes among ungulates range from the seven-ton male African elephant to the rabbit-sized dik-dik antelope. Most ungulates are herbivores. Ungulates are also the only mammals with horns or antlers, although not all of them have this bony headgear. They are native to all earth's continents except Australia and Antarctica.

Key Concepts
- **antlers:** branched, temporary horns made of solid bone, shed and regrown yearly
- **carnivore:** any animal that eats only the flesh of other animals
- **gestation:** the term of pregnancy
- **herbivore:** an animal that eats only plants
- **nocturnal:** active at night
- **omnivore:** an animal that eats both plants and other animals
- **true horns:** straight, permanent, hollow bone horns

DESCRIBING UNGULATES

Ungulates are the hoofed mammals, belonging to the phylum *Chordata*. The word "ungulate" comes from Latin *ungula*, meaning "hoof." Ungulates are a large group of dissimilar vertebrate animals, grouped together because their outermost toe joints are encased in hooves. There are four ungulate orders. Those ungulates having an odd number of toes belong to the order *Perissodactyla*. This includes horses (one-toed), rhinoceroses (three-toed), and tapirs (four-toed on the front feet and three-toed on the back feet). Entirely even-toed ungulates belong to the order *Artiodactyla*. This includes pigs (four-toed) and two-toed ruminants such as camels, giraffes, antelope, deer, cattle, sheep, and goats.

The two other orders are *Proboscidea* (elephants) and *Hyracoidea* (rabbitlike hyraxes). The size

Cattle, colloquially referred to as cows, are domesticated ungulates, a member of the subfamily Bovinae of the family Bovidae. As shown: Simmental is a brown-white cattle breed originating in the Simme valley in Switzerland.

extremes among ungulates range from the seven-ton male African elephant to the rabbit-sized dik-dik antelope. Most ungulates are herbivores. Ungulates are also the only mammals with horns or antlers, although not all of them have this bony headgear. They are native to all earth's continents except Australia and Antarctica.

WILD UNGULATES

Antelope, elephants, hippopotamuses (hippos), and deer are some common wild ungulates. Horses, sheep, goats, cattle, and pigs are mostly domestic ungulates. Ungulate appearance varies widely, but there are common physical and digestive characteristics. Most are artiodactyls, which walk on two toes. Their ancestors had five toes, but evolution deleted the first toe and made the second and fifth toes vestigial. The third and fourth toes provide support, and end in protective hoofs. Hippopotamuses, unique among artiodactyls, have four toes of equal dimensions.

Most ungulates are herbivorous ruminants. They eat only plants, and have specialized digestive tracts with three or four chambers in their stomachs. They chew and swallow vegetation, which, after partial digestion, is regurgitated, chewed again, and reenters the stomach for more digestion. This leads to maximum nutrient uptake from vegetable food. Ungulates usually lack upper incisor and canine teeth. They have hard pads in their upper jaws, which help the lower teeth to grind food.

Deer and antelope are swift-running, hoofed ruminants. Male deer have solid, branched antlers (temporary horns) made of bone, which are shed and regrown yearly. Antelope of both genders have unbranched, permanent, hollow bone horns (true horns). Deer inhabit Asia, Europe, the Americas, and North Africa. Antelope inhabit Africa, Asia, and Europe. Both deer and antelope live in woods, prairies, marshes, mountains, and tundra. Their sizes range from huge moose and elands to rabbit-sized species. Deer and antelope eat twigs, leaves, bark, and grass. The largest antelope are ox-sized.

Giraffes and hippos are very unusual African ungulates. Giraffes live south of the Sahara desert. They have very long legs and necks. Males are over sixteen feet tall and both sexes have short horns. Their flexible tongues and upper lips pull leaves—their main food source—from trees. The two-ton animals can go months without drinking, getting most of their water from the leaves they eat. The three- to four-ton hippos walk on all four toes of each foot. They have short legs, large heads, no horns, small eyes and ears, and nostrils that close underwater. Hippos have long, sharp incisors and canines in both jaws. A hippo can be fifteen feet long and five feet high at the shoulder. They spend most of their time submerged, eating aquatic plants.

Brazilian tapirs inhabit South American forests from Colombia and Venezuela to Paraguay and Brazil. They look a bit like elephants and pigs, but are related to horses. The tapirs have stout bodies and short necks and legs, well adapted for pushing through dense forests, and have short, rigid manes, which protect them from predators. Each tapir has a short trunk with a flexible "finger" at its tip. Like elephants, they use the finger to pull leaves into the mouth. They are dark brown to reddish colored, about 2 meters long, 80 centimeters tall at shoulder height, and weigh up to 300 kilograms. Tapirs are nocturnal herbivores, spending much of the night eating grass, grasses, aquatic vegetation, leaves, buds, soft twigs, fruits, and plant shoots. The tapirs roam the forest and can climb river banks and mountains. Excellent swimmers, they spend a lot of time in the water, eating and cooling off.

DOMESTICATED UNGULATES

Bovids—cattle—are domesticated ungulates. Most have true horns. Bovid horns are spiral, straight, tall, or grow from the sides of the head and then up. Most are herbivorous ruminants. Cattle are raised to provide meat, milk, and leather. Modern cattle come from European, African, and Asian imports. Breeding modern cattle began in mid-eighteenth century Europe; today there are three hundred breeds. Dairy cattle such as Holsteins make milk, and beef cattle such as Angus yield meat.

Sheep and goats are also domesticated, ruminant ungulates. Sheep were domesticated eleven thousand years ago from Asiatic mouflons. They have paired, spiral true horns, largest in males. Adults reach lengths of 2.5 meters and weights from 100 to 200 kilogramss. The eight hundred domesticated breeds provide wool for clothing, meat, and milk. Goats, closely related to sheep, have shorter tails, different horn shape, and beards. They eat grass, leaves, and branches. Numerous breeds are domesticated for meat and milk. Angora goats yield silky mohair. Goat milk is as nutritious as cow milk.

The horse, donkey, zebra (HDZ) family are perissodactyls. They live in habitats ranging from grassland to desert. They eat grasses, bark, leaves, buds, fruits, and roots, spending most of their waking hours foraging and eating. Wild specimens inhabit East Africa and the Near East. Domesticated horses and donkeys are used for food, meat, and leather. Zebras are too savage to domesticate.

Members of the HDZ family lack horns. They have long heads and necks, slender legs, manes on their necks, and long tails. They have good wide-angle day and night vision and a keen sense of smell. The smallest family members are African wild donkeys, 1.25 meters tall, 2 meters long, and weighing nearly 225 kilograms. The largest, Grevy's zebras, are 1.75 meters tall, 3 meters long, and weigh up to 400 kilograms. Zebras have black or brown and white, vertically striped coats. The other HDZ family members are brown, black, gray, white, and mixtures of these colors.

THE LIFESTYLES OF UNGULATES

The lifestyles of ungulates are very different. Many of them are very sociable and live in large herds, including many bovids, horses, deer, antelope, and zebras. In other cases, the animals live in smaller family groups, or are solitary, coming together only to breed.

Wild donkeys and horses live in herds made up of a male and his mates. Young stay in the herd until two or four years old for females and males, respectively. Males then join other bachelors until winning a herd. Females join other herds. Goats and sheep are also herd animals. Young goats, sheep, and cattle join herds or live in solitary fashion after they are weaned.

Moose are quite different. Males are solitary until they fight for mates and breed in the fall. A successful male often leads several females and babies all winter. In the spring he returns to the solitary life. Giraffes and male elephants are solitary. Female elephants and young form herds whose members breed with visiting males, protect each other, and raise young.

Ungulates are of great importance to humans and to the world. First of all, in the wild state they are food for many carnivores and omnivores. Domesticated, they provide meat, milk, hides, and sinew for human use. Furthermore, elephants, horses, and reindeer have long been used as beasts of burden. In addition, ungulates are biologically important because as herbivores, they prevent overgrowth of all sorts of plants by eating them.

—*Sanford S. Singer*

FURTHER READING

Arnold, Caroline, and Richard Hewitt. *Zebras*. William Morrow, 1987.
Gerlach, Duane, Sally Atwater, and Judith Schnell. *Deer*. Stackpole, 1994.
Rath, Sara. *The Complete Cow*. Voyageur Press, 1990.
Sherr, Lynn. *Tall Blondes: A Book About Giraffes*. Andrews and McMeel, 1997.
Shoshani, Jeheskel, and Frank Knight. *Elephants: Majestic Creatures of the Wild*. Rodale Press, 1992.
Walker, Sally M., and Gerry Ellis. *Hippos*. Carolrhoda Books, 1998.

URBAN AND SUBURBAN WILDLIFE

FIELDS OF STUDY

Ecology; Environmental Studies; Evolutionary Biology; Ornithology; Herpetology; Mammalogy

ABSTRACT
Animals and plants of cities and suburbs are categorized as urban wildlife. As cities and suburbs grow ever larger and displace natural habitats, many city and suburban landscapes have become more attractive for certain kinds of wildlife, or at least urban wildlife has become more noticeable. Urban wildlife consists of an eclectic and unlikely mix of escaped pets (mostly exotics and caged birds), feral animals, furtive and temporary intruders from adjacent natural habitats, and species whose natural ecology and behavior enables them to fit within human-modified landscapes and tolerate living in close proximity to humans.

Key Concepts

anthropogenic: originating from human sources, such as aerosols and other pollutants

biodiversity: variety of life found in a community or ecosystem; includes both species richness and the relative number of individuals of each species

exotics: organisms, usually animals, that have been deliberately or inadvertently introduced into a new habitat, such as monk parakeets in New England, or brown snakes on Guam

feral animals: domestic animals that have reverted to a wild or semiwild condition, such as cats, dogs, or caged birds that have been released or escaped and now survive in the wild

morphology: development, structure, and function of form in organisms

open space: natural or partly natural areas in and around cities and suburbs, such as woodlots, greenbelts, parks, and cemeteries

urban wildlife: generally the nondomestic invertebrates and vertebrates of urban, suburban, and urbanizing areas; may include domestic animals that have escaped and are subsequently feral

URBANE CITY-DWELLERS

Animals and plants of cities and suburbs are categorized as urban wildlife. As cities and suburbs grow ever larger and displace natural habitats, many city and suburban landscapes have become more attractive for certain kinds of wildlife, or at least urban wildlife has become more noticeable. Urban wildlife consists of an eclectic and unlikely mix of escaped pets (mostly exotics and caged birds), feral animals, furtive and temporary intruders from adjacent natural habitats, and species whose natural ecology and behavior enables them to fit within human-modified landscapes and tolerate living in close proximity to humans.

Urban landscapes present a seemingly stark and forbidding environment for wildlife. The horizontal pavement of streets and sidewalks is punctuated by rising angles and arches of concrete and steel which in turn are topped by wood and metal rooftops. Overhead, a maze of telephone, power, and cable lines limits vertical movement, while vehicle and foot traffic pose a constant threat to surface movement. All of these edifices and connecting corridors and lines result in a complex, vertically structured environment within which some animals find difficult

Pigeons in Amsterdam.

to maneuver yet to which other animals quickly adapt. In addition to monotonous and often dangerous structural diversity, urban wildlife is subject to elevated and often almost continuous noise and disturbance and is constantly exposed to an enormous variety of residential wastes (garbage, litter, excess water, salts, sewage), vehicular pollutants (lubricants, greases, gasoline, hydrocarbons, nitrogen oxides), and chemical wastes (pesticides, paints, lead, mercury, contaminants).

Despite the forbidding features offered by urban habitats, a surprising variety of wildlife manages to exist on a more-or-less permanent status. In fact, some kinds of wildlife can be found even in the midst of the most degraded forms of urban blight. *Ailthanthus*, which is also commonly called tree-of-heaven, is but one of many opportunistic trees and shrubs that can take root and grow given a bare minimum of soil and nutrients. A simple linear crack in the pavement of a sidewalk, a little-used roadway, an unused parking area or vacant lot can trap enough windswept dirt to offer a growing substrate for *Ailanthus* and similar hardy plants. Each *Ailanthus*, in turn, provides food and shelter for equally tough and adaptable wildlife, ranging from the variety of invertebrates that colonize and feed upon *Ailanthus* to birds and mammals that take shelter or find food in its branches and foliage. Similarly, every invading sprig of grass, wildflower, shrub, or tree, however large or small, creates its own suite of microhabitats which, in turn, offer colonization opportunities for other plants and animals, the whole ultimately contributing to an overall increase in urban biodiversity.

CHARACTERISTICS OF URBAN WILDLIFE

Ailthanthus is an example of those plants and animals able to tolerate the most extreme urban conditions, but in reality most urban wildlife derives a number of benefits by living within the confines of cities and suburbs. Far from being homogenous expanses of concrete, most urban centers are a patchwork of different habitats—residential, commercial, and industrial buildings, warehouses, power stations, vacant lots, parks or green spaces, detached gardens, rooftop gardens, and alleyways—that each offers innumerable opportunities for wildlife. Many urban areas also have a number of limited access areas that animals are quick to adopt for shelter and breeding places; these include fenced-in lots and boarded-up buildings, along with a rabbit warren of underground tunnels, ducts, steam and water pipes, basements, and access ways.

City lights extend foraging time and opportunities, allowing wildlife to hunt for food not only throughout the day but also during much of the night, as needed. Urban nooks and crannies offer an extensive variety of microhabitats that differ fundamentally in size, microclimate, and other structural features. These microhabitats serve primarily as shelters and breeding sites for city wildlife. Many birds, such as house sparrows (*Passer domesticus*) and Eurasian starlings (*Sturnis vulgaris*) nest in innumerable crevices, cracks, nooks, niches, and sheltered rooftops. Pigeons (*Columbia livia*) and starlings hide in sheltered enclaves offered by bridge abutments and supports, archways, and other edifices.

The most adaptable wildlife are quick to find and take advantage of subtle advantages offered in urban habitats: Many birds cluster around chimneys and roof reflectors or in shelters afforded by lee sides of rooftops during harsh cold and windstorms. Others are equally quick to obtain warmth by sitting on poles, rooftops, or other elevated perches to orient toward sunlight, while at ground level animals gather near gratings, vents, and underground heating pipes.

Urban wildlife quickly concentrates in areas where potential food is made available, for instance, during trash pickup, then just as quickly disperses to find new food sources. Most urban wildlife forage opportunistically as scavengers, specializing in finding and consuming all bits of discarded food, raiding trash cans, and concentrating at waste collection and disposal centers. Thus, the rubbish dumps, found in or immediately adjacent to every city of the world, attract an amazing diversity of small mammals and birds. Feeding on the scavenged food of urban areas and bird feeders is much more efficient because it requires less energy to find or catch and is usually available throughout the year.

Because of the need to find and exploit temporary food resources, some of the most successful urban animals forage in loose groupings or flocks—the more eyes, the more searching, and the more feeding opportunities can be identified and exploited. Solitary and nonsocial species often do less well in urban environments simply because they lack the collective power of the group to find food and shelter, and avoid enemies.

The availability of a year-round food supply—however tenuous and temporary—along with the presence of an enormous variety of safe shelters and breeding sites promotes a higher life expectancy, which partly or mostly balances the higher vehicle-related death rates to which urban wildlife is continuously subject.

Parks and open space provide the only true refuges of natural habitats set deep within urban and suburban landscapes. Such open-space habitats function as ecological islands in a sea of urbanism. Most are necessarily managed habitats rather than entirely natural and, like the urban environment that surrounds them, are usually subject to constant disturbance from adjacent traffic, noise, and other forms of pollution. Economically, since most open-space parks are set aside and maintained for a variety of recreational purposes rather than as natural habitats, the wildlife that colonizes these unnatural natural habitats must have an unusually high tolerance for human presence and recreational activities of all kinds.

SOURCES AND TYPES OF URBAN WILDLIFE

For some urban wildlife, the urban landscape is merely a manmade version of their natural environment. Thus, for pigeons the ledges, cracks, and crevices of buildings and bridges represent an urban version of the cracks and crevices of cliffs and rock outcrops that they use for roosting and nesting in the native habitats. Similarly, the short-eared owls (*Asio flammeus*) and snowy owls (*Nyctea scandiaca*) that show up in winter to stand as silent sentinels at airports, golf courses, and other open areas are simply substituting these managed short-grass habitats for the tundra habitats preferred by snowy owls and the coastal marshes hunted by short-eared owls. Their summer replacements include a host of grassland nesting species such as grasshopper sparrows, kildeer, and upland sandpipers, which all find these managed habitats to be ideal substitutes for the native grasslands which they displaced or replaced.

Many bird inhabitants of urban and suburban environments are exotics which were deliberately or inadvertently introduced into urban areas. Certainly the three birds with the widest urban distribution in North America, the pigeon or rock dove, European starling, and house sparrow or English sparrow, all fit within this category. The introduction of the European starling into North American cities and suburbs resulted from the dedicated efforts of the American Acclimitization Society of the late 1800s. The goal of this society was the successful introduction of all birds mentioned in the works of Shakespeare into North America. Unfortunately for North Americans, the character of Hotspur in *Henry IV* makes brief note of the starling, so the society repeatedly attempted to introduce the starling into Central Park until they were finally successful. Since then, the starling has become the scourge of cities and suburbs throughout much of North America and the rest of the civilized world. Starlings damage and despoil crops, and dirty buildings with their droppings.

The association between house sparrows and urban centers is apparently very old. Evidence suggests that they abandoned their migratory ways to become permanent occupants of some of the earliest settlements along the Nile and Fertile Crescent, a trend that has continued to this day. Sparrows and starlings both share certain characteristics that enable them effectively to exploit urban and suburban habitats; both are aggressive colonizers and competitors, able to feed opportunistically on grains, crops, discarded bits of garbage, and other food supplies.

Avian occupants also include an increasing diversity of released caged pets. Thus, urban locales in Florida, Southern California, and along the Gulf Coast support an ever increasing diversity of parakeets, parrots, finches, and lovebirds, all stemming from caged pet birds either deliberately released or lost as escapees.

Feral animals, mostly dogs (*Canidae*) and cats (*Felidae*), represent another important source and component of urban wildlife. Feral dogs revert to primal adaptive behaviors, gathering in loose packs that usually forage and take shelter together, but have limited success because almost all cities in developed countries have ongoing measures to control and remove them whenever found. Feral cats are often more successful because they are secretive, mostly nocturnal, and can clearly better exploit available urban food sources. The role of other feral animals as urban wildlife, mostly escaped pets, is not well known.

HUMANS AND URBAN WILDLIFE

The attitude of urban dwellers toward urban wildlife varies greatly. For many humans, urban wildlife offers a welcome respite from their otherwise dreary and mundane surroundings. Urban wildlife in all of its forms and colors can be aesthetically attractive, even beautiful, and is also compellingly interesting. For example, the nesting of a pair of red-tailed hawks (*Buteo jamaicensis*) in New York City's Central Park sparked a remarkable interest in birdwatching in the city and a heightened awareness of exactly how exciting wildlife watching can actually be. All facets of the pair's courtship and nesting were watched and reported in newsprint, novellas, and even a book, *Red-Tails in Love*. Other animals, while not nearly as large, conspicuous, and glamorous in their color and disposition, also elicit interest. Urban wildlife adds lively color and contrast to the otherwise monotonous gray and grime of streets and sidewalks. Part of the attraction is that urban birds are usually already sufficiently tolerant to be semitame in spirit, easily seen and observed, and, in some instances, easily attracted. Strategically placed bird feeders and bird houses also attract these birds.

Public attitude toward urban predators varies considerably. Some people find them attractive and interesting and even put out food for them. Others consider them pests or potentially dangerous and avoid them. During rabies outbreaks or public scares, most urban wildlife is targeted by various control programs to remove unwanted animals.

SUBURBAN WILDLIFE HABITATS

The vast sprawl of suburbs across the landscape offers many types of wildlife yet another habitat opportunity to exploit, either as residences or as temporary components of the search for food or shelter. Like urban areas, suburbs offer a range of differing habitats. The simplest suburbs are merely extensions of urban row houses with minimal yardscapes, but there is an increasing progression toward more open and natural yards in outlying suburbs that merge with rural areas and natural habitats. The larger and more diverse yards at the edges of suburbs often help blur the distinction and diversity between human landscapes and natural landscapes.

Ornamental trees, shrubs, flowers, gardens, and lawns that characterize almost all suburban habitats provide a series of artificial habitats that can actually increase wildlife diversity. Again, the chief wildlife benefactors are species that can best ecologically exploit the unnatural blend of woodland, edge, and meadow that suburban landscapes offer. It is no accident that some of the most common components of suburban wildlife include thrushes such as robins, finches, and cardinals, titmice, blue jays, crows, and many other similar birds. All of these species are actually responding to the structural components of the suburban landscape, which provide suitable substitutes for their natural landscapes.

The blend of ornamental and garden vegetation offered by most suburban landscapes offers food for a diversity of what were once considered less tolerant wildlife. Deer, wild turkey, grouse, and a host of other animals, large and small, make periodic forays into suburbs in search of foods. Crepuscular and nocturnal wildlife is much more likely effectively to exploit food sources offered by suburban landscapes than diurnal wildlife, which is more at risk because of its high visibility during daylight hours.

Well-wooded suburban habitats that attract a variety of wildlife also attract an increasing number of predators. American kestrels (*Falco sparverius*), Cooper's hawks (*Accipiter cooperi*), barn owls (*Tyto alba*), screech owls (*Otus* spp.), and little owls (*Athene noctua*) provide but a small sampling of birds of prey that nest deep within urban and suburban environments, taking advantage of open-space habitats deep within cities and quickly exploiting unused areas within most suburbs. Terrestrial predators are almost equally common, but most are nocturnal or nearly so; consequently, their contacts with humans are quite limited. Many urban predators are, in fact, mistaken for neighborhood pets and left alone or avoided: Coyotes (*Canis latrans*) are often mistaken for dogs, especially when seen in twilight. The wily coyote is equally at home in the suburbs of Los Angeles, California, and the urban parks of New Haven, Connecticut, joining a host of small and medium-sized mammal predators such as foxes (*Vulpes* spp.) and scavengers such as opossums (*Didelphis marsupalis*), raccoons (*Procyon lotor*), and skunks (*Mephitis mephitis*). These urban predators have many behavioral attributes in common. All are omnivorous and able to feed on a wide variety of natural foods such as fruit, small birds and mammals, insects, and invertebrates such as beetles, grasshoppers, and earthworms.

Foraging and food habits of urban predators sometimes conflict with human concerns. Urban foxes hunt and kill cats, especially kittens, if given the opportunity, while the larger and stronger urban coyote will often not hesitate to kill and eat cats and dogs, to the pet owners' dismay.

CONSERVATION AND MANAGEMENT OF URBAN AND SUBURBAN WILDLIFE

Urban wildlife must be much more closely managed than wildlife of natural environments because urban and suburban habitats attract an enormous number of pest species as well as interesting and beneficial species. Introduced species such as starlings may also transmit histoplasmosis, a fungal disease that attacks human lungs. Other birds may also be harbingers, carriers, and vectors of various diseases, the most notable of which are the parrots and parakeets, which transmit parrot fever or psittocosis. Rats and mice (Rodentia) carry and spread disease and despoil both residential and public buildings and other structures.

The growing interest in urban wildlife has stimulated innumerable programs to promote beneficial wildlife. Both public and private organizations and agencies have embarked on a variety of programs aimed at remodeling existing habitats and even creating new habitats for urban wildlife.

Programs aimed at creating new or modifying existing urban habitats come in a variety of categories, such as linear parks, greenways, urban wildlife acres programs, backyard gardens, and treescaping streets and roadways, all of which create biodiversity, which in turn provides attractive habitats for colonization by additional animals and plants. Modification of existing habitats to increase animal biodiversity includes "critter crossings," roadside habitats, backyard gardens, and arbor plantings, all of which provide refuges, shelters, breeding sites, connecting corridors, and safe havens that promote the welfare of urban and suburban wildlife.

Many existing open-space habitats are also being modified. Many urban renewal commissions have placed new and more restrictive regulations on the use of pesticides and fertilizers on golf courses, which not only reduces the incidence and intensity of nonpoint pollution from the golf courses but also reduces the incidence of wildlife poisoning. These steps cannot help but increase the biotic potential of golf courses for supporting local biodiversity.

—*Dwight G. Smith*

FURTHER READING

Adams, Lowell W. *Urban Wildlife Habitats: A Landscape Perspective.* University of Minnesota Press, 1994.

Bird, David, Daniel Varland, and Juan Josè Negro, eds. *Raptors in Human Landscapes.* Academic Press, 1996.

Forman, Richard, and Michel Godron. *Landscape Ecology.* John Wiley & Sons, 1986.

Gill, Don, and Penelope Bonnett. *Nature in the Urban Landscape: A Study of City Ecosystems.* York Press, 1973.

McDonnell, Mark J., and Steward T. A. Pickett, eds. *Humans as Components of Ecosystems.* Springer-Verlag, 1993.

Vertebrates

Fields of Study

Animal Anatomy; Animal Physiology; Evolutionary Biology; Paleontology

Abstract

The six vertebrate classes are lampreys, true fish, frogs and toads, reptiles, birds, and mammals. A vertebrate animal has a spinal column made of bone or cartilage, and a brain case. Vertebrates also have two pairs of limbs, and a bilaterally paired muscular system. The backbone is a group of small bones or cartilage pieces with articulating surfaces. Ribs and bones that support the limbs are attached to the backbone. The ribs protect the heart, lungs, and other internal organs, and can expand and contract. The earliest vertebrate fossils occur in rock from the Paleozoic era.

Key Concepts

articulate: interconnect by joints
bone: dense, semirigid, calcified connective tissue, the main component of skeletons of adult vertebrates
cartilage: elastic, fibrous connective tissue, the main component of fetal vertebrate skeletons, it turns into bone
collagen: a fibrous substance plentiful in bone, cartilage, and other connective tissue
connective tissue: fibrous tissue that connects or supports organs
notochord: a flexible, rodlike structure in lower chordates and vertebrate embryos, it supports like the backbone of a vertebrate

WHAT MAKES A VERTEBRATE A VERTEBRATE

The *Vertebrata* are a subphylum of the phylum *Chordata*. The six vertebrate classes are lampreys, true fish, frogs and toads, reptiles, birds, and mammals. A vertebrate animal has a spinal column (backbone) made of bone or cartilage, and a brain case (skull). Vertebrates also have two pairs of limbs (though some have lost limbs through evolution), and a bilaterally paired muscular system. The backbone that gives the subphylum its name is a group of small bones or cartilage pieces with articulating surfaces (vertebrae). Ribs and bones that support the limbs are attached to the backbone. The ribs protect the heart, lungs, and other internal organs, and can expand and contract. The earliest vertebrate fossils occur in rock from the Paleozoic era.

The vertebrae serve to encase and protect the spinal cord, a major part of the vertebrate nervous system. The central nervous system also has an enlarged, highly differentiated upper portion, the brain. The bony skull of a vertebrate also serves to encase and protect that brain, as well as providing a base for the vulnerable sensory organs of eyes, ears, nose, and mouth, which are thus efficiently located close to the brain.

The more primitive members of the phylum Chordata have notochords, solid or segmented columns that cover the nervous system. Vertebrates resemble chordates in having notochords in their embryonic state. The vertebrae develop around the notochord. The trunk of a vertebrate is a hollow cavity in which the heart, lungs, and digestive tract are suspended. The central nervous system branches out from the spinal column to reach all internal organs, muscles, and skin. The support offered to the brain and nervous system by the vertebrae has allowed vertebrates to evolve increasingly large brains, which in turn has allowed vertebrates to become increasingly intelligent and responsive to their environment.

BONE AND BONES

Bone, the material that forms the vertebrae, first evolved about 500,000,000 years ago. Bone is the hard supportive framework of all vertebrates. The

elastic, due to its collagen fibrils. The fibrils provide mechanical stability and high tensile strength, while allowing nutrients to enter chondrocytes. Blood vessels around cartilage supply nutrients and remove wastes.

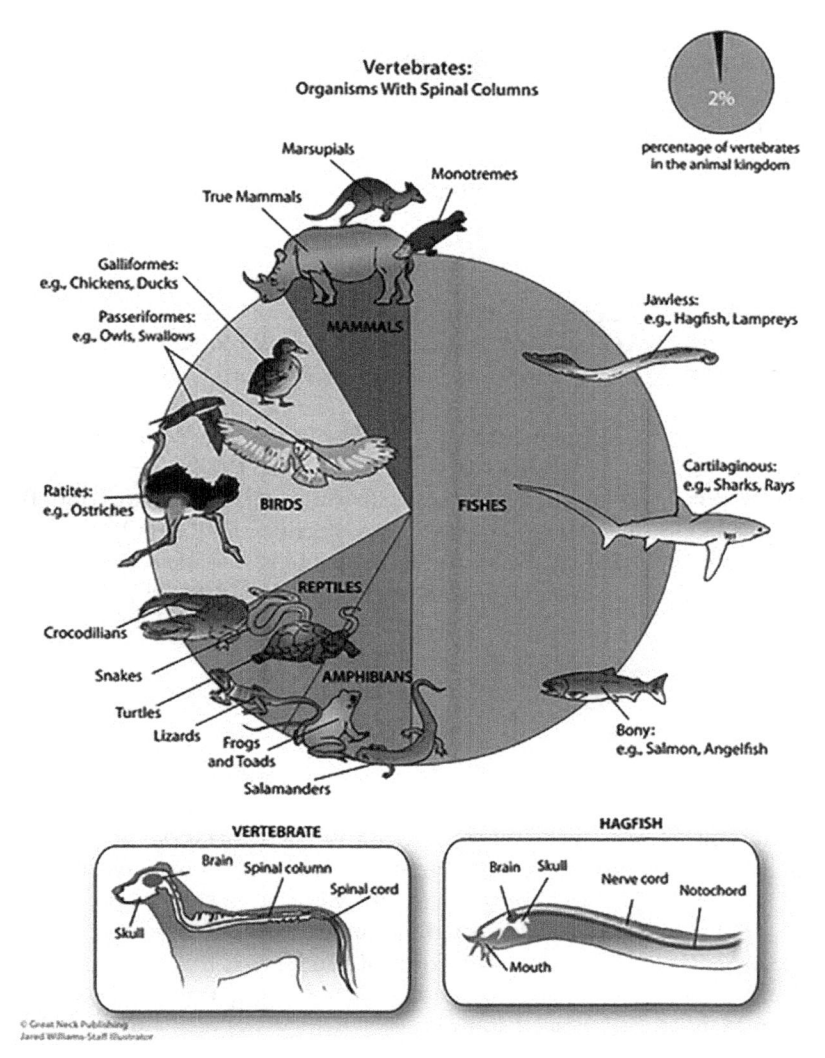

Cartilage-containing skeletons of newborn vertebrates become bone by a process of calcification, chondrocyte destruction, and replacement by bone cells. In young vertebrates, cartilage is the site of growth and calcification that lengthens bone to attain adult size.

VERTEBRAE

In higher vertebrates, each vertebra consists of a lower part, called the centrum, and an upper, Y-shaped part, called the neural arch. The arch has a downward and backward projection, which can be felt as the bumps along a vertebrate's back, and two sideways projections, where muscles and ligaments can attach. The space between the arms of the Y on the neural arch and the centrum create an opening called the vertebral foramen, through which the spinal cord passes. Intervertebral disks, made of cartilage, separate the centrums and serve as shock absorbers.

The vertebral column has five regions: the cervical region (neck), thoracic (chest), lumbar (lower back), sacral (pelvic girdle), and caudal (tail). The top two cervical vertebrae, called the atlas and axis vertebrae, make a joint to attach with the skull. The number of vertebrae varies by species.

VERTEBRATE EVOLUTION

Vertebrates, all designed on the same general plan, flourish on land, in the air, and in both fresh and salt water. Vertebrates first evolved in the Silurian period, around 438 to 408 million years ago. In the intervening millennia, according to need, evolution produced valuable morphological changes to optimize vertebrate biofunctions. For example, whales, once

Vertebrates are living things that possess a spinal column, or backbone. There are five groups of vertebrates including mammals, fishes, amphibians, reptiles, and birds.

framework, a skeleton, has hundreds of separate parts; for instance, there are 206 bones in a human skeleton. Bones protect delicate organs, such as the brain and lungs. Muscles, attached to bones, enable walking, flying, swimming, and all other means of motion. Bones provide body calcium needs and contain sites for making blood cells.

Much bone in adult vertebrates arises from cartilage, an elastic, fibrous connective tissue, and the main component of fetal vertebrate skeletons. Such bone is cartilage bone. Cartilage is an extracellular matrix made by chondrocyte cells. It is firm and

625

land dwellers, evolved into ocean dwellers with lungs and sonar to help them navigate and find food. Birds developed wings to ride the air, and mammals proliferated in forms that fit varied habitats worldwide. Fish were the first vertebrates to appear, around 480 million years ago. Amphibians and reptiles appeared around 360 million years ago, while birds and mammals begin to appear around 205 million years ago.

Morphological change led to a balance of nature, where herbivores ate plants, preventing their overgrowth, and carnivores ate herbivores, preventing their superabundance. Then the ultimate vertebrates, humans, developed civilization. Humans domesticated animals for food, clothing, transportation, and pets. In so doing, many species have been eradicated, endangered, or put at risk. For example, blue whales were endangered because their blubber was useful. Wolves and tigers have been eradicated locally or endangered globally by the quest for hunting trophies and to keep them from eating livestock. Mustelids were treated similarly because their pelts made attractive fur garments. Fortunately, human tolerance for this treatment of animals is decreasing, lest many species be seen only in zoos or faded photographs.

—*Sanford S. Singer*

Further Reading

Chatterjee, Sankar. *The Rise of Birds: 225 Million Years of Evolution.* The Johns Hopkins University Press, 1997.

Crispens, Charles G. *The Vertebrates: Their Forms and Functions.* Charles C Thomas, 1978.

Pough, F. Harvey, Christine M. Janis, and John B. Heiser. *Vertebrate Life.* Prentice Hall, 1999.

Rosen, Vicki, and R. Scott Theis. *The Molecular Basis of Bone Formation and Repair.* R. G. Landes, 1995.

Vocalizations

Fields of Study

Ethology; Neurophysiology; Animal Anatomy; Animal Physiology; Population Biology

Abstract

A vocal mechanism typically consists of lungs to provide an air stream, a trachea to conduct the air to a larynx or syrinx, a pharynx, and the associated oral and nasal cavities. Vocalization, any sound produced by the respiratory system, implies that air flowing from the lungs has been converted into an oscillating air stream. The sound can be melodious or noisy; when the vocal folds vibrate the sound is termed phonation.

Key Concepts

larynx: the vocal mechanism of mammals, consisting of a structure of cartilage at the upper end of the trachea containing the vocal folds

pharynx: lower part of vocal tract, connecting the mouth and nasal cavities to the larynx

syrinx: the vocal mechanism of birds, consisting of one or more membranous structures at the lower end of the trachea, where the windpipe divides into two bronchial tubes leading to the lungs; the membranes vibrate due to pressure differences when air streams across their surfaces

trachea: a cartilaginous tube that transports air from the lungs to the pharynx

vocal folds: small, laminated sheets of muscle which meet at the front of the larynx; they are open for breathing, and are brought together to vibrate for voiced sounds

ARE YOU TALKING TO ME?

A vocal mechanism typically consists of lungs to provide an air stream, a trachea to conduct the air to a larynx or syrinx, a pharynx, and the associated oral and nasal cavities. Vocalization, any sound produced by the respiratory system, implies that air flowing from the lungs has been converted into an oscillating air stream. The sound can be melodious or noisy; when the vocal folds vibrate the sound is termed phonation.

Animal vocalizations evolved with hearing because of the many evolutionary advantages of sound communication. Sound can be varied in pitch, duration, tonality, and repetition rate, making it possible to communicate considerable amounts of detailed information quickly. Animals may vocalize while keeping their limbs free or while hiding. Because sound waves pass readily through vegetation and around obstacles, vocalization is used among insects, frogs, and birds to indicate sexual receptivity.

Vocalizations are also an important component in the behavioral displays of reptiles, birds, and mammals. Although animals typically employ body language and nonvocal noises in their displays, vocalization provides an impressive elaboration impossible to achieve otherwise. The fearsome sight of a gorilla beating its chest and stomping the ground is enhanced considerably by its bloodcurdling roar.

There are two different types of sound-generating mechanism used for animal vocalizations. The first requires a vibrating structure, such as the voiced sounds of human speech produced by vibrating vocal folds. The second is aerodynamically excited, such as for unvoiced sounds or whistled tones. The vibrating element of voiced sounds is a flow valve, which vibrates when pressure is applied from the lungs. Two different systems occur in the animal kingdom. The first is the larynx, common among mammals; when pressure is applied from the lungs, the vocal folds swing outward, stretching muscular ligaments which tend to restore them. These alternating forces induce vibration in the form of air pulses, which propagate through the trachea to be emitted by the mouth or nose. The second is the avian syrinx, which is blown open by excess pressure on either side of the mechanism and restored by forces supplied by pressurized air sacs surrounding the syrinx. In both cases, pitch can be varied by the muscles associated with the valve mechanism. Airflow through the valve is nonlinear, generating a complete set of harmonics in the radiated sound.

Whistled sounds are created when an air jet impinges on a sharp edge or an aperture, creating a sinuous instability in the air stream. The sound is considerably louder when the air stream is acoustically coupled to a resonating oral or nasal cavity, the size and shape of which determines the resulting pitch.

In any acoustic communication system, signals must compete with noise in the environment. The environment has a preponderance of natural low-frequency noise, but high frequencies are produced more easily by small animals. Most vertebrates, however, do not communicate with high frequency sounds, because they are more readily attenuated and thus do not travel far. The optimum frequencies for auditory communication thus depend on the desired communication range. High frequencies are best for small animals communicating over short distances, while low frequencies better serve large animals communicating over longer distances.

REPTILES AND FROGS

Although the vocal apparatus of alligators is rather primitive, they can produce noise-excited roaring and hissing vocalizations when provoked by low-frequency sounds, such as a horn or a cannon. Crocodilians live as lone individuals and establish individual territories defined by their loud vibrant roars. To roar, crocodiles tense their body muscles and raise their heads and tails high above the water. The emitted sound vibrates the animal's flanks so violently that water is sprayed into the air. Crocodiles are also capable of deep grunting sounds, used during courtship. Geckos are small nocturnal lizards with soft skin. Their voice varies, by species, from faint chirps to loud squawks.

Frog vocalizations encode several pieces of information; the species, the sex, and whether or not it has mated. Male mating calls attract females and indicate the number of other males nearby, critical information for females who wish to deposit their eggs in the most receptive habitat. The vocalizations are produced by primitive vocal cords, consisting of a pair of slits at the throat opening in the floor of the mouth. When the frog forces air from the lungs, the cords vibrate to produce sound. Many species also

Lioness (Panthera leo) roaring. Photo was taken at the Louisville Zoo.

have a vocal sac, an inflatable chamber located in the throat region of males, which swells to a large size when calls are produced. Air vibrates the vocal cords while passing back and forth between the lungs and the sac, while the vocal pouch acts as a resonating chamber which amplifies the sound. The frog's mouth remains closed while vocalizing, so it can call even while under water.

BIRDS

In birds, the voice is well developed, having such distinctive sound patterns that many species are named onomatopoeically, such as the whippoorwill and the killdeer. Although birds use different calls for different purposes, each species has a primary song, often repeated incessantly, used for species recognition. Male birds also use vocalizations to mark their territory and to attract mates. Some species can even identify their mates by sound. During the breeding season, the male emperor penguin leaves for several days to forage for food; when he returns he is able to locate his mate, out of a pack of hundreds of birds, from the calls emitted.

The green-backed sparrow utters a hoarse scream when attack or escape is likely to occur, and medium hoarse notes when the bird's indecision between the two courses of action make it unlikely that either will occur. To a family of migrating geese, the sounds of other geese on the ground conveys the information that there is probably food and a safe shelter.

Cuckoos are a highly vocal species; they use a variety of contact calls, alarm notes, and melodious songs used to define territory or attract mates. The male's song is characterized by a repetition of loud, short notes on a descending scale. The common cuckoo found throughout Europe, Asia, and Africa emits the well-known two-note call imitated in cuckoo clocks.

Owls produce a variety of vocalizations with a pitch, timbre, and rhythm unique to each species. Most vocalize at dusk and dawn before beginning to hunt. Their songs vary from the deep hoots of large species to the chirps and warbles of small owls. When its nest is threatened, the nestling burrowing owls emit a buzzing noise, resembling the warning sound of the rattlesnakes that frequently inhabit rodent burrows. North American screech owls begin mating when a special song, commenced by the male, is answered by a distant female. After fifteen minutes of antiphonal singing while gradually approaching, the couple meets, sings a duet with a different song pattern, and mates. Other calls of the screech owl include sounds to prompt the young to reveal their location, a food-soliciting call by the young, and barking calls used to eject the matured young from the parents' territory.

The "voicebox" used by birds to produce birdsong is the syrinx, located where the windpipe divides into the two bronchial tubes leading to the lungs. The syrinx varies considerably among different birds. In the ordinary chicken it is quite simple, consisting of four uncomplicated membranes, which produce the characteristic clucking sounds when activated. An asymmetric chamber at the base of the ducks' trachea adds a noise component to its vocalization, which humans hear as "quacking." The trachea of trumpeter swans enters the sternum, flexes twice into bony pockets, then coils back to the lungs, somewhat analogous to bass orchestral wind instruments. This long resonator implements the production of its clarion, trumpetlike call.

The human brain can perceive speech in sounds having only the remotest resemblance to speech if the rhythm and intonation matches that of a simple sentence. Mynah birds use this phenomenon to deceive us into believing they can speak. They have a syrinx valve on each bronchial tube which can be independently controlled to produce two simultaneous wavering tones, which we perceive as speech when they mimic the rhythm of a sentence.

MAMMALS

Among mammals, vocalization is used first for survival. Infants vocalize to express hunger or pain, or to be located when lost. Other cries, such as the lion's roar or the trumpeting of an elephant, mandate caution. Animal vocalizations of this type, often accompanied by an offensive posture, are used to startle or intimidate an opponent. There is a direct correlation between vocal anatomy and behavior among mammals. Social animals that readily vocalize have larynges which open less widely when they breathe, thus reducing breathing efficiency. The vocal folds must close to start phonation (wailing of cats, howling of wolves). For breathing, they must open wide so as to not obstruct air flow to the lungs. Horses and animals whose survival depends on running long distances while breathing aerobically have simple vocal folds that can open wide to offer an unobstructed air passage, but which consequently cannot be effective phonators. The giraffe's vocal folds are so poorly developed that the animal was long thought to be mute. In actuality, giraffes can phonate to a limited extent; they groan when injured and call their young when they stray. The more highly developed vocal folds of primates enhance phonation, but at the expense of a more constricted airway.

Vocalization is an important aspect of mammal communication. When the wild dogs of India (dholes) hunt, the leader coordinates the pack's motions with a series of sharp yelps. The black-tailed prairie dog combines a visual and vocal display consisting of jumping into the air with its nose straight up while emitting an abrupt two-part vocalization. This display indicates that some behavior is about to be interrupted or prevented by fleeing, which usually occurs immediately thereafter. The display is only employed when an alternative to flight also exists. Hyenas have no organized social behavior but often cooperate while hunting. Their cries suggest human laughter—a low-pitched, hysterical chuckling that rises to higher tones. The female deer emits a sharp, staccato bark to warn its young when it senses danger, and lions coordinate a hunt by grunting while stalking prey.

Elephants use their trunks for communication by trumpeting, humming, roaring, piping, purring, and rumbling. At least three dozen distinct elephant vocalizations have been documented, including an assortment of trumpeting sounds ranging from outright blasts to a low groan that males use to indicate that a jousting session is finished. Elephant screams range from expressions of social agitation to the pulsating bellow emitted by a female pursued by an unwanted suitor. Babies scream when they want milk; the scream gets progressively louder until their hunger is satisfied.

Elephants also communicate with infrasonic frequencies (below the range of human hearing), which humans detect as an air pulsation accompanied by low-frequency rumblings. Rumbles constitute the majority of elephant vocalizations and explain the uncanny ability of widely separated groups to coordinate their activities. There are rumbles of reassurance, rumbles to say "Let's go," rumbles to maintain contact, rumbles to cry "I'm lost," courtship and mating rumbles, and a humming rumble produced by mothers for newborn calves. Rumbles also coordinate activities within a given group when preparing to fight a dominance battle with another group, and mothers use a special rumble to reassemble the younger members of her family. About fifteen of the known rumbles have an infrasonic component, which enables elephants to maintain contact over long distances. Because low frequencies dissipate less rapidly in air, they can travel up to five miles. Elephants also emit infrasound to alert others to listen carefully for faint, higher-frequency sounds containing more detailed information.

Highly territorial mammals, such as lions, coyotes, and wolves, vocalize extensively at dawn and dusk to establish and maintain territory. Some species vocalize to attract mates and to intimidate rivals. Male moose give hoarse, bellowing cries during mating season to locate cows; the cows respond with a softer, somewhat longer lowing sound. The male elk, in an effort to collect as large a harem as possible, challenges competitors by emitting a buglelike sound. This vibrant call begins in the low register, ascends to high pitch, then abruptly drops in a scream. Bull seals, arriving at breeding grounds before the females, attempt to obtain as many cows as possible for their harems by frightening away competitors with loud roaring.

Bats and dolphins utilize high-frequency ultrasonic vocalization, or echolocation, as a type of animal sonar to find prey and to navigate in the dark. High frequencies are desirable to locate small targets, as high frequency waves are more directional, and a wave cannot "see" an object smaller than its wavelength.

PRIMATES

Primates communicate by various vocal sounds as well as by facial expressions. Apes and monkeys use growls, grunts, twitterings, chirpings, whispers, barks, screams, and cries to warn of danger, indicate alarm or distress, and keep the members of a clan together. Vervet monkeys are known to have three different alarm calls: One warns of eagles, one of snakes, one of leopards. Howling monkeys emit loud, disconcerting, barking roars. The sound is produced by air passing through a resonating chamber in the throat. Rival groups fighting over territory engage in a duel of roaring until one group retreats. While roaring, all other activity, such as feeding, playing, or exploring comes to a halt. A female howler may also wail in distress when one of her young falls from a tree, while the youngster emits diminutive cries to indicate its position.

Primates lower the natural resonant frequencies of their vocal tracts when faced by danger in order to project the sonic aura of a larger animal. Apes and monkeys achieve this by protruding and partly closing their lips to generate low-pitched, aggressive sounds.

Among the great apes, gibbons are the most vociferous. Their raucous cries, especially boisterous at sunrise and sunset, can carry more than a mile. Apes only vocalize to express an emotional state. The ability to produce vocal sounds not linked to instinct or emotion is the primary difference between human speech and ape calls. The changing sound patterns of human speech represent abstract concepts, while apes produce simple melodies tied to their mood. The seat of human language is the cerebral cortex, while ape vocalizations are controlled by the subcortical neural structures involved in emotion. Emotionally based human vocalizations, such as sobbing, crying, laughing, giggling, or shouting in pain, are also controlled subcortically.

—*George R. Plitnik*

FURTHER READING

Brackenbury, J. H. "The Structural Basis of Voice Production and Its Relationship to Sound Characteristics." In *Acoustic Communication in Birds*, edited by D. E. Kroodsma and E. H. Miller. Academic Press, 1982.

Chadwick, Douglas. *The Fate of the Elephant.* Sierra Club, 1992.

Fletcher, Neville. *Acoustic Systems in Biology.* Oxford University Press, 1992.

Lewis, B., ed. *Bioacoustics: A Comparative Approach.* Academic Press, 1983.

Stebbins, W. C. *The Acoustic Sense of Animals.* Harvard University Press, 1983.

Strong, William, and George Plitnik. *Music, Speech, Audio.* 3rd ed. Soundprint, 1992.

Water Balance in Vertebrates

Fields of Study
Animal Physiology; Animal Anatomy; Dermatology; Endocrinology

Abstract
Water, which makes up about 60% of the body weight of vertebrates, may be the most neglected nutrient. Drinking and eating are the obvious ways of obtaining water, but metabolism, the processes of synthesis and breakdown within the body's cells, also provides water for organisms. In fact, diet and metabolism provide 40% of the water necessary for human life.

Key Concepts

apocrine gland: a type of sweat gland that becomes active at puberty and responds to emotional stress; the glands are found at the armpits, groin, and nipples

eccrine gland: a type of sweat gland that helps maintain body temperature; the glands are located on the palms and soles, forehead, neck, and back

extracellular fluid: the fluid outside cell membranes, including fluid in spaces between cells (interstitial), in blood vessels (plasma), in lymph vessels (lymph), and in the central nervous system (cerebrospinal)

homeostasis: the dynamic balance between body functions, needs, and environmental factors which results in internal constancy

hyperosmotic: a solution with a higher osmotic pressure and more osmotically active particles than the solution with which it is being compared

hypoosmotic: a solution with a lower osmotic pressure and fewer osmotically active particles than another solution with which it is being compared

isosmotic: a solution with the same osmotic pressure and the same number of osmotically active particles as the solution with which it is being compared

UBIQUITOUS, VITALLY IMPORTANT WATER

Water, which makes up about 60% of the body weight of vertebrates, may be the most neglected nutrient. Drinking and eating are the obvious ways of obtaining water, but metabolism, the processes of synthesis and breakdown within the body's cells, also provides water for organisms. In fact, diet and metabolism provide 40% of the water necessary for human life.

Drinking water is limited by the environment. Areas such as deserts, which have little rainfall, have little potable groundwater available; even plants have to develop some means of conserving the little water their roots can find or the dew that settles on exposed surfaces during the cool desert night. With no surface water and few plants as sources of water, some desert mammals, such as the kangaroo rat, get most of their water from metabolism. They often do not drink even when a supply of water is nearby.

Water balance varies both daily and seasonally as environmental factors such as temperature, humidity, and wind vary and as activity levels change. The body must maintain nearly constant volume and composition of the extracellular fluid despite fluctuations caused by drinking, eating, metabolism, activity, and environment.

WATER LOSS FROM KIDNEYS AND LUNGS

There are four sites of water loss: kidneys, lungs, skin, and intestines. Control of water balance depends on the efficiency of water retention compared with the necessity of water loss in the normal functioning of both terrestrial and marine vertebrates. For freshwater vertebrates, for example, excretion of excess water without losing salts is vital.

The kidney can produce urine that can be highly concentrated or very dilute. For humans, urine can be four times more concentrated than body fluids and contain 1,200 milliosmoles of solute. Other animals, particularly some desert species, can produce urine five times more concentrated than that. The more concentrated the urine, the more water is retained during the excretion of waste materials.

Generally, increasing length of the loop of Henle in the nephron is associated with increased concentrating ability of the kidney in mammals. Although some of their nephrons contain loops of Henle, birds cannot match the mammals' concentrating ability. The maximum urine-to-plasma concentration ratio in birds is only a little more than five. The reason for this is that mammals excrete osmotically active urea, whereas birds excrete precipitates of uric acid and uric acid salts that do not contribute to osmotic pressure. The osmotic pressure of birds' urine primarily comes from sodium chloride. Birds also allow their plasma to become twice as concentrated as that of mammals during dehydration.

When water must be conserved, urine is concentrated to a greater degree than when water is plentiful. As water becomes scarce, the concentrations of solutes in the body fluids increase. Osmoreceptors in the hypothalamus sense the increase and stimulate the release of antidiuretic hormone (ADH). Antidiuretic hormone is a small peptide that varies somewhat in composition among vertebrate classes. It affects kidney cells, increasing the permeability of the distal tubule of the nephron and the collecting duct to water. More water is reabsorbed and, therefore, retained by the body. In some species, ADH also affects the numbers of nephrons filtering.

The earliest water-balance problem that the vertebrates faced in their evolutionary history was an excess of fresh water and a scarcity of sodium. The hormone aldosterone evolved in the fishes to cope with this. Aldosterone is secreted by the adrenal gland and increases the reabsorption of sodium by the kidney. Some potassium is lost from the body in exchange for sodium, but since plants are rich in potassium, its loss could be made up in the diet.

Water is also lost or gained when respiring; all gas-exchange surfaces must be moist. Aquatic organisms take in water through their gills in fresh water and lose water through the gills in marine environments. Terrestrial organisms have lungs, and the moist exchange surface is inside the body. Even if the atmosphere is dry, air entering the lungs is moistened to nearly 100% relative humidity during its passage through the airways. Expired air loses water to those same airway walls as it leaves the body. Some moisture, however, remains in the air and is lost to the body. The cloud of vapor seen as animals exhale in cold weather is this lost moisture condensing in the cold air.

Breathing rate and volume influence the amount of water lost from the lungs. As breathing rate increases, air moves out of the body more quickly, and there is less time for moisture to condense on the cool airway walls before it is exhaled. The movement of greater volumes of air also increases loss of water vapor from the respiratory tract, because less comes in contact with the airway walls. Since heavy exertion often involves breathing through the mouth rather than through the moisture-conserving nasal cavity, further water is lost. As much as 25% of the moisture present in expired air may be lost in a dry environment.

WATER LOSS FROM SKIN AND INTESTINES

The skin is another site through which water is lost. The sweating which occurs in warm weather is an obvious example. Even during the winter, when the air is cool, "insensible perspiration" occurs as water diffuses through the skin. Insensible perspiration occurs at all times. The perspiration that can be sensed comes from eccrine glands. They secrete a watery fluid that cools the skin by using body heat to evaporate the liquid. The amount of salts and nitrogenous wastes in perspiration is not large; however, when one is working in a hot environment, the loss may be significant. Primarily, though, it is water that is lost and must be replaced.

The apocrine sweat glands are located in the armpits and groin. They become active during emotional stimulation. These are the glands associated with the musky odor that some animals exude. These glands are not important in regulating body temperature, and their evaporation of water is not a major component of water balance mechanisms. The losses of water from the respiratory tract and the skin are obligatory, usually amounting to about 850 milliliters a day.

The intestinal tract is a source of water gain, as it ingests both liquids and food (with its associated water). Some water, however, is also lost because the copious intestinal secretions contain water. In fact, one day's intestinal secretions may amount to twice the body's plasma volume (from which it is derived). Not all water is reabsorbed in the passage through the stomach, small intestine, and large intestine: About one hundred milliliters are lost in the feces each day.

In a normal human diet, ingested fluid tends to exceed the minimum required by about one liter. Whatever excess is not used in evaporative cooling and lost from expired air or in feces is excreted by the kidney. The minimum water uptake required for balance is defined as that required to provide minimum urine volume without weight loss. The stomach and the small intestine reabsorb most of the ingested and secreted water. Only 35% reaches the large intestine. The large intestine is specialized to absorb water and produce semisolid feces for excretion. The maximum rate of water absorption by the intestines lies well above what is normally required.

The body fluid compartments provide an excellent example of the steady-state system characteristic of living things. Intracellular fluid must maintain a composition that promotes chemical reactions and diffusion despite the changes that those reactions bring. Extracellular compartments must retain their individual characteristics even though they communicate with one another. Hormones and nerves coordinate the interactions of the digestive, respiratory, integumentary, and urinary systems, which contribute to the constant conditions of volume and composition of the body fluid compartments.

STUDYING WATER BALANCE

The study of water balance is often difficult because so many body systems and physical factors are involved. Gross methods include measuring moisture in respired air, feces, and urine, as well as in ingested foods. In other studies, the weight of water in bodies or organs may be obtained by drying: The amount of water is the difference between the wet and dry weights. These methods are crude and give rough estimates of the fluid volume or fluid balance in the body or an individual compartment or organ.

To obtain more precise information, dilution techniques are used. One method is the injection or infusion of an indicator or test substance. This substance must be distributed only in the fluid compartment being measured. It must be safe for the organism, while not being metabolized or synthesized in the body. If the substance can be excreted, it must be measurable in the excreta. Radioactive tracers are also used to determine dilution of an injected or infused sample. The isotope chosen must not influence the mechanisms governing the size and composition of the compartment being measured. Moreover, the body must not be able to distinguish between the radioactive molecule and the unlabeled isotope.

The concentration of the test substance in a sample of the blood, or lymph, or cerebrospinal fluid gives an indication of the dilution caused by the volume of the fluid within the compartment. There is no perfect test substance; each is associated with problems affecting the accuracy of the measurement. Inulin is used to determine the volume of interstitial fluid, but inulin diffuses slowly through dense connective tissue. Radioactive sodium enters most compartments, but it binds to the crystalline structure of bone. The dye Evans blue, which binds to plasma proteins, and radio-iodinated serum albumin are used to measure plasma volume, but these substances move out of capillaries.

All the fluid compartments communicate, but some, such as bone and cartilage, communicate more slowly than others, such as lymph and plasma. The resistance between compartments is often supposed to be at the interface between compartments. There is evidence, however, that the rate-limiting factors may not only be the permeability of the cell membrane alone but also the physical state of some of the water within the compartment. For example, in red blood cells, some water is bound to protein and is not accessible to solutes. A portion of the water in mitochondria is not free to participate in osmotic processes. For these reasons, dilution techniques may underestimate the amount of water in the compartment being measured. Although dilution techniques use sophisticated technology, the measurements are often extrapolations and not exact. These techniques do provide a general picture of the distribution of body fluids, but since water balance is a continuing, dynamic process, the values are not stable.

REGULATING WATER LOSS

Loss of water through the skin and respiratory tract is obligatory. All respiratory surfaces must be moist so that gases can pass through them. Amphibians are limited in their geographical distribution primarily because their skin is a respiratory surface and, therefore, water loss from it cannot be curtailed. The water losses vary with environmental factors such as temperature, humidity, and wind. Because of this, amphibians must remain near a source of water which their skin can imbibe.

Loss of body water through "insensible perspiration" is not controllable; it is obligatory. The sweating that helps regulate body temperature is facultative, and it varies with weather and exercise. If sweating is prevented when the ambient temperature is high, the body temperature can rise explosively. This will cause death as surely as the dehydration which was prevented by not sweating would have.

Mammals and birds can minimize water loss by modifying the depth and rate of breathing. On exertion, the rate and depth increases, and correspondingly, the loss of water through the airways and across the skin surface increases. Unless the organism intends to quit breathing and allows its body temperature to rise, some water must be lost in this way.

Losses through the digestive tract are often involuntary. Diarrhea and vomiting accompany many illnesses, and since these uncontrolled losses are from the digestive system, which secretes a volume twice that of the plasma each day, their unreclaimed losses are massive. Dehydration and electrolyte imbalance follow quickly if these losses are not made up. This is particularly crucial for infants, for whom the daily diet makes up 25% of the total body water. With dehydration, the volume of the circulatory system decreases and circulatory failure results. In addition, since the extracellular fluid compartments are continuous with the intracellular compartments, the fluid inside the cells becomes hyperosmotic, and metabolic reactions cannot take place.

The retention of water depends upon several conditions. The most important is the sources of water available. If fresh water is not available, a human will die after eleven to twenty days, depending upon the rapidity of onset of dehydration. By that time, the person will have lost 15% to 20% of initial body weight. The excretion of wastes, the act of breathing, and insensible perspiration (even in moderate temperatures with shade available) are accompanied by obligatory water losses that cannot be reduced.

For other organisms, water is present in excess and becomes a problem. Freshwater fish take in water through their respiratory surface, the gills. They must release great quantities of urine without losing the salts necessary to maintain proper internal osmotic conditions. The hormone aldosterone promotes that salt absorption from the nephrons.

For marine creatures, on the other hand, the entry of salt is a problem. They must eliminate the excess salt without losing too much of their precious body water. Because kidney function always involves water loss as well as the loss of ions, and because fish and reptiles do not concentrate urine efficiently, many of these vertebrates have evolved salt glands. The salt glands use metabolic energy to excrete sodium chloride with very little water. Each environment presents its own water-balance problems to an organism. Yet even in the world's harshest, driest conditions—in the Antarctic—tiny mites and spiders, penguins, and predatory birds have found ways to live and obtain all the water they need in a land where water is solid most of the time.

—*Judith O. Rebach*

FURTHER READING

Deetjen, Peter, John W. Boylan, and Kurt Kramer. *Physiology of the Kidney and of Water Balance.* Springer-Verlag, 1975.

Raven, Peter H., and George B. Johnson. *Understanding Biology.* 3rd ed. Wm. C. Brown, 1995.

Slonim, N. Balfour, and Lyle H. Hamilton. *Respiratory Physiology.* 5th ed. C. V. Mosby, 1987.

Weisburd, Stefi. "Death-Defying Dehydration: Sugars Sweeten Survival for Dried-Out Animals, Membranes, and Cells." *Science News* 133 (February 13, 1988): 107–110.

Widdicombe, John, and Andrew Davies. *Respiratory Physiology.* 2nd ed. Edward Arnold, 1991.

Withers, Philip C. *Comparative Animal Physiology.* W. B. Saunders, 1992.

Wildlife Management

Fields of Study

Population Biology; Ecology; Population Genetics; Environmental Studies

Abstract

Wildlife management, also known as game management, is often compared with farming or forestry, because one of its goals is to ensure annual "crops" of wild animals. Animals considered to be game included deer and animals such as coyotes that do damage to domestic animals or crops. Now, however, the term "wildlife" has replaced game, and virtually all living organisms, including invertebrates and plants, are included in management considerations.

Key Concepts

carrying capacity: the number of individual animals that a habitat can support
community: all the living organisms existing in an area at a particular time
furbearers: mammals that are harvested for their fur, such as muskrat, mink, and beaver
game: economically important animals, usually birds or mammals; it includes those taken for recreation or products and those that damage human property
habitat: a specific type of environment or physical place where an animal lives; it usually emphasizes the vegetation of an area
home range: the physical area that an animal uses in its daily activities to get all its needs, such as food and water
succession: change in a plant or animal community over time, with one kind of organism or plant being replaced by others in a more or less predictable pattern
sustained annual yield: the harvest of no more animals than are produced, so that the total population remains the same
wildlife: traditionally, the term included only mammals and birds that were hunted or considered economically important; today, it includes all living organisms

WILDLIFE, NOT GAME

Wildlife management, also known as game management, is often compared with farming or forestry, because one of its goals is to ensure annual "crops" of wild animals. Aldo Leopold, in 1933, defined "game management" as the art of making land produce sustained annual crops of wild game for recreational use. At that time, animals considered to be game included deer and animals such as coyotes that do damage to domestic animals or crops. Now, however, the term "wildlife" has replaced game, and virtually all living organisms, including invertebrates and plants, are included in management considerations.

APPROACHES TO WILDLIFE MANAGEMENT

The process of wildlife management has moved through a sequence of six approaches: the restriction of harvest (by law); predator control; the establishment of refuges, reserves, and parks; the artificial stocking of native species and introduction of exotic ones; environmental controls, or management of habitat; and education of the general public. All six are used in modern wildlife management programs, but most emphasis is placed on habitat management and control of harvest.

Primitive man practiced a form of wildlife management simply by setting fires. These fires stimulated the growth of new grasses that lured grazing animals to the areas near tribal camps. It was then easier to kill the animals for food. Tribal taboos often regulated

Churchill Wildlife Management Area welcome sign.

the use of animals, but the first written wildlife law is probably contained in the biblical Mosaic law (in Deuteronomy 22:6). The Egyptians hunted for sport, and the Romans and Greeks had game laws. The first comprehensive wildlife management program was in the Mongol Empire: Marco Polo reported in the thirteenth century that the Great Khan protected animals from hunters between March and October and provided food for animals during the winter.

During feudal times in Europe, wildlife belonged to the royal family; today it legally belongs to the landowner. In most other countries of the world, wildlife belongs to the state, province, or federal government. In the United States, wildlife belongs to the state, as originally granted legally by the Magna Carta, in 1215, to the people of England—the right was transferred to the state as part of the Common Law when independence was won by the colonies from England.

Only in the last century has the philosophy of wildlife management been not only to preserve but also to increase wildlife abundance. All fifty states of the United States have departments responsible for wildlife conservation, as do all Canadian provinces and territories. An appointed board of directors or commission oversees the actions of the departments. Groups for wildlife law enforcement, research, management, and information and education make recommendations to the board of directors regarding wildlife management actions. The federal government of the United States also has many agencies that manage wildlife on public lands. The US Fish and Wildlife Service is involved with animals that cross state lines, including migratory birds such as waterfowl, marine mammals, and any plants and animals listed as rare or endangered by the National Environmental Protection Act. Other agencies, such as the US Forest Service, Bureau of Land Management, Soil Conservation Service, and the US National Park Service, do extensive wildlife work. Many private organizations, such as the National Wildlife Federation, the Audubon Society, and the Sierra Club actively promote wildlife conservation.

Wildlife management decisions involve the entire range of biological, sociological, political, and economic considerations of human society. Today, the wildlife resource in the United States is managed primarily either for consumptive use (such as sport hunting) or for nonconsumptive use (such as bird watching). Virtually all wildlife management problems are related to the large human population of the earth. Some specific problems are habitat losses (for example, the destruction of tropical rain forests), pollution, diseases introduced by domestic animals into wildlife populations, and the illegal killing of animals for their parts, such as the poaching of elephants for their ivory.

THE PROCESS OF MANAGEMENT

A wildlife manager must first determine the physical and biological conditions of the organism or organisms being managed. Issues include what the best habitat for the animal is and how many animals this habitat can support. The stage of ecological succession determines the presence or absence of particular animals in an area. All animals need food, water, and protection from weather and predators. Special needs, such as a hollow tree in which to raise young, for example, must be fulfilled within the animal's home range. Wildlife managers attempt to remove or provide items that are most limiting to a population of animals. In many respects, solving wildlife management problems is an art; it is similar to medicine in that it often must deal with symptoms (birds dying, for example) and imprecise information.

The stage of ecological succession may be maintained by plowing lands, spraying unwanted plants with a chemical to kill them, or using fire, under controlled conditions, to burn an area to improve the habitat for a certain wildlife group. Refuges and preserves may be set aside to assure that some of the needed habitat is available; nest boxes and water supplies may even be provided.

Periodic surveys of the number of animals in a population provide guidelines for their protection. If animals are more abundant than the lowest carrying capacity, a controlled harvest may be allowed. Sustained annual yield assures that no more than the population surplus is taken. Wildlife laws protect the animals, provide for public safety, often set ethical guides for sporting harvest, and attempt to provide all hunters with an equitable chance of obtaining certain animals (for example, by setting bag limits). If proper wildlife management procedures are followed, no animal need become rare or endangered by sport hunting. Market hunting, the taking of animals for the sale of their products, such as meat or hides, has been stopped in the United States since the 1920s and is also illegal in most other areas of the world. There are almost no societies left that are true subsistence hunters—that is, living exclusively on the materials produced by the wildlife resource.

THE NEED FOR WILDLIFE MANAGEMENT

The proper management of wildlife resources, based on sound ecological principles, is essential to the well-being of humans. All domestic plants and animals came from wild stock, and this genetic reservoir must be maintained. Maintaining the web of life that includes these organisms is necessary for man's survival. Wildlife resources are used by at least 60% of the citizens of the United States each year, and about 6% are sport hunters. Wildlife provides considerable commercial value from products, such as meat; it also offers aesthetic values of immeasurable worth. Seeking and observing wildlife provides needed relief from the everyday tensions of human life. Moreover, by observing wildlife reactions to environmental quality, investigators can monitor the status of the biological system within which man lives. Wildlife populations serve as a crucial index of environmental quality.

Wildlife management is a dynamic force that, to be effective, must reflect an understanding of and respect for the natural world. It cannot be practiced in a vacuum but must encompass the realm of complex human interactions that often have conflicting goals and values. Aldo Leopold once defined conservation as man living in harmony with the land; successful wildlife management will help assure that this occurs.

—*David L. Chesemore*

FURTHER READING

Anderson, S. H. *Managing Our Wildlife Resources.* 3rd ed. Prentice Hall, 1999.
Bailey, James A. *Principles of Wildlife Management.* John Wiley & Sons, 1984.
Bissonette, John A., ed. *Wildlife and Landscape Ecology: Effects of Pattern and Scale.* Springer-Verlag, 1997.
Cooperrider, Allen Y., R. J. Boyd, H. R. Stuart, and Shirley L. McCulloch. *Inventory and Monitoring of Wildlife Habitat.* US Government Printing Office, 1986.
Dasmann, R. F. *Wildlife Biology.* John Wiley & Sons, 1981.
Di Silvestro, Roger, ed. *Audubon Wildlife Report, 1986.* National Audubon Society, 1986.
Giles, R. H., Jr. *Wildlife Management.* W. H. Freeman, 1978.
Leopold, Aldo. *Game Management.* New York: Charles Scribner's Sons, 1939. Reprint. University of Wisconsin Press, 1986.
Matthiessen, Peter. *Wildlife in America.* Viking Press, 1987.
Robinson, W. L., and E. G. Bolen. *Wildlife Ecology and Management.* 4th ed. Prentice Hall, 1999.

WINGS

FIELDS OF STUDY

Ornithology; Animal Anatomy; Animal Physiology; Evolutionary Biology

ABSTRACT

It is believed that insect wings evolved from bilateral, dorsal flaps called gill plates on thoracic and abdominal segments. These gill plates are seen on some fossilized early insects. During the evolution of gill plates into winglike appendages, the appendages grew larger and became restricted to the second and third thoracic segments. Some early insects used these winglike appendages to sail on the surface of ponds or glide in gentle updrafts and air currents.

KEY CONCEPTS

camber: the degree to which wings are convex on their top and concave on their underside
lift: the upward force that is developed by moving wings, which opposes the pull of gravity
slotting: the separation of primary feathers at the tip of the wings

thrust: the forward force that is developed by engines, rotors, or moving wings, which pushes planes, helicopters, or flying animals and which opposes drag

GIVE ME WINGS THAT I MAY FLY

It is believed that insect wings evolved from bilateral, dorsal flaps called gill plates on thoracic and abdominal segments. These gill plates are seen on some fossilized early insects. During the evolution of gill plates into winglike appendages, the appendages grew larger and became restricted to the second and third thoracic segments. Some early insects used these winglike appendages to sail on the surface of ponds or glide in gentle updrafts and air currents.

FLY WINGS

In the fruit fly *Drosophila melanogaster*, each wing develops from a packet of epithelial cells called the imaginal disk. The imaginal disk grows into a wide, baglike structure that flattens out to become a transparent wing only a few cells thick. Hemolymph-filled veins along the anterior wing margin and within the wing support the fragile wing and supply the wing cells with nutrients and water.

Drosophila powers its single pair of wings quite differently than the more primitive dragonflies and grasshoppers that have two pairs of wings. Flies have internal dorsoventral (vertical) muscles on each side of the body that pull the top of the thorax down when they contract. This moves each of the wings up because the wing bases are inserted into the top of the thorax. Contracting longitudinal muscles (running along the length of the thorax) force the top of the thorax back to its original position. This causes each of the wings to move down. Wing movement in most advanced flying insects is associated with the extremely rapid deformation of the thorax. Houseflies can flap their wings up to two hundred times per second, whereas gnats can flap up to one thousand times per second.

The wings of true flies not only flap up and down but they also alter their angle of attack while moving forward during the down stroke and moving backward during the upstroke. These complex movements provide both lift and thrust and require a number of highly evolved muscle groups.

LIFT

For many insects, simple wing flapping cannot generate sufficient lift for flight or for hovering. This is especially true of tiny insects or for insects that have small wing size to body-weight ratios, such as bumblebees. Some flying insects are less than one millimeter long and have extremely small wings that have difficulty moving through air. The viscosity of the air prevents the development of sufficiently small leading-edge vortices. So how do insects produce sufficient lift for flight and hovering?

Beginning with wings above the body, the downstroke moves the wings down and forward through the air at a high attack angle that produces lift. Leading and lagging edge vortices that lower the air pressure above the wing cause lift. As the wing moves down and forward, it continues to rotate slightly, creating larger vortices above the wing that increase the lift. The leading edge vortex begins near the base of the wing and develops quickly toward the tip. Air entering the vortex moves from the base of the wing to the tip. The upstroke provides insects with additional lift. In the upstroke, the wing rotates so that the leading edge is still ahead of the trailing edge. The wing moves up and backward through the air. During the upstroke, lift is created by air reentering the wake created by the down stroke. This moving air pushes against the wing and provides lift. Lift is the sum of "delayed stall" during the downstroke and "wake capture" on the upstroke. In most insects, wing curvature also plays a small role in the development of lift. Most insect wings do not remain flat or rigid during flight or hovering.

Black-legged kittiwakes fly at Cape Hay in the High Arctic. (Hans Hillewaert).

Unlike airplanes, which use wings only for lift and have propellers or jet engines for thrust, insects use their wings to generate both lift and thrust. Although lift is generated during both the downstroke and upstroke, thrust is generated only on the upstroke. In fact, during the downstroke there is a force on the insect that slows it or causes it to hover.

PTEROSAUR WINGS

The first pterosaurs appeared in the fossil record about 215 million years ago. They were a very successful group of reptiles, surviving until the mass extinction that marks the Cretaceous-Tertiary boundary 65 million years ago. These flying animals evolved from a population of tetrapod, carnivorous reptiles. Their wings consisted of leathery skin supported by forelimbs, body trunk, and hind legs. The leathery skin of the wings was reinforced by stiff fibers that ran from the front of the wing to its trailing margin.

The forelimbs of these flying reptiles consisted of an upper arm bone (humerus), two forearm bones (radius and ulna), wrist bones (carpals), palm bones (metacarpals), bones of three claw-tipped fingers (phalanges), and the bones of a fourth, very long, clawless finger (phalanges). The first, second, and third finger bones are the thumb, the index finger, and the middle finger. The greatly elongated palm and fourth finger bones along the distal front edge of the wing was longer than the upper and lower arm combined and supported about three-quarters of the wing.

The wings of all pterosaurs were long and sickle-shaped, yet they could fold up like a fan along the sides of the body. Pterosaurs and their wings varied in size. The smallest pterosaurs were about the size of a starling, with a wing span of about 30 centimeters, whereas the largest animal, belonging to the genus *Quetzalcoatlus*, was about 10 meters in length from bill to trailing feet, with a wing span of nearly 13 meters. The elongated palm and fourth finger bones were over 5 meters. *Quetzalcoatlus* was the largest flying animal ever, weighing up to 100 kilogramss. The long, thin, swept-back, sickle-shaped wings suggest that the pterosaurs were rapid flyers and consummate gliders.

BIRD WINGS

Approximately 155 million years ago, birds began to evolve from a population of feathered, bipedal, meat-eating dinosaurs. Their wings consisted of feathers supported by the forelimbs. The forelimbs of these early, birdlike animals consisted of an upper arm bone (humerus), two forearm bones (radius and ulna), wrist bones (carpals), palm bones (metacarpals), and bones of the three claw-tipped fingers (phalanges).

In modern birds, the wing bones of the palm and fingers have changed significantly. The palm and first finger bones have been fused and greatly shortened so that a single, small thumb bone protrudes near the wrist. The feather or feathers originating in the skin at the end of the thumb is known as the alula. The palm bones associated with the second and third fingers have partially fused at their ends to become a bone that roughly resembles a fused miniature radius and ulna. The second finger that supports the feathers forming the tip of the wing has become shorter. The third finger has been reduced to a single, tiny bone. The long feathers protruding from the skin covering the palm and second finger are called primaries, whereas the long feathers attached to forearm are known as secondaries. Feathers attached to the upper arm are called tertiary feathers.

Wing flapping is powered by muscles that stretch from the base of the humerus to the large, keel-like sternum. Muscles attached to the top of the humerus base and to the sternum pull the wings up, whereas muscles attached to the bottom of the humerus base and the sternum pull the wings down.

WING SHAPES FOR SPECIAL TASKS

A bird's lifestyle and habitat have selected for wings that may be divided into a number of major categories. Birds that live in forested or densely wooded habitats or birds that prey on flying insects and other fast-moving animals have short, broad wings, sometimes described as elliptical wings. Elliptical wings allow birds to carry out rapid, intricate maneuvers. These wings have a high degree of camber and extensive slotting. Highly cambered wings provide greater lift than more flattened wings. The separation of primary feathers at the wing tip provides most of the needed thrust by acting as miniwings. In general, the primary feathers that constitute the wing tip generate most of the thrust and even some lift.

The first finger (thumb) has one or more feathers, parallel to the anterior wing margin, which can be

639

raised above the front of the wing to eliminate air turbulence above the wing. While turbulence significantly decreases lift, the raising of the alula decreases turbulence and increases lift.

Sparrows, finches, cuckoos, barn owls, warblers, grouse, and similar birds have elliptical wings that develop a high degree of camber and slotting. Because of these wing characteristics, birds with elliptical wings are capable of intricate maneuvers. Pigeons have broad wings for maneuverability but also elongated, pointed tips to increase their speed. To muffle the noise generated when barn owls swoop down on prey, their flight feathers have developed fringed edges and their coverts have become soft and downy.

Large birds of prey, such as vultures, buzzards, eagles, condors, swans, and storks, have high-lift wings. Their wings are wide and highly cambered with extensive slotting at the tips. The slightest updraft of air provides lift for these heavy birds. The inner flight feathers provide most of the lift within a thermal, whereas the long primary feathers that resemble fingers are used for maneuvering and creating thrust as well as additional lift. The larger the bird, the fewer wing-beats per second. Large vultures flap their wings about once each second.

Many marine birds that live along the shore and fly long distances to find food, as well as many migrating birds, have nearly flat, narrow, long, pointed wings. These wings are associated with rapid flying and, in some cases, with hovering. Peregrine falcons, gulls, geese, ducks, swifts, and swallows are examples of fast-flying birds. Long, narrow, swept-back wings are found in some birds that develop high speeds, such as swifts, swallows, and falcons. The common swift flies up to five hundred miles each day, gathering insects for its chicks. The Arctic tern is known for its long migration from the North American Arctic via Europe and Africa to Antarctica, approximately 17,500 kilometerss. The peregrine falcon is an extremely rapid flyer, sometimes reaching speeds of 280 kilometers per hour when it dives in pursuit of other birds.

Kingfishers, hummingbirds, and kestrels are both fast and capable of hovering. Hummingbirds beat their wings forty to eighty times each second, depending on their size. In addition, they can hover and fly backward for short distances. To achieve these feats, the wing bones evolved quite differently than in other birds. Hummingbird wing bones are mostly hand bones, as the forearm and upper arm became extremely short. The wrist joint and elbow became rigid, so that the wing only rotates at the shoulder. Their wings are thin, flat, pointed and can become highly slotted. When these birds hover or fly backward, they obtain lift on both the up- and downstrokes of the wing, somewhat like insects. Larger birds usually only generate lift on the downstrokes.

Accomplished gliders, such as the albatrosses and frigate birds, have nearly flat, extremely long, narrow wings. These wings efficiently create lift from updrafts and surface air movements over water and land. These marine birds, which fly long distances when feeding or during migrations, have very long, slender, pointed wings. Each of these birds catches updrafts and is able to glide long distances without using much energy flapping their wings.

BAT WINGS

Bats evolved from tetrapod, insectivorous mammals some time after the Cretaceous-Tertiary mass extinction 65 million years ago. Their wings consist of elastic, leathery skin supported by forelimbs, body trunk, and hind legs. The upper arm and forearm support the proximal half of the extended wing, while the elongated hand and digit bones support the distal half of the wing like the struts of an umbrella. Greatly elongated palm bones and digits two, three, four, and five provide the outer wing struts. The thumb, midway along the anterior margin of the wing, is a mere stump. Depending upon the bat, sometimes the thumb has a claw the bat uses to hold onto its perch.

Bats are extremely agile fliers because they can alter the camber and shape of their wings. The skin between the body and the fifth finger and the tail membrane generate most of the lift. The skin between the second and fifth fingers produces forward thrust. Highly maneuverable bats have relatively short, broad wings, whereas migrating bats have exceptionally long, narrow, pointed wings. These long wings increase lift and allow for extended gliding, but bats with long wings are not as agile as those with broader wings. Tail membranes are continuations of the wings and are used for sudden turns and changes of direction. The tail membrane is controlled by the back legs. The tail membrane is also used during landings to brake and help flip the bat upside-down so it can attach to its roost. The bat uses both its wings and its tail membrane to stall as it lands.

Some species of bats are strong enough to take off from the ground, but most initiate their flight from an elevated roost. Although bats do not have a keeled sternum like birds, some are able to take off from the ground and most are very agile fliers.

—*Jaime Stanley Colomé*

FURTHER READING

Dalton, Stephen. *Borne on the Wind: The Extraordinary World of Insects in Flight.* Reader's Digest Press, 1975.

Dickinson, Michael H., Fritz-Olaf Lehmann, and Sanjay P. Sane. "Wing Rotation and the Aerodynamic Basis of Insect Flight." *Science* 284, no. 5422 (June 18, 1999): 1954–1960.

Dowswell, Paul, John Malam, Paul Mason, and Steve Parker. *The Ultimate Book of Dinosaurs.* Parragon, 2000.

Frey, Eberhard, Hans-Dieter Sues, and Wolfgang Munk. "Gliding Mechanism in the Late Permian Reptile *Coelurosauravus.*" *Science* 275, no. 5305 (March 7, 1997): 1450–1452.

Graham, Gary L. *Bats of the World.* St. Martin's Press, 1994.

Padian, Kevin, and Luis M. Chiappe "The Origin of Birds and Their Flight." *Scientific American* 278, no. 2 (1998): 38–47.

Sereno, Paul C. "The Evolution of Dinosaurs." *Science* 284, no. 5423 (June 25, 1999): 2137–2146.

Welty, Joel Carl, and Luis Baptista. *The Life of Birds.* 4th ed. Saunders College Publishing, 1988.

Zoology

Fields of Study

Zoology

Abstract

Attempts at animal classification are known from documents in the collection of the Greek physician, Hippocrates, as early as 400 BCE. However, the Greek philosopher Aristotle (384–322 BCE) was the first to devise a system of classifying animals that recognized commonalities among diverse organisms. The Swedish botanist Carolus Linnaeus developed the system of nomenclature still in use today, referred to as the binomial system of genus and species. Linnaeus also established taxonomy as a discipline. During the twentieth century, zoology has become more diversified and less confined to such traditional issues as classification and anatomy.

Key Concepts

comparative anatomy: the branch of natural science dealing with the structural organization of living things
ecology: the study of the interactions between animals and their environment
embryology: the study of the development of individual animals
evolutionary zoology: the study of the mechanisms of evolutionary change and the evolutionary history of animal groups
morphology: the study of structure; includes gross morphology, which examines entire structures or systems, such as muscles or bones; histology, which examines body tissues; and cytology, which focuses on cells and their components
phylogenetics: the study of the developmental history of groups of animals
physiology: the study of the functions, activities, and processes of living organisms
systematics: the delineation and description of animal species and their arrangement into a classification
taxonomy: the classification of organisms in an ordered system that indicates natural relationships
zoogeography: the study of the distribution of animals over the earth

ORIGINS OF ZOOLOGY

Attempts at animal classification are known from documents in the collection of the Greek physician, Hippocrates, as early as 400 BCE. However, the Greek philosopher Aristotle (384–322 BCE) was the first to devise a system of classifying animals that recognized commonalities among diverse organisms. Aristotle arranged groups of animals according to mode of reproduction and habitat. After observing the development of selected animal groups, he noted that general structures appear before specialized ones, and he also distinguished between sexual and asexual reproduction. Aristotle was also interested in form and structure, and concluded that different animals can have similar embryological origins and different structures can have similar functions.

In Roman times, Pliny the Elder (23–79 CE) compiled four volumes on zoology widely read during the Middle Ages. Some scholars have deemed those volumes little more than a collection of folklore, myth, and superstition. One of the more influential figures in the history of physiology, the Greek physician Galen (c. 130-ca. 201 CE), dissected farm animals, monkeys, and other mammals and described many features accurately, although scholars have noted that some of these features were then wrongly applied to the human body. His misconceptions, especially with regard to the movement of blood, remained virtually unchanged for hundreds of years. In the seventeenth century, the English physician William Harvey established the true mechanism of blood circulation.

THE FOUNDATIONS OF ZOOLOGY

Until the Middle Ages, zoology was little more than a collection of folklore and superstition. However, during the twelfth century, zoology began to emerge as a science. The thirteenth century German scholar and naturalist St. Albertus Magnus refuted many of the superstitions associated with biology and reintroduced the work of Aristotle. The anatomical studies of Leonardo da Vinci in the fifteenth century have been noted as being far ahead of their time. His dissections and comparisons of the structure of humans and other animals led him to several important conclusions. For example, Leonardo noted that the arrangement of joints and bones in the leg are similar in both horses and humans, thus embracing the concept of homology, or the similarity of corresponding parts in different kinds of animals, suggesting a common grouping. A Flemish physician of the sixteenth century, Andreas Vesalius, is considered the father of anatomy for establishing the principles of comparative anatomy.

Throughout most of the seventeenth and eighteenth centuries, classification dominated zoology. The Swedish botanist Carolus Linnaeus developed the system of nomenclature still in use today, referred to as the binomial system of genus and species. Linnaeus also established taxonomy as a discipline. His work was built on that of the English naturalist John Ray and relied upon the form of teeth and toes to differentiate mammals and upon beak shape to classify birds. Another leading figure in systematic development of this era was the French biologist Comte Georges-Louis Leclerc de Buffon. The study of comparative anatomy was further developed by men such as Georges Cuvier, who devised a systematic organization of animals based on specimens sent to him from all over the world.

A cell is the smallest structural unit of an organism capable of independent functioning. Although the word "cell" was introduced in the seventeenth century by the English scientist Robert Hooke, it was not until 1839 that two Germans, Matthias Schleiden and Theodor Schwann, proved that the cell is the common structural unit of living things. The concept of the cell provided impetus for progress in embryology and animal physiology, including the concept of homeostasis, referring to the stability of the body's internal environment.

The formation of scientific expeditions in the eighteenth and nineteenth centuries gave scientists the opportunity to study plant and animal life throughout the world. The most famous scientific expedition was the voyage of the HMS *Beagle* in the early 1830s. During this voyage, Charles Darwin observed the plant and animal life of South America and Australia and developed his theory of evolution by natural selection. Although Darwin recognized the importance of heredity in understanding the evolutionary process, he was unaware of the work of a contemporary, the Austrian monk Gregor Mendel, who first formulated the concept of particulate hereditary factors, later called genes. Mendel's work was not widely disseminated until 1900.

MODERN ZOOLOGY

During the twentieth century, zoology has become more diversified and less confined to such traditional

Pierre Belon systematically compared the skeletons of birds and humans in his Book of Birds *(1555).*

issues as classification and anatomy. Zoology, broadening its span to include such areas of study as genetics, ecology, and biochemistry, has become an interdisciplinary field applying a wide variety of techniques to obtain knowledge about the animal kingdom. The current study of zoology has two main focuses, taxonomic groups, and the structures and processes common to these groups. Studies of taxonomy concentrate on the different divisions of animal life. Invertebrate zoology deals with multicellular animals without backbones; its subdivisions include entomology (the study of insects) and malacology (the study of mollusks). Vertebrate zoology, the study of animals with backbones, is divided into ichthyology (the study of fish), herpetology (amphibians and reptiles), ornithology (birds), and mammalogy (mammals). Taxonomic groups also subdivide paleontology, the study of fossils. In each of these fields, researchers investigate the classification, distribution, life cycle, and evolutionary history of the particular animal or group of animals under study. Most zoologists are also specialists in one or more of the related disciplines of morphology, physiology, embryology, and ecology.

Animal behavioral studies have developed along two lines. The first of these, animal psychology, is primarily concerned with physiological psychology and has traditionally concentrated on laboratory techniques such as conditioning. The second, ethology, had its origins in observations of animals under natural conditions, concentrating on courtship, flocking, and other social contacts. One of the important recent developments in the field is the focus on sociobiology, which is concerned with the behavior, ecology, and evolution of social animals such as bees, ants, schooling fish, flocking birds, and humans.

—Mary E. Carey

Further Reading

Allaby, Michael, ed. *The Concise Oxford Dictionary of Zoology*. 2nd ed. Oxford University Press, 1999.

_____, ed. *The Dictionary of Zoology*. 2nd ed. Oxford University Press, 1999.

Anderson, Donald Thomas. *Invertebrate Zoology*. Oxford University Press, 1999.

Griffin, Donald R. *Animal Minds*. University of Chicago Press, 1992.

Proctor, Noble S., and Patrick J. Lynch. *Manual of Ornithology: Avian Structure and Function*. Yale University Press, 1993.

Taylor, Barbara. *Animal Encyclopedia*. Dorling Kindersley, 2000.

Wernert, Susan J., ed. *Reader's Digest North American Wildlife: An Illustrated Guide to Two Thousand Plants and Animals*. Reader's Digest Association, 1998.

Whitaker, John O. *National Audubon Society Field Guide to North American Mammals*. 2nd ed. Alfred A. Knopf, 1996.

Wilson, Edward O. *Sociobiology: The New Synthesis*. The Belknap Press of Harvard University Press, 2000.

Zoos

Fields of Study

Zoology; Population Genetics; Genetics; Ethology

Abstract

Ancient rulers maintained wild animal collections beginning about 3000 BCE in Mesopotamia (in the area that is now Iraq), China, and possibly the Indus Valley (now Pakistan). Renaissance collections first developed in Italy, where the importation of animals from Asia and Africa was already well established, and then spread throughout Europe as the continental nation-states developed their own trade routes. Many new zoos and aquariums were built in the decades after World War II, and many older zoos and aquariums were renovated around this time.

Key Concepts

- **aquarium:** self-contained aquatic environmental exhibits, maintained either independently or in association with a zoo
- ***ex situ:*** conservation and research out of the animal's natural environment (at a zoo)

frozen zoo: frozen tissue bank maintained at a zoo that contains wild animal tissue and reproductive samples for use in future breeding programs

in situ: conservation and research in the animal's natural habitat

menagerie: a French word for zoos first used in the early 1700s to describe the keeping (and later exhibition) of animals

vivaria: a Latin word for a structure housing living animals, first used by the Romans to describe the places holding elephants and other animals for their shows

zoo: an English abbreviation of the term "zoological garden"; first used in the early 1800s to describe early British zoos, it tended to replace the word menagerie in the 1900s

EARLIEST ZOOS

Ancient rulers maintained wild animal collections beginning about 3000 BCE in Mesopotamia (in the area that is now Iraq), China, and possibly the Indus Valley (now Pakistan). Collections of native and exotic wildlife were kept in royal parks. These parks were a combination protomenagerie, hunting reserve, and garden park. They were used for falconry, hunting wild beasts, entertaining guests, and personal pleasure. Animals kept in these parks often included elephants, wild bulls, lions, apes, ostriches, deer, gazelle, and ibex. Kings and other wealthy individuals also had fishponds, flight cages for birds, and household pets. Keepers and veterinarians were employed to care for these animals.

From around 2700 BCE, Egyptian pharaohs and wealthy Egyptians had animal collections as well. Native species were caught locally and exotic species were obtained through commerce and tribute. Animals known to have been kept included lions, leopards, hyena, gazelle, ibex, baboons, giraffe, bears, and elephants. Exotic animals came from the so-called Divine Land (Palestine, Syria, and Mesopotamia), Nubia (on Egypt's southern border), and Punt (the Ethiopia area).

In China each ruling dynasty created animal reserves, beginning with the Shang dynasty, the first to unify the region, from about 1520 BCE. These royal parks were large, walled-in natural areas, where wild animals roamed freely and were maintained by park administrators, keepers, and veterinarians. Animals in these parks included fish, turtles, alligators, birds, camels, horses, yaks, deer, elephants, rhinoceroses, and possibly giant pandas.

ANCIENT AND MEDIEVAL ANIMAL COLLECTIONS

The Greco-Roman societies, from 1100 BCE to 476 CE, were the next to evolve urban centers, overtaking the Mesopotamian and Egyptian civilizations and their animal collections, while the Chinese society and collections continued separately from the Western world. Greek curiosity, travel, and trade were favorable for the development of animal collections; however, their ruling city-states did not have the wealth or influence to develop large collections. There were pets, temple collections with animals used in processionals, showmen and animal trainers with animals used in itinerant entertainment acts, and small collections featuring native animals, including the large ones such as bears and lions, that could still be found in the area.

Chimpanzees at the Bronx Zoo.

The Roman Republic (509–27 BCE) provided its rulers and wealthy citizens with the opportunity to maintain small collections of native species. Among the first exotic animals seen in the Republic were Indian elephants taken in battle (280 BCE). Hunts and processionals, which began during the Republic period, evolved into increasingly elaborate spectacles during the Empire period (27 BCE–476 CE). Much has been made of these shows because of the large number of animals that were displayed and killed, but it is their appearances in these spectacles that reveal the introduction of exotic species to Rome. Unfortunately, little is known about the collections in which these animals were kept. Large collections kept in vivaria by emperors, municipalities, and military units likely supplied animals for the spectacles. Public entertainment also included itinerant performing animal acts. Private collections of native and exotic animals were kept in villa gardens, fresh and salt water ponds, bird enclosures, cages, parks, and hunting reserves.

Persian and Arabic societies between about 546 BCE and 1492 CE also had gardens containing wild animals. During this period, Persian and Arabic collections extended throughout the Middle East, India, Asia, northern Africa, and Spain. Meanwhile, medieval Europe (476–1453 CE) saw the fall of the Roman infrastructure and the rise of monarchies, monasteries, and municipalities. These centers of administrative, religious, and social life were also centers for medieval animal collections. Kings and wealthy barons kept elephants, lions, bears, camels, monkeys, and birds, especially falcons. Monasteries maintained modest collections for economic and aesthetic reasons. Towns had animals in moats (when no longer needed for defensive purposes), pits, and cages; they often kept animals used in their coat of arms, such as deer, lions, bears, and eagles.

Collections even existed in the Americas, although it is not known precisely when they began. Both the Aztecs of Central America and the Incas of South America had extensive animal collections. Montezuma's estates, in what is now Mexico, included a large bird building, which included freshwater and saltwater ponds, as well as a staff of three hundred keepers. A separate collection combined birds of prey, reptiles, and mammals, along with its own staff of three hundred keepers. The Incas also had animal collections, along with domesticated herds of guanacos and vicuñas.

EARLY MODERN MENAGERIES

Europe emerged into its Renaissance period as the sixteenth century dawned. The accompanying age of exploration brought to light many new species from distant lands. The proliferation of sixteenth-century European animal collections was greatly influenced by these discoveries, increasing the collections in both size and number. Renaissance collections first developed in Italy, where the importation of animals from Asia and Africa was already well established, and then spread throughout Europe as the continental nation-states developed their own trade routes. Living animal collections, an essential part of royal courts, became commonplace among wealthy collectors. However, obtaining, shipping, and caring for wild, living animals was expensive and difficult, especially since little was known about the needs of the newly discovered species. Nevertheless, these new animals were status symbols and were of immense interest, and so the collections grew.

During the seventeenth and eighteenth centuries these collections, known as menageries, developed throughout Europe and the European colonies. The earliest colonial menageries began as acclimatization farms and animal holding areas at colonial botanical stations located at the more important colonial posts. A tremendous increase in these stations occurred during the nineteenth century, as the exchange of plants and animals became commonplace.

As the nineteenth century advanced, menageries could be found worldwide, due to the intense interest in the many new species and the exotic places from which they came. Knowledge about these animals increased and their transport was greatly improved. As a result, menageries evolved into zoological gardens during the early 1800s and aquariums developed in the mid-1800s.

EARLY ZOOS AND AQUARIUMS

The establishment of the Zoological Society of London's zoological garden at Regent's Park in 1828 was a significant event in the history of animal collecting and may be considered a transition in the evolution of zoos. This collection was intended from its inception to surpass any then in existence, with an emphasis on education and research. Private European menageries evolved into public zoological gardens, going from collections for the few (royalty and wealthy collectors) to zoological gardens for all citizens. In the United States, exotic animals were first introduced in 1716 when a lion arrived at Boston. Other species were gradually introduced over the remainder of the eighteenth century, and menageries containing many species appeared toward the end of this century. Traveling and circus menageries became popular in the early 1800s.

A few small urban menageries were already established when the Zoological Society of Philadelphia (chartered in 1859) opened its zoological garden in 1874. Most of these urban menageries eventually closed, but a few, such as those at Lincoln Park in Chicago and Central Park in New York City, continued and eventually became modern zoological gardens. Throughout the rest of the world, menageries were developing into zoological gardens, although often hindered by local economies and politics.

Robert Warington and Philip H. Gosse first developed the modern aquarium during the early 1850s in England. Gosse also worked with the Zoological Society of London to establish the first public aquarium at the London Zoo in 1853. Other public aquariums, along with the home aquarium craze, swept Europe and the United States soon after.

The turn of the century brought about a tremendous increase in the number of zoos and, to a lesser extent, aquariums that lasted up through World War I. It was a period during which zoos and aquariums improved their programs in conservation, animal husbandry, research, and education. Beginning in the 1890s, conservation of wildlife became an important concern in the United States and other former and current European colonies. Europe had been dealing with conservation issues for many centuries, but these other regions saw their seemingly limitless resources quickly disappear. The New York Zoological Park (the Bronx Zoo), along with other United States zoos, played a significant role in conserving the American bison, which had become extinct in the wild. European zoos did likewise, saving the European bison (wisent), Père David's deer, and Przewalski's horse from extinction.

Animal husbandry research improved when the Penrose Research Laboratory opened at the Philadelphia Zoo around 1901. *In situ* field research began at the New York Zoological Park with the inception of its Department of Tropical Research in 1916. Exhibition techniques received further attention when Hagenbeck's Tierpark (in Stellingen, Germany) opened in 1907. This was an important event in the trend toward moated, open exhibits (rather than buildings), ecological exhibits (rather than systematic arrangements), and mixed species exhibits (rather than single-species exhibits). Hagenbeck's Tierpark featured panorama exhibits based on zoogeographic themes, with a series of back-to-back moated displays designed in such a way that, from the visitor's perspective, the animals appeared to be together in one space.

World War I affected zoos because of a loss of employees to the war effort, a loss of revenues to operate the facilities, a loss of paying visitors, difficulty in finding food for the animals, and a loss of some animals. After recovering from the social and economic impacts of this war there was another surge in the number of zoos and aquariums up through World War II. As zoos and aquariums increased in numbers, the professional management of these institutions improved and became more organized.

During the 1930s, a studbook for the European wisent was established, the first of many species-specific wild animal studbooks. By 1997, there were 150 mammal, bird, and reptile international studbooks and world registers. In 1887, the Verband Deutscher ZooDirektoren (Association of German Zoo Directors) formed in Germany, the first of several early professional associations. In the United States, the American Association of Zoological Parks and Aquariums (later the Association of Zoos and Aquariums) formed in 1924.

World War II repeated the problems faced during the previous war but to a much greater extent, because in addition to the previous problems, there was more physical destruction of European zoos. The physical, social, and economic damage from this war took longer to correct than did the first war's. Improvements began in earnest as soon as the war ended, however, and the postwar period became a time of scientific advances, improved technology, better animal husbandry, new exhibit designs, improved education, more intense conservation programs, and professionalism.

MODERN ZOOS AND AQUARIUMS

Many new zoos and aquariums were built in the decades after World War II, and many older zoos and aquariums were renovated around this time. In addition, many advances were made in areas important to animal husbandry, such as veterinary medicine, chemical immobilization and transportation, animal nutrition, reproductive biology, conservation techniques, biotechnology, materials and exhibit design, and water management technology for aquariums.

Zoo staff training became more formal and systematic beginning in the 1950s. The work had become a profession, and the required knowledge had increased, making formal training a necessity. In 1959, a zoo school was established at the Wroclaw Zoo, Poland, as well as at several German zoos in the 1960s and 1970s. The Association of Zoos and Aquariums published a zookeeper training manual in 1968 and began a series of training classes in 1975. In 1972, the first of several academic programs in zookeeper training began at the Santa Fe Community College in Gainesville, Florida.

Professional associations increased as well. The American Association of Zoo Veterinarians formed in 1946, the American Association of Zoo Keepers formed in 1967, and the Association of British Wild Animal Keepers was founded in 1972. Since then, several other segments of the zoo and aquarium work force have formed professional associations.

Conservation efforts intensified as the seriousness of the endangered species problem increased. More attention was paid to endangered species and environmental problems in the 1960s and 1970s, particularly after activities were held at many zoos and aquariums for the first Earth Day on April 22, 1970. Government laws and regulations concerning endangered species and wildlife importation increased significantly during these decades as well. Conservation efforts at zoos and aquariums included the establishment of international and regional studbooks, regional species survival programs, taxon advisory groups, conservation assessment and management plans, International Union for Conservation of Nature and Natural Resources (IUCN) species survival commission action plans, fauna interest groups, breeding consortia, scientific advisory groups, species survival plans, and conservation research studies at American zoos and aquariums.

More recent trends include the development of more naturalistic exhibits, participation in in situ conservation research, development of frozen zoos, and the use of biotechnology to bring back extinct species. Increasing scientific recognition of animal consciousness and self-awareness has drawn criticism to the practice of keeping animals in captivity for human entertainment and education; however, this scientific understanding of animal consciousness has also brought major improvements to exhibition designs and training techniques.

Many zoos place a new emphasis on promoting the psychological welfare of animals in captivity in addition to ensuring their physical well-being. Many zoos have invested in enrichment programs, which provide stimulation and entertainment to animals in captivity. However, despite improved efforts to replicate animals' natural environments, zoo exhibits often cover only a tiny fraction of animals' range in the wild. A greater understanding and appreciation of animal psychology has led to changes in breeding programs, in which zoo animals are moved cross-country to other zoos, in order to avoid the disruption of families or packs. Zoos and aquariums have evolved over the past five thousand years and continue to evolve to improve conservation efforts.

PROS AND CONS OF ZOOS

Zoos, particularly in the late twentieth and early twenty-first centuries, have often been controversial, drawing both support and criticism from various sources. Proponents of zoos claim that they serve multiple purposes, including valuable scientific research, conservation and protection of rare and endangered species, education and outreach, and public awareness and appreciation for animals. Indeed, many prominent zoos around the world champion wildlife research and environmentalism efforts and captive animal specimens have proved important to breeding programs for threatened species. Most large zoos are highly regulated and have adopted policies to provide animals with as natural a habitat as possible while providing humane treatment according to ethical guidelines. Unlike in former times, zoos generally maintain a population of captive-bred animals rather than individuals taken from the wild, reducing the negative impact on natural ecosystems.

However, opponents of zoos counter many of these claims. Some argue that it is unethical to keep any animal captive and all zoos and aquariums should be disbanded. Others suggest that, despite even the best intentions, zoos are typically driven by economic concerns centered on the entertainment value of animals and that conservation and research programs are only secondary. This conflict of interest could allegedly keep zoos from truly serving the best interests of the captive animals. Even when zoos install open-range habitats and take other measures to approximate life in the wild, the space and conditions are never identical to an animal's native habitat (especially when the zoo is located in a climate radically different from the native habitat), and critics claim that this has a negative effect on many species. Studies have suggested that various health problems including psychological stress can impact captive animals and cause altered behavior and even shortened life spans. Critics of zoos especially target small roadside zoos, which often lack well-equipped facilities, as well as zoos in countries that lack strong regulations and animal rights protections.

The mid-2010s saw something of a shift in public opinion against animal captivity, and though zoos remained popular attractions several organizations responding with major policy changes. In 2016 Seaworld, a prominent aquarium, announced it would stop breeding captive orcas, or killer whales, for performances after the 2013 documentary film *Blackfish* drew widespread attention to the consequences of keeping orcas in captivity. The same year the US Department of Agriculture banned the practice of allowing visitors to interact directly with lion, tiger, and other big cat cubs. Though popular and a significant source of income for roadside zoos, the practice was found to negatively affect the cubs' development. Also in 2016, a three-year-old child climbed into an enclosure at the Cincinnati Zoo and Botanical Garden, where he was subsequently picked up and guarded by a Western lowland gorilla named Harambe. A zookeeper then shot and killed Harambe in order to rescue the child. The incident was captured on video that was widely distributed on social media and news outlets, drawing sharp criticism of the zoo's enclosure design in particular and of the dangers and ethics of keeping wild animals in close proximity to humans in general.

—*Vernon N. Kisling, Jr.*

FURTHER READING

Bekoff, Marc. "Why Was Harambe the Gorilla in a Zoo in the First Place?" *Scientific American Guest Blog*. Scientific American, 31 May 2016. Web. 14 Oct. 2016.

Bell, Catharine, ed. *Encyclopedia of the World's Zoos*. Dearborn, 2001.

Croke, Vicki. *The Modern Ark: The Story of Zoos, Past, Present, and Future*. Scribner's, 1997.

Downes, Azzedine, and Jane Goodall. "Jane Goodall, Azzedine Downs Together Offer Thoughts on Tragic Harambe Killing." *International Fund for Animal Welfare*. IFAW, 19 June 2016. Web. 14 Oct. 2016.

Hakberstadt, Alex. "Zoo Animals and Their Discontents." *New York Times*. New York Times, 3 July 2014. Web. 23 June 2015.

Hoage, R. J., and William A. Deiss, eds. *New Worlds, New Animals: From Menagerie to Zoological Park in the Nineteenth Century*. Johns Hopkins University Press, 1996.

Kisling, Vernon N., Jr., ed. *Zoo and Aquarium History: Ancient Animal Collections to Zoological Gardens*. CRC, 2001.

Milman, Oliver. "Sanctuaries or Showbiz? What's the Future of Zoos." *Guardian*. Guardian News and Media, 23 Mar. 2016. Web. 5 Apr. 2016.

Milman, Oliver. "US Government Cracks Down on Letting Zoo Visitors Play with Lion and Tiger Cubs." *Guardian*. Guardian News and Media, 5 Apr. 2016. Web. 5 Apr. 2016.

Norton, Bryan G., Michael Hutchins, Elizabeth F. Stevens, and Terry L. Maple, eds. *Ethics on the Ark: Zoos, Animal Welfare, and Wildlife Conservation*. Smithsonian Institute Press, 1995.

Owen, David. "Bears Do It: But Pandas in Captivity Often Won't." *New Yorker*. Condé Nast, 2 Sept. 2013. Web. 23 June 2015.

Smith, Laura. "Zoos Drive Animals Crazy." *Slate*. Slate Media Group, 20 June 2014. Web. 23 June 2015.

Wemmer, Christen M., ed. *The Ark Evolving: Zoos and Aquariums in Transition*. Smithsonian Inst. Conservation and Research Center, 1995.

APPENDIXES

BRANCHES OF ZOOLOGY

Acarology—study of mites and ticks

Anthrozoology—study of interaction between humans and other animals

Arachnology—study of spiders and related animals such as scorpions, pseudoscorpions, and harvestmen, collectively called arachnids

Entomology—study of insects

Myrmecology—study of ants

Ethology—study of animal behavior, usually with a focus on behavior under natural conditions, and viewing behaviour as an evolutionarily adaptive trait

Herpetology—study of amphibians and reptiles

Batrachology—study of amphibians including frogs and toads, salamanders, newts, and caecilians

Cheloniology—study of turtles and tortoises

Ichthyology—study of fish

Malacology—study of mollusks

Mammalogy—study of mammals

Neuroethology—study of animal behavior and its underlying mechanistic control by the nervous system

Ornithology—study of birds

Paleozoology—study of deals with the recovery and identification of multicellular animal remains from geological (or even archeological) contexts, and the use of these fossils in the reconstruction of prehistoric environments and ancient ecosystems

Parasitology—study of parasites, their hosts, and the relationship between them

Helminthology – parasitic worms (helminths)

Planktology—study of plankton, various small drifting plants, animals and microorganisms that inhabit bodies of water

Primatology—study of primates

Protozoology—study of study of protozoan, the "animal-like" (i.e., motile and heterotrophic) protists

Endocrinology—study of endocrine systems

Nematology—study of nematodes (roundworms)

Key Figures in Conservation and Zoology

Abbey, Edward (1927–89): American environmental activist and author. The originality of Abbey's ideas regarding the preservation of nature, expressed with great eloquence in his writings, helped to increase awareness of environmental issues and inspired a radical environmental movement.

Amory, Cleveland (1917–98): American author and animal rights activist. Amory's decades of activism for animal rights and animal protection saved thousands of animals from extermination and helped bring the issue of cruelty to animals into the public spotlight.

Attenborough, Sir David (b. 1926): English naturalist and television broadcaster. Attenborough is an esteemed presenter of nature documentaries in Great Britain. His first success was with the series *Life on Earth* in 1979. *The Living Planet* followed in 1984, and in 1990 *The Trials of Life* looked at animal behavior. His 2006 two-part program *Are We Changing Planet Earth?* addressed the issue of global warming.

Audubon, John James (1785–1851): French American naturalist and wildlife artist. Through his unique paintings and his writings, Audubon demonstrated ecological relationships among organisms and set new standards for field observation. By illustrating the beauty of birds and animals, he helped to lay the foundation for a national environmental consciousness in the United States.

Bailey, Michael (b. 1954): Canadian conservationist and film and video producer. Bailey was an early member of Greenpeace and is especially interested in saving dolphins and whales. Through his involvement with the Climate Summit, a consortium begun by former US vice president Al Gore, he uses video technology to educate the public on climate change and global warming. In 1997 Bailey was instrumental in pressuring the Japanese government to release dolphins in connection with the Japan-based ELSA Nature Conservancy.

Bancroft, Tom (b. 1951): American ecologist. Bancroft was named chief scientist at the National Audubon Society in 2007. In 2008 he testified before the US Congress regarding the decline in bird populations around the world, citing environmental causes for the decline that include global warming, habitat destruction as the result of agriculture and encroachment of human settlements, and introduction of nonnative species.

Berry, Wendell (b. 1934): American author of books on conservation and agrarianism. Berry's integrated professions of farmer, writer, and critic of industrial development have placed him among the major figures of the twentieth century in both conservation and literature.

Brown, Lester (b. 1934): American agricultural scientist and author. Brown founded the Worldwatch Institute, an environmental think tank the mission of which is to analyze the state of the earth and to act as "a global early warning system."

Brundtland, Gro Harlem (b. 1939): Norwegian politician, physician, and environmental advocate. Brundtland has been called the "Green Goddess" because of the innovative environmental programs she initiated during her career as prime minister of Norway.

Burroughs, John (1837–1921): American nature writer. Through his best-selling books, Burroughs raised Americans' awareness of the beauty of nature and the importance of preserving it.

Carson, Rachel (1907–1964): American author and environmentalist. As the author of *Silent Spring* (1962) and other best-selling books, Carson helped to spark the modern environmental movement.

Cousteau, Jacques (1910–97): French explorer, conservationist, and filmmaker. Cousteau, one of the twentieth century's best-known explorers and conservationists, gained widespread attention for environmental issues, particularly those concerning the world's oceans.

Darwin, Charles (1809–82): English naturalist. Darwin's theory of evolution through natural selection, the dominant paradigm of the biological sciences, underlies the study of ecosystems.

Davy, Sir Humphry (1778–1829): A chemist, teacher, and inventor born in England, Davy began conducting scientific experiments as a child. As a teen he worked as a surgeon's apprentice and became addicted to nitrous oxide. He later researched galvanism and electrolysis, discovered the elements sodium, chlorine, and potassium, and contributed to the discovery of iodine. He invented the Davy safety lamp for use in coal mines, was a founder of the Zoological Society of London, and served as president of the Royal Society.

Fossey, Dian (1932-85): American zoologist and author. Fossey influenced views of animal behavior and the need for animal protection through her writings about the mountain gorillas of Central Africa and her passionate attempts to save the gorillas from poachers. She was murdered in 1985 many believe in relatiation for her activism.

Francis of Assisi, Saint (c. 1181-1226): Italian monk. In modern times Saint Francis of Assisi has become the patron saint of environmentalists because of his love of nature. Among his works admired by environmentalists is his "Canticle of the Creatures," in which he uses the expressions "Brother Sun" and "Sister Moon."

Freudenburg, William R. (b. 1951-2010): American environmental researcher and educator. Freudenburg, who taught environmental studies at the University of California, Santa Barbara, was known for his work on the relationship between environment and society. His areas of research included resource-dependent communities, the social impacts of environmental and technological change, and risk analysis. He examined such topics as the social impacts of US oil dependence and the polarizing nature of debates over protection of the northern spotted owl from logging in old-growth forest habitat.

Frisch, Karl von (1886-1982): The Vienna-born son of a surgeon-university professor, von Frisch initially studied medicine before switching to zoology and comparative anatomy. Working as a teacher and researcher out of Munich, Rostock, Breslau, and Graz universities, he focused his research on the European honeybee. He made many discoveries about the insect's sense of smell, optical perception, flight patterns, and methods of communication that have since proved invaluable in the fields of apiology and botany. He was awarded the 1973 Nobel Prize in Physiology of Medicine in recognition of his pioneering work.

Gibbons, Euell (1911-75): American ethnobotanist and nature writer. Gibbons improved the public image of wild food foraging and thus of environmentalism in general, as his staid, avuncular image made environmentalism acceptable to Americans who had tended to perceive environmental activism as subversive.

Gibbs, Lois (b. 1951): American environmental activist. Gibbs united her community by forming the Love Canal Homeowners Association and leading efforts to compel state and federal officials to relocate residents in her neighborhood whose homes were compromised by exposure to toxic waste.

Gill, Frank B. (b. 1941): American ornithologist. From 1996 to 2004 Gill was senior vice president and director of science for the National Audubon Society. He is especially known as the author of the book *Ornithology*, which is considered the leading textbook in the field; several editions of the work have appeared since the first was published in 1990.

Goethe, Johann Wolfgang von (1749-1832): German philosopher, scientist, dramatist, poet, and novelist. Goethe held a holistic view of nature—that is, he believed that although humans are the crowning achievement of nature, they are a part of nature like any other part. He had a passionate respect for and even veneration of the natural world, and this holistic view penetrated every aspect of his literary as well as his scientific work.

Hansen, James E. (b. 1941): American climate change scientist. As a prominent climate scientist and activist, Hansen has been an important contributor to increased public awareness of global warming.

Hardin, Garrett (1915-2003): American ecologist. Through his writings—in particular his widely read 1968 article "The Tragedy of the Commons"—Hardin raised awareness of the environmental problems caused by human overpopulation and overexploitation of resources.

Hemenway, Harriet (1858-1960): American socialite and activist for protection of birds. Hemenway began the Massachusetts Audubon Society in 1896 after she realized that many thousands of birds were being slaughtered to provide feathers for women's hats. She convinced women to give up wearing feathers and campaigned for milliners to design featherless hats.

By 1905 fifteen other states had formed Audubon Societies, which eventually combined to form the National Audubon Society.

Kress, Stephen W. (b. 1945): American ornithologist. Kress is an expert in seabird conservation and, as vice president for bird conservation for the National Audubon Society, has had extraordinary success leading that organization's Project Puffin seabird restoration program in Maine. He has also published several books that provide expert tips for creating bird-friendly habitats.

Leopold, Aldo (1887–1948): American wilderness conservationist and environmental philosopher. Leopold, who has been called the father of modern wildlife management and ecology, applied his insightful concepts of ethics and philosophy to conservation strategies and thus helped raise awareness of environmental issues.

Linnaeus, Carolus (1707–1778): A Swedish botanist, zoologist, physician, and teacher, Linnaeus began studying plants as a child. As an adult, he embarked on expeditions throughout Europe observing and collecting specimens of plants and animals and wrote numerous works about his findings. He devised the binomial nomenclature system of classification for living and fossil organisms—called taxonomy—still used in modern science, which provides concise Latin names of genus and species for each example. Linnaeus also cofounded the Royal Swedish Academy of Science.

Lovelock, James (b. 1919): English environmentalist and inventor. Lovelock is best known for his Gaia hypothesis, which suggests that the earth itself is the source of life and that all living things on the planet have coevolved and therefore are inextricably intertwined.

Manabe, Syukuro (b. 1931): Japanese meteorological scientist. Manabe's research using computer modeling has improved humankind's understanding of the role that the oceans play in the global climate.

Marsh, George Perkins (1801–82): American statesman, diplomat, and author. Marsh's widely read book *Man and Nature: Or, Physical Geography as Modified by Human Action* (1864), a treatise on environmental history, became one foundation for the conservation and environmental movements of the twentieth century.

Mendes, Chico (1944–88): Brazilian rubber tapper and trade union leader. Mendes spent his entire life working against the forces of environmental destruction in the Amazon forest in order to sustain a way of life for his fellow rubber tappers and other indigenous peoples of western Brazil. He earned international recognition as a defender of the Amazon ecosystem.

Muir, John (1838–1914): Scottish American naturalist, preservationist, and writer. Muir, one of America's most notable preservationists and a founder of the Sierra Club, introduced Americans to California's Sierra Nevada and worked hard to protect much of the region's wilderness, including Yosemite, against development.

Naess, Arne (1912–2009): Norwegian philosopher. Naess's ideas, in particular his introduction of the concept of deep ecology, have had a great deal of influence on environmental philosophy and activism.

Osborn, Henry Fairfield, Jr. (1887–1969): American naturalist and conservationist. Through his work with the New York Zoological Society and his writings, Osborn promoted the preservation of endangered species and their habitats and also raised public awareness of the dangers of human overpopulation.

Owen, Richard (1804–1892): An English biologist, taxonomist, anti-Darwinist, and comparative anatomist, Owen founded and directed the natural history department at the British Museum. He originated the concept of homology, a similarity of structures in different species that have the same function, such as the human hand, the wing of a bat, and the paw of an animal. He also cataloged many living and fossil specimens, contributed numerous discoveries to zoology, and coined the term "dinosaur." Owen advanced the theory that giant flightless birds once inhabited New Zealand long before their remains were found there.

Passmore, John Arthur (1914–2004): Australian philosopher. Passmore argued that humans cannot continue to exploit the environment, but he also believed that those who say that nature has intrinsic value or that nature has rights of its own are irrational. According to Passmore, the value in nature lies in what it contributes to living things, including humans.

Pinchot, Gifford (1865–1946): American conservationist and forester. As the first head of the US Forest Service, Pinchot influenced national policy making concerning the conservation of natural resources as well as their management for human use.

Powell, John Wesley (1834–1902): American geologist and explorer. Powell contributed significantly to scientific knowledge of the American West in the mid-nineteenth century, and his ideas regarding environmental policy are recognized as being ahead of their time.

Roddick, Anita (1942–2007): English businesswoman and environmental activist. In 1976 Roddick founded the Body Shop, a company that sells cosmetics, soaps, and lotions made from natural ingredients and not tested on animals. Roddick designed the Body Shop stores to do more than sell merchandise—they would also serve as centers for education about social justice and environmental issues.

Roosevelt, Theodore (1858–1919): American politician and conservationist who served as governor of New York and president of the United States. During his years as US president, 1901–09, Roosevelt did more to boost conservation efforts in the United States than any other president before him.

Schumacher, E. F. (1911–77): German British economist. Schumacher's promotion of nonmaterialist values and his writings emphasizing the importance of protecting resources while attending to the needs of humans had a great influence on the environmental movement in the 1970s and 1980s.

Schweitzer, Albert (1875–1965): German philosopher and physician. In 1913 Schweitzer established a hospital in west central Africa, in what is now Gabon; in doing so, he took care to preserve the surrounding forest and rejected the use of any technology that would degrade the environment. He believed that all life is precious and that although killing may at times be necessary, it is never ethical. Schweitzer won the Nobel Peace Prize for his work in equatorial Africa, and Rachel Carson dedicated *Silent Spring* (1962) to him.

Seattle (c. 1780–1866): Native American chief. Chief Seattle is known for the peaceful negotiations he carried out with white settlers in what is now the state of Washington. His words have become a source of inspiration for many environmentalists, particularly a speech in favor of ecological responsibility that has been attributed to him. In this speech he asked, "How can you buy or sell the sky—the warmth of the land? The idea is strange to us."

Singer, Peter (b. 1946): Australian philosopher and bioethicist. Singer, author of the 1975 book *Animal Liberation* and numerous other works in applied ethics, is considered by many to have launched the modern animal liberation movement.

Suzuki, David (b. 1936): Canadian genetics scientist and environmentalist. Suzuki is known for his activism regarding climate change and for his television and radio programs that have addressed various issues related to science and the environment. He became the host of the Canadian Broadcasting Corporation television series *The Nature of Things* in 1979; this widely viewed program has focused on the topics of nature, wildlife, and sustainable human societies. Suzuki was awarded the United Nations Environment Programme Medal for his 1985 series *A Planet for the Taking*, in which he called for a change in the way human beings relate to nature.

Tansley, Arthur G. (1871–1955): English botanist. Tansley, who coined the term "ecosystem," published scholarly articles and books on natural processes that have become central to ecological theory.

Thoreau, Henry David (1817–1862): American naturalist and philosopher. Best remembered as a persuasive advocate of nonviolent civil disobedience to protest unjust laws, Thoreau was also an early advocate of environmentalism.

Udall, Stewart L. (1920–2010): American politician who served as US secretary of the interior. During his term as secretary of the interior (1961–69), Udall acquired for the federal government 1.56 million hectares (3.85 million acres) of new lands, including four national parks and six national monuments. He was an early supporter of Rachel Carson and was instrumental in the signing of the Radiation Exposure Compensation Act of 1990, which provided benefits for persons sickened by radiation as the result of American nuclear weapons testing.

Watson, Paul (b. 1950): Canadian animal rights and environmental activist. A dissident Greenpeace member and experienced sailor, Watson founded the Sea Shepherd Conservation Society, one of the world's most aggressive environmental organizations. He and his organization have mounted vigilante (but deliberately nonlethal) attacks against the efforts of seal hunters, whalers, and drift-net fishers.

Weyler, Rex (b. 1947): American Canadian author, journalist, and ecologist. During the 1970s, Weyler served as director of the original Greenpeace Foundation. He was a cofounder of Greenpeace International in 1979 and is the author of *Blood of the Land: The Government and Corporate War Against the American Indian Movement* (1982), a book about Native Americans' rights, and *Greenpeace: How a Group of Ecologists, Journalists, and Visionaries Changed the World* (2004).

Wilmut, Ian (b. 1944): English reproductive biologist. Wilmut, one of the world's foremost authorities on biotechnology and genetic engineering, conducted a landmark cloning experiment in 1996 that produced Dolly the sheep, the first mammal clone ever produced from adult cells.

Wilson, Edward O. (b. 1929): American evolutionary biologist and author. An evolutionary biologist with extensive field experience, especially in studying ants, Wilson became a political target during the 1970s because of his application of sociobiology to humans. More recently, he has championed biological diversity and has worked to save species from extinction.

Zahniser, Howard Clinton (1906–64): American conservationist and nature writer. Zahniser was an influential figure in the wilderness preservation movement of the mid-twentieth century. In addition to serving as executive secretary of the Wilderness Society for more than twenty years, he authored the landmark Wilderness Act of 1964.

—*Winifred O. Whelan*

FURTHER READING

Allitt, Patrick. *A Climate of Crisis: America in the Age of Environmentalism.* Penguin, 2015.

Dunlap, Riley E., and Angela G. Mertig. *American Environmentalism: The US Environmental Movement, 1970–1990.* Taylor & Francis, 2014.

Frome, Michael. *Rediscovering National Parks in the Spirit of John Muir.* U of Utah P, 2015.

McNeill, John Robert, and Erin Stewart Mauldin, editors. *A Companion to Global Environmental History.* John Wiley & Sons, 2014.

Milton, Kay. *Environmentalism and Cultural Theory: Exploring ahe Role of Anthropology in Environmental Discourse.* Routledge, 2013.

INDEX

A

Acarina, 108
accessory organs, 552–553
acid(s), 486
acid-base disturbances, 486–487
acid rain, 505
actin, 444
actin myofibrils, 444
Actinopterygii, 290
adaptations, animal, 11
   environment and survival, 12–13
   function of, 14–15
   mutation and natural selection, 12
   physiological *versus* behavioral adaptation, 13
   theory and practice, 13–14
adaptive radiation
   adaptation, 1
   evidence and implications, 3
   evolution, 2–3
   natural selection, principles of, 1–2
   studying, 3
adenosine triphosphate (ATP), 470
adrenocorticotropic hormone (ACTH), 244
age-specific approach, animal demographics, 32
aggregation, animal
   animal associations, 15–16
   communal foraging and hunting, 17–18
   division of labor, 18–19
   reproduction and rearing young, 16
aggression, territoriality and, 607
aggressive mimicry, 431
aging
   causes of death, 6
   change and, 4–5
   effects of, 5–6
Agnatha, 525
agricultural pollution, 507
agriculture, animal, 198
Ailthanthus, 620
air pollution, 506–507
Aistopoda, 8
Allen's rule, 360
allopatric speciation model, 540
alpacas, animal domestication, 41
altricial species, 472–473

Amblypygi, 107
Ambystomatidae, 8
American robin, 483
American Sign Language (ASL), 534
amino acids, 470
amoebocytes, 205
amoeboid motion, 80
Amphibia, 7, 492
amphibians, 597
   caecilians, 9
   characterization, bases of, 10
   circulatory systems, of vertebrates, 163
   courtship, 26–27
   estivation in, 252
   evolution of, 7
   fossil amphibians, taxonomy of, 7–8
   frogs, 9–10
   life, 6–7
   marshes, 74–75
   metamorphosis, 426
   salamanders, 8–9
ancestral multicellular organisms, 440
anemones, animal kingdom, 70
anglerfish, 412
animal(s)
   adaptations, 11
      environment and survival, 12–13
      function of, 14–15
      mutation and natural selection, 12
      physiological *versus* behavioral adaptation, 13
      theory and practice, 13–14
   aggregation
      animal associations, 15–16
      communal foraging and hunting, 17–18
      division of labor, 18–19
      reproduction and rearing young, 16
   agriculture, 198
   animal kingdom, 69–70
      arthropods, mollusks and echinoderms, 71–72
      phylogeny, 491
      sponges, hydra, anemones and jellyfish, 70
      tunicates, lancelets and vertebrates, 72–73
      worms, diverse forms of, 70–71
   behavior, sense organs, 574–575

661

animal(s) (*continued*)
   bioluminescence
      functions of, 20
      in nature, 19–20
   cells
      and life, 21
      nucleus and contents, 22–23
      shapes and types, 24
      size and function, 23–24
      structure, 21–22
      studying cells, 24–25
   communication, 396
   courtship
      amphibian and reptile courtship, 26–27
      bird, 27–28
      courtship and ritualistic behavior, 25–26
      fish, 26
      mammals, 28–29
   demographics
      age-specific approach, 32
      age structures and sex ratios, 32
      birth, death and, 29–30
      parameters, 30
      patterns of reproduction, 31–32
      patterns of survival, 30–31
      time-specific approach, 33
      uses of, 33
   dentition, 598
   development
      biogenetic law, nineteenth century, 35
      description *versus* experimentation, 36
      interdisciplinary studies, 37
      ontogeny and phylogeny, 34–35
      ramifications beyond science, 37
      recapitulation, 35–36
   domestication
      characteristics of, 38–39
      dogs and reindeer, 40
      history of, 39–40
      sheep, pigs and cattle, 40
      taming *versus*, 38
   embryology, 42
      fertilization and, 43–44
      gametogenesis, 43
   emotions
      anthropomorphism, 44
      biology of, 46–47
      defining and communicating emotions, 44–45
      primary and secondary emotions, 45–46
   evolution
      Darwinism acceptance, 49–50
      Darwin's revolution, 49
      evolutionary theory, historical context of, 50–51
      evolution of evolution, 47–48
      struggle to conceptualization, 48–49
      studies, 50
   growth
      and development, 52
      developmental biology, 54–55
      postnatal development, 53
      studies, 53–54
      witnessing development, 52
   immune systems
      antigen presentation and receptors, 60–61
      immunity, 60
      nonspecific defenses, 59–60
      primary and secondary immune responses, 61
      protection, 59
   instincts
      ethology, fieldwork of, 64–65
      instincts and genes, 65
      nature *versus* nurture debate, 63–64
      sequences of behavior, 64
      study of, 62–63
   intelligence
      brilliance, flashes of, 66
      cognitive ethology, theories of, 66–67
      primates and sign language, 67–68
   life spans
      and life expectancy, 76
      limiters and extenders, 76–77
      size and, 77–78
      theories of, 78–79
   mating
      courtship and reproduction, 83–84
      hermaphroditism, 85
      mating systems, 84–85
   pH maintenance in, 485–489
   pregnancy and prenatal development in, 519–523
   reproductive strategies in, 547–550
   reptiles and, 561–562
   respiratory systems in, 562–565
Animalia, 491
Annelida, 71
annelids, 139, 161
   muscles, invertebrates, 447

anosmia, 588
Anseriformes, 130
ant, 376–377
antelope, 617
Anthracosauria, 8
anthropomorphism, 44
antidiuretic hormone (ADH), 245, 265
antigen molecules, 60
antigen presentation, 60–61
Anura, 9
apes
    to hominids, 101–105
    vocalizations, 629
aphotic zone, 407
apical ectodermal ridge (AER), 435
Aplysia, 58
apocrine sweat glands, 632
Apodiformes, 131
aposematic display, 214
aquatic arthropods, 110
aquatic biomes, 321
aquatic herbivores, 331–332
aquatic vertebrate animals, locomotion for, 286
Arachnids
    Acarina, 108
    Aranae, 107
    characteristics, 106
    Opiliones, 107–108
    scorpions, 106
    Uropygi, Schizomida, Amblypygi and *Palpigradi,* 107
arachnids, 501–502
Aranae, 107
Archaeology, 39–40
Archaeopteryx, 528–529
Ardipithecus, 350
Argentinosaurus, 210
arthropod(s), 71–72, 109, 383
    external structure and function, 109–110
    hydrostatic skeletons, 364–365
    internal body plan, 110
    muscles, invertebrates, 447–448
    phylogeny, 111
    shells of, 582–583
Arthropoda, 382, 491
arthropodic exoskeletons, 266
arthropodization, 266
artificial breeding, 120
artificial nesting, 459
artiodactyls, 331

asexual reproduction
    parthenogenesis, 112–113
        sex differences, 259
    in simple life forms, 112
Asiatic buffaloes, animal domestication, 40
assortative mating, 466–467
asteroids, 272–273
asynchronous muscles, 445
atrial natriuretic peptide (ANP), 245
auditory communication, 176–177
Australopithecus anamensis, 258
automatic strategy, reproductive strategies, 547–548
autonomic nervous system (ANS), 454
autotrophic protozoa, 538
Aves, 492

B
bacterioplankton, 408
basal bone, 287
basal metabolic rate (BMR), 423
Batesian mimicry, 430–431
bats, 615
    wings, 640–641
bears
    carnivores, 151
    scavengers, 572–573
beefalo, 362
bee societies, 376–377
beetles, scavengers, 572
behavior
    animal, sense organs, 574–575
    biological rhythms and, 121–122
        chronobiology, 123
        circadian and circannual rhythms, 122
        endocrine and nervous system rhythms, 124
        implications of, 124–125
        marine rhythms, 122–123
        studies, 123–124
    crypsis, 147–148
    determinism, 395–396
    hormonal triggers of, 356–357
    hormones, 359
    marsupials, 420
    reproductive strategies, animals, 548
behavioral adaptation, 13
behaviorism, 45
beneficial rodents, 567–568
benthos, 410–411
bicarbonate buffer syst, 486

Index

bills and beaks, 114
    feeding, modifications, 116
    structure of, 114–115
    variations, bill structure, 115–116
bindins, 284
binomial system, 643
biodiversity, 322–323
    definition, 117–118
    habitat loss and degradation, 118–119
    increasing awareness of, 120–121
    research, 119–120
    threats to, 118
biogenetic law, animal development, 35
biological clock, 358
biological rhythms, 121–122
    chronobiology, 123
    circadian and circannual rhythms, 122
    endocrine and nervous system rhythms, 124
    implications of, 124–125
    marine rhythms, 122–123
    studies, 123–124
biology, 125–127
bioluminescence, animal
    functions of, 20
    in nature, 19–20
biomes, 319–320
    biosphere, 321
    biosphere and biodiversity, 322–323
    North American biomes, 320–321
    studies, 321–322
biosphere, 321, 322–323
biotechnology, 126
birds
    aquatic birds and fowl, 129–130
    avian physiology, 128
    circulatory systems, of vertebrates, 163–164
    claws, 166
    courtship, 27–28
    estivation in, 252
    feathers, 127–128
    flightless and perching, 129
    homes, 342–343
    lab, studying, 131–132
    nest, 458
    ornithology, 132
    pair-bonding in, 483–484
    predator birds, 130–131
    scavengers, 573
    stones, 614
    swamps and marshes, 75
    tracking and studying, 131
    types of, 129
    vocalizations, 628
    wings, 639
birth
    oviparous, 134
    ovoviviparous, 134
    viviparous, 133–134
bivalves, 583
    hydrostatic skeletons, 364–365
blastocoel, 232
blastula, 43, 231
body fluids, hydrogen ion concentration of, 485–486
body size, 423
body symmetry, 544
body temperature
    pH maintenance, 487
    regulation, 608–610
Bombyx mori, 427
bones
    composition, development and remodeling, 137
    of contention, 135–136
    physical characteristics of, 136
bony fishes, fish, 289
bovids, 332
bovine spongiform encephalopathy (BSE), 198
bowerbirds, 28
box jellyfish, 602
Brachiosaurus, 210
brain
    evolutionary development of, 139–141
    invertebrates with and without, 138–139
    primate, 141
    reproductive system, male mammals, 555
    structure, 138
breathing air, respiratory systems, 564–565
breathing water, respiratory systems, 564–565
breeding programs
    breeding domesticated animals, 143–144
    breeding wild animals, 144–145
    domestic origins, 142–143
bugs, 380

C

caecilians, 9
calcium, 245
calcium ions, 582
calcium phosphate, 471

calorie, 424
camouflage, 13, 146–148
cannibalism, 148–150
Caprimulgiformes, 130
carbohydrates, 470
carbonate ions, 582
carbon dioxide, 303, 478
cardiac muscle, 450
carnivores, 150–151, 475
    bears, cats and civets, 151
    panda, 152
    raccoons, 152
    social organization, 405
    weasels, 152–153
carrying capacity, 548–549
cartilage, 135, 249
Catarrhini, 532
catastrophe, extinction, 270–271
cats
    animal domestication, 41
    carnivores, 151
    Felidae, 280–282
cattle, 332
    animal domestication, 40
cells, animal, 643
    and life, 21
    nucleus and contents, 22–23
    shapes and types, 24
    size and function, 23–24
    structure, 21–22
    studying cells, 24–25
cell-cell hybridization, 54
cell determination
    genes, role of, 154–155
    genetic recombination and transcription, 155
    methodology, 156–157
    specialization, process of, 153–154
    studying embryology, 156
cell-mediated immunity, 60
cell metabolism, fertilization, 285
cellular biology, 125
cellulolytic protozoa, 538
central nervous system (CNS), 453–454
centrifugation, 451
cephalopods, 383
cephalothorax, 195
cerebral cortex, 455
cervix, 552
change, and aging, 4–5

Chelonia, 559
chemicals
    genetic mutations, 309
    pollution effects, on animal life of, 505–506
    warfare, 595
chickens, animal domestication, 41
chimpanzees, 534
    tool use, 614
chipmunks, 340
chitin, 268
chlorofluorocarbon (CFC), 505
choanocytes, 204
cholecystokininpancreozymin (CCKP), 246
Chordata, 492
chromatography, 379
chromosomes, genetics, 313–314
chronobiology, 123
chyme, 208
Ciliophora, 537
circadian rhythms, 122
circannual rhythms, 122
circulation, invertebrates, 383
circulatory systems, 158
    closed and open, 159–160, 161–162
    components of, 158
    fluid transport systems, 159
    pressure patterns, in heart action, 160
    studying invertebrate circulation, 160–161
    of vertebrates
        birds and mammals, 163–164
        blood and cells, 162–163
        blood volume and blood vessels, 164
        fish, amphibians and reptiles, 163
        studies, 164–165
civets, carnivores, 151
Cladistei, 290
claws, 165–166
cleavage, embryonic development, 231–232
climate, and extinction, 271
clock model, 36
cloning, extinct/endangered species, 168–170
Cnidaria, 70, 382
cnidarian polyps, 112
cnidarians, 383
cnidoblasts, 388
Cnidospora, 537
coastal swamps, 73–74
cockroaches, regeneration, 544–545
coelenterates, 205, 601–602

coevolution, 171
    coevolutionary alliances, 172–173
    coevolutionary warfare, 172
    Gaia, 171
    maintaining balances, 174
    symbioses, commensalisms and parasitisms, 173
    unraveling, tricacies, 173–174
cognitive ethology, theories of, 66–67
colligative properties, 477
colonial nesting, 16
colostrum, 391
commensalisms, coevolution, 173
commensals, 593
communication
    animal, 396
    auditory, 176–177
    insects, 377–378
    pheromones, 175
    studies, 177–178
    tactile and electrical, 177
    talking to animals, 178
    visual signals, 176
communities, 319–320
    community structure, mechanisms of, 181
    complex systems, 182
    components of, 179–180
    dominant and keystone species, 180
    long-term dynamics, 181–182
    natural disturbances, 181
    nature of, 180
    patterns of responses, 182–183
comparative physiology, 91, 92
competition, 181, 183–184
    within niches, 186
    observation, 186–187
    pecking orders, 185
    between within species, 185
    struggle to survive, 184
conditioned response (CR), 398
conditioned stimulus (CS), 398
conjugation, 536
connective cells, 24
consumers, 224, 438
continuous evolution, 540–542
continuous feeders, 201
contractile molecules, 444
convergence, 8
convergent and divergent evolution
    evolutionary convergence, 190

fossil record, tracing, 189–190
Galápagos finches, 189
observation, comparison and classification, 190–191
species and niches, 188–189
cooperative hunting, 16, 18, 482
copulation
    copulatory behaviors, 193
    copulatory organs, 192–193
    coupling constant, 192
coral, 295
countershading, 146, 147
courtship, 25–26, 484
    animal
        amphibian and reptile courtship, 26–27
        bird, 27–28
        courtship and ritualistic behavior, 25–26
        fish, 26
        mammals, 28–29
    animal mating, 83–84
    of ring doves, 356
crabs, Crustaceans, 195–196
crocodiles, 559–560
cro-magnons, 351–352
crossbills, 186
crosscurrent flow, 302
Crossopterygii, 289–290
crustacean arthropods, 161
Crustaceans
    Crustaceae, dominance of, 193, 194
    life cycle of, 196
    physical characteristics, 194–195
    shrimp, lobsters and crabs, 195–196
crypsis, 146
    behavior and ecology, 147–148
Cryptobranchidae, 8
cuckoos, 628
cultural learning, 400
cuticle, 110
cylindrical organisms, 382
cytoplasm, 22
cytotoxic T cells, 61

D
death
    causes of, 6
    life spans, variability of, 197
    organized animal fighting, animal farming and sport hunting, 198–199

666

pollutants, 197–198
scientific experimentation, 199
viral and bacterial disease, 198
decomposers, 224
deep-bodied fishes, 290
deer, 331, 617
defensive mutualism, 498
deformities, 234
degradation, 118–119
demographics, animal
age-specific approach, 32
age structures and sex ratios, 32
birth, death and, 29–30
parameters, 30
patterns of reproduction, 31–32
patterns of survival, 30–31
time-specific approach, 33
uses of, 33
dendrobatids, 9
dentition, animal, 598
deoxyribonucleic acid (DNA), 23, 307
genetic mutations, 308–309
neutral mutations, 460
descriptive ecology, 225
descriptive embryology, 36
destructive rodents, 567–568
determination, 153
development
animal
biogenetic law, nineteenth century, 35
description *versus* experimentation, 36
interdisciplinary studies, 37
ontogeny and phylogeny, 34–35
ramifications beyond science, 37
recapitulation, 35–36
invertebrates, 383–384
marsupials, 419–420
developmental biology, 126
dichlorodiphenyltrichloroethane (DDT), 118–119, 294, 505–506
diet, insects, 374–375
digestion
continuous and noncontinuous feeders, 201
digestive specialization, 201–202
enzymes, 202
food, glorious food, 200
sequence, 200–201
studies, 202–203

and survival, 203
digestive hormones, endocrine systems, 245–246
digestive tract, 204
of complex animals, 205–206
mucous layers, 206–207
nerve and muscle layers, 207
of simple animals, 204–205
small intestine, 208
studies, 208–209
digitigrade locomotion, 81
dinoflagellates, 538
dinosaur, 559–560
extinction theories, 211–212
life-earth interaction, 213
ornithischians, 211
saurischians, 210–211
study of, 212–213
and times, 209–210
Diplodocus, 210
Dipnoi, 289
Discoglossidae, 9
displays
interpretation, 215
modality, 215–216
types of, 214
DNA-DNA hybridization, 462, 463
dogs
animal domestication, 40
vocalization, 629
dolphins, 577, 615
domesticated artiodactyls, 332
domesticated ruminants, 569
domesticated ungulates, 617–618
domestication, animal, 144
characteristics of, 38–39
dogs and reindeer, 40
history of, 39–40
sheep, pigs and cattle, 40
taming *versus*, 38
dominance social behavior, 39
dominant species, 180
dorsal lip, 232
drinking water, 631
Drosophila melanogaster, 545, 638
Dryopithecus, 102
Dryopithecus major, 102
dual-process habituation-sensitization theory, 56
duodenum, 208

E

Eberhardt method, 417
ecdysis, 110
ecdysone, 242, 427, 428
echinoderms, 71–72
    hydrostatic skeletons, 364–365
ecological isolation, 385
ecological niches, 217
    and community, 220–221
    field research, 219–220
    niche overlap, 219
    organisms, interrelationships among, 218–219
    species specialization, 219
    trophic levels, 217–218
ecological pyramids, 228
ecological theory, 78
ecology
    all-encompassing nature, 225–226
    demography and population regulation, 223
    descriptive, experimental and mathematical ecology, 225
    ecology, 224
    ecosystems, 224
    environment and natural selection, 222–223
    nature of, 222
    of sleep, 584
    species, interactions between, 223–224
ecosystems, 319–320
    definition, 227
    development of, 227–228
    disturbance, responding to, 230
    energy production and transmission, 229
    measuring ecosystem productivity, 229–230
    in twentieth century thought, 228
ectotherms, 610
eel-like fishes, 290
egg, gametogenesis, 298–299
egg-polarity genes, 345, 347
electrical communication, 177
electron microscope (EM), 250
elephants, 601
    vocalization, 629
embryology, animal, 42, 232
    fertilization and, 43–44
    gametogenesis, 43
embryonic development, 231
    cleavage, 231–232
    gastrulation, 232–233
    neurulation, 233–234

emotions, animal
    anthropomorphism, 44
    biology of, 46–47
    defining and communicating emotions, 44–45
    primary and secondary emotions, 45–46
endangered species
    deaths of species, 236
    greed and ignorance, 236–237
    international trade bans, 238
    restoring, successes in, 238–239
    species endangerment, causes of, 235–236
    United States' Endangered Species Preservation Act, 237–238
    vanishing wildlife, 235
Endangered Species Act, 238
endocannibalism, 150
endocrine system
    hormones, 359
    of invertebrate, 240–243
    rhythms, 124
    of vertebrates, 244–247
endocuticle, 268
endoplasmic reticulum (ER), 23
endoreceptors, sense organs, 576–577
endoskeletons
    axial and appendicular skeletons, 248
    development of, 247–248
    endoskeletal research, 250
    histology, 249
    late twentieth century advances, 250–251
    study, 247
    vertebrate evolution, 249–250
endotherms, 610–611
energy budget, 223
energy transfer, 294–295
enterocytes, 208
environment, 12–13
    ecology, 222–223
epidermis, 268
epididymus, reproductive system, male mammals, 556
epinephrine, 245
episodic movements, 87–88
epithelial cells, 24
epithelio-chorial placenta, 495
estivation
    invertebrate, 251
    lungfish estivation, 251–252
    in mammals and birds, 252
    in reptiles and amphibians, 252

estrogen, 520, 552
estrus
    evolution and ethology, 254
    pregnancy, 253
    variations between and within species, 253–354
estuaries, 411
ethics, 126
ethological isolation, 385
ethology, 126
    estrus, 254
Euglena, 536
Euornithopoda, 528
evolution, 2–3, 255
    animal
        Darwinism acceptance, 49–50
        Darwin's revolution, 49
        evolutionary theory, historical context of, 50–51
        evolution of evolution, 47–48
        struggle to conceptualization, 48–49
        studies, 50
    animal life, 255–256
    environment, adaptation to, 257
    estrus, 254
    existing and extinct human species and australopithecines, 258
    and extinction, 271
    of fins, 286
    genes and, 256–257
    marsupials, 420–421
    patterns of, 541–542
    time frame, 257–258
evolutionary clocks, 461
evolutionary consequences, infanticide, 367–368
evolutionary convergence, 190
evolutionary explosions, extinctions and, 275–278
evolutionary theory, of animal evolution, 50–51
excretory organs, excretory system, 263–264
excretory system
    cell state maintainance, 263
    excretory organs, 263–264
    fluid homeostasis, 263
    kidney, 264
    loop of Henle, 265
    studies, 265
exocannibalism, 150
exoskeleton, 270
    and arthropod success, 266–267
    drawbacks of, 268–269
    in fossil record, 269–270
    natural body armor, 266
    parts of, 267–268
    studies, 269
exotic species, 118
experimental ecology, 225
extensors, 445
external gills, 526
exteroreceptors, sense organs, 576–577
extinction
    asteroids, 272–273
    catastrophe *versus* uniformity, 270–271
    climate, evolution and, 271
    and evolutionary explosions, 275–278
    genetic factors in, 271–272
    interdisciplinary study, 273–274
    mass extinctions, 272
extinction theories, dinosaurs, 211–212

F

facile husbandry, 39
Falconiformes, 130
fangs, 599–600
feeding relationships, in community, 293–294
Felidae
    cats, appearance of, 280
    classification, 280–281
    feline anatomy, 281
    feline behavior, 281–282
Felinae, 281
female animals, reproductive system of, 551–554
fertilization
    and animal embryology, 43–44
    cell metabolism and meiosis, 285
    eggs and sperm assuring, 283
    invertebrates, 384
    penetration, 284
fetal hormones, 357
fetal loss, 521–522
filter feeding, 369
fins
    aquatic vertebrate animals, locomotion for, 286
    of bony fishes, 287–288
    evolution of, 286
    flippers, 288
    uses of, 286–287
fish
    bony fishes and jawless fishes, 289
    circulatory systems, of vertebrates, 163

fish (*continued*)
    courtship, 26
    ichthyology, 291–292
    marshes, 74–75
    pair-bonding in, 484
    prehistoric animals, 525
    reproduction, 291
    shapes and habits, 290–291
    smell in, 587
    subclasses, 289–290
fitness, 12
fixed action patterns (FAPs), 64
flamingos, 343
flatworms, 70, 205
flexors, 445
flippers, 288
fluid-fluid boundary, 302
fluid homeostasis, excretory system, 263
fluid transport systems, circulatory systems, 159
fly wings, 638
follicle stimulating hormone (FSH), 244, 300, 552
food chains, 293–296
food webs, 293–296
foraminifers, 538
forebrain, nervous systems of vertebrates, 455–456
fossil amphibians, taxonomy of, 7–8
freshwater environments, 480
freshwater organisms, 478
frogs, 9–10
    vocalizations, 627–628
fruit flies, regeneration, 544–545
Fungi, 491

G
Gaia, 171
Galápagos finches, 14, 189
gallbladder, 208
game theory, 607
gametogenesis, 297
    animal embryology, 43
    eggs, 298–299
    RNA, 299
    sperm, 298
    studies, 299–300
gap genes, 345
gardener/Vogelkop bowerbird, 28
gas diffusion, 301–302
gas exchange, 301
    diffusion, 301–302
    fluid-fluid boundary, 302

    measurement, 303
    organs, 301
    partial pressure difference measurement, 304
    of respiratory systems, in animals, 563–564
    studies, 304–305
    ventilation, 302–303
    ventilation measurement, 304
gastropods, 583
gastrulation, 52
    embryonic development, 232–233
Gaviiformes, 130
gene exchange, gene flow, 306
gene flow
    genes and gene exchange, 306
    hybridization, 306–307
    hybrid zones, 360
    pre-and post-darwin science, 305–306
    studies, 307–308
genes
    gene flow, 306
    genetics, 311–312, 313–314
genetic(s)
    early developments, 312
    the genes, 311–312
    genes and chromosomes, 313–314
    genetic engineering, 314–315
    Mendelian genetics, 312–313
    molecular genetics, 314
genetically modified organisms (GMOs), 315
genetic drift, 467–468
genetic engineering, 314–315
genetic equilibrium, Hardy-Weinberg law of, 323–327
genetic mutations
    chemicals, 309
    development and evolution, 310
    DNA and proteins, 308–309
    high-energy electromagnetic radiation and particles, 309–310
    homeotic genes, 310–311
    spontaneous mutations, cause of, 309
    usefulness of, 310
genetic recombination, cell determination, 155
genetic variability, 83, 468
genotypes, 326–327, 512
gestation
    mammals, 519–520
    studies, 522–523
giant squid, 603
gibbons, 484
    vocalizations, 630

Gigantopithecus, 103
gills, 563
giraffes, 331, 617
glucagon, 245
glycerinated muscle, 446
glycosylation, 78
goats, 332
    domesticated ungulates, 617–618
Golgi apparatus, 23
gorillas, 343
grass frog clines, 361
gray matter, 454
grief, 46
grooming, 316
    as cleaning behavior, 316
    interspecies grooming, 318
    as social behavior, 316–317
groundhogs, 340
ground squirrels, 459
growth, animal
    and development, 52
    developmental biology, 54–55
    postnatal development, 53
    studies, 53–54
    witnessing development, 52
growth hormone, 244
guinea pig, 38, 41
gut, arthropod, 110

H

habitat, 319–320
    destruction, 221
    loss, 118–119
    marsupials, 421
habituation
    learning, 398
    neurology, stimulus response, 56–57
    neurotransmitters, 57
    sensitization, 56
    stimulus and response, 56
    studies, 57–58
    survival, 58–59
    and training, 55–56
hadrosaurs, 211
Hadrosaurus, 528
Haeckel's method, 37
hair cells, 328
hairlessness, 395
hamadryas baboon, 404

Haplorhini, 531–532
Hardy-Weinberg law, of genetic equilibrium, 323–327
Haversian canal, 136
hearing
    sensory receptors and nerve impulses, 327–328
    sound detection, 329–330
    vertebrate ear, 328–329
heavy sleepers, 341
hemolymph, 374
herbivores, 475
    aquatic herbivores, 331–332
    domesticated artiodactyls, 332
    wild herbivores, 330–331
herbivorous species, 473
herds
    to avoid predation, 333–334
    formations, 333
    migratory and social herds, 334–335
hermaphroditism, 85
    sex change, triggers for, 336
    sexual dimorphism, 335–336
    simultaneous *versus* sequential, 336
hermit crabs, 593
heterochrony, 35
    paedomorphosis and peramorphosis, 337–338
    studies, 338–339
    Von Baer and Haeckel, contributions of, 338
hibernation, 339–341
high-energy electromagnetic radiation, genetic mutations, 309–310
hippos, 331, 617
Hirudinida, 364
home building
    insect and bird, 342–343
    lifestyles and forms, 344
    mammal, 343
    primate homes, 343–344
homeobox, 347
homeodomain protein, 346
homeosis
    body plans, genetic control of, 345
    homeotic research, implications of, 347–348
    homeotic selector genes, 345–346
    mutants, 344–345
    studies, 346–347
homeostasis, 488–489, 608
homeotic genes, 310–311, 347
homeotic selector genes, 345–346
Hominidae, 532

hominids, apes to, 101–105
hominin, 349
Homo erectus, 350
Homo sapiens, 258, 349
    animal evolution, human evolution, 354–355
    cro-magnons, 351–352
    earliest *Homo sapiens,* 350–351
    emergence of, 350
    modern humans, emergence of, 352–353
    neanderthals, 351
    paleoanthroplogy and physical anthropology, 353
    sciences and social sciences, 353–354
homozygous recessive, 313
honeybees, animal domestication, 41
hooves, 165–166
hormones, 207, 544
    endocrine system, nervous system and behavior, 359
    experimental endocrinology, 358–359
    fetal hormones, 357
    invertebrate, 241–242
    and mating behavior, 357–358
    ring doves, courtship of, 356
horse, donkey, zebra (HDZ) family, 618
horses, animal domestication, 41
hosts, and symbiosis, 592–593
Human Genome Project, 315
human immunodeficiency virus (HIV), 61
humoral immunity, 60
hybridization
    gene flow, 306–307
    hybrid zones, 361–362
hybrid swarms, 362
hybrid zones
    gene flow *versus* natural selection, 360
    grass frog clines, 361
    hybridization, 361–362
    introgression, 362
    north-south clines, 360–361
hydra
    animal kingdom, 70
    tool use, 615
hydrochloric acid, 206
hydrostatic skeletons, 266
    arthropods, bivalves and echinoderms, 364–365
    power of water under pressure, 363
    vertebrate hydrostatic skeletons, 365
    worm, 363–364

hyenas
    scavengers, 572–573
    vocalization, 629
Hymenoptera, 377, 378
Hynobildae, 8–9
hyperventilation, 97
Hypoplectrus, 291
hyposmia, 588
hypothalamo-hypophysial portal vessel, 244
hypoxia, 97
Hypsilophodon, 211

I
ichthyology, fish, 291–292
Ichthyostega, 526
immune systems, animal, 5
    antigen presentation and receptors, 60–61
    immunity, 60
    nonspecific defenses, 59–60
    primary and secondary immune responses, 61
    protection, 59
immunity, 60
immunology, 595
inbreeding, 468–469
indirect calorimetry, 424
infanticide, 366
    by direct competitors, 366–367
    evolutionary consequences, 367–368
    by genetic relatives, 366
    sexually selected infanticide, 367
ingestion, 368
    large item ingestion, 370
    liquid, 370–371
    necessity of food, 372
    small particle ingestion, 368–370
    studies, 371–372
Insecta, 376
insects, 74, 373–374, 501–502
    anatomy, 374
    ant, wasp and bee societies, 376–377
    behavior, 375–376
    communication, 377–378
    cooperation, benefits of, 378–379
    diet, 374–375
    environmental significance of, 380
    homes, 342–343
    metamorphosis, 426–427

nests, 458
smell in, 587
social, 380
societies, 376
studies, 379
termite society, 377
in situ hybridization, 347
inspiration, 565
instincts, animal
    ethology, fieldwork of, 64–65
    instincts and genes, 65
    nature *versus* nurture debate, 63–64
    sequences of behavior, 64
    study of, 62–63
intelligence, animal
    brilliance, flashes of, 66
    cognitive ethology, theories of, 66–67
    primates and sign language, 67–68
International Union for the Conservation of Nature and Natural Resources (IUCN), 238
interneurons, 454
interspecies grooming, 318
intertidal zone, 407
intestinal tract, 632
intestines, water loss from, 632–633
introgression, hybrid zones, 362
introgressive hybridization, 307
invertebrates
    brain, 138–141
    circulatory systems of, 157–161
    claws, 166
    endocrine systems of, 240–243
    estivation, 251
    feeding, 382–383
    muscles in, 444–448
    reproduction and development, 383–384
    respiration and circulation, 383
    shapes, 382
    tails in, 597
in vitro cell culture, 24
ion exchange, pH maintenance, 487
isolating mechanisms, evolution, 385
    enhancing reproductive efficiency, 387
    field studies and experimental studies, 386–387
    postmating mechanisms, 386
    premating mechanisms, 385
isozymic polymorphism, 460

J
jackals, scavengers, 572–573
jawless fishes, 72
    fish, 289
jellyfish, 20, 370, 388
    animal kingdom, 70
    feeding strategy of, 388–389
    reproduction, 389
    tool use, 615
jumping mouse, 341
juvenile hormone (JH), 242, 428
juvocimines, 500

K
kangaroo, 418
kangaroo rats, 459
Kaufman, Thomas, 36
keystone predators, 413–414
keystone species, 180
kidneys
    excretory system, 264
    water loss, 631–632
koalas, 39
K selection, 548–549

L
lactation, 390
    mammary glands, 390–391
    milk production, 391
    study of, 392–393
    suckling, 391–392
ladybugs, 380
lancelets, 72–73
language, 393–394
    animal communication, 396
    behavior determinism, 395–396
    comparative language research, 396–397
    protolanguage theory, 394–395
    talking animals, 396
learning, 397–398
    classical and operant conditioning, 398–399
    ethological and psychological approaches, 400–401
    habituation and sensitization, 398
    insight, 399–400
    process, 401
    programmed learning, 399
leeches, 371

Leiopelmatidae, 10
lekking, 16
Lepidosauria, 560
lepidotrichia, 287
Leydig cells, 556
life spans, animal
    and life expectancy, 76
    limiters and extenders, 76–77
    size and, 77–78
    theories of, 78–79
lift, wings, 638–639
light sleepers, 340
limbic system, 455
limb morphogenesis, 435–436
Lincoln-Petersen index, 416
Liopleurodon, 527
lipids, 470
liquid ingestion, 370–371
lizard, 166, 597
Llamas, animal domestication, 41
lobsters, Crustaceans, 195–196
local resource competition, 366–367
locomotion, 79, 560
    for aquatic vertebrate animals, 286
    physiology of, 79–80
    swimming, gliding and flying, 81–82
long-term dynamics, 181–182
loop of Henle, 265
love, 46
low oxygen, 97–100
luciferin, 20
lungfish estivation, 251–252
lungs, water loss, 631–632
luteinizing hormone (LH), 244, 300, 552
lysosomes, 23

M

macronutrients, 294
mad cow disease, 198
major histocompatibility complex (MHC), 60–61
male mammals, reproductive system of, 554–558
Mammalia, 492
mammalian brains, 140
mammalian social systems
    carnivore social organization, 405
    fieldwork and laboratory studies, 405–406
    lifestyles, 403–404
    primate social organization, 404
    rodent social organization, 405
    social organization and food, 406
    ungulate social organization, 404–405
mammal-like reptiles, 559
mammals
    circulatory systems, of vertebrates, 163–164
    courtship, 28–29
    estivation in, 252
    gestation, 519–520
    homes, 343
    nests, 458–459
    pair-bonding in, 484
    swamps and marshes, 75
    vocalizations, 628–629
mammary glands, lactation, 390–391
mangrove swamps, 74
marine animals, 407
    benthos, 410–411
    nekton, 409–410
    plankton, 408–409
    zones, 407–408
marine biology, 411–412
    open ocean, organisms of, 412–413
    organisms of bottom, 413
    trophic cascades and keystone predators, 413–414
marine ecosystem, food chains and food webs in, 295–296
marine rhythms, 122–123
marked cells, 436
marking, 415–416
marshes, 73–74
    fish, reptiles and amphibians, 74–75
    human destruction, 75
    insects and invertebrates, 74
marsupials, 418–419
    behavior, 420
    classification and physiology, 419
    evolution, 420–421
    habitat, 421
    reproduction and development, 419–420
mass extinctions
    causes of, 275
    historical, 276
    post-extinction recoveries, 276–277
Mastodonsaurus giganteus, 526
maternal care, 474
maternal egg-polarity genes, 345
mathematical ecology, 225

mating, animal
    courtship and reproduction, 83–84
    hermaphroditism, 85
    mating systems, 84–85
mating behavior, hormones and, 357–358
meiosis, 43
    fertilization, 285
meiosis-inducing substance (MIS), 241
melanocytestimulating hormone (MSH), 244
Mendelian genetics, 312–313
menstrual cycle, reproductive system, 553
message/conveyance cells, 24
messenger ibonucleic acid (mRNA), 299
metabolic acidosis, 487
metabolic alkalosis, 487
metabolic alteration, respiration, 98–99
metabolic rates, animals, 422
    factors, 423
    measuring metabolism, 424
metamorphosis, 9, 330, 425, 545
    amphibian metamorphosis, 426
    hormonal regulation of, 427–428
    insect metamorphosis, 426–427
    laboratory studies, 428–429
    silkworm metamorphosis, 427
microbiology, 126
micronutrients, 294
microsauria, 8
midbrain, nervous systems of vertebrates, 455–456
migration, animal
    excursions, and dispersals, 86
    importance of, 89–90
    means and reasons, 86–87
    programmed and episodic movements, 87–88
    studies, 88–89
migratory herds, 334–335
mimicry, 429–430
    aggressive mimicry, 431
    Batesian and Müllerian mimicry, 430–431
mineral calcium, 249
mitochondria, 22–23
modern embryology, 36
modifiable action patterns (MAPs)
Mola mola, 291
molecular biology, 125
molecular clock, 463
    concept of, 462
    hypothesis of, 461
molecular genetics, 314

mollusks, 71–72
    muscles, invertebrates, 447
    shells of, 583
    tentacles, 602
monkeys, 343
    aging in, 5–6
    vocalizations, 629
monogamy, 28, 84
Monoplacophora, 583
monotremes, 493–494
morphogenesis, 52, 434–435
    limb, 435–436
    rules of, 435
    studies, 436–437
morula, 494
mosquitoes, 371, 587
motilin, 246
movement cells, 24
mucous layers, digestive tract, 206–207
Müllerian-inhibiting substance (MIS), 579, 580
Müllerian mimicry, 430–431
multicellularity
    advantages and origins, 438–439
    eukaryotic organisms, rise of, 439
    in evolutionary process, 442–443
    metazoan origin, theories of, 440–441
    multicellular animals, rise of, 439–440
    multicellular kingdoms, 438
    multicellular life, theories of origin, 440
    plants, animals and fungi, 437–438
    rocks, evidence from, 441–442
multicellular kingdoms, 438
multinucleate sphere unknown (MSX), 480
muscles
    fiber, 444
    invertebrates
        annelids and mollusks, 447
        arthropods, 447–448
        mechanism of movement, 444
        muscle response rates, 445
        simple and complex arrangements, 446–447
        specifics of, 444–445
        studies, 446
        variations of, 445–446
    vertebrates, 448–449
        cardiac muscle, 450
        importance of, 451–452
        skeletal muscle, 449–450
        smooth muscle, 450–451
        studies, 451

muscle response rates, 445
mutation, 306, 467–468
    animal adaptations, 12
mutual grooming, 317
mutualism, symbiosis, 593
mynah birds, 628
Myobatrachidae, 10
myosin, 444

N

nails, 165–166
National Environmental Protection Act, 636
natural disturbances, 181
natural selection
    animal adaptations, 12
    ecology, 222–223
    gene flow, 305
    hybrid zones, 360
    principles of, 1–2
neanderthals, 351
Nectridea, 8
nekton, 409–410
neoteny, 36
nephron, 264
nerve impulses, 327–328
nervous system
    hormones, 359
    rhythms, 124
    of vertebrates
        complex structure, 456–457
        midbrain and forebrain, 455–456
        pattern of, 453–454
        studies, 456
        vertebrate brain, 454–455
nesting, 457–458
    artificial nesting, 459
    bird and insect nests, 458
    mammal nests, 458–459
net production, 223
neurosecretory cells, endocrine systems, 240–241
neurotransmitters, 57
neurulation, 232, 233–234
neutral mutations
    advantages and disadvantages of, 462
    DNA and chance changes, 460
    hypothesis, testing, 462–463
    molecular clock, hypothesis of, 461
    population genetics, 513–514
    selection *versus* chance, 460–461
    theory, uses of, 463

newborn, 32
newts, regeneration, 544–545
niche overlap, 219
nitrous acid, 309
nocturnal animals, 464–465
nomads, 86
noncontinuous feeders, 201
nonrandom mating, 466
nonrapid eye movement sleep (NREM), 585
nonsymbiotic mutualism, 498–499
North American biomes, 320–321
north-south clines, 360–361
notochord, 232
nutrient cycle, 294–295
nutrient molecules, 470–471
nutrient requirements of animals, 469–471

O

octopuses, 614–615
octopus tentacle, 602–603
offspring care
    maternal care *versus* paternal care, 474
    parental behavior, spectrum of, 474–475
    precocial species *versus* altricial species, 472–473
    prey species *versus* predatory species, 473
    quantity *versus* quality parenting, 473–474
olfaction, 587
olfactory receptor cells, 588
olive baboon, 404
omnivores, 475–476
one-humped dromedary, animal domestication, 41
ontogenetic migrations, 89
ontogeny, 34–35
oogonia, 298–299
opercular pumping, 564
Opiliones, 107–108
optical microscopes, 269
optimality theory, 607
orangutans, tool use, 614
organelles, 438
organogenesis, 52
ornithischians, 211
osmoconformers, 478
osmoregulation
    challenges of, 477–478
    freshwater *versus* saltwater environments, 480
    measurement, 479–480
    osmosis and osmotic pressure, 477

osmotic pressure, hormonal regulation of, 479
    salt loss, 478–479
osmosis, 477
osmoticity, 477
osmotic pressure, 477, 479
osteoblast, 137
ovaries, reproductive system, 552
oviparous birth, 134
oviparous mammals, 493–494
ovoviviparous birth, 134
ovoviviparous mammals, 493–494
owls, 628
oxidation, respiratory systems, 563
oxygen partial pressure, 565
oxygen uptake, 303
oxytocin, 391

P

packs, 481
    and predation, 481–482
    and rearing of young, 482
paedomorphosis, 337–338
pair-bonding
    in birds, 483–484
    displays, 215
    in fish, 484
    formation and maintenance, 484–485
    in mammals, 484
    partnering up, 483
pair-rule genes, 345
paleoanthroplogy, 353
paleocortex, 140
Palpigradi, 107
pancreatic acinar cell, 23
pancreatic hormones insulin, 245
panda, carnivores, 152
Panthera, 280
paramecium, 203
paramyosin, 445
Parasaurolophus, 528
parasitic insects, 375
parasitisms
    coevolution, 173
    symbiosis, 594
parathyroid hormone (PTH), 245
parental behavior, spectrum of, 474–475
parrots, 597
parthenogenesis, 383
partial pressure difference measurement, 304

paternal care, 474
pecking orders, competition, 185
pectoral girdle, 288
Pelecaniformes, 129, 130
pelvic bones, 526
pelvic girdles, 288
penis, reproductive system, 556–557
Pentastoma, 491
peramorphosis, 337–338
periosteum, 135, 136
peripheral nervous system (PNS), 453–454
persistent organic pollutants (POPs), 198
phagocytosis, 204–205
phase-contrast microscope, cells, 24
phasic receptors, 576
pheromones, 175, 216, 375, 377
pH maintenance, 486
    acid-base disturbances, 486–487
    body temperature and ion exchange, 487
    and homeostasis, 488–489
    studies, 487–488
phoreses, 593
phosphate, 245
phosphate buffer system, 486
photosynthetic organisms, 372
Photuris, 375
phyla, 491
    protozoa, 536–537
phylogeny, 7, 34–35
    animal kingdom, 491
    classification
        benefits of, 492–493
        system of, 489–490
    classifying earth's species, 490
    life-forms, history of, 491–492
    non-animal kingdoms, 490–491
    phylum chordata, 492
phylum chordata, 492
physical anthropology, 353
physiological adaptation, 13
physiological buffering, 486
physiology, animal
    experimentation, 92
    science of, 90–91
    studies, 91
pigs, 40, 521
pinniped herds, 335
placental mammals
    gestation periods, 495–496

placental mammals (*continued*)
    oviparous, viviparous and ovoviviparous mammals, 493–494
    structure, 494–495
    types, 495
plankton, 408–409
plant and animal interactions, 497
    defensive mutualism, 498
    ecology, interaction, 500
    experiments with, 499–500
    mutualism, 497–498
    nonsymbiotic mutualism, 498–499
plant-animal mutualism, 497–498
plant-eating animals, 570–571
planuloids, 441
Plasmodium, 537
Platyhelminthes, 491
Platyrrhini, 532
Plesiadapis, 531
Plethodontidae, 9
poisonous amphibians, 503
poisonous animals, 501–502
poisonous fish, 503
poisonous lizards, 501–502
poisonous snakes, 502
poison tentacles, 602
polar coordinate model, 546
pollination, 498
pollutants, death and dying, 197–198
pollution effects, on animal life
    air pollution, 506–507
    of chemicals, 505–506
    remedies, 507
    sources and types of pollution, 504–505
    ticking time bomb, 504
polygamy, 484
    animal mating, 85
Polyplacophora, 583
polyspermy, 284
Pongidae, 532
population analysis
    density effects, population growth, 509–510
    modeling animal populations, 508–509
    sex ratio and age structure, population growth, 510
    size determination, 508
population genetics, 323–324, 466–467, 511
    average effects, argument against, 513
    explaining, quantifying and predicting evolution, 514–515
    forces against change, 512–513
    neutral mutation and sexual reproduction, 513–514
    sources of variability, 511–512
    theory, observation and experiment, 514
Porifera, 382
power of water, 363
precambrian metazoan fossils, 442
precocial species, 472–473
predation
    packs and, 481–482
    predator-prey interactions, 516
    and prey, 516
    prey, choice of, 517
    relationships and population fluctuation, 517–518
    studies, 518
predators, savannas, 571
predatory species, 473
pregnancy
    course of, 520–521
    fetal loss, 521–522
    fostering optimal pregnancies, 523
    gestation, mammals, 519–520
    gestation studies, 522–523
    uterine environment, 521
prehistoric animals, 524–525
    Archaeopteryx, 528–529
    dinosaurs, 527–528
    first terrestrial vertebrates, 525–526
    fish evolution, 525
    flying reptiles, 527
    reptile diversity, 526–527
    reptiles returned to sea, 527
pressure patterns, circulatory systems, 160
prey
    choice of, 517
    predation and, 516
    predator-prey interactions, 516–517
    species, 473
primary producers, 224, 438
primates, 67–68, 101–102
    ancient fossils and, 104
    brain, 141
    branches of, 530
    classification, 531

genetic comparisons, 533
Haplorhini, 531–532
higher primate fossil record, 102
homes, 343–344
humans and, 533–534
social organization, 404
Strepsirhini, 532
study types, 533
vision, 530–531
vocalizations, 629–630
primordial soup, 443
Proanura, 8
Proconsul, 102
progesterone, 520
programmed death theories, 78
programmed learning, 399
programmed movements, 87–88
prolactin, 244, 426
promiscuity, 84
proteins, 424, 470
genetic mutations, 308–309
prothoracicotropic hormone (PTTH), 427, 429
Protoctista, 491
protolanguage theory, 394–395
Protopterus, 252
protozoa, 535, 539
diversity, 535–536
food chain, base of, 538
infections, 538–539
phyla, 536–537
protozoology, 537–538
reproduction, 536
shells, 582
Pseudidae, 9
Pseudocoelomate, 364
pseudopregnancy, 520
Pteranodon, 527
pterosaur wings, 639
punctuated equilibrium, 540–542
implications of, 541

Q
quality parenting, 473–474
quantity parenting, 473–474

R
rabbits, animal domestication, 41
raccoons, 340
carnivores, 152
scavengers, 572–573

radial bone, 287
radioimmunoassay (RIA) techniques, 554
Ramapithecus, 103, 349
ram ventilation, 564
random damage theories, 78
random mutation, 188
rapid eye movement (REM) sleep, 585
rattlesnakes, 465
recapitulation, 35–36
receptor cells, sense organs, 575–576
recessive allele, 313
recombinant deoxyribonucleic acid (DNA) technology, 54
reefs, 295
refractory period, 358
regeneration
cockroaches, newts and fruit flies, 544–545
fix, repair/replace, 543
models of, 545–546
principles of, 544
studies, 546
reindeer, animal domestication, 40
Reissner's membrane, 329
relaxation, 446
reproduction
animal
beginnings of, 93
competition and reproductive success, 95
field research and laboratory studies, 95–96
fighting for mates, 94–95
mate competition and mate choice, 94
reproductive success, 93–94
animal mating, 83–84
fish, 291
invertebrates, 383–384
marsupials, 419–420
patterns of, 31–32
protozoa, 536
reproductive strategies, animals, 550
reproductive system, male mammals, 555
reproductive hormones, endocrine systems, 246
reproductive strategies, animals
automatic strategy, 547–548
criticism of model, 549
r and K selection, 548–549
reproduction and survival, 550
studies, 549–550
traits and behaviors, 548

reproductive system
   arthropod, 110
   female animals, 551–552
      accessory organs, 552–553
      menstrual cycle, 553
      ovaries, 552
      sexual maturity, 553
      studies, 553–554
   male mammals
      brain and reproduction, 555
      controlling reproduction, 557–558
      epididymus and vas deferens, 556
      penis, 556–557
      studies, 557
      testes, 555–556
Reptantia, 195
reptile courtship, 26–27
reptiles, 559
   and animals, 561–562
   circulatory systems, of vertebrates, 163
   estivation in, 252
   Lepidosauria and *Squamata*, 560
   marshes, 74–75
   snakes, 560–561
   studies, 561
   turtles and crocodiles, 559–560
   vocalizations, 627–628
Reptilia, 492
respiration
   animal
      blood flow and oxygen delivery, 98
      evolution, metabolism and ecology, 100
      hypoxia, 97
      metabolic alteration, 98–99
      respiratory and circulatory systems, 99–100
      ventilation, increase in, 97–98
   invertebrates, 383
respiratory acidosis, 487, 488
respiratory alkalosis, 487
respiratory systems, in animals
   air and water environments, 564
   breathing water and breathing air, 564–565
   gas exchange, organs of, 563–564
   oxidation, 563
   setting breathing rate, 565
respirometry, 424
Rhacophoridae, 9–10
ribonucleic acid (RNA), 22, 299
ribosomal RNA (rRNA), 299

ring doves, courtship of, 356
ritualistic behavior, 25–26
Rodentia, 566
rodents
   destructive and beneficial rodents, 567–568
   lives of, 567
   physical characteristics of, 566–567
   Rodentia, 566
   social organization, 405
rotifers, 369, 382
roundworms, 71
Roux, Wilhelm, 42
r selection, 360, 548–549
ruminants, 568–569

S

Sahelanthropus tchadensis, 349, 350
salamanders, 8–10
Salamandridae, 9
salt loss, 478–479
saltwater environments, 480
sampling effect, 467
sarcomere, 444
saurischians, 210–211
Sauropoda, 528
sauropods, 210
savannas
   plant-eating animals, 570–571
   predators and scavengers, 571
   tropical grasslands, 570
scanning electron microscopy (SEM), 269, 436
Scaphopoda, 583
scavengers, 571
   bears, jackals, raccoons and hyenas, 572–573
   benefits of, 574
   birds, 573
   nature's clean-up crew, 571–572
   sharks, 573–574
   worms and beetles, 572
Schizomida, 107
Schnabel method, 416
scientific agriculture, 144
scorpions, 71
   Arachnids, 106
sedentary sea anemones, 446
segmentation genes, 345
self-fertilization, 85
sense organs
   animal behavior, 574–575

exteroreceptors and endoreceptors, 576–577
receptor cells and transduction, 575–576
sensory data, responding to, 577–578
sensitization
    habituation, 56
    learning, 398
sensory hair cells, 329
sensory receptors, 327–328
sequential hermaphroditism, 336
sex change, trigger for, 336
sex differences
    evolutionary origin of
        asexual reproduction, 259
        sexual reproduction, 259–260
        sexual reproduction, cost of, 261
        sexual selection, 260
        studies, 261
        testing theories, 262
    origin of, 579–580
sex hormones, 579
sexual development
    sex differentiation, origin of, 579–580
    sexual reproduction, evolutionary advantage of, 580–581
    species, males and females of, 579
    studies, 580
sexual dimorphism, 149, 335–336
sexually selected infanticide, 367
sexual maturity, reproductive system, 553
sexual reproduction
    evolutionary advantage of, 580–581
    population genetics, 513–514
    process, 26
    sex differences, 259–260
    of sex differences, evolutionary origin, 261
sexual selection, sex differences, 260
sharks, scavengers, 573–574
shedding, 432–434
shedding hormone, 241
sheep, 332
    animal domestication, 40
    domesticated ungulates, 617–618
shells
    of arthropods, 582–583
    calcium and carbonate ions, formation from, 582
    of mollusks, 583
    protozoa, 582
shrimp, Crustaceans, 195–196
sickle-cell anemia, 511

silkworm metamorphosis, 427
simultaneous hermaphroditism, 336
skeletal muscle, 449–450
skeletonized faunas, 440
skin, water loss from, 632–633
sleep, 584
    across-species and developmental patterns of, 585
    ecology of, 584
    physiology of, 584–585
    possible functions of, 585–586
small intestine, 208
small particle ingestion, 368–370
smell, 586–587
    in fish, 587
    in insects, 587
    in terrestrial vertebrates, 587–588
smooth muscle, 450–451
snakes, 599
    reptiles, 560–561
social behavior, grooming as, 316–317
social herds, 334–335
social hierarchies, 589
    characteristics of, 589–590
    dominance hierarchies, special properties of, 590
    dominants and subordinates, 590–591
social insects, 342, 376, 380
social organization, 406
social roles, 5
solid food eaters, 203
somatic cells, 23
somatic mutations theory, 78
somatotropin (STH), 244
sonar, 615
song learning, 399
sound production, 375
sound waves, 329
specimens, 269
spermatids, 43
sperm, gametogenesis, 298
spiders, 71, 371
    tool use, 615
Spirostomum, 536
sponges, 138, 383
    animal kingdom, 70
spontaneous mutations, cause of, 309
Sporozoa, 537
spotted hyenas, 343
Squamata, 560
squid tentacles, 602–603

squirrel, 597
starfish, 382
status badge, 214
Stegosaurus, 211
steroid hormones, 245
sticks, 614
stimulus response, neurology of, 56–57
stones, 614
Strepsirhini, 532
Strigiformes, 130
structural adaptations, 13
Styracosaurus, 528
suburban wildlife, habitats, 621–622
suckling, lactation, 391–392
surface-oriented fishes, 290
survival
    animal adaptations, 12–13
    animal aggregation, 16–17
    digestion and, 203
    reproductive strategies, animals, 550
survivorship, 31, 32
swamps, 73–74
    fish, reptiles and amphibians, 74–75
    human destruction, 75
    insects and invertebrates, 74
symbiosis
    animal interactions, 592
    coevolution, 173
    hosts and, 592–593
    mutualism, 593
    parasitism, 594
    species, interrelationship of, 595
    studies, 594–595
symbiotic grooming, 318
symphilism, 593

T

tactile communication, 177
tails
    in invertebrates, 597
    uses of, 596
    in vertebrates, 597–598
talking animals, 396
taming domestication, 38
tapeworms, 382
tautomeric shifts, 309
teeth, 599
teleost fishes, 290–291
Temnodontosaurus, 527

Temnospondyli, 7–8
temperature compensation, 609
Tennessee Valley Authority, 238
tentacles, 601–602
    octopus and squid tentacles, 602–603
    poison tentacles, 602
    worm and mollusk tentacles, 602
termite, 342
termite society, 377
terrestrial ecosystem, food chains and food webs, 295
terrestrial organisms, 480
terrestrial vertebrates, smell in, 587–588
terrible lizard, 209
territorial defense, 606
territoriality, 185
    and aggression, 607
    causes of, 604–605
    in field, 606–607
    functions of, 605
    roles, 605–606
    and species, 603–604
testes, reproductive system, 555–556
testosterone, 556
Testudines, 559
thermoregulation
    applications of, 612
    body-temperature regulation, 608–610
    ectotherms, 610
    endotherms, 610–611
    studies, 611–615
theropods, 210
Thomson's model, 598
thyroid-stimulating hormone (TSH), 244
thyroxine, 426
time-specific approach, animal demographics, 33
Titanosaurus, 528
tonic receptors, 576
tool use, 613–614
    sonar, 615
    sticks and stones, 614
tooth-bearing animals, 599
traits, reproductive strategies, 548
transcription, cell determination, 155
transduction, sense organs, 575–576
transmembrane proteins, 22
transmission electron microscopy (TEM), 24, 436
transverse tubules, 444
Triceratops, 528
trimethylaminoxide (TMAO), 478

trophic cascades, 413–414
tropomyosin, 444
troponin, 445
T tubules, 444
tunicates, 72–73
turtles, 559–560
tusks, 598, 600–601
two-humped Bactrian, animal domestication, 41
tylopods, 568–569

U

ultraviolet (UV) radiation, 309, 505
unconditioned response (UCR), 398
unconditioned stimulus (US), 398
ungulates, 616–617
    domesticated ungulates, 617–618
    hooves, 166
    lifestyles of, 618
    social organization, 404–405
    wild, 617
uniformity, extinction, 270–271
United States' Endangered Species Preservation Act, 237–238
urbane city-dwellers, 619
urban wildlife
    characteristics of, 620
    conservation and management, 622–623
    humans and, 621
    sources and types of, 620–621
    suburban wildlife habitats, 621–622
    urbane city-dwellers, 619
urea, 478
Uropygi, 107
uterine milk, 520

V

variability, population genetics, 511–512
vas deferens, reproductive system, 556
vasoactive intestinal polypeptide, 246
venom injection, 561
ventilation, 97–98
    gas exchange, 302–303
vertebrae, 625
vertebrates, 72–73, 624
    bone and bones, 624–625
    brain, nervous systems of vertebrates, 454–455
    ear, 328–329
    endocrine systems of, 244–247
    hydrostatic skeletons, 365
    muscles in, 448–452
    tails in, 597–598
    vertebrae, 625
    vertebrate evolution, 625–626
    water balance in, 631–634
vipers, 600
visual signals, 176
vitamins, 471
viviparous birth, 133–134
viviparous mammals, 493–494
vocalizations, 626–627
    birds, 628
    mammals, 628–629
    primates, 629–630
    reptiles and frogs, 627–628
voicebox, 628
von Baer, Karl Ernst, 34, 42
vultures, 343

W

Wallace, Alfred Russel, 13
wasp, 376–377
water balance, 631
    control of, 245
    kidneys and lungs, water loss from, 631–632
    skin and intestines, water loss from, 632–633
    studies, 633
    ubiquitous, vitally important water, 631
    water loss regulation, 633–634
water vascular system (WVS), 383
weasels, carnivores, 152–153
white matter, 454
wild herbivores, 330–331
wildlife management
    approaches, 635–636
    need for, 637
    process of, 636–637
wild ruminant species, 569
wild ungulates, 617
wings, 638
    bat, 640–641
    bird, 639
    fly, 638
    lift, 638–639
    pterosaur wings, 639
    special tasks, shapes for, 639–640
winter lethargy, 340
Wolffian ducts, 580
wolves, 343

woodchucks, 341
woodpecker, 2
worms
    diverse forms of, 70–71
    hydrostatic skeletons, 363–364
    scavengers, 572
    tentacles, 602

Z

zone of polarizing activity (ZPA), 435
zookeepers, 468
zoology
    foundations of, 643
    modern zoology, 643–644
    origins of, 642
zooplankton, 408, 412
zoos
    ancient and medieval animal collections, 645–646
    earliest zoos, 645
    early modern menageries, 646
    early zoos and aquariums, 646–647
    modern zoos and aquariums, 648
    pros and cons of, 648–649
zygote, 26
    animal mating, 83